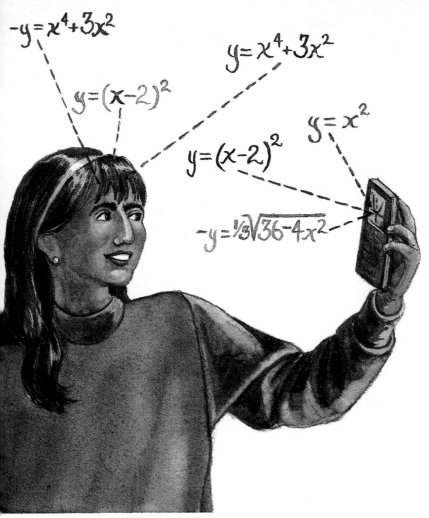

$$-y = x^4 + 3x^2$$

$$y = x^4 + 3x^2$$

$$y = (x-2)^2$$

$$y = (x-2)^2$$

$$y = x^2$$

$$-y = \tfrac{1}{3}\sqrt{36 - 4x^2}$$

When I started this program, right off the bat scores were better.

Ellen Hook
Granby High School
Norfolk, VA

I honestly believe this is the first time some of my students have had success with word or story problems.

William Putnam
John Marshall High School
Rochester, MN

The grades on the final exam were better than those from previous years when we reviewed extensively.

Robert E. Smith
Southside High School
Elmira, NY

Students who have not done as well in the past do as well as or better than those who are good at the symbolic mathematics.

Anne Thompson
Washington High School
Sioux Falls, SD

The use of a graphing utility — whether a hand-held graphing calculator or computer graphing software — is a necessary part of this program. Technology allows the focus of the course to be on problem solving and exploration, while building a deeper understanding of algebraic techniques.

SUCCESS THROUGH DISCOVERY AND PROBLEM SOLVING

I feel successful when the kids are motivated by word problems or real-world problems. Hey, word problems are what we do in life!

William Leonard
Shawnee Mission West High School
Overland Park, KS

It opens up a lot more discovery. What students discover, they remember. You really have to see it to believe it!

Linda George
Richardson Pearce High School
Richardson, TX

We're not just teaching them how to pull the rabbit out of the hat .. but where to find the rabbit!

Milton Norman
Granby High School
Norfolk, VA

As a natural outgrowth of this excitement, students complete the course with a better understanding of mathematics and a solid, intuitive foundation for calculus.

EXPLORE WITH A GRAPHING UTILITY

Graph the following functions together in the same viewing rectangle:

$$y = 2^x, \qquad y = 3^x, \quad \text{and} \quad y = 5^x.$$

1. Find a viewing rectangle that clearly distinguishes the complete graphs of these three functions.

2. For what values of x is it true that $2^x > 3^x > 5^x$?

3. For what values of x is it true that $2^x < 3^x < 5^x$?

Answer the same three questions for these functions:

$$y = 2^{-x}, \qquad y = 3^{-x}, \quad \text{and} \quad y = 5^{-x}.$$

Generalize How do the graphs of $y = a^x$ and $y = a^{-x}$ compare? How do the graphs of $y = a^x$ compare for various values of a?

> Kids used to be afraid of word problems; it was such a bug-a-bear for them. Now it's the core of what we do. The approach is problem oriented, not just memorizing separate math skills.

Gayle Garrison
Heritage High School
Conyers, GA

> The use of the graphing calculator opens up so many more doors to the type of problems we can do.

Kathy Layton
Beverly Hills High School
Beverly Hills, CA

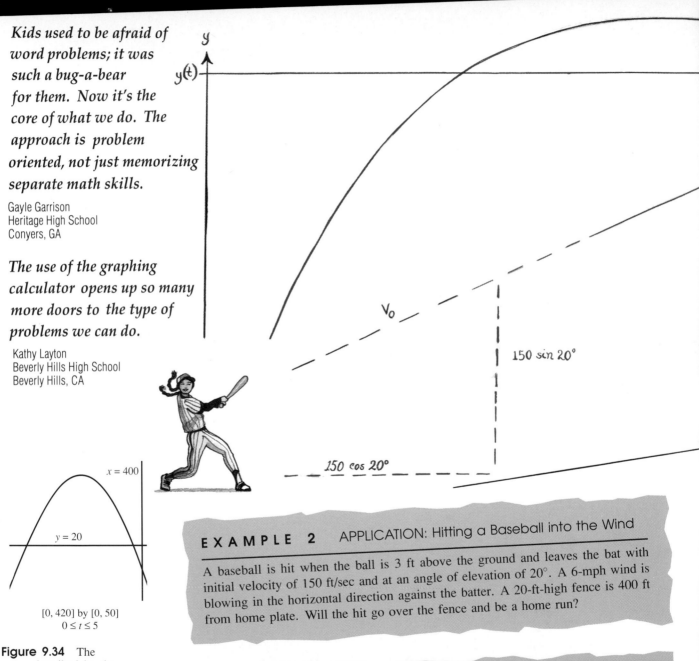

Figure 9.34 The curve described by the system of equations
$x(t) = 150t \cos 20° - 8.8t$ and
$y(t) = 150t \sin 20° + 3 - 16t^2$.

[0, 420] by [0, 50]
$0 \le t \le 5$

EXAMPLE 2 APPLICATION: Hitting a Baseball into the Wind

A baseball is hit when the ball is 3 ft above the ground and leaves the bat with initial velocity of 150 ft/sec and at an angle of elevation of 20°. A 6-mph wind is blowing in the horizontal direction against the batter. A 20-ft-high fence is 400 ft from home plate. Will the hit go over the fence and be a home run?

Figure 9.34 shows a graph of this pair of parametric equations. Clearly, when $x = 400$, $y < 20$ and the ball will *not* clear the fence. Therefore the hit will not be a home run.

THE REAL WORLD INTO THE CLASSROOM WITH OTHER DISCIPLINES

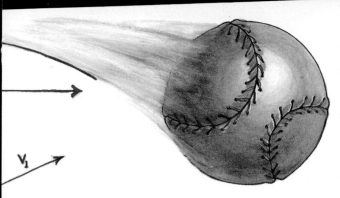

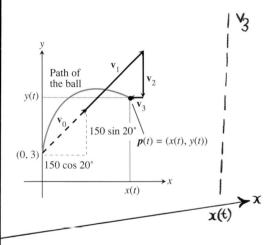

Students get a better handle on the big picture. They understand what a function is... and that they naturally occur in other disciplines, such as science and social science.

John Diehl
Hinsdale Central High School
Hinsdale, IL

Students are developing their thought processes so they can transfer knowledge.

Judy West
Temple High School
Temple, TX

They learn mathematics in a connected sense.

Ruth Casey
Franklin County High School
Frankfort, KY

The text blends great with a physics course.

Tim Tilton
Winton Woods High School
Cincinnati, OH

Technology eliminates the need for contrived problems and opens the door for realistic and interesting applications.

Planet	P Period (Days)	x Semimajor Axes (Miles)
Earth	365	92,600,000
Mercury	88	36,000,000
Venus	225	67,100,000
Mars	687	141,700,000
Jupiter	4330	483,000,000
Saturn	10,750	886,100,000

57. Plot $\ln P$ against $\ln x$. That is, plot the pairs $(\ln x, \ln P)$. Verify that the relationship is linear and estimate the slope m and y-intercept.

58. Use Exercise 57 to show that $P = ax^m$ for constants a and m. This result was discovered by Kepler in the early 1700s and is known as *Kepler's third law of planetary motion*.

What will change?

I'm having the best year of the 21 I've taught!

Tim Tilton
Winton Woods High School
Cincinnati, OH

I find myself answering questions as opposed to giving prepared lectures.

Gene Olmstead
Elmira Free Academy
Elmira, NY

The enthusiasm displayed by those teachers using graphing calculators cannot be overemphasized! You've got kids and teachers thinking and talking mathematics.

Robert E. Smith
Southside High School
Elmira, NY

65. Writing to Learn Write a paragraph that explains how solving the equation $\sin y = 0.5$ is different from finding the value $\sin^{-1}(0.5)$.

What will remain the same?

I can still do the same rich mathematics that I need to do.

Ruth Casey
Franklin County High School
Frankfort, KY

The content is exactly the same. The difference is that with true understanding, the students require less drill. So the trade-off is less rote learning for more time for investigation.

Gayle Garrison
Heritage High School
Conyers, GA

The mathematics classroom is transformed into a mathematics laboratory, with a new interactive instructional approach that focuses on problem solving.

What training is needed?

A lot of people think you need months of training on this approach before you can teach it, but it's not true. Teachers can use the approach without the training session ... especially because it's so easy to get excited about!

William Leonard
Shawnee Mission West High School
Overland Park, KS

Bert says, "Be sure that students understand the difference between screen coordinates and Cartesian coordinates of points on the graph."

Bert, Frank, and Stan are with you every step of the way in the new Teacher's Edition, which includes

- Lesson objectives
- Key ideas for each lesson
- Teaching notes for each lesson
- Common errors
- Notes on examples
- Notes on exercises
- Assignment guides
- Special notes from Bert, Frank, and Stan

In addition, students are provided with sidelight boxes, which include graphing calculator hints and reminders.

FINDING THE VIEWING RECTANGLE

When finding a complete graph of a polynomial, you must choose a viewing rectangle that includes all the zeros. If c is an upper bound and d is a lower bound for zeros, then you know to choose a viewing rectangle so that $X\min < d < c < X\max$.

GRAPHER SKILLS UPDATE

The square root function can also be accessed by entering it in the exponential form:

$$x \wedge (1/2) \text{ or } x \wedge 0.5.$$

The parentheses around the 1/2 are needed because computers follow the conventional rules for order of operations. Entering $x \wedge 1/2$ produces the expression $x/2$.

GRAPHER UPDATE

Scientific calculators and graphing calculators have both a degree mode and a radian mode. Ensure the mode setting for angle measure is correct.

AND MUCH MORE...
AND MUCH MORE...

The Teacher's Resource Package makes PRECALCULUS *even easier to use*

Graphing Technology Manual

- Worksheets and a user's guide for the most common graphing utilities

Tests

- Chapter and semester tests including alternative assessment

Quizzes

- Two quizzes for each chapter

Inservice Seminar Video

Frank and Bert team up to show you how to bring the POWER OF VISUALIZATION to your classroom.

Also Available

- Solutions Manual
- MasterGrapher software (Apple/IBM/Macintosh)
- OmniTest software (IBM/Macintosh)

Workshops, conferences, and newsletters are provided for *Calculator and Computer Precalculus* (C^2PC) and *Teachers Teaching with Technology* (T^3). Contact your regional sales office for current information.

Addison-Wesley Publishing Company

2725 Sand Hill Road, Menlo Park, CA 94025, (415) 853-2541

Northeastern Region
Jacob Way
Reading, MA 01867
1 (800) 521-0011

International Division
(617) 944-3700 x2517

Midwestern Region
1843 Hicks Road
Rolling Meadows, IL 60008
1 (800) 535-4391

Southeastern Region
1100 Ashwood Parkway
Suite 145
Atlanta, GA 30338
(404) 394-7233
1 (800) 241-3532

Southwestern Region
1815 Monetary Lane
Carrollton, TX 75006
1 (800) 441-1438 (Texas)
1 (800) 527-2701(outside Texas)

Western Region
200 F Twin Dolphin Drive
Redwood City, CA 94065
1 (800) 533-4075 (California)
1 (800) 548-4885(outside California)

0-201-95868-6 Printed in USA.

PRECALCULUS MATHEMATICS
A Graphing Approach   **Third Edition**

Franklin Demana and Bert K. Waits *The Ohio State University*

Stanley R. Clemens *Bluffton College*

with the assistance of

Tommy Eads *North Lamar Independent School District, Paris, Texas*

Gregory D. Foley *Sam Houston State University*

Pamela Giles and Diana Taggert *Jordan School District, Sandy, Utah*

Addison-Wesley Publishing Company

Menlo Park, California • Reading, Massachusetts • New York • Don Mills, Ontario
Wokingham, England • Amsterdam • Bonn • Sydney • Singapore • Tokyo
Madrid • San Juan • Paris • Seoul, Korea • Milan • Mexico City • Taipei, Taiwan

This book was produced with T$_E$X.

Library of Congress Cataloging-in-Publication Data

Demana, Franklin D. , 1938-
    Precalculus mathematics: a graphing approach / Franklin Demana,
Bert K. Waits, Stanley R. Clemens with the assistance of Tommy Eads,
Gregory D. Foley, Pamela Giles, Diana Taggert.—3rd ed., annotated
instructor's ed.
      p. cm.
   Includes index.
   ISBN 0-201-52905-X
   1. Algebra.   2. Trigonometry.   I. Waits, Bert K.   II. Clemens,
Stanley R.   III. Title.
QA154,2.D444   1993
512'.13'028541—dc20   92-44754
                                                  CIP

4 5 6 7 8 9 10–CRW–979695

Preface

Precalculus Mathematics: A Graphing Approach, grew out of our strong conviction that incorporating graphing technology into the precalculus curriculum better prepares students for further study in mathematics and science. Our own research at The Ohio State University and at dozens of other high school and college test sites shows that the use of a calculator- or computer-based graphing approach dramatically changes results in the classroom. Instead of being bored and discouraged by conventional contrived problems, students suddenly grow excited by their ability to explore problems that arise from real world situations and learn from their experiences. The mathematics classroom is transformed into a mathematics laboratory, with a new interactive instructional approach that focuses on problem solving. As a natural outgrowth of this excitement, students complete the course with a better understanding of mathematics and a solid intuitive foundation for calculus.

The Graphing Approach

As in the previous editions, this text is designed to be used in a one or two semester precalculus course. We take advantage of the power and speed of modern technology to apply a graphing approach to the course. The characteristics of this approach are described below.

Integration of Technology Use of a graphing utility—whether a hand-held graphing calculator or computer graphing software—is not optional. Technology allows the focus of the course to be on problem solving and exploration, while building a deeper understanding of algebraic techniques. Students are expected to have regular and frequent access to a graphing utility for class activities as well as homework.

Problem Solving   The ultimate power of mathematics is that it can be used to solve problems. Technology removes the need for contrived problems and opens the door for realistic and interesting applications. Throughout this text, we focus on what we call problem situations—situations from the physical world, from our social environment, or from the quantitative world of mathematics. Using real life situations makes the math understandable to the students, and students come to value mathematics because they appreciate its power.

Throughout this text, we use a three step problem solving process. Students will be asked to:

1. Find an algebraic representation of the problem;

2. Find a complete graph of the algebraic representation; and

3. Find a complete graph of the problem solving situation.

These three steps prepare the student to find either a graphical or an algebraic solution to the problem. Problem situations are highlighted in the exercise sets, and we encourage students to complete all the exercises which deal with that problem. See page 74, exercises 143–147.

Multiple Representations   A quantitative mathematical problem can often be approached using multiple representations. In a traditional precalculus course, problems are analyzed using an algebraic representation, and perhaps a numerical representation. However, modern technology allows us to take full advantage of a graphical, or geometric, representation of a problem. Our understanding of the problem is enriched by exploring it numerically, algebraically, and graphically. See pages 84 and 85.

Exploration   We believe that a technology-based approach enriches the student's mathematical intuition through exploration. With modern technology, accurate graphs can be obtained quickly and used to study the properties of functions. Students learn to decide for themselves what technique should be used. The speed and power of graphing technology allows an emphasis on exploration. See page 45.

Geometric Transformations   The exploratory nature of graphing helps students learn how to transform a graph geometrically by horizontal or vertical shifts, horizontal or vertical stretches and shrinks, and reflection with respect to the axes. This develops students' abilities so that they can sketch graphs of functions quickly and understand the behavior of graphs. See page 152.

Foreshadowing Calculus   We foreshadow important concepts of calculus through an emphasis on graphs. Using graphs, students can find maxima and minima of functions, and intervals where functions are increasing or decreasing and limiting behavior of functions are determined graphically. We do not borrow the techniques of calculus—rather we lay the foundation for the later study by providing students with rich intuitions about functions and graphs. See page 147.

Approximate Answers   Technology allows a proper balance between exact answers that are rarely needed in the real world and accurate approximations. Graphing techniques such as zoom-in provide an excellent geometric vehicle for discussion about error in answers. Students can read answers from graphs with accuracy up to the limits of machine precision. See page 80.

Visualization   Graphing helps students to gain an understanding of the properties of graphs and makes the addition of geometric representations to the usual numeric and algebraic representations very natural. Exploring the connections between graphical representations and problem situations deepens student understanding about mathematical concepts and helps them appreciate the role of mathematics.

About the Third Edition

This third edition of *Precalculus Mathematics: A Graphing Approach* grew out of the experiences of hundreds of classrooms. We have carefully listened to comments and suggestions of both teachers and students, and incorporated them fully into the text. The entire text has been extensively revised and rewritten.

Development of Functions   In this edition a presentation of functions including operations on functions, composition of functions, graphs of functions, and transformations applied to graphs of functions is the major focus of Chapter 1. Finding zeros of functions as a way of solving equations and solving applied problems is the focus of Chapter 2. Parametric equations are also introduced in Chapter 2 as a way of defining a relation. Parametric equations are then used to define inverse functions as a special type of relation. Polynomial functions and their graphs remain in a separate chapter (Chapter 3).

Rational Functions   Coverage of rational functions and functions involving radicals (Chapter 4) has been streamlined from seven sections to four.

Trigonometry Chapters   The three chapters on trigonometry (Chapters 6–8) have been developed to make concepts even more accessible, and now contain an even greater emphasis on graphing. The trigonometric functions are introduced in terms of the unit circle rather than in terms of right triangles as in the previous editions. A more complete development of trigonometric identities and solving trigonometric equations has been included since students need to practice these skills to be successful in calculus.

Parametric Equations and Polar Coordinates   These topics are now covered in a separate chapter (Chapter 9) together with conic sections. The material on conics is treated in two sections. Topics from this chapter may be incorporated earlier if the instructor chooses.

Systems of Equations   Sections on solving systems of equations algebraically and graphically are now combined with material on matrices. The concept of solving systems of equations graphically is first introduced in Chapter 3. Chapter 10 focuses on systems of equations and the use of matrices to solve them. Also in

Chapter 10 matrices are used to describe rotations and rotation images of conic sections.

Topics in Discrete Mathematics   Chapter 11 focuses on a variety of topics from discrete mathematics-sequences and series, the binomial theorem, mathematical induction, and permutations and combinations. It also includes a discussion of some topics from three dimensional geometry.

Topics in Statistics   This third edition includes new coverage on a variety of statistics topics. Chapter 12 includes a discussion of probability, stem-and-leaf tables, histograms, box-and-whisker plots, scatter plots, and line graphs, mean, mode, median, variance, and standard deviation, and curves of best fit.

Algebra Skills   A new appendix was added that provides a brief review of concepts from intermediate algebra. This algebra review illustrates how a grapher can be used to support algebraic skills.

Features

New pedagogical features have been incorporated into this text. It is our hope that these features will make the text a stronger teaching and learning tool. The pedagogy now includes:

Explore with a Graphing Utility   This recurring box places the student in the role of participant in the development of the mathematics. By introducing topics through this experience-based process, the book literally interacts with the student. Students develop their critical thinking skills, and form generalizations about the behavior of functions.

Sidelight Boxes   Shaded boxes placed in the margin provide commentary on the mathematical development, and include problem solving tips, calculator hints, and reminders.

Color   A functional use of color has been introduced to help the reader better navigate through the text. In addition to using color to mark beginnings and ends of examples, and to identify definitions and theorems, the text uses color in the artwork to help the student correctly identify the concept being illustrated.

Artwork   We have made a visual distinction between graphs generated with a graphing utility and hand-sketched art. Art which has been derived from a grapher is outlined with a colored box; while the graphs have been drawn more smoothly we still try to emulate what the student sees on the grapher. Traditional art uses color within the graphs but is not boxed. The distinction between types of artwork underscores the difference between a sketch and a grapher-drawn complete graph.

Examples   As in the previous editions, we have included many examples to develop the concepts. Titled examples help the student focus on the purpose of the example.

Exercises   We have closely focused on correlating end of section exercises to examples, and added many new exercises. Writing to Learn and Discussion exercises have also been included in nearly every exercise set.

Acknowledgments

We would like to thank the many wonderful teachers who participated in the development of this text. Their dedication and enthusiasm made this revision possible. Creative suggestions for this edition came from so many of our family of teachers that it would be impossible to name them all here. However, we particularly want to thank the reviewers of this edition and the participants of focus groups:

Carl Arvendsen, *Grand Valley State University*; Elinor Berger, *Columbus College*; David Jaspers, *University of South Carolina—Aiken*; Samuel Buchanan, *University of Central Arkansas*; Donald Davis, *Ohio State University—Newark*; Phil DeMarois, *William Rainey Harper College*; Yuanan Diao, *Kennesaw State College*; Patricia Ernst, *St. Cloud State University*; John Fink, *Kalamazoo College*; Marge Friar, *Grand Valley State University*; Jelena Gill, *Michigan State University*; Margaret Greene, *Florida Community College at Jacksonville*; Dwight Horan, *Wentworth Institute*; Keith Kuchar, *Northern Illinois University*; Mercedes McGowen, *William Rainey Harper College*; Michael Perkowski, *University of Missouri at Columbia*; Hal Schoen, *University of Iowa*; Eddie Warren, *University of Texas*; Steven Wilkinson, *Northern Kentucky University*; Howard Wilson, *Oregon State University*.

We thank our colleagues, Greg Foley and Alan Osborne, for their constructive suggestions. Special thanks are due to Penny Dunham, who coordinated the revision of the answer section and managed a team of teachers to solve and check the answers. We would like to thank the entire team for their valued contribution: Arne Engebretson, Babs Merkert, Tommy Eads, Pam Giles, Ray Barton, Karen Longhart, Betty Roseborough, and Jeri Nichols. We also thank Gerald White, of Western Illinois University, and Gloria Dion, of Penn State—Ogontz Campus, for their assistance in checking the mathematical accuracy of the manuscript.

Bert K. Waits
Franklin Demana
Stan Clemens

Contents

CHAPTER 1

Functions and Graphs

1.1

Real Numbers and the Coordinate Plane

The set of numbers used most frequently in algebra is known as the **real numbers**. Real numbers are either **rational** or **irrational**. The set of **integers** is a subset of the set of rational numbers, since for each integer a, $a = a/1$ (see Fig. 1.1). Following are some examples:

$$\text{integers:} \qquad 3, \qquad -5, \qquad 48$$

$$\text{rational numbers:} \qquad \frac{3}{8}, \qquad -\frac{2}{3}, \qquad \frac{22}{7}$$

$$\text{irrational numbers:} \qquad \pi, \qquad \sqrt{2}, \qquad \sqrt{17}$$

When represented as decimal numbers, integers have all zeros to the right of the decimal point, rational numbers always have a block of digits that repeat, and irrational numbers have no repeating blocks of digits. Some examples of rational and irrational numbers are

$$\pi = 3.141592654\cdots, \quad \tfrac{5}{8} = 0.625, \quad \sqrt{2} = 1.414213562\cdots, \quad \tfrac{1}{3} = 0.33333\cdots.$$

Because a calculator or computer display of a decimal number can show only a finite number of digits, usually 7 to 10, many displays represent only approxima-

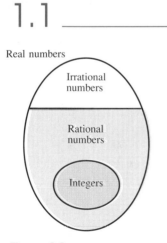

Real numbers

Figure 1.1 All integers are rational numbers, and all rational numbers are real numbers. A real number is either rational or irrational but not both.

π

3.141592654

Figure 1.2   When an irrational number is keyed into a calculator, the digits to the right of the ten digits displayed and the three additional check digits are usually interpreted to be zero. Thus the number displayed is a rational number approximation to the exact value of the number. What does your calculator give for π?

Objective

Students will be able to make connections between algebraic concepts related to the real number line, the coordinate plane, and geometric graphs by using technology.

Key Ideas

Properties of real numbers
Real number line
Absolute value
Distance between points on a line
Cartesian coordinate system
Graphical representation
Graphing utility
Viewing rectangle
Pixel
Screen coordinates
Distance between points in the coordinate plane

tions to the number keyed in. However, they are very accurate approximations (see Fig. 1.2).

Arithmetic Operations

There are four binary operations for numbers: addition, subtraction, multiplication, and division, represented by the symbols $+$, $-$, $\times$ (or $\cdot$), $\div$, respectively.

Addition and multiplication satisfy a set of properties that can be used to change the form of a mathematical expression into an equivalent form. These properties are summarized as follows.

Real-number Properties of Addition and Multiplication

Let a, b, and c represent real numbers. Then all of the following are true:

	Addition	*Multiplication*
Closure:	$a + b$ is real.	$a \cdot b$ is real.
Commutative:	$a + b = b + a$	$a \cdot b = b \cdot a$
Associative:	$a + (b + c) = (a + b) + c$	$a \cdot (b \cdot c) = (a \cdot b) \cdot c$
Identity:	$a + 0 = 0 + a = a$	$a \cdot 1 = 1 \cdot a = a$
Inverse:	$a + (-a) = (-a) + a = 0$	$a \cdot \left(\dfrac{1}{a}\right) = \left(\dfrac{1}{a}\right) \cdot a$ $= 1 \quad (a \neq 0)$
Distributive:	$a(b + c) = ab + ac$	

An **equation** is a statement of equality between two expressions. For example, $3 + 4$ and $2 + 5$ both represent 7. Therefore we can write the equation $3 + 4 = 2 + 5$. We can use the commutative property of addition to conclude that for any real number x, $x + 5 = 5 + x$. In solving problems involving equations, three important properties of equality will be useful.

Real-number Properties of Equality

Let a, b, and c be real numbers. Then the following properties are true:

Reflexive: $a = a$.
Symmetric: If $a = b$, then $b = a$.
Transitive: If $a = b$ and $b = c$, then $a = c$.

E X A M P L E   1   Using Properties of Real Numbers

If x is any real number, show that $2 \cdot (x + 3) = 6 + 2 \cdot x$.

Solution

$$2 \cdot (x + 3) = 2 \cdot x + 2 \cdot 3 \quad \text{Distributive property}$$

$$2 \cdot x + 2 \cdot 3 = 6 + 2 \cdot x \qquad \text{Commutative property}$$

so $\quad\quad 2 \cdot (x + 3) = 6 + 2 \cdot x. \qquad \text{Transitive property of equality}$ ≣

Real-number Line

The set of all real numbers is often represented as points on a line (see Fig. 1.3). To construct a **coordinate system** on a line, draw a line and label one point 0. This point is called the **origin**. Then mark equally spaced points on each side of 0. Label points to the right of zero 1, 2, 3, ... and to the left of zero $-1, -2, -3, \ldots$. The properties of *order* in Definition 1.1 describe the placement of all other numbers on the line. Figure 1.3 is called a **real-number line**.

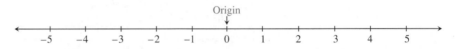

Figure 1.3   Each real number corresponds to one and only one point on the number line, and each point on the number line corresponds to one and only one real number.

The number associated with a point P is called the **coordinate** of point P.

Figure 1.4 a is less than b.

Definition 1.1   Order on the Real-number Line

If a and b are any two real numbers, then **a is less than b** if $b - a$ is a positive number. In this case, a is to the left of b on the real-number line (see Fig. 1.4). This order relation is denoted by the **inequality $a < b$**. In all, there are four inequality symbols that express order relationships, as follows:

$$a < b \quad a \text{ is less than } b.$$

$$a \leq b \quad a \text{ is less than or equal to } b.$$

$$a > b \quad a \text{ is greater than } b.$$

$$a \geq b \quad a \text{ is greater than or equal to } b$$

E X A M P L E   2   Solving Inequalities

On a real-number line, draw all numbers that are solutions to the following inequalities:

a) $x \leq 3$

b) $-3 \leq x < 2$

Solution

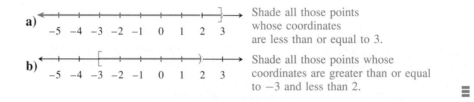

a) Shade all those points whose coordinates are less than or equal to 3.

b) Shade all those points whose coordinates are greater than or equal to -3 and less than 2.

Note in the number lines above the use of a square bracket to show inclusions and a rounded parenthesis to show exclusions of particular endpoints.

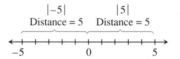

Figure 1.5   Both -5 and 5 are a distance of 5 from 0.

The numbers 5 and -5 are the same distance from zero (0) on the number line (see Fig. 1.5). So are π and $-\pi$ and $\sqrt{2}$ and $-\sqrt{2}$. To communicate this idea of distance from the origin of the coordinate line, we introduce a symbol. The **absolute value** of a number c, denoted $|c|$, represents the distance of c from zero on the real-number line. For example, $|-3| = 3$. A more formal definition follows.

Definition 1.2   Absolute Value

If a is a real number, then the **absolute value of a** is given by

$$|a| = \begin{cases} a & \text{if } a \geq 0 \\ -a & \text{if } a < 0. \end{cases}$$

Teaching Note

It is sometimes helpful for students to see the distance between two points by plotting the points on a number line. As in Example 3, students will plot the point for 2 on the number line and then plot the point representing $\sqrt{3}$, that is, somewhere between 1 and 2. If a and b are the two points to be plotted on the number line, then the distance always can be written without absolute value notation as $b - a$ if $b > a$. Since 2 is to the right of $\sqrt{3}$ on the number line, $2 - \sqrt{3}$ is the correct representation.

Distance $= |-2 - (-9)| = 7$

$$-9\ -8\ -7\ -6\ -5\ -4\ -3\ -2$$

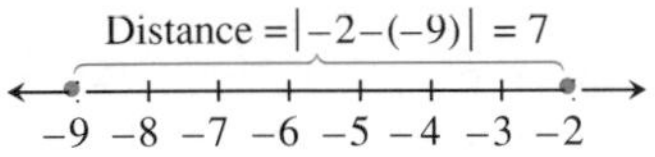

Figure 1.6 Distance between -2 and -9.

EXAMPLE 3 Evaluating Absolute Value

Write the expression $|\sqrt{3} - 2|$ without using absolute value notation.

Solution

$$|\sqrt{3} - 2| = -(\sqrt{3} - 2) \quad \text{Because } \sqrt{3} \text{ is less than 2, } \sqrt{3} - 2 \text{ is negative}.$$

We can use absolute value notation to describe the distance between any two points on the real-number line. Notice that the distance between the points with coordinates 3 and 8 is $|3 - 8|$.

Definition 1.3 Distance between Points on a Line

Suppose A and B are two points on a real-number line with coordinates a and b, respectively. The **distance between A and B**, denoted $d(A, B)$, is given by

$$d(A, B) = |a - b|.$$

EXAMPLE 4 Finding the Distance between Points on a Real-number Line

Find the distance between the points with the coordinates -2 and -9.

Solution (See Fig. 1.6.)

$$d(-2, -9) = |-2 - (-9)| = |7| = 7 \quad \text{Notice that } d(-2, -9) = d(-9, -2).$$

Notice in the solution to Example 4 that -2 and -9 are used as both the coordinates and the names of points on the number line. It is common to simplify notation this way.

Cartesian Coordinate System

Just as each point on a real-number line is associated with a real number, each point in a plane is associated with an ordered pair of real numbers. To determine which pairs of numbers get associated with which points, we use the **Cartesian coordinate system** (also called the **rectangular coordinate system**). To construct a coordinate system, draw a pair of perpendicular real-number lines, one horizontal and the other vertical, with the lines intersecting at their respective origins (see Fig. 1.7). The horizontal line is usually called the **x-axis** and the vertical line is usually called the **y-axis**. The positive direction on the x-axis is to the right, and the positive direction on the y-axis is up.

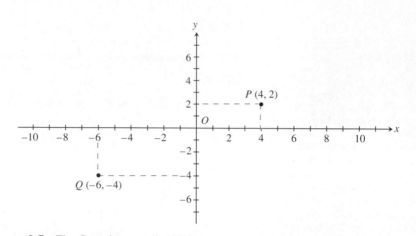

Figure 1.7   The Cartesian coordinate plane.

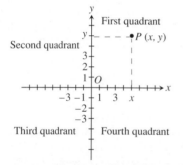

Figure 1.8   The four quadrants.

The intersection of the two coordinate axes, point O, is called the **origin**. The coordinate axes divide the plane into four **quadrants**, as shown in Fig. 1.8.   Each point P of the plane is associated with an **ordered pair** (x, y) of real numbers called the **coordinates** of the point. The x-**coordinate** represents the directed distance from the y-axis to P, and the y-**coordinate** represents the directed distance from the x-axis to P.

Graphing calculators and graphing software for personal computers can be used to generate **graphical representations** of the Cartesian coordinate plane. In this text, the terms **grapher** and **graphing utility** will be used interchangeably to refer to a graphing calculator or a desktop computer with graphing software.

Bert says, "Be sure that students understand the difference between screen coordinates and Cartesian coordinates of points on the graph."

E X A M P L E   5   Displaying the Coordinate Plane on a Grapher

Display the coordinate plane on your grapher.

Solution   Use the keystrokes appropriate to a grapher to show the coordinate plane. Use appropriate keys to move the cursor up and down and to the right and left. As the cursor moves around, note that the displayed **screen coordinates** of the point represented by the cursor also change.

It is important to understand how graphers work. Only a portion of the entire plane can be displayed. The portion that is displayed at any one time is called a **viewing rectangle**. In particular, the viewing rectangle [Xmin, Xmax] by [Ymin, Ymax] is the set of all points in the plane satisfying the relationships $X\text{min} \leq x \leq X\text{max}$ and $Y\text{min} \leq y \leq Y\text{max}$ (see Fig. 1.9).

Frank says, "The mathematics of the viewing screen itself is very rich and students can uncover mathematical concepts through exploration of the process by which a graphing utility produces the graph of a function."

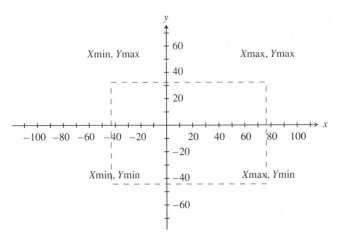

The viewing rectangle [Xmin, Xmax] by [Ymin, Ymax].

Figure 1.9

You can change the size of a viewing rectangle by giving the grapher the appropriate commands. The coordinates of the points (x, y) in a particular viewing rectangle satisfy the following inequalities:

$$X\text{min} \leq x \leq X\text{max} \qquad \text{and} \qquad Y\text{min} \leq y \leq Y\text{max}.$$

Notice the large "gaps" in the screen coordinates as the cursor moves point by point. On one model grapher, the origin $(0, 0)$ has screen coordinates $x = 0.10526316$ and $y = 0.15873016$ for the viewing rectangle $[-10, 10]$ by $[-10, 10]$. If the viewing rectangle is changed, the screen coordinates of the origin change. Can you figure out why?

GAPS IN SCREEN COORDINATES

On a graphing utility display, a **pixel** is the smallest dot that can be lighted. One model grapher uses 96 columns of pixels and 64 rows of pixels. That means that for the $[-10, 10]$ by $[-10, 10]$ viewing rectangle, the "gap" in the horizontal screen coordinates will be $(10 - (-10)) \div 95 = 0.2105$ and the "gap" in the vertical screen coordinates will be $(10 - (-10)) \div 63 = 0.3175$. (In a numerical situation, it is understood that $=$ means approximately equals.)

EXAMPLE 6   Changing the Viewing Rectangle

Set your grapher for the viewing rectangle: $[-20, 20]$ by $[-20, 20]$. Display the coordinate plane and move the cursor to the bottom-left and top-right corners of the screen to verify the coordinates of these points are correct. Approximate the widths of the "gaps" in both of the screen x- and y-coordinates for the viewing rectangle $[-20, 20]$ by $[-20, 20]$ on your grapher.

Solution   Set the range screen of a graphing utility as follows:

$$X\text{min} = -20$$

$$X\text{max} = 20$$

Notes on Exercises

Ex. 22–25 focus upon the reasonable spacing of tick marks. Have students observe the changes in the coordinate readout as they trace from pixel to pixel on their grapher screens. The concept developed in these problems is useful in Ex. 46–48 and 51.

$$Xscl = 1$$

$$Ymin = -20$$

$$Ymax = 20$$

$$Yscl = 1$$

For the viewing rectangle $[-20, 20]$ by $[-20, 20]$ on one model grapher, the x-gap $= 0.42$ and the y-gap $= 0.63$.   ▤

E X A M P L E   7   Adjusting the *X*scl and *Y*scl values

How does your grapher's viewing rectangle change if you change the range values used in Example 6 so that $Xscl = 5$ and $Yscl = 5$?

Solution   The size of the viewing rectangle does not change. The tick marks on the coordinate axes are five units apart instead of one unit apart.   ▤

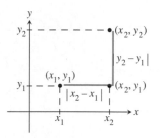

Figure 1.10 Finding the distance between points on a vertical or a horizontal line.

Distance between Two Points in the Plane

To find the distance between two points on a vertical or horizontal line in the coordinate plane, use the method illustrated in Fig. 1.10 and described in Example 4.

E X A M P L E   8   Finding the Distance between Points on a Horizontal Line

Find the distance between the following pairs of points:

a) $P(1, 3)$ and $Q(5, 3)$

b) $P(1, 3)$ and $R(1, 4)$

Solution   (See Fig. 1.10.)

a) $d(P, Q) = |5 - 1|$ Because the y-coordinates of these points are equal, find this distance using only the x-coordinates.

$\qquad\qquad\quad = 4$

b) $d(P, R) = |3 - 4|$ Because the x-coordinates of these points are equal, find this distance using only the y-coordinates.

$\qquad\qquad\quad = 1$ ▤

Consider the general situation of two points $P(x_1, y_1)$ and $Q(x_2, y_2)$, and let the point $R(x_2, y_1)$ be the vertex of a right triangle, $\triangle PRQ$. Then $d(P, R) = |x_1 - x_2|$ and $d(R, Q) = |y_1 - y_2|$ (see Fig. 1.11).

Common Errors

Students tend to think that setting the scale on a graphing utility affects the way a graph will be represented on the viewing screen. For Example 7, have students set the Xscl and Yscl values to 1. Then, when they press *graph* they will see two "fat" axes. Point out that the Xscl and Yscl values tell you how far the tick marks will be apart. Use the example of a fence that is 40 feet long. If fence posts are placed five feet apart in a straight line along the 40 feet of fence, then only 9 posts will be needed. On the grapher, the "fence posts" will be far enough apart to distinguish the tick marks clearly.

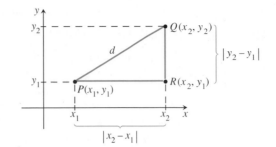

Figure 1.11 $d(P, Q)^2 = d(P, R)^2 + d(R, Q)^2$.

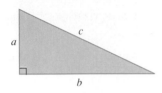

Figure 1.12   Recall the **Pythagorean theorem**: $a^2 + b^2 = c^2$.

Apply the Pythagorean theorem (see Fig. 1.12) to find the distance $d(P, Q)$.

$$d(P, Q) = \sqrt{|x_1 - x_2|^2 + |y_1 - y_2|^2}$$

By replacing $|x_1 - x_2|^2$ and $|y_1 - y_2|^2$ with the equivalent expressions $(x_1 - x_2)^2$ and $(y_1 - y_2)^2$, we arrive at the following general rule.

Distance Formula

The distance $d(P, Q)$ between points $P(x_1, y_1)$ and $Q(x_2, y_2)$ in the coordinate plane is

$$d(P, Q) = \sqrt{(x_1 - x_2)^2 + (y_1 - y_2)^2}.$$

E X A M P L E   9   *Finding the Distance between Two Points in a Plane*

Find the distance between the points $P(1, 5)$ and $Q(6, 2)$.

Solution

$$
\begin{aligned}
d(P, Q) &= \sqrt{(1 - 6)^2 + (5 - 2)^2} \quad \text{Apply the distance formula.}\\
&= \sqrt{(-5)^2 + 3^2}\\
&= \sqrt{25 + 9}\\
&= \sqrt{34}
\end{aligned}
$$

Midpoint

The coordinates of the midpoint of a segment can be found if the coordinates of the endpoints are known.

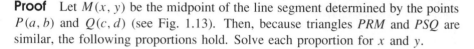

> ### Midpoint Formula
>
> The **midpoint** of the line segment with endpoints (a, b) and (c, d) is the point with coordinates $\left(\dfrac{a+c}{2}, \dfrac{b+d}{2}\right)$.

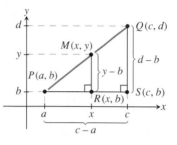

Figure 1.13 Proof of the midpoint formula.

Proof   Let $M(x, y)$ be the midpoint of the line segment determined by the points $P(a, b)$ and $Q(c, d)$ (see Fig. 1.13). Then, because triangles *PRM* and *PSQ* are similar, the following proportions hold. Solve each proportion for x and y.

$$\frac{x - a}{c - a} = \frac{1}{2} \qquad\qquad \frac{y - b}{d - b} = \frac{1}{2}$$

$$x - a = \frac{c - a}{2} \qquad\qquad y - b = \frac{d - b}{2}$$

$$x = \frac{a + c}{2} \qquad\qquad y = \frac{b + d}{2} \qquad\qquad \blacksquare$$

An easy way to remember the midpoint formula is to observe that the coordinates of the midpoint are found by averaging the x- and y-coordinates of the two points.

E X A M P L E   10   Finding Midpoint Coordinates

Find the midpoint of the segment with endpoints $(-5, 2)$ and $(3, 7)$.

Solution   From the midpoint formula, we have

$$x = \frac{-5 + 3}{2} \qquad\qquad y = \frac{2 + 7}{2}$$

$$x = -1 \qquad\qquad y = 4.5.$$

The midpoint is $(-1, 4.5)$. $\qquad\qquad\blacksquare$

E X A M P L E   11   Showing That Diagonals of a Parallelogram Bisect Each Other

Show that the diagonals of a parallelogram bisect each other.

Solution   Draw a diagram of a parallelogram. Recall from your study of geometry that opposite sides of a parallelogram are parallel. Use this fact to assign a coordinate system to your figure so that the vertices of the parallelogram are $(0, 0)$, (a, b), $(a + c, b)$, and $(c, 0)$ (see Fig. 1.14).

Notes on Exercises

Ex. 35–37 are "show" and "prove" problems. A coordinate approach to proof is used here rather than a synthetic approach as in many geometry books. You may want students to work in cooperative groups on these problems.

Ex. 51 is the first in a series of Writing to Learn exercises. Some teachers have students keep journals, and these exercises can be used for journal writing. Be sure to call upon a few students to read their answers to this exercise.

Assignment Guide

Day 1: Ex. 2, 3, 8, 11, 14, 19, 28, 31, 34–37, 39, 43, 46, 51

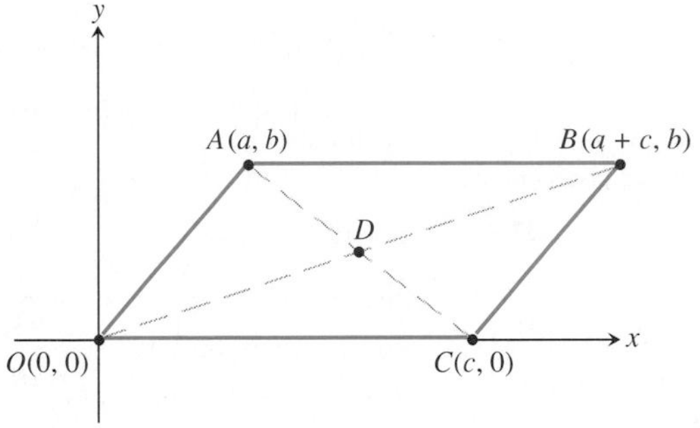

Figure 1.14 Parallelogram $OABC$.

Then
$$\text{midpoint of } OB = \left(\frac{0+a+c}{2}, \frac{0+b}{2}\right) = \left(\frac{a+c}{2}, \frac{b}{2}\right),$$
$$\text{midpoint of } AC = \left(\frac{a+c}{2}, \frac{b+0}{2}\right) = \left(\frac{a+c}{2}, \frac{b}{2}\right).$$

Because these midpoints are the same point, conclude that the diagonals of parallelogram $OABC$ bisect each other. ▤

Exercises for Section 1.1

In Exercises 1–4, identify the real-number properties that explain why each are valid equations.

1. $(3 + \sqrt{2}) + 7 = 3 + (\sqrt{2} + 7)$

2. $2(\sqrt{3} + 4) = 2 \cdot \sqrt{3} + 8$

3. $2 + (x + 5) = (5 + x) + 2$

4. $\frac{1}{2} \cdot (4 \cdot x) = \left(\frac{1}{2} \cdot 4\right) \cdot x = 2x$

In Exercises 5–10, draw on the real-number line all numbers that are solutions to the inequalities.

5. $-2 \le x < 3$ **6.** $4 \le x \le 5$

7. $5 \ge x$ **8.** $-3 < x < -1$

9. $x \le \sqrt{2}$ **10.** $-4 \le x \le 5$

In Exercises 11–14, write an exact value for each expression without using absolute value bars.

11. $|-8|$ **12.** $|4 - 9|$

13. $|5 - \sqrt{3}|$ **14.** $|2 - \sqrt{5}|$

In Exercises 15–20, find the distance between the points on the real-number line with each of the given coordinates.

15. $5, \ 17$ **16.** $-3, \ 4$

17. $-7, \ -3$ **18.** $2, \ -6$

19. $x, \ 4$ **20.** $x, \ -3$

21. Graph the points in (a) to (f) on the same coordinate plane. Identify the quadrant containing each point.

    a) $(2, 4)$ **b)** $(0, 3)$ **c)** $(3, 0)$

    d) $(-1, -4)$ **e)** $(-2, 3)$ **f)** $\left(-\frac{1}{2}, -5\right)$

In Exercises 22–25, set your grapher for the given viewing rectangles. In each case, move the cursor to the bottom-left and top-right corners of the viewing rectangle to confirm that you have the correct rectangle. Also, set the Xscl and Yscl on the range menu so that the tick marks are reasonably spaced on each axis.

22. $[-10, 20]$ by $[-20, 10]$

23. $[-5, 5]$ by $[-3, 4]$

24. $[-100, 100]$ by $[-50, 50]$

25. $[-20, 20]$ by $[-2, 2]$

In Exercises 26–29, find the distance between the points in the plane with the following coordinates:

26. $(0, 0)$; $(3, 4)$ **27.** $(-1, 2)$; $(2, 3)$

28. $(-3, -2)$; $(6, -2)$ **29.** $(-4, 6)$; $(2, 3)$

30. Show that the points $(1, 3)$, $(4, 7)$, and $(8, 4)$ are the vertices of an isosceles triangle.

In Exercises 31–34, find the midpoint of the line segment determined by the pair of points.

31. $(-1, 3)$; $(5, 9)$ **32.** $(2, -3)$; $(-1, -1)$

33. $(3, \sqrt{2})$; $(6, 2)$ **34.** (a, b); $(3, -6)$

35. Let $A = (3, 0)$, $B = (-1, 2)$, and $C = (5, 4)$. Prove that the triangle determined by A, B, and C is isosceles but not equilateral.

36. Show that the midpoint of the hypotenuse of a right triangle with vertices $(0, 0)$, $(5, 0)$, and $(0, 7)$ is equidistant from all three vertices.

37. Prove that the midpoint of the hypotenuse of a right triangle is the same distance from each of the three vertices.

38. Let $A = (-1, 3)$ and $B = (9, 6)$. Find the coordinates of the two points on the line AB that divide the segment AB into three equal segments.

39. Let $A = (-1, 3)$ and $B = (9, 6)$. Find the coordinates of the three points on the line AB that divide the segment AB into four equal segments.

In Exercises 40–43, write an expression that gives the distance between the given points in a coordinate plane.

40. (a, b), $(2, 3)$ **41.** $(0, y)$, $(4, 0)$

42. $(0, y)$, $(-4, 0)$ **43.** $(x, 0)$, $(-1, 3)$

In Exercises 44 and 45, find x so that the distance between the given points is 10.

44. $(x, 5)$, $(1, -3)$ **45.** $(2, -1)$, $(x, 7)$

46. Suppose the viewing rectangle on a grapher is $[-20, x]$ by $[-20, y]$. What are the values of x and y if each pixel represents "one unit"? (Answers may vary depending on the type of grapher being used.)

47. Find a viewing rectangle for your grapher for which a single move of the cursor causes a change, or gap, of two units.

48. Set the viewing window of your grapher so that the coordinate axes are centered (or nearly centered) and such that the cursor, as you move it around the screen, always has integer values.

49. Show that $|a| \geq 0$.

50. Show that $|a \cdot b| = |a| \cdot |b|$.

51. **Writing to Learn.** Write several sentences that explain the differences between the *coordinates of points* in the Cartesian coordinate plane (math coordinates) and the *screen coordinates* in a computer graphical representation of the coordinate plane. Give an example of a point that lies in a particular viewing rectangle whose (x, y) math coordinates do not appear as screen coordinates. (*Hint:* First consider Exercises 46 and 47.)

1.2 _____ Graphing Utilities and Complete Graphs

Objective

Students will be able to apply problem-solving strategies in problem situations and use algebraic and geometric models as tools to solve problems.

Problem Solving

Mathematics is used frequently as a tool for solving scientific and economic problems.

George Pólya is sometimes called the father of modern problem solving because of his significant writing and analysis of the mathematical problem-solving process. Born in Hungary in 1887, Pólya completed his Ph.D. at the University of

Budapest. In 1940 he went to Brown University and then in 1942, joined the faculty at Stanford University. He died at the age of 97 in 1985. His four-step process continues to be valid and helpful in this age of computers and graphing utilities.

In Step 1 of this four-step process, you might need to do all of the following:

- Read the problem as stated, several times if necessary.
- Restate the problem in your own words.
- Write down a statement or variable name that identifies what the problem asks you to find or solve for.
- Identify clearly what information is given and what you need to find. Many people find it helpful to discuss their understanding of a problem out loud.

In Step 2, you devise a plan for solving the problem. Your plan will usually consist of identifying which strategy or combination of strategies you can use to solve the problem. Each of the strategies listed in the margin suggests a **problem-solving process**. Successful problem solvers learn how to match a strategy or several strategies with the problem that needs to be solved.

In Step 3, you carry out the plan that has been devised while completing Steps 1 and 2. Sometimes Step 3 will take less time than a careful completion of Steps 1 and 2. *Most beginning problem solvers skip over Steps 1 and 2 and begin pushing a pencil or punching a calculator too soon. Taking time to follow Steps 1 and 2 will generally lead to more success with Step 3.*

In Step 4, you look back and analyze the process you have completed. By reflecting on the relationship between this problem and others you have solved in the past, you may see a pattern. Then you identify which characteristics of the problem might provide clues for strategies most likely to succeed as you encounter similar problems in the future.

EXAMPLE 1   Observing the Pattern and Generalizing

PROBLEM-SOLVING STRATEGIES

- Use a variable
- Complete a table
- Consider a special case
- Look for a pattern
- Guess and test
- Draw a picture
- Draw a diagram
- Make a list
- Solve a simpler, related problem
- Use reasoning
- Solve an equivalent problem
- Work backward
- Solve an equation
- Look for a formula
- Use coordinates

A Problem Situation Consider this economic situation. Quality Rent-a-Car charges $15 plus $0.20/mi to rent a car. One can ask many questions about this situation that pose a mathematical problem. Consequently, we call this situation a **problem situation**.

Quality Rent-a-Car charges $15 plus $0.20/mi to rent a car. How much does Quality Rent-a-Car charge if a rented car is driven

a) 50 mi?

b) 75 mi?

c) 100 mi?

d) 200 mi?

e) x mi?

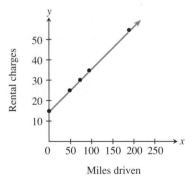

Stan says, "All of us have different problem-solving styles. I have found that most students are either *impulsive* or *reflective* problem solvers. The impulsive problem solver is a brainstormer and a risk taker. The reflective problem solver seems more intent on giving the right answer and may withdraw from class discussions. In using technology in the classroom, it is essential that the teacher observe the problem-solving style of each student and encourage all students to participate in class discussions."

Notes on Exercises

Ex. 1–6 provide opportunities for students to relate tabular, graphical, and algebraic representations.

Solution   Use the problem-solving strategies of "Look for a pattern" and "Use a variable."

a) $0.20(50) + 15 = \$25.00$ Cost/mi times 50 mi plus \$15

b) $0.20(75) + 15 = \$30.00$

c) $0.20(100) + 15 = \$35.00$

d) $0.20(200) + 15 = \$55.00$

Generalization   To answer part (e), for x miles the cost y is found by using the equation

$$y = 0.20 \cdot x + 15.$$

The equation $y = 0.2x + 15$ is called an **algebraic representation** or a **mathematical model** of the problem situation. The power of mathematics is that this model equation can now be used to find the cost (y) for any other number of miles driven (x).

Drawing a graph is one way to picture this relationship. Draw a pair of axes and label them as in Fig. 1.15. (Because distance is represented by positive values, only the positive axes appear.) Then plot the five points whose coordinates are the **data pairs** listed in the following table. (This kind of table is often called a **table of values** or **data table**.) These points satisfy the equation $y = 0.2x + 15$.

Figure 1.15   Relationship between miles driven and rental charges: $y = 0.2x + 15$.

x	0	50	75	100	200	x
y	15	25	30	35	55	$0.2x + 15$

Note that these points appear to lie on a line. Draw a ray beginning at point $(0, 15)$ and continuing through the other four points. This ray is a graphical representation of the Quality Rent-a-Car problem situation. To summarize, in a given problem situation there may be both an algebraic representation and a graphical representation of the problem situation.

In Example 1, we collected data and found an algebraic representation for the data. Next, we consider the converse situation, where an equation is used to generate data.

EXAMPLE 2   Using an Equation to Find Data

Referring back to Example 1, suppose Sarah is charged $50 for renting a car from Quality Rent-a-Car. How many miles did she drive the car?

Solution   **Algebraic Method** $y = 0.2x + 15$ and the cost y is 50.

$$50 = 0.2x + 15 \qquad \text{Substitute 50 for } y \text{ and solve the resulting equation for } x.$$

$$50 - 15 = 0.2x$$

$$\frac{35}{0.2} = x$$

so

$$x = 175 \qquad \text{Sarah drove 175 mi.}$$

Graphical Method   Apply the following steps:

1. Draw a graph of the model $y = 0.2x + 15$ (see Fig. 1.16).

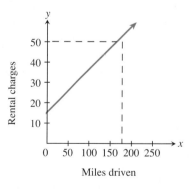

Figure 1.16   Finding the number of miles for a $50 fee.

2. Begin at the point $(0, 50)$ on the y-axis and move horizontally to the graph.

3. Read the x-coordinate of this point to find an *estimate* of the problem's solution.

Solutions found graphically will be approximations with the accuracy depending on the viewing rectangle used.

Bert says, "The beauty of using a graphing utility to solve problems is that we are not restricted to four or five values from the *x*-column of a table, and that we can actually see ordered pairs displayed at the bottom of a screen which change simply with the move of a cursor."

🔎 EXPLORE WITH A GRAPHING UTILITY

Use your graphing utility to graph $y = 0.2x + 15$.

Questions

1. Are you able to see the graph with your current viewing rectangle?
2. Find an appropriate viewing window and move the cursor as close as possible to the point $(0, 50)$. Are the coordinates of this point exactly $(0, 50)$?
3. After positioning the cursor near $(0, 50)$, move the cross hairs to the right until the cursor coincides with the graph. What approximate solution do you find?

When doing this Exploration, you may have discovered that the first viewing rectangle you chose was not large enough to include the point $(175, 50)$, which is needed to solve the problem graphically. The viewing rectangle $[0, 200]$ by $[0, 100]$ will contain the point $(175, 50)$. What viewing rectangle did you use?

This situation illustrates the importance of choosing an appropriate viewing rectangle when graphing an equation. What is appropriate in one situation may not be so in the next situation. Deciding which viewing rectangle to use is a judgment skill you need to use a graphing utility.

Related to this discussion is the concept of complete graph. A graph of an equation is a **complete graph** if it suggests all points of the graph and all of the important features of the graph. Once you find a viewing rectangle that shows a complete graph, a slightly larger or smaller viewing rectangle usually produces a complete graph also. Thus a given equation generally has *many* complete graphs.

The next example illustrates several of the kinds of things that can happen if the selected viewing rectangle does not show a complete graph.

Teaching Note

Have students discuss the notion of what is a *complete graph* after viewing several examples of linear functions, quadratic functions, cubic functions, and so on. After the class has agreed upon the attributes of a complete graph, have students write a definition of a complete graph in their notebooks or as a journal entry.

EXAMPLE 3 Finding a Complete Graph with a Grapher

Graph $y = x^2 + 10$ in each of the following viewing rectangles:

a) $[-5, 5]$ by $[-5, 5]$

b) $[-10, 10]$ by $[-10, 10]$

c) $[-10, 10]$ by $[-50, 50]$

Choose which viewing rectangle gives a complete graph.

Solution The best viewing rectangle is $[-10, 10]$ by $[-50, 50]$ (see Fig. 1.17). The viewing rectangle $[-5, 5]$ by $[-5, 5]$ includes no points of a graph, and the $[-10, 10]$ by $[-10, 10]$ rectangle includes only one point of a graph.

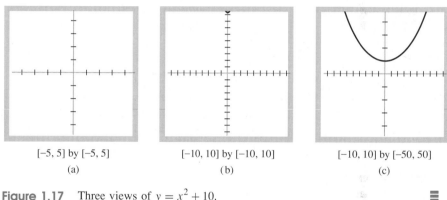

| $[-5, 5]$ by $[-5, 5]$ | $[-10, 10]$ by $[-10, 10]$ | $[-10, 10]$ by $[-50, 50]$ |
| (a) | (b) | (c) |

Figure 1.17 Three views of $y = x^2 + 10$.

STANDARD VIEWING RECTANGLE

The **standard viewing rectangle** is the rectangle $[-10, 10]$ by $[-10, 10]$. As you attempt to find a complete graph of an equation, begin with the standard viewing rectangle and change to an alternate viewing rectangle as needed.

The curve given by a complete graph of $y = x^2 + 10$ is called a **parabola**.

Notice that in Fig. 1.17(a, b), the scale marks on both coordinate axes are one unit apart. However, in Fig. 1.17(c), the scale marks on the x-axis are one unit apart, while the scale marks on the y-axis are 10 units apart. When you change the viewing rectangle on a grapher, it is often helpful also to change the scale marks on the two axes.

Many keys on your calculator can be used to generate a table of (x, y) data pairs that can be graphed. Consider for example the $\sqrt{x}$ key, where $\sqrt{x}$ stands for the nonnegative square root of x. This function key can be used to generate the pairs $(x, \sqrt{x})$. A picture of the graph of all these pairs can be obtained by graphing the equation $y = \sqrt{x}$.

EXAMPLE 4   Finding a Complete Graph

Find the graph of the equation $y = \sqrt{x + 20}$. Which of these rectangles shows a complete graph?

a) $[-10, 10]$ by $[-10, 10]$

b) $[-50, 50]$ by $[-10, 10]$

c) $[-10, 10]$ by $[-50, 50]$

GRAPHER SKILLS UPDATE

You should be able to

- graph an equation that begins $y =$,
- experiment with viewing rectangles until a complete graph can be determined, and
- use the trace feature to move along the graph.

Solution   (See Fig. 1.18.)

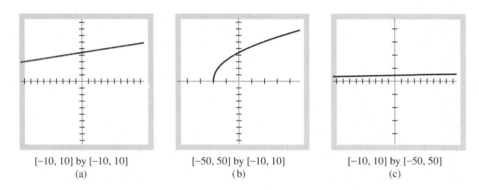

$[-10, 10]$ by $[-10, 10]$	$[-50, 50]$ by $[-10, 10]$	$[-10, 10]$ by $[-50, 50]$
(a)	(b)	(c)

Figure 1.18 Three views of $y = \sqrt{x + 20}$.

The viewing rectangle in Fig. 1.18(b) shows a complete graph, while those in Fig. 1.18(a,c) do not show a significant portion of the graph or properly indicate the behavior of the graph.   ≡

Examples 3 and 4 illustrate that finding a complete graph depends on choosing an appropriate viewing rectangle. It is important to realize that a given function does not have a unique, complete graph. Consequently, we speak of a complete graph rather than the complete graph of a function.

As you progress in this course, you will encounter occasions when you cannot find a complete graph in a single viewing rectangle. You may need to use a large rectangle to obtain a global view and then to zoom in on some small piece of the graph to see necessary detail.

On the other hand, when you are collecting and plotting data points for a hand-drawn graph, the task of finding a complete graph is different. Here, the question may be, what is the correct way to connect the data points?

Because the four points graphed in the Quality Rent-a-Car problem fell on a straight line, it was relatively easy to determine the graph. When the points do not lie on a straight line, a hand-drawn graph is less reliable.

FINDING AN APPROPRIATE VIEWING RECTANGLE

Choosing an appropriate viewing rectangle is sometimes tricky. Here's a suggestion. Notice that when the trace cursor moves off the selected screen, the x- and y-coordinates of points on the graph continue. Use this feature to help select an appropriate viewing rectangle.

Frank says, "Explorations of problem situations using a graphing utility have led many of our students to pursue analytical approaches using pencil-and-paper methods."

Special Agreement about Graphs

This text illustrates graphs produced by hand and on a grapher. Note that figures of grapher-produced graphs (for example, Figs. 1.17 and 1.18) have a color border, while those representing hand-drawn graphs (such as Fig. 1.19) do not.

In the text and exercises, the words *draw* and *sketch* refer to hand-drawn graphs, and the word *find* indicates use of the grapher. Sometimes you will be asked to "draw a complete graph," as in Example 5. Usually, however, you will be asked to "find a complete graph."

Notes on Exercises

Ex. 11–17 may need to be hand-sketched on quadrille graph paper so that students can get a feeling for the size of the viewing window needed. Ex. 18–31 allow students to explore the effects of changing viewing rectangles. This gives them an opportunity to explore key concepts that relate to scale and the notion of a complete graph.

EXAMPLE 5   Drawing Sketches of Complete Graphs

The following data table shows the (x, y) pairs for some relationship.

x	-3	-2.5	-2	-1.5	-1	-0.5	0	0.5	1	1.5	2	2.5	3
y	-15	-5.625	0	2.625	3	1.875	0	-1.875	-3	-2.625	0	5.625	15

a) On a rectangular coordinate system, graph the data from this table.

b) Draw three different complete graphs of this relationship between x and y. (Recall the agreement about the use of the word *draw*.)

Solution   A complete graph must contain the data from the table. The graph also must suggest the *pattern* of the data.

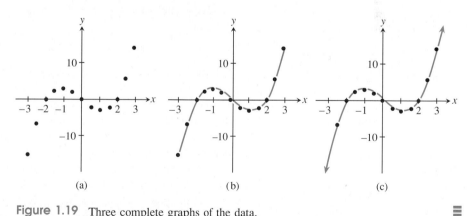

Figure 1.19   Three complete graphs of the data.

Figure 1.19(a) is strictly the graph of the data and so by definition is a complete graph. However, it is not the only possible complete graph; Fig. 1.19(b, c) show two other complete graphs.

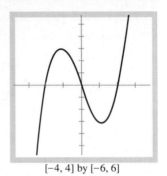

[−4, 4] by [−6, 6]

Figure 1.20 Complete graph of $y = x^3 - 4x$.

E X A M P L E   6   Finding an Equation to Model the Data

Show that the equation $y = x^3 - 4x$ is a model for the data table given in Example 5.

Solution

$$(-3)^3 - 4(-3) = -15 \qquad \text{Substitute } -3 \text{ for } x \text{ and solve for } y.$$

$$(-2.5)^3 - 4(-2.5) = -5.625 \qquad \text{Substitute } -2.5 \text{ for } x.$$

$$\vdots$$

$$(2.5)^3 - 4(2.5) = 5.625$$

$$(3)^3 - 4(3) = 15$$

Figure 1.20 shows a complete graph of the equation $y = x^3 - 4x$. ■

One possible complete graph through the data points found in Example 5 is precisely the graph of $y = x^3 - 4x$. In a real sense, it is also the best complete graph because it is a smooth curve that contains the given data. If more data points were given following the same pattern, this choice of model equation would continue to be appropriate.

Graphers work by creating a finite table of values from a given equation (model), plotting the data points, and connecting the points with a smooth curve. One model grapher uses 96 data points from Xmin to Xmax to create its graphs. Some computer graphing utilities use 500 or more points between Xmin and Xmax. Graphers do the tedious computation and allow us to view graphs today that in the past, would have been too time consuming to produce.

Exercises for Section 1.2

In Exercises 1 and 2, graph on a rectangular coordinate system the (x, y) pairs from the given data. In each case, estimate a complete graph.

1.

x	−4	−1	1	4	9
y	−2	1	3	6	11

2.

x	1	2	3	4	5
y	1	4	9	16	25

3. Use a graphing utility to graph $y = 100\sqrt{x}$. Then use the technique you used in the Exploration on page 16 to complete the following table of values:

x	?	?	?	?	?
y	10	20	30	40	50

4. Show that the equation $y = x + 2$ is a model for the data in Exercise 1.

5. Show that the equation $y = x^2$ is a model for the data in Exercise 2.

6. Which of the following equations is a model for the following set of data?

x	1	2	3	4
y	0	4	10	18

a) $y = x^2$ **b)** $y = x^2 + x - 2$ **c)** $y = x^3$

Exercises 7–10, refer to the following problem situation: A school club pays \$35.75 to rent a video cassette player to show a movie for a fund-raising project. The club charges \$0.25 per ticket for the movie.

7. How much profit does the club make if it sells
    a) 100 tickets?   **b)** 175 tickets?
    c) 250 tickets?   **d)** 425 tickets?

8. Find the pattern and generalize. What is the profit y if the club sells x tickets?

9. Draw a complete graph of the equation that describes the profit y for selling x tickets.

10. Use the graph you completed in Exercise 9 to determine the minimum number of tickets that must be sold for the club to do the following:
    a) Break even   **b)** Realize a profit

Which of the following points lie in the viewing rectangle $[-2, 3]$ by $[1, 10]$?

11. $(0, 0)$ **12.** $(0, 5)$ **13.** $(5, 0)$

14. $(3, 2)$ **15.** $(1, 6)$ **16.** $(1, -1)$

17. Choose a viewing rectangle $[X\min, X\max]$ by $[Y\min, Y\max]$ that includes all of the indicated points:
    a) $(-9, 10)$, $(2, 8)$, and $(3, 12)$
    b) $(20, -3)$, $(13, 56)$, and $(-11, 2)$

18. Which viewing rectangles show a complete graph of $y = x^2 - 30x + 225$?
    a) $[-5, 5]$ by $[-5, 5]$ **b)** $[-10, 10]$ by $[-10, 10]$
    c) $[15, 30]$ by $[-50, 50]$ **d)** $[5, 25]$ by $[-100, 100]$
    e) $[10, 20]$ by $[-1000, 1000]$

19. Which viewing rectangles show a complete graph of $y = 20 - 30x + x^3$?
    a) $[-10, 10]$ by $[-10, 10]$ **b)** $[-50, 50]$ by $[-10, 10]$
    c) $[-10, 10]$ by $[-50, 50]$ **d)** $[0, 10]$ by $[-100, 100]$
    e) $[-10, 10]$ by $[-50, 100]$

For Exercises 20–29, use your grapher to draw a complete graph of each equation. Begin with the standard viewing rectangle and modify it until you arrive at a satisfactory viewing rectangle.

20. $y = 6 - x$ **21.** $y = 2x^2 + 1$

22. $y = x^3 + x$ **23.** $y = 2x^2 - 3x + 5$

24. $y = 4 - x - x^3$ **25.** $y = \dfrac{(x + 3)^2}{5} + 2x + 4$

26. $y = 200x - 10$ **27.** $y = 10x^2 + 500$

28. $y = \sqrt{x - 2}$ **29.** $y = \sqrt{2x + 10}$

In Exercises 30 and 31, find a complete graph of each equation on the same coordinate system.

30. $y = x^2$, $\ y = (x - 1)^2$, $\ y = (x - 2)^2$, $\ y = (x + 3)^2$

31. $y = -1 + x^2$, $\ y = 2 + x^2$, $\ y = -2 + x^2$, $\ y = x^2 - 4$

In Exercises 32–34, complete a table of (x, y) pairs for the given values of x. Round your y-values to the second decimal place.

32. $y = \sqrt{x + 4}$ for $x = -3, -2, -1, 0, 1, 2, 3, 4, 5, 6$

33. $y = \log x$ for $x = 0.25, 0.5, 1, 1.5, 2, 3, 4, 5$

34. $y = \sin x$ for $x = 0, 0.52, 1.05, 1.57, 2.09, 2.62, 3.14,$ $3.67, 4.19, 4.71$

35. Graph the (x, y) pairs from the data table you completed for Exercise 32 and sketch your best estimate of a complete graph for this data.

36. Graph the (x, y) pairs from the data table you completed for Exercise 33 and sketch your best estimate of a complete graph for this data.

37. Graph the (x, y) pairs from the data table you completed for Exercise 34 and sketch your best estimate of a complete graph for this data.

38. If the point (a, b) is on the graph of $y = 100/x$, find the product ab.

39. **Collect Data** Find an algebraic and a graphical representation that shows the relationship between gallons of gasoline used and miles driven for your personal or family (or a friend's) car.

40. **Writing to Learn** Write a paragraph that explains your answer to the following questions: The two equations $y = x + 1$ and $y = x + 1.01$ have what appear to be identical graphs in the standard viewing rectangle even though their equations are not identical. (Check that this is true.) Why does this happen? Does this happen for all viewing rectangles?

41. **Writing to Learn** Why does the graph of the straight line $y = 0.5x$ appear jagged when using a grapher?

42. **Writing to Learn** Why does the graph of $y = \sqrt{x + 20}$ appear to be a straight line in some viewing rectangles (see Example 4)?

1.3 _____

Objective

Students will be able
to represent functions
algebraically and geometrically
and will be able to determine
the domain and range for
functions in problem-solving
situations.

Key Ideas

Function
Function notation
Domain of a function
Graph of a function
Vertical line test for a function
The absolute value function
The greatest integer function
Step function
Projectile motion
Piecewise-defined functions

Stan says, "It is very
important preparation for
calculus that students have
a firm grip on the concept
of function. The graphing
calculator enables students to
understand this concept in a
way that never before existed."

Functions

Example 1 in Section 1.2 refers to the Quality Rent-a-Car problem situation. The equation $y = 0.2x + 15$ was found to be an algebraic representation for this problem, and we learned that given any value of x, which represents miles, the value of y, which represents the cost, can be calculated. Think of the x as an "input" value and the y as the "output" value.

input	x	0	50	75	100	200	x
output	y	15	25	30	35	55	$0.2x + 15$

A significant property of the above algebraic representation is that each input value has *only one* output value. When that is true, we call the algebraic representation a **function**.

Definition 1.4   Function

A **function** f **of** x is a correspondence that associates each x in a set X with exactly one y in a set Y. If x is the input value and y is the output value, we call y the **image of** x under f.

The **domain** of f is the set X and the **range** consists of the set of all images of elements in X.

```
log 2
                .3010299957
```

Figure 1.21 The input value 2 has the output value $\log 2$.

We encounter functions routinely in our daily lives in tabular form: height and weight charts, time and temperature charts, price and sales tax charts, income tax tables, and so forth.

That functions are important in scientific applications is evident by the fact that functions abound on a scientific calculator. For example, the calculator keys

$$[\sqrt{\ }], \qquad [x^2], \qquad [\sin], \qquad \text{and} \qquad [\log]$$

each represent a function because for each *input* value x, there is only one *output* value—the one shown on the screen—which represents the y-value. Figure 1.21 shows that for the function $y = \log x$, the input value 2 has the output value 0.3010299957. We will study many of the functions on a scientific calculator more thoroughly later in this book.

Function Notation

If x is an element of the domain of a function, the output value that corresponds to x is usually denoted by $f(x)$. The notation and terminology used for the function $f(x) = 4x^2 + x - 2$ is summarized as follows:

Input	Output	Function Rule
x	$f(x)$	$f(x) = 4x^2 + x - 2$

The symbol $f(x)$ is read "f of x" and is called the **value of f at x**. This value $f(x)$ is the y-value that corresponds to the value of x, so $y = f(x)$. To find the value of f at $x = -1$, substitute -1 for x. Thus

$$f(-1) = 4(-1)^2 + (-1) - 2$$
$$= 4(1) - 1 - 2$$
$$= 4 - 3$$
$$= 1.$$

In a similar fashion, we find that $f(3) = 37$.

E X A M P L E 1   Finding Functional Values

Let f be the function defined by the equation $f(x) = x^2 + 1$. Find $f(-2)$, $f(3)$, $f(a)$, and $f(x - 3)$.

Solution

$$f(-2) = (-2)^2 + 1 = 5 \qquad \text{Replace } x \text{ with } -2.$$
$$f(3) = (3)^2 + 1 = 10$$
$$f(a) = a^2 + 1 \qquad \text{Replace } x \text{ with } a.$$
$$f(x - 3) = (x - 3)^2 + 1 = x^2 + 2x(-3) + (-3)^2 + 1 \qquad \text{Replace } x \text{ with } x - 3 \text{ and simplify.}$$
$$= x^2 - 6x + 10 \qquad \blacksquare$$

Domain of a Function

Let $y = f(x)$ be a function. The **understood domain of f** is the largest set of real numbers for which the equation $y = f(x)$ makes sense. That is, the real number a is in the understood domain of f if and only if $f(a)$ is a real number. When the

Bert says, "Students should be exposed to many different algebraic representations of functions. The graphing calculator allows us to explore many real-world phenomena that can be represented geometrically using this technology."

Notes on Exercises

Ex. 1–10 give students practice in linking the algebraic and numerical representations of a function.

Teaching Note

Many students have difficulty with the concepts of domain and range. Be sure they have opportunities to discuss and write down the domains and ranges of many functions. The examples and exercises in this section lend themselves to using cooperative groups in the classroom. Working in groups allows students to help each other and to take advantage of a variety of exploratory activities.

domain of a function is not explicitly stated, you should assume that its domain is its understood domain.

EXAMPLE 2   Finding Domain and Range

Find a complete graph of the function $f(x) = \sqrt{x - 3}$. Also find its domain and range.

Solution   (See Fig. 1.22 for a complete graph.)

$$x - 3 \geq 0 \qquad \text{The expression under a square root must be nonnegative.}$$

$$x \geq 3$$

The domain of f is the interval $[3, \infty)$. We read from the graph that the range is all numbers y such that $y \geq 0$. In other words, the range of f is the interval $[0, \infty)$. ▬

The **graph of a function** consists of all points in the rectangular coordinate plane whose x-coordinate represents an input value and whose y-coordinate represents the corresponding output value of the function. In other words, the

$$\text{graph of } f = \{(x, y) \quad \text{such that} \quad y = f(x)\}.$$

EXAMPLE 3   Finding Domain and Range

Compare complete graphs of $f(x) = x^2 - 4$ and $g(x) = |x^2 - 4|$. Determine the domain and range of each function.

Solution   Figure 1.23 shows complete graphs. For each function, x can represent any real number. Therefore the

$$\text{domain of } f = (-\infty, \infty) \qquad \text{and the} \qquad \text{domain of } g = (-\infty, \infty).$$

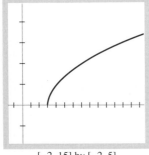

$[-2, 15]$ by $[-2, 5]$

Figure 1.22 $f(x) = \sqrt{x - 3}$.

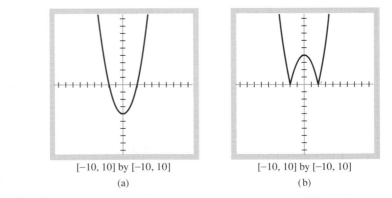

$[-10, 10]$ by $[-10, 10]$ $[-10, 10]$ by $[-10, 10]$
(a)   (b)

Figure 1.23 $f(x) = x^2 - 4$ and $g(x) = |x^2 - 4|$. ▬

GRAPHER UPDATE

A graphing calculator uses the notation $y = x^3 - 3x + 2$ to represent the function $f(x) = x^3 - 3x + 2$.

Notes on Exercises

Ex. 14–23 are of varying difficulty. Students can use graphs to help determine domains and ranges but must be careful to make use of algebraic knowledge and expectations. Determining the range of a function sometimes requires solving an extreme-value problem.

The range of f is the set of all numbers y such that $y \geq -4$; in other words, the range of f is the interval $[-4, \infty)$. The range of g is the interval $[0, \infty)$.

Notice that because the absolute value of an expression is always nonnegative, the graph of $g(x) = |x^2 - 4|$ can never go below the x-axis. The absolute value symbols in g have the effect of bringing the part of the graph of f that lies below the x-axis to above the x-axis.

This example leads to the following general principle.

Graph of Absolute Value

To obtain a complete graph of $y = |f(x)|$ from a complete graph of $y = f(x)$, reflect the portion of the graph of $y = f(x)$ that lies below the x-axis about the x-axis and join this reflection with the portion of the graph of $y = f(x)$ that lies above the x-axis.

Not all equations in variables x and y determine a function of x. The equation in Example 4 does not determine a function.

E X A M P L E 4   Determining When an Equation Is Not a Function of x

Show that $y^2 = x$ does not determine a function of x.

Solution

x	0	0.5	1	1.5	2	2.5	3	4	5
y	0	±0.7	±1	±1.2	±1.4	±1.6	±1.7	±2	±2.2

Because there are two y-values corresponding to some x-values, this table of values does not meet the condition for being a function of x. ▤

Notice that restricting the table of values in Example 4 to the positive y-values results in pairs for the function $y = \sqrt{x}$:

x	0	0.5	1	1.5	2	2.5	3	4	5
y	0	0.7	1	1.2	1.4	1.6	1.7	2	2.2

We see that the graph of $y^2 = x$ is not a function because the vertical line $x = 1$ intersects the graph at the two points $(1, 1)$ and $(1, -1)$ (see Fig. 1.24). In general, an equation does not determine a function of x if there exists one

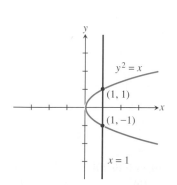

Figure 1.24 Testing whether $y^2 = x$ is a function using $x = 1$.

Notes on Exercises

Ex. 11–13 let students focus on the graphical representations of would-be functions in the context of the vertical line test.

Common Errors

Some calculators have a built-in integer truncation function that differs from the greatest integer function in the way it treats negative inputs. Graphing utilities can vary in terms of their plotting mode. Some operate in a point-plot mode, while others operate in a line-plot (connected) mode. Some models offer a choice of the mode. Students should be aware that the selection of a particular mode may impose limitations on graphs or may produce graphs that can lead them to make false conclusions.

GRAPHER SKILLS UPDATE

The square root function can also be accessed by entering it in the exponential form:

$$x \wedge (1/2) \text{ or } x \wedge 0.5.$$

The parentheses around the 1/2 are needed because computers follow the conventional rules for order of operations. Entering $x \wedge 1/2$ produces the expression $x/2$.

vertical line that intersects the graph of the equation in more than one point. This observation leads to the **vertical line test** for functions.

Definition 1.5   Vertical Line Test for a Function

If every vertical line intersects the graph of an equation in at most one point, then the equation determines a function of x.

E X A M P L E   5   Using the Vertical Line Test

Which of the relations represented by the following graphs are functions?

Solution   All four graphs satisfy the vertical line test except part (b). Therefore only part (b) is not the graph of a function of x.   ≡

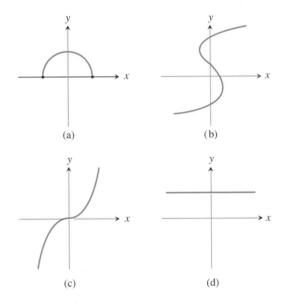

(a)   (b)

(c)   (d)

Built-in Functions

Computers and graphing calculators have special built-in functions that can be accessed by pushing a special key or typing a special code. For example, in graphing the equation in Example 3, the syntax of a graphing utility usually requires that $f(x) = |x^2 - 4|$ be written in the form $f(x) = \text{ABS}(x^2 - 4)$. Also, typing $\text{SQR}(x)$—or $\text{SQRT}(x)$ on some computers—will usually produce the **square root function** $f(x) = \sqrt{x}$.

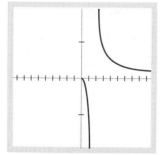

[-10, 10] by [-2, 2]

Figure 1.25
$f(x) = \sqrt{x}/(2x - 4)$.

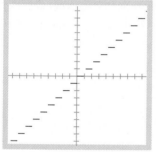

[-10, 10] by [-10, 10]

Figure 1.26 $f(x) = \text{INT}(x)$.

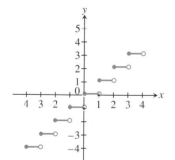

Figure 1.27   Sketch of
$f(x) = \text{INT}(x)$.

E X A M P L E   6   *Finding Domain and Range*

Find a complete graph of $f(x) = \sqrt{x}/(2x - 4)$ and determine its domain and range.

Solution   Figure 1.25 shows a complete graph of the function. Notice that 2 is not in the domain because the denominator $2x - 4$ cannot equal zero. Also, x cannot be less than zero because of the $\sqrt{x}$ in the numerator. Thus the domain of f is $[0, 2) \cup (2, \infty)$.

Notice that $f(0) = 0$, so 0 is in the range of f. Investigate further to conclude from the graph that the range of f is the set of all real numbers.   ▤

The **greatest integer function** is another function that is built in on many graphers. It is denoted by $[\![x]\!]$ or $\text{INT}(x)$ where $[\![x]\!]$ is defined to be the greatest integer less than or equal to x. For example, $[\![3.7]\!] = 3$, $[\![-1.2]\!] = -2$, and $[\![6]\!] = 6$.

E X A M P L E   7   *Graphing the Greatest Integer Function*

Find a complete graph of $f(x) = \text{INT}(x)$ and determine its domain and range.

Solution   (See Fig. 1.26.)

Domain of $f = (-\infty, \infty)$ There is a unique largest integer less than or equal to any number x.

Range of $f = \{\ldots, -3, -2, -1, 0, 1, 2, 3, \ldots\}$ The value of the function is always an integer.   ▤

Notice that the graph of $f(x) = \text{INT}(x)$ coincides with the x-axis when x is between 0 and 1 since $\text{INT}(x) = 0$ in that interval. Moreover, a grapher with the standard viewing rectangle does not show whether INT(3) is 2 or 3 (see Fig. 1.26). The detail at each integer value is shown better in a hand-drawn sketch (Fig. 1.27). A solid circle is placed at the left end of each segment to show the function value for each integer, and an open circle is placed at the right of each segment to show that the endpoint is missing on the right of each line segment. This illustrates, for example, that INT(1) = 1 and INT(3) = 3.

The function $f(x) = \text{INT}(x)$ is sometimes called a **step function** because its completed graph looks like a series of steps.

Applications of Functions

Functions can be used to model many problem situations. Often a graphing utility can be used to help visualize and interpret an algebraic representation of the problem situation. It is important that you use an appropriate viewing rectangle whenever using the grapher.

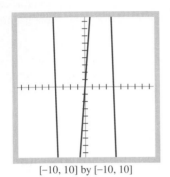

[−10, 10] by [−10, 10]

Figure 1.28 $f(x) = 16x - x^3$.

E X A M P L E   8   Finding an Appropriate Viewing Rectangle

Find a complete graph of the function $f(x) = 16x - x^3$ and determine the domain and range of f.

Solution   Begin with the standard viewing rectangle [−10, 10] by [−10, 10]. Figure 1.28 illustrates that the graph of $f(x) = 16x - x^3$ in the standard rectangle goes off the screen vertically. That means that the values for Ymin and Ymax must be increased. Experiment until you find an acceptable viewing rectangle. See Fig. 1.29. Conclude that the domain and range are $(-\infty, \infty)$, the set of real numbers.   ∎

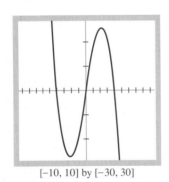

[−10, 10] by [−30, 30]

Figure 1.29 $f(x) = 16x - x^3$.

Example 9 considers the area of a sidewalk around a pool.

E X A M P L E   9   APPLICATION: Determining the Area of a Sidewalk

A rectangular swimming pool with dimensions 30 by 50 ft is surrounded by a sidewalk of uniform width x (see Fig. 1.30). What is the width of the sidewalk when the sidewalk area is 600 sq ft?

Solution

1. First find the area of the sidewalk, A, as a function of x.

$$\text{total length} = 50 + 2x$$

$$\text{total width} = 30 + 2x$$

$$\text{total area} = (50 + 2x)(30 + 2x)$$

$$\text{swimming pool area} = 50(30)$$

$$A(x) = (50 + 2x)(30 + 2x) - (50)(30)$$

$$= 4x^2 + 160x$$

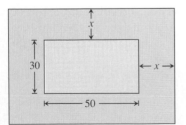

Figure 1.30   Diagram of a swimming pool with a sidewalk of uniform width x.

2. Next, determine what portion of the complete graph of $A(x) = 4x^2 + 160x$ (see Fig. 1.31) represents the problem situation. Only the positive values of x make sense in this situation, so the domain of $A(x)$ for it is $(0, \infty)$.

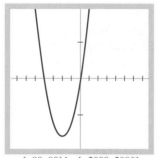

[−80, 80] by [−2000, 2000]

Figure 1.31 $A(x) = 4x^2 + 160x$.

3. Graph both $A(x) = 4x^2 + 160x$ and $y = 600$ in the viewing rectangle [0, 10] by [0, 1000] (see Fig. 1.32). The x-coordinate a of the point of intersection is approximately 3.5. Thus when the area of the sidewalk is 600 sq ft, the width is about 3.5 ft. ≡

E X A M P L E 10   APPLICATION: Determining Projectile Motion

A model rocket is shot straight up from ground level with an initial velocity of 64 ft/sec.

a) At what time will the object be 50 ft above the ground?

b) What is the maximum height reached by this object, and how long does it take to reach this height?

Solution   The height $s(t)$ after t seconds is given by $s(t) = -16t^2 + v_0 t + s_0$. So in this problem, $s(t) = -16t^2 + 64t$.

a) Graph the functions $s(t) = -16t^2 + 64t$ and $s = 50$ (see Fig. 1.33). Why is [0, 4] the appropriate domain for this problem situation? By zooming in on the points of intersection, you will discover that the object is 50 ft above the ground at 1.06 sec and 2.94 sec after the object is thrown.

b) Zoom-in on the highest point and confirm that it is (2, 64). This means that the object reaches a maximum height of 64 ft after 2 sec. ≡

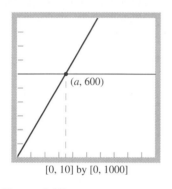

(a, 600)

[0, 10] by [0, 1000]

Figure 1.32 $A(x) = 4x^2 + 160x$.

Piecewise-defined Functions

In a **piecewise-defined function**, the domain of the function is divided into several parts and a different function rule is applied to each part. For example, consider the following function:

$$f(x) = \begin{cases} 3 - x^2 & \text{if } x < 1 \\ x^3 - 4x & \text{if } x \geq 1. \end{cases} \tag{1}$$

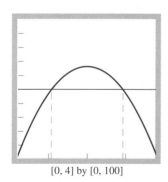

[0, 4] by [0, 100]

Figure 1.33 $s(t) = -16t^2 + 64t$.

The rule $f(x) = 3 - x^2$ applies only when x falls in the interval $(-\infty, 1)$, and the rule $f(x) = x^3 - 4x$ applies only when x falls in the interval $[1, \infty)$. To find a complete graph of this function, piece together the unshaded portion of the graph in Fig. 1.34(a) with the unshaded portion of the graph in Fig. 1.34(b).

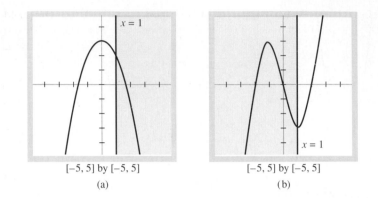

[−5, 5] by [−5, 5]   [−5, 5] by [−5, 5]
(a)   (b)

Figure 1.34 (a) $y = 3 - x^2$; (b) $y = x^3 - 4x$. Combining the unshaded portions results in the piecewise function in Eq. (1).

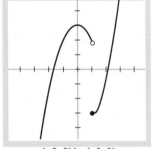

[−5, 5] by [−5, 5]

Figure 1.35 The graph of the piecewise function defined in Example 11.

E X A M P L E   11   Piecewise-defined Function

Sketch a complete graph of the function

$$f(x) = \begin{cases} 3 - x^2 & \text{if } x < 1 \\ x^3 - 4x & \text{if } x \geq 1. \end{cases}$$

Solution   (See Fig. 1.35.)

Exercises for Section 1.3

The functions $f(x) = x^2 - 1$ and $g(x) = 1/(x + 1)$ apply to Exercises 1–6. Find:

1. $f(0)$ and $f(1)$ **2.** $f(3)$ and $f(-5)$

3. $g(0)$ and $g(1)$ **4.** $g(3)$ and $g(-5)$

5. $g(2/t)$ **6.** $f(x + 2)$

In Exercises 7–10, consider the graph of the function $y = f(x)$ given in the accompanying figure. Estimate the indicated values. (Each tick mark represents a value of 1.)

7. $f(0)$, $f(-1)$, and $f(4)$ **8.** x if $f(x) = 0$

9. x if $f(x) = 1$ **10.** x if $f(x) = -8$

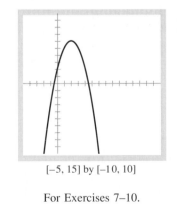

[−5, 15] by [−10, 10]

For Exercises 7–10.

In Exercises 11–13, use the vertical line test to decide which are graphs of functions of x.

11.

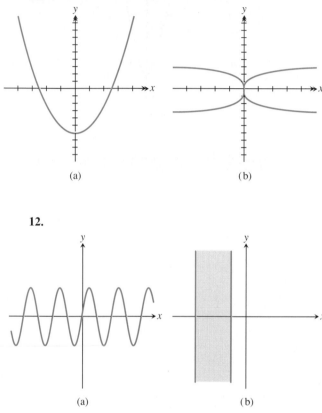

(a)

(b)

12.

(a)

(b)

13.

(a)

(b)

In Exercises 14–23, determine the domain and range of each function. Use a grapher if necessary.

14. $f(x) = \sqrt{x}$

15. $f(x) = \dfrac{2}{x - 3}$

16. $f(x) = x^2 - 3$

17. $g(x) = |x + 5|$

18. $h(x) = \dfrac{x}{|x + 4|}$

19. $f(x) = \sqrt{8 - x}$

20. $g(x) = \dfrac{x - 2}{x^2 + 5}$

21. $h(x) = \sqrt{\dfrac{1}{x + 1}}$

22. $h(x) = \sqrt{x^3 - 8x}$

23. $f(x) = x^2 - 5x - 10$

24. Which viewing rectangle gives the best complete graph of $f(x) = x^3 + 10x + 100$?
   a) $[-10, 10]$ by $[-10, 10]$
   b) $[-10, 10]$ by $[-100, 100]$
   c) $[-10, 10]$ by $[0, 100]$
   d) $[-10, 10]$ by $[-100, 200]$
   e) $[-10, 10]$ by $[-1000, 1000]$

25. Which viewing rectangle gives the best complete graph of $f(x) = 12x^2 - 5x + 19$?
   a) $[-10, 10]$ by $[-10, 10]$
   b) $[0, 10]$ by $[-100, 100]$
   c) $[-50, 50]$ by $[-5, 5]$
   d) $[-5, 5]$ by $[-50, 50]$
   e) $[-10, 0]$ by $[0, 50]$

26. Which viewing rectangle gives the best complete graph of $f(x) = 0.005x^3 - 5x^2 - 60$?
   a) $[-10, 10]$ by $[-100, 100]$
   b) $[-100, 100]$ by $[-100,000, 100,000]$
   c) $[-1000, 2000]$ by $[-1,000,000, 1,000,000]$
   d) $[0, 2000]$ by $[0, 1,000,000]$
   e) $[-1000, 1000]$ by $[-1000, 1000]$

In Exercises 27–42, find a complete graph of each function.

27. $g(x) = \dfrac{x - 4}{5}$

28. $f(x) = x^2 - 4$

29. $f(x) = 2x^2 - 3x + 5$

30. $f(x) = x^3 - x + 1$

31. $g(x) = (x + 3)^2 + 2x + 4$

32. $h(x) = 4 - x - x^3$

33. $h(x) = \sqrt{x + 1}$

34. $g(x) = -1 - \sqrt{x - 1}$

35. $k(x) = 1 + |x - 2|$

36. $f(x) = -|2x^3|$

37. $g(x) = \text{INT}(x + 2)$

38. $f(x) = (x - 30)(x + 20)$

39. $k(x) = |x^2 - 6x - 12|$

40. $h(x) = 2x^3 - x + 3$

41. $f(x) = x^3 - 8x$

42. $g(x) = 10x^3 - 20x^2 + 5x - 30$

In Exercises 43 and 44, find a complete graph of each function. Record each series of graphs on the *same* coordinate system.

43. $f(x) = x^2$, $g(x) = (x-1)^2$, $h(x) = (x-2)^2$, $k(x) = (x+3)^2$

44. $f(x) = -1+x^2$, $g(x) = 2+x^2$, $h(x) = 4-x^2$, $k(x) = x^2 - 4$

In Exercises 45–47, sketch a complete graph of each piecewise-defined function. Support your work with a graphing utility.

45. $f(x) = \begin{cases} x^2 & \text{if } x < 3 \\ x - 4 & \text{if } x \geq 3 \end{cases}$

46. $g(x) = \begin{cases} -x & \text{if } 0 \leq x < 4 \\ \sqrt{x-3} & \text{if } x \geq 4 \end{cases}$

47. $h(x) = \begin{cases} 4 - x + x^2 & \text{if } x < 2 \\ 1 - 2x + 3x^2 & \text{if } x \geq 2 \end{cases}$

48. Jerry runs at a constant rate of 4.125 mph. Describe the distance that Jerry runs as a function $y = d(t)$, and find a complete graph of this function.

49. Use the graph in Exercise 48 to determine the time it takes Jerry to run a 26-mi marathon.

Exercises 50–52 refer to the following **problem situation:** A 4- by 6.5-in. picture pasted on a cardboard sheet is surrounded by a border of uniform width x and area A.

50. Determine the area A of the border region as a function $y = A(x)$, and find a complete graph of this function.

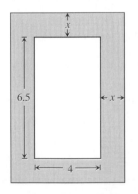

For Exercises 50–52.

51. In the function $A(x)$ found in Exercise 50, what values of x are possible in this problem situation?

52. Use the graph in Exercise 50 to determine the width of the border if its area is 20 sq in.

Exercises 53 and 54 refer to the following **problem situation:** 80 ft of fencing are cut into two unequal lengths, and each piece is used to make a square enclosure. Let x be the perimeter of the smaller enclosure.

53. Express the total area A of both enclosures as a function $y = A(x)$. What domain values are possible in this problem situation?

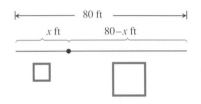

For Exercises 53 and 54.

54. Find a complete graph of the function found in Exercise 53 and use it to approximate the smaller length of fence when $A = 300$ sq ft.

Exercises 55–57 refer to the following **problem situation:** The regular long-distance telephone charges C from Columbus, Ohio, to Lancaster, Pennsylvania, are \$0.48 for the first minute or fraction of a minute, and \$0.28 for each additional minute.

55. Express the cost C of calling Lancaster from Columbus as a function $y = C(t)$, and sketch a complete graph. What are the possible range values of this function?

56. What part of the domain of the function in Exercise 55 represents this problem situation?

57. Use the graph of the function in Exercise 55 to determine how long Bill talked to his girlfriend if he spent \$1.88.

Exercises 58–60 refer to a projectile motion **problem situation** in which an object 10 ft above ground level is shot straight up with an initial velocity of 150 ft/sec. See the description of the projectile motion problem situation in Example 10.

58. Determine the height s of the object above the ground as a function $s = f(t)$, and find a complete graph.

59. What portion of the graph of the function found in Exercise 58 represents the problem situation?

60. At what time t will the object be 300 ft above the ground?

1.4 _____

Linear Functions and Inequalities

Objective

Students will be able to graph
and solve problems using
equations and inequalities in
two variables.

Key Ideas

Linear function
Slope
Slope-Intercept form of a line
Point-Slope form
General form
Linear inequality

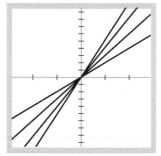

[−3, 3] by [−10, 10]

Figure 1.36 $f_1(x) = 2x$,
$f_2(x) = 3x$, $f_3(x) = 4x$, and
$f_4(x) = 5x$.

Teaching Note

Be sure to do the "Explore
with a Graphing Utility"
with students. Reviewing
the familiar concepts of
slope and y-intercept using
a graphing utility gives
students an opportunity to
make generalizations about the
behavior of linear functions.

A primary reason for studying mathematics is to solve everyday problems. Section 1.2 introduced the Quality Rent-a-Car problem. The function $f(x) = 0.2x + 15$, which was a mathematical model for that problem, is an example of a linear function.

🔎 EXPLORE WITH A GRAPHING UTILITY

We want to explore the similarities and differences among the graphs of the four functions $f_1(x) = 2x$, $f_2(x) = 3x$, $f_3(x) = 4x$, and $f_4(x) = 5x$. Graph on a graphing utility these four functions, one after another without erasing (see Fig. 1.36).

Questions

1. How are the four graphs alike? As the coefficient of x changes from 2 to 3 to 4 to 5, how do they differ?

2. Predict how the graphs of $f_1(x) = -2x$, $f_2(x) = -3x$, $f_3(x) = -4x$, and $f_4(x) = -5x$ will look. Check your predictions.

Generalize   How would you describe the graph of $f(x) = mx$, where m is a constant?

You have probably discovered that the graph of $f(x) = mx$ is a straight line for every value of m. When m is a positive number, the line rises as it moves from left to right, and when m is negative, the line falls as it moves from left to right.

The equation $y = 2x + 3$ is also a line. Experiment with other equations of the form $y = mx + b$, where m and b are constants.

Definition 1.6   Linear Function

A **linear function** is a function that can be written in the form $f(x) = mx + b$, where m and b are real numbers and $m \neq 0$.

Many of the functions used for business and economic models are linear functions. Example 1 illustrates one such situation.

E X A M P L E   1   APPLICATION: Determining the Value of a House

A house was purchased 10 years ago for $60,000. This year it was appraised at $85,000. Assume the value V of the house is $V(t) = mt + b$ for some real numbers m and b where t represents the time in years. When was the house worth $71,250?

Solution   In this problem situation, we assume that the problem is modeled by a linear function. We want to find the specific values of m and b and use the model to determine when the house was worth $71,250.

1. First find the value of b by noticing that the point $(0, 60{,}000)$ is a point on the line.

$$V = mt + b$$

$$60{,}000 = m(0) + b \quad \text{When } t = 0, V = 60{,}000.$$

$$60{,}000 = b$$

Therefore $V = mt + 60{,}000$ for some value of m. Next, find the value of m.

$$85{,}000 = m(10) + 60{,}000 \quad \begin{matrix} \text{Substitute } t = 10 \text{ and } V = 85{,}000 \\ \text{into the equation } V = mt + 60{,}000. \end{matrix}$$

$$85{,}000 - 60{,}000 = 10m + 60{,}000 - 60{,}000$$

$$25{,}000 = 10m$$

$$2500 = m$$

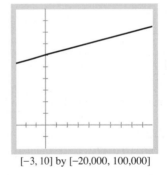

[−3, 10] by [−20,000, 100,000]

Figure 1.37 $V(t) = 2500t + 60{,}000.$

The algebraic representation of this problem situation is $V(t) = 2500t + 60{,}000$ (see Fig. 1.37).

2. To find t when $V = 71{,}250$, substitute $V = 71{,}250$ and solve for t.

$$V = 2500t + 60{,}000$$

$$71{,}250 = 2500t + 60{,}000$$

$$11{,}250 = 2500t$$

$$4.5 = t$$

Therefore the house was worth $71,250, 4.5 years after its purchase.   ≡

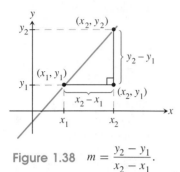

Figure 1.38 $m = \dfrac{y_2 - y_1}{x_2 - x_1}$.

When the equation for a line is written as a linear equation, the numbers m and b give useful information about the graph. First let's consider m.

Suppose that (x_1, y_1) and (x_2, y_2) are two distinct points on the graph of $y = mx + b$ (see Fig. 1.38). Then $y_1 = mx_1 + b$ and $y_2 = mx_2 + b$. (Why?) We can isolate m by subtracting one equation from the other.

$$y_2 - y_1 = (mx_2 + b) - (mx_1 + b)$$
$$= m(x_2 - x_1)$$
$$m = \frac{y_2 - y_1}{x_2 - x_1}$$

The number m thus describes the steepness, or slope, of the line.

Definition 1.7   Slope

The **slope** of the graph of $y = mx + b$ is the quotient

$$m = \frac{y_2 - y_1}{x_2 - x_1},$$

where (x_1, y_1) and (x_2, y_2) are any two points on the line $y = mx + b$ such that $x_1 \neq x_2$.

The slope tells us two things about the line. The sign of m determines whether the line rises or falls as it moves from left to right: When m is positive, the line rises, and when m is negative, the line falls. The absolute value of m determines the steepness of that rise or fall.

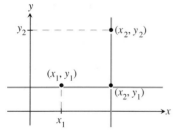

Figure 1.39 Vertical line $x = x_2$ and horizontal line $y = y_1$.

E X A M P L E 2   *Calculating the Slope*

Find the slope of the line passing through the points $(-1, 2)$ and $(4, -2)$.

Solution   Let $(x_1, y_1) = (-1, 2)$ and $(x_2, y_2) = (4, -2)$. Then

$$m = \frac{y_2 - y_1}{x_2 - x_1} = \frac{-2 - 2}{4 - (-1)} = -\frac{4}{5}.$$
Substitute into the slope definition the values of the coordinates of the given points.

Figure 1.39 shows that the slope of the horizontal line is $(y_1 - y_1)/(x_2 - x_1) = 0$. The slope of any horizontal line is zero because the numerator in the slope fraction is always $y_1 - y_1 = 0$. *The slope of a vertical line is not defined* because in the slope calculation, the denominator is zero.

Example 3 asks that we find the equation of a horizontal and vertical line.

E X A M P L E 3   Finding Horizontal and Vertical Lines

Write equations for the horizontal and vertical lines through the point (2, 3).

Solution   (See Fig. 1.40.) The horizontal line is $y = 3$, and the vertical line is $x = 2$.   ▤

Slope-intercept Form

Slope is one characteristic of the equation of a line. Another characteristic relates to where the line crosses the y-axis. If the graph of an equation crosses the y-axis at $(0, b)$, then $(0, b)$ (or simply b) is the y-intercept of the graph. We conclude with the following generalization.

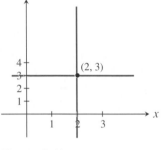

Figure 1.40 $x = 2$ and $y = 3$.

Slope-intercept Form

The graph of $y = mx + b$ is a straight line with slope m and y-intercept b. This form is called the **slope-intercept form** of the line.

Sometimes it is advantageous to change an equation of a line and write it in the slope-intercept form. For example, $2x + 3y = 6$ can be written in the equivalent form $y = (-2/3)x + 2$. In this form, we conclude that the slope of the line is $-2/3$ and that its y-intercept is 2.

E X A M P L E 4   Finding Slope and y-Intercept

Change the equation $3y - \sqrt{20}x = 9$ to slope-intercept form and find the slope and the y-intercept for the line. Sketch a complete graph.

Solution

$$3y - \sqrt{20}x = 9$$
$$3y = \sqrt{20}x + 9$$
$$y = \frac{\sqrt{20}}{3}x + 3 \quad \text{Solve for } y.$$

▤

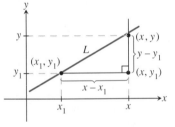

Figure 1.41 $y = (\sqrt{20}/3)x + 3$.

The slope of this line is $\sqrt{20}/3$, and its y-intercept is 3. Use this information to sketch a correct graph (see Fig. 1.41).

Point-slope Form

Suppose we know that the slope of a line L is m and we are given a point (x_1, y_1) on L that is not necessarily the y-intercept. How do we find an equation of that line?

To figure this out, suppose that (x, y) is *any other* point on line L (see Fig. 1.42). Then the slope m of line L satisfies the following equations:

$$m = \frac{y - y_1}{x - x_1}$$

$$y - y_1 = m(x - x_1)$$

This observation is summarized in Theorem 1.1.

Figure 1.42 Finding the equation of a line given the slope and a point (x_1, y_1) on the line.

Theorem 1.1 Point-slope Form

If L is the line through the point (x_1, y_1) with slope m, then the equation

$$y - y_1 = m(x - x_1)$$

is an equation of line L. An equation in this form is called the **point-slope form** for the equation of line L.

E X A M P L E 5 Point-slope Form

Write an equation for the line that has a slope of 0.5 and contains the point $(-2, 3)$.

Solution $m = 0.5$ and $(x_1, y_1) = (-2, 3)$. Therefore

$$y - y_1 = m(x - x_1)$$
$$y - 3 = 0.5(x + 2) \quad \text{Substitute into the slope-intercept form.}$$
$$y = 0.5x + 1 + 3$$
$$y = 0.5x + 4.$$

E X A M P L E 6 Writing an Equation Given Two Points

Write an equation for the line that contains the two points $(-3, 1)$ and $(4, -3)$.

Solution

$$m = \frac{-3 - 1}{4 - (-3)} = -\frac{4}{7} \qquad \text{First find the slope and}$$
then use the point-slope form.

$$y - 1 = -\frac{4}{7}(x + 3) \qquad \text{Substitute } (-3, 1) \text{ into the point-slope form.}$$

$$y = -\frac{4}{7}x - \frac{12}{7} + 1$$

$$y = -\frac{4}{7}x - \frac{5}{7}$$

$$4x + 7y = -5$$

If we had chosen to use the point $(4, -3)$, we would have obtained the same equation. Notice that $4(4) + 7(-3) = -5$, which demonstrates that $(4, -3)$ is on this line.

≡

The equation of the line in Example 6 can be written as $4x + 7y + 5 = 0$. This form $Ax + By + C = 0$ is called the **general form** of a linear equation, where A and B are not both zero. Every line has an equation that can be written in the slope-intercept form.

Following is a summary of the several forms for equations of lines that have been discussed.

Notes on Exercises

Ex. 5–36 include many traditional problems from the analytic geometry of lines.

Forms for Linear Equations

General form:	$Ax + By + C = 0$, A and B not both zero.
Vertical line:	$x = a$
Horizontal line:	$y = b$
Slope-intercept form:	$y = mx + b$
Point-slope form:	$(y - y_1) = m(x - x_1)$

Parallel and Perpendicular Lines

If two lines are parallel, then either they are both vertical (slope undefined) or they have equal slopes. For example, the lines with equations $y = 3x - 4$ and $y = 3x + 7$ both have a slope of 3, so the lines are parallel.

The next Exploration concerns perpendicular lines.

Notes on Exercises

Ex. 37–41 require a graphing utility for exploration.

🔎 EXPLORE WITH A GRAPHING UTILITY

Using the viewing rectangle $[-15, 15]$ by $[-10, 10]$, graph each pair of equations. Do they appear to be perpendicular?

1. $y = x - 3$ and $y = -x + 2$

2. $y = \frac{1}{3}x + 7$ and $y = -3x + 2$

3. $y = 2x - 3$ and $y = \frac{1}{2}x + 1$

4. $y = \frac{5}{4}x + 1$ and $y = -\frac{4}{5}x - 3$

5. $y = -2x + 1$ and $y = \frac{1}{5}x - 2$

6. $y = -\frac{2}{3}x + 2$ and $\frac{3}{2}x - 3$

Generalize How are the slopes of lines related when they are perpendicular?

The following theorem is given without proof.

Theorem 1.2 Parallel and Perpendicular Lines

Let ℓ_1 be a line with equation $y = m_1 x + b_1$ and ℓ_2 be a line with equation $y = m_2 x + b_2$. Then

- the lines ℓ_1 and ℓ_2 are parallel if and only if $m_1 = m_2$, and

- the lines ℓ_1 and ℓ_2 are perpendicular if and only if $m_1 = -\dfrac{1}{m_2}$.

E X A M P L E 7   Determining a Line

Write equations for lines passing through $(1, -2)$ and parallel and perpendicular to the line $3x - 2y = 1$.

Solution First, change the equation to the slope-intercept form to find its slope.

$$3x - 2y = 1$$

$$-2y = -3x + 1$$

$$y = \frac{3}{2}x - \frac{1}{2}$$

Next, use the point-slope form twice, once with $m = \frac{3}{2}$ and once with $m = -\frac{2}{3}$, and the point $(1, -2)$.

$$y + 2 = \tfrac{3}{2}(x - 1) \qquad \text{or} \qquad 3x - 2y = 7$$
$$y + 2 = -\tfrac{2}{3}(x - 1) \qquad \text{or} \qquad 2x + 3y = -4$$

$\blacksquare$

Linear Inequalities

Notes on Exercises

Ex. 52 uses the concept of inequality and then requires that a complete graph be found.

Suppose that a linear equation or a linear function is changed to an inequality. How do the graphs of the equality and the inequality relate?

$\mathcal{P}$ **EXPLORE WITH A GRAPHING UTILITY**

Find a complete graph of $y = 2x + 3$ in a square viewing rectangle. Move the cursor to a point on the graph. Then move the cursor up and down.

Questions

1. Does the x-coordinate of the moving point change as you move the cursor up and down?

2. What happens to the y-coordinate of the moving point?

3. Would the coordinates of this moving point satisfy the inequality $y < 2x + 3$ or the inequality $y > 2x + 3$? How do you know?

Generalize If you would shade in all points that satisfy $y > 2x + 3$, what points would be shaded?

This Exploration has motivated the following definition.

Definition 1.8   Linear Inequality

A **linear inequality** in the two variables x and y is an inequality that can be written in one of the following forms:

$$y < mx + b, \qquad y \leq mx + b, \qquad y > mx + b, \qquad \text{or} \qquad y \geq mx + b.$$

Consider the linear equation $y = 2x + 3$. The point $(1, 5)$ is on this line since $5 = 2(1) + 3$. Suppose c is a number such that $c < 2(1) + 3$. Will the point $(1, c)$ lie below or above the line $y = 2x + 3$? For example, $(1, 3)$ and $(1, -1)$ both lie below the line. These points also satisfy the inequality $y < 2x + 3$.

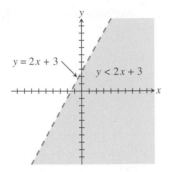

Figure 1.43 $y = 2x + 3$ (dashed line) and $y < 2x + 3$ (shaded area). The line is dashed to indicate it is *not* part of the solution to $y < 2x + 3$.

The graph of an inequality consists of all points whose coordinates satisfy the inequality. So, among the points on the vertical line $x = 1$, the points below the graph of $y = 2x + 3$ are solutions to $y < 2x + 3$, and the points above the graph of $y = 2x + 3$ are *not* solutions to $y < 2x + 3$. Since a similar argument can be made for each vertical line, a complete graph of $y < 2x + 3$ consists of all the points below the graph of $y = 2x + 3$ (see Fig. 1.43).

EXAMPLE 8   Graphing an Inequality

Draw the graph of $y \geq 2x + 3$.

Solution

1. Plot the line $y = 2x + 3$. Make the line solid to indicate that it will be part of the solution to $y \geq 2x + 3$.

2. Shade the region above the line.

The result is shown in Fig. 1.44.

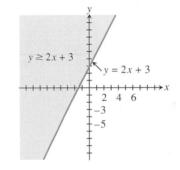

Figure 1.44 $y \geq 2x + 3$.

Notes on Examples

Examples 8 and 9 are very important in establishing the notion of a linear inequality. In Example 9, the inequality is rewritten to obtain an equivalent inequality in slope-intercept form.

Common Errors

Before students assume that in graphing any linear inequality they only need to look at the inequality sign to determine whether to shade above or below the line, discuss an example using the general form of a line. For instance, before graphing $2x + 3y < 4$, ask whether the graph will be shaded above the line $2x + 3y = 4$ or below it. Can students generalize after looking at several inequalities?

EXAMPLE 9   Graphing Another Inequality

Draw the graph of $2x + 3y < 4$.

Solution

$$2x + 3y < 4$$
$$3y < -2x + 4$$
$$y < -\frac{2}{3}x + \frac{4}{3} \qquad \text{Solve for } y.$$

Draw the graph of $y = -\frac{2}{3}x + \frac{4}{3}$ and shade the region below this line (see Fig. 1.45).

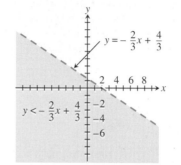

Figure 1.45 $y < -\frac{2}{3}x + \frac{4}{3}$. Why is the line dashed?

Exercises for Section 1.4

In Exercises 1–4, determine which equations have graphs that rise (going from left to right) and which have graphs that fall. Support your answers by graphing each equation on a graphing utility.

1. $y = 3x, \quad y = \frac{1}{3}x, \quad y = -3x$

2. $y = -3x, \quad y = 2x, \quad y = -\frac{1}{4}x$

3. $y = -2x - 1, \quad y = -4x + 3, \quad y = 2x + 5$

4. $y = 3 - 2x, \quad y = -2x + 3, \quad y = 5 + 3x$

In Exercises 5–8, find the slope of the line through each pair of points.

5. $(-1, 2); \quad (2, -6)$ **6.** $(0, -6); \quad (-4, 7)$

7. $(-1, -3); \quad (4, 1)$ **8.** $(-3, 1); \quad (-1, -5)$

In Exercises 9–14, determine the slope and y-intercept of the graph of each equation.

9. $y = 2x - 4$ **10.** $y = 1 - 3x$

11. $2x + 3y = 2$ **12.** $-x + 3y + 2 = 0$

13. $x + 2y = 3$ **14.** $-x - 3y = 8$

In Exercises 15–18, draw the line that contains point P and has slope m.

15. $P(1, 2); \quad m = 2$ **16.** $P(2, -1); \quad m = \frac{1}{3}$

17. $P(0, 3); \quad m = 0$ **18.** $P(3, -1); \quad m = \text{undefined}$

In Exercises 19–22, write an equation for the line with the given slope m that contains the given point.

19. $m = \frac{3}{2}; \quad (-1, 0)$ **20.** $m = -\frac{2}{3}; \quad (2, -4)$

21. $m = 0; \quad (2, 5)$ **22.** $m = 10; \quad (-1, 3)$

In Exercises 23–26, write an equation for the line with the given slope m and y-intercept b.

23. $m = \frac{3}{2}; \quad b = 2$ **24.** $m = -\frac{3}{4}; \quad b = -3$

25. $m = \frac{1}{5}; \quad b = -\frac{2}{3}$ **26.** $m = 6; \quad b = \frac{1}{2}$

In Exercises 27 and 28, write an equation of both the vertical and the horizontal line through each point.

27. $P(-2, 3)$ **28.** $Q(0, -2)$

In Exercises 29–36, find an equation of the line through the given pair of points.

29. $(0, 0); \quad (2, 3)$ **30.** $(2, -1); \quad (-3, 4)$

31. $(-1, 0); \quad (3, 1)$ **32.** $(8, 1); \quad (8, -4)$

33. $(0, 3); \quad (2, 6)$ **34.** $(1, 1); \quad (0, 2)$

35. $(0, 300); \quad (10, 365)$ **36.** $(5, 200); \quad (35, 1050)$

37. When the x- and y-axes have the same scale, a line with a slope of 1 appears to be a $45°$ line. Find a *square viewing* *rectangle*, that is, a viewing rectangle on which $y = x$ looks like a 45° line.

38. Find a graph of the following linear equations. Which one appears steeper? Which one has the greater slope?
   a) $y = 3x + 1$, viewing rectangle of $[-10, 10]$ by $[-10, 10]$
   b) $y = 5x - 1$, viewing rectangle of $[-10, 10]$ by $[-50, 50]$

39. Suppose the equation $y = 5x - 2$ is graphed using the following viewing rectangles. Which graph will appear steeper? Is the slope the same or different in the two cases?
   a) $[-10, 10]$ by $[-10, 10]$
   b) $[-2, 2]$ by $[-10, 10]$

40. Suppose the equation $y = 5x - 2$ is graphed using the following viewing rectangles. Which graph will appear steeper? Is the slope the same or different in the two cases?
   a) $[-10, 10]$ by $[-10, 10]$
   b) $[-25, 25]$ by $[-10, 10]$

41. The graph of the function $f(x) = x$ drawn with a viewing rectangle $[-10, 10]$ by $[-10, 10]$ will appear identical to the graph of $g(x) = 5x$ on which viewing rectangle?

In Exercises 42 and 43, write an equation for the perpendicular bisector of the line segment determined by each pair of points.

42. $(3, -5); (-6, 10)$ **43.** $(-1, 3); (5, -3)$

In Exercises 44–47, write an equation for the line determined by the given conditions.

44. It contains the point $(4, -1)$ and is perpendicular to the line $2x - y = 4$.

45. It contains the point $(-2, 4)$ and is parallel to the line $x - 4y = 8$.

46. It contains the point $(-2, 0)$ and is parallel to the line $x = 4$.

47. It contains the point $(0, 2)$ and is parallel to the line $y = 8$.

48. Use a graphing utility to complete Example 1, part 2. Change the viewing rectangle until you are able to get an approximation for the solution to within the nearest hundredth.

49. Show that the triangle with vertices $(-1, 2)$, $(-6, -2)$, and $(2, -12)$ is a right triangle.

50. What do a and b represent in the equation $x/a + y/b = 1$? (*Hint:* Draw a complete graph of the equation for a few specific values of a and b.)

51. Explain why either point can be labeled (x_1, y_1) or (x_2, y_2) when applying the slope formula.

52. Candy worth \$0.95/lb is mixed with candy worth \$1.35/lb. Determine an inequality that expresses the requirement that the value of the new mixture not exceed \$1.15/lb. Find its complete graph.

53. Use the concept of slope to determine whether the three points $(-1, 2)$, $(2, 4)$, and $(6, 9)$ are *collinear*, that is, whether they all lie on the same line.

54. Determine D so that $A = (1, 2)$, $B = (3, 5)$, $C = (7, 7)$, and D are vertices of a parallelogram. (There are three answers.)

55. Do the midpoints of the sides of the quadrilateral with vertices $(-1, 2)$, $(2, 6)$, $(-3, 7)$, and $(8, -2)$ form a rectangle?

In Exercise 56 let $A = (3, 0)$, $B = (-1, 2)$, and $C = (5, 4)$.

56. Prove that the line through A and the midpoint of BC is perpendicular to BC.

57. Triangle ABC has vertices $A(0, 0)$, $B(2, 6)$, and $C(5, 0)$. Write an equation for each of the three lines determined by a vertex and the midpoint of the opposite side of the triangle.

58. Show that in any triangle, the line segment joining the midpoints of two sides is parallel to the third side.

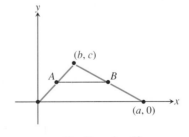

For Exercise 58.

59. Show that if the midpoints of consecutive sides of any quadrilateral are connected, the result is a parallelogram.

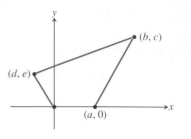

For Exercise 59.

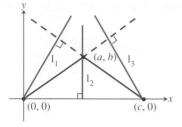

For Exercise 67.

60. Prove that the line segments joining the midpoints of the opposite sides of any quadrilateral bisect each other.

61. Prove that the diagonals of a rectangle are equal in length.

62. Prove that if the diagonals of a rectangle are perpendicular, then the rectangle is a square.

63. Prove that the diagonals of a rhombus are perpendicular.

For Exercises 64–66, refer to the following figure.

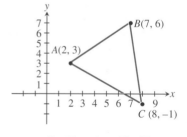

For Exercises 64–66.

64. Find an equation for the line through A and B.

65. Find an equation for the line through A and the midpoint of BC. This segment is called a **median**.

66. Find an equation for the line through A and perpendicular to BC. This segment is called an **altitude**.

67. Show that the altitudes of a triangle are *coincident* (that is, they intersect in a common point). *Hint:* Show that the three altitudes ℓ_1, ℓ_2, and ℓ_3 intersect at the point $(a, a(c - a)/b)$.

68. Prove that if $c \neq d$ and a and b are not both zero, then $ax + by = c$ and $ax + by = d$ are parallel lines.

Exercises 69–71 refer to the following **problem situation**: A house was purchased 8 years ago for $42,000. This year it was appraised at $67,500. Assume that the value V of the house changes linearly with time (t).

69. Find a linear equation that models this problem situation.

70. Use a graph to estimate when this house will be worth $90,000.

71. Determine algebraically when this house will be worth $90,000.

Exercises 72–74 refer to the following **problem situation**: Mary Ellen intends to invest $18,000, putting part of the money in one savings account, which enables her to withdraw her money without penalty, that pays 5% annually. The rest she will invest in another account that pays 8% annually.

72. Write an equation that describes the total interest I received at the end of 1 yr in terms of the amount A invested at 8%.

73. Find a complete graph of the equation in Exercise 72.

74. If Mary Ellen's annual interest is $1020, how much of her original $18,000 did she invest at 8%?

75. The equation describing home appreciation in Example 3 was $V = 2500t + 60{,}000$. The percent appreciation in year 1 is

$$\frac{\text{change in value during year}}{\text{value at beginning of year}} = \frac{2500}{60{,}000} = 4.17\%.$$

What is the percent appreciation
a) During year 2?
b) During year 3?
c) During year 4?
d) During year 5?

In Exercises 76–83, find a complete graph of each equation or inequality.

76. $y = 3x - 6$

77. $y < x - 2$

78. $x - 2y \geq 6$

79. $2x + y = 8$

80. $2x - 6y = 8$

81. $y \geq x$

82. $5x - 2y < 8$

83. $2x + 3y - 5 \geq 0$

84. Complete these four steps to show that the distance d from point $P(x_0, y_0)$ to line L with equation $Ax + By + C = 0$ is given by

$$d = \frac{|Ax_0 + By_0 + C|}{\sqrt{A^2 + B^2}}.$$

a) Find the equation of the line through P and perpendicular to L.

b) Find the coordinates of the point Q of intersection of line L and the line found in part (a).

c) Use the distance formula to find the distance d between points P and Q.

d) Express the coordinates of Q in terms of x_0 and y_0 to obtain the desired equation.

85. Use the formula developed in Exercise 84 to find the length of each altitude of the triangle used for Exercises 64–66.

86. Is there a line that cannot be graphed with a function grapher? Are there any lines whose equation cannot be written in the form $y = mx + D$?

87. Tickets to a concert cost $2.50 for students and $3.75 for adults. Suppose x represents the number of student tickets sold and y represents the number of adult tickets sold. Write an inequality that describes the condition that total receipts must exceed $1500.

88. Discussion Form groups of two or three students and discuss with each other the following situation. Suppose you plan to graph the inequality in Exercise 87. If this graph represents the problem situation of Exercise 87, discuss whether your graph should include all the points above a certain line or only a square grid of points above a certain line. Defend your decision.

89. Discussion Do you think a linear model of economic growth is a good one to explain economic activity as you have experienced it? For example, consider the value of the house in Section 1.4, Example 1. Is house value likely to change linearly? Consider Exercise 75.

90. Writing to Learn Write several sentences that explain how the solution to the inequality $0 < 3x + 5$ differs from the solution to the inequality $y < 3x + 5$.

91. Writing to Learn Read in a reference manual about aspect ratio. Then write a paragraph explaining how aspect ratio is related to Exercise 38.

1.5 _____ Quadratic Functions and Geometric Transformations

This section focuses on a family of functions known as the quadratic functions.

Definition 1.9 Quadratic Function

A function f is a **quadratic function** if it can be written in the form $f(x) = ax^2 + bx + c$, where $a, b,$ and c are real numbers, called **coefficients,** and $a \neq 0$.

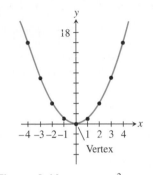

Figure 1.46 $f(x) = x^2$.

The simplest and most basic quadratic function is $f(x) = x^2$ (see Fig. 1.46). Observe that $f(-x) = f(x)$, which means that if (x, y) is a point of the graph, $(-x, y)$ is also. The y-axis is a line of **symmetry**. Recall that the graph of this function is called a parabola. The point $(0, 0)$ is the lowest point of the graph and is called the **vertex** of the parabola.

Transforming the Graph of $y = x^2$

We shall show that the graph of any quadratic function can be obtained from $y = x^2$ by using the transformations of stretching or shrinking, shifting the whole graph horizontally or vertically, reflecting the graph through the x-axis, or some combination of these transformations.

Frank says, "Graphing utilities are great tools for showing the transformations applied to parent functions such as $y = x^2$."

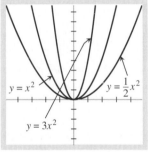

[−5, 5] by [−5, 10]

Figure 1.47   Three parabolas.

Notes on Exercises

Ex. 1–6 require the use of a graphing utility. In each exercise, three related graphs are drawn on the same coordinate system. Have students describe how one graph can be obtained from the other or conjecture the position of the new graph before using a graphing utility.

🔍 EXPLORE WITH A GRAPHING UTILITY

Find graphs for the following four functions, one after another without erasing: $f_1(x) = x^2$, $f_2(x) = \frac{1}{3}x^2$, $f_3(x) = 2x^2$, and $f_4(x) = 4x^2$.

Questions:

1. Which graphs lie below the graph of $f(x) = x^2$?
2. Which graphs lie above the graph of $f(x) = x^2$?
3. Propose an equation of a quadratic function whose graph is between the graphs of $f(x) = 2x^2$ and $f(x) = 4x^2$. Support your guess graphically.

Generalize   Describe the graph of $f(x) = ax^2$ for various positive values of the coefficient a.

Conjecture   How will these graphs change if a is negative?

The graphs in Fig. 1.47 add additional evidence to support your generalizations. Notice that all these graphs pass through the point $(0, 0)$, that all of them lie above the x-axis, and that the y-axis is a line of symmetry for all of them.

If x is any nonzero number, then $\frac{1}{2}x^2 < x^2$ and the graph of $y = \frac{1}{2}x^2$ lies below the graph of $y = x^2$. Similarly, $x^2 < 3x^2$ and the graph of $y = x^2$ lies below the graph of $y = 3x^2$. These general relationships can be summarized as follows.

Value of a	Comparison of Graphs
$a > 1$	$y = ax^2$ is above $y = x^2$.
$0 < a < 1$	$y = ax^2$ is below $y = x^2$.

Teaching Note

You can illustrate vertical stretch and shrink by taking a thin piece of stiff wire and bending it into the form of a parabola. Have a student hold the "vertex" of the wire firmly in place with an index finger on the top of a flat surface. Have another student hold the ends of the wire firmly with each hand and then pull upward (stretch the parabola) to show that a coefficient of 6, for example, in front of the x-squared term will produce a stretched graph.

Notes on Examples

For Example 1, be sure that the graphing utility is set in a sequential graphing mode rather than a simultaneous mode. This will allow students to see the transformations appear one at a time in the same order they were entered into the calculator.

The graph of $y = x^2$ can be *stretched up* to the graph of $y = 3x^2$ or *shrunk down* to the graph of $y = \frac{1}{2}x^2$.

Definition 1.10   Vertical Stretch and Shrink

If $a > 1$, the graph of $f(x) = ax^2$ can be obtained from the graph of $y = x^2$ with a **vertical stretch** of the graph of $y = x^2$ by a factor of a.

   If $0 < a < 1$, the graph of $f(x) = ax^2$ can be obtained from the graph of $y = x^2$ with a **vertical shrink** of the graph of $y = x^2$ by the factor a.

   The graph of f has vertex $(0, 0)$, and the y-axis is a line of symmetry.

The next example confirms whether your conjecture in the Exploration was correct.

E X A M P L E   1   Obtaining a Graph by Reflection

Find complete graphs of the functions $y = -x^2$, $y = -3x^2$, and $y = -\frac{1}{2}x^2$, one after the other in the same viewing rectangle. How do these graphs differ from those in Fig. 1.47?

Solution   Figure 1.48 shows that all the graphs are parabolas that open downward, whereas in Fig. 1.47 all the parabolas open upward.   ≡

If the coordinate plane in Fig. 1.47 is folded along the x-axis, the resulting graphs would coincide with those in Fig. 1.48.

   In general, any graph of the form $y = -ax^2$ can be obtained by reflecting the graph of $y = ax^2$ through the x-axis.

E X A M P L E   2   Combining Two Transformations

Describe how a complete graph of $y = -5x^2$ can be obtained from the graph of $y = x^2$ by stretching, shrinking, and/or reflection.

Solution   There are two solutions. You can vertically stretch the graph of $y = x^2$ by a factor of 5 and then reflect the result through the x-axis. Or you can reflect the graph of $y = x^2$ through the x-axis and then stretch the graph of $y = -x^2$ by a factor of 5.   ≡

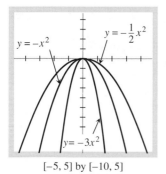

$[-5, 5]$ by $[-10, 5]$

Figure 1.48   Three downward-opening parabolas.

Common Errors

Students sometimes make errors in the order in which they apply transformations to a graph. Warn students of the danger of reversing the order of some of these transformations. Point out that, in general, they are not commutative. Explore which transformations are commutative.

Teaching Note

At first, have students operate within a fixed viewing rectangle. Then have them vary the viewing rectangle, keeping the functions fixed. This should lead to an interesting discussion of issues relating to transformations, particularly the effect of the parameter a on the graph of $y = ax^2$.

In Example 2, the order in which the two transformations are performed does not matter. That will not always be the case, however. Let's summarize the transformations explored so far.

Transformations of Stretching/Shrinking and Reflection

Condition on coefficient a	To obtain $y = ax^2$ from $y = x^2$
$a > 1$	Stretch by a factor of a.
$0 < a < 1$	Shrink by a factor of a.
$-1 < a < 0$	Shrink by factor $|a|$ and reflect through the x-axis.
$a < -1$	Stretch by factor $|a|$ and reflect through the x-axis.

E X A M P L E   3   Obtaining a Graph by a Vertical Shift

Describe how complete graphs of $y = x^2 + 2$ and $y = x^2 - 3$ can be obtained from the graph of $y = x^2$.

Solution   The graph of $y = x^2 + 2$ can be obtained from the graph of $y = x^2$ by shifting it up 2 units (see Fig. 1.49a).

Similarly, the graph of $y = x^2 - 3$ can be obtained from the graph of $y = x^2$ by shifting it down 3 units (see Fig. 1.49b). When shifting a graph up or down, the y-axis remains a line of symmetry.

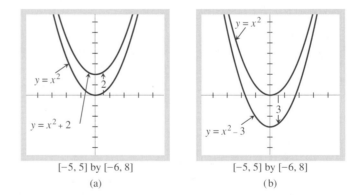

[−5, 5] by [−6, 8]   [−5, 5] by [−6, 8]
(a)   (b)

Figure 1.49   Vertical-shift transformations of $y = x^2$: (a) to $y = x^2 + 2$; (b) to $y = x^2 - 3$.

Notes on Exercises

Ex. 7–13 give students a further opportunity to make and test conjectures concerning geometric transformations. Ex. 14–31 give students a chance to use their knowledge of the parts of a parabola and transformations.

Example 3 can be generalized to the following definition.

Definition 1.11   Vertical Shift

The graph of $f(x) = x^2 + k$ is said to be obtained from the graph of $y = x^2$ by a **vertical shift**.

- If $k > 0$, the shift is *up* k units.
- If $k < 0$, the shift is *down* $|k|$ units.

For any k, the graph of f has the vertex $(0, k)$ and the y-axis is the line of symmetry.

The vertical-shift transformation is a rigid motion because it preserves both the size and shape of the graph. In contrast, the transformations vertical stretch and vertical shrink are not rigid motion transformations.

E X A M P L E 4   Obtaining a Graph by a Horizontal Shift

Describe how complete graphs of $y = (x - 3)^2$ and $y = (x + 2)^2$ can be obtained from the graph of $y = x^2$.

Solution   The graph of $y = (x - 3)^2$ can be obtained from the graph of $y = x^2$ by shifting it right 3 units (see Fig. 1.50a). Similarly, the graph of $y = (x + 2)^2$ can be obtained from the graph of $y = x^2$ by shifting it left 2 units (see Fig. 1.50b). When a graph is shifted right or left, the line of symmetry of the graph slides in the same direction.

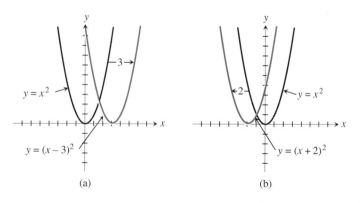

(a)   (b)

Figure 1.50 Two horizontal-shift transformations of $y = x^2$: (a) to $y = (x - 3)^2$; (b) to $y = (x + 2)^2$.

From Example 4, we can generalize the following definition.

> ### Definition 1.12   Horizontal Shift
>
> The graph of $f(x) = (x-h)^2$ is said to be obtained from the graph of $y = x^2$ by a **horizontal shift**.
>
> - If $h > 0$, the shift is *right* h units.
> - If $h < 0$, the shift is *left* $|h|$ units.
>
> For any h, the graph of f has the vertex $(h, 0)$ and the line of symmetry $x = h$.

In summary, we have considered four types of transformations:

1. Vertical stretch or shrink

2. Vertical shift

3. Horizontal shift

4. Reflection through the x-axis

As these transformations are performed on the parabola $y = x^2$, it is often helpful to keep track of the vertex and the line of symmetry.

Examples 5–7 illustrate how these transformations can be combined.

E X A M P L E   5   Combining Two Shift Transformations

Find the vertex and line of symmetry of the graph of $f(x) = (x+2)^2 - 3$. Describe how its complete graph can be obtained from the graph of $y = x^2$.

Solution   Study the algebraic form of the equation $y = (x + 2)^2 - 3$. Rewriting the factor $(x + 2)^2$ as $(x - (-2))^2$ shows that a horizontal shift left 2 units is required. The "-3" indicates a shift down 3 units (see Fig. 1.51). The new vertex is $(-2, -3)$, and the new line of symmetry is $x = -2$.

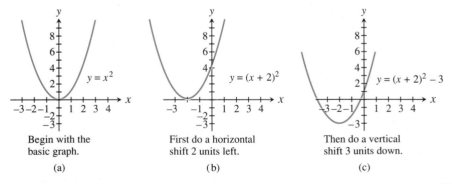

Begin with the basic graph.	First do a horizontal shift 2 units left.	Then do a vertical shift 3 units down.
(a)	(b)	(c)

Figure 1.51   Transforming $y = x^2$ to $y = (x + 2)^2 - 3$.

A look at the graphs in Fig. 1.51 shows that the order of the transformations in the solution could have been reversed. The vertical shift could have been completed first, followed by the horizontal shift.

E X A M P L E   6   Combining Stretch and Shift Transformations

Describe how the complete graph of $f(x) = 5x^2 + 4$ can be obtained from the graph of $y = x^2$. Find the vertex and line of symmetry of f.

Solution   (See Fig. 1.52.)

1.   Begin with the basic graph of $y = x^2$.
2.   Stretch vertically 5 units to get $y = 5x^2$.
3.   Then shift up 4 units to get $y = 5x^2 + 4$.

The vertex is $(0, 4)$, and the line of symmetry is $x = 0$.

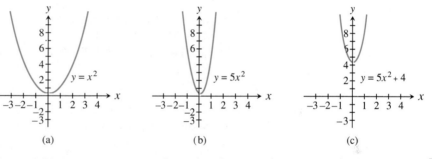

(a)   (b)   (c)

Figure 1.52   Transforming $y = x^2$ to $y = 5x^2 + 4$.

Notice that in Example 6, a vertical stretch followed by a vertical shift does not result in the same graph as the reverse order of transformations. To shift up first and then do a vertical stretch would yield the following sequence of equations:

$$y = x^2, \qquad y = x^2 + 4, \qquad \text{and} \qquad y = 5(x^2 + 4) = 5x^2 + 20.$$

Note that this final equation is not the same as $y = 5x^2 + 4$.

E X A M P L E   7   Combining Several Transformations

Describe how the complete graph of $f(x) = -3(x - 1)^2 + 4$ can be obtained from the graph of $y = x^2$ and find its vertex and line of symmetry.

Solution

1. Begin with the basic graph of $y = x^2$.

2. Shift right 1 unit to find $y = (x - 1)^2$.

3. Stretch vertically by a factor of 3 to get $y = 3(x - 1)^2$.

4. Reflect through the x-axis to arrive at $y = -3(x - 1)^2$.

5. Shift up 4 units to get $y = -3(x - 1)^2 + 4$.

The vertex is $(1, 4)$, and the line of symmetry is $x = 1$.   ▤

Thus far in this section, all the quadratic functions have been written in a particular form called standard form (see Definition 1.13). Doing this has been helpful for finding the right combination of transformations asked for in the examples. Later in the section, we will see how to convert other quadratic equation forms into standard form in order to find the right transformations.

Notes on Exercises

Ex. 32–39 involve the use of more than one transformation. Students must again consider the question of the commutativity of transformations.

Definition 1.13   Standard Form of a Quadratic Function

The **standard form of a quadratic function** $y = f(x)$ is
$f(x) = a(x - h)^2 + k$, where a, h, and k are real-number constants.

We summarize how the graph of a quadratic function in standard form can be obtained by transforming the graph of the function $y = x^2$.

Transforming $y = x^2$ to $f(x) = a(x - h)^2 + k$

The graph of a quadratic function in standard form $f(x) = a(x - h)^2 + k$ can be obtained from the graph of $y = x^2$ through the following sequence of transformations, performed in the indicated order:

1. *Horizontal shift h units:* Shift right if $h > 0$ and left if $h < 0$.

2. *Vertical stretch/shrink by factor $|a|$.* Stretch if $|a| > 1$ and shrink if $0 < |a| < 1$.

3. *Reflect in x-axis only if $a < 0$.*

4. *Vertical shift k units:* Shift up if $k > 0$ and down if $k < 0$.

The vertex of the graph of f is the point (h, k), and its line of symmetry is $x = h$.

Finding the Graph of $f(x) = ax^2 + bx + c$

To find the vertex and line of symmetry of a general quadratic function and to identify the transformations used to transform $y = x^2$, use the technique of completing the square to change the function to standard form.

E X A M P L E   8   Writing a Quadratic Function in Standard Form

Describe how the complete graph of $f(x) = -2x^2 - 12x - 13$ can be obtained from the graph of $y = x^2$. Find the vertex and line of symmetry of f.

Solution

$$f(x) = -2x^2 - 12x - 13 \quad \text{First complete the square.}$$

$$= -2(x^2 + 6x) - 13$$

$$= -2(x^2 + 6x + 9) - 13 + 18$$

$$= -2(x + 3)^2 + 5$$

The graph of this function (see Fig. 1.53) can be obtained from the graph of $y = x^2$ as follows: Shift left 3 units, stretch vertically by a factor of 2, reflect through the x-axis, and finally shift up 5 units.

The vertex is $(-3, 5)$, and the line of symmetry is $x = -3$.

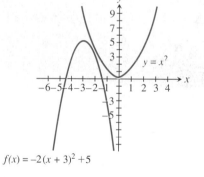

$f(x) = -2(x + 3)^2 + 5$

Figure 1.53   Notice how much easier it is to graph $f(x) = -2x^2 - 12x - 13$ once it is written in the form $f(x) = -2(x + 3)^2 + 5$.   ≡

This following theorem describes the coordinates of the vertex and the line of symmetry for a general quadratic function.

Theorem 1.3   Symmetry and Vertex of $f(x) = ax^2 + bx + c$

For the graph of the function $f(x) = ax^2 + bx + c$,

- the vertex is $\left(-\dfrac{b}{2a},\, c - \dfrac{b^2}{4a}\right)$, and

- the line of symmetry is $x = -\dfrac{b}{2a}$.

Proof

$$f(x) = ax^2 + bx + c$$

$$= a\left(x^2 + \frac{b}{a}x\right) + c$$

$$= a\left(x^2 + \frac{b}{a}x + \frac{b^2}{4a^2}\right) + c - \frac{b^2}{4a}$$

$$= a\left(x + \frac{b}{2a}\right)^2 + \left(c - \frac{b^2}{4a}\right)$$

So the line of symmetry is $x = -b/2a$. This means that the x-coordinate of the vertex is also $-b/2a$ and the y-coordinate is $f(-b/2a) = c - b^2/4a$. ■

Zeros of a Quadratic Function

A number k is a zero for a function f if $f(k) = 0$. Consequently, a number k is a zero for a quadratic function $f(x) = ax^2 + bx + c$ if, and only if, it is a solution to the equation $ax^2 + bx + c = 0$. This means that $(k, 0)$ is an x-intercept of the graph of f. A quadratic function can have zero, one, or two zeros (see Fig. 1.54). The discriminant $b^2 - 4ac$ can be used to determine the number of real-number zeros for a quadratic function.

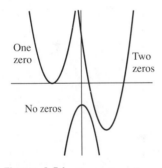

One zero

Two zeros

No zeros

Figure 1.54   Examples of quadratic graphs with different numbers of zeros.

Theorem 1.4   Discriminant and Zeros

Let $f(x) = ax^2 + bx + c$.

- If $b^2 - 4ac < 0$, then f has no real-number zeros, and the graph of f does not cross the x-axis.

- If $b^2 - 4ac = 0$, then f has exactly one real-number zero, and the graph of f is tangent to the x-axis.

- If $b^2 - 4ac > 0$, then f has two real-number zeros, and the graph of f crosses the x-axis twice.

Proof   Recall that zeros to the quadratic equation $ax^2 + bx + c = 0$ are given by the quadratic formula

$$x = \frac{-b \pm \sqrt{b^2 - 4ac}}{2a}.$$

If $b^2 - 4ac < 0$, this expression is not defined and the equation has no real solutions; if $b^2 - 4ac = 0$, the quadratic equation has only one real solution; and if $b^2 - 4ac > 0$, there are both positive and negative square roots, which means the equation has two real solutions.   ≡

Notes on Exercises

Ex. 58–65 offer some classical applications of quadratic functions. The Writing to Learn exercises can be used to analyze students' understanding of the concepts.

E X A M P L E   9   APPLICATION: Determining Maximum Enclosed Area

If 200 ft of fence are used to enclose a rectangular plot of land using an existing wall as one side of the plot, find the dimensions of the rectangle with maximum enclosed area.

Solution   Let x represent the width of the rectangle. Then $200 - 2x$ represents its length, and $A(x) = x(200 - 2x)$ is an algebraic representation for the area of the rectangular plot. Figure 1.55(a) suggests that the maximum area is about 5000 when x is about 50.

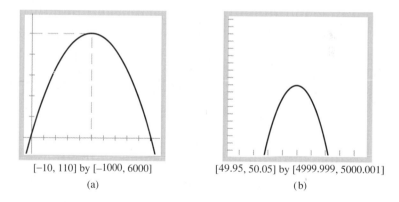

[-10, 110] by [-1000, 6000]   [49.95, 50.05] by [4999.999, 5000.001]

(a)   (b)

Figure 1.55   Two views of $A(x) = x(200 - 2x)$.   ≡

Assignment Guide

Day 1: Ex. 1–53 odd
Day 2: Ex. 55–57, 62, 63, 65

The zeros of $A(x) = x(200 - 2x)$ are 0 and 100. Thus, by symmetry, the x-coordinate of the vertex is 50 and the y-coordinate of the vertex is $A(50) = 5000$. This confirms that the dimensions of the maximum enclosed area are 50 ft by 100 ft.

Exercises for Section 1.5

In Exercises 1–6, find complete graphs of the three functions in the same viewing rectangle.

1. $y = 3x^2$, $y = 3(x-2)^2$, $y = 3(x+2)^2$

2. $y = -x^2$, $y = -(x+4)^2$, $y = -(x-4)^2$

3. $y = 2x^2$, $y = 2(x-1)^2$, $y = 2(x+1)^2$

4. $y = -x^2$, $y = -x^2 - 4$, $y = -x^2 + 4$

5. $y = x^2 + 3$, $y = (x+2)^2 + 3$, $y = (x-3)^2 + 3$

6. $y = x^2 - 2$, $y = 3x^2 - 2$, $y = -2x^2 - 2$

In Exercises 7–9, when both functions are graphed in the same viewing rectangle, which graph is "above" the other?

7. $y = 3x^2$ or $y = 4x^2$ **8.** $y = \frac{2}{3}x^2$ or $y = \frac{3}{4}x^2$

9. $y = 3.21x^2$ or $y = 3.021x^2$

In Exercises 10–13, describe a transformation that can be used to transform the graph of the first function to the graph of the second function.

10. $y = (x-3)^2$ and $y = (x+3)^2$

11. $y = (x-2)^2$ and $y = (x-3)^2$

12. $y = (x+5)^2$ and $y = (x+2)^2$

13. $y = (x+3.2)^2$ and $y = (x+0.2)^2$

In Exercises 14–18, describe how each graph can be obtained from the graph of $y = x^2$. Support your work with a grapher.

14. $f(x) = 4x^2$ **15.** $f(x) = -3x^2$

16. $f(x) = (x-5)^2$ **17.** $f(x) = (x+1)^2$

18. $f(x) = 2x^2 - 3$ **19.** $f(x) = -3x^2 + 2$

In Exercises 20–25, find the vertex and the line of symmetry for each parabola.

20. $f(x) = 3(x-1)^2 + 5$ **21.** $f(x) = -3(x+2)^2 - 3$

22. $f(x) = 5(x-3)^2 - 7$ **23.** $f(x) = 2(x-\sqrt{3})^2 + 4$

24. $f(x) = (x-5)^2 + \sqrt{2}$ **25.** $f(x) = \sqrt{5}(x+4)^2 + 3$

In Exercises 26–31, draw a complete graph of each function. Support your sketch with a grapher.

26. $f(x) = (x-4)^2 + 3$ **27.** $f(x) = -(x+3)^2 - 2$

28. $f(x) = 2(x-1)^2 + 3$ **29.** $f(x) = -3(x+4)^2 - 5$

30. $f(x) = 3(x-4)^2 + 7$ **31.** $f(x) = (x-\sqrt{2})^2 - \sqrt{3}$

32. Write an equation whose graph can be obtained from the graph of $y = x^2$ by vertically stretching by a factor of 3 and then shifting right 4 units.

33. Write an equation whose graph can be obtained from the graph of $y = x^2$ by shifting right 4 units and then vertically stretching by a factor of 3.

34. Are the graphs of the equations found in Exercises 32 and 33 the same? Explain any differences.

35. Write an equation whose graph can be obtained from the graph of $y = x^2$ by vertically stretching by a factor of 3 and then vertically shifting up 4 units.

36. Write an equation whose graph can be obtained from the graph of $y = x^2$ with a vertical shift up 4 units followed by a vertical stretch by a factor of 3.

37. Are the graphs of the equations found in Exercises 35 and 36 the same? Explain any differences.

38. Write an equation whose graph can be obtained from the graph of $y = x^2$ by shifting left 2 units, then vertically stretching by a factor of 3, and finally shifting down 4 units.

39. Write an equation whose graph is obtained from the graph of $y = x^2$ by shifting right 4 units, then vertically stretching by a factor of 2, followed by a reflection through the x-axis, and ending with a shift up 3 units.

In Exercises 40–43, complete the square for each function. Then write a sequence of transformations that will produce its graph from the graph of $y = x^2$. In each case, find the vertex and the line of symmetry of the parabola. Support your work with a grapher.

40. $f(x) = x^2 - 4x + 6$ **41.** $f(x) = x^2 - 6x + 12$

42. $f(x) = 2x^2 - 8x + 20$ **43.** $f(x) = 10 - 16x - x^2$

In Exercises 44–49, use the discriminant to determine how many real-number zeros each quadratic function has. Support your work with a grapher.

44. $f(x) = 2x^2 + 5x + 1$ **45.** $f(x) = x^2 - 2x + 1$

46. $f(x) = x^2 + x + 1$ **47.** $f(x) = 2x^2 - 4x + 1$

48. $f(x) = 3x^2 - 7x - 3$ **49.** $f(x) = 2x^2 - x + 3$

50. Find the midpoint of the segment whose endpoints are the real zeros of $f(x) = (x-8)(x+2)$.

51. Find the midpoint of the segment whose endpoints are the real zeros of $f(x) = x^2 - x - 1$.

52. Show that the line of symmetry for the graph of $f(x) = (x-a)(x-b)$ is $x = (a+b)/2$, where a and b are any positive real numbers.

53. Find the vertex of $f(x) = (x - a)(x - b)$, where a and b are any positive real numbers.

54. Let f be the function given by the following graph. Determine the points on the graph of $y = 2 + 3f(x + 1)$ corresponding to the points $(-2, f(-2))$, $(0, f(0))$, and $(4, f(4))$.

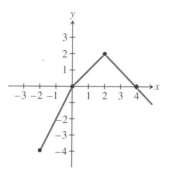

For Exercise 54.

In Exercises 55–57, $f(x) = |x|$.

55. Complete the equation $f(x-2) = ?$ Explain how the graph of $f(x - 2)$ can be obtained from the graph of $f(x)$.

56. Complete the equation $3f(x) = ?$ Explain how the graph of $3f(x)$ can be obtained from the equation of $f(x)$.

57. Complete the equation $2f(x+3) - 1 = ?$ Explain how the graph of $2f(x+3) - 1$ can be obtained from the graph of $f(x)$.

58. A rectangle is 3 ft longer than it is wide. If each side is increased by 1 ft, the area of the new rectangle is 208 sq ft. Find the dimensions of the original rectangle.

59. A rectangular pool with dimensions of 25 by 40 ft is surrounded by a walk with a uniform width. If the area of the walk is 504 sq ft, find the width of the walk.

60. Among the rectangles with perimeters of 100 ft, find the dimensions of the one with the maximum area.

61. A piece of wire 20 ft long is cut into two pieces so that the sum of the squares of the length of each piece is 202 sq ft. Find the length of each piece.

62. A long rectangular sheet of metal 10 in. wide is to be made into a gutter by turning up sides of equal length perpendicular to the sheet. Find the length that must be turned up to produce a gutter with maximum cross-sectional area.

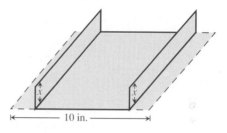

For Exercise 62.

63. A rectangular fence is to be constructed around a field so that one side of the field is bounded by the wall of a large building. Determine the maximum area that can be enclosed if the total length of fencing to be used is 500 ft.

64. **Writing to Learn** Suppose the graph of a function f is obtained from the graph of a function g by performing a sequence of transformations. Write several sentences that explain when the sequence of transformations can be performed in any order versus when the order makes a difference.

65. **Writing to Learn** Find a complete graph of $y = x^2$ in $[-10, 10]$ by $[-10, 10]$ and $y = 3x^2$ in $[-10, 10]$ by $[-50, 50]$. Write several paragraphs that describe which graph appears to be the steepest and describe what you mean by that phrase. But which graph really is the steepest and why?

1.6 _____ Operations on Functions and Composition of Functions

A function consists of a domain and a rule. The rule is often described by using an algebraic formula. A new function can be defined by combining several other functions using addition, subtraction, multiplication, or division. For example, consider the functions $f(x) = x^2$ and $g(x) = \sqrt{x+1}$.

$$f(x) + g(x) = x^2 + \sqrt{x+1} \qquad \text{Sum of } f \text{ and } g$$

$$f(x) - g(x) = x^2 - \sqrt{x+1} \qquad \text{Difference of } f \text{ and } g$$

$$f(x)g(x) = x^2\sqrt{x+1} \qquad \text{Product of } f \text{ and } g$$

$$\frac{f(x)}{g(x)} = \frac{x^2}{\sqrt{x+1}} \quad (x \neq -1) \qquad \text{Quotient of } f \text{ and } g$$

The domain of the new function consists of all numbers x that belong to the domains of *both* f and g. The quotient's domain also requires that $g(x) \neq 0$.

Definition 1.14 Sum, Difference, Product, and Quotient of Functions

Let f and g be two functions. We define the **sum, difference, product**, and **quotient** of f and g to be the functions whose domains are the set of all numbers common to the domains of f and g and that are defined as follows:

$$(f + g)(x) = f(x) + g(x)$$

$$(f - g)(x) = f(x) - g(x)$$

$$(fg)(x) = f(x)g(x)$$

$$\left(\frac{f}{g}\right)(x) = \frac{f(x)}{g(x)} \qquad (g(x) \neq 0)$$

Notice that in the quotient function, the denominator cannot be zero.

EXAMPLE 1   Combining Functions

Let $f(x) = \sqrt{x+3}$ and $g(x) = \sqrt{x-2}$. Find the domains and rules for $f + g$ and f/g.

Solution

$$(f + g)(x) = f(x) + g(x)$$

$$= \sqrt{x+3} + \sqrt{x-2} \quad x+3 \geq 0 \text{ and } x-2 \geq 0$$

$$\left(\frac{f}{g}\right)(x) = \frac{f(x)}{g(x)} = \frac{\sqrt{x+3}}{\sqrt{x-2}} \quad x+3 \geq 0 \text{ and } x-2 > 0$$

The domain of $f = [-3, \infty)$, and the domain of $g = [2, \infty)$. Therefore the domain of $f + g$ is $[2, \infty) \cap [-3, \infty) = [2, \infty)$.

Since the denominator of (f/g) cannot be zero, its domain is $(2, \infty)$. ≡

E X A M P L E   2   Combining Functions

Let $f(x) = |x|$ and $g(x) = x$. Find a complete graph of $(f/g)(x) = f(x)/g(x)$.

Solution   Figure 1.56 shows the graph found with a graphing utility. However, the grapher does not show the fact that $f(0)$ is not defined. This fact can be conveyed when sketching this graph by adding small open circles to the endpoints of the horizontal pieces of the graph (see Fig. 1.57).   ≡

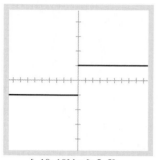

[−10, 10] by [−5, 5]

Figure 1.56 $f(x) = |x|/x$. Note that $f(0)$ is not defined, but the grapher cannot indicate this fact.

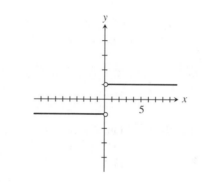

Figure 1.57 Complete graph of $|x|/x$.

Notes on Exercises

Ex. 8–11 establish the notation for the composition of functions.

Composition of Functions

The function $h(x) = \sqrt{x^2 + 3}$ can be thought of as built from two previously defined functions. Suppose that $g(x) = \sqrt{x}$ and $f(x) = x^2 + 3$. Observe that

$$g(f(x)) = g(x^2 + 3) \quad \text{Replace } f(x) \text{ by } x^2 + 3.$$

$$= \sqrt{x^2 + 3}.$$

With this process a new function, called the composition of f by g, is formed from two given functions f and g. For example, we say that $h(x) = \sqrt{x^2 + 3}$ is the composition of $x^2 + 3$ by $\sqrt{x}$.

Common Errors

Students must understand how the composition notation works. The composition notation $f \circ g(x)$ should be compared to the $f(g(x))$ notation so that students can move from one notation to the other with facility. You may wish to read $f \circ g$, "f following g," to emphasize the order in which the functions are applied.

Definition 1.15   Composition of Functions f by g

The **composition of functions f by g**, denoted $g \circ f$, is given by

$$g \circ f(x) = g(f(x)).$$

The domain of $g \circ f$ is the set of all x in the domain of f such that $f(x)$ is in the domain of g.

Domain of f Range of f

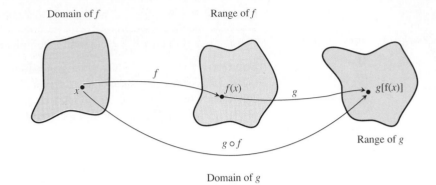

Domain of g

Figure 1.58   The composition of functions f by g.

POTENTIAL MISUNDERSTAND-ING

Since we read from left to right, it is easy to make the error of thinking that in the composition $g \circ f$, function g is performed first. In fact, the opposite is true (see Fig. 1.58).

Notice that in order for a value of x to be in the domain of $g \circ f$, two conditions must be met:

a) x must be in the domain of f, and

b) $f(x)$ must be in the domain of g.

In other words, the range of f must be a subset of the domain of g.

The next several examples illustrate forming the composition.

E X A M P L E   3   Forming the Composition

Let $f(x) = x + 1$ and $g(x) = \sqrt{x}$. Find the following:

a) $f(g(9))$

b) $g(f(3))$

c) $f(g(x))$

d) $g(f(x))$

Solution

a) $f(g(9)) = f(\sqrt{9}) = f(3) = 4$ Substitute $\sqrt{9}$ for $g(9)$ and evaluate $\sqrt{9}$.

b) $g(f(3)) = g(4) = \sqrt{4} = 2$

c) $f(g(x)) = f(\sqrt{x}) = \sqrt{x} + 1$ Substitute $\sqrt{x}$ for $g(x)$.

d) $g(f(x)) = g(x + 1) = \sqrt{x + 1}$ ▤

Example 3 illustrates that $f(g(x))$ and $g(f(x))$ are generally not equal functions.

E X A M P L E 4   Finding the Domain of the Composition

Let $f(x) = x^2 - 1$ and $g(x) = \sqrt{x}$. Find $(g \circ f)(x)$ and $(f \circ g)(x)$ and the domain of each.

Solution   See Figs. 1.59 and 1.60 to see the domain and range of functions f and g.

$$(g \circ f)(x) = g(f(x))$$
$$= g(x^2 - 1)$$
$$= \sqrt{x^2 - 1}$$

$$(f \circ g)(x) = f(g(x))$$
$$= f(\sqrt{x}), \qquad x \geq 0$$
$$= (\sqrt{x})^2 - 1$$
$$= x - 1$$

Domain of $g \circ f$

$$(g \circ f)(x) = g(x^2 - 1)$$

$$x^2 - 1 \geq 0 \quad x^2 - 1 \text{ must be in the domain of } g.$$

$$x^2 \geq 1$$

So $x \leq -1$ or $x \geq 1$.   Therefore the domain of $g \circ f = (-\infty, -1] \cup [1, \infty)$.

Domain of $f \circ g$

$$(f \circ g)(x) = f(\sqrt{x})$$

$$x \geq 0. \quad \sqrt{x} \text{ must be defined for } x \text{ to be in the domain of } f \circ g.$$

Therefore the domain of $f \circ g = [0, \infty)$.   ▤

We see from Example 4 that in general, the domains of $f \circ g$ and $g \circ f$ are not equal. This example also underscores the fact that x is in the domain of $(g \circ f)(x)$ only if $f(x)$ is in the domain of g.

A similar consideration must be made when determining the range. In Example 4, we found that $(f \circ g)(x) = x - 1$ when $x \geq 0$. We must be cautious and not simply look at $x - 1$ and conclude that the range of $f \circ g$ is all real numbers. In fact, $(f \circ g)(x)$ is in the range of $f \circ g$ only if x is in the domain of $f \circ g$.

E X A M P L E 5   Domain and Range of a Composition

Let $f(x) = \sqrt{x}$ and $g(x) = x - 3$. Find $g \circ f$ and $f \circ g$ and the domain and range of each.

Solution   (See Fig. 1.61.)

$$(g \circ f)(x) = g(f(x)) = g(\sqrt{x}) \qquad (f \circ g)(x) = f(g(x)) = f(x - 3)$$

$$= \sqrt{x} - 3 \quad \text{So } x \geq 0. \qquad \qquad = \sqrt{x - 3} \quad \text{So } x \geq 3$$

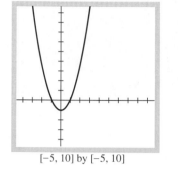

[−5, 10] by [−5, 10]

Figure 1.59 $f(x) = x^2 - 1$.

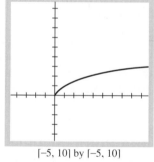

[−5, 10] by [−5, 10]

Figure 1.60 $g(x) = \sqrt{x}$.

domain of $g \circ f = [0, \infty)$, range of $g \circ f = [-3, \infty)$.

domain of $f \circ g = [3, \infty)$, range of $f \circ g = [0, \infty)$.

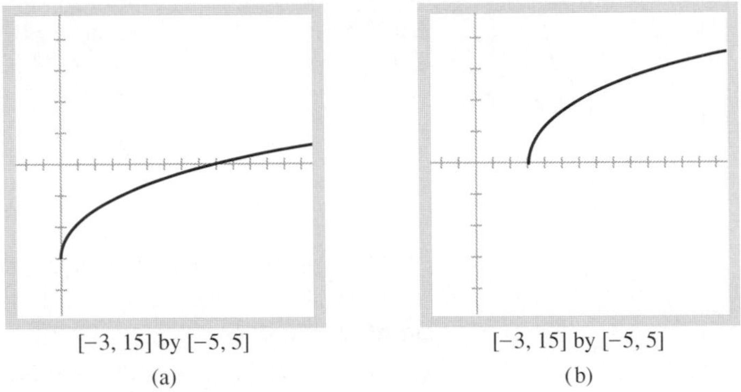
[−3, 15] by [−5, 5]
(a)
[−3, 15] by [−5, 5]
(b)

Figure 1.61 (a) $(g \circ f)(x) = \sqrt{x} - 3$; (b) $(f \circ g)(x) = \sqrt{x-3}$.

Transformations as Compositions of Two Functions

Teaching Note

If your students understand transformations, you can shorten the treatment of transformations as compositions to the point of seeing whether they can operate with the new composition notation. Check to see whether students have trouble in ascertaining the order in which the transformations are applied. Some students may need help in thinking this through. A good activity to use at the end of this section is to ask your students to prepare a summary of transformational properties in terms of composition.

In Section 1.5, the graphs of functions $h_1(x) = ax^2$, $h_2(x) = x^2 + k$, and $h_3(x) = (x - h)^2$ were found to be transformations of the graph of $f(x) = x^2$. The next example shows that for each of these functions, it is possible to find a function g so that these functions can be represented as composites of f and g.

EXAMPLE 6 Transformations as Compositions of Two Functions

Let $f(x) = x^2$.
Find the functions $g(x)$ such that

a) $h_1(x) = ax^2 = g(f(x))$,

b) $h_2(x) = x^2 + k = g(f(x))$, and

c) $h_3(x) = (x - h)^2 = f(g(x))$.

Solution

a) If $g(x) = ax$, then $(g \circ f)(x) = g(f(x)) = g(x^2) = ax^2 = h_1(x)$.

b) If $g(x) = x + k$, then $(g \circ f)(x) = g(f(x)) = g(x^2) = x^2 + k = h_2(x)$.

c) If $g(x) = x - h$, then $(f \circ g)(x) = f(g(x)) = f(x - h) = (x - h)^2 = h_3(x)$.

Reflection in the y-axis

The last transformation that we study is the **reflection in the y-axis**. Suppose that $f(x) = (x + 3)^2$ and that $g(x) = f(-x)$. How does the graph of g compare to the graph of f? Notice that

$$g(x) = f(-x) = (-x + 3)^2$$
$$= [(-1)(x - 3)]^2$$
$$= (x - 3)^2.$$

We see in Fig. 1.62 that the graph of g is obtained from the graph of f by a reflection in the y-axis. Therefore the graph of $f(-x)$ can be obtained from the graph of $f(x)$ by a reflection in the y-axis.

So far in this chapter, we have applied transformation to $y = x^2$. However, these transformations can be performed on a general function $y = f(x)$ as well as on the function $y = x^2$. Here are the summary results.

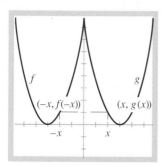

Figure 1.62 $f(x) = (x + 3)^2$ and $g(x) = (x - 3)^2$.

Stan says, "It is important for students to see how transformations applied to quadratic functions in the previous section can be applied to all functions in the form $y = f(x)$."

Transformation Equations for $y = f(x)$

Suppose that $a, k,$ and h are positive real numbers. Then transformations of the graph of $y = f(x)$ are represented as follows:

Transformation Performed on $y = f(x)$	Transformation Function
Vertical stretch or shrink	$h(x) = af(x)$
Vertical shift k units up	$h(x) = f(x) + k$
Vertical shift k units down	$h(x) = f(x) - k$
Horizontal shift h units right	$h(x) = f(x - h)$
Horizontal shift h units left	$h(x) = f(x + h)$
Reflection in the x-axis	$h(x) = -f(x)$
Reflection in the y-axis	$h(x) = f(-x)$

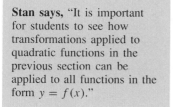

Figure 1.63   Graph of function for Example 7.

E X A M P L E   7   Transforming the Graph of $f(x)$

A graph of a function f is given in Fig. 1.63. Sketch the graph of $y = 2f(x+1) - 3$.

Solution   First, do a vertical stretch by a factor of 2 to get $y = 2f(x)$ (see Fig. 1.64a). Then, do a horizontal shift left 1 unit to get $y = 2f(x+1)$ (see Fig. 1.64b). Finally, do a vertical shift down 3 units to get $y = 2f(x + 1) - 3$ (see Fig. 1.64c).

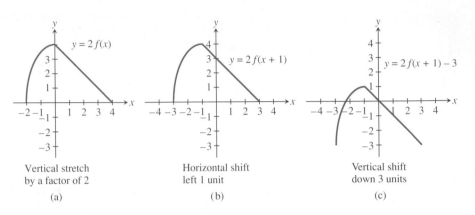

Vertical stretch   Horizontal shift   Vertical shift
by a factor of 2   left 1 unit   down 3 units

(a)   (b)   (c)

Figure 1.64   Transformations performed on the graph of $y = f(x)$ in Fig. 1.63 to get $y = 2f(x+1) - 3$.

Composition of functions is important in applications. Suppose a spherical balloon is being inflated. This means that both the radius r and volume V of the sphere change as time changes. That is, r and V are both functions of t. Recall that $V = \frac{4}{3}\pi r^3$. Thus if $r = f(t)$, then $V(t)$ can be expressed as the composition

$$V(t) = \frac{4}{3}\pi (f(t))^3.$$

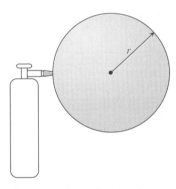

Figure 1.65   Inflation of a balloon with radius r.

E X A M P L E   8   APPLICATION: Finding the Volume of a Balloon

A spherically shaped balloon is being inflated so that the radius r is changing at the constant rate of 2 in./sec (see Fig. 1.65). Assume that $r = 0$ at time $t = 0$. Find an algebraic representation $V(t)$ for the volume as a function of t and determine the volume of the balloon after 5 sec.

Solution

$$r = 2t \qquad \text{\small r changes at 2 in./sec, so after t sec $r = 2t$ in.}$$

$$V = \frac{4}{3}\pi r^3$$

$$= \frac{4}{3}\pi (2t)^3 \qquad \text{\small This equation expresses V as a composition of functions.}$$

$$V(t) = \frac{32}{3}\pi t^3$$

Substitute $t = 5$ into the algebraic representation $V(t) = \frac{32}{3}\pi t^3$.

$$V(5) = \frac{32}{3}\pi (5)^3$$

$$= \frac{4000}{3}\pi \qquad \text{Completing this calculation on a calculator}$$
$$\qquad\qquad \text{yields an approximation for } V.$$

The volume V is approximately 4188.79 cu in. after 5 sec of inflation. ▤

E X A M P L E   9   APPLICATION: Determining the Length of a
Shadow

Anita is 5 ft tall and walks at the rate of 4 ft/sec away from a street light with its
lamp 12 ft above level ground. Find an algebraic representation for the length of
Anita's shadow as a function of time t, and find the length of the shadow after
7 sec.

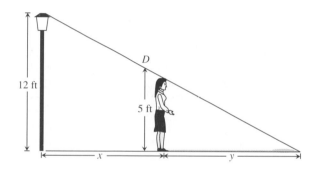

Figure 1.66   A 12-ft street light casts a shadow y feet long as Anita walks away from the
street light.

Solution   (See Fig. 1.66.)
Given:
  y: length of Anita's shadow
  x: distance from Anita to the street light
  $x = 4t$: distance from Anita to the street light after t seconds
Find: y as a function of t
    To obtain the following proportion, use the fact that the two right triangles
are similar:

$$\frac{12}{5} = \frac{x+y}{y} \qquad \text{This proportion works because the}$$
$$\qquad\qquad\qquad \text{two right triangles are similar.}$$

$$12y = 5x + 5y$$

$$7y = 5x$$

Assignment Guide

Day 1: Ex. 1, 4, 6, 8, 10–12, 14–17
Day 2: Ex. 19, 21, 22, 25, 26, 29, 32, 33, 38–41, 46–51
Day 3: Ex. 52, 55, 58, 61, 62, 65, 66, 69, 72, 74–77

$$y = \frac{5}{7}x \qquad \text{Substitute } x = 4t \text{ into this equation.}$$

$$y = \frac{5}{7}(4t)$$

$$y = \frac{20}{7}t$$

Substitute $t = 7$ into the algebraic representation $y = \frac{20}{7}t$ to find that the length of the shadow after 7 sec is 20 ft. ▤

E X A M P L E   10   APPLICATION: Finding the Area of a Rectangle

The initial dimensions of a rectangle are 3 by 4 cm, and the length and width of the rectangle are increasing at the rate of 1 cm/sec. How long will it take for the area to be at least 10 times its initial size?

Solution

length ℓ:	$\ell = 3 + t$	Length after t seconds
width w:	$w = 4 + t$	Width after t seconds
area:	$A(t) = (3 + t)(4 + t)$	Area after t seconds

When the area is 10 times its initial size, $A = 10(3)(4) = 120$. Thus we want to solve the inequality $A(t) \geq 120$ or $(3 + t)(4 + t) \geq 120$. Find complete graphs of the algebraic representations $y = 120$ and $A(t) = (3 + t)(4 + t)$ in the same viewing rectangle (see Fig. 1.67). Notice the value of t at this point of intersection is about 7. In Exercise 19, you will show that the exact solution is

$$t \geq \frac{-7 + \sqrt{481}}{2}.$$

The area will be at least ten times its initial size when $t \geq 7.47$. ▤

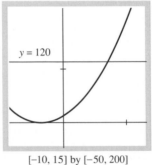

$y = 120$

[–10, 15] by [–50, 200]

Figure 1.67 $A(t) = (3 + t)(4 + t).$

Exercises for Section 1.6

In Exercises 1–5, determine the domain and a rule for $f + g$, $f - g$, fg, and f/g.

1. $f(x) = 2x - 1;\ \ g(x) = x^2$

2. $f(x) = (x - 1)^2;\ \ g(x) = 3 - x$

3. $f(x) = x^2;\ \ g(x) = 2x$

4. $f(x) = \sqrt{x};\ \ g(x) = x - 2$

5. $f(x) = x + 3;\ \ g(x) = \dfrac{2x - 1}{3}$

In Exercises 6 and 7, determine the domain and range, and find a complete graph.

6. $f(x) = \dfrac{|x - 5|}{x - 5}$ **7.** $f(x) = \dfrac{|x + 4|}{x + 4}$

In Exercises 8–11, find $(f \circ g)(3)$ and $(g \circ f)(-2)$.

8. $f(x) = 2x - 3;\ \ g(x) = x + 1$

9. $f(x) = x^2 - 1;\ \ g(x) = 2x - 3$

10. $f(x) = x^2;\ \ g(x) = \sqrt{x - 1}$

11. $f(x) = 2x - 3$; $g(x) = x^2 - 2x + 3$

In Exercises 12–18, find both $f \circ g$ and $g \circ f$ and the domain and range for each composite.

12. $f(x) = 3x + 2$; $g(x) = x - 1$

13. $f(x) = x^2 - 1$; $g(x) = \dfrac{1}{x - 1}$

14. $f(x) = 2x - 5$; $g(x) = \dfrac{x + 3}{2}$

15. $f(x) = x^2 - 2$; $g(x) = \sqrt{x + 1}$

16. $f(x) = \dfrac{1}{x - 1}$; $g(x) = (x + 1)^2$

17. $f(x) = x^2 - 3$; $g(x) = \sqrt{x + 2}$

18. $f(x) = 3x - 1$; $g(x) = \dfrac{x + 1}{3}$

19. Use the quadratic formula to show that the exact solution to Example 10 is $t \geq \dfrac{-7 + \sqrt{481}}{2}$.

In Exercises 20 and 21, find a complete graph of $f, g, f \circ g$, and $g \circ f$ in the same viewing rectangle. Add the line $y = x$. What symmetry do you see?

20. $f(x) = \dfrac{2x - 3}{4}$; $g(x) = \dfrac{4x + 3}{2}$

21. $f(x) = \sqrt{x + 3}$; $g(x) = x^2 - 3$, $x \geq 0$

In Exercises 22 and 23, let $f(x) = x^2$. Find $g(x)$ so that $g \circ f$ produces the specified transformation.

22. Shift the graph of f up 6 units.

23. Shift the graph of f down 3 units.

In Exercise 24–29, determine functions g and h so that $f(x) = h(g(x))$.

24. $f(x) = (x + 3)^2$ **25.** $f(x) = \left(\dfrac{1}{x + 1}\right)^3$

26. $f(x) = \sqrt{x + 3}$ **27.** $f(x) = \dfrac{2}{(x - 3)^2}$

28. $f(x) = (x - 3)^4 - 2$ **29.** $f(x) = 3 - \sqrt{x}$

In Exercises 30 and 31, let $f(x) = x^2$. Find $g(x)$ so that $f \circ g$ produces the specified transformation.

30. Shift the graph of f right 4 units.

31. Shift the graph of f left 8 units.

In Exercises 32 and 33, express each function as a composition of $f(x) = x^2$, $g(x) = 3x$, $h(x) = x + 4$, and $k(x) = x - 2$.

32. $y = 3(x + 4)^2 - 2$ **33.** $y = 3(x - 2)^2 + 4$

34. Verify that $(f \circ g) \circ h = f \circ (g \circ h)$ for the functions $f(x) = x^2 - 1$, $g(x) = 1/x$, and $h(x) = \sqrt{x - 2}$.

For Exercises 35–38, let f be given by the following figure. Sketch a complete graph of each composite function.

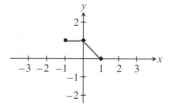

For Exercises 35–38.

35. $y = -1 + f(x)$ **36.** $y = f(x + 1)$

37. $y = -2f(x)$ **38.** $y = 2f(x - 1)$

In Exercises 39–42, let f be given by the following figure. Sketch a complete graph of each composite function.

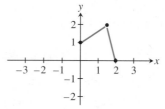

For Exercises 39–42.

39. $y = 1 + f(x)$ **40.** $y = -2 + f(x)$

41. $y = 2f(x)$ **42.** $y = -2f(x)$

In Exercises 43–46, let f be given by the following figure. Sketch a complete graph of each composite function.

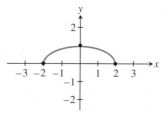

For Exercises 43–46.

43. $g(x) = 3 + f(x)$ **44.** $g(x) = 3f(x)$

45. $g(x) = f(x - 3)$ **46.** $g(x) = f(x + 3)$

In Exercises 47–52, let f be given by the following figure. Sketch a complete graph of each composite function.

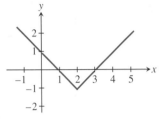

For Exercises 47–52.

47. $g(x) = -3f(x)$ **48.** $g(x) = -1 + 2f(x)$

49. $g(x) = 1 - 2f(x - 1)$ **50.** $g(x) = 1 + 2f(-x)$

51. $y = f(x + 2)$.   **52.** $y = -1 + 2f(x - 1)$.

In Exercises 53–56, list the transformations that can be applied, in order, to obtain the graph of $f(x)$ from the graph of $y = |x|$. Draw a complete graph of the given function without using a graphing utility.

53. $f(x) = |x - 3|$ **54.** $f(x) = -2|x + 3|$

55. $f(x) = 2 + 3|x - 4|$ **56.** $f(x) = 3 - |x - 2|$

In Exercises 57–60, assume that the point $(3, 4)$ is on the graph of $y = f(x)$. What point is on the graph of the following composites?

57. $y = 2f(x)$ **58.** $y = 4f(x)$

59. $y = 2 + f(x)$ **60.** $y = -2 - 3f(x)$

61. Assume that point $(4, 3)$ is on the graph of $y = f(x)$. Find b so that $(2, b)$ is on the graph of $y = f(x + 2)$.

62. Assume that point $(-1, 5)$ is on the graph of $y = f(x)$. Find b so that $(2, b)$ is on the graph of $y = f(x - 3)$.

In Exercises 63–66, assume that points $(2, 5)$ and $(0, 3)$ are on the graph of $y = f(x)$. Name two points that are on the graph of each function.

63. $y = f(x - 2)$ **64.** $y = f(x) + 2$

65. $y = -3f(x - 2)$ **66.** $y = 3 + 2f(x - 2)$

67. Suppose the balloon in Example 8 will burst when its volume is 10,000 cu in. When will the balloon burst?

68. Express the distance D between the street light's lamp and the tip of Anita's shadow (in Example 9) as a function of t. When will that distance D be 100 ft?

69. The initial dimensions of a rectangle are 5 by 7 cm, and the length and width of the rectangle are increasing at the rate of 2 cm/sec. How long will it take for the area to be at least 5 times its initial size?

70. Leon is 6 ft 8 in. tall and walks at the rate of 5 ft/sec away from a street light with a lamp 15 ft above ground level. Determine an algebraic representation for the length of Leon's shadow, and find the length of the shadow after 5 sec.

71. A spherical balloon is inflated so that the radius r is increasing at the rate of 3 in./sec such that $r = 0$ at time $t = 0$. Determine an algebraic representation for the volume of the balloon, and find the volume when $t = 3$ sec.

72. A rock is tossed into a pond. The radius of the first circular ripple (wave) increases at the rate of 2.3 ft/sec. Determine an algebraic representation for the area of the ripple, and find the area when $t = 6$ sec.

73. The surface area of a sphere of radius r is given by $S = 4\pi r^2$. A hard candy ball of radius 1.6 cm is dropped into a glass of water. Its radius decreases at the rate of 0.0027 cm/sec. Determine an algebraic representation for the surface area of the candy ball as a function of t. When will the candy be completely dissolved?

74. The initial dimensions of a computer image of a box are 5 by 7 by 3 cm. If each of the three side lengths is increasing at a rate of 2 cm/sec, how long will it take for the volume of the box to be at least 5 times its initial size?

Exercises 75–77 refer to the following **problem situation**: A certain long-distance phone company charges the following amounts for calls placed after 6 P.M. and before 8 A.M. from Columbus, Ohio, to San Francisco, California. For the first 5 min (up to but less than 5 min), the charge is $0.72 per minute or fraction of a minute. For the next 10 min (beginning at 5 min but less than 15 min), the charge is $0.63 per minute or fraction of a minute. After 15 min the charge is $0.51 per minute or fraction of a minute.

75. Determine a piecewise-defined function that is an algebraic representation of the cost of a call placed from 6 P.M. to 8 A.M.

76. Determine the domain and range of the function in Exercise 75.

77. Sketch a graph of this problem situation without using a graphing utility.

Chapter 1 Review

KEY TERMS (The number following each key term indicates the page of its introduction.)

absolute value, 4
algebraic representation, 14
associative property, 2
Cartesian coordinate
   system, 5
closure property, 2
commutative property, 2
complete graph, 16
composition of f by g, 59
coordinate of a point, 3
coordinate system, 3
difference of f and g, 58
discriminant, 54
distance, 5
distributive property, 2
domain of a function, 22
equation, 2
evaluating a function, 23
function, 22
function notation, 23
general form of
   a linear equation, 38

graph of a function, 24
grapher, 6
graphical representation, 6
graphing utility, 6
horizontal shift, 50
identity property, 2
inequality, 3
integers, 1
inverse property, 2
irrational numbers, 1
less than, 3
linear function, 33
linear inequality, 40
mathematical model, 14
midpoint, 10
number line, 3
ordered pair, 6
origin, 3
parallel lines, 39
perpendicular bisector, 39
perpendicular lines, 39
piecewise-defined function, 29

point-slope form, 37
problem-solving
   process, 13
product of f and g, 58
Pythagorean theorem, 9
quadrants, 6
quadratic function, 45
quotient of f and g, 58
range of a function, 24
rational numbers, 1
real numbers, 1
rectangular coordinate
   system, 5
reflection, 25
reflection in y-axis, 63
reflexive property, 2
rule of a function, 23
screen coordinate, 6
slope, 35
slope-intercept form, 36
standard form of a
   quadratic, 52

standard viewing
   rectangle, 17
step function, 27
sum of f and g, 58
symmetric property, 2
symmetry of a
   quadratic, 46
transitive property, 2
understood domain, 23
vertex, 46
vertical line test, 26
vertical shift (slide), 49
vertical shrink or
   stretch, 47
viewing rectangle, 6
x-axis, 5
x-intercept, 54
y-axis, 5
y-intercept, 36
zeros of a quadratic
   function, 54

REVIEW EXERCISES

Identify the real-number properties that are used to explain why each of the following are valid equations.

1. $2(3 + \sqrt{5}) = 2 \cdot \sqrt{5} + 2 \cdot 3$

2. $x + (2 + y) = (x + y) + 2$

In Exercises 3–8, draw on a number line all numbers that are solutions to these inequalities.

3. $-3 \leq x < 4$

4. $1 \leq x \leq 4$

5. $-5 < x \leq -3$

6. $x \leq 2$

7. $5 < x$

8. $x < 3$

In Exercises 9–16, evaluate each numerical expression.

9. $|-2.5|$

10. 2^{-3}

11. $27^{2/3}$

12. $8^{-2/3}$

13. $|-(5 - 2)|$

14. $|2 - 6|$

15. $16^{3/4}$

16. $4^2 \cdot 8^3$

17. Write the exact value of the indicated expression without using absolute-value notation.
   a) $|\sqrt{7} - 2.6|$
   b) $|2.6 - \sqrt{7}|$
   c) $|\pi - 3|$
   d) $|x - 5|$ where $x > 5$

In Exercises 18 and 19, consider the points $P(-2, 4)$ and $Q(3, -1)$ in the coordinate plane.

18. Find the coordinates of the point in the first quadrant that is on the vertical line through Q and on the horizontal line through P.

19. Find the coordinates of the point in the third quadrant that is on the horizontal line through Q and on the vertical line through P.

20. Find the distance between the two points on the number line with the given coordinates.

   a) $-2, 4$ **b)** $5, -2$ **c)** $-\pi, -1$

In Exercises 21–24, find the distance between the points in the plane with the given coordinates.

21. $(1, 1), (3, 4)$ **22.** $(-3, -8), (2, 3)$

23. $(2, -1), (1, -3)$ **24.** $(-3, -1), (-5, -4)$

For Exercises 25 and 26, write an expression that gives the distance in the plane between each pair of points.

25. $(0, y), (-2, 3)$ **26.** $(x, 0), (0, -4)$

27. Let $A = (-2, 3)$, $B = (3, -4)$, and $C = (-4, 1)$.
   a) Prove that triangle ABC is isosceles but not equilateral.
   b) Prove that the line passing through B and the midpoint of AC is perpendicular to AC.

28. Let $A = (0, 0)$, $B = (4, 2)$, and $C = (6, -2)$.
   a) Write an equation for each of the three lines determined by a vertex and the midpoint of the opposite side of triangle ABC.
   b) Prove that the three lines in part (a) intersect in one point.

29. Prove that the diagonals of a square are perpendicular.

For Exercises 30–32, consider the triangle ABC shown in the following figure:

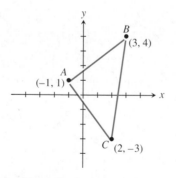

For Exercises 30–32.

30. Find an equation of the line through B and C.

31. Find an equation of the line through A and the midpoint of BC (median).

32. Find an equation of the line through A and perpendicular to BC (altitude).

For Exercises 33 and 34, the accompanying graph shows a line that goes through the origin and the upper right corner of the display. For the given viewing rectangles select the equation whose graph has this characteristic.

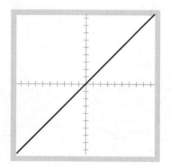

For Exercises 33 and 34.

33. For viewing rectangle $[-10, 10]$ by $[-10, 10]$, the figure shows a complete graph of which equation?
   a) $y = x$
   b) $y = 5x$
   c) $y = \frac{1}{2}x$

34. For viewing rectangle $[-10, 10]$ by $[-50, 50]$, the figure shows a complete graph of which equation?
   a) $y = x$
   b) $y = 5x$
   c) $y = 50x$

In Exercises 35–38, find a complete graph of each equation without using a graphing utility. Choose an appropriate scale for each axis.

35. $y - 5 = \frac{1}{2}x$ **36.** $x - y = 6$

37. $y = 200x - 10$ **38.** $y = 2x^2 + 1000$

39. Which of the following viewing rectangles gives the best complete graph of $y = x^2 - 30x - 100$?
   a) $[-10, 5]$ by $[-10, 5]$
   b) $[-10, 10]$ by $[-10, 10]$
   c) $[25, 50]$ by $[-500, 1000]$
   d) $[-10, 50]$ by $[-400, 800]$
   e) $[-100, 100]$ by $[-10, 10]$

40. Which of the following viewing rectangles gives the best complete graph of $y = 200 - 10x - x^3$?
   a) $[-10, 10]$ by $[-10, 10]$
   b) $[-50, 50]$ by $[-10, 10]$
   c) $[-10, 10]$ by $[-50, 50]$
   d) $[-10, 10]$ by $[-1000, 1000]$
   e) $[-100, 100]$ by $[-10, 10]$

In Exercises 41 and 42, choose a viewing rectangle $[X\text{min}, X\text{max}]$ by $[Y\text{min}, Y\text{max}]$ that will include all of the indicated points.

41. $(-9, 12)$, $(2, -8)$, $(3, 20)$

42. $(20, -11)$, $(18, 156)$, $(-11, 2)$

In Exercises 43–50, find a complete graph of each of the following functions. Begin with the standard viewing rectangle, and modify until an appropriate viewing rectangle is found. Transfer your graph to paper and record the viewing rectangle and scale used.

43. $f(x) = 5 - 3x^2$ **44.** $f(x) = x^3 - 4x^2 + 3x - 2$

45. $f(x) = 17x^2 - 48$ **46.** $f(x) = \sqrt{x + 21}$

47. $f(x) = 13x - 5x^2$ **48.** $f(x) = 2x + x^3 - 28$

49. $f(x) = 5(x - 12)^2 - 17$

50. $f(x) = -(9x - 12)$

51. Let $f(x) = x^2 + 2$. Find $f(0)$, $f(1)$, $f(3)$, $f(-5)$, $f(t)$, $f(-t)$, $f(-1/t)$, $f(a + h)$, and $[f(a + h) - f(a)]/h$.

52. Let $g(x) = 1/(2 - x)$. Find $g(2)$, $g(0)$, $g(-2)$, $g(a)$, $g(1/a)$, and $g(a + h)$.

53. Consider the graph of the function $y = f(x)$ in the following figure and estimate the indicated values.
   a) $f(0)$, $f(-1)$, $f(2)$
   b) x if $f(x) = 0$
   c) x if $f(x) = 2$

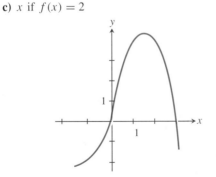

For Exercise 53.

In Exercises 54–59, find a complete graph of each function. Explain why the graph is a complete graph.

54. $g(x) = x^2 - 40$ **55.** $f(x) = 10x^3 - 400x$

56. $h(x) = 2 - x$ **57.** $f(x) = \sqrt{x - 3}$

58. $t(x) = \left(\dfrac{x}{4}\right)^2 + \left(\dfrac{200 - x}{4}\right)^2$

59. $g(x) = -2 - \sqrt{x + 3}$

60. Determine the domain and range of the function in Exercise 57.

61. Determine the domain and range of the function in Exercise 58.

62. Determine the domain and range of the function in Exercise 59.

In Exercises 63–68, sketch a complete graph of the function without using a graphing utility. Determine the domain and range.

63. $f(x) = \begin{cases} x + 7, & x \le -3 \\ x^2 + 1, & x > -3 \end{cases}$

64. $f(x) = \begin{cases} 4 - x^2, & x < 1 \\ \sqrt{x + 2}, & x \ge 1 \end{cases}$

65. $f(x) = \begin{cases} -2x, & x < 1 \\ \sqrt{x + 5}, & x \ge 1 \end{cases}$

66. $f(x) = \begin{cases} 3 - x, & x < -3 \\ (x - 2)^2, & -3 \le x < 2 \\ x^3, & x \ge 2 \end{cases}$

67. $f(x) = \dfrac{2|x - 3|}{x - 3}$

68. $g(x) = \begin{cases} 2x + x^5, & x < 2 \\ x^3 - x - 1, & x \ge 2 \end{cases}$

69. Draw a line through the point $(3, 4)$ with a slope of $-\frac{1}{2}$.

70. Draw a line through the point $(-1, 3)$ with a slope of 2.

71. Draw graphs of the functions $y = 2x^2 + 3$, $y = 2x^2 - 1$, and $y = 2x^2 - 3$ on the same coordinate system. Do not use a graphing utility.

72. Determine the slope of the line through the points $(0, 3)$ and $(-2, 6)$.

73. Determine the y-intercept of the graph of the function $3x - 2y + 8 = 0$. Graph the function.

In Exercises 74 and 75, determine a formula for the linear function f that satisfies the given conditions.

74. $f(2) = 0$, $f(0) = 4$ **75.** $f(0) = 250$, $f(-10) = 200$

In Exercises 76–79, sketch a complete graph of each equation or inequality.

76. $2x + y \leq 3$

77. $3x - 4y = 12$

78. $x + 3y = 9$

79. $y \leq x$

80. Write an equation for the line with a slope of $-\frac{2}{3}$ and a y-intercept at $(0, 4)$.

81. Write an equation for the line with a slope of $\frac{3}{4}$ that contains the point $(1, 2)$.

82. Write an equation for the line determined by the points $(-3, 4)$ and $(2, 5)$.

83. Given the point $A = (5, 7)$, write an equation for the following:
   a) the vertical line through A
   b) the horizontal line through A.

84. Find the midpoint of the line segment determined by the two points $(2, -3)$ and $(-4, 6)$.

85. Write an equation for the perpendicular bisector of the line segment determined by the two points $(1, 3)$ and $(-3, 7)$.

In Exercises 86 and 87, write an equation for the line determined by the given conditions.

86. The line contains the point $(6, 1)$ and is perpendicular to the line $3x - 2y = 4$.

87. The line contains the point $(5, -8)$ and is perpendicular to the line $y = 31$.

In Exercises 88–95, sketch a complete graph of $y = f(x)$ without using a graphing utility. Describe how each graph is obtained from the graph of $y = x^2$.

88. $y = -x^2$

89. $y = (x - 3)^2$

90. $y = -2(x + 3)^2 + 4$

91. $y = (x + 1)^2 - 4$

92. $y = 4(x - 2)^2$

93. $y = 4(x + 4)^2$

94. $y = -2(x + 3)^2$

95. $y = 2(x - 3)^2$

In Exercises 96 and 97, find a complete graph of $y = f(x)$. Describe how each graph is obtained from the graph of $y = |x|$.

96. $y = 2|x - 3|$

97. $y = -3 - |x + 4|$

98. Assume the point $(4, 3)$ is on the graph of $y = f(x)$. Find b so that $(1, b)$ is on the graph of $y = f(x + 3)$.

99. Assume the points $(-3, 2)$ and $(1, 0)$ are on the graph of $y = f(x)$. Find b so that
   a) $(-1, b)$ is on the graph of $y = f(x - 2)$
   b) $(3, b)$ is on the graph of $y = f(x - 2)$

c) $(-1, b)$ is on the graph of $y = -3f(x - 2)$
d) $(3, b)$ is on the graph of $y = 2 + 2f(x - 2)$

100. Consider the graph of $y = x^2$.
   a) Vertically stretch this graph by a factor of 2 followed by a vertical shift up 1 unit. What is an equation of this new graph? Sketch a complete graph.
   b) What is an equation of the transformed graph if the order of the two transformations in part (a) is reversed?
   c) Are the two graphs the same? Explain the effect of reversing the order of the transformations.

In Exercises 101 and 102, write a sequence of transformations that will produce a complete graph of each function from the graph of $y = x^2$. Specify an order in which the transformations should be applied.

101. $f(x) = 2x^2 - 12x + 4$ **102.** $f(x) = 14 - 6x - x^2$

In Exercises 103 and 104, list the transformations that will produce a complete graph of each function from the graph of $y = x^3$. Specify the order in which the transformations should be applied, and then sketch a complete graph of the function without using a graphing utility. Use a graphing utility to support your work.

103. $y = (x + 2)^3$ **104.** $y = -2(x - 3)^3 - 5$

In Exercises 105–107, consider the complete graphs of $y = f(x)$ and $y = g(x)$ shown in the following figure:

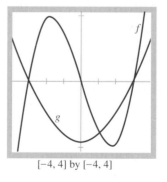

$[-4, 4]$ by $[-4, 4]$

For Exercises 105–107.

105. Compare the following:
   a) $f(3)$ and -3
   b) $g(4)$ and 1

106. Compare $f(a)$ and $g(a)$
   a) if $a = -2$
   b) if $a = 2$

107. Is $x = -1$ a solution to
   a) $f < g$?
   b) $f = g$?
   c) $f > g$?

108. List the transformations that will produce a complete graph of the function $y = (x + 3)^4 + 2$ from the graph of $y = x^4$. Specify the order in which the transformations should be applied, and sketch a complete graph of the function. Then use a graphing utility to check your answer.

109. Consider $f(x) = 1/(x - 1)$ and $g(x) = (x + 2)^2$. Determine the domain and a rule for $f + g$, $f - g$, fg, and f/g.

In Exercises 110 and 111, determine $(f \circ g)(-3)$ and $(g \circ f)(2)$.

110. $f(x) = 5x + 7$; $g(x) = x - 4$

111. $f(x) = x^2 + 4$; $g(x) = \sqrt{1 - x}$

112. Determine $f \circ g$ and $g \circ f$ if $f(x) = x^2 + 2$ and $g(x) = (x + 3)/2$.

113. Find the domain and range of f, g, $f \circ g$, and $g \circ f$ if $f(x) = (2x - 3)/4$ and $g(x) = x^2 - 2$.

114. Let $f(x) = x^2$. Find $g(x)$ so that $g \circ f$ produces the specified transformation.
   a) Vertically shift the graph of f down 5 units.
   b) Vertically shift the graph of f up 4 units.

115. Let $f(x) = x^2$. Find $g(x)$ so that $f \circ g$ produces the specified transformation.
   a) Horizontally shift the graph of f left 5 units.
   b) Horizontally shift the graph of f right 6 units.

116. Express $y = 2(x + 3)^2 - 4$ as a composition of $f(x) = x^2$, $g(x) = 2x$, $h(x) = x + 3$, and $k(x) = x - 4$.

In Exercises 117 and 118, find functions g and h so that $f(x) = (h \circ g)(x) = h(g(x))$.

117. $f(x) = (x + 5)^2$ **118.** $f(x) = \dfrac{3}{(x - 2)^2}$

In Exercises 119 and 120, sketch a complete graph of $y = f(x)$. Then check your answer with a graphing utility.

119. $f(x) = 2(x - 2)^2 + 3$ **120.** $f(x) = 4 - 3\sqrt{x + 1}$

121. The perimeter P of a rectangle is given by the equation $P = 2L + 2W$, where L is the length and W is the width. If the width is 220 units, then write an equation for the perimeter P as a function of the length. Find a complete graph showing how P varies with length.

122. The area A of a rectangle is given by the equation $A = \ell w$. If the length is 125 units, write the area A in terms of the width. Find a complete graph of A showing how A varies with width.

123. Jerry runs at a constant speed of 4.25 mph. Find the algebraic representation for distance d in terms of time t. Use a complete graph of this algebraic representation to find how long it takes Jerry to run a 26-mile marathon.

124. $A = \pi r^2$ is an equation for the area of a circle with a radius of r units. Find a complete graph of this algebraic representation, and use it to find the radius of a circle whose area is 150 square units.

Exercises 125–128 refer to this **problem situation**. A school club buys a scientific calculator for $18.25 to use as a raffle prize for a fund-raising project. The club charges $0.50 per raffle ticket.

125. Find an algebraic representation that gives the club's profit or loss, P, as a function of the number of tickets sold, n.

126. Find a complete graph of the algebraic representation. What part of this graph represents the problem situation? *Hint:* You cannot sell half a ticket.

127. What are the domain and range of the algebraic representation?

128. Use your graph from Exercise 126 to determine the minimum number of tickets that must be sold for the club to realize a profit.

Exercises 129–131 refer to the following **problem situation**. A bike manufacturer determines the annual cost C of making x bikes to be $85 per bike plus $75,000 in fixed overhead costs.

129. Find an algebraic representation of the total annual cost as a function of the number of bikes made. Find a complete graph of this algebraic representation.

130. What portion of your graph from Exercise 129 represents possible total annual costs of producing the bikes?

131. Use the graph in Exercise 129 to determine the number of bikes made if the total cost is $143,000.

Exercises 132–133 refer to the following **problem situation**. Consider all rectangles whose length plus width equals 150 in., and let x be the length of such a rectangle.

132. Show that $A = x(150 - x)$ is an algebraic representation that gives the area A of a rectangle in the collection in terms of its length.

133. Find a complete graph of the algebraic representation in Exercise 132.

134. What values of x make sense in this problem situation? Find a graph of the problem situation.

For Exercises 135–137, consider the following **problem situation.** The annual profit P of a baby food manufacturer is determined by the formula $P = R - C$, where R is the total revenue generated from selling x jars of baby food and C is the total cost of making and selling x jars of baby food. Each jar sells for $0.60 and costs $0.45 to make. The fixed costs of making and selling the baby food are $83,000 annually.

135. Find the algebraic representation for the company's annual profit as a function of x.

136. How many jars of baby food must be sold for the company to break even?

137. How many jars of baby food must be sold for the company to make a profit of $10,000$?

Exercises 138–140 refer to this **problem situation**. A trucker averages 48 mph on a cross-country trip from Boston to Seattle. Let t be the time since the trucker left Boston.

138. Find a complete graph of an algebraic representation for the distance the trucker travels as a function of time.

139. What part of the graph in Exercise 138 represents a graph of the problem situation?

140. Use the graph in Exercise 138 to find how many hours have elapsed since the trucker left Boston if the distance traveled is 1200 mi.

141. A fast-food restaurant makes $0.40 profit on a medium-sized soft drink, $0.15 profit on a taco, and breaks even on the other items. If the weekly overhead is $700, write an inequality that expresses the requirement that the weekly profits must exceed $500. Sketch its complete graph.

142. Ethanol worth $0.75/gal is mixed with gasoline worth $1.10/gal. The value of the new mixture cannot exceed $0.999/gal. (At a gas station this would be written as $0.99\frac{9}{10}$/gal.) Write an inequality expressing this requirement, and sketch its complete graph.

Exercises 143–147 refer to the following **problem situation:** An object is shot straight up (launched) from the top of a building 200 ft tall with an initial velocity of 70 ft/sec. Let t be the time in seconds since the object was launched.

143. Determine an algebraic representation that gives the height of the object above ground level as a function of t.

144. Draw a complete graph of the algebraic representation.

145. What portion of the graph of the algebraic representation represents this problem situation?

146. Determine the time when the object is 225 ft above the ground.

147. Determine the time when the object is more than 225 ft above the ground.

CARLOS DE SIGUENZA Y GÓNGORA

The Mayan civilization, which flourished in the Yucatan Peninsula of Mexico from about A.D. 300 to 800, is known for its advances in mathematics. The Maya were the first to use zero as a place holder. They were also adept astronomers and created an accurate calendar. Following the Spanish conquest, Mexico continued to be a center for the study of mathematically related fields. The Royal University of Mexico, established in 1553, soon became a gathering place for astronomers, surveyors, and cartographers as gifted as any in Europe at that time.

   Carlos de Siguenza y Góngora (1645–1700) was an astronomer, mathematician, geographer, and writer. He advocated the theories of René Descartes, and he showed a remarkable intellectual curiosity in science as well as in literature. He is best known as the author of *Infortunios de Alfonso Ramirez (The Misadventures of Alfonso Ramirez)*, considered a forerunner of the Mexican novel.

2

Functions and Their Zeros

2.1 _____

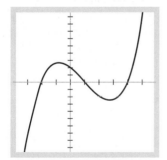

[−2, 3] by [−10, 10]

Figure 2.1
$f(x) = 2x^3 - 3x^2 - 3x + 2$.

Zeros of Polynomial Functions

The function $f(x) = 2x^3 - 3x^2 - 3x + 2$ is an example of a polynomial function. In this first Exploration, you will be studying this function.

🔍 **EXPLORE WITH A GRAPHING UTILITY**

Graph $f(x) = 2x^3 - 3x^2 - 3x + 2$ using the viewing rectangle $[-2, 3]$ by $[-10, 10]$. See Fig. 2.1.

Questions

1. How many times does the graph cross the x-axis?

2. What seems to be true about the y-coordinate of the points where the graph crosses the x-axis? (Use the trace key on your grapher.)

3. Show that $(x + 1)(2x - 1)(x - 2) = 2x^3 - 3x^2 - 3x + 2$.

4. What are the exact solutions to $2x^3 - 3x^2 - 3x + 2 = 0$?

5. How do the x-coordinates of the points where the graph crosses the x-axis compare with the solutions to the equation $2x^3 - 3x^2 - 3x + 2 = 0$?

Generalize   Make some general statements that you think will always be true about the coordinates of the points where the graph crosses the x-axis.

Summary of Your Experience   From the Exploration, you should have come to these conclusions:

1. A solution to $2x^3 - 3x^2 - 3x + 2 = 0$ corresponds to the x-coordinate of a point where the graph of the function $f(x) = 2x^3 - 3x^2 - 3x + 2$ crosses the x-axis.

2. If $x = a$ is a real solution to $2x^3 - 3x^2 - 3x + 2 = 0$, then $(a, 0)$ is a point on the graph of the function $f(x) = 2x^3 - 3x^2 - 3x + 2$.

   Such a point $(a, 0)$ is called an ***x*-intercept** of $f(x) = 2x^3 - 3x^2 - 3x + 2$. The value a is also called a **root** or a **zero** of the function f because it is an x-value that makes $f(x)$ equal to zero. Sometimes the x-intercept is called a rather than $(a, 0)$.

Linear Equations

The linear equation $ax + b = 0$ has a unique solution, namely $-b/a$, (where $a \neq 0$). The graph of the equation $y = ax + b$ is a straight line that crosses the x-axis at the point $(-b/a, 0)$.

E X A M P L E   1   Solving a Linear Equation Graphically

Solve $2x + 3 = 0$.

Solution   (See Fig. 2.2.) The x-intercept appears to be $-\frac{3}{2}$, so reading from the graph gives an approximate solution of -1.5. The exactness of this approximation can be confirmed by substitution.   ≡

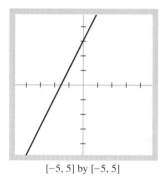

[−5, 5] by [−5, 5]

Figure 2.2 $y = 2x + 3$.

Quadratic Equations

The graphs of quadratic equations of the form $f(x) = ax^2 + bx + c$ are called **parabolas**. In Chapter 3, we will study these graphs in more detail, but for now

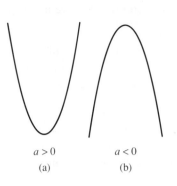

Figure 2.3 The two possible graphs of $f(x) = ax^2 + bx + c$.

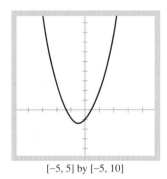

$[-5, 5]$ by $[-5, 10]$

Figure 2.5 $y = 2x^2 + 2x - 1$.

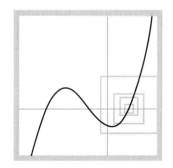

Figure 2.6 Zoom-in creates a nested sequence of viewing rectangles decreasing in size.

assume that a complete graph of $f(x) = ax^2 + bx + c$ looks like one of the two parabolas in Fig. 2.3.

Figure 2.4 illustrates that the number of x-intercepts for a quadratic equation can be zero, one, or two depending on whether the discriminant $b^2 - 4ac$ is negative, zero, or positive, respectively.

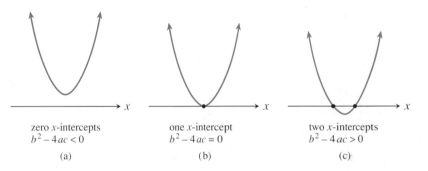

zero x-intercepts
$b^2 - 4ac < 0$

(a)

one x-intercept
$b^2 - 4ac = 0$

(b)

two x-intercepts
$b^2 - 4ac > 0$

(c)

Figure 2.4 Graphs of $f(x) = ax^2 + bx + c, (a > 0)$. Notice how the sign of the discriminant affects the number of x-intercepts.

E X A M P L E   2   Solving a Quadratic Equation Graphically

Solve $2x^2 = 1 - 2x$.

Solution

1. Rewrite the equation $2x^2 = 1 - 2x$ to the equivalent form $2x^2 + 2x - 1 = 0$.

2. Graph the function $y = 2x^2 + 2x - 1$ in the viewing rectangle $[-5, 5]$ by $[-5, 10]$ (see Fig. 2.5).

3. The x-intercepts appear to be approximately 0.4 and -1.4. ≡

Finding Solutions Graphically Using Zoom-in

A graphing utility can be used to find solutions to a high degree of accuracy with a procedure called **zoom-in**. This is accomplished by "trapping" the x-intercept in a sequence of viewing rectangles, each new one contained within the previous one (see Fig. 2.6). Zoom-in continues until the viewing rectangle has enlarged a small enough portion of the graph that the user can read the value of x to the level of accuracy desired. However, it is important to note that solutions

cannot be read more accurately than the graphing utility permits. Most computers and graphing calculators allow answers to be read to at least 9 or 10 significant digits.

EXAMPLE 3   Solving Graphically with Zoom-in

Use a complete graph to show that $x^3 + 2x = 1$ has only one solution, then find an approximation of that solution.

Solution   Rewrite the equation as $x^3 + 2x - 1 = 0$, and complete the following steps:

1. Graph the function $y = x^3 + 2x - 1$ in the standard viewing rectangle (see Fig. 2.7a). The graph crosses the x-axis between $x = 0$ and $x = 1$ (scale marks are 1 unit apart).
2. Find a new graph using the viewing rectangle [0, 1] by [−1, 1] with scale marks 0.1 unit apart (see Fig. 2.7b). The graph crosses the x-axis between $x = 0.4$ and $x = 0.5$.
3. Find a new graph using the viewing rectangle [0.4, 0.5] by [−0.1, 0.1] with scale marks 0.01 unit apart (see Fig. 2.7c).

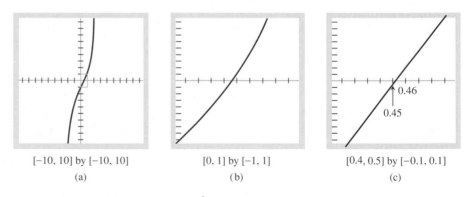

| [−10, 10] by [−10, 10] | [0, 1] by [−1, 1] | [0.4, 0.5] by [−0.1, 0.1] |
| (a) | (b) | (c) |

Figure 2.7   Three views of $f(x) = x^3 + 2x - 1$.

Read the graph and report that an approximate solution is 0.453.   ≡

Analysis of Error

Study Example 3 in more detail. The (exact) solution of $x^3 + 2x - 1 = 0$ is between 0.45 and 0.46, a fact that can be read from the graph in Fig. 2.7(c). Furthermore, since the difference between 0.45 and 0.46 is 0.01, *any* number in the interval [0.45, 0.46] is the solution with an error of at most 0.01.

Which number should be reported as the solution with an error of at most 0.01? One reader might read the graph in Fig. 2.7(c) as crossing the *x*-axis about $\frac{3}{10}$ of the way between two tick marks and could report 0.453 as a solution (with an error of at most 0.01). A second reader might read the graph as crossing the *x*-axis about $\frac{4}{10}$ of the way between two tick marks and could report the solution as 0.454 (with an error of at most 0.01). Both conclusions would be correct.

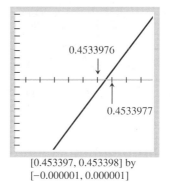

Figure 2.8   A typical graph.

Definition 2.1   Error of a Solution

Suppose a graph crosses the *x*-axis between two consecutive scale marks *a* and *b* and $b - a = r$ (see Fig. 2.8). If *c* is any number in the interval $[a, b]$ determined by the scale marks, *c* is a **solution with error of at most *r***.

Suppose the graph of an equation crosses the *x*-axis in viewing rectangle [*X*min, *X*max] by [*Y*min, *Y*max] in which no scale marks are displayed. Then any number between *X*min and *X*max would be a solution with an error of at most $r = X\text{max} - X\text{min}$.

E X A M P L E   4   *Using Zoom-in for a Very Accurate Solution*

Solve the equation $x^3 + 2x = 1$ with an error of at most 0.0000001.

Solution

1. Continue the process begun in Example 3 by graphing the equation using viewing rectangle [0.45, 0.46] by [−0.01, 0.01] with scale marks 0.001 unit apart.

2. Using smaller and smaller viewing rectangles, zoom-in until you see viewing rectangle [0.453397, 0.453398] by [−0.000001, 0.000001] (see Fig. 2.9).

3. Notice that the graph crosses the *x*-axis about halfway between $x = 0.4533976$ and $x = 0.4533977$. Thus the solution is $x = 0.4533976 + 0.00000005 = 0.45339765$ with an error of at most $0.4533977 - 0.4533976 = 0.0000001$. ≡

0.4533976

0.4533977

[0.453397, 0.453398] by [−0.000001, 0.000001]

Figure 2.9 $y = x^3 + 2x - 1$.

Ordinarily it isn't necessary to approximate solutions as accurately as in Example 4. However, the example illustrates that accuracy is limited only by machine precision.

Stan says, "Exact solutions are important mostly to mathematicians. In the real world of mathematics, very few problems yield exact answers. Realistically, real-world problems are solved within parameters of a certain degree of accuracy."

Teaching Note

Be sure to make clear the distinction between exact and approximate answers. Many students think that $\sqrt{3}$ (exact) and 1.732 (approximate) are exactly the same.

Accuracy Agreement

Throughout the rest of this text, we shall adhere to the following convention: Unless stated to the contrary, to *solve an equation* means "to approximate all real solutions of the equation with an error of at most 0.01" or "to state the exact solution." For example, when it has been determined that a solution is between 2.03 and 2.04, you can report either of these numbers as the solution with an error of at most 0.01.

Assume that numbers provided in examples and exercises are exact unless otherwise specified.

Third-degree Polynomial Equations

A **third-degree polynomial equation** has the form

$$ax^3 + bx^2 + cx + d = 0.$$

Solving this equation graphically requires first finding the complete graph of the function $y = ax^3 + bx^2 + cx + d$ and then using zoom-in as illustrated in Examples 3 and 4.

It is essential to begin with a complete graph to determine how many real solutions exist. For example, the graph of $f(x) = x^3 + 2x - 1$ in Example 3 crossed the x-axis once; therefore it has only one real solution.

When completing exercises, assume that a complete graph of a third-degree polynomial looks like one of the graphs in Fig. 2.10. These forms will be confirmed in Chapter 3.

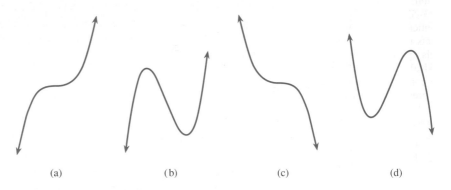

(a)   (b)   (c)   (d)

Figure 2.10   The graph of $y = ax^3 + bx^2 + cx + d$ will always have one of these shapes.

E X A M P L E 5   Finding the Number of Solutions

How many real-number solutions are there to the equation $x^3 - 5x^2 + 6x - 1 = 0$?

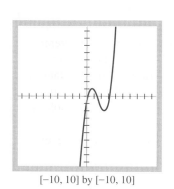

[−10, 10] by [−10, 10]

Figure 2.11
$y = x^3 - 5x^2 + 6x - 1$.

Notes on Exercises

Ex. 27–42 are follow-up problems to Example 6. Do your students set up the zoom-in sequence of viewing rectangles to get the solutions expeditiously? In Ex. 36, students may want to use the distance interpretation of absolute value to solve the equation. Ex. 43–53 offer a variety of applications. Are your students using the language of models appropriately? Ex. 54 allows students to make connections among the ideas presented in this section.

Assignment Guide

Day 1: Ex. 1–43 odd
Day 2: Ex. 44–47, 52, 53, 55, 56

Solution   Find the complete graph of the function $f(x) = x^3 - 5x^2 + 6x - 1$ as shown in Fig. 2.11. Because the graph crosses the x-axis three times, the function f has three zeros. Therefore there are three real-number solutions. ≡

Using the power of a graphing utility, we can find the zeros of a function even when they are not whole numbers. The next example illustrates this fact.

E X A M P L E   6   Approximating One of Several Solutions

Find the middle of the three real-number solutions to $x^3 - 5x^2 + 6x - 1 = 0$ with an error of at most 0.01.

Solution

1. Graph the function $f(x) = x^3 - 5x^2 + 6x - 1$ in the standard viewing rectangle (see Fig. 2.11). Observe that the middle solution is between $x = 1$ and $x = 2$.

2. Use zoom-in several times to arrive at viewing rectangle [1.5, 1.6] by [−0.1, 0.1] with scale marks 0.01 units apart.

3. Note that the solution is $x = 1.555$. ≡

The equations in Examples 4, 5, and 6 can be solved with very complicated algebraic techniques that go beyond the scope of this book. With a grapher, the solutions are manageable. Furthermore, for polynomial equations of degrees 5 and greater, an algebraic method is usually impossible, whereas the graphical method is always possible.

Summary of the Graphical Method for Solving Equations

- Write an equation in the form $y = \ldots$ whose x-intercepts are the solutions to the original equation.

- Find a complete graph of the equation in the previous step so that all solutions are visible. Two or more viewing rectangles may be necessary.

- Use zoom-in to approximate each solution to an error at most 0.01 unless otherwise stated.

Exercises for Section 2.1

In Exercises 1–6, use an algebraic method to solve each equation. Support your answer with a graphing utility.

1. $(x - 3)(x + 2) = 0$
2. $x^2 = 14$
3. $x^2 - 3x + 2 = 0$
4. $x^2 - 2x + 3 = 0$
5. $x^3 - x = 0$
6. $|x - 2| = 6$

In Exercises 7–14, find a complete graph whose x-intercepts are solutions to the given equation. In each case, state how many real solutions there are and record the viewing rectangle that you used.

7. $x^2 - 3x - 10 = 0$
8. $x^2 + 10x - 119 = 0$
9. $1000 - 15x - x^2 = 0$
10. $x^2 - 14x - 10 = 3$
11. $x^3 - 25x = 0$
12. $x^3 + 2x^2 - 109x - 110 = 0$
13. $x^3 - 2x^2 + 3x - 5 = 0$
14. $x^3 - 65x + 10 = 0$

In Exercises 15 and 16, find a complete graph whose x-intercepts are solutions to the given equation. Explain why you may need to use both a complete graph and zoom-in to determine how many solutions there are to these equations.

15. $\frac{1}{8}x^4 - 5x^2 + 2 = 0$
16. $\frac{1}{2}x^3 - 7x^2 = -3$

In Exercises 17–24, assume that the graph of an equation crosses the x-axis in the interval given. Find the value of r such that any number in this interval is a solution to the corresponding equation with an error of at most r.

17. $[3.25, 3.26]$
18. $[4.8, 4.9]$
19. $[-3.5, -3.0]$
20. $[6.213, 6.214]$
21. $[1.32, 1.33]$
22. $[-2.008, -2.007]$
23. $[0.036, 0.362]$
24. $[1.5, 1.55]$

In Exercises 25 and 26, find a sequence of four viewing rectangles containing each solution. Choose each sequence to permit the solutions to be read with errors of at most 0.1, 0.01, 0.001, and finally, 0.0001.

25. $x^3 - x^2 + x - 3 = 0$
26. $x^3 - 2x + 3 = 0$

In Exercises 27–30, find one positive solution to each equation with an error of at most 0.01.

27. $x^3 - x - 2 = 0$
28. $\frac{1}{100}x^3 - x - 2 = 0$
29. $\frac{1}{10}x^3 - x^2 - 2 = 0$
30. $x^3 - \frac{1}{10}x = 0$

In Exercises 31–36, solve the equation.

31. $3x^2 - 15x + 8 = 0$
32. $x^3 - 2x^2 + 3x - 1 = 0$
33. $x(x - 25)(x - 35) = 3000$
34. $x^4 - 5x^3 + x^2 - 3x = 2$
35. $|x| + |x - 3| = 6$
36. $|x| - |x - 6| = 0$

37. Find three distinct approximations to the one real-number solution to $x^3 - 10 = 0$ with an error of at most 0.01. What is the exact solution?

38. Find three distinct approximations to the one real-number solution to $x^3 + 4 = 0$ with an error of at most 0.01. What is the exact solution?

In Exercises 39–42, solve for x in the given interval with error of at most 0.01.

39. $x^4 - 3x^3 - 6x + 5 = 0$, where $0 \le x \le 10$
40. $\dfrac{x^3 - 10x^2 + x + 50}{x - 2} = 0$, where $-10 \le x \le 10$
41. $3\sin(x - 5) = 0$, where $0 \le x \le 10$
42. $\sqrt[3]{x^2 - 2x + 3} = 0$, where $-10 \le x \le 10$

43. The owner of the Olde Time Ice Cream Shoppe pays $1000 per month for fixed expenses such as rent, electricity, and wages. Ice cream cones are sold at $0.75 each, of which $0.40 goes for ice cream, cone, and napkin. How many cones must be sold to break even?

Exercises 44–47 refer to the following **problem situation**: There are many rectangles whose perimeters are 320 in. Consider all these rectangles.

44. Illustrate the problem-solving strategy "Draw a picture" by drawing a picture of this problem situation. Label the length x. What values of x make sense in this problem situation?

45. Find an algebraic representation of the area of the rectangles in this collection.

46. Find a graphical representation of the area of the rectangles in this collection.

47. Verify that $(-40, -8000)$ is a point on a complete graph of the algebraic representation of this problem situation. What meaning do these coordinates have?

48. For a certain car, it has been determined that $D = r + r^2/19.85$ is an algebraic representation that approximates the stopping distance D (in feet) when a car is traveling at a speed of r mph. Use a graphical representation of this problem situation to estimate the speed of a car if the stopping distance is 300 ft.

49. The rate at which a blood cell flows depends on the distance of the cell from the center of the artery. Research has determined that a mathematical model of this problem situation is the equation $v = 1.19 - (1.85 \times 10^4)r^2$, where r is the distance (in centimeters) of the blood cell from the center of the artery and v is the velocity (in centimeters per second). If a blood cell is traveling at 0.975 cm/sec, use a graphical representation of the model equation to estimate the distance of the blood cell from the center of the artery.

Exercises 50 and 51 refer to the following **problem situation**: A graphic artist designs pages $8\frac{1}{2}$ by 11 in. with a picture centered on the paper. Suppose the distance from the outer edge of the paper to the picture is x inches on all sides.

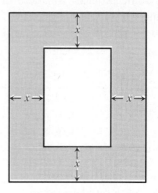

For Exercises 50 and 51.

50. Find an algebraic representation that describes the area A of the picture when the width of the border is x inches on all sides.

51. Use a graphical representation on a grapher to estimate the width of the uniform border if the area of the picture is 50 sq in.

Exercises 52 and 53 refer to the following **problem situation**: A single-commodity open market is driven by the supply-and-demand principle. Economists have determined that supply curves usually increase (that is, as the price increases, the sellers increase production) and demand curves usually decrease (as the price increases, the consumer buys less).

52. Suppose $p = 15 - 0.023x$. Find the price p if the production level x is 120 units, and find the production level x if the price p is 2.30.

53. Suppose $p = 100 - 0.0015x^2$. Find the price p if the production level x is 120 units, and find the production level x if the price p is 2.30.

54. Writing to Learn A student writes that $-1, 2$, and 3 are solutions to the equation $x^3 + 2x^2 - 5x - 6 = 0$. Write several paragraphs explaining whether you agree. Include in your response how you could determine if this equation has any real solutions and how graphing can help determine all real solutions.

Two or more viewing windows may be needed to show a complete graph of an equation. Take that into account in these next two exercises.

55. Find a complete graph of $y = x^3 - 8x^2 + 12.99x - 5.94$.

56. Solve $x^3 - 8x^2 + 12.99x - 5.94 = 0$.

57. Solve Example 10 from Section 1.6 by using zoom-in.

2.2 _____ Applications and Finding Complete Graphs of Problem Situations

Recall the Quality Rent-a-Car problem situation studied in Section 1.2. Here, the function $y = 0.2x + 15$ is an algebraic representation of the problem situation, and Fig. 2.12 is a graphical representation of the problem situation. The equation, its graph, or both together are often referred to as a mathematical model of the problem situation because the data given in the problem satisfy this equation.

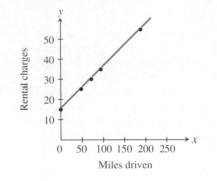

Figure 2.12   Graph of the Quality Rent-a-Car Problem Situation: $y = 0.2x + 15$.

When solving a problem posed for some problem situation, first make sure you understand the problem situation and the problem. Then find an algebraic or graphical representation of the problem situation. Determine what values of x make sense in the problem situation.

Strategy: Analyzing a Problem Situation

Understand the problem situation and the problem before trying to develop a model. Then follow this two-step process:

- Find an algebraic representation of the problem situation.
- Find and analyze a complete graph of the problem situation.

E X A M P L E   1   Finding an Algebraic Representation

Bill invests $20,000, a portion of this at 6.75% simple interest and the remainder at 8.6% simple interest. Find an algebraic representation for the total interest earned assuming that Bill invests x dollars at 6.75% and the rest at 8.6%.

Solution   Let x represent the amount invested in the 6.75% account and y represent the total interest Bill earns in 1 yr.

The problem-solving strategy "Completing a table" (see Section 1.2) is helpful for this problem:

Interest Rate	Dollars Invested	Interest Earned
6.75%	x	$0.0675x$
8.6%	$20,000 - x$	$0.086(20,000 - x)$

Total interest = interest from the 6.75% account + interest from the 8.6% account.

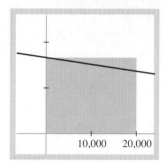

Figure 2.13
$y = -0.0185x + 1720$. Shaded area shows viewing rectangle [0, 20,000] by [0, 1720].

[−6000, 24,000] by [−600, 2,400]

Thus an algebraic representation is

$$y = 0.0675x + 0.086(20,000 - x).$$

This equation can be simplified as follows:

$$y = 0.0675x + 0.086(20,000 - x)$$
$$= 0.0675x + 1720 - 0.086x$$
$$= -0.0185x + 1720.$$

Complete Graph of an Algebraic Representation vs. Complete Graph of a Problem Situation

In Example 1, the algebraic representation of the problem situation was $y = -0.0185x + 1720$. The complete graph of this equation is called the **complete graph of the algebraic representation**.

However, in the context of the problem situation, perhaps only a part of such a complete graph represents the problem. For example, in the context of the problem in Example 1, the only values of x that make sense are *nonnegative values that do not exceed* 20,000. Consequently the graph of $y = -0.0185x + 1720$ in the viewing rectangle [0, 20,000] by [0, 1720] is a **complete graph of the problem situation** (shaded portion of Fig. 2.13).

In a different context from the problem situation in Example 1, a graph of the equation $y = -0.0185x + 1720$ could include values for x that are negative or greater than 20,000. Thus the whole graph shown in Fig. 2.13 (ignoring the shading) is called a complete graph of the algebraic representation $y = 0.0675x + 0.086(20,000 - x)$.

In other words, the complete graph of a problem situation is often *only a part* of a complete graph of the algebraic representation.

E X A M P L E 2   Solving the Problem: Graphically

Use a graphing utility and the problem situation in Example 1 to estimate the amount invested at each rate if Bill receives $1509.10 in interest in 1 yr.

Solution

1. Find a graph of $y = -0.0185x + 1720$ in the viewing rectangle [0, 20,000] by [0, 1720] (see Fig. 2.13). This is a complete graph of the problem situation.

2. Locate the point on the graph whose y-coordinate is closest to $1509.10.

3. Read the x-coordinate of the point.

   The x-coordinate of the point indicates that Bill invests $11,400 at 6.75% and $20,000 - $11,400 = $8,600 at 8.6%.

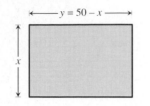

Figure 2.14   Rectangle with a perimeter of 100 in., where x = width.

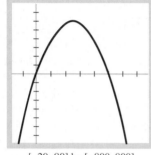

[−20, 80] by [−800, 800]

Figure 2.15 $A = x(50 - x)$: complete graph of the *algebraic representation*.

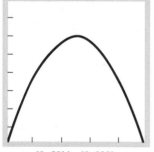

[0, 50] by [0, 800]

Figure 2.16 $A = x(50 - x)$: complete graph for the *problem situation* in Example 4. Compare with Fig. 2.15.

EXAMPLE 3   Finding an Algebraic Representation

Consider the collection of all rectangles having a perimeter of 100 in. Find an algebraic representation that describes the area of these rectangles in terms of their width.

Solution   The problem-solving strategies "Drawing a picture" and "Using variables" are helpful in this problem situation (see Fig. 2.14). Let x = width and y = length in inches. Then

$$100 = 2x + 2y \quad \text{or} \quad 50 = x + y. \quad \text{The perimeter is 100 in.}$$

$$\text{Area} = xy \quad \text{where} \quad y = 50 - x.$$

Therefore the equation $A = x(50 - x)$ is an algebraic representation of the problem situation. ∎

EXAMPLE 4   APPLICATION: Finding the Area of a Rectangle

Find a complete graph of the problem situation of Example 3.

Solution   The variable x in the algebraic representation $A = x(50 - x)$, independent of the problem situation, can represent any real number. Therefore the graph in Fig. 2.15 is a complete graph of the algebraic representation. However, in the context of the problem situation, x is positive and is less than 50 because the area of a rectangle is always positive. Thus the complete graph of the problem situation includes *only* the portion of the graph of $A = x(50 - x)$ between $x = 0$ and $x = 50$, as shown in Fig. 2.16. ∎

EXAMPLE 5   APPLICATION: Finding Manufacturing Costs

A Problem Situation   Companies that produce goods have fixed costs and variable costs. Fixed costs include salaries, benefits, equipment maintenance, utilities, and so forth. Quick Manufacturing Company produces T-shirts and has fixed annual costs of $200,000. The variable cost to produce one T-shirt is $1.50, and each shirt sells for $4.

Find complete graphs of the algebraic representations of the total cost C and the revenue R in this problem situation.

Solution

Total cost = variable cost × number produced + fixed costs.
Let x be the number of T-shirts produced in 1 yr. Then

$$C = 1.5x + 200,000$$ 1.5x is the cost of producing x T-shirts.
Add the fixed cost of $200,000.

$$R = 4x.$$ 4x is the revenue generated by selling x T-shirts.

Complete graphs are shown in Fig. 2.17.   ≣

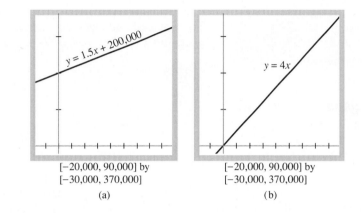

[−20,000, 90,000] by   [−20,000, 90,000] by
[−30,000, 370,000]   [−30,000, 370,000]
     (a)   (b)

Figure 2.17   (a) $y = 1.5x + 200,000$; (b) $y = 4x$.

Note that Fig. 2.17 shows complete graphs of the algebraic representations. Because the only values of x that make sense in this problem situation are whole numbers, complete graphs of the problem situation would consist of only the points on the lines in Fig. 2.17 whose x-coordinates are positive whole numbers.

E X A M P L E   6   APPLICATION: Determining the Break-even Point

In order to break even, total revenue R from a product must equal the total cost C of production. How many T-shirts must Quick Manufacturing Company sell so that their total revenue equals their total cost?

Solution

Graphical Method

1. Find complete graphs of both $R = 4x$ and $C = 1.5x + 200,000$ in the same viewing rectangle [−20,000, 90,000] by [−30,000, 370,000] (see Fig. 2.18).
2. To the left of the point where the two graphs intersect, R is less than C. To the right, R is greater than C. Therefore cost and revenue are equal when x equals the x-coordinate of this point of intersection.

Notes on Examples

In Example 6, anticipate that some students will observe that there is a difference between shirts made and shirts sold. Ask students to identify the assumptions made about total revenue and selling all shirts made.

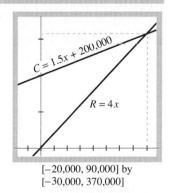

[−20,000, 90,000] by
[−30,000, 370,000]

Figure 2.18   Complete graphs of C and R for Example 5.

Notes on Exercises

Ex. 9–24 are problems that one finds in a second-year algebra course. Since the settings are familiar, focus attention on the mathematical models and the relationships between representations.

Assignment Guide

Day 1: Ex. 1–16
Day 2: Ex. 18–23, 25–30
Day 3: Ex. 31–34, 35–39

3. Use trace to estimate the x-coordinate of this point as 80,000.

Algebraic Method   To solve this problem algebraically, express C and R in terms of x and then write and solve the equation $R = C$.

$$R = C$$

$$4x = 1.5x + 200,000$$

$$2.5x = 200,000$$

$$x = 80,000$$

The graphical and algebraic methods agree. The break-even point for Quick Manufacturing Company is $x = 80,000$ shirts.   ▬

Exercises for Section 2.2

In Exercises 1–4, assume that the equations are algebraic representations of some problem situation. Find a complete graph of each algebraic representation.

1. $P = 2N + 50$ **2.** $V = 5L^2$

3. $S = 1105 + 1105(0.08)N$ **4.** $E = 1000C^2$

Exercises 5–8 use the following information. The area A of a rectangle is given by the equation $A = LW$, where L is the length and W is the width.

5. If the width is 50 units, then write the area A in terms of the length L.

6. If the length is 200 units, then write the area A in terms of the width W.

7. Find a complete graph of the algebraic representation in Exercise 5.

8. Find a complete graph of the algebraic representation in Exercise 6.

Exercises 9–12 use the following information: The perimeter P of a rectangle is given by the equation $P = 2L + 2W$, where L is the length and W is the width.

9. If the width is 100 units, then write the perimeter P in terms of the length L.

10. If the length is 40 units, then write the perimeter P in terms of the width W.

11. Find a complete graph of the algebraic representation in Exercise 9.

12. Find a complete graph of the algebraic representation in Exercise 10.

In Exercises 13–16, consider the collection of all rectangles having a length of twice their width.

13. Write the area A in terms of the width W of the rectangle.

14. Write the area A in terms of the length L of the rectangle.

15. Write the perimeter P in terms of the width W of the rectangle.

16. Write the perimeter P in terms of the length L of the rectangle.

17. Refer to the problem situation in Example 3. Use a graph to find the dimensions of one of the rectangles if its area is 500 sq in.

For Exercises 18–20, consider the collection of all rectangles the sum of whose length and width is 75 in. Let x be the length of such a rectangle and consider the area of these rectangles.

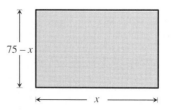

For Exercises 18–20.

18. Show that $A = x(75 - x)$ is an algebraic representation for the area of one of these rectangles in terms of its length x.

19. Verify graphically that $(-10, -850)$ is a point on the graph of the algebraic representation in Exercise 18. What meaning do these coordinates have in this problem situation?

20. Is the graph used in Exercise 19 a complete graph of the algebraic representation or a complete graph of the problem situation?

For Exercises 21–23, consider the collection of all rectangles whose perimeter is 360 in. Let x be the length of such a rectangle, and consider the area A of all these rectangles.

21. Find an algebraic representation of this problem situation.

22. What values of x make sense in this problem situation? Why?

23. What is the smallest viewing rectangle that will show a complete graph of the problem situation? Find the complete graph of the problem situation.

24. $A = \pi x^2$ is an equation that gives the area A of a circle with radius x. How might a complete graph of this algebraic representation differ from a complete graph of this problem situation?

For Exercises 25–27, consider this **problem situation**: Janet is making a rectangular end table for her living room. She has decided the tabletop should have a surface area of 625 sq in. Let L be the length of the tabletop in inches and W be the width in inches.

25. Write an equation that relates the length L and the width W.

26. Solve the equation you found in Exercise 25 for L, and find a complete graph of this algebraic representation.

27. What portion of the graph in Exercise 26 represents a complete graph of the problem situation?

Exercises 28–30 refer to the following **problem situation**: A bag of lawn fertilizer can cover a rectangular lawn having an area of 6000 sq ft. Let x be the length of the lawn (in feet) and y the width (in feet).

28. Write an equation that relates the length and the width of the lawn.

29. Find a complete graph of this problem situation.

30. Use the graph in Exercise 29 to find the length of the lawn covered by fertilizer if the width is 45 ft.

Exercises 31–34 refer to the following **problem situation**: Reggie invests $12,000, part at 7% simple interest and the remainder at 8.5% simple interest.

31. Determine the total interest Reggie receives in one year in terms of the amount he invests at 7%.

32. Find a complete graph of the algebraic representation in Exercise 31.

33. Find a complete graph of the problem situation in Exercise 32.

34. Use the graph in Exercise 32 to find the amount invested at each rate if Reggie receives $900 interest in one year.

Exercises 35–39 refer to the following **problem situation**: A shoe manufacturer determines that the annual cost C of making x pairs of shoes is $30 per pair plus $100,000 in fixed overhead costs. Each pair of shoes that is manufactured is sold for $50 per pair.

35. Write an algebraic representation for the total cost C in terms of the number of pairs x produced.

36. Find a complete graph of this problem situation.

37. Use the graph in Exercise 36 to find the number of shoes produced if the total cost is $340,000.

38. Write an algebraic representation for revenue R in terms of the number of pairs x produced.

39. How many shoes must be sold before revenue exceeds annual cost?

40. Discussion Exercises 1–4 asked you to find complete graphs of algebraic representations of problem situations without knowing any details about them. Could you also find the complete graphs of the problem situations without knowing more about them?

41. Writing to Learn Write several paragraphs that explain the difference between a complete graph of a problem situation and a complete graph of an algebraic representation of the problem situation.

42. Solve Example 2 algebraically.

43. Solve Exercise 34 algebraically.

2.3 _____ Solving Linear and Quadratic Equations Algebraically

Objective

Students will be able to find solutions to equations using algebraic techniques.

Key Ideas

Identity
Conditional equation
Solution to an equation
Linear equations
Equivalent equations
Quadratic equations
Linear equations with
     absolute value
Quadratic formula
Discriminant

Common Errors

Many students readily accept almost any solution found by using equivalent equations until the form "$x = $ a number" is produced. For this reason, it is important that students understand that the *number* is the solution, not the equation. Also, students should check their solutions in the original equation.

Teaching Note

Encourage students to check answers for two reasons: (1) to check for computational or calculator errors, and (2) to establish the expectation of checking approximate answers that are found using a graphing utility. Not all examples have checks provided in the textbook. These are left for students to do.

The concept of problem situation was introduced in Section 1.2 and revisited in Section 2.2. A problem situation can be described using both an algebraic and a graphical representation, and so we often use both in finding a solution to the problem. Become proficient in analyzing the connections among the model (often an equation or an inequality), its associated graphs, and the problem situation.

We have seen in earlier sections that finding a solution to a problem situation often requires that we find the zero of a function. This is equivalent to saying that we want to find a solution to an equation $f(x) = 0$. Sometimes it is quicker and often more precise to find a solution to such an equation using an algebraic method.

Recall that an equation is a statement equating two expressions. If at least one of the expressions is an algebraic expression in a variable x, the statement is called **an equation in x**. To **solve** an equation in x means that you find all the values of x for which the equation is true.

There are two types of equations. An **identity** is an equation that is true for all values of the variable for which the expressions are defined. On the other hand, an equation in x that is true for only certain values of x is called a **conditional equation**. Here are some examples of each:

$$\text{identities:} \qquad x + 3 = 3 + x, \qquad x(x-3) = x^2 - 3x$$

$$\text{conditional equations:} \qquad x + 2 = 5, \qquad\qquad x^2 = 25$$

In this chapter, you will study certain types of conditional equations.

A number a is a **solution to an equation in x** if a true statement results when x is replaced by a everywhere in the equation.

EXAMPLE 1   Verifying a Solution

Show that $x = -2$ is a solution to the equation $x^3 - x + 6 = 0$.

Solution

$$x^3 - x + 6 = 0 \qquad\qquad \text{Begin with the equation.}$$

$$(-2)^3 - (-2) + 6 = -8 + 2 + 6 \qquad \text{Substitute } -2 \text{ for } x.$$

$$= 0 \qquad\qquad\qquad ≡$$

Solution Agreement   Once all the real-number solutions for a given equation have been found, the equation is considered solved. In this section, the focus is on algebraic techniques for solving equations.

Perhaps the simplest type of conditional equation is a linear one.

Definition 2.2   Linear Equation in x

A **linear equation in** x is an equation that can be written in the form $ax + b = 0$, where a and b are real numbers and $a \neq 0$.

Solving Linear Equations by Writing Equivalent Equations

Equivalent equations are two or more equations with exactly the same set of solutions. One important method of solving a linear equation requires transforming it into an equivalent equation whose solutions are obvious. For example, the equations $2x - 4 = 0$, $x - 2 = 0$, and $x = 2$ are all equivalent, and the solution to the last one is obvious.

Obtaining Equivalent Equations

An equivalent equation is obtained if any one of the following operations is performed:

	Given Equation	Equivalent Equation
Combine like terms or reduce fractions.	$2x + x = \frac{3}{9}$	$3x = \frac{1}{3}$
Add or subtract the same real number to each side of the equation.	$x + 3 = 7$	$x = 4$
Add or subtract the same polynomial to each side of the equation.	$4x^2 - 2x = x$	$4x^2 - 3x = 0$
Multiply or divide each side of the equation by the same nonzero number.	$3x = 12$	$x = 4$

We will use these methods of obtaining equivalent equations in the next several examples.

E X A M P L E   2   Combining Like Terms

Solve $2(2x - 3) + 3(x + 1) = 5x + 2$.

Solution

$$2(2x - 3) + 3(x + 1) = 5x + 2$$

$$4x - 6 + 3x + 3 = 5x + 2 \qquad \text{Use the distributive property.}$$

$$7x - 3 = 5x + 2 \qquad \text{Combine like terms.}$$

$$2x - 3 = 2 \qquad \begin{array}{l}\text{Subtract } 5x \text{ from both sides} \\ \text{of the equation.}\end{array}$$

$$2x = 5 \qquad \text{Add 3 to both sides.}$$

$$x = 2.5 \qquad \text{Divide each side by 2.}$$

Check the Solution

$$2\big(2(2.5) - 3\big) + 3(2.5 + 1) \stackrel{?}{=} 5(2.5) + 2 \qquad \begin{array}{l}\text{Substitute } x = 2.5 \text{ into both} \\ \text{sides of the equation.}\end{array}$$

$$2(2) + 3(3.5) \stackrel{?}{=} 14.5$$

$$14.5 = 14.5 \qquad \begin{array}{l}\text{It checks. The solution } x = 2.5 \\ \text{is correct.}\end{array} \quad \blacksquare$$

This method can be extended to equations that involve expressions of the form $|ax + b|$. Recall that $|a|$ represents the distance on the real-number line from the origin to the point with coordinate a.

E X A M P L E 3 Solving a Linear Equation with Absolute Value

Solve $|2x + 3| = 5$.

Solution The expression inside the absolute value symbols must be either 5 or -5, so $2x + 3$ equals 5 or -5. Solving for both values of the expression yields two values for x.

$$\begin{array}{ll} 2x + 3 = -5 & 2x + 3 = 5 \\ 2x = -8 & 2x = 2 \\ x = -4 & x = 1 \end{array}$$

Check the Solution

$$|2(-4) + 3| = 5 \qquad \text{and} \qquad |2(1) + 3| = 5$$

The solutions to $|2x + 3| = 5$ are -4 and 1. $\qquad \blacksquare$

E X A M P L E 4 APPLICATION: Solving a Mixture Problem

Sparks Drug Store keeps two acid solutions on hand to fill orders for its customers. One solution is 10% acid and the other is 25% acid. An order is received for 15 l

---

Notes on Examples

Example 3 is useful for reviewing the algebraic process for solving an absolute value equation. In addition to Example 3, solve $|2y - 1| + 3 = 0$ in class to check students' reasoning abilities for working with absolute value equations. For an absolute value equation, the solution process to produce an equivalent equation is not simply a matter of operating on both sides of the equation. Students need to use the definition of absolute value. Some students may find the following alternative approach viable. Since $|2y - 1| + 3$ is the sum of a nonnegative and a positive number, how can this sum be zero? Or, alternatively, since $|2y - 1| = -3$, how can this equation be true?

of 12% acid solution. How much 10% acid solution and how much 25% acid solution should be combined to fill this order?

Solution Let x represent the number of liters of 10% acid solution needed to make the 12% mixture. There will be 15 l of the mixture, so $15 - x$ represents the number of liters of 25% acid solution.

An acid solution is made by taking pure acid and diluting it with water. A 10% solution means that 10% of the mixture is pure acid. Therefore

$$0.1x = \text{amount of pure acid in the 10\% solution}$$

$$0.25(15 - x) = \text{amount of pure acid in the 25\% solution}$$

$$0.12(15) = \text{amount of pure acid in the final 15-l solution.}$$

The total amount of acid in the mixture is the sum of the acid in the 10% and 25% solutions. The equation is

$$0.1x + 0.25(15 - x) = 0.12(15)$$

$$0.1x + 3.75 - 0.25x = 1.8$$

$$-0.15x + 3.75 = 1.8$$

$$-0.15x = -1.95$$

$$x = 13. \qquad \text{13 l of 10\% solution are combined}$$
$$\text{with 2 l of 25\% solution.}$$

Check the Solution

$$\frac{0.1(13) + 0.25(2)}{15} = 0.12$$

≡

Many problem situations can be solved by using linear equations. Another important class of equations is the quadratic equation.

Definition 2.3   Quadratic Equations in x

A **quadratic equation in x** is an equation that can be written in the form $ax^2 + bx + c = 0$, where a, b, and c are real numbers and $a \neq 0$.

Solving Quadratic Equations by Factoring

Linear equations are solved by finding equivalent equations that have obvious solutions. In a similar fashion, quadratic equations can be solved by using the method of factoring to find equivalent equations.

E X A M P L E   5   Solving a Quadratic Equation by Factoring

Solve $2x^2 + 5x - 3 = 0$.

Solution

$$2x^2 + 5x - 3 = 0$$

$$(2x - 1)(x + 3) = 0$$

$$2x - 1 = 0 \quad \text{or} \quad x + 3 = 0 \qquad \text{A product is zero only if at least one factor is zero.}$$

The solutions are $x = \frac{1}{2} = 0.5$ and $x = -3$.   ≡

Often, the linear factors of a quadratic equation cannot be quickly seen. In that case, you can solve the equation by using the quadratic formula.

Quadratic Formula

The solutions to a quadratic equation in x in the standard form $ax^2 + bx + c = 0$, $(a \neq 0)$ are given by the **quadratic formula**

$$x = \frac{-b \pm \sqrt{b^2 - 4ac}}{2a}.$$

Consider the general quadratic equation $ax^2 + bx + c = 0$. The quadratic formula is derived from this equation using a technique called **completing the square**. This technique calls for changing the equation to a form that includes a perfect square—an expression of the form $(x + k)^2$.

$$ax^2 + bx + c = 0$$

$$ax^2 + bx = -c \qquad \text{Subtract } c \text{ from each side.}$$

$$x^2 + \frac{b}{a}x = -\frac{c}{a} \qquad \text{Divide each side by } a.$$

$$x^2 + \frac{b}{a}x + \left(\frac{b}{2a}\right)^2 = -\frac{c}{a} + \left(\frac{b}{2a}\right)^2 \qquad \text{Add } (b/2a)^2 \text{ to each side.}$$

$$\left(x + \frac{b}{2a}\right)^2 = -\frac{c}{a} + \left(\frac{b}{2a}\right)^2 \qquad \text{Factor the left-hand side to get a perfect square.}$$

$$\left(x + \frac{b}{2a}\right)^2 = \frac{b^2 - 4ac}{4a^2} \qquad \text{Combine fractions on the right-hand side.}$$

$$x + \frac{b}{2a} = \pm\sqrt{\frac{b^2 - 4ac}{4a^2}} \qquad \text{Take the square root of both sides.}$$

$$x = \frac{-b \pm \sqrt{b^2 - 4ac}}{2a} \qquad \text{Solve for } x.$$

E X A M P L E   6   Using the Quadratic Formula

Solve $2x^2 + 2x - 1 = 0$ using the quadratic formula.

Solution   In this case, $a = 2$, $b = 2$, and $c = -1$. Thus

$$x = \frac{-b \pm \sqrt{b^2 - 4ac}}{2a} \qquad \text{Quadratic formula}$$

$$= \frac{-2 \pm \sqrt{2^2 - 4(2)(-1)}}{2(2)} \qquad \begin{array}{l}\text{Substitute into the quadratic formula}\\ a = 2, b = 2, \text{ and } c = -1.\end{array}$$

$$= \frac{-2 \pm \sqrt{12}}{4}$$

$$= \frac{-1 \pm \sqrt{3}}{2}. \qquad \text{Recall that } \sqrt{12} = \sqrt{4 \cdot (3)} = 2\sqrt{3}.$$

Verify with a calculator that

$$x = \frac{-1 + \sqrt{3}}{2} = 0.3660254038 \qquad \text{and} \qquad x = \frac{-1 - \sqrt{3}}{2} = -1.366025404$$

are accurate to the number of displayed decimal places.   ☰

EXPRESSING IRRATIONAL NUMBERS AS DECIMALS

In this text, we will write $\sqrt{3} = 1.732050808$ (instead of $\sqrt{3} \approx 1.732050808$) with the understanding that the right side is a decimal approximation for the irrational number on the left side, accurate to the number of decimal places displayed.

E X A M P L E   7   Using the Quadratic Formula

Solve $3x - x^2 = 1$.

Solution   Rewrite the given equation into the following standard form: $-x^2 + 3x - 1 = 0$. Then it is clear that $a = -1$, $b = 3$, and $c = -1$. The quadratic formula then yields

$$x = \frac{-b \pm \sqrt{b^2 - 4ac}}{2a}$$

$$= \frac{-3 \pm \sqrt{3^2 - 4(-1)(-1)}}{2(-1)}$$

$$= \frac{-3 \pm \sqrt{9 - 4}}{-2}$$

$$= \frac{3 \mp \sqrt{5}}{2}. \qquad \text{☰}$$

REMINDER

When a quadratic equation is not in standard form, be sure you choose the correct values for a, b, and c.

The expression under the **radical** (square root sign) in the quadratic formula, $b^2 - 4ac$, is called the **discriminant** of the quadratic equation $ax^2 + bx + c = 0$. The discriminant is important because it determines whether there are any real-number solutions and if so, how many. If $b^2 - 4ac$ is positive, there are two real solutions. If it is zero, the equation has exactly one solution. If it is negative, there are no real-number solutions. (Recall that the square root of a negative number is not a real number.)

EXAMPLE 8   APPLICATION: Finding Box Dimensions

A Problem Situation   Squares with a side length of 5 in. are cut from each corner of a rectangular piece of cardboard with width w and length ℓ (see Fig. 2.19). By folding along the dashed lines in Fig. 2.19(a), a box is formed whose height is 5 in., width is $(w - 10)$ in., and length is $(\ell - 10)$ in.

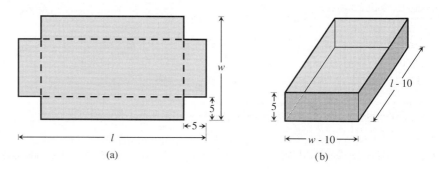

(a)   (b)

Figure 2.19   Box dimensions and completed box for the problem situation.

Suppose the length of the cardboard in Fig. 2.19 is twice its width. Find the dimensions of the cardboard if the volume of the resulting box is 2040 cu in.

Solution   Let V be the volume of the box. Since $V = \ell w h$,

$$5(w - 10)(\ell - 10) = V$$

$$5(w - 10)(2w - 10) = 2040 \qquad \ell = 2w \text{ and } V = 2040.$$

$$(w - 10)(w - 5) = 204 \qquad \begin{array}{l}\text{Factor 2 from } 2w - 10 \text{ and divide} \\ \text{both sides by 10.}\end{array}$$

$$w^2 - 15w - 154 = 0 \qquad \begin{array}{l}\text{Multiply the left-hand side} \\ \text{and combine terms.}\end{array}$$

$$(w - 22)(w + 7) = 0. \qquad \begin{array}{l}\text{Use the quadratic formula} \\ \text{if this factorization is not evident.}\end{array}$$

Note that -7 and 22 are both solutions to this equation; however, because width must be positive, only 22 is a solution to the problem. The dimensions of the cardboard are 22 by 44 in.   ▰

Exercises for Section 2.3

In Exercises 1–6, determine whether each equation is a *conditional equation* or an *identity*.

1. $x + 3 = -1$ **2.** $x + 3 = -1 + x + 4$

3. $3(x + 2) = 3x + 6$ **4.** $5/x = 35$

5. $5 - x = 18$ **6.** $x^2 + 4x + 3 = (x + 3)(x + 1)$

7. Is $x = 2$ a solution to $x^3 - 3x - 2 = 0$?

8. Is $x = 3$ a solution to $x^5 - 4x^3 - 6x + 117 = 0$?

In Exercises 9 and 10, decide which of the given values of x are solutions to the equation.

9. $x^3 + 2x^2 - 5x - 6 = 0$
    a) $x = -3$ **b)** $x = -2$ **c)** $x = -1$

10. $x^3 + 2x^2 - 5x - 6 = 0$
    a) $x = 1$ **b)** $x = -3$ **c)** $x = 2$

In Exercises 11–21, use an algebraic method to solve each equation. Check by substitution into the original equation.

11. $2x - 3 = 4x - 5$ **12.** $4(x - 2) = 5x$

13. $\frac{1}{2}x - \frac{2}{3} = 2x + 7$ **14.** $-3x + 4 = 2(x + 4)$

15. $2(3 - 4x) - 5(2x + 3) = x - 17$

16. $\dfrac{t - 1}{3.5} = 8$ **17.** $|x + 1| = 4$

18. $|x - 2| = -5$ **19.** $|2t - 3| - 1/2 = 0$

20. $|3 - 5x| = |-4|$ **21.** $|x - 3| = |2x + 1|$

In Exercises 22–29, solve each equation by factoring. Support with a graphing utility.

22. $x^2 + x - 2 = 0$ **23.** $x^2 - 5x + 6 = 0$

24. $x^2 - x - 20 = 0$ **25.** $x^2 - 4x + 3 = 0$

26. $2x^2 + 5x - 3 = 0$ **27.** $4x^2 - 8x + 3 = 0$

28. $x^2 - 8x = -15$ **29.** $x^2 + 4x - 3 - 2$

In Exercises 30–33, solve each equation by completing the square.

30. $x^2 + 7x - 2 = 0$ **31.** $x^2 + 6x = 16$

32. $x^2 - 4x = 7$ **33.** $x^2 - 5x - 3 = 0$

In Exercises 34–41, solve each equation by using the quadratic formula. Support your answer with a graphing utility.

34. $x^2 + x - 1 = 0$ **35.** $x^2 - 4x + 2 = 0$

36. $x^2 + 8x - 2 = 0$ **37.** $2x^2 - 3x + 1 = 0$

38. $x^2 - 2x = 7$ **39.** $3x + 4 = x^2$

40. $5 - x^2 = 8x$ **41.** $x^2 - 5 = \sqrt{3}x$

In Exercises 42–45, use any methods to solve the equation.

42. $x^2 - 5x + 6 = 30$ **43.** $3y^3 - 2y^2 + y = 0$

44. $|x^2 + 4| = 8$ **45.** $x^4 + 4x^2 - 5 = 0$

In Exercises 46–49, determine the number of real solutions to each equation.

46. $x^2 + 3x + 2 = 0$ **47.** $2x^2 - 3x + 2 = 0$

48. $x^2 - \pi x + \sqrt{3} = 0$ **49.** $x^2 - 1 = 0$

Exercises 50 and 51 refer to the formula $F = \frac{9}{5}C + 32$, which gives the Fahrenheit temperature F as a function of the Celsius temperature C.

50. Normal body temperature is $98.6°$F. How many degrees Celsius is this?

51. Solve the formula for C.

Exercises 52–54 refer to the following **problem situation**: The perimeter of a rectangle is 360 in. and its width is 20 in.

52. Write an equation that includes the variable l representing length.

53. Find the area of the rectangle.

54. If the width of the rectangle is changed so that its length is twice its width, what is its length and width?

Exercises 55–57 refer to the following **problem situation**: A storage box with a rectangular base has a height of 30 cm and a volume of 5400 cm^3.

55. If x represents the length of the box in centimeters and y represents the width in centimeters, write an equation in the form $y = ?$ where the right-hand side of the equation is an expression in x. What does y represent?

56. Find the width of the box if its length is 24 cm.

57. Find the length of the box if its width is 12 cm.

Exercises 58 and 59 refer to the following **problem situation**: A laboratory keeps two acid solutions on hand. One is 20% acid and the other is 35% acid.

58. How much 20% acid solution and how much 35% acid solution should be used to fill an order for 25 l of a 26% acid solution?

59. How many liters of distilled water should be added to a liter of the 35% acid solution in order to dilute it to a 20% acid solution?

Exercises 60 and 61 refer to the following **problem situation**: An investment pays simple interest. If P dollars are invested at an interest rate r per year, the value S after n years is given by the algebraic representation $S = P(1 + rn)$.

60. Solve this equation for n.

61. How many years are required for an investment earning 8% simple interest to triple in value?

62. The formula for the area of a trapezoid is $A = \frac{1}{2}h(b_1 + b_2)$. Solve this equation for b_1.

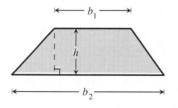

For Exercise 62.

63. A semicircle is placed on one side of a square so that its diameter coincides with a side of the square. Find the side length of the square if the total area of the square plus the semicircle is 200 square units.

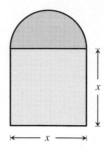

For Exercise 63.

Exercises 64–66 refer to the following **problem situation**: A single-commodity open market is driven by the supply-and-demand principle. Economists have determined that supply curves are usually increasing; that is, as the price increases, the sellers increase production. Use the algebraic representation relating the price p and the number of units produced x.

64. If $p = 12 + 0.025x$ is an algebraic representation of this problem situation, find the price if the production level is 120 units.

65. Using the same algebraic representation as in Exercise 64, find the production if the price is $23.00.

66. Repeat Exercises 64 and 65 assuming the algebraic representation is $p = 5 + 0.01x^2$.

Exercises 67–69 refer to the following **problem situation**: A real estate company has two agents. Company profits P depend on the number of weekly listings x and are described by the algebraic representation $P = -5x^2 + 100x + 20$.

67. How many weekly listings are necessary to realize a profit of $300?

68. Is there a maximum possible profit in this situation?

69. What can you say about the weekly profit as the number of listings increases? Explain why this might be reasonable.

2.4 _____ Solving Linear Inequalities

Objective

Students will be able to solve problems involving inequalities using algebraic and graphical techniques.

Inequalities were defined in Section 1.1 in the context of order on the real-number line. Definition 1.1 says: For real numbers a and b, $a < b$ if $b - a$ is a positive number. Similarly, we defined the relations represented by the symbols $\leq$, $>$, and $\geq$. The symbols $<$, $\leq$, $>$, and $\geq$ are inequality signs, and the expressions $a < b$, $a \leq b$, $a > b$, and $a \geq b$ are called inequalities.

Inequalities may include variables just as equations do. This Exploration provides experience with inequalities.

🔍 EXPLORE WITH A GRAPHING UTILITY

1. Find the graphs of both $y = -x^2 + 5x + 2$ and $y = 6$ in the viewing rectangle $[-2, 8]$ by$[-2, 10]$.

2. Are there any values of x for which $-x^2 + 5x + 2 \geq 6$? What characteristic of these graphs gives you an answer?

3. Trace the graph to estimate values of x that satisfy this inequality.

4. Graph $y = -x^2 + 5x - 4$ on the same viewing rectangle. How do the solutions to the inequality $-x^2 + 5x + 2 \geq 6$ compare with the interval associated with where this third graph lies above the x-axis?

Generalize   Can you formulate a procedure for solving an inequality by a graphical method that would always work?

A number a is called a solution to an inequality if replacing the variable x with a results in a true statement. Sometimes it is convenient to combine two inequalities into one statement. For example, if $-1 \leq x$ and $x < 3$, we can write this information as the single statement

$$-1 \leq x < 3.$$

We call the set of all solutions to this double inequality an **interval** on the real-number line. Notice that -1 is included in the interval but 3 is not. This interval is said to be "closed on the left and open on the right" and is thus called a half-open interval. There are four types of **bounded intervals,** each with its own notation.

Interval Notation for Bounded Intervals

Notation	Interval Type	Inequality	Graph
$[a, b]$	Closed	$a \leq x \leq b$	
(a, b)	Open	$a < x < b$	
$[a, b)$	Half-open: closed-left, open-right	$a \leq x < b$	
$(a, b]$	Half-open: open-left, closed-right	$a < x \leq b$	

An **unbounded interval** is indicated by the symbols $-\infty$ and ∞ in interval notation. For example, $(-\infty, 3]$ stands for all the numbers less than or equal to

3. In other words, $(-\infty, 3]$ represents the set of all solutions to the inequality $x \leq 3$. It is unbounded on the left and closed on the right. There are four kinds of unbounded intervals.

REMINDER

It is not correct to write the two inequalities $x \leq -1$ and $x > 3$ in the single statement

$$-1 \geq x > 3$$

because no real number is less than or equal to -1 and at the same time greater than 3.

Interval Notation for Unbounded Intervals

Notation	Interval Type	Inequality	Graph
$(-\infty, b]$	Unbounded-left, closed	$x \leq b$	
$(-\infty, b)$	Unbounded-left, open	$x < b$	
$[a, \infty)$	Closed, unbounded-right	$a \leq x$	
(a, ∞)	Open, unbounded-right	$a < x$	

E X A M P L E   1   Writing Inequalities

Write an inequality represented by each of the following intervals:

a) $(-3, 5]$

b) $[-4, \infty)$

c) $[-8, -3]$

Solution

a) $(-3, 5]$ represents $-3 < x \leq 5$.

b) $[-4, \infty)$ represents $-4 \leq x$.

c) $[-8, -3]$ represents $-8 \leq x \leq -3$.

E X A M P L E   2   Drawing Number-line Graphs of Intervals

The graphs shown in the previous two boxes on interval notation are called number-line graphs. Draw number-line graphs of the solutions to the following inequalities: (a) $x \geq 2$, (b) $x < 5$, and (c) $-1 \leq x < 3$.

Solution

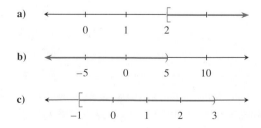

Algebraic Method for Solving an Inequality

When asked to solve an inequality, you are asked to find all numbers that are solutions to the inequality. Two inequalities are said to be **equivalent** if they have exactly the same set of solutions. To solve an inequality algebraically, replace the given inequality with an equivalent one that is simpler. Use any of the properties listed in Theorem 2.1 to change a given inequality to an equivalent one.

Theorem 2.1   Properties of Inequalities

Let a, b, and c be real numbers.

Addition of a number   If $a < b$, then $a + c < b + c$.
Multiplication by a positive number   If $a < b$ and $c > 0$, then $ac < bc$.
Multiplication by a negative number   If $a < b$ and $c < 0$, then $ac > bc$.
Transitive property   If $a < b$ and $b < c$, then $a < c$.

These four properties remain true when $<$ is replaced by any of the other three inequality symbols: $\leq$, $>$, or $\geq$.

Proof   We shall prove multiplication by a negative number.

$a < b$ implies that $b - a > 0$ Definition of *less than*

$(b - a)c < 0$ The product of a positive and a negative is negative.

$bc - ac < 0$ Distributive property

$ac > bc$ Definition of *less than*   ≡

Proofs of the other three properties are similar and are left as an exercise.

Definition 2.4   Linear Inequality in *x*

A **linear inequality in** x is an equality that can be written in the form

$$ax + b > 0, \qquad ax + b \geq 0, \qquad ax + b < 0, \qquad \text{or} \qquad ax + b \leq 0,$$

where a and b are real numbers and $a \neq 0$.

Example 3 illustrates that often an inequality should be rewritten into an equivalent form using the properties of Theorem 2.1 as a part of finding a solution.

E X A M P L E   3   Solving Inequalities Algebraically

Solve the inequality $3(x - 1) + 2 \leq 5x + 6$ algebraically and draw a number-line graph of the solution.

Solution

$$3(x - 1) + 2 \leq 5x + 6$$

$$3x - 3 + 2 \leq 5x + 6 \qquad \text{Distributive property}$$

$$3x - 1 \leq 5x + 6$$

$$-2x \leq 7 \qquad \text{Subtract } 5x \text{ from both sides.}$$

$$x \geq -\tfrac{7}{2} \qquad \text{Divide both sides by } -2.$$

The solution interval is $[-\tfrac{7}{2}, \infty)$, which contains all real numbers greater than or equal to $-\tfrac{7}{2}$. Figure 2.20 is the number-line graph for the solution inequality.

Figure 2.20 $[-3.5, \infty)$

Most linear inequalities can be solved by using the algebraic method illustrated in Examples 2 and 3.

Graphical Method for Solving an Inequality

An inequality can always be solved using a graphing method on a grapher. For more complicated inequalities, the graphical method is preferred. Examples 4 and 5 illustrate two different approaches to a graphical solution. Method 1 requires rewriting the inequality so that only a zero remains on one side of the inequality. For example, rewrite $4x - 1 < 2$ as follows:

$$4x - 1 < 2 \iff 4x - 3 < 0.$$

Now graph the equation $y = 4x - 3$ and determine where the graph is *below* the x-axis.

To solve the inequality $4x - 3 > 0$, determine where the graph is *above* the x-axis.

E X A M P L E   4   Solving Inequalities Graphically: Method 1

Solve $4x - 1 < 2$ graphically.

Solution

1. Rewrite $4x - 1 < 2$ as $4x - 3 < 0$.
2. Find the graph of $y = 4x - 3$ in the viewing rectangle $[-5, 5]$ by $[-5, 5]$ (see Fig. 2.21).

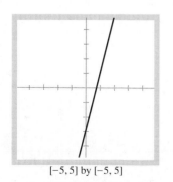

$[-5, 5]$ by $[-5, 5]$

Figure 2.21 $y = 4x - 3$.

3. Determine the values of x when the graph of $y = 4x - 3$ is below the x-axis.

4. Notice that the graph of $y = 4x - 3$ appears to cross the x-axis at about $x = \frac{3}{4}$. This is easily confirmed algebraically: $4x - 3 = 0$, so $x = \frac{3}{4}$.

5. Thus the graph of $y = 4x - 3$ lies below the x-axis when $x < \frac{3}{4}$.

The solution is the interval $(-\infty, 3/4)$.   ≡

Example 5 illustrates a second method of solving the same inequality graphically.

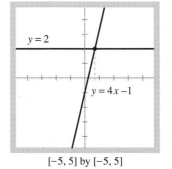

$y = 2$

$y = 4x - 1$

[−5, 5] by [−5, 5]

Figure 2.22 $y = 2$ and $y = 4x - 1$.

E X A M P L E   5   Solving Inequalities Graphically: Method 2

Solve $4x - 1 < 2$ graphically by graphing both $y = 4x - 1$ and $y = 2$ on the same coordinate system.

Solution

1. Draw graphs of both $y = 4x - 1$ and $y = 2$ in the same viewing rectangle [−5, 5] by [−5, 5] (see Fig. 2.22).

2. Identify the x-coordinates of points on the graph of $y = 4x - 1$ that lie *below* the graph of $y = 2$.

3. Note that the two graphs appear to intersect at point $(0.75, 2)$.

4. Thus the values of x where the graph of $y = 4x - 1$ is below the graph of $y = 2$ is when $x < \frac{3}{4}$.

The solution interval is $(-\infty, \frac{3}{4})$.   ≡

Example 6 illustrates that a double inequality can also be solved graphically.

E X A M P L E   6   Solving Double Inequalities

Solve

$$-3 < \frac{2x + 5}{3} \le 5$$

both algebraically and graphically.

Solution   The algebraic solution is found by applying the properties of real numbers and of inequalities until the variable remains alone in the middle expression.

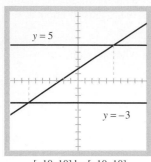

[−10, 10] by [−10, 10]

Figure 2.23 $y = (2x + 5)/3$, $y = 5$, and $y = -3$.

Note that each property or operation must be applied to all three expressions in a double inequality, not just to the left-hand and right-hand sides.

$$-3 < \frac{2x + 5}{3} \le 5 \qquad \text{Multiply each expression by 3.}$$

$$-9 < 2x + 5 \le 15 \qquad \text{Add } -5 \text{ in each expression.}$$

$$-14 < 2x \le 10 \qquad \text{Multiply each expression by 1/2.}$$

$$-7 < x \le 5$$

A graphical solution can be found by reproducing and analyzing the graph in Fig. 2.23. However, the fact that the endpoint $x = 5$ is a solution cannot be deduced from a graph.   ≡

E X A M P L E   7   APPLICATION: Finding a Rectangle Perimeter

Consider the set of all rectangles whose length is one unit more than twice its width.

a) Find an algebraic representation of those rectangles whose perimeters are less than 100 in.

b) Find the possible widths of these rectangles.

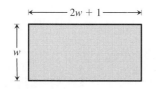

Figure 2.24   Rectangles with a length 1 unit more than twice the width w.

Solution   If w is the width of a rectangle, then its length is $2w + 1$ (see Fig. 2.24).

a) The perimeter is $2w + 2(2w + 1)$, and an algebraic representation of the problem situation is

$$2w + 2(2w + 1) < 100.$$

b) Solve the algebraic representation to get the width.

$$2w + 4w + 2 < 100$$

$$6w + 2 < 100$$

$$6w < 98$$

$$w < \frac{49}{3}$$

The width can be any positive number less than $\frac{49}{3}$ in. That is, it can be any number in the open interval $(0, \frac{49}{3})$.

Notes on Examples

An application that requires the use of inequalities is developed in Example 7. Inequalities that represent the real-world constraints determine the portion of the model that applies to the problem situation.

A graphical solution of this problem can be obtained by finding a complete graph of the problem situation, such as a graph of $y = 2x + 2(2x + 1)$ and $y = 100$ in the same viewing rectangle [0, 25] by [0, 150].

Inequalities with Absolute Value

A real number x is a solution to the inequality $|x| \leq 2$ if, and only if, the distance from the point to the origin is less than or equal to 2 (see Fig. 2.25). This set of numbers is precisely all those between (and including) -2 and 2.

Figure 2.25 $|x| \leq 2$.

It is convenient to use the symbol $\cup$, called the **union** symbol. The notation $(a, b) \cup (c, d)$ indicates the collection of all real numbers that belong to (a, b) or (c, d) or both.

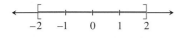

Inequalities with Absolute Value

Inequality	Solution Described by an Inequality	Solution Described in Interval Notation
$\lvert x\rvert < a$	$-a < x < a$	$(-a, a)$
$\lvert x\rvert \leq a$	$-a \leq x \leq a$	$[-a, a]$
$\lvert x\rvert > a$	$x < -a$ or $x > a$	$(-\infty, -a) \cup (a, \infty)$
$\lvert x\rvert \geq a$	$x \leq -a$ or $x \geq a$	$(-\infty, -a] \cup [a, \infty)$

Communication in mathematics occurs with words, symbols, and pictures. Often a single idea is communicated in all three ways. The examples in this section provide practice in recognizing when words, symbols, and pictures are all communicating the idea of interval.

Notes on Examples

Students will be better prepared to interpret absolute value expressions after completing Example 8, especially parts (b) and (c). Building a complete understanding of the absolute value function is important before students face complicated equations and inequalities.

EXAMPLE 8 Changing Inequalities to Graphs

On a real-number line, draw the interval represented by each inequality: (a) $-3 < x < 8$, (b) $|x| \geq 2$, and (c) $|x - 3| < 3$.

Solution

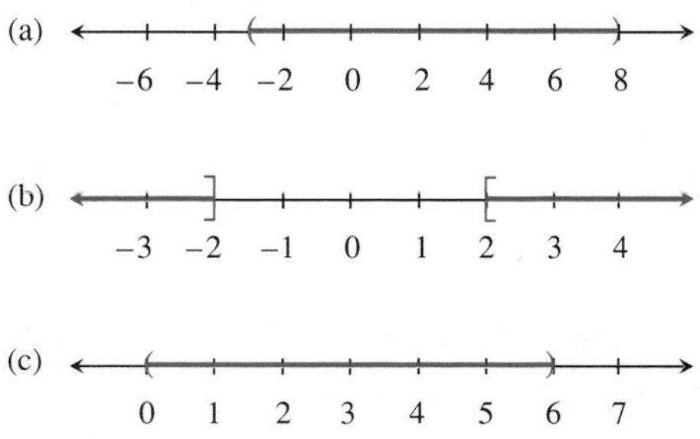

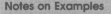

E X A M P L E 9   *Translating from Words into Math*

Use absolute value to write an inequality that represents each of the following statements:

a) x is within 4 units of the origin on the real-number line.

b) x is less than 3 units from the point 2 on the real-number line.

c) x is at least 5 units from the point -3 on the real-number line.

Solution

a) $|x| < 4$

b) $|x - 2| < 3$ $|x - 2|$ describes the distance between
                     x and 2 on the real-number line.

c) $|x - (-3)| \geq 5$ $|x - (-3)|$ is the distance
                     between x and -3.   ▤

Example 10 illustrates how to solve an inequality using either an algebraic or a graphical method. A goal is to learn how to judge when an algebraic method is preferable to a graphical one and vice versa.

E X A M P L E 10   *Comparing Algebraic and Graphical Methods*

Solve $|3x - 2| > 1$ both algebraically and graphically.

Solution

Algebraic Method

$$|3x - 2| > 1$$

$$3x - 2 < -1 \quad \text{or} \quad 3x - 2 > 1$$
$$3x < 1 \quad \text{or} \quad 3x > 3$$
$$x < \tfrac{1}{3} \quad \text{or} \quad x > 1$$

The solution is $(-\infty, \tfrac{1}{3}) \cup (1, \infty)$.

Graphical Method

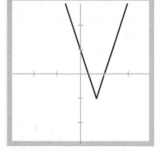

Figure 2.26 $y = |3x - 2| - 1$.

1. Since $|3x - 2| > 1$ is equivalent to $|3x - 2| - 1 > 0$, find a complete graph of $y = |3x - 2| - 1$ (see Fig. 2.26).

2. Notice that the graph appears to be above the x-axis when $x < \tfrac{1}{3}$ and also when $x > 1$.

Notes on Examples

Examples 12 and 13 are more complicated inequalities and may give some students difficulty if they try to solve them algebraically. If the expression on each side of the inequality sign can be written in the form $y = f(x)$, a graphical solution is possible. These examples provide good material for discussion. Students should be encouraged to talk about mathematical expressions and discuss what they know (or can guess) about the behavior of a function before seeing it graphed.

3. Confirm by direct substitution that the graph crosses the x-axis at $x = \frac{1}{3}$ and $x = 1$.

Again, the solution is $(-\infty, \frac{1}{3}) \cup (1, \infty)$.   ■

E X A M P L E 11   Choosing the Method

Solve $|1 - 2x| \le 4$, and draw a number-line graph of the solution.

Solution   Example 6 should convince you that the algebraic method is easier for this type of inequality.

$$|1 - 2x| \le 4$$
$$-4 \le 1 - 2x \le 4$$
$$-5 \le -2x \le 3$$
$$\frac{5}{2} \ge x \ge -\frac{3}{2} \qquad \text{Multiplying an inequality by a negative number reverses the inequality.}$$

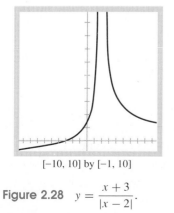

Figure 2.27 $-\frac{3}{2} \le x \le \frac{5}{2}$.

The solution is $[-\frac{3}{2}, \frac{5}{2}]$, and a number-line graph is shown in Fig. 2.27.

Although the double inequality is correct as expressed above, it is customary and helpful to reverse the signs:

$$-\frac{3}{2} \le x \le \frac{5}{2}.$$

Now the left-hand side corresponds to the left of the two numbers given in the solution interval and also to the left-hand side of the number-line graph. It is easier to avoid mistakes in both reading and constructing graphs if left- and right-hand sides of the algebraic and graphic representations match.   ■

When faced with an unfamiliar or difficult inequality, it may not be clear how to proceed with an algebraic approach. In these cases, the graphical approach may be easier. Examples 12 and 13 illustrate two, more complex inequalities.

E X A M P L E 12   Solving a More Difficult Inequality

Solve $\dfrac{x + 3}{|x - 2|} > 0$.

Solution   For this inequality, it may be easier to begin with a graphical approach.

Graphical Method

$[-10, 10]$ by $[-1, 10]$

Figure 2.28 $y = \dfrac{x + 3}{|x - 2|}$.

1. Graph $y = (x + 3)/|x - 2|$ in the $[-10, 10]$ by $[-1, 10]$ viewing rectangle (see Fig. 2.28).

Notes on Exercises

Ex. 60–74 present applied problems that students can solve algebraically or graphically using their own discretion. In Ex. 64, the assumption that the electrician rounds off her time to the nearest quarter hour means that she rounds her time up to the nearest quarter hour.

2. Study the graph to determine where it lies *above* the x-axis. Notice that $x = 2$ results in a zero in the denominator of the inequality, so $x = 2$ cannot be a part of the solution.

3. Notice that the graph appears to cross the x-axis at $x = -3$. Confirm this by substituting $x = -3$ into the inequality. The grapher does not visually distinguish between $>$ and $\geq$; this substitution does that.

The solution consists of all values of x greater than -3 except $x = 2$. Write this solution in interval notation as $(-3, 2) \cup (2, \infty)$.

Algebraic Method Since the denominator of $(x + 3)/|x - 2|$ is always nonnegative, the quotient is greater than zero whenever the numerator is greater than zero and the denominator is not zero. That is, the solution to the inequality consists of all values of x such that

$$x + 3 > 0 \qquad \text{and} \qquad x - 2 \neq 0,$$

or, $x > -3, x \neq 2$. ≡

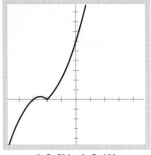

Figure 2.29 $y = (x + 3)|x + 2|$.

E X A M P L E   13   Using a Graphical Method

Solve $(x + 3)|x + 2| \geq 0$.

Solution

1. Find a complete graph of $y = (x + 3)|x + 2|$ in the $[-5, 5]$ by $[-5, 10]$ viewing rectangle (see Fig. 2.29).

2. Notice that the graph appears to be above the x-axis for $-3 < x < -2$ and for $x > -2$.

3. Verify by substitution that $x = -3$ and $x = -2$ are solutions to the inequality. The solution to the inequality is $[-3, \infty)$. ≡

In this section, we have seen that some inequalities can be solved by either an algebraic method or a graphical method. Even when the algebraic method is easier, the graphical method may provide additional insight.

Assignment Guide

Day 1: Ex. 1–32, 35, 38, 40
Day 2: Ex. 41–45, 47, 48, 50, 53–55, 57, 60, 66–68, 71, 73, 74

Exercises for Section 2.4

In Exercises 1–4, use the problem-solving strategy "Guessing and testing" to find three specific values of x that are solutions to each inequality.

1. $2x - 3 < 7$

2. $3 - x \geq 5$

3. $5 \geq x - 3 \geq -1$

4. $12 < 4x + 3 \leq 25$

In Exercises 5–8, write an inequality that represents each interval.

5. $[-3, 5)$

6. $(-3, 7)$

7. $(-\infty, 4)$ **8.** $[3, \infty)$

In Exercises 9–12, graph each interval on a real-number line.

9. $x \le 2$ or $x > 5$ **10.** $x > -3$ or $x \le -6$

11. $x \ge -3$ and $x < 5$ **12.** $x \le -1$ and $x > -3$

In Exercises 13–15, use interval notation to name the intervals depicted graphically.

13.

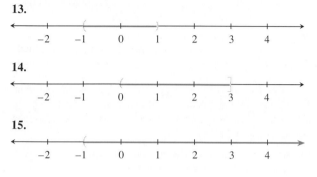

14.

15.

In Exercises 16 and 17, use interval notation to describe the intervals depicted graphically. You may need to use the $\cup$ symbol.

16.

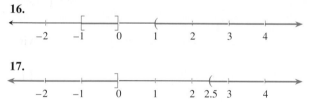

17.

In Exercises 18–21, graph each interval on a number line.

18. $(3.5, 7]$ **19.** $(-\infty, 5)$

20. $[\frac{5}{4}, 7]$ **21.** $[-2, \infty)$

In Exercises 22 and 23, use absolute value notation to describe the intervals depicted graphically. *Hint:* Find numbers a and b so that the solutions to $|x - a| < b$ (or similar expressions using $\le$, $>$, or $\ge$) are the indicated interval or intervals.

22.

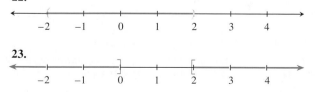

23.

In Exercises 24–31, use an algebraic method to solve the inequality.

24. $2x - 1 > 4x + 3$

25. $\frac{1}{2}(x - 4) - 2x \le 5(3 - x)$

26. $\frac{3x - 2}{5} > -1$

27. $\frac{1}{2}(x + 3) + 2(x - 4) < \frac{1}{3}(x - 3)$

28. $\frac{3 - x}{2} + \frac{5x - 2}{3} < -1$

29. $2 \le x + 6 < 9$

30. $-1 < 3x - 2 < 7$

31. $4 \ge \frac{2x - 5}{3} \ge -2$

In Exercises 32–35, use either an algebraic or a graphical method to solve the inequality.

32. $2x - 10 < -\frac{1}{2}x - 1 < x - 4$

33. $2x < 4 \le x + 7$

34. $-x < 2x + 3 < -8x + 3$

35. $-x + 4 \le -3 \le 3x$

In Exercises 36–39, translate each English phrase into interval notation and inequality notation. Use absolute value notation if possible.

36. The set of all real numbers less than or equal to 5 and greater than 2

37. The set of all numbers greater than 3 and less than 7

38. x is less than 2 units from 3 on the real-number line

39. x is within 4 units of 0 on the real-number line

In Exercises 40–43, draw the solution to each inequality on a real-number line.

40. $|x| \ge 2$ **41.** $|x| < 3$

42. $|x - 1| > 2$ **43.** $|x - 2| < 5$

In Exercises 44–47, use both algebraic and graphical methods to solve the inequalities.

44. $|x - 3| < 2$ **45.** $|x + 3| \le 5$

46. $\frac{3x - 8}{2} > 6$ **47.** $3|x| - 4 > 0$

In Exercises 48–51, choose between an algebraic method and a graphical method, and solve the inequality.

48. $x|x - 2| > 0$

49. $\dfrac{x - 3}{|x + 2|} < 0$

50. $\left|\dfrac{1}{x}\right| < 3$

51. $|x| < |x - 3|$

52. Solve the inequality $3x + 4 < 19$ graphically using Method 1 (see Example 6). That is, graph $y = 3x - 15$ and find the values of x for which the graph lies below the x-axis.

53. Solve the inequality $2x - 1 > 6$ graphically using Method 2 (see Example 7). That is, graph $y = 2x - 1$ and $y = 6$ in the same viewing rectangle. Find the values for x for which the graph of $y = 2x - 1$ is *above* the graph of $y = 6$.

In Exercises 54–59, use an algebraic method to solve the inequality. Write your answer in interval notation. Support your solutions with a graphing utility.

54. $\dfrac{1}{2} < \dfrac{5x - 2}{6} \leq \dfrac{8}{3}$

55. $\dfrac{3}{x - 2} > 0$

56. $-\dfrac{3}{4} < \dfrac{3 - x}{2} < 8$

57. $-2(4 - \dfrac{x}{3}) < 3 + 5x$

58. $\dfrac{x + 5}{3} < 2$

59. $0 < \dfrac{2}{x + 5} < 6$

Exercises 60–62 refer to the following **problem situation**. Sarah has $45 to spend and wishes to take as many friends as possible to a concert. Parking is $5.75 and concert tickets are $7.50 each.

60. Let x represent the number of friends Sarah takes to the concert. Write an inequality that is an algebraic representation for this problem situation.

61. Solve the inequality in Exercise 60.

62. How many friends can Sarah take to the concert?

63. Barb wants to drive to a city 105 mi from her home in no more than 2 hrs. What average speed must she drive?

64. An electrician charges $18/hr plus $25 per service call for home repair work. How long did she work if her charges were less than $100? Assume she rounds off her time to the nearest quarter hour.

65. Consider the collection of all rectangles that have a length 2 in. less than twice the width (in inches). Find the possible widths of these rectangles if their perimeters are less than 200 in. Solve this problem algebraically and with a graphing utility.

66. A candy company finds that the cost of making a certain candy bar is $0.23/bar plus fixed costs of $2000/wk. If each bar sells for $0.25, find the minimum number of candy bars that must be made and sold in order for the company to make a profit.

67. *Boyle's law* for a certain gas states that $PV = 400$, where P is pressure and V is volume. If $20 \leq V \leq 40$, what is the corresponding range for P?

68. A company has current assets (cash, property, inventory, and accounts receivable) of $200,000 and current liabilities (taxes, loans, accounts payable) of $50,000. How much can it borrow if it wants its ratio of assets to liabilities to be less than 2? Assume the amount borrowed is added to both current assets and current liabilities.

69. Complete a formal definition of the relation *greater than*. *Hint:* Model your definition after the definition given for *less than* in Section 1.1.

Exercises 70 and 71 refer to the following **problem situation**: The Celsius-to-Fahrenheit temperature conversion formula is $F = \frac{9}{5}C + 32$. Water boils when its temperature is greater than or equal to $212°$F.

70. Write an inequality that describes algebraically the temperature Celsius at which water will boil.

71. Solve the inequality found in Exercise 70.

Exercises 72–74 refer to the following **problem situation**: The annual profit P of a candy manufacturer is determined by the formula $P = R - C$, where R is the total revenue generated from selling x pounds of candy, and C is the total cost of making and distributing x pounds of candy. Each pound of candy sells for $1.80 and costs $1.38 to make. The fixed costs of making and distributing the candy are $20,000 annually.

72. Write an algebraic representation of the company's annual profit in terms of x.

73. Write an algebraic representation for the number of pounds of candy that must be produced and sold for the company to make a profit.

74. Use a graphing utility to find the production level that will yield a profit for the year.

In Exercises 75–77, use any method to solve the inequality.

75. $|x| - |8 - x| > 0$

76. $|x + 3| < |x|$

77. $|x + 5| > |x|$

Exercises 78–81 refer to the equation $y = 3x - 5$.

78. Write intervals equivalent to the following inequalities:
a) $|x - 2| < 0.1$ **b)** $|y - 1| < 0.3$

79. On the x-axis, draw a number-line graph of the solution to $|x - 2| < 0.1$.

80. On the y-axis, draw a number-line graph of the solution to $|y - 1| < 0.3$.

81. Find all values of $d > 0$ such that if $|x - 2| < d$, then $|y - 1| < 0.01$.

82. Prove the first, second, and fourth parts of Theorem 2.1.

83. Write several paragraphs that explain how to use graphical Methods 1 and 2 to solve the inequality $3(x - 1) + 2 \leq 5x + 6$ from Example 3.

2.5 _____   Solving Higher-order Inequalities Algebraically and Graphically

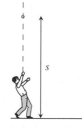

Figure 2.30   Distance s of an object above ground level at time t.

Objective

Students will be able to solve higher-order inqualities algebraically and graphically.

Key Ideas

Projectile motion
Sign pattern

Notes on Examples

Example 1 is an important but sometimes difficult problem for many students. Discuss it in detail. Notice that feet and seconds are used as units of measure in this problem. Some students may be more familiar with meters than feet and thus use $-4.9t^2$ instead of $-16t^2$.

In this section, we illustrate that using a graphing method with a graphing utility is nearly always more reasonable than using an algebraic method for solving higher-order inequalities.

Projectile motion is a frequently studied problem situation in physical science. Consider the following common problem situation and its algebraic representation.

Projectile Motion

An object is thrown straight up from a point s_0 feet above ground level with an initial velocity of v_0. If s represents the distance in feet of the object above ground level t seconds after it is thrown, then we know from physics that $s = -16t^2 + v_0 t + s_0$ (see Fig. 2.30).

E X A M P L E   1   Solving Projectile Motion Problems Graphically

Suppose a baseball is thrown straight up from ground level with an initial velocity of 80 ft/sec. When will the baseball be at least 64 ft above the ground?

Solution   Because the ball is thrown from the ground, $s_0 = 0$. The initial velocity is $v_0 = 80$. The algebraic representation of this problem is

$$s = -16t^2 + 80t.$$

Because we want to find out when $s \geq 64$, we must solve the inequality $-16t^2 + 80t \geq 64$.

1. Graph $s = -16t^2 + 80t$ and $s = 64$ in the same viewing rectangle and determine the values of t for which the graph of $s = -16t^2 + 80t$ lies above the graph of $s = 64$.

2. Figure 2.31(a) shows complete graphs of both algebraic representations of s. Figure 2.31(b) shows a complete graph of the problem situation, since t is nonnegative and t cannot be greater than 5. Can you explain why t cannot be greater than 5?

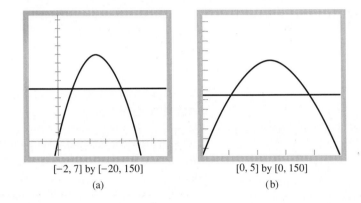

[−2, 7] by [−20, 150]   [0, 5] by [0, 150]

(a)   (b)

Figure 2.31   Two views of $s = -16t^2 + 80t$ and $s = 64$.

3. It appears that $s \geq 64$ (in other words, that the baseball is at least 64 ft above the ground) when $1 \leq t \leq 4$.

4. We can verify by direct substitution that $s = 64$ when $t = 1$ or $t = 4$.

$$\text{For } t = 1: \quad -16(1)^2 + 80(1) = -16 + 80 = 64$$

$$\text{For } t = 4: \quad -16(4)^2 + 80(4) = -16(16) + 80(4)$$

$$= -256 + 320 = 64$$

For the sake of comparing methods, Example 2 uses an algebraic method.

E X A M P L E 2   Solving Projectile Motion Problems Algebraically

Use an algebraic method to answer these questions:

a) How long does the baseball in Example 1 stay in the air?

b) When is the ball at least 64 ft off the ground?

Solution

a) $$-16t^2 + 80t \geq 0$$

$$-16t(t - 5) \geq 0$$

$t \geq 0$ and $t - 5 \leq 0$ From the context of the problem $t > 0$, in order for the product to be positive, $t - 5 < 0$.

The baseball is in the air when $0 < t < 5$, so it is in the air 5 sec.

b) $-16t^2 + 80t \geq 64$

$-16t^2 + 80t - 64 \geq 0$

$t^2 - 5t + 4 \leq 0$ Multiplying the inequality by $-\frac{1}{16}$
reverses the inequality.

$(t - 1)(t - 4) \leq 0$ This product is less than or equal to zero whenever
it is not positive, that is, whenever $t \leq 4$ and $t \geq 1$.
So the solution to this inequality is [1, 4].

If $1 \leq t \leq 4$, the baseball is at least 64 ft off the ground   ≡

Notes on Examples

Some type of sign pattern approach may be familiar to you from your previous work. If so, read Examples 3, 4, and 5 quickly so that your classroom presentation will match the text. If not, examine Examples 3, 4, and 5 carefully.

Sign Pattern

The problem-solving strategy "Study the pattern" can be used advantageously to solve algebraically an inequality like $(t - 1)(t - 4) \leq 0$. Every linear factor is zero for exactly one value. For example, $t - 1$ is zero for $t = 1$. So when $t < 1$, the factor $t - 1 < 0$ is negative and when $t > 1$, the factor $t - 1 > 0$ is positive.

It is convenient to record this information in a **sign-pattern picture**:

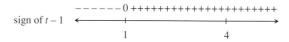

In a similar fashion, we determine the sign of $t - 4$. Because the product $(t - 1)(t - 4)$ is positive when both factors are positive or both negative, we can see with a glance at the following graphic when $(t - 1)(t - 4)$ is positive. Similarly, we can see when the product is negative.

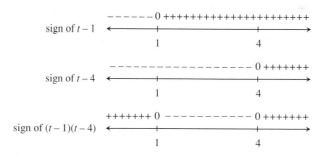

Notes on Exercises

Ex. 6–15 preview a skill that is essential for sign patterns. Ex. 20–27 provide a significant opportunity to see if students understand how to use sign patterns.

We see from this sign-pattern picture that $(t - 1)(t - 4) \leq 0$ when $1 \leq t \leq 4$. We say that a solution to the inequality is the interval [1, 4]. A sign-pattern picture like the one illustrated here can be used *whenever an expression can be factored into factors with known signs*.

The next three examples illustrate the use of sign-pattern pictures to solve an inequality algebraically.

E X A M P L E   3   Using a Sign Pattern for Two Factors

Use a sign-pattern picture to solve $2x^2 + 9x - 5 > 0$. Support your solution graphically.

Solution $2x^2 + 9x - 5 = (2x - 1)(x + 5) = 0$ if, and only if, $x = \frac{1}{2}$ or $x = -5$.

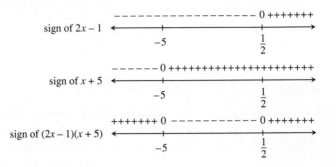

The solution to the inequality $2x^2 + 9x - 5 > 0$ is $(-\infty, -5) \cup (\frac{1}{2}, \infty)$. This solution is supported graphically by finding a complete graph of $y = 2x^2 + 9x - 5$ (see Fig. 2.32).   ▤

This method of using sign-pattern pictures also can be used when there are more than two factors. As the number of factors increases, the number of intervals in the picture increases.

E X A M P L E   4   Using a Sign Pattern for Three Factors

Solve $x(x + 3)(x - 1) \geq 0$ algebraically. Check your answer graphically.

Solution $x(x + 3)(x - 1) = 0$ when $x = 0, -3$, or 1.

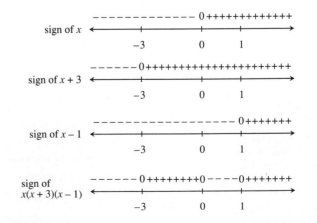

[–10, 10] by [–50, 100]

Figure 2.32 $y = 2x^2 + 9x - 5$.

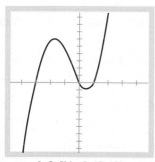

[−5, 5] by [−10, 10]

Figure 2.33 $y = x(x+3)(x-1)$.

The solution to $x(x + 3)(x - 1) \geq 0$ is $[-3, 0] \cup [1, \infty)$. This conclusion is supported graphically by finding a complete graph of $y = x(x + 3)(x - 1)$ (see Fig. 2.33). The graph is on or above the x-axis when $-3 \leq x \leq 0$ or $x \geq 1$. ▤

The next example illustrates that a sign-pattern picture can be used for quotients of polynomials also.

E X A M P L E   5   Using a Sign Pattern for a Quotient

Use a sign-pattern picture to solve

$$\frac{x - 4}{2x + 5} \leq 0.$$

Solution   The numerator equals zero when $x = 4$, and the denominator equals zero when $x = -\frac{5}{2}$.

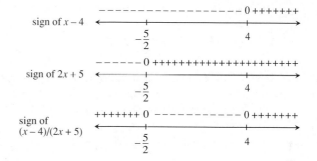

The solution to the inequality is $(-2.5, 4]$. Notice that -2.5 is not included in the solution. Why? ▤

This next example is an important one in the section, since the expression cannot be factored over the rational numbers. It illustrates both the power of the graphical method and the limitation of an algebraic method. In solving real-world problems, it is extremely rare to find an inequality in which the expression can be factored over the rational numbers; hence there will be few opportunities to apply the sign-pattern method. The graphical method is the most universal means of solving inequalities.

Example 6 uses zoom-in to determine the endpoints of the intervals in the solution. Recall that a grapher cannot distinguish between the signs $>$ and $\geq$ when solving an inequality. Therefore it is important to include the endpoints of intervals when solving a $\geq$ inequality and to omit the endpoints when solving a $>$ inequality.

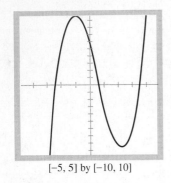

[−5, 5] by [−10, 10]

Figure 2.34 $y = x^3 - 2x^2 - 7x + 6$.

Assignment Guide

Day 1: Ex. 5, 6, 8, 9, 11, 12, 14, 17, 18
Day 2: Ex. 18, 21, 22, 24, 26, 28, 31, 34, 37, 39, 45–47, 54–56

E X A M P L E   6   Using a Graphical Method

Solve $x^3 - 2x^2 - 5x + 7 \geq 2x + 1$.

Solution

1. The inequalities

$$x^3 - 2x^2 - 5x + 7 \geq 2x + 1 \qquad \text{and} \qquad x^3 - 2x^2 - 7x + 6 \geq 0$$

   are equivalent. So find a complete graph of $y = x^3 - 2x^2 - 7x + 6$ (see Fig. 2.34).

2. Use zoom-in to find the three x-intercepts, which are $-2.264, 0.756$, and 3.508. The solutions to the equation with an error of at most 0.01 are $-2.26, 0.76$, and 3.51.

3. Solutions to the inequality coincide with intervals on which the graph is above the x-axis. Therefore the solution is $[-2.26, 0.76] \cup [3.51, \infty)$. ▇

Exercises for Section 2.5

Exercises 1–3 refer to the baseball throw described in Example 1, whose algebraic representation is $s = -16t^2 + 80t$.

1. For how many seconds will the ball be at least 64 ft above the ground?

2. Write an equation that could be used to find how long (in seconds) the ball is in the air.

3. Write an inequality that could be solved to find when the ball is
   a) more than 10 ft off the ground;
   b) at least 10 ft off the ground.

4. If a baseball is thrown straight up from ground level at an initial velocity of 48 ft/sec, what is the algebraic representation for the distance above the ground?

5. If a baseball is thrown straight up from a platform that is 120 ft above the ground with an initial velocity of 32 ft/sec, what is the algebraic representation for the distance above the ground?

In Exercises 6–15, write the intervals that should be used when completing a sign-pattern picture for each inequality.

6. $(x - 3)(x + 2) \geq 0$ 7. $(x + 5)(x + 6) < 0$

8. $x^2 + 4x + 3 > 0$ 9. $(x + 3)(x - 2)(x + 5) < 0$

10. $(x - 4)(x^2 - x - 6) \leq 0$ 11. $x^2 - 6x + 10 \geq 2$

12. $(3 - x)(x + 2)(x - 5) \geq 0$

13. $(x^2 - 1)(x^2 - 4) > 0$

14. $(x - 2)(x + 3)^2 < 0$

15. $(x - 3)(x^2 + 2x - 1) \geq 0$

In Equations 16 to 19, write an equation that you could graph in order to solve the given inequality graphically.

16. $x^2 - 3x + 2 \geq -3$

17. $7x^3 - 2x^2 > 5x - 3$

18. $x^2 - 5x - 17 \geq -2x + 1$

19. $x^5 - 4x^2 > 2x^2 + x + 3$

In Exercises 20–27, use a sign-pattern picture to solve each inequality. Write the solution in interval notation.

20. $(x - 1)(x + 2) < 0$ 21. $x(x - 3) \leq 0$

22. $x^2 - 5x + 6 \geq 0$ 23. $x^2 - 4x - 21 < 0$

24. $2x^2 + 5x \geq 3$ 25. $x^2 < x$

26. $\dfrac{x+3}{x-1} \geq 0$

27. $x(x-4)(x+2) \geq 0$

In Exercises 28–31, use a graphical method to solve each inequality.

28. $3x^3 - 8x^2 - 5x + 6 \leq 0$

29. $x^3 - 9x^2 + 6x + 55 < 0$

30. $|x^2 - 1|(x-2) \geq 3$

31. $5 \sin x < 2$ for $0 \leq x \leq 10$

In Exercises 32–43, decide whether you prefer the sign-pattern method or a graphical method and solve each inequality.

32. $(x-2)^3 > 0$

33. $(x+1)(x-2)(x+4) < 0$

34. $x^4 + 3x^2 < 4$

35. $\dfrac{x+3}{x-1} < 0$

36. $x^2 - 5x + 3 > 0$

37. $2 - 8x + 3x^2 \leq 0$

38. $3x^2 - x + 6 \leq 2x^3 - x^2 + 3x - 4$

39. $2x + 1 > x^3 + 2x^2 - 3x + 5$

40. $\dfrac{x-2}{x+5} > 0$

41. $\dfrac{x^2+4}{x-2} < 0$

42. $\dfrac{x^2 - 2x - 3}{x+1} > 0$

43. $\dfrac{x-1}{(x-3)(x+1)} \geq 0$

Exercises 44–47 refer to the following projectile-motion **problem situation**, whose algebraic representation is $s = -16t^2 + v_0 t + s_0$. Suppose an object is propelled upward from a tower 200 ft tall with an initial velocity of 100 ft/sec.

44. Write an equation that is an algebraic representation for the question, "When will the object hit the ground?"

45. Find a complete graph of the problem situation. How does this graph compare with a complete graph of the algebraic representation in Exercise 44?

46. For how long will the object be above the ground?

47. When will the object hit the ground?

48. Answer the questions in Exercises 44–47 if the object is propelled straight upward from a tower 155 ft tall with an initial velocity of 275 ft/sec.

Exercises 49–51 refer to the following **problem situation**: A swimming pool with dimensions of 20 by 30 ft is surrounded by a sidewalk of uniform width x. Find the possible widths of the sidewalk if the total area of the sidewalk is to be greater than 200 sq ft but less than 360 sq ft.

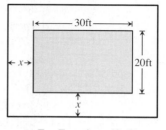

For Exercises 49–51.

49. Write an equation that gives the area of the sidewalk in terms of x.

50. Write an inequality that is an algebraic representation of this problem situation.

51. Choose a method and find a solution to the problem situation.

Exercises 52 and 53 refer to the following **problem situation**: An investor can earn interest using the simple interest formula $S = P(1 + rn)$, where r is the interest rate, n is the number of years, P is the amount invested, and S is the total amount in the account. The investor wants to accumulate at least \$30,000 in 18 yr by investing \$5000 today. What interest rate is required for the investor to meet his goal?

52. Write an inequality that describes this problem situation.

53. Find a complete graph of each side of the inequality in Exercise 52 in the same viewing rectangle and determine a solution to the problem situation.

54. Equal squares are removed from the four corners of a 22-by 29-in. rectangular sheet of cardboard. The sides are turned up to make a box with no lid. Find the possible lengths of the sides of the removed squares if the volume of the box is to be less than 2000 cu in.

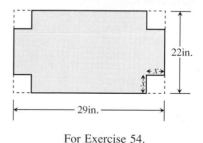

For Exercise 54.

Exercises 55 and 56 refer to the following **problem situation**: Sam makes two initial investments of \$1000 each.

The first earns 10% simple interest—after n years its value is $S = 1000(1 + 0.10n)$. The second investment earns interest using 5% *simple discount*—after n years its value is $S = 1000/(1 - 0.05n)$.

55. Find a complete graph of each algebraic representation in the viewing rectangle $[0, 30]$ by $[0, 6000]$ and determine whether there is a time when both investments are equal.

56. Is there a period of time when the value of the simple-interest investment is less than the value of the simple-

discount investment?

In Exercises 57–59, decide if each inequality is true for (a) some values of x, (b) all values of x, or (c) no values of x.

57. $|x + (5)| \leq |x| + |5|$

58. $|x + (-3)| \leq |x| + |-3|$

59. $|x + (-7)| \leq |x| + |-7|$

60. Generalize   Decide whether $|a + b| \leq |a| + |b|$ is true for all real-number values of a and b.

2.6 _____ Relations and Parametric Equations

Any set of ordered pairs (x, y) in the coordinate plane determines a relation. In this section, we discuss two methods for specifying a relation. The first method uses an equation or inequality in x and y and the second uses parametric equations.

Defining a Relation with an Equation in x and y

The equation $x^2 + y^2 = 25$ defines a relation. If a pair (a, b) satisfies the equation, we say that a and b are **related** or that (a, b) is **in the relation**.

To show that $(-3, 4)$ is in the relation $x^2 + y^2 = 25$, substitute -3 for x and 4 for y.

$$x^2 + y^2 = 25$$
$$(-3)^2 + (4)^2 = ?$$
$$9 + 16 = 25$$

Because the pair $(-3, 4)$ satisfies the equation, we say that $(-3, 4)$ is in the relation.

To show that $(2, 7)$ is not in the relation, substitute 2 for x and 7 for y.

$$x^2 + y^2 = 25$$
$$(2)^2 + (7)^2 = ?$$
$$4 + 49 \neq 25$$

Because the pair $(2, 7)$ does not satisfy the equation, we say that $(2, 7)$ is not in the relation.

EXAMPLE 1   Checking Number Pairs in a Relation

For $x^2 y + y^2 = 5$, show (a) that $(2, -5)$ is in the relation and (b) that $(1, 3)$ is not in the relation.

Solution

a) Check $(2, -5)$:

$$x^2 y + y^2 = 5$$

$$(2^2)(-5) + (-5)^2 \stackrel{?}{=} 5 \qquad \text{Substitute } x = 2 \text{ for } x \text{ and } y = -5 \text{ for } y.$$

$$-20 + 25 = 5. \qquad \text{The sum is 5. It checks.}$$

b) Check $(1, 3)$:

$$x^2 y + y^2 = 5$$

$$(1)^2(3) + (3)^2 \stackrel{?}{=} 5 \qquad \text{Substitute } x = 1 \text{ and } y = 3 \text{ into the equation.} \\ \text{Check whether the equation is true.}$$

$$3 + 9 \neq 5.$$

Thus $(2, -5)$ is in the relation, but $(1, 3)$ is not.   ▇

Defining a Relation with Parametric Equations for *x* and *y*

A second method for defining a relation in x and y employs two functions in which the input value for each function is a variable t and the output values are x and y. The variable t is called a **parameter** and the two equations are called **parametric equations**.

For example, consider the parametric equations

$$x(t) = t^2 - 2$$

$$y(t) = 3t.$$

Often we drop the "(t)" and write

$$x = t^2 - 2$$

$$y = 3t.$$

As the parameter t varies through some set of numbers, a set of ordered pairs (x, y) is determined. This set of pairs (x, y) determines a relation in x and y.

The following Exploration will help you become familiar with parametric equations. As you complete this Exploration, change the values of Tmin and Tmax to discover the impact of that change on the graph.

Bert says, "Teachers should introduce parametric graphing by using a linear function or a quadratic function for which the t-values in the range match the x-values in the range."

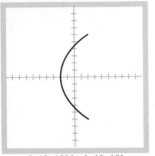

$[-10, 10]$ by $[-10, 10]$

Figure 2.35 $X_1(T) = T^2 - 2$ and $Y_1(T) = 3T$ for $-2 \leq T \leq 2$.

Stan says, "The parametric graphing capabilities of modern technology will greatly enhance a student's ability to visualize the graph of a relation."

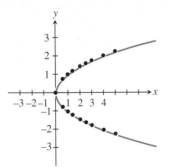

Figure 2.36 The relation $y^2 = x$ plotted from a table of (x, y)-values.

$\mathcal{P}$ EXPLORE WITH A GRAPHING UTILITY

Set a graphing calculator as follows:

- Rad mode,   Param mode
- Range: $T\min = -2$, $T\max = 2$, $T\text{step} = 0.1$
- Viewing rectangle: $[-10, 10]$ by $[-10, 10]$
- Define: $X_1(T) = T^2 - 2$ and $Y_1(T) = 3T$

Describe what happens when you press the **graph** key (see Fig. 2.35).

1. How is the curve different if $T\max = 2$ is changed to $T\max = 3$? Or if $T\min = -2$ is changed to $T\min = -3$?

2. As you move the **trace** cursor along the curve, how does the value of T relate to the values of X and Y on the grapher screen? How are these values related to the defining equations?

Graph of a Relation

The **graph of a relation** consists of all points in the rectangular coordinate plane whose coordinates satisfy the condition that defines the relation.

To graph a relation by the traditional method of point plotting, complete a table of number pairs that satisfy the equation. Then graph these number pairs and connect the points with a smooth curve.

For example, consider the equation $y^2 = x$. Substitute values for x and find approximate corresponding y-values, as shown here:

x	0	0.5	1	1.5	2	2.5	3	4	5
y	0	± 0.7	± 1	± 1.2	± 1.4	± 1.6	± 1.7	± 2	± 2.2

Then graph these points and connect them as shown in Fig. 2.36. The technology of graphing utilities reduces the importance of this traditional graphing method.

The relation $y^2 = x$ can also be described using the following parametric equations:

$$x = t^2$$
$$y = t$$

Teaching Note

If students are not familiar with parametric graphing, it might be helpful to show them the graph of the linear function $f(x) = 3x - 2$ and compare it to one defined parametrically as $f(t) = 3t - 2$, using a trace key to show how t, x, and y are related.

We verify this claim by noticing that if $y = t$, then $y^2 = t^2$. Therefore $x = t^2 = y^2$. Figure 2.37 shows several familiar relations and their graphs.

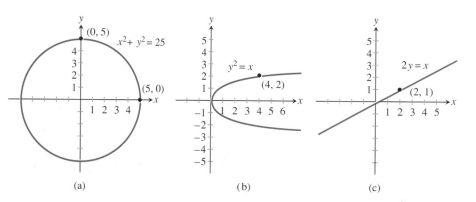

Figure 2.37 (a) A circle, (b) a parabola, and (c) a straight line.

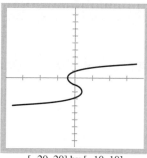

[−20, 20] by [−10, 10]

Figure 2.38 $x(t) = t^3 + 3t^2 - 2$ and $y(t) = t$ with $-4 \le t \le 2$.

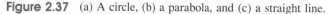

E X A M P L E   2   Finding a Graph of Parametric Equations

Find a complete graph of the parametric equations $x = t^3 + 3t^2 - 2$ and $y = t$ for values of the parameter $-4 \le t \le 2$.

Solution   Use parametric mode as you learned to do when completing the Exploration. The complete graph shown in Fig. 2.38 has the viewing rectangle $[-20, 20]$ by $[-10, 10]$. When the parameter has value $t = -4$, the (x, y) point is

$$x(-4) = (-4)^3 + 3(-4)^2 - 2 = -18 \quad \text{and} \quad y(-4) = -4.$$

When the parameter has value $t = 2$, the (x, y) point is

$$x(2) = (2)^3 + 3(2)^2 - 2 = 18 \quad \text{and} \quad y(2) = 2.$$

So as the parameter varies from $t = -4$ to $t = 2$, the complete graph goes from point $(-18, -4)$ to point $(18, 2)$. ▤

GRAPHER UPDATE

1. To graph $y^2 = x$, graph the pair of equations $y = \sqrt{x}$ and $y = -\sqrt{x}$ in the same viewing rectangle.

2. To graph $x^2 + y^2 = 25$, graph $y = \sqrt{25 - x^2}$ and $y = -\sqrt{25 - x^2}$ together.

Intercepts of a Graph

Chapter 1 introduced the concept of x-intercept. An **intercept** of the graph of an equation is a point that lies on both the graph and one of the coordinate axes. Intercepts are among the easiest points on a graph to identify. Following is a more formal definition than the one given in Chapter 1.

Notes on Exercises

Ex. 19–24 deal with intercepts. Students should be able to solve these by inspection. They can be used as oral exercises in class.

Definition 2.5   Intercepts

A point $(a, 0)$ is called an **x-intercept** of the graph if the number pair $(a, 0)$ satisfies the equation.

A point $(0, b)$ is called a **y-intercept** of the graph if the number pair $(0, b)$ satisfies the equation.

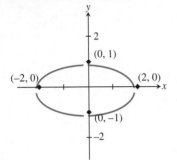

Figure 2.39 $x^2 + 4y^2 = 4$.
Note the x- and y-intercepts.

In Fig. 2.39, the intercepts are highlighted. Strictly speaking the x- and y-intercepts refer to the *points* $(a, 0)$ and $(0, b)$, as given in Definition 2.5. However, it is convenient to abbreviate and speak of the x-intercept as a and the y-intercept as b. We will refer to intercepts both ways in this text.

E X A M P L E 3   Finding Intercepts

The graph of the relation $x^2 + 4y^2 = 4$ is shown in Fig. 2.39. Find the x- and y-intercepts of the graph.

Solution

a) Let $y = 0$ and solve for x.

$$x^2 + 4(0)^2 = 4$$
$$x^2 = 4$$
$$x = \pm 2$$

The points $(-2, 0)$ and $(2, 0)$ are x-intercepts.

b) Let $x = 0$ and solve for y.

$$(0)^2 + 4y^2 = 4$$
$$y = \pm 1$$

The points $(0, -1)$ and $(0, 1)$ are y-intercepts.

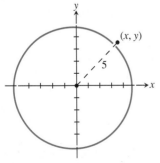

Figure 2.40   Circle with center $(0, 0)$ and a radius of 5.

Equations of Circles

Suppose that a point (x, y) in the coordinate plane has a distance 5 units from the origin (see Fig. 2.40). From the distance formula, we see that

$$\sqrt{(x - 0)^2 + (y - 0)^2} = 5$$
$$x^2 + y^2 = 25.$$

Consequently, each point 5 units from the origin has coordinates (x, y), which are in the relation $x^2 + y^2 = 25$. So the graph of the relation $x^2 + y^2 = 25$ is the circle centered at the origin with a radius of 5.

Suppose the point (h, k) is the center of a circle of radius r (see Fig. 2.41). If (x, y) is a typical point on the circle, then the distance formula yields the equation

$$\sqrt{(x - h)^2 + (y - k)^2} = r.$$

By squaring both sides, we obtain the following:

$$(x - h)^2 + (y - k)^2 = r^2.$$

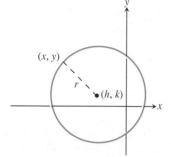

Figure 2.41 Circle with center (h, k) and radius r.

Standard Form of the Equation of a Circle

For a circle with center (h, k) and radius r, the standard equation is

$$(x - h)^2 + (y - k)^2 = r^2.$$

E X A M P L E   4   Finding the Standard Equation of a Circle

Find an equation of the circle with center $(-4, 1)$ and a radius of 8.

Solution

$$(x - h)^2 + (y - k)^2 = r^2$$
$$(x - (-4))^2 + (y - 1)^2 = 8^2 \qquad \text{To get the standard form,}$$
$$\text{substitute } h = -4, k = 1, \text{ and } r = 8.$$
$$(x + 4)^2 + (y - 1)^2 = 64$$

Symmetry of Graphs

The graphs of some relations possess **symmetry**. One type of symmetry is called *reflectional symmetry*. This means that one half of the graph is a mirror image of the other half. Often a mirror image can be found by mentally folding the coordinate plane so that one half of the graph folds to exactly coincide with the other half. The line where this fold occurs is called a **line of symmetry,** and the process of rotation about this line is called **reflection**.

A graph with reflectional symmetry may have only one line of reflectional symmetry, or it may have two or more. For example, in Fig. 2.36 the x-axis is the line of symmetry, in Fig. 2.39 the x- and y-axes are both lines of symmetry, and in Fig. 2.40 all the lines passing through the origin are lines of symmetry.

The graph in Fig. 2.42(a) is symmetric about the y-axis because the point $(-x, y)$ lies on the graph of $y = x^2$ whenever (x, y) does. Similarly, Fig. 2.42(b) is symmetric about the x-axis, because the point $(x, -y)$ lies on the graph of $y^2 = x$ whenever (x, y) does.

Figure 2.42(c) is an interesting case. Notice that there are no lines of symmetry, that is, there is no way to fold the coordinate plane and have one half of the graph exactly coincide with the other half. However, there is another type of symmetry called *symmetry about the origin.* Algebraically we say that Fig. 2.42(c) is symmetric about the origin because the point $(-x, -y)$ lies on the graph of $y = x^3 + x$ whenever (x, y) does.

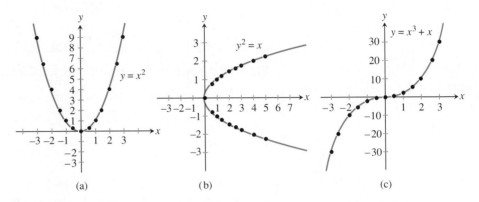

Figure 2.42   Three kinds of symmetry: (a) about the y-axis; (b) about the x-axis; (c) about the origin.

The three examples in Fig. 2.42 motivate the following algebraic definitions of symmetry:

Definition 2.6   Symmetry about Coordinate Axes and Origin

- A graph is **symmetric about the x-axis** if $(x, -y)$ is on the graph whenever (x, y) is on the graph.

- A graph is **symmetric about the y-axis** if $(-x, y)$ is on the graph whenever (x, y) is on the graph.

- A graph is **symmetric about the origin** if $(-x, -y)$ is on the graph whenever (x, y) is on the graph.

Because these three types of symmetry have been formulated in algebraic terms, it is possible to use an algebraic method to determine whether a relation has a symmetric graph. For example, to show that the graph of $y = x^4 + 3x^2$

is symmetric about the y-axis, verify that the point $(-x, y)$ satisfies the equation whenever (x, y) does.

EXAMPLE 5   Checking for Symmetry

Show that the graph of $y^2 = x^4 + 3x^2$ is symmetric about both the x-axis and y-axis.

Solution

***x*-axis symmetry**   Replace y with $-y$ in the equation, and show that the result is an equivalent equation.

$$(-y)^2 = x^4 + 3x^2$$

$$y^2 = x^4 + 3x^2$$

***y*-axis symmetry**   Replace x with $-x$ in the equation $y^2 = x^4 + 3x^2$ and show that the result is an equivalent equation.

$$y^2 = (-x)^4 + 3(-x)^2$$

$$y^2 = x^4 + 3x^2 \quad \begin{array}{l}(-x)^k = x^k \text{ whenever } k \\ \text{is an even integer.}\end{array}$$

≡

There is a relationship between symmetry about the x- and y-axes and symmetry about the origin that is summarized in this theorem.

Theorem 2.2   Symmetry about the Origin

If a graph is symmetric about both the x-axis and the y-axis, then it also is symmetric about the origin.

Analysis of Example 5 shows that the graph of $y^2 = x^4 + 3x^2$ is symmetric about the origin.

EXAMPLE 6   Checking for Symmetry (Continued)

Determine the symmetry of the graph of the equation $y = 2x^3 + 5x$.

Solution

***y*-axis symmetry**

$$y = 2(-x)^3 + 5(-x) \qquad \begin{array}{l}\text{Replace } x \text{ with } -x \text{ and check whether} \\ \text{the equations are equivalent.}\end{array}$$

$$y = -2x^3 - 5x$$

$$y = -(2x^3 + 5x)$$

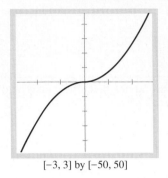

[−3, 3] by [−50, 50]

Figure 2.43 $y = 2x^3 + 5x$.

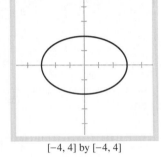

[−4, 4] by [−4, 4]

Figure 2.44   Graph of $y = \frac{1}{3}\sqrt{36 - 4x^2}$ and $y = -\frac{1}{3}\sqrt{36 - 4x^2}$. (See solution to Example 6.)

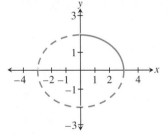

Figure 2.45   Graph of $4x^2 + 9y^2 = 36$ plotted in the first quadrant and then reflected through the x- and y-axes.

The equations $y = 2x^3 + 5x$ and $y = -(2x^3 + 5x)$ are not equivalent, so the graph is not symmetric with respect to the y-axis.

Next, check for x-axis symmetry. In a similar fashion to the solution above, we can show that the graph is not symmetric with respect to the x-axis.

Symmetry with respect to the origin

$$(-y) = 2(-x)^3 + 5(-x) \qquad \text{Replace } x \text{ with } -x \text{ and } y \text{ with } -y.$$

$$-y = -2x^3 - 5x$$

$$-y = -(2x^3 + 5x) \qquad \text{Multiply both sides of the equation by } -1 \text{ to obtain an equivalent equation.}$$

$$y = 2x^3 + 5x$$

The graph is symmetric about the origin. The graph in Fig. 2.43 supports the algebraic analysis.

Observations about the symmetry of a graph can be used when you are trying to draw the graph. For example, once you know that a graph is symmetric about both the x- and y-axes, knowing the graph in the first quadrant is sufficient to determine the complete graph.

EXAMPLE 7   *Finding and Drawing a Symmetric Graph*

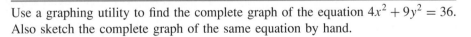

Use a graphing utility to find the complete graph of the equation $4x^2 + 9y^2 = 36$. Also sketch the complete graph of the same equation by hand.

Solution

1. Verify that the graph is symmetric about both the x- and y-axes by using the algebraic methods illustrated in Examples 5 and 6.

2. Solve $4x^2 + 9y^2 = 36$ for y.

3. Graph $y = \frac{1}{3}\sqrt{36 - 4x^2}$ and $y = -\frac{1}{3}\sqrt{36 - 4x^2}$ (see Fig. 2.44).

4. To sketch this graph, first plot the graph in quadrant I and then reflect it in one axis to obtain half the ellipse. Then reflect this half ellipse in the other axis to obtain the complete ellipse (see Fig. 2.45).   ≡

Exercises for Section 2.6

1. Show that the point $(3, 8)$ is in the relation $2x - y = -2$ and that $(4, 2)$ is not.

2. Show that the point $(2, 5)$ is in the relation $2x^2 - y = 3$ and that $(5, 2)$ is not.

3. Show that the point $(-3, 6)$ is in the relation $x + y^2 = 33$ and that $(-2, 4)$ is not.

4. Show that the point $(1, 2)$ is in the relation $5x^3 - 2y = 1$ and that $(3, 2)$ is not.

In Exercises 5–12, determine whether the given number pairs are in the given relation.

5. $(3, 5)$; $x^2 + y^2 = 8$ **6.** $(7, 2)$; $x - 5y = -3$

7. $(1, 2)$; $x^3 + 3y = -1$ **8.** $(3, 1)$; $x - 4y^3 = 2$

9. $(2, 3)$; $x^2 - y = 3$ **10.** $(3, 4)$; $x^2 + y^2 = 25$

11. $(2, \sqrt{2})$; $x^3 - 3y^2 = 2$ **12.** $(\sqrt{3}, 2)$; $4x - 3y^2 = 5$

In Exercises 13–18, find a complete graph of the given parametric equations with the parameter t satisfying $-2 \le t \le 2$.

13. $x = 2t$ and $y = 3t - 1$ **14.** $x = 5t + 1$ and $y = 3t - 2$

15. $x = t^2$ and $y = t + 1$ **16.** $x = t$ and $y = t^2 - 3$

17. $x = t$ and $y = t^3 - 2t + 3$

18. $x = t^3 - 2t + 3$ and $y = t$

In Exercises 19–24, find the x- and y-intercepts of the graph of each equation.

19. $x^2 + 4y^2 = 4$ **20.** $4x^2 + 9y^2 = 36$

21. $4x^2 - y^2 = 4$ **22.** $x - 4 = y^2$

23. $3x^2 + 5y^2 = 15$ **24.** $x + 5 = -y^2$

In Exercises 25–28, complete the square to find the center and radius of the circle whose equation is given.

25. $x^2 - 6x + y^2 + 8y = -9$

26. $x^2 - 2x + y^2 + 6y = 39$

27. $x^2 - 4x + y^2 + 6y + 11 = 0$

28. $x^2 - 14x + y^2 - 8y + 46 = 0$

In Exercises 29–32, find an equation of the circle with the given center and radius.

29. $(1, 2)$; $r = 5$ **30.** $(-3, 2)$; $r = 1$

31. $(-1, -4)$; $r = 3$ **32.** $(5, -3)$; $r = 8$

33. Find the center and radius of the circle whose equation is $(x - 3)^2 + (y - 1)^2 = 36$.

34. Find the center and radius of the circle whose equation is $(x + 4)^2 + (y - 2)^2 = 121$.

35. Suppose that the point $(2, 3)$ is on a graph. Determine a second point on the graph if the graph is symmetric with respect to
a) the x-axis,   **b)** the y-axis,   **c)** the origin.

36. Suppose that the point $(a, -b)$ is on a graph. Determine a second point on the graph if the graph is symmetric about
a) the x-axis,   **b)** the y-axis,   **c)** the origin.

37. Determine whether the following graphs are symmetric about the x-axis, the y-axis, or the origin:

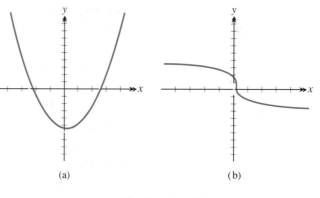

(a)   (b)

For Exercise 37.

38. Determine whether the following graphs are symmetric about the x-axis, the y-axis, or the origin:

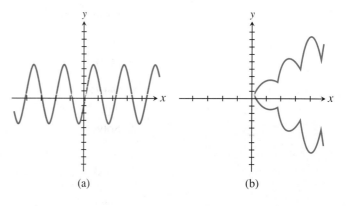

(a)   (b)

For Exercise 38.

In Exercises 39–42, complete the following graph if you know that a complete graph is symmetric about the following:

39. x-axis only

40. y-axis only

41. origin only

42. y-axis and origin

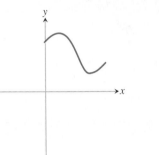

For Exercises 39–42.

In Exercises 43–50, determine whether the graph of each equation is symmetric about the x-axis, the y-axis, or the origin.

43. $y = x^2 + 1$

44. $y^2 = x + 1$

45. $y^3 = x$

46. $x^2 y = 1$

47. $x^2 - y^4 = 8$

48. $y = (x - 2)^2$

49. $x^2 + y^3 = 1$

50. $y^2 + y = x$

In Exercises 51 and 52, use a graphing utility to determine whether the graph is symmetric about the x-axis, the y-axis, or the origin.

51. $y = -\sin x$

52. $y = 2\cos x$

53. Draw a complete graph of $y = (x - 2)^2$. Does the graph have a reflectional line of symmetry? If so, what is an equation of that line?

54. Draw a complete graph of $y = x^3 + 2$. Is the graph symmetric about some point? If so, what is the point of symmetry?

In Exercises 55–58, give an equation for a line whose graph is a line of symmetry for the graph of each function.

55. $y = |x + 2|$

56. $y = |x - 5|$

57. $y = |3 - x|$

58. $y = |x| + 2$

59. Use a graphing utility to find a complete graph of $y^2 = 2x^2 + 1$ by solving for y and then finding two complete graphs in the same viewing rectangle.

60. Use a graphing utility to find a complete graph of $y^2 = 3x + x^2$ by solving for y and then finding two complete graphs in the same viewing rectangle.

61. Graph the relation $y^2 = x^4 + 3x^2$ on a graphing utility by graphing $y = \sqrt{x^4 + 3x^2}$ and $y = -\sqrt{x^4 + 3x^2}$.

2.7 Inverse Functions

Roughly speaking, *inverse* means "undoing." Suppose that we know the effect of a certain process on an object. For example, if heat is removed from water at 32°F, the water becomes ice. The inverse process, if there is one, is what has to be done to return to the starting point. In our example, the inverse process would be to add heat to convert the ice back to water.

Cubing is a process that is undone by taking the principal cube root.

$$2 \quad \overset{cubing}{\underset{(do)}{\longrightarrow}} \quad 8 \quad \overset{cube\ root}{\underset{(undo)}{\longrightarrow}} \quad 2$$

In Section 2.6, we saw that a relation can be defined by an equation and that an ordered pair (x, y) is in the relation if it satisfies the equation.

Using equations, as done in Section 2.6, we can say that

(2, 8) is in the relation $y = x^3$, since $8 = 2^3$, and

(8, 2) is in the relation $x = y^3$, since $8 = 2^3$.

In other words, (2, 8) is in the relation "cubing" and (8, 2) is in the relation "cube root." Notice that the relation $x = y^3$ and $\sqrt[3]{x} = y$ are identical.

Inverse Relations

Suppose that R is a relation consisting of all ordered pairs (x, y) given by a pair of parametric equations $x = f(t)$ and $y = g(t)$.

Whenever a relation R is defined, a second relation, called its inverse, is also defined.

Teaching Note

Some students may need to be reminded that in considering inverse relations, it is important to consider what happens to points. Thus, examining (a, b) as an element of a relation R and (b, a) as an element of the relation R^{-1} is critical to thinking about inverse relationships. The concept of symmetry drives much of the thinking about inverses.

REMINDER

The -1 in R^{-1} is not to be interpreted as an exponent. That is, $R^{-1} \neq 1/R$. R^{-1} is simply a symbol used to name the inverse relation of R.

Definition 2.7   Inverse Relation

Suppose that a relation R is given by an equation. The **inverse relation of R**, denoted R^{-1}, consists of all those ordered pairs (b, a) for which (a, b) belong to R.

In other words, (a, b) is in the relation R if, and only if, (b, a) is in the relation R^{-1}.

Example 1 shows how to find a complete graph of both a relation and its inverse that are defined by parametric equations.

E X A M P L E   1   Finding a Complete Graph of a Relation and Its Inverse

Find a complete graph of both the relation (x, y) determined by the parametric equations

$$x = t \qquad y = t^3 - 4t^2 + 3 \qquad \text{for} -2 \le t \le 5$$

and its inverse relation.

Solution   Set your grapher to parametric mode with the t parameter varying from -2 to 5. Figure 2.46 shows the complete graph with viewing rectangle $[-10, 10]$ by $[-10, 10]$. Note that this relation is a function because it satisfies the vertical line test.

We see from Definition 2.7 that if (a, b) is in relation R, then (b, a) is in relation R^{-1}. Therefore the inverse of the relation $y = t^3 - 4t^2 + 3$, $x = t$ for

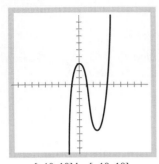

[−10, 10] by [−10, 10]

Figure 2.46 $x = t$ and $y = t^3 - 4t^2 + 3$.

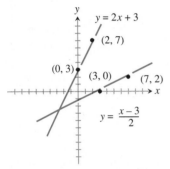

Figure 2.48 A function and its inverse relation.

$-2 \le t \le 5$ is the relation

$$x = t^3 - 4t^2 + 3 \qquad y = t \qquad \text{for } -2 \le t \le 5.$$

A complete graph of this inverse relation is shown in Fig. 2.47.

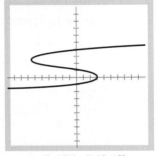

[−10, 10] by [−10, 10]

Figure 2.47 $x = t^3 - 4t^2 + 3$ and $y = t$.

Notice that the relation in Fig. 2.46 is a function; however, its inverse relation, shown in Fig. 2.47, is not a function because it does not satisfy the vertical line test. Example 2 illustrates that the inverse relation of some functions is again a function.

E X A M P L E   2   Verifying Inverse Relations

Show that $y = 2x + 3$ and $y = (x - 3)/2$ are inverse relations.

Solution   (See Fig. 2.48.) We shall show that if (a, b) belongs to the first relation, then (b, a) belongs to the second relation.

$$y = 2x + 3 \qquad \text{This equation defines a relation } R.$$

$$b = 2a + 3 \qquad \text{Suppose } (a, b) \text{ is in this relation.}$$

$$b - 3 = 2a$$

$$a = \frac{b - 3}{2} \qquad \begin{array}{l}\text{This equation shows that } (b, a) \text{ satisfies} \\ \text{the equation } y = (x - 3)/2.\end{array}$$

Because these steps can be reversed, we conclude that $y = 2x + 3$ and $y = (x - 3)/2$ are inverse relations.

Notice that both $y = 2x + 3$ and its inverse relation $y = (x - 3)/2$ are functions.

Notes on Exercises

Ex. 1–12 can be used to illustrate the relationship between a function and its inverse. In Ex. 30–35, students should be encouraged to make conjectures about the exercises and then support their conjectures graphically.

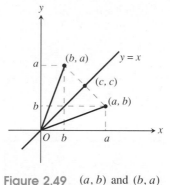

Figure 2.49 (a, b) and (b, a) are symmetric with respect to the line $y = x$.

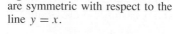

The Graph of a Relation and Its Inverse

The points (a, b) and (b, a) are symmetric with respect to the line $y = x$ (see Fig. 2.49). So the graph of a function and its inverse relation are symmetric with respect to this line. This observation is confirmed by Examples 1 and 2 and leads to the following theorem.

Theorem 2.3 Graph of a Relation and Its Inverse

The graph of the inverse of a relation can be obtained by reflecting the graph of the relation through the line $y = x$.

Notes on Examples

In discussing Example 3, it is important to look at the graph of the inverse relation to determine if it is a function. Students must track the domain and range and think about whether an inverse is indeed a function. This provides the rationale for the concept of one-to-one function. The horizontal line test operationalizes the one-to-one idea to make it easier for students to understand.

E X A M P L E 3   Verifying and Graphing Inverse Relations

Show that relations $y = x^2$ and $x = y^2$ are inverse relations and find a complete graph of each.

Solution

$y = x^2$ This equation defines a relation R. Suppose (a, b) is in this relation.

$b = a^2$ This equation shows that (b, a) satisfies the equation $x = y^2$.

This shows that if (a, b) is in the relation $y = x^2$, then (b, a) is in the relation $x = y^2$. By reversing these arguments, we could show that the converse is also true. So $y = x^2$ and $x = y^2$ are inverse relations (see Fig. 2.50). ≡

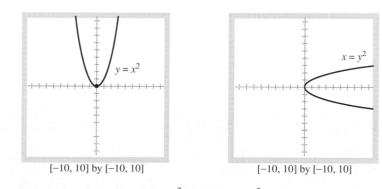

[−10, 10] by [−10, 10]   [−10, 10] by [−10, 10]

Figure 2.50 Inverse relations (a) $y = x^2$ and (b) $x = y^2$.

Because only one y-value is associated with each x-value, the relation $y = x^2$ is a function. Notice that the vertical line test of Definition 3.4 is not met for the inverse relation $x = y^2$. So the inverse relation $x = y^2$ is not a function.

Inverse Functions

From Examples 1 and 2, we see that some functions have inverse relations that are also functions and others have inverse relations that are *not* functions.

Definition 2.8   Inverse Function

If the inverse relation of a function f is also a function, it is called the **inverse function of f**, denoted f^{-1}. A function and its inverse are related by the following two equations:

$$f(f^{-1}(x)) = x \quad \text{for all values of } x \text{ in the domain of } f^{-1}$$

$$f^{-1}(f(x)) = x \quad \text{for all values of } x \text{ in the domain of } f$$

An important relationship between the domain and range of a function and its inverse is that the following sets are equal:

$$\text{domain of } f = \text{range of } f^{-1} \quad \text{and} \quad \text{domain of } f^{-1} = \text{range of } f.$$

If either $f(f^{-1}(x)) = x$ or $f^{-1}(f(x)) = x$ fails for some value of x, the inverse relation of f is not a function.

E X A M P L E   4   Demonstrating Inverse Functions

Show that functions $f(x) = x^2$, where $x \geq 0$, and $f^{-1}(x) = \sqrt{x}$ are inverse functions. Find complete graphs of both.

Solution   We must show that $f(f^{-1}(x)) = x$ and $f^{-1}(f(x)) = x$ for all $x \geq 0$.

$$f(f^{-1}(x)) = f(\sqrt{x})$$
$$= (\sqrt{x})^2$$
$$= x \quad \text{All } x\text{-values are non-negative.}$$
$$f^{-1}(f(x)) = \sqrt{f(x)}$$
$$= \sqrt{x^2}$$
$$= x \quad \text{All } x\text{-values are non-negative.}$$

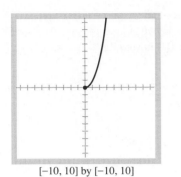

[−10, 10] by [−10, 10]

Figure 2.51 $f(x) = x^2$, $x \geq 0$.

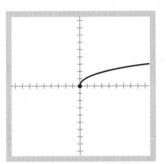

[−10, 10] by [−10, 10]

Figure 2.52 $f^{-1}(x) = \sqrt{x}$.

Figures 2.51 and 2.52 are complete graphs of f and f^{-1}.

The remainder of this section deals with two questions:

1. When is the inverse relation of the function $y = f(x)$ also a function?

2. Given a function whose inverse is also a function, how can we find a rule for the inverse?

The function $y = x^2$ with domain $(-\infty, \infty)$ is graphed in Fig. 2.50, and $y = x^2$ with domain $[0, \infty)$ is graphed in Fig. 2.51. There is a major difference in these two examples. In the first case, horizontal lines intersect the graph of $y = x^2$ in more than one point. This means the *inverse* relation of $y = x^2$ is not a function because its graph does not satisfy the vertical line test. However, every horizontal line intersects the graph of $y = x^2$ with the restricted domain in at most one point. That means that its inverse satisfies the vertical line test and is a function (see Fig. 2.52).

Theorem 2.4   Horizontal Line Test and f^{-1}

The inverse relation R^{-1} of a function $y = f(x)$ is also a function if, and only if, f passes the **horizontal line test**; that is, every horizontal line intersects the graph of f in at most one point.

There is another way to describe those functions whose inverses are also functions. A function f is said to be **one-to-one** if for every pair of distinct values in the domain, x_1 and x_2, it is also true that $f(x_1)$ and $f(x_2)$ are distinct. In other words, if $x_1 \neq x_2$, then $f(x_1) \neq f(x_2)$. This condition is satisfied only if the horizontal line test is met.

Theorem 2.5   A One-to-one Function and f^{-1}

The inverse f^{-1} of f is a function if, and only if, f is a one-to-one function. Furthermore, f is one-to-one if, and only if, every horizontal line intersects the graph of f in at most one point.

E X A M P L E   5   Determining One-to-one Functions

Determine whether the following functions are one-to-one: (a) $f(x) = -3x + 4$; (b) $g(x) = x^3 - 4x$.

Solution Find complete graphs of each function. The graph of $f(x) = -3x + 4$ (see Fig. 2.53a) meets the horizontal line test and hence is one-to-one. The graph of $g(x) = x^3 - 4x$ (see Fig. 2.53b) does not meet the horizontal line test and so is not one-to-one.

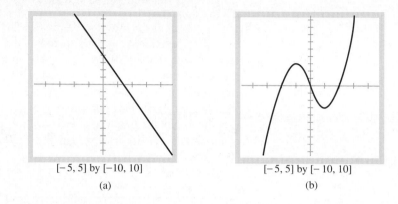

[−5, 5] by [−10, 10]   [−5, 5] by [−10, 10]

(a)   (b)

Figure 2.53 (a) $f(x) = -3x + 4$; (b) $g(x) = x^3 - 4x$. The inverse of f is a function while the inverse of g is not.

Finding Inverse Functions

In addition to determining when an inverse relation is also a function, it is often necessary to find a rule that describes the inverse function.

Finding an Inverse Function

Find the inverse function f^{-1} of a function f by following these steps:

1. Show that f is one-to-one.
2. Interchange x and y in the equation $y = f(x)$.
3. Solve for y.
4. Confirm that the domain of f^{-1} is equal to the range of f.

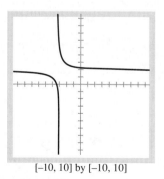

[−10, 10] by [−10, 10]

Figure 2.54
$f(x) = (2x + 7)/(x + 3)$.

E X A M P L E   6   Finding an Inverse Function

Show that $f(x) = (2x + 7)/(x + 3)$ is one-to-one and find a rule for its inverse.

Solution f is one-to-one by the horizontal line test (see Fig. 2.54).

$$y = \frac{2x + 7}{x + 3}$$

$$x = \frac{2y + 7}{y + 3} \qquad \text{Interchange } x \text{ and } y.$$

$$x(y + 3) = 2y + 7 \qquad \text{Solve for } y.$$

$$xy + 3x = 2y + 7$$

$$xy - 2y = -3x + 7$$

$$y(x - 2) = -3x + 7$$

$$y = \frac{-3x + 7}{x - 2}$$

The inverse function is $f^{-1}(x) = (-3x + 7)/(x - 2)$ (see Fig. 2.55).

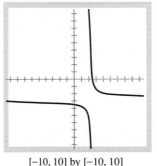

[−10, 10] by [−10, 10]

Figure 2.55 $f^{-1}(x) = (-3x + 7)/(x - 2)$.

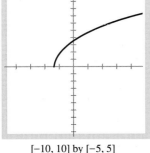

[−10, 10] by [−5, 5]

Figure 2.56 $y = \sqrt{x + 3}$.

EXAMPLE 7 Finding an Inverse Function (Continued)

Let $f(x) = \sqrt{x + 3}$. Show that the inverse of f is a function, find a rule for f^{-1}, and determine the domain and range of f^{-1}.

Solution Figure 2.56 shows that f satisfies the horizontal line test. Consequently, f^{-1} is a function.

$$y = \sqrt{x + 3}$$

$$x = \sqrt{y + 3} \qquad \text{Interchange } x \text{ and } y.$$

$$x^2 = y + 3 \qquad \begin{array}{l}\text{Square both sides in order to solve for } y.\\ \text{This step may introduce extraneous solutions.}\end{array}$$

$$y = x^2 - 3$$

The rule $y = x^2 - 3$ does not represent f^{-1} unless x is restricted to the interval $[0, \infty)$. This restriction must be done because the squaring process introduced extraneous solutions.

Assignment Guide

Day 1: Ex. 2, 3, 5, 6, 8, 9, 11, 12, 14, 15, 17, 20, 21
Day 2: Ex. 23, 24, 26, 27, 29, 30, 32, 35, 37, 38, 40

Exercises for Section 2.7

In Exercises 1–4, write a pair of parametric equations that define the inverse relation for the given parametric equations.

1. $x = 2t - 3$ and $y = t^2 - 2$

2. $x = t^3 - 2$ and $y = 2t$

3. $x = 3^t$ and $y = t$

4. $x = t^2$ and $y = t^3$

In Exercises 5–8, find a complete graph of both the given relation and its inverse.

5. $x = 2t$ and $y = t^2$

6. $x = t^2 - 3t + 2$ and $y = 2t - 3$

7. $x = t^2 - 1$ and $y = t^3 + 3$

8. $x = t - t^3$ and $y = 2t$

In Exercises 9–12, verify that each pair of equations are inverse relations.

9. $y = x^3$; $x = y^3$

10. $y = x^2 + 1$; $x = y^2 + 1$

11. $y = x^2 - 4$; $x = y^2 - 4$

12. $y = x^3 + x^2 - 6x$; $x = y^3 + y^2 - 6y$

In Exercises 13–16, show that the given functions f and g are inverses of each other by demonstrating that $f(g(x)) = x$ and that $g(f(x)) = x$ for all x in the respective domains.

13. $f(x) = 3x - 2$; $g(x) = \dfrac{1}{3}(x + 2)$

14. $f(x) = \dfrac{x + 3}{4}$; $g(x) = 4x - 3$

15. $f(x) = x^3 + 1$; $g(x) = (x - 1)^{1/3}$

16. $f(x) = \dfrac{1}{x}$; $g(x) = \dfrac{1}{x}$

In Exercises 17–22, find a complete graph of each function and decide whether it is one-to-one.

17. $y = x^2 - 5$ **18.** $y = x^3 - 4x + 5$

19. $y = x^4 - 5x^2 + 1$ **20.** $y = 0.001x^3$

21. $f(x) = 4 - x^3$ **22.** $g(x) = \sqrt{x - 4}$

In Exercises 23–25, consider the relation $y = 2x + 4$.

23. Is point $(0, 4)$ on the graph of the inverse of the relation?

24. Is point $(4, 0)$ on the graph of the inverse of the relation?

25. Find a point (a, b) that is on both the graph of the relation and the graph of its inverse.

In Exercises 26 and 27, sketch a complete graph of the inverse of each relation.

26.

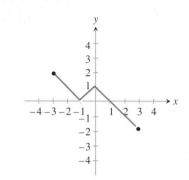

27.

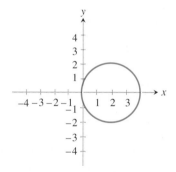

28. Let the domain of f be $(-\infty, 0]$ and let $f(x) = x^2$. Find a rule for f^{-1}.

29. Find a rule for the inverse of $f(x) = \sqrt{x - 2}$. Determine the domain and range of both f and f^{-1}.

In Exercises 30–35, show that each function f is one-to-one and find a rule for f^{-1}.

30. $f(x) = 3x - 6$

31. $f(x) = 2x + 5$

32. $f(x) = x^3$

33. $f(x) = \dfrac{x + 3}{x - 2}$

34. $f(x) = \dfrac{2x - 3}{x + 1}$

35. $f(x) = \sqrt{x + 2}$

In Exercises 36–40, use a graphing utility to find a complete graph of each function. Sketch the inverse relation of each function without the aid of a graphing utility. Identify those functions whose inverses are also functions.

36. $f(x) = x^3 - 8x$

37. $f(x) = x^4 - 2x + 3$

38. $f(x) = \dfrac{x^2 - 2x + 3}{x + 2}$

39. $f(x) = x^3 - 2x - 6$

40. $f(x) = x^3 + 2x + 2$

Chapter 2 Review

KEY TERMS   (The number following each key term indicates the page of its introduction.)

bounded interval, 99
circle, 123
closed interval, 99
complete graph of
  the algebraic
  representation, 85
complete graph of
  the problem
  situation, 85
completing the square, 94
conditional equation, 90
discriminant, 96
equivalent equations, 91
equivalent inequalities, 101
graph of a relation, 120

horizontal line test, 133
identity, 90
inequality, 98
intercept, 121
interval, 99
inverse function, 128
inverse function of f, 132
inverse relation of R, 129
linear equation in x, 91
linear function, 76
linear inequality, 101
line of symmetry, 123
one-to-one function, 133
open interval, 99
parabola, 76

parametric equations, 119
perfect square, 94
polynomial equation, 75
projectile motion, 111
quadratic equation in x, 93
quadratic formula, 94
radical, 96
relation, 118
root, 76
sign-pattern picture, 113
solution to an equation
  in x, 90
solution to an inequality, 99
solution with error of
  at most r, 79

symmetry, 123
  about the x-axis, 123
  about the y-axis, 123
  about the origin, 124
  of graphs, 123
transformation, 91
third-degree polynomial, 80
unbounded interval, 99
union, 105
vertical line test, 129
x-intercept, 76
y-intercept, 122
zero of a polynomial
  function, 76
zoom-in, 77

REVIEW EXERCISES

In Exercises 1–6, use an algebraic method to solve each equation. Check your answers by substitution into the original equation.

1. $\frac{1}{3}x + 2 = 3x + \frac{1}{5}$

2. $3(2 - x) - 2(3x + 7) = x + 2$

3. $x^2 - 4x - 21 = 0$ **4.** $x^2 - 3x - 28 = 0$

5. $2x^4 - 4x^2 - 6 = 0$ **6.** $|2 - 3x| = 7$

In Exercises 7–10, use a number line to graph each interval.

7. $(-3, 5]$ **8.** $[-5, -2]$

9. $-2 \le x < 7$ **10.** $(-\infty, -3)$

In Exercises 11–18, use a number line to graph solutions to each inequality.

11. $|x - 3| < 5$ **12.** $|4 - x| < 7$

13. $|2x - 3| < 1$ **14.** $|3x + 2| \le 3$

15. $|x + 4| \ge 2$ **16.** $|2x - 3| > 5$

17. $|5x - 1| < |-3|$ **18.** $|2x - \sqrt{3}| < \sqrt{7}$

In Exercises 19–22, use an algebraic method to solve each equation. Support each answer with a graphing utility.

19. $x^2 = 6$ **20.** $x^2 - 2x - 8 = 0$

21. $2x^2 - 3x - 7 = 0$ **22.** $x^2 + 3x - 8 = 0$

In Exercises 23–26, use a graphical method to solve each equation.

23. $4x^2 - 5x - 3 = 0$ **24.** $2x - 3x^2 + 15 = 0$

25. $x^2 - 3 - \sqrt{x} = 0$ **26.** $3x^2 - 17x - 39 = 0$

In Exercises 27–36, solve each equation. Use whatever method is preferred.

27. $3x^2 - 4x + 2 = 0$ **28.** $2x^3 - 4x + 7 = 0$

29. $4x^3 - 3x + 1 = 0$ **30.** $2x^3 - 6x^2 + 7x - 5 = 0$

31. $2x^3 - 4x^2 + 3x - 9 = 0$

32. $x^3 - 31x + 2 = 0$ **33.** $x^3 - 5x = 1/x$

34. $x^2 - 3x = \sqrt{x}$ **35.** $x^3 + 5x - 3 = 2\sqrt{x}$

36. $4x^3 + 60x^2 - 103x - 65 = 0$

In Exercises 37–42, use an algebraic method to solve each inequality. Write your answers in interval notation, and support your answers with a graphing utility.

37. $3x + 5 \leq x - 4$ **38.** $\dfrac{x-1}{2} - \dfrac{2x+1}{5} > 1$

39. $\dfrac{1}{4} \leq \dfrac{4x-1}{12} < \dfrac{11}{6}$ **40.** $|3x - 2| \geq 7$

41. $5|x| + 3 < 2$ **42.** $\dfrac{3x+6}{|x+1|} \geq 0$

In Exercises 43–54, use a sign-pattern picture to solve each inequality. Write the solution in interval notation, and support your answers with a graphing utility.

43. $(x - 4)(x + 1) \geq 0$ **44.** $(x - 2)(x + 3) < 0$

45. $(x + 3)(x + 7) \geq 0$ **46.** $(x - 8)(x + 5) > 0$

47. $x^2 - 5x + 6 \geq 0$ **48.** $x^2 + 20 < x$

49. $x(x + 2)(2x - 3) > 0$ **50.** $\dfrac{4x+8}{x+5} > 0$

51. $10 + x - 2x^2 < 0$ **52.** $|x - 1|(2x - 4) \leq 0$

53. $\dfrac{4x-8}{x+5} < 0$ **54.** $\dfrac{5-3x}{x-1} \geq 0$

In Equations 55–58, solve each inequality. Use whatever method is preferred.

55. $3x^2 - 5x + 1 \geq 0$

56. $x^3 - 4x^2 - 2x + 3 < 0$

57. $2x^2 + x - 6 \leq x^3 - x^2 + 4x$

58. $\sqrt{x+2} > 1 - x$

59. Consider the equation $3x^3 + 8x^2 + 24x - 9 = 0$. It has one real-number solution. Find a sequence of four viewing rectangles containing the solution. The first viewing rectangle should permit the solution to be read with an error of at most 0.1, the second 0.01, the third 0.001, and the last 0.0001.

60. The owner of Christine's Crewel Craft Center pays $1500/mo for fixed expenses such as rent, lights, and wages. Craft kits are sold at $22 each, and $15 is required for material, floss, needle, and instructions for each kit. How many kits must be sold to break even?

61. A storage box with a height of 25 cm has a volume of 10,500 cu cm. Let x be the length of the box in centimeters and y the width in centimeters. Find the algebraic representation for the width y in terms of the length x. What is the length when the width is 12 cm?

62. Three hundred and twenty five tickets were sold for a movie. There were two ticket prices: adults at $5.50 and children at $3.00. How many tickets of each type were sold if the total proceeds from the sale of the tickets were $1500?

63. The total value of 23 coins consisting of pennies, nickels, and dimes is $1.51. If there are twice as many dimes as pennies, determine the number of pennies, nickels, and dimes.

64. Sandy can swim 1 mi upstream (against the current) in 20 min. She can swim the same distance downstream in 9 min. Find Sandy's swimming speed and the speed of the current if both speeds are constant.

65. The length of a certain rectangle is 5 in. greater than its width. Find the possible widths of the rectangle if its perimeter is less than 300 in.

66. A single-commodity open market is driven by the supply-and-demand principle. Economists have determined that supply curves are usually increasing (that is, as the price increases, the sellers increase production). For the supply equation $p = 0.003x + 2$, find the price p if the production level is 2000 units.

Exercises 67–69 refer to the following **problem situation**. A farmer has 530 yd of fence to use to make a rectangular corral.

67. If x is the length of such a rectangle, show that $A = x(265 - x)$ is an algebraic representation of the area in terms of x. Find a complete graph of this model.

68. Verify that $(100, 16,500)$ is a point on the graph found in Exercise 67. What meaning do these coordinates have in this problem situation?

69. Verify that $(-100, -36,500)$ is a point on the graph in Exercise 67. What meaning do these coordinates have in this problem situation?

Exercises 70–72 refer to the following **problem situation**. A picture frame's outside dimensions measure 16 by 20 in. The picture will be surrounded by a mat with a uniform border. The distance from the edge of the frame to the picture is x in. on all sides.

70. Find an algebraic representation for the area A of the picture in terms of x, and find a complete graph of this algebraic representation.

71. What values of x make sense in this problem situation?

72. Use a graphing utility to estimate the width of the uniform border if the area of the picture is 250 sq in.

Exercises 73 and 74 refer to the following **problem situation**: Chris has $36 to spend on a pizza party for herself and her friends. Each pizza costs $8.50 and serves two.

73. Let x represent the number of friends Chris can invite. Write an inequality that is an algebraic representation for this problem situation.

74. Solve the inequality found in Exercise 73 and find how many friends Chris can invite.

Exercises 75–77 refer to the following **problem situation**: A publishing company prints x books each week at a total cost $C = (x - 100)^2 + 400$ (in dollars). The company receives $15.25 for each book sold. Assume all books printed are sold.

75. Find an algebraic representation expressing the weekly profit of the company as a function of x and find a complete graph of this algebraic model.

76. What portion of the graph in Exercise 75 is a graph of the problem situation?

77. Use the graph in Exercise 75 to determine the level of weekly production the company must maintain to break even.

Exercises 78–80 refer to the following **problem situation**: The annual profit P of a candy manufacturer is determined by the formula $P = R - C$, where R is the total revenue generated from selling x pounds of candy and C is the total cost of making and selling x pounds of candy. Each pound of candy sells for $5.15 and costs $1.26 to make. The fixed costs of making and selling the candy are $34,000 annually.

78. Find an algebraic representation for the company's annual profit in terms of x, the number of pounds of candy made and sold.

79. Write an inequality that represents a company profit of at least $42,000/year. Solve this problem both algebraically and graphically.

80. How many pounds of candy must be produced and sold for the company to make a profit of at least $42,000 for the year?

Exercises 81–83 refer to the following **problem situation**: A park with dimensions of 125 by 230 ft is surrounded by a sidewalk of uniform width x.

81. Find an algebraic representation for the area of the sidewalk in terms of the width x of the sidewalk.

82. Write an inequality that describes the situation in which the area of the sidewalk is greater than 2900 sq ft but less than 3900 sq ft.

83. Solve the inequality in Exercise 82 graphically to determine possible sidewalk widths that meet the conditions of this problem situation.

Exercises 84–86 refer to the following **problem situation**: An object 30 ft above level ground is shot straight up with an initial velocity of 250 ft/sec.

84. Find an algebraic representation for the height of the object above the ground in terms of t.

85. Draw a complete graph of the algebraic representation in Exercise 84 and indicate what portion of the graph represents the problem situation.

86. At what time t will the object be 550 ft above the ground?

87. Use the distance-formula representation of absolute values to solve $|x - 6| - |x| > 0$.

In Exercises 88–93, determine whether the graph of the equation is symmetric with respect to the x-axis, the y-axis, the origin, or none of these.

88. $x^2 y = 1$ **89.** $x^2 - xy^4 = 2$

90. $x^2 y^3 + 3xy = 1$ **91.** $xy^3 + x^3 y = 25$

92. $y^2 = x^2 + 4$ **93.** $y = x^3 - 5x$

94. Find a complete graph of $f(x) = (x + 5)^2$. Is the graph symmetric about some line? If so, what is an equation of that line of symmetry?

In Exercises 95–100, sketch a complete graph of each equation and specify its domain and range. Give an equation of the line of symmetry, and indicate which equations define y as a function of x.

95. $y = |x - 3|$ **96.** $y = 5 - |x|$

97. $|y - 4| = x$ **98.** $y = |x^3 - 8x|$

99. $x|y| = 3$ **100.** $y = \left| \dfrac{3}{x - 2} \right|$

101. Complete the following graph if a complete graph is symmetric with respect to (a) the x-axis, (b) the y-axis, and (c) the origin.

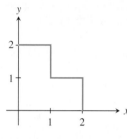

For Exercise 101.

In Exercises 102 and 103, draw a complete graph of each relation and its inverse without using a graphing utility. Determine an equation for the inverse and specify whether it is a function.

102. $y = 5 - 3x$ **103.** $y = x^2 - 2$

In Exercises 104–105, determine whether each function is one-to-one, and find an equation for the inverse relation of each function. Is the inverse a function? If so, find a function rule $y = f^{-1}(x)$ for the inverse. Draw a graph of f and its inverse relation. Then support your answer with a graphing utility.

104. $f(x) = (x + 2)^2$ **105.** $f(x) = 2\sqrt{x - 4}$

106. Determine the domain and range of both the function and its inverse in Exercise 104.

107. Determine whether each graph in the following figure is symmetric with respect to the x-axis, the y-axis, or the origin. Justify your answers.

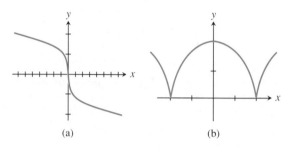

(a)   (b)

For Exercise 107.

SOPHIE GERMAIN

Sophie Germain (1776–1831) was a middle-class woman from a liberal, educated family in France at the time of the French Revolution. She fought fiercely against the prejudices of the time to become an accomplished mathematician. Starting at age 13, Germain studied secretly at night by candlelight. Although not allowed to attend the university, she obtained lecture notes for many courses and submitted comments to instructors under a male pseudonym.

Even though Germain was denied the stimulation of learning alongside other scientists, she did exchange letters with several prominent mathematicians, including Carl Friedrich Gauss and Joseph-Louis Lagrange. Germain is best known for her work in the areas of number theory and the theory of elasticity. She competed or collaborated with many of the eminent mathematicians and physicists of her day, and she was proud to work at the frontier of nineteenth-century science.

CHAPTER 3

Polynomial Functions

3.1 — Graphs of Polynomial Functions

Polynomial functions are related to the polynomial equations studied in Chapter 2. We define them here.

Objective

Students will be able to find local maximum and local minimum points on the graph of a polynomial function and will be able to determine the intervals in which a polynomial function is increasing or decreasing.

Key Ideas

Polynomial function
Degree of a polynomial function
Local maximum value
Local minimum value
Complete graphs of fourth and fifth-degree polynomial functions
Increasing and decreasing functions
Increasing on an interval
Decreasing on an interval

> **Definition 3.1 Polynomial Function**
>
> A **polynomial function** is one that can be written in the form
>
> $$f(x) = a_n x^n + a_{n-1} x^{n-1} + \ldots + a_1 x + a_0,$$
>
> where n is a nonnegative integer and the coefficients $a_0, a_1, \ldots, a_n$ are real numbers. If $a_n \neq 0$, then n is the **degree of the polynomial function**.

Linear and quadratic functions are special cases of polynomial functions. They are polynomial functions of degree 1 and 2, respectively.

Functions f, g, and h are polynomial functions, whereas k is not a polynomial function.

$$f(x) = 3x - 2, \qquad\qquad g(x) = 5x^2 - 3x + 1$$

$$h(x) = 3x^4 + 5x^2 - 3x + 2, \quad k(x) = \sqrt{x^2 + 3}$$

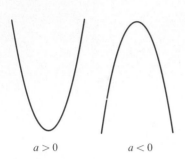

$a > 0$ $a < 0$

Figure 3.1   (a) Upward parabola; (b) downward parabola.

If $f(x)$ is a polynomial function, x can represent any value. So the domain of a polynomial function is the set of all real numbers. The range varies and depends on the characteristics of the specific polynomial.

This chapter focuses on two major themes concerning the behavior of polynomial functions.

1. What are the main characteristics of the graphs of polynomial functions?
2. How many zeros does a given polynomial function have and how can they be found?

In Chapter 2, we saw that the graph of a degree 2 function $f(x) = ax^2 + bx + c$ is called a parabola and has one of the shapes represented in Fig. 3.1. A third-degree polynomial function $f(x) = ax^3 + bx^2 + cx + d$ is sometimes called a **cubic** function. Graphs of cubic polynomial functions fall into one of the four types illustrated in Fig. 3.2.

Figure 3.2   The graphs of $f(x) = ax^3 + bx^2 + cx + d$.

Frank says, "A polynomial is a curve with a few wiggles."

The visual information found in Fig. 3.2 helps as you try to find a complete graph of a cubic function.

🔎 EXPLORE WITH A GRAPHING UTILITY

Graph each of these functions in the standard viewing rectangle. Then continue to modify the viewing rectangle until you have found a complete graph. Record the sequence of viewing rectangles you tried.

1. $f(x) = x^3 + 13x^2 + 10x - 4$
2. $g(x) = x^3 + x^2 - 132x$

Write a paragraph explaining how the information from Fig. 3.2 helped you find complete graphs of these functions.

Fourth- and Fifth-degree Polynomial Functions

Graphs of degree 4 polynomial functions fall into one of four types as indicated in Fig. 3.3.

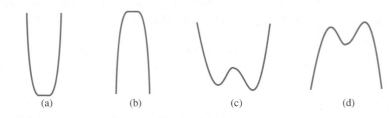

(a)   (b)   (c)   (d)

Figure 3.3   A graph of $f(x) = ax^4 + bx^3 + cx^2 + dx + e$ is one of four types.

E X A M P L E 1   Graphing a Degree 4 Polynomial Function

Find a complete graph of $f(x) = x^4 + 5x^3 + 2x^2 - 8x + 1$. Which one of the four graphs in Fig. 3.3 most closely matches this one? Describe the domain and range of f.

Solution   Begin by graphing f in the standard viewing rectangle. Notice that the resulting graph covers a fairly narrow portion of the viewing rectangle.

Try a smaller interval for x and a larger interval for y, as shown in Fig. 3.4. This graph is like Fig. 3.3(c).

Domain of f $(-\infty, \infty)$

Range of f   All $y \geq k$, where k appears to be approximately -12 ≡

In the same fashion, to find a complete graph of a degree 5 polynomial function it is important to know the general shapes of these graphs. Figure 3.5 shows the six types of graphs for degree 5 polynomial functions. With calculus it can be shown that these shapes are the only possible types of degree 5 polynomial functions.

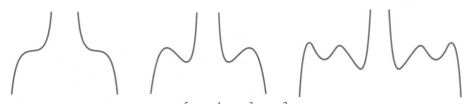

Figure 3.5   A graph $f(x) = ax^5 + bx^4 + cx^3 + dx^2 + ex + k$ is one of six types.

Local Maximum and Minimum Values

Suppose that Ryne Sandberg hits an infield fly ball straight up in the Houston Astrodome with an initial velocity of 84 ft/sec. Will the baseball hit the ceiling of

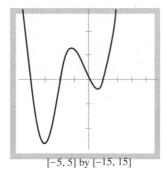

[−5, 5] by [−15, 15]

Figure 3.4 $f(x) = x^4 + 5x^3 + 2x^2 - 8x + 1$.

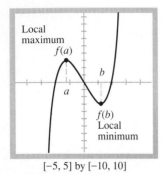

[−5, 5] by [−10, 10]

Figure 3.6 $f(x) = x^3 - 4x$.

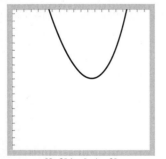

[0, 2] by [−4, −2]

Figure 3.7 $f(x) = x^3 - 4x$.

the dome? To answer this question, we need to know how high the ball will fly, that is, its maximum height. One way to answer this question would be to graph an algebraic representation $y = f(x)$ and find the maximum value of y.

Study Fig. 3.6. Notice that often you can identify a portion of a complete graph of a polynomial function that could be characterized as a peak or a valley. The mathematical terminology is local maximum value or local minimum value. Figure 3.6 shows a complete graph of the cubic function $f(x) = x^3 - 4x$ on which the local maximum and local minimum have been labeled.

Definition 3.2   Local Maximum and Local Minimum

A value $f(a)$ is a **local maximum** of f if there is an open interval (c, d) containing a such that $f(x) \leq f(a)$ for all values of x in (c, d).
A value $f(b)$ is a **local minimum** of f if there is an open interval (c, d) containing b such that $f(x) \geq f(b)$ for all values of x in (c, d).

Local maximum and minimum values together are called **local extremum values**, or simply, **local extrema**.

Notice that the maximum and minimum values of a function, if they exist, are "output" values of the function. In Fig. 3.6, the local maximum value is $f(a)$, not a.

EXAMPLE 2   Finding a Local Minimum

Find the local minimum value of the function $f(x) = x^3 - 4x$ and where it occurs.

Solution

1. The graph in Fig. 3.6 suggests that the local minimum occurs between $x = 0$ and $x = 2$. Find the graph in the viewing rectangle [0, 2] by [−4, −2] shown in Fig. 3.7.

2. Estimate that the low point in Fig. 3.7 has its x-coordinate between 1.1 and 1.2 and its y-coordinate between −3.1 and −3.0. Find the graph in the viewing rectangle [1.1, 1.2] by [−3.1, −3.0] (see Fig. 3.8a).

3. Move the cursor to the valley point (see Fig. 3.8a). The y-coordinate of this point is −3.079, but the x-coordinate is difficult to identify because the graph is too flat to determine when the cursor is at the lowest point. To alleviate this problem, change the viewing rectangle so that its width is 10 times its height (see Fig. 3.8b).

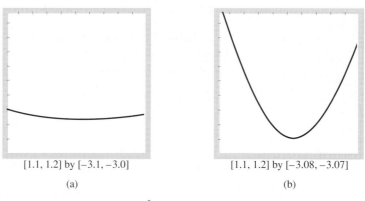

$[1.1, 1.2]$ by $[-3.1, -3.0]$ (a) $[1.1, 1.2]$ by $[-3.08, -3.07]$ (b)

Figure 3.8 Two views of $f(x) = x^3 - 4x$.

4. Using the viewing rectangle $[1.1, 1.2]$ by $[-3.08, -3.07]$, move the cursor to the low point. Its coordinates are $(1.155, -3.0792)$.

Therefore f achieves a local minimum value of -3.0792 when x is 1.155. The error of the x-coordinate is at most 0.01 and of the y-coordinate is at most 0.001. The y-coordinate was found to a greater precision than the x-coordinate so that the error in x would be at most 0.01. ≡

E X A M P L E   3   Finding the Local Maximum Value

Find the local maximum of $f(x) = x^3 - 4x$.

Solution   Use information from Example 2 together with the fact that the graph of f is symmetric about the origin.

1. The graph of f is symmetric about the origin.

$$f(-x) = (-x)^3 - 4(-x)$$
$$= -x^3 + 4x$$
$$= -(x^3 - 4x) = -f(x)$$

2. In Example 2, we found that the valley point of the graph is $(1.155, -3.0792)$. Because the graph of f is symmetric about the origin, we can conclude that the peak point of the graph is the point $(-1.155, 3.0792)$.

Therefore f achieves a local maximum value of 3.0792 when x is -1.155. ≡

Often we are able to use the method of Example 2 to solve applied problems. Here are several examples.

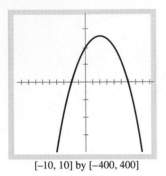

[−10, 10] by [−400, 400]

Figure 3.9 $s = -16t^2 + 64t + 200.$

Notes on Examples

Examples 4 and 5 will be familiar to teachers of calculus. These examples are excellent models of polynomial functions introduced in this chapter.

E X A M P L E   4   APPLICATION: Finding the Maximum Height

A baseball is thrown straight up from the top of a building 200 ft tall with an initial velocity of 64 ft/sec. When does the baseball reach its maximum height above ground, and what is this maximum height?

Solution   This is a projectile motion problem situation. The height of the object t seconds after launch is

$$s = -16t^2 + 64t + 200.$$

Review the Projectile Motion Problem Situation in Section 2.5, where we learned that $y = -16t^2 + v_0 t + s_0$.

Graphical Solution   Find the graph of $s = -16t^2 + 64t + 200$ (see Fig. 3.9) and use zoom-in to find the coordinates of the highest point. The highest point is $(2, 264)$ with an error of at most 0.01. The maximum height that the object achieves is 264 ft, as shown by the algebraic solution.

Algebraic Solution   Complete the square of $s = -16t^2 + 64t + 200$ and obtain the form $y = -16(t - 2)^2 + 264$. We know that the coordinates of the vertex are exactly $(2, 264)$. The maximum height that the baseball achieves is exactly 264 ft. This maximum height occurs exactly 2 sec after the baseball is thrown.   ▤

E X A M P L E   5   APPLICATION: Finding the Maximum Volume

Squares are cut from the corners of a 20- by 25-in. piece of cardboard, and a box is made by folding up the flaps (see Fig. 3.10). Determine the graph of the problem situation and find the dimensions of the squares so that the resulting box has the maximum possible volume.

Solution

x = width of square and height of the resulting box in inches

$20 - 2x$ = width of the base in inches

$25 - 2x$ = length of the base in inches

$V(x) = x(20 - 2x)(25 - 2x)$ Volume is equal to length × width × height.

Find a complete graph of $V(x) = x(20 - 2x)(25 - 2x)$ (see Fig. 3.11a). In this problem situation, $0 < x < 10$, so the complete graph of the problem situation is shown in Fig. 3.11(b).

Use zoom-in to find the coordinates of the local maximum (see Fig. 3.11c). Read the coordinates of the high point as $(3.68, 820.5282)$ with an error of at most 0.01. Thus the maximum volume is 820.5282 cu in. and this volume occurs when $x = 3.68$ in.   ▤

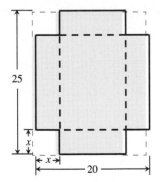

Figure 3.10   Squares cut from a piece of cardboard measuring 20 by 25 in.

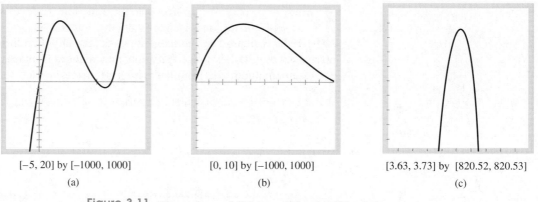

[−5, 20] by [−1000, 1000]   [0, 10] by [−1000, 1000]   [3.63, 3.73] by [820.52, 820.53]

(a)   (b)   (c)

Figure 3.11   Three views of $V(x) = x(20 − 2x)(25 − 2x)$.

Notes on Exercises

Ex. 37–56 provide applied problem situations for polynomial functions. You may want to spend time in class deriving one or two of the models. Be sure to use the terminology introduced in previous sections (model, problem situation, graph of the problem situation, algebraic representation).

Increasing and Decreasing Functions

This next Exploration introduces the concepts *increasing function* and *decreasing function*.

🔎 **EXPLORE WITH A GRAPHING UTILITY**

Find a complete graph of each of the following functions:

1. $f(x) = 1 − x^3$
2. $f(x) = x^3 − x^2 − 6x$
3. $f(x) = x^3 − 4$
4. $f(x) = −x^3 − 2x^2 + 8x + 2$

Note that as the trace cursor moves along the graph from the extreme left to the extreme right, the *x-coordinate of the cursor continually increases*. In each case ask yourself, does the *y*-coordinate

a) continually increase?

b) increase for a while, then decrease, and then increase?

c) decrease for a while, then increase, and then decrease?

d) continually decrease?

Generalize   What do you think it means to call a function *an increasing function on an interval,* and what do you think it means to call a function *a decreasing function on an interval*?

Study Fig. 3.12(a). Notice that the function f decreases on the intervals $(-\infty, a)$ and (b, ∞) and increases on the interval (a, b). The function g in Fig. 3.12(b) increases on the intervals $(-\infty, a)$ and (b, ∞) and decreases on the interval (a, b). Further notice that a function changes from increasing to decreasing at a local maximum and from decreasing to increasing at a local minimum.

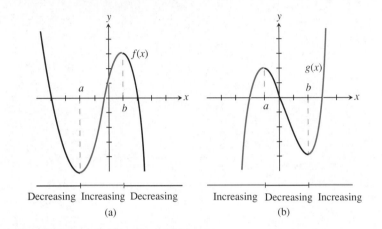

Figure 3.12   (a) The local maximum and minimum of f are $f(b)$ and $f(a)$, respectively.
(b) The local maximum and minimum of g are $g(a)$ and $g(b)$, respectively.

Definition 3.3   Increasing and Decreasing Functions

A function f is **increasing on an interval J** if the values $f(x)$ increase as x increases in J. A function f is **decreasing on an interval J** if the values $f(x)$ decrease as x increases in J.

This intuitive definition is stated with more precision in a calculus course.

E X A M P L E   6   Finding Increasing and Decreasing Intervals

Find the intervals on which $f(x) = x^3 - 4x$ is increasing and decreasing.

Solution   In Examples 2 and 3, we found the following information:

$$(-1.155, 3.0792) = \text{point where a local maximum occurs}$$

$$(1.155, -3.0792) = \text{point where a local minimum occurs}$$

[−5, 5] by [−10, 20]

Figure 3.13
$f(x) = x^3 - 4x^2 - 4x + 16.$

Use this information to conclude the following about this function:

- f is increasing on the intervals $(-\infty, -1.155)$ and $(1.155, \infty)$.
- f is decreasing on the interval $(-1.155, 1.155)$.

EXAMPLE 7   *Finding Increasing and Decreasing Intervals*

Find the intervals on which $f(x) = x^3 - 4x^2 - 4x + 16$ is increasing and decreasing.

Solution

1. Find a complete graph of f (see Fig. 3.13).
2. Use zoom-in to show that a local maximum occurs when $x = -0.43$ and that a local minimum occurs when $x = 3.10$.

The function f is increasing on the intervals $(-\infty, -0.43)$ and $(3.10, \infty)$. The function f is decreasing on the interval $(-0.43, 3.10)$.

Assignment Guide

Day 1: Ex. 3–29 odd, 32
Day 2: Ex. 38–40, 45–49, 54–57

Exercises for Section 3.1

In Exercises 1–8, find a complete graph of each polynomial function. Specify the viewing rectangle you used.

1. $f(x) = x^2 - x + 3$
2. $y = 12x^2 - 20x + 60$
3. $A = \left(\dfrac{x}{4}\right)^2 + \left(\dfrac{500 - x}{4}\right)^2$
4. $A = x(8 - x^2)$
5. $V = x(30 - 2x)(40 - 2x)$
6. $g(x) = 2x^3 - 6x^2 + 3x - 5$
7. $V(x) = 200 - 20x^2 + 0.01x^3$
8. $f(x) = 2x^5 - 10x^4 + 3x^3 + 2x^2 - 10x + 5$

In Exercises 9–12, determine the *exact* values of all local maximum and minimum values and the corresponding values of x that give these values.

9. $y = x^2 - 4x + 5$
10. $f(x) = 30 - 3x + 7x^2$
11. $g(x) = |10 - 7x + x^2|$
12. $f(x) = 3 - 2|4x + 7|$

In Exercises 13–24, draw a complete graph of the function. Find the points where the graphs cross the x-axis. Determine all local maximum and minimum values and the corresponding values of x that give these local extrema.

13. $y = 10 - x^2$
14. $f(x) = 2x^2 - 6x + 10$
15. $g(x) = x^3 - 10x^2$
16. $y = 2x^3 - x^2 + x - 4$
17. $f(x) = x^4 - 4x^2 - 3x + 12$
18. $y = |2x^3 - 4x^2 + x - 3|$
19. $f(x) = x^3 - 2x^2 + x - 30$
20. $V(x) = x(34 - 2x)(53 - 2x)$
21. $T(x) = |20x^3 + 2x^2 - 10x + 5|$
22. $V(x) = x(22 - 2x)(8 - 2x)$
23. $f(x) = 12 - x + 3x^2 - 2x^3$
24. $y = x^5 - 3x^2 + 3x - 6$

Notes on Exercises

Ex. 15–18 need to be discussed thoroughly since they highlight the intermediate value property. First, students should check that f is continuous on the interval $[a, b]$. Then they should find $f(a)$ and $f(h)$ and check that the designated number is between $f(a)$ and $f(h)$. Finally, they should solve for the number c.

Because -3.3 is not between -2 and 3, $(-3 + \sqrt{13})/2$ is the only number between -2 and 3 such that

$$f\left(\frac{-3 + \sqrt{13}}{2}\right) = -3.$$

≡

The intermediate value property is important because it guarantees the existence of a real number that may be difficult or impossible to compute. It is perhaps most commonly used to locate where a function value is zero. In this case the intermediate value property is sometimes known as the "Location Theorem".

E X A M P L E   5   Approximating the Zero of a Function

Show that $f(x) = x^5 + 2x - 1$ has a zero in the interval $[0.25, 0.75]$.

Solution   We want to find c so that $f(c) = 0$ where c is in $[0.25, 0.75]$.

$$f(0.25) = (0.25)^5 + 2(0.25) - 1 = -0.499$$

$$f(0.75) = (0.75)^5 + 2(0.75) - 1 = 0.737$$

Because f is continuous and $f(0.25) < 0 < f(0.75)$, we can conclude from the intermediate value property that there is a value c in $[0.25, 0.75]$ such that $f(c) = 0$, that is, there is a zero between 0.25 and 0.75.

≡

Example 5 provides the basis for a numerical zoom-in procedure for estimating a zero of a function. The **bisection method** involves continually bisecting the interval in the previous viewing rectangle considered. At each new midpoint, the function is evaluated to see whether it is positive or negative. By using the intermediate value property, you can then determine in which of the two intervals resulting from the bisection that the graph of the function crosses the x-axis.

Example 5 shows that $f(0.25)$ is negative and $f(0.75)$ is positive. Theorem 3.2 tells us there is a zero between 0.25 and 0.75. Bisecting the interval $(0.25, 0.75)$ results in the midpoint 0.5. The new interval will thus have 0.5 as one endpoint and one of the two previous endpoints as the other endpoint. How is the other endpoint chosen? It must be the point whose sign is opposite to that of the midpoint. Since both $f(0.5)$ and $f(0.75)$ are positive, we do not choose 0.75 as the new endpoint. However, $f(0.5)$ and $f(0.25)$ have opposite signs, so the new interval for bisection is $(0.25, 0.5)$.

This second interval is bisected to get a second midpoint. The process is repeated until an interval small enough to provide the desired accuracy is reached. In this case, we continue the method for two more rounds until the zero is found to be in the interval $(0.4375, 0.5)$. The following table records the process; notice that the midpoints are boldfaced. See also Fig. 3.21.

	Starting Interval	Step 1	Step 2	Step 3
Positive values	$f(0.75)$	$\boldsymbol{f(0.5)}$	$f(0.5)$	$f(0.5)$
Negative values	$f(0.25)$	$f(0.25)$	$\boldsymbol{f(0.375)}$	$\boldsymbol{f(0.4375)}$

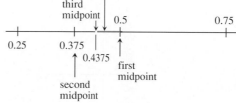

Figure 3.21   Bisection method for approximating a zero of $f(x) = x^5 + 2x - 1$.

End Behavior of a Polynomial Function

What happens to the graph of a function on the extreme right and left ends of the x-axis? This is the same as asking about the **end behavior** of the function: What happens to the $f(x)$ values when $|x|$ is large?

Consider the graph of $f(x) = x^3$ (see Fig. 3.22).

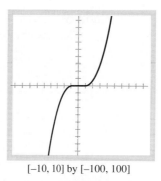

[−10, 10] by [−100, 100]

Figure 3.22 $f(x) = x^3$ has end behavior type $(\swarrow, \nearrow)$.

We say: As x approaches infinity, $f(x)$ approaches infinity.

We write: As $x \to \infty$, $f(x) \to \infty$.

We mean: As x gets very large, $f(x)$ also gets very large.

Similarly, we write: As $x \to -\infty$, $f(x) \to -\infty$. When $f(x) \to +\infty$ at the extreme right of the graph and $f(x) \to -\infty$ at the extreme left of the graph, we describe the **end behavior type** of f as $(\swarrow, \nearrow)$. So $(\swarrow, \nearrow)$ is a notational abbreviation for the following two statements:

$$\text{As } x \to -\infty, f(x) \to -\infty, \quad \text{and} \quad \text{as } x \to \infty, f(x) \to \infty.$$

There are four end behavior types for polynomial functions. These four types are $(\nwarrow, \nearrow), (\nwarrow, \searrow), (\swarrow, \searrow), (\swarrow, \nearrow)$ and are illustrated in Fig. 3.23.

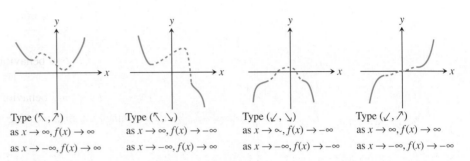

Type $(\nwarrow, \nearrow)$
as $x \to \infty, f(x) \to \infty$
as $x \to -\infty, f(x) \to \infty$

Type $(\nwarrow, \searrow)$
as $x \to \infty, f(x) \to -\infty$
as $x \to -\infty, f(x) \to \infty$

Type $(\swarrow, \searrow)$
as $x \to \infty, f(x) \to -\infty$
as $x \to -\infty, f(x) \to -\infty$

Type $(\swarrow, \nearrow)$
as $x \to \infty, f(x) \to \infty$
as $x \to -\infty, f(x) \to -\infty$

Figure 3.23 The four possible end behaviors for polynomial functions.

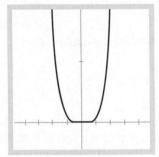

[−5, 5] by [−5, 20]

Figure 3.24 $f(x) = x^4$.

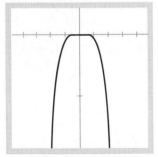

[−5, 5] by [−20, 5]

Figure 3.25 $g(x) = -x^4$.

E X A M P L E 6 Identifying the End Behavior Type

Identify the end behavior type for $f(x) = x^4$ and $g(x) = -x^4$.

Solution Study Figs. 3.24 and 3.25.

The end behavior type of f is $(\nwarrow, \nearrow)$, and the end behavior type of g is $(\swarrow, \searrow)$. ≡

🔎 **EXPLORE WITH A GRAPHING UTILITY**

Find the end behavior type for each of these functions. Record your results.

$$f(x) = 3x^4 \qquad f(x) = 2x^3$$
$$f(x) = -2x^4 \qquad f(x) = -4x^5$$
$$f(x) = -5x^2 \qquad f(x) = -2x^3$$
$$f(x) = 2x^6 \qquad f(x) = 2x^7$$

Generalize Describe the end behavior type of a function $f(x) = a_n x^n$.

The following box summarizes the generalization that you may have discovered in this Exploration.

End Behavior of $f(x) = a_n x^n$

The polynomial functions of the form $f(x) = a_n x^n (n \geq 1)$ have the following types of end behavior:

	n **even**	n **odd**
$a_n > 0$	$(\nwarrow, \nearrow)$ end behavior	$(\swarrow, \nearrow)$ end behavior
$a_n < 0$	$(\swarrow, \searrow)$ end behavior	$(\nwarrow, \searrow)$ end behavior

---

Teaching Note

Have students compare values in the columns in the table on this page when discussing Example 7. Students should think about what it means for the fraction $\dfrac{f(x)}{2x^3}$ to approach 1 as x becomes large. It may be helpful for some students to write $f(x)$ as $2x^3 - 7x^2 - 8x + 16$. The third column entries in the table (headed $-7x^2 - 8x + 16$) should be examined in terms of how small the absolute value of each entry is in comparison to the entries in the second and fourth columns. A thorough discussion of this aspect of end behavior should lead students to the same conclusions made in comparing different viewing rectangles.

Using Zoom-out to Find a Model of End Behavior

Each polynomial function belongs to one of four categories of end behavior type as shown in Fig. 3.23. For example, the polynomial $f(x) = 2x^3 - 7x^2 - 8x + 16$ has end behavior type $(\swarrow, \nearrow)$. In particular, as $x \to \infty$, $f(x) \to \infty$ and as $x \to -\infty$, $f(x) \to -\infty$. But how fast does $f(x)$ increase as x increases?

The term having the highest degree in a polynomial is called the **leading term**. For $f(x) = 2x^3 - 7x^2 - 8x + 16$, the leading term is $2x^3$. In the following table, notice that the values of $2x^3$ are very large in absolute value compared with the absolute values of $-7x^2 - 8x + 16$. We describe this by saying that the leading term dominates the polynomial f.

Comparing the Leading Term $2x^3$ and $f(x)$

x	$2x^3$	$-7x^2 - 8x + 16$	$f(x)$
20	16,000	2,944	13,056
50	250,000	−17,884	232,116
100	2,000,000	−70,784	1,929,216
−20	−16,000	−2,624	−18,624
−50	−250,000	−17,084	−267,084
−100	−2,000,000	−69,184	−2,069,184

Another way of saying it is that $2x^3$ models how rapidly $f(x) = 2x^3 - 7x^2 - 8x + 16$ increases; that is, $2x^3$ is the end behavior model of $f(x) = 2x^3 - 7x^2 - 8x + 16$. This leads to the following definition.

> ### Definition 3.4   End Behavior Model
>
> The **end behavior model** of a polynomial
> $$f(x) = a_n x^n + a_{n-1} x^{n-1} + \cdots + a_1 x + a_0 \quad (a_n \neq 0)$$
> is the polynomial $g(x) = a_n x^n$.

A grapher can provide visual evidence to support this numerical evidence of end behavior. The process of viewing a graph in increasingly larger viewing rectangles is called **zoom-out**. We can use zoom-out to find functions that model the end behavior of a polynomial.

EXAMPLE 7   Visualizing End Behavior

Graph both $f(x) = 2x^3 - 7x^2 - 8x + 16$ and $g(x) = 2x^3$ in viewing rectangle $[-8, 8]$ by $[-100, 100]$. Then zoom out to the viewing rectangle $[-40, 40]$ by $[-10,000, 10,000]$.

Solution   In Fig. 3.26(a), the differences between the two functions are evident. In Fig. 3.26(b), which is a zoom-out version of Fig 3.26(a), the end behaviors of the two functions appear alike.

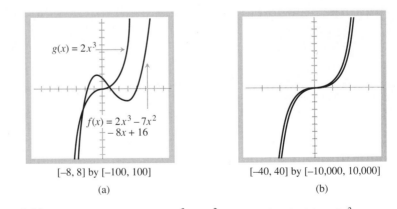

[–8, 8] by [–100, 100]

(a)

[–40, 40] by [–10,000, 10,000]

(b)

Figure 3.26   Two views of $f(x) = 2x^3 - 7x^2 - 8x + 16$ and $g(x) = 2x^3$.

Example 7 gives a visual justification for the concept *end behavior model*. We see that $2x^3$ does model the function $f(x) = 2x^3 - 7x^2 - 8x + 16$ when $|x|$ is large.

Notes on Examples

When determining the end behavior model for a polynomial function, it is helpful to set the scale factors for both x and y equal to zero so that no tick marks will appear on the screen. When visualizing end behavior, as in Example 7, a good technique to use for zooming-out on the TI-81 calculator is to set the zoom factor for x at 10 and the zoom factor for y at 1000. After determining a viewing rectangle in which both the polynomial function and the end behavior model are clearly distinguishable, begin the zoom-out procedure until the graphs appear to be the same. For a fourth-degree polynomial, such as Example 8, the zoom factor for x would be 10 and the zoom factor for y would be 10,000.

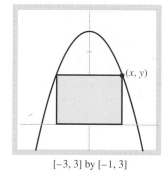

[−3, 3] by [−1, 3]

Figure 3.27 $y = 4 - x^2$.

EXAMPLE 8   Finding the End Behavior Model

Find the end behavior model for $f(x) = -0.5x^4 + 5x^3 - 10x + 1$. Also determine the end behavior type of f.

Solution   The function $g(x) = -0.5x^4$ is the end behavior model of $f(x) = -0.5x^4 + 5x^3 - 10x + 1$. The end behavior type of f is $(\nearrow, \searrow)$.   ▤

EXAMPLE 9   Finding the Maximum Possible Area

Find the maximum possible area of a rectangle that has its base on the x-axis and its upper two vertices on the graph of the equation $y = 4 - x^2$ (see Fig. 3.27).

Solution

$$\text{width of rectangle:} \quad 2x$$

$$\text{height of rectangle:} \quad y = 4 - x^2$$

$$\text{area of rectangle:} \quad A(x) = 2x(4 - x^2)$$

The only values of x that make sense in the problem situation are those such that $0 < x < 2$. Why?

Find a complete graph of $A(x)$ and observe that there is a local maximum between 0 and 2. Use zoom-in to find the local maximum.

The value of x, with an error of at most 0.01, producing the largest value for the area of the rectangle, is 1.155. The area for $x = 1.155$ is 6.16 square units.   ▤

Exercises for Section 3.2

In Exercises 1–8, assume that each graph is complete. Identify the function as continuous or discontinuous. If it is discontinuous, name two intervals on which it is continuous and two intervals on which it is discontinuous.

1.

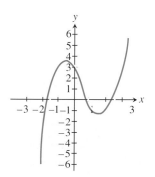

2.

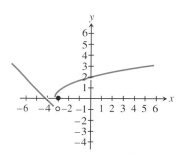

3.

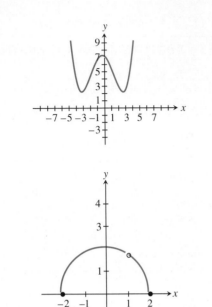

4.

5.

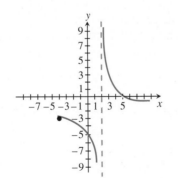

6.

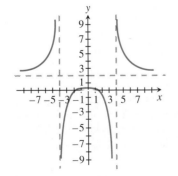

7.

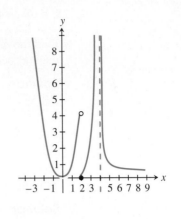

8.

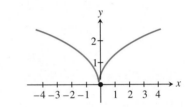

In Exercises 9–14, determine the points of discontinuity of each function.

9. $f(x) = \text{INT}(x + 2)$

10. $g(x) = 2x^2 - 3x + 6$

11. $V(x) = 4x^3 - 200x^2 + 30x$

12. $f(x) = \begin{cases} -\dfrac{1}{x} & \text{for } x < 0 \\ x^2 + 1 & \text{for } x \geq 0 \end{cases}$

13. $y = \dfrac{2}{x - 3}$

14. $f(x) = \begin{cases} |x - 2| & \text{for } x < -1 \\ 2x & \text{for } x \geq -1 \end{cases}$

15. Consider the function $f(x) = 3x - 12$. Verify that 5 is a number between $f(-10)$ and $f(10)$ and determine a value of c so that $f(c) = 5$ and $-10 \leq c \leq 10$.

16. Consider the function $f(x) = 4x - 7$. Verify that 0 is a number between $f(1)$ and $f(2)$ and determine a value of c so that $f(c) = 0$ and $1 \leq c \leq 2$.

17. Consider the quadratic function $f(x) = 2x^2 + 4x - 10$. Verify that 50 is a number between $f(0)$ and $f(10)$. Determine a value of c so that $f(c) = 50$ and $0 \leq c \leq 10$.

18. Consider the linear function $f(x) = ax + b$. Let L be a number between $f(-10)$ and $f(10)$ and determine a value of c so that $f(c) = L$ and $-10 \le c \le 10$.

For Exercises 19 and 20, assume the graph of $y = f(x)$ is complete, and complete each statement.

19. As $x \to \infty$, $f(x) \to$?   **20.** As $x \to -\infty$, $f(x) \to$?

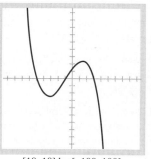

[10, 10] by [−100, 100]

For Exercises 19 and 20.

In Exercises 21 and 22, assume the graph of $y = f(x)$ is complete, and complete each statement.

21. As $x \to \infty$, $f(x) \to$?   **22.** As $x \to -\infty$, $f(x) \to$?

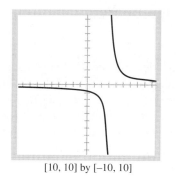

[10, 10] by [−10, 10]

For Exercises 21 and 22.

In Exercises 23–28, find the the end behavior type of the function.

23. $f(x) = x^3 - 4x + 5$

24. $f(x) = 2x^4 - x^3 + 3x^2 + 1$

25. $f(x) = x^2 - x^3 - 4x + 1$

26. $f(x) = 3x^{15} - 2x^5 + x^2 - 2$

27. $f(x) = x^3 + 2x^5 - 4x + 1$

28. $f(x) = 7 - x + x^2 - x^3$

In Exercises 29–32, find the end behavior model for the given function.

29. $f(x) = 3x^4 - 4x^2 + 2$

30. $f(x) = 0.05x^7 - 4x^3 + x^2 + 3$

31. $f(x) = 6x^2 + 4x^3 - 17$

32. $f(x) = 2(x - 3)^5$

33. Find a viewing rectangle that demonstrates visually that $y = 3x^2$ is the end behavior model for $f(x) = 3x^2 - 4x + 6$.

34. Find a viewing rectangle that demonstrates visually that $y = 2x^3$ is the end behavior model of $f(x) = 2x^3 + 3x^2 - 4x + 5$.

Exercises 35–37 refer to the following **problem situation**: The total daily revenue of Chris's Cookie Shop is given by the equation $R = xp$, where x is the number of pounds of cookies sold and p is the price of 1 lb of cookies. Assume the price per pound of cookies is given by the supply equation $p = 0.2 + 0.01x - 0.00001x^2$.

35. What values of x make sense in this problem situation?

36. Find a complete graph of the revenue function R and determine for what values of x the revenue function is increasing.

37. Find the number of pounds of cookies Chris's Cookie Shop needs to sell to achieve maximum revenue. What is this maximum revenue?

38. A box is made by cutting equal squares from the four corners of a 16- by 28-in. piece of material. Determine the size of the square that must be cut out to produce a box with maximum volume. What is the corresponding maximum volume?

39. Consider the following figure, which represents a rectangular piece of fabric 3 ft wide. Suppose one corner E is folded over along the line DF until the corner E touches the opposite edge at B. Let A be the area of the shaded triangle BCD. Let x be the distance between the vertex C and point B. It can be shown that the area of triangle BCD is $A = \frac{3}{4}x - \frac{1}{12}x^3$. What is the largest possible area A?

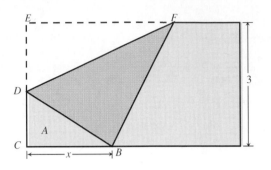

For Exercise 39.

Exercises 40 and 41 refer to the following algebraic situation: Given $n + 1$ points in the coordinate plane, it can be shown that there is a unique polynomial of a degree less than or equal to n whose graph passes through these $n + 1$ points. Such a polynomial is said to *interpolate* these data points. (Such

interpolating polynomials are important tools in understanding data generated from many kinds of scientific experiments and statistical analyses.)

40. Show that $L(x) = 2x^3 - 11x^2 + 12x + 5$ interpolates the data points $(-1, -20)$, $(0, 5)$, $(1, 8)$, and $(2, 1)$.

41. Assume that $L(x) = 2x^3 - 11x^2 + 12x + 5$ interpolates a set of data points from some experiment. Suppose that $(1.5, 6)$ is another data point generated by the same experiment. What is the predicted outcome if $x = 1.5$ is used in the interpolating polynomial $L(x)$?

42. Writing to Learn   Write several paragraphs that explain the difference between the concepts *end behavior type* and *end behavior model*.

43. Writing to Learn   Write several paragraphs describing a few motions or activities that can be represented by one quantity varying continuously as another variable changes.

3.3

Real Zeros of Polynomials: The Factor Theorem

The words *zero* and *root*, introduced in Section 2.1, are used interchangeably when discussing functions. The zeros or roots of a function f are the solutions to the equation $f(x) = 0$. A number a is a zero or a root if, and only if, $f(a) = 0$.

Section 2.1 showed how to find exact zeros for linear and quadratic functions. In this section and Section 4.4, we focus on special techniques that can be used to find real zeros algebraically for polynomial functions of degrees greater than or equal to 3.

Factors and Long Division

The zeros of a polynomial function can be found by finding its factors. For example, if $x - 2$ is a factor of f, long division can be used to find a quotient $q(x)$. We can then write $f(x)$ as

$$f(x) = (x - 2)q(x).$$

Because

$$f(2) = (2 - 2)q(2) = 0,$$

2 is a zero of f. So if a linear factor can be found, a zero has also been found.

Whether $x - c$ is a factor of a polynomial can be determined by using long division to divide the polynomial by $x - c$ and checking to see if the remainder is zero.

E X A M P L E 1   Completing Polynomial Long Division

Divide the polynomial $x^3 - 8x - 3$ by $x - 3$ and find a zero of $x^3 - 8x - 3$.

Solution   When dividing $x - 3$ into $x^3 - 8x - 3$, first find a partial quotient of x^2 by dividing x into the first term x^3 to obtain x^2. Then multiply x^2 by $x - 3$ and subtract the product $x^3 - 3x^2$ from the dividend. Repeat the process until you reach a remainder.

$$
\begin{array}{r}
x^2 + 3x + 1 \quad \longleftarrow \text{ Quotient} \\
x - 3 \overline{\smash{)}\, x^3 + 0x^2 - 8x - 3} \quad \longleftarrow \text{ Dividend} \\
\underline{x^3 - 3x^2} \qquad\qquad\ \longleftarrow x^2 \text{ multiplied by } (x - 3) \\
3x^2 - 8x - 3 \quad \longleftarrow \text{ Result of first subtraction} \\
\underline{3x^2 - 9x} \qquad\quad \longleftarrow 3x \text{ multiplied by } (x - 3) \\
x - 3 \\
\underline{x - 3} \\
0 \quad \longleftarrow \text{ Remainder}
\end{array}
$$

Divisor $\longrightarrow$

Because the remainder is zero, we can conclude that

$$x^3 - 8x - 3 = (x - 3)(x^2 + 3x + 1)$$

and 3 is a zero of $x^3 - 8x - 3$.   ≡

Notice that the terms *dividend, divisor, quotient*, and *remainder* are used here exactly as they are in the division of integers.

E X A M P L E 2   Completing Polynomial Long Division (Continued)

Find the quotient and remainder when $2x^4 - x^3 - 2$ is divided by $2x^2 + x + 1$.

Solution

$$\begin{array}{r} x^2 - x \qquad\qquad \leftarrow \text{Quotient} \\ 2x^2 + x + 1 \overline{\smash{\big)}\ 2x^4 - x^3 + 0x^2 + 0x - 2} \\ \underline{2x^4 + x^3 + x^2} \qquad\qquad \\ -2x^3 - x^2 + 0x \qquad \\ \underline{-2x^3 - x^2 - x} \qquad \\ x - 2 \quad \leftarrow \text{Remainder} \end{array}$$

Notice that the degree of the remainder is less than the degree of the divisor. ▬

We can write the dividend as the product of the divisor and the quotient plus the remainder.

$$\underset{\text{Dividend}}{} \quad = \quad \underset{\text{Divisor}}{} \times \underset{\text{Quotient}}{} + \underset{\text{Remainder}}{}$$

$$2x^4 - x^3 - 2 = (2x^2 + x + 1)(x^2 - x) + (x - 2)$$

Examples 1 and 2 illustrate the general process called the **division algorithm for polynomials**, which can be stated more formally as follows.

Division Algorithm for Polynomials

If $f(x)$ and $h(x)$ are polynomials, then there are polynomials $q(x)$ and $r(x)$, called the **quotient** and **remainder**, such that

$$f(x) = h(x)q(x) + r(x).$$

Either $r(x) = 0$ or the degree of $r(x)$ is less than the degree of $h(x)$.

Factors and Zeros

The division algorithm tells us that the degree of the remainder is less than the degree of the divisor. This means that whenever a given polynomial is divided by a linear polynomial (degree 1), the remainder must be a zero-degree polynomial, in other words, a constant.

So if $f(x)$ is divided by $x - c$, it is true for all values of x that

$$f(x) = (x - c)q(x) + r,$$

where r is a constant. In particular, for $x = c$,

$$f(c) = (c - c)q(c) + r$$

$$f(c) = r.$$

This proves the following theorem.

Theorem 3.3 Remainder Theorem

If a polynomial $f(x)$ is divided by $x - c$, then the remainder is $f(c)$. Thus

$$f(x) = (x - c)q(x) + f(c),$$

where $q(x)$ is the quotient.

E X A M P L E   3   Applying the Remainder Theorem

Find the remainder when the polynomial $f(x) = x^3 - 2x^2 + x - 5$ is divided by
(a) $x - 3$ and (b) $x + 1$.

Solution

a) $f(x) = x^3 - 2x^2 + x - 5$

$$f(3) = (3)^3 - 2(3)^2 + 3 - 5$$

The remainder theorem says that the remainder is $f(3)$ when $f(x)$ is divided by $x - 3$.

$$= 7$$

The remainder when $f(x)$ is divided by $x - 3$ is $f(3) = 7$.

b) $f(-1) = (-1)^3 - 2(-1)^2 + (-1) - 5$

The remainder theorem says that the remainder is $f(-1)$ when $f(x)$ is divided by $x + 1$.

$$= -9$$

The remainder when $f(x)$ is divided by $x + 1$ is $f(-1) = -9$. ≡

Suppose that a polynomial function $f(x)$ is divided by $x - c$. Then the remainder theorem says that for all values of x,

$$f(x) = (x - c)q(x) + f(c).$$

Suppose that $x - c$ is a factor of $f(x)$. Then the remainder is zero and $f(c) = 0$. This means that c is a zero of f.

Conversely, suppose that c is a zero of f. Then $f(c) = 0$ and by the remainder theorem, we can conclude that

$$f(x) = (x - c)q(x).$$

Thus $x - c$ is a factor. This discussion proves the following theorem.

Notes on Exercises

Ex. 1–8 address the understanding of the division algorithm, which is necessary to understand the factor theorem.

Theorem 3.4   Factor Theorem

Let $f(x)$ be a polynomial. Then $x - c$ is a factor of $f(x)$ if, and only if, c is a zero of $f(x)$.

E X A M P L E   4   Applying the Factor Theorem

Use the factor theorem to show that $x + 2$ is a factor of $x^3 + 5x^2 + 5x - 2$.

Solution   Show that -2 is a zero of $f(x) = x^3 + 5x^2 + 5x - 2$.

$$f(-2) = (-2)^3 + 5(-2)^2 + 5(-2) - 2$$
$$= -8 + (20) + (-10) - 2$$
$$= 0$$

So -2 is a zero, which means that $x - (-2) = x + 2$ is a factor.   ≡

There are many equivalent ways to formulate a statement using the terms *factor, root,* and *intercept.* These various formulations are based on equivalent statements about roots and zeros.

Equivalent Statements about Roots and Zeros

The following statements are equivalent for a polynomial $f(x)$ and a real number c:

- c is a solution to the equation $f(x) = 0$.
- c is a zero of $f(x)$.
- c is a root of $f(x)$.
- $x - c$ is a factor of $f(x)$.
- When $f(x)$ is divided by $x - c$, the remainder is 0.
- c is an x-intercept of the graph of $y = f(x)$.

These equivalent statements show that when you know a root of a polynomial function, you also know a factor. Conversely, when you know a factor, you also know a root.

E X A M P L E   5   Finding a Polynomial with Certain Zeros

Find a polynomial of degree 3 whose coefficients are real numbers and whose zeros are $-2, 1, 3$.

Solution

-2 is a zero of f $\Longleftrightarrow$ $x + 2$ is a factor of f.

1 is a zero of f $\Longleftrightarrow$ $x - 1$ is a factor of f.   These statements are true by the factor theorem.

3 is a zero of f $\Longleftrightarrow$ $x - 3$ is a factor of f.

Therefore $(x + 2)(x - 1)(x - 3) = x^3 - 2x^2 - 5x + 6$ is a polynomial of degree 3 with zeros $-2, 1,$ and 3.   ≡

When the factors of a polynomial are recognized or can otherwise be found, the factors can be used to find zeros of the polynomial algebraically. Example 6 illustrates this method.

E X A M P L E   6   Factoring to Find Zeros of a Polynomial

Use factoring to find the real zeros of $f(x) = x^3 + 2x^2 - 4x - 8$.

Solution   Factor by grouping the first two terms and the last two terms together.

$$x^3 + 2x^2 - 4x - 8 = x^2(x + 2) - 4(x + 2)$$

$$= (x + 2)(x^2 - 4)$$ Factor out the common factor $x + 2$.

$$- (x + 2)(x - 2)(x + 2)$$

$$= (x + 2)^2(x - 2)$$

The real zeros of f are 2 and -2.   ≡

Notes on Examples

Example 7 shows how a graph can be used to think about the factor and remainder theorems to identify zeros of functions. Ease of graphing gives practical power to finding the zeros of a function via the factor or remainder theorems. Give some attention to the fact that the discussion and examples in this section focus on the real zeros of the functions.

Although factoring can be used to find the zeros of a polynomial, it is often difficult to recognize the factors of a polynomial, particularly if the polynomial is of degree 3 or greater. Sometimes a grapher can be used to help find one zero and hence a corresponding linear factor.

E X A M P L E   7   Using a Grapher to Find Factors

Use the graph of $f(x) = 2x^3 - 4x^2 + x - 2$ to find the real zeros of $f(x)$.

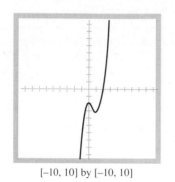

[–10, 10] by [–10, 10]

Figure 3.28
$f(x) = 2x^3 - 4x^2 + x - 2.$

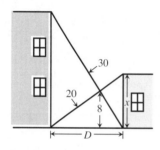

Figure 3.29   Diagram for Example 8.

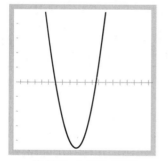

[0, 20] by [–5000, 5000]

Figure 3.30 $f(x) = x^4 -$
$16x^3 + 500x^2 - 8000x + 32,000.$
Knowing that the y-intercept is
32,000 and that $y = x^4$ is a model
of end behavior are clues that a
large viewing rectangle is needed.

Solution   The graph shown in Fig. 3.28 suggests that f has a zero at $x = 2$. Indeed,

$$f(2) = 2(2^3) - 4(2^2) + 2 - 2 = 0.$$ Conclude from the factor theorem that $x - 2$ is a factor.

To find the other factor, complete the following long division.

$$
\begin{array}{r}
2x^2 + 1 \\
x - 2 \overline{\smash{)}\ 2x^3 - 4x^2 + x - 2} \\
\underline{2x^3 - 4x^2} \\
x - 2 \\
\underline{x - 2} \\
0
\end{array}
$$

Thus $f(x) = (x - 2)(2x^2 + 1)$. From this, conclude that 2 is the only real zero of f. (Why?)   ▤

Even with the help of a graph to suggest roots, in most cases the zeros of a polynomial function cannot be found by algebraic means. In that case, the zoom-in method is used to find a numerical root.

To illustrate this situation, consider the following application.

E X A M P L E   8   APPLICATION: Crossing Ladders

Suppose two ladders, one 20 ft long and one 30 ft long, are placed between two buildings as shown in Fig. 3.29. How far apart are the buildings if the ladders cross at a height of 8 ft?

Solution   In Exercise 40, you will verify that the value of x in Fig. 3.29 satisfies the equation $x^4 - 16x^3 + 500x^2 - 8000x + 32,000 = 0$. Once x is known, D can be found because $20^2 = D^2 + x^2$.

To find a solution to this fourth-degree equation, find a complete graph of $f(x) = x^4 - 16x^3 + 500x^2 - 8000x + 32,000$ (see Fig. 3.30).

There is one solution near 6 and another near 12. Use zoom-in to find that the zeros are 5.945 and 11.712 with an error of at most 0.01.

Studying Fig. 3.29 shows that in this problem situation, the value of x must be between 8 and 20. (Why?) So let $x = 11.712$ in the equation $20^2 = D^2 + x^2$.

Then

$$D^2 = 20^2 - (11.712)^2$$

$$D = \sqrt{(20)^2 - (11.712)^2}$$

$$= 16.212.$$

Assignment Guide

Day 1: Ex. 1, 2, 4, 5, 7, 8, 10, 11, 13, 16, 18, 19, 22
Day 2: Ex. 24, 25, 28, 31, 32, 38–40

The buildings are 16.21 ft apart.

Exercises for Section 3.3

In Exercises 1–4, find the quotient $q(x)$ and remainder $r(x)$ when $f(x)$ is divided by $h(x)$. Then compute $q(x)h(x) + r(x)$ and compare the result with $f(x)$.

1. $f(x) = x^2 - 2x + 3$; $h(x) = x - 1$

2. $f(x) = x^3 - 1$; $h(x) = x + 1$

3. $f(x) = 4x^3 - 8x^2 + 2x - 1$; $h(x) = 2x + 1$

4. $f(x) = x^4 - 2x^3 + 3x^2 - 4x + 6$; $h(x) = x^2 + 2x - 1$

In Exercises 5–8, use the remainder theorem to determine the remainder when $f(x)$ is divided by $x - c$. Check by using long division.

5. $f(x) = 2x^2 - 3x + 1$; $c = 2$

6. $f(x) = x^4 - 5$; $c = 1$

7. $f(x) = 2x^3 - 3x^2 + 4x - 7$; $c = 2$

8. $f(x) = x^5 - 2x^4 + 3x^2 - 20x + 3$; $c = -1$

In Exercises 9–12, use the factor theorem to determine whether the first polynomial is a factor of the second polynomial.

9. $x + 2$; $x^2 - 4$ **10.** $x - 1$; $x^3 - x^2 + x - 1$

11. $x - 3$; $x^3 - x^2 - x - 15$

12. $x + 1$; $2x^{10} - x^9 + x^8 + x^7 + 2x^6 - 3$

In Exercises 13–18, find all real zeros of each polynomial by factoring. Use a graphing utility to support your answer.

13. $f(x) = x^2 - 5x + 6$ **14.** $f(x) = 6x^2 + 8x - 8$

15. $f(x) = x^3 - 9x$ **16.** $g(x) = x^3 - 1$

17. $T(x) = 2x^4 + x^2 - 15$ **18.** $g(x) = x^3 - x^2 - 2x + 2$

In Exercises 19–22, draw a complete graph of each function. Use the graph as an aid in factoring the polynomial, and then determine all real zeros.

19. $f(x) = 5x^2 - 2x - 51$

20. $f(x) = x^3 - 2x^2 - 3x + 6$

21. $f(x) = x^3 - 11x^2 + x - 11$

22. $g(x) = x^4 - 16$

In Exercises 23–26, determine all real zeros.

23. $f(x) = x^3 + 3x^2 - 10x - 1$

24. $g(x) = 2 - 15x + 11x^2 - 3x^3$

25. $f(x) = 100x^3 - 403x^2 + 406x + 1$

26. $f(x) = 100x^3 - 403x^2 + 406x - 1$

27. A toy rocket is shot straight up into the air with an initial velocity of 48 ft/sec. Its height $s(t)$ above the ground after t seconds is given by $s(t) = -16t^2 + 48t$. What is the maximum height attained by the rocket?

28. An object is shot straight up from the top of a 260-ft tower with an initial velocity of 35 ft/sec.
   a) When will the object hit the ground?
   b) When does the object reach its maximum height above the ground, and what is this maximum height?

29. Draw the graph of $f(x) = 100x^3 - 203x^2 + 103x - 1$ in the $[-5, 5]$ by $[-100, 100]$ viewing rectangle.
   a) How many real zeros are evident from this graph?
   b) How many actual real zeros exist in this case?
   c) Find all real zeros.

In Exercises 30–33, find a polynomial with real coefficients satisfying the given conditions.

30. Degree 2, with 3 and -4 as zeros.

31. Degree 2, with 2 as the only real zero.

32. Degree 3, with -2, 1, and 4 as zeros.

33. Degree 3, with -1 as the only real zero.

Exercises 34 and 35 refer to the following **problem situation** NICE-CALC Company manufactures the only calculator with the ZAP feature. Use the monthly supply function $P = S(x)$ and monthly demand function $P = D(x)$ to find the equilibrium price (break-even price) and the associated production level needed to achieve equilibrium.

34. $S(x) = 6 + 0.001x^3$ and x is in thousands; $D(x) = 80 - 0.02x^2$ and x is in thousands.

35. $S(x) = 20 - 0.1x + 0.00007x^4$ and x is in thousands; $D(x) = 150 - 0.004x^3$ and x is in thousands.

36. If $f(x) = 2x^3 - 3kx^2 + kx - 1$, find a number k so that the graph of f contains the point $(1, 9)$.

37. Determine a linear factor of $f(x) = (x - 1)^6 - 64$.

38. Determine the remainder when
a) $x^{40} - 3$ is divided by $x + 1$;
b) $x^{63} - 17$ is divided by $x - 1$.

39. The hypotenuse of a right triangle is 2 in. longer than one of its legs and the triangle has an area of 50 sq in.
a) Show that if x denotes the length of this leg, then

$$10{,}000 - 4x^3 - 4x^2 = 0.$$

b) Determine the solutions to the equation in part (a).

40. Challenge Derive the fourth-degree polynomial that gives the solution to the ladder problem of Example 8. (*Hint:* Consider the following figure. Notice that $x^2 + z^2 = 20^2$ and $y^2 + z^2 = 30^2$. Write one equation that involves only x and y. Now show that the length of $AF = 8z/x$ and the length of $BF = 8z/y$. (Supply only the reasons.) Next, use the fact that the length of AF plus the length of BF is equal to z to determine a second equation in terms of x and y only. Solve the second equation for y and substitute it into the first equation.)

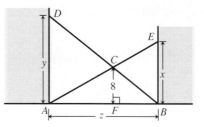

For Exercise 40, $AE = 20$, $DB = 30$.

3.4 _____ Rational Zeros and Horner's Algorithm

Objective

Students will be able to use the rational zeros theorem and Horner's algorithm to explore the upper and lower bounds of the real zeros of a function and to write the linear factors of a polynomial function.

Some polynomial functions have zeros that are irrational numbers. For example,

$$f(x) = x^2 - 2 = (x - \sqrt{2})(x + \sqrt{2}).$$

The factor theorem shows that the function $f(x) = x^2 - 2$ has zeros $\sqrt{2}$ and $-\sqrt{2}$. Because these numbers are irrational, the function f is said to have **irrational zeros**. On the other hand, the function $g(x) = 4x^2 - 9$ has zeros $\frac{3}{2}$ and $-\frac{3}{2}$. Because these numbers are rational, g has **rational zeros**.

A polynomial function can have rational or irrational zeros or it can have some of each. In this section, we learn some techniques that can be used to find rational zeros if they exist.

We begin with a theorem that tells what the possible rational zeros are for polynomial functions whose coefficients are integers.

Teaching Note

The ideas of Section 3.3 are revisited in this section to consolidate the relationship among roots, zeros of a function, and graphs. However, the ideas are extended via the rational zeros theorem. This theorem provides the means of dealing with the leading coefficient of a polynomial that is other than one. If students are progressing easily through this material, you may want to develop a proof of the theorem for purposes of enrichment.

Common Errors

The point of the rational zeros theorem is to be able to make a prediction of what the zeros are by examining the coefficients. It does require that all coefficients be integers. This theorem is powerful for examining roots. Students should be cautioned about using the rational zeros theorem with a polynomial function that may have noninteger coefficients.

Notes on Examples

Example 2 shows how the rational zeros theorem can be used to determine the character of roots that are not rational.

> **Theorem 3.5   Rational Zeros Theorem**
>
> Suppose all the coefficients in the polynomial
> $$f(x) = a_n x^n + a_{n-1} x^{n-1} + \cdots + a_0 \qquad (a_n \neq 0, a_0 \neq 0)$$
> are integers.
>
> If $x = p/q$ is a rational zero, where p and q have no common factors, then
>
> - p is a factor of the constant term a_0, and
> - q is a factor of the leading coefficient a_n.

E X A M P L E   1   Finding All the Rational Zeros

Make a complete list of possible rational number zeros of $f(x) = 10x^5 - 3x^2 + x - 6$.

Solution   Suppose p/q is a rational zero. Then

$$p: \pm 1, \pm 2, \pm 3, \pm 6 \qquad \text{\small p must be a factor of the constant term -6.}$$

$$q: \pm 1, \pm 2, \pm 5, \pm 10. \qquad \text{\small q must be a factor of the leading coefficient 10.}$$

To list all possible rational zeros, choose a numerator from the p list and a denominator from the q list and write the rational number in reduced form. The complete list is

$$\frac{p}{q}: \pm 1, \pm 2, \pm 3, \pm 6, \pm \frac{1}{2}, \pm \frac{1}{5}, \pm \frac{1}{10}, \pm \frac{2}{5}, \pm \frac{3}{2}, \pm \frac{3}{5}, \pm \frac{3}{10}, \pm \frac{6}{5}. \qquad \blacksquare$$

The complete list in Example 1 includes 24 rational numbers. To verify algebraically which ones, if any, are zeros you could use one of these techniques:

1. Calculate $f(p/q)$ in each case to see whether the remainder is zero.
2. Divide $10x^5 - 3x^2 + x - 6$ by $(x - p/q)$ in each case to see whether the remainder is zero.

Either method is time consuming and tedious, even if a calculator is used. Once again, the work can be shortened by using a graphing utility.

E X A M P L E   2   Using a Grapher to Support an
Algebraic Approach

Show that $f(x) = 10x^5 - 3x^2 + x - 6$ has exactly one real number root and that it is an irrational number.

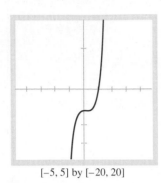

[−5, 5] by [−20, 20]

Figure 3.31
$f(x) = 10x^5 - 3x^2 + x - 6.$

Solution Find a complete graph of f and observe that there is only one real number zero and that it appears to be 1 (see Fig. 3.31).

$$f(1) = 10(1)^5 - 3(1)^2 + 1 - 6 = 2 \neq 0 \qquad \text{Conclude from the remainder theorem that } x = 1 \text{ is not a zero.}$$

Use zoom-in to observe that the zero is approximately $x = 0.95$. Using information from Example 1, notice that the nearest rational zero less than 1 is $x = 3/5$. Then

$$f\left(\frac{3}{5}\right) = 10\left(\frac{3}{5}\right)^5 - 3\left(\frac{3}{5}\right)^2 + \frac{3}{5} - 6 \neq 0. \qquad \text{Conclude from the remainder theorem that } x = 3/5 \text{ is not a zero.}$$

Thus there are no rational zeros and there is one irrational zero. ▤

Because a grapher cannot usually distinguish between rational and irrational zeros, it is often necessary to use algebraic methods if you specifically want to find the rational zeros.

Horner's Algorithm (Synthetic Division)

Suppose you want to evaluate $f(x) = 4x^3 - 3x^2 + x - 4$ for $x = 2$. It appears that 2 must be raised to powers since

$$f(2) = 4(2)^3 - 3(2)^2 + 2 - 4.$$

However, Horner's algorithm provides a method of evaluating polynomials that uses only multiplication and addition. Consider this expression:

$$f(x) = 4x^3 - 3x^2 + x - 4$$
$$= x[4x^2 - 3x + 1] - 4$$
$$= x[x(4x - 3) + 1] - 4.$$

The basis for Horner's algorithm is the last line of nested parentheses. For $f(2)$ this expression becomes

6. Add −4
4. Add 1
2. Add −3
↓

$$f(2) = 2[2(4 \cdot 2 - 3) + 1] - 4.$$

1. Multiply by 4
3. Multiply by 2
5. Multiply by 2

We evaluate from the innermost set of parentheses and work towards the outside. Multiply 4×2, add −3, multiply by 2, add 1, multiply by 2, and add −4.

This computation can be arranged in a 3-row format as follows. List the coefficients in the first row. Next, obtain each number in the second row by multiplying the entry in row 3 of the previous column by 2. Last, obtain each number in the third row by adding the corresponding entries in rows 1 and 2. This procedure is known as **Horner's algorithm**.*

$$(x = 2) \quad \underline{2} \begin{array}{rrrr} 4 & -3 & 1 & -4 \\ & 8 & 10 & 22 \\ \hline 4 & 5 & 11 & 18 \end{array} \qquad \longleftarrow \text{ Coefficients of } f(x)$$

Coefficients of quotient

$f(2)$

Notice that the last entry in the last row is precisely $f(2)$. So $f(2) = 18$. It can be shown that the remaining numbers in row 3 are the coefficients of the polynomial that results when $4x^3 - 3x^2 + x - 4$ is divided by $x - 2$. In other words,

$$4x^3 - 3x^2 + x - 4 = (x - 2)(4x^2 + 5x + 11) + 18.$$

If $x = 2$ were a zero of this polynomial, then the last entry in row 3 would be zero and a factorization would have been obtained.

Stan says, "Possible rational zeros may be directly substituted into a polynomial function using a graphing utility. For Example 3, one could store $\frac{3}{2}$ or 1.5 in X on the blank screen. If the function is defined on the Y screen as Y1, then by pressing Y1 and ENTER, the remainder will be displayed on the screen. A remainder of zero would indicate that $\frac{3}{2}$ is a zero of the function."

E X A M P L E   3   Using Horner's Algorithm

Show that 3/2 is a zero of $f(x) = 2x^3 - 5x^2 + x + 3$, and write $2x^3 - 5x^2 + x + 3$ in factored form.

Solution   First use Horner's algorithm.

$$\frac{3}{2} \begin{array}{rrrr} 2 & -5 & 1 & 3 \\ & 3 & -3 & -3 \\ \hline 2 & -2 & -2 & 0 \end{array}$$

Because the last entry in the last row is 0, $f(\frac{3}{2}) = 0$ and $\frac{3}{2}$ is thus a zero.

Also, $2x^2 - 2x - 2$ must be a factor, so

$$2x^3 - 5x^2 + x + 3 = \left(x - \tfrac{3}{2}\right)(2x^2 - 2x - 2)$$

$$= \left(x - \tfrac{3}{2}\right)2(x^2 - x - 1)$$

$$= (2x - 3)(x^2 - x - 1). \qquad \blacksquare$$

---

* William Horner, 1786–1837, was an English mathematician who developed this method for finding zeros of polynomials. This procedure is often referred to in algebra books as *synthetic division* or *synthetic substitution*. However, in recent years computer scientists have revived this method and use it as the basis of a recursive algorithm.

We can find the remaining zeros of f by applying the quadratic formula to $x^2 - x - 1$ to obtain the irrational zeros

$$x = \frac{1 + \sqrt{5}}{2}, \qquad x = \frac{1 - \sqrt{5}}{2}.$$

Notice that the factor theorem tells us that $x^2 - x - 1$ can be factored as

$$x^2 - x - 1 = \left(x - \frac{1 + \sqrt{5}}{2}\right)\left(x - \frac{1 - \sqrt{5}}{2}\right).$$

The complete factorization of $2x^3 - 5x^2 + x + 3$ is

$$2x^3 - 5x^2 + x + 3 = (2x - 3)\left(x - \frac{1 + \sqrt{5}}{2}\right)\left(x - \frac{1 - \sqrt{5}}{2}\right).$$

However, we will not ordinarily factor expressions into linear terms that involve irrational numbers. The exception is when we specifically want to display the irrational zeros.

Teaching Note

Students may also wish to program their graphing utility to do synthetic division using Horner's algorithm. The program should allow the student to indicate the degree of the polynomial and then input the coefficients. The program should output the coefficients of the reduced polynomial and the remainder.

E X A M P L E   4   Using Horner's Algorithm (Continued)

Use Horner's algorithm to find the quotient and remainder when $3x^4 + 7x^3 + x - 11$ is divided by $x + 3$.

Solution

$$
\begin{array}{r|rrrrr}
-3 & 3 & 7 & 0 & 1 & -11 \\
   &   & -9 & 6 & -18 & 51 \\
\hline
   & 3 & -2 & 6 & -17 & 40 \\
\end{array}
$$

$$\underset{3x^3 \quad -2x^2 + \; 6x \quad -17}{\downarrow \qquad \downarrow \qquad \downarrow \qquad \downarrow}$$

The quotient is $3x^3 - 2x^2 + 6x - 17$, and the remainder is 40. ☰

E X A M P L E   5   Using Horner's Algorithm (Continued)

Use Horner's algorithm to show that 3 is *not* a zero of $f(x) = 2x^3 + 5x^2 - 2x - 3$.

Solution

$$
\begin{array}{r|rrrr}
3 & 2 & 5 & -2 & -3 \\
  &   & 6 & 33 & 93 \\
\hline
  & 2 & 11 & 31 & 90 \\
\end{array}
$$

Because $f(3) = 90$, 3 is not a zero of f. ☰

Finding Zeros of a Polynomial Function

By combining the methods of graphing, the rational zeros theorem, Horner's algorithm, and factoring, we can find the rational zeros of polynomials. We summarize the procedure.

Strategy for Finding Zeros

Given a polynomial function

$$f(x) = a_n x^n + \cdots + a_0 \qquad (a_n \neq 0)$$

with integer coefficients, the real zeros can be found by following these steps:

1. List all potential rational zeros by using the rational zeros theorem.
2. Find a complete graph of f with a graphing utility to decide which rational numbers in this list are most likely to be zeros. Use Horner's algorithm and the remainder theorem to determine whether any of these rational numbers is a zero.

   a) If one is a zero, the numbers in the last row of Horner's algorithm are the coefficients of the degree $(n-1)$ factor. Repeat Steps 1 and 2 with that factor.

   b) If no rational zero is found, use the graphing utility to find zeros to whatever degree of precision is desired.

E X A M P L E   6   Finding All Rational Zeros

Find all rational zeros of $f(x) = x^3 - 3x - 2$ and find a complete graph.

Solution   (See Fig. 3.32.)

$$p: \ \pm 1, \pm 2 \qquad p \text{ must be a factor of } -2.$$

$$q: \ \pm 1 \qquad\quad q \text{ must be a factor of } 1.$$

The only possible rational zeros are ± 1 and ± 2. The graph in Fig. 3.32 suggests that -1 and 2 are the most likely zeros. Check -1 using Horner's algorithm.

$$
\begin{array}{r|rrrr}
-1 & 1 & 0 & -3 & -2 \\
   &   & -1 & 1 & 2 \\
\hline
   & 1 & -1 & -2 & 0
\end{array}
$$

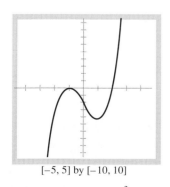

[−5, 5] by [−10, 10]

Figure 3.32 $f(x) = x^3 - 3x - 2.$

Note that $x^2 - x - 2$ is a factor of f. Therefore

$$x^3 - 3x - 2 = (x + 1)(x^2 - x - 2)$$
$$= (x + 1)(x + 1)(x - 2)$$
$$= (x + 1)^2(x - 2).$$

The zeros of f are -1 and 2. ▤

E X A M P L E 7 Finding All Rational Zeros (Continued)

Find all rational zeros of $f(x) = 6x^3 - 5x^2 + 3x - 1$.

Solution

$$p: \pm 1$$

$$q: \pm 1, \pm 2, \pm 3, \pm 6$$

The only possible rational zeros are $\pm 1, \pm \frac{1}{2}, \pm \frac{1}{3}, \pm \frac{1}{6}$. The graph of f in Fig. 3.33 suggests that we check $\frac{1}{2}$.

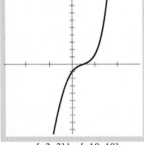

[−3, 3] by [−10, 10]

Figure 3.33
$f(x) = 6x^3 - 5x^2 + 3x - 1.$

$$\begin{array}{r|rrrr} \frac{1}{2} & 6 & -5 & 3 & -1 \\ & & 3 & -1 & 1 \\ \hline & 6 & -2 & 2 & 0 \end{array}$$

$$6x^3 - 5x^2 + 3x - 1 = (x - \tfrac{1}{2})(6x^2 - 2x + 2)$$
$$= (x - \tfrac{1}{2})(2)(3x^2 - x + 1)$$
$$= (2x - 1)(3x^2 - x + 1)$$

Check the discriminant $b^2 - 4ac$ for $(3x^2 - x + 1)$:

$$b^2 - 4ac = (-1)^2 - 4(3)(1) = -11.$$

Because the discriminant is less than zero, $3x^2 - x + 1$ does not have real zeros, and $\frac{1}{2}$ is the only real zero for f. ▤

E X A M P L E 8 Finding and Classifying Zeros

Find all real number zeros of $f(x) = 3x^3 - 8x^2 + x + 2$. Classify them as integer, noninteger, rational, or irrational.

Solution The possible rational zeros are as follows:

$$\frac{p}{q}: \pm \frac{1}{3}, \pm \frac{2}{3}, \pm 1, \pm 2.$$

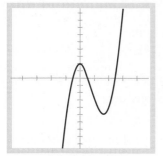

[−5, 5] by [−10, 10]

Figure 3.34
$f(x) = 3x^3 - 8x^2 + x + 2.$

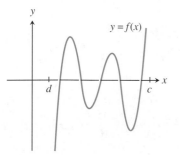

Figure 3.35 c is an upper bound and d is a lower bound for the set of zeros.

The graph in Fig. 3.34 suggests that there are three real zeros, one close to $\frac{2}{3}$; use Horner's algorithm to confirm this.

$$\begin{array}{c|cccc} \frac{2}{3} & 3 & -8 & 1 & 2 \\ & & 2 & -4 & -2 \\ \hline & 3 & -6 & -3 & 0 \end{array}$$

Therefore

$$3x^3 - 8x^2 + x + 2 = (x - \frac{2}{3})(3x^2 - 6x - 3)$$

$$= (3x - 2)(x^2 - 2x - 1)$$

Use the quadratic formula to find the zeros of $x^2 - 2x - 1$.

$$x = \frac{2 \pm \sqrt{(-2)^2 - 4(-1)}}{2} \quad \text{or} \quad 1 \pm \sqrt{2}.$$

Thus $\frac{2}{3}$ is a noninteger rational zero and $1 + \sqrt{2}$ and $1 - \sqrt{2}$ are irrational zeros.

Upper and Lower Bounds for Real Zeros

A number c is an **upper bound** for the set of real zeros of a function f if $f(x) \neq 0$ whenever $x > c$. Likewise, a number d is a **lower bound** for the set of real zeros of f if $f(x) \neq 0$ whenever $x < d$ (see Fig. 3.35).

Horner's algorithm can be used to find upper and lower bounds as described in the next two theorems.

Theorem 3.6 Upper Bound for Zeros

Let f be a polynomial with a $\left\{ \begin{array}{c} \text{positive} \\ \text{negative} \end{array} \right\}$ leading coefficient, and let c be a positive number. If the last line of Horner's algorithm contains no $\left\{ \begin{array}{c} \text{negative} \\ \text{positive} \end{array} \right\}$ numbers using c as a potential zero, then c is an upper bound.

EXAMPLE 9 Checking for Upper Bounds

Show that 3 is an upper bound for the real number zeros of $f(x) = x^3 - 3x^2 + x - 1$.

Solution

$$\begin{array}{c|cccc} 3 & 1 & -3 & 1 & -1 \\ & & 3 & 0 & 3 \\ \hline & 1 & 0 & 1 & 2 \end{array}$$

Teaching Note

The idea of an upper or lower bound is important and provides an equivalent means of analyzing limit behavior of functions in the study of calculus. The discussion of upper bound should focus students' thinking on what happens to the $f(x)$ values in terms of changes in the x-values. Students need to work through the feature of the sign of the leading coefficient and the results of dividing synthetically by positive or negative numbers.

Notes on Examples

Example 9 uses synthetic division to identify the zero of the function. Caution students against becoming so preoccupied with the technique of synthetic division that they miss the reasoning associated with the bounded behavior. Visually representing the curve on an overhead projector should enhance the discussion of the bound definition and provide a means for stating results in terms of an open interval.

The last line of Horner's algorithm has no negative numbers, so 3 is an upper bound.

Here's why. Horner's algorithm gives the factorization

$$f(x) = x^3 - 3x^2 + x - 1 = (x - 3)(x^2 + 1) + 2.$$

Note that $x^2 + 1$ is a positive number for each value of x. So, when $x > 3$, then $x - 3 > 0$ and $f(x) > 2$. Therefore f has no zero greater than 3. ▰

To find lower bounds, we convert the problem to one of finding upper bounds. Figure 3.36 shows the graph of a function f and its reflection about the y-axis. Notice that a negative number c is a lower bound for the set of zeros of $f(x)$ if the positive number $-c$ is an upper bound for the zeros of the function $f(-x)$. This observation leads to Theorem 3.7.

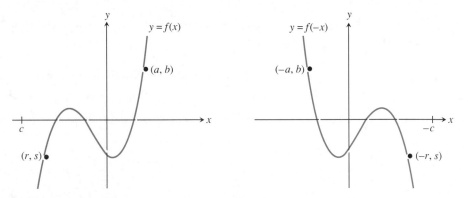

Figure 3.36   A function $f(x)$ and the related function $f(-x)$. These graphs are related by symmetry with respect to the y-axis. Notice that whenever (a, b) is on the graph of $f(x)$, $(-a, b)$ is on the graph of $f(-x)$.

FINDING THE VIEWING RECTANGLE

When finding a complete graph of a polynomial, you must choose a viewing rectangle that includes all the zeros. If c is an upper bound and d is a lower bound for zeros, then you know to choose a viewing rectangle so that $X\min < d < c < X\max$.

Theorem 3.7   Lower Bound for Zeros

The negative number d is a lower bound for the real zeros of $f(x)$ if, and only if, the positive number $-d$ is an upper bound for the real zeros of $f(-x)$.

E X A M P L E   10   Checking for Lower Bounds

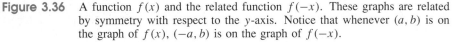

Show that -6 is a lower bound for $f(x) = x^3 + 2x^2 - 21x + 16$.

Solution According to Theorem 3.7, this problem is equivalent to showing that 6 is an upper bound for $f(-x)$.

$$f(-x) = (-x)^3 + 2(-x)^2 - 21(-x) + 16$$

$$= -x^3 + 2x^2 + 21x + 16$$

$$
\begin{array}{r|rrrr}
6 & -1 & 2 & 21 & 16 \\
 & & -6 & -24 & -18 \\
\hline
 & -1 & -4 & -3 & -2
\end{array}
$$

Because all the numbers in the third row are negative, we see that 6 must be an upper bound for $f(-x)$. Therefore -6 is a lower bound for $f(x)$. ≡

After using Theorems 3.6 and 3.7 to find upper and lower bounds for the zeros of a polynomial, you can gain information about the number of positive and negative real zeros for $f(x) = 0$ by counting the number of sign variations of $f(x)$ and $f(-x)$. Consider $f(x) = x^3 - 3x^2 + 4x - 5$, which has 3 sign variations, whereas $f(-x) = -x^3 - 3x^2 - 4x - 5$ has zero sign variations.

Theorem 3.8   Descartes's Rule of Signs

Let $f(x)$ be a polynomial with real coefficients and a nonzero constant term. If the polynomial is arranged with descending powers of the variable, then:

1. the number of positive real zeros of $f(x) = 0$ equals the number of variations in sign of the terms of $f(x)$, or is less than this number by an even integer; and
2. the number of negative real zeros of $f(x) = 0$ equals the number of variations in sign of the terms of $f(-x)$, or is less than this number by an even integer.

E X A M P L E   11   Finding the Number of Zeros

Find the number of positive and negative zeros of $f(x) = x^3 + 3x + 5$.

Solution

$$f(x) = x^3 + 3x + 5 \qquad$$ There is a positive sign between each term.
There are zero sign variations.

$$f(-x) = -x^3 - 3x + 5 \qquad$$ There is one sign variation.

There can be no positive zeros and at most one negative zero. We know by end behavior characteristics that there is at least one zero. So there is exactly one zero and it is negative. This result can be supported with a graph. ▤

We end this section with an application.

E X A M P L E   12   APPLICATION: Finding the Sink Depth of a Floating Buoy

Suppose a spherical floating buoy has a radius of 1 m and a density $\frac{1}{4}$ that of sea water. Find the depth that the buoy sinks in sea water (see Fig. 3.37).

Solution   By Archimedes' law, the volume of the displaced water is equal to the volume of the submerged portion of the buoy. Choose variables:

$x =$ depth that the buoy sinks
$r =$ radius of the submerged spherical segment
$V =$ volume of the submerged spherical segment

Then

$$V = \frac{\pi}{6}x(3r^2 + x^2) \qquad \text{This formula comes from solid geometry.}$$

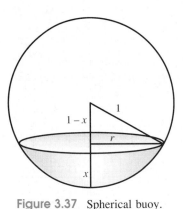

Figure 3.37 Spherical buoy.

$$(1 - x)^2 + r^2 = 1. \qquad \begin{array}{l}\text{Apply the Pythagorean theorem to obtain}\\ \text{a relationship between } x \text{ and } r.\end{array}$$

Solve for r^2 and substitute into the volume formula to obtain

$$V = \frac{\pi}{6}x(3r^2 + x^2)$$

$$= \frac{\pi}{6}x[3(1 - (1 - x)^2) + x^2]$$

$$= \frac{\pi}{3}(3x^2 - x^3).$$

On the other hand, since the density of the buoy is $\frac{1}{4}$ that of water, $\frac{1}{4}$ of the volume of the buoy is submerged. The volume of this sphere is $4\pi/3$ (since the volume of a sphere is $4\pi r^3/3$). Therefore the volume of the submerged spherical segment is $\pi/3$.

Setting these two volumes equal to each other, we obtain

$$\frac{\pi}{3} = \frac{\pi}{3}(3x^2 - x^3)$$

$$1 = 3x^2 - x^3$$

$$0 = x^3 - 3x^2 + 1.$$

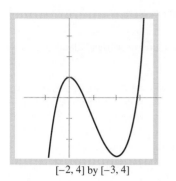

$[-2, 4]$ by $[-3, 4]$

Figure 3.38
$f(x) = x^3 - 3x^2 + 1$.

The problem will be solved by finding a zero for $f(x) = x^3 - 3x^2 + 1$ that is between 0 and 1.

Exercise 60 asks you to show that there are no rational zeros to this function, so a graphical method must be used.

Figure 3.38 shows a complete graph of f. Notice that $x < 1$. Using zoom-in, show that $x = 0.653$ is a solution with an error of less than 0.01. The buoy sinks 0.653 m into the sea. ≡

Exercises for Section 3.4

In Exercises 1–4, use the rational zeros theorem to list all possible rational zeros of each function.

1. $f(x) = x^3 - 2x^2 + 3x - 4$

2. $g(x) = 2x^3 - x + 1$

3. $f(x) = 2x^3 - 4x^2 - 5x + 10$

4. $g(x) = 4x^3 - x^2 + 2x - 6$

In Exercises 5–8, show that each function has no rational zeros.

5. $f(x) = x^3 + 2x^2 + x - 1$

6. $f(x) = x^3 - x + 2$

7. $f(x) = 2x^4 - x^3 + 1$

8. $f(x) = 3x^3 + x^2 - 2x + 1$

In Exercises 9–12, use Horner's algorithm to show that the indicated number is a zero of $f(x)$. Then factor f.

9. $f(x) = x^3 + x^2 - 10x + 8$; 2

10. $f(x) = x^4 + 3x^3 - 4x^2 - 11x + 3$; -3

11. $f(x) = 2x^3 + x^2 - 6x - 3$; $-\frac{1}{2}$

12. $f(x) = 3x^3 - 7x^2 + 10x - 8$; $\frac{4}{3}$

In Exercises 13–16, use Horner's algorithm to find the quotient and remainder when the first polynomial is divided by the second polynomial.

13. $x^3 + 2x^2 - 3x + 1$; $x + 2$

14. $-2x^3 - x^2 - 4$; $x - 1$

15. $x^4 - 2x^3 + x^2 - x + 2$; $x + 1$

16. $2x^4 - 3x^2 + x + 5$; $x + 3$

In Exercises 17–20, find all rational zeros of each function using an algebraic method.

17. $f(x) = x^2 - 3x + 4$

18. $g(x) = x^3 - 4x$

19. $g(x) = x^3 - 5x$

20. $f(x) = x^3 + x$

In Exercises 21–24, draw a complete graph of each polynomial. Confirm that there is at least one rational root in each case. Use Horner's algorithm to find the other quadratic factor and then determine all real zeros. Classify each real-number zero as rational or irrational.

21. $f(x) = x^3 + 4x^2 - 4x - 1$

22. $f(x) = 6x^3 - 5x - 1$

23. $f(x) = 3x^3 - 7x^2 + 6x - 14$

24. $f(x) = 2x^3 - x^2 - 9x + 9$

In Exercises 25–30, determine the real-number zeros of each polynomial and classify each as rational or irrational. Carefully outline why you are sure of your results.

25. $f(x) = 2x^4 - 7x^3 - 2x^2 - 7x - 4$

26. $f(x) = 3x^4 - 2x^3 + 3x^2 + x - 2$

27. $f(x) = 2x^3 - x^2 - 18x + 9$

28. $f(x) = 3x^3 - x^2 + 27x - 9$

29. $f(x) = x^4 + x^3 - 3x^2 - 4x - 4$

30. $f(x) = x^4 + 3x^2 + 2$

In Exercises 31–34, use Horner's algorithm to find upper and lower bounds for the real-number zeros of each polynomial function.

31. $f(x) = x^2 - 2x - 5$

32. $f(x) = x^3 - x^2 + 2x - 5$

33. $g(x) = x^3 + 2x^2 - 3x - 1$

34. $f(x) = x^4 + 2x^3 - 7x^2 - 8x + 12$

35. Does $f(x) = 3x^4 - 2x^2 + x - 1$ have any rational zeros? Any irrational zeros? If so, how many of each? Give both graphical and algebraic reasons for your answer.

36. The function $f(x) = 6x^3 + x^2 - 3x + 12$ has 24 possible rational roots. How many of them can be eliminated as candidates by showing that 1 is an upper bound of all the zeros?

In Exercises 37–40, use Descartes's Rule of Signs to algebraically determine the number of *possible* positive real and negative real zeros for each polynomial. Use a graphing utility to determine the *exact* number of positive and negative real zeros.

37. $f(x) = x^3 + x^2 - 1$

38. $f(x) = x^3 - x^2 - x - 1$

39. $f(x) = 2x^3 - x^2 + x - 1$

40. $f(x) = x^3 + x^2 + x + 1$

Exercises 41–43 refer to the following **problem situation** A 172-ft-long steel beam is anchored at one end of a piling 20 ft above the ground. It is known that the steel beam bends s feet (vertical distance) if a 200-lb object is placed d feet from the anchored end and that s is given by

$$s = (3 \times 10^{-7})d^2(550 - d).$$

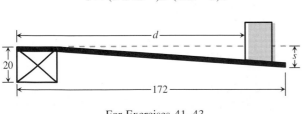

For Exercises 41–43.

41. Draw a complete graph of the algebraic representation of this problem situation. What is the domain of this model of the problem? What values of d make sense in this problem situation?

42. How far is the 200-lb object from the anchored end if the vertical deflection is 1.25 ft?

43. What is the greatest amount of vertical deflection, and where does it occur?

Exercises 44–47 refer to the following **problem situation** Biologists have determined that the polynomial function $P(t) = -0.00001t^3 + 0.002t^2 + 1.5t + 100$ approximates the population t days later of a certain group of wild turkeys left to reproduce on their own with no predators.

44. Draw a complete graph of the algebraic model $y = P(t)$ of this problem situation.

45. Find the maximum turkey population and when it occurs.

46. When will this turkey population be extinct?

47. Create a scenario that could explain the "growth" exhibited by this turkey population.

Exercises 48–53 refer to the complete graph of $y = f(x)$ given in the following figure. Draw a complete graph of each function and describe the transformations needed to obtain its graph from the graph of $y = f(x)$.

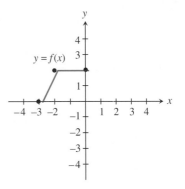

For Exercises 48–53.

48. $y = 1 + 2f(x)$ **49.** $y = -f(x)$

50. $y = -\frac{1}{2}f(x)$ **51.** $y = -1 + 2f(-x)$

52. $y = -2f(x - 2)$ **53.** $y = 2 - f(x + 1)$

Exercises 54–59 refer to the complete graph of $y = f(x)$ given in the following figure. Draw a complete graph of each function and describe the transformations needed to obtain its graph from the graph of $y = f(x)$.

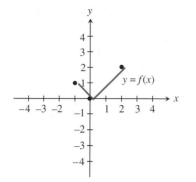

For Exercises 54–59.

54. $y = f(-x)$ **55.** $y = -f(x)$

56. $y = -f(x - 3)$ **57.** $y = -f(3 - x)$

58. $y = 2 - 3f(x + 1)$ **59.** $y = -1 + 2f(1 - x)$

Exercises 60–63 refer to the floating-buoy **problem situation** of Example 12.

60. Show that the equation solved in Example 12 does not have any rational roots.

61. Use zoom-in to show that $x = 0.653$ is a solution to Example 12.

62. Find the depth the buoy sinks in sea water if its density is $\frac{1}{3}$ that of sea water.

63. Find the depth the buoy sinks in sea water if its density is $\frac{1}{5}$ that of sea water.

64. **Writing to Learn** Write several paragraphs that describe how the zeros of $f(x) = \frac{1}{3}x^3 + x^2 + 2x - 3$ are related to the zeros of $g(x) = x^3 + 3x^2 + 6x - 9$. In what ways does this example illustrate how the rational zeros theorem can be generalized to include all polynomials with rational number coefficients?

3.5  Complex Numbers as Zeros

In the last two sections, we focused on the task of finding zeros of polynomial functions. We saw that a real number c is a zero of a polynomial function f if $f(c) = 0$. This is equivalent to saying that $(c, 0)$ is an x-intercept of the graph of f, which implies that if the graph of a polynomial function does not cross the x-axis, it cannot have any real-number zeros. Figure 3.39 shows that the function $f(x) = x^2 + 1$ does not have any real-number zeros. Equivalently, $x^2 + 1 = 0$ has no solutions that are real numbers. There are applications in mechanics, electricity, and other areas of physics in which it is helpful for each polynomial equation to have a solution.

A new symbol i, called the **imaginary unit**, is defined as a solution to the equation $x^2 + 1 = 0$. Accordingly, the imaginary unit i satisfies these properties:

$$i^2 + 1 = 0$$

$$i^2 = -1$$

$$i = \sqrt{-1}.$$

[−5, 5] by [−3, 10]

Figure 3.39 $f(x) = x^2 + 1$.

Objective

Students will be able to apply the fundamental theorem of algebra, to use operations with complex numbers, and to find complex zeros of a polynomial function.

Key Ideas

Imaginary unit
Complex number
Real part
Imaginary part
Complex conjugate
Complex conjugate zeros
Polynomial function of odd
    degree
Roots of unity
Fundamental theorem of
    algebra
Linear factorization theorem
Multiplicity

Because there is no real number equal to $\sqrt{-1}$, i cannot be a real number. Notice that

$$(-i)^2 + 1 = -1 + 1 = 0,$$

which means that $-i$ is also a solution to $x^2 + 1 = 0$.

A new system of numbers, called the complex numbers, is based on the symbol i. This new system includes real-number multiples of i as well as sums of real numbers and real-number multiples of i. The following expressions are all complex numbers:

$$3i \qquad 5i \qquad -7i \qquad \frac{5}{2}i \qquad 7.3i$$

$$2 + 3i \qquad 5 - i \qquad \frac{1}{3} + \frac{4}{5}i$$

Definition 3.5   Complex Numbers

For real numbers a and b, the expression

$$a + bi$$

is called a **complex number**. The real number a is called the **real part** and the real number b is called the **imaginary part** of the complex number $a + bi$.

Common Errors

Some students, upon looking at an equation such as $x^2 - 4 = 0$, will say there is no complex number that is a solution. Point out that $2 = 2 + 0i$ and $-2 = -2 + 0i$.

Note that all real numbers are also complex numbers, since any real number a can be expressed in the form $a + bi$ where $b = 0$.

Two complex numbers $a + bi$ and $c + di$ are equal if their real parts are equal and their imaginary parts are equal. That is

$$a + bi = c + di$$

if, and only if,

$$a = c \text{ and } b = d.$$

Operations with Complex Numbers

Adding two complex numbers involves adding the real parts and the imaginary parts separately. Two complex numbers can be subtracted in a similar fashion.

> **Definition 3.6   Addition and Subtraction of Complex Numbers**
>
> Let $a + bi$ and $c + di$ be any two complex numbers. The sum and the difference of these two complex numbers are given as follows:
>
> **Addition:** $(a + bi) + (c + di) = (a + c) + (b + d)i$
> **Subtraction:** $(a + bi) - (c + di) = (a - c) + (b - d)i$

E X A M P L E   1   *Adding and Subtracting Complex Numbers*

Complete each of these operations: a) $(7 - 3i) + (4 + 5i)$ and b) $(4 - 2i) - (6 + i)$.

Solution

a) $(7 - 3i) + (4 + 5i) = (7 + 4) + (-3 + 5)i$

$$= 11 + 2i$$

b) $(4 - 2i) - (6 + i) = (4 - 6) + (-2 - 1)i$

$$= -2 + (-3)i$$

$$= -2 - 3i$$

Two complex numbers can be multiplied by thinking of them as binomials and multiplying them. To simplify the resulting expression, use the fact that $i^2 = -1$.

E X A M P L E   2   *Multiplying Complex Numbers*

Multiply $2 + 3i$ and $5 - i$.

Solution

$$(2 + 3i)(5 - i) = 2(5) + 2(-i) + 5(3i) + (3i)(-i)$$

$$= 10 - 2i + 15i - 3i^2$$

$$= 10 + 13i + 3 \quad \text{Use the relation } i^2 = -1.$$

$$= 13 + 13i$$

To find a general rule for the product of two complex numbers, multiply $a + bi$ and $c + di$ using the same method as in Example 2.

$$(a + bi)(c + di) = a(c + di) + (bi)(c + di) \qquad \text{Use the distributive property.}$$

$$= ac + (ad)i + (bc)i + (bd)i^2 \qquad \text{Use the distributive property.}$$

$$= ac + (ad + bc)i + (bd)(-1) \qquad \text{Use } i^2 = -1 \text{ and factor } i \text{ from the middle two terms.}$$

$$= (ac - bd) + (ad + bc)i$$

Teaching Note

Check to see if students are familiar with the concept of a complex conjugate because it is a key idea in thinking about zeros of functions. Students should observe from their experience with quadratic equations that roots did appear as conjugate pairs. In Example 4, lead students to this observation, but show the function on the graphing utility so that students can see that the curve does not touch or cross the x-axis.

Complex Conjugates

Understanding the division of complex numbers requires introducing a new concept. The complex number pairs $a + bi$ and $a - bi$ are **complex conjugates**. Notice that the product of complex conjugates $a + bi$ and $a - bi$ is the real number $a^2 + b^2$.

$$(a + bi)(a - bi) = a^2 - abi + abi - b^2 i^2$$
$$= a^2 + b^2$$

Complex conjugates are used to complete division of complex numbers, as illustrated in this next example.

E X A M P L E 3   Dividing Complex Numbers

Find $(5 + i) \div (2 - 3i)$ and write the quotient in the form $a + bi$.

Solution

$$\frac{5 + i}{2 - 3i} = \frac{(5 + i)(2 + 3i)}{(2 - 3i)(2 + 3i)} \quad \text{Multiply the numerator and denominator} \atop \text{by the complex conjugate of the denominator.}$$

$$= \frac{10 + 15i + 2i - 3}{4 + 9}$$

$$= \frac{7}{13} + \frac{17}{13}i \qquad\qquad\qquad \equiv$$

These last two examples illustrate how complex numbers are multiplied and divided. A general definition of these two operations follows.

Notes on Exercises

Ex. 1–18 provide a check on the proficiency of students in computing with complex numbers. If students are capable, do not assign all of the exercises. Rather, progress to the section on the zeros of functions that are complex numbers.

> **Definition 3.7   Multiplication and Division of Complex Numbers**
>
> Let $a + bi$ and $c + di$ be any two complex numbers.
>
> **Multiplication:** $\qquad\qquad\qquad (a + bi)(c + di) = (ac - bd) + (ad + bc)i$
>
> **Division:** $\qquad\qquad \dfrac{a + bi}{c + di} = \dfrac{ac + bd}{c^2 + d^2} + \dfrac{bc - ad}{c^2 + d^2}i \quad (c^2 + d^2 \neq 0)$

Finding Complex Zeros of a Polynomial Function

Some polynomials have only real-number zeros. Other polynomials have both real-number zeros and nonreal complex zeros, and still others have only nonreal complex zeros. We begin with an example.

Notes on Exercises

Ex. 19–28 allow students to put together all of the ideas concerning complex numbers as zeros of functions. They must use the key skills and ideas in order to work these exercises.

E X A M P L E   4   Complex Zeros

Find the zeros of $f(x) = x^2 + x + 1$.

Solution   Use the quadratic formula to solve the equation $x^2 + x + 1 = 0$.

$$x = \frac{-1 \pm \sqrt{1-4}}{2} \qquad \text{Simplify by using } \sqrt{-3} = \sqrt{-1}\sqrt{3} = i\sqrt{3}.$$

$$= \frac{-1}{2} \pm \frac{\sqrt{3}}{2}i$$

Thus the two zeros are $-\dfrac{1}{2} + \dfrac{\sqrt{3}}{2}i$ and $-\dfrac{1}{2} - \dfrac{\sqrt{3}}{2}i$.   ≡

Notice that the two zeros of $x^2 + x + 1$ are complex conjugates. It can be shown that complex zeros always occur in conjugate pairs for polynomials with real coefficients. The following theorem is stated without proof.

Teaching Note

Theorem 3.9 plays a key role in thinking of complex numbers as zeros of functions. It says that if you know one root of a polynomial is a complex number $a + bi$, then you know one other root, $a - bi$.

Teaching Note

Ask students how 5 can be the single zero of $f(x) = (x - 5)$ since 5 is a complex number and zeros appear as conjugate pairs. Where is the other member of the pair? Theorem 3.9 examines this result in terms of factors of a polynomial function. Example 5 shows how this can be used as an aid in factoring higher-degree polynomials. Example 6 focuses attention on the roots of unity but will receive more attention later after trigonometric concepts are developed through DeMoivre's theorem.

> **Theorem 3.9   Complex Conjugate Zeros**
>
> If $a + bi$ with $b \neq 0$ is a zero of a polynomial function f with real coefficients, then its complex conjugate $a - bi$ is also a zero of f.

It is important to observe that the remainder and factor theorems hold for both real and nonreal complex roots. Consequently Horner's algorithm can also be used with nonreal complex zeros, as illustrated in the next example.

E X A M P L E   5   Finding Complex Zeros

Show that $1 - 2i$ is a zero of $f(x) = 4x^4 + 17x^2 + 14x + 65$ and find all other zeros of f.

Solution   Use Horner's algorithm to show that $f(1 - 2i) = 0$.

$1 - 2i$	4	0	17	14	65
		$4 - 8i$	$-12 - 16i$	$-27 - 26i$	-65
	4	$4 - 8i$	$5 - 16i$	$-13 - 26i$	0

Thus $1 - 2i$ is a zero of f. By Theorem 3.9, $1 + 2i$ is also a zero. Use Horner's algorithm again to find the remaining quadratic factor.

$1 + 2i$	4	$4 - 8i$	$5 - 16i$	$-13 - 26i$
		$4 + 8i$	$8 + 16i$	$13 + 26i$
	4	8	13	0

Therefore

$$f(x) = \left[x - (1 - 2i)\right]\left[x - (1 + 2i)\right](4x^2 + 8x + 13).$$

Finally, use the quadratic formula to find the two zeros of $4x^2 + 8x + 13$.

$$x = \frac{-8 \pm \sqrt{64 - 208}}{8}$$

$$= \frac{-8 \pm \sqrt{-144}}{8}$$

$$= \frac{-8 \pm 12i}{8}$$

$$= -1 \pm \frac{3}{2}i$$

Thus the four zeros of f are $1 - 2i$, $1 + 2i$, $-1 + \frac{3}{2}i$, and $-1 - \frac{3}{2}i$.   ▤

A complete graph of the fourth-degree polynomial $f(x) = 4x^4 + 17x^2 + 14x + 65$ in Example 5 would show that it does not cross the x-axis anywhere. This is visual evidence that there are no real roots.

Consider the case of an odd-degree polynomial. Any odd-degree polynomial has end behavior type either ($\swarrow$, $\nearrow$) or ($\nwarrow$, $\searrow$). It follows from the intermediate value theorem that such a polynomial has at least one real zero.

Theorem 3.10   Polynomial Function of Odd Degree

A polynomial function f of odd degree has at least one real-number zero. The domain and range of f are each $(-\infty, \infty)$.

Finding a complete graph of f will help identify when a polynomial has complex zeros.

E X A M P L E   6   *Finding Complex Zeros*

Find a complete graph of $f(x) = x^3 - 1$ to show that it has only one real-number zero. Find all zeros.

Solution   The complete graph of $f(x) = x^3 - 1$ in Fig. 3.40 suggests that 1 is a real zero of $x^3 - 1$. Use Horner's algorithm to confirm that 1 is a zero and that f can be factored as follows:

$$x^3 - 1 = (x - 1)(x^2 + x + 1).$$

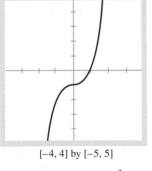

[−4, 4] by [−5, 5]

Figure 3.40 $f(x) = x^3 - 1$.

In Example 4, we found that $-\frac{1}{2} + (\sqrt{3}/2)i$ and $-\frac{1}{2} - (\sqrt{3}/2)i$ are the two zeros of $x^2 + x + 1$. Thus $x^3 - 1$ has three zeros:

$$1, \qquad -\frac{1}{2} + \frac{\sqrt{3}}{2}i, \qquad -\frac{1}{2} - \frac{\sqrt{3}}{2}i.$$

Notice that the zeros of $x^3 - 1$ are solutions to the equivalent equations

$$x^3 - 1 = 0 \iff x^3 = 1.$$

So these zeros are often called the cube roots of 1 or the cube roots of unity. In general, a solution to the equation $x^n - 1 = 0$ is called an **nth root of unity**.

Notes on Exercises

Ex. 46 and 47 focus on the roots of unity. Interestingly, a few students may not realize the connection between the word *unity* and the 1 (one) in the polynomial.

EXAMPLE 7   Finding Roots of Unity

Find the fourth roots of unity.

Solution   Find solutions to $x^4 - 1 = 0$, or equivalently, find the zeros of $f(x) = x^4 - 1$.

$$x^4 - 1 = (x^2 - 1)(x^2 + 1) \qquad \text{Factor the difference of squares.}$$

$$x^4 - 1 = (x - 1)(x + 1)(x^2 + 1)$$

$$= (x - 1)(x + 1)(x - i)(x + i) \qquad i \text{ and } -i \text{ are zeros of } x^2 + 1.$$

The four fourth roots of unity are 1, -1, i, and $-i$. ▰

Teaching Note

The fundamental theorem of algebra is a major idea that students encounter in high school mathematics. Give it the attention it deserves. It is equivalent to saying that an nth degree polynomial has n roots. Moreover, we do not need another type of number beyond the complex numbers in order to solve a polynomial equation or to find the zeros of a polynomial function. Make an issue about complex and real zeros and the relation of the graph to the x-axis. Multiplicity is important when considering the number of zeros. You might like to graph examples such as $f(x) = (2x - 5)^4$.

Notice that in Examples 4 through 7, the number of zeros, real plus complex, equals the degree of the given polynomial. The next two theorems and the following discussion will clarify this observation.

Theorem 3.11   Fundamental Theorem of Algebra

If f is a polynomial function of degree $n > 0$ with real coefficients, then f has at least one zero in the set of complex numbers.

Note that the zero mentioned in Theorem 3.11 may be a real number since any real number r can be expressed as the complex number $r + 0i$.

The fundamental theorem of algebra is used in the proof of this theorem.

> **Theorem 3.12 Linear Factorization Theorem**
>
> If $f(x)$ is a polynomial of degree $n > 0$ with real coefficients, then f has precisely n linear factors
>
> $$f(x) = a(x - c_1)(x - c_2) \cdots (x - c_n),$$
>
> where $c_1, c_2, \ldots, c_n$ are complex numbers and a is the leading coefficient of $f(x)$.

Here is the idea behind the proof for the case that the roots are all real numbers. If c_1 is a real-number zero of f, then $x - c_1$ is a factor of f; that is,

$$f(x) = (x - c_1)f_1(x),$$

where $f_1(x)$ is a polynomial of degree $n - 1$. Repeat this argument for $f_1(x)$ to obtain the factorization

$$f(x) = (x - c_1)(x - c_2)f_2(x),$$

where $f_2(x)$ is a polynomial of degree $n - 2$. Continue this process to eventually obtain a factorization of linear factors. This argument can be modified when there are nonreal complex roots.

It is important to observe that the c_i's in Theorem 3.12 need not be distinct. For example, we could have

$$f(x) = (x - 2)(x - 2)(x - 2)(x + 1)(x + 1)$$
$$= (x - 2)^3(x + 1)^2$$

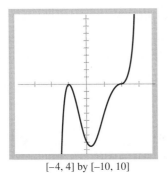

Figure 3.41
$f(x) = (x - 2)^3(x + 1)^2.$

(see Fig. 3.41). In this case, we say that for f, 2 is a **zero of multiplicity 3** and -1 is a **zero of multiplicity 2**. Counting multiplicities, we can conclude from Theorem 3.12 that a polynomial of degree n has n zeros.

E X A M P L E   8   Finding Linear Factors

Write $f(x) = 3x^5 - 2x^4 + 6x^3 - 4x^2 - 24x + 16$ as a product of linear factors. Classify the zeros of f as rational, irrational, or nonreal complex.

Solution From the rational zeros theorem, the only possible rational zeros are

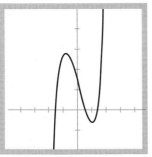

[−5, 5] by [−20, 50]

Figure 3.42 $f(x) = 3x^5 - 2x^4 + 6x^3 - 4x^2 - 24x + 16.$

$$\pm 1, \pm 2, \pm 4, \pm 8, \pm 16, \pm\frac{1}{3}, \pm\frac{2}{3}, \pm\frac{4}{3}, \pm\frac{8}{3}, \pm\frac{16}{3}.$$

The complete graph of f in the $[-5, 5]$ by $[-20, 50]$ viewing rectangle (see Fig. 3.42) suggests that the list of possible rational zeros can be reduced to $-\frac{4}{3}, \frac{2}{3}, \frac{4}{3}$.

Use Horner's algorithm to check $\frac{2}{3}$.

$$\frac{2}{3} \begin{array}{|rrrrrr} 3 & -2 & 6 & -4 & -24 & 16 \\ & 2 & 0 & 4 & 0 & -16 \\ \hline 3 & 0 & 6 & 0 & -24 & 0 \end{array}$$

Thus $f(\frac{2}{3}) = 0$ and f factors as follows:

$$f(x) = (x - \tfrac{2}{3})(3x^4 + 6x^2 - 24)$$

$$= (x - \tfrac{2}{3})(3)(x^4 + 2x^2 - 8)$$

$$= (3x - 2)(x^4 + 2x^2 - 8)$$

$$= (3x - 2)(x^2 - 2)(x^2 + 4)$$

The zeros of $x^2 - 2$ are $\pm\sqrt{2}$, and the zeros of $x^2 + 4$ are $\pm 2i$. Therefore

$$f(x) = (3x - 2)(x - \sqrt{2})(x + \sqrt{2})(x - 2i)(x + 2i).$$

The two real zeros $\sqrt{2}$ and $-\sqrt{2}$ are irrational, and the real zero $\frac{2}{3}$ is rational. Finally, the zeros $2i$ and $-2i$ are nonreal complex numbers. ▀

Exercises for Section 3.5

In Exercises 1–8, write each expression in the form $a + bi$ where a and b are real numbers.

1. $2 - 3i + 6$

2. $2 - 3i + 6 - 4i$

3. $(2 + 3i)(2 - i)$

4. $(2 - i)(1 + 3i)$

5. $2(1 + i) - 1 - i$

6. $3(6 - i) - 2(-1 - 3i)$

7. $3(2 + i)^2 - 4i$

8. $(1 - i)^3$

In Exercises 9–12, determine the complex conjugate of each complex number.

9. $2 - 3i$

10. $-6i$

11. $-3 + 4i$

12. $-1 - \sqrt{2}i$

In Exercises 13–18, write each expression in the form $a + bi$ where a and b are real numbers.

13. $\dfrac{1}{2 + i}$

14. $\dfrac{i}{2 - i}$

15. $\dfrac{2 + i}{2 - i}$

16. $\dfrac{2 + i}{3i}$

17. $\dfrac{(2 + i)^2(-i)}{1 + i}$

18. $\dfrac{(2 - i)(1 + 2i)}{5 + 2i}$

In Exercises 19–24, determine the number of real-number zeros and the number of nonreal complex zeros of each function.

19. $f(x) = x^2 - 2x + 7$

20. $f(x) = x^3 - 3x^2 + x + 1$

21. $f(x) = x^3 - x + 3$

22. $f(x) = x^4 - 2x^2 + 3x - 4$

23. $f(x) = x^4 - 5x^3 + x^2 - 3x + 6$

24. $f(x) = x^5 - 2x^2 - 3x + 6$

In Exercises 25–28, find all the zeros of each polynomial. Classify each zero as integer, noninteger rational, irrational, or nonreal complex.

25. $f(x) = x^3 + 4x - 5$

26. $f(x) = x^3 - 10x^2 + 44x - 69$

27. $f(x) = x^4 + x^3 + 5x^2 - x - 6$

28. $f(x) = 3x^4 + 8x^3 + 6x^2 + 3x - 2$

29. Show that $1 + i$ is a zero of $f(x) = 3x^3 - 7x^2 + 8x - 2$ and find all other zeros of f.

30. Show that $3 - 2i$ is a zero of $f(x) = x^4 - 6x^3 + 11x^2 + 12x - 26$ and find all the other zeros of f.

In Exercises 31–34, write each polynomial as a product $f(x) = k(x - c_1)(x - c_2) \cdots (x - c_n)$, where n is the degree of the polynomial, k is a real number, and each c_i is a zero of f.

31. $f(x) = x^3 - x^2 + x - 1$

32. $f(x) = x^4 - 6x^2 + 5$

33. $f(x) = 2x^3 - x^2 + 3x - 4$

34. $f(x) = x^4 + 6x^3 + 7x^2 - 12x - 18$

In Exercises 35–38, find a polynomial with real-number coefficients satisfying the given conditions.

35. degree 2; zero $2 - 3i$ **36.** degree 3; zeros 1 and i

37. degree 3; zeros 3 and $1 - i$

38. degree 4; zeros $-2 + i$ and $1 - i$

39. Can you find a polynomial of degree 3 with real-number coefficients that has -2 as its only real-number zero?

40. Can you find a polynomial of degree 3 with real-number coefficients that has $2i$ as its only nonreal zero?

41. Find a polynomial $f(x)$ of degree 4 with real-number coefficients that has -3, $1 + i$, and $1 - i$ as its only zeros.

42. Find a polynomial $f(x)$ of degree 4 with real-number coefficients that has -3, $1 + i$, and $1 - i$ as its only zeros

and that also satisfies $f(0) = 1$.

Exercises 43–45 refer to the following **problem situation** *Archimedes' law* states that when a sphere of radius r with density d_S is placed in a liquid of density $d_L = 62.5$ lb/ft^3, it will sink to a depth h where

$$\frac{\pi}{3}(3rh^2 - h^3)d_L = \frac{4}{3}\pi r^3 d_S.$$

43. If $r = 5$ ft and $d_S = 20$ lb/ft^3, use zoom-in to determine h with an error of less than 0.01.

44. If $r = 5$ ft and $d_S = 45$ lb/ft^3, use zoom-in to determine h with an error of less than 0.01.

45. If $r = 5$ ft and $d_S = 70$ lb/ft^3, use zoom-in to determine h with an error of less than 0.01.

46. Find the two square roots of unity; that is, solve the equation $x^2 = 1$.

47. Find the three cube roots of 8; that is, solve the equation $x^3 = 8$.

48. We denote $\bar{z}$ as the conjugate of $z = a + bi$. Prove that $z + \bar{z}$ is a real number for any complex number z.

49. Prove that $z \cdot \bar{z}$ is a real number for any complex number z.

50. Verify that the complex number i is a zero of the polynomial $f(x) = x^3 - ix^2 + 2ix + 2$.

3.6 _____ Polynomial Functions and Inequalities

Recall that the solution to an inequality is usually reported in interval notation as one or more intervals. In this section, we are interested in solving inequalities like $x^4 \geq x^2$. Notice that solving $x^4 \geq x^2$ is equivalent to solving $x^4 - x^2 \geq 0$.

When using an algebraic method of solution, the form $x^4 - x^2 \geq 0$ is the preferred form. We can solve this form in a manner similar to the one used in finding the zeros of $f(x) = x^4 - x^2$ in the last several sections.

Chapter 2 showed that a graphical approach to solving equations and inequalities is often quicker and more general than an algebraic approach. Consequently, we will use a graphical method in this section. The graphical method allows either the form $x^4 \geq x^2$ or the form $x^4 - x^2 \geq 0$ to be used. When using the form $x^4 \geq x^2$, we are making comparisons between the graphs of $y = x^n$ for various values of n.

Your experiences in the following Exploration will help you understand how the graphs of $y = x^2$, $y = x^3$, $y = x^4$, and $y = x^5$ compare for values of x near 1.

Teaching Note

The zoom-in procedure may be used to examine closely the behavior of the functions $f(x) = x^2$ and $g(x) = x^4$ near the points of intersection.

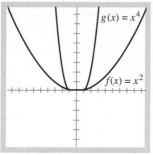

$[-10, 10]$ by $[-60, 100]$

Figure 3.43 $f(x) = x^2$ and $g(x) = x^4$.

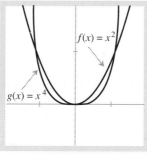

$[-2, 2]$ by $[-1, 2]$

Figure 3.44 $f(x) = x^2$ and $g(x) = x^4$ in a smaller viewing rectangle.

🔍 EXPLORE WITH A CALCULATOR

What happens when you raise a number to a power? Does it get larger or smaller? Use a calculator to find all values for the following table:

x	x^2	x^3	x^4	x^5
0.5	0.25	0.125	0.0625	
0.9				
0.99				
1				
1.01				
1.1				

Complete a Generalization For what values of x do the powers of x get larger? For what values of x do the powers of x get smaller?

Graphs of $y = x^n$ When n Is Even

We begin by comparing the graphs of $y = x^n$ when n is even. For example, compare $f(x) = x^2$ and $g(x) = x^4$. Figure 3.43 shows the graphs in the viewing rectangle $[-10, 10]$ by $[-60, 100]$. The graph of g seems to be *above* the graph of f. But is it? Recall from the Exploration that if $0 < x < 1$, powers of x get smaller. So for a complete comparison, we must also consider the viewing rectangle $[-2, 2]$ by $[-1, 2]$ shown in Fig. 3.44.

The points of intersection of these graphs seem to be $(-1, 1)$, $(0, 0)$, and $(1, 1)$. To confirm this claim algebraically, consider the following equations:

$$x^4 = x^2$$

$$x^4 - x^2 = 0$$

$$x^2(x^2 - 1) = 0$$

$$x^2(x - 1)(x + 1) = 0$$

The solutions are $x = -1$, $x = 0$, and $x = 1$.

To summarize what we have learned:

$$x^4 < x^2 \iff x \text{ is in intervals } (-1, 0) \cup (0, 1);$$

$$x^4 = x^2 \iff x = -1, x = 0, \text{ or } x = 1;$$

$$x^4 > x^2 \iff x \text{ is in intervals } (-\infty, -1) \cup (1, \infty).$$

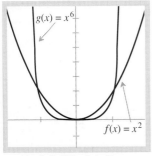

[−2, 2] by [−1, 4]

Figure 3.45 $f(x) = x^2$ and $g(x) = x^6$.

Notes on Examples

Example 1 illustrates a comparison of two inequalities. Point out that the graph of a polynomial function $y = x^n$, where n is even, flattens near zero as n increases, and rises more sharply for values of x where $|x| > 1$. Compare the graphs of $y = x^6$ and $y = x^2$ by using zoom-in.

Notes on Examples

Example 2 illustrates graphs of $y = x^n$ where n is odd. In all the examples of this section, the strategy for solving an inequality of the form $f(x) - g(x) > 0$ (or $<, \leq, \geq$) is to rewrite it as $f(x) > g(x)$ (or $<, \leq, \geq$). The graphs of the functions $f(x)$ and $g(x)$ are compared directly. Be sure to discuss the similarities and differences of the graphs of $y = x^n$ where n is even and where n is odd.

E X A M P L E 1   Solving Inequalities

Solve the inequality $x^6 \leq x^2$.

Solution Let $f(x) = x^2$ and $g(x) = x^6$. For what values of x does $g(x) = f(x)$, and when is $g(x) < f(x)$? Consider the graphs of $f(x) = x^2$ and $g(x) = x^6$ shown in Fig. 3.45.

$$x^6 \leq x^2 \quad \Longleftrightarrow \quad x \text{ is in the interval } [-1, 1]$$

Notice the use of square brackets to indicate that the endpoints are included in the interval (that is, the solution is a closed interval). ■

Using the method of Example 1, we can solve any inequalities of the form $x^n \leq x^m$, where n and m are even integers.

Graphs of $y = x^n$ When n Is Odd

We now turn our attention to comparing graphs of $y = x^n$ when n is odd. For example, compare $f(x) = x^3$ and $g(x) = x^5$ as shown in Fig. 3.46.

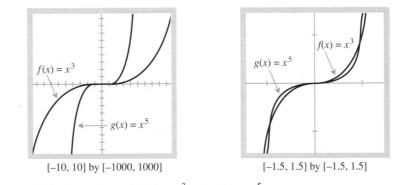

[−10, 10] by [−1000, 1000]   [−1.5, 1.5] by [−1.5, 1.5]

Figure 3.46   Two views of $f(x) = x^3$ and $g(x) = x^5$.

At least two viewing rectangles must be considered to support the behavior of these two graphs. These two graphs appear to intersect at the points $(-1, -1)$, $(0, 0)$, and $(1, 1)$. The graph of $f(x) = x^3$ is above the graph of $g(x) = x^5$ when x is in the intervals $(-\infty, -1)$ and $(0, 1)$ and below the graph of $g(x) = x^5$ when x is in the intervals $(-1, 0)$ and $(1, \infty)$.

E X A M P L E 2   Solving Inequalities

Solve the inequality $x^5 - x^3 > 0$ algebraically and graphically.

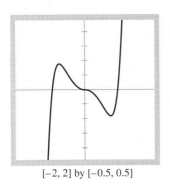

[−2, 2] by [−0.5, 0.5]

Figure 3.47 $h(x) = x^5 - x^3$.

Solution **Method 1** Rewrite the inequality in the form $x^5 > x^3$. Read the solution from the graphs in Fig. 3.46 by observing where the graph of $g(x) = x^5$ is above the graph of $f(x) = x^3$.

$$x^5 > x^3 \iff x \text{ is in } (-1, 0) \cup (1, \infty).$$

Method 2 Find a complete graph of $h(x) = x^5 - x^3$, find the zeros, and determine where the graph is above the x-axis (see Fig. 3.47).

$$x^5 - x^3 = 0$$

$$x^3(x^2 - 1) = 0$$

$$x^3(x - 1)(x + 1) = 0$$

The zeros of h are $-1, 0$, and 1.

$$x^5 - x^3 > 0 \iff x \text{ is in } (-1, 0) \cup (1, \infty). \qquad \blacksquare$$

In Example 2, it was possible to determine the zeros algebraically since $x^5 - x^3$ could be factored. Usually the algebraic technique is very hard or impossible. In that case, the zeros can be determined graphically by using zoom-in. This next example illustrates the zoom-in method.

E X A M P L E 3 Solving Inequalities

Solve the inequality $x^3 < 4x - 1$.

Solution Write the inequality in the form $x^3 - 4x + 1 < 0$ and find the values of x for which the complete graph of $f(x) = x^3 - 4x + 1$ lies below the x-axis. Use zoom-in to find

$$\text{zeros of } f : \quad 2.11, 0.25, 1.86$$

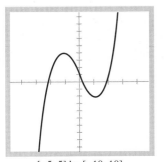

[−5, 5] by [−10, 10]

Figure 3.48 $f(x) = x^3 - 4x + 1$.

with an error of less than 0.01 (see Fig. 3.48).
Therefore

$$x^3 - 4x + 1 < 0 \iff x \text{ is in } (-\infty, -2.11) \cup (0.25, 1.86). \qquad \blacksquare$$

Endpoint Agreement

Even though the endpoints reported in solutions are approximations, we treat them as if they are exact. Consequently, we will continue to use parentheses ( ) to communicate the open intervals for the relations $<$ and $>$, and square brackets [ ] to communicate the closed intervals for the relations $\leq$ and $\geq$.

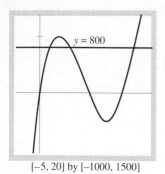

[−5, 20] by [−1000, 1500]

Figure 3.49 $y = 800$ and $V(x) = x(20 − 2x)(30 − 2x)$.

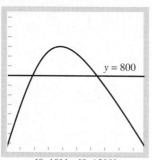

[0, 10] by [0, 1500]

Figure 3.50 $y = V(x)$ and $y = 800$.

Assignment Guide

Day 1: Ex. 1–15, 18, 22, 25
Day 2: Ex. 26–30, 34, 35

EXAMPLE 4   APPLICATION: Finding Minimum Box Volume

A box is formed by removing squares of side length x from each corner of a rectangular piece of cardboard 20 in. wide by 30 in. long. Determine x so that the volume of the resulting box is at least 800 cu in.

Solution   Let x = width of square = height of the resulting box, and $V(x)$ = the volume of the resulting box. Then

$$20 − 2x = \text{width of the base}$$

$$30 − 2x = \text{length of the base}$$

$$V(x) = x(20 − 2x)(30 − 2x).\quad \text{Volume is equal to length} \times \text{width} \times \text{height.}$$

Figure 3.49 shows a complete graph of $V(x)$ and $y = 800$. However, the problem situation allows only values of x between $x = 0$ and $x = 10$. (Why?) Consequently, the graph in Fig. 3.50 is a graphical representation of the problem situation.

We must find the values of x for which the graph of V is on or above the graph of $y = 800$. Using zoom-in, we find that the x-coordinates of the points of intersection are

$$x = 1.88, x = 6.36$$

with an error of less than 0.01. Therefore the resulting box has a volume of at least 800 cu in., provided x is in the interval [1.88, 6.36].   ■

Exercises for Section 3.6

In Exercises 1–14, solve the inequality.

1. $x^4 < 2x^2$ **2.** $x^3 > 3x$

3. $x^6 < x^4$ **4.** $5x^2 \geq x^4$

5. $x^2 > x^3$ **6.** $x^7 < x^9$

7. $x^3 < x^5$ **8.** $2x^3 > 4x^5$

9. $x^3 + 8 > 0$ **10.** $x^3 − 8x + 1 \leq 0$

11. $x^3 − 2x^2 + 3 > 0$

12. $2x^3 − 3x^2 + 2x − 5 \geq 0$

13. $x^3 − 2x + 3 < 2x^2 − 3x + 5$

14. $x^4 \geq x^3 − 1$

In Exercises 15–20, use the sign-pattern method shown in Section 2.6 to solve each inequality algebraically. Support your answer with a graphing utility.

15. $x^2(x − 1) > 0$ **16.** $x(x − 1)(x + 3) \leq 0$

17. $x^3 > x$ **18.** $x^3 − x^2 − 2x > 0$

19. $x^3 − 4x^2 − x + 4 \geq 0$ **20.** $|x^3 − 8x| \leq 0$

In Exercises 21 and 22, draw a complete graph of each inequality.

21. $y < x^3 - 5x^2 - 3x + 15$

22. $y \geq |x^2 + 2x - 3|$

23. Squares of side length x are removed from a 15- by 60-in. piece of cardboard, and a box with no top is formed. Determine x so that the volume of the resulting box is at most 450 cu in.

24. Squares of side length x are removed from a 10- by 25-cm piece of cardboard, and a box with no top is formed. Determine x so that the volume of the resulting box is at least 175 cu cm.

25. The function $V = 2666x - 210x^2 + 4x^3$ represents the volume of a certain box that has been made by removing equal squares of side length x from each corner of a rectangular sheet of material and then folding up the sides. What are the possible values of x for the height of the box?

Exercises 26–28 refer to the following   **problem situation** A rectangular area, with one side against an existing wall, is to be enclosed by three sides of fencing totaling 335 ft in length. Let x be the length of the side of the fence perpendicular to the existing wall.

26. Find an algebraic representation that gives the area enclosed as a function of x.

27. Find a complete graph of this problem situation. (Recall that the complete graph of a problem situation is only a portion of the complete graph of the algebraic representation.)

28. Find x so that the area is less than or equal to 11,750 sq ft.

Exercises 29 and 30 refer to the following   **problem situation**

The profit P of a business is determined by the formula $P = R - C$, where R is the total revenue generated by the business and C is the total cost of operating the business. Suppose $R(x) = 0.0125x^2 + 412x$ is the total annual revenue of the business where x is the number of customers patronizing the business. Further, suppose $C(x) = 12{,}225 + 0.00135x^3$ is the total annual cost of doing business.

29. Find the number of customers that the business must have in 1 yr for it to make a profit.

30. How many customers must the business have for it to realize an annual profit of $60,000?

Exercises 31 and 32 relate to finding hidden behavior of a function.

31. Draw the graph of $f(x) = 33x^3 - 100x^2 + 101x + 5$ in each viewing rectangle.
   a) $[-10, 10]$ by $[-1000, 1000]$
   b) $[-1, 1]$ by $[-100, 100]$
   c) $[0.9, 1.1]$ by $[20, 50]$

32. The function f in Exercise 31 appears to be increasing on any interval. *It is not.* There is *hidden behavior* that can be determined by zooming in near the point $(1, f(1))$. Find a viewing rectangle that exhibits the hidden behavior.

In Exercises 33–36, draw a complete graph. Determine all local maximum and minimum values. Determine all real-number zeros. Determine the intervals on which the function is increasing and those on which the function is decreasing.

33. $f(x) = 2 - 3x + x^2 - x^3$

34. $f(x) = (x - 1)^2 x^3$

35. $g(x) = 3x^4 - 5x^3 + 2x^2 - 3x + 6$

36. $T(x) = 2x^5 - 3x^4 + 2x^3 - 3x^2 + 7x - 4$

3.7 _____   Solving Systems of Equations Graphically

Sometimes a problem situation requires that we solve several different equations simultaneously. For example, suppose that both of the equations

$$2x + 3y = 5$$

$$-3x + 5y = 21$$

Key Ideas

System of linear equations
Solution to a system of
  equations
Error agreement for solutions
  to systems of equations

Teaching Note

The graph of an equation
in two variables is the set
of solutions to the equation
considered as points in the
plane. It is important that the
connection between *solutions
to the equation* and *points in
the plane* be emphasized.

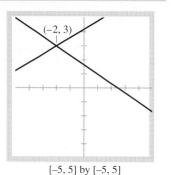

[−5, 5] by [−5, 5]

Figure 3.51 Graphs of
$y = -\frac{2}{3}x + \frac{5}{3}$ and $y = \frac{3}{5}x + \frac{21}{5}$.

Teaching Note

Pay particular attention to the
definitions at the beginning of
the section. Ask students:
What is a system of equations?
What does it mean to have
simultaneous solutions to a
system of equations? Point
out to students that if there
are two variables, then two
nonequivalent equations
are needed to have a finite
number of solutions. This
should reinforce the idea of
equivalent equations.

must be true at the same time. We call this pair of equations a **system of linear equations.**

Definition 3.8   Solution to a System of Equations

A pair of real numbers is a **solution to a system of equations** in two variables if, and only if, the pair of numbers is a solution to each equation. When we have found all solutions to the system of equations, we say that we have **solved** the system of equations.

When a system of linear equations includes only two equations, each with the same two variables, it is possible to solve the system graphically. You simply graph the equations in the same viewing rectangle and find the coordinates of the point where the graphs cross.

E X A M P L E   1   Solving a System of Linear Equations

Find a graphical solution to the system of linear equations:

$$2x + 3y = 5$$
$$-3x + 5y = 21$$

Solution   First change these equations to the equivalent forms.

$$y = -\frac{2}{3}x + \frac{5}{3}$$
$$y = \frac{3}{5}x + \frac{21}{5}$$

Next find graphs of these equations as shown in Fig. 3.51.

After zooming in several times, we find the point of intersection $x = -2.001$ and $y = 2.998$. We suspect that $x = -2$ and $y = 3$ is the exact solution. To confirm this, substitute into the equations.

$$2(-2) + 3(3) = 5$$
$$-3(-2) + 5(3) = 21$$

In Example 1, the equations in the system were linear. The same graphical technique also can be used when the equations are not linear.

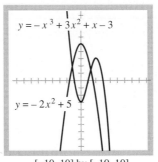

$y = -x^3 + 3x^2 + x - 3$

$y = -2x^2 + 5$

[−10, 10] by [−10, 10]

Figure 3.52 In this viewing rectangle, there appear to be two points of intersection to system (1).

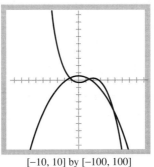

[−10, 10] by [−100, 100]

Figure 3.53 System (1) seen in a different viewing rectangle.

[−0.1, 0.1] by [−0.1, 0.1]

Figure 3.54 The points suggest a grid of squares, each with a length and width of 0.01.

E X A M P L E   2   Finding the Number of Solutions to a System

Find the number of solutions to the system

$$y = -x^3 + 3x^2 + x - 3$$
$$y = -2x^2 + 5.$$

(1)

Solution Begin by finding the graph of each function in the standard viewing rectangle (see Fig. 3.52). There appear to be two solutions. However, we should be suspicious enough to investigate the fourth quadrant more thoroughly: Notice that as x increases, the two graphs seem to be approaching each other and may intersect "offscreen." So it is appropriate to investigate the end behavior models of these two functions.

1. The end behavior model of $y = -x^3 + 3x^2 + x - 3$ is $y = -x^3$.

2. The end behavior model of $y = -2x^2 + 5$ is $y = -2x^2$.

Knowing that $y = -x^3$ decreases more rapidly than $y = -2x^2$ for large values of x confirms the suspicion that there may be a point of intersection in the fourth quadrant. Find graphs of the functions in system (1) in a viewing rectangle with greater vertical range, say $[-10, 10]$ by $[-100, 100]$. Figure 3.53 reveals that there is indeed a third point of intersection. Can you be sure there are no more? Why?   ▤

Throughout the text, we have been following the convention that graphical solutions be found with an error of at most 0.01. For systems of equations, that means the solution for each variable must be found with an error of at most 0.01.

There is a geometric interpretation of this agreement. Consider the viewing rectangle $[-0.1, 0.1]$ by $[-0.1, 0.1]$ with the grid of small squares shown in Fig. 3.54. The length and width of each square is 0.01. This grid suggests the geometric interpretation to our error agreement.

Error Agreement for Solutions to Systems of Equations

If the solution of a system of equations falls within one of the squares of the grid shown in Fig. 3.54, then any ordered pair (a, b) that falls within the same small square is reported as a *solution with an error of at most* 0.01.

Example 1 explains that system (1) has three solutions. There is a difference between the two solutions that appear in the standard viewing rectangle and the third solution. So we consider them separately and explain the difference.

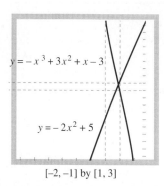

[−2, −1] by [1, 3]

Figure 3.55 One solution to system (1) lies in the viewing rectangle [−1.3, −1.2] by [2.0, 2.1], which is indicated by the dashed lines.

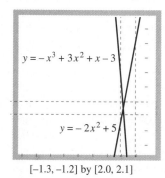

[−1.3, −1.2] by [2.0, 2.1]

Figure 3.56 Zooming in to [−1.22, −1.21] by [2.03, 2.04] (dashed lines) will generate a solution to system (1) with the desired degree of accuracy.

Notes on Examples

Use the GRID ON feature of your graphing utility when solving systems of equations, as in Example 3, and for finding solutions with an error of at most 0.01, as in Example 4.

E X A M P L E   3   Solving a System of Equations: Two Points

Find all solutions, with errors of at most 0.01, that fall within the viewing rectangle [−10, 10] by [−10, 10] for the system

$$y = -x^3 + 3x^2 + x - 3$$

$$y = -2x^2 + 5.$$

Solution   We can estimate from Fig. 3.52 that there is one solution in the viewing rectangle [−2, −1] by [1, 3] and a second one in the viewing rectangle [1, 2] by [1, 2]. We consider these solutions one at a time.

Case 1 Figure 3.55 shows the graphs in the viewing rectangle [−2, −1] by [1, 3]. The vertical dashed lines in this figure are $x = -1.3$ and $x = -1.2$, and the horizontal dashed lines are $y = 2.0$ and $y = 2.1$. These values determine the viewing rectangle [−1.3, −1.2] by [2.0, 2.1] shown in Fig. 3.56. Notice that any point in the viewing rectangle [−1.22, −1.21] by [2.03, 2.04] will be a solution with an error of at most 0.01.

We report the solution as the point (−1.218, 2.035).

Case 2 By using a technique similar to that in Case 1, we can show that the second solution is the point (1.350, 1.356) (not shown in a figure). ▤

An important characteristic to note about the two solutions found in Example 3 is that we were able to have the scale marks represent the same horizontal and vertical intervals throughout the process of zooming in. This will *not* be possible for finding the remaining point of intersection.

E X A M P L E   4   Solving a System of Equations: Third Point

Find the solution, with an error of at most 0.01, that falls in the viewing rectangle [4, 5] by [−50, −40] for system (1) in Example 2.

Solution   Figure 3.57 shows the graphs of system (1) in the viewing rectangle [4, 5] by [−50, −40]. To find the third solution, continue to zoom in so that the length and width of each succeeding viewing rectangle is reduced by a factor of 1/10 from the previous one. Successive viewing rectangles will be [4.8, 4.9] by [−43, −42] and then [4.86, 4.87] by [−42.4, −42.3].

Finally, Fig. 3.58 shows the graphs in the viewing rectangle [4.864, 4.869] by [−42.42, −42.36]. We read the solution to be (4.868, −42.39) with an error of at most 0.01. ▤

Notice in Example 3 that we need an error of at most 0.001 in the x-coordinate in order to obtain an error of at most 0.01 in the y-direction. This is because the horizontal and vertical scales are not the same. Observe that in Fig. 3.57, the

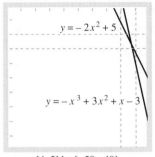

[4, 5] by [−50,−40]

Figure 3.57   The third solution to system (1) appears to be in the rectangle [4.8, 4.9] by [−43, −42].

Common Errors

Many students do not realize that finding a solution with an error of at most 0.01 applies to both the x- and y-coordinates of the plane. Students must understand the error agreement for solutions to systems of equations.

horizontal scale marks represent 0.1 unit and the vertical scale marks represent 1 unit. In each of the successive viewing rectangles, the horizontal and vertical scale marks represent different units. This is in contrast to the solutions in Example 2.

   To understand visually why the horizontal and vertical scales are chosen differently, consider the viewing rectangle [4.8, 4.9] by [−42.4, −42.3] as shown in Fig. 3.59. Even though the solution appears to fall in this rectangle, clearly this rectangle is not a good one to use because the coordinates of the point of intersection will be hard to determine.

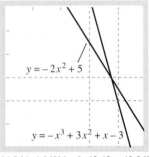

[4.864, 4.869] by [−42.42, −42.36]

Figure 3.58   The third solution is (4.868, −42.39) with an error of at most 0.01. This point must fall within the rectangle indicated by the dashed lines in order to have the desired accuracy.

Box Problem Revisited

In Section 3.1, Example 5, we considered the volume of a box made by cutting a square of side length x from each corner of a cardboard rectangle 20 by 25 in. (see Fig. 3.60). That problem asked us to find the value of x that would maximize the volume of the box. The next example asks a slightly different question.

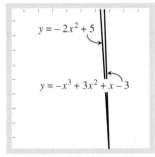

[4.8, 4.9] by [−42.4, −42.3]

Figure 3.59   The solution to the third point of intersection of system (1) cannot be easily determined using this viewing rectangle.

E X A M P L E   **5**   APPLICATION: Determining Box Dimensions

Find the side length of the square that must be cut from a 20- by 25-in. piece of cardboard to form a box with a volume of 500 cu in.

Solution   Use Fig. 3.60 to find that the volume of the box is described by $V(x) = x(20 - 2x)(25 - 2x)$. To solve the problem, set $y = V(x)$ and solve the following system of equations:

$$y = x(20 - 2x)(25 - 2x)$$
$$y = 500$$

(2)

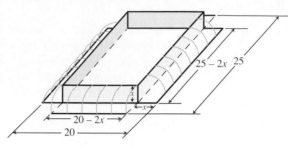

Figure 3.60   Squares cut from a piece of cardboard measuring 20 by 25 in. Fold along the dotted lines to form a box.

Because the width of the rectangle is 20, x must be restricted to $0 < x < 10$. (Why?) Therefore consider the graph of this system in the viewing rectangle $[0, 10]$ by $[0, 1000]$ (see Fig. 3.61a). There are two solutions.

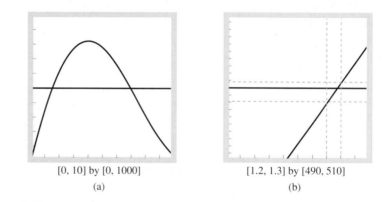

$[0, 10]$ by $[0, 1000]$	$[1.2, 1.3]$ by $[490, 510]$
(a)	(b)

Figure 3.61 $y = x(20 - 2x)(25 - 2x)$ and $y = 500$. (a) This graph is a complete graph for the problem situation in Example 5. (b) Zoom in to a smaller viewing rectangle. The dashed lines indicate the viewing rectangle $[1.27, 1.28]$ by $[498, 501]$, which will yield a solution with an error of at most 0.01.

Zoom in to the viewing rectangle $[1.2, 1.3]$ by $[490, 510]$ (see Fig. 3.61b). Now the graph of $V(x)$ appears to be a straight line from the bottom-left corner to the top-right corner of the viewing rectangle $[1.27, 1.28]$ by $[498, 501]$ and crosses the graph of $y = 500$ in this rectangle.

Because the width of this rectangle is 0.01, any point in this rectangle represents a solution whose x-coordinate has an error of at most 0.01. The length of x is 1.277 with an error of at most 0.01 when the volume is 500 cu in.   ≡

Notice that in Example 4, we did not find a solution for *both* x and $y = V(x)$ accurate to at least 0.01. While the x-coordinate had an error of at most 0.01, the value of y was found with an error of at most 3 (see Fig. 3.61b).

To find the value of y with greater accuracy than an error of at most 3, the accuracy of x must be increased also. To see this, try to increase the accuracy of $y = V(x)$ while keeping the accuracy of x the same. For example, consider the [1.2, 1.3] by [499.9, 500.1] viewing rectangle. Figure 3.62(a) shows that this viewing rectangle does not work because the graph of $y = V(x)$ is too near vertical. Therefore in order to increase the accuracy of $V(x)$, we must increase the accuracy of x. Figure 3.62(b) shows that $V = 500$ with an error of at most 0.02 when $1.2767 \leq x \leq 1.2768$.

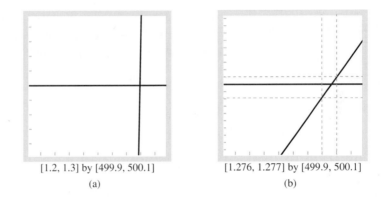

[1.2, 1.3] by [499.9, 500.1]   [1.276, 1.277] by [499.9, 500.1]

(a)   (b)

Figure 3.62 Two attempts to increase the accuracy of the y-coordinate of the solution to system (2). In (b), the horizontal dashed lines show that y has an error of at most 0.03.

We must find x with an error of at most 0.0001 in order to find $y = V(x)$ with an error of at most 0.03.

Although most nonlinear systems of equations with two variables are very difficult or impossible to solve algebraically, it may be possible to do so with some systems. The next example illustrates a system whose exact solutions can be found algebraically.

E X A M P L E   6   Solving a System Graphically and Algebraically

Find the simultaneous solutions to the system

$$y = x^3 - x \tag{3a}$$

$$y = 3x. \tag{3b}$$

Solution **Graphical Method** Find the complete graph of this system in the viewing rectangle [−5, 5] by [−10, 10] (see Fig. 3.63). The figure suggests that $(−2, −6)$, $(0, 0)$, and $(2, 6)$ are solutions to this system. Verify directly by substitution that these three pairs are solutions to the system.

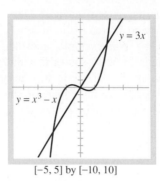

[−5, 5] by [−10, 10]

Figure 3.63 $y = x^3 - x$ and $y = 3x$.

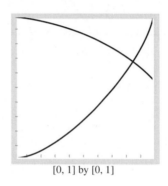

[0, 1] by [0, 1]

Figure 3.64 $y = \cos x$ and $y = x^2$.

Algebraic Method   Substitute the value of y from Eq. (3a) into Eq. (3b) and factor to find the zeros.

$$x^3 - x = 3x$$

$$x^3 - 4x = 0$$

$$x(x^2 - 4) = 0$$

$$x(x - 2)(x + 2) = 0$$

The x-values are -2, 0, and 2. The corresponding values of y are -6, 0, and 6, respectively. Thus the three solutions are $(-2, -6)$, $(0, 0)$, and $(2, 6)$.   ■

Note that an algebraic approach, although possible in Example 6, generally will not work. However, a graphical approach always will, provided the equations can be graphed with a graphing utility, as illustrated in the next example.

E X A M P L E   7   Solving a System Graphically

Find the simultaneous solution to the system

$$y = \cos x$$

$$y = x^2$$

such that $0 \le x \le 1$.

Solution   Since x is between 0 and 1, find the graph of the system in the viewing rectangle [0, 1] by [0, 1]. Then zoom in to find the unique solution to be the pair $x = 0.82$ and $y = 0.68$ (see Fig. 3.64).   ■

Exercises for Section 3.7

In Exercises 1 and 2, graph the equations E_1, E_2, and E_3 in the standard viewing rectangle $[-10, 10]$ by $[-10, 10]$.

$$x - 2y = 4 \qquad \qquad (\mathbf{E_1})$$

$$4y - 2x = 12 \qquad \qquad (\mathbf{E_2})$$

$$4y - 2x = -8 \qquad \qquad (\mathbf{E_3})$$

1. Describe geometrically the simultaneous solution to the system of equations E_1 and E_2.

2. Describe geometrically the simultaneous solution to the system of equations E_1 and E_3.

In Exercises 3–6, use an algebraic method to determine the simultaneous solution to each system. Use a graphing utility and zoom-in to support your answer.

3. $\begin{cases} x - y = \ \ 4 \\ 2x + y = \ \ 14 \end{cases}$ 4. $\begin{cases} 2x - 3y = 33 \\ 5x + 2y = 35 \end{cases}$

5. $\begin{cases} 2x + 2y = \ \ 40 \\ \qquad xy = 100 \end{cases}$ 6. $\begin{cases} y = x^3 - x \\ y = x \end{cases}$

7. Consider the system of equations

$$y = x^3 - 2x^2 + 3x - 5$$

$$y = 5x.$$

Find each simultaneous solution (x, y) to the system and determine what maximum error in x will ensure that the maximum error in y is at most 0.01. For each simultaneous solution, specify a viewing rectangle for which the solution can be read with an error of at most 0.01.

In Exercises 8–17, solve each system. (*Remember:* Both x and y must have an error of at most 0.01.)

8. $\begin{cases} 2x - 5y = 6 \\ y = 6 - x^2 \end{cases}$

9. $\begin{cases} y = x^2 - 4 \\ y = 6 - x^2 \end{cases}$

10. $\begin{cases} y = x^3 \\ y = 4 - x^2 \end{cases}$

11. $\begin{cases} y = 2x^2 - 3x - 10 \\ y = \dfrac{1}{x} \end{cases}$

12. $\begin{cases} y = 16x - x^3 \\ y = 10 - x^2 \end{cases}$

13. $\begin{cases} y = 2x + 10 - x^2 \\ y = 2x^3 + 13x^2 - 9x + 1 \end{cases}$

14. $\begin{cases} y = \sin x \\ y = 2 - x^2 \end{cases}$

15. $\begin{cases} y = 2 - \cos x \\ y = x^3 \end{cases}$

16. $\begin{cases} y = \sin x \\ y = x^3 - x \end{cases}$

17. $\begin{cases} y = \dfrac{1}{x} \\ y = 1 + \cos x \end{cases}$

18. Let $H(x) = ax + b$ be a polynomial function of degree 1. If $H(2) = -3$ and $H(-4) = -5$, determine a and b.

19. A function T describing the behavior of a certain physical phenomenon is known to be a polynomial function of degree 2; that is, $T(x) = ax^2 + bx + c$. It is also known that $T(0) = 5$, $T(1) = -2$, and $T(3) = 6$. Determine the function T.

20. A scholarship fund earns income from three investments at simple interest rates of 5%, 6%, and 10%, respectively. The annual income available for scholarships from the three investments is $1000. If the amount invested at 5% is 5 times the amount invested at 6% and the total amount invested at 5% and 6% is equal to the amount invested at 10%, determine how much is invested at each rate.

21. The total value of 17 coins consisting of nickels, dimes, and quarters is $2.95. There are twice as many quarters as nickels. Determine the number of nickels, dimes, and quarters.

22. The sum of the digits of a certain two-digit number is 16. If the digits are reversed, the original number is increased by 18. What is the original number?

For Exercises 23–25, sketch the graphs of $x^2 + y^2 = 16$ and $x + y = 2$ on the same coordinate system.

23. Determine the number of simultaneous solutions to the system

$$x^2 + y^2 = 16$$
$$x + y = 2.$$

24. Determine the solutions with errors of at most 0.01 using zoom-in with a graphing utility. (*Hint:* Write $x^2 + y^2 = 16$ as $y = \sqrt{16 - x^2}$ or $y = -\sqrt{16 - x^2}$.)

25. Determine the simultaneous solutions algebraically.

For Exercises 26–28, sketch the graphs of $x^2 + y^2 = 25$ and $y = x^2 - 16$ on the same coordinate system.

26. Determine the number of simultaneous solutions to the system

$$x^2 + y^2 = 25$$
$$y = x^2 - 16.$$

27. Determine the solutions with errors of at most 0.01 using zoom-in with a graphing utility.

28. Determine the simultaneous solutions algebraically.

Exercises 29–31 refer to the following **problem situation** : Let x be the side length of the square that must be cut out from each corner of a 30- by 40-in. piece of cardboard to form a box with no top.

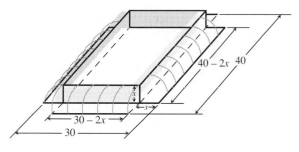

For Exercises 29–31.

29. Express the volume $V(x)$ of the box as a function of x.

30. Write a system of equations to solve in order to find the size of the square that must be cut from each corner of the piece of cardboard to produce a box with a volume of 1200 cu in. Find a complete graph of the system on a grapher. How many solutions are there to this problem situation?

31. Use zoom-in to determine all values of x that produce a box whose volume is 1200 cu in. State the maximum error

in x necessary to produce an error in the volume of at most 0.01 cu in.

Exercises 32 and 33 refer to the following economic **problem situation:** Economists have determined that supply curves are usually increasing (as the price increases, sellers increase production) and the demand curves are usually decreasing (as the price increases, consumers buy less). Suppose that a certain single-commodity market situation is modeled by the following system:

$$\text{Supply:} \quad P = 10 + 0.1x^2$$

$$\text{Demand:} \quad P = \frac{50}{1 + 0.1x}$$

32. Find the graph of both the supply and demand equations in the *first* quadrant; P is the price (vertical axis), and x represents the number of units of the commodity produced.

33. Determine the equilibrium price, that is, the price at which supply is equal to demand.

34. Repeat Exercises 32 and 33 for a single-commodity market situation modeled by the following system:

$$\text{Supply:} \quad P = 5 + 0.014x^2$$

$$\text{Demand:} \quad P = \frac{133}{1 + 0.0251x}$$

Exercises 35–37 refer to the following investment **problem situation:** Sam makes two initial investments of $1000 each. The first earns 10% simple interest, and the second earns interest using 5% *simple discount*. After n years, the value of the first investment is $S = 1000(1 + 0.10n)$ and the value of the second investment is $S = 1000/(1 - 0.05n)$.

35. Find the graph of each of the above functions in the viewing rectangle [0, 30] by [0, 6000].

36. Is there a time when the future values of both investments are equal?

37. What happens to the value of the simple discount investment at the end of
a) 19 yr?
b) 19.5 yr?
c) 19.9 yr?
d) 19.99 yr?
e) Exactly 20 yr?
Could your results explain why the simple discount model is used only for short periods of time in actual practice? (*Note:* Normally, interest is paid at the end of a term. In the case of simple discount, interest is paid in advance, at the beginning of the term.)

Chapter 3 Review

KEY TERMS (The number following each key term indicates the page of its introduction.)

REVIEW EXERCISES

In Exercises 1–4, graph the two functions in the same viewing rectangle.

1. $y = (x + 1)(x - 3)$; $y = -2(x + 1)(x - 3) + 4$

2. $y = (x - 1)(x + 2)(x - 3)$; $y = 3(x - 1)(x + 2)(x - 3) - 5$

3. $y = (x + 2)(x - 3)^2$; $y = 3(x + 2)(x - 3)^2 - 5$

4. $y = (x - 1)(x - 2)(x - 3)(x - 4)$;
$y = 3(x - 1)(x - 2)(x - 3)(x - 4) - 8$

In Exercises 5–12, draw a complete graph for each function. Find all the real-number zeros and all local extrema.

5. $f(x) = x^3 - 3x - 2$

6. $f(x) = x^3 - 6x^2 + 9x + 1$

7. $f(x) = 3x^4 - 4x^3 + 3$

8. $f(x) = x^4 + 8x^3 - 270x^2 + 10$

9. $f(x) = x^5 + 2x^4 - 6x^3 + 2x - 3$

10. $f(x) = 1000x^3 + 780x^2 - 5428x + 3696$

11. $f(x) = x^4 - 5x^3 + 2x - 10$

12. $f(x) = 100x^5 - 270x^4 + 36x^3 + 400x^2 - 280x + 460$

In Exercises 13–16, draw a complete graph of the function $y = f(x)$ in an appropriate viewing rectangle. Then find two values for c for which $f - c$ has three real zeros.

13. $f(x) = x^3 - 6x^2 + 9x + 6$

14. $f(x) = x^3 + 2x^2 + 2x - 5$

15. $f(x) = x^3 + x^2 - 4x - 9$

16. $f(x) = -x^3 - x^2 + 2x - 10$

In Exercises 17 and 18, draw a complete graph of the function $y = f(x)$ in an appropriate viewing rectangle. Then find two values for c for which $f - c$ has four real zeros.

17. $f(x) = x^4 - 4x^3 + 3x^2 + 2$

18. $f(x) = x^4 - 2x^2 - 4$

In Exercises 19–21, draw a complete graph for each function. Find all the real-number zeros and all the local extrema.

19. $f(x) = 2x^3 - 3x^2 - 72x + 76$

20. $f(x) = 3x^4 - 4x^3 - 12x^2 + 13$

21. $f(x) = x^5 + 2x^4 - 6x^3 + 2x + 3$

In Exercises 22–25, determine the domain, range, and the points of discontinuity of each function.

22. $g(x) = \begin{cases} x^2 - 1 & x < 2 \\ \dfrac{3x}{2} & x \geq 2 \end{cases}$

23. $V(x) = 5x^3 - 25x^2 + 30x + 9$

24. $y = \dfrac{1}{x + 5}$

25. $f(x) = \text{INT}(x + 2)$

In Exercises 26 and 27, determine the domain, range, and values of x for which the function given by the graph is discontinuous. Assume the graph is complete.

26.

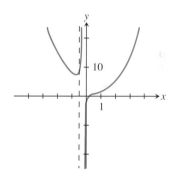

27.

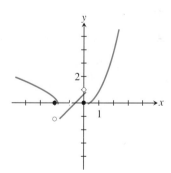

28. Graph the polynomial functions $f(x) = 5x^5 + 2x^2 - 4x + 7$ and $g(x) = 5x^5$ in a viewing rectangle that graphically illustrates that the end behavior of the two functions are the same, that is, a viewing rectangle in which the two graphs are nearly identical.

In Exercises 29 and 30, assume the graph of $y = f(x)$ is complete.

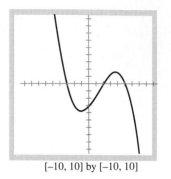

[–10, 10] by [–10, 10]

For Exercises 29 and 30.

29. As $x \to \infty$, $f(x) \to$?

30. As $x \to -\infty$, $f(x) \to$?

In Exercises 31 and 32, assume the graph of $y = f(x)$ is complete.

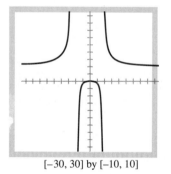

[–30, 30] by [–10, 10]

For Exercises 31 and 32.

31. As $x \to \infty$, $f(x) \to$?

32. As $x \to -\infty$, $f(x) \to$?

In Exercises 33 and 34, find the quotient $q(x)$ and remainder $r(x)$ when $f(x)$ is divided by $h(x)$.

33. $f(x) = 3x^2 - 2x + 7$; $h(x) = x + 2$

34. $f(x) = 6x^3 - x^2 + 9x + 1$; $h(x) = 3x + 2$

35. Use the remainder theorem to determine the remainder when $f(x) = 3x^3 - 2x + 17$ is divided by $x - 1$. Check by using division.

36. Determine the remainder when $f(x) = 3x^{38} - 2x^{15} + 17x^2 - 3x + 12$ is divided by $x - 1$.

37. Use the factor theorem to determine whether $x - 3$ is a factor of $x^3 - 2x^2 - 4x + 3$.

38. Determine the coordinates of a point (a, b) that must be added to the graph of

$$f(x) = \frac{x^2 - 4}{x + 2}$$

to make the function continuous for every real number.

In Exercises 39 and 40, find all real-number zeros of each polynomial by factoring. Use a graphing utility to support your answer.

39. $f(x) = 8x - x^3$ **40.** $T(x) = x^4 - 13x^2 + 36$

41. Draw a complete graph of $f(x) = x^3 - x - 6$. Use the graph as an aid in factoring the polynomial and then determine all real-number zeros.

42. Determine *all* the zeros, real and nonreal, of the polynomial in Exercise 41.

43. Determine all the real-number zeros of $f(x) = x^5 - 5x^4 - x^3 + 5x^2 + 16x - 80$.

44. Draw the graph of $f(x) = 800x^3 + 780x^2 - 21x - 1$ in the $[-5, 5]$ by $[-150, 150]$ viewing rectangle.
 a) How many real-number zeros are evident from this graph?
 b) How many actual real-number zeros exist in this case?
 c) Find all real-number zeros.

45. How many real-number zeros does $f(x) = 2x^4 + 2x^3 + 3x^2 + 2x + 1$ have? Why?

46. Draw the graph of $f(x) = 33x^3 - 300.5x^2 + 903x + 500$ in the $[-10, 10]$ by $[-3000, 3000]$ viewing rectangle.
 a) Where does the function appear to be increasing based on this graph?
 b) How many local extrema does the function have based on this graph?
 c) Determine the actual number of local extrema of the function.

47. Draw a complete graph of $f(x) = 4x^3 + mx$ for $m = -3$, -2, 0, 2, and 4. For each value of m, determine
 a) the number of real-number zeros of f,
 b) the number of local extrema of f,
 c) the intervals on which f is increasing, and
 d) the intervals on which f is decreasing.

48. Determine a polynomial of degree 3 with real-number co-efficients that has
  a) 2 and -3 as zeros,
  b) 2 and -3 as the *only* zeros, and
  c) 2 and $2 - i$ as zeros.

49. If $f(x) = 3x^3 + 2kx^2 - kx + 2$, find a number k so that the graph of f contains the point $(1, 3)$.

In Exercises 50 and 51, determine all real-number zeros of each function using algebraic methods. Classify them as integer, noninteger rational, or irrational. Support your work with a graphing utility.

50. $f(x) = x^3 + 3x^2 - 5x - 15$

51. $f(x) = x^4 - 8x^2 - 9$

In Exercises 52 and 53, use the rational zeros theorem to list all possible rational zeros of each function. Find the graph of each function in an appropriate viewing rectangle and indicate which of the possible rational zeros are actual zeros. Determine all real-number zeros (both rational and irrational) of each function.

52. $f(x) = x^3 - 5x^2 + 2x - 10$

53. $f(x) = 3x^3 + 3x^2 - 4x - 2$

54. Use synthetic division to find the quotient and remainder when $2x^3 + x^2 - 4x + 5$ is divided by $x + 5$.

55. Draw a complete graph of $f(x) = x^3 + 7x^2 + 6x - 8$. Confirm that there is at least one rational root. Use synthetic division to find the other quadratic factor and then determine all real-number zeros. Classify each real-number zero as rational or irrational.

56. Determine the real-number zeros of $f(x) = x^4 + 3x^3 - 3x^2 - 3x + 2$. Classify them as rational or irrational. Carefully outline why you are sure of your results.

In Exercises 57 and 58, use synthetic division to find upper and lower bounds for the zeros of each polynomial function.

57. $f(x) = x^3 + 3x - 1$

58. $g(x) = x^3 - x^2 - 2x + 1$

59. Show that $x + 3$ is a factor of $g(x) = x^6 - 729$.

In Exercises 60–63, write each expression in the form $a + bi$ where a and b are real numbers.

60. $(5 - 2i)(1 - i)$

61. $3(2 + 3i) - 4 + 2i$

62. $\dfrac{3 + i}{5 + 4i}$

63. $(i^{93})^4$

In Exercises 64 and 65, determine the number of real zeros and the number of nonreal complex zeros of each function.

64. $f(x) = 2x^3 + 3x^2 + 3x + 1$

65. $f(x) = x^3 - 2x^2 + 2x - 2$

66. Use computer graphing, the rational zeros theorem, synthetic division, and the fundamental theorem of algebra to find all the zeros of $f(x) = x^4 - 3x^3 - 12x - 16$. Classify each zero as integer, noninteger rational, irrational, or nonreal complex.

In Exercises 67 and 68, find a polynomial with real-number coefficients satisfying the given condition.

67. Degree 2; zero $1 - 2i$

68. Degree 3; zeros $2 + 5i$ and 2

69. Find a polynomial $f(x)$ of degree 4 with real-number coefficients that has 2, $3 + i$, and $3 - i$ as its only zeros.

70. Find a polynomial $f(x)$ of degree 4 with real-number coefficients that has 2, $3 + i$, and $3 - i$ as its only zeros and that also satisfies $f(0) = 1$.

Exercises 71–75 refer to the following **problem situation**: The temperature in a certain town for a 24-hr period is given by the algebraic representation $f(t) = 0.04t(t - 24)(t - 22) + 30$. Assume $t = 0$ is 6 A.M.

71. Without using a graphing utility, sketch a complete graph of the model. What is the domain of the model? What values of t make sense in this problem situation?

72. Use a graphing utility to draw a complete graph of the model and compare it with your sketch in Exercise 71.

73. When will the temperature be $45°$F?

74. What is the highest temperature, and when is it achieved?

75. What is the lowest temperature, and when is it achieved?

A good deal of real-world mathematics involves interpreting concrete data generated from experiments, observations, or computer simulations. Polynomials are frequently used to create functions that approximate the observed data and then used as algebraic representations of the real-world problem. Exercises 76–80 explore the nature of *Chebyshev polynomials*, which are useful in this type of investigation.

76. Draw a complete graph of the third-degree Chebyshev polynomial $C_3(x) = 4x^3 - 3x$.

77. Draw a complete graph of the fourth-degree Chebyshev polynomial $C_4(x) = 8x^4 - 8x^2 + 1$.

78. Draw a complete graph of the fifth-degree Chebyshev polynomial $C_5(x) = 16x^5 - 20x^3 + 5x$.

79. What are the local extrema of each of the Chebyshev polynomials in Exercises 76 to 78?

80. Can you predict the graph of the sixth-degree Chebyshev polynomial?

81. The polynomial function $M(x) = x^m(1-x)^{n-m}$ is used in applied statistics to determine *maximum likelihood estimates*. For the indicated values of m and n, draw the graph of the function $y = M(x)$ in the $[-2, 2]$ by $[-2, 2]$ viewing rectangle.
a) $m = 2$, $n = 5$
b) $m = 4$, $n = 8$
c) $m = 3$, $n = 10$

Exercises 82–85 refer to the following **problem situation**: A 255-ft-long steel beam is anchored between two pilings 50 ft above the ground. An algebraic representation for the amount of vertical deflection s caused by a 250-lb object placed d feet from the west piling is $s = (8.5 \times 10^{-7})d^2(255 - d)$.

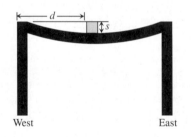

West   East

For Exercises 82–85.

82. Draw a complete graph of the model. What is the domain of the model? What values of d make sense in this problem situation?

83. Where is the object placed if the amount of vertical deflection is 1 ft?

84. What is the greatest amount of vertical deflection, and where does it occur?

85. Give a possible scenario explaining why the solution to Exercise 84 does not occur at the halfway point.

Exercises 86–89 refer to the following **problem situation**: A liquid storage container on a truck is in the shape of a cylinder with hemispheres on each end. The cylinder and hemispheres

have the same radius. The total length of the container is 140 ft.

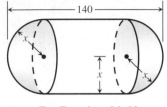

For Exercises 86–89.

86. Determine the volume V of the container as a function of the radius x.

87. Draw a complete graph of the volume function $y = V(x)$.

88. What are the possible values of the radius determined by this problem situation?

89. What is the radius of the container with largest possible volume?

Exercises 90–93 refer to the following **problem situation**: The total daily revenue of Henrietta's Hamburger Haven is given by the equation $R = x \cdot p$, where x is the number of hamburgers sold and p is the price of one hamburger. Assume the price per hamburger is given by the supply equation $p = 0.3x + 0.05x^2 + 0.0007x^3$.

90. Determine for what positive values of x the supply function is increasing.

91. Draw a complete graph of the revenue function R.

92. What values of x make sense in the problem situation?

93. Determine the maximum possible daily revenue of Henrietta's Hamburger Haven and the number of hamburgers she needs to sell to achieve maximum revenue.

Exercises 94–97 refer to the following **problem situation**: Biologists have determined that the polynomial function $P(t) = -0.00005t^3 + 0.003t^2 + 1.2t + 80$ approximates the population t days later of a certain group of wild pheasants left to reproduce on their own with no predators.

94. Draw a complete graph of $y = P(t)$.

95. Find the maximum pheasant population and when it occurs.

96. When will this pheasant population be extinct?

97. Create a scenario that could explain the "growth" exhibited by this pheasant population.

In Equations 98–107, use a graphical method to solve each system.

98. $\begin{cases} 2x - 3y = 2 \\ 4x + y = 5 \end{cases}$

99. $\begin{cases} 5x - 17y = 42 \\ 9x + 6y = 19 \end{cases}$

100. $\begin{cases} 0.32x + 1.8y = 3.2 \\ 2.4x - 0.08y = 5.2 \end{cases}$

In Exercises 101–107, solve each system.

101. $\begin{cases} 3x + 2y = 6 \\ y = x^2 - 2x + 5 \end{cases}$

102. $\begin{cases} y = |2x + 5| \\ y = 3x^2 - 2x + 1 \end{cases}$

103. $\begin{cases} x^2 - 2y - 4x + 10 = 0 \\ 5x - 4y + 24 = 0 \end{cases}$

104. $\begin{cases} x^2 - 2y - 4x + 10 = 0 \\ 5x - y + 24 = 0 \end{cases}$

105. $\begin{cases} y = 16 - x^2 \\ y = 9x - x^3 \end{cases}$

106. $\begin{cases} y = 6 - x^2 \\ y = 3\sin(x - 4) \end{cases}$

107. $\begin{cases} y = 115x - 3x^3 \\ y = 50\cos x \end{cases}$

MARIA AGNESI

Maria Gaetana Agnesi (1718–1799) was recognized as a genius at an early age. At age 9, her interests included logic, mechanics, chemistry, botany, zoology, mineralogy, and analytic geometry. By age 11, she was fluent in seven languages. Agnesi's first concern was helping the poor and underprivileged people of her native Italy, but her father persuaded her to concentrate on studying mathematics.

Agnesi's masterpiece, *Analytical Institutions for the Use of Italian Youth,* included discoveries in algebra, calculus, and differential equations. Her name is associated with a bell-shaped curve, the *versiera* of Agnesi, which is still used in mathematics and physics. Agnesi was appointed to the chair of mathematics and natural philosophy at the University of Bologna, but she never actually lectured there. After her father died, she withdrew from her studies to devote herself to charitable work and religious studies.

CHAPTER 4

Rational Functions and Functions Involving Radicals

4.1

Rational Functions and Asymptotes

Chapter 3 examined the class of functions known as the polynomials. Recall that polynomial functions are continuous functions that often have local maximums and minimums. Also recall the exploration of end behavior models and zeros of polynomial functions.

Your knowledge of these characteristics of polynomial functions will be useful when learning about a class of functions known as rational functions—functions that are a quotient of two polynomial functions.

Definition 4.1 Rational Functions

A **rational function** is one that can be written in the form

$$f(x) = \frac{p(x)}{q(x)},$$

where $p(x)$ and $q(x)$ are polynomial functions, $q(x) \neq 0$.

The following are examples of rational functions:

$$f(x) = \frac{x^2 + 1}{x - 3} \qquad g(x) = \frac{3x^2 - 4x + 1}{x^2 + x - 2} \qquad h(x) = \frac{4x^3 + 7x^2 - 3}{(x - 2)(x + 4)}$$

The domain of a rational function is the set of all real numbers except those for which the denominator is zero.

EXAMPLE 1 Finding Domains of Rational Functions

Find the domains of the following rational functions:

a) $f(x) = \dfrac{x^2 + 1}{x - 3}$

b) $g(x) = \dfrac{3x^2 - 4x + 1}{x^2 + x - 2}$

c) $h(x) = \dfrac{4x^3 + 7x^2 - 3}{(x - 2)(x + 4)}$

Solution

a) Domain of $f = (-\infty, 3) \cup (3, \infty)$

b) Domain of $g = (-\infty, -2) \cup (-2, 1) \cup (1, \infty)$ $x^2 + x - 2 = (x + 2)(x - 1)$

c) Domain of $h = (-\infty, -4) \cup (-4, 2) \cup (2, \infty)$ ▤

Horizontal Asymptotes

Perhaps the simplest rational function is the **reciprocal function**, $f(x) = 1/x$, which is graphed in Fig. 4.1. The following table gives numerical evidence that as $|x|$ gets larger, $f(x)$ gets closer and closer to zero.

x	± 1	± 2	± 3	± 4	$\cdots$	± 1000
$f(x)$	± 1	$\pm 1/2$	$\pm 1/3$	$\pm 1/4$	$\cdots$	$\pm 1/1000$

The following Exploration will give visual evidence of the same thing.

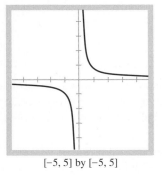

[−5, 5] by [−5, 5]

Figure 4.1 $f(x) = \dfrac{1}{x}$.

🔍 EXPLORE WITH A GRAPHING UTILITY

Find the graph of $f(x) = 1/x$ in each of the following viewing rectangles. In each case, observe the values of the x- and y-coordinates as you move the cursor along the graph to the right using the trace key.

a) $[-10, 10]$ by $[-2, 2]$

b) $[-100, 100]$ by $[-0.1, 0.1]$

c) $[-1000, 1000]$ by $[-0.01, 0.01]$

MATHEMATICAL NOTATION

Another way of writing:

as $x \to \infty$, $f(x) \to 0$

is: $\lim\limits_{x \to \infty} f(x) = 0$.

This notation is read "the limit of $f(x)$ as x approaches infinity is zero."

For the function $f(x) = 1/x$ used in the Exploration, we say, "as x approaches positive or negative infinity, $f(x)$ approaches 0." We write

as $x \to \infty$ or as $x \to -\infty$, $f(x) \to 0$.

The horizontal line $y = 0$ is called a horizontal asymptote of $f(x) = 1/x$.

Figure 4.2 shows several other cases of horizontal asymptotes. In each case, the graph approaches the horizontal line $y = L$ as $|x| \to \infty$.

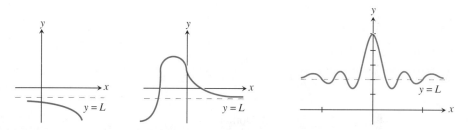

Figure 4.2 Examples of graphs and their horizontal asymptotes.

Definition 4.2 Horizontal Asymptote

The horizontal line $y = L$ is a **horizontal asymptote** of a function f if

$$f(x) \to L \text{ as } x \to \infty \qquad or \qquad f(x) \to L \text{ as } x \to -\infty.$$

E X A M P L E   2   Finding Horizontal Asymptotes

Find the horizontal asymptotes of $g(x) = 1/x + 2$ and $h(x) = 1/x - 3$.

Solution

$$\frac{1}{x} \to 0 \text{ as } |x| \to \infty$$

Therefore

$$g(x) \to 2 \qquad \text{and} \qquad h(x) \to -3, \text{ as } |x| \to \infty.$$

The line $y = 2$ is a horizontal asymptote for g, and the line $y = -3$ is a horizontal asymptote for h.

The graphs of g and h are shown in the standard viewing rectangle in Figs. 4.3 and 4.4. Try other viewing rectangles where $|X\text{max}|$ and $|X\text{min}|$ are large.

The graphs of g and h provide visual support for these conclusions. These graphs can be found either by using a graphing utility or by using the horizontal and vertical shift transformations studied in Section 3.6. These transformations are illustrated in Example 3.

E X A M P L E   3   Sketching Graphs Using Transformations

Sketch the graphs of functions g and h defined in Example 2 by applying transformations studied in Section 3.6 to the graph of $f(x) = 1/x$.

Solution

$$g(x) = \frac{1}{x} + 2 = f(x) + 2$$

$$h(x) = \frac{1}{x} - 3 = f(x) - 3$$

Therefore the graph of g is obtained from the graph of f by a vertical shift up 2 units. Similarly, the graph of h is obtained from the graph of f by a vertical shift down 3 units.   ▤

Example 4 shows that a function may have two horizontal asymptotes.

E X A M P L E   4   Two Horizontal Asymptotes

Find a complete graph of

$$f(x) = \frac{x - 2}{|x| + 3}$$

and show algebraically that f has two horizontal asymptotes.

Solution   A complete graph appears in Fig. 4.5.

Because $|x| \geq 0$, we see that $|x| + 3 > 0$. Consider two cases:

Case 1   Suppose $x \geq 0$. Then $|x| = x$. Using division, f can be written as

$$f(x) = \frac{x - 2}{x + 3} = 1 - \frac{5}{x + 3} \qquad (x \geq 0).$$

As $x \to \infty$, $f(x) \to 1$, which means that the line $y = 1$ is a horizontal asymptote.

Case 2   Suppose $x < 0$. Then $|x| = -x$. Using division, f can be written as

$$f(x) = \frac{x - 2}{-x + 3} = -1 + \frac{1}{-x + 3} \qquad (x < 0).$$

As $x \to -\infty$, $f(x) \to -1$, which means that the line $y = -1$ is a horizontal asymptote.   ▤

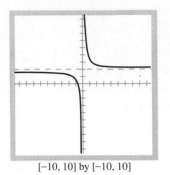

[−10, 10] by [−10, 10]

Figure 4.3 $g(x) = 1/x + 2$.

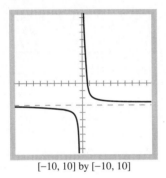

[−10, 10] by [−10, 10]

Figure 4.4 $h(x) = 1/x - 3$.

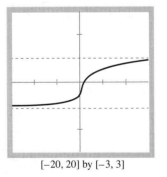

[−20, 20] by [−3, 3]

Figure 4.5 $f(x) = \dfrac{x - 2}{|x| + 3}$.

Vertical Asymptotes

Again consider the reciprocal function $f(x) = 1/x$ shown in Fig. 4.1. The following table shows that as x approaches 0 from the positive side, $f(x)$ increases without bound. That is,

$$f(x) \to +\infty \quad \text{as} \quad x \to 0^+.$$

The plus sign in 0^+ means that x assumes only values greater than zero. Likewise, 0^- means x assumes only values less than zero.

x	1	1/2	1/3	1/4	$\cdots$	1/1000	$x \to 0^+$
$f(x)$	1	2	3	4	$\cdots$	1000	$f(x) \to \infty$

A similar table with x approaching 0 through values less than zero would show that

$$f(x) \to -\infty \text{ as } x \to 0^-.$$

We say that the vertical line $x = 0$ is a vertical asymptote. Figure 4.6 shows some additional examples of vertical asymptotes.

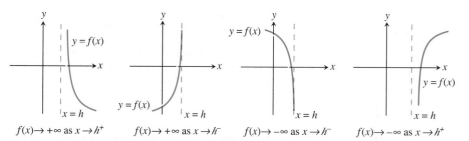

Figure 4.6 Examples of graphs and their vertical asymptotes.

Definition 4.3 Vertical Asymptotes

The vertical line $x = h$ is called a **vertical asymptote** of a function f if

$$f(x) \to +\infty \qquad \text{or} \qquad f(x) \to -\infty$$

as $x \to h$ from the right or from the left.

If the numerator and denominator of a rational function have no common factors, a vertical asymptote will occur whenever the denominator is zero.

E X A M P L E 5   Finding Vertical Asymptotes

Find the vertical asymptotes of (a) $f(x) = 1/(x-2)$ and (b) $g(x) = 1/(x+3) - 1$.

Solution

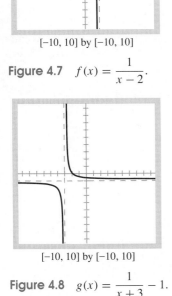

[−10, 10] by [−10, 10]

Figure 4.7 $f(x) = \dfrac{1}{x-2}$.

a) Consider where the denominators are zero. The function f is undefined if $x - 2 = 0$.

$$x - 2 = 0 \Longleftrightarrow x = 2$$

Therefore the vertical line $x = 2$ is a vertical asymptote for f (see Fig. 4.7).

b) The function g is undefined if $x + 3 = 0$.

$$x + 3 = 0 \Longleftrightarrow x = -3$$

Therefore the vertical line $x = -3$ is a vertical asymptote for g. ≡

[−10, 10] by [−10, 10]

Figure 4.8 $g(x) = \dfrac{1}{x+3} - 1$.

E X A M P L E 6   Sketching Graphs Using Transformations

Sketch the graphs of (a) $f(x) = 1/(x-2)$ and (b) $g(x) = 1/(x+3) - 1$ by applying transformations to the graph of $h(x) = 1/x$.

Solution

a)
$$f(x) = h(x-2) = \frac{1}{x-2}$$

Therefore the graph of f is obtained from the graph of $1/x$ by a horizontal shift 2 units right (see Fig. 4.7).

b)
$$g(x) = \frac{1}{x+3} - 1 = h(x+3) - 1$$

The graph of g is obtained from the graph of $1/x$ by a horizontal shift left 3 units followed by a vertical shift down 1 unit (see Fig. 4.8). ≡

REMINDER

Notice that there are no vertical asymptotes for

$$f(x) = \frac{4 - 2x}{x - 2}$$

because this numerator and denominator have a common factor of $x - 2$, leaving $f(x) = -2$.

E X A M P L E 7   Sketching Graphs Using Long Division

Determine the horizontal and vertical asymptotes for

$$f(x) = \frac{2x - 13}{x - 5}.$$

Also describe how a graph of f can be obtained from $y = 1/x$.

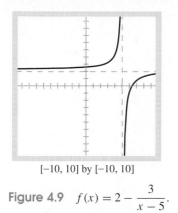

[−10, 10] by [−10, 10]

Figure 4.9 $f(x) = 2 - \dfrac{3}{x-5}$.

Notes on Examples

Example 7 shows students how to transform a rational function of the form $f(x) = \dfrac{(ax+b)}{(cx+d)}$ into $f(x) = \dfrac{r}{(x-h)} + k$ by dividing the numerator by the denominator. The latter form provides critical information to determine the attributes of the graph.

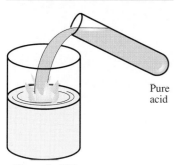

50 oz. of a 35% acid solution

Figure 4.10   Mixing pure and diluted acid solutions.

Assignment Guide

Day 1: Ex. 1, 7, 10, 13, 16, 19, 22
Day 2: Ex. 25, 28, 31, 34, 37, 40, 42, 43
Day 3: Ex. 46–51

Solution   Begin by finding a vertical asymptote:

$$x - 5 = 0 \iff x = 5 \qquad \text{and} \qquad 2x - 13 \neq 0 \text{ when } x = 5.$$

Because the numerator is not also zero when $x = 5$, the vertical line $x = 5$ is a vertical asymptote for f. Because the numerator is not a constant, complete the following long division to find that the quotient is 2 and the remainder is -3:

$$
\begin{array}{r}
2 \\
x - 5 \overline{)\, 2x - 13} \\
2x - 10 \\
\hline
-\,3
\end{array}
$$

Therefore

$$f(x) = \frac{2x - 13}{x - 5} = 2 - \frac{3}{x - 5}.$$

The horizontal line $y = 2$ is a horizontal asymptote.

The graph of f shown in Fig. 4.9 can be obtained from $y = 1/x$ by applying the following transformations:

1. Horizontal shift 5 units right;
2. Vertical stretch by a factor of 3;
3. Reflection through the x-axis; and
4. Vertical shift up 2 units.   ▤

Rational functions often occur as algebraic models of problem situations like the mixture problem in the next example. Recall from an earlier mixture problem that if you have 50 oz of a 35% acid solution, there are $50(0.35) = 17.5$ oz of pure acid in the solution.

If 10 oz of pure acid are added to this 50-oz solution, the resulting 60-oz solution contains 27.5 oz of pure acid. In this mixture, the ratio of pure acid to total solution is $27.5/60 = 0.458$, resulting in a 45.8% solution.

E X A M P L E   8   APPLICATION: Solving a Mixture Problem

Suppose that x ounces of pure acid is added to 50 oz of a 35% acid solution (see Fig. 4.10).

a) Find a complete graph of the algebraic representation $C(x)$ that represents the concentration of the new mixture as a function of x.

b) Find a complete graph of the problem situation. (That is, what values of x make sense?)

c) How much pure acid should be added to the 35% acid solution to produce a mixture that is at least 75% acid?

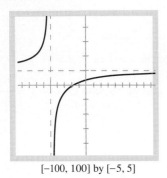

[−100, 100] by [−5, 5]

Figure 4.11   A complete graph of the algebraic representation of $C(x)$.

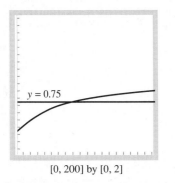

[0, 200] by [0, 2]

Figure 4.12   A complete graph of the problem situation in Example 8.

Solution

a) Let $x =$ ounces of acid added. Then

$$x + 50(0.35) = \text{ounces of pure acid in the new solution}$$

$$x + 50 = \text{ounces of new solution}$$

$$\frac{\text{ounces of pure acid}}{\text{ounces of total solution}} = \text{concentration of acid in the new solution}$$

$$C(x) = \frac{x + 17.5}{x + 50}.$$

Notice that the domain of the algebraic representation $C(x)$ is all the real numbers except -50. However, in this problem only $x \geq 0$ makes sense.

A complete graph of $C(x)$ is found in Fig. 4.11. Notice that the line $x = -50$ is a vertical asymptote. It appears that line $y = 1$ is a horizontal asymptote.

b) The portion of the graph in Fig. 4.11 where $x \geq 0$ is a complete graph of the problem situation. This portion is shown in Fig. 4.12.

c) Because the mixture is a 75% solution when $C(x) = 0.75$, the line $y = 0.75$ is also shown in Fig. 4.12. Use zoom-in to see that the two graphs intersect at the point $(80, 0.75)$. At least 80 oz of pure acid must be added.   ▤

Exercises for Section 4.1

In Exercises 1–10, find the domain of each rational function. Support your answer with a graphing utility.

1. $f(x) = \dfrac{x}{(x-2)(x+1)}$

2. $g(x) = \dfrac{5}{x^2 - 1}$

3. $h(x) = \dfrac{x+2}{(x+3)(x-1)}$

4. $f(x) = \dfrac{2}{x^2 + 4x + 3}$

5. $f(x) = \dfrac{x^2 + 1}{x^2 - 1}$

6. $t(x) = \dfrac{2}{2x^2 - x - 3}$

7. $f(x) = \dfrac{x^2 - 4}{x^2 - 4x - 1}$

8. $P(x) = \dfrac{4x - 2}{x^2 + 5x + 8}$

9. $f(x) = \dfrac{x^4 - 3x^2 - 5}{x^5 - x}$

10. $f(x) = \dfrac{2x^2 + 5}{x^3 - 2x^2 + x}$

In Exercises 11–16, find a horizontal asymptote for each of these functions.

11. $f(x) = -\dfrac{1}{x}$

12. $g(x) = \dfrac{1}{x} + 3$

13. $h(x) = \dfrac{1}{x} - 4$

14. $f(x) = 2 - \dfrac{1}{x}$

15. $g(x) = -12 + \dfrac{1}{x}$

16. $h(x) = 17 + \dfrac{1}{x}$

In Exercises 17–22, find the horizontal and vertical asymptotes for each of these functions.

17. $y = \dfrac{2}{x+1}$ **18.** $y = -\dfrac{1}{x-2}$

19. $y = 4 + \dfrac{1}{x+1}$ **20.** $y = -2 + \dfrac{2}{x+3}$

21. $y = \dfrac{2}{4-x}$ **22.** $f(x) = -1 - \dfrac{3}{x+1}$

In Exercises 23–30, list the geometric transformations needed to produce a complete graph of the function f from a complete graph of $y = 1/x$. Specify the order in which the transformations should be applied. In each case, draw a graph of the function without the aid of a graphing utility and write the equations of any vertical or horizontal asymptotes. Support your answer with a graphing utility.

23. $f(x) = \dfrac{1}{x-3}$ **24.** $f(x) = \dfrac{2}{x+2}$

25. $y = -\dfrac{2}{x+5}$ **26.** $y = \dfrac{1}{x+3}$

27. $y = -3 + \dfrac{1}{x+1}$ **28.** $y = -2 + \dfrac{1}{x+3}$

29. $y = \dfrac{5}{1-x}$ **30.** $f(x) = -1 - \dfrac{1}{x+1}$

In Exercises 31–36, find the horizontal and vertical asymptotes of each function. Use long division first if necessary.

31. $g(x) = \dfrac{3x-1}{x+2}$ **32.** $f(x) = \dfrac{8x+6}{2x-4}$

33. $f(x) = \dfrac{2x+4}{x-3}$ **34.** $h(x) = \dfrac{x-3}{2x+5}$

35. $t(x) = \dfrac{x-1}{x+4}$ **36.** $k(x) = \dfrac{2x-3}{x+2}$

Exercises 37 and 38 are based on the following computer-generated graph of $y = f(x)$.

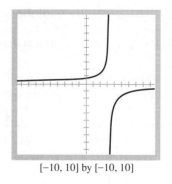

[−10, 10] by [−10, 10]

For Exercises 37 and 38.

37. As $x \to \infty$, $f(x) \to$?   **38.** As $x \to 3^-$, $f(x) \to$?

Exercises 39 and 40 are based on the following computer-generated graph of $y = g(x)$ with a vertical asymptote at $x = -3$ and a horizontal asymptote at $y = 4$.

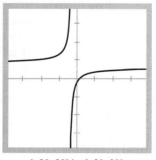

[−20, 20] by [−20, 20]

For Exercises 39 and 40.

39. As $x \to -\infty$, $g(x) \to$?   **40.** As $x \to -3^+$, $g(x) \to$?

In Exercises 41 and 42, draw a complete graph of each function. Give the domain and range and write equations of any asymptotes.

41. $y = \dfrac{x-1}{2x-2}$ **42.** $y = \dfrac{3x-9}{6-2x}$

43. Describe how the graph of $y = 3 - 2/(x-1)$ is obtained from the graph of $y = 1/x$.

44. Describe how the graph of $f(x) = r/(x-h) + k$ is obtained from the graph of $y = 1/x$.

45. Consider the function $f(x)$ with $c \neq 0$ and $bc > ad$:

$$f(x) = \frac{ax+b}{cx+d}.$$

List the transformations that will produce a complete graph of f from a complete graph of $y = 1/x$ and specify an order in which they should be applied. (*Hint:* Use long division.)

46. Consider the following **problem situation**: The area of a rectangle is 300 square units. Determine an algebraic representation that gives the length of this rectangle as a function of its width x. Sketch a complete graph of both this algebraic representation and of this problem situation. What is the width of the rectangle if the length is 2000 units?

47. Solve Example 8 algebraically.

48. Consider the following **problem situation**: x ounces of pure acid are added to 125 oz of a 60% acid solution.
   a) Find a complete graph of this problem situation where

$C(x)$ represents the concentration of the new mixture as a function of x.
   b) Use a computer-drawn graph to determine how much pure acid should be added to the 60% solution to produce a new mixture that is at least 83% acid.
   c) Solve part (b) algebraically.

49. Sally wishes to obtain 100 oz of a 40% acid solution by combining a 60% acid solution and a 10% acid solution. How much of each solution should Sally use? Solve both algebraically and graphically.

50. Find the domain and range of $f(x) = r/(x - h) + k$.

51. Find the domain and range of $f(x)$ where $c \neq 0, d \neq 0$:

$$f(x) = \frac{ax + b}{cx + d}.$$

4.2

Rational Functions and Their Graphs

Objective

Students will investigate the limit concept by writing end behavior models and asymptotes for rational functions.

Key Ideas

Removable discontinuity
End behavior model
End behavior asymptote
Slant asymptote
Hidden behavior

Frank says, "A rational function is a polynomial with a few bad places and a polynomial is a line with a few wiggles."

All the rational functions in Section 4.1 met the special condition that the numerator and denominator were both linear polynomials.

In the rational functions in this section, either the numerator or the denominator (or both) are polynomials of degree 2 or higher.

Numerator and Denominator with Common Factors

When the numerator and denominator of a rational function have a common factor, the function has a discontinuity that may not be noticed when using a graphing utility. Consider the following two rational functions:

$$f(x) = \frac{x - 1}{x - 1} \quad \text{and} \quad g(x) = \frac{(x - 1)^2}{x - 1}.$$

The domain of each function consists of all the real numbers except $x = 1$. These functions can be simplified algebraically as follows:

$$f(x) = \frac{x - 1}{x - 1} = 1 \qquad (x \neq 1)$$

$$g(x) = \frac{(x - 1)^2}{x - 1} = x - 1 \qquad (x \neq 1).$$

E X A M P L E   1   Common Factors in Rational Functions

Sketch complete graphs of the functions:

$$f(x) = \frac{x-1}{x-1} \quad \text{and} \quad g(x) = \frac{(x-1)^2}{x-1}.$$

Solution   The graph of f and the line $y = 1$ are identical except at $x = 1$. Because f is undefined at $x = 1$, the graph of f is the horizontal line $y = 1$ with a missing point at $(1,1)$ (see Fig. 4.13).

The graph of g and the line $y = x - 1$ are identical except for $x = 1$. So the graph of g is the line $y = x - 1$ with a missing point at $(1,0)$ (see Fig. 4.14). (*Note:* The open circles in Figs. 4.13 and 4.14 may not appear on a graphing utility.)   ■

There are several important observations to make about the graphs in Example 1:

1. Even though the denominator is zero for $x = 1$, there is no vertical asymptote because the numerator is also zero for $x = 1$.

2. Both functions are discontinuous since their graphs have a missing point. This discontinuity is called a **removable discontinuity** because the discontinuity can be removed by redefining the function at $x = 1$.

3. It is usually impossible to see a single-point discontinuity on graphs produced by graphing utilities. Be alert to the possibility of these discontinuities and find them algebraically.

Because of these conditions, the rational functions in the remainder of this section, except $g(x)$ in Example 2, have been chosen to be without common factors in the numerator and the denominator. No loss of generality in the discussion will occur in making this assumption.

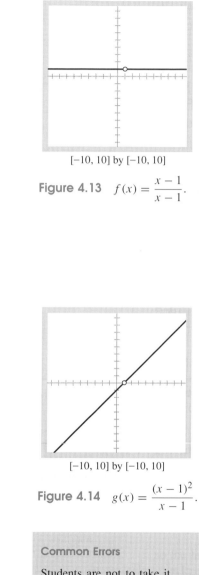

[−10, 10] by [−10, 10]

Figure 4.13 $f(x) = \dfrac{x-1}{x-1}$.

[−10, 10] by [−10, 10]

Figure 4.14 $g(x) = \dfrac{(x-1)^2}{x-1}$.

Common Errors

Students are not to take it upon themselves to "remove" discontinuities. The idea in this section is that the discontinuity *could be* removed by redefining the function at only one point, the point of discontinuity.

End Behavior Models for Rational Functions

Suppose that f is a rational function defined as follows:

$$f(x) = \frac{a_n x^n + a_{n-1}x^{n-1} + \cdots + a_0}{b_m x^m + b_{m-1}x^{m-1} + \cdots + b_0}.$$

Recall these two facts studied in Section 3.2 about end behavior models for polynomials:

1. $a_n x^n$ is the end behavior model for $a_n x^n + a_{n-1}x^{n-1} + \cdots + a_0$ ($a_n \neq 0$).
2. $b_m x^m$ is the end behavior model for $b_m x^m + b_{m-1}x^{m-1} + \cdots + b_0$ ($b_m \neq 0$).

226 22

Definition 4.5 End Behavior Asymptote

Suppose $h(x)$ and $g(x)$ are polynomial functions that are the numerator and denominator, respectively, of the rational function

$$f(x) = \frac{h(x)}{g(x)}.$$

Suppose $q(x)$ and $r(x)$ are the quotient and remainder resulting from this division and that

$$f(x) = q(x) + \frac{r(x)}{g(x)}.$$

Then the function $q(x)$ is the **end behavior asymptote of** f.

There are several points to note about this concept. The first is the distinction between end behavior model and end behavior asymptote: An end behavior *model* is found by using the leading terms of both numerator and denominator (see Definition 4.4), whereas the end behavior *asymptote* is the quotient $q(x)$ as given in Definition 4.5.

Second, the end behavior model and the end behavior asymptote are sometimes the same and sometimes not. In Example 5 and in the next example, the end behavior model and the end behavior asymptote of f are both $2x^2$. Example 7 will examine a function in which the end behavior model differs from the end behavior asymptote.

And third, until now all asymptotes have been defined as straight lines. For rational functions where $n \leq m$ (which we explored earlier in this section), this continues to be true: The end behavior asymptote is simply the horizontal asymptote. However, for cases where $n > m$, the end behavior asymptote may be a curve. In Examples 5 and 6, for instance, it is a parabola.

Section 3.2 showed how to use zoom-out as a way of supporting a claim of an end behavior model. The same can be done for rational functions.

E X A M P L E   6   *Using Zoom-out to Support End Behavior*

Use zoom-out to support a claim that $g(x) = 2x^2$ is an end behavior model of

$$f(x) = \frac{2x^3 - 4x^2 + 3}{x - 2}.$$

Solution   Find the graph of f in the viewing rectangle $[-10, 10]$ by $[-10, 30]$ (see Fig. 4.19). There appears to be a vertical asymptote at $x = 2$, which is just as expected with a denominator of $x - 2$.

Next, zoom out for the graphs of both f and $g(x) = 2x^2$ in the viewing rectangle $[-100, 100]$ by $[-1000, 1000]$ (see Fig. 4.20); you can no longer see

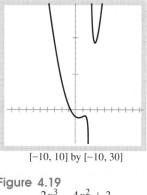

$[-10, 10]$ by $[-10, 30]$

Figure 4.19
$$f(x) = \frac{2x^3 - 4x^2 + 3}{x - 2}.$$

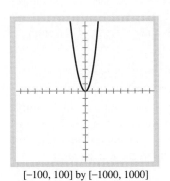

[−100, 100] by [−1000, 1000]

Figure 4.20 $g(x) = 2x^2$ and $f(x) = \dfrac{2x^3 - 4x^2 + 3}{x - 2}$.

that f has a vertical asymptote at $x = 2$. However, the graphs of f and g look identical, which supports the claim that $2x^2$ is the end behavior model of f. Example 5 shows that $y = 2x^2$ is an end behavior asymptote of f. ▤

The information gathered about end behavior in Examples 5 and 6 can be used to complete a rough sketch of the graph of f, as shown in Fig. 4.21.

Follow these three steps:

1. Sketch a graph of $y = 2x^2$, the end behavior asymptote of f, and then erase a portion near $x = 2$.

2. Sketch the vertical asymptote $x = 2$.

3. Complete the graph using the facts that $f(x) \to \infty$ as $x \to 2^+$ and $f(x) \to -\infty$ as $x \to 2^-$.

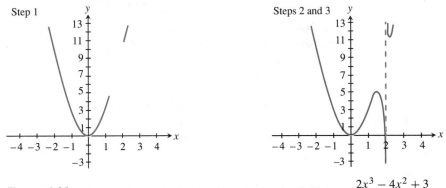

Figure 4.21 Three-step process for sketching the graph of $f(x) = \dfrac{2x^3 - 4x^2 + 3}{x - 2}$.

E X A M P L E   7   Graphing a Rational Function

Determine the end behavior asymptote and the vertical asymptotes and sketch a complete graph of

$$f(x) = \frac{2x^3 + 7x^2 - 4}{x^2 + 2x - 3}.$$

Solution   Complete long division to find the end behavior asymptote.

$$
\begin{array}{r}
2x + 3 \\
x^2 + 2x - 3 \,\overline{\big)\, 2x^3 + 7x^2 - 4} \\
\underline{2x^3 + 4x^2 - 6x} \\
3x^2 + 6x - 4 \\
\underline{3x^2 + 6x - 9} \\
5
\end{array}
$$

The end behavior asymptote is thus the line $y = 2x + 3$. Such an asymptote is sometimes called a *slant asymptote*.

The vertical asymptotes are the lines $x = -3$ and $x = 1$. This claim is confirmed by the following equation:

$$x^2 + 2x - 3 = (x + 3)(x - 1).$$ Factor the denominator to look for vertical asymptotes.

To sketch the complete graph, follow the same three steps as in Example 6. Sketch a graph in which the end behavior asymptote is drawn and then draw the vertical asymptotes (see Fig. 4.22a). Finally, use a calculator to gather the following information about the behavior of f near the vertical asymptotes:

$$f(x) \to \infty \quad \text{as } x \to -3^-, \qquad f(x) \to -\infty \quad \text{as } x \to 1^-;$$

$$f(x) \to -\infty \quad \text{as } x \to -3^+, \qquad f(x) \to \infty \quad \text{as } x \to 1^+.$$

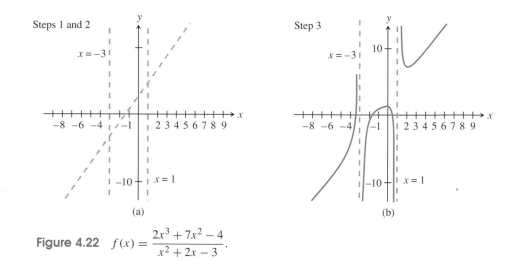

Figure 4.22 $\quad f(x) = \dfrac{2x^3 + 7x^2 - 4}{x^2 + 2x - 3}.$

Use all of the above information to complete a sketch (see Fig. 4.22b). A graphing utility graph will support your work (see Fig. 4.23).

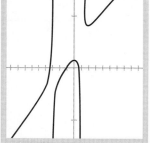

$[-10, 10]$ by $[-15, 15]$

Figure 4.23
$f(x) = \dfrac{2x^3 + 7x^2 - 4}{x^2 + 2x - 3}.$

Hidden Behavior of Rational Functions

The following rational function is more complicated than the others that have been graphed so far:

$$f(x) = \frac{2x^4 + 7x^3 + 7x^2 + 2x}{x^3 - x + 50}.$$

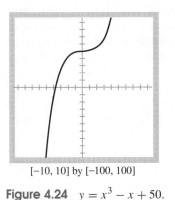

[−10, 10] by [−100, 100]

Figure 4.24 $y = x^3 - x + 50$.

In particular, the zeros of the denominator cannot be found by factoring. However, a complete graph of the denominator $x^3 - x + 50$ (see Fig. 4.24) shows that there is one zero; hence f has one vertical asymptote between $x = -3$ and $x = -4$.

Knowing that the denominator of f has one zero helps in understanding the complete graph of f, although Example 8 will show there are some surprises. In fact, sometimes more than one viewing rectangle is needed to gain a complete understanding of the nature of a rational function graph.

EXAMPLE 8   Finding Hidden Behavior

Find a complete graph of

$$f(x) = \frac{2x^4 + 7x^3 + 7x^2 + 2x}{x^3 - x + 50},$$

including the hidden behavior between $x = -2$ and $x = 1$.

Solution   The end behavior model for $f(x)$ is

$$g(x) = \frac{2x^4}{x^3} = 2x.$$

IMPORTANT OBSERVATION

The theory of calculus can be used to predict where hidden behavior will occur. A graphing utility cannot. It must be found by an inquiring mind, as illustrated in Example 8.

This observation may contribute to a belief that the graph in Fig. 4.25(a) is a complete graph of f. However, this graph looks suspiciously flat around the origin. Use zoom-in to investigate the graph near the origin (see Fig. 4.25b).   ■

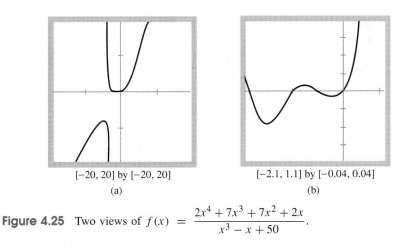

[−20, 20] by [−20, 20]   [−2.1, 1.1] by [−0.04, 0.04]
(a)   (b)

Figure 4.25 Two views of $f(x) = \dfrac{2x^4 + 7x^3 + 7x^2 + 2x}{x^3 - x + 50}$.

We close this section with a problem situation that has a rational function for an algebraic model.

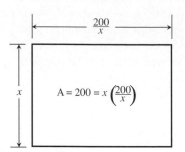

Figure 4.26
$P(x) = 2x + 400/x.$

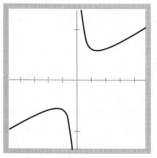

[−50, 50] by [−150, 150]

Figure 4.27
$P(x) = 2x + 400/x.$

E X A M P L E   9   APPLICATION: Finding a Minimum Perimeter

Problem Situation   An architect might be interested in finding what shape of rectangle has the minimum perimeter for a given area. A minimum perimeter would reduce construction costs.

Consider all rectangles with an area of 200 sq ft. Let x be the width in feet of one such rectangle. Find an algebraic representation and a complete graph for $P(x)$, the perimeter as a function of x. Find the dimensions of the rectangle with the least possible perimeter.

Solution   If the width of a rectangle with an area of 200 is x feet, then its length must be $200/x$ feet (see Fig. 4.26). Therefore

$$P(x) = 2x + 2\left(\frac{200}{x}\right) = 2x + 400/x.$$

Notice that $y = 2x$ is the end behavior asymptote of $P(x)$ and that line $x = 0$ is a vertical asymptote of $P(x)$. A complete graph of P is shown in Fig. 4.27. Also observe that only the portion of the graph for $x > 0$ applies to this problem situation. Using zoom-in, you will find the local minimum near $x = 15$ to be $x = 14.14$. Therefore the dimensions of a rectangle with minimum perimeter are

$$x = 14.14 \qquad \text{and} \qquad P(x) = \frac{200}{14.14} = 14.14.$$

The minimum perimeter occurs when the rectangle is a square.   ■

Exercises for Section 4.2

In Exercises 1–8, find the domain of each function algebraically. Support your answer with a graphing utility.

1. $f(x) = \dfrac{x^2 + 1}{x^2 - 1}$

2. $f(x) = \dfrac{x^2 - 4}{x^2 - 4x - 1}$

3. $f(x) = \dfrac{x^2 + 1}{x}$

4. $f(x) = \dfrac{x^2 - 1}{x}$

5. $f(x) = \dfrac{x^3 + 1}{x^2 - 1}$

6. $f(x) = \dfrac{x^3 - 1}{x^2 + 1}$

7. $t(x) = \dfrac{2}{2x^2 - x - 3}$

8. $P(x) = \dfrac{4x - 2}{x^2 + 5x + 8}$

In Exercises 9–14, find the end behavior model for each rational function.

9. $f(x) = \dfrac{x^4 - 3x^2 - 5}{x^5 - x}$

10. $f(x) = \dfrac{2x^2 + 5}{x^3 - 2x^2 + x}$

11. $f(x) = \dfrac{3x^2 - 2x + 4}{x^2 - 4x + 1}$

12. $f(x) = \dfrac{4x^2 + 2x}{x^2 - 4x + 8}$

13. $f(x) = \dfrac{x^3 - 1}{x^2 + 4}$

14. $f(x) = \dfrac{x^3 - 1}{x^2 - 4}$

In Exercises 15–18, use algebraic means to sketch the graph of each rational function. Find all vertical and horizontal asymptotes. Determine the end behavior model for each function. Support your answer with a graphing utility.

15. $f(x) = \dfrac{2}{x - 3}$

16. $g(x) = \dfrac{x - 2}{x^2 - 2x - 3}$

17. $f(x) = \dfrac{x-1}{x^2+3}$

18. $h(x) = \dfrac{3x^2-12}{4-x^2}$

For Exercises 19 and 20,

$$f(x) = \frac{x-1}{x^2-x-6},$$

the function from Example 3. Use a calculator to complete each of the following tables:

19.

x	-1	-10	-100	-1000	$-10,000$
$f(x)$	?	?	?	?	?

20.

x	3	2.9	2.99	2.999	2.9999
$f(x)$	?	?	?	?	?

21. Use the table in Exercise 19 to determine numerically what happens to $f(x)$ as $x \to -\infty$.

22. Use the table in Exercise 20 to determine numerically what happens to $f(x)$ as $x \to 3^-$.

In Exercises 23–28, find a complete graph of each function. Also find the domain, range, asymptotes, and end behavior model for each.

23. $f(x) = \dfrac{x+1}{x^2}$

24. $f(x) = \dfrac{1}{x-1} + \dfrac{2}{x-3}$

25. $g(x) = \dfrac{x+1}{x^2-1}$

26. $f(x) = \dfrac{x^2+1}{x^4+1} + \dfrac{2}{x-3}$

27. $T(x) = \dfrac{x-1}{x^2+3x-1}$

28. $g(x) = \dfrac{2x}{(x-3)(x+2)}$

In Exercises 29–32, let $g(x) = 3/(x-2)$ and

$$f(x) = 2x^2 + \frac{3}{x-2} = \frac{2x^3 - 4x^2 + 3}{x-2},$$

the function of Example 6. Use a calculator to complete each of the following tables:

29.

x	3	2.1	2.01	2.001	2.0001
$f(x)$	?	?	?	?	?
$g(x)$	?	?	?	?	?

30.

x	1	1.9	1.99	1.999	1.9999
$f(x)$	?	?	?	?	?
$g(x)$	?	?	?	?	?

31.

x	1	10	100	1000	10, 000
$f(x)$	?	?	?	?	?
$g(x)$	?	?	?	?	?

32.

x	-1	-10	-100	-1000	$-10,000$
$f(x)$	?	?	?	?	?
$g(x)$	?	?	?	?	?

33. Use the table in Exercise 29 to determine numerically what happens to $f(x)$ as $x \to 2^+$.

34. Use the table in Exercise 30 to determine numerically what happens to $f(x)$ as $x \to 2^-$.

35. Use the table in Exercise 31 to determine numerically what happens to $f(x)$ as $x \to \infty$.

36. Use the table in Exercise 32 to determine numerically what happens to $f(x)$ as $x \to -\infty$.

In Exercises 37–44, use long division to find the end behavior asymptote of each function. Sketch a complete graph including all asymptotes. Support your answer with a graphing utility.

37. $f(x) = \dfrac{x^2-2x+3}{x+2}$

38. $g(x) = \dfrac{3x^2-x+5}{x^2-4}$

39. $f(x) = \dfrac{x^3-1}{x-1}$

40. $h(x) = \dfrac{x^3+1}{x^2+1}$

41. $f(x) = \dfrac{x^2-3x-7}{x+3}$

42. $g(x) = \dfrac{2x^3-2x^2-x+5}{x-2}$

43. $f(x) = \dfrac{2x^4-x^3-16x^2+17x-5}{2x-5}$

44. $g(x) = \dfrac{2x^5-3x^3+2x-4}{x-1}$

In Exercises 45–50, use a graphing utility to find a complete graph of each function. Find the domain, range, asymptotes, and end behavior model for each.

45. $f(x) = \dfrac{x^2-2x+5}{x+1}$

46. $g(x) = \dfrac{x^2-2x+1}{x-2}$

47. $f(x) = \dfrac{2x^3-x^2+3x-2}{x^3+3}$

48. $g(x) = \dfrac{x^3-2x+1}{x-2}$

49. $f(x) = \dfrac{x^4-2x^2-x+3}{x^2+4}$

50. $g(x) = \dfrac{x^4 - 2x^2 - x + 3}{x^2 - 4}$

In Exercises 51–54, determine the intervals on which each function is increasing and those on which each function is decreasing.

51. $f(x) = \dfrac{x^2 + 1}{x}$ **52.** $g(x) = \dfrac{2x^3 - 3x + 1}{x^2 + 4}$

53. $f(x) = \dfrac{x^3 - 2x - 1}{3x + 5}$ **54.** $g(x) = \dfrac{2x^3 - x^2 + 1}{2 - x}$

In Exercises 55–59, determine all local extrema, intervals when the function is increasing and decreasing, and all zeros of the function. Determine the end behavior model and look for any hidden behavior.

55. $f(x) = \dfrac{x^3 - 2x^2 + x - 1}{2x - 1}$

56. $f(x) = \dfrac{2x^3 + x^2 - 24x - 12}{x^2 + x - 12}$

57. $f(x) = \dfrac{2x^4 + 3x^3 + x^2 + 2}{x^3 - x^2 + 20}$

58. $f(x) = \dfrac{x^3 + 1}{x}$

59. $g(x) = \dfrac{2x^4 - 2x^2 + x + 5}{x^2 - 3x - 4}$

60. Let $f(x) = \dfrac{2x^4 - 3x^2 + 1}{3x^4 - x^2 + x - 1}$. Find the graph of f in the standard viewing rectangle. Is a vertical asymptote appar-

ent? Now find the graph in the $[-0.5, 1.0]$ by $[-1.5, 1.5]$ viewing rectangle. What changes do you see?

61. Use zoom-out to find the end behavior model for

$$f(x) = \dfrac{3x^5 - 2x^4 + 3x^2 + 5x - 6}{x^2 - 3x + 6}$$

and find the end behavior asymptote of f.

62. Consider all the rectangles with an area of 182 square units. Let x be the width of one such rectangle. Find an algebraic representation and a complete graph of $P(x)$, the perimeter as a function of x. Find the dimensions of a rectangle that has the least possible perimeter.

63. Consider all rectangles with an area of 375 sq ft. Let x be the width of such a rectangle. Find an algebraic representation and a complete graph of $P(x)$, the perimeter as a function of x. Find the dimensions of a rectangle that has the least possible perimeter.

64. Consider the following **problem situation**: Pure acid is added to 78 oz of a 63% acid solution. Let x be the amount (in ounces) of pure acid added. Find an algebraic representation for $C(x)$, the concentration of acid as a function of x. Also find a complete graph of this problem situation on a graphing utility. Use this graph to determine how much pure acid should be added to the 63% solution to produce a new mixture that is at least 83% acid.

65. A certain amount x of a 100% pure barium solution is added to 135 oz of a 35% barium solution to obtain a 63% barium solution. Use a graphing utility graph and zoom in to determine x.

4.3

Equations and Inequalities with Rational Functions

Objective

Students will be able to solve problems by using equations and inequalities that involve rational functions and by using both algebraic and graphical techniques.

Key Ideas

Extraneous solution
Equation-solving principle

This section uses both graphical and algebraic techniques to solve equations and inequalities that involve rational functions. We also shall investigate problem situations that have such equations and inequalities as models.

E X A M P L E   1   APPLICATION: Solving the Time-rate Problem

A Problem Situation   Sue drove 30 mi to a train station and then completed her trip by train. In all, she traveled 120 mi. The average rate of the train was 20 mph faster than the average rate of the car.

Find an algebraic representation that gives the total time T required to complete the trip as a function of the rate x of the car.

Solution Choose variables as suggested by Fig. 4.28.

$$t_1 = \text{time in hours to travel distance } d_1 = 30 \text{ mi}$$

$$t_2 = \text{time in hours to travel distance } d_2 = 90 \text{ mi}$$

$$t_1 + t_2 = \text{total time in hours to travel the 120 mi}$$

$$d = rt \quad \text{Remember that } r \text{ represents average rate.}$$

Let x be the average rate in mph of the car; therefore $x + 20$ is the average rate of the train.

$$d_1 = xt_1 \quad \text{and} \quad d_2 = (x + 20)t_2$$

$$\frac{d_1}{x} = t_1 \quad \text{and} \quad \frac{d_2}{x + 20} = t_2$$

$$T(x) = t_1 + t_2 = \frac{d_1}{x} + \frac{d_2}{x + 20}$$

$$T(x) = \frac{30}{x} + \frac{90}{x + 20}$$

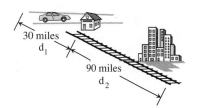

30 miles
d₁

90 miles
d₂

Figure 4.28 Diagram for Example 1.

E X A M P L E 2   Analyzing an Algebraic Representation

Find the real zeros, the vertical asymptotes, and the end behavior model for

$$T(x) = \frac{30}{x} + \frac{90}{x + 20}.$$

Also determine a complete graph.

Solution   First rewrite $T(x)$ by finding a common denominator.

$$T(x) = \frac{30}{x} + \frac{90}{x + 20}$$

$$= \frac{30(x + 20) + 90x}{x(x + 20)}$$

$$= \frac{120x + 600}{x(x + 20)}$$

The vertical asymptotes are thus $x = 0$ and $x = -20$. To find the x-intercept, determine where the numerator is zero.

$$120x + 600 = 0 \iff x = \frac{-600}{120} = -5$$

A rational function is zero only when the numerator is zero, provided the denominator is not zero for the same value.

The end behavior model is

$$y = \frac{120x}{x^2} = \frac{120}{x},$$

and the horizontal asymptote is $y = 0$.

All this information is combined in the complete graph shown in Fig. 4.29.

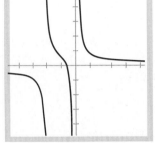

[−50, 50] by [−50, 50]

Figure 4.29

$T(x) = \dfrac{120x + 600}{x(x + 20)}.$

With this analysis of the time-rate problem situation, we are ready to solve the following problem.

E X A M P L E 3   Solving the Time-rate Problem (Continued)

Sue has 2 hr to complete the 120-mi trip described in Example 1. Use both a graphical and an algebraic method to find the rate she must travel by car.

Solution   Examples 1 and 2 showed that

$$T(x) = \frac{30}{x} + \frac{90}{x + 20} = \frac{120x + 600}{x(x + 20)}.$$

We need to find a value of x such that $T(x) = 2$.

Graphical Method   Figure 4.30 shows a graphical representation of the problem situation together with the line $y = 2$. Use zoom-in to show that the x-coordinate of the point of intersection is $x = 46.46$ with an error of at most 0.01. This is the value of x that solves the problem.

Algebraic Method   Solve the equation

$$\frac{120x + 600}{x(x + 20)} = 2.$$

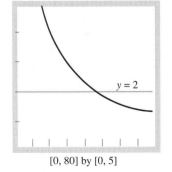

$y = 2$

[0, 80] by [0, 5]

Figure 4.30

$T(x) = \dfrac{120x + 600}{x(x + 20)}.$

Begin by multiplying both sides of the equation by $x(x + 20)$ and simplifying.

$$120x + 600 = 2x(x + 20)$$

$$120x + 600 = 2x^2 + 40x$$

$$0 = 2x^2 - 80x - 600$$

$$0 = x^2 - 40x - 300$$

Use the quadratic formula to obtain

$$x = 20 + 10\sqrt{7} \qquad \text{and} \qquad x = 20 - 10\sqrt{7}.$$

Because $x = 20 - 10\sqrt{7}$ is a negative number, the only solution relevant to the problem situation is $20 + 10\sqrt{7} = 46.46$. Sue must travel an average of 46.46 mph.

Analysis of This Algebraic Solution

In the algebraic solution to the problem in Example 3, both sides of the original equation were multiplied by an expression containing the variable x. Whenever this is done, the resulting equation may not be equivalent to the original equation; that is, there may be solutions to the resulting equation that were not solutions to the original equation.

Any solution to the resulting equation that is not a solution of the original equation is called an **extraneous solution.**

Consider the following simple example:

$$x - 1 = 0. \qquad \text{This equation has one solution: } x = 1.$$

Multiply both sides of the equation by x.

$$x(x - 1) = x \cdot 0 = 0 \qquad \begin{array}{l} \text{This equation has two solutions,} \\ x = 0 \text{ and } x = 1. \end{array}$$

Therefore $x = 0$ is an extraneous solution to the original equation $x - 1 = 0$.

PARTIAL FRACTIONS

Ex. 44–47 describe a method called **partial fractions** that can be used to split up a rational expression into a sum of rational expressions.

> **Equation-solving Principle**
>
> When each side of an equation is multiplied by an expression containing the variable, each solution of the resulting equation must be checked to ensure it is a solution of the original equation.

E X A M P L E 4 Solving a Rational Equation

Solve the following rational equation for x:

$$\frac{2x}{x - 1} + \frac{1}{x - 3} = \frac{2}{x^2 - 4x + 3}.$$

Teaching Note

Have students graph several steps in the algebraic solution of one of the equations where an extraneous solution is introduced. See if they can identify the step that causes the extraneous solution to be introduced. Ask students to explain what is special about this step. Emphasize that extraneous solutions reaffirm the importance of checking solutions.

Solution

Algebraic Method Notice that $x^2 - 4x + 3 = (x - 1)(x - 3)$. Thus the least common denominator of the three fractions in the equation is $(x - 1)(x - 3)$. Multiply both sides of the equation by this common denominator.

$$(x - 1)(x - 3)\left[\frac{2x}{x - 1} + \frac{1}{x - 3}\right] = (x - 1)(x - 3)\frac{2}{x^2 - 4x + 3}$$

$$\frac{2x(x - 1)(x - 3)}{x - 1} + \frac{(x - 1)(x - 3)}{x - 3} = \frac{2(x - 1)(x - 3)}{x^2 - 4x + 3} \qquad \text{Distributive property}$$

$$2x(x - 3) + (x - 1) = 2 \qquad \begin{array}{l}\text{Simplify numerator} \\ \text{and denominator} \\ \text{that have equal factors}\end{array}$$

$$2x^2 - 6x + x - 1 = 2$$

$$2x^2 - 5x - 3 = 0$$

$$(2x + 1)(x - 3) = 0$$

$$x = 3, \qquad x = -\frac{1}{2}$$

Check each solution: Substitute $x = 3$ and $x = -\frac{1}{2}$ into the original equation.

$$\frac{2(3)}{3 - 1} + \frac{1}{3 - 3} \neq \frac{2}{3^2 - 4 \cdot (3) + 3} \qquad \begin{array}{l}\text{These expressions are not defined} \\ \text{because they involve division by zero.}\end{array}$$

$$\frac{2(-\frac{1}{2})}{(-\frac{1}{2}) - 1} + \frac{1}{(-\frac{1}{2}) - 3} = \frac{2}{(-\frac{1}{2})^2 - 4(-\frac{1}{2}) + 3}$$

$$\frac{8}{21} = \frac{8}{21}$$

So we see that $x = 3$ is not a solution, whereas $x = -\frac{1}{2}$ is.

Graphical Method Solve the equation graphically by finding the zeros of the function

$$f(x) = \frac{2x}{x - 1} + \frac{1}{x - 3} - \frac{2}{x^2 - 4x + 3}.$$

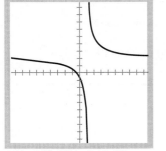

[–10, 10] by [–10, 10]

Figure 4.31 $f(x) = \frac{2x}{x - 1} + \frac{1}{x - 3} - \frac{2}{x^2 - 4x + 3}.$

A complete graph of f that supports the $-\frac{1}{2}$ solution found algebraically appears in Fig. 4.31. Notice that line $x = 3$ is not a vertical asymptote even though the function is not defined for $x = 3$. We see why this is true algebraically by combining the three fractions of f to show that

$$f(x) = \frac{2x^2 - 5x - 3}{(x - 1)(x - 3)} = \frac{(2x + 1)(x - 3)}{(x - 1)(x - 3)}.$$

So f has a removable discontinuity at $x = 3$.

Stan says, "The notion of an extraneous solution provides an opportunity for students to investigate the interplay between graphical and algebraic representations of equations."

Notes on Examples

Example 6 provides students with an opportunity to review the idea of a solution being read from a graph when the graph is below the x-axis.

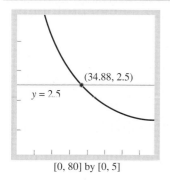

[0, 80] by [0, 5]

Figure 4.32
$T(x) = \dfrac{30}{x} + \dfrac{90}{x+20}$
and $y = 2.5$.

Notes on Exercises

Ex. 28–43 include many opportunities for developing students' problem-solving skills.

The next example illustrates that *all* solutions of a rational equation may turn out to be extraneous solutions.

E X A M P L E   5   Solving a Rational Equation

Find an algebraic solution to the equation

$$\frac{x-3}{x} + \frac{3}{x+2} + \frac{6}{x^2+2x} = 0.$$

Solution   Multiply both sides of the equation by the least common denominator of $x(x+2)$, and simplify.

$$\frac{x-3}{x} + \frac{3}{x+2} + \frac{6}{x^2+2x} = 0$$

$$(x-3)(x+2) + 3x + 6 = 0$$

$$x^2 - x - 6 + 3x + 6 = 0$$

$$x^2 + 2x = 0$$

$$x(x+2) = 0$$

$$x = 0, \qquad x = -2$$

Notice that neither solution is a solution of the original equation since both values result in division by zero. ◼

Knowing that multiplying both sides of an equation sometimes introduces extraneous solutions, we return to the problem of Sue's trip by car and train.

E X A M P L E   6   APPLICATION: Revisiting the Time-rate Problem

At what possible rates could Sue travel by car to ensure that the total time for her to complete the trip is less than 2.5 hr?

Solution   We must find the positive values of x for which $T(x) < 2.5$, where

$$T(x) = \frac{30}{x} + \frac{90}{x+20}.$$

Figure 4.32 shows a graph of the line $y = 2.5$ and the complete graph of Sue's trip. To solve the inequality $T < 2.5$, find the positive values of x for which the graph of T lies *below* the graph of $y = 2.5$.

Use zoom-in to find that the coordinates of the point of intersection are $(34.88, 2.5)$ with an error of at most 0.01. Therefore when $x > 34.88$, the graph of $T(x)$ lies below the line $y = 2.5$.

Thus, to complete her trip in time, Sue must travel at an average speed greater than 34.88 mph. ◼

We have seen in Examples 3 and 4 that equations involving rational expressions can often be solved by either algebraic or graphical methods. This is also true for inequalities involving rational expressions.

The next example illustrates that the graphical method, which is a more general method, also has the potential to hide information.

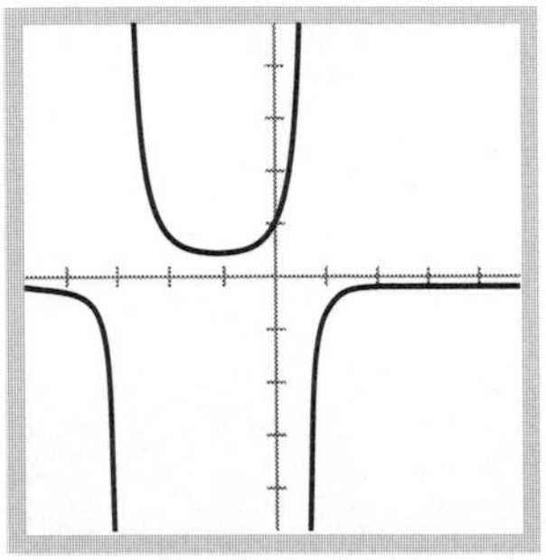

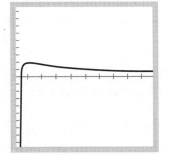

[−5, 5] by [−10, 10]

Figure 4.33
$$f(x) = \frac{2x^2 + 6x - 8}{2x^2 + 5x - 3} - 1.$$

EXAMPLE 7   Solving a Rational Inequality

Use both graphical and algebraic methods to solve

$$\frac{2x^2 + 6x - 8}{2x^2 + 5x - 3} < 1.$$

Solution

Graphical Method   Begin by graphing

$$f(x) = \frac{2x^2 + 6x - 8}{2x^2 + 5x - 3} - 1$$

in a standard viewing rectangle (see Fig. 4.33). Notice that f has vertical asymptotes at $x = -3$ and $x = \frac{1}{2}$, which can be confirmed algebraically by factoring the denominator $2x^2 + 5x - 3 = (2x - 1)(x + 3)$.

It appears that the graph is below the x-axis over the interval $(-\infty, -3)$. We might be tempted to conclude the same for when x is positive. However, there is hidden behavior for $x > 0$.

To see this, zoom in along the y-axis where $x > 0$. Figure 4.34 shows the graph in the viewing rectangle $[0, 100]$ by $[-0.1, 0.1]$. In this viewing rectangle, it becomes clear that the graph of f crosses the x-axis at $x = 5$. Thus the interval $(\frac{1}{2}, 5)$ is part of the solution.

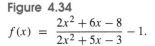

[0, 100] by [−0.1, 0.1]

Figure 4.34
$$f(x) = \frac{2x^2 + 6x - 8}{2x^2 + 5x - 3} - 1.$$

Algebraic Method   In Fig. 4.33, the graph appears to be below the x-axis when $0 < x < 5$. However, this is not correct because there is a vertical asymptote at $x = \frac{1}{2}$.

To proceed with an algebraic solution to the inequality, subtract 1 from each side of the inequality, combine terms, and simplify.

$$\frac{2x^2 + 6x - 8}{2x^2 + 5x - 3} - 1 < 0$$

$$\frac{2x^2 + 6x - 8}{2x^2 + 5x - 3} - \frac{2x^2 + 5x - 3}{2x^2 + 5x - 3} < 0 \qquad \text{Find a common denominator and subtract.}$$

$$\frac{2x^2 + 6x - 8 - 2x^2 - 5x + 3}{2x^2 + 5x - 3} < 0$$

$$\frac{x-5}{2x^2+5x-3} < 0 \qquad \text{Simplify the numerator.}$$

$$\frac{x-5}{(2x-1)(x+3)} < 0 \qquad \text{Factor the denominator.}$$

Because the numerator and denominator polynomials can be factored, a sign-pattern method can be used for solving this inequality.

We see that the three linear factors are zero for $x = 5$, $x = \frac{1}{2}$, and $x = -3$. These three numbers divide the real-number line into four intervals, as shown in the following sign-pattern pictures. Each of the three factors $(x-5)$, $(2x-1)$, and $(x+3)$ is either positive or negative throughout each interval.

Because $f(x) < 0$ when an odd number of factors is negative (in this case all three factors or only one), the complete solution to

$$f(x) = \frac{2x^2+6x-8}{2x^2+5x-3} - 1 < 0$$

is $(-\infty, -3) \cup (\frac{1}{2}, 5)$. A graphical analysis supports this solution.   ≡

That the graphical method is more general than the algebraic method is underscored with the next example.

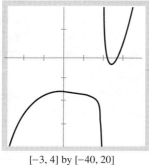

[−3, 4] by [−40, 20]

Figure 4.35 A complete graph of $f(x) = \dfrac{x^4 - 3x^3 + 2x^2 + 2}{x - 2} - 15.$

E X A M P L E 8   Solving Another Rational Inequality

Solve the following inequality for x:

$$\frac{x^4 - 3x^3 + 2x^2 + 2}{x - 2} \geq 15.$$

Solution   Find a complete graph of

$$f(x) = \frac{x^4 - 3x^3 + 2x^2 + 2}{x - 2} - 15.$$

Assignment Guide

Day 1: Ex. 1–10, 13, 16, 19, 21, 26
Day 2: Ex. 28–31, 38–47

(See Fig. 4.35.) Look for values of x corresponding to where the graph of f lies above the x-axis. Note from the denominator of f that $x = 2$ is a vertical asymptote.

Use zoom-in to determine that the two x-intercepts are 2.223 and 2.679 with an error of at most 0.01.

The solution to the inequality is $(2, 2.223] \cup [2.679, \infty)$. ▤

Exercises for Section 4.3

In Exercises 1–10, solve each equation algebraically. Support your answer with a graphing utility.

1. $\dfrac{x-1}{x+2} = 3$

2. $\dfrac{2}{x-1} + x = 5$

3. $\dfrac{1}{x} - \dfrac{2}{x-3} = 4$

4. $\dfrac{3}{x-1} + \dfrac{2}{x} = 8$

5. $\dfrac{x^2 - 2x + 1}{x + 5} = 0$

6. $\dfrac{x^2 - 6x + 5}{x^2 - 2} = 3$

7. $\dfrac{x^3 - x}{x^2 + 1} = 0$

8. $\dfrac{2x^4 + x^2 - 1}{x^2 - 9} = 0$

9. $\dfrac{3x}{x+2} + \dfrac{2}{x-1} = \dfrac{5}{x^2 + x - 2}$

10. $\dfrac{x-3}{x} - \dfrac{3}{x+1} + \dfrac{3}{x^2 + x} = 0$

In Exercises 11–19, solve each inequality algebraically using the sign-pattern picture method. Support your answer with a graphing utility.

11. $\dfrac{1}{x-3} + 4 > 0$

12. $\dfrac{x-3}{x+5} - 6 \le 3$

13. $\dfrac{x+3}{2x-7} < 5$

14. $\dfrac{x-1}{x+4} < 0$

15. $\dfrac{x-1}{x+4} > 3$

16. $\dfrac{x-1}{x^2 - 4} < 0$

17. $\dfrac{x^2 - 1}{x^2 + 1} \ge 0$

18. $\dfrac{x^2 - 3x - 2}{x^2 - 4x + 4} \le 0$

19. $\dfrac{x^2 - x + 1}{x + 2} < 3$

In Exercises 20–23, solve the equation or the inequality.

20. $\dfrac{x^4 - 2x^2 - 3}{x^2 - 5} = -3$

21. $\dfrac{x^6 - x^4 + x^3 - 2x - 4}{x^2 - 2x + 1} = 5$

22. $\dfrac{x^5 - 2x^2 + 4}{x^3 + 5} < 3$

23. $\dfrac{x^4 - 3x^2 + 4x - 2}{x - 3} \ge 2$

In Exercises 24–27, find the zeros, y-intercepts, vertical asymptotes, and end behavior asymptotes of each rational function. Find a complete graph that shows all of this behavior. (More than one viewing rectangle may be needed.)

24. $f(x) = \dfrac{x^3 - 2x + 3}{x^2 - x + 4}$

25. $f(x) = \dfrac{x^4 - 2x^2 + 1}{x^2 + x - 1}$

26. $f(x) = \dfrac{x^4 - 2x^3 + 3x^2 + x - 4}{x^2 - 5x + 4}$

27. $f(x) = \dfrac{x^5 - x^3 + x - 5}{x^2 - 3x - 1}$

Exercises 28–31 refer to the following **problem situation**: Josh rode his bike 17 mi from his home to Columbus, Ohio, and then completed a 53-mi trip by car from Columbus to Mansfield, Ohio. Assume the average rate of the car was 43 mph faster than the average rate of the bike.

28. Find an algebraic representation for the total time T required to complete the 70-mi trip (bike and car) as a function of the average rate x of the bike.

29. Find a complete graph of the algebraic representation of this problem situation, indicating any vertical and horizontal asymptotes and any zeros.

30. Make a complete graph of this problem situation showing only the values of x that make sense in this situation.

31. Use a graphical method to find the bike's rate if the total time of the trip was 1 hr and 40 min.

Exercises 32–34 refer to the following **problem situation**: Drains A and B are used to empty a swimming pool. Drain A alone can empty the pool in 4.75 hr. Let t be the time it takes for drain B alone to empty the pool. (Assume the pool drains at a constant rate.)

32. Find an algebraic representation that gives the part of the drainage that can be done in 1 hr with both drains open at the same time as a function of t.

33. Find a complete graph of this problem situation.

34. Use a graphical method to find the time it takes for drain B alone to empty the pool if both drains, when open at the same time, can empty the pool in 2.60 hr.

Exercises 35–37 refer to the following **problem situation**: The total electrical resistance R of two resistors connected in parallel with resistance R_1 and R_2 is given by

$$R = \frac{R_1 R_2}{R_1 + R_2}.$$

One resistor has a resistance of 2.3 ohms. Let x be the resistance of the second resistor.

35. Find an algebraic representation for $R(x)$, the total resistance of the pair of resistors connected in parallel.

36. Draw a complete graph of this problem situation.

37. Use a graphical method to find the resistance of the second resistor if the total resistance of the pair is 1.7 ohms.

Exercises 38–40 refer to the following **problem situation**: A cylindrical soda pop can of radius r and height h is to hold exactly 355 ml (milliliters) of liquid when completely full. A manufacturer wishes to find the dimensions of the can with the minimum surface area.

For Exercises 38–40.

38. Find the algebraic representation for the surface area S as a function of r. (*Hint:* Show that the surface area of the can is $S = 2\pi r^2 + 2\pi r h$. Then use the fact that the volume of the can is $355 = \pi r^2 h$.)

39. What are the restrictions on r for this problem situation? Find a complete graph of this problem situation.

40. Use zoom-in to determine the radius r and height h that yield a can of minimal surface area. What is the minimal surface area?

Exercises 41–43 refer to the following **problem situation**: A single-story house with a rectangular base is to contain 900 sq ft of living area. Local building codes require that both the length L and the width W of the base of the house be greater than 20 ft. To minimize the cost of the foundation, the builder wants to minimize the perimeter of the foundation.

For Exercises 41–43.

41. Find an algebraic representation for the perimeter P as a function of L, the length of the base.

42. Find a complete graph of this problem situation showing only the values of L that make sense.

43. Use a graphical method to find the value of L that minimizes the perimeter. What is the minimum perimeter?

Exercises 44–47 use the method known as finding **partial fractions**. Often, an algebraic solution required that two fractions be simplified. In Example 2, we see that

$$\frac{30}{x} + \frac{90}{x + 20} = \frac{120x + 600}{x(x + 20)}.$$

For some applications, it is necessary to reverse that process, that is, to split a fraction into a sum. This method is known as **finding partial fractions**.

44. Find constants A and B that make the following equation valid:

$$\frac{2}{(x-1)(x+1)} = \frac{A}{x-1} + \frac{B}{x+1}.$$

(*Hint:* Multiply both sides of the equation by $(x-1)(x+1)$ and solve the resulting equation by equating coefficients to find that $A = 1$ and $B = -1$.)

45. Find constants A and B that make the following equation valid:

$$\frac{3}{(x-2)(x+3)} = \frac{A}{x-2} + \frac{B}{x+3}.$$

46. Find constants A and B that make the following equation valid:

$$\frac{x+1}{(x-2)(x+4)} = \frac{A}{x-2} + \frac{B}{x+4}.$$

47. Find constants A and B that make the following equation valid:

$$\frac{2x-1}{(x-1)(x+3)} = \frac{A}{x-1} + \frac{B}{x+3}.$$

4.4 _____   Radical Functions

In this section, we investigate the graphs of functions that involve radicals; for example,

$$y = \sqrt{x-2} \qquad \text{and} \qquad y = \sqrt[3]{x^2-1} + x.$$

Before proceeding further, it will be useful to review what the radical symbol means.

Radical Symbol

Any expression of the form $\sqrt[n]{a}$, where n is an integer greater than 1, is called a radical. The integer n is the **index** of the radical, and a is the **radicand**. The symbol $\sqrt{}$ is the **radical sign**. If no index appears over the radical sign, n is understood to equal 2.

Radical expressions can also be written using fractional exponents, as illustrated by the following expressions:

$$\sqrt{x-2} = (x-2)^{1/2}, \qquad \sqrt[3]{x^2-1} = (x^2-1)^{1/3}, \qquad \text{and} \qquad \sqrt[5]{x^4} = x^{4/5}.$$

We restate more formally what the radical symbol means.

Definition 4.6   Radical Symbol

Let n be an integer greater than 1 and let a be a real number. Then the **principal nth root of a**, denoted $\sqrt[n]{a}$, is

1. the real solution to $x^n = a$, if n is odd or
2. the nonnegative real solution to $x^n = a$, if n is even and $a \geq 0$.

When n is even, a must be nonnegative because an even power of x is always a nonnegative number.

The principal square root of a is $\sqrt{a}$, and the principal cube root of a is $\sqrt[3]{a}$.

🔍 EXPLORE WITH A CALCULATOR

What happens when you find the root of a number? Is the root larger or smaller than the number? Use a calculator to complete the following table of values:

x	0.5	0.9	0.99	1	1.01	1.1
$\sqrt{x}$						
$\sqrt[3]{x}$						
$\sqrt[4]{x}$						
$\sqrt[5]{x}$						

Complete a Generalization For what values of x is the root of x larger than x? For what values of x is the root of x smaller than x?

Graphs of $y = \sqrt[n]{x}$ When n Is Even

We can see from the above Exploration that when $0 < x < 1$, $\sqrt[n]{x}$ is larger than x, and that when $x > 1$, $\sqrt[n]{x}$ is smaller than x. What impact does this observation have on the graphs of $y = \sqrt[n]{x}$?

EXAMPLE 1   Comparing Graphs of Radical Functions

Find and compare complete graphs of $y = \sqrt{x}$, $y = \sqrt[4]{x}$, and $y = \sqrt[6]{x}$. Determine the domain and range of each function.

Solution   The domain of each function is $[0, \infty)$. The graphs of all three functions are seen using viewing rectangle $[-5, 30]$ by $[-5, 10]$ (see Fig. 4.36a). These graphs appear to overlap near the origin. Choose the smaller viewing rectangle shown in Fig. 4.36(b) to investigate the detail near the origin.

The range of each function is the interval $[0, \infty)$.   ≡

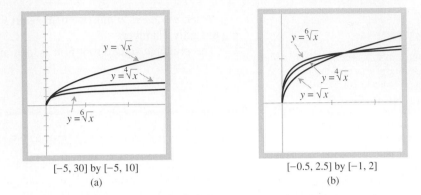

$[-5, 30]$ by $[-5, 10]$ $[-0.5, 2.5]$ by $[-1, 2]$

(a)   (b)

Figure 4.36   Three radical functions with even indexes.

Figure 4.36 suggests the following summarizing points about the graphs of radical functions with even indexes:

- Points common to all graphs: $(0, 0)$ and $(1, 1)$.
- For values of x in the interval $(0, 1)$, the graph of $y = \sqrt{x}$ lies below the graph of $y = \sqrt[4]{x}$, which in turn lies below the graph of $y = \sqrt[6]{x}$. That is,

$$\sqrt{x} < \sqrt[4]{x} < \sqrt[6]{x} < \cdots < \sqrt[2n]{x} \qquad (0 < x < 1).$$

- For values of x in interval $(1, \infty)$, the graph of $y = \sqrt{x}$ lies above the graph of $y = \sqrt[4]{x}$, which in turn falls above the graph of $y = \sqrt[6]{x}$. That is,

$$\sqrt{x} > \sqrt[4]{x} > \sqrt[6]{x} > \cdots > \sqrt[2n]{x} \qquad (x > 1).$$

- The domain of each function is $[0, \infty)$.
- The range of each function is $[0, \infty)$. On some graphing utilities, the range may appear to be $[a, \infty)$ for some $a > 0$ because of the graph's steepness near the origin.

Graphs of $y = \sqrt[n]{x}$ When n is Odd

We begin with an example that compares the graphs of radical functions with odd indexes.

E X A M P L E   2   *Comparing Graphs of Radical Functions*

Find and compare complete graphs of $y = \sqrt[3]{x}$, $y = \sqrt[5]{x}$, and $y = \sqrt[7]{x}$. Determine the domain and range of each function.

Solution All domains are $(-\infty, \infty)$. All ranges as observed from the graphs in Fig. 4.37 are also $(-\infty, \infty)$.

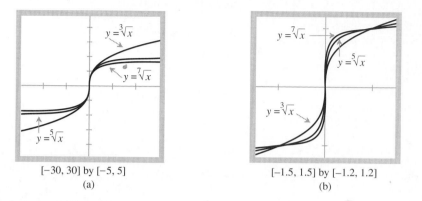

[−30, 30] by [−5, 5]
(a)

[−1.5, 1.5] by [−1.2, 1.2]
(b)

Figure 4.37 Three radical functions with odd indexes.

Both the domain and range for all these functions is $(-\infty, \infty)$. Figure 4.37 suggests the following summarizing points about radical functions with odd indexes:

- Points common to all graphs: $(-1, -1)$, $(0, 0)$, and $(1, 1)$.
- For values of x in the interval $(-\infty, -1) \cup (0, 1)$, the graph of $\sqrt[3]{x}$ lies below the graph of $\sqrt[5]{x}$, which in turn is below the graph of $\sqrt[7]{x}$. That is,

$$\sqrt[3]{x} < \sqrt[5]{x} < \sqrt[7]{x} < \cdots < \sqrt[2n+1]{x} \qquad (x < -1 \text{ or } 0 < x < 1).$$

- For values of x in the interval $(-1, 0) \cup (1, \infty)$, the graph of $\sqrt[3]{x}$ falls above the graph of $\sqrt[5]{x}$, which is above the graph of $\sqrt[7]{x}$. That is,

$$\sqrt[3]{x} > \sqrt[5]{x} > \sqrt[7]{x} > \cdots > \sqrt[2n+1]{x} \qquad (-1 < x < 0 \text{ or } x > 1).$$

Transformations of $y = \sqrt[n]{x}$

In earlier sections of this text, we graphically transformed $y = x^2$ into the graph of $f(x) = 2(x - 3)^2 - 4$ and the graph of $y = 1/x$ into the graph of $f(x) = 1/(x + 3) - 2$.

In a similar spirit, we examine in this section how the graph of $y = a\sqrt[n]{bx + c} + d$ can be obtained from the graph of $y = \sqrt[n]{x}$. As before, the answer involves transformations of vertical and horizontal shifts and stretches and reflection through the x- or y-axis.

E X A M P L E   3   Using Transformations with Radicals

Explain how to sketch each of the following functions from the graph of $f(x) = \sqrt{x}$ using transformations:

a) $g(x) = \sqrt{x - 2}$

b) $h(x) = \sqrt{x + 3}$

c) $k(x) = \sqrt{3 - x}$

Solution   (See Fig. 4.38.)

a) $g(x) = \sqrt{x - 2} = f(x - 2)$ A horizontal shift 2 units right

b) $h(x) = \sqrt{x + 3} = f(x + 3)$ A horizontal shift 3 units left

c) $k(x) = \sqrt{(3 - x)} = \sqrt{-(x - 3)}$

$\qquad\qquad = f(-(x - 3))$ A reflection through the y-axis followed by a horizontal shift 3 units right

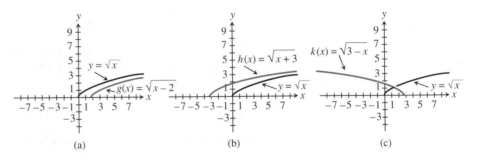

Figure 4.38   Transformations applied to the graph of $y = \sqrt{x}$.

Applying Transformations to $y = \sqrt[n]{x}$

It can be shown that the graph of any radical function of the form

$$y = a\sqrt[n]{bx + c} + d$$

can be obtained by applying transformations to the graph of $y = \sqrt[n]{x}$.

Knowing that the graph of $f(x) = a\sqrt[n]{bx + c} + d$ can be obtained as the result of applying transformations to the graph of $y = \sqrt[n]{x}$ tells us what the overall shape of f is going to be. The location and general orientation of the graph of f can be found by plotting two or three points. When doing so, always be careful to include the point of $f(x) = a\sqrt[n]{bx + c} + d$ that corresponds to the point $(0, 0)$ of $y = \sqrt[n]{x}$. Examples 4 and 5 illustrate this method.

EXAMPLE 4   Sketching a Graph by Plotting Several Points

Sketch a complete graph of $f(x) = 2\sqrt[3]{3 - 2x} + 1$.

Solution   If $x = \frac{3}{2}$, then $\sqrt[3]{3 - 2x} = 0$. Therefore $f\left(\frac{3}{2}\right) = 1$. So the point $\left(\frac{3}{2}, 1\right)$ corresponds to the point $(0, 0)$ of $y = \sqrt[3]{x}$ under the transformations that convert $y = \sqrt[3]{x}$ to $f(x) = 2\sqrt[3]{3 - 2x} + 1$.

Find a point on the graph of f to the left and to the right of $\left(\frac{3}{2}, 1\right)$.

$$f\left(\frac{11}{2}\right) = 2\sqrt[3]{3 - (11)} + 1 \qquad f\left(-\frac{5}{2}\right) = 2\sqrt[3]{3 - (-5)} + 1$$

$$= 2\sqrt[3]{-8} + 1 \qquad\qquad = 2\sqrt[3]{8} + 1$$

$$= -4 + 1 = -3 \qquad\qquad = 4 + 1 = 5$$

Points $\left(\frac{11}{2}, -3\right)$ and $\left(-\frac{5}{2}, 5\right)$ are on the graph of f. These three points are sufficient to determine the graph of f (see Fig. 4.39).   ≡

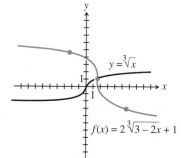

Figure 4.39
$f(x) = 2\sqrt[3]{3 - 2x} + 1$.

EXAMPLE 5   Sketching a Graph by Plotting Two Points

Sketch a complete graph of $f(x) = 3\sqrt[4]{4 - 3x} - 2$.

Solution   If $x = \frac{4}{3}$, then $\sqrt[4]{4 - 3x} = 0$. Therefore $f\left(\frac{4}{3}\right) = -2$. So the point $\left(\frac{4}{3}, -2\right)$ corresponds to the point $(0, 0)$ of $y = \sqrt[4]{x}$ under the transformations that convert $y = \sqrt[4]{x}$ to $f(x) = 3\sqrt[4]{4 - 3x} - 2$.

$$f(-4) = 3\sqrt[4]{4 - (-12)} - 2$$

$$= 3\sqrt[4]{16} - 2$$

$$= 6 - 2 = 4$$

Thus the points $(-4, 4)$ and $\left(\frac{4}{3}, -2\right)$ are on the graph of f. A complete graph is shown in Fig. 4.40.   ≡

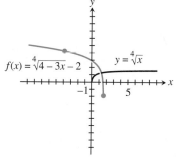
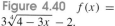

Figure 4.40 $f(x) = 3\sqrt[4]{4 - 3x} - 2$.

Solving Equations and Inequalities Involving Radicals

As before, you can sometimes use either a graphical or an algebraic method to solve an equation that involves a radical.

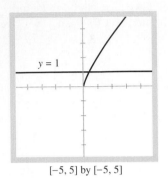

[−5, 5] by [−5, 5]

Figure 4.41 $y = x + \sqrt{x}$ and $y = 1$.

Notes on Examples

Examples 6–9 provide a review of familiar graphical techniques for solving equations and inequalities. Discuss the algebraic techniques for solving radical equations, especially the notion of introducing extraneous roots. Use the techniques of the previous section for illustrating extraneous solutions.

E X A M P L E   6   Solving an Equation Involving a Square Root

A real number plus its principal square root equals 1. Find all such numbers using both a graphical and an algebraic method.

Solution   If x is such a number, then $x + \sqrt{x} = 1$. To find all solutions to this equation, begin with the easier graphical method.

Graphical Method   Find a complete graph of $y = x + \sqrt{x}$ and $y = 1$ (see Fig. 4.41). We see that there is exactly one solution and it is between 0 and 1. Use zoom-in to find that the solution is $x = 0.38$ with an error of at most 0.01.

Algebraic Method

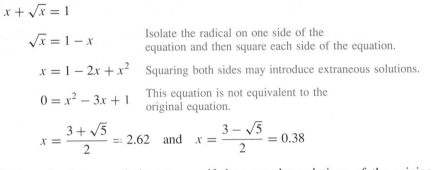

$x + \sqrt{x} = 1$

$\sqrt{x} = 1 - x$ Isolate the radical on one side of the equation and then square each side of the equation.

$x = 1 - 2x + x^2$ Squaring both sides may introduce extraneous solutions.

$0 = x^2 - 3x + 1$ This equation is not equivalent to the original equation.

$x = \dfrac{3 + \sqrt{5}}{2} = 2.62$ and $x = \dfrac{3 - \sqrt{5}}{2} = 0.38$

Check each of these solutions to see if they are also solutions of the original equation.

$2.62 + \sqrt{2.62} = 4.24$ 2.62 is not a solution.

$0.38 + \sqrt{0.38} = 1.00$ ▤

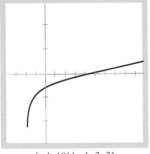

[−4, 10] by [−3, 3]

Figure 4.42
$f(x) = \sqrt{6x + 12} - \sqrt{4x + 9} - 1$.

E X A M P L E   7   Solving an Equation Involving a Square Root

Find all real numbers that are solutions to $\sqrt{6x + 12} - \sqrt{4x + 9} = 1$ using both graphical and algebraic methods.

Solution

Graphical Method   Find the x-intercepts of a complete graph of $f(x) = \sqrt{6x + 12} - \sqrt{4x + 9} - 1$ (see Fig. 4.42). It appears that $x = 4$ is the only solution.

Algebraic Method

$$\sqrt{6x + 12} - \sqrt{4x + 9} = 1$$

$$\sqrt{6x + 12} = \sqrt{4x + 9} + 1 \qquad \text{Prepare to square both sides of the equation.}$$

$$6x + 12 = 4x + 9 + 2\sqrt{4x + 9} + 1 \qquad \text{Square both sides.}$$

$$2x + 2 = 2\sqrt{4x + 9} \qquad \text{Collect terms and simplify.}$$

$$x + 1 = \sqrt{4x + 9}$$

$$x^2 + 2x + 1 = 4x + 9 \qquad \text{Square both sides of the equation a second time.}$$

$$x^2 - 2x - 8 = 0$$

$$(x - 4)(x + 2) = 0$$

So either $x = 4$ or $x = -2$. Checking for extraneous solutions shows that $x = 4$ is a solution whereas $x = -2$ is not. ≡

If radicals other than square roots are involved, then an algebraic solution requires that each side of the equation be raised to higher powers.

EXAMPLE 8   A Radical with an Index Greater Than 2

Find all real numbers x that are solutions to $\sqrt[3]{x^2 - 2x + 2} = x$.

Solution

Graphical Method   Find the zeros of the function $y = \sqrt[3]{x^2 - 2x + 2} - x$ or find where the graphs of $y = \sqrt[3]{x^2 - 2x + 2}$ and $y = x$ intersect. Either method works; we choose the latter. Find a complete graph (see Fig. 4.43). Use zoom-in to find that the solution is 1.00 with an error of at most 0.01.

Algebraic Method   Cube both sides of the equation to eliminate the radical.

$$x = \sqrt[3]{x^2 - 2x + 2}$$

$$x^3 = x^2 - 2x + 2 \qquad \text{Cube both sides.}$$

$$x^3 - x^2 + 2x - 2 = 0$$

$$x^2(x - 1) + 2(x - 1) = 0$$

$$(x - 1)(x^2 + 2) = 0 \qquad \begin{array}{l}x = 1 \text{ is the only real solution.} \\ \text{Check that this is true.}\end{array}$$

The solution is $x = 1$. ≡

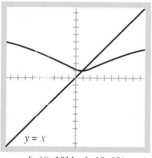

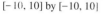

$y = x$

[−10, 10] by [−10, 10]

Figure 4.43

$y = \sqrt[3]{x^2 - 2x + 2}$ and $y = x$.

Examples 6 through 8 show that an equation can often be solved using either a graphical method or an algebraic method. The graphical method is more general and usually easier. On the other hand, some things, like proving there is a vertical asymptote, require an algebraic method for verification. It is essential to be comfortable using both methods.

The next example requires a graphical method since it cannot be solved algebraically.

EXAMPLE 9   Solving an Inequality with a Radical

Solve the following inequality for x:

$$\sqrt[5]{9 - x^2} \geq x^2 + 1.$$

Solution   There are two approaches to a graphical method:

1. Find the values of x for which the graph of $h(x) = \sqrt[5]{9 - x^2} - x^2 - 1$ lies on or above the x-axis.
2. Find the values of x for which the graph of $f(x) = \sqrt[5]{9 - x^2}$ lies above or intersects the graph of $g(x) = x^2 + 1$.

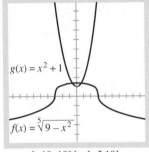

$[-10, 10]$ by $[-5, 10]$

Figure 4.44 $f(x) = \sqrt[5]{9 - x^2}$ and $g(x) = x^2 + 1$.

With the second approach, the problem can be visualized better since complete graphs of both functions are shown. Figure 4.44 shows both graphs. Use zoom-in to find that the x-coordinates of the points of intersection are -0.73 and 0.73 with an error of at most 0.01.

The solution to the inequality is the interval $[-0.73, 0.73]$.   ▤

EXAMPLE 10   APPLICATION: Finding Maximum Illumination

A street light is situated on a pole x feet above the ground. The intensity of illumination I at a point P that is 24 ft from the base of the pole is known to vary according to the equation $I = kx/d^3$, where d is the distance from P to the light and k is some positive real number (see Fig. 4.45). How high above the ground should a security light be placed to provide maximum illumination at the point P on the ground.

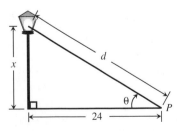

Figure 4.45 Diagram for the maximum illumination problem situation.

Solution   Describe d in terms of x and substitute the expression into the equation $I = kx/d^3$.

$$d = \sqrt{x^2 + 24^2} = \sqrt{x^2 + 576}$$

Therefore

$$I(x) = \frac{kx}{(x^2 + 576)^{3/2}}.$$

Figure 4.46 shows a complete graph of

$$I(x) = \frac{x}{(x^2 + 576)^{3/2}}.$$

Notice we set $k = 1$. The graph of I for any other value of k is a vertical stretching or shrinking of

$$y = \frac{x}{(x^2 + 576)^{3/2}}$$

by the factor k.

The value of x that gives the maximum value of I (when k is positive) is the same no matter what the actual value of k is. (Why?)

Figure 4.46 shows a complete graph of the algebraic representation. Use zoom-in to see that $x = 16.97$ with an error of at most 0.01 gives the maximum value of I.

For maximum illumination, the light should be placed on the pole 16.97 ft above the ground. ∎

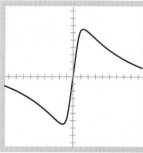

[−100, 100] by [−0.001, 0.001]

Figure 4.46

$$I(x) = \frac{x}{(x^2 + 576)^{3/2}}.$$

Assignment Guide

Day 1: Ex. 1–5, 8, 10, 11, 14, 17, 19, 22, 23, 26, 27
Day 2: Ex. 29, 31, 34, 35, 43, 44–51

Exercises for Section 4.4

In Exercises 1–4, use a graphing utility to solve each inequality for x.

1. $\sqrt[3]{x} > \sqrt[4]{x}$

2. $\sqrt[4]{x} > \sqrt[2]{x}$

3. $\sqrt{x} < 2\sqrt[4]{x}$

4. $\sqrt{x} > 4\sqrt[3]{x}$

In Exercises 5–10, list the transformations that can be applied to the graph of $y = \sqrt{x}$, $y = \sqrt[3]{x}$, or $y = \sqrt[4]{x}$ to obtain sketches of each function.

5. $y = -2\sqrt{x + 3}$

6. $y = 4 + (x + 2)^{1/2}$

7. $y = -3 + \sqrt{x - 5}$

8. $y = 3 - 2(x - 3)^{1/3}$

9. $y = 2\sqrt[3]{3x - 5}$

10. $y = -2 + 3(2 - 5x)^{1/3}$

In Exercises 11–18, sketch a complete graph of each function. Support your answer with a graphing utility.

11. $y = 3\sqrt{x - 2}$

12. $y = -1 + (x + 5)^{1/2}$

13. $y = -1 + \sqrt{x + 2}$

14. $y = 3 - 2(x + 1)^{1/3}$

15. $y = 2\sqrt[3]{x + 3}$

16. $y = 2 - 3(1 + 2x)^{1/3}$

17. $y = 1 - \sqrt[4]{2 - 4x}$

18. $y = 2 + 3\sqrt[4]{2x + 8}$

In Exercises 19–28, solve each equation algebraically. Support your answer with a graphing utility.

19. $\sqrt{x - 2} = 6$

20. $(2x - 1)^{1/2} = 2$

21. $(2x - 1)^{1/3} = 2$

22. $2\sqrt{3 - x} = -1$

23. $\sqrt[3]{x^2 - 1} = 3$

24. $(x^2 - 1)^{1/3} = -\frac{1}{2}$

25. $x - \sqrt{x} = 1$

26. $\sqrt{x - 1} = \frac{x}{5} + 1$

27. $\sqrt{x - 3} - 3\sqrt{x + 12} = -11$

28. $\sqrt{x - 5} - \sqrt{x + 3} = -2$

In Exercises 29 and 30, solve each inequality algebraically. Support your answer with a graphing utility.

29. $\sqrt{x + 3} > 6$

30. $\sqrt{x^2 - 4x - 5} > x + 2$

In Exercises 31–34, solve each equation or inequality.

31. $\sqrt{x^3 + 2} = 5$ **32.** $\sqrt[3]{x^2 - 2x + 1} = 3x$

33. $\sqrt{9 - x^2} > x^2 + 1$ **34.** $2x + 5 < 10 + 4\sqrt{3x - 4}$

In Exercises 35 and 36, sketch (without using a graphing utility) a complete graph of the region satisfying the inequality. Then support the boundary of your answer using a graphing utility.

35. $y < \sqrt{x - 2}$ **36.** $y > 4 - \sqrt[3]{x + 3}$

37. Draw a graph of $y = |x|/x$ without using a graphing utility. (*Hint:* Consider two cases, $x \geq 0$ and $x < 0$. What are the domain and range of each?)

38. Show how to determine the number of real-number solutions to $\sqrt[3]{x - 1} = 4 - x^2$. Do not use a graphing utility.

39. Use zoom-out to determine the end behavior of $f(x) = \sqrt{x^2 - 4x - 5}$.

40. Determine the end behavior of $f(x) = \sqrt{ax^2 + bx + c}$.

41. A number plus twice its square root equals 2. Find the initial and resulting numbers using an algebraic method. Support your solution with a graphing utility.

42. This **problem situation** is similar to that in Example 10. A pole is situated 30 ft from the front of a store. How high above the ground should a security light bulb be placed in order to provide maximum illumination at point P at the base of the storefront?

43. Determine the minimum *vertical* distance between the parabola $y = \frac{1}{10}(x - 4)^2 + 28$ and the line $y = 2x - 11$. At what point on the parabola does it occur?

Exercises 44–47 refer to the following **problem situation**: Penny is boating 20 mi offshore. She wishes to reach a coastal city 60 mi further down the shore by steering the boat to a point P along the shore and then driving the remaining distance.

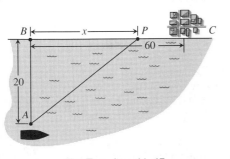

For Exercises 44–47.

Let A denote the position of the boat, C that of the city, and AB the perpendicular to the shoreline. Furthermore, let x be the distance between B and P, and let T be the total time (in hours) for the trip.

44. If Penny's boat speed is 30 mph and her driving speed is 50 mph, where should she steer her boat to arrive at the city in the least amount of time?

45. Express T as a function of x and compute $T(0)$ and $T(60)$. Interpret these values in terms of the problem situation.

46. Find a complete graph that includes only those values of x that make sense in the problem situation.

47. What value of x results in the least amount of time for the trip?

Exercises 48–51 refer to the following **problem situation**: The surface area S of a (right circular) cone excluding the base is given by the equation $S = \pi r \sqrt{r^2 + h^2}$, where r is the radius and h is the height. The volume of the cone is $V = \frac{1}{3}\pi r^2 h$.

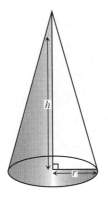

For Exercises 48–51.

48. Suppose the height of the cone is 21 ft. Find an algebraic representation and a complete graph of the surface area S as a function of the radius r.

49. If the height of the cone is 21 ft, what radius produces a surface area of 155 sq ft?

50. Suppose the volume is 380 cu ft. Find an algebraic representation and a complete graph of S as a function of r. (*Hint:* Solve for h in the volume equation and substitute the resulting expression into the surface area equation.)

51. Find the dimensions of a cone with volume 380 cu ft that has the minimum surface area.

Chapter 4 Review

KEY TERMS   (The number following each key term indicates the page of its introduction.)

end behavior asymptote
  for a rational
  function, 229
end behavior model for a
  rational function, 226

extraneous solution, 239
horizontal asymptote, 217
index, 246
principal nth root of a, 246
radical function, 246

radical sign, 246
radicand, 246
rational function, 215
reciprocal function, 216
removable discontinuity, 225

vertical asymptote, 219

REVIEW EXERCISES

In Exercises 1–6, find the domain of each function by algebraic means. Then support your answer with a graphing utility.

1. $f(x) = \dfrac{x + 1}{(x - 1)(x + 2)}$

2. $g(x) = \dfrac{(x + 3)(x - 2)}{x + 1}$

3. $h(x) = \dfrac{3}{x^2 - 2x - 3}$

4. $f(x) = \dfrac{3x - 5}{x^2 + x - 6}$

5. $g(x) = \dfrac{5}{x^2 - 3x + 1}$

6. $f(x) = \dfrac{x^3 + 1}{x^2 + 5}$

In Exercises 7–10, sketch a complete graph of each function without using a graphing utility. Find all the vertical and horizontal asymptotes. Then support your answer with a graphing utility.

7. $f(x) = \dfrac{7}{x + 5}$

8. $g(x) = \dfrac{2x^2 - 6}{3 - x^2}$

9. $y = \dfrac{5x}{x - 3}$

10. $y = \dfrac{2 + |x|}{x + 1}$

11. Find the domain and range of the function in Exercise 9.

12. Write the function in Exercise 10 as a piecewise-defined function without using absolute value symbols.

Exercises 13 and 14 refer to the following complete graph of $y = f(x)$.

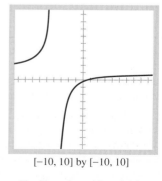

[−10, 10] by [−10, 10]

For Exercises 13 and 14.

13. As $x \to \infty$, $f(x) \to$?

14. As $x \to -4^+$, $f(x) \to$?

In Exercises 15–18, list the geometric transformations needed to produce the graph of f from the graph of $y = 1/x$ or $y = x^2$ or $y = \sqrt{x}$. Specify the order in which the transformations should be applied. In each case, sketch a complete graph of the function without the aid of a graphing utility. Write an equation for any vertical or horizontal asymptotes. Then support your answer with a graphing utility.

15. $f(x) = \dfrac{5}{x - 2}$

16. $f(x) = 2 - \dfrac{3}{x + 5}$

17. $g(x) = 2(x - 2)^2 + 3$

18. $h(x) = 4 - 3\sqrt{x + 1}$

In Exercises 19–21, draw a complete graph of each function and determine its domain and range. Find all the vertical asymptotes and the end behavior asymptote. Find all the zeros of the function.

19. $g(x) = \dfrac{3x^2}{(x - 5)(x + 4)}$

20. $f(x) = \dfrac{x^2 - 4}{x^2 + 4}$

21. $g(x) = \dfrac{2x^3 + 3x^2 - 6x - 1}{x^3 + 2}$

In Exercises 22 and 23, determine the intervals on which each function is increasing and the intervals on which each function is decreasing.

22. $f(x) = \dfrac{x^3 + 1}{x^2}$

23. $g(x) = \dfrac{-1}{x^3 + x}$

In Exercises 24 and 25, draw a complete graph of each function. Determine all local minima and maxima.

24. $f(x) = \dfrac{3x^2 + 10}{x}$

25. $f(x) = \dfrac{x^4 + 3x^3 + 2x^2 - 7}{x^2 + 2x - 1}$

In Exercises 26 and 27, sketch a complete graph. Determine the end behavior model. Find all vertical and horizontal asymptotes. Determine all zeros. Determine all extrema and the intervals where the function is increasing and decreasing.

26. $g(x) = \dfrac{x^3 + 1}{x^2 - 9}$ **27.** $f(x) = \dfrac{x^2 - 4x + 13}{x + 2}$

In Exercises 28–31, find the end behavior model and the end behavior asymptote of each function. Support your answer with a graphing utility.

28. $f(x) = \dfrac{x^3 - 5x^2 - 7x - 1}{x + 2}$

29. $g(x) = \dfrac{x^3 - 8}{x - 2}$ **30.** $f(x) = \dfrac{x^2 - x - 2}{x^3 + 1}$

31. $f(x) = \dfrac{x^4 - 2x^3 - 8x^2 + 2x + 3}{x + 1}$

32. Let $g(x) = 5/(x + 1)$ and

$$f(x) = x^2 + \frac{5}{x + 1} = \frac{x^3 + x^2 + 5}{x + 1}.$$

Complete the following table using a calculator:

x	-0.9	-0.99	-0.999	-0.9999
$f(x)$	?	?	?	?
$g(x)$	?	?	?	?

33. Use the table in Exercise 32 to numerically determine what happens to the function values $f(x)$ as $x \to -1^+$.

For Exercises 34 and 35, let

$$f(x) = \frac{x + 5}{x^2 - 2x - 8}.$$

Use a calculator to create tables of values that can be used for these exercises.

34. Determine numerically what happens to the values of the function $f(x)$ as $x \to -\infty$.

35. Determine numerically what happens to the values of the function $f(x)$ as $x \to 4^-$.

In Exercises 36–39, solve each equation algebraically. Support your answer with a graphing utility.

36. $\dfrac{2}{x} - \dfrac{3}{x + 5} = 0$ **37.** $\dfrac{1}{x - 3} + \dfrac{5}{x} = 2$

38. $\dfrac{x^3 + x}{x^2 + 1} = 0$ **39.** $\dfrac{2}{x} - \dfrac{3}{x - 1} = 6$

In Exercises 40–43, solve each inequality algebraically using the sign-pattern picture method. Support your answer with a graphing utility.

40. $\dfrac{x - 2}{x + 5} < 0$ **41.** $\dfrac{x^2 - 4}{x + 4} \geq 0$

42. $\dfrac{2}{x - 3} + 6 < 0$ **43.** $\dfrac{x + 3}{x - 2} < 5$

In Exercises 44–47, solve each equation or inequality.

44. $\dfrac{x^2 - x - 7}{x^2 + 1} = -1$ **45.** $\dfrac{x^4 + x^2 - 12}{x^2 - 3} = -5$

46. $\dfrac{x^4 + 3x^2 - x - 1}{x^2 + 3} < 0$ **47.** $\dfrac{x^5 - 3x^2 + 1}{x^3 + 7} < 3$

48. Find the zeros, y-intercept, vertical asymptotes, and end behavior asymptote of the following function:

$$f(x) = \frac{x^4 - 2x^3 + 3x^2 + x - 4}{x^2 - 5x - 14}.$$

Find a complete graph that shows all of this behavior. (More than one viewing rectangle may be needed.)

In Exercises 49–52, solve each equation algebraically. Support your answer with a graphing utility.

49. $2(x - 1)^{1/2} = 2$ **50.** $\sqrt[3]{x^2 + 3x + 1} = 2$

51. $\sqrt{3x - 5} = 2$ **52.** $(x^2 + 2x - 3)^{1/3} = 5$

53. Use a graphical argument to show that if $x < 5$, then x is *not* a solution to the inequality $\sqrt{x + 5} > 5 - x/3$. Solve the inequality.

In Exercises 54 and 55, sketch a complete graph of each function. Then support your answer with a graphing utility. List the transformations that can be applied to $y = \sqrt{x}$ or $y = \sqrt[3]{x}$ to obtain the given graph. Specify the order in which the transformations should be applied.

54. $f(x) = \sqrt[3]{x - 3}$ **55.** $g(x) = -3 + 2\sqrt{5 - x}$

56. The area of a triangle is to be 150 square units.
  a) Express the height of this triangle as a function of the length x of its base and draw a complete graph of the function.
  b) Find the domain and range of the function found in part (a). What values of x make sense in the context of this problem?
  c) Use a graphical method to find the length of the base of the triangle if the height is 800 units.

57. The area of a rectangle is to be 500 sq ft. Let x be the width of such a rectangle.
  a) Express the perimeter P as a function of x.
  b) Determine the vertical asymptotes and the end behavior asymptote and find a complete graph of the function in part (a).

c) Find the dimensions of a rectangle with an area of 500 sq ft and the least possible perimeter. What is this perimeter?

Exercises 58–61 refer to the following **problem situation**: Judy is 5 ft 6 in. tall and walks at the rate of 4 ft/sec away from a street light with a lamp 14.5 ft above level ground.

58. Find an algebraic representation of the length of Judy's shadow.

59. At what rate is the length of Judy's shadow increasing?

60. Express the distance D between the lamp and the tip of Judy's shadow as a function of time t.

61. When will the distance D be 100 ft?

Exercises 62–64 refer to the following **problem situation**: A balloon in the shape of a sphere is being inflated. Assume the radius r of the balloon is increasing at the rate of 3 in./sec and is zero when $t = 0$.

62. Express the volume V of the balloon as a function of time t.

63. Determine the volume of the balloon at $t = 5$ sec.

64. Suppose the balloon will burst when its volume is 15,000 cu in. When will the balloon burst?

Exercises 65–67 refer to the following **problem situation**: Pure acid is added to 150 oz of a 50% acid solution. Let x be the number of ounces of pure acid added.

65. Determine an algebraic representation of the acid concentration of the mixture.

66. Draw a complete graph of both the algebraic representation and of the problem situation.

67. Use a graphical method to find how much pure acid should be added to the 50% solution to produce a new mixture that is at least 78% acid.

68. Lisa wishes to obtain 85 oz of a 40% acid solution by combining a 72% acid solution with a 25% acid solution. How much of each solution should Lisa use?

Exercises 69–72 refer to the following **problem situation**: Eric is boating 15 mi offshore. He wishes to reach a coastal city 55 mi further down the shore by steering the boat to a point P along the shore and then driving the remaining distance.

Assume that his boat speed is 25 mph and his driving speed is 40 mph. Let A denote the position of the boat, C that of the city, and AB the perpendicular to the shoreline. Furthermore, let x be the distance between B and P and T be the total time (in hours) for the trip.

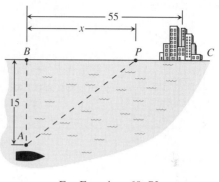

For Exercises 69–72.

69. Determine an algebraic representation for the total time T of the trip. Find $T(0)$ and $T(55)$ and interpret these values in the problem situation.

70. Draw a complete graph of the algebraic representation function T. Also find the graph of the problem situation.

71. Where should he steer his boat in order to arrive at the city in the least amount of time?

72. What is the least amount of time?

Exercises 73–75 refer to the following **problem situation**: Jeri rode her bike 11 mi from her home to Columbus, Ohio, and then completed a 45-mi trip by car from Columbus to Marysville, Ohio. Assume the average rate of the car was 41 mph faster than the average rate of the bike.

73. Find an algebraic representation for the total time T required to complete the trip (bike and car) as a function of the rate x of the bike.

74. What values of x make sense in this problem situation?

75. Use a graphical method to find the bike's rate if the total time of the trip was 55 min.

AL-KHWARIZMI

Mohammad ibn Musa al-Khwarizmi (ca. A.D. 780–850) was born in what is present-day Iran. He became a professor at the Arab University in Baghdad during the first golden age of Islamic science. His greatest contribution to mathematics was his algebra book, *Kitab al-jabr w'al muqabalah* (*The Book of Integration and Equation*). The book established algebra as an independent discipline for the first time. It showed how to solve equations, manipulate terms, and use the quadratic formula. Al-Khwarizmi also taught the use of Hindu-Arabic numerals, rather than the clumsier Roman numerals then used in Europe.

Al-Khwarizmi's impact on mathematics is reflected in our language to this day. The Arabic word *al-jabr* eventually became the modern word *algebra*. Also, the name Al-Khwarizmi later took on the meaning of "a mathematical procedure," becoming the modern word *algorithm*, by way of Latin.

THE AZTECS

The advanced Aztec civilization in modern-day Mexico was notable for its art, architecture, commerce, and system of justice. To help keep accurate accounts, the Aztecs developed an efficient system of numerals and arithmetic. They borrowed the concepts of place value and a base-20 number system from the earlier Mayan and Olmec cultures, but they had a set of distinctive Aztec numerals. For example, a small ear of corn was used for the zero symbol.

The Aztecs applied mathematics to the records that registered land ownership. These records showed the boundaries, area, and market value of property, and were used to calculate the taxes that the owners had to pay. The measurements included in the Aztec records were more accurate than the measurements of the same farms drawn up by the Spanish conquerors.

5

Exponential and Logarithmic Functions

5.1

Exponential Functions

Objective

Students will be able to explore the characteristics of exponential functions, make generalizations about the properties of exponential functions, and use exponential functions, including the number e, in applications.

So far in this text, three categories of functions have been discussed: polynomial functions, rational functions, and functions with radicals. This chapter introduces two more categories of functions: exponential and logarithmic functions.

We begin our study of exponential functions by comparing two notational conventions that represent two different functions. Both x^2 and 2^x have a **base number** and an **exponent**.

x^2: The base is x, and the exponent is 2.

2^x: The base is 2, and the exponent is x.

In the first case, the variable x is the base, and in the second case, x is the exponent. When the variable is in the base position and the exponent is a nonnegative integer, as in x^2, the function is a polynomial, as we saw earlier. However, when the variable is in the exponent position and the base is a positive real number $\neq 1$, as in 2^x, the function is called an exponential function.

Definition 5.1 Exponential Function

Let a be a positive real number other than 1. The function $f(x) = a^x$, whose domain is the set of all real numbers, is the **exponential function with base a**.

The following are examples of exponential functions:

$$f(x) = 10^x, \qquad g(v) = 4^v, \qquad h(x) = \pi^x, \quad \text{and} \quad y = 2^x.$$

These functions are not exponential functions:

$$f(x) = x^3, \qquad g(x) = \frac{x^4}{2}, \qquad y = \frac{\sqrt{x}}{x^2 + 3}, \quad \text{and} \quad y = 3^5.$$

When studying calculus, you will learn a more complete definition of exponential functions. In this text, you will become familiar with exponential functions by studying their graphs using a graphing utility.

Teaching Note

Students should exercise care in determining appropriate viewing rectangles for complete graphs in this section. Attention to domain and range should be emphasized in class discussion.

🔍 **EXPLORE WITH A GRAPHING UTILITY**

Graph the following functions together in the same viewing rectangle:

$$y = 2^x, \qquad y = 3^x, \quad \text{and} \quad y = 5^x.$$

1. Find a viewing rectangle that clearly distinguishes the complete graphs of these three functions.
2. For what values of x is it true that $2^x > 3^x > 5^x$?
3. For what values of x is it true that $2^x < 3^x < 5^x$?

    Answer the same three questions for these functions:

$$y = 2^{-x}, \qquad y = 3^{-x}, \quad \text{and} \quad y = 5^{-x}.$$

Generalize How do the graphs of $y = a^x$ and $y = a^{-x}$ compare? How do the graphs of $y = a^x$ compare for various values of a?

Graphs of Exponential Functions

Using properties of exponents, we know that $a^{-1} = 1/a$. Therefore

$$a^{-x} = \left(a^{-1}\right)^x = \left(\frac{1}{a}\right)^x.$$

Teaching Note

The shapes of exponential graphs and transformations of exponential graphs are an important background topic for calculus.

So, in particular,

$$2^{-x} = \left(\frac{1}{2}\right)^x = 0.5^x.$$

While completing the Exploration, you may have made the following observation about the behavior of exponential functions.

Generalization about Exponential Functions

- If $a > 1$, the graph of $f(x) = a^x$ has a shape like the graph of $y = 2^x$.
- If $0 < a < 1$, the graph of $f(x) = a^x$ has a shape like the graph of $y = 0.5^x$.
- For all base values $a > 0$, $f(0) = 1$.

E X A M P L E   1   Graphing Exponential Functions

Find and compare complete graphs of $f(x) = 2^x$ and $g(x) = 0.5^x = 2^{-x}$. Find the domain and range of each function.

Solution   Observe the following properties of f and g by studying Figs. 5.1 and 5.2:

$$\text{Domain for both:} \quad (-\infty, \infty).$$
$$\text{Range for both:} \quad (0, \infty). \qquad\qquad \blacksquare$$

Several other general properties of the functions f and g are illustrated in Example 1: The x-axis is a horizontal asymptote for each; also, f is an increasing function, while g is a decreasing function.

We can make two additional generalizations about the end behavior of f and g.

$$f(x) \to \infty \text{ as } x \to \infty \qquad\qquad f(x) \to 0 \text{ as } x \to -\infty$$
$$\text{and}$$
$$g(x) \to 0 \text{ as } x \to \infty \qquad\qquad g(x) \to \infty \text{ as } x \to -\infty$$

Finally, it is important to observe that

$$f(-x) = 2^{-x} = g(x) \qquad \text{and} \qquad g(-x) = 2^{-(-x)} = f(x),$$

which means that the graph of each function is the reflection about the y-axis of the other.

As with other categories of functions, exponential functions may be transformed by vertical and horizontal shifts and by vertical stretches.

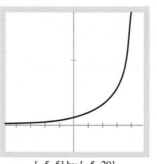

[−5, 5] by [−5, 20]

Figure 5.1 $f(x) = 2^x$.

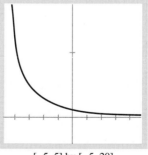

[−5, 5] by [−5, 20]

Figure 5.2
$g(x) = 0.5^x = 2^{-x}$.

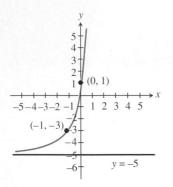

Figure 5.3 $g(x) = 2(3^{x+1}) - 5$.

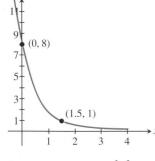

Figure 5.4 $f(x) = 2^{3-2x}$ and $g(x) = 0.25^{x-1.5}$.

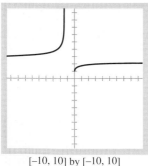

$[-10, 10]$ by $[-10, 10]$

Figure 5.5 $f(x) = \left(1 + \dfrac{1}{x}\right)^{x}$.

E X A M P L E   2   Transforming Exponential Functions

List the transformations used to obtain a complete graph of

$$g(x) = 2(3^{x+1}) - 5$$

from $f(x) = 3^x$. Sketch a complete graph of g.

Solution   Apply the following transformations, in order, to $f(x) = 3^x$:

1. Vertical stretch by a factor of 2 to obtain $y = 2(3^x)$,
2. Horizontal shift left 1 unit to obtain $y = 2(3^{x+1})$, and
3. Vertical shift down 5 units to obtain $y = 2(3^{x+1}) - 5$.

(See Fig. 5.3.)   ▤

Often the properties of exponents are used to change the form of an exponential function.

E X A M P L E   3   Using Properties of Exponents

Use the properties of exponents to show that $f(x) = 2^{3-2x}$ and $g(x) = 0.25^{x-1.5}$ are the same function. Sketch the graph of each.

Solution

$$2^{3-2x} = 2^{-2(x-3/2)} = (2^{-2})^{x-1.5} = 0.25^{x-1.5}$$

Because $0 < 0.25 < 1$, g is a decreasing function and can be obtained from the graph of $y = 0.25^x$ by a horizontal shift right 1.5 units (see Fig. 5.4).   ▤

Natural Base *e*

For the exponential functions of the form $f(x) = a^x$ considered so far, a has had integer or rational number values such as $2, 3, 5, 10, 0.5$, or 0.25.

One particular exponential function is especially important because it models many phenomena in both the natural world and the world of manufacturing. We begin the development of this special exponential function by considering the graph of the function $f(x) = (1 + 1/x)^x$.

Recall that the base of an exponential function must be positive. Therefore

$$1 + \frac{1}{x} > 0 \Longleftrightarrow x > 0 \text{ or } x < -1.$$

In other words, the domain of the function f is $(-\infty, -1) \cup (0, \infty)$.

More importantly, however, the graph of f in Fig. 5.5 suggests that there is a horizontal asymptote. You can approximate the y-coordinate of the horizontal

Bert says, "Using a graphing utility to explore the concept of the number e is one of the more powerful visualizations in mathematics."

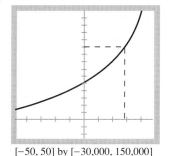

[−50, 50] by [−30,000, 150,000]

Figure 5.6
$f(t) = (50,000)(1 + 0.025)^t$.
Notice that when
$y = f(t) = 100,000$, the value for t is about 28.

asymptote by a careful selection of viewing rectangles. It turns out that this asymptote is the line $y = k$ where k is an irrational number that can be approximated accurately to nine decimal places as $k = 2.718281828 \cdots$.

Recall that the Greek letter π is used to represent the irrational ratio C/d of a circle (where C is the circumference and d is the diameter). Similarly, the special number $k = 2.718281828 \cdots$ is represented by a particular symbol, e, which is the base of the natural exponential function.

Definition 5.2   Natural Exponential Function

The **natural exponential function** is the function

$$f(x) = e^x,$$

where $e = $ the irrational number $2.718281828 \cdots$.

Applications of Exponential Functions to Growth and Decay Problems

Suppose that the population P of a certain town is increasing at constant rate r each year, where r is in decimal form.

$$P + Pr = P(1 + r) \qquad \text{Population one yr later}$$

$$P(1 + r) + P(1 + r)r = P(1 + r)^2 \qquad \text{Population two yr later}$$

$$\vdots$$

$$P(1 + r)^{t-1} + P(1 + r)^{t-1}r = P(1 + r)^t \qquad \text{In general, the population } t \text{ years later}$$

If the population is *decreasing*, the population at the end of t years would be $P(1 - r)^t$. These are examples of problems of **exponential growth** and **exponential decay**, which are found in biology, chemistry, business, and other social and physical sciences.

E X A M P L E   4   APPLICATION: Tracking Population Growth
_____

A town has a population of 50,000 that is increasing at the rate of 2.5% each year. Use a complete graph of $f(t) = (50,000)(1 + 0.025)^t$ to find when the population of the town will be 100,000.

Solution   Begin by finding a complete graph of f (see Fig. 5.6).

You need to find the value of t that makes $f(t) = 100,000$. Use zoom-in to find that $t = 28.07$ with an error of at most 0.01. ▰

Is Fig. 5.6 a complete graph of the problem situation? It depends on how the situation is interpreted. If $t = 0$ represents present time and the 2.5% growth rate extends only into the future, then the complete graph would include only the portion where $t \geq 0$, or points to the right of the y-axis. On the other hand, if the 2.5% rate also applies to prior years, then the graph to the left of the y-axis describes the problem situation in prior years.

The word *population* is used broadly. In addition to humans, it can refer to some species of animal or plant, to bacteria or molecules. It could even refer to currency such as the dollar.

For example, assume that the number of bacteria in a certain bacterial culture doubles every hour and that there are 100 present initially. Study the following pattern and generalize to find a description of the population P after t hours:

$$200 = 100(2) \qquad \text{Total bacteria after 1 hr}$$

$$400 = 100(2^2) \qquad \text{Total bacteria after 2 hr}$$

$$800 = 100(2^3) \qquad \text{Total bacteria after 3 hr}$$

$$\vdots$$

$$P = 100(2^t) \qquad \text{Total bacteria after } t \text{ hr}$$

E X A M P L E   5   Graphical Method for Biological Growth

Suppose a culture of 100 bacteria are put in a petri dish and the culture doubles every hour. Find when the number of bacteria will be 350,000.

Solution   From the discussion prior to this example, we see that the population P of the petri dish after t hours is $P(t) = 100(2^t)$. Find a complete graph of this function.

We must find an appropriate viewing rectangle. Observe that $100(2^{13}) = 819,200$. Consider the graph of $P(t) = 100(2^t)$ in the viewing rectangle $[0, 13]$ by $[0, 450,000]$ (see Fig. 5.7).

The solution is found at the point of intersection of the graphs of $P(t) = 100(2^t)$ and $y = 350,000$. The graph of f shows that t at the point of intersection is a little less than 12. Use zoom-in to find that t is 11.77 with an error of at most 0.01.

Thus after 11 hr and 46.2 min, the number of bacteria is approximately 350,000.   ▤

Exponential functions also model radioactive decay. The **half-life** of a radioactive substance is the amount of time it takes for half of the substance to decay. Suppose the half-life of a certain radioactive substance is 20 days and that there are 5 g initially. Study the following pattern and generalize to find a

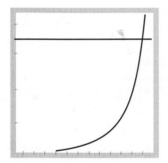

[0, 13] by [0, 450,000]

Figure 5.7 $P(t) = 100(2^t)$ and $y = 350,000$.

Notes on Exercises

Ex. 47–53 provide a rich opportunity for class discussion.
For Ex. 54–61, be sure to explore questions other than those asked in the textbook.

description of the number of grams $f(t)$ present after t days:

$$\frac{5}{2} = 5\left(\frac{1}{2}\right) \qquad \text{Grams remaining after 20 days}$$

$$\frac{5}{4} = 5\left(\frac{1}{2}\right)^2 \qquad \text{Grams remaining after 40 days}$$

$$\vdots$$

$$f(t) = 5\left(\frac{1}{2}\right)^{t/20} \qquad \text{Grams present after } t \text{ days}$$

We use this generalization in Example 6.

E X A M P L E   6   APPLICATION: Calculating Radioactive Decay

Suppose the half-life of a certain radioactive substance is 20 days and there are 5 g present initially. Draw a complete graph of an algebraic representation of this problem situation and find when there will be less than 1 g of the substance remaining.

Solution   Based on the discussion prior to the example, an algebraic representation of this problem situation is

$$f(t) = 5\left(\frac{1}{2}\right)^{t/20}.$$

Because $f(0) = 5$ and we want to find t such that $f(t) = 1$, a viewing rectangle with $Y\min = -5$ and $Y\max = 10$ should be reasonable for the initial view (see Fig. 5.8). Use zoom-in to find that two graphs intersect at $t = 46.44$.

Thus there will be less than 1 g of the radioactive substance left after 46.44 days.

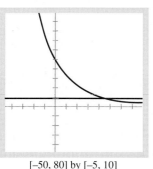

[−50, 80] by [−5, 10]

Figure 5.8 $f(t) = 5\left(\frac{1}{2}\right)^{t/20}$ and $y = 1$.

We close this section by investigating an important graph that occurs in the study of statistics: the **normal distribution curve**, sometimes called a **bell curve** because of its shape.

E X A M P L E   7   APPLICATION: Graphing a Normal Distribution Curve

Let

$$f(x) = \frac{1}{\sqrt{2\pi}} e^{-x^2/2}.$$

Find a complete graph of f and find its maximum value.

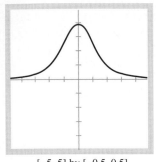

[−5, 5] by [−0.5, 0.5]

Figure 5.9

$f(x) = \dfrac{1}{\sqrt{2\pi}} e^{-x^2/2}.$

Solution   No graph is evident in the standard viewing rectangle. So the graph either lies outside the standard viewing rectangle or is hidden within it.

$$\frac{x^2}{2} \geq 0 \Longrightarrow 0 < e^{-(x^2/2)} \leq 1$$

$$\Longrightarrow 0 < \frac{1}{\sqrt{2\pi}} e^{-(x^2/2)} < \frac{1}{\sqrt{2\pi}}$$

$$0 < f(x) \leq \frac{1}{\sqrt{2\pi}} \approx 0.399$$

Try the viewing rectangle $[-5, 5]$ by $[-0.5, 0.5]$ (see Fig. **??**).
The maximum value is $f(0) = \dfrac{1}{\sqrt{2\pi}} = 0.399$ with an error of at most 0.01.

∎

Exercises for Section 5.1

1. Find complete graphs of
$$y_1 = 3^x, \quad y_2 = 5^x, \quad \text{and} \quad y_3 = 10^x$$
in the same viewing rectangle and show that all three pass through the point $(0, 1)$.

2. The graph of $y = 3^{-x}$ is symmetric about the y-axis with which one of the following curves?

a) $y_1 = -3^x$ b) $y_2 = 3^x$ c) $y_3 = \dfrac{1}{3^x}$

In Exercises 3 and 4, determine without using a graphing utility which of the three functions are decreasing functions. Then support your answer with a graphing utility.

3. a) $y_1 = 4^{-x}$ b) $y_2 = 6^x$ c) $y_3 = 0.5^x$
4. a) $y_1 = 0.25^x$ b) $y_2 = 0.8^x$ c) $y_3 = 1.3^x$

In Exercises 5–10, list the transformations that can be used to obtain a complete graph of the given function from the graph of $y = 2^x$.

5. $y = 2^x - 4$ 6. $y = 3(2^x)$
7. $y = 2^{x-3}$ 8. $y = -2^{-x}$
9. $y = 2^{x+1} + 7$ 10. $y = -2 \cdot 2^{x+2} - 1$

In Exercises 11–14, draw the graphs of $y = \left(\frac{1}{4}\right)^x$, $y = \left(\frac{1}{3}\right)^x$, and $y = \left(\frac{1}{2}\right)^x$ in the same $[-2, 4]$ by $[-1, 2]$ viewing rectan-

gle. Solve each inequality.

11. $\left(\dfrac{1}{4}\right)^x > \left(\dfrac{1}{3}\right)^x$ 12. $\left(\dfrac{1}{3}\right)^x > \left(\dfrac{1}{2}\right)^x$

13. $\left(\dfrac{1}{4}\right)^x < \left(\dfrac{1}{3}\right)^x$ 14. $\left(\dfrac{1}{3}\right)^x < \left(\dfrac{1}{2}\right)^x$

In Exercises 15–18, use properties of exponents to select the pair of functions that are equal.

15. a) $y_1 = 3^{2x+4}$ b) $y_2 = 3^{2x} + 4$ c) $y_3 = 9^{x+2}$

16. a) $y_1 = 2^{3x-1}$ b) $y_2 = 8^{x-1/2}$ c) $y_3 = 2^{3x-3/2}$

17. a) $y_1 = 4^{3x-2}$ b) $y_2 = 2 \cdot 2^{3x-2}$ c) $y_3 = 2^{3x-1}$

18. a) $y_1 = 2^{-(x-4)}$ b) $y_2 = (0.5)^{x-4}$ c) $y_3 = (0.25)^{2x-8}$

In Exercises 19–24, sketch a complete graph of each function without using a graphing utility. Then support your answer with a graphing utility.

19. $f(x) = 1 + 2^x$ 20. $f(x) = 1 - 3^x$
21. $f(x) = 2^{x-3}$ 22. $y = 3 \cdot 2^{x+2}$
23. $g(x) = -2 - 2^{x+3}$ 24. $D(x) = -2 + \left(\dfrac{1}{2}\right)^{x+1}$

25. Determine the domain and range of the function in Exercise 19.

26. Determine the domain and range of the function in Exercise 20.

In Exercises 27–30, use the rules of exponents to solve each equation.

27. $2^x = 4^2$ **28.** $x^4 = 16$

29. $8^{x/2} = 4^{x+1}$ **30.** $(-8)^{5/3} = 2(4^{x/2})$

In Exercises 31–34, sketch a complete graph and determine the domain of each function.

31. $f(x) = e^{2x-1}$ **32.** $g(x) = 2^{x^2-1}$

33. $f(x) = x(3^x)$ **34.** $f(x) = xe^{-x}$

In Exercises 35–40, determine where each function is increasing and decreasing. Determine all local maximum and minimum values and the range of each function.

35. $f(x) = e^{2x-1}$ **36.** $g(x) = 2^{x^2-1}$

37. $f(x) = x(3^x)$ **38.** $f(x) = xe^{-x}$

39. $f(x) = x(2^{-x^2})$ **40.** $g(x) = -x(10^{x^2/50})$

In Exercises 41 and 42, sketch a complete graph of each inequality.

41. $y < e^x - 1$ **42.** $y \geq 2e^{x-3}$

43. Find the points that satisfy the equations $y = x^2$ and $y = 2^x$ simultaneously. Use both of the following methods and compare them. Which do you prefer?

Method 1 Find a complete graph of each equation in the same viewing rectangle; then use zoom-in to find the points where the two graphs intersect.

Method 2 Find the zeros of the function $y = 2^x - x^2$. Use the zeros to find where the graphs cross.

44. Use the two methods outlined in Exercise 43 to find the points that satisfy each of these equations simultaneously: $y = 3^x$ and $y = x^3$.

45. Investigate graphically the end behavior of $y = (1+2/x)^x$ by drawing the graph of $y = (1+2/x)^x$ and the line $y = e^2$ in the $[0, 100]$ by $[0, 10]$ viewing rectangle. Repeat in the $[100, 1000]$ by $[7, 8]$ viewing rectangle. Find an equation for a horizontal asymptote.

46. Investigate graphically the end behavior of $y = (1+3/x)^x$ by drawing the graph of $y = (1+3/x)^x$ and the line $y = e^3$ in the $[0, 100]$ by $[0, 25]$ viewing rectangle. Repeat in the $[100, 1000]$ by $[19, 21]$ viewing rectangle. Find an equation for a horizontal asymptote.

Exercises 47–49 refer to the following **problem situation**: The number of bacteria in a petri dish culture doubles every 3 hr. Initially there are 2500 bacteria present.

47. Find the number of bacteria present after 3 hr. After 6 hr. After t hours.

48. Find a complete graph that shows the number of bacteria present during the first 24-hr period.

49. Find when the number will be 100,000.

Exercises 50–53 refer to the following **problem situation**: The number of rabbits in a certain population doubles every month, and there are 20 rabbits present initially.

50. How many rabbits are present after 1 yr? After 5 yr?

51. Find a graph that shows the number of rabbits present during the first year.

52. When will the number be 10,000?

53. **Writing to Learn** Write a paragraph that explains why this exponential growth model is not a good model for rabbit population growth over a long period of time. What factors influence population growth?

54. The population P of a town is 475,000 and is increasing at the rate of 3.75% each year. Find an algebraic representation for P as a function of time. When will the population be 1 million?

55. The population of a small town in the year 1890 was 6250. Assuming it increased at the rate of 3.75% per year, what was the population in 1915? In 1940?

56. The population P of a town is 123,000 and is decreasing at the rate of 2.375% each year. Find an algebraic representation for P as a function of time. Determine when the population will be 50,000.

57. The half-life of a certain radioactive substance is 14 days, and there are 6.58 g present initially. Find an algebraic representation for P as a function of time. When will there be less than 1 g remaining?

58. The half-life of a certain radioactive substance is 65 days, and there are 3.5 g present initially. Draw a complete graph of this problem situation where A is the amount and is a function of t. When will there be less than 1 g remaining?

Exercises 59–61 refer to the following **problem situation**: The half-life of a certain radioactive substance is 1.5 sec, and S represents the amount of the substance initially (in grams).

59. How much of the substance is left after 1.5 sec? After 3 sec? After t seconds?

60. Sketch a complete graph of an algebraic representation if there are 2 g of the substance initially.

61. What is the initial amount of the substance needed if there is to be 1 g left after 1 min?

5.2 _____

Simple and Compound Interest

Many people pay rent for the use of an apartment that is owned by someone else.
Interest paid on a savings deposit or charged for a loan works essentially in the
same way. It is "rent" for the use of someone else's money. Interest can be
calculated in two ways: simple and compound. We will see that interest is an
economic application of exponential functions.

Simple Interest

Suppose you deposit \$200 in a bank savings account that pays 7% for 1 yr. The
interest earned (the rent the bank pays you for the use of your money) is \$200(0.07),
or \$14. We say that \$200 is invested at a simple annual interest rate of 7%.

Definition 5.3   Simple Interest

Suppose P dollars are invested at a **simple interest rate** r, where r is a
decimal, then P is called the **principal** and $P \cdot r$ is the **interest** received at
the end of one interest period.

The following pattern, which results from applying Definition 5.3, leads to a
general formula for the total amount after n interest periods:

$$P + Pr \qquad \text{Principal plus the interest for one period}$$

$$P + 2Pr \qquad \text{Total amount after two interest periods}$$

$$P + 3Pr \qquad \text{Total amount after three interest periods}$$

$$\vdots$$

$$P + Pnr = P(1 + nr) \qquad \text{Total amount after } n \text{ interest periods}$$

Simple Interest Formula

Suppose P dollars are invested at a simple interest rate r, then the **simple
interest formula** for the total amount T after n interest periods is

$$T = P(1 + nr).$$

EXAMPLE 1   Calculating Simple Interest

Silvia deposits $500 in an account that pays 7% simple annual interest. How much will she have saved after 10 yr?

Solution   Let $P = 500$, $n = 10$, and $r = 7\%$ (or 0.07); let T be the total saved after 10 yr. Applying the simple interest formula, we get

$$T = P(1 + nr)$$

$$T = 500\big[1 + 10(0.07)\big] \quad \text{Substitute values for } P, r, \text{ and } n.$$

$$= 850.$$

The value of the investment after 10 yr will be $850.   ≡

Teaching Note

Developing the concepts of compound interest and continuous interest can be enhanced by using the large-screen display of a graphing calculator along with its recursive, replay, and editing capabilities. Begin by setting the decimal display to two places to show answers in dollars and cents.

Compound Interest

Financial institutions commonly allow interest to compound; that is, they pay *interest on the interest*. Suppose P dollars are invested at rate r with interest compounded at the end of each year. Then P_n is the total amount after year n.

$$P_1 = P(1 + r)$$

$$P_2 = P_1(1 + r) = P(1 + r)(1 + r) = P(1 + r)^2$$

$$P_3 = P_2(1 + r) = P(1 + r)^2(1 + r) = P(1 + r)^3$$

$$\vdots$$

$$P_n = P(1 + r)^n$$

Interest accumulated in this way is called **compound interest**, and the total amount S at the end of n interest periods is given by the formula

$$S = P(1 + r)^n.$$

When the interest period is 1 yr, we say that the interest is **compounded annually**.

Notice that the compound interest formula is an exponential function with a base of $1 + r$. Compare this formula with the model of exponential growth in Section 5.1.

Teaching Note

To illustrate the growth that leads to the answer in Example 2, enter 500 on the computation screen of your calculator (key in the number followed by ENTER or EXE). The calculator should return the answer of 500.00. Then key in ANS X (1.07) followed by ENTER or EXE ten times, one press for each year of growth. Each time you press the ENTER or EXE key, a new line of text should appear. The screen should eventually show a table of values that culminates in 983.58. This progression of values is a geometric sequence with first term 500.00 and common ratio 1.07.

EXAMPLE 2   Calculating Compound Interest

Suppose $500 is invested at 7% interest compounded annually. Find the value of the investment after 10 yr.

Frank says, "Calculating compound and continuous interest using the power of the blank screen of a graphing utility is one of the main reasons that I believe a graphing calculator is the best scientific calculator money can buy."

Solution   For the compound interest formula $S = P\left(1 + \dfrac{r}{k}\right)^n$, we are given that $r = 0.08$, $k = 365$, $n = 365 \cdot 5 = 1825$, and $S = 5000$. We need to find P.

$$S = P\left(1 + \frac{r}{k}\right)^n$$

$$5000 = P\left(1 + \frac{0.08}{365}\right)^{1825} \qquad \begin{array}{l}\text{Substitute values for } S, r, k, \text{ and } n, \text{ and}\\ \text{solve for } P.\end{array}$$

$$P = \frac{5000}{(1 + 0.08/365)^{1825}}$$

$$P = 3351.75$$

An investment of \$3351.75 at 8% APR compounded daily will have a value of \$5000 after 5 yr.   ▇

The next example assumes that the original investment amount and the final goal are both known. We ask what APR is needed to meet the goal. It also illustrates that some interest-rate problems can be solved either graphically or algebraically.

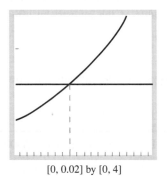

[0, 0.02] by [0, 4]

Figure 5.12 $y = (1 + x)^{84}$ and $y = 2$.

E X A M P L E   6   Calculating Compound Interest

What APR compounded monthly is required for a \$2000 investment to accumulate to \$4000 in 7 yr?

Solution   We need to solve the compound interest formula for r.

$$4000 = 2000\left(1 + \frac{r}{12}\right)^{84}$$

$$2 = (1 + x)^{84} \qquad \text{Let } x = r/12.$$

Graphical Method   Find where complete graphs of $y = (1 + x)^{84}$ and $y = 2$ intersect (see Fig. 5.12). Use zoom-in to find that $x = 0.00828$.

Algebraic Method   Solve the following equation for x:

$$2 = (1 + x)^{84}$$

$$2^{1/84} = \left[(1 + x)^{84}\right]^{1/84} \qquad \text{Take the 84th root of both sides of the equation.}$$

$$2^{1/84} = 1 + x \qquad \text{Solve the equation for } x.$$

$$x = 2^{1/84} - 1 \qquad \text{Evaluate this expression with a calculator.}$$

$$x = 0.0082858917$$

Finally, solve for r.

$$x = \frac{r}{12}$$

$$r = 12x$$

$$r = 12(0.0082858917) = 0.099430703$$

Rounding, we get $r = 0.09943$.

An APR of 9.943% compounded monthly will produce $4000 in 7 yr.   ▬

Suppose $1000 is invested at 8%. How much difference is there in the total value of the investment after 1 yr based on different compounding periods? The following calculations show that this value increases as the number of compounding periods increases. The value appears to be approaching a little less than $1083.30.

Compounding Period	Total Value after 1 Year	
Annually	$S = 1000(1 + 0.08)$	$= \$1080.00$
Quarterly	$S = 1000\left(1 + \dfrac{0.08}{4}\right)^{4}$	$= \$1082.43$
Monthly	$S = 1000\left(1 + \dfrac{0.08}{12}\right)^{12}$	$= \$1083.00$
Weekly	$S = 1000\left(1 + \dfrac{0.08}{52}\right)^{52}$	$= \$1083.22$
Daily	$S = 1000\left(1 + \dfrac{0.08}{365}\right)^{365}$	$= \$1083.28$
Hourly	$S = 1000\left(1 + \dfrac{0.08}{8760}\right)^{8760}$	$= \$1083.29$

Notice that what we really want to know is the end behavior of the function

$$f(x) = 1000\left(1 + \frac{0.08}{x}\right)^{x}.$$

It can be shown that

$$\left(1 + \frac{0.08}{n}\right)^{n} \longrightarrow e^{0.08} \text{ as } n \to \infty.$$

See Fig. 5.13. In general, if P dollars are invested at compound interest rate r for t years and the number of compounded periods approaches infinity, then the value of the investment approaches Pe^{rt}. This is called **continuous interest**.

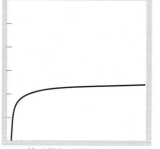

[0, 15] by [1070, 1100]

Figure 5.13

$$y = 1000\left(1 + \frac{0.08}{x}\right)^{x}.$$

Notes on Exercises

These exercises serve as a good foundation for future sections of this chapter. Carefully select problems to be done as homework and for students to present in class. This is a good opportunity to use cooperative groups. Assign representative problems of varying types to different groups.

Ex. 46–49 allow students to see how *inflation* and *purchasing power* are related to exponential growth.

Assignment Guide

Day 1: Ex. 1, 4, 5, 8, 11, 13–16, 19, 22, 25
Day 2: Ex. 32–36, 38, 41, 44, 46–50

Continuous Interest Formula

If P dollars are invested at APR r (in decimal form) and **compounded continuously**, then the value of the investment after t years is given by $S = Pe^{rt}$, where e is 2.718281828 (calculated to 9 places).

E X A M P L E   7   Calculating Continuous Interest

Suppose Noah invests $1000 at an 8% APR compounded continuously. After 5 yr, what will his account be worth?

Solution   Let $P = 1000$, $r = 0.08$, and $t = 5$. Then

$$S = Pe^{rt}$$

$$S = 1000e^{0.08(5)}$$

Most calculators have e built in so that this value can be computed with a calculator. You can also enter 2.718281828 for e and compute the value of S.

$$= 1491.82.$$

The value of the investment after 5 yr is $1491.82.   ≡

Exercises for Section 5.2

1. Suppose you invest $1250 at 8% simple interest for 2 yr. How much interest will your investment earn? What will be its value after 2 yr?

2. Suppose you invest $5000 at 9% simple interest. At the end of 5 yr, you use part of the investment to make a $7000 down payment for a house. How much of your investment remains?

3. Suppose you borrow $575 at 11.5% simple interest. The note needs to be repaid in a single payment at the end of 1 yr. What is the total payoff?

4. A certain investment earns 7% simple interest. If you invest $500, when will your investment be worth $600?

In Exercises 5–11, find the value of the investment for each initial amount, invested at the given rate, after the stated elapsed time.

5. $700, 7% compounded annually, 5 yr

6. $1200, 9% compounded annually, 8 yr

7. $7000, 11% compounded annually, 15 yr

8. $5000, 8% compounded quarterly, 5 yr

9. $20,000, 7.5% compounded monthly, 8 yr

10. $8000, 8.3% compounded daily, 4 yr

11. $3540, 9% compounded monthly, 12 yr

12. Would you rather invest at 11.1% compounded monthly or 11.2% compounded quarterly? Why?

13. Would you rather invest money for 10 yr at 11% simple interest or 8% compounded annually?

14. Draw a complete graph in the same viewing rectangle for $0 \le x \le 10$ of both

$$y = 1000(1 + 0.06x) \quad \text{and} \quad y = 1000(1 + 0.09x).$$

What value of Ymax did you use to make these graphs

clearly distinguishable?

15. Draw a complete graph in the same viewing rectangle for $0 \le x \le 10$ of both

$$y = 1000(1 + 0.05x) \quad \text{and} \quad y = 1000(1.05)^x.$$

What value of Ymax did you use to make these graphs clearly distinguishable?

16. The functions in Exercise 14 are algebraic representations for simple interest. What is the annual percentage rate (APR) in each case?

17. One function in Exercise 15 is an algebraic representation for simple interest and the other one is an algebraic representation for compound interest. Which one is which, and what is the APR in each case?

18. Draw a complete graph in the same viewing rectangle for $0 \le x \le 10$ of both $y = 1000(1.06)^x$ and $y = 1000(1.09)^x$. What value of Ymax did you use to make these graphs clearly distinguishable?

Exercises 19–24 involve equations that occur in either simple or compound interest problems. Solve each of these equations algebraically.

19. Find r if $2500 = 1000(1 + 12r)$.

20. Find t if $2500 = 1000(1 + 0.07t)$.

21. Determine P if $2500 = P[1 + 0.08(16)]$.

22. Find r if $2500 = 1000(1 + r)^{12}$.

23. Find P if $2500 = P(1.08)^{16}$.

24. Find S if $S = \dfrac{(1 + i)^{120} - 1}{i}$ and $i = 0.06$.

25. Use a graphical method to solve the equation $2500 = 1000(1.07)^n$ for n.

26. Find S if $S = 500\dfrac{(1 + i)^{48} - 1}{i}$ and $i = 0.08$.

Exercises 27–31 refer to the following **problem situation**: Sally deposits \$500 in a bank that pays 6% annual interest. Assume she makes no other deposits or withdrawals. Determine how much her investment will be worth after 5 yr if the bank pays interest compounded as follows:

27. Annually   **28.** Quarterly

29. Monthly   **30.** Daily

31. Continuously

Exercises 32–36 refer to the following **problem situation**: Escobar deposits \$4500 in a bank that pays 7% annual interest.

Assume that he makes no other deposits or withdrawals. Find out how much his investment is worth after 8 yr if the bank pays interest compounded as follows:

32. Annually   **33.** Quarterly

34. Monthly   **35.** Daily

36. Continuously

37. Find when an investment of \$2300 accumulates to a value of \$4150 if the investment earns 9% interest compounded quarterly.

38. Find when an investment of \$1500 accumulates to a value of \$3750 if the investment earns 8% interest compounded monthly.

39. A \$1580 investment earns interest compounded annually. Determine the annual interest rate if the value of the investment is \$3000 after 8 yr.

40. A \$22,000 investment earns interest compounded monthly. Determine the annual interest rate if the value of the investment is \$36,500 after 5 yr.

41. What is the value of an initial investment of \$2575 at 8% interest compounded continuously for 6 yr?

42. Find when an investment doubles in value at 6% interest compounded continuously.

43. Find r if $2300 = 1500\, e^{10r}$. Make up a problem situation for which the equation is a model.

44. Determine how long it will take for an investment to double in value at 5.75% interest compounded quarterly.

45. Determine how long it will take for an investment to triple in value at 6.25% interest compounded monthly.

Exercises 46–49 refer to the following **problem situation**: The formula $S = C(1+r)^n$ is frequently used to model inflation. In such a case, r is the annual inflation rate, C is the value today, and S is the *inflated* value n years from now. The *purchasing power* of \$1 n years from now (assuming an annual inflation rate of r) is $C = 1/(1 + r)^n$.

46. What is the purchasing power of \$1 in 10 yr if the annual inflation rate is 3%?

47. What is the purchasing power of \$1 in 10 yr if the annual inflation rate is 8%?

48. What is the purchasing power of \$1 in 10 yr if the annual inflation rate is 15%?

49. Assume an inflation rate of 8%. Use a graph to determine the value of a \$55,000 house in 7 yr. (Assume no other factors affect the value of the house.)

50. Writing to Learn   Write a paragraph explaining the difference between simple interest and compound interest. Include in your discussion a calculation that shows the simple interest rate required to yield the same as 8% compounded quarterly for 10 yr.

5.3 _____

Effective Rate and Annuities

Effective Annual Rate

Objective

Students will be able to apply the formulas for finding the present and future value of an annuity and will be able to calculate the effective annual rate of an investment.

Key Ideas

Effective annual rate
Effective yield
Ordinary annuity
Future value of an annuity
Present value of an annuity

How does one compare investment options when there are so many ways interest can be compounded? For example, would you be better off investing in an account that pays 8.75% APR compounded quarterly or one that pays 8.7% compounded monthly?

One way to decide is to compare a given method of compounding to the simple interest rate that would yield the same balance at the end of 1 yr. This simple interest rate is called the effective yield or the effective annual rate (i_{eff}).

Definition 5.4   Effective Annual Rate

The **effective annual rate i_{eff}** of APR r compounded k times per year is given by the equation

$$i_{\text{eff}} = \left(1 + \frac{r}{k}\right)^k - 1.$$

Another name for effective annual rate is **effective yield**.

Example 1 illustrates how the effective-annual-rate formula can be used.

Notes on Examples

Example 1 can be made more meaningful for students if local financial institutions can provide current quotes on interest paid on investments. Have students call local banks and savings and loans and ask for quotes on rates and periods.

E X A M P L E   1   Comparing Effective Annual Rates

Compare the effective annual rates of an account paying 8.75% compounded quarterly with an account paying 8.75% compounded monthly.

Solution   For 8.75% compounded quarterly, the effective annual rate is

$$i_{\text{eff}} = \left(1 + \frac{r}{k}\right)^k - 1$$

$$= \left(1 + \frac{0.0875}{4}\right)^4 - 1$$

$$= 0.0904,$$

or 9.04%. For 8.75% compounded monthly, the effective annual rate is

$$i_{\text{eff}} = \left(1 + \frac{r}{k}\right)^k - 1$$

$$= \left(1 + \frac{0.0875}{12}\right)^{12} - 1$$

$$= 0.0911,$$

or 9.11%, which is just slightly better than the effective rate of 8.75% compounded quarterly. ≡

Ordinary Annuity

So far in Sections 5.2 and 5.3, we have focused on finding the value of a single investment as a result of either simple or compound interest. In this section, we consider investments that involve regular equal deposits into the investment over regular time intervals. An investment program of this type is called an ordinary annuity.

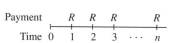

Figure 5.14 In an ordinary annuity, equal payments of R dollars are invested each payment period.

Definition 5.5   Ordinary Annuity

An **ordinary annuity** is a sequence of equal regular periodic payments to be made in the future (see Fig. 5.14).

E X A M P L E   2   Calculating Ordinary Annuity

Sarah makes quarterly $500 payments into a retirement account that pays quarterly interest. If the account pays 8% compounded quarterly, how much will be in Sarah's account at the end of the first year?

Solution   At the end of the first quarter, the value is equal only to the original $500 payment. At the end of the second quarter, the value includes the first payment plus interest, 500(1.02), plus a second $500 payment. The pattern continues as shown in this table.

	Total Value
End of 1st quarter	$500
End of 2nd quarter	500(1.02) + 500 = $1010
End of 3rd quarter	1010(1.02) + 500 = $1530.20
End of 4th quarter	1530.20(1.02) + 500 = $2060.80

At the end of the year, the account's value will be $2060.80. ≡

When you calculate the value S of an account in the future, you have found what is called the future value of S. In Example 2, we found the future value of Sarah's account at the end of 1 yr.

It is not practical to try to find the future value of Sarah's account at the end of 35 yr using the step-by-step solution method of Example 2. Instead, we will use a formula that will be verified in Chapter 11.

Future Value of an Ordinary Annuity

The **future value S of an ordinary annuity** consisting of n equal payments of R dollars, each with the interest rate i per period (payment interval), is given by

$$S = R\frac{(1+i)^n - 1}{i}. \tag{1}$$

E X A M P L E   3   Finding the Future Value of an Ordinary Annuity

Suppose you make monthly $25 deposits into a retirement account that pays an APR of 9%, or 0.75% monthly. What is the value S of this annuity at the end of 47 yr?

Solution   Let $R = \$25$, $i = 0.0075$, and $n = 12(47) = 564$. Use Eq. (1) to find the value S.

$$S = R\frac{(1+i)^n - 1}{i}$$

$$= 25\left(\frac{(1.0075)^{564} - 1}{0.0075}\right)$$

$$= \$222,137.13$$

The value S of the annuity after 47 yr is $222,137.13.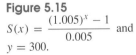

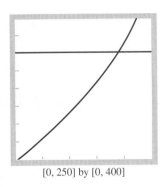

[0, 250] by [0, 400]

Figure 5.15
$S(x) = \dfrac{(1.005)^x - 1}{0.005}$ and $y = 300$.

Notice that 47 yr is approximately the length of time between high school graduation and retirement. Example 3 shows that investing $25 monthly will yield a good start at building a retirement fund.

If we replace the variable n in Eq. (1) with x, we have a function $S(x)$ that describes the future value of an annuity x payment intervals into the future. The function in Fig. 5.15,

$$S(x) = \frac{(1.005)^x - 1}{0.005},$$

describes the value of depositing $1 each month into an account that pays 6% APR.

payments. Find that monthly payment
term of 30 yr.

18. An $86,000 mortgage loan at 12% APR
payments. Find that monthly payment
term of 15 yr.

19. A $100,000 mortgage requires monthly
yr at 12% APR. How much is each payr

20. An $86,000 mortgage for 30 yr at 12
monthly payments of $884.61. Suppos
make monthly payments of $1050.00.
mortgage loan be completely paid?

21. Suppose you make payments of $884.61
mortgage in Exercise 20 for 10 yr and the
of $1050 until the loan is paid. When v
loan be completely paid under these circ

In Exercises 22–25, each function is an alge
tion for the present value of a certain annui
determine from the form of the function th
rate and the number of payments per year.

22. $f(x) = 200 \dfrac{1 - (1 + 0.08/12)^{-x}}{0.08/12}$

23. $g(x) = 200 \dfrac{1 - (1 + 0.11/4)^{-x}}{0.11/4}$

24. $f(x) = 200 \dfrac{1 - (1.01)^{-x}}{0.01}$

25. $g(x) = 500 \dfrac{1 - (1.0075)^{-x}}{0.0075}$

26. Draw a complete graph of the following
same viewing rectangle for $0 \leq x \leq 60$:

a) $y = 200 \dfrac{1 - (1 + 0.08/12)^{-x}}{0.08/12}$

b) $y = 200 \dfrac{1 - (1.01)^{-x}}{0.01}$

27. Explain how the graphs in Exercise 26
nancing a car requiring monthly payment

28. Consider the following equation:

$$86,000 = R \dfrac{1 - (1.01)^{-x}}{0.01}$$

a) Solve for R.
b) Draw a complete graph of $R = f(x)$ f
c) Explain how the graph in part (b) relat
mortgage loan at 12% APR.

29. Solve the following equation for t using a

Teaching Note

Theorem 5.1 is tl
for developing th
logarithms. In Se
function $y = a^x$ i
on $(-\infty, \infty)$ if a
decreasing on $(-$
$0 < a < 1$. Thus,
of the one-to-one
$y = a^x$ is a funct
proof of Theorem
the ideas of one-t
increasing and de
functions to estab
existence of the l
functions from ex
functions. Domai
range ideas are es
ascertain the poss
of logarithms of r
The examples of
exponential functi
their logarithmic i
given in Fig. 5.17
logarithmic functi
associated bases o
graphical footing.

Teaching Note

Show students how an
investment in an annuity can
accumulate over a 40-year
period if they were to begin
making monthly deposits of
$50, beginning the first month
after graduating from high
school.

EXAMPLE 4   Finding the Future Value of an Ordinary Annuity

Use the graph shown in Fig. 5.15 to find how many months it will take to accumulate $300 when $1 is deposited each month at an interest rate of 0.5%.

Solution   Find the x-coordinate of the point of intersection of the graphs of

$$y = \frac{(1.005)^x - 1}{0.005} \quad \text{and} \quad y = 300.$$

Use zoom-in to show that $x = 183.72$ with an error of at most 0.01. The value of the annuity will exceed $300 for the first time after 184 mo.   ▪

In Example 3, we knew the amount of each equal periodic payment and we used the future-value-of-an-annuity formula to project the accumulated value into the future.

Next, consider this situation: Ellie has received an inheritance. How much of it should she save today in order for this one-time investment to accumulate at retirement the same amount as Carlos, who is saving $50 each month? The same question can be asked in general terms: What is the total amount A that should be invested today as a single lump sum in order to accumulate the same amount as an annuity of equal periodic payments, where each has the same interest rate?

In other words, we want to solve the following equation for A:

$$A(1 + i)^n = R \frac{(1 + i)^n - 1}{i}$$

$$A = R \frac{1 - (1 + i)^{-n}}{i} \quad \text{Divide both sides of the equation by } (1 + i)^n.$$

We call this amount A the present value of an ordinary annuity.

Present Value of an Ordinary Annuity

The **present value A of an ordinary annuity** consisting of n equal payments or deposits of R dollars, each with an interest rate i per payment interval, is given by the formula

$$A = R \frac{1 - (1 + i)^{-n}}{i}.$$

Loan payments to a bank are regular periodic payments characteristic of an annuity. If a bank lends you $9000 for the purchase of a car, the monthly payments represent the equal payments of the annuity and the $9000 represents the present value of the annuity.

Notes on Examples

Using Example 5, enter the future value formula on the computation screen of your graphing utility. The formula would be entered with values as follows: $9000((1 - (1 + 0.125/12)^{\wedge} - 60)/(0.125/12))$ When you press ENTER, you will get a monthly payment of 202.48. Now recall the formula from the memory buffer and change the interest rate to 14% (change 0.125 to 0.14) and see what effect this has on the monthly payment.

Assignment Guide

Day 1: Ex. 1–5, 8, 11, 13, 14, 17, 20
Day 2: Ex. 22, 25–27, 30, 36–38

Exercises for Section 5.3

In Exercises 1–4, determine which investm est return.

1. 6% compounded quarterly or 5.75% c
2. 8.25% compounded monthly or 8% cc
3. 7% compounded quarterly or 7.20% c
4. 8.5% compounded quarterly or 8.40% c
5. Amy contributes $50 monthly into an I tirement Account) annuity for 25 yr. earns 6.25% annual interest, what is IRA account after 25 yr?
6. Frank contributes $50 monthly into an yr. Assuming the IRA earns 5.5% ann the value of Frank's IRA account afte
7. Betsy contributes to a retirement an earns 8.5% annual interest compound wants to accumulate $125,000 by the much should she invest each quarter?

In Exercises 8–11, each function is an alge for the future value of a certain annuity. mine from the form of the function the ann

---

E X

Wh
con

Sol
(W

Sub

Use
mo

38. What is the r

In Exercises 39 method is closely mining appreciat

5.4

Objective

Students will be the properties o to develop skill with the graphs and natural log exponential fun

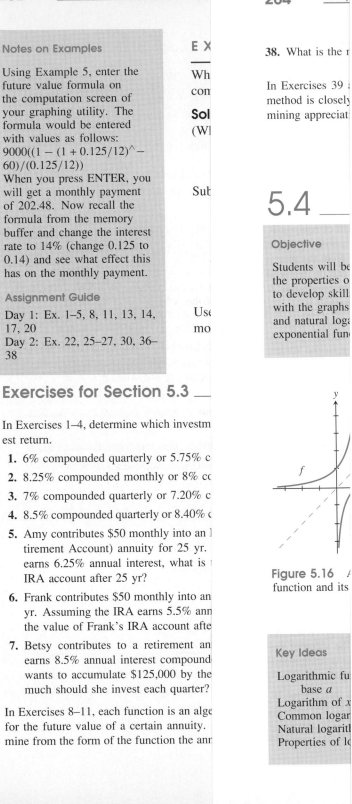

Figure 5.16
function and its

Key Ideas

Logarithmic fu
  base a
Logarithm of x
Common logar
Natural logarith
Properties of l

USING A PARAMETRIC GRAPHING UTILITY TO FIND INVERSES

Try this:

1. Select parametric graphing mode.
2. Let $X_1(T) = T$, $Y_1(T) = e^T$, $X_2(T) = e^T$, $Y_2(T) = T$.
3. Find the graph in the standard viewing rectangle for $-10 \leq T \leq 10$.
4. Use the trace key to see the function and inverse function values.
5. Add $X_3(T) = T$ and $Y_3(T) = T$ and explain your visualization.

Solution

$$8 = 2^3 \implies (3, 8) \text{ is a solution to } y = 2^x$$
$$\implies (8, 3) \text{ is a solution to the inverse } y = \log_2 x$$

Therefore $3 = \log_2 8$.

The reasoning used in Example 1 can be used for logarithmic functions of all bases. If we replace the pair $(3, 8)$ with the pair (x, y), we obtain the following generalization.

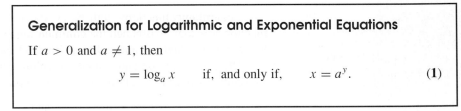

Generalization for Logarithmic and Exponential Equations

If $a > 0$ and $a \neq 1$, then

$$y = \log_a x \qquad \text{if, and only if,} \qquad x = a^y. \qquad (1)$$

Notice that $\log_a x$ is an exponent. It is the exponent y such that $a^y = x$. We use this generalization in the following examples.

E X A M P L E   2   Evaluating a Logarithm

Use the above generalization to compute $\log_5\left(\frac{1}{25}\right)$.

Solution

$$\log_5 \frac{1}{25} = x \iff 5^x = \frac{1}{25}$$

Notice that $x = -2$ is a solution to this equation since $5^{-2} = \left(\frac{1}{25}\right)$.

$$= \frac{1}{5^2}$$
$$= 5^{-2}$$

Therefore $x = -2$ and $\log_5\left(\frac{1}{25}\right) = -2$.

This same generalization can be used to solve equations involving logarithms.

E X A M P L E   3   Solving Logarithmic Equations

Use exponents to solve the following equations for x:

a) $\log_5 x = 1.5$

b) $\log_2(x - 3) = 4$

c) $\log_x 3 = 2$

d) $\log_x(\frac{1}{8}) = -3$

Solution

a) $\log_5 x = 1.5 \iff x = 5^{1.5}$ Use a calculator to evaluate x accurate to hundredths.

$$x = 11.18$$

b) $\log_2(x - 3) = 4 \iff x - 3 = 2^4$

$$x = 19$$

c) $\log_x 3 = 2 \iff x^2 = 3$

$$x = \sqrt{3}$$ The base of a logarithm must be positive.

d) $\log_x\left(\dfrac{1}{8}\right) = -3 \iff x^{-3} = \left(\dfrac{1}{8}\right) = 2^{-3}$

$$x = 2$$

■

Properties of Logarithms

The graphs of the logarithmic functions give us visual information about the nature of these functions. But information about the nature of these functions can also be gained by exploring some of the properties that they satisfy. The next Exploration provides a visual way to discover whether two expressions are equal.

🔍 EXPLORE WITH A GRAPHING UTILITY

Written below are equations that *may or may not* be true. Find the graph of each side of each equation in the viewing rectangle $[-3, 10]$ by $[-2, 3]$ and decide whether you think the equation is true for all values of x.

1. $\log(x + 2) = \log x + \log 2$

2. $\log x^2 = (\log x)^2$

3. $\log \dfrac{x}{4} = \dfrac{\log x}{\log 4}$

Which of these equations are true?

In this Exploration, you may have discovered several expressions you guessed to be true that turned out to be false. It is also important to learn those properties that *are* true. The next two theorems describe such properties. A graphing utility can be used to support and demonstrate them.

Theorem 5.2 Properties of Logarithms

Let $a > 0$ and $a \neq 1$. Then, the following statements are true:

a) $a^{\log_a x} = x$ for every positive real number x

b) $\log_a a = 1$

c) $\log_a 1 = 0$

Proof

a) $y = \log_a x \iff x = a^y$ Use statement (1).

$x = a^y = a^{\log_a x}$

b) $a^1 = a \iff \log_a a = 1$ This directly follows statement (1).

c) $a^0 = 1 \iff 0 = \log_a 1$ ▤

Theorem 5.3 Properties of Logarithms

Let a, r, and s be positive real numbers such that $a \neq 1$. Then the following statements are true:

a) $\log_a rs = \log_a r + \log_a s$

b) $\log_a \dfrac{r}{s} = \log_a r - \log_a s$

c) $\log_a r^c = c \log_a r$ for every real number c

d) $\log_a a^x = x$ for every real number x

Proof   Let $\log_a r = u$ and $\log_a s = v$. Then, using the exponential form, we have $a^u = r$ and $a^v = s$.

a) $rs = a^u a^v = a^{u+v}$. The logarithmic form of $rs = a^{u+v}$ is $\log_a rs = u + v$. Substituting for u and v results in the relation

$$\log_a rs = u + v = \log_a r + \log_a s \,.$$

b) $r/s = a^u/a^v = a^{u-v}$. The logarithmic form of the equation $r/s = a^{u-v}$ is $\log_a(r/s) = u - v$. Substituting for u and v, we find that

$$\log_a \frac{r}{s} = \log_a r - \log_a s \,.$$

c) $r^c = (a^u)^c = a^{uc}$. Thus $\log_a r^c = uc = cu$. Substituting for u yields

$$\log_a r^c = c \log_a r \,.$$

d) Substitute $r = a$ and $c = x$ in Theorem 5.3(c) to obtain $\log_a a^x = x \log_a a$. By Theorem 5.2(b), we know that $\log_a a = 1$. Thus $\log_a a^x = x$. ▬

Theorem 5.3 is used to express the logarithm of a complicated expression in terms of the logarithms of its simpler components. The basic idea is illustrated in the next example.

E X A M P L E 4   Using the Properties of Logarithms

Express $\log_a[(x^3 y^{3/2})/\sqrt{z}]$ in terms of $\log_a x$, $\log_a y$, and $\log_a z$.

Solution Use Theorem 5.3 to obtain the following:

$$\log_a \frac{x^3 y^{3/2}}{\sqrt{z}} = \log_a (x^3 y^{3/2}) - \log_a \sqrt{z} \qquad \text{Theorem 5.3(b)}$$

$$= \log_a x^3 + \log_a y^{3/2} - \log_a z^{1/2} \qquad \text{Theorem 5.3(a)}$$

$$= 3 \log_a x + (3/2) \log_a y - (1/2) \log_a z. \quad \text{Theorem 5.3(c)} \quad ▬$$

Example 4 shows that the logarithm of an expression involving products, quotients, and exponents can be expressed as sums and differences of products of logarithms and numbers. On the other hand, Example 5 shows that an exponential function with an exponent involving sums and differences can be expressed as a product and quotient of exponential expressions.

E X A M P L E 5   Using the Properties of Exponents

Express $a^{2x+3y-z}$ in terms of a^x, a^y, and a^z.

Solution Use properties of exponents to obtain the following:

$$a^{2x+3y-z} = a^{2x} a^{3y} a^{-z}$$

$$= \frac{(a^x)^2 (a^y)^3}{a^z}. \qquad ▬$$

In Example 4 of Section 5.1, a graphical method was used to find how long it would take a town with a population of 50,000, increasing at the rate of 2.5% yearly, to reach a population of 100,000. In this next example, we see how logarithms can be used to solve the same problem algebraically.

E X A M P L E   6   Problem Solving with Logarithms

Use an algebraic method to find how long it would take a town with a population of 50,000, increasing at the rate of 2.5% yearly, to reach a population of 100,000.

Solution   The population of the town at any time t is $P(t) = 50{,}000(1.025)^t$. Thus we need to solve the following equation for t:

$$50{,}000(1.025)^t = 100{,}000$$

$$1.025^t = 2$$

Divide both sides of the equation by 50,000.

$$\log(1.025^t) = \log 2$$

For any function f, if $a = b$, then $f(a) = f(b)$. In particular, $\log a = \log b$.

$$t \log(1.025) = \log 2$$

$$t = \frac{\log 2}{\log 1.025} = 28.071.$$

Thus $t = 28.07$ is accurate to 0.01.

The population of the town will reach 100,000 in 28 yr and a few days.   ≡

Data Analysis

Scientists working in a laboratory collect enough experimental data to establish that a given rule is plausible. The data points are graphed, and the scientist tries to find the **curve of best fit**. If the relationship is linear, the data points will nearly lie on a line, taking into account experimental error. This line is called the **line of best fit** and is relatively easy to find.

However, if the relationship is not linear, the data points may look something like the graph is Fig. 5.18, and the curve of best fit may not be obvious.

Often it is possible to use a logarithmic function to convert the data from a nonlinear relationship to a linear one, making it easier to find the experimental relationship. For example, suppose the actual relationship is described by a function of the form $y = ax^m$, where a is a positive constant and x is a positive independent variable. Such a function is called a **power function**. Then the following equations are equivalent:

$$y = ax^m$$

$$\ln y = \ln ax^m$$

$$\ln y = \ln a + \ln x^m$$

$$\ln y = \ln a + m \ln x.$$

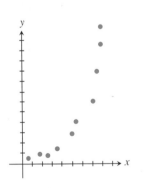

Figure 5.18   What is the best-fit curve for these data points.

Notice that $\ln y$ is the *linear* function of $\ln x$ with slope m and y-intercept $\ln a$.

> ### Theorem 5.4 Converting a Power Function Relationship to a Linear Function Relationship
>
> Let a, x, and y be positive. Then $y = ax^m$ if, and only if, $\ln y$ is a linear function of $\ln x$.

Notes on Examples

A graphing calculator can be used to find a smooth curve, using the power rule model for y in terms of x for the data given in Example 7. Consult your graphing calculator manual for instructions on how to accomplish this. Be sure to compare the values for a and b generated by a graphing utility with the calculated values found in Example 7. Compare the power of technology to visualize these models with the old method of plotting the points on log-log graph paper.

Proof We have already shown that when $y = ax^m$, $\ln y = \ln a + m \ln x$. Thus $\ln y$ is a linear function of $\ln x$.

Conversely, suppose $\ln y$ is a linear function of $\ln x$. Then there are real numbers m and c so that $\ln y = m \ln x + c$. The natural logarithm function is one-to-one and has the range $(-\infty, \infty)$. Thus there is a unique positive real number a such that $\ln a = c$. We substitute this form for c in the expression for $\ln y$.

$$\ln y = m \ln x + c$$
$$\ln y = m \ln x + \ln a$$
$$\ln y = \ln x^m + \ln a$$
$$\ln y = \ln ax^m$$
$$y = ax^m$$

This next example applies Theorem 5.4 to a set of data in order to conclude that the data satisfy a power function relationship.

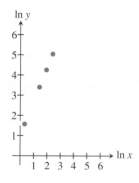

Figure 5.19 Plot of the data points $(\ln x, \ln y)$ for Example 7.

EXAMPLE 7 Finding a Power Function Relationship for Data

Determine a power rule model for y in terms of x for the following data:

x	1	3	5	8
y	5	31.2	73.1	160

Solution According to Theorem 5.4, we can show that y is a power function of x by showing that $\ln y$ is a linear function of $\ln x$. Calculate and graph the data points $(\ln x, \ln y)$ (see Fig. 5.19).

$\ln x$	0	1.10	1.61	2.08
$\ln y$	1.61	3.44	4.29	5.08

The plotted data appear to have a linear relationship. Use two of the data points to find an equation of a line. This line should be close to the line of best fit. Use the points $(\ln 1, \ln 5)$ and $(\ln 8, \ln 160)$ to determine m and b in $\ln y = m \ln x + b$.

$$\ln 5 = m \ln 1 + b \qquad \text{Substitute } (1, 5) \text{ into the equation } \ln y = m \ln x + b.$$

$$\ln 5 = m \cdot 0 + b$$

$$\ln 5 = b$$

$$\ln 160 = m \ln 8 + \ln 5 \qquad \text{Substitute } (8, 160) \text{ and } b = \ln 5 \text{ into the equation } \ln y = m \ln x + b.$$

$$\ln 160 = \ln(8^m)(5) \qquad \text{Use properties from Theorem 5.3.}$$

$$160 = 5(8^m)$$

$$32 = 8^m \qquad \text{Express both sides of the equation as a power of 2.}$$

$$2^5 = 2^{3m}$$

$$m = \frac{5}{3}$$

We can use the values of m and b to write y as a power function of x.

$$\ln y = m \ln x + b$$

$$\ln y = \frac{5}{3} \ln x + \ln 5$$

$$\ln y = \ln(5x^{5/3})$$

$$y = 5x^{5/3}$$

See Fig. 5.20.

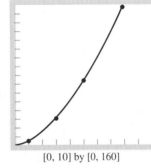

[0, 10] by [0, 160]

Figure 5.20 $y = 5x^{5/3}$.

Assignment Guide

Day 1: Ex. 1–9, 12–29, 32, 34
Day 2: Ex. 35, 38, 39, 42, 47, 48, 50, 51, 57–59

Even though in Example 7 the exact values for b and m were found, often we would simply approximate their values. In the past, before calculators became common, the values would be plotted on log-log graph paper.

Exercises for Section 5.4

Compute Exercises 1–8 without using a calculator.

1. $\log_4 16$

2. $\log_{10} 10$

3. $\log_2 8$

4. $\log_3 81$

5. $\log_{1/2} 16$

6. $\ln 1$

7. $\log_2(-8)$

8. $\log_4(-2)^4$

In Exercises 9–12, sketch a complete graph without using a graphing utility. Use the inverse property.

9. $f(x) = \log_3 x$

10. $g(x) = \log_5(x)$

11. $y = \log_{1/4} x$

12. $y = \log_2(-x)$

In Exercises 13–20, solve each equation without using a calculator.

13. $\log_9 x = 2$

14. $\log_2 x = 5$

15. $\log_x(\frac{1}{125}) = -3$

16. $\log_3(x + 1) = 2$

17. $\log_x 81 = 9$

18. $\log_2 x^2 = -2$

19. $\log_3 |x| = 1$

20. $\log_6(x^2 - 2x + 1) = 0$

Express Exercises 21–24 in logarithmic form.

21. $y = 7^x$

22. $xy = 3^4$

23. $x + y = 2^8$

24. $(1 + r)^n = P$

Express Exercises 25–28 in exponential form.

25. $\log_3 x = 5$

26. $\log_2 x = y$

27. $\log_3 \dfrac{x}{y} = -2$

28. $\dfrac{\ln P}{\ln(1 + r)} = n$

In Exercises 29–34, use properties of logarithms to write the expression as a sum, difference, or product of simple logarithms (that is, logarithms without sums, products, quotients, or exponents).

29. $\log_2(x^3 y^2)$

30. $\ln(xy^3)$

31. $\log_a\left(\dfrac{x^2}{y^3}\right)$

32. $\log_{10} 1000x^4$

33. $\log[5000x(1 + r)^{360}]$

34. $\log(\sqrt[5]{216z^3})$

In Exercises 35–38, find an appropriate viewing rectangle that shows a complete graph of each function.

35. $f(x) = e^{1/x}$

36. $f(x) = \dfrac{\ln x}{x}$

37. $f(x) = x^{1/x}$

38. $f(x) = \dfrac{x}{\ln x}$

In Exercises 39–42, use a calculator to investigate numerically the end behavior of each function.

39. $f(x) = e^{2/x}$

40. $f(x) = \dfrac{\ln x^2}{x}$

41. $f(x) = x^{1/x}$

42. $f(x) = \dfrac{3x}{\ln x}$

In Exercises 43–47, use logarithms, if needed, to solve each problem algebraically.

43. The population of a town is 475,000 and is increasing at the annual rate of 3.75%. Determine when the population of the town will be 1 million.

44. The population of a town is 123,000 and is decreasing at the annual rate of 2.375%. Determine when the population of the town will be 50,000.

45. The population of a small town in the year 1890 was 6250. Assume the population increased at the annual rate

of 3.75%. Determine the town's population in 1915 and in 1940.

46. The half-life of a certain radioactive substance is 14 days, and there are 6.58 g present initially. When will there be less than 1 g of the substance remaining?

47. The half-life of a certain radioactive substance is 65 days, and there are 3.5 g present initially. When will there be less than 1 g of the substance remaining?

48. Determine a specific power rule algebraic representation for the data in this table.

x	4	6.5	8.5	10
y	2816	31908	122,019	275,000

49. Consider the two sets of data given below. Which one has an algebraic representation that is a power rule? Determine the specific power rule.

x	2	3	7.5	7.7
y	7.48	7.14	6.43	6.41

x	8	12	15	40
y	23.84	58.2	113.69	162.13

In Exercises 50–52, use an algebraic method to solve the equation for t. Support your answer with a graphing utility.

50. $2500 = 1000(1.08)^t$

51. $6000 = 4600(1.05)^t$

52. $10,000 = 3500(1.07)^t$

53. Explain how Exercises 50–52 relate to interest-bearing savings accounts.

54. Solve for t: $S = P(1 + r)^t$.

55. Use an algebraic method to determine how long it will take for an initial deposit of $1250 to double in value at 7% interest compounded monthly.

56. Use an algebraic method to determine how long it will take for an initial deposit of $1250 to triple in value at 7% interest compounded monthly.

Exercises 57–59 refer to the following **problem situation** The table below gives the period of one complete revolution about the sun for the listed planets, along with the length of their semimajor orbit axes. Let P be the orbit period and x the length of the semimajor axis.

Planet	Period (Days)	Semimajor Axes (Miles)
Earth	365	92,600,000
Mercury	88	36,000,000
Venus	225	67,100,000
Mars	687	141,700,000
Jupiter	4330	483,000,000
Saturn	10,750	886,100,000

57. Plot $\ln P$ against $\ln x$. That is, plot the pairs $(\ln x, \ln P)$. Verify that the relationship is linear and estimate the slope m and y-intercept.

58. Use Exercise 57 to show that $P = ax^m$ for constants a and m. This result was discovered by Kepler in the early 1700s and is known as *Kepler's third law of planetary motion.*

59. Predict the period of Pluto's orbit if its semimajor orbit axis length is 3,660,000,000 mi.

Exercises 60 and 61 refer to the following **problem situation** The *at-rest blood pressure P* and weight x of various primates were measured as shown in this table:

Weight, x (lbs)	20	50	80	110	125	140
Blood pressure P	106	133	150	162	167	172

60. Plot $\ln P$ against $\ln x$, that is, plot the pairs $(\ln x, \ln P)$. Verify that the relationship is linear. Determine the slope m and y-intercept.

61. Use Exercise 60 to find a formula expressing the blood pressure as a function of weight.

5.5

Graphs of Logarithmic Functions

Objective

Students will be able to use the properties of logarithms of different bases and develop skills associated with the geometric transformations of graphs of logarithms.

Key Ideas

Change-of-base formula
Given base
Desired base

Teaching Note

Students may have difficulty remembering the change-of-base formula. When evaluating $\log_5 8$, for example, suggest that they place the "base in the basement." This will help students to remember that $\log 8$ is the numerator and $\log 5$ is the denominator. At the blank text screen, enter $\log 8 / \log 5$ and press ENTER to get a result of 1.29029674.

In Section 5.4, the only logarithmic functions graphed with a graphing utility were the *common logarithmic function* $y = \log x$ and the *natural logarithmic function* $y = \ln x$. These may be graphed using the built-in function keys for $y = \log x$ and $y = \ln x$ found in most graphing utilities.

However, to graph other logarithmic functions such as $y = \log_2 x$, $y = \log_5 x$, and $y = \log_8 x$, a formula called the **change-of-base formula** is used to transform the function. You can use this formula to compute logarithms with any base using a calculator and to obtain a complete graph of any logarithmic function using a graphing utility.

Theorem 5.5   Change-of-base Formula

Let a and b be positive real numbers, with $a \neq 1, b \neq 1$. Then

a) $\log_b x = \dfrac{\log_a x}{\log_a b}$, and

b) $\log_b a = \dfrac{1}{\log_a b}$.

In these formulas, a is sometimes called the **given base** and b the **desired base**.

Teaching Note

Theorem 5.5 is used to establish the relationship between logarithmic functions with different bases. Theorem 5.5(a) is often called the change-of-base formula. The proof examines this relationship in terms of the functions. The application of the change-of-base formula is shown in the computational examples just prior to Example 1. In Example 2, the theorem is applied in a graphical setting.

Proof

a)
$$y = \log_b x \iff x = b^y$$

Apply the logarithm base a to both sides of the equation $x = b^y$.

$$\log_a x = \log_a b^y$$

$$\log_a x = y \log_a b$$

$$y = \frac{\log_a x}{\log_a b}$$

b)
$$\log_b x = \frac{\log_a x}{\log_a b}$$ Substitute $x = a$ into Theorem 5.5 (a).

$$\log_b a = \frac{\log_a a}{\log_a b}$$ Substitute $\log_a a = 1$ (Theorem 5.2b).

$$\log_b a = \frac{1}{\log_a b}$$

To apply the change-of-base formula, let $a = 10$ or $a = e$. Then the logarithm of any base b can be represented in terms of either common logarithms or natural logarithms; for example,

$$\log_3 x = \frac{\log x}{\log 3}, \qquad \log_7 x = \frac{\ln x}{\ln 7}, \qquad \log_5 x = \frac{\log x}{\log 5}, \quad \text{and} \quad \log_9 x = \frac{\ln x}{\ln 9}.$$

E X A M P L E   1   Graphing a Logarithmic Function

Use a graphing utility to find a complete graph of $f(x) = \log_3 x$. Also find its domain and range.

Solution   Use the change-of-base formula found in Theorem 5.5(a).

$$f(x) = \log_3 x = \frac{\log x}{\log 3} = \frac{\ln x}{\ln 3}$$

To find the complete graph of f, graph either $y = \log x / \log 3$ or $y = \ln x / \ln 3$ (see Fig. 5.21). The domain and range of f are the same as for logarithmic functions with other bases; that is, the domain is $(0, \infty)$ and the range is $(-\infty, \infty)$.

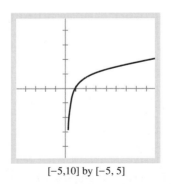

[−5,10] by [−5, 5]

Figure 5.21 $f(x) = \log_3 x$.

The method shown in Example 1 can be used to find the graphs of logarithmic functions with other bases. The standard transformations of vertical and horizontal shift, vertical stretch, and reflection about the x- and y-axes can be used to find the graphs of logarithms.

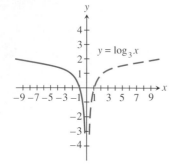

Figure 5.22 $g(x) = \log_3(-x)$ reflecting the graph of $f(x) = \log_3 x$ about the y-axis results in $g(x) = \log_3(-x)$.

Notes on Examples

Notice that Example 2 deals with the reflection of a graph about the y-axis.

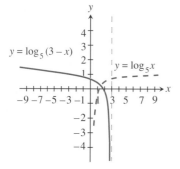

Figure 5.23 $f(x) = \log_5(3-x)$.

Applying Transformations to $y = \log_b x$

The graph of any logarithmic function of the form $y = a \log_b(cx+d)+k$ can be obtained by applying geometric transformations to the graph of $y = \log_b x$.

E X A M P L E   2   Using Transformations to Graph Logarithms

Explain how a sketch of the graph of $g(x) = \log_3(-x)$ can be obtained from the graph of $f(x) = \log_3 x$. Also determine the domain, range, and vertical asymptote of g.

Solution   (See Fig. 5.22.)

$$g(x) = \log_3(-x) = f(-x)$$

A graph of g can be obtained from a graph of f by reflection through the y-axis.

$$-x > 0 \iff x < 0$$

This means that the domain of g is $(-\infty, 0)$.

Domain of g $(-\infty, 0)$.

Range of g $(-\infty, \infty)$.

Vertical Asymptote of g $x = 0$.   ≡

E X A M P L E   3   Using Transformations to Graph Logarithms

Explain how a sketch of the graph of $g(x) = \log_5(3 - x)$ can be obtained from a sketch of $f(x) = \log_5 x$. Also, find the domain, range, and asymptotes of g.

Solution

$$g(x) = \log_5(3 - x) = \log_5[(-1)(x - 3)]$$

$$= f[-(x - 3)]$$

Recall that the -1 signifies a reflection through the y-axis.

The graph of f can be obtained from the graph of $y = \log_5 x$ through the following sequence of transformations:

1. A reflection about the y-axis.

2. A horizontal shift right 3 units (see Fig. 5.23). The vertical asymptote of f also shifts 3 units right.

$$3 - x > 0 \iff x < 3$$

This means that the domain of f is $(-\infty, 3)$.

Domain of g $(-\infty, 3)$.

Range of g $(-\infty, \infty)$.

Vertical Asymptote of g $x = 3$.

■

The next example illustrates that because we know the general shape of the graph of a logarithmic function, we can find its position and orientation by plotting a few specific points. At the same time, we investigate the effect of the coefficient a in $y = \log_b(ax + c)$.

E X A M P L E   4   Sketching a Graph by Plotting Several Points

Sketch a complete graph of $g(x) = \log_2(3x + 5)$ by plotting several points and taking into account the transformations that can be applied to the graph of $f(x) = \log_2 x$. Also find its domain, range, and asymptotes.

Solution   Suppose we rewrite $g(x)$ as follows:

$$g(x) = \log_2(3x + 5)$$

$$= \log_2[3(x + \tfrac{5}{3})]$$

$$= \log_2 3 + \log_2(x + \tfrac{5}{3})$$

$$= f(x + \tfrac{5}{3}) + \log_2 3. \quad \text{The graph of } y = \log_2 x \text{ shifts left } \tfrac{5}{3} \text{ units,} \\ \text{then up } \log_2 3 \text{ units.}$$

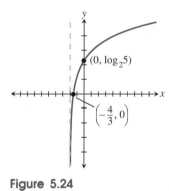

Figure 5.24
$g(x) = \log_2(3x + 5)$.

To help locate the position of the graph, find its x- and y-intercepts.

$$\log_2(3x + 5) = 0 \iff 3x + 5 = 2^0 = 1$$

$$x = -\frac{4}{3}$$

$$g(0) = \log_2 5$$

The points $(-\tfrac{4}{3}, 0)$ and $(0, \log_2 5)$ are on the graph of g (see Fig. 5.24).

$$3x + 5 > 0 \iff 3x > -5 \quad \begin{array}{l} 3x + 5 \text{ must be positive to be} \\ \text{in the domain of } g. \end{array}$$

$$\iff x > -\frac{5}{3} \quad \begin{array}{l} \text{This means that the domain} \\ \text{of } g \text{ is } \left(-\tfrac{5}{3}, \infty\right). \end{array}$$

Domain of g $\left(-\tfrac{5}{3}, \infty\right)$.

Range of g $(-\infty, \infty)$.

Vertical Asymptote of g $x = -\tfrac{5}{3}$.

■

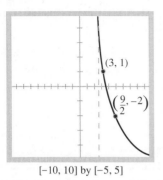

[-10, 10] by [-5, 5]

Figure 5.25
$g(x) = -3\log_4(2x - 5) + 1.$

EXAMPLE 5   Combining Methods to Find a Graph

Find a complete graph of $g(x) = -3\log_4(2x - 5) + 1$. Determine its domain, range, and asymptote.

Solution   Combine all the methods of the last several examples.

To find a complete graph of g with a graphing utility, enter g in the form

$$g(x) = -3\frac{\log(2x - 5)}{\log 4} + 1.$$

(See Fig. 5.25.)

Using properties of logarithms, we obtain the following:

$$g(x) = -3\log_4(2x - 5) + 1$$

$$= -3\log_4[2(x - \tfrac{5}{2})] + 1$$

$$= -3[\log_4 2 + \log_4(x - \tfrac{5}{2})] + 1$$

$$= -3\log_4 2 - 3\log_4(x - \tfrac{5}{2}) + 1$$

$$= -3\log_4(x - \tfrac{5}{2}) + (1 - 3\log_4 2)$$

$$= -3f(x - \tfrac{5}{2}) + (1 - 3\log_4 2)$$
The graph of $f(x) = \log_4 x$ shifts right $\tfrac{5}{2}$ units, reflects through the x-axis, stretches by a factor of 3, and then shifts up $(1 - 3\log_4 2)$ units.

$$2x - 5 > 0 \iff x > \frac{5}{2}.$$
The domain of g is $(\tfrac{5}{2}, \infty)$ and the vertical asymptote is $x = \tfrac{5}{2}$.

Confirm that the points $(3, 1)$ and $\left(\tfrac{9}{2}, -2\right)$ are on the graph of g.

Domain of g $\left(\tfrac{5}{2}, \infty\right)$.

Range of g $(-\infty, \infty)$.

Vertical Asymptote of g $x = \tfrac{5}{2}$.   ≡

We close this section by applying our knowledge of logarithms to a couple of applications. Example 6 examines the effect of the monthly payment on the term (length) of a loan, and Example 7 considers the Richter scale, a common means to measure earthquakes.

EXAMPLE 6   APPLICATION: Calculating Loan Repayment

A Problem Situation

Suppose that to buy a house, Alicia takes out an $86,000 loan at APR 12% compounded monthly. She wants to pay off the loan as quickly as possible. The interest payment required the first month is $0.01(86{,}000) = \$860$. Thus monthly payments must exceed \$860 in order to pay off the loan. Assume the interest rate and payment are fixed.

a) Find a complete graph of an algebraic representation that shows how the number of months t for the loan depends on the amount of the monthly payment x.

b) What are the monthly payments if the loan is paid in 25 yr?

c) If the amount in part (b) is increased by \$50, how long will it take to pay off the loan?

Solution This problem is a present-value-of-an-annuity problem. Use the formula introduced in Section 5.3,

$$A = R\frac{1 - (1+i)^{-n}}{i},$$

and let $A = \$86{,}000$ and $i = 0.01$. Let x represent the unknown monthly payment and t represent the time of the loan. Then

a)
$$86{,}000 = x\frac{1 - 1.01^{-t}}{0.01}.$$

Next, solve this equation for t.

$$86{,}000 = x\frac{1 - 1.01^{-t}}{0.01}$$

$$860 = x(1 - 1.01^{-t})$$

$$\frac{860}{x} = 1 - 1.01^{-t}$$

$$1.01^{-t} = 1 - \frac{860}{x}$$

$$1.01^{-t} = \frac{x - 860}{x}$$

$$1.01^{t} = \frac{x}{x - 860}$$

$$\ln(1.01^{t}) = \ln\left(\frac{x}{x - 860}\right)$$

$$t = \frac{1}{\ln 1.01}\ln\left(\frac{x}{x - 860}\right)$$

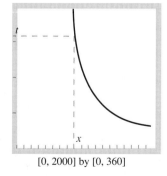

[0, 2000] by [0, 360]

Figure 5.26

$$t = \frac{1}{\ln 1.01}\ln\left(\frac{x}{x - 860}\right).$$

This is an algebraic representation for time t (in months) as a function of the amount of the monthly payment x. Figure 5.26 shows a complete graph of this representation.

b) Let $t = 300$ mo and use zoom-in to find the point of intersection of the graph of t and the horizontal line $t = 300$. The x-coordinate of that point of intersection

Notes on Exercises

Ex. 52–54 are Richter scale problem situations which use an application of logarithmic functions.
Ex. 55–58 are an application of the Beer-Lambert law of light absorption, a classical use of logarithms in the display and interpretation of data in problem solving.
Ex. 59–62 give an example of the use of semilogarithmic and logarithmic graph paper. This approach helped to capture a considerable amount of data efficiently and thereby reduced many logarithmic-based problems to situations that could be interpreted with linear functions. While this use of logarithms is no longer as common as it once was, some applications in engineering, the life sciences, business, and economics still require it.

is 905.77, which means that to pay off the loan in 25 yr, the monthly payments would need to be $905.77.

c) Substitute $x = 955.77$ into the algebraic representation of part (a) to find t.

$$t = \frac{1}{\ln 1.01} \ln \left(\frac{955.77}{955.77 - 860} \right) = 231.21$$

Thus $t = 231.21$ mo, which is equal to 19.27 yr. If the payments are increased by $50, the loan will be paid off about 5.73 yr sooner. ▤

The intensity of an earthquake is rated by its measurement on the Richter scale. This scale is actually the logarithmic function

$$R = \log \left(\frac{a}{T} \right) + B,$$

where R is the magnitude of the intensity, a is the amplitude (in micrometers) of the vertical ground motion at the receiving station, T is the period of the seismic wave (in seconds), and B is a factor that accounts for the weakening of the seismic wave with increasing distance from the epicenter of the earthquake.

EXAMPLE 7   APPLICATION: Measuring Earthquakes Using the Richter Scale

At the early stages of a particular earthquake, the amplitude of vertical ground motion was a, and at later stages the amplitude became $10a$. How much did the intensity R change as the earthquake progressed?

Solution   Let R_1 be the intensity when the amplitude was a and R_2 be the intensity when the amplitude was $10a$. So $R_1 = \log(a/T) + B$ and $R_2 = \log(10a/T) + B$. Combine these two equations and solve for either R_1 or R_2.

$$R_2 - R_1 = \left[\log \left(\frac{10a}{T} \right) + B \right] - \left[\log \left(\frac{a}{T} \right) + B \right]$$

$$= \log \left(\frac{10a}{T} \right) - \log \left(\frac{a}{T} \right)$$

$$= (\log 10a - \log T) - (\log a - \log T)$$

$$= \log 10a - \log a$$

$$= \log \frac{10a}{a} = \log 10 = 1$$

$$R_2 = R_1 + 1$$

The intensity increased by 1 Richter scale unit. ▤

Assignment Guide

Day 1: Ex. 2, 5, 8, 11, 14, 17, 20, 23, 26, 29, 32, 35, 36, 39
Day 2: Ex. 42, 45, 47, 50, 52–58, 63, 64

Thus, for example, an earthquake with a Richter scale intensity of 6 is 10 times greater than one with an intensity of 5.

Exercises for Section 5.5

In Exercises 1–6, use a graphing utility to find a complete graph for each function. State the viewing rectangle that you used for each and each function's domain and range.

1. $f(x) = \log_4 x$

2. $g(x) = \log_8 x$

3. $h(x) = \log_5 x + 3$

4. $f(x) = \log_2 x - 10$

5. $k(x) = \log_3 x + \log_2 64$

6. $f(x) = \log_7 x + 25$

7. Use a graphing utility to draw complete graphs of $y = \log_2 x$, $y = \ln x$, $y = \log_3 x$, $y = \log_5 x$, and $y = \log x$ in the same viewing rectangle.

In Exercises 8–15, explain how a sketch of a complete graph of each function can be obtained from the graph of $y = \log_a x$ for the appropriate value of a. In each case, identify the vertical asymptote.

8. $f(x) = \log_3(x - 2)$

9. $g(x) = \log_4(x + 5) - 1$

10. $g(x) = \log(5 - x)$

11. $f(x) = \log_5(2 - x)$

12. $k(x) = \log_2(3x - 4)$

13. $j(x) = \ln(2x + 3) + 2$

14. $f(x) = -\log(x + 3) - 2$

15. $g(x) = -\log(4 - x) - 2$

In Exercises 16–19, sketch a complete graph of each function by plotting a pair of points. State the domain and range of each function.

16. $f(x) = \log_2(x + 3)$

17. $k(x) = \log_3(2x - 1)$

18. $g(x) = 2\log(-x)$

19. $f(x) = \log(3 - x)$

In Exercises 20–27, use any method to sketch a complete graph of each function. Support your answer with a graphing utility.

20. $f(x) = 2\ln(x + 3)$

21. $g(x) = -1 - \ln(x - 1)$

22. $h(x) = 1 + \log_3(x - 2)$

23. $f(x) = 2 - \log_2(2x + 6)$

24. $k(x) = \log_5 \sqrt{x - 3}$

25. $f(x) = \log_2(x + 3)^2$

26. $g(x) = \ln(5 - 2x)$

27. $f(x) = 4 + \log_5 \sqrt{x - 3}$

In Exercises 28–31, solve each inequality graphically.

28. $\log_2 x < \log_3 x$

29. $\log_2 x > \log_3 x$

30. $\ln x < \log_3 x$

31. $\ln x > \log_3 x$

In Exercises 32–35, determine the end behavior of each function; that is, determine what $f(x)$ approaches as $|x| \to \infty$.

32. $f(x) = x\ln x$

33. $f(x) = x^2 \ln x$

34. $f(x) = x^2 \ln |x|$

35. $f(x) = \dfrac{\ln x}{x}$

In Exercises 36–39, find a complete graph of each function. Determine where the function is increasing and decreasing and all local maximum and minimum values.

36. $f(x) = x\ln x$

37. $g(x) = x^2 \ln x$

38. $f(x) = x^2 \ln |x|$

39. $g(x) = \dfrac{\ln x}{x}$

In Exercises 40 and 41, sketch the graph of each inequality without using a graphing utility.

40. $y < \ln(3 - x)$

41. $y \geq 2 - \ln(x + 2)$

In Exercises 42 and 43, find the points of intersection of each equation.

42. $y = 6 - x$ and $y = \ln(x - 2)$

43. $y = x^2 - 2$ and $y - 1 = \ln|x + 5|$

44. Show that $y = \frac{1}{2}(e^x - e^{-x})$ and $y = \ln(x + \sqrt{x^2 + 1})$ are inverses of each other. Give a graphical argument.

In Exercises 45–50, for each pair of functions f and g, find the solutions to $f(x) > g(x)$, $f(x) < g(x)$, and $f(x) = g(x)$ graphically.

45. $f(x) = 3^x$; $g(x) = x^3$

46. $f(x) = e^x$; $g(x) = x^e$

47. $f(x) = 10^x$; $g(x) = x^x$

48. $f(x) = e^x$; $g(x) = (\ln x)^x$

49. $f(x) = \ln x$; $g(x) = \ln(\ln x)$

50. $f(x) = \ln x$; $g(x) = x^{1/3}$

51. How would you graph $y = \log_x 4$ using a graphing utility?

Exercises 52–54 refer to the Richter scale **problem situation** in Example 7.

52. What is the magnitude on the Richter scale of an earthquake if $a = 250$, $T = 2$, and $B = 4.250$?

53. Explain why an earthquake of magnitude 6 on the Richter scale is 100 times more intense than one with the same epicenter of magnitude 4 on the Richter scale. Assume that T and B in the formula for the Richter scale are constant. The change in the vertical ground motion (amplitude) is directly related to the *intensity* of the earthquake.

54. Draw a complete graph of the Richter scale model (magnitude R as a function of amplitude a). Assume that $T = 2$ and $B = 4.250$. What are the values of a that make sense in this problem situation?

Exercises 55–58 refer to the following **problem situation**: The *Beer-Lambert law of light absorption* is given by

$$\log\left(\frac{I}{I_0}\right) = Kx,$$

where I_0 and I denote the intensity of light of a particular type before and after the light passes through a body of material, respectively, and x denotes the length of the path followed by the beam of light passing through the material. K is a constant.

55. Let $I_0 = 12$ lumens and assume that $K = -0.00235$. Express I as a function of x.

56. Draw a complete graph of $I = f(x)$.

57. Suppose I_0 represents the intensity of light measured at the surface of a lake. The Beer-Lambert law can be used to determine the intensity I of the light measured at a depth of x feet from the surface. Assume the constants given in Exercise 55 are correct for Lake Erie. What is the intensity of the light at a depth of 30 ft?

58. Suppose that for Lake Superior, $K = -0.0125$ and the surface intensity of the light is 12 lumens. What is the intensity at a depth of 30 ft?

Exercises 59–62 refer to the table of data given below. Exponential relations appear to be linear when plotted on *semilogarithmic graph paper*. Before calculators and computers were readily available, techniques for data analysis frequently involved using this type of paper. The same analysis can be carried out using ordinary graph paper and a scientific or graphing calculator.

x	2.3	4	5.5	7	9
y	43.8	283.5	1473.1	7654.5	68,890.5

59. Plot the data pairs $(x, \log y)$ on regular graph paper using the values for x and y given above. Verify that the points are *collinear* (fall on the same line).

60. Compute the slope m and the y-intercept y_0 of the line in Exercise 59. Verify that the data can be represented by the exponential relationship $y = ab^x$, where $m = \log b$ and $\log a = y_0$.

61. Prove that if f is the exponential function $f(x) = ab^x$, then $\log f(x)$ varies linearly with x; that is, show that $\log f(x) = mx + y_0$ for some constants m and y_0.

62. Determine the actual exponential function that provides a model of the data in Exercise 59. (*Hint:* Use Exercise 61.)

63. Consider a $75,000 mortgage loan with interest at 10.50% APR compounded monthly.
a) Draw a graph that shows how the term of the loan depends on the monthly payment amount.
b) What is the monthly payment amount required to pay off the loan in 25 yr?
c) If the amount in part (b) is increased by $50, how long will it take to pay off the loan?

64. Consider a $110,000 mortgage loan with interest at 9% APR compounded monthly.
a) Draw a graph that shows how the term of the loan depends on the amount of the monthly payment.
b) What is the monthly payment amount required to pay off the loan in 25 yr?
c) If the amount in part (b) is increased by $50, how long will it take to pay off the loan?

5.6 — Equations, Inequalities, and Extreme Value Problems

Objective

Students will be able to use exponential and logarithmic functions to solve equations and inequalities by algebraic and graphical methods.

Key Ideas

Principle about extraneous solutions
Normal distribution functions

In this section, we solve equations and inequalities that involve exponential or logarithmic functions. Both graphical and algebraic methods will be used. The algebraic method generally involves moving from the exponential form to the logarithmic form and vice versa. Sometimes the solution of the problem requires that we find a local maximum or minimum of a function.

E X A M P L E   1   Solving a Logarithmic Equation with a Calculator

Solve $\log x = -2.5$.

Solution

$$\log x = -2.5 \iff x = 10^{-2.5} \quad \text{Apply statement (1) from Section 5.4.}$$

Use the exponentiation key to show that $x = 0.0031623$.   ▀

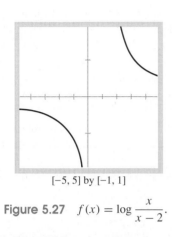

[−5, 5] by [−1, 1]

Figure 5.27 $f(x) = \log \dfrac{x}{x-2}$.

Teaching Note

The graphs in Figs. 5.27 and 5.28 should be examined carefully for purposes of class discussion. Students should be encouraged to make conjectures about log graphs. What are permissible replacements for x? Some students need to be prompted frequently to encourage speculation.

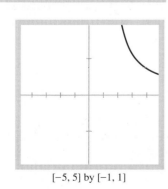

[−5, 5] by [−1, 1]

Figure 5.28
$g(x) = \log x - \log(x - 2)$.

Using Properties of Logarithms to Solve Equations

Suppose we want to solve the equation

$$\log x - \log(x - 2) = -2 \qquad \textbf{(1)}$$

using algebraic methods. The properties summarized in Theorem 5.3 can be used.

Caution: Extraneous solutions are sometimes introduced when both sides of an equation are squared, as seen in Section 4.3. The same thing can happen when using the properties of logarithms. For example, suppose Theorem 5.3(b) is used to conclude that

$$\log x - \log(x - 2) = \log \frac{x}{x - 2}.$$

The fact that extraneous roots are introduced can be seen by comparing the domains of $f(x) = \log[x/(x - 2)]$ and $g(x) = \log x - \log(x - 2)$, as follows:

$$f(x) = \log \frac{x}{x - 2} \text{ is defined} \iff \frac{x}{x - 2} > 0;$$

$$\iff x \text{ is in } (-\infty, 0) \cup (2, \infty).$$

$$g(x) = \log x - \log(x - 2) \text{ is defined} \iff x > 0 \text{ and } x - 2 > 0;$$

$$\iff x \text{ is in } (2, \infty).$$

This illustrates that the domains are not identical. It is evident from the graphs of these two functions (see Figs. 5.27 and 5.28) that

$$\log \frac{x}{x - 2} = -2$$

has a solution but that

$$\log x - \log(x - 2) = -2$$

does not have a solution.

This comparison suggests the following **principle of extraneous solutions.**

Frank says, "Developing the conceptual consciousness of domain and range restrictions throughout this section will simplify your instruction in future sections concerned with inverses and the trigonometric functions."

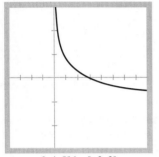

[–4, 8] by [–3, 3]

Figure 5.29
$y = \frac{1}{2}\log_5(x+6) - \log_5 x.$

Principle of Extraneous Solutions

To find all solutions, we can replace

$$\log_a f(x) - \log_a g(x) \quad \text{by} \quad \log_a \frac{f(x)}{g(x)}$$

or

$$\log_a f(x) + \log_a g(x) \quad \text{by} \quad \log_a\big[f(x) \cdot g(x)\big].$$

However, extraneous solutions may be introduced through these replacements.

The algebraic solution in this next example illustrates how to use the principle on extraneous solutions.

EXAMPLE 2   Solving a Logarithmic Equation

Solve $\frac{1}{2}\log_5(x+6) - \log_5 x = 0$.

Solution

Graphical Method   A complete graph of $y = \frac{1}{2}\log_5(x+6) - \log_5 x$ (see Fig. 5.29) suggests that $x = 3$ is the one zero of this function. Use zoom-in or direct numerical substitution to support this claim or use the following algebraic solution.

Algebraic Method   Use the properties of Theorem 5.3:

$$\frac{1}{2}\log_5(x+6) - \log_5 x = 0$$

$$\log_5(x+6)^{1/2} - \log_5 x = 0$$

$$\log_5 \sqrt{x+6} - \log_5 x = 0$$

$$\log_5 \frac{\sqrt{x+6}}{x} = 0 \qquad \text{This step might introduce extraneous solutions.}$$

$$\frac{\sqrt{x+6}}{x} = 5^0$$

$$\frac{\sqrt{x+6}}{x} = 1.$$

Next, solve $\sqrt{x+6}/x = 1$ in the usual way:

$$\frac{\sqrt{x+6}}{x} = 1$$

$$\sqrt{x+6} = x$$

$$x + 6 = x^2$$
$$0 = x^2 - x - 6$$
$$0 = (x - 3)(x + 2)$$
$$x = 3, \qquad x = -2.$$

Check   Let $x = 3$. Then

$$\frac{1}{2} \log_5(3 + 6) - \log_5 3 = 0. \quad \text{Use a calculator to evaluate this expression.}$$

The expression $\log_5(-2)$ is not defined, so $x = -2$ is not a solution.   ≡

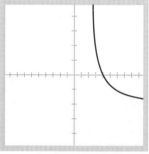

[−10,10] by [−5, 5]

Figure 5.30 $f(x) = \log_2(2x + 1) - \log_2(x - 3) - 2\log_2 3$.

E X A M P L E 3   Solving a Logarithmic Equation

Solve $\log_2(2x + 1) - \log_2(x - 3) = 2\log_2 3$.

Solution

Graphical Method   A complete graph of $f(x) = \log_2(2x + 1) - \log_2(x - 3) - 2\log_2 3$ (see Fig. 5.30) suggests that $x = 4$ is the one zero of f. Use zoom-in to support this claim or use the following algebraic solution.

Algebraic Method

$$\log_2(2x + 1) - \log_2(x - 3) = 2\log_2 3$$

$$\log_2 \frac{2x + 1}{x - 3} = \log_2 9$$

$$\frac{2x + 1}{x - 3} = 9 \quad \text{Because } f(x) = \log_2 x \text{ is a one-to-one function, } \log a = \log b \text{ implies that } a = b.$$

$$9x - 27 = 2x + 1$$

$$7x - 28$$

$$x = 4.$$

Check   Let $x = 4$. Then

$$\log_2(2 \cdot 4 + 1) - \log_2(4 - 3) = 2 \, \log_2 3$$

$$\log_2 9 - \log_2 1 = 2 \, \log_2 3$$

$$\log_2 3^2 - 0 = 2 \, \log_2 3$$

$$2 \, \log_2 3 = 2 \, \log_2 3.$$

So $x = 4$ is a solution to the equation.   ≡

E X A M P L E 4   Solving an Exponential Equation

When will your money triple in value at 8% interest compounded annually?

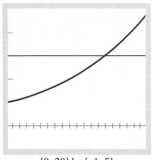

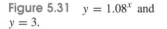

[0, 20] by [−1, 5]

Figure 5.31 $y = 1.08^x$ and $y = 3$.

Solution   To answer this question, solve the equation $3P = P(1 + 0.08)^x$, which is equivalent to $3 = 1.08^x$.

$$1.08^x = 3 \qquad \text{Take the base 10 logarithm of both sides of the equation.}$$

$$\log 1.08^x = \log 3$$

$$x \log 1.08 = \log 3$$

$$x = \frac{\log 3}{\log 1.08} \qquad \text{Use a calculator to evaluate this expression.}$$

$$x = 14.27$$

You can use zoom-in to support this solution by graphing $y = 1.08^x$ and $y = 3$ (see Fig. 5.31).   ▤

E X A M P L E   5   Solving an Exponential Equation

Solve the equation $(3^x - 3^{-x})/2 = 5$.

Solution

Graphical Method   Figure 5.32 shows a complete graph of $y = 5$ and $y = (3^x - 3^{-x})/2$. Because these graphs are complete, there appears to be one solution. Use zoom-in to find the x-coordinate of the point of intersection, or use the following algebraic method.

Algebraic Method

$$\frac{3^x - 3^{-x}}{2} = 5$$

$$3^x - 3^{-x} = 10$$

$$(3^x)^2 - 1 = (10)3^x \qquad \text{Multiply each side by } 3^x.$$

$$(3^x)^2 - (10)3^x - 1 = 0$$

This equation is a quadratic in 3^x. Therefore the quadratic formula yields

$$3^x = \frac{10 \pm \sqrt{104}}{2} = 5 \pm \sqrt{26}.$$

$5 - \sqrt{26}$ is negative and so does not result in a solution.

$$3^x = 5 + \sqrt{26}$$

$$\log_3 3^x = \log_3(5 + \sqrt{26}) \qquad \text{Use a calculator to evaluate this expression.}$$

$$x = 2.10$$ ▤

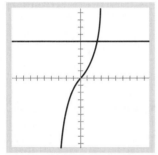

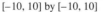

[−10, 10] by [−10, 10]

Figure 5.32 $y = \dfrac{3^x - 3^{-x}}{2}$ and $y = 5$.

Teaching Note

Help students improve their conceptual development by graphing functions that lead to errors. Present some examples for which the graph reveals key behaviors to help students think in terms of graphs and the restrictions on domains.

The algebraic technique needed to solve an inequality involving logarithms is similar to the technique used to solve equations. The resulting inequalities are not necessarily equivalent, so solutions must be checked in the original inequality.

E X A M P L E 6   Solving an Inequality

Solve $3\log_4 x - 1 < 0$.

Solution Notice that we must have $x > 0$. (Why?) Rewrite the inequality as $\log_4 x < \frac{1}{3}$. Figure 5.33 shows the graph of $y = \log_4 x$ and $y = \frac{1}{3}$. Use zoom-in to show that the solution is $0 < x < 1.59$.

≡

The next example asks you to find any local extrema.

E X A M P L E 7   Finding Maximum and Minimum Values

Find the domain, range, and local extrema of $f(x) = \ln\big[x(3-x)\big]$ and draw a complete graph.

Solution f is defined if

$$x(3-x) > 0 \iff x \text{ is in } (0, 3).$$

Figure 5.34 shows a complete graph of f. Use zoom-in to show that the local maximum value is 0.81 with an error of at most 0.01.

Domain of f $(0, 3)$.

Range of f $(-\infty, 0.81]$.

≡

The next example introduces another application with an exponential function for a model.

E X A M P L E 8   APPLICATION: Calculating Atmospheric Pressure

Scientists have established that standard atmospheric pressure of 14.7 pounds per square inch is reduced by half for each 3.6 mi of vertical ascent. This rule for atmospheric pressure holds for altitudes up to 50 mi. Express the atmospheric pressure P as a function of altitude and determine when the atmospheric pressure will be 4 lb/in^2.

Solution Use the given information to conclude that

$$P = 14.7(0.5)^{h/3.6}$$

REMINDER

To find the graph of $y = \log_4 x$ on a grapher, use the fact that

$$\log_4 x = \frac{\log x}{\log 4}.$$

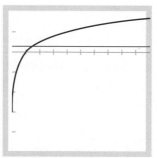

[0, 10] by [−5, 2]

Figure 5.33 $y = \log_4 x$.

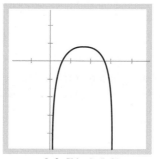

[−2, 5] by [−5, 3]

Figure 5.34
$f(x) = \ln[x(3-x)]$.

Notes on Examples

Using the trace feature is especially useful in this example to verify the domain, range, and extreme values.

Notes on Examples

The normal distribution function is explored in Example 8. This is a follow-up to the example of the normal curve which was given in a previous section.

Notes on Exercises

Ex. 1–45 provide students with the opportunity to explore concepts that have been themes throughout this course—domain, range, extrema, and symmetry.

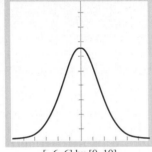

[−6, 6] by [0, 10]

Figure 5.35 $f(x) = \left(\dfrac{25}{\sqrt{2\pi \cdot (2.35)}}\right)e^{-x^2/4.7}$.

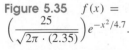

Assignment Guide

Day 1: Ex. 1–15, 18, 19–25
Day 2: Ex. 26, 29, 32, 35, 38, 41–44, 49–52

is an algebraic representation of atmospheric pressure as a function of h. Solve this equation for h with $P = 4$.

$$14.7(0.5)^{h/3.6} = 4$$

$$(0.5)^{h/3.6} = \frac{4}{14.7}$$

$$\frac{h}{3.6}\log(0.5) = \log\left(\frac{4}{14.7}\right)$$

$$h = \frac{3.6\log(4/14.7)}{\log(0.5)} \qquad \text{Use a calculator to evaluate this expression.}$$

$$= 6.7598793$$

Thus the atmospheric pressure will be 4 lb/in^2 at an altitude of approximately 6.76 mi. ▤

EXAMPLE 9 APPLICATION: Using the Probability Theory

An important function in probability theory is the **normal distribution function**

$$f(x) = \left(\frac{c}{r\sqrt{2\pi}}\right)e^{-(x-u)^2/2r^2},$$

where u represents the mean and r^2 is the variance. If $u = 0$, $c = 25$, and $r^2 = 2.35$, what is the maximum of the function f?

Solution If $u = 0$, $c = 25$, and $r^2 = 2.35$, the function $f(x)$ becomes

$$f(x) = \left(\frac{25}{\sqrt{2\pi \cdot (2.35)}}\right)e^{-x^2/4.7}.$$

Figure 5.35 shows a complete graph of f. Use zoom-in to find that the maximum is 6.51 with an error of at most 0.01. ▤

Exercises for Section 5.6

In Exercises 1–14, solve each equation algebraically. Support your answer with a graphing utility.

1. $\log x = 4$

2. $\ln x = -1$

3. $\log_4(x - 5) = -1$

4. $\log_x(1 - x) = 1$

5. $\log(x + 3) = 2$

6. $\log(x - 1) = 3$

7. $\log_2 x = 4$

8. $\log_3(x + 7) = 12$

9. $\log_x(12 - x) = 2$

10. $3\log_2 x = 4$

11. $\ln x + \ln 2 = 3$

12. $\frac{1}{2}\log_3(x + 1) = 2$

13. $\log_2 x + \log_2(x + 3) = 2$

14. $\log_4(x + 1) - \log_4 x = 1$

In Exercises 15–18, compare the domain and range of each pair of functions.

15. $y = 2\log x$; $y = \log x^2$

16. $y = \log x + \log(x + 1)$; $y = \log[x(x + 1)]$

17. $y = \log_5 \sqrt{x+6} - \log_5 x$; $y = \log_5 \dfrac{\sqrt{x+6}}{x}$

18. $y = \log_2 x - \log_2(x+1)$; $y = \log_2 \left(\dfrac{x}{x+1}\right)$

In Exercises 19–24, find the exact solutions to each equation.

19. $(1.06)^x = 4.1$

20. $(0.98)^x = 1.6$

21. $(1.09)^x = 18.4$

22. $(1.12)^x = 3.2$

23. $\dfrac{2^x - 2^{-x}}{3} = 4$

24. $\dfrac{2^x + 2^{-x}}{2} = 3$

25. Use a calculator to find a decimal approximation to each solution in Exercises 19–24.

In Exercises 26–31, find a complete graph of each function. Determine the domain and range, find where the function is increasing or decreasing, and locate all local extrema.

26. $y = xe^x$

27. $y = x^2 e^{-x}$

28. $y = \dfrac{3^x - 3^{-x}}{3}$

29. $y = \dfrac{3^x + 3^{-x}}{3}$

30. $y = \ln(x^2 + 2x)$

31. $y = \ln\left(\dfrac{x}{2+x}\right)$

In Exercises 32 and 33, solve each inequality.

32. $2\log_2 x - 4\log_2 3 > 0$ **33.** $2\log_3(x+1) - 2\log_3 6 < 0$

In Exercises 34–37, solve each equation or inequality.

34. $e^x + x = 5$

35. $\ln|x| - e^{2x} \geq 3$

36. $e^x < 5 + \ln x$

37. $e^{2x} - 8x + 1 = 0$

In Exercises 38 and 39, solve each equation using a method of your choice.

38. $\dfrac{\ln x}{x} = \dfrac{1}{10} \ln 2$

39. $2^x = x^{10}$

40. Explain why Exercises 38 and 39 have the same *positive* solutions.

Exercises 41–44 refer to the normal probability density function problem situation introduced in Example 9. Let a mean of $u = 11.6$, a variance $r^2 = 2.35$, and $c = 25$.

41. Find a complete graph of f.

42. Determine the domain of f.

43. Show that the graph of f is symmetric with respect to the line $x = u$.

44. Determine all x where $f(x) = 2$.

45. Using the model for atmospheric pressure from Example 8, find the pressure at an altitude of 50 mi.

Exercises 46–48 refer to the following **problem situation**: In a certain molecular structure, the total *potential energy E* be-

tween two ions is given by

$$E = \dfrac{-5.6}{r} + 10e^{-r/3},$$

where r is the distance separating the nuclei.

46. Draw a complete graph of E as a function of r. Determine the domain of E.

47. What values of r make sense in the problem situation?

48. Find any relative extrema of the total potential energy on $(0, \infty)$ and the number r for which they occur.

For the mortgage interest problems in Exercises 49 and 50, there is no known method to determine an exact, explicit solution. Use a graphing zoom-in method.

49. Carlos and Maria can afford to make monthly mortgage payments of $800. They want to purchase a home requiring a 30-yr, $92,000 mortgage. What APR will be required?

50. Jill and Benny can afford to make monthly mortgage payments of $1200. They want to purchase a home requiring a 30-yr $130,000 mortgage. What APR will be required?

Exercises 51 and 52 refer to the following **problem situation**: From the time an item is new (original cost C) until it reaches the end of its useful life (salvage value S), it depreciates. A continuous model for book value B at any time t using the *sinking fund method of depreciation* is

$$B = C - \left[\dfrac{C - S}{[(1+i)^n - 1]/i}\right]\left[\dfrac{(1+i)^t - 1}{i}\right],$$

where n is the useful life, $0 \leq t \leq n$ and i is an annual interest rate. In this case, the annual depreciation charge is assumed to earn interest at the rate i per year, and the annual depreciation charges accumulate to the total depreciation $D = C - S$.

51. A machine costs $17,000, has a useful life of 6 yr, and has a salvage value of $1200. Draw a complete graph of the book-value model $y = B(t)$. What values of t make sense in the problem situation? Assume $i = 5\%$.

52. Consider the machine in Exercise 51. What is the book value of the machine in 4 yr, 3 mo?

The exponential function $y = e^x$ is an important function because it models such a variety of phenomena. A significant theory in calculus shows that $y = e^x$ can be approximated by polynomials. Consider this list of polynomial functions of increasing degree.

$$f_1(x) = x + 1$$

$$f_2(x) = \frac{x^2}{2!} + x + 1$$

$$f_3(x) = \frac{x^3}{3!} + \frac{x^2}{2!} + x + 1$$

$$f_4(x) = \frac{x^4}{4!} + \frac{x^3}{3!} + \frac{x^2}{2!} + x + 1$$

$$\vdots$$

$$f_n(x) = \frac{x^n}{n!} + \frac{x^{n-1}}{(n-1)!} + \ldots + x + 1$$

$$\vdots$$

$$n! = n(n-1)(n-2)\cdots 3 \cdot 2 \cdot 1$$

Exercises 53–56 give a glimpse of that result.

53. Show algebraically that $f_n(x) = \dfrac{x^n}{n!} + f_{n-1}(x)$.

54. Draw the graphs of $y = f_1(x)$, $y = f_2(x)$, $y = f_3(x)$, $y = f_4(x)$, and $y = e^x$ in the $[-5, 5]$ by $[-25, 25]$ viewing rectangle.

55. For what values of x does $f_3(x)$ approximate e^x with an error of less than 0.1?

56. For what values of x does $f_5(x)$ approximate e^x with an error of less than 0.1?

Chapter 5 Review

KEY TERMS (The number following each key term indicates the page of its introduction.)

annual percentage rate (APR), 272
base number, 261
bell curve, 267
change-of-base formula, 294
common logarithmic function, 285
compounded annually, 271
compounded continuously, 276
compound interest, 271
compound interest formula, 272

continuous interest, 275
continuous interest formula, 276
desired base, 294
effective annual rate i_{eff}, 278
effective yield, 278
exponent, 261
exponential function with base a, 262
exponential decay, 265
exponential growth, 265
given base, 294

future value S of an annuity, 280
half-life, 266
interest, 270
interest rate per period, 272
logarithm of x with base a, 285
logarithmic function with base a, 285
natural exponential function, 265

natural logarithmic function, 285
normal distribution curve, 267
normal distribution function, 308
ordinary annuity, 279
present value A of an annuity, 281
principal, 270
simple interest formula, 270
simple interest rate r, 270

REVIEW EXERCISES

In Exercises 1–6, sketch a complete graph of each function without using a graphing utility. Then support your answer with a graphing utility.

1. $f(x) = 2 + 3^x$

2. $f(x) = 3 - 2^{x+2}$

3. $y = 2 + \ln(x - 3)$

4. $y = \ln(x - 2) - 1$

5. $y = 3 \ln x$

6. $y = 2\ln(x+4)$

7. Use rules of exponents to solve $2^{x+1} = 4^3$.

In Exercises 8–12, use a graphing utility to find a complete graph of each function. Determine the domain and range. Find all existing local extrema.

8. $f(x) = (8x)2^{-x}$

9. $g(x) = \dfrac{\ln\sqrt{x+2}}{x-2}$

10. $h(x) = \log_3(x-5)^2$

11. $f(x) = \ln\left(\dfrac{x}{x-3}\right)$

12. $g(x) = \dfrac{2^x + 2^{-x}}{3}$

In Exercises 13 and 14, determine where each function is increasing and decreasing and find all local extrema.

13. $f(x) = e^{-2x}$ **14.** $g(x) = (8x)2^{-x}$

In Exercises 15 and 16, solve each equation without using a calculator.

15. $\log_5 125$ **16.** $\log_3(-9)$

In Exercises 17 and 18, solve each equation without using a calculator.

17. $\log_x \frac{1}{9} = -2$ **18.** $\log_4 x = 3$

19. Use the properties of logarithms to write the expression $\log[(1+r)^{12}/r]$ as a sum, difference, and/or product of simple logarithms.

20. Use the properties of logarithms to solve $z = (23+x)^M$ for M.

21. Use a calculator to investigate numerically the end behavior of the function $f(x) = x^2/\ln x$. That is, find what $f(x)$ approaches as $x \to \infty$.

22. Use a calculator to investigate the behavior of $g(x) = x/\ln(x-2)$ as $x \to 2^+$.

23. Use a calculator to investigate the behavior of $f(x) = x/\ln x^2$ as $x \to 1^-$.

24. Solve $\ln x > \log_2 x$ graphically.

25. Determine what $f(x) = x^3 \ln x$ approaches as $x \to \infty$.

26. Draw a complete graph of $y = x^3 \ln x$. Determine where the function is increasing and decreasing and all local extrema.

In Exercises 27 and 28, solve each equation algebraically. Support your answer with a graphing utility.

27. $5\log_3 x = 2$

28. $\log_5 x + \log_5(x-4) = 1$

29. Compare the domain and range of the functions $y = \log_3 x - \log_3(x+2)$ and $y = \log_3[x/(x+2)]$.

30. Find the exact solution to the equation $(1.5)^x = 0.90$.

31. Use a calculator to find a decimal approximation to the solution in Exercise 30. Then use a graphing utility to check the solution.

Solve Exercises 32–34 algebraically. Check with a graphing utility.

32. $3\log_5(x-1) - 2\log_5 4 > 0$

33. $\log_2(x-1) + \log_2(x+2) = 2$

34. $\log_3(x+5) + 2\log_3 x > 1$

In Exercises 35–38, use a graphing utility to solve each equation or inequality.

35. $e^x + \ln x = 5$

36. $2\log x - e^x = -3$

37. $e^{2x} + 3e^x \le 10$

38. $75\log x - e^{2x} \ge -10$

39. Find the simultaneous solution to the system

$$\begin{cases} y = x^2 \\ y = 2^x \end{cases};$$

that is, solve $x^2 = 2^x$. For what values of x is $2^x > x^2$? For what values of x is $x^2 > 2^x$?

40. Find the simultaneous solution to the system

$$\begin{cases} y = x^3 \\ y = 3^x \end{cases};$$

that is, solve $x^3 = 3^x$. For what values of x is $3^x > x^3$? For what values of x is $x^3 > 3^x$?

Exercises 41–46 refer to the following **problem situation**: Linda deposits $500 in a bank that pays 7% interest annually. Assume she makes no other deposits or withdrawals. How much will she accumulate after 5 yr if the bank pays interest compounded as follows:

41. Annually   **42.** Quarterly

43. Monthly   **44.** Daily

45. Semiannually **46.** Continuously

47. Determine when an investment of $1500 accumulates to a value of $2280 if the investment earns interest at the rate of 7% interest compounded monthly.

48. Determine how long it will take for an investment to double in value if interest is earned at 6.25% compounded quarterly.

49. Draw a graph that shows the value at any time t of a $500 investment that earns 9% simple interest. In the same viewing rectangle, do the same for a $500 investment that earns 9% interest compounded monthly.

Exercises 50–52 refer to the following **problem situation**: The population of a town is 625,000 and is increasing at the annual rate of 4.05%.

50. Find an algebraic representation of the population P as a function of time.

51. Draw a complete graph of the function P in Exercise 50. What portion of the graph represents the problem situation?

52. Determine when the population will be 1 million.

Exercises 53 and 54 are the basis for what is called the *rule of 72*.

53. Graph $y = r$ and $y = \ln(1 + r)$ for $0 \le r \le 1$ in the same viewing rectangle. When is $\ln(1 + r)$ approximately equal to r?

54. Use 0.72 as an approximate value for $\ln 2$ and show how you can derive an easy-to-remember "rule" that says money invested at annual rate r compounded annually will double in value after $72/100r$ yr. For example, at 6% interest compounded annually, you would expect money to double in value after $72/[100(0.06)] = 72/6 = 12$ yr. Is this a good rule of thumb?

Exercises 55–57 refer to the following **problem situation**: The half-life of a certain radioactive substance is 21 days and there are 4.62 g present initially.

55. Find an algebraic representation for the amount A of substance remaining as a function of time.

56. Find a complete graph of the function in Exercise 55.

57. When will there be less than 1 g of the substance remaining?

Exercises 58 and 59 refer to the Richter scale problem situation described in Section 5.5.

58. What is the magnitude on the Richter scale of an earthquake if $a = 275$, $T = 2.5$, and $B = 4.250$?

59. Draw a complete graph of the Richter scale model (magnitude R as a function of amplitude a). Assume $T = 2.5$ and $B = 4.250$. What are the values of a that make sense in this problem situation?

Exercises 60 and 61 refer to the following **problem situation**: It is known from the theory of electricity that in an *RL circuit*, the current I in the circuit is given by

$$I = \frac{V}{R}(1 - e^{-(Rt)/L})$$

as a function of time t. Here V, R, and L are constants for voltage, resistance, and self-inductance, respectively.

60. Assume $V = 10$ volts, $R = 3$ ohms, and $L = 0.03$ henries. Draw a complete graph of $I = f(t)$.

61. Show that $I \to V/R = 10/3$ as $t \to \infty$. Give a graphical argument.

62. Find the present value of an ordinary annuity for $100 payments in each of 30 payment intervals where the interest rate is 4% per payment interval.

63. Use a graphing utility to solve

$$10{,}000 = 250[1 - (1 + i)^{-60}]/i$$

for i. (*Note:* Think about solving this equation algebraically. There is no known method of explicitly solving this equation for i!)

64. Explain why the formula

$$B(n) = R\frac{1 - (1 + i)^{-(360-n)}}{i}$$

gives the outstanding loan balance of a 30-yr mortgage loan requiring monthly payments of R dollars with an APR of $12i$ as a function of the number n of payments made.

65. Let $f(x) = ab^x$. Show that

$$\log b = \frac{\log f(x_2) - \log f(x_1)}{x_2 - x_1}$$

for any pair of real numbers x_1 and x_2 such that $x_1 \ne x_2$. Assume that $a > 0$ and $b > 1$.

66. Consider the equation

$$y = \$884.61\frac{1 - (1.01)^{-(360-x)}}{0.01}.$$

a) Draw a complete graph.

b) Explain how this graph relates to the outstanding loan balance of an $86,000 mortgage loan with a 30-yr term requiring monthly payments of $884.61.

67. Some mortgage loans are available that require payments every 2 wk (26 times each year). Consider a mortgage loan of $80,000 for a 30-yr term with interest at 10% APR and monthly payments.

a) Determine the monthly payment.

b) Suppose one half of the monthly payment was made every 2 wk. When would the mortgage loan be completely paid? Assume the interest rate per 2-wk interval is one half the monthly interest rate.

CHU SHIH-CHIEH

Chu Shih-chieh (ca. 1280–1303) was the last and greatest mathematician of the golden age of Chinese mathematics during the Sung Dynasty. Very little is known about his personal life. He probably lived in Yen-shan, near modern-day Beijing, but he spent 20 years traveling extensively in China as a renowned mathematician and teacher.

His book *Introduction to Mathematical Studies* was lost for a time, but later became an influential textbook in Korea and Japan. Another book, *Precious Mirror of the Four Elements*, strongly influenced the development of the theory of equations. It discusses four "elements"—heaven, earth, man, and matter—which represent four unknowns in a single equation. Although the book introduces two concepts later named for Western mathematicians (the Horner method and Pascal's triangle), Chu Shih-chieh's contributions to mathematics remain largely anonymous.

CHAPTER 6

The Trigonometric Functions

6.1 _____ Angles and Their Measure

So far in this book, we have studied several types of functions: polynomial, rational, exponential and logarithmic, and those involving radicals. In this chapter, you will learn of another collection of functions called the **trigonometric functions**.

Trigonometric functions are important because they model a wide variety of cyclical phenomena—from pendulum swings to alternating electrical current. In this text, we discuss both historical and modern approaches to this subject.

In geometry, an **angle** is defined as the union of two rays (half lines) with a common endpoint called the **vertex of the angle**. Trigonometry takes a more dynamic view by describing an angle in terms of a rotating ray. The beginning ray, called the **initial side of the angle,** is rotated about its endpoint. The final position is called the **terminal side of the angle,** and the endpoint of the ray is again called the vertex of the angle (see Fig. 6.1). Notice that the initial and terminal sides of an angle divide a plane into two different angles, one clockwise and the other counterclockwise.

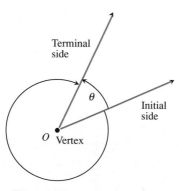

Figure 6.1 A clockwise and a counterclockwise angle.

315

Objective

Students will be able to draw angles in standard position using a rotational approach, to convert between radians and degrees, and to find arc length.

Key Ideas

Angle
Initial side
Terminal side
Standard position
Measure of an angle
Coterminal angles
Supplementary angles
Radian
Central angle
Intercepted arc
Arc length
Angular speed
Linear speed

An angle is usually named with a single Greek letter, such as α (alpha), β (beta), γ (gamma), or θ (theta), or by an uppercase letter such as A, B, or C. We also use the symbol $\angle$ (angle), so that $\angle A$ is read "angle A."

An angle is called an **angle in standard position** when it is positioned on a rectangular coordinate system with its vertex at the origin and its initial side on the positive x-axis. Figure 6.2 shows two examples of angles in standard position.

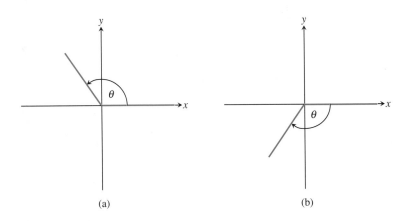

(a)   (b)

Figure 6.2   Two angles in standard position: (a) counterclockwise; (b) clockwise.

Teaching Note

Many students will find the rotational approach to angle measure new. Focus attention on the basic idea of rotation and how angle measure can be viewed in terms of rotation.

Teaching Note

Many new terms are introduced in the section. Make sure students understand the vocabulary. A brief vocabulary quiz may even be in order.

Degree Measure

Angles are often measured using units called **degrees**. The measure of an angle determined by one complete counterclockwise rotation is defined to be 360 degrees, denoted $360°$. So the measure of one-half and one-quarter of a complete counterclockwise rotation would be $180°$ and $90°$, respectively (see Fig. 6.3). When the angle is rotated counterclockwise, the measure in degrees is positive, and when the angle is rotated clockwise, its measure in degrees is negative. Figure 6.4 shows both positive and negative angles. Notice that an angle will be more than $360°$ if it makes more than one complete revolution (Fig 6.4a,c).

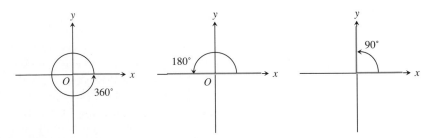

Figure 6.3   Angles with positive measure.

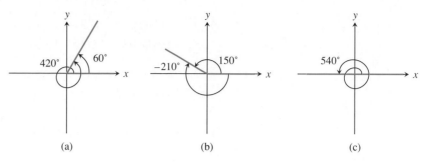

(a)   (b)   (c)

Figure 6.4   Examples of positive, negative, and coterminal angles.

Two angles are **coterminal** if they have the same initial side and the same terminal side. Coterminal angles can be found by adding or subtracting an integer multiple of 360°. For example, Fig. 6.4 (a) shows that a 60° angle and a 420° angle are coterminal angles; notice that $420 - 360 = 60$. Similarly, angles of $-210°$ and $150°$ are also coterminal: $-210 + 360 = 150$ (see Fig. 6.4b).

E X A M P L E   1   Finding Coterminal Angles

Sketch each of the following angles in standard position and find one negative and one positive angle coterminal with each:

a) $\theta = 30°$

b) $\theta = 145°$

c) $\theta = -45°$

Solution   (See Fig. 6.5.)

Notes on Exercises

See Ex. 1–20.

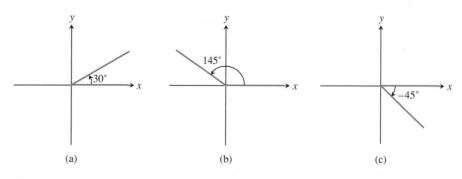

(a)   (b)   (c)

Figure 6.5   Three angles in standard position.

a) $\theta = 30°$ is coterminal with

$$\alpha = 30° + 360° = 390° \qquad \text{and} \qquad \beta = 30° + (-360°) = -330°.$$

b) $\theta = 145°$ is coterminal with

$$\alpha = 145° + 360° = 505° \quad \text{and} \quad \beta = 145° + (-360°) = -215°.$$

c) $\theta = -45°$ is coterminal with

$$\alpha = -45° + 360° = 315° \quad \text{and} \quad \beta = -45° + (-360°) = -405°.$$

Because a coterminal angle can be obtained by adding or subtracting an integer multiple of $360°$, there are an infinite number of correct answers. ≡

Certain angles are given special names because they occur so frequently in trigonometric relations, as we shall see later in the chapter. A $90°$ angle is a **right angle**. An angle θ is an **acute angle** when $0° < \theta < 90°$, and an angle α is an **obtuse angle** when $90° < \alpha < 180°$. Two positive angles α and β are **supplementary angles** when their sum is $180°$ and **complementary angles** when their sum is $90°$. Negative angles do not have complements or supplements; positive angles greater than $90°$ do not have complements.

EXAMPLE 2   Finding Complements and Supplements

Find the supplement and the complement of each of the following angles:

Notes on Exercises

Ex. 21–28 incorporate the
ideas of Example 2.

a) $48°$

b) $135°$

c) $-18°$

d) $x°$ where $0 < x < 90$

Solution   To find the complement of the angle, subtract the angle from $90°$; to find its supplement, subtract the angle from $180°$.

a) $\theta = 48°$: $180° - \theta = 132°$ (supplement); $90° - \theta = 42°$ (complement)

b) $\theta = 135°$: $180° - \theta = 45°$ (supplement); the complement of θ does not exist.

c) $\theta = -18°$ has no supplement or complement.

d) $\theta = x°$: $180° - x°$ (supplement); $90° - x°$ (complement)   ≡

Teaching Note

You will need to help students
think in terms of radians as
indicating the measure of an
angle. Have students explore
the geometric basis for this
new measure of angles. In
Section 6.3, students will
need to think of real numbers
as the domain elements of
trigonometric functions, and to
see real numbers as lengths of
arcs on the unit circle.

Radian Measure

In addition to the degree, there is a second unit of angle measure called the radian. Radian measure is the unit of angle measure used frequently in trigonometry, calculus, and more advanced mathematics. It is important to understand radian measure and its relationship to degree measure.

If the vertex of an angle θ is the center of a circle, θ is called a **central angle** of the circle. If the sides of the angle intersect the circle at points A and B, the angle has an **intercepted arc** AB (see Fig. 6.6). If the length of this arc AB is equal to the radius r of the circle, then θ has a measure of 1 radian.

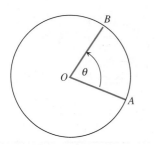

Figure 6.6 Central angle θ with intercepted arc AB.

Definition 6.1 Radian Measure

One **radian** (rad.) is the measure of a central angle θ whose intercepted arc has a length equal to the circle's radius. (See Fig. 6.7.)

To explore the relationship of radian measure to degree measure, we begin by considering a circle with radius $r = 1$. Such a circle is called a **unit circle**.

The radian measure of a central angle of a unit circle is equal to the length of the intercepted arc. Because the degree measure of a straight angle is 180° and half the circumference of a unit circle is π, we conclude that an angle of $180° = \pi$ radians (see Fig. 6.8). This fact can be used to convert radians to degrees and vice versa.

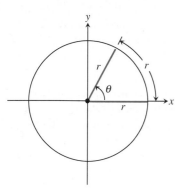

Figure 6.7 An angle of 1 radian.

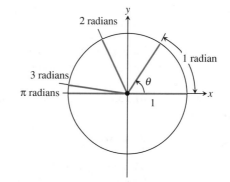

Figure 6.8 π radians $= 180°$.

RADIAN MEASURE CONVENTION

Hereafter, we omit the word *radian* when giving the measure of an angle in radians. For example, an angle of 2 radians will be reported as an angle whose measure is 2. We continue to use the established abbreviation for degree measure; that is, 2° stands for an angle that measures 2 degrees.

Conversion between Radians and Degrees

Because π radians $= 180°$, it follows that

$$1 \text{ rad.} = \frac{180°}{\pi} \qquad \text{and} \qquad 1° = \frac{\pi}{180} \text{ rad.}$$

E X A M P L E   3   Converting from Degrees to Radians

Change each of the following common angles from degrees to radians: 30°, 45°, 60°, and 90°.

Solution

$$30° = 30° \left(\frac{\pi \text{ rad.}}{180°} \right) = \frac{\pi}{6} \text{ rad.} \qquad 45° = 45° \left(\frac{\pi \text{ rad.}}{180°} \right) = \frac{\pi}{4} \text{ rad.}$$

$$60° = 60° \left(\frac{\pi \text{ rad.}}{180°} \right) = \frac{\pi}{3} \text{ rad.} \qquad 90° = 90° \left(\frac{\pi \text{ rad.}}{180°} \right) = \frac{\pi}{2} \text{ rad.} \qquad \blacksquare$$

E X A M P L E   4   Converting from Radians to Degrees

Change each of the following angles from radians to degrees: $3\pi/2$, $2\pi/3$, $5\pi/4$, and $7\pi/12$.

Solution

$$\frac{3\pi}{2} = 3 \left(\frac{\pi}{2} \right) = 3(90°) = 270° \qquad \frac{2\pi}{3} = 2 \left(\frac{\pi}{3} \right) = 2(60°) = 120°$$

$$\frac{5\pi}{4} = 5 \left(\frac{\pi}{4} \right) = 5(45°) = 225° \qquad \frac{7\pi}{12} = \frac{7\pi}{12} \left(\frac{180°}{\pi} \right) = 105° \qquad \blacksquare$$

The angles $30°$, $45°$, $60°$, and $90°$ are used frequently in trigonometry. Because they are factors of $180°$, their radian measure equivalents, $\pi/6$, $\pi/4$, $\pi/3$, and $\pi/2$, and their integer multiples, usually are expressed in this fractional form. The radian measure of other angles commonly are expressed in decimal form. Using a grapher or a scientific calculator, you can find very accurate approximations to 10 digits.

E X A M P L E   5   Converting from Degrees to Radians

Find the radian measure of an angle whose measure is the following:

a) $500°$

b) $-200°$

Solution

a) $500° = 500° \left(\frac{\pi}{180°} \right) = 8.72664626$

b) $-200° = -200° \left(\frac{\pi}{180°} \right) = -3.490658504$ $\qquad \blacksquare$

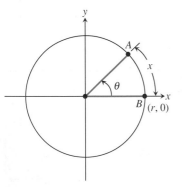

Figure 6.9 Length x of arc A/B subtends the central angle θ.

In a unit circle, the length of an arc and its subtended central angle measured in radians have the same measure. However, this is not true when the radius of the circle is not 1. Figure 6.9 shows a central angle with radian measure θ. The length x of arc AB determined by this angle is a fraction of the circumference $C = 2\pi r$

of the circle and can be determined from the following proportion:

$$\frac{\theta}{2\pi} = \frac{x}{2\pi r}$$

$$x = \frac{\theta}{2\pi}(2\pi r) = \theta r.$$

In this proportion, angle θ must be measured in radians.

Theorem 6.1   Arc Length

Given a circle of radius r, if x is the length of the intercepted arc of a central angle θ, then

$$x = r\theta.$$

Notice that the formula in Theorem 6.1 is correct only for θ expressed in radians.

E X A M P L E   6   Finding Arc Length

Find the length x of the arc of a circle of radius 2 intercepted by a central angle measuring $\pi/4$.

Solution   Use Theorem 6.1.

$$x = r\theta = 2\left(\frac{\pi}{4}\right)$$

$$= \frac{\pi}{2}$$ ≡

Applications

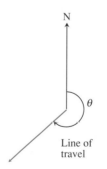

Figure 6.10   A certain angle west of north.

Many real-world applications use degrees for angle measure, while many scientific applications use radians.

In navigation, the **course**, or **path**, of an object is sometimes given as the measure of the clockwise angle θ that the **line of travel** makes with due north (see Fig. 6.10). That is, the due-north ray is the initial side, and the line of travel is the terminal side of the angle of the path. By convention, the angle measure θ is positive even though the angle rotation is clockwise.

We also say that the line of travel has **bearing θ**.

E X A M P L E   7   APPLICATON: Determining a Boat's Bearing

The captain of a boat steers a course 35° from home port. Draw the boat's path.

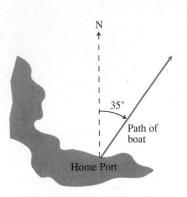

Figure 6.11 East of north.

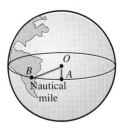

Figure 6.12 If $\angle AOB$ is $1/60°$, then the length of AB is 1 nautical mile.

Solution Draw a ray from home port in the due-north direction. Rotate the north ray 35° clockwise (see Fig. 6.11). The boat travels in the direction of this second ray. We say that the bearing of the boat is 35°. ▤

In navigation, it's often necessary to measure angles smaller than 1°. The degree is divided into 60 equal parts, called **minutes** (denoted by ′), and each minute is divided into 60 equal parts, called **seconds** (denoted by ″). So an angle of 23° 15′ 45″ is equal to

$$23 + \frac{15}{60} + \frac{45}{3600} = 23 + \frac{900}{3600} + \frac{45}{3600}$$

One second is 1/60th of 1/60th, or 1/3600th, of a degree.

$$= 23.2625°.$$

The measure of distance used by navigators is the nautical mile. It is defined in terms of Earth itself with point O located at the center of the planet. If $\angle AOB$ has measure 1 minute (1/60 of a degree), then the distance between A and B on the circumference of the circle is 1 **nautical mile** (see Fig. 6.12). The diameter of Earth averages 7912.176 statute miles, where a **statute mile** is a standard mile of 5280 ft.

E X A M P L E   8   APPLICATON: Finding the Number of Feet in a Nautical Mile

How many feet are in 1 nautical mile?

Solution We need to find the length x of arc AB in Fig. 6.12.

$$\text{radius } r \text{ of Earth} = \frac{7912.176}{2} \text{ mi and 5280 ft/mi}$$

$$x = r\theta$$

$$= \frac{7912.176}{2}\left(\frac{1}{60}\right)^{\circ}(5280)$$

$$= \frac{7912.176}{2}\left(\frac{\pi}{60 \cdot 180}\right)5280$$

$$= 6076.115 \text{ ft}$$

There are 6076.115 ft in 1 nautical mile. ▤

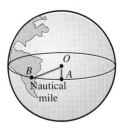

Figure 6.13 A rotating wheel.

Radian measure also can be used to analyze the motion of a point moving at a constant speed along a circular path. Suppose for example that a wheel of radius r is rotating at a constant rate (see Fig. 6.13). Let P be a point on the circumference of the wheel. The **angular speed** of the wheel, in radians per second, is the angle swept out in 1 sec by the line segment from the center of the wheel to the point P on the wheel's circumference. The **linear speed** of the point P, in feet per second, is the distance P travels in 1 sec.

Angular speed: $\quad \theta$ rad./sec $= \dfrac{s}{r}$ rad./sec

Linear speed: $\quad s$ ft/sec $= r\theta$ ft/sec

Notes on Examples

Use craft hoops, an old spare tire, a lazy Susan, or some other circular object to help students see the relationship between angular speed and linear speed.

EXAMPLE 9 APPLICATION: Determining Angular Speed

A wheel with a radius of 18 in. is rotating at 850 rpm (revolutions per minute). Determine the following:

a) The angular speed of the wheel in radians per second

b) The linear speed in feet per second of a point on the circumference of the wheel (See Fig. 6.13.) (*Hint:* 1 rpm $= 2\pi$ rad./min.)

Solution

a) First, change revolutions per minute to radians per second.

$$850 \text{ rpm} = \frac{850(2\pi)}{60} \text{ rad./sec} = \left(\frac{85}{3}\right)\pi \text{ rad./sec}$$

The angular speed of the wheel is $(85/3)\pi$ rad./sec.

b) From Theorem 6.1, we know that if x represents arc length $x = r\theta$ and from part (a), that $\theta = \left(\frac{85}{3}\right)\pi$.

$$x = r\theta$$

$$= 1.5 \left(\frac{85}{3}\right)\pi \quad 18 \text{ in.} = 1.5 \text{ ft}$$

$$= 133.52 \text{ ft/sec}$$

Assignment Guide

Day 1: Ex. 2, 5, 8, 10, 11, 14, 16, 19, 22, 25, 28, 29, 32, 35, 38, 40
Day 2: Ex. 41–55, 58, 60, 61, 63, 64, 68, 74–76

The linear speed of a point on the wheel is 133.52 ft/sec. ▤

Exercises for Section 6.1

In Exercises 1–6, determine the quadrant of the terminal side of each angle in standard position.

1. $-160°$ **2.** $280°$ **3.** $452°$

4. $-827°$ **5.** $1150°$ **6.** $-455°$

In Exercises 7–10, determine the measure of an angle θ coterminal with the given angle that satisfies the specified condition.

7. $48°$; $360° \le \theta \le 720°$ **8.** $110°$; $-360° \le \theta \le 0°$

9. $-15°$; $180° \le \theta \le 540°$ **10.** $-250°$; $360° \le \theta \le 720°$

In Exercises 11–20, sketch each angle and determine four different coterminal angles, two with positive and two with negative measures.

11. $55°$ **12.** $-22°$ **13.** $410°$

14. $-150°$ **15.** $\dfrac{\pi}{4}$ **16.** $\dfrac{3\pi}{2}$

17. $\dfrac{\pi}{6}$ **18.** $\dfrac{5\pi}{6}$ **19.** $-\dfrac{\pi}{3}$

20. $-\dfrac{7\pi}{4}$

In Exercises 21–28, find the complement and supplement of each angle. Use the same unit of angle measure as in the given angle.

21. $35°$ **22.** $23°$ **23.** $68°$

24. $12°$ **25.** $\pi/3$ **26.** $\pi/12$

27. $\dfrac{5\pi}{13}$ **28.** $\dfrac{3\pi}{7}$

In Exercises 29–34, assume the given point is on the terminal side of an angle θ in standard position, where $0° \le \theta \le 360°$. Determine θ in both degrees and radians.

29. $(-1, 0)$ **30.** $(0, 5)$ **31.** $(3, 3)$

32. $(-2, 2)$ **33.** $(5, -5)$ **34.** $(10, 0)$

In Exercises 35–40, determine the measure in both degrees and radians and draw the angle in standard position.

35. One-half counterclockwise rotation

36. One-third clockwise rotation

37. Four-thirds counterclockwise rotation

38. Five-thirds clockwise rotation

39. 2.5 counterclockwise rotations

40. 3.5 clockwise rotations

In Exercises 41–46, find the equivalent radian measure for each angle.

41. $45°$ **42.** $60°$ **43.** $135°$

44. $210°$ **45.** $120°$ **46.** $330°$

In Exercises 47–50, find the equivalent radian measure for each angle.

47. $23°$ **48.** $118°$

49. $72°$ **50.** $249°$

In Exercises 51–54, find the equivalent degree measure for each angle.

51. $\dfrac{5\pi}{12}$ **52.** 4.5

53. 3.75 **54.** 12.9

In Exercises 55–58, determine the length of the arc of a circle with the specified radius subtended by a central angle of the given measure.

55. $r = 2$; $\theta = 30°$ **56.** $r = 3.75$; $\theta = 122°$

57. $r = 5.76$; $\theta = 155°$ **58.** $r = 20.55$; $\theta = 72°$

In Exercises 59–62, let θ be a central angle of a circle of radius r and s the length of the subtended arc. Find the arc length s.

59. $r = 15$ in.; $\theta = 22°$ **60.** $r = 5.6$ ft; $\theta = 3°$

61. $r = 3.5$ cm; $\theta = 1.750$ **62.** $r = 5.1$ in.; $\theta = \dfrac{\pi}{12}$

63. The captain of a boat follows a $38°$ course for 2 mi and then changes to a $47°$ course, which he follows for the next 4 mi. Sketch the path of this trip.

64. Find the angles that describe the following compass directions:
a) NE (northeast)
b) NNE (north-northeast)
c) ENE (east-northeast)

65. A ship's captain traveling east discovers he is traveling on a course that is $5°$ north of the correct direction of his destination. If the captain makes a single midcourse correction when he is exactly halfway to his destination, will he make a course correction of $5°$, $10°$, $-5°$, or $-10°$?

66. Points A and B are 257 nautical miles apart. How far apart are A and B in *statute* miles?

67. Diagram the navigational courses of $35°$, $128°$, and $310°$.

68. At a certain time, an airplane is between two signal towers that lie on an east–west line. The bearing from the plane to each tower is $340°$ and $37°$, respectively. Draw the exact location of the plane.

69. Let A be the origin of a coordinate plane. Diagram the location of point B if the bearing from A to B is the following:
a) $22°$
b) $185°$
c) $292°$

Exercises 70 and 71 refer to the following **problem situation:** A wheel with a radius of 5 ft is rotating at 1200 rpm.

70. Determine the angular speed of the wheel.

71. Determine the linear speed of a point on the wheel's circumference.

Exercises 72 and 73 refer to the following **problem situation:** A wheel with a radius of 2.8 feet is rotating at 600 rpm.

72. Determine the angular speed of the wheel.

73. Determine the linear speed of a point on the wheel's circumference.

74. Elena can reach 42 mph on her exercise bike in high gear. The bike wheels are 30 in. in diameter, the pedal sprocket is 16 in. in diameter, and the wheel sprocket is 5 in. in diameter (in high gear). Find the angular speed of the wheel and of both sprockets. *(Note:* The linear speed of a point on the wheel's circumference also is 42 mph.)

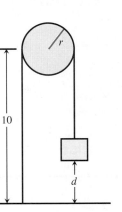

Exercises 75 and 76 refer to the following **problem situation:** A simple winch used to lift heavy objects is positioned 10 ft above ground level. Assume the radius of the winch is r feet. For the given radius r and winch rotation θ, determine the distance the object is lifted above ground.

For Exercises 75 and 76.

75. $r = 4$ in.; $\theta = 720°$ **76.** $r = 1$ ft; $\theta = 720°$

6.2 _____

The Trigonometric Functions and the Unit Circle

In Section 6.1, the radian unit of angular measure was defined. The six trigonometric functions defined in this section are defined using the unit circle and radian measure. We begin with a discussion of a relationship between the real-number line and the unit circle.

Real-number Line and the Unit Circle

Recall that the unit circle is the circle with a radius of 1, as defined by the relation $x^2 + y^2 = 1$ (see Fig. 6.14). Imagine that a flexible string with infinite length

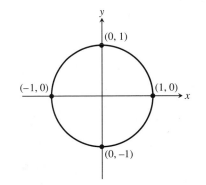

Figure 6.14 The unit circle.

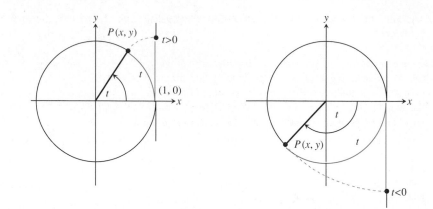

Figure 6.15   Wrapping the real-number line around the unit circle.

represents the real-number line. The origin of this number line is attached to the circle at point $(1, 0)$ and wraps around the circle such that the positive real numbers wrap counterclockwise and the negative ones wrap clockwise (see Fig. 6.15).

As the string wraps around the unit circle, each real number t on the string is associated with a point $P(x, y)$ on the circle. Thus the real-number line from 0 to t makes an arc of length t beginning on the circle at $(1, 0)$ and ending at a point $P(t) = P(x, y)$ on the unit circle. Notice that this arc **subtends an angle** in standard position whose radian measure is t. To find the point on the unit circle associated with t requires that we locate the coordinates of the endpoint of an arc of length t that begins at $(1, 0)$.

Angles in standard position whose measures are multiples of $\pi/4$ radians or multiples of $\pi/6$ are often called the **familiar angles.** They determine eight equally spaced points and 12 equally spaced points, respectively, around the unit circle, as shown in Fig. 6.16.

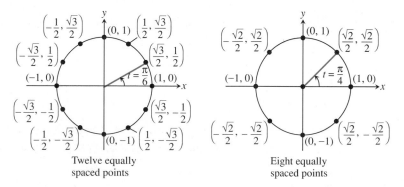

Twelve equally
spaced points

Eight equally
spaced points

Figure 6.16   Multiples of $\pi/6$ and $\pi/4$.

As you read Example 1, refer to the coordinates of the points in Fig. 6.16.

E X A M P L E 1   Finding $P(t)$

Find the point $P(t)$ on the unit circle that matches each of the following numbers in the wrapping process:

a) $t = \pi/2$

b) $t = 3\pi/4$

c) $t = -2\pi/3$

d) $t = 9\pi/4$

Solution   Find the coordinates of P by finding the values of x and y for the right triangles in Fig. 6.17.

a) An arc of length $\pi/2$ beginning at $(1, 0)$ completes a one-quarter counterclockwise rotation ending at $(0, 1)$. Therefore

$$P\left(\frac{\pi}{2}\right) = (0, 1).$$

b) An arc of length $3\pi/4$ beginning at $(1, 0)$ completes a three-eighths counterclockwise rotation ending at $\left(-\sqrt{2}/2, \sqrt{2}/2\right)$. Therefore

$$P\left(\frac{3\pi}{4}\right) = \left(-\frac{\sqrt{2}}{2}, \frac{\sqrt{2}}{2}\right).$$

c) An arc of length $-2\pi/3$ beginning at $(1, 0)$ completes a one-third clockwise rotation ending at $\left(-1/2, -\sqrt{3}/2\right)$. Therefore

$$P\left(-\frac{2\pi}{3}\right) = \left(-\frac{1}{2}, -\frac{\sqrt{3}}{2}\right).$$

d) An arc of length $9\pi/4$ beginning at $(1, 0)$ completes a one-and-one-eighth counterclockwise rotation ending at $\left(\sqrt{2}/2, \sqrt{2}/2\right)$. Therefore

$$P\left(\frac{9\pi}{4}\right) = \left(\frac{\sqrt{2}}{2}, \frac{\sqrt{2}}{2}\right).$$

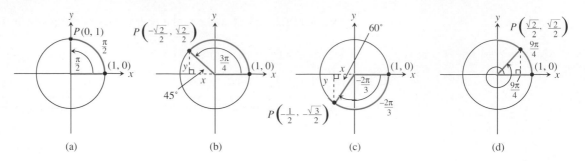

Figure 6.17   The four values for $P(t) = P(x, y)$ in Example 1.

Definitions of the Trigonometric Functions

The six trigonometric functions of sine, cosine, tangent, cotangent, secant, and cosecant are defined in terms of the coordinates of points on the unit circle. Notice that abbreviations for these six functions are used in the following definition.

Definition 6.2   Trigonometric Functions in Radian Measure

Let $P(t) = (x, y)$ be the point on the unit circle corresponding to the real number t. Then for any angle t, the following is true:

$$\text{sine } t = y \qquad\qquad \text{cotangent } t = \frac{x}{y} \quad (y \neq 0)$$

$$\text{cosine } t = x \qquad\qquad \text{secant } t = \frac{1}{x} \quad (x \neq 0)$$

$$\text{tangent } t = \frac{y}{x} \quad (x \neq 0) \qquad \text{cosecant } t = \frac{1}{y} \quad (y \neq 0)$$

We often use the abbreviation *trig* for *trigonometry* or *trigonometric*. Also, the following are standard abbreviations for the trig functions:

$$\text{sine } \theta = \sin \theta \qquad\qquad \text{cosecant } \theta = \csc \theta$$

$$\text{cosine } \theta = \cos \theta \qquad\qquad \text{secant } \theta = \sec \theta$$

$$\text{tangent } \theta = \tan \theta \qquad\qquad \text{cotangent } \theta = \cot \theta$$

In Examples 2 and 3, Definition 6.2 is applied to find the values of the trigonometric functions.

Notes on Exercises

Ex. 13–20 support the material found in Example 2.

REMINDER

- $\cos t$ is the x-coordinate of the point $P(x, y)$ associated with t.

- $\sin t$ is the y-coordinate of the point $P(x, y)$ associated with t.

EXAMPLE 2   Using the Unit Circle for Sine and Cosine Values

Let $f(t) = \sin t$ and $g(t) = \cos t$. Use the unit circle to find the value of the functions f and g for the following:

a) $t = 5\pi/6$

b) $t = 5\pi/3$

Solution   (See Fig. 6.18.)

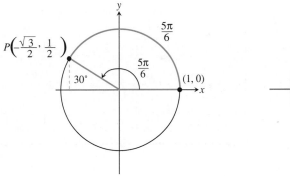
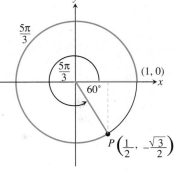

Figure 6.18   Two angles in standard position on a unit circle.

a) $f\left(\dfrac{5\pi}{6}\right) = \sin\dfrac{5\pi}{6} = \dfrac{1}{2}$ $g\left(\dfrac{5\pi}{6}\right) = \cos\dfrac{5\pi}{6} = -\dfrac{\sqrt{3}}{2}$

b) $f\left(\dfrac{5\pi}{3}\right) = \sin\dfrac{5\pi}{3} = -\dfrac{\sqrt{3}}{2}$ $g\left(\dfrac{5\pi}{3}\right) = \cos\dfrac{5\pi}{3} = \dfrac{1}{2}$

EXAMPLE 3   Using the Unit Circle for Other Trig Values

CONVENTION REGARDING FUNCTION NOTATION

When trig functions are written using function notation $f(t) = \sin t$, $g(t) = \cos t$, and $h(t) = \tan t$, it is understood that radian measure is being used.

Let $f(t) = \tan t$ and $g(t) = \sec t$. Use the unit circle to find the following:

a) $f(5\pi/6)$

b) $g(5\pi/3)$

Solution   (See Fig. 6.18.)

a) $f\left(\dfrac{5\pi}{6}\right) = \tan\dfrac{5\pi}{6} = \dfrac{y}{x}$ Substitute the coordinates of the point P.

$$= \dfrac{\frac{1}{2}}{-\frac{\sqrt{3}}{2}} = -\dfrac{1}{\sqrt{3}} = -\dfrac{\sqrt{3}}{3}$$

b) $g\left(\dfrac{5\pi}{3}\right) = \sec\dfrac{5\pi}{3} = \dfrac{1}{x} = \dfrac{1}{\frac{1}{2}} = 2$

When we write a trig function of a number, for example $\sin(\pi/6)$, the number represents the radian measure of an angle.

The sine, cosine, and tangent values for angles with radian measure $\pi/6$, $\pi/4$, and $\pi/3$ are summarized in the next table. The values for these familiar angles can be found using Fig. 6.16.

Sine, Cosine, and Tangent for Angles of $\dfrac{\pi}{6}$, $\dfrac{\pi}{4}$, and $\dfrac{\pi}{3}$ Radians

$$\sin\frac{\pi}{6} = \frac{1}{2} \qquad \sin\frac{\pi}{4} = \frac{\sqrt{2}}{2} \qquad \sin\frac{\pi}{3} = \frac{\sqrt{3}}{2}$$

$$\cos\frac{\pi}{6} = \frac{\sqrt{3}}{2} \qquad \cos\frac{\pi}{4} = \frac{\sqrt{2}}{2} \qquad \cos\frac{\pi}{3} = \frac{1}{2}$$

$$\tan\frac{\pi}{6} = \frac{1}{\sqrt{3}} \qquad \tan\frac{\pi}{4} = 1 \qquad \tan\frac{\pi}{3} = \frac{\sqrt{3}}{1}$$

If an angle is not a multiple of one of the familiar angles, the values of the trig functions of that angle can be found using a scientific calculator.

USING A CALCULATOR

Ensure the mode setting is correct when you use a calculator with trig functions. Check the owner's manual of your calculator to see how the trig keys work.

E X A M P L E   4   Finding Trig Values with a Calculator

Find $\sin\theta$, $\cos\theta$, and $\tan\theta$ for the following:

a) $\theta = 42°$

b) $\theta = \pi/12$ rad.

Solution   Use a scientific calculator to find 10-digit approximations.

a) $\sin 42° = 0.6691306064$ **b)** $\sin\dfrac{\pi}{12} = 0.2588190451$

$\cos 42° = 0.7431448255$ $\cos\dfrac{\pi}{12} = 0.9659258263$

$\tan 42° = 0.9004040443$ $\tan\dfrac{\pi}{12} = 0.2679491924$

Notes on Exercises

Ex. 29–44 are follow-up problems for Examples 4 and 5.

Most calculators have function keys for $\sin t$, $\cos t$, and $\tan t$ but not for $\cot t$, $\sec t$, or $\csc t$. To use a calculator to find values for these latter three functions, use the following reciprocal trigonometric relationships. They can be verified directly from Definition 6.2 and are valid for all angles t. Consequently, these equations are called **identities**.

Reciprocal Identities

If θ is an acute angle, then

$$\csc t = \frac{1}{\sin t}, \qquad \sec t = \frac{1}{\cos t}, \quad \text{and} \quad \cot t = \frac{1}{\tan t}.$$

E X A M P L E   5   Finding Trig Values with a Calculator

Find the following:

a) cot 0.29   **b)** sec 0.29   **c)** csc 0.29

(Ensure the calculator is set in radian mode.)

Solution   Use the reciprocal identities and a scientific calculator in radian mode.

a) $\cot 0.29 = \dfrac{1}{\tan 0.29} = 3.35106284$ **b)** $\sec 0.29 = \dfrac{1}{\cos 0.29} = 1.043575676$

c) $\csc 0.29 = \dfrac{1}{\sin 0.29} = 3.497087668$ ▤

Properties of sin *t* and cos *t*

The following Exploration relates the definitions of sine and cosine and allows you to experience some of the properties of these functions.

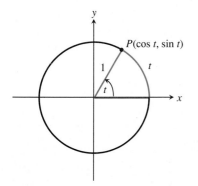

Figure 6.19   A unit circle.

🔍 **EXPLORE WITH A GRAPHING UTILITY**

Set a graphing calculator as follows:

- Radian mode,   parametric mode
- Range: $T\min = 0$, $T\max = 6.28$, $T\text{step} = 0.0314$
- Viewing rectangle: $[-1.5, 1.5]$ by $[-1, 1]$
- $X_1(T) = \cos T$ $Y_1(T) = \sin T$

1. Describe what you see when you press ⎡GRAPH⎤.

2. As you use the ⎡TRACE⎤ key, what is the meaning of the variables T, x, and y?

3. How does Fig. 6.19 relate to what you are seeing?

When you press the $\boxed{\text{TRACE}}$ key in this Exploration, the T variable represents the arc length from $(1, 0)$ counterclockwise around the circle to the cursor. The (x, y) coordinates of the cursor are $(\cos t, \sin t)$.

Experiment some more. Notice how the $\sin t$ and $\cos t$ functions vary as t varies.

Consider these observations and the resulting properties that follow:

1. Because the circle has a radius of 1, the coordinates of the cursor always fall between -1 and 1 inclusive. Therefore $\sin t$ and $\cos t$ are also both between -1 and 1 for all values of t. (See Fig. 6.19.)

2. If the cursor continues going around and around, an additional arc length of 2π is added with each revolution. The coordinates of $P(\cos t, \sin t)$ repeat their values.

Properties of sin _t_ and cos _t_

1. The domain of $f(t) = \sin t$ and of $g(t) = \cos t$ is the set of all real numbers.
2. For each real number t,

$$-1 \leq \sin t \leq 1 \qquad \text{and} \qquad -1 \leq \cos t \leq 1.$$

3. For each real number t and for each integer n,

$$\sin t = \sin(t \pm 2\pi n) \qquad \text{and} \qquad \cos t = \cos(t \pm 2\pi n).$$

Definition 6.2 and the Exploration explain why the trig functions are often called **circular functions.**

Property 3 above says that the values of the sine and cosine repeat each time the cursor goes around the circle. This is called the **periodic property of sine and cosine.** Later we will define the concept *periodic function*.

E X A M P L E   6   Using the Periodic Property

Find $f(t) = \sin t$ and $g(t) = \cos t$ for the following:

a) $t = 7\pi/3$

b) $t = 25\pi/6$

using the periodic properties of these functions:

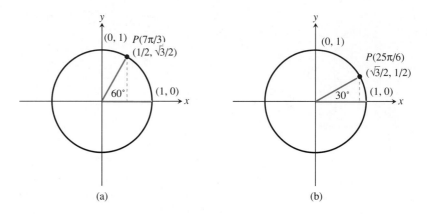

Figure 6.20 Using a unit circle definition in Example 4.

Solution (See Fig. 6.20.)

a)
$$\frac{7\pi}{3} = 2\pi + \frac{\pi}{3}$$

$$f\left(\frac{7\pi}{3}\right) = \sin\frac{7\pi}{3} = \sin\left(2\pi + \frac{\pi}{3}\right) = \sin\frac{\pi}{3} = \frac{\sqrt{3}}{2} \qquad \text{Use the periodic property of } \sin t.$$

$$g\left(\frac{7\pi}{3}\right) = \cos\frac{7\pi}{3} = \cos\left(2\pi + \frac{\pi}{3}\right) = \cos\frac{\pi}{3} = \frac{1}{2}$$

b)
$$\frac{25\pi}{6} = \frac{24\pi}{6} + \frac{\pi}{6} = 2\cdot 2\pi + \frac{\pi}{6}$$

$$f\left(\frac{25\pi}{6}\right) = \sin\frac{25\pi}{6} = \sin\left(2\cdot 2\pi + \frac{\pi}{6}\right) = \sin\frac{\pi}{6} = \frac{1}{2} \qquad \text{Use the periodic property of } \sin t.$$

$$g\left(\frac{25\pi}{6}\right) = \cos\frac{25\pi}{6} = \cos\left(2\cdot 2\pi + \frac{\pi}{6}\right) = \cos\frac{\pi}{6} = \frac{\sqrt{3}}{2}$$

The results of Example 6 can be confirmed using a grapher.

Fundamental Trigonometric Identities

Section 2.3 described the difference between a conditional equation and an identity. The equation $x + 3 = 5$ is a conditional equation since the equation is true only if x meets a certain condition. On the other hand, $x + 3 = 3 + x$ is an identity since the equation is true for all permissible values of x.

An equation that includes trigonometric functions and is true for all permissible values of the variable is called a **trigonometric identity**.

Bert says, "A graphing utility is an excellent tool for visually confirming a trigonometric identity. We can say that there is strong visual evidence that the problem in question is an identity, but we cannot say that the visual representation replaces an analytical proof."

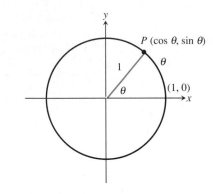

Figure 6.21 Unit circle.

(The reciprocal identities were introduced after Example 4 in this section.)

For any value t, the point $P(\cos t, \sin t)$ lies on the unit circle (see Fig. 6.21) and satisfies the equation $x^2 + y^2 = 1$, which defines the unit circle. That is,

$$x^2 + y^2 = 1 \quad \text{The defining equation for the unit circle}$$

$$(\cos t)^2 + (\sin t)^2 = 1 \quad \text{Substitute } x = \cos t \text{ and } y = \sin t.$$

$$\cos^2 t + \sin^2 t = 1.$$

There are 11 **fundamental trig identities**, summarized below. The last two Pythagorean identities can be verified by dividing both sides of the equation $\cos^2 \theta + \sin^2 \theta = 1$ by either $\sin^2 \theta$ or $\cos^2 \theta$.

REMINDER

To save writing parentheses, we agree that $(\sin t)^2 = \sin^2 t$ and $\sin(t^2) = \sin t^2$. In particular, $\sin^2 t \neq \sin t^2$.

Teaching Note

Trigonometric identities are introduced more formally here. Point out the utility of identities as equations that represent the interrelationships among the six functions. Student-generated proofs of identities are delayed until later sections.

Fundamental Trig Identities

Reciprocal Identities

$$\sin t = \frac{1}{\csc t} \qquad \cos t = \frac{1}{\sec t} \qquad \tan t = \frac{1}{\cot t}$$

$$\csc t = \frac{1}{\sin t} \qquad \sec t = \frac{1}{\cos t} \qquad \cot t = \frac{1}{\tan t}$$

Tangent and Cotangent Identities

$$\tan t = \frac{\sin t}{\cos t} \qquad \cot t = \frac{\cos t}{\sin t}$$

Pythagorean Identities

$$\sin^2 t + \cos^2 t = 1 \qquad 1 + \tan^2 t = \sec^2 t \qquad 1 + \cot^2 t = \csc^2 t$$

These fundamental trig identities are used to verify the equivalence of expressions involving trigonometric functions. We illustrate this practice in Example 7.

E X A M P L E 7   Using Trig Identities

Assignment Guide

Day 1: Ex. 3, 6, 9, 12, 14, 17, 20, 21, 24, 27, 29, 32, 35, 36, 38, 39, 42, 43
Day 2: Ex. 45, 48–54, 56, 57, 59–63

Show by an algebraic argument that $\cos^2 t - \sin^2 t = 1 - 2\sin^2 t$.

Solution   Subtracting $\sin^2 t$ from both sides of $\sin^2 t + \cos^2 t = 1$ results in the equation $\cos^2 t = 1 - \sin^2 t$. Therefore

$$\cos^2 t - \sin^2 t = (1 - \sin^2 t) - \sin^2 t \quad \text{Substitute } 1 - \sin^2 t \text{ for } \cos^2 t.$$

$$= 1 - 2\sin^2 t.$$

Exercises for Section 6.2

In Exercises 1–12, find the point $P(t) = P(x, y)$ on the unit circle that corresponds to each value of t.

1. $t = \dfrac{\pi}{2}$ **2.** $t = \dfrac{\pi}{4}$ **3.** $t = -\dfrac{\pi}{4}$

4. $t = \dfrac{5\pi}{6}$ **5.** $t = -\dfrac{3\pi}{2}$ **6.** $t = \dfrac{5\pi}{3}$

7. $t = -\dfrac{3\pi}{4}$ **8.** $t = \pi$ **9.** $t = \dfrac{11\pi}{6}$

10. $t = \dfrac{11\pi}{4}$ **11.** $t = \dfrac{8\pi}{3}$ **12.** $t = -\dfrac{5\pi}{6}$

In Exercises 13–20, let $f(t) = \sin t$ and $g(t) = \cos t$. Use the unit circle to find $f(t)$ and $g(t)$, where t equals the given values.

13. $t = \dfrac{\pi}{6}$ **14.** $t = \dfrac{\pi}{4}$ **15.** $t = \dfrac{\pi}{3}$

16. $t = \dfrac{2\pi}{3}$ **17.** $t = \dfrac{3\pi}{4}$ **18.** $t = \dfrac{5\pi}{6}$

19. $t = -\dfrac{2\pi}{3}$ **20.** $t = -\dfrac{\pi}{6}$

In Exercises 21–28, use the unit circle to find $\tan t$, $\cot t$, $\sec t$, and $\csc t$, where t equals the given values.

21. $t = \dfrac{\pi}{6}$ **22.** $t = \dfrac{\pi}{4}$ **23.** $t = \dfrac{\pi}{3}$

24. $t = \dfrac{2\pi}{3}$ **25.** $t = \dfrac{3\pi}{4}$ **26.** $t = \dfrac{5\pi}{6}$

27. $t = -\dfrac{2\pi}{3}$ **28.** $t = -\dfrac{\pi}{6}$

In Exercises 29–36, use a calculator to find the values of $\sin\theta$, $\cos\theta$, and $\tan\theta$ for each acute angle. In each case, decide whether to use the degree or radian mode.

29. $\theta = 21°$ **30.** $\theta = 49°$

31. $\theta = 1.23$ **32.** $\theta = 0.78$

33. $\theta = 82°$ **34.** $\theta = 19°$

35. $\theta = 0.27$ **36.** $\theta = 0.95$

In Exercises 37–44, use a calculator to find the values of $\cot\theta$, $\sec\theta$, and $\csc\theta$ for each acute angle. In each case, decide whether to use the degree or radian mode.

37. $\theta = 38°$ **38.** $\theta = 72°$

39. $\theta = 0.83$ **40.** $\theta = 0.12$

41. $\theta = 46°$ **42.** $\theta = 62°$

43. $\theta = 1.35$ **44.** $\theta = 1.03$

In Exercises 45–48, use the periodic properties to find $f(x) = \sin x$ and $g(x) = \cos x$.

45. $x = \dfrac{13\pi}{4}$ **46.** $x = \dfrac{13\pi}{6}$

47. $x = \dfrac{8\pi}{3}$ **48.** $x = \dfrac{23\pi}{6}$

49. Prove that $1 + \tan^2\theta = \sec^2\theta$ by dividing both sides of $\sin^2\theta + \cos^2\theta = 1$ by $\cos^2\theta$.

50. Prove that $1 + \cot^2 \theta = \csc^2 \theta$ by dividing both sides of $\sin^2 \theta + \cos^2 \theta = 1$ by $\sin^2 \theta$.

For Exercises 51–54, complete the following table by adding a $+$ or $-$ to record intervals on which $\sin t$, $\cos t$, and $\tan t$ are positive and negative:

	sin t	cos t	tan t
51. $0 < t < \dfrac{\pi}{2}$	$+$		
52. $\dfrac{\pi}{2} < t < \pi$	$-$		
53. $\pi < t < \dfrac{3\pi}{2}$			
54. $\dfrac{3\pi}{2} < t < 2\pi$			

In Exercises 55–60, find $\sin \theta$, $\cos \theta$, and $\tan \theta$ for each angle, without using a calculator when possible.

55. $\theta = \dfrac{7\pi}{3}$ **56.** $\theta = \dfrac{31\pi}{6}$ **57.** $\theta = \dfrac{5\pi}{12}$

58. $\theta = \dfrac{8\pi}{12}$ **59.** $\theta = \dfrac{11\pi}{24}$ **60.** $\theta = \dfrac{7\pi}{36}$

In Exercises 61–63 use the fundamental identities to prove the equation is valid.

61. $\cos^2 t - \sin^2 t = 2\cos^2 t - 1$

62. $\sec^2 t - \tan^2 t = 1$

63. $1 + \dfrac{\sin^2 t}{\cos^2 t} = \sec^2 t$

6.3 _____ Graphs of sin x and cos x

Objective

Students will be able to generate the graphs for the sine and cosine functions and explore various transformations upon these graphs.

Key Ideas

Graph of sine
Graph of cosine
Symmetry
Amplitude

Common Errors

Be sure the exploration activity is using the simultaneous graphing option.

🔍 EXPLORE WITH A GRAPHING UTILITY

Set a graphing calculator as follows:

- Radian mode,   parametric mode,   simultaneous mode
- Range: $T\min = 0$, $T\max = 6.28$, $T\text{step} = 0.1$
- Viewing rectangle: $[-2, 6.28]$ by $[-2.5, 2.5]$
- $X_1(T) = -1 + \cos T$ $Y_1(T) = \sin T$
  $X_2(T) = T$ $Y_2(T) = \sin T$

1. Describe what you see when you press $\boxed{\text{GRAPH}}$. How is this graph related to that in the Exploration in Section 6.2?

2. As you move the $\boxed{\text{TRACE}}$ cursor from one curve to the other by using the $\boxed{\text{UP}}$ or $\boxed{\text{DOWN}}$ key, how do the y-coordinates of the cursor compare?

This Exploration provides a dynamic simulation of $y = \sin x$. A textbook, however, can provide only static representations, as we provide in the next several figures.

Consider the arc of the unit circle from $(1, 0)$ to P whose arc length is x. The value $f(x) = \sin x$, then, is the y-coordinate of P. (See Fig. 6.22.)

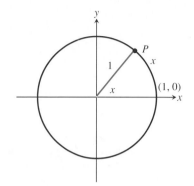

Figure 6.22 If an arc from $(1, 0)$ to P has length x, then the coordinates of P are $(\cos x, \sin x)$.

As x increases along the x-axis, the value of $f(x)$ is the y-coordinate of the corresponding point $P(x)$ on the unit circle (see Fig. 6.23a). As x varies from 0 to $\pi/2$, $\sin x$ varies from 0 to 1; as x varies from $\pi/2$ to π, $\sin x$ varies from 1 to 0 (see Fig. 6.23b).

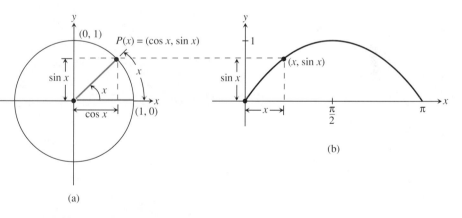

Figure 6.23 $f(x) = \sin x$ for $0 \le x \le \pi$: (a) unit circle definition; (b) graph of $f(x) = \sin x$.

In a similar fashion, we see that as x varies from 0 to $-\pi$, $\sin x$ varies from 0 to -1 and back to 0, as shown in Fig. 6.24. In particular,

$$f(-\pi) = 0, \quad f\left(-\frac{\pi}{2}\right) = -1, \quad f(0) = 0, \quad f\left(\frac{\pi}{2}\right) = 1, \quad \text{and} \quad f(\pi) = 0.$$

Frank says, "Parametric graphing allows students to explore trigonometry in a way that was not possible before. Seeing the unit circle unwrap on the graphing screen is a powerful visualization that must not be ignored."

Teaching Note

Discuss with students the connection between the opening exploration activity and Fig. 6.23. A central part of your discussion should be the relationship between the two parts of Fig. 6.23 (and the corresponding parts of the graphing utility image). Have students identify the arc length of the circle and discuss how it relates to thinking about length on the x-axis.

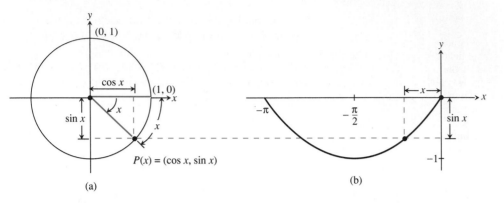

Figure 6.24 $f(x) = \sin x$ for $-\pi \le x \le 0$: (a) unit circle definition; (b) graph of $f(x)$.

The graphs in Figs. 6.23 and 6.24 are pieced together to form the graph of $f(x) = \sin x$ on the interval $[-\pi, \pi]$, shown in Fig. 6.25.

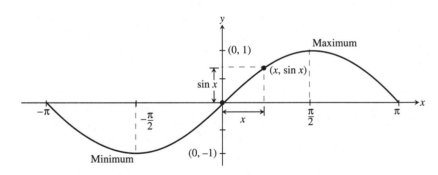

Figure 6.25 Graph of $y = \sin x$ for $-\pi \le x \le \pi$.

If $x > \pi$ or $x < -\pi$, the values that $\sin x$ has over the interval $[-\pi, \pi]$ repeat in what is called a **periodic function**.

Definition 6.3 Periodic Function

A function f is said to be a **periodic function** if there is a positive real number h such that $f(x + h) = f(x)$ for every value of x in the domain of f. The smallest such positive number h is called the **period of f**. The period corresponds to the length of one cycle on the graph of f.

One property of $\sin x$ (see Section 6.2) is that $\sin x = \sin(x \pm 2n\pi)$, which suggests that the period of $\sin x$ is 2π. The graph in Fig. 6.25 also suggests that the period of $y = \sin x$ is 2π. In addition, from the graph of $y = \sin x$ over the interval $[-3\pi, 3\pi]$ (see Fig. 6.26), it is evident that the graph begins repeating itself as it extends beyond the interval $[0, 2\pi]$, thus confirming that the period of $y = \sin x$ is 2π.

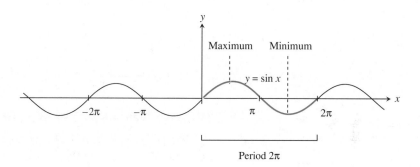

Figure 6.26 Complete graph of $f(x) = \sin x$.

As point P on the unit circle in Fig. 6.27 rotates, the x-coordinate of P goes through the same cycle of values as the y-coordinate does; they simply begin at a different point in the cycle. Confirm this by checking the values $\cos(-\pi)$, $\cos(-\pi/2)$, $\cos 0$, $\cos(\pi/2)$, and $\cos \pi$. The graph of $y = \cos x$ is shown in Fig. 6.28.

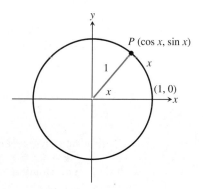

Figure 6.27 Unit circle definition.

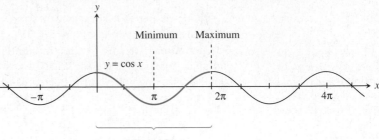

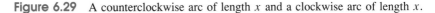

Figure 6.28 Complete graph of $y = \cos x$.

Symmetry of the Graphs of Sine and Cosine

Figure 6.29 shows that if (a, b) is the point on the unit circle associated with $P(x)$, then $(a, -b)$ is the point associated with $P(-x)$. This means that

$$\sin(-x) = -\sin x$$

for all real numbers x.

In other words, the graph of $f(x) = \sin x$ is symmetric about the origin.

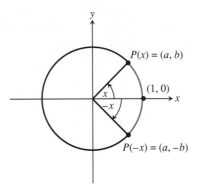

Figure 6.29 A counterclockwise arc of length x and a clockwise arc of length x.

In a similar manner, it can be shown that

$$\cos(-x) = \cos x$$

for all real numbers x.

In other words, the graph of $f(x) = \cos x$ is symmetric about the y-axis.

Amplitude of Sine and Cosine

A graphing utility can be used to discover properties of the trig functions experimentally.

🔎 EXPLORE WITH A GRAPHING UTILITY

Find a complete graph of each of the following functions in the same viewing rectangle:

- $y = 3 \sin x$
- $y = \sin x$
- $y = 2 \sin x$

1. How are these graphs the same? How do they differ?

2. What are the maximum and minimum values for each function?

Generalize What are the maximum and minimum values of the function $y = a \sin x$, where a is any positive real number?

This Exploration leads us to the following definition.

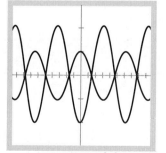

[−10, 10] by [−3, 3]

Figure 6.30 $y = -2 \cos x$ and $y = \cos x$.

Definition 6.4 Amplitude of Sine and Cosine

The **amplitude** of functions $f(x) = a \sin x$ and $g(x) = a \cos x$ is the maximum value of y, where a is any real number; **amplitude** $= |a|$.

E X A M P L E   1   Finding the Amplitude of a Cosine Function

Find a complete graph of $y = -2 \cos x$. Also find its amplitude and describe how the negative factor affects the graph.

Solution   (See Fig. 6.30.)
Amplitude: $|-2| = 2$.

   We have seen in Chapter 1 that the graph of $y = -f(x)$ is obtained by reflecting the graph of $y = f(x)$ about the x-axis. Therefore the graph of $y = -2 \cos x$ is obtained from the graph of $y = 2 \cos x$ by reflection about the x-axis.

≡

Teaching Note

Students should be expected to apply previously examined transformations to the graphs of sine and cosine. Give attention to translation, reflection, stretching, and shrinking. A brief review on how $f(x) = x^2$ is related to $f(x) = ax^2$ and to $f(x) = (x - h)^2$ and relating this discussion to transformations applied to trigonometric functions will be valuable.

Horizontal Stretch or Shrink for Sine and Cosine

The period for $y = \sin x$ and for $y = \cos x$ is 2π. Now consider the functions $y = \sin bx$ and $y = \cos bx$, where b is a positive constant coefficient of x. As in the graphs of earlier functions, b horizontally stretches or shrinks the graph, depending on the value of b. When b is a positive integer, the period will be some fraction of 2π.

🔍 EXPLORE WITH A GRAPHING UTILITY

Find a graph of each of the following functions in the viewing rectangle $[-3.14, 3.14]$ by $[-2, 2]$:

- $y = \sin 2x$
- $y = \sin x$
- $y = \sin 3x$
- $y = \sin 0.5x$

How many complete periods occur in the given viewing rectangle for each function?

Generalize   What is the period of $y = \sin bx$?

This Exploration might have led you to determine the effect of a constant on the period of a sine function. It also is possible to determine this relationship in a more algebraic way, as illustrated in the next example.

E X A M P L E   2   Finding the Period of $y = \cos 5x$

Find the period of the function $y = \cos 5x$.

Solution

Graphical Method   Observe in Fig. 6.31 that $y = \cos 5x$ completes five full cycles over the interval $[-\pi, \pi]$, which means the period is $2\pi \div 5 = 2\pi/5$.

Algebraic Method   The function $y = \cos t$ completes one full cycle as t varies from 0 to 2π. Likewise, as $5x$ varies from 0 to 2π, $y = \cos 5x$ completes one cycle.

$$\text{Cycle begins: } 5x = 0 \quad\quad \text{or} \quad\quad x = 0$$

$$\text{Cycle ends: } 5x = 2\pi \quad\quad \text{or} \quad\quad x = 2\pi/5 \quad\quad ▣$$

[−3.14, 3.14] by [−3, 3]

Figure 6.31 $y = \cos 5x$.

This example can be generalized to any functions $y = \sin bx$ and $y = \cos bx$, as summarized next.

Theorem 6.2   Period of Sine and Cosine Functions

The period for $y = \sin bx$ or $y = \cos bx$ is $2\pi/|b|$.

Proof   If the period for $y = \sin bx$ or $y = \cos bx$ begins at $x = 0$, then it ends at $bx = 2\pi$, or $x = 2\pi/b$; therefore the period, or length of one cycle, is $|2\pi/b| = 2\pi/|b|$ for each function.   ≡

E X A M P L E   3   Using a Graphing Utility

Find a complete graph of $f(x) = 4\sin 3x$ in the viewing rectangle $[-3.14, 3.14]$ by $[-5, 5]$. Find the domain, range, amplitude, and period of $f(x)$.

Solution

Graphical Method   (See Fig. 6.32.) Using the $\boxed{\text{TRACE}}$ key, observe that the maximum value of f is shown on the screen display as 3.99 and the minimum value as -3.99. The graph supports the conclusion that the amplitude is 4 and the period is $2\pi/3$. This claim is confirmed by Definition 6.6 and Theorem 6.2.
Domain of f: $(-\infty, \infty)$.
Range of f: $[-4, 4]$.   ≡

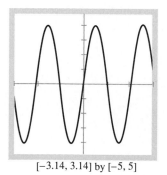

[−3.14, 3.14] by [−5, 5]

Figure 6.32 $f(x) = 4\sin 3x$.

Horizontal Shift Identities

Figure 6.33 shows graphically that $f(x) = \cos x$ (dashed line) can be obtained from $g(x) = \sin x$ by a horizontal shift left $\pi/2$ units. In other words, it appears from the graphs that $f(x) = g(x + \pi/2)$.

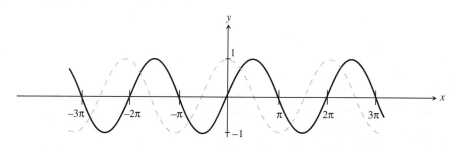

Figure 6.33 $f(x) = \cos x$ and $g(x) = \sin x$.

The following horizontal shift identities can be proved.

> ### Theorem 6.3   Horizontal Shift Identities
>
> For all values of x,
>
> $$\cos x = \sin\left(x + \frac{\pi}{2}\right) \qquad \text{and} \qquad \sin x = \cos\left(x - \frac{\pi}{2}\right)$$

Proof   We shall prove the first of these two identities. (See Fig. 6.34.)

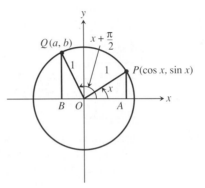

Figure 6.34   Angles x and $x + \pi/2$ in standard position.

$P(\cos x, \sin x)$ is the point associated with angle x, and $Q(a, b)$ is the point associated with angle $x + \pi/2$. Because $\triangle OAP$ is congruent with $\triangle QBO$, we see that $QB = OA$. This means that $b = \cos x$. But b, the y-coordinate of Q, is equal to $\sin(x + \pi/2)$, that is,

$$\cos x = \sin\left(x + \frac{\pi}{2}\right).$$

The graph of $f(x + \pi/2)$ is obtained from the graph of $f(x)$ by a horizontal shift left $\pi/2$ units.   ▆

Section 7.4 will develop what it means to verify a trigonometric identity. In the meantime, we will use only identities.

E X A M P L E   4   Using an Identity to Verify a Trig Equation

Verify that $\sin x = \cos(x - \pi/2)$.

Solution   Let $x = u + \pi/2$. Then

$$\sin x = \sin\left(u + \frac{\pi}{2}\right)$$

$$= \cos u \qquad \text{Use the first identity of Theorem 6.3.}$$

$$= \cos\left(x - \frac{\pi}{2}\right). \qquad \text{Substitute } (x - \pi/2) \text{ for } u. \qquad ▰$$

Solving Trig Equations

Acute-angle solutions to trig equations like $\sin x = 0.8$ can be solved by evaluating $x = \sin^{-1}(0.8)$, a technique that we will use in later sections. A graphical method can be used that is identical to that used with functions in previous chapters.

EXAMPLE 5   Solving a Trig Equation

Find both solutions of $\sin x = 0.4$ in the interval $0 \le x < 2\pi$.

Solution

Graphical Method   Find a complete graph of $f(x) = \sin x$ and $g(x) = 0.4$ and use zoom-in to find the solutions $x = 0.41$ and $x = 2.73$ (see Fig. 6.35). ▰

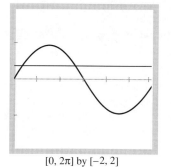

$[0, 2\pi]$ by $[-2, 2]$

Figure 6.35 $f(x) = \sin x$ and $g(x) = 0.4$.

A calculator computation can be used to support the solutions in Example 5. For example, $x = \sin^{-1}(0.4) = 0.4115168461$. The other solution is $\pi - 0.4115168461 = 2.730075808$.

EXAMPLE 6   Finding Period and Amplitude

Find a complete graph of $f(x) = 3\sin(x - 1)$. Determine the domain, range, period, and amplitude.

Solution   Compare the graph of f to the graph of $g(x) = 3\sin x$.

$$g(x - 1) = 3\sin(x - 1)$$

From this, we see that the graph of f is obtained from the graph of g by a horizontal shift right 1 unit (see Fig. 6.36).

Domain: $(-\infty, \infty)$.
Range: $[-3, 3]$.
Amplitude: 3.
Period: 2π. ▰

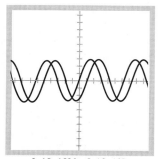

$[-10, 10]$ by $[-10, 10]$

Figure 6.36 $f(x) = 3\sin(x-1)$ is shifted 1 unit right of $g(x) = 3\sin x$, which passes through the origin.

Exercises for Section 6.3

In Exercises 1–6, find a complete graph, the maximum value and the amplitude of each function.

1. $y = 4 \sin x$

2. $y = \cos 3x$

3. $y = 15 \cos 2x$

4. $y = -3 \sin x$

5. $y = -5 \cos x$

6. $y = -12 \sin 2x$

In Exercises 7–12, find a complete graph of each function. Also find the domain, range, and period of each.

7. $y = \cos 3x$

8. $y = \sin 7x$

9. $y = 4 \sin 5x$

10. $y = -2 \cos 9x$

11. $y = -3 \cos 2x$

12. $y = 6 \sin 9x$

In Exercises 13–20, sketch a complete graph of each function without using a graphing utility. State the domain, range, and period. Check with a grapher.

13. $y = \sin 3x$

14. $y = 2 + \sin x$

15. $y = 2 + \sin(x - 4)$

16. $y = -1 - \cos(x - \pi)$

17. $y = -1 + 3 \cos(x - 2)$

18. $y = 2 - 2 \cos(x + 3)$

19. $y = -1 + 3 \sin(x - 2)$

20. $y = 2 - 2 \sin(x + 1)$

In Exercises 21–26, find a viewing rectangle for each function that displays exactly one period of the function.

21. $y = \sin 2x$

22. $y = \cos \frac{1}{2}x$

23. $y = 2 \cos \frac{1}{2}x$

24. $y = \sin \frac{1}{3}x$

25. $y = 3 \cos \frac{1}{5}x$

26. $y = 3 \sin \frac{1}{4}x$

In Exercises 27–34, solve for $0 \le x < 2\pi$. Use both a graphical and a calculator method and confirm that the two methods agree.

27. $\sin x = 0.5$

28. $\sin x = -0.6$

29. $\sin x = 0.8$

30. $\sin x = 0.3$

31. $\cos x = 0.6$

32. $\cos x = -0.4$

33. $3 \sin x = 1.7$

34. $2 \cos \frac{1}{2}x = 1.5$

In Exercises 35–38, solve for $0 \le x < 2\pi$ using a graphical method.

35. $\sin x = \cos x$

36. $\sin \frac{1}{2}x = \cos x$

37. $3 \cos \frac{1}{2}x = 2 \sin x$

38. $\sin x = -\cos x$

In Exercises 39–42, solve the inequalities for $0 \le x < 2\pi$.

39. $\sin x < 0.6$

40. $3x \cos x < 4$

41. $2 \cos x > -1.3$

42. $\sin x > 0.05$

43. Use identities from this section to verify that $\tan(-x) = -\tan x$.

44. Is applying a vertical stretch factor of 2 to the graph of $y = \sin x$ followed by a vertical shift up 3 units the same or different as applying a vertical shift up 3 units to the graph of $y = \sin x$ followed by a vertical stretch factor of 2?

45. Determine the two equations of the resulting graphs from Exercise 44. Confirm that they are different graphs.

46. A signal buoy bobs up and down so that at time t (in seconds), it is $\sin t$ feet above the average water level. A bell rings whenever the buoy is 0.5 ft above the average water level. What are the time intervals between bell rings in a sequence of six rings?

47. A ferris wheel 50 ft in diameter makes one revolution every 2 min. If the center of the wheel is 30 ft above the ground, how long after reaching the low point is a rider 50 ft above the ground?

48. Writing to Learn In a certain video game, a cursor bounces back and forth across the screen at a constant rate. Its distance d from the center of the screen varies with time t and hence can be described as a function of t. Write several sentences and draw a graph to explain that this horizontal distance d from the center of the screen *does not* vary according to an equation $d = a \sin bt$, where t represents seconds.

6.4 Graphs of the Other Trig Functions

In this section, we will develop the graphs of the functions $y = \tan x$, $y = \cot x$, $y = \sec x$, and $y = \csc x$.

Objective

Students will be able to generate the graphs for the tangent, cotangent, secant, and cosecant functions and to explore various transformations upon their graphs.

Key Ideas

Graph of tangent
Graph of cotangent
Graph of secant
Graph of cosecant

Common Errors

Remind students to pay close attention to the radian and degree mode settings on their calculators. They should always check to see if the calculator mode is set properly.

Graphs of Tangent and Cotangent Functions

Exercise 43 of Section 6.3 established that $\tan(-x) = -\tan x$. This means that the graph of $y = \tan x$ is symmetric about the origin.

Recall from Section 6.2 the tangent identity

$$\tan x = \frac{\sin x}{\cos x}.$$

This identity allows us to make the following observations:

1. $\tan x = 0$ whenever $\sin x = 0$. That is,

$$\tan x = 0 \iff x = \ldots, -2\pi, -\pi, 0, \pi, 2\pi, \ldots.$$

2. $\tan x$ is undefined whenever $\cos x = 0$. Consequently, the graph of $y = \tan x$ is expected to have a vertical asymptote whenever

$$x = \ldots, -\frac{3\pi}{2}, -\frac{\pi}{2}, \frac{\pi}{2}, \frac{3\pi}{2}, \ldots.$$

3. When x is very close to $\pi/2$ on the left, $\sin x$ is close to 1 and $\cos x$ is close to zero. *Therefore the quotient $\tan x$ is a large positive number.* Similarly, if x is very close to $\pi/2$ from the right, $\tan x$ is a large negative number. Figure 6.37(a) shows the graph of $y = \tan x$.

The identity $\cot x = 1/\tan x$ contributes to our understanding of the graph of $y = \cot x$. In particular,

$$\tan x = 0 \implies \cot x \text{ is undefined}$$

$$\tan x \to \pm\infty \implies \cot x \to 0.$$

So we see that there are vertical asymptotes whenever $\tan x = 0$. (See Fig. 6.37b.)

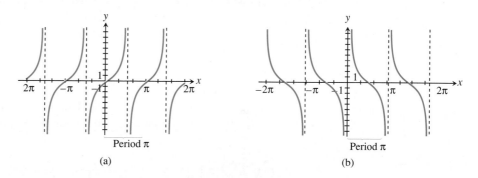

(a)   (b)

Figure 6.37 (a) $y = \tan x$. (Note that $\tan x \to \infty$ as $x \to (\pi/2)^-$ and $\tan x \to -\infty$ as $x \to (\pi/2)^+$); (b) $y = \cot x$. (Note that $\cot x \to \infty$ as $x \to \pi^+$, and $\cot x \to -\infty$ as $x \to \pi^-$.)

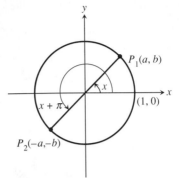

Figure 6.38 Unit circle showing the angles x and $x + \pi$.

Period of Tangent and Cotangent

The graph in Fig. 6.37 suggests that the period of $y = \tan x$ is π. In other words, $\tan(x + \pi) = \tan x$ for all values of x for which $\tan x$ is defined. To verify this equation analytically, consider Fig. 6.38. Here, P_1 is the point on the unit circle associated with angle x and P_2 is the point associated with angle $x + \pi$. Therefore

$$\tan x = \frac{b}{a} \quad \text{and} \quad \tan(x + \pi) = \frac{-b}{-a} = \frac{b}{a}.$$

A similar argument yields an identity for $y = \cot x$.

Identities Showing Period of Tangent and Cotangent

For all values of x,

$$\tan x = \tan(x + \pi) \quad \text{and} \quad \cot x = \cot(x + \pi).$$

We see from Fig. 6.37 that the periods of $y = \tan x$ and $y = \cot x$ are π. If a is a positive constant coefficient of x, a affects the period of the tangent function just as it does the period of $y = \sin ax$, that is, by horizontally stretching or shrinking the graph. It is helpful to recognize that one period in any tangent or cotangent function represents the width of the graph between adjacent vertical asymptotes.

E X A M P L E   1   Finding the Period of a Tangent Function

Find the period of $y = \tan 2x$. Also find a complete graph in a viewing rectangle that shows three complete periods.

Solution   It is convenient to consider a period of the tangent function from $-\pi/2$ to $\pi/2$.

One period begins:	$2x = -\pi/2$	or	$x = -\pi/4$
One period ends:	$2x = \pi/2$	or	$x = \pi/4$

The period of $y = \tan 2x$ is $\pi/2$. Use the viewing rectangle $[-3\pi/4, 3\pi/4]$ by $[-10, 10]$ to find a complete graph with three complete periods. (See Fig. 6.39.)

≡

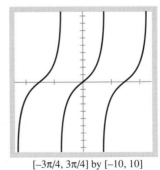

$[-3\pi/4, 3\pi/4]$ by $[-10, 10]$

Figure 6.39 $y = \tan 2x$.

E X A M P L E   2   Comparing Graphs of Tangent Functions

Identify which graph in Fig. 6.40 is the graph of $y = 2\tan x$ and which is the graph of $y = 5\tan x$.

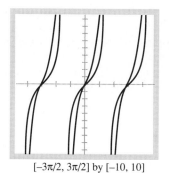

[−3π/2, 3π/2] by [−10, 10]

Figure 6.40 $y = 2\tan x$ and $y = 5\tan x$.

Solution As a grapher traces out these graphs, it will be obvious which is which. The graph of $y = 2\tan x$ is obtained from $y = \tan x$ by stretching it vertically by a factor of 2, and the graph of $y = 5\tan x$ is obtained from $y = \tan x$ by stretching it vertically by a factor of 5.

Graphs of Secant and Cosecant Functions

The reciprocal identities

$$\sec x = \frac{1}{\cos x} \quad \text{and} \quad \csc x = \frac{1}{\sin x}$$

contribute to our understanding of the graphs of $y = \sec x$ and $y = \csc x$. Because $y = \sin x$ and $y = \cos x$ are periodic functions, it follows that $y = \sec x$ and $y = \csc x$ are also. When $\cos x$ is nearly zero, $|\sec x|$ is large; in particular, the graph of $y = \sec x$ has a vertical asymptote whenever $\cos x = 0$. When $\cos x = 1$, then $\sec x$ is also 1. (See Fig. 6.41a.)

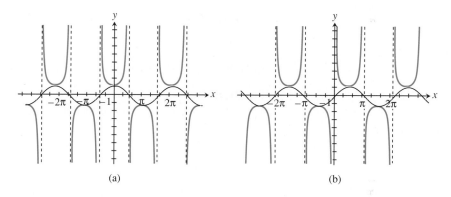

(a)   (b)

Figure 6.41 (a) $y = \sec x$ (color) and $y = \cos x$; (b) $y = \csc x$ (color) and $y = \sin x$.

EXAMPLE 3   Exploring the Secant Function

Find the domain, range, period, and vertical asymptotes of the function $f(x) = \sec x$.

Solution The following information can be found in Fig. 6.41(a):
Domain: All x except $x = \pi/2 + k\pi$ for any integer k.
Range: $(-\infty, -1] \cup [1, \infty)$.
Period: 2π.
Vertical asymptotes: $x = \pi/2 + k\pi$ for any integer k.

The graph of $y = \csc x$ looks like the graph of $y = \sec x$ except that it has been shifted $\pi/2$ units right. (See Fig. 6.41b.)

E X A M P L E   4   Exploring the Cosecant Function

Find the domain, range, period, and vertical asymptotes of the function $f(x) = \csc x$.

Solution   The following information can be found in Fig. 6.41(b):
Domain: All x except $x = k\pi$ for any integer k.
Range: $(-\infty, -1] \cup [1, \infty)$.
Period: 2π.
Vertical asymptotes: $x = k\pi$ for any integer k.

Finding the Period of a Function

Because the four trig functions $y = \tan x$, $y = \cot x$, $y = \sec x$, and $y = \csc x$ can be described in terms of $\sin x$ and $\cos x$, it's not surprising to find that the periods of these four functions are related also to the periods of $\sin x$ and $\cos x$.

E X A M P L E   5   Finding the Period of a Cosecant Function

Find the period of the function $y = \csc 3x$.

Solution   It is convenient to consider a period of $y = \csc x$ that begins at the vertical asymptote $x = -\pi$ and ends at $x = \pi$. In a similar fashion, analyze $y = \csc 3x$.

$$\text{One period begins:} \qquad 3x = -\pi \qquad \text{or} \qquad x = -\pi/3$$

$$\text{One period ends:} \qquad 3x = \pi \qquad \text{or} \qquad x = \pi/3$$

Therefore the period of $y = \csc 3x$ is $\pi/3 + \pi/3 = 2\pi/3$.

Notes on Exercises

Ex. 29–54 may be solved numerically, graphically, or even algebraically. The algebraic approach will be covered in a subsequent chapter. Help students make the connection between the different methods of solution.

Assignment Guide

Day 1: Ex. 1–6, 8, 9, 11–13, 16, 17, 20, 21
Day 2: Ex. 24, 25, 28, 29, 32, 33, 36, 37, 40–44, 46, 47, 50, 51, 54

> **Theorem 6.4   Period of Tangent, Cotangent, Secant, and Cosecant Functions**
>
> For $y = \tan bx$ or $y = \cot bx$,
>
> $$\text{period} = \frac{\pi}{|b|}.$$
>
> For $y = \sec bx$ or $y = \csc bx$,
>
> $$\text{period} = \frac{2\pi}{|b|}.$$

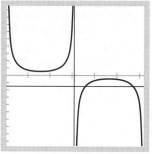

[0, 2π] by [−10, 10]

Figure 6.42 $y = \csc x$ and $y = -1.6$.

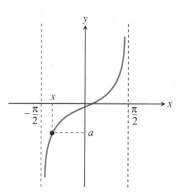

Figure 6.43 $y = \tan x$. $\tan x = a$ has a unique solution over $(-\pi/2, \pi/2)$.

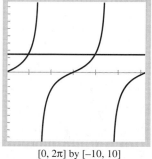

[0, 2π] by [−10, 10]

Figure 6.44 $y = \tan x$ and $y = 2.5$.

Solving Equations with Trig Functions

An equation $\sin x = a$ or $\cos x = a$ (where a is a constant) has either two solutions or no solutions over any interval of length 2π, since the period of each function is 2π. The same is true for the equations $\sec x = a$ and $\csc x = a$. To find all solutions to an equation of any one of these four types, first find the two solutions over the interval of length 2π. Then add $2k\pi$ to each value for all integer values of k.

EXAMPLE 6 Solving csc x = a Graphically

Find all solutions to $\csc x = -1.6$.

Solution Find a complete graph of $y = \csc x$ and $y = -1.6$ in the viewing rectangle $[0, 2\pi]$ by $[-10, 10]$ (see Fig. 6.42).

Use zoom-in to find the two solutions $x = 3.82$ and 5.61 in the interval $[0, 2\pi]$. Now add $2k\pi$ to each of these values for each integer value of k. Therefore

$$x = 3.82 + 2k\pi \qquad \text{and} \qquad x = 5.61 + 2k\pi$$

for all integer values of k. ≡

An equation of the form $\tan x = a$ has a unique solution over the interval $(-\pi/2, \pi/2)$, as indicated in Fig. 6.43. After finding that unique solution, we can find all other solutions of the equation by adding $k\pi$ for all integer values of k.

EXAMPLE 7 Solving tan x = a Graphically

Find all solutions to $\tan x = 2.5$.

Solution Find a complete graph of $y = \tan x$ and $y = 2.5$ in the same viewing rectangle, as shown in Fig. 6.44. Use zoom-in to find the solution $x = 1.19$. Therefore

$$x = 1.19 + k\pi$$

for all integer values of k. ≡

Exercises for Section 6.4

1. Use a graphing utility to draw the graph of $y = \tan x$ in each of the following viewing rectangles. (Use $\pi = 3.14159$.) What generalization do you observe?

a) $\left[-\dfrac{\pi}{2}, \dfrac{\pi}{2}\right]$ by $[-10, 10]$

b) $\left[\dfrac{7\pi}{2}, \dfrac{9\pi}{2}\right]$ by $[-10, 10]$

c) $\left[-\dfrac{11\pi}{2}, -\dfrac{9\pi}{2}\right]$ by $[-10, 10]$

d) $\left[\dfrac{71\pi}{2}, \dfrac{73\pi}{2}\right]$ by $[-10, 10]$

2. The following figure shows one period of the graphs of both $y = \csc x$ and $y = 2\csc x$. Identify each function.

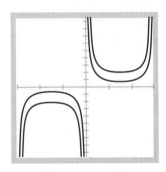

For Exercise 2.

3. The following figure shows two periods of the graphs of both $y = 0.5\tan x$ and $y = 5\tan x$. Identify each function.

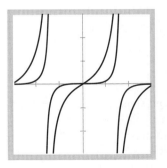

For Exercise 3.

4. Identify three viewing rectangles that illustrate that portions of the graph of $y = \cot x$ are identical to each other, as illustrated in Exercise 1 for $y = \tan x$.

5. Identify three viewing rectangles that illustrate that portions of the graph of $y = \sec x$ are identical to each other, as illustrated in Exercise 1 for $y = \tan x$.

6. The following figure shows graphs of two trig functions. Which of these pairs correctly identifies them?

a) $y = \tan x$ and $y = \sec x$

b) $y = \cot x$ and $y = \csc x$

c) $y = \tan x$ and $y = \csc x$

d) $y = \cot x$ and $y = \sec x$

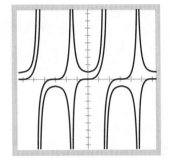

For Exercise 6.

In Exercises 7–12, find for each function its period and a complete graph in a viewing rectangle that shows exactly three periods.

7. $y = \tan 2x$ **8.** $y = \cot 3x$

9. $y = 2\sec 2x$ **10.** $y = \csc \dfrac{x}{2}$

11. $y = \sec 3x$ **12.** $y = 3\tan \dfrac{x}{2}$

In Exercises 13–28, sketch a graph of each function. Determine the domain, range, period, and asymptotes (if any).

13. $y = 3\tan x$ **14.** $y = -\tan x$

15. $y = \dfrac{1}{2}\sec x$ **16.** $y = \sec(-x)$

17. $y = 3\csc x$ **18.** $y = 2\tan x$

19. $y = -3\tan \dfrac{1}{2}x$ **20.** $y = 2\cot \dfrac{1}{2}x$

21. $y = 2\csc x$ **22.** $y = -2\sec \dfrac{1}{2}x$

23. $y = 2 \tan 3x$

24. $y = \sec\left(-\frac{1}{2}x\right)$

25. $y = -\tan\frac{\pi}{2}x$

26. $y = 2 \tan \pi x$

27. $y = 3 \sec 2x$

28. $y = 4 \csc\frac{1}{3}x$

In Exercises 29–32, solve each equation for $-2\pi \le x \le 2\pi$.

29. $\tan x = 3.25$

30. $\cot x = -5.6$

31. $\sec x = 1.2$

32. $-\csc x = 3.1$

In Exercises 33–36, solve each equation for $-1 \le x \le 1$.

33. $\tan x = \cos x$

34. $\sin x = \cos x$

35. $\cot x = \sin x$

36. $\tan x = \cos 2x$

In Exercises 37–40, solve each inequality for $-\pi/2 < x < \pi/2$.

37. $5 \tan x > \sec x$

38. $\tan x < \cot x$

39. $\tan x \ge \cos x$

40. $\sec x < \csc x$

Exercises 41–44 concern the equation $\sin x = 0.75$. All real solutions are given by $x + 2k\pi$, $k = 0, \pm1, \pm2, \ldots$, where x is 0.85 or 2.29. Use zoom-in to confirm this statement for each value of k.

41. $k = 0$

42. $k = 2$

43. $k = -3$

44. $k = 5$

In Exercises 45–54, find all real solutions.

45. $\sin x = 0.25$

46. $\tan x = 4$

47. $\cos x = 0.42$

48. $\sec x = 3$

49. $\sin x < 0.15$

50. $\tan x > 2$

51. $\sin 3x = 0.55$

52. $\cos 2x = 0.85$

53. $\csc\frac{1}{2}x \le 4$

54. $3 \sin\frac{1}{2}x > 2 \cos x$

6.5   Trigonometric Functions of an Acute Angle

The Greeks developed trigonometry over 2000 years ago to measure angles and sides of triangles, particularly for land measurement. In fact, the word *trigonometry* derives from two Greek words that mean "triangle measurement." Because the Greeks based their calculations on the lengths of sides of a right triangle, the historical development of trigonometry is based on the right triangle. The more modern development, based on the unit circle, was introduced in Section 6.2.

We begin with some terminology regarding right triangles.

For triangle ABC, denoted $\triangle ABC$, in Fig. 6.45, side a is opposite $\angle A$, side b is opposite $\angle B$, and side c is opposite $\angle C$. The side c opposite the right angle is the **hypotenuse,** and a and b are the **legs.**

Side a is called the **side opposite** $\angle A$ and b is called the **side adjacent to** $\angle A$. Likewise, side b is opposite $\angle B$ and side a is adjacent to $\angle B$.

Historically, the following definitions of the six trigonometric functions of an angle θ were the definitions used to introduce the trig functions. These definitions are in terms of the lengths of the sides of a right triangle with an acute angle θ. It can be shown that these definitions agree with the unit circle definition found in Definition 6.2.

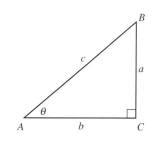

Figure 6.45 In a right triangle, one angle (here $\angle C$) is a right angle and angles A and B are acute.

Definition 6.5 Trigonometric Functions

The six trigonometric functions of any angle $0° < \theta < 90°$ are defined as follows. (See Fig. 6.45.) The abbreviations opp, adj, and hyp represent the lengths of the sides opposite and adjacent to angle θ and the length of the hypotenuse.

$$\text{sine } \theta = \frac{a}{c} = \frac{\text{opp}}{\text{hyp}} \qquad \text{cosecant } \theta = \frac{c}{a} = \frac{\text{hyp}}{\text{opp}}$$

$$\text{cosine } \theta = \frac{b}{c} = \frac{\text{adj}}{\text{hyp}} \qquad \text{secant } \theta = \frac{c}{b} = \frac{\text{hyp}}{\text{adj}}$$

$$\text{tangent } \theta = \frac{a}{b} = \frac{\text{opp}}{\text{adj}} \qquad \text{cotangent } \theta = \frac{b}{a} = \frac{\text{adj}}{\text{opp}}$$

For the trig functions to be well defined, their values for a given acute angle θ must be the same no matter which right triangle with acute angle θ is used in the definition.

Recall from your study of geometry that **similar triangles** are triangles whose corresponding angles are **congruent**. Figure 6.46 shows that any two right triangles with the same acute angle θ are similar. (Why?) Consequently, corresponding ratios of the two triangles are equal. From this we see that the trig functions have the same value no matter what right triangle is used with $\angle\theta$.

Because the angles $30°, 45°$, and $60°$ occur frequently, it is convenient to know their trig values without having to use a calculator. Examples 1 and 2 give you these values, which you can compare with Fig. 6.16.

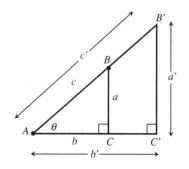

Figure 6.46 $\triangle ABC$ and $\triangle AB'C'$ are similar, so ratios of corresponding sides are equal.

EXAMPLE 1   Finding Trig Values of 30° and 60° Angles

Find the values of the trig functions at $30°$ and $60°$.

Solution In the equilateral triangle ABC shown in Fig. 6.47, altitude BD is the angle bisector of $\angle B$ and also is the perpendicular bisector of AC. It follows that $\triangle ABD$ has angles of $30°$, $60°$, and $90°$, so if $AB = 2$, then $AD = 1$ and $BD = \sqrt{3}$. Apply the following definitions of the trig functions to the triangles in Fig. 6.47:

$$\sin 30° = \frac{1}{2} \qquad \csc 30° = \frac{2}{1} \qquad \sin 60° = \frac{\sqrt{3}}{2} \qquad \csc 60° = \frac{2}{\sqrt{3}}$$

$$\cos 30° = \frac{\sqrt{3}}{2} \qquad \sec 30° = \frac{2}{\sqrt{3}} \qquad \cos 60° = \frac{1}{2} \qquad \sec 60° = \frac{2}{1}$$

$$\tan 30° = \frac{1}{\sqrt{3}} \qquad \cot 30° = \frac{\sqrt{3}}{1} \qquad \tan 60° = \frac{\sqrt{3}}{1} \qquad \cot 60° = \frac{1}{\sqrt{3}}$$

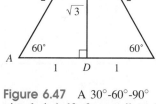

Figure 6.47 A $30°$-$60°$-$90°$ triangle is half of an equilateral triangle.

E X A M P L E   2   Finding Trig Values of a 45° Angle

Find the values of the trig functions at $45°$.

Solution Consider an isosceles right triangle with a $45°$ angle and legs of length 1 (see Fig. 6.48).

$$\sin 45° = \frac{1}{\sqrt{2}} = \frac{\sqrt{2}}{2} \qquad \csc 45° = \frac{\sqrt{2}}{1} = \sqrt{2}$$

$$\cos 45° = \frac{1}{\sqrt{2}} = \frac{\sqrt{2}}{2} \qquad \sec 45° = \frac{\sqrt{2}}{1} = \sqrt{2}$$

$$\tan 45° = \frac{1}{1} = 1 \qquad \cot 45° = \frac{1}{1} = 1$$

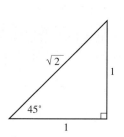

Figure 6.48 A right triangle with a $45°$ angle.

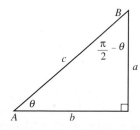

Figure 6.49 The two acute angles in a right triangle are complementary.

Sine and cosine are called **cofunctions,** as are tangent and cotangent, and secant and cosecant. Basic identities between each of the cofunctions are summarized next.

Notice that in Fig. 6.49, angles A and B are complementary.

$$\sin \theta = \sin A = \frac{a}{c}$$

$$\cos \left(\frac{\pi}{2} - \theta \right) = \cos B = \frac{a}{c}$$

Therefore $\sin \theta = \cos(\pi/2 - \theta)$. The identities for the other cofunctions can be established in a similar fashion.

Cofunctions of Complementary Angles

If θ is any acute angle, a trig function value of θ is equal to the cofunction of the complement of θ, as follows:

$$\sin \theta = \cos \left(\frac{\pi}{2} - \theta\right) \qquad \cot \theta = \tan \left(\frac{\pi}{2} - \theta\right)$$

$$\cos \theta = \sin \left(\frac{\pi}{2} - \theta\right) \qquad \sec \theta = \csc \left(\frac{\pi}{2} - \theta\right)$$

$$\tan \theta = \cot \left(\frac{\pi}{2} - \theta\right) \qquad \csc \theta = \sec \left(\frac{\pi}{2} - \theta\right)$$

If you know the value of one trig function for an acute angle θ, you can find the values for the other five functions, as illustrated in Example 3.

E X A M P L E   3   Finding Trig Values

Let θ be an acute angle such that $\sin \theta = \frac{5}{6}$. Find the values of all the trig functions of θ.

Solution   Using the information $\sin \theta = \frac{5}{6}$, label a right triangle as shown in Fig. 6.50. From the Pythagorean theorem, it follows that

$$b^2 + 5^2 = 6^2$$

$$b^2 = 6^2 - 5^2 = 11$$

$$b = \sqrt{11}.$$

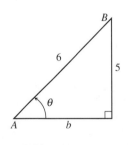

Figure 6.50   In Example 3, the hypotenuse and one leg are known.

Use the value $b = \sqrt{11}$ to find the remaining trig functions.

$$\sin \theta = \frac{5}{6} \qquad \csc \theta = \frac{6}{5}$$

$$\cos \theta = \frac{\sqrt{11}}{6} \qquad \sec \theta = \frac{6}{\sqrt{11}}$$

$$\tan \theta = \frac{5}{\sqrt{11}} \qquad \cot \theta = \frac{\sqrt{11}}{5}$$

Solving Right Triangles

One of the earliest uses of trigonometry was finding unknown parts of a triangle. The ancient Greeks established that if the measures of two sides of a right triangle or of one side and one angle are known, then the measures of the other sides or angles of the right triangle can be found. Determining the measures of the missing parts of a right triangle is often referred to as **solving a right triangle**, as illustrated in the next two examples.

E X A M P L E 4   Solving a Right Triangle (Part 1)

One angle of a right triangle measures $37°$, and the hypotenuse has a length of 8. Find the measures of the remaining sides of the right triangle (see Fig. 6.51).

Solution

$$\sin 37° = \frac{a}{8} \qquad\qquad \cos 37° = \frac{b}{8}$$

$$a = 8 \sin 37° \qquad\qquad b = 8 \cos 37°$$

$$= 4.814520185 \qquad\qquad = 6.38908408$$

Rounded to the nearest hundredths, $a = 4.81$ and $b = 6.39$.   ≡

We can use the Pythagorean theorem as a check on the trig calculations in Example 4. In Fig. 6.51, $a^2 + b^2 = 8^2$, or $8 = \sqrt{a^2 + b^2}$.

$$\sqrt{a^2 + b^2} = \sqrt{(4.81)^2 + (6.39)^2} \qquad \text{Substitute } a = 4.81 \text{ and } b = 6.39.$$

$$= \sqrt{23.1361 + 40.8321}$$

$$= 7.998012253$$

Because a and b were rounded to the nearest hundredth, this calculation will not be exactly 8. However, the result is close enough to 8 to be confident that 4.81 and 6.39 are correct. If the complete expressions for $8 \sin 37°$ and $8 \cos 37°$ had been used, the result would be 8. Support with a calculator that

$$\sqrt{a^2 + b^2} = \sqrt{(8 \sin 37°)^2 + (8 \cos 37°)^2}$$

$$= 8.$$

E X A M P L E 5   Solving a Right Triangle (Part 2)

The hypotenuse and one leg of a right triangle measure 12.7 and 6.1, respectively. Find the measure of the angle θ formed by these two sides.

Solution   From Fig. 6.52, notice that

$$\cos \theta = \frac{6.1}{12.7} = 0.4803149606.$$

We want to find an angle (in degrees) whose cosine is 0.4803149606; that is, we want to find $\cos^{-1}(6.1/12.7)$. (See the Agreement on Notation.)

$$\cos^{-1}\left(\frac{6.1}{12.7}\right) = \boxed{\text{2nd}}\ \equiv \boxed{\text{COS}}\left(\frac{6.1}{12.7}\right) = 61.29°$$ ≡

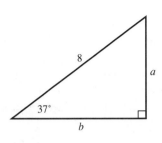

Figure 6.51 In Example 4, the hypotenuse and one acute angle are known.

Teaching Note

Some attention should be paid to showing students keying sequences that yield efficient computations for solving right triangles.

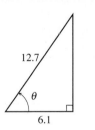

Figure 6.52 In Example 5, only the lengths of two sides are known.

AGREEMENT ON NOTATION

We use the notation

$$\cos^{-1}(6.1/12.7)$$

to denote the angle θ (in degree mode) for which $\cos\theta = 6.1/12.7$. We use $\cos^{-1}$ to denote the calculator keys $\boxed{\text{INV}}\,\boxed{\text{COS}}$, or $\boxed{\text{2nd}}$ $\boxed{\text{COS}}$, or $\boxed{\text{SHIFT}}\,\boxed{\text{COS}}$. The symbols $\sin^{-1}$ and $\tan^{-1}$ are used similarly.

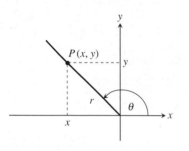

Figure 6.53 An angle in standard position.

The answer to Example 5 can be degrees whose cosine is 0.4803149606, confirmed by evaluating the trig value on a calculator. For example, cos 61.29° = 0.48037, which is sufficiently close to 6.1/12.7 to confirm the solution.

Defining Trig Functions for Angles in Standard Position

Up to this point, we have defined each trig function as the ratio of the lengths of two sides of a right triangle. In this section, we extend the definitions of the trig functions as follows.

Suppose θ is any angle in standard position and that point $P(x, y)$ is on the terminal side of the angle (see Fig. 6.53). Notice that θ can have any value either positive or negative and angles are no longer restricted to acute angles. The trig functions will now be defined in terms of the coordinates of P and the distance r between P and the origin.

Definition 6.6 Trigonometric Functions of Any Angle

Let θ be an angle in standard position, $P(x, y)$ a point other than the origin on the terminal side of P, and $r = \sqrt{x^2 + y^2}$. (See Fig. 6.53.) The six trig function values of θ are defined as follows:

$$\sin\theta = \frac{y}{r} \qquad\qquad \csc\theta = \frac{r}{y} \quad (y \neq 0)$$

$$\cos\theta = \frac{x}{r} \qquad\qquad \sec\theta = \frac{r}{x} \quad (x \neq 0)$$

$$\tan\theta = \frac{y}{x} \quad (x \neq 0) \qquad \cot\theta = \frac{x}{y} \quad (y \neq 0)$$

Notice that the ratios in this definition do not depend on the choice of the point P on the terminal side of the angle. Figure 6.54 shows that if point P_2 rather than P_1 is chosen, the two triangles formed are similar and the ratios of corresponding sides are equal. Also, these definitions are consistent with all previous definitions in this chapter.

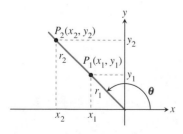

Figure 6.54 Choosing different points on the same terminal side results in the same ratios.

EXAMPLE 6   Finding Trig Values

Find the values of the trig functions at an angle θ in standard position with point $P(2, 3)$ on its terminal side (see Fig. 6.55).

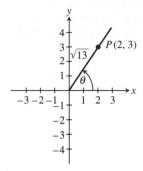

Figure 6.55   Angle in standard position with point $P(2, 3)$ on its terminal side.

Solution   Apply Definition 6.3, where $x = 2$, $y = 3$, and $r = \sqrt{2^2 + 3^2} = \sqrt{13}$.

$$\sin \theta = \frac{3}{\sqrt{13}} \qquad \csc \theta = \frac{\sqrt{13}}{3}$$

$$\cos \theta = \frac{2}{\sqrt{13}} \qquad \sec \theta = \frac{\sqrt{13}}{2}$$

$$\tan \theta = \frac{3}{2} \qquad \cot \theta = \frac{2}{3}$$

E X A M P L E   7   Finding Angle Measure

Find the measure of the angle described in Example 6, assuming it is an acute angle.

Solution   We have seen in Example 6 that $\tan \theta = 1.5$. Therefore

$$\theta = \tan^{-1}(1.5) = 56.31°.$$

Recall the agreements on inverse notation and accuracy that followed Example 5.

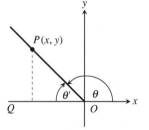

Figure 6.56   Angle θ in standard position with reference angle θ'.

When the terminal side of an angle falls in quadrants II, III, or IV, the angle is not an acute angle, as was assumed in Example 7. For example, in Fig. 6.56 the angle θ satisfies the inequality $90° < \theta < 180°$. Whenever point P is selected on the terminal side of a nonacute angle, a perpendicular dropped from P to the x-axis forms a right triangle ($\triangle POQ$ in Fig. 6.56). This triangle is called the **reference triangle** of angle θ, and the acute angle θ' in this reference triangle is called the **reference angle**.

An acute angle can serve as the reference angle for four different terminal sides, as shown in Fig. 6.57.

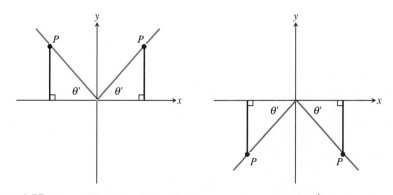

Figure 6.57   Four different angles with the same reference angle θ'.

E X A M P L E   8   Finding Trig Values and Angle Measure

Find the values of all the trig functions and the angle measure of angle θ, where $P(-2, 3)$ is a point on the terminal side of θ.

Solution $r = \sqrt{(-2)^2 + 3^2} = \sqrt{13}$

$$\sin\theta = \frac{3}{\sqrt{13}} \qquad \csc\theta = \frac{\sqrt{13}}{3}$$

$$\cos\theta = -\frac{2}{\sqrt{13}} \qquad \sec\theta = -\frac{\sqrt{13}}{2}$$

$$\tan\theta = -\frac{3}{2} \qquad \cot\theta = -\frac{2}{3}$$

To find angle θ, first find the reference angle $\theta' = \tan^{-1}(1.5) = 56.31°$ (see Fig. 6.56). Therefore

$$\theta = 180° - 56.31° = 123.69°.$$

Values of the Trig Functions for Quadrantal Angles

A **quadrantal angle** is an angle in standard position whose terminal side coincides with a coordinate axis. The following are quadrantal angles:

$$t = \ldots, -2\pi, 0, 2\pi, 4\pi, \ldots, 2n\pi, \ldots$$

$$t = \ldots, -\frac{3\pi}{2}, \frac{\pi}{2}, \frac{\pi}{2} + 2\pi, \frac{\pi}{2} + 4\pi, \ldots, \frac{\pi}{2} + 2n\pi, \ldots$$

$$t = \ldots, -\pi, \pi, \pi + 2\pi, \pi + 4\pi, \ldots, \pi + 2n\pi, \ldots$$

$$t = \ldots, -\frac{\pi}{2}, \frac{3\pi}{2}, \frac{3\pi}{2} + 2\pi, \frac{3\pi}{2} + 4\pi, \ldots, \frac{3\pi}{2} + 2n\pi, \ldots$$

Using Definition 6.1 and Fig. 6.58, we can see at a glance that for the quadrantal angles, the trig functions are either $1, -1, 0, 1/0$ (and hence undefined), or $-1/0$ (undefined).

E X A M P L E   9   Finding Trig Values for Quadrantal Angles

Find the values of the trig functions for the following quadrantal angles:
(a) 0   **(b)** $\pi/2$ **(c)** π **(d)** $3\pi/2$

Solution   Use Definition 6.2 and the coordinates of the points $(0, 0)$, $(0, 1)$, $(-1, 0)$, and $(0, -1)$ (see Fig. 6.58).

t	$\sin t$	$\cos t$	$\tan t$	$\cot t$	$\sec t$	$\csc t$
0	0	1	0	undefined	1	undefined
$\dfrac{\pi}{2}$	1	0	undefined	0	undefined	1
π	0	−1	0	undefined	−1	undefined
$\dfrac{3\pi}{2}$	−1	0	undefined	0	undefined	−1

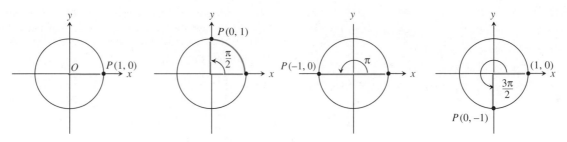

Figure 6.58   Quadrantal angles.

Exercises for Section 6.5

In Exercises 1–6, consider the following right triangle and complete each equation.

1. $\sin A = ?$ **2.** $\cos A = ?$ **3.** $\tan A = ?$

4. $\cot A = ?$ **5.** $\sec A = ?$ **6.** $\csc A = ?$

For Exercises 1–7.

7. Complete the six equations given in Exercises 1–6 for $\angle B$.

In Exercises 8–14, consider the following right triangle and complete each equation.

8. $\sin A = ?$ **9.** $\cos A = ?$ **10.** $\tan A = ?$

11. $\cot A = ?$ **12.** $\sec A = ?$ **13.** $\csc A = ?$

For Exercises 8–14.

14. Complete the same six equations for $\angle B$.

In Exercises 15–20, use Definition 6.6 to find the values $\sin\theta$, $\cos\theta$, $\tan\theta$, $\cot\theta$, $\sec\theta$, and $\csc\theta$.

15. $\theta = 30°$ **16.** $\theta = 45°$ **17.** $\theta = \dfrac{\pi}{3}$

18. $\theta = \dfrac{\pi}{6}$ **19.** $\theta = 60°$ **20.** $\theta = \dfrac{\pi}{4}$

In Exercises 21–24, use a right triangle to determine the values of all the trig functions at θ.

21. $\sin\theta = \dfrac{3}{5}$ **22.** $\tan\theta = \dfrac{1}{3}$

23. $\cos\theta = \dfrac{\sqrt{3}}{2}$ **24.** $\sin\theta = \dfrac{12}{13}$

In Exercises 25–38, use a calculator to determine θ. Report answers in both degrees and radians.

25. $\sin\theta = \dfrac{1}{2}$ **26.** $\sin\theta = 0.8245$

27. $\cos\theta = \dfrac{4}{5}$ **28.** $\cos\theta = 0.125$

29. $\tan\theta = 1$ **30.** $\tan\theta = 3$

31. $\tan\theta = 0.423$ **32.** $\tan\theta = 2.80$

33. $\csc\theta = 2$ **34.** $\sec\theta = 3.81$

35. $\cot\theta = \dfrac{3}{5}$ **36.** $\cot\theta = 1.875$

37. $\theta = \sin^{-1}\left(\dfrac{2}{3}\right)$ **38.** $\theta = \tan^{-1}(2)$

In Exercises 39–48, solve the following right triangle:

39. $\angle A = 20°$; $a = 12.3$ **40.** $a = 3$; $b = 4$

41. $\angle A = 41°$; $c = 10$ **42.** $\angle A = 55°$; $b = 15.58$

43. $b = 5$; $c = 7$

44. $a = 20.2$; $c = 50.75$

45. $a = 2$; $b = 9.25$

46. $a = 5$; $\angle B = 59°$

47. $c = 12.89$; $\angle B = 12.55°$

48. $\angle A = 10.2°$; $c = 14.5$

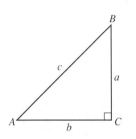

For Exercises 39–48.

For Exercises 49–56, let θ be an angle in standard position with the point P on the terminal side of θ. Sketch the reference triangle and the reference angle θ' determined by each angle θ. Find the values of the six trig functions at θ and determine the measures of θ' and θ.

49. $P(-3, 0)$

50. $P(0, 5)$

51. $P(4, 3)$

52. $P(-1, 2)$

53. $P(-4, -6)$

54. $P(5, -2)$

55. $P(22, -22)$

56. $P(-8, -1)$

In Exercises 57–60, sketch the angle θ, the reference angle θ', and a reference triangle. Determine the trig function values of θ and θ'.

57. $156°$

58. $-305°$

59. $614°$

60. $213°$

In Exercises 61–66, determine which trig functions are undefined for each quadrantal angle t.

61. $t = \dfrac{\pi}{2}$

62. $t = \pi$

63. $t = \dfrac{3\pi}{2}$

64. $t = 5\pi$

65. $t = \dfrac{5\pi}{2}$

66. $t = \dfrac{7\pi}{2}$

6.6

Applying Trig Functions

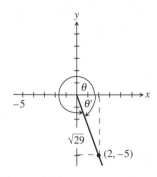

Figure 6.59 $y = -2.5x$.

Our development of the trig functions has included three different approaches. Section 6.2 introduced the trig functions in terms of a unit circle. Section 6.5 defined them in terms of ratios of sides of right triangles. By the end of Section 6.5, the trig functions were defined in terms of the coordinates of any point on the terminal side of an angle in standard position. It is important to emphasize that these three definitions agree with each other.

Now that all three approaches have been developed, our task is to choose the one that seems most appropriate for the problem being analyzed. This section begins by considering the problem of how to find the angle that a line makes with the x-axis and how to find the angle of intersection of two lines.

E X A M P L E   1   Finding Trig Function Values

The terminal side of θ is in the fourth quadrant and lies on the line $y = -2.5x$. Find all trig function values of θ and the measures of both angle θ and the reference angle θ'.

Solution   Any point on the line $y = -2.5x$ that falls in quadrant IV will be on the terminal side of θ and generate a reference triangle containing θ'. We choose the point $(2, -5)$ (see Fig. 6.59). Find the distance r from the origin to $(2, -5)$.

$$r = \sqrt{2^2 + (-5)^2} = \sqrt{29}$$

Apply Definition 6.3 for $x = 2$, $y = -5$, and $r = \sqrt{29}$.

$$\sin \theta = -\frac{5}{\sqrt{29}} \qquad \csc \theta = -\frac{\sqrt{29}}{5}$$

$$\cos \theta = \frac{2}{\sqrt{29}} \qquad \sec \theta = \frac{\sqrt{29}}{2}$$

$$\tan \theta = -\frac{5}{2} \qquad \cot \theta = -\frac{2}{5}$$

Because $\tan \theta = -\dfrac{5}{2}$, we use a calculator to determine that the reference angle $\theta' = 68.2°$. Therefore

$$\theta = 360° - 68.2° = 291.8°.$$

Angle between Intersecting Lines

When two lines intersect, they form four angles (see Fig. 6.60). The **angle between** ℓ_1 **and** ℓ_2 is the smaller, acute angle between the lines. We can use the methods applied in Example 1 to find the angle measure between any two lines in the plane.

E X A M P L E   2   Finding Angles between Lines

Determine the angle between the x-axis and the line $y = -2x + 3$.

Solution   The line $y = -2x$ is parallel to $y = -2x + 3$ (see Fig. 6.61). We will use $y = -2x$ in this solution since it goes through the origin and thus determines an angle in standard position.

Choose a point on the terminal side of θ: $P(-1, 2)$. Because $\tan \theta = -2$, we see that angle θ in Fig. 6.61 is $116.57°$ and the angle between the x-axis and graph of $y = 2x + 3$ is $66.43°$.

The angle between the x-axis in the positive direction and a line is called **the angle of inclination** of the line. In Fig. 6.61, θ is the angle of inclination of the line $y = -2x$.

Applications for Solving Right Triangles

Since the time of the ancient Greeks, people have used the principles of solving a right triangle to measure and survey the landscape. Determining these measurements requires finding the measures of sides or angles of the right triangle that

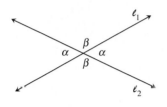

Figure 6.60   The intersection of two lines always creates two pairs of equal angles.

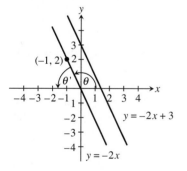

Figure 6.61   A pair of parallel lines.

model the problem. The terms **angle of elevation** or **angle of depression** are used to describe the angle between the line of sight and the horizontal (see Fig. 6.62).

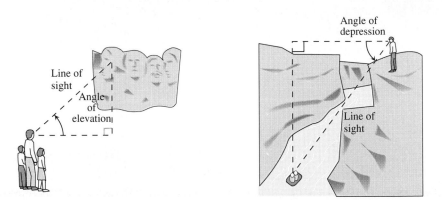

Figure 6.62   A right triangle is used in figuring the angle of elevation or the angle of depression.

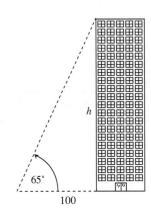

Figure 6.63   Sketch for Example 3.

E X A M P L E   3   APPLICATION: Determining an Unknown Height

The angle of elevation of the top of a building from a point 100 ft away from the building on level ground is $65°$ (see Fig. 6.63). Find the height h of the building.

Solution   Notice from Fig. 6.63 that

$$\tan 65° = \frac{h}{100}$$

$$h = 100 \tan 65° \quad \text{Enter the right-hand side on your calculator.}$$

$$= 214.45.$$

The height of the building is 214.45 ft.   ≡

E X A M P L E   4   APPLICATION: Finding the Angle of Elevation

A guy wire 75 ft long runs from a radio tower to a point on level ground 10 ft from the center of the tower's base (see Fig. 6.64). Determine the angle α the guy wire makes with the horizontal, the angle β the guy wire makes with the tower, and the distance h between the ground and the point B where the guy wire is attached to the tower.

Solution

$$\cos \alpha = \frac{10}{75} \qquad \text{or} \qquad \alpha = \cos^{-1}\left(\frac{10}{75}\right)$$

Use the calculator keys $\boxed{\text{INV}}$ $\boxed{\text{COS}}$, $\boxed{\text{2nd}}$ $\boxed{\text{COS}}$, or $\boxed{\text{SHIFT}}$ $\boxed{\text{COS}}$ to determine that

$$\alpha = 82.34° \qquad \text{and} \qquad \beta = 90° - 82.34° = 7.66°.$$

To find h, use either the Pythagorean theorem or the tangent function, as follows:

$$75^2 = 10^2 + h^2 \qquad\qquad \tan \alpha = \frac{h}{10}$$

$$h^2 = 5625 - 100 \qquad\qquad h = 10 \tan 82.34°$$

$$h = \sqrt{5525} = 74.33 \qquad\qquad = 74.35$$

Thus the guy wire makes a 82.34° angle with the horizontal and a 7.66° angle with the tower, and the guy wire is attached to a point 74.33 ft above ground. ≡

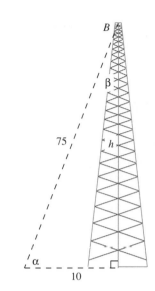

Figure 6.64   Sketch for Example 4.

What explains the discrepancy in the two solutions for h in Example 4? Which value for h is the most reliable? (See Exercise 41 in the Exercises for Section 6.6.)

E X A M P L E   5   APPLICATION: Determining the Angle of Depression

The angle of depression of a buoy from a point on a lighthouse 130 ft above the surface of the water is 6°. Find the distance x from the base of the lighthouse to the buoy.

Solution   (See Fig. 6.65.)

$$\tan \theta = \tan 6° = \frac{130}{x}$$

$$x = \frac{130}{\tan 6°} = 1236.87$$

The distance from the base of the lighthouse to the buoy is 1236.87 ft. ≡

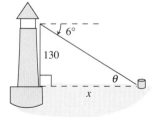

Figure 6.65   Diagram of the problem situation in Example 5.

E X A M P L E   6   APPLICATION: Calculating Indirect Measurement

From the top of a 100-ft building, a man observes a car moving toward him. If the angle of depression of the car changes from 22° to 46° during the period of observation, how far does the car travel?

Solution (See Fig. 6.66.) $\alpha = 22°$ and $\beta = 46°$

$$\tan \alpha = \frac{100}{x + d} \qquad \tan \beta = \frac{100}{d}$$

$$x + d = \frac{100}{\tan 22°} \qquad d = \frac{100}{\tan 46°}$$

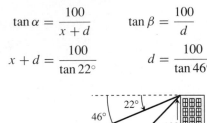

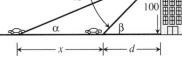

Figure 6.66 Diagram for Example 6.

Therefore

$$x = \frac{100}{\tan 22°} - d$$

$$= \frac{100}{\tan 22°} - \frac{100}{\tan 46°}$$

$$= 150.94.$$

The car travels 150.94 ft.

≡

E X A M P L E 7   APPLICATION: Determining a Boat's Bearing

A boat travels at 35 mph for 2 hr on a course 53° east of north; then it changes
to a course 143° east of north and travels for another 3 hr. Determine the distance
from the boat to its home port (Fig. 6.67). What is the bearing from the boat's
home port to its location at the end of 5 hr?

Solution

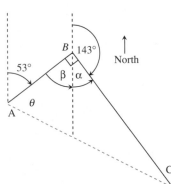

Figure 6.67 Diagram for
Example 7.

A: boat's home port
B: position when boat changes course
C: position at end of trip
$\alpha = 37°$, $\qquad \beta = 53°$, $\qquad$ segment AC is the hypotenuse of right $\triangle ABC$

$AB = (35 \text{ mi/hr})(2 \text{ hr}) = 70 \text{ mi}$
$BC = (35 \text{ mi/hr})(3 \text{ hr}) = 105 \text{ mi}$
$AC = \sqrt{70^2 + 105^2} = 126.19$ Use the Pythagorean theorem.
Bearing from A to C: $53° + \theta$, where

$$\tan \theta = \frac{BC}{AB} = \frac{105}{70} \qquad \text{or} \qquad \theta = \tan^{-1}(1.5)$$

The bearing from A to C is $53° + \tan^{-1}(1.5) = 109.31°$.

≡

Exercises for Section 6.6

In Exercises 1–8, let θ be an angle in standard position with its terminal side as given by the equation. Find the values of the six trig functions at θ.

1. $y = x$, quadrant III

2. $y = 2x$, quadrant I

3. $y = x$, quadrant I

4. $y = 2x$, quadrant III

5. $y = \frac{3}{4}x$, quadrant I

6. $y = -x$, quadrant II

7. $y = \frac{2}{3}x$, quadrant III

8. $y = -\frac{3}{5}x$, quadrant IV

In Exercises 9–12, determine the angle between the x-axis and the line.

9. $y = 2x$

10. $2x - 3y = 4$

11. $y = -3x + 1$

12. $3x - 5y = -2$

In Exercises 13–16, find the quadrant containing the terminal side of θ, which is an angle in standard position.

13. $\sin\theta < 0$ and $\tan\theta > 0$

14. $\cos\theta > 0$ and $\tan\theta < 0$

15. $\tan\theta > 0$ and $\sec\theta < 0$

16. $\sin\theta > 0$ and $\cos\theta < 0$

For Exercises 17–39, sketch a picture of each **problem situation** and solve the problem.

17. The angle of elevation of the top of a building from a point 100 ft away from the building on level ground is 45°. Determine the building's height.

18. The angle of elevation of the top of a building from a point 80 ft away from the building on level ground is 70°. Determine the building's height.

19. The angle of depression of a buoy from a point on a lighthouse 120 ft above the water's surface is 10°. Find the distance from the lighthouse to the buoy.

20. The angle of depression of a buoy from a point on a lighthouse 100 ft above the water's surface is 3°. Find the distance from the lighthouse to the buoy.

21. The angle of elevation of the top of a building from a point 250 ft away from the building on level ground is 23°. Determine the building's height.

22. A guy wire 30 m long runs from an antenna to a point on level ground 5 m from the antenna's base. Determine the angle the guy wire makes with the horizontal, the angle the guy wire makes with the antenna, and the distance between the ground and the point where the guy wire is attached to the antenna.

23. A building casts a shadow 130 ft long when the angle of elevation of the sun (measured from the horizon) is 38°. How tall is the building?

24. A wire stretches from the top of a vertical pole to a point on level ground 16 ft from the pole's base. If the wire makes an angle of 62° with the ground, determine the height of the pole and the length of the wire.

25. A lighthouse L stands 3 mi from the nearest point P on the shore. Point Q is located down the shoreline and $\overline{PQ} \perp \overline{PL}$. Determine the distance from P to a point Q along the shore if $\angle PQL = 35°$.

26. Using a sextant, a surveyor determines that the angle of elevation of a mountain peak is 35°. Moving 1000 ft further away from the mountain, the surveyor determines the angle of elevation to be 30°. What is the mountain's height?

27. An observer on the ground is 1 mi from a building 1200 ft tall. What is the angle of elevation from the observer to the top of the building?

28. The angle of elevation from an observer to the base of the roof of a tower located 200 ft from the observer is 30°. The angle of elevation from the observer to the top of the same roof is 40°. What is the height AB? (Assume points A, B, and C are on the same perpendicular line to the ground.)

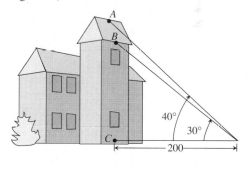

For Exercise 28.

29. From the top of a 100-ft building, a man observes a car moving toward him. If the angle of depression of the car changes from 15° to 33° during the period of observation, how far does the car travel?

30. A boat travels at 30 mph from its home port on a course of 95° for 2 hr and then changes to a course of 185° for 2 hr. Determine the distance from the boat to its home port and the bearing from the home port to the boat.

31. A boat travels at 40 mph from its home base on a course of 65° for 2 hr and then changes to a course of 155° for 4 hr. Determine the distance from the boat to its home port and the bearing from the home port to the boat.

32. A point on the north rim of the Grand Canyon is 7256 ft above sea level. A point on the south rim directly across from that point is 6159 ft above sea level. The canyon is 3180 ft wide (horizontal distance) between the two points. What is the angle of depression from the north-rim point to the south-rim point?

33. A ranger spots a fire from a 73-ft tower in Yellowstone National Park. She measures the angle of depression to be 1°20′. How far is the fire from the tower?

34. A footbridge is to be constructed along an east–west line across a river gorge. The bearing of a line of sight from a point 325 ft due north of the west end of the bridge to the east end of the bridge is 117°. What is the length ℓ of the bridge?

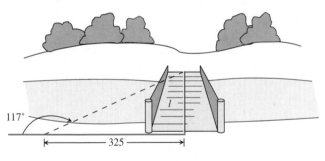

For Exercise 34.

35. The angle of elevation of a space shuttle launched from Cape Canaveral is measured to be 17° relative to the point of launch when it is directly over a ship 12 mi down range. What is the altitude of the shuttle when it is directly over the ship?

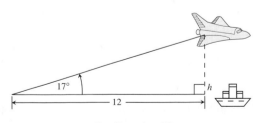

For Exercise 35.

36. A truss for a barn roof is constructed as shown in the following figure. What is the height of the vertical center span?

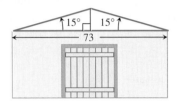

For Exercise 36.

37. A hot-air balloon over Park City, Utah, is 760 ft above the ground. The angle of depression from the balloon to a small lake is 5.25°. How far is the lake from a point on the ground directly under the balloon?

38. A shoreline runs north–south and a boat is due east of the shoreline. The bearings of lines of sight from two points on the shore to the boat are 110° and 100°. Assume the two points are 550 ft apart. How far is the boat from the nearest point on shore?

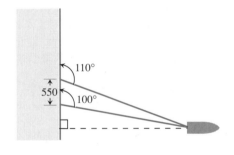

For Exercise 38.

39. Boats A and B leave from ports on opposite sides of a large lake. The ports are on an east–west line. Boat A steers a course of 105° and boat B steers a course of 195°. Boat A averages 23 mph and collides with boat B (it was a foggy night). What was boat B's average speed?

40. **Writing to Learn**   Write several paragraphs that explain how the three approaches to the trig functions represented by Definitions 6.2, 6.3, and 7.4 are consistent with one another.

41. **Writing to Learn**   Write a paragraph that explains the discrepancy in the two solutions for h in Example 4 of Section 6.6. In particular, which value do you think is the most reliable: $h = 74.33$ or $h = 74.35$?

Chapter 6 Review

KEY TERMS   (The number following each key term indicates the page of its introduction.)

acute angle, 318
amplitude, 341
angle, 315
angle between ℓ_1 and ℓ_2, 363
angle in standard position, 316
angle of depression, 364
angle of elevation, 364
angular speed, 322
bearing, 321
central angle, 318
circular function, 332
cofunctions, 355

complementary angles, 318
coterminal angles, 317
course, 321
degree, 316
fundamental trig identity, 334
hypotenuse, 353
initial side of an angle, 315
intercepted arc AB, 319
linear speed, 322
line of travel, 321
minutes, 322
nautical mile, 322

obtuse angle, 318
path, 321
periodic function, 338
periodic property of sine
   and cosine, 332
period of f, 338
quadrantal angle, 360
radian, 319
reference angle, 359
reference triangle, 359
right angle, 318
seconds, 322

side adjacent to an angle, 353
side opposite an angle, 353
similar triangles, 354
solving a right triangle, 356
subtends an angle, 326
supplementary angles, 318
statute mile, 322
terminal side of an angle, 315
trigonometric function, 315
trigonometric identity, 333
unit circle, 319
vertex of an angle, 315

REVIEW EXERCISES

In Exercises 1–8, determine the quadrant of the terminal side of each angle in standard position. If the angle is in degrees, give its radian equivalent; if it is in radians, give its degree equivalent.

1. $135°$

2. $-45°$

3. $\dfrac{5\pi}{2}$

4. $\dfrac{3\pi}{4}$

5. $78°$

6. $112°$

7. $\dfrac{\pi}{12}$

8. $\dfrac{7\pi}{10}$

In Exercises 9 and 10, determine the angle measure in both degrees and radians. If the terminal side of the angle is obtained as follows, draw the angle in standard position:

9. A three-quarters' counterclockwise rotation

10. Two and one-half counterclockwise rotations

In Exercises 11–16, the given point is on the terminal side of an angle in standard position. Give the smallest positive angle in both degrees and radians for this angle.

11. $(2, 4)$

12. $(-1, 1)$

13. $(-1, \sqrt{3})$

14. $(-3, -3)$

15. $(6, -12)$

16. $(\sqrt{3}, 1)$

In Exercises 17–28, find the exact value of each trig function.

17. $\sin 30°$

18. $\cos 270°$

19. $\tan 135°$

20. $\sin \dfrac{5\pi}{6}$

21. $\csc \dfrac{2\pi}{3}$

22. $\sec 135°$

23. $\csc \dfrac{\pi}{3}$

24. $\csc 210°$

25. $\cos \dfrac{2\pi}{3}$

26. $\sec 330°$

27. $\cot 135°$

28. $\cot \dfrac{5\pi}{4}$

In Exercises 29–38, use a grapher to find each value.

29. $\sin \dfrac{\pi}{9}$

30. $\cos \dfrac{\pi}{5}$

31. $\tan \dfrac{\pi}{6}$

32. $\sin \dfrac{\pi}{8}$

33. $\cot \dfrac{\pi}{5}$

34. $\sec \dfrac{\pi}{12}$

35. $\sin 121°$

36. $\cos 72°$

37. $\tan 83°$

38. $\sin 19°$

In Exercises 39–44, state the period and amplitude, if it exists, for each function.

39. $f(x) = \sin 3x$

40. $g(x) = 5 \cos 2x + 1$

41. $h(x) = 3 \tan 2x$

42. $f(x) = 6 \cos 4x$

43. $g(x) = 2 \sec 3x$

44. $h(x) = 2 \cos 3\pi x$

In Exercises 45–50, use properties of a 30°-60° right triangle and a 45° right triangle to determine the exact values for all six trig functions for these angles.

45. 0

46. $-\dfrac{\pi}{6}$

47. $\dfrac{3\pi}{4}$

48. $60°$

49. $-135°$

50. $300°$

In Exercises 51–58, sketch a complete graph of each function without using a graphing utility. Check your answers with a graphing utility.

51. $y = \sin(x + 45°)$

52. $y = 3 + 2\cos x$

53. $y = -\sin(x + \pi/2)$

54. $y = -2 - 3\sin(x - \pi)$

55. $y = \tan 2x$

56. $y = 3\cos 2x$

57. $y = -2\sin 3x$

58. $y = \csc \pi x$

59. Find all six trig functions of $\angle A$ in $\triangle ABC$ of the following figure:

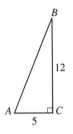

For Exercise 59.

60. Use a right triangle to determine the values of all trig functions at θ with $\cos \theta = \frac{5}{7}$.

61. Use a right triangle to determine the values of all trig functions at θ with $\cos \theta = \frac{12}{13}$.

62. Use a calculator to solve $\cos \theta = \frac{3}{7}$ if $0° \le \theta \le 90°$. Ensure your calculator is set in degree mode. (Recall the meaning and use of $\cos^{-1}$.)

63. Use a calculator to solve $\cot \theta = \frac{1}{3}$ if $0° \le \theta \le 90°$.

64. Use a calculator to solve $\tan x = 1.35$ if $0 \le x \le \pi/2$. Ensure your calculator is set in radian mode.

65. Use a calculator to solve $\sin x = 0.218$ if $0 \le x \le \dfrac{\pi}{2}$.

In Exercises 66–71, solve the right triangle ABC in the following figure for each of the given data.

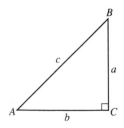

For Exercises 66–71.

66. $\angle A = 35°; c = 15$

67. $b = 8; c = 10$

68. $\angle B = 48°; a = 7$

69. $\angle A = 28°; c = 8$

70. $b = 5; c = 7$

71. $a = 2.5; b = 7.3$

In Exercises 72–75, find the quadrant of x if x is an angle in standard position and $0 \le x \le 2\pi$.

72. $\sin x < 0$ and $\tan x > 0$

73. $\cos x < 0$ and $\csc x > 0$

74. $\tan x < 0$ and $\sin x < 0$

75. $\csc x > 0$ and $\sec x < 0$

76. Show that

$$\cot \theta = \frac{\cos \theta}{\sin \theta}$$

for all acute angles θ.

In Exercises 77 and 78, use the fundamental trig identities to complete each function. Check your answers with a calculator.

77. $\tan \theta$ if $\sin \theta = 0.58$

78. $\tan \theta$ if $\cos \theta = 0.82$

In Exercises 79–82, let θ be an angle in standard position with the point P on the terminal side of θ. Find the values of each of all six trig functions if $0 < \theta \le 2\pi$ in each case.

79. $(-3, 6)$

80. $(12, 7)$

81. $(-5, -3)$

82. $(4, 9)$

In Exercises 83 and 84, investigate graphically what happens to the values of $f(x)$ as $x \to 0$.

83. $f(x) = \dfrac{1}{x} \sin\left(\dfrac{1}{x}\right)$

84. $f(x) = \dfrac{\sin x}{x^2}$

In Exercises 85–88, use a calculator to evaluate each expression. Give your answers in both degrees and as real numbers.

85. $\sin^{-1}(0.766)$

86. $\sin^{-1}(0.479)$

87. $\sin^{-1}\left(\dfrac{\sqrt{3}}{2}\right)$

88. $\tan^{-1} 1$

In Exercises 89–92, use a calculator to evaluate each expression and give your answer as a real number.

89. $\sin[\sin^{-1}(0.25)]$

90. $\cos^{-1}[\tan(0.2)]$

91. $\tan(\sin^{-1} 2)$

92. $\sin^{-1}[\sin(\sqrt{3}\cos 3)]$

93. Solve the inequality $3\cos x < x + 2$.

94. The angle of elevation of the top of a building from a point 100 m away from the building on level ground is $78°$. Determine the building's height.

95. A tree casts a shadow 51 ft long when the angle of elevation of the sun (measured from the horizon) is 25°. How tall is the tree?

96. From the top of a 150-ft building, Flora observes a car moving toward her. If the angle of depression of the car changes from 18° to 42° during the observation, how far does the car travel?

97. A lighthouse L stands 4 mi from the closest point P along a straight shore. Find the distance from P to a point Q along the shore if $\angle PLQ = 22°$.

98. An airplane at a certain time is between two signal towers positioned on a line with bearing 0° (that is, along a north–south line). The bearings between the plane and the north and south towers are 23° and 128°, respectively. Draw the exact location of the plane.

99. A shoreline runs north and south and a boat is due east of the shoreline. The bearings of lines of sight from a boat to two points on shore are 115° and 123°. Assume the two points are 855 ft apart. How far is the boat from the nearest point on shore if the shore is a straight line?

THE ANCIENT EGYPTIANS

About 5,000 years ago, African artists in Egypt developed the idea of proportional drawing for creating immense paintings. A small drawing was divided into a grid and transferred, square by square, from the small sketch to a large surface. Angles and curves are kept the same, and the lengths in the large figure are proportional to the lengths in the small figure. This technique is still used today.

Another Egyptian discovery was the base-10 system of numerals. Their numbers were written with pictures called *hieroglyphics*. Egyptian mathematicians also developed a type of algebra to solve practical problems. They were aware of the need to check the solutions they obtained by substituting the solutions into the original equations. This is significant in the history of mathematics as one of the earliest instances of proof.

7

Analytic Trigonometry

—————— **Transformations and Trigonometric Graphs**

Students will be able to
apply transformations to
obtain the graph of the form
$y = af(bx + c) + d$ with
emphasis on sinusoidal graphs.

In Chapter 6, trig functions were introduced as ratios of sides of right triangles. This chapter focuses on graphs of the trig functions, equation solving, and trig identities. These aspects of trigonometry, which are important in calculus, are often referred to as **analytic trigonometry**.

In earlier chapters, you learned to transform the graph of a function by horizontal and vertical shifts, by vertical stretching and shrinking, and by reflection through the x-axis. Following is a review of these transformations:

Review of Transformations

Suppose that $a, c,$ and d are positive real numbers.

To obtain the graph of:	**From the graph of $y = f(x)$:**
$y = af(x)$	Vertical stretch or shrink by a factor of a units
$y = -f(x)$	Reflect through the x-axis
$y = f(x + c)$	Shift left c units
$y = f(x - c)$	Shift right c units
$y = f(x) + d$	Shift up d units
$y = f(x) - d$	Shift down d units
$y = f(-x)$	Reflect through the y-axis

These same transformations can be applied to the graphs of any of the trigonometric functions. These transformations are illustrated for the functions $y = \sin x$ and $y = \cos x$.

E X A M P L E   1   Finding Sine Function Transformations

Describe how a complete graph of the following functions can be obtained from $y = \sin x$. Support your answer with a graphing utility.

a) $f(x) = 5 \sin x$

b) $g(x) = \sin x - 2$

Solution

a) A complete graph of $f(x) = 5 \sin x$ can be obtained from a graph of $y = \sin x$ by a vertical stretch by a factor of 5 (see Fig. 7.1a).

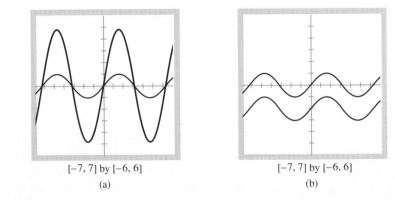

$[-7, 7]$ by $[-6, 6]$	$[-7, 7]$ by $[-6, 6]$
(a)	(b)

Figure 7.1   (a) $y = \sin x$ and $f(x) = 5 \sin x$; (b) $y = \sin x$ and $g(x) = \sin x - 2$.

Teaching Note

Within this setting, students are expected to apply previously examined transformations. Attention must be given to translation, stretching, and shrinking. Review how $f(x) = x^2$ is related to $f(x) = ax^2$ or $f(x) = (x - c)^2$ and make the connection to transformations applied to a trig function, such as the cosine function.

Notes on Examples

A discussion of Example 2 provides an opportunity to remind students that the cosine curve is a sine curve out of phase by $\frac{\pi}{2}$ radians.

b) A complete graph of $g(x) = \sin x - 2$ can be obtained from a graph of $y = \sin x$ by a vertical shift down 2 units (see Fig. 7.1b). ▤

Recall that the vertical stretch factor 5 in Example 1 is called the amplitude of $f(x)$. The amplitude of a sine or cosine function determines the local maximum and local minimum of that function.

This next example reviews how horizontal shift transformations can be used with cosine functions.

E X A M P L E   2   Finding Cosine Function Transformations

Describe how a complete graph of the following functions can be obtained from $y = \cos x$. Support your answer with a graphing utility.

a) $f(x) = \cos(x - 1.5)$

b) $g(x) = \cos(x + 0.5)$

Solution

a) A complete graph of $f(x) = \cos(x - 1.5)$ can be obtained from a graph of $y = \cos x$ by a shift right 1.5 units (see Fig. 7.2a).

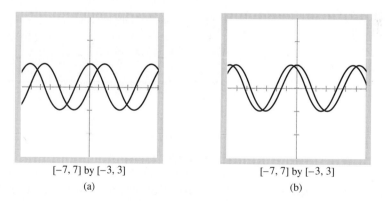

[−7, 7] by [−3, 3]   [−7, 7] by [−3, 3]
(a)   (b)

Figure 7.2   (a) $y = \cos x$ and $f(x) = \cos(x - 1.5)$; (b) $y = \cos x$ and $g(x) = \cos(x + 0.5)$.

b) A complete graph of $g(x) = \cos(x + 0.5)$ can be obtained from a graph of $y = \cos x$ by a shift left 0.5 units (see Fig. 7.2b). ▤

Trigonometry is important in electronics because the sine and cosine functions model alternating current. Physicists have developed their own terminology for this application. They would say that the graphs of $f(x) = \cos(x - 1.5)$ and $g(x) = \cos(x + 0.5)$ are obtained from the graph of $y = \cos x$ by a **phase shift.** Example 2(a) illustrated a phase shift to the right 1.5 units.

To summarize, so far we have considered the transformations of vertical and horizontal shifting and vertical stretching. The graph of $y = a\sin(x + c) + d$ can be obtained from the graph of $y = \sin x$ by using these transformations.

The following Exploration focuses on what impact the constant b has on the graph of $y = \sin bx$.

🔍 EXPLORE WITH A GRAPHING UTILITY

Graph each pair in the viewing rectangle $[-10, 10]$ by $[-4, 4]$. (The $+2$ and -2 are included so that the two graphs will not overlap.) Would the graph of the second function be obtained by compressing (that is, horizontally shrinking) or stretching the graph of the first function?

1. $y = \sin x + 2;$ $\quad y = \sin 2x - 2$
2. $y = \sin x + 2;$ $\quad y = \sin 3x - 2$
3. $y = \sin x + 2;$ $\quad y = \sin \frac{1}{2}x - 2$

Try several other examples of your choice.

Generalize Under what conditions do you think that the graph of $y = \sin bx$ can be obtained from the graph of $y = \sin x$ by a horizontal stretch? By a horizontal shrink?

Graphs of $y = \sin(bx + c)$

Consider the graphs of functions of the form $y = \sin(bx + c)$, where b is a positive real number and c is any real number, positive or negative. We saw in Chapter 6 that the trigonometric functions are periodic. The period of $f(x) = \sin bx$ is $2\pi / |b|$, so the constant b determines the period of f.

How do the constants b and c affect the graph of $y = \sin(bx + c)$? Figure 7.3 shows the beginning and end of one period of $y = \sin(bx + c)$. Because the sine function is a periodic function with a period of 2π, the beginning of the period shown occurs where $bx + c = 0$ and the end of the period occurs where $bx + c = 2\pi$.

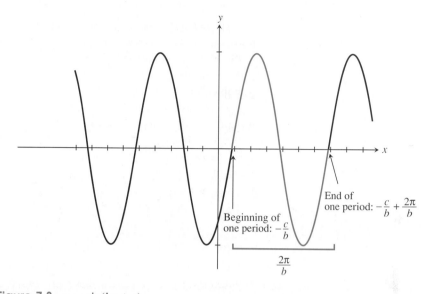

Figure 7.3 $y = \sin(bx + c)$.

We can use this information to find the x-intercepts at the beginning and the end of the period.

Beginning of Period	**End of Period**
$bx + c = 0$	$bx + c = 2\pi$
$bx = -c$	$bx = 2\pi - c$
$x = -\dfrac{c}{b}$	$x = \dfrac{2\pi}{b} - \dfrac{c}{b}$

The constants b and c affect both the period and phase shift of the graph. The graph of $y = \sin(bx + c)$ has a phase shift of $-c/b$ and a period of $2\pi/|b|$. The same is true for any periodic function, a fact that leads to this generalization.

> **Theorem 7.1 Phase Shift and Period of** $y = f(bx + c)$
>
> Suppose that $y = f(x)$ is a trigonometric function with period P and that b is a positive number. Then
>
> **a)** the number $-c/b$ is the phase shift of $y = f(bx + c)$, and
> **b)** the period of $y = f(bx + c)$ is $P/|b|$.
>
> We say that the graph of $y = f(bx)$ can be obtained from the graph of $y = f(x)$ by a horizontal shrink if $b > 1$ and by a horizontal stretch if $0 < b < 1$.

EXAMPLE 3   Analyzing a Complete Graph

Find a complete graph and the domain, range, period, and phase shift of $f(x) = 2\sin(3x + 4) - 1$.

Solution

Domain: x can represent any real number, so the domain of f is $(-\infty, \infty)$.

Range: Consider the graph of $y = 2\sin(3x + 4) = f(x) + 1$. The amplitude of y is 2, which means that $-2 \le y \le 2$. Therefore $-2 - 1 \le f(x) \le 2 - 1$ and the range of f is the interval $[-3, 1]$.

Phase shift: It follows from Theorem 7.1 that the phase shift is $-\frac{4}{3}$.

Period: It follows from Theorem 7.1 that the period is $2\pi/3$. A complete graph is shown in Fig. 7.4. ▬

We now summarize the characteristics of amplitude, period, phase shift, and maximum and minimum values for the completely general sine function $f(x) = a\sin(bx + c) + d$.

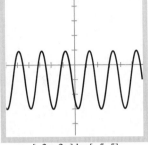

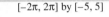

[−2π, 2π] by [−5, 5]

Figure 7.4
$f(x) = 2\sin(3x + 4) - 1$.

Summary: $f(x) = a\sin(bx + c) + d$

The function $f(x) = a\sin(bx + c) + d$, where $a, b, c,$ and d are any real numbers, is a periodic function with the following characteristics:

Amplitude: $|a|$ Period: $\dfrac{2\pi}{|b|}$ Phase shift: $-\dfrac{c}{b}$

Local maximum of $f(x)$: $d + |a|$

Local minimum of $f(x)$: $d - |a|$

The functions $f(x) = a\cos(bx + c) + d$, $f(x) = a\sec(bx + c) + d$, and $f(x) = a\csc(bx + c) + d$ have the same characteristics, except that the secant and cosecant functions do not have amplitudes and the local maximum and local minimum values are reversed from the cosine and sine functions.

Example 4 supports the claim that the functions $f(x) = a\tan(bx + c) + d$ and $f(x) = a\cot(bx + c) + d$ do not have amplitudes or a maximum or minimum and that their period is $\pi/|b|$.

EXAMPLE 4   Analyzing a Complete Graph

Describe how the function $f(x) = -4\cot(2x + \pi/3)$ can be obtained from $y = \cot x$ by applying transformations.

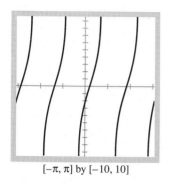

$[-\pi, \pi]$ by $[-10, 10]$

Figure 7.5
$f(x) = -4\cot(2x + \pi/3)$

Solution   Notice that $f(x)$ can be rewritten as

$$f(x) = -4\cot\left(2x + \frac{\pi}{3}\right) = -4\cot\left[2\left(x + \frac{\pi}{6}\right)\right].$$

Apply the following transformations to the graph of $y = \cot x$ to obtain the graph of f as expressed by this equation:

1. A vertical stretch by a factor of 4 to obtain $y = 4\cot x$
2. A reflection through the x-axis to obtain $y = -4\cot x$
3. A horizontal shrink by a factor of $\frac{1}{2}$ to obtain $y = -4\cot 2x$
4. A horizontal shift left $\pi/6$ units to obtain $f(x) = -4\cot\left[2\left(x + \frac{\pi}{6}\right)\right]$

A complete graph is shown in Fig. 7.5.   ≣

Sinusoids

The functions discussed in Examples 1 through 3 belong to the category of functions called sinusoids.

Definition 7.1   Sinusoids

A **sinusoid** is a function that can be written in the form

$$f(x) = a\sin(bx + c) + d,$$

where a, b, c, and d are real numbers.

Notes on Exercises

In Ex. 35–42, use zoom-in to identify all solutions with an error of at most 0.01.
Ex. 43–52 are problems for which students should make conjectures about the complete graphs of each sinusoid.
Ex. 53–69 offer a variety of sinusoidal applications to problem situations that lend themselves to a cooperative learning approach.

Notice that the graph of a sinusoid can be obtained from the graph of $y = \sin x$ by a combination of horizontal stretching or shrinking, horizontal shifting, vertical stretching or shrinking, and vertical shifting. In the next few Explorations, you will investigate what happens when you add or subtract sine and cosine functions.

A sum of the form $a\sin cx + b\cos dx$ where a, b, c, and d are constants is called a **linear combination** of $\sin cx$ and $\cos dx$. In the next several Explorations, you will investigate the conditions under which a linear combination of $\sin cx$ and $\cos dx$ is also a sinusoid.

Bert says, "It is appropriate to focus attention on sinusoidal curves because of the richness of their applications."

🔍 **EXPLORE WITH A GRAPHING UTILITY**

Experiment by graphing these functions. Which appear to be sinusoidal?

$$y = 3\sin x + 2\cos x \qquad y = 2\sin x - 3\cos x$$

$$y = 2\sin 3x + 4\cos 2x \qquad y = 3\sin 5x - 5\cos 5x$$

$$y = 4\sin x - 2\cos x \qquad y = 3\cos 2x + 2\sin 3x$$

Experiment with other sums and differences of sine and cosine functions.

Write a Conjecture Under what conditions will these sums and differences be sinusoidal?

In this Exploration, you have discovered that not all of the functions are sinusoidal. Your experience may have led you to guess that any linear combination of $\sin x$ and $\cos x$ is a sinusoid. This is the subject of a theorem.

Theorem 7.2 Sums That Are Sinusoidal

For all real numbers a and b, the function

$$f(x) = a\sin x + b\cos x$$

is a sinusoid. In particular, there exist real numbers A and α such that

$$a\sin x + b\cos x = A\sin(x + \alpha),$$

where $|A|$ is the amplitude and α is the phase shift.

E X A M P L E   5   Finding the Amplitude and Phase Shift

Show that $f(x) = 2\sin x + 5\cos x$ is a sinusoid. Also approximate A and α so that $A\sin(x + \alpha) = 2\sin x + 5\cos x$.

Solution By Theorem 7.2, the function f is sinusoidal, which is confirmed by the graph in Fig. 7.6. Using the trace key and/or zoom-in, estimate the following values:

Amplitude: 5.4, or $A = 5.4$.

Phase shift: 1.2 units to the left, or $\alpha = 1.2$.

Period: 2π.

[−2π, 2π] by [−10, 10]

Figure 7.6 $f(x) = 2\sin x + 5\cos x$ has the same graph as $y = 5.4\sin(x + 1.2)$.

Support these estimates by overlaying the graph of $y = 5.4 \sin(x + 1.2)$ with the graph of $f(x) = 2 \sin x + 5 \cos x$ (see Fig. 7.6). (Example 5 of Section 7.5 gives an algebraic way to do this exercise.)   ≡

Theorem 7.2 and Example 5 were about linear combinations of $\sin x$ and $\cos x$. The next Exploration broadens the investigation to include linear combinations of functions with phase shifts and periods other than 2π.

🔍 **EXPLORE WITH A GRAPHING UTILITY**

Experiment and decide which of these functions appear to be sinusoids:

$y = 2 \sin(3x + 1) - 5 \cos(3x - 2)$ $y = 3 \sin(2x - 0.5) + \cos(2x + 1)$
$y = \sin(3x - 1) + 3 \cos(3x + 2)$ $y = 2 \sin(x - 2) + 3 \cos(4x + 1)$
$y = 3 \sin(4x + 1) - 2 \cos(2x - 3)$ $y = 2 \sin(3x - 2) + 3 \cos(3x + 4)$

Experiment with other sums and differences of sine and cosine functions of your own choice.

Write a Conjecture   Under what conditions will these linear combinations be sinusoidal?

This Exploration supports the following theorem.

Theorem 7.3   Sums That Are Sinusoidal

For all real numbers $a, b, d, h,$ and k, the function

$$f(x) = a \sin(bx + h) + d \cos(bx + k)$$

is a sinusoid. In particular, there exist real numbers A and α such that

$$a \sin(bx + h) + d \cos(bx + k) = A \sin(bx + \alpha).$$

E X A M P L E   6   Finding the Amplitude and Phase Shift

Show that $f(x) = 3 \sin(2x - 1) + 4 \cos(2x + 3)$ is a sinusoid. Also estimate A and α so that $A \sin(2x + \alpha) = 3 \sin(2x - 1) + 4 \cos(2x + 3)$.

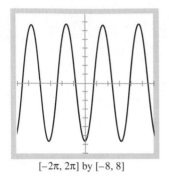

[−2π, 2π] by [−8, 8]

Figure 7.7 $f(x) =$
$3 \sin(2x − 1) + 4 \cos(2x + 3)$
has the same graph as
$y = 6.6 \sin(2x − 1.42)$.

Notes on Exercises

Ex. 43–52 require students to
apply Theorems 7.2 and 7.3.
You may wish to do a class
demonstration of Examples 5
and 6, or investigate some of
these sinusoidal exercises in
class, before students are asked
to work them independently.

Solution From Theorem 7.3, we know that f is sinusoidal, which is supported
by the graph in Fig. 7.7. Using the trace key and/or zoom-in, estimate the following
values:

Amplitude: 6.6.

Phase shift: 0.71 units right or $\alpha = −1.42$.

Period: π.

Support these estimates by overlaying the graph of $y = 6.6 \sin[2(x − 0.71)]$
with the graph of $f(x) = 3 \sin(2x − 1) + 4 \cos(2x + 3)$ in Fig. 7.7. ≡

Notice that in Theorem 7.2 as well as Theorem 7.3, both the sine and cosine
terms in the linear combination have the same period. If the periods of the sine
and cosine are different, then the linear combination is not a sinusoid.

A value $f(c)$ is an **absolute maximum** for a function f on an interval $[a, b]$
if $f(c) \geq f(x)$ for all x in $[a, b]$. Similarly, $f(c)$ is an **absolute minimum** if
$f(c) \leq f(x)$ for all x in $[a, b]$.

A value $f(c)$ is an **absolute extremum** if it is either an absolute maximum
or an absolute minimum.

E X A M P L E   7   Showing a Sum Is Not a Sinusoid

Show that the function $f(x) = \sin 2x + \cos 3x$ is not sinusoidal. Also find the
domain, range, and period of f.

Solution Figure 7.8(a) shows a complete graph of f. From this graph, we
conclude that f is not sinusoidal.

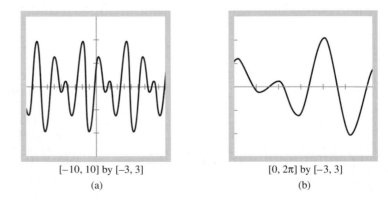

| [−10, 10] by [−3, 3] | [0, 2π] by [−3, 3] |
| (a) | (b) |

Figure 7.8 Two views of $f(x) = \sin 2x + \cos 3x$.

Frank says, "Examples 8 and 9 are problems which, until now, could only be solved by using calculus. Today, they can be used to foreshadow calculus because precalculus students can solve them using technology."

Domain: $(-\infty, \infty)$. (Why?)

Period: The period of $\sin 2x$ is π, and the period of $\cos 3x$ is $2\pi/3$. The period of f must be a common multiple of these two periods, and the least common multiple is 2π.

Range: In order to determine the range of f, we investigate a graph of the single period of f shown in the $[0, 2\pi]$ by $[-3, 3]$ viewing rectangle (see Fig. 7.8b).

Use zoom-in to determine that the absolute maximum of f in $[0, 2\pi]$ is 1.91 and occurs when $x = 4.11$. The absolute minimum is -1.91 and occurs when $x = 5.32$. So the range is $[-1.91, 1.91]$. ▤

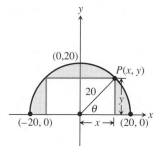

Figure 7.9 A semicircular tunnel has walls and ceiling forming a rectangular cross section.

Applications of Trigonometric Functions

Analyzing the graphs of trigonometric functions is often the key to solving an applied problem. We illustrate with several examples.

E X A M P L E 8   APPLICATION: Building a Tunnel

The cross section of a tunnel is a semicircle with a radius of 20 ft. The interior walls of the tunnel form a rectangle as illustrated in Fig. 7.9. An engineer is asked to find the width and height of the tunnel opening with the maximum cross-sectional area.

Solution We want to find an algebraic representation of the tunnel's cross-sectional area. Let $P(x, y)$ be the upper right corner point of the rectangle and θ the angle that OP makes with the positive x-axis. Then $\cos\theta = x/20$ and $\sin\theta = y/20$, or $x = 20\cos\theta$ and $y = 20\sin\theta$.

Because the area A of the rectangle is the quantity to be maximized, express A as a function of θ.

$$A = 2xy$$
$$= 2(20\cos\theta)(20\sin\theta)$$
$$= 800\sin\theta\cos\theta$$

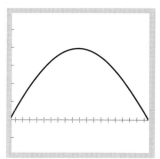

$[0, \pi/2]$ by $[-200, 600]$

Figure 7.10 $A = 800\sin\theta\cos\theta$.

The context of the problem situation indicates that θ must be in the interval $[0, \pi/2]$. Figure 7.10 shows a graph of $A = 800\sin\theta\cos\theta$ in $[0, \pi/2]$. Use zoom-in to find that the coordinates of the highest point of the graph are $(0.785, 400)$ with an error at most 0.01.

The maximum area occurs when $\theta = 0.785$, so we can see that

$$x = 20\cos 0.785 = 14.14 \text{ ft} \quad \text{and} \quad y = 20\sin(0.785) = 14.14 \text{ ft}. \quad ▤$$

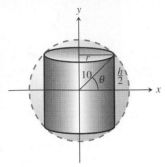

Figure 7.11   A hole is cut through the center of a sphere.

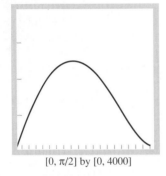

$[0, \pi/2]$ by $[0, 4000]$

Figure 7.12 $V = 2000\pi \sin\theta \cos^2\theta$.

The next problem situation deals with a sphere with a hole drilled through it. From Fig. 7.11, it is evident that as the radius of the hole increases, the height of the hole decreases.

E X A M P L E   9   APPLICATION: Drilling a Hole in a Sphere

A hole is drilled through the center of a sphere of radius 10. The portion of the sphere drilled out is a right circular cylinder and two spherical caps. Find the dimensions of the cylindrical hole with the maximum volume that can be drilled out of a sphere.

Solution   Using the reference triangle in Fig. 7.11, we see that $h/2 = 10\sin\theta$ and $r = 10\cos\theta$. Because volume is the quantity to be maximized, express V as a function of θ.

$$V = \pi r^2 h$$
$$= \pi(10\cos\theta)^2(20\sin\theta)$$
$$= 2000\pi \sin\theta \cos^2\theta$$

Notice that in this problem situation, θ must be in the interval $[0, \pi/2]$. Fig. 7.12 shows a complete graph of the problem situation. Use zoom-in to determine that the coordinates of the highest point of the graph of V are $(0.615, 2418.399)$ with an error at most 0.01.

The maximum volume is 2418.399 cubic units, and it occurs when $\theta = 0.615$ rad. Substituting this value of θ into the equations for h and r, we find that the maximum volume occurs when $r = 8.17$ and $h = 11.54$.   ≡

Exercises for Section 7.1

In Exercises 1–4, list the transformations in the order of their application to transform $y = \sin x$ or $y = \tan x$ into the indicated graph.

1. $y = 3 - 4\sin(2x - \pi)$ **2.** $y = 2 + 3\sin\left(\frac{1}{2}x + \frac{\pi}{2}\right)$

3. $y = 2 - \tan(x - \pi)$ **4.** $y = 1 + 2\tan(2x - \pi)$

In Exercises 5–14, sketch a complete graph of each function without using a graphing utility. Support your answer with a graphing utility if necessary.

5. $y = 2 + \sin\left(x - \frac{\pi}{4}\right)$ **6.** $y = -1 - \cos(x - \pi)$

7. $y = -1 + 3\sin(x - 2)$ **8.** $y = 2 - 2\sin(x + 3)$

9. $f(x) = 2 - 3\sin(2x)$ **10.** $g(x) = -2 + \tan\left(\frac{1}{2}x\right)$

11. $y = 2 + 3\cos\left(\frac{x - 1}{2}\right)$ **12.** $y = -2 - 4\sin(2x + 6)$

13. $T(x) = -3 + 2\sin(2x - \pi)$

14. $y = 2 - 3\cos(4x - 2\pi)$

In Exercises 15–18, determine the domain, range, period, and asymptotes (if any) of the function in each exercise.

15. Exercise 3   **16.** Exercise 5

17. Exercise 6   **18.** Exercise 8

In Exercises 19–24, find the period and phase shift for each function.

19. $f(x) = 3\sin(2x - \pi)$ **20.** $g(x) = \cos\left(5x - \dfrac{\pi}{2}\right)$

21. $k(x) = \tan\left(x - \dfrac{\pi}{4}\right)$ **22.** $h(x) = \cot(3x - 1.5)$

23. $f(x) = \sec(2x - 0.5)$ **24.** $g(x) = \cos\left(\dfrac{x}{2} - 3\right)$

In Exercises 25–28, write an equation of a sinusoidal function with the given amplitude A, phase shift B, and period P. Draw a complete graph.

25. $A = 3, B = \dfrac{\pi}{2}, P = \pi$

26. $A = \dfrac{1}{3}, B = 2, P = 4$

27. $A = 2, B = -\dfrac{\pi}{4}, P = 4\pi$

28. $A = 5, B = -1, P = 1$

In Exercises 29–34, find a complete graph and the domain, range, period, and phase shift of each function.

29. $y = 3\sin 2x$ **30.** $y = 1 + 2\cos(x - \pi)$

31. $y = -\cot(x - \pi)$ **32.** $y = 2\sec\left(x + \dfrac{\pi}{2}\right)$

33. $y = 2 - 3\sin(4x - \pi)$

34. $y = -3 + 2\cos\left(\dfrac{1}{3}x - \dfrac{\pi}{6}\right)$

In Exercises 35–38, find all real-number solutions.

35. $2 = 4\sin 3x$ **36.** $3 = 2\sin 3x$

37. $\cos\dfrac{1}{2}x = 0.24$ **38.** $4\sin 2x < x$

In Exercises 39–42, find all real-number solutions for x in $[0, 5]$.

39. $5\sin(x - \pi) = 3\cos\left(x + \dfrac{\pi}{2}\right)$

40. $\tan(x - \pi) = 3\sin 2x$

41. $3\sin x > 2\cos(x - 1)$

42. $3\sin^2 x > 2.65$

In Exercises 43–46, find a complete graph of each function. Show that each is a sinusoid in the form $y = A\sin(bx + \alpha)$ by estimating A, b, and α. Overlay the complete graph with the graph of $y = A\sin(bx + \alpha)$ to check your estimated values.

43. $y = 3\sin x + 2\cos x$ **44.** $y = -5\sin x + 3\cos x$

45. $y = -\sin(x + \pi) + \cos x$

46. $y = 4\sin 2x - 3\cos 2x$

In Exercises 47–52, find a complete graph of each function. Find the domain, range, and period of each function. Determine all local and absolute extrema in the interval $[0, 2\pi]$.

47. $f(x) = \sin x + \cos 2x$

48. $g(x) = 2\sin x + 3\cos 2x$

49. $f(x) = \sin 3x + \cos x$ **50.** $t(x) = 3\sin 2x - \cos x$

51. $h(x) = 2\sin x + 5\cos x$ **52.** $k(x) = \sin\dfrac{x}{2} + \sin\dfrac{x}{3}$

Exercises 53–56 refer to the following **problem situation**: The *frequency* of a periodic function is the reciprocal of its period. One *cycle* of a periodic function is the graph in one period. It follows that the frequency represents the number of cycles of a periodic function that occur over an interval whose length is 1 unit. Determine the period and frequency.

53. $y = 2\sin 3x$ **54.** $y = -4\cos\dfrac{1}{2}x$

55. $y = 3\sin\dfrac{\pi}{2}x$ **56.** $y = 2\cos 2\pi x$

Exercises 57 and 58 refer to the following **problem situation**: A belt rotates around two wheels having radii r_1 and r_2 as shown.

57. If $r_1 = 22$ in. and $r_2 = 15$ in. and the centers of the wheels are 50 in. apart, find the length of the belt. (*Hint:* Compute d, then compute $\angle\alpha$).

58. Determine the length of a belt if $r_1 = 18$ in. and $r_2 = 11$ in. and the centers of the wheels are 36 in. apart.

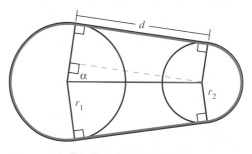

For Exercises 57 and 58.

59. A rocket is launched straight up from ground level at a rate of 200 ft/sec. An observation post is located at a point P located 2055 ft from the launch point. Describe θ as a function of t. Determine the angle θ 15 sec after launch.

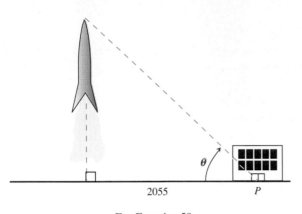

For Exercise 59.

Exercises 60 and 61 refer to the following **problem situation**: The sinusoidal equation

$$T = A + B \sin \left\{ \frac{\pi}{6} \left[360 - (360 - t)/30 \right] + C \right\}$$

is a model for the mean daily temperature T as a function of time t measured in days. A is the mean yearly temperature, B is $\frac{1}{2}$ of the total mean temperature variation, and C is a phase-shift factor for the beginning of a temperature cycle.

60. In Columbus, Ohio, the mean yearly temperature is $58°F$ and the total mean temperature variation is $56°$. Assume the temperature on April 1 is the mean yearly temperature, so $C = 0$, and t is the number of days past April 1. Find a complete graph of this problem situation.

61. Determine approximately the hottest month and day and the coldest month and day during the year in Columbus.

Exercises 62–64 refer to the following **problem situation**: The function $y = A \cos Bt$ is a sinusoidal model for *simple harmonic motion*, which describes the bouncing motion of an object hung from the end of a spring. Introduce a coordinate system so the vertical axis of the spring coincides with the y-axis and the rest position of the object is at the origin. Suppose that at $t = 0$, the object has been stretched to a position $(0, A)$, where $A < 0$, and released (when $t > 0$). Let y be the vertical position of the object as a function of time t. (Assume there is no friction to stop the motion once it has started.)

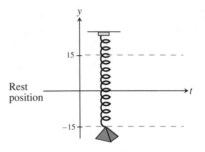

For Exercises 62 to 65.

62. Suppose a spring oscillates between $y = -15$ and $y = 15$ and the frequency of oscillation is $\frac{1}{4}$ cycle per second. Determine the time interval of one complete cycle (that is, the period). Then determine A and B of the algebraic representation for the vertical position y at time t.

63. Find a complete graph of the algebraic representation in Exercise 62 and describe which part of the graph represents the problem situation.

64. At what times is the object in Exercise 62 5 units from the rest position?

65. Suppose an object on a spring oscillates between $y = -28$ and $y = 28$ and the *frequency* of oscillation is 3 cycles per second. Find the algebraic representation for the vertical position of this object and determine the times the object is 5 units from the rest position.

66. A right circular cylinder is inscribed in a sphere of radius 20 (see Example 9). Find a complete graph of the problem situation for the volume V of the cylinder in terms of the angle θ. Find the dimensions of the cylinder of maximum volume that can be inscribed in the sphere.

Exercises 67 and 68 refer to the following **problem situation**: Actual oscillatory motion of an object on a spring is affected by friction, so eventually the object returns to an at-rest position. This behavior is called *damped motion*.

67. Find a complete graph of $y = (20 - x) \cos(x - 3)$.

68. **Writing to Learn**   Write a paragraph that explains how a portion of the graph in Exercise 67 could be a model for the motion of an object hung from a stretched spring.

69. **Writing to Learn**   Write a paragraph that explains when a linear combination of sine and cosine functions is a sinusoid.

7.2 _____

Inverse Trigonometric Functions

In Section 2.7, you learned that if (x, y) belongs to a relation R, then (y, x) belongs to the inverse relation R^{-1}. Theorem 2.3 established that the graph of an inverse relation can be obtained by reflecting the graph of the relation about the line $y = x$. Applying this result to the relation $y = \sin x$, we conclude that the graph in Fig. 7.14 is the inverse of the relation $y = \sin x$ shown in Fig. 7.13.

It is evident that the inverse relation of $y = \sin x$ (see Fig. 7.14) does not satisfy the vertical line test and hence is not a function. This is what we expect to find, given that the graph of $y = \sin x$ does not satisfy the horizontal line test. (Recall that the horizontal line test states that the inverse of a function f is also a function if, and only if, every horizontal line intersects the graph of f at most once.) Therefore the inverse of the sine function over $(-\infty, \infty)$ is not a function.

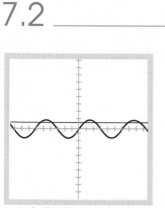

[–10, 10] by [–10, 10]

Figure 7.13 $y = \sin x$ intersects an arbitrary horizontal line $y = c$ more than once, thus indicating that the inverse of $y = \sin x$ will not be a function.

Defining the Inverse Sine Function

Although there is no inverse function for the sine when it is defined over its whole domain, we can define the inverse sine function if we restrict the domain of $y = \sin x$ so that it satisfies the horizontal line test. Consider the function $f(x) = \sin x$ over the interval $[-\pi/2, \pi/2]$. The graph of this function f, shown in Fig. 7.15(a), is a small portion of the complete graph of $y = \sin x$.

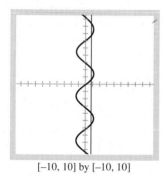

[–10, 10] by [–10, 10]

Figure 7.14 The inverse of $y = \sin x$ is not a function since it fails the vertical line test.

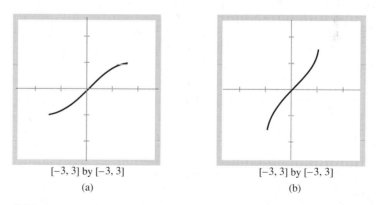

[–3, 3] by [–3, 3]

(a)

[–3, 3] by [–3, 3]

(b)

Figure 7.15 (a) $f(x) = \sin x$ for $-\pi/2 \le x \le \pi/2$; (b) inverse of f.

Notice that the graph of the function f in Fig. 7.15(a) satisfies the horizontal line test. Thus the inverse of f shown in Fig. 7.15(b) is also a function. The inverse sine function can now be defined.

Objective

Students will be able to relate the concept of inverse function to trigonometric functions.

Definition 7.2 Inverse Sine Function

The **inverse sine function**, denoted $y = \sin^{-1} x$ or $y = \arcsin x$, is the function with a domain of $[-1, 1]$ and a range of $-\pi/2 \leq y \leq \pi/2$ that satisfies the relation $\sin y = x$.

The following theorem, stated without proof, confirms that the restricted sine and arcsine are inverse functions.

Theorem 7.4 Inverse Functions

Let $f(x) = \sin x$ such that $-\pi/2 \leq x \leq \pi/2$. Then $f^{-1}(x) = \sin^{-1} x$ and

a) $(f^{-1} \circ f)(x) = x$ [or $\sin^{-1}(\sin x) = x$] for all x in $[-\pi/2, \pi/2]$, and
b) $(f \circ f^{-1})(x) = x$ [or $\sin(\sin^{-1} x) = x$] for all x in $[-1, 1]$.

Common Errors

Students often have difficulty with the symbol used to represent the concept of inverse function. The symbol used to denote inverse functions, f^{-1}, frequently causes confusion because the same symbolism is used for the reciprocal of a number, x^{-1}. The symbol x^{-1} is defined to be $\frac{1}{x}$, but $f^{-1}(x) \neq \frac{1}{f(x)}$.

Defining the Inverse Cosine and Tangent Functions

The inverse cosine function, denoted $y = \cos^{-1} x$, and the inverse tangent function, denoted $y = \tan^{-1} x$, can be defined by following a procedure similar to the one used for defining $y = \sin^{-1} x$. The complete graphs of $y = \cos x$ and $y = \tan x$ (see Fig. 7.16) show that neither relation satisfies the horizontal line test.

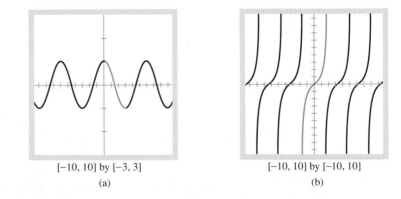

[−10, 10] by [−3, 3]   [−10, 10] by [−10, 10]
(a)   (b)

Figure 7.16 (a) $y = \cos x$; (b) $y = \tan x$. The colored portion of each curve represents the portion of the function used to define the inverse.

However, if the domain of each is restricted to include only the colored portion of each graph, the graph satisfies the horizontal line test. The inverse cosine function and inverse tangent function can be defined when thus restricted.

Frank says, "Be sure students make a clear distinction between inverse *function* and inverse *relation*. Domain and range restrictions must be considered."

Definition 7.3   Inverse Cosine Function

The **inverse cosine function**, denoted by $y = \cos^{-1} x$ or $y = \arccos x$, is the function with a domain of $[-1, 1]$ and a range of $[0, \pi]$ that satisfies the relation $\cos y = x$.

Teaching Note

Graphing calculators can graph the function $y = \sin^{-1} x$, but they cannot graph the relation $\sin y = x$; hence, the name *function grapher*. In Chapter 6, you saw how to use parametric graphing for trigonometry. These same techniques can be used here to graph the inverse of $y = \sin x$.

Definition 7.4   Inverse Tangent Function

The **inverse tangent function**, denoted by $y = \tan^{-1} x$ or $y = \arctan x$, is the function with a domain of $(-\infty, \infty)$ and a range of $(-\pi/2, \pi/2)$ that satisfies the relation $\tan y = x$.

Complete graphs of $y = \sin^{-1} x$, $y = \cos^{-1} x$, and $y = \tan^{-1} x$ are shown in Fig. 7.17.

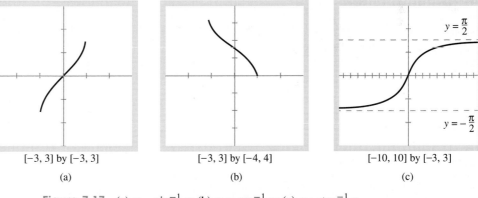

[−3, 3] by [−3, 3]	[−3, 3] by [−4, 4]	[−10, 10] by [−3, 3]
(a)	(b)	(c)

Figure 7.17   (a) $y = \sin^{-1} x$; (b) $y = \cos^{-1} x$; (c) $y = \tan^{-1} x$.

The lines $y = -\pi/2$ and $y = \pi/2$ are horizontal asymptotes of $y = \tan^{-1} x$. The inverse cotangent, secant, and cosecant functions are so rarely used that definitions will not be given for them here.

Teaching Note

A graphing utility automatically restricts the domain, thus graphing the function $y = \arcsin x$. If you wish to demonstrate the arcsin relation, use the parametric mode in your graphing utility. For example,

$x_1 t = T$
$y_1 t = \sin T$
$t \min = -2\pi$
$t \max = 2\pi$.

Notes on Examples

Encourage students to generalize the effects of the coefficient b in a transformed inverse function of the type $y = f^{-1}(bx)$. As shown in Example 1, b has the effect of a horizontal stretch or shrink which affects the domain. But are the results the same for the arctan? What is affected, the domain or the range?

Transformations of Inverse Trig Functions

The graph of the inverse sine function can be transformed by stretch factors and vertical and horizontal shift constants just like other functions we have studied.

E X A M P L E 1   Finding Graphs of Inverse Sine Functions

Find the domain and range and a complete graph of the following:

a) $f(x) = \sin^{-1}(2x)$ **b)** $g(x) = \sin^{-1}\left(\frac{1}{3}x\right)$

Solution   Because the domain of the inverse sine function is $[-1, 1]$, conclude the initial inequality for both parts (a) and (b).

a) For f:

$$-1 \le 2x \le 1$$

$$-\frac{1}{2} \le x \le \frac{1}{2}$$

The domain of f is the interval $[-\frac{1}{2}, \frac{1}{2}]$. We say that f has been obtained from $y = \sin^{-1} x$ by a horizontal shrink by a factor of $\frac{1}{2}$.

b) For g:

$$-1 \le \frac{1}{3}x \le 1$$

$$-3 \le x \le 3$$

The domain of g is the interval $[-3, 3]$. We say that g has been obtained from $y = \sin^{-1} x$ by a horizontal stretch by a factor of 3.

The range of both f and g is $[-\pi/2, \pi/2]$. The complete graphs of f and g are seen in Fig. 7.18.

[−3, 3] by [−3, 3]	[−4, 4] by [−4, 4]
(a)	(b)

Figure 7.18   (a) $f(x) = \sin^{-1}(2x)$; (b) $g(x) = \sin^{-1}\left(\frac{1}{3}x\right)$.

The graph in Fig. 7.18(a) can be obtained from $y = \sin^{-1} x$ by a horizontal shrink by a factor of $\frac{1}{2}$, and the graph in Fig. 7.18(b) can be obtained from $y = \sin^{-1} x$ by a horizontal stretch by a factor of 3.

E X A M P L E   2   Identifying Transformations

Identify transformations that can be used to obtain the graph of

$$f(x) = 3 \sin^{-1}\left[\frac{1}{4}(x+8)\right] - 5$$

from the graph of $y = \sin^{-1} x$. Then find the domain and range and sketch a complete graph of f.

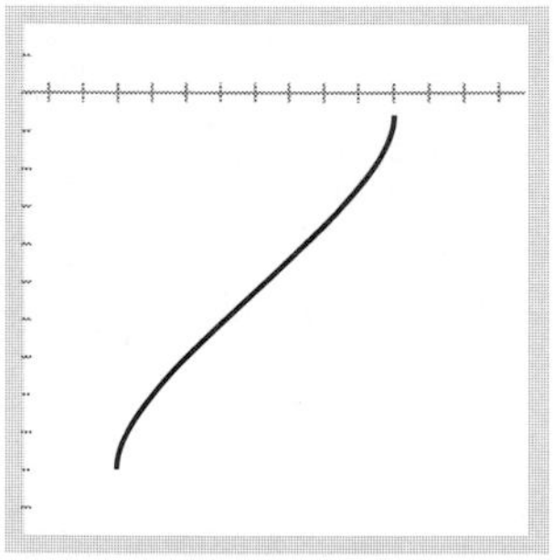

Solution   The graph of f (see Fig. 7.19) can be obtained from the graph of $y = \sin^{-1} x$ by applying, in order, the following transformations:

1. A horizontal stretch by a factor of 4

2. A horizontal shift left 8 units

3. A vertical stretch by a factor of 3

4. A vertical shift down 5 units

The first two transformations change the domain of $y = \sin^{-1} x$ from $[-1, 1]$ to $[-12, -4]$, and the third and fourth transformations change the range of $y = \sin^{-1} x$ from $[-\pi/2, \pi/2]$ to $[-3\pi/2 - 5, 3\pi/2 - 5]$. ▤

[−15, 0] by [−12, 2]

Figure 7.19
$f(x) = 3 \sin^{-1}\left[\frac{1}{4}(x+8)\right] - 5.$

Evaluating Inverse Trigonometric Functions

In Section 6.2 when we were solving an equation like $\cos x = 0.28$, the notation $\cos^{-1} x$ was introduced to mean the angle whose cosine is x. The calculator keys $\boxed{\text{INV}}\ \boxed{\text{COS}}$, $\boxed{\text{2nd}}\ \boxed{\text{COS}}$, or $\boxed{\text{SHIFT}}\ \boxed{\text{COS}}$ were used to find a unique angle solution to this equation. Similar calculator function keys can be used to solve other trig equations. For example, $\boxed{\text{INV}}\ \boxed{\text{SIN}}$, $\boxed{\text{2nd}}\ \boxed{\text{SIN}}$, or $\boxed{\text{SHIFT}}\ \boxed{\text{SIN}}$ can be used for evaluating $\sin^{-1} x$. Thus we have implicitly been using inverse functions when solving some trig equations even before the concept of inverse function was formally defined.

Notes on Exercises

Ex. 27–56 include many types of problems traditionally associated with the topic of inverse trigonometric functions.

E X A M P L E   3   Evaluating the Inverse Sine Function

Evaluate each of the following inverse sine functions at the given value:

a) $\sin^{-1}(0.5)$

b) $\sin^{-1}(-0.7)$

c) $\sin^{-1}(1.2)$

Solution   The basis for doing these problems is the definition

$$y = \sin^{-1} x \qquad \text{if, and only if,} \qquad x = \sin y \text{ and } -\frac{\pi}{2} \le y \le \frac{\pi}{2}.$$

a) $y = \sin^{-1}(0.5)$ if, and only if, $\sin y = 0.5$ and $-\pi/2 \le y \le \pi/2$. The sine of what angle y is 0.5? Recall that $\sin \pi/6 = \sin 30° = 0.5$ and therefore

$$\sin^{-1}(0.5) = \pi/6. \qquad \text{\footnotesize If you do not remember this value, you also can use a calculator.}$$

b) Use a calculator set in radian mode to get $\sin^{-1}(-0.7) = -0.7753974966$.

c) A calculator gives an error message for $\sin^{-1}(1.2)$. The error message occurs because there is no angle whose sine equals 1.2. The sine of any angle or real number is a number in the interval $[-1, 1]$.   ▤

Example 3 illustrates how to use a calculator to find numerical values of the inverse trig functions. In other situations, an algebraic expression may be needed. Example 4 shows that when evaluating an expression like $\cos(\sin^{-1} v)$, it is helpful to think of $\sin^{-1} v$ as an angle.

E X A M P L E   4   Evaluating a Trig Expression

Write an algebraic expression in terms of v for $\cos(\sin^{-1} v)$ and $\tan(\sin^{-1} v)$.

Solution   It follows from the definition for $\sin^{-1} v$ that

$$\sin^{-1} v \text{ is an angle } \theta \text{ such that } \sin \theta = v \qquad \text{and} \qquad -\frac{\pi}{2} \le \theta \le \frac{\pi}{2}.$$

Notice that θ can be in either the first or fourth quadrant depending on whether v is positive or negative. Figure 7.20 shows that $v > 0$ if, and only if, θ is in the first quadrant. Likewise, $v < 0$ if, and only if, θ is in the fourth quadrant. Use the Pythagorean theorem to find the length of the side adjacent to θ to be $\sqrt{1 - v^2}$. Thus

$$\cos(\sin^{-1} v) = \sqrt{1 - v^2} \qquad \text{and} \qquad \tan(\sin^{-1} v) = \frac{v}{\sqrt{1 - v^2}}.$$

Notice that if $v > 0$, $\tan \theta > 0$, and if $v < 0$, $\tan \theta < 0$. However, the $\cos \theta$ is positive whether v is positive or negative.   ▤

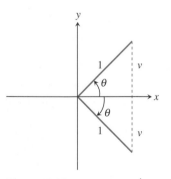

Figure 7.20   Picture $\sin^{-1} v$ as an angle.

E X A M P L E   5   Finding Values without a Calculator

Without using a calculator, find the exact value of each expression:

a) $\sin^{-1}(\tan 3\pi/4)$

b) $\cos[\tan^{-1}(\frac{1}{2})]$

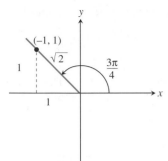

Figure 7.21 Reference triangle for Example 5, part (a): $\theta = 3\pi/4$.

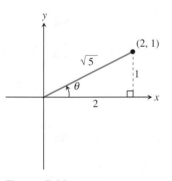

Figure 7.22 Reference triangle for Example 5, part (b): $\tan \theta = \frac{1}{2}$.

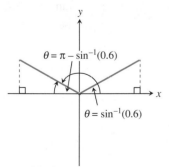

Figure 7.23 Reference triangles for the two solutions to Exercise 6, part (a): $\sin x = 0.6$.

Solution

a) The point $(-1, 1)$ is on the terminal side of angle $3\pi/4$ (see Fig. 7.21). Therefore

$$\tan \frac{3\pi}{4} = -1 \qquad \text{and} \qquad \sin^{-1}\left(\tan \frac{3\pi}{4}\right) = \sin^{-1}(-1).$$

Figure 7.15(b) is a reminder that the domain of $y = \sin^{-1} x$ is restricted to $[-1, 1]$ and the range is $[-\pi/2, \pi/2]$. So $\sin^{-1}(-1) = -\pi/2$.

b) Let θ be an angle between $-\pi/2$ and $\pi/2$ whose tangent is $\frac{1}{2}$. Then θ must be in the first quadrant since the tangent is negative in the fourth quadrant (see Fig. 7.22).

$$\theta = \tan^{-1}\left(\frac{1}{2}\right)$$

$$\cos \theta = \cos\left[\tan^{-1}\left(\frac{1}{2}\right)\right] \qquad \text{Use Fig. 7.22.}$$

$$= \frac{2}{\sqrt{5}}$$

We end this section by illustrating how to use a calculator to solve certain trigonometric equations.

E X A M P L E   6   Solving Equations

Solve each of the following equations for x:

a) $\sin x = 0.6$

b) $\cot x = 2.5$

Solution   Because of the periodic nature of the trig functions, there are many solutions to these equations. The inverse trig functions can be used to find the solution in one period. The arguments that follow show how to find the rest.

a) One solution to $\sin x = 0.6$ can be found by using a calculator to determine

$$\sin^{-1}(0.6) = 0.6435011.$$

A second solution is the angle in the second quadrant $\pi - 0.6435011 = 2.4980915$ (see Fig. 7.23). All other solutions are angles θ that are coterminal with one of these two solutions. Thus the solutions to $\sin \theta = 0.6$ are

$$0.6435011 + 2k\pi \qquad \text{and} \qquad 2.4980915 + 2k\pi,$$

where k is an integer.

Notes on Exercises

This set of exercises is routine. Check that in using calculators to produce solutions, students are aware of the restrictions for inverses that affect how many solutions they are producing. Proficiency in using the technology is one of the goals of this section. Focus students' attention on what happens to domain and range.

In Ex. 13–20, students should be cautioned to check the mode setting (degrees versus radians).
Ex. 37–42 are computed without the use of a calculator.

Assignment Guide

Day 1: Ex. 1, 3, 5–9, 13– 26 odd, 27–38, 40, 45, 47, 49
Day 2: Ex. 51, 52, 61—63

b) Because calculators do not have $\cot x$ as a built-in function, we can change this equation to one involving $\tan x$.

$$\cot x = 2.5 \qquad \text{if, and only if,} \qquad \tan x = \frac{1}{\cot x} = \frac{1}{2.5} = 0.4$$

$$\tan^{-1}(0.4) = 0.3805064$$

The tangent function is positive in both first and third quadrants; therefore the other solutions to $\tan x = 0.4$ are

$$0.3805064 + k\pi,$$

where k is any integer. ■

E X A M P L E   7   Verifying an Identity

Show that $\sin^{-1} x + \cos^{-1} x = \pi/2$ for all x in $[-1, 1]$.

Solution   Let $\theta = \sin^{-1} x$. Then

$$x = \sin\theta \qquad \text{and} \qquad -\frac{\pi}{2} \le \theta \le \frac{\pi}{2}.$$

By applying the concept of horizontal shift identities explained in Section 6.4 and because $\cos(\theta - \pi/2) = \cos(\pi/2 - \theta)$, we see that

$$x = \sin\theta = \cos\left(\frac{\pi}{2} - \theta\right) \qquad \text{and} \qquad 0 \le \frac{\pi}{2} - \theta \le \pi.$$

It follows that

$$\cos^{-1} x = \frac{\pi}{2} - \theta$$

$$\theta + \cos^{-1} x = \frac{\pi}{2}$$

$$\sin^{-1} x + \cos^{-1} x = \frac{\pi}{2}.$$

■

Exercises for Section 7.2

In Exercises 1–6, find the domain, range, and a complete graph of each function.

1. $f(x) = \sin^{-1}(3x)$

2. $g(x) = \cos^{-1}(x) - \dfrac{\pi}{2}$

3. $h(x) = \arcsin(x + 1)$

4. $k(x) = 2\sin^{-1}\left(\dfrac{x}{3}\right)$

5. $f(x) = 3\arccos(2x - 4)$

6. $g(x) = \tan^{-1}(x - 1) + \pi$

In Exercises 7–12, identify transformations that can be used to draw the graph of each function. Then sketch a complete graph and find the domain and range of each.

7. $y = 1 - \arcsin x$

8. $y = 3 + \cos^{-1}(x - 2)$

9. $y = -0.25\tan^{-1}(x - \pi)$

10. $g(x) = 3\arccos\left(\dfrac{1}{2}x - \pi\right)$

11. $y = \sin(\sin^{-1} x)$

12. $y = \sin^{-1}(\sin x)$

In Exercises 13–16, use a calculator to evaluate each expression. Express your answer in degrees.

13. $\sin^{-1}(0.362)$

14. $\arcsin(-1.67)$

15. $\tan^{-1}(0.125)$ **16.** $\tan^{-1}(-2.8)$

In Exercises 17–20, use a calculator to evaluate each expression. Express your answer in radians.

17. $\sin^{-1}(0.46)$ **18.** $\cos^{-1}(-0.853)$

19. $\tan^{-1}(2.37)$ **20.** $\tan^{-1}(-22.8)$

In Exercises 21–26, compute the exact value in radians without using a calculator.

21. $\sin^{-1} 1$ **22.** $\tan^{-1} \sqrt{3}$

23. $\sin^{-1} \dfrac{\sqrt{2}}{2}$ **24.** $\cos^{-1}\left(-\dfrac{\sqrt{3}}{2}\right)$

25. $\tan^{-1}(-\sqrt{3})$ **26.** $\sin^{-1}(-1)$

In Exercises 27–36, use a calculator to evaluate each expression. Express your answer as a real number.

27. $\sin[\sin^{-1}(0.36)]$ **28.** $\sin[\arccos(0.568)]$

29. $\sin^{-1}(\cos 20)$ **30.** $\cos[\sin^{-1}(-0.125)]$

31. $\sin[\sin^{-1}(1.2)]$ **32.** $\sin^{-1}(\sin 1.2)$

33. $\sin[\sin^{-1}(2)]$ **34.** $\sin^{-1}(\sin 2)$

35. $\tan^{-1}(\sin 2)$ **36.** $\tan(\arctan 3)$

In Exercises 37–42, compute the exact value without using a calculator.

37. $\cos\left[\sin^{-1} \frac{1}{2}\right]$ **38.** $\sin[\tan^{-1} 1]$

39. $\cos[(2 \sin^{-1} \frac{1}{2})]$

40. $\cos[\tan^{-1}(\sqrt{3})] - \sin[\tan^{-1}(0)]$

41. $\cos[\sin^{-1}(0.6)]$ **42.** $\sin[\tan^{-1}(2)]$

In Exercises 43–46, solve for x. Find the exact solution(s) in each case.

43. $\sin(\sin^{-1} x) = 1$ **44.** $\cos^{-1}(\cos x) = 1$

45. $2 \sin^{-1} x = 1$ **46.** $\tan^{-1} x = -1$

In Exercises 47–50, find an equivalent algebraic expression not involving trig functions.

47. $\sin(\tan^{-1} x)$ **48.** $\cos(\tan^{-1} x)$

49. $\tan(\sin^{-1} x)$ **50.** $\cot(\cos^{-1} x)$

In Exercises 51–54, solve for x with an error of at most 0.01. Find *all* solutions.

51. $\tan x = 2.3$ **52.** $\sin x = -0.75$

53. $\sec x = 3$ **54.** $\cot x = -5$

In Exercises 55 and 56, use a graphical method to find the solutions to each inequality over the interval $[-\pi, \pi]$.

55. $(\sin x)(\tan^{-1} x) \geq 0$ **56.** $\dfrac{\sin^{-1}(2x)}{\sin x} \geq 2$

In Exercises 57–60, verify each identity.

57. $\sin^{-1}(-x) = -\sin^{-1} x$ for $|x| \leq 1$

58. $\sin^{-1} x = \tan^{-1} \dfrac{x}{\sqrt{1-x^2}}$ for $|x| < 1$

59. $\arccos x + \arcsin x = 90°$

60. $\cos(\sin^{-1} x) = \sqrt{1 - x^2}$ for $|x| \leq 1$

61. The length L of the shadow cast by a tower 50 ft tall depends on θ, the angle of elevation of the sun (measured from the horizontal). Express θ as a function of L. Draw a complete graph of the function and describe which portion of the graph represents the problem situation.

62. Consider the following **problem situation**: A revolving light beacon L stands 3 mi from the closest point P along a straight shoreline. Express angle $PLQ = \theta$ as a function of the distance x from P to Q. Draw a complete graph of this function and describe which portion of the graph represents the problem situation.

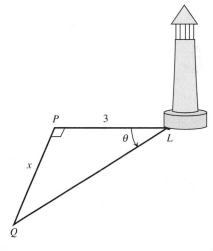

For Exercise 62.

63. Suppose triangle ABC is isosceles with $a = b$ and $\angle C = \theta$. Show that the area of the triangle is given by $A = \frac{1}{2}a^2 \sin\theta$.

64. Use a graphical method to find all values of x such that

$$\sin^{-1}\left(\frac{2x}{x^2+1}\right) = 2\tan^{-1} x.$$

65. Writing to Learn   Write a paragraph that explains how solving the equation $\sin y = 0.5$ is different from finding the value $\sin^{-1}(0.5)$.

7.3

Solving Trigonometric Equations and Inequalities Graphically

Even though trigonometric functions differ in many respects from polynomial, logarithmic, and exponential functions, the graphical methods for solving equations used in previous chapters apply to trigonometric functions as well.

Solving Trigonometric Equations

Example 6 in Section 7.2 shows that there are an infinite number of solutions to trigonometric equations of the type that will be solved in this section. The procedure used in this section to solve trig equations is as follows. First, use a graphical method to find all solutions for one period (excluding the right endpoint) of the trig function found in the equation. Then, find all other solutions by taking into account the periodicity of the functions.

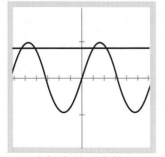

[$-2\pi, 2\pi$] by [$-2, 2$]

Figure 7.24 $y = \sin x$ and $y = 0.8$.

E X A M P L E   1   Solving an Equation with a Single Trig Function

Find all solutions to the equation $\sin x = 0.8$.

Solution   Complete graphs of functions $y = \sin x$ and $y = 0.8$ are shown in Fig. 7.24. Notice that these two graphs will have an infinite number of solutions in the domain $(-\infty, \infty)$.

Our method is to first find all solutions in one period of $y = \sin x$. Then we find all other solutions.

Solutions in $[0, 2\pi)$ From Fig. 7.24, estimate that $x = 0.9$ and $x = 2.2$ are solutions. The value $\sin^{-1}(0.8) = 0.927$, which by definition of the function $y = \sin^{-1} x$ is in the first quadrant.

The second-quadrant solution is $\pi - \sin^{-1}(0.8) = 2.214$. (Why?)

Complete Solution   The solution set consists of all real numbers of the form $0.927 + 2k\pi$ or $2.214 + 2k\pi$, where k is any integer.

If $y = f(x)$ is a trigonometric function, then solving the equation $f(x) = 0$ is equivalent to finding the zeros of the function $y = f(x)$. The method for finding all zeros of a trigonometric function is summarized next.

Zeros of Periodic Functions

If $y = f(x)$ is a periodic function with period P, the equation $f(x) = 0$ is called a **periodic equation**.

Suppose $x_1, x_2, \ldots, x_n$ are all the solutions to $f(x) = 0$ over an interval of length P. Then the set of all solutions to $f(x) = 0$ over $(-\infty, \infty)$ consists of the numbers $x_i + kP$ where $i = 1, 2, \ldots, n$ and k is any integer.

This method of finding zeros of a trigonometric equation is illustrated in Examples 2 through 4.

E X A M P L E   2   Solving an Equation with Several Trig Functions

Find all solutions to the equation $2 \cos^2 t + \sin t + 1 = 0$.

Solution   A complete graph of $y = 2 \cos^2 t + \sin t + 1$ is given in Fig. 7.25. We see that this graph shows two complete periods and that there is only one solution in the single period interval $[0, 2\pi)$. Use zoom-in to determine this solution to be $x = 4.71$ with an error at most 0.01.

Complete Solution   The solution set consists of all real numbers of the form $4.71 + 2k\pi$, where k is any integer.   ≡

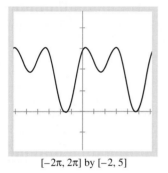

$[-2\pi, 2\pi]$ by $[-2, 5]$

Figure 7.25 $y = 2 \cos^2 t + \sin t + 1$.

Section 7.6 will show that the solutions to the equation in Example 2 can be derived using an algebraic method. However, because the graphical method can be used when the algebraic methods fail, it is more general. Examples 3 and 4 deal with such equations.

E X A M P L E   3   Solving an Equation with Several Trig Functions

Find all the solutions to the equation $\tan x = 3 \cos x$.

Solution   We need to find the zeros of the function $f(x) = \tan x - 3 \cos x$. The period of $y = \tan x$ is π, and the period of $3 \cos x$ is 2π. So we first consider solutions in the interval $[0, 2\pi)$.

Solutions in $[0, 2\pi)$ Solutions to the equation $\tan x = 3 \cos x$ can be found by finding zeros to the function $f(x) = \tan x - 3 \cos x$. Figure 7.26 shows that there are two zeros in the interval $[0, 2\pi)$.

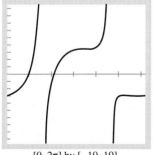

$[0, 2\pi]$ by $[-10, 10]$

Figure 7.26   A complete graph of $f(x) = \tan x - 3 \cos x$.

Use zoom-in to determine that the two solutions in this interval are $x = 1.01$ and $x = 2.13$ with errors of at most 0.01.

Complete Solution The solution set consists of all real numbers of the form $1.01 + 2k\pi$ and $2.13 + 2k\pi$, where k is any integer. ▤

E X A M P L E   4   Solving an Equation with Several Trig Functions

Find all solutions to the equation $3 \sin^3 2x - 2 \cos x = 0$.

Solution We first find all zeros in one period of the function $f(x) = 3 \sin^3 2x - 2 \cos x$. It can be shown that the period of $f(x) = 3 \sin^3 2x - 2 \cos x$ is 2π, and so we begin by finding all solutions in the interval $[0, 2\pi)$.

Solutions in $[0, 2\pi)$ Figure 7.27 shows that $f(x) = 3 \sin^3 2x - 2 \cos x$ has six zeros in the interval $[0, 2\pi)$. Use zoom-in to show that they are 0.50, 1.25, 1.57, 1.89, 2.65, and 4.71 with errors of at most 0.01.

Complete Solution The solution set consists of all real numbers of the form $0.50 + 2k\pi$, $1.25 + 2k\pi$, $1.57 + 2k\pi$, $1.89 + 2k\pi$, $2.65 + 2k\pi$, and $4.71 + 2k\pi$. ▤

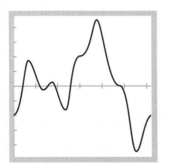

$[0, 2\pi]$ by $[-5, 5]$

Figure 7.27 A complete graph of $f(x) = 3 \sin^3 2x - 2 \cos x$.

Solving Inequalities

Examples 1 to 4 illustrated how trigonometric equations can be solved using a graphical method. The technique used was to first find all solutions in one period of the function and then find all other solutions by using the fact that functions are periodic. Inequalities can be solved graphically in much the same way. First we will find solutions to the corresponding equation and then take into account the inequality.

Notes on Examples

In Example 5, some students may enter $y_1 = \sin x < \cot x$ to determine the solution. This will give them the correct interval if their graphing utility has a Boolean Algebra operator. However, the graph will be different from Fig. 7.28.

E X A M P L E   5   Solving a Trigonometric Inequality

Solve $\sin x < \cot x$.

Solution The inequality $\sin x < \cot x$ is equivalent to the inequality $\sin x - \cot x < 0$. Consider the graph of $f(x) = \sin x - \cot x$ (see Fig. 7.28) and determine what portion lies below the x-axis.

   Because it can be shown that the period of f is 2π, it is sufficient to consider the solution in the interval $[0, 2\pi)$. Furthermore, f has vertical asymptotes at $x = 0$, $x = \pi$, and $x = 2\pi$.

   Use zoom-in to determine the two zeros of f in $[0, 2\pi)$ to be $x = 0.90$ and $x = 5.38$ with errors of at most 0.01.

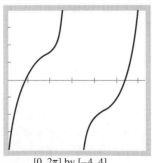

$[0, 2\pi]$ by $[-4, 4]$

Figure 7.28 A complete graph of $f(x) = \sin x - \cot x$.

Solution in $[0, 2\pi)$ The solution to $\sin x < \cot x$ in $[0, 2\pi)$ consists of all numbers whose x-values lie in the intervals $(0, 0.90)$ or $(\pi, 5.38)$.

Complete Solution The solution set consists of $R = (0, 0.90) \cup (\pi, 5.38)$ together with all horizontal shifts of R right and left through all integer multiples of 2π. ☰

The power of the graphical method is illustrated perhaps most dramatically by using it to solve an equation or an inequality that combines trigonometric functions with nontrigonometric functions. Algebraic methods are generally not useful for this type of equation or inequality.

EXAMPLE 6 Solving a Trig and Polynomial Inequality

Find all solutions to $\sin 2x \geq x^2 - 1$.

Solution Figure 7.29 shows the complete graphs of both $y = \sin 2x$ and $y = x^2 - 1$. Notice that the graph of $y = \sin 2x$ lies above the graph of $y = x^2 - 1$ for only one interval. Use zoom-in to find that the two curves intersect when $x = -0.46$ and $x = 1.26$ with errors of at most 0.01.

The solution to $\sin 2x \geq x^2 - 1$ is the interval $[-0.46, 1.26]$. Notice that the endpoints are included since the inequality is $\geq$ rather than $>$. ☰

We end this section by considering several functions that have interesting and unusual behaviors.

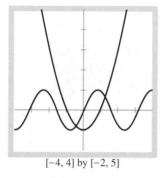

[−4, 4] by [−2, 5]

Figure 7.29 $y = \sin 2x$ and $y = x^2 - 1$.

🔍 EXPLORE WITH A GRAPHING UTILITY

Find a graph of $f(x) = \sin(1/x)$. Experiment with various viewing rectangles to observe the behavior of $f(x)$.

1) As $x \to \infty$, what does $f(x)$ do? That is, what is its end behavior?

2) As $x \to 0$, how does $f(x)$ behave?

The experience from this Exploration should help you understand the next two examples.

EXAMPLE 7 Solving a Special Trig Equation

How many solutions are there to the equation $\sin(1/x) = 0$?

Solution An exploration of the graph of $f(x) = \sin(1/x)$ in three different viewing rectangles reveals some interesting behavior.

Figure 7.30(a) shows the end behavior of the function. In particular, the x-axis is an asymptote and there are no solutions to the equation when $|x|$ is large. Figure 7.30(b) shows interesting behavior near $x = 0$, while Fig. 7.30(c) supports a claim that there are an infinite number of solutions to the equation.

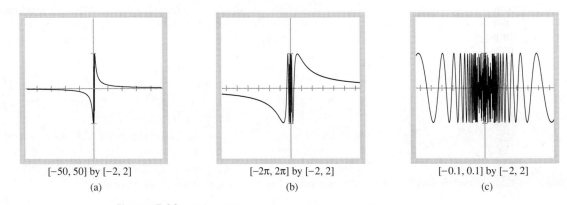

| $[-50, 50]$ by $[-2, 2]$ | $[-2\pi, 2\pi]$ by $[-2, 2]$ | $[-0.1, 0.1]$ by $[-2, 2]$ |
| (a) | (b) | (c) |

Figure 7.30   Three different views of $f(x) = \sin(1/x)$.

TRY THIS

1. Set your grapher mode to *simultaneous*.

2. Graph $y = x$, $y = x \sin x$, and $y = -x$ simultaneously in the viewing rectangle $[-30, 30]$ by $[-30, 30]$.

To analyze this behavior near $x = 0$ algebraically, observe that $\sin \alpha = 0$ whenever $\alpha = k\pi$ (where k is a nonzero integer). Therefore

$$\sin \frac{1}{x} = 0 \quad \text{if, and only if,} \quad \frac{1}{x} = k\pi$$

$$\text{if, and only if,} \quad x = \frac{1}{k\pi}.$$

As $|k|$ gets larger, each value $1/k\pi$ is closer to zero. Therefore we see that $f(x) = \sin(1/x)$ has an infinite number of zeros, all in the interval $[-1, 1]$. ▬

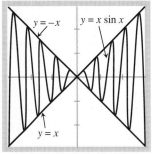

$[-30, 30]$ by $[-30, 30]$

Figure 7.31 $y = x$, $y = -x$, and $y = x \sin x$.

E X A M P L E   8   Solving a Double Inequality

Use a graphical method to describe solutions for the following inequalities:
a) $-x \leq x \sin x \leq x$ **b)** $-|x| \leq x \sin x \leq |x|$

Solution   Typically, to solve an inequality, we would graph the function on each side of the inequality sign in the same viewing rectangle. This method will work with double inequalities also. Find graphs of the three functions $y = x$, $y = -x$, and $y = x \sin x$ in the viewing rectangle $[-30, 30]$ by $[-30, 30]$ (see Fig. 7.31). Notice that the graph of $y = x \sin x$ oscillates back and forth and always lies between the lines $y = x$ and $y = -x$.

At first glance, you may be inclined to say that the two inequalities are satisfied for all values of x. However, that is not correct.

a) Notice that the graph of $y = x \sin x$ lies on or below the graph of $y = x$ for all x-values to the right of the y-axis. However, only for certain specific negative values of x is it true that $x \sin x \leq x$.

Also, the graph of $y = x \sin x$ lies on or above the graph of $y = -x$ to the right of the y-axis. Again, only for certain specific negative values of x is it true that $x \sin x \geq -x$.

The negative values of x for which $x \sin x = x$ are not the same values of x for which $x \sin x = -x$, so we conclude that the solution to the inequality $-x \leq x \sin x \leq x$ is the interval $[0, \infty)$.

b) The graph of $y = x \sin x$ lies on or below the graph of $y = |x|$ for all values of x and lies on or above the graph of $y = -|x|$ for all values of x.

The solution to $-|x| \leq x \sin x \leq |x|$ is the interval $(-\infty, \infty)$. ≣

Exercises for Section 7.3

In Exercises 1–4, use a graphing utility.

1. Solve the equation in Example 2 by considering the graphs of $y = 2\cos^2 t$ and $y = -\sin t - 1$.

2. Support the solution to the equation in Example 2 by comparing the graphs of $y = 2\cos^2 t + \sin t$ and $y = -1$. Does the solution agree with Example 2 and Exercise 1?

3. Find the solutions to the equation $\tan x = 3\cos x$ in Example 3 by finding the points of intersection of the graphs of $y = \tan x$ and $y = 3\cos x$.

4. Solve the equation in Example 4 by finding the points of intersection of the graphs of $y = 3\sin^3 2x$ and $y = 2\cos x$.

In Exercises 5–8, solve each equation.

5. $\sin x = 0.7$

6. $\cos x = 0.9$

7. $\sin x = 1.3$

8. $\tan x = 2.75$

In Exercises 9–18, solve each equation over the interval $[0, 2\pi)$.

9. $\sin 2x = 1$

10. $\sin 3t = 1$

11. $2\sin x = 1$

12. $2\sin 3x = 1$

13. $3\cos 2x = 1$

14. $\sin 2t = \sin t$

15. $\sin^2 x = 0$

16. $\sin x \tan x + \sin x = 0$

17. $(\cos x)(\sin x - 1) = 0$

18. $\sin^2 x - 1 = 0$

In Exercises 19–28, find the complete solution for each equation.

19. $\sin^2 \theta - 2\sin \theta = 0$

20. $2\cos 2t = 1$

21. $3\sin t = 2\cos^2 t$

22. $\cos(\sin x) = 1$

23. $1 = \csc x - \cot x$

24. $\sin 2t = \sin t$

25. $2\sin^2 x + 3\sin x - 2 = 0$

26. $2\cos^2 x + \cos x - 1 = 0$

27. $\cos 2x + \cos x = 0$

28. $\tan^2 x \cos x + 5\cos x = 0$

In Exercises 29–32, solve each inequality.

29. $\sin x < \dfrac{1}{2}$ for $0 \leq x \leq 2\pi$

30. $|\sin x| < \dfrac{1}{2}$ for $0 \leq x \leq 2\pi$

31. $\sin 2x < \cos 2x$

32. $\tan x > 0$

In Exercises 33–40, find the complete solution to each of these equations or inequalities.

33. $3\sin x = x$

34. $\tan x = x$ for $-\pi \leq x \leq \pi$

35. $3\sin 2x - x^2 = 0$

36. $\sin x < \cos(x - \pi)$

37. $5 \sin 2x < 3 \cos x$

38. $x \sin x \geq 1$ for $-10 \leq x \leq 10$

39. $\sin 3x + \cos x = 0$

40. $3 \sin 2x = \cos x$

In Exercises 41–44, find the graph in the indicated interval. Determine all local extrema over the interval.

41. $g(x) = 2 - 3 \sin \left(\dfrac{1}{2} x - \dfrac{\pi}{2} \right)$ for $0 \leq x < 4\pi$

42. $f(x) = x^2 \sin x$ for $-\pi \leq x < \pi$

43. $g(x) = -1 + 2 \cos(\pi x - 2)$ for $0 \leq x \leq 3$

44. $f(x) = e^{-x/2} \sin 2x$ for $0 \leq x \leq 8$

45. Show that the function $f(x) = x \sin x$ is not a periodic function.

46. Prove or disprove that $g(x) = \sin 1/x$ is a periodic function.

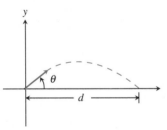

For Exercises 47 to 52.

Exercises 47–49 refer to the following **problem situation**: A cannon shell is fired with an initial velocity V (in feet per second). If the cannon barrel makes an angle θ with the ground, then the horizontal distance d the shell travels before it hits the ground is given by $d = \dfrac{V^2}{16} \sin \theta \cos \theta$.

47. Suppose $V = 500$ ft/sec. Draw a complete graph of d as a function of θ. What portion of the graph represents the

problem situation?

48. If the initial velocity is 500 ft/sec, at what angle should the cannon be aimed to hit a target 2350 ft away?

49. If the initial velocity is 500 ft/sec, what is the maximum distance the shell travels?

Exercises 50–52 refer to the following **problem situation** as shown by the preceding figure: The range d of a projectile shot at an angle of elevation θ and with an initial velocity v is given by

$$d(\theta) = \frac{v^2}{g} \sin 2\theta,$$

where g is the acceleration due to gravity (32 ft/sec^2). Assume $v = 85$ ft/sec.

50. Draw a complete graph of the trigonometric representation d and discuss the values of θ that make sense in this problem situation.

51. At what angle should the projectile be aimed in order to hit a target that is 215 ft down range?

52. For what value of θ is the range maximum?

53. **Writing to Learn**   Find a complete graph of $f(x) = x \sin 1/x$ and then write a paragraph that describes the end behavior of f.

54. **Writing to Learn**   Find a complete graph of $f(x) = x^{\sin x}$ and then write a paragraph that describes the end behavior of f.

55. **Writing to Learn**   Write a paragraph or two that describes at least two graphical methods for solving the inequality $f(x) \leq g(x)$ where f and g are any two functions.

56. **Writing to Learn**   Explain why in Example 1, when solutions in one period are found, the right endpoint in $[0, 2\pi)$ is excluded.

7.4 _____   Trigonometric Identities

You will recall from earlier work in solving equations that an identity is an equation that is true for all values of the variable for which the expressions are defined. We restate here for convenience the fundamental identities introduced in Section 6.2.

Objective

Students will be able to use the common identities to simplify trigonometric expressions.

Key Ideas

Fundamental identities
Reciprocal identities
Tangent and cotangent
 identities
Pythagorean identities
Verifying trigonometric
 identities

Teaching Note

The identities and proofs presented in this section are fairly standard. Although the role of identities is different in today's technologically oriented mathematics, you should feel free to explore fundamental identities using graphing utilities and to design other approaches to teaching about these identities.

Fundamental Trig Identities

Reciprocal Identities

$$\sin\theta = \frac{1}{\csc\theta} \qquad \cos\theta = \frac{1}{\sec\theta} \qquad \tan\theta = \frac{1}{\cot\theta}$$

$$\csc\theta = \frac{1}{\sin\theta} \qquad \sec\theta = \frac{1}{\cos\theta} \qquad \cot\theta = \frac{1}{\tan\theta}$$

Tangent and Cotangent Identities

$$\tan\theta = \frac{\sin\theta}{\cos\theta} \qquad\qquad \cot\theta = \frac{\cos\theta}{\sin\theta}$$

Pythagorean Identities

$$\sin^2\theta + \cos^2\theta = 1 \qquad 1 + \tan^2\theta = \sec^2\theta \qquad 1 + \cot^2\theta = \csc^2\theta$$

Trig identities have a variety of uses. Example 4 in Section 6.2 illustrates that the reciprocal identities are applied whenever the cotangent, secant, or cosecant functions are evaluated using a calculator.

Often it is necessary to change the form of an equation before a solution can be found. In calculus, trig identities often will be used to change a trigonometric expression to a form more appropriate for solving the given problem.

Another important reason for learning to simplify trigonometric expressions is that the simplified expressions are helpful for interpreting certain computer output.

E X A M P L E 1   Simplifying a Trig Expression

Simplify the expression $(\cot\theta)/(\csc\theta)$.

Solution

$$\frac{\cot\theta}{\csc\theta} = \cot\theta \cdot \frac{1}{\csc\theta}$$

$$= \frac{\cos\theta}{\sin\theta} \cdot \frac{\sin\theta}{1} = \cos\theta \qquad \text{Use the fundamental identity for } \cot\theta \text{ and simplify algebraically.}$$ ▬

The algebra in the Example 1 solution shows that $(\cot\theta)/(\csc\theta) = \cos\theta$ is a trig identity; that is, this equation is true for all values of θ for which the expressions are defined.

An equation is not an identity if the equation fails to be true for at least one value of the variable for which the expressions are defined. Graphing the function represented by each side of the equation visually supports the conclusion that an

SUGGESTION

Often a trigonometric expression can be simplified if all functions are changed to expressions involving $\sin\theta$ and $\cos\theta$. That is the method used in the solution to Example 1.

equation is or is not an identity. This use of a graphing utility is illustrated in Example 2.

E X A M P L E   2   Demonstrating a Trig Nonidentity

Show that

$$\frac{1 + \sin^2 \theta}{\cos \theta} = \cos \theta$$

is not an identity.

Solution   There are some solutions for this equation; in particular, $\theta = k\pi$ is a solution to the equation for any integer value of k.

Because some solutions exist, the question occurs, is this equation an identity? We see that the graphs of $y = (1 + \sin^2 \theta)/\cos \theta$ and $y = \cos \theta$ in Fig. 7.32 show that the two sides of the equation are not identical, that is, this equation is not an identity. ≡

[$-2\pi, 2\pi$] by [$-5, 5$]

Figure 7.32 $y = \dfrac{1 + \sin^2 \theta}{\cos \theta}$ and $y = \cos \theta$.

Verifying Trigonometric Identities

To verify that a trig equation is an identity, we must show that the equation is true for all values of the variable for which all expressions in the equation are defined. Follow one of these steps.

Verifying an Identity

To verify that an equation is an identity, follow one of these two approaches:

1. Simplify the more complicated side until it is identical to the other side.
2. Simplify each side separately until both sides are identical.

E X A M P L E   3   Verifying an Identity

Verify that

$$\sin \theta = \frac{\sec \theta}{\tan \theta + \cot \theta}$$

is an identity.

Solution   We simplify the right side of the equation until it is identical to the left side.

$$\frac{\sec \theta}{\tan \theta + \cot \theta} = \frac{1}{\cos \theta} \left(\frac{\sin \theta}{\cos \theta} + \frac{\cos \theta}{\sin \theta} \right)^{-1}$$

$$= \frac{1}{\cos\theta}\left(\frac{\sin^2\theta + \cos^2\theta}{\sin\theta\cos\theta}\right)^{-1}$$

$$= \frac{1}{\cos\theta} \cdot \frac{\sin\theta\cos\theta}{1}$$

$$= \frac{\sin\theta\cos\theta}{\cos\theta}$$

$$= \sin\theta$$

Stan says, "Graphing utilities are powerful tools. They will quickly convince students about the validity of identities. Seeing is believing."

When verifying that an equation is an identity, it is important not to perform a step that already assumes the equation is valid for all values of the variable. In particular, *it is not valid to multiply both sides of an equation by an expression.*

≡

EXAMPLE 4 Multiplying Numerator and Denominator

Verify the identity

$$\frac{\cos t}{1 - \sin t} = \frac{1 + \sin t}{\cos t}.$$

Solution   It may be tempting to begin by multiplying both sides of the equation by $\cos t$ and by $1 - \sin t$. However, that method presumes the equation is an identity, or that it is true before it has been verified.

A valid method is to simplify the right side of the equation until it is identical to the left.

$$\frac{1 + \sin t}{\cos t} = \frac{1 + \sin t}{\cos t} \cdot \frac{1 - \sin t}{1 - \sin t} \qquad \text{Multiply numerator and denominator by the conjugate of the numerator.}$$

$$= \frac{1 - \sin^2 t}{(\cos t)(1 - \sin t)}$$

$$= \frac{\cos^2 t}{(\cos t)(1 - \sin t)}$$

$$= \frac{\cos t}{1 - \sin t}$$

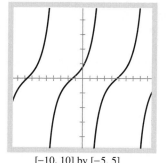

[−10, 10] by [−5, 5]

Figure 7.33 $y = \dfrac{\cos t}{1 - \sin t}$ and $y = \dfrac{1 + \sin t}{\cos t}$.

Therefore

$$\frac{\cos t}{1 - \sin t} = \frac{1 + \sin t}{\cos t}$$

is an identity.

≡

The graph in Fig. 7.33 supports this algebraic verification.

In some cases, it is easier to verify an identity by working with both sides of the equation. Work on one side until it has become simpler, and then work on the other side until it becomes identical.

E X A M P L E   5   Working on Each Side

Verify the identity

$$\frac{\tan^2 x - 1}{\sec^2 x} = \frac{\tan x - \cot x}{\tan x + \cot x}.$$

Solution   First change the right side.

$$\frac{\tan x - \cot x}{\tan x + \cot x} = \frac{\tan x - \dfrac{1}{\tan x}}{\tan x + \dfrac{1}{\tan x}}$$

$$= \frac{(\tan x)\left(\tan x - \dfrac{1}{\tan x}\right)}{(\tan x)\left(\tan x + \dfrac{1}{\tan x}\right)}$$

$$= \frac{\tan^2 x - 1}{\tan^2 x + 1}$$

Then change the left side.

$$\frac{\tan^2 x - 1}{\sec^2 x} = \frac{\tan^2 x - 1}{\tan^2 x + 1}$$

The right side was simplified to $(\tan^2 x - 1)/(\tan^2 x + 1)$, and the left side was simplified to $(\tan^2 x - 1)/(\tan^2 x + 1)$. ≡

A graphing utility can be used to support a claim that an equation is an identity. After finding a complete graph of the function on the left side of the equation, overlay a graph of the right side of the equation in the same viewing rectangle. Check that the graphs appear to be identical.

E X A M P L E   6   Providing Visual Support for an Identity

Use a grapher to provide visual support for the claim that the equation

$$(\cos x)(1 + \tan x)^2 = \frac{1}{\cos x} + 2\sin x$$

is an identity.

SUGGESTIONS FOR SUPPORTING IDENTITIES GRAPHICALLY

The graphs of $y_1 = \dfrac{\cos t}{1 - \sin t}$ and $y_2 = \dfrac{1 + \sin t}{\cos t}$ appear to be identical. Can you see the graph of $y_1 - y_2$? Why? Try graphing $y_1 - y_2 + 1 = y$.

Notes on Exercises

For Ex. 1–16, attention should be focused on algebraic skills; however, feel free to incorporate the use of a graphing utility to support answers, or for further exploration.
Ex. 17–38 allow students an opportunity to pick the best method of verification. Algebraic skills can be enhanced by doing these exercises with pencil and paper. Graphing may be used to confirm partial steps of the identity.

Assignment Guide

Day 1: 1–39 odd
Day 2: 2–40 even

[−6, 6] by [−5, 5]

Figure 7.34 $y = (\cos x)(1 + \tan x)^2$ and $y = \dfrac{1}{\cos x} + 2\sin x$.

Solution Figure 7.34 shows a complete graph of $y = (\cos x)(1 + \tan x)^2$. The graph of

$$y = \frac{1}{\cos x} + 2\sin x$$

is overlaid and appears to be identical. ▤

Notice that the visual support provided in Example 6 is not a verification that the equation is an identity. Formal verification can only be done algebraically.

Exercises for Section 7.4

In Exercises 1–10, simplify each expression using the fundamental identities.

1. $\tan\theta\cos\theta$

2. $\cot x \tan x$

3. $\sec x \cos x$

4. $\cot\theta\sin\theta$

5. $\dfrac{1 + \tan^2 x}{\csc^2 x}$

6. $\dfrac{1 - \cos^2\theta}{\sin\theta}$

7. $(\sec^2 x + \csc^2 x) - (\tan^2 x + \cot^2 x)$

8. $\dfrac{\tan x \csc x}{\sec x}$

9. $\dfrac{1 + \tan x}{1 + \cot x}$

10. $\dfrac{\sec x + \tan x}{\sec x + \tan x - \cos x}$

In Exercises 11–16, reduce each expression to an equivalent expression involving only sines and cosines.

11. $\tan x + \cot x$

12. $\sin\theta + \tan\theta\cos\theta$

13. $(\sec t + \csc t)^2 \cot t$

14. $(\csc\theta - \sec\theta)\sin\theta\cos\theta$

15. $\dfrac{1}{\csc^2 x} + \dfrac{1}{\sec^2 x}$

16. $\dfrac{\sec x \csc x}{\sec^2 x + \csc^2 x}$

In Exercises 17–20, use a grapher to identify whether you think the equation is an identity. Notice that this method can be used to develop conjectures but not to verify proofs.

17. $\dfrac{\sin t - \cos t}{\cos t} + 1 = \tan t$

18. $\csc(x + \pi) = -\sec x$

19. $\cos(3\pi + x) = -\cos x$

20. $\dfrac{1}{\tan t} + \dfrac{\sin t}{\cos t - 1} = -\csc t$

In Exercises 21–24, use a grapher to provide visual support for a claim that each equation is an identity. Then verify the identity algebraically.

21. $1 + \cot^2\theta = \csc^2\theta$

22. $1 - 2\sin^2 x = 2\cos^2 x - 1$

23. $\cos^2\theta + 1 = 2\cos^2\theta + \sin^2\theta$

24. $\sin\theta + \cos\theta\cot\theta = \csc\theta$

In Exercises 25–38, verify that each of the following is an identity:

25. $\dfrac{\sin x}{\tan x} = \cos x$

26. $\sec^2\theta(1 - \sin^2\theta) = 1$

27. $(\cos t - \sin t)^2 + (\cos t + \sin t)^2 = 2$

28. $\sin^2\alpha - \cos^2\alpha = 1 - 2\cos^2\alpha$

29. $\dfrac{1 + \tan^2 x}{\sin^2 x + \cos^2 x} = \sec^2 x$

30. $\dfrac{1}{\tan\beta} + \tan\beta = \sec\beta\csc\beta$

31. $\dfrac{1 - \cos\theta}{\sin\theta} = \dfrac{\sin\theta}{1 + \cos\theta}$

32. $\dfrac{\tan x}{\sec x - 1} = \dfrac{\sec x + 1}{\tan x}$

33. $\dfrac{\sin t - \cos t}{\sin t + \cos t} = \dfrac{2\sin^2 t - 1}{1 + 2\sin t \cos t}$

34. $\dfrac{1 + \cos x}{1 - \cos x} = \dfrac{\sec x + 1}{\sec x - 1}$

35. $(x\sin\alpha + y\cos\alpha)^2 + (x\cos\alpha - y\sin\alpha)^2 = x^2 + y^2$

36. $\dfrac{\sin t}{1 - \cos t} + \dfrac{1 + \cos t}{\sin t} = \dfrac{2(1 + \cos t)}{\sin t}$

37. $\dfrac{\sin\theta}{1 + \cos\theta} + \dfrac{1 + \cos\theta}{\sin\theta} = 2\csc\theta$

38. $\dfrac{\sin A \cos B + \cos A \sin B}{\cos A \cos B - \sin A \sin B} = \dfrac{\tan A + \tan B}{1 - \tan A \tan B}$

39. Writing to Learn   Explain why even though $\sin^2 x + \cos^2 x = 1$ is an identity, $\sin x = \sqrt{1 - \cos^2 x}$ is not.

40. Writing to Learn   Graph both $y = \sin x$ and $y = x$ in the viewing rectangle $[-0.1, 0.1]$ by $[-0.1, 0.1]$. Write several paragraphs to explain whether this experience with the grapher is proof that $\sin x = x$.

7.5 ————— Sum and Difference Identities

Objective

Students will be able to use traditional algebraic and trigonometric methods that apply to identities.

Key Ideas

Sum and difference identities for sine, cosine, and tangent.
Double-angle identities
Half-angle identities

COMMON ERROR

Beginning trigonometry students often make the error of assuming that you can think of the symbol "sin" as a variable and apply the distributive property to equations like those in the Exploration. In particular, $\sin(x + 4) \neq \sin x + \sin 4$.

Example 6 in Section 7.4 reminds us that graphical methods do not provide proof that an equation is really an identity. Algebraic and trigonometric methods continue to be important. Sections 7.5 and 7.6 develop some of the more important trig identities.

We begin this section with an Exploration.

🔍 EXPLORE WITH A GRAPHING UTILITY

Experiment with a graphing utility and decide which of these equations might be an identity:

1. $\sin(x + 3) = \sin x + \sin 3$
2. $\cos(x - 1) = \cos x - \cos 1$
3. $\tan(2 + x) = \tan 2 + \tan x$

You may have been tempted in this Exploration to apply the distributive property of algebra to trig functions and conclude that the equations in the Exploration are identities. In fact, all the equations in the Exploration fail to be true for most values of x. The distributive property cannot be applied as suggested by the Exploration.

Next, we shall establish trig identities about sums and differences of angles.

Difference and Sum Formulas

This section introduces equations known as the **sum and difference identities for sine and cosine**; these contain the expressions $\cos(\alpha - \beta)$, $\cos(\alpha + \beta)$, $\sin(\alpha - \beta)$, and $\sin(\alpha + \beta)$. We first develop an identity for $\cos(\alpha - \beta)$.

Figure 7.35(a) shows angles α and β in standard position, whereas the difference $\alpha - \beta$ is not; Fig. 7.35(b) shows the angle $\alpha - \beta$ in standard position. These two positions allow us to establish an equation and conclude some algebraic relationships.

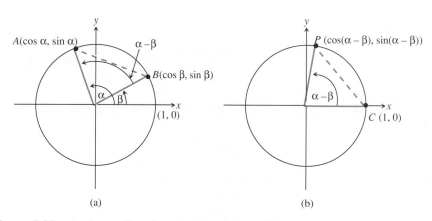

Figure 7.35   Angle $\alpha - \beta$ on the unit circle. (a) α and β are both in standard position but $\alpha - \beta$ is not; (b) $\alpha - \beta$ is in standard position.

Notice that triangle AOB in Fig. 7.35(a) is congruent to triangle POC in Fig. 7.35(b), which means that $d(A, B) = d(C, P)$. Therefore $d^2(C, P) = d^2(A, B)$. First, we convert the left side of this equation into an equation involving trig functions.

$$[d(A, B)]^2 = (\cos \alpha - \cos \beta)^2 + (\sin \alpha - \sin \beta)^2$$

$$= \cos^2 \alpha - 2 \cos \alpha \cos \beta + \cos^2 \beta + \sin^2 \alpha - 2 \sin \alpha \sin \beta + \sin^2 \beta$$

$$= (\sin^2 \alpha + \cos^2 \alpha) + (\sin^2 \beta + \cos^2 \beta) - 2 \cos \alpha \cos \beta - 2 \sin \alpha \sin \beta$$

$$= 2 - 2 \cos \alpha \cos \beta - 2 \sin \alpha \sin \beta$$

Next, we find a trig equation for $d^2(C, P)$.

$$[d(C, P)]^2 = [1 - \cos(\alpha - \beta)]^2 + [0 - \sin(\alpha - \beta)]^2$$

$$= 1 - 2 \cos(\alpha - \beta) + \cos^2(\alpha - \beta) + \sin^2(\alpha - \beta)$$

$$= 2 - 2 \cos(\alpha - \beta)$$

Finally, we simplify the equation $[d(C, P)]^2 = [d(A, B)]^2$.

$$2 - 2\cos(\alpha - \beta) = 2 - 2\cos\alpha\cos\beta - 2\sin\alpha\sin\beta$$

$$-2\cos(\alpha - \beta) = -2(\cos\alpha\cos\beta + \sin\alpha\sin\beta)$$

$$\cos(\alpha - \beta) = \cos\alpha\cos\beta + \sin\alpha\sin\beta$$

This development verifies the identity for $\cos(\alpha - \beta)$ stated in the following theorem.

Theorem 7.5 Formula for Cosine of a Difference

For all angles α and β,

$$\cos(\alpha - \beta) = \cos\alpha\cos\beta + \sin\alpha\sin\beta.$$

Example 1 illustrates how to use this identity to find an exact value for $\cos 15°$.

E X A M P L E 1 Using the Identity for $\cos(\alpha - \beta)$

Find an exact value for $\cos 15°$.

Solution

$$\cos 15° = \cos(45° - 30°)$$
$$= \cos 45° \cos 30° + \sin 45° \sin 30°$$
$$= \frac{\sqrt{2}}{2} \cdot \frac{\sqrt{3}}{2} + \frac{\sqrt{2}}{2} \cdot \frac{1}{2}$$
$$= \frac{\sqrt{6} + \sqrt{2}}{4}$$

Theorem 7.5 is also important because it can be used to verify other trig identities, as Example 2 illustrates.

E X A M P L E 2 Verifying Identities with the $\cos(\alpha - \beta)$ Identity

Let θ be any angle. Use Theorem 7.5 to verify the following identities. Then use a grapher to provide visual support for your work.

a) $\cos(\pi/2 - \theta) = \sin\theta$ **b)** $\cos(\pi - \theta) = -\cos\theta$

Solution

a) Use Theorem 7.5 with $\alpha = \pi/2$ and $\beta = \theta$.

$$\cos\left(\frac{\pi}{2} - \theta\right) = \cos\frac{\pi}{2}\cos\theta + \sin\frac{\pi}{2}\sin\theta$$

$$= 0\cos\theta + 1\sin\theta$$

$$= \sin\theta$$

Figure 7.36 provides visual support for this identity. Both $y = \cos(\pi/2 - \theta)$ and $y = \sin\theta$ are graphed in Fig. 7.36 and they coincide.

b) Use Theorem 7.5 with $\alpha = \pi$ and $\beta = \theta$.

$$\cos(\pi - \theta) = \cos\pi\cos\theta + \sin\pi\sin\theta$$

$$= (-1)\cos\theta + (0)\sin\theta$$

$$= -\cos\theta.$$

Figure 7.37 provides visual support for this identity. ▤

The identity for the cosine of a difference is used to verify an identity for the cosine of a sum, which is the subject of Theorem 7.6.

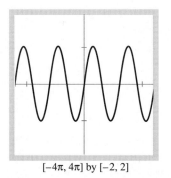

[−4π, 4π] by [−2, 2]

Figure 7.36 $y = \cos(\pi/2 - \theta)$ and $y = \sin\theta$.

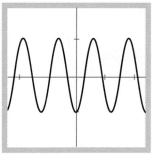

[−4π, 4π] by [−2, 2]

Figure 7.37 $y = \cos(\pi - \theta)$ and $y = -\cos\theta$.

Theorem 7.6   Formula for Cosine of a Sum

For all angles α and β,

$$\cos(\alpha + \beta) = \cos\alpha\cos\beta - \sin\alpha\sin\beta.$$

Notes on Exercises

Ex. 9–16 are traditional problems found in most trigonometry textbooks. However, using graphing technology to confirm results by overlaying graphs is a powerful technique.

Frank says, "Do analytically, support graphically."

Proof   We use the fact that $\cos(-x) = \cos x$ and $\sin(-x) = -\sin x$ for all values of x. Then,

$$\cos(\alpha + \beta) = \cos[\alpha - (-\beta)]$$

$$= \cos\alpha\cos(-\beta) + \sin\alpha\sin(-\beta) \qquad \text{Recall that } \cos(-x) = \cos x$$
$$\text{and } \sin(-x) = -\sin x.$$

$$= \cos\alpha\cos\beta - \sin\alpha\sin\beta. \qquad \qquad ▤$$

E X A M P L E   3   Using the Identity for $\cos(\alpha + \beta)$

Find the exact value of $\cos 75°$.

Solution

$$\cos 75° = \cos(45° + 30°)$$

$$= \cos 45° \cos 30° - \sin 45° \sin 30°$$

$$= \frac{\sqrt{2}}{2} \cdot \frac{\sqrt{3}}{2} - \frac{\sqrt{2}}{2} \cdot \frac{1}{2}$$

$$= \frac{\sqrt{6} - \sqrt{2}}{4}$$

≡

Because we could approximate the value of $\cos 75°$ directly with a calculator, we might expect to use this method. However, the identity remains important for solving theoretical applications, as illustrated in Example 2 and Theorem 7.7. Numerical examples, like Example 3, help you become familiar with these identities.

Theorem 7.7   Formulas for Sine of a Sum and Difference

For all angles α and β,

$$\sin(\alpha + \beta) = \sin \alpha \cos \beta + \cos \alpha \sin \beta, \text{ and}$$

$$\sin(\alpha - \beta) = \sin \alpha \cos \beta - \cos \alpha \sin \beta.$$

Proof

$$\sin(\alpha + \beta) = \cos\left[\frac{\pi}{2} - (\alpha + \beta)\right] \quad \text{See Example 2(a).}$$

$$= \cos\left[\left(\frac{\pi}{2} - \alpha\right) - \beta\right]$$

$$= \cos\left(\frac{\pi}{2} - \alpha\right) \cos \beta + \sin\left(\frac{\pi}{2} - \alpha\right) \sin \beta$$

$$= \sin \alpha \cos \beta + \cos \alpha \sin \beta$$

Proof of the second identity is left as an exercise.

≡

It was mentioned earlier that the identities for the sine and cosine of the sum and difference are needed in calculus. In particular, if $f(x) = \sin x$, the quotient

$$\frac{f(x + h) - f(x)}{h},$$

which is used in calculus in the definition of the derivative of f, uses the identity for the sine of a sum. Example 4 illustrates this application.

EXAMPLE 4   Using the Identity for $\sin(\alpha + \beta)$

Let $f(x) = \sin x$ and $h \neq 0$. Show that

$$\frac{f(x+h) - f(x)}{h} = \sin x \left(\frac{\cos h - 1}{h} \right) + \cos x \left(\frac{\sin h}{h} \right).$$

Solution

$$\frac{f(x+h) - f(x)}{h} = \frac{\sin(x+h) - \sin x}{h}$$

$$= \frac{\sin x \cos h + \cos x \sin h - \sin x}{h}$$

$$= \frac{\sin x (\cos h - 1) + \cos x \sin h}{h}$$

$$= \sin x \left(\frac{\cos h - 1}{h} \right) + \cos x \left(\frac{\sin h}{h} \right) \qquad\qquad \blacksquare$$

Theorems 7.2 and 7.3 stated that the sum of a sine function and a cosine function is a sinusoid. Example 5 shows how to find the period and phase-shift constants of the resulting sinusoid. The solution in Example 5 can be generalized to provide a proof of Theorem 7.3.

EXAMPLE 5   Finding a Sinusoid

Find constants A and α so that for every value of x, $2 \cos 3x + 5 \sin 3x = A \sin(3x + \alpha)$.

Solution

$$2 \cos 3x + 5 \sin 3x = A \sin(3x + \alpha)$$

$$= A \sin 3x \cos \alpha + A \cos 3x \sin \alpha$$

We conclude that

$$A \cos \alpha = 5 \qquad \text{or} \qquad \cos \alpha = 5/A, \text{ and}$$

$$A \sin \alpha = 2 \qquad \text{or} \qquad \sin \alpha = 2/A.$$

Therefore

$$\tan \alpha = \frac{\sin \alpha}{\cos \alpha} = \frac{2/A}{5/A} = \frac{2}{5} \qquad \text{or} \qquad \alpha = \tan^{-1}\left(\frac{2}{5} \right).$$

We can conclude from Fig. 7.38 that $A = \sqrt{2^2 + 5^2} = \sqrt{29}$.

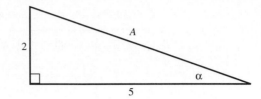

Figure 7.38 $\cos C = 5/A$ and $\sin C = 2/A$.

Graph $y = \sqrt{29}\sin(3x + \tan^{-1}(\frac{2}{5}))$ and overlay the graph of $y = 2\cos 3x + 5\sin 3x$.

Finally, there are two **sum and difference identities for tangent**, stated here in Theorem 7.8. The proofs are saved for Exercises 47 and 48.

> **Theorem 7.8   Tangent of a Sum and a Difference**
>
> For all angles α and β for which the expressions are defined,
>
> $$\tan(\alpha + \beta) = \frac{\tan\alpha + \tan\beta}{1 - \tan\alpha\tan\beta}, \text{ and}$$
>
> $$\tan(\alpha - \beta) = \frac{\tan\alpha - \tan\beta}{1 + \tan\alpha\tan\beta}.$$

Double-angle Formulas

In this section, we discuss identities for $\sin 2x$, $\cos 2x$, and $\tan 2x$. It is tempting to claim that the 2 can be factored out to give $\sin 2x = 2\sin x$. Students who manipulate symbols without thinking about what they mean often make this assertion. Is it true? The following Exploration should help you decide.

> ### 🔎 EXPLORE WITH A GRAPHING UTILITY
>
> Experiment with a graphing utility and decide which of the following equations might be an identity:
>
> **1.** $\sin 2x = 2\sin x$
>
> **2.** $\cos 3x = 3\cos x$
>
> **3.** $\tan 2x = 2\tan x$

You should have discovered that none of the equations in the Exploration are valid. Notice that $\sin 2\theta$ is equal to $\sin(\theta + \theta)$. Consequently, the identity for $\sin(\alpha + \beta)$ can be used to develop an identity for $\sin 2\theta$, called a **double-angle identity**.

Theorem 7.9   Double-angle Identities

For all angles θ for which the expressions are defined,

$$\sin 2\theta = 2\sin\theta\cos\theta,$$

$$\cos 2\theta = \cos^2\theta - \sin^2\theta = 1 - 2\sin^2\theta = 2\cos^2\theta - 1, \text{ and}$$

$$\tan 2\theta = \frac{2\tan\theta}{1 - \tan^2\theta}.$$

E X A M P L E   6   Verifying a Double-angle Identity

Verify that $\sin 2\theta = 2\sin\theta\cos\theta$.

Solution   Use the identity for $\sin(\alpha + \beta)$, where $\alpha = \beta = \theta$. Then,

$$\sin 2\theta = \sin(\theta + \theta) \qquad \text{Use the identity for } \sin(\alpha + \beta), \text{ where } \alpha \text{ and } \beta \text{ are both } \theta.$$

$$= \sin\theta\cos\theta + \cos\theta\sin\theta$$

$$= 2\sin\theta\cos\theta. \qquad\qquad \blacksquare$$

Proofs of the identities for $\cos 2x$ and $\tan 2x$ are saved for Exercises 49 through 51.

E X A M P L E   7   Using Double-angle Identities

Write $2\cos x + \sin 2x$ as an expression involving only $\sin x$ and $\cos x$.

Solution

$$2\cos x + \sin 2x = 2\cos x + 2\sin x\cos x \qquad \text{Use the identity for } \sin 2x.$$

$$= (2\cos x)(1 + \sin x) \qquad\qquad \blacksquare$$

E X A M P L E   8   Finding Zeros in a Double-angle Function

Use both graphical and algebraic methods to find all the zeros of the function $f(x) = \sin 2x + \cos 3x$ over the interval $[0, 2\pi)$.

Solution

Graphical Method Figure 7.39 shows a complete graph of f for the interval $[0, 2\pi]$. Use zoom-in to show that the zeros of f are 0.94, 1.57, 2.20, 3.46, 4.71, and 5.97.

Algebraic Method To solve the equation $\sin 2x + \cos 3x = 0$ algebraically, rewrite the equation in terms of $\sin x$ and $\cos x$.

$$
\begin{aligned}
0 = \sin 2x + \cos 3x &= \sin 2x + \cos(2x + x) \\
&= \sin 2x + \cos 2x \cos x - \sin 2x \sin x \\
&= 2\sin x \cos x + (1 - 2\sin^2 x)\cos x - 2\sin x \cos x \sin x \\
&= 2\sin x \cos x + \cos x - 2\sin^2 x \cos x - 2\sin^2 x \cos x \\
&= (\cos x)(2\sin x + 1 - 4\sin^2 x)
\end{aligned}
$$

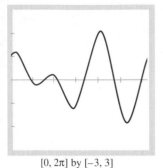

$[0, 2\pi]$ by $[-3, 3]$

Figure 7.39 A complete graph of $f(x) = \sin 2x + \cos 3x$.

Therefore either $\cos x = 0$ or $2\sin x + 1 - 4\sin^2 x = 0$. We know that the solutions of $\cos x = 0$ in $[0, 2\pi)$ are $\pi/2$ and $3\pi/2$.

To find the solutions to $4\sin^2 x - 2\sin x - 1 = 0$, use the fact that the equation is a quadratic in $\sin x$ to obtain the following results:

$$
\sin x = \frac{2 \pm \sqrt{4 + 16}}{8} = \frac{1 \pm \sqrt{5}}{4}.
$$

The solutions to $\sin x = (1 + \sqrt{5})/4$ in $[0, 2\pi)$ are 0.94 and 2.20. The solutions to $\sin x = (1 - \sqrt{5})/4$ in $[0, 2\pi)$ are 3.46 and 5.97.

Therefore the zeros of $f(x)$ are 0.94, 1.57, 2.20, 3.46, 4.71, and 5.97. ∎

Half-angle Formulas

The double-angle identities can be used to obtain formulas that permit us to write $\sin(\theta/2)$, $\cos(\theta/2)$, and $\tan(\theta/2)$ in terms of $\sin\theta$, $\cos\theta$, and $\tan\theta$. For example, applying Theorem 7.9, for angle α

$$
\cos 2\alpha = 2\cos^2\alpha - 1,
$$

which is equivalent to

$$
\cos^2\alpha = \frac{1 + \cos 2\alpha}{2}.
$$

Let $2\alpha = \theta$, so that $\alpha = \theta/2$, and obtain

$$
\cos\frac{\theta}{2} = \pm\sqrt{\frac{1 + \cos\theta}{2}}.
$$

This identity is known as a **half-angle identity**.

Common Errors

Students are often confused as to which of the $+$, $-$ signs is correct in the sine and cosine half-angle formulas. They must be shown that the signs are chosen according to the quadrant of $\frac{\theta}{2}$. It is also interesting to have students discuss why $\tan\frac{\theta}{2}$ is always positive (because the signs on $\sin\theta$ and $\tan\frac{\theta}{2}$ are the same) while the sign on $(1 + \cos\theta)$ is always nonnegative.

Notes on Exercises

Ex. 27–30 demonstrate the limitations of algebraic methods. Have students take full advantage of the power of technology to explore these exercises.

Assignment Guide

Day 1: Ex. 1–16, 17–25 odd
Day 2: Ex. 27–33 odd, 34, 37–43
Day 3: Ex. 47, 49, 50, 52, 55

Theorem 7.10   Half-angle Identities

For all angles θ,

$$\sin\frac{\theta}{2} = \pm\sqrt{\frac{1 - \cos\theta}{2}},$$

$$\cos\frac{\theta}{2} = \pm\sqrt{\frac{1 + \cos\theta}{2}}, \text{ and}$$

$$\tan\frac{\theta}{2} = \frac{\sin\theta}{1 + \cos\theta}.$$

Exercises for Section 7.5

In Exercises 1–8, use the sum and difference identities to find exact values of each of the following:

1. $\sin 15°$ **2.** $\tan 15°$ **3.** $\sin 75°$

4. $\tan 75°$ **5.** $\cos\dfrac{\pi}{12}$ **6.** $\sin\dfrac{5\pi}{12}$

7. $\tan\dfrac{5\pi}{12}$ **8.** $\tan\dfrac{\pi}{12}$

In Exercises 9–12, express each of the following in terms of $\sin x$, $\cos x$, $\tan x$, or $\cot x$:

9. $\sin(x + 90°)$ **10.** $\tan(180° + x)$

11. $\cos\left(\dfrac{\pi}{2} + x\right)$ **12.** $\cot(x + 2\pi)$

In Exercises 13–16, prove each identity. Use a grapher to provide visual support for your work.

13. $\sin\left(\theta + \dfrac{\pi}{2}\right) = \cos\theta$

14. $\cos\left(\theta - \dfrac{\pi}{4}\right) = \dfrac{\sqrt{2}}{2}(\cos\theta + \sin\theta)$

15. $\tan\left(\theta + \dfrac{\pi}{4}\right) = \dfrac{1 + \tan\theta}{1 - \tan\theta}$

16. $\cos\left(\theta + \dfrac{\pi}{2}\right) = -\sin\theta$

In Exercises 17–22, write each of the following as an expression involving only $\sin\theta$ and $\cos\theta$:

17. $\sin 2\theta + \cos\theta$ **18.** $\sin 2\theta + \cos 2\theta$

19. $\sin 2\theta + \cos 3\theta$ **20.** $\sin 3\theta + \cos 2\theta$

21. $\sin 4\theta + \cos 3\theta$ **22.** $\tan 2\theta + \tan\theta$

In Exercises 23–26, use a grapher to find a complete graph of each equation. If it is a sinusoid, give the period, amplitude, and phase shift.

23. $y = \cos 3x + 2\sin 3x$

24. $y = 3\cos 2x - 2\sin 2x$

25. $y = 3\sin x + 5\sin(x + 2)$

26. $y = 3\sin(2x - 1) + 5\sin(2x + 3)$

In Exercises 27–30, find the domain, range, and a complete graph of each function.

27. $f(x) = \sec x^2$ **28.** $g(x) = x^2\sin\left(x - \dfrac{\pi}{2}\right)$

29. $g(x) = 3\sin^2 x$ **30.** $f(x) = \dfrac{\tan x}{x^2}$

In Exercises 31 and 32, let $\sin x = \dfrac{2}{3}$ and $\pi/2 < x < \pi$ and find the value of each expression.

31. a) $\cos x$ **b)** $\csc x$

32. a) $\tan 2x$ **b)** $\cos\dfrac{x}{2}$

In Exercises 33–36, let $\cos x = -\dfrac{1}{2}$ and $\pi < x < 3\pi/2$ and find the value of each expression.

33. $\sec x$ **34.** $\sin x$ **35.** $\tan 2x$ **36.** $\cos\dfrac{x}{2}$

In Exercises 37–44, solve each equation for x in the interval $[0, 2\pi)$. Find exact answers when possible. Check your answer using a graphing utility.

37. $\sin x = \dfrac{1}{2}$

38. $\cos x = 0$

39. $\sin x = \dfrac{\sqrt{3}}{2}$

40. $\tan x = 1$

41. $\sin^2 x - 1 = 0$

42. $2\sin^2 x + \sin x - 1 = 0$

43. $\sin^2 x - 2\sin x = 0$

44. $\cos 2x + \sin 3x = 0$

45. Use Example 2(a) to verify that $\sin(\pi/2 - \alpha) = \cos \alpha$ for all angles α.

46. Verify the identity
$$\sin(\alpha - \beta) = \sin \alpha \cos \beta - \cos \alpha \sin \beta.$$

47. Verify the identity
$$\tan(\alpha + \beta) = \frac{\tan \alpha + \tan \beta}{1 - \tan \alpha \tan \beta}.$$

48. Verify the identity $\tan(\alpha - \beta) = \dfrac{\tan \alpha - \tan \beta}{1 + \tan \alpha \tan \beta}$.

49. Verify the identity $\cos 2\theta = \cos^2 \theta - \sin^2 \theta$.

50. Verify the two alternate identities for $\cos 2\theta$.

a) $\cos 2\theta = 1 - 2\sin^2 \theta$ **b)** $\cos 2\theta = 2\cos^2 \theta - 1$

51. Verify the identity
$$\tan 2\theta = \frac{2\tan \theta}{1 - \tan^2 \theta}.$$

52. Verify the identity
$$\sin \frac{\theta}{2} = \pm \sqrt{\frac{1 - \cos \theta}{2}}.$$

53. Verify the identity
$$\tan \frac{\theta}{2} = \frac{\sin \theta}{1 + \cos \theta}.$$

54. Use identities to verify that $f(x) = 2\sin x \cos x$ and $g(x) = 1 - 2\sin^2 x$ are both sinusoidal functions.

55. **Writing to Learn**   Explain how the graphs of $y = \cos(x + 4)$ and $y = \cos x + 4$ can be obtained from $y = \cos x$. Write a paragraph explaining how this graphical approach can be used to show that $\cos(x + 4) \neq \cos x + 4$.

56. Use the fact that $\cos(\pi/2 - \theta) = \sin \theta$ to show that $\sin(\pi/2 - \alpha) = \cos \alpha$. (*Hint:* Let $\alpha = \pi/2 - \theta$.)

57. Verify the second identity stated in Theorem 7.7.

7.6

Solving Trigonometric Equations and Inequalities Analytically

In Section 7.3, graphical methods were used to solve a variety of trig equations and inequalities. In this section, we revisit equation and inequality solving using algebraic techniques. These methods are useful in situations where exact solutions are desired.

Example 1 deals with the equation $2\cos^2 t + \sin t + 1 = 0$, which we solved graphically in Example 2 of Section 7.3. Now we see how trig identities can be combined with general algebraic properties to solve the equation analytically.

E X A M P L E   1   Solving a Trig Equation with Factoring

Find all solutions to $2\cos^2 t + \sin t + 1 = 0$ in the interval $[0, 2\pi)$.

Teaching Note

Throughout this section stress the algebraic method, and use the graphing method to support the analytic solutions.

Solution   Rewrite the equation into an equivalent form that involves powers of $\sin x$ instead of combinations of $\sin x$ and $\cos x$.

$$2\cos^2 t + \sin t + 1 = 0$$

$$2(1 - \sin^2 t) + \sin t + 1 = 0$$

$$2 - 2\sin^2 t + \sin t + 1 = 0$$

$$-2\sin^2 t + \sin t + 3 = 0$$

$$2\sin^2 t - \sin t - 3 = 0 \quad \text{This quadratic equation in } \sin t \text{ can now be factored.}$$

$$(2\sin t - 3)(\sin t + 1) = 0$$

So t is a solution to $2\cos^2 t + \sin t + 1 = 0$ if, and only if,

$$2\sin t - 3 = 0 \qquad\qquad \sin t + 1 = 0$$
$$\text{or}$$
$$\sin t = \frac{3}{2} \qquad\qquad \sin t = -1.$$

There are no solutions for $\sin t = \frac{3}{2}$. (Why?) Thus it is sufficient to consider $\sin t = -1$. The solution to $\sin t = -1$ is $t = 3\pi/2$.

Therefore the solution to $2\cos^2 t + \sin t + 1 = 0$ in $[0, 2\pi)$ is $t = 3\pi/2$. Figure 7.40 shows this solution graphically.

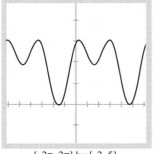

[−2π, 2π] by [−2, 5]

Figure 7.40 $y = 2\cos^2 t + \sin t + 1$.

As discussed in Section 7.3, the complete set of solutions to an equation with period P can be found by adding multiples of P (where k is an integer) to the solutions found over an interval of length P. In this section, we ordinarily will not list the complete set of solutions.

In the next example, we again use identities to change the original equation to a quadratic in $\sin x$ in order to solve the equation. The quadratic is factored if possible; otherwise, we shall use the quadratic formula.

Notes on Examples

Once students observe in Example 1 that the period is 2π, they can focus their attention on the interval $(0, 2\pi)$, and thus simplify both the algebraic and graphical methods of finding the solution. Although students are not asked to list the complete set of solutions in this section, it may be advisable to work some examples with them so they can see what it means to find *all* solutions.

E X A M P L E   2   Using the Quadratic Formula

Find all solutions to $\tan x = 3\cos x$ over the interval $[0, 2\pi)$.

Solution   It is often helpful to rewrite a trig equation in terms of the sine and cosine and then use identities and algebraic properties as necessary to change the form to a quadratic equation in $\sin x$.

$$\tan x = 3\cos x$$

$$\frac{\sin x}{\cos x} = 3\cos x$$

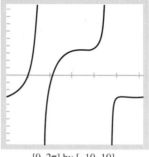

[0, 2π] by [−10, 10]

Figure 7.41 $f(x) = \tan x - 3\cos x$.

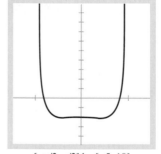

[−π/2, π/2] by [−5, 10]

Figure 7.42 $f(x) = 3\tan^4 x - 1 - \sec^2 x$.

$$\sin x = 3\cos^2 x$$

Because both sides of the equation are multiplied by $\cos x$, extraneous solutions may be introduced.

$$\sin x = 3(1 - \sin^2 x)$$

$$3\sin^2 x + \sin x - 3 = 0$$

It is not evident how to factor this expression, so use the quadratic formula to get

$$\sin x = \frac{-1 \pm \sqrt{37}}{6}.$$

The equation

$$\sin x = \frac{-1 - \sqrt{37}}{6}$$

has no solutions. However, the equation

$$\sin x = \frac{-1 + \sqrt{37}}{6}$$

has two solutions, one in the first quadrant and one in the second quadrant. The solution in the first quadrant is

$$\sin^{-1}\left(\frac{-1 + \sqrt{37}}{6}\right) = 1.01,$$

and the solution in the second quadrant is $\pi - 1.01 = 2.13$.

The graphical support seen in Fig. 7.41 shows that these solutions are not extraneous. Check that these values are solutions to the original equation. The solutions to $\tan x = 3\cos x$ in $[0, 2\pi)$ are $x = 1.01$ and $x = 2.13$. ≡

E X A M P L E   3   Solving a Trig Equation with Factoring

Find all solutions in one period of the periodic equation $3\tan^4 x = 1 + \sec^2 x$.

Solution   Because $\sec^2(x + \pi) = \sec^2 x$, $\sec^2 x$ has period π and because $\tan x$ also has period π, the period of $\sec^2 x$ is a divisor of π. The graph in Fig. 7.42 shows that the period is at least π, so we conclude that the period is π. We shall find all solutions in the interval $[-\pi/2, \pi/2)$.

$$3\tan^4 x = 1 + \sec^2 x$$

$$= 1 + 1 + \tan^2 x$$

$$3\tan^4 x - \tan^2 x - 2 = 0$$

$$(3\tan^2 x + 2)(\tan^2 x - 1) = 0$$

So we need to solve the following equations:

$$3\tan^2 x + 2 = 0 \qquad\qquad \tan^2 x - 1 = 0$$

$$\tan^2 x = -\tfrac{2}{3} \qquad\qquad \tan^2 x = 1$$

$$\tan x = \pm 1$$

However, the equation $\tan^2 x = -\frac{2}{3}$ has no real-number solutions.

A solution to $\tan x = 1$ is $\tan^{-1}(1) = \pi/4$, and a solution to $\tan x = -1$ is $\tan^{-1}(-1) = -\pi/4$.

It is clear from the graph in Fig. 7.42 that there are only two zeros to $f(x) = 3\tan^4 x - 1 - \sec^2 x$ in $[-\pi/2, \pi/2)$.

So the solutions to $3\tan^4 x = 1 + \sec^2 x$ in $[-\pi/2, \pi/2)$ are $x = \pi/4$ and $x = -\pi/4$. ≡

EXAMPLE 4   Factoring by Grouping

Find all solutions in one period of the periodic equation $2\cot x \cos x - 3\cos x + 6\cot x - 9 = 0$.

Solution   Because $\cos x$ has a period of 2π and $\cot x$ has a period of π, the period of $f(x) = 2\cot x \cos x - 3\cos x + 6\cot x - 9$ is a divisor of 2π. It is evident from the graph in Fig. 7.43 that the period is at least 2π. So we look for all solutions in the interval $[0, 2\pi)$.

Group terms and factor as shown.

$$(2\cot x \cos x - 3\cos x) + (6\cot x - 9) = 0$$

$$(\cos x)(2\cot x - 3) + 3(2\cot x - 3) = 0$$

$$(\cos x + 3)(2\cot x - 3) = 0$$

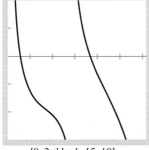

[0, 2π] by [−15, 10]

Figure 7.43 $f(x) = 2\cot x \cos x - 3\cos x + 6\cot x - 9.$

The equation $\cos x + 3 = 0$ has no solution. So consider the equation $2\cot x - 3 = 0$.

$$2\cot x - 3 = 0 \qquad \text{is equivalent to} \qquad \cot x = \tfrac{3}{2}$$

$$\cot x = \tfrac{3}{2} \qquad \text{is equivalent to} \qquad \tan x = \tfrac{2}{3}$$

One solution to $\tan = \frac{2}{3}$ is $\tan^{-1}(\frac{2}{3}) = 0.59$, an angle in the first quadrant. Because the tangent function is positive in the third quadrant, $0.59 + \pi$ is also a solution.

The solutions to $2\cot x \cos x - 3\cos x + 6\cot x - 9 = 0$ in $[0, 2\pi)$ are $x = 0.59$ and $x = 0.59 + \pi = 3.73$. ≡

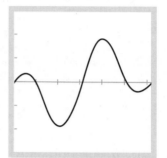

[0, 2π] by [−3, 3]

Figure 7.44 $f(x) = \sin 2x - \sin x$.

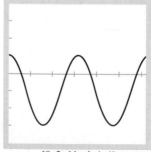

[0, 2π] by [−4, 4]

Figure 7.45 $y = \cos 2x - 2\sin^2 x$.

Sometimes an analytical method for solving a trig equation depends on a double-angle identity.

EXAMPLE 5   Using a Double-angle Identity

Find all solutions in one period of the periodic equation $\sin 2x = \sin x$.

Solution   The period of the function $f(x) = \sin 2x - \sin x$ is 2π, so we shall find all solutions in $[0, 2\pi)$ (see Fig. 7.44).

$$\sin 2x - \sin x = 0$$

$$2\cos x \sin x - \sin x = 0$$

$$(\sin x)(2\cos x - 1) = 0$$

So we must solve the equations

$$\sin x = 0 \qquad \text{and} \qquad \cos x = \tfrac{1}{2}.$$

The solutions to $\sin x = 0$ are 0 and π, and the solutions to $\cos x = \tfrac{1}{2}$ are $\pi/3$ and $5\pi/3$. The solutions to $\sin 2x = \sin x$ in $[0, 2\pi)$ are 0, $\pi/3$, π, and $5\pi/3$. ▇

EXAMPLE 6   Using Fundamental and Double-angle Identities

Find all solutions in one period of the periodic equation $\cos 2x - 2\sin^2 x = 0$.

Solution   The period of this periodic equation is π. Therefore we shall find all solutions in $[0, \pi)$.

$$\cos 2x - 2\sin^2 x = 0$$

$$(1 - 2\sin^2 x) - 2\sin^2 x = 0 \quad \text{Use the identity for } \cos 2x.$$

$$1 - 4\sin^2 x = 0$$

$$\sin^2 x = \tfrac{1}{4}$$

$$\sin x = \pm\tfrac{1}{2}$$

The solutions to $\sin x = \tfrac{1}{2}$ are $\pi/6$ and $5\pi/6$; $\sin x = -\tfrac{1}{2}$ has no solution in $[0, \pi]$.

The solutions to $\cos 2x - 2\sin^2 x = 0$ over $[0, \pi)$ are $x = \pi/6$ and $x = 5\pi/6$. Figure 7.45 provides visual support for these solutions. ▇

Exercises for Section 7.6

In Exercises 1–6, find all solutions in one period of each periodic equation.

1. $\sin x = 0.3$ **2.** $\cos x = 0.75$

3. $\tan x = 1.5$ **4.** $\sin x = 1.25$

5. $\cos 2t = \frac{1}{2}$ **6.** $\tan 2\theta = 1$

In Exercises 7–10, find all solutions in one period of each periodic equation.

7. $(\sin x - 0.5)(\cos x + 0.3) = 0$

8. $(\sin x + 0.2)(\cos 2x - 1) = 0$

9. $(\cos x + 0.8)(\cos x - 0.5) = 0$

10. $(\sin x + 1.4)(\cos x - 0.1) = 0$

In Exercises 11–21, find all solutions in one period of each periodic equation.

11. $(\sin x)(\tan x - 1)(\cos x - 1) = 0$

12. $(\tan x - 1.5)(2 \sin x + 1)(2 \cos x - 1) = 0$

13. $(\sin^2 x - 1)(\cos x - 1) = 0$

14. $2 \sin x \cos x = \sin x$ **15.** $2 \sin^2 x = 1 + \cos x$

16. $4 \sin^2 x = 3$ **17.** $2 \tan^2 x = \sec x - 1$

18. $\sin 2x = \cos x$ **19.** $2 \tan^2 x + 3 \sec x = 0$

20. $2 \cos^2 x = 1 - \sin x$ **21.** $2 \sin^2 t - \sin t = 1$

In Exercises 22 and 23, find all solutions over the interval $[0, 2\pi)$ for each equation.

22. $\cos 2x - \sin x = \dfrac{1}{2}$

23. $\cos 2x + \cos x = 0$

Exercises 24–26 develop a formula for the area $\mathcal{A}$ of a triangle based on the figure below.

24. Show that

$$\mathcal{A} = \frac{1}{2}x_1 h + \frac{1}{2}x_2 h.$$

25. Use the formula in Exercise 24 to show that

$$\mathcal{A} = \frac{1}{2}ab \sin\theta_1 \cos\theta_2 + \frac{1}{2}ab \cos\theta_1 \sin\theta_2.$$

26. Use the formula in Exercise 25 to show that

$$\mathcal{A} = \frac{1}{2}ab \sin C.$$

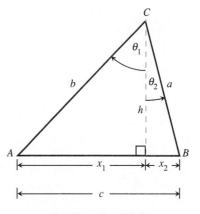

For Exercises 24–26.

Exercises 27–30 refer to the following **problem situation**: When light passes through a medium, it changes direction, as shown in the figure. This bending is referred to as *light refraction*. If α is the angle of incidence and β is the angle of refraction, *Snell's law* states that for a given medium,

$$\frac{\sin\alpha}{\sin\beta} = C,$$

where the constant C is called the *index of refraction*.

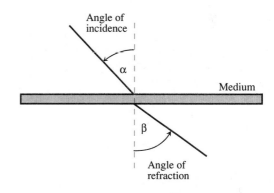

For Exercises 27–30.

27. Suppose the index of refraction for a diamond is 2.42. Show that the trigonometric representation for β as a function of α is $\beta = \sin^{-1}(0.413 \sin\alpha)$. Find a complete graph of this problem situation and state the domain.

28. What is the maximum value for the angle of refraction for this diamond medium?

29. What angle of incidence should you choose if you want the angle of refraction to be 0.38 radians?

30. Find the angle of refraction when the angle of incidence is $\pi/4$.

31. Writing to Learn Try to solve the equation $\tan x = \sec x$ both algebraically and graphically. Write a paragraph that explains which method is most convincing for you on this problem and why.

Chapter 7 Review

KEY TERMS (The number following each key term indicates the page of its introduction.)

analytic trigonometry, 373
double-angle identity, 415
half-angle identity, 416
inverse cosine function, 389

inverse sine function, 388
inverse tangent function, 389
linear combination, 379
periodic equation, 397

phase shift, 376
sinusoid, 379
sum and difference
  identities for sine

and cosine, 409
sum and difference
  identities for tangent, 414

REVIEW EXERCISES

In Exercises 1–6, determine the amplitude, period, phase shift, domain, and range for each function.

1. $f(x) = 2\sin 3x$

2. $g(x) = 3\cos 4x$

3. $f(x) = 1.5\sin\left(2x - \pi/4\right)$ **4.** $g(x) = 2\sin(3x + 1)$

5. $f(x) = 4\sin(2x - 1)$ **6.** $g(x) = -2\sin\left(3x - \pi/3\right)$

In Exercises 7–12, describe the transformations that can be used to obtain a graph of the given function from the graph $y = \sin x$.

7. $f(x) = 3\sin(2x - 1) + 7$

8. $g(x) = -2\sin(2x - 0.5) - 3$

9. $h(x) = 5\sin(3x + 1.5) + 2.3$

10. $f(x) = -3\sin(\pi x - 2) + 4$

11. $g(x) = 2\sin(2\pi x + 3) - 2$

12. $f(x) = -\sin(x - \pi/2) - 1$

In Exercises 13–16, sketch a graph of each function.

13. $f(x) = 2\cos(3x - \pi) + 1$

14. $h(x) = 3\sin\left(2\pi x - \dfrac{\pi}{4}\right)$

15. $g(x) = -2\cos(x + 1)$

16. $h(x) = 5\cos\left(4x + \dfrac{\pi}{3}\right) - 2$

In Exercises 17–20, find a complete graph of each function. Show that each is a sinusoid of the form $y = A\sin(bx + \alpha)$ and estimate the value of $A, b,$ and α.

17. $y = 2\sin x - 4\cos x$ **18.** $y = 3\cos 2x - 2\sin 2x$

19. $y = 3\sin(2x - 1) + 5\cos(2x + 3)$

20. $y = 2\sin(3x + 1) + 3\cos(3x - 2)$

In Exercises 21–24, estimate values of A and C that solve each equation.

21. $2\cos 2x + 3\sin 2x = A\sin(2x + C)$

22. $3\sin 3x + 4\cos 3x = A\cos(3x + C)$

23. $5\sin 2x - 3\cos 2x = A\sin(2x + C)$

24. $7\cos 3x + 4\sin 3x = A\sin(3x + C)$

In Exercises 25–32, sketch a complete graph of each function. What transformations did you use, if any, in completing this graph?

25. $y = \sin^{-1} x$ **26.** $y = \cos^{-1} x$

27. $y = \sin^{-1} 2x$ **28.** $y = \tan^{-1} x$

29. $y = \tan^{-1} 2x$ **30.** $y = \sin^{-1}(2x - 1)$

31. $y = \sin^{-1}(3x - 1) + 2$ **32.** $y = \cos^{-1}(2x + 1) - 3$

In Exercises 33–42, use a calculator if necessary to find all solutions for each of these trigonometric equations.

33. $\sin 2x = 0.5$ **34.** $\cos x = \dfrac{\sqrt{3}}{2}$

35. $\tan x = -1$ **36.** $\sin x = 0.7$

37. $\cos 2x = 0.13$ **38.** $\cot x = 1.5$

39. $3\sin x = 0.9$ **40.** $2\cos 3x = 0.45$

41. $2\sin^{-1} x = \sqrt{2}$ **42.** $\tan^{-1} x = 1$

In Exercises 43–50, find a complete graph of each function and determine the domain and range of each.

43. $y = 3x\cos x + \sin 2x$ **44.** $y = 2\sin 3x - 5\cos 2x$

45. $y = \dfrac{2x - 1}{\tan x}$ **46.** $f(x) = x^2 \sin x$

47. $g(x) = 5\cos^2 2x$ **48.** $y = \arccos x - 1$

49. $f(x) = (\sin^{-1} x)^2$ **50.** $g(x) = (\tan^{-1} x)^2$

In Exercises 51–58, prove each identity algebraically. Support these identities by graphing each side of the equation and comparing their graphs.

51. $\cos 3x = 4\cos^3 x - 3\cos x$

52. $\cos^2 2x - \cos^2 x = \sin^2 x - \sin^2 2x$

53. $\tan^2 x - \sin^2 x = \sin^2 x \tan^2 x$

54. $2\sin\theta\cos^3\theta + 2\sin^3\theta\cos\theta = \sin 2\theta$

55. $\csc x - \cos x\cot x = \sin x$

56. $\dfrac{\tan\theta + \sin\theta}{2\tan\theta} = \cos^2\left(\dfrac{\theta}{2}\right)$

57. $\dfrac{1 + \tan\theta}{1 - \tan\theta} + \dfrac{1 + \cot\theta}{1 - \cot\theta} = 0$

58. $\sin 3\theta = 3\cos^2\theta\sin\theta - \sin^3\theta$

In Exercises 59 and 60, use a graphing utility to investigate whether the equation is an identity. If not, determine a *counterexample*.

59. $\sec x - \sin x\tan x = \cos x$

60. $(\sin^2\alpha - \cos^2\alpha)(\tan^2\alpha + 1) = \tan^2\alpha - 1$

In Exercises 61–64, write each expression in terms of $\sin x$ and $\cos x$ only.

61. $\sin 3x + \cos 3x$ **62.** $\sin 2x + \cos 3x$

63. $\cos^2 2x - \sin 2x$ **64.** $\sin 3x - 3\sin 2x$

In Exercises 65–68, find an equivalent algebraic expression that does not involve trig functions.

65. $\sin(\cos^{-1} x)$ **66.** $\sin(\tan^{-1} x)$

67. $\cos(\sin^{-1} x)$ **68.** $\sin(2\cos^{-1} x)$

In Exercises 69–72, solve each equation graphically.

69. $\sin^2 x - 3\cos x = -0.5$

70. $\cos^3 x - 2\sin x - 0.7 = 0$

71. $\sin^4 x + x^2 = 2$

72. $\sin 2x = x^3 - 5x^2 + 5x + 1$

In Exercises 73–78, solve each equation algebraically over the interval $[0, 2\pi)$. Support your conclusions with a graphing utility.

73. $2\sin 2x = 1$ **74.** $2\cos x = 1$

75. $\sin 3x = \sin x$ **76.** $\sin^2 x - 2\sin x - 3 = 0$

77. $\cos 2t = \cos t$ **78.** $\sin(\cos x) = 1$

In Exercises 79–82, solve each inequality algebraically. Support your conclusions with a graphing utility.

79. $2\cos 2x > 1$ for $0 \le x < 2\pi$

80. $\sin 2x > 2\cos x$ for $0 < x \le 2\pi$

81. $2\cos x < 1$ for $0 \le x < 2\pi$

82. $\tan x < \sin x$ for $-\dfrac{\pi}{2} < x < \dfrac{\pi}{2}$

In Exercises 83 and 84, draw a complete graph and determine the end behavior of the function.

83. $y = \dfrac{1}{x}\sin\dfrac{1}{x}$ **84.** $y = x^3\cos x$

Exercises 85–88 refer to the following **problem situation**: A single cell in a beehive is a regular hexagonal prism open at the front with a trihedral top at the back. It can be shown that the surface area of a cell is given by

$$S(\theta) = 6ab + \frac{3}{2}b^2\left(-\cot\theta + \frac{\sqrt{3}}{\sin\theta}\right),$$

where θ is the trihedral angle, a is the depth of the cell, and $2b$ is the length of the line segment through the center connecting opposite vertices of the hexagonal front. Assume $a = 1.75$ (inches) and $b = 0.65$ (inches).

85. Draw a complete graph of the trigonometric representation for the surface area.

86. What values of θ make sense in the problem situation?

87. What value of θ gives the minimum surface area? (*Remark:* This answer is quite close to the observed angle in nature.)

88. What is the minimum surface area?

ERATOSTHENES

Eratosthenes (274–194 B.C.) was born in Cyrene, in present-day Libya, and educated in Athens, Greece. He was a teacher, librarian, and poet at the great university in Alexandria, Egypt. His best-known contribution in mathematics was the prime-number "sieve." He was able to compile a list of prime numbers by listing consecutive whole numbers greater than 1 and then systematically eliminating the multiples of 2, then the multiples of 3, and so on, until only the prime numbers remained in the list.

Eratosthenes was also interested in geometry. He discussed a famous problem known as the "duplication of the cube." In this problem, one attempts to find the appropriate edge length to create a cube with twice the volume of the given cube.

HYPATIA

Hypatia (A.D. 370–415) was the only known woman professor at the famous university in Alexandria, Egypt. Among her research subjects was the geometry of conic sections. These are the figures formed by the intersection of a plane and a cone. Depending on the angle of the plane, the figure formed is either a circle, an ellipse, a parabola, or a hyperbola.

Hypatia was also interested in science. She wrote descriptions of plans for building an instrument called an astrolabe, which is used to measure the positions of the stars and planets. She also invented several pieces of equipment for working with liquids. Hypatia lived during a time of great social upheaval in Egypt; she was murdered by a mob of fanatics because of her religious convictions. Some historians believe that her death marks the end of ancient mathematics and science.

CHAPTER 8

More Applications of Trigonometry

8.1 _____ Law of Sines

Objective

Students will be able to understand the proof for the law of sines and use the computational applications of the law of sines to solve a variety of problems.

Key Ideas

Oblique triangles
Solving triangles
Law of sines

Section 6.5 demonstrated how to use trigonometric functions to solve right triangles when an acute angle and one side are known, or when two sides are known. Recall that solving a triangle involves using known angle and side measures to find the remaining angles and side lengths.

In this section, we generalize the technique for solving **oblique triangles**, that is, triangles that are not right triangles. Oblique triangles include **acute triangles**, in which all the angles measure less than $90°$, and **obtuse triangles**, in which one angle is greater than $90°$ (see Fig. 8.1). The following notation will be used throughout this chapter. The vertices of a triangle will be denoted A, B, and C, and the angles at A, B, and C will be denoted α, β, and γ, respectively. The lengths of the sides opposite A, B, and C are denoted a, b, and c, respectively. So aside from its vertices, a triangle has six "parts": three sides and three angles (see Fig. 8.1).

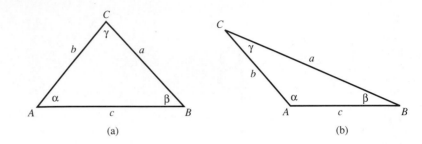

Figure 8.1   Oblique triangles and their notation: (a) an acute triangle; (b) an obtuse triangle.

An important theorem that can be used to solve triangles is the **law of sines**. We begin by stating and proving this theorem.

Theorem 8.1   Law of Sines

In any triangle ABC,

$$\frac{\sin \alpha}{a} = \frac{\sin \beta}{b} = \frac{\sin \gamma}{c}.$$

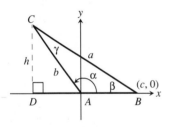

Figure 8.2   A triangle on the rectangular coordinate system used for proving the law of sines.

Proof   Without loss of generality, assume that triangle ABC is an obtuse triangle. Position ABC on a rectangular coordinate system in such a way that A is at the origin and B lies on the positive x-axis at $(c, 0)$. Let D be the foot of the perpendicular from C to the x-axis (see Fig. 8.2).

Because α is in standard position, $\sin \alpha = h/b$ or $h = b \sin \alpha$. Likewise, $\sin \beta = h/a$ or $h = a \sin \beta$. Therefore

$$b \sin \alpha = a \sin \beta$$

$$\frac{\sin \alpha}{a} = \frac{\sin \beta}{b}.$$

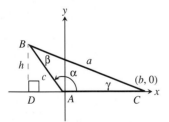

Figure 8.3   Reposition the triangle in Fig. 8.2 so C is now on the x-axis.

Next, reposition triangle ABC with A at the origin and vertex C on the x-axis at $(b, 0)$ (see Fig. 8.3) and repeat the same argument. From doing this, we can conclude that

$$\frac{\sin \alpha}{a} = \frac{\sin \gamma}{c}.$$

Therefore

$$\frac{\sin \alpha}{a} = \frac{\sin \beta}{b} = \frac{\sin \gamma}{c}.$$

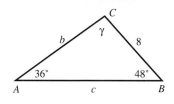

Figure 8.4   Triangle for Example 1.

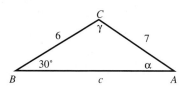

Figure 8.5   Triangle for Example 2.

Because this argument can be repeated in the case where α is acute, the proof is complete.   ▤

Two Angles and a Side Specified

The law of sines allows us to solve a triangle when two angles and a side are specified. Example 1 illustrates how we can find the length of the other two sides and the measure of the third angle using the law of sines.

EXAMPLE 1   Finding Two Sides and One Angle

In triangle ABC, $\alpha = 36°$, $\beta = 48°$, and $a = 8$. Find the remaining sides and angle.

Solution   We need to find b, c, and γ (see Fig. 8.4).
$$\gamma = 180° - (36° + 48°) = 96°$$

$$\frac{\sin \alpha}{a} = \frac{\sin \beta}{b} \qquad \frac{\sin \alpha}{a} = \frac{\sin \gamma}{c} \qquad \text{By the law of sines}$$

$$\frac{\sin 36°}{8} = \frac{\sin 48°}{b} \qquad \frac{\sin 36°}{8} = \frac{\sin 96°}{c}$$

$$b = \frac{8 \sin 48°}{\sin 36°} \qquad c = \frac{8 \sin 96°}{\sin 36°}$$

$$b = 10.11 \qquad c = 13.54 \qquad ▤$$

The method shown in Example 1 will work whenever you are given two angles whose sum is less than 180° and a side. It is generally true that a unique triangle is determined when two angles and a side are given.

Two Sides and an Angle Specified

The next example looks very similar to Example 1 except that the lengths of two sides and one angle are given.

EXAMPLE 2   Finding Two Angles and One Side

In triangle ABC, $\beta = 30°$, $a = 6$, and $b = 7$. Find the unknown side and angles.

Solution   Study Fig. 8.5. As illustrated by Example 1, to use the law of sines to find the length of a side, the angle opposite that side must be given. Therefore, before finding length c, you need to find γ. Because $\alpha + \beta + \gamma = 180°$, γ can be found once α is known. Use the law of sines to find α.

$$\frac{\sin \alpha}{a} = \frac{\sin \beta}{b}$$

$$\frac{\sin \alpha}{6} = \frac{\sin 30°}{7}$$

$$\sin \alpha = \frac{6 \sin 30°}{7}$$

$$\sin \alpha = \frac{3}{7}$$

Two values of α in the interval $0 < \alpha < 180°$ solve $\sin \alpha = \frac{3}{7}$, namely, $\alpha = \sin^{-1}\left(\frac{3}{7}\right) = 25.38°$ and $\alpha = 180° - 25.38° = 154.62°$. However, if $\alpha = 154.62°$, then $\alpha + \beta = 184.62°$, which is not possible. Therefore

$$\alpha = 25.38° \qquad \text{and} \qquad \gamma = 180° - (25.38° + 30°) = 124.62°.$$

Now use the law of sines to find c.

$$\frac{\sin \beta}{b} = \frac{\sin \gamma}{c}$$

$$\frac{\sin 30°}{7} = \frac{\sin 124.62°}{c}$$

$$c = \frac{7 \sin 124.62°}{\sin 30°}$$

$$c = 11.52$$

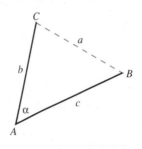

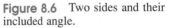

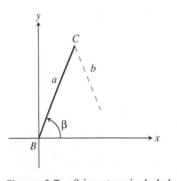

Figure 8.6   Two sides and their included angle.

Example 2 illustrates two important points. One is that the law of sines solution method worked because the given angle was *opposite* one of the given sides. If the given angle had been *included between* the given sides (see Fig. 8.6), the law of sines would not have been helpful. The law of cosines, to be introduced in Section 8.2, can be used to solve that situation.

A second aspect of Example 2 that must be discussed is that a unique triangle exists with the given angle $\beta = 30°$ and given sides, $a = 6$ and $b = 7$. But when two sides and a noninincluded angle are given, there are several other possibilities.

For example, suppose that angle β and side a are given and that β is placed on a coordinate system in standard position as shown in Fig. 8.7. There are several ways of completing a triangle by drawing side b from vertex C down to the x-axis, and these are shown in Fig. 8.8.

Figure 8.7 β is not an included angle.

1. If b is too short, no triangle is formed (see Fig. 8.8a).

Teaching Note

It may be helpful to have students construct triangles using compass and ruler, manipulatives, or geometric software to develop an intuitive sense for Fig. 8.8.

Notes on Examples

Example 3 is best taught by referring to Fig. 8.8 since items a–d are the same in both. In Examples 3 and 4 some students may need help making the connection between the right triangle relationship of $\sin B = \frac{h}{a}$ ($h = a \sin B$) as the altitude of the triangle in Fig. 8.8, and $6 \sin 30°$ as its value for the specific cases in Example 4. Help students to visualize geometrically and analytically how the height can be used to identify the four ways to define a triangle.

Teaching Note

The law of sines and the law of cosines (Section 8.2) provide an opportunity for students with a graphing calculator to write programs to solve the various cases of oblique triangles. A program just to run the ambiguous case of SSA offers an excellent analytical challenge for students.

Notes on Examples

Because Example 4 illustrates the four cases of Example 3, it may be best to do both of these examples in one class period.

2. If b represents the perpendicular distance from C to the x-axis, a right triangle is formed (see Fig. 8.8b).

3. If b is just a bit longer than in Fig. 8.8(b) but still shorter than a, two triangles can be formed (see Fig. 8.8c).

4. If b is as long as or longer than a, a unique triangle is formed (see Fig. 8.8d).

Notice that Example 2 falls in the fourth category shown in Fig. 8.8(d).

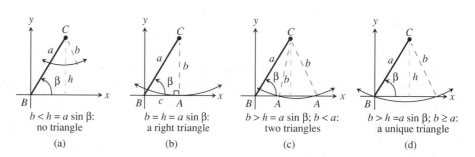

Figure 8.8   Four ways to complete a triangle given an acute angle in standard position and one adjacent side.

EXAMPLE 3   Finding the Number of Triangles

Suppose in triangle ABC that $\beta = 30°$ and $a = 6$. How many triangles are formed if b equals the following?

a) $b = 2$ **b)** $b = 3$
c) $b = 5$ **d)** $b = 7$

Solution

a) $b = 2 < 6 \sin 30° = a \sin \beta$: No triangle is formed.

b) $b = 3 = 6 \sin 30° = a \sin \beta$: A unique right triangle is formed.

c) $b = 5$ is greater than $a \sin \beta = 6 \sin 30° = 3$ and less than $a = 6$: Two triangles are formed.

d) $b = 7 \geq 6 = a$: A unique triangle is formed (see Example 2).　　▤

When two sides and a nonincluded angle of a triangle are given, it is important to identify to which of the four cases shown in Fig. 8.8 the triangle belongs. Doing this determines whether the triangle has no solution, a unique solution, or two solutions.

E X A M P L E   4   Solving Triangles with Nonincluded Angles

In triangle ABC, $\beta = 30°$ and $a = 6$. Find the unknown sides and angles if

a) $b = 3$,

b) $b = 5$.

Solution

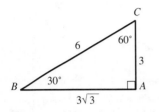

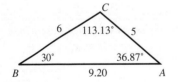

Figure 8.9   Triangle for Example 4, part (a).

a) Because $b = 3 = a \sin \beta$, Example 3 shows that there is a unique right triangle formed (see Fig. 8.9). Therefore $\gamma = 90° - 30° = 60°$, and by the Pythagorean theorem $c = \sqrt{6^2 - 3^2} = \sqrt{36 - 9} = 3\sqrt{3}$.

b) Because $b = 5 > 6 \sin 30° = a \sin \beta$ and $b = 5 < 6 = a$, Example 3 and Fig. 8.8 show that there are two possible values of α.

$$\frac{\sin \alpha}{a} = \frac{\sin \beta}{b} \quad \text{Law of sines}$$

$$\frac{\sin \alpha}{6} = \frac{\sin 30°}{5}$$

$$\sin \alpha = \frac{6 \sin 30°}{5}$$

$$\sin \alpha = \frac{3}{5}$$

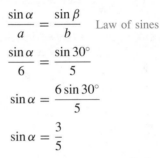

Figure 8.10   One solution to Example 4, part (b).

The two solutions to this equation are $\alpha = \sin^{-1}\left(\frac{3}{5}\right) = 36.87°$ and $\alpha = 180° - 36.87° = 143.13°$.

If $\alpha = 36.87°$, then $\gamma = 180° - (36.87° + 30°) = 113.13°$, and we find c.

$$\frac{\sin \beta}{b} = \frac{\sin \gamma}{c} \quad \text{Law of sines}$$

$$\frac{\sin 30°}{5} = \frac{\sin 113.13°}{c}$$

$$c = 9.20$$

One solution is $\alpha = 36.87°$, $\gamma = 113.13°$, and $c = 9.20$ (see Fig. 8.10).

If $\alpha = 143.13°$, then $\gamma = 180° - (143.13° + 30°) = 6.87°$, and we find c.

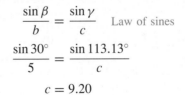

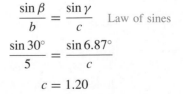

Figure 8.11   A second solution to Example 4, part (b).

$$\frac{\sin \beta}{b} = \frac{\sin \gamma}{c} \quad \text{Law of sines}$$

$$\frac{\sin 30°}{5} = \frac{\sin 6.87°}{c}$$

$$c = 1.20$$

The second solution is $\alpha = 143.13°$ and $c = 1.20$ (see Fig. 8.11).   ≡

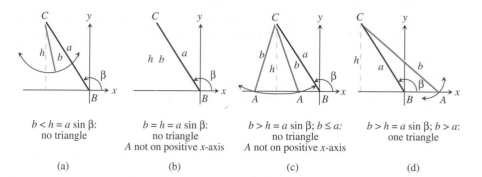

$$b < h = a \sin \beta:$$
no triangle

$$b = h = a \sin \beta:$$
no triangle
A not on positive x-axis

$$b > h = a \sin \beta; b \le a:$$
no triangle
A not on positive x-axis

$$b > h = a \sin \beta; b > a:$$
one triangle

(a)   (b)   (c)   (d)

Figure 8.12 Ways to complete a triangle given an obtuse angle in standard position and one adjacent side.

In Example 4, β was an acute angle. What happens if β is an obtuse angle? Figure 8.12 shows the four cases considered in Fig. 8.8 except that now β is obtuse. However, in this case there is either exactly one solution or no solution since A is on the positive x-axis.

Applications

Applications of the law of sines are illustrated in the next two examples.

EXAMPLE 5   APPLICATION: Locating a Fire

A forest ranger at an observation point A along a straight road sights a fire in the direction $32°$ east of north. Another ranger at a second observation point B 10 mi due east of A on the road sights the same fire $48°$ west of north. Find the distance from each observation point to the fire and find the shortest distance CD from the road to the fire.

Solution   In Fig. 8.13, vertex C represents the location of the fire. We are asked to find a, b, and h, the distance to the road.

The fire is $32°$ east of north, so $\alpha = 90° - 32° = 58°$. Similarly, $\beta = 90° - 48° = 42°$. So $\gamma = 180° - (58° + 42°) = 80°$.

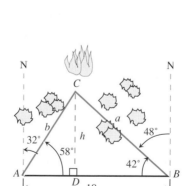

Figure 8.13   Triangle for Example 5.

$$\frac{\sin \alpha}{a} = \frac{\sin \gamma}{c} \qquad\qquad \frac{\sin \beta}{b} = \frac{\sin \gamma}{c}$$

$$\frac{\sin 58°}{a} = \frac{\sin 80°}{10} \qquad\qquad \frac{\sin 42°}{b} = \frac{\sin 80°}{10}$$

$$a = \frac{10 \sin 58°}{\sin 80°} \qquad\qquad b = \frac{10 \sin 42°}{\sin 80°}$$

$$a = 8.6113 \qquad\qquad b = 6.7945$$

The distance h is $b \sin 58° = a \sin 42° = 5.76$.

Thus the fire is 6.79 mi from A and 8.61 mi from B, and the shortest distance from the road AB is 5.76 mi. ☰

EXAMPLE 6   APPLICATION: Determining the Height of a Pole

A road slopes at a $10°$ angle with the horizontal. A vertical telephone pole stands by the road. When the angle of elevation of the sun is $62°$, the telephone pole casts a 14.5-ft shadow downhill parallel with the road. Find the height of the telephone pole.

Solution   It helps to begin by drawing and labeling a figure (see Fig. 8.14). Let BC represent the pole and AB the pole's shadow. Because the angle of elevation of the sun is $62°$, $\gamma = 90° - 62° = 28°$. The road makes a $10°$ angle with the horizontal, so apply geometry to Fig. 8.14 to conclude that $\beta = 100°$ and $\alpha = 52°$.

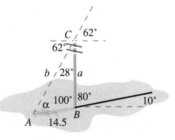

Figure 8.14   Diagram for Example 6.

Notes on Exercises

For Ex. 18–23, students are given a drawing of the problem situation. You may wish to supplement these exercises with some additional problems that require students to draw their own diagrams to solve the problem.
Ex. 22 is a good problem for cooperative learning.

Assignment Guide

Day 1: Ex. 1–15 odd
Day 2: Ex. 17–23

The length of the shadow is $c = 14.5$. Now the law of sines can be used to find the height, a, of the telephone pole.

$$\frac{\sin \alpha}{a} = \frac{\sin \gamma}{c}$$

$$\frac{\sin 52°}{a} = \frac{\sin 28°}{14.5}$$

$$a = \frac{14.5 \sin 52°}{\sin 28°}$$

$$a = 24.34$$

Thus the height of the telephone pole is 24.34 ft.   ☰

Exercises for Section 8.1

1. Give the side lengths a, b, and c for two different triangles ABC whose interior angles are $\alpha = 32°$, $\beta = 75°$, and $\gamma = 73°$.

2. Give the side lengths a, b, and c of two different triangles ABC whose interior angles are $\alpha = 54°$ and $\gamma = 16°$.

Solve each triangle specified by the angles and sides given in Exercises 3–16.

3. $\alpha = 40°$, $\beta = 30°$, $b = 10$

4. $\alpha = 60°$, $a = 3$, $b = 4$

5. $\beta = 30°$, $a = 12$, $b = 6$

6. $\alpha = 50°$, $\beta = 62°$, $a = 4$

7. $\alpha = 33°$, $\beta = 79°$, $b = 7$

8. $\beta = 85°$, $a = 4$, $b = 6$

9. $\alpha = 50°$, $a = 4$, $b = 5$

10. $\beta = 38°$, $a = 16$, $b = 20$

11. $\beta = 38°$, $a = 16$, $b = 12$

12. $\beta = 116°$, $a = 11$, $b = 13$

13. $\beta = 116°$, $a = 11$, $b = 10$

14. $\beta = 116°$, $a = 11$, $b = 8$

15. $\beta = 152°$, $a = 8$, $b = 10$

16. $\gamma = 103°$, $\beta = 16°$, $c = 12$

17. If $a = 10$ and $\beta = 42°$, determine the values of b for which α has the following:

   a) Two values   **b)** One value   **c)** No value

18. Two markers A and B on the same side of a canyon rim are 56 ft apart, and a third marker is located across the rim at point C. A surveyor determines that $\angle BAC = 72°$ and $\angle ABC = 53°$.

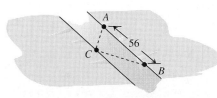

For Exercise 18.

a) What is the distance between C and point A?

b) What is the distance between the two canyon rims? (Assume they are parallel.)

19. A weather forecaster at an observation point A along a straight road sights a tornado in the direction $38°$ east of north. Another forecaster, at an observation point B on the road and 25 mi due east of A, sights the same tornado at $53°$ west of north. The tornado is moving due south. Find the distance from each observation point to the tornado. Also find the distance between the tornado and the road.

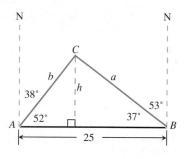

For Exercise 19.

20. A vertical flagpole stands by the side of a road that slopes at an angle of $15°$ with the horizontal. When the angle of elevation of the sun is $62°$, the flagpole casts a 16-ft shadow downhill parallel to the road. Find the height of the flagpole.

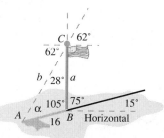

For Exercise 20.

21. A hot-air balloon is seen over Park City, Utah, simultaneously by two observers at points A and B 2.32 mi

apart. Assume the balloon and the observers are in the same vertical plane and that the angles of elevation and the distances are as shown in the following figure. How high above ground is the balloon?

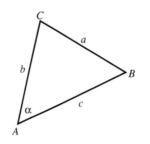

For Exercise 21.

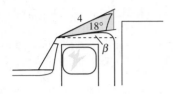

For Exercise 22.

23. Solve triangle ABC shown here, given that $a = 5$, $b = 8$, and $\gamma = 22°$. (*Hint:* Draw a perpendicular from A to the line through B and C.)

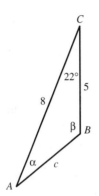

For Exercise 23.

22. A 4-ft airfoil is attached to the cab of a truck to reduce wind resistance, as shown in the following figure. The angle between the airfoil and the cab top is $18°$. If angle β is $10°$, what is the length of a vertical brace positioned as shown in the figure?

8.2 _____

Law of Cosines

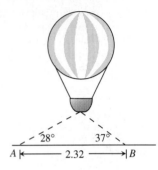

Figure 8.15 Two sides and the included angle.

In Section 8.1, the law of sines was used to find unknown parts of a triangle if two angles and a side were known or if two sides and a nonincluded angle were known. However, in the case where two sides and the included angle are given, as shown in Fig. 8.15, the law of sines does not apply. Instead, we apply the **law of cosines** to find the length of the side opposite the given angle.

Theorem 8.2   Law of Cosines

For any triangle ABC labeled in the usual way (see Fig. 8.15),

$$a^2 = b^2 + c^2 - 2bc \cos \alpha,$$

$$b^2 = a^2 + c^2 - 2ac \cos \beta, \text{ and}$$

$$c^2 = a^2 + b^2 - 2ab \cos \gamma.$$

Proof Without loss of generality, we place triangle ABC so that vertex A is at the origin and vertex B lies on the positive x-axis at $(c, 0)$. Choose D on the x-axis at $(d, 0)$ so that segment CD is perpendicular to the x-axis. If h is the altitude from C, then the coordinates of C are (d, h) (see Fig. 8.16). Because α is in standard position,

$$\cos \alpha = \frac{d}{b} \qquad \text{and} \qquad \sin \alpha = \frac{h}{b}$$

$$d = b \cos \alpha \qquad\qquad\qquad h = b \sin \alpha.$$

Using the Pythagorean theorem and substituting these values of d and h, we have

$$a^2 = (c - d)^2 + (0 - h)^2$$

$$a^2 = (c - d)^2 + h^2$$

$$= (c - b \cos \alpha)^2 + (b \sin \alpha)^2$$

$$= c^2 - 2bc \cos \alpha + b^2 \cos^2 \alpha + b^2 \sin^2 \alpha$$

$$= c^2 - 2bc \cos \alpha + b^2 (\cos^2 \alpha + \sin^2 \alpha)$$

$$a^2 = b^2 + c^2 - 2bc \cos \alpha.$$

The other two equations follow by beginning with the angle β or γ in standard position and repeating the above argument.

Two Sides and the Included Angle Specified

If two sides and an angle opposite one of those sides are given, we found in Section 8.1 that zero, one, or two triangles were possible. In the case where two sides and an included angle of measure between $0°$ and $180°$ are specified, then precisely one triangle is determined. We use the law of cosines to find the remaining side and angles, as illustrated in Example 1.

E X A M P L E   1   Finding Two Angles and One Side

Solve triangle ABC if $a = 4$, $b = 7$, and $\gamma = 42°$.

Solution First use the law of cosines to find the third side.

$$c^2 = a^2 + b^2 - 2ab \cos \gamma$$

$$c^2 = 4^2 + 7^2 - 2(4)(7) \cos 42°$$

$$c^2 = 23.383890$$

$$c = 4.8357$$

The value of c was not rounded to hundredths because it will be used again.

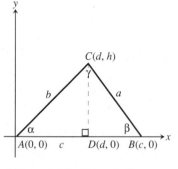

Figure 8.16 Triangle for Theorem 8.2.

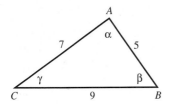

Figure 8.17   Triangle for Example 1.

Next, use the law of sines to determine α.

$$\frac{\sin\alpha}{a} = \frac{\sin\gamma}{c}$$

$$\frac{\sin\alpha}{4} = \frac{\sin 42°}{4.8357}$$

$$\sin\alpha = \frac{4\sin 42°}{4.8357}$$

$$\alpha = \sin^{-1}\left(\frac{4\sin 42°}{4.8357}\right)$$

$$\alpha = 33.61°$$

Thus $\alpha = 33.61°$, $\beta = 104.39°$, $c = 4.84$, and we have determined the triangle in Fig. 8.17.

Three Sides Specified

Suppose that the measures of the three sides of a triangle are specified. In order for there actually to be such a triangle, the sum of the measures of any two sides must exceed the measure of the third side. (Why?) The law of cosines can be applied in this case to find the measures of the angles.

E X A M P L E 2   Finding Three Angles

Solve triangle ABC if $a = 9$, $b = 7$, and $c = 5$.

Solution   First, use the law of cosines to find angle α (see Fig. 8.18).

$$a^2 = b^2 + c^2 - 2bc\cos\alpha$$

$$9^2 = 7^2 + 5^2 - 2(7)(5)\cos\alpha$$

$$\cos\alpha = \frac{7^2 + 5^2 - 9^2}{2(7)(5)}$$

$$\cos\alpha = -0.1$$

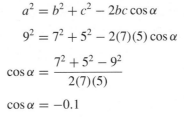

Figure 8.18   Triangle for Example 2.

The only solution to $\cos\alpha = -0.1$ where $0° < \alpha < 180°$ is $\cos^{-1}(-0.1) = 95.7392°$. Now use the law of cosines to find angle β.

$$b^2 = a^2 + c^2 - 2ac\cos\beta$$

$$7^2 = 9^2 + 5^2 - 2(9)(5)\cos\beta$$

$$\cos\beta = \frac{9^2 + 5^2 - 7^2}{2(9)(5)}$$

Thus $\beta = 50.7035°$. It follows that $\gamma = 180° - (95.7392° + 50.7035°) = 33.5573°$. So the three angles are $\alpha = 95.74°$, $\beta = 50.70°$, and $\gamma = 33.56°$.

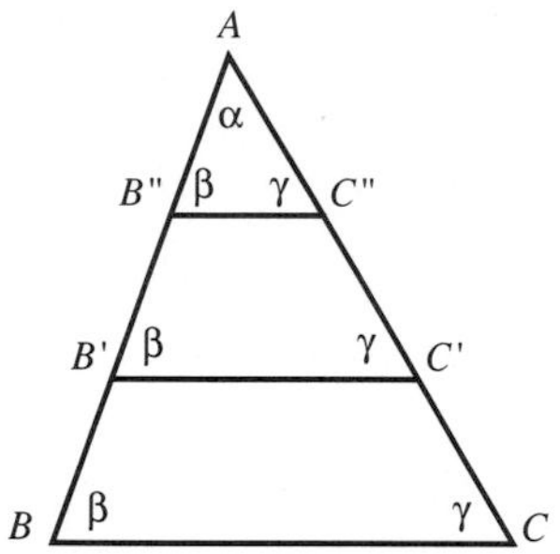

Figure 8.19 Three of the infinite number of triangles possible when three angles and no sides are specified.

Three Angles Specified

We have considered all the possibilities for three parts of a triangle to be specified, except for one. Suppose all three angles are specified. Figure 8.19 illustrates that there are infinitely many different triangles with the same three angles. Consequently, it is not possible to solve for the three sides. They are not uniquely determined.

Following is a summary of the number of triangles determined when three parts of a triangle are given.

Teaching Note

For students who are still unsure of case 3 in this summary, you may wish to refer them to Section 8.1, Fig. 8.8 on page 431. It may be helpful to add a third column to this summary that identifies likely computational processes or relationships to apply in each of the four cases.

Notes on Exercises

Students can be encouraged to solve Ex. 1–10 by writing a program that applies the law of cosines, and then submit their programs and the resulting measurements for the assigned problems. Ex. 11 and 12 give students an opportunity to apply their knowledge of solving the ambiguous case 3 of SSA.

Summary: Solving an Oblique Triangle

Parts Given	Number of Possible Triangles
1. Three angles (sum equals 180°)	Infinitely many
2. Two angles (sum less than 180°), one side	One
3. One angle, two sides	Zero, one, or two
4. Three sides (sum of any two greater than the third)	One

Area of Triangles

The law of cosines allows us to calculate the length of one side of a triangle given the opposite angle and the other two sides. Therefore it is possible to find the area of the resulting triangle also.

Theorem 8.3 Area of Triangles

Let ABC be a triangle labeled in the usual way. Then the area $\mathcal{A}$ of the triangle is given by

$$\mathcal{A} = \frac{1}{2}bc\sin\alpha = \frac{1}{2}ac\sin\beta = \frac{1}{2}ab\sin\gamma.$$

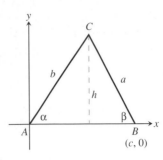

Figure 8.20 Triangle for Theorem 8.3.

Proof Place triangle ABC in standard position and let h be an altitude of triangle ABC (see Fig. 8.20). Notice that $h = b \sin \alpha$. Then

$$\mathcal{A} = \frac{1}{2} ch$$

$$= \frac{1}{2} cb \sin \alpha$$

$$= \frac{1}{2} bc \sin \alpha.$$

The other two forms of the theorem follow by a similar argument. Notice that if $\beta = 90°$, then $\mathcal{A} = \frac{1}{2} ac \sin 90° = \frac{1}{2} ac$, so as expected, $\mathcal{A} = \frac{1}{2}$ base × height.

≡

EXAMPLE 3   Finding the Area of a Triangle

Find the area of triangle ABC if $a = 8$, $b = 5$, and $\gamma = 52°$.

Solution   Theorem 8.3 says that the area is given by

$$\mathcal{A} = \frac{1}{2} ab \sin \gamma$$

$$= \frac{1}{2} (8)(5) \sin 52°$$

$$= 15.76.$$

Thus the area is 15.76 square units.

≡

If we know two sides and the included angle, we can find a base and altitude and thus the area of a triangle. Theorem 8.4 gives **Heron's formula**, which can be used to find the area of a triangle when all three sides are known.

Theorem 8.4   Heron's Formula

Let triangle ABC be labeled in the usual way. Then the area $\mathcal{A}$ of the triangle is given by

$$\mathcal{A} = \sqrt{s(s - a)(s - b)(s - c)},$$

where $s = \frac{1}{2}(a + b + c)$ is one half of the perimeter, or the **semiperimeter**.

Proof   We begin by rewriting the formula for area found in Theorem 8.3 in such a way that the law of cosines can be used to obtain an equation involving only

Notes on Exercises

Ex. 13–18 are area problems. Ex. 17 and 18 can be solved using Heron's formula, or by using the law of cosines and Theorem 8.3, if you choose not to teach Heron's formula.

sides a, b, and c:

$$\mathcal{A} = \frac{1}{2}ab \sin \gamma$$

$$2\mathcal{A} = ab \sin \gamma$$

$$4\mathcal{A}^2 = a^2 b^2 \sin^2 \gamma$$

$$= a^2 b^2 (1 - \cos^2 \gamma)$$

$$= a^2 b^2 - a^2 b^2 \cos^2 \gamma \qquad (1)$$

Next, we multiply each side of Eq. (1) by 4 again and then use the law of cosines to replace $2ab \cos \gamma$ by $a^2 + b^2 - c^2$:

$$16\mathcal{A}^2 = 4a^2 b^2 - (2ab \cos \gamma)^2$$

$$= 4a^2 b^2 - (a^2 + b^2 - c^2)^2 \qquad (2)$$

It can be verified that the right-hand side of Eq. (2) is the same as $(a + b + c)(a + b - c)(b + c - a)(c + a - b)$. Thus

$$16\mathcal{A}^2 = (a + b + c)(a + b - c)(b + c - a)(c + a - b).$$

Notice that $a + b + c = 2s$, $a + b - c = 2s - 2c$, $b + c - a = 2s - 2a$, and $a + c - b = 2s - 2b$. Substituting these values, we obtain

$$16\mathcal{A}^2 = 2s(2s - 2c)(2s - 2a)(2s - 2b)$$

$$16\mathcal{A}^2 = 16s(s - c)(s - a)(s - b)$$

$$\mathcal{A} = \sqrt{s(s - a)(s - b)(s - c)}. \qquad \blacksquare$$

EXAMPLE 4 Using Heron's Formula

Find the area of triangle ABC if $a = 9$, $b = 7$, and $c = 5$.

Solution First note that $s = \frac{1}{2}(a + b + c) = 10.5$. Then compute the area using Heron's formula.

$$\mathcal{A} = \sqrt{s(s - a)(s - b)(s - c)}$$

$$= \sqrt{10.5(1.5)(3.5)(5.5)}$$

$$= 17.41 \qquad \blacksquare$$

Applications of the Law of Cosines

EXAMPLE 5 APPLICATION: Measuring Distance Indirectly

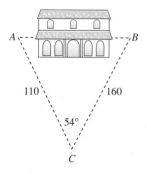

Figure 8.21 Points A and B are on opposite sides of a building.

A surveyor wants to find the distance between two points A and B on opposite sides of a building (see Fig. 8.21). The surveyor locates a point C that is 110 ft from A and 160 ft from B. If $\gamma = 54°$, find distance AB.

Notes on Exercises

Ex. 20–23 are good application problems for baseball fans. Ex. 19–29 offer a variety of applications, including some which should be assigned because students can draw their own figures to represent the problem situations. Ex. 30–33 are interesting and worthwhile problems from geometry.

Solution The distance from A to B in triangle ABC is c. Using the law of cosines,

$$c^2 = a^2 + b^2 - 2ab \cos \gamma$$

$$c^2 = 160^2 + 110^2 - 2(160)(110) \cos 54°$$

$$c = 130.42.$$

Thus the distance between A and B is 130.42 ft.

E X A M P L E   6   APPLICATION: Finding Distances on a Baseball Diamond

In major league baseball, the four bases form a square whose sides are 90 ft long. The front edge of the pitching rubber on which the pitcher stands is 60.5 ft from home plate. Find the distance from the front edge of the pitching rubber to first base.

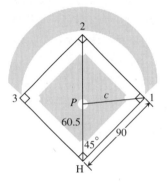

Figure 8.22 Baseball diamond for Example 6.

Solution In Fig. 8.22, let P be the middle of the front edge of the pitching rubber, H be home plate, and c the distance from P to the corner of first base. Because $\gamma = 45°$, distance c can be calculated as follows:

$$c^2 = (60.5)^2 + (90)^2 - 2(60.5)(90)\cos 45°$$

$$c = 63.72$$

Thus the distance from the front edge of the pitching rubber to first base is 63.72 ft.

Recall that in Section 6.6, we defined the angle between two nonperpendicular lines to be the acute angle determined by the lines. Example 7 shows how to apply the law of cosines to find the angle between two lines.

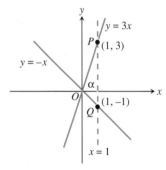

Figure 8.23 Two intersecting lines for Example 7.

E X A M P L E   7   Finding the Angle between Two Lines

Find the angle between the lines $y = 3x$ and $y = -x$.

Solution Using the distance formula, we see that in Fig. 8.23 segments $OP = \sqrt{10}$, $OQ = \sqrt{2}$, and $PQ = 4$. Applying the law of cosines to triangle POQ with $\alpha = \angle POQ$ yields the following:

$$4^2 = (\sqrt{10})^2 + (\sqrt{2})^2 - 2\sqrt{2}\sqrt{10} \cos \alpha$$

$$16 = 10 + 2 - 4\sqrt{5} \cos \alpha$$

$$-\frac{1}{\sqrt{5}} = \cos\alpha$$

$$\alpha = \cos^{-1}\left(-\frac{1}{\sqrt{5}}\right) = 116.57°.$$

Therefore the acute angle between these two lines is $180° - 116.57° = 63.43°$.

≡

Assignment Guide

Day 1: Ex. 1–17 odd
Day 2: Ex. 19, 24–26, 29

Exercises for Section 8.2

In Exercises 1–10, solve each triangle ABC.

1. $a = 1$, $b = 5$, $c = 4$

2. $a = 1$, $b = 5$, $c = 8$

3. $a = 3.2$, $b = 7.6$, $c = 6.4$

4. $\alpha = 21°$, $\beta = 17°$, $c = 15$

5. $\alpha = 55°$, $b = 12$, $c = 7$

6. $\beta = 125°$, $a = 25$, $c = 41$

7. $\beta = 103°$, $b = 13$, $a = 18$

8. $\alpha = 36°$, $b = 17$, $a = 14$

9. $\beta = 110°$, $b = 13$, $c = 15$

10. $a = 5$, $b = 7$, $c = 6$

11. If $a = 8$ and $\beta = 58°$, determine the values of b for which α has
   a) two values,
   b) one value, or
   c) no value.

12. If $a = 12$ and $\beta = 32°$, determine the values of b for which α has
   a) two values,
   b) one value, or
   c) no value.

In Exercises 13–18, find the area of each triangle ABC determined by the given information.

13. $b = 6$, $c = 8$, $\alpha = 47°$

14. $a = 17$, $c = 14$, $\beta = 103°$

15. $\alpha = 15°$, $\beta = 65°$, $a = 8$

16. $\alpha = 10°$, $\gamma = 110°$, $c = 12.3$

17. $a = 2$, $b = 6$, $c = 7$

18. $a = 20$, $b = 36$, $c = 50$

19. To determine the distance between two points A and B on opposite sides of a lake, a surveyor chooses a point C that is 860 ft from A and 175 ft from B. If the measure of the angle at C is 78°, find the distance between A and B.

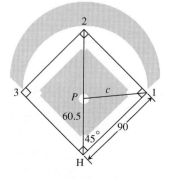

For Exercise 19.

20. Find the distance from the center of the front edge of the pitching rubber to second base.

21. Find the angle created by home plate, the center of the front edge of the pitching rubber, and first base.

22. Find the angle formed by the center of the front edge of the pitching rubber, first base, and home plate.

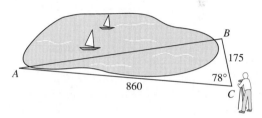

For Exercises 20–22.

23. In softball, the four bases form a square whose sides are 60 ft in length. The front edge of the pitching rubber is 40 ft from home plate. Find the distance from the front edge of the pitching rubber to first base.

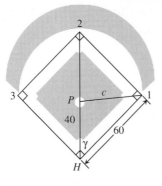

For Exercise 23.

24. Find the radian measure of the largest angle in the triangle whose sides have lengths of 4, 5, and 6.

25. The sides of a parallelogram have lengths of 18 and 26 ft and one angle of 39°. Find the length of the longer diagonal.

26. Two observers are 600 ft apart on opposite sides of a flagpole. The angles of elevation from the observers to the top of the pole are 19° and 21°. Find the height of the flagpole.

27. A blimp is sighted simultaneously by two observers: A at the top of a 650-ft tower and B at the base of the tower. Find the distance of the blimp from observer A if the angle of elevation (from the horizontal) as viewed by A is 32° and the angle of elevation as viewed by B is 56°. How high is the blimp?

28. In a parallelogram, two adjacent sides meet at an angle of 35° and are 3 and 8 ft in length. What is the length of the shorter diagonal of the parallelogram?

29. Two observers are 400 ft apart on opposite sides of a tree. If the angles of elevation from the observers to the top of the tree are 15° and 20°, how tall is the tree?

30. Suppose that ℓ_1 and ℓ_2 are two lines with slopes m_1 and m_2, respectively. Show that the angle α between the two lines is either

$$\alpha = \cos^{-1}\left(\frac{1 + m_1 m_2}{\sqrt{(1 + m_1^2)(1 + m_2^2)}}\right) \qquad \text{if } \alpha < 90°$$

$$\text{or } 180° - \alpha \qquad \text{if } \alpha > 90°.$$

Exercises 31–33 refer to the following **problem situation**: Let AOB be a sector of a circle of radius r with central angle θ.

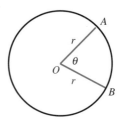

For Exercises 31 to 33.

31. Show that the area $\mathcal{A}$ of the sector is $r^2\theta/2$.

32. Assume the perimeter $P = 2r + \theta r$ of the sector is fixed. What is the maximum area of the sector?

33. Explain how the function

$$f(x) = \frac{x}{2(2 + x)^2}$$

is related to the solution of Exercise 32.

8.3 _____ Trigonometric Form of Complex Numbers

Complex numbers were introduced in Section 3.5. Recall that a complex number is written in the form $a + bi$, where a and b are real numbers. Also recall that

$$(a + bi) + (c + di) = (a + c) + (b + d)i.$$

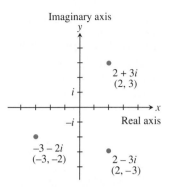

Figure 8.24 Complex numbers represented as points in the coordinate plane.

In this section, we introduce two additional ways to represent complex numbers. The first is geometric representation and the second is trigonometric.

Geometric Representation of Complex Numbers

There is a natural one-to-one correspondence between the set of complex numbers and the points in the coordinate plane. The complex number $a + bi$ corresponds to the point $P(a, b)$. When the coordinate plane is viewed as a representation of the complex numbers, it is called the **complex plane** or the **Gaussian plane**. Figure 8.24 shows the complex numbers $2+3i$, $-3-2i$, and $2-3i$ on the complex plane. The x-axis of the complex plane is called the **real axis** and the y-axis is called the **imaginary axis.**

If we draw a segment from the origin to the point representing $a + bi$, the length of this segment is $\sqrt{a^2 + b^2}$ (see Fig. 8.25). This length is the concept defined in the following definition.

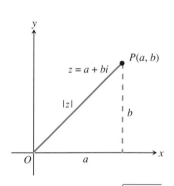

Figure 8.25 $|z| = \sqrt{a^2 + b^2}$.

Definition 8.1   Absolute Value of a Complex Number

The **absolute value** or **modulus** of the complex number $z = a + bi$ is given by

$$|a + bi| = \sqrt{a^2 + b^2}.$$

The absolute value of $a + bi$ is the length of the segment from the origin O to $a + bi$.

Complex numbers have many of the algebraic properties of real numbers. In particular, they can be added and subtracted. A close association exists between complex number addition and vector addition that will be defined in Section 8.5. Consequently, we use a geometric representation to introduce complex number addition.

If $z_1 = a + bi$ is represented by point $P(a, b)$ and $z_2 = c + di$ is represented by point $Q(c, d)$, then the complex number $z_1 + z_2$ is represented by the point $R(a+c, b+d)$. (See Fig. 8.26.) It can be shown that segment OR is the diagonal of the parallelogram determined by segments OP and OQ.

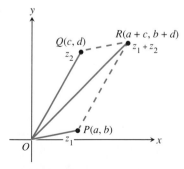

Figure 8.26 Geometric representation of complex addition.

E X A M P L E   1   Finding Geometric Representations

Show a geometric representation for

a) $(2 + i) + (1 + 3i)$, and

b) $(3 + 5i) - (2 + i)$.

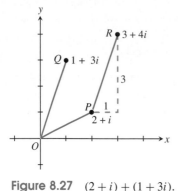

Figure 8.27 $(2 + i) + (1 + 3i)$.

Solution

a) $(2 + i) + (1 + 3i)$ corresponds to the vertex R of the parallelogram $OPRQ$ in Fig. 8.27. To locate point R, begin at P(2,1) and move right 1 unit and up 3 units to the point with coordinates $(3, 4)$. The complex number that corresponds to the point with coordinate $(3, 4)$ is

$$(1 + 3i) + (2 + i) = (3 + 4i).$$

b) $(3 + 5i) - (2 + i) = a + bi$, where $(a + bi) + (2 + i) = 3 + 5i$. Because $a + 2 = 3$ and $b + 1 = 5$, it is evident that $a = 1$ and $b = 4$. Therefore the complex number $a + bi$ is associated with point $(1, 4)$. (See Fig. 8.28.) ∎

Trigonometric Form for Complex Numbers

A common form for expressing complex numbers involves the trig functions $\sin \theta$ and $\cos \theta$. To develop this form we will use a geometric representation of complex numbers.

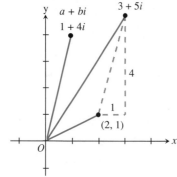

Figure 8.28 Finding $(3 + 5i) - (2 + i)$.

The complex number $a + bi$ corresponds to the point $P(a, b)$ in the complex plane. In Fig. 8.29, we see that for the right triangle determined by $z = a + bi$, the lengths of the three sides a, b, and r are given by

$$r = \sqrt{a^2 + b^2}, \qquad \cos \theta = \frac{a}{r}, \qquad \sin \theta = \frac{b}{r}.$$

Therefore we can write

$$a + bi = (r \cos \theta) + (r \sin \theta)i$$
$$= r(\cos \theta + i \sin \theta).$$

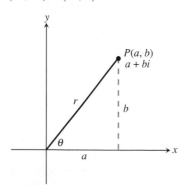

Figure 8.29 The complex number $a + bi$ determines a right triangle.

Definition 8.2   Trigonometric Form of a Complex Number

Let $z = a + bi \neq 0$ be a complex number. A **trigonometric form of z** is

$$z = r(\cos \theta + i \sin \theta),$$

where $a = r \cos \theta$, $b = r \sin \theta$, and $r = |z| = \sqrt{a^2 + b^2}$.

The angle θ can always be chosen so that $0 \le \theta < 2\pi$, although any angle coterminal with θ could be used. Consequently, the angle θ is not unique and hence the trigonometric representation of z is not unique.

E X A M P L E   2   Finding Trigonometric Representations

Find the trigonometric representation with $0 \le \theta < 2\pi$ for the following:

a) $-2 + 2i$

b) $3 - 3\sqrt{3}i$

c) $-3 - 4i$

Solution   First calculate r and then use that calculated value to find θ.

a) $r = |-2 + 2i| = \sqrt{(-2)^2 + 2^2} = \sqrt{8} = 2\sqrt{2}$. The angle θ formed by the positive x-axis and $-2 + 2i$ is $3\pi/4$ (see Fig. 8.30a). Therefore

$$-2 + 2i = 2\sqrt{2}\left(\cos\frac{3\pi}{4} + i\sin\frac{3\pi}{4}\right).$$

b) $r = |3 - 3\sqrt{3}i| = \sqrt{9 + 27} = 6$. The angle θ formed by the positive x-axis and $3 - 3\sqrt{3}i$ is $5\pi/3$ (see Fig. 8.30b). (Why?) Therefore

$$3 - 3\sqrt{3}i = 6\left(\frac{1}{2} - i\frac{\sqrt{3}}{2}\right) = 6\left(\cos\frac{5\pi}{3} + i\sin\frac{5\pi}{3}\right).$$

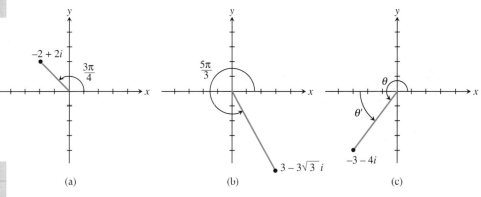

(a)   (b)   (c)

Figure 8.30 Graphs of the trigonometric representations for Example 2.

c) $r = |-3 - 4i| = \sqrt{(-3)^2 + (-4)^2} = 5$. Let θ be the angle that $-3 - 4i$ makes with the positive x-axis, and let θ' be the reference angle (see Fig. 8.30c). Since $\tan\theta = (-4)/(-3)$,

$$\theta' = \tan^{-1}\left(\tfrac{4}{3}\right) = 0.9273.$$

It follows that $\theta = \pi + 0.9273 = 4.0689$. Therefore

$$-3 - 4i = 5(\cos 4.07 + i \sin 4.07).$$ ≡

The trigonometric representation of complex numbers is particularly convenient to use when multiplying or dividing complex numbers. The following theorem describes this method.

> ### Theorem 8.5   Product and Quotient of Trig Representation
>
> Let $z_1 = r_1(\cos\theta_1 + i\sin\theta_1)$ and $z_2 = r_2(\cos\theta_2 + i\sin\theta_2)$. Then
>
> **a)** $z_1 z_2 = r_1 r_2 \big[\cos(\theta_1 + \theta_2) + i\sin(\theta_1 + \theta_2)\big]$, and
>
> **b)** $\dfrac{z_1}{z_2} = \dfrac{r_1}{r_2}\big[\cos(\theta_1 - \theta_2) + i\sin(\theta_1 - \theta_2)\big]$ $(r_2 \neq 0)$.

Proof

a) Multiply z_1 and z_2 in the usual way and apply the identities for $\sin(\theta_1 + \theta_2)$ and $\cos(\theta_1 + \theta_2)$.

$$z_1 z_2 = [r_1(\cos\theta_1 + i\sin\theta_1)][r_2(\cos\theta_2 + i\sin\theta_2)]$$

$$= r_1 r_2[(\cos\theta_1\cos\theta_2 - \sin\theta_1\sin\theta_2) + i(\sin\theta_1\cos\theta_2 + \cos\theta_1\sin\theta_2)]$$

$$= r_1 r_2\big[\cos(\theta_1 + \theta_2) + i\sin(\theta_1 + \theta_2)\big]$$

b) To divide z_1 by z_2, multiply the numerator and denominator by the complex conjugate of the portion of the denominator inside the parentheses.

$$\frac{z_1}{z_2} = \frac{r_1(\cos\theta_1 + i\sin\theta_1)}{r_2(\cos\theta_2 + i\sin\theta_2)}$$

$$= \frac{r_1}{r_2} \cdot \frac{(\cos\theta_1 + i\sin\theta_1)}{(\cos\theta_2 + i\sin\theta_2)} \cdot \frac{(\cos\theta_2 - i\sin\theta_2)}{(\cos\theta_2 - i\sin\theta_2)}$$

$$= \frac{r_1}{r_2} \cdot \frac{(\cos\theta_1\cos\theta_2 + \sin\theta_1\sin\theta_2) + i(\sin\theta_1\cos\theta_2 - \cos\theta_1\sin\theta_2)}{\cos^2\theta_2 + \sin^2\theta_2}$$

$$= \frac{r_1}{r_2}\big[\cos(\theta_1 - \theta_2) + i\sin(\theta_1 - \theta_2)\big] \quad \text{Use identities for the sine and cosine of the difference.}$$ ≡

E X A M P L E   3   Multiplying Complex Numbers

Suppose $z_1 = -2 + 2i$ and $z_2 = 3 - 3\sqrt{3}i$. Use the trigonometric form to find $z_1 z_2$ and the definition of product to confirm that your answer agrees with the product.

Solution   In Example 2, we found that

$$-2 + 2i = 2\sqrt{2}\left(\cos\frac{3\pi}{4} + i\sin\frac{3\pi}{4}\right) \quad \text{and} \quad 3 - 3\sqrt{3}i = 6\left(\cos\frac{5\pi}{3} + i\sin\frac{5\pi}{3}\right).$$

We now compute $z_1 z_2$.

$$
\begin{aligned}
z_1 z_2 &= (-2 + 2i)(3 - 3\sqrt{3}i) \\
&= \left[2\sqrt{2}\left(\cos\frac{3\pi}{4} + i\sin\frac{3\pi}{4}\right)\right]\left[6\left(\cos\frac{5\pi}{3} + i\sin\frac{5\pi}{3}\right)\right] && \text{Substitute the} \\
&&& \text{trig form.} \\
&= 12\sqrt{2}\left[\cos\left(\frac{3\pi}{4} + \frac{5\pi}{3}\right) + i\sin\left(\frac{3\pi}{4} + \frac{5\pi}{3}\right)\right] && \text{Apply} \\
&&& \text{Theorem 8.5.} \\
&= 12\sqrt{2}\left(\cos\frac{29\pi}{12} + i\sin\frac{29\pi}{12}\right) \\
&= 12\sqrt{2}\left(\cos\frac{5\pi}{12} + i\sin\frac{5\pi}{12}\right)
\end{aligned}
$$

Use the definition for $z_1 z_2$ to obtain

$$(-2 + 2i)(3 - 3\sqrt{3}i) = (-6 + 6\sqrt{3}) + (6 + 6\sqrt{3})i.$$

Next, use a calculator to confirm that $-6 + 6\sqrt{3}$ and $12\sqrt{2}\cos(5\pi/12)$ are equal and that $6 + 6\sqrt{3}$ and $12\sqrt{2}\sin(5\pi/12)$ are equal. Therefore

$$(-6 + 6\sqrt{3}) + (6 + 6\sqrt{3})i = 12\sqrt{2}\left(\cos\frac{5\pi}{12} + i\sin\frac{5\pi}{12}\right). \qquad \blacksquare$$

E X A M P L E   4   Dividing Complex Numbers

Suppose $z_1 = -2 + 2i$ and $z_2 = 3 - 3\sqrt{3}i$. Use the trigonometric form to find z_1/z_2 and confirm that your answer agrees with the quotient using the definition of quotient.

Solution   Use Theorem 8.5 to find z_1/z_2.

$$
\begin{aligned}
\frac{z_1}{z_2} &= \frac{-2 + 2i}{3 - 3\sqrt{3}i} \\
&= \frac{2\sqrt{2}\left(\cos\frac{3\pi}{4} + i\sin\frac{3\pi}{4}\right)}{6\left(\cos\frac{5\pi}{3} + i\sin\frac{5\pi}{3}\right)} \\
&= \frac{\sqrt{2}}{3}\left[\cos\left(\frac{3\pi}{4} - \frac{5\pi}{3}\right) + i\sin\left(\frac{3\pi}{4} - \frac{5\pi}{3}\right)\right]
\end{aligned}
$$

Notes on Exercises

Ex. 7–16 are students' first encounter with the geometric and trigonometric aspects of complex numbers. These exercises, therefore, warrant careful attention to be sure that students are off to a good start. Ex. 17–32 provide an opportunity for students to move back and forth between the trigonometric and rectangular forms of complex numbers. Ex. 33–46 provide an opportunity to discuss how much Theorem 8.5 can simplify computations when applicable. A thorough treatment of these problems will prepare students for the next section.

Assignment Guide

Day 1: Ex. 1–11, 13, 15, 16, 17–27 odd
Day 2: Ex. 18, 26, 28–30, 33, 34, 36, 37, 39, 41, 43, 45–48.

$$= \frac{\sqrt{2}}{3} \left[\cos\left(-\frac{11\pi}{12}\right) + i \sin\left(-\frac{11\pi}{12}\right) \right]$$

$$= \frac{\sqrt{2}}{3} \left(\cos\frac{11\pi}{12} - i \sin\frac{11\pi}{12} \right) \qquad \text{Recall that } \cos(-x) = \cos x \text{ and } \sin(-x) = -\sin x.$$

Another way to find the quotient is to multiply numerator and denominator by the conjugate of the denominator. Then

$$\frac{-2+2i}{3-3\sqrt{3}i} = \frac{-2+2i}{3-3\sqrt{3}i} \cdot \frac{3+3\sqrt{3}i}{3+3\sqrt{3}i}$$

$$= -\frac{1+\sqrt{3}}{6} + i\frac{1-\sqrt{3}}{6}.$$

Use a calculator to conclude that $-(1+\sqrt{3})/6$ and $(\sqrt{2}/3)\cos(11\pi/12)$ are equal and that $(1-\sqrt{3})/6$ and $-(\sqrt{2}/3)\sin(11\pi/12)$ are equal. Therefore

$$-\frac{1+\sqrt{3}}{6} + \frac{1-\sqrt{3}}{6}i = \frac{\sqrt{2}}{3}\left(\cos\frac{11\pi}{12} - i\sin\frac{11\pi}{12}\right). \qquad \equiv$$

Exercises for Section 8.3

In Exercises 1–6, simplify and write in the form $a + bi$.

1. $(-1+2i)+(3+5i)$ **2.** $3(2-3i)$

3. $(2-4i)(3+2i)$ **4.** $(-1+2i)^2$

5. $\dfrac{1+i}{1-i}$ **6.** $\dfrac{2-3i}{4+5i}$

In Exercises 7–10, consider the complex numbers z_1 and z_2 as represented in the figure that follows.

7. If $z_1 = a + bi$, determine a, b, and $|z_1|$.

8. If $z_2 = a + bi$, determine a, b, and $|z_2|$.

9. Determine the trig form of z_1.

10. Determine the trig form of z_2.

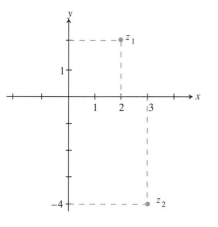

For Exercises 7 to 10.

In Exercises 11–16, plot each of the following in the complex plane.

11. 3   **12.** $3i$

13. $4 - 4i$ **14.** $-4 + 3i$

15. $2(\cos 30° + i \sin 30°)$ **16.** $3\left(\cos \dfrac{\pi}{4} + i \sin \dfrac{\pi}{4}\right)$

In Exercises 17–28, find the trigonometric representation for each complex number such that $0 \le \theta < 2\pi$.

17. $1 + i$ **18.** $-1 + \sqrt{3}i$

19. $-2\sqrt{3} - 2i$ **20.** $3 - 4i$

21. $1 + 3i$ **22.** $-2 + 3i$

23. $4 - 3i$ **24.** $2i$

25. $2 + 3i$ **26.** $-4 + 4i$

27. $\sqrt{2} - 8i$ **28.** $-1 - 2i$

In Exercises 29–32, write the complex number in the form $a + bi$.

29. $2(\cos 60° + i \sin 60°)$ **30.** $5\left(\cos \dfrac{3\pi}{4} + i \sin \dfrac{3\pi}{4}\right)$

31. $3.4(\cos 5 + i \sin 5)$ **32.** $10(\cos 6\pi + i \sin 6\pi)$

In Exercises 33–36, complete each operation.

33. $2(\cos 30° + i \sin 30°) \cdot 3(\cos 60° + i \sin 60°)$

34. $\dfrac{2(\cos 30° + i \sin 30°)}{3(\cos 60° + i \sin 60°)}$

35. $\sqrt{2}(\cos 3 + i \sin 3) \cdot \sqrt{3}(\cos 5 + i \sin 5)$

36. $(2 - 3i)(-3 + 4i)$

In Exercises 37–40, use the trigonometric form to find the product and quotient of each pair of complex numbers.

37. $\dfrac{\sqrt{2}}{2} + \dfrac{\sqrt{2}}{2}i$ and $\dfrac{\sqrt{3}}{2} + \dfrac{1}{2}i$

38. $\dfrac{\sqrt{2}}{2} - \dfrac{\sqrt{2}}{2}i$ and $\dfrac{1}{2} + \dfrac{\sqrt{3}}{2}i$

39. $-3 + 3i$ and $\sqrt{3} - i$ **40.** $4 + 3i$ and $\sqrt{3} - i$

In Exercises 41–46, let $z_1 = -2 + 3i$, $z_2 = 3 + 4i$, and $z_3 = 2 - 5i$. Perform the indicated computation and write in the form $a + bi$.

41. $z_1 + z_2$ **42.** $4z_3$ **43.** $z_1 z_3$

44. $|z_1 - z_2|$ **45.** $|z_1 z_2|$ **46.** $\dfrac{z_2}{z_3}$

47. Determine $(1 + 3i) + (4 + 4i)$ using a geometric representation.

48. Determine $(-3 + i) - (4 + 2i)$ using a geometric representation.

8.4 _____   De Moivre's Theorem and *n*th Roots

Theorem 8.5 can be used to raise a complex number to a power. For example, let $z = r(\cos \theta + i \sin \theta)$. Then

$$z^2 = r(\cos \theta + i \sin \theta) \cdot r(\cos \theta + i \sin \theta)$$

$$= r^2[\cos(\theta + \theta) + i \sin(\theta + \theta)]$$

$$= r^2(\cos 2\theta + i \sin 2\theta)$$

Use this result to find z^3.

$$z^3 = z \cdot z^2 = r(\cos \theta + i \sin \theta) \cdot r^2(\cos 2\theta + i \sin 2\theta)$$

$$= r^3\left[\cos(\theta + 2\theta) + i \sin(\theta + 2\theta)\right]$$

$$= r^3(\cos 3\theta + i \sin 3\theta)$$

In a similar fashion, we could find $z^4, z^5, \ldots$ and arrive at the generalization known as **De Moivre's theorem**.

Theorem 8.6   De Moivre's Theorem

Let n be any integer. Then

$$[r(\cos\theta + i\sin\theta)]^n = r^n(\cos n\theta + i\sin n\theta).$$

It is helpful to interpret De Moivre's theorem geometrically. When a complex number is squared, the angle in its trig form is doubled and its absolute value is squared. We see in Fig. 8.31 that drawing z^2 involves rotating z through an angle θ and stretching or shrinking it to its appropriate length. If z is outside the unit circle, then its absolute value is greater than 1 and z^2 is further away from the origin than z is (see Fig. 8.31a). If z has an absolute value of 1 and is on the unit circle, then so is z^2 or any higher power of z (see Fig. 8.31b). If z is inside the unit circle, then its absolute value is less than 1 and z^2 is closer to the origin than z is.

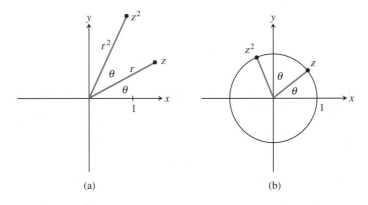

Figure 8.31   To find z^2, double the angle θ and square the modulus of z: (a) $|z| > 1$; (b) $|z| = 1$.

E X A M P L E   1   Using De Moivre's Theorem

Find $(1 + \sqrt{3}i)^3$ using De Moivre's theorem.

Solution Use the technique of Example 2 of Section 8.3 to write $1 + \sqrt{3}i$ in the trigonometric form $2[\cos(\pi/3) + i\sin(\pi/3)]$.

$$(1 + \sqrt{3}i)^3 = \left[2\left(\cos\frac{\pi}{3} + i\sin\frac{\pi}{3}\right)\right]^3$$

$$= 2^3\left[\cos\left(3 \cdot \frac{\pi}{3}\right) + i\sin\left(3 \cdot \frac{\pi}{3}\right)\right]$$

$$= 8(\cos\pi + i\sin\pi)$$

$$= 8(-1 + 0i) = -8.$$

≣

Notes on Examples

Problems like Example 2 are found on math contest tests. They are easy if a student knows De Moivre's theorem.

E X A M P L E   2   Using De Moivre's Theorem

Find $[-\sqrt{2}/2 + (\sqrt{2}/2)i]^8$ using De Moivre's theorem.

Solution Begin by writing $-\sqrt{2}/2 + (\sqrt{2}/2i)$ in the trigonometric form $\cos(3\pi/4) + i\sin(3\pi/4)$. Then

$$\left(-\frac{\sqrt{2}}{2} + \frac{\sqrt{2}}{2}i\right)^8 = \left(\cos\frac{3\pi}{4} + i\sin\frac{3\pi}{4}\right)^8$$

$$= \cos\left(8 \cdot \frac{3\pi}{4}\right) + i\sin\left(8 \cdot \frac{3\pi}{4}\right)$$

$$= \cos 6\pi + i\sin 6\pi$$

$$= \cos 0 + i\sin 0$$

$$= 1 + 0i = 1.$$

≣

Think how difficult it would be to compute $\left[-\sqrt{2}/2 + (\sqrt{2}/2)i\right]^8$ without using De Moivre's theorem.

Example 1 shows a complex number that satisfies the equation $z^3 = -8$, and Example 2 shows a complex number that satisfies the equation $z^8 = 1$. These equations suggest employing the terminology of *roots* used with real numbers.

Definition 8.3   *n*th Roots

Let a be a complex number and n be an integer greater than 1. An **nth root of a** is any complex number z satisfying the relation

$$z^n = a.$$

If z is a number satisfying $z^n = 1$, then z is called an **nth root of unity**.

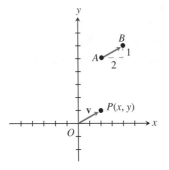

Figure 8.40 Equal vectors for Example 3, part (a).

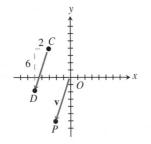

Figure 8.41 Equal vectors for Example 3, part (b).

Solution

a) The terminal point B of $\overrightarrow{AB}$ is 2 units right and 1 unit up from the initial point A. So the terminal point P of $\overrightarrow{OP}$ must also be 2 units right and 1 unit up from the initial point $(0, 0)$. Thus $P = (2, 1)$ and $\mathbf{v} = (2, 1)$ (see Fig. 8.40).

b) P must be 2 units left and 6 units down from the origin, so $P = (-2, -6)$. Therefore $\mathbf{v} = (-2, -6)$ (see Fig. 8.41).

Notice in Example 3, part (a), that the coordinates of vector $\mathbf{v} = (2, 1)$ can be obtained by subtracting the coordinates of point $A = (2, 5)$ from the coordinates of point $B = (4, 6)$. That is, $\mathbf{v} = (2, 1) = (4 - 2, 6 - 5)$. Similarly, for the vector of Example 3, part (b), we have $\mathbf{v} = (-2, -6) = (-5 - (-3), -2 - 4)$. These are special cases of the following theorem.

Theorem 8.10   Finding the Terminal Point of a Vector

Let $A = (a_1, a_2)$ and $B = (b_1, b_2)$ be two points and $\mathbf{v} = (x, y)$ be the vector that is equal to $\overrightarrow{AB}$. Then

$$x = b_1 - a_1 \text{ and } y = b_2 - a_2.$$

That is, $\mathbf{v} = (x, y) = (b_1 - a_1, b_2 - a_2)$.

Addition and Subtraction of Vectors

Because vectors model force, vector addition should be defined so that it models the way forces add together. And forces add according to a parallelogram property.

The sum of vectors $\mathbf{u}$ and $\mathbf{v}$ is represented by the diagonal of the parallelogram whose sides are $\mathbf{u}$ and $\mathbf{v}$. Figure 8.42 shows that $\mathbf{u} + \mathbf{v}$ is equal to $\mathbf{v} + \mathbf{u}$.

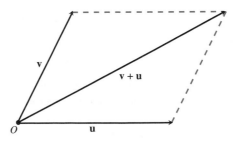

Figure 8.42 $\mathbf{u} + \mathbf{v} = \mathbf{v} + \mathbf{u}$.

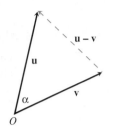

Figure 8.43 $\mathbf{v} + (\mathbf{u} - \mathbf{v})$.

The difference of two vectors $\mathbf{u}$ and $\mathbf{v}$, denoted $\mathbf{u} - \mathbf{v}$, is the vector that when added to $\mathbf{v}$ gives $\mathbf{u}$ (see Fig. 8.43), that is, $\mathbf{v} + (\mathbf{u} - \mathbf{v}) = \mathbf{u}$. Notice that the initial point of vector $\mathbf{u} - \mathbf{v}$ is the terminal point of $\mathbf{v}$ and that the terminal point of $\mathbf{u} - \mathbf{v}$ is the terminal point of $\mathbf{u}$.

Vector addition and subtraction can also be defined in terms of vector components.

Definition 8.4   Vector Addition and Subtraction

Suppose $\mathbf{u} = (a, b)$ and $\mathbf{v} = (c, d)$. Then

$$\mathbf{u} + \mathbf{v} = (a, b) + (c, d) = (a + c, b + d)$$

$$\mathbf{u} - \mathbf{v} = (a, b) - (c, d) = (a - c, b - d).$$

Teaching Note

If students are having difficulty, encourage them to draw pictures and analyze the geometry of the situation.

Notice that the vectors $\mathbf{u} = (a, b)$ and $\mathbf{v} = (c, d)$ can also be represented by the complex numbers $z_1 = a + bi$ and $z_2 = c + di$. Figure 8.26 was used in Section 8.3 to represent addition of complex numbers. We duplicate it here as Fig. 8.44 to show that it also illustrates the vector sum $\mathbf{u} + \mathbf{v}$, where $\mathbf{u} = (a, b)$ and $\mathbf{v} = (c, d)$.

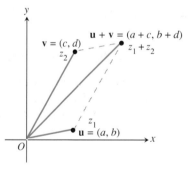

Figure 8.44   Complex number addition and vector addition.

E X A M P L E   4   Doing Vector Addition

Let $\mathbf{u} = (2, 5)$ and $\mathbf{v} = (4, 3)$. Find $\mathbf{u} + \mathbf{v}$ using

a) component addition, and

b) a geometric definition.

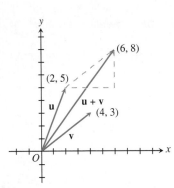

Figure 8.45   Figure for Example 4(b).

Solution

a) (See Fig. 8.45.)

$$\mathbf{u} + \mathbf{v} = (2, 5) + (4, 3)$$
$$= (2 + 4, 5 + 3)$$
$$= (6, 8)$$

b) Consider the vector equal to $\mathbf{v} = (4, 3)$ with initial point $(2, 5)$. Its terminal point is $(6, 8)$. Therefore

$$\mathbf{u} + \mathbf{v} = (6, 8). \qquad \blacksquare$$

Multiplication of a Vector by a Scalar

A vector can be multiplied by a real number, called a **scalar**, to obtain another vector. A geometric model for multiplication of a vector by a scalar is the following. If a vector $\mathbf{v}$ is multiplied by a positive scalar a, then $\mathbf{v}$ is stretched (if $a > 1$) or shrunk (if $0 < a < 1$) by the factor a. If $\mathbf{v}$ is multiplied by a negative scalar a, then the direction of $\mathbf{v}$ is reversed as well as being stretched or shrunk (see Fig. 8.46).

The multiplication of a vector by a scalar can also be defined in terms of the components of the vector.

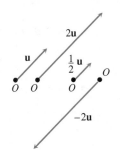

Figure 8.46   Examples of vector-scalar multiplication.

Definition 8.5   Multiplication of a Vector by a Scalar

Suppose that r is any real number and that $\mathbf{v} = (a, b)$ is any vector. Then the **scalar product** $r\mathbf{v}$ is defined to be

$$r\mathbf{v} = r(a, b) = (ra, rb).$$

Notice that a scalar product always yields a new *vector*, not a scalar.

EXAMPLE 5   Completing Vector-scalar Computations

Let $r = 2$, $s = 3$, $\mathbf{u} = (-1, 3)$, and $\mathbf{v} = (2, -4)$. Find $r\mathbf{u} + s\mathbf{v}$ and $|r\mathbf{u} + s\mathbf{v}|$.

Solution

a) $r\mathbf{u} + s\mathbf{v} = 2(-1, 3) + 3(2, -4)$
$$= (-2, 6) + (6, -12)$$
$$= [-2 + 6, 6 + (-12)]$$
$$= (4, -6)$$

b) $|r\mathbf{u} + s\mathbf{v}| = |(4, -6)|$

$$= \sqrt{4^2 + (-6)^2}$$

$$= \sqrt{16 + 36}$$

$$= \sqrt{52} = 2\sqrt{13}$$

≡

Teaching Note

The presentation of linear combinations is driven by the notion of independent basis elements for a vector space. It is limited to vectors in the plane and only considers orthogonal elements. The linear combination that provides a unique description should be stressed throughout.

Linear Combinations of Vectors

The vector $r\mathbf{u} + s\mathbf{v}$ in Example 5 is an example of a linear combination of vectors $\mathbf{u}$ and $\mathbf{v}$.

Definition 8.6   Linear Combination

The vector $\mathbf{w}$ is said to be a **linear combination of the vectors $\mathbf{u}$ and $\mathbf{v}$** if there are scalars a and b such that

$$\mathbf{w} = a\mathbf{u} + b\mathbf{v}.$$

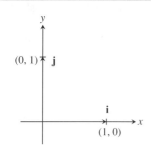

Figure 8.47   Unit vectors.

The two vectors $\mathbf{i} = (1, 0)$ and $\mathbf{j} = (0, 1)$ have magnitudes of 1 and consequently are called **unit vectors** (see Fig. 8.47). Every vector can be written as a linear combination of these two unit vectors in a very natural way. For example, let $\mathbf{v} = (a, b)$ be any vector. Then

$$\mathbf{v} = (a, b)$$

$$= (a, 0) + (0, b)$$

$$= a(1, 0) + b(0, 1)$$

$$= a\mathbf{i} + b\mathbf{j}.$$

Notice that the scalars in this case are the x- and y-components of the vector $\mathbf{v} = (a, b)$ (see Fig. 8.48).

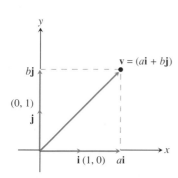

Figure 8.48 $\mathbf{v}$ as a linear combination of $\mathbf{i}$ and $\mathbf{j}$.

Relationship between Vector Components and Vector Direction

The next theorem explains how the components of a vector $\mathbf{v}$ can be found if we know both the angle that $\mathbf{v}$ forms with the positive x-axis and its absolute value.

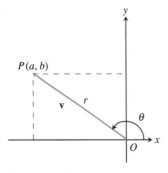

Figure 8.49   Finding the components of $\mathbf{v} = (a, b)$.

> **Theorem 8.11   Finding Vector Components**
>
> Let $\mathbf{v} = (a, b)$ be any vector and P be the point with coordinates (a, b). If θ is an angle in standard position with terminal side OP, then
>
> $$a = |\mathbf{v}| \cos\theta \quad\text{and}\quad b = |\mathbf{v}| \sin\theta.$$

Proof   We see from Fig. 8.49 that

$$|\mathbf{v}| = \sqrt{a^2 + b^2} = r, \qquad \cos\theta = \frac{a}{r}, \qquad \sin\theta = \frac{b}{r}.$$

Therefore

$$a = |\mathbf{v}| \cos\theta \quad\text{and}\quad b = |\mathbf{v}| \sin\theta.$$ ▤

E X A M P L E   6   *Finding the Components of a Vector*

Find the components of the vector shown in Fig. 8.50.

Solution   We apply Theorem 8.11.

$$a = 6\cos 115° = -2.54 \quad\text{and}\quad b = 6\sin 115° = 5.44$$

Thus $\mathbf{v} = (-2.54, 5.44)$. ▤

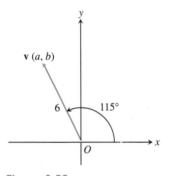

Figure 8.50   Vector $\mathbf{v}$ given for Example 6.

In Example 6, we were given both the direction and magnitude of a vector and asked to find its components. Example 7 gives us the components of two vectors and asks us to find their directions.

E X A M P L E   7   *Finding the Direction of a Vector*

Let $\mathbf{u} = (3, 2)$ and $\mathbf{v} = (-2, 5)$. Find the following:

a) The angle each vector makes with the positive x-axis

b) The angle with initial side $\mathbf{u}$ and terminal side $\mathbf{v}$

Solution   Study Fig. 8.51.

a) $\tan\alpha = \frac{2}{3}$ and therefore $\alpha = \tan^{-1}\left(\frac{2}{3}\right) = 33.6901°$

$$\tan\beta = -\left(\frac{5}{2}\right) \quad\text{and}\quad \tan^{-1}\left(-\frac{5}{2}\right) = -68.1986°$$

Teaching Note

Remind students about their work with complex numbers, that is, rectangular to trigonometric form. It will provide them with a familiar base for thinking about vectors.

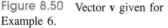

The angle β is in the second quadrant, and $\tan^{-1}\left(-\frac{5}{2}\right) = -68.1986°$ is in the fourth quadrant. Therefore $\beta = 180° + \tan^{-1}\left(-\frac{5}{2}\right) = 111.8014°$.

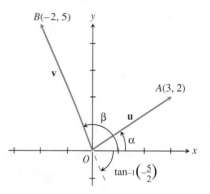

Figure 8.51 Solving **u** and **v** for Example 7.

b) Notice from Fig. 8.51 that the angle θ with initial side **u** and terminal side **v** is $\beta - \alpha$, or $78.11°$.

Using the Law of Cosines

In Example 7, part(b), the angle between two vectors was found. Example 8 shows an alternate way to complete this calculation.

EXAMPLE 8   Finding the Angle between Two Vectors

Find the angle $\beta - \alpha$ from vector $\mathbf{u} = (3, 2)$ to vector $\mathbf{v} = (-2, 5)$ shown in Fig. 8.51.

Solution   Apply the law of cosines to triangle AOB.

$$d^2(A, B) = |\mathbf{u}|^2 + |\mathbf{v}|^2 - 2|\mathbf{u}||\mathbf{v}| \cos(\beta - \alpha)$$
$$34 = 13 + 29 - 2\sqrt{13}\sqrt{29} \cos(\beta - \alpha)$$
$$\beta - \alpha = \cos^{-1}\left(\frac{13 + 29 - 34}{2\sqrt{13}\sqrt{29}}\right)$$
$$\beta - \alpha = 78.11°$$

Applications of Vectors

We have already pointed out that vectors model forces in physics. The **velocity** of a moving object is another physical situation modeled by vectors because velocity has both magnitude and direction. Usually the magnitude of velocity is called **speed**.

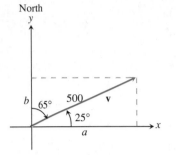

North
y

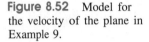

Figure 8.52 Model for the velocity of the plane in Example 9.

E X A M P L E 9 APPLICATION: Calculating a Plane's Velocity

A plane is flying on a bearing $65°$ east of north at 500 mph. Express the velocity of the plane as a vector.

Solution The vector **v** in Fig. 8.52 represents the velocity of the plane.

Direction: $65°$ east of north is $25°$ north of east.

Magnitude: $|\mathbf{v}| = 500$ since the plane is flying 500 mph.
Apply Theorem 8.11 to get:

$$\mathbf{v} = (500\cos 25°, 500\sin 25°)$$

$$= (453.15, 211.31)$$

The next example illustrates the kind of problem that a flight engineer must solve on a daily basis.

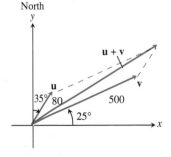

North
y

Figure 8.53 Model of the velocity of the plane described in Example 10.

E X A M P L E 10 APPLICATION: Calculating Actual Speed and Velocity

A plane is flying on a bearing of $65°$ east of north at 500 mph. A tail wind that adds to the plane's velocity is blowing in the direction $35°$ east of north at 80 mph. Express the actual velocity of the plane as a vector. Determine the actual speed of the plane.

Solution Let **v** be the velocity of the plane and **u** the velocity of the wind (see Fig. 8.53).
From Example 9, we know that $\mathbf{v} = (453.15, 211.31)$.

Because **u** has a direction $55°$ north of east,

$$\mathbf{u} = (80\cos 55°, 80\sin 55°) = (45.89, 65.53).$$

The actual velocity of the plane is $\mathbf{u} + \mathbf{v}$:

$$\mathbf{u} + \mathbf{v} = (45.89, 65.53) + (453.15, 211.31)$$

$$= (499.04, 276.84)$$

The speed of the plane is $|\mathbf{u} + \mathbf{v}| = \sqrt{499.04^2 + 276.84^2} = 570.69$ mph.

Assignment Guide

Day 1: Ex. 1, 3, 5–13, 15–19, 21, 23.
Day 2: Ex. 27, 29–36, 38–41
Day 3: Ex. 42–58

Usually pilots face a slightly different problem. The actual speed and direction **u** + **v** are known as well as the tail-wind or head-wind vector **u**. The flight bearing must be established. This problem is explored in Exercises 43 and 44.

Exercises for Section 8.5

In Exercises 1–4, find the magnitude of vector $\overrightarrow{AB}$ and the angle that $\overrightarrow{AB}$ makes with the positive x-axis.

1. $A = (1, 2), B = (4, 5)$ **2.** $A = (2, 3), B = (-3, 2)$

3. $A = (-2, 5), B = (1, 4)$ **4.** $A = (1, 3), B = (2, 5)$

In Exercises 5–12, let **u** $= (-1, 3)$, **v** $= (2, 4)$, and **w** $= (2, -5)$ be vectors. Calculate the following:

5. **u** + **v**   **6.** **u** − **v**

7. **u** − **w**   **8.** 3**v**

9. 2**u** + 3**w**   **10.** 2**u** − 4**v**

11. |**u** + **v**| **12.** |**u** − **v**|

In Exercises 13 and 14, use the geometric definitions of addition and subtraction of vectors to find the sum **u** + **v** and the difference **u** − **v**. Compare your answers with the results obtained using the respective component-wise definitions.

13. **u** $= (1, 3)$, **v** $= (3, 6)$ **14.** **u** $= (-1, 2)$, **v** $= (4, -2)$

In Exercises 15–20, let $A = (-1, 2), B = (3, 4), C = (-2, 5)$, and $D = (2, -8)$. Write each of the following in the form **v** $= (x, y)$.

15. $\overrightarrow{AB}$ **16.** $\overrightarrow{CD}$

17. $3\overrightarrow{AB}$ **18.** $2\overrightarrow{AB} - \overrightarrow{BA}$

19. $\overrightarrow{AB} + \overrightarrow{CD}$ **20.** $2\overrightarrow{AB} + \overrightarrow{CD}$

In Exercises 21–24, compute the following:

21. $|\overrightarrow{AB}|$ **22.** $|2\overrightarrow{AB}|$

23. $|\overrightarrow{AB} + \overrightarrow{BA}|$ **24.** $|\overrightarrow{AB} + \overrightarrow{CD}|$

In Exercises 25 and 26, show geometrically that $\overrightarrow{AB} = \overrightarrow{OB} - \overrightarrow{OA}$.

25. $A = (2, 3), B = (5, 2)$

26. $A = (-2, 4), B = (2, 6)$

In Exercises 27–30, express $\overrightarrow{AB}$ as a linear combination of **i** $= (1, 0)$ and **j** $= (0, 1)$.

27. $A = (3, 0), B = (0, 6)$ **28.** $A = (2, 1), B = (5, 0)$

29. $A = (3, -2), B = (2, 6)$

30. $A = (-2, 1), B = (-3, -5)$

In Exercises 31–34, find the angle each vector makes with the positive x-axis. Then determine the angle between the vectors.

31. **u** $= (3, 4)$, **v** $= (1, 0)$ **32.** **u** $= (-1, 2)$, **v** $= (3, 2)$

33. **u** $= (-1, 2)$, **v** $= (3, -4)$

34. **u** $= (2, -3)$, **v** $= (-3, -5)$

35. Let **r** $= (1, 2)$ and **s** $= (2, -1)$. Show that the vector **v** $= (5, 7)$ can be expressed as a linear combination of **r** and **s**. Explain what this means geometrically.

36. Let **r** $= (1, 2)$ and **s** $= (2, -1)$. Show that *any* vector **v** $= (x, y)$ can be expressed as a linear combination of **r** and **s**.

37. Let **r** $= (a, b)$ and **s** $= (c, d)$. Show that any vector **v** $= (x, y)$ can be expressed as a linear combination of **r** and **s** for almost all values of a, b, c, and d. For what values of a, b, c, and d is it *impossible* to express **v** as a linear combination of **r** and **s**?

38. A plane is flying on a bearing $25°$ west of north at 530 mph. Express the velocity of the plane as a vector.

39. A plane is flying on a bearing $10°$ east of south at 460 mph. Express the velocity of the plane as a vector.

40. A plane is flying on a bearing $20°$ west of north at 325 mph. A tail wind is blowing in the direction $40°$ west of north at 40 mph. Express the actual velocity of the plane as a vector. Determine the actual speed and direction of the plane.

41. A plane is flying on a bearing $10°$ east of south at 460 mph. A tail wind is blowing in the direction $20°$ west of south at 80 mph. Express the actual velocity of the plane as a vector. Determine the actual speed and direction of the plane.

42. A sailboat under auxiliary power is proceeding on a bearing 25° north of west at 6.25 mph in still water. Then a tail wind blowing 15 mph in the direction 35° south of west alters the course of the sailboat. Express the actual velocity of the sailboat as a vector. Determine the actual speed and direction of the boat.

43. A pilot must actually fly due west at a constant speed of 382 mph against a head wind of 55 mph blowing in the direction 22° south of east. What direction and speed must the pilot maintain to keep on course (due west)?

44. A jet fighter pilot must actually fly due north at a constant speed of 680 mph against a head wind of 80 mph blowing in the direction 10° west of south. What direction and speed must the pilot maintain to keep on course (due north)?

45. If possible, find a way to illustrate the vector sum $(1, 2) + (3, 5) = (4, 7)$ on your graphing utility.

Exercises 46–49 refer to the following **problem situation**: A ball is tossed up in the air at an angle of 70° with the horizontal and with an initial velocity of 36 ft/sec. Neglect air resistance and gravity.

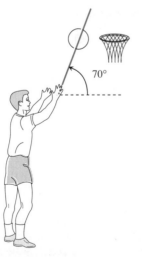

For Exercises 46 to 49.

46. Specify the position of the ball as a vector 1, 2.5, and 4 sec after the ball is released.

47. Express the position of the ball at time t, where t is the number of seconds after the ball is released.

48. Model this problem situation on a hand-held graphing computer using parametric graphs.

49. Explain why the description of this problem situation does not provide a realistic model of a thrown ball. For example, when does the ball return to the ground?

Exercises 50 and 51 refer to the following **problem situation**: Consider an object dropped with initial velocity *zero* (a so-called freely falling body) from a tower at point A. Let point B be the position of the object t seconds after it is released (neglecting air resistance and wind effects). It is a fact from elementary physics that on Earth, the vector $\overrightarrow{AB}$ is equal to the *effect-of-gravity vector* $\mathbf{g} = (0, -16t^2)$. Assume point A is 220 ft above the ground.

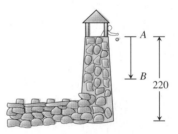

For Exercises 50 and 51.

50. Determine a vector that describes the position of the object after 2, 3, and 4 sec. Specify the coordinates of point B.

51. When will the object hit the ground?

Exercises 52–54 refer to the following **problem situation**: Consider the thrown ball described for Exercises 46–49 and modify that problem situation by accounting for the *effect of gravity*. (*Hint:* Let **u** be the vector describing the positions in Exercise 46 and **g** be the effect-of-gravity vector described for Exercises 50 and 51, and then find $\mathbf{u} + \mathbf{g}$.)

52. Describe a vector **v** that gives the position of the ball at time t.

53. When will the ball hit the ground?

54. What maximum height above ground will the ball reach? How far will the ball travel in the horizontal direction?

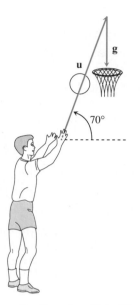

For Exercises 52 to 54.

55. Suppose that β is the angle between vectors **u** and **v**. Show that

$$\beta = \cos^{-1}\left(\frac{|\mathbf{u}|^2 + |\mathbf{v}|^2 - d^2}{2|\mathbf{u}||\mathbf{v}|}\right),$$

where d is the distance between the terminal points of **u** and **v**.

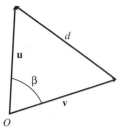

For Exercise 55.

56. Use the method of Example 8 to show that each pair of vectors is perpendicular.
   a) $\mathbf{u} = (2, 3)$ and $\mathbf{v} = (-3, 2)$
   b) $\mathbf{u} = (1, -1)$ and $\mathbf{v} = (-1, -1)$
   c) $\mathbf{u} = (5, -3)$ and $\mathbf{v} = (-3, -5)$

If vectors **u** and **v** have components (a, b) and (c, d) respectively, then the **dot product of u and v**, denoted $\mathbf{u} \cdot \mathbf{v}$, is defined by

$$\mathbf{u} \cdot \mathbf{v} = a \cdot c + b \cdot d$$

57. Calculate the **dot product** of each vector pair in Exercise 56. What generalization do you make?

58. Use your generalization about the dot product from Exercise 57 to find several vectors that are perpendicular to $\mathbf{u} = (4, -1)$.

Chapter 8 Review

KEY TERMS   (The number following each key term indicates the page of its introduction.)

REVIEW EXERCISES

In Exercises 1–8, solve the following triangle.

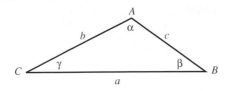

For Exercises 1 to 9.

1. $\alpha = 79°, \beta = 33°, a = 7$ **2.** $a = 5, b = 8, \beta = 110°$
3. $a = 14.7, \alpha = 29.3°, \gamma = 37.2°$
4. $a = 8, b = 3, \beta = 30°$ **5.** $\alpha = 34°, \beta = 74°, c = 5$
6. $c = 41, \alpha = 22.9°, \gamma = 55.1°$
7. $a = 5, b = 7, c = 6$ **8.** $\alpha = 85°, a = 6, b = 4$
9. Refer to the figure above. If $a = 12$ and $\beta = 28°$, determine the values of b for which α has the following:
a) Two values **b)** One value **c)** No values

In Exercises 10 and 11, find the area of triangle ABC.

10. $a = 3, b = 5, c = 6$ **11.** $a = 10, b = 6, \gamma = 50°$

12. Two markers A and B on the same side of a canyon rim are 80 ft apart. A hiker is located across the rim at point C. A surveyor determines that $\angle BAC = 70°$ and $\angle ABC = 65°$.
a) What is the distance between the hiker and point A?
b) What is the distance between the two canyon rims (assume they are parallel)?

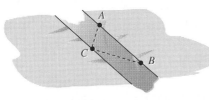

For Exercise 12.

13. A hot-air balloon is seen over Tucson, Arizona, simultaneously by two observers at points A and B 1.75 mi apart. Assume the balloon and the observers are in the same vertical plane and the angles of elevation are as shown here. How high above ground is the balloon?

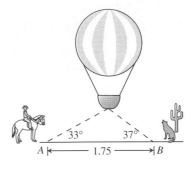

For Exercise 13.

14. To determine the distance between two points A and B on opposite sides of a lake, a surveyor chooses a point C that is 900 ft from A and 225 ft from B (see the figure). If the measure of the angle at C is 70°, find the distance between A and B.

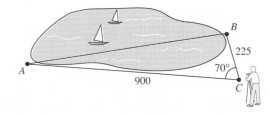

For Exercise 14.

15. Find the radian measure of the largest angle of the triangle whose sides have lengths of 9, 8, and 10.

In Exercises 16–21, convert each expression to the form $a + bi$ where a and b are real numbers.

16. $(4 + 3i) - (2 - i)$ **17.** $(3 + 2i) + (-5 - 7i)$
18. $(1 - i)(3 + 2i)$ **19.** $(2 - 4i)(3 + i)$
20. $\dfrac{-2 + 4i}{1 - i}$ **21.** $\dfrac{-1 + 2i}{3 - 7i}$

Graph Exercises 22–25 in the complex plane.

22. $-5 + 2i$ **23.** $4 - 3i$
24. $2\left(\cos\dfrac{\pi}{3} + i\sin\dfrac{\pi}{3}\right)$
25. $3\left[\cos\left(-\dfrac{\pi}{6}\right) + i\sin\left(-\dfrac{\pi}{6}\right)\right]$

For Exercises 26 and 27, consider the complex number z_1 as represented in this figure.

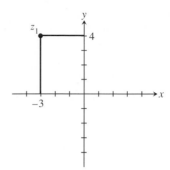

For Exercises 26 and 27.

26. If $z_1 = a + bi$, determine $a, b,$ and $|z_1|$.

27. Determine the trig form of z_1.

In Exercises 28–31, let $z_1 = -1 + 4i$, $z_2 = 3 + 4i$, and $z_3 = 5 - 2i$. Perform the indicated computation and write it in the form $a + bi$.

28. $z_1 - z_2$

29. $z_2 z_3$

30. $|z_1 z_3|$

31. $z_1(z_2 + z_3)$

In Exercises 32–35, write each complex number in the form $a + bi$.

32. $6(\cos 30° + i \sin 30°)$

33. $3(\cos 150° + i \sin 150°)$

34. $2.5 \left(\cos \dfrac{4\pi}{3} + i \sin \dfrac{4\pi}{3} \right)$

35. $4(\cos 2.5 + i \sin 2.5)$

Write Exercises 36–39 in trig form where $0 \le \theta \le 2\pi$. Then give all possible trig forms.

36. $3 - 3i$

37. $-1 + \sqrt{2}i$

38. $3 - 5i$

39. $-2 - 2i$

In Exercises 40–43, use Theorem 8.5 to compute each expression.

40. $[3(\cos 30° + i \sin 30°)][4(\cos 60° + i \sin 60°)]$

41. $[2(\cos 30° + i \sin 30°)][5(\cos 30° + i \sin 30°)]$

42. $\dfrac{3(\cos 30° + i \sin 30°)}{4(\cos 60° + i \sin 60°)}$

43. $\dfrac{5(\cos 75° + i \sin 75°)}{2(\cos 30° + i \sin 30°)}$

44. Write the two complex numbers in Exercise 42 in the form $a+bi$, perform the division, and verify that the result is the same as that obtained using the trig form and Theorem 8.5.

45. Write the two complex numbers in Exercise 43 in the form $a+bi$, perform the division, and verify that the result is the same as that obtained using the trig form and Theorem 8.5.

In Exercises 46–49, use De Moivre's theorem to compute each of the following. Express the answer in the form $a + bi$.

46. $\left[3 \left(\cos \dfrac{\pi}{4} + i \sin \dfrac{\pi}{4} \right) \right]^5$

47. $\left[2 \left(\cos \dfrac{\pi}{12} + i \sin \dfrac{\pi}{12} \right) \right]^8$

48. $\left[5 \left(\cos \dfrac{5\pi}{3} + i \sin \dfrac{5\pi}{3} \right) \right]^3$

49. $\left[7 \left(\cos \dfrac{\pi}{24} + i \sin \dfrac{\pi}{24} \right) \right]^6$

In Exercises 50 and 51, find and graph the nth roots of each complex number for the specified value of n.

50. $3 + 3i$, $n = 4$

51. 8, $n = 3$

52. Determine z and the three cube roots of z if one cube root is $2 + 2i$.

53. Solve $z^4 = 4 - 4i$ and graph the solutions.

In Exercises 54–57, let $\mathbf{u} = (2, -1)$, $\mathbf{v} = (4, 2)$ and $\mathbf{w} = (1, -3)$ be vectors and determine each expression.

54. $\mathbf{u} - \mathbf{v}$

55. $2\mathbf{u} - 3\mathbf{w}$

56. $|\mathbf{u} + \mathbf{v}|$

57. $|\mathbf{w} - 2\mathbf{u}|$

58. Use the geometric definitions of addition and subtraction of vectors to find $\mathbf{u} + \mathbf{v}$ and $\mathbf{u} - \mathbf{v}$. Then compare these results with those obtained using the respective component-wise definitions for $\mathbf{u} = (2, 4)$ and $\mathbf{v} = (-1, 3)$.

In Exercises 59–62, let $A = (2, -1)$, $B = (3, 1)$, $C = (-4, 2)$, and $D = (1, -5)$. Write each expression as a vector whose initial point is at the origin. Also find the indicated magnitudes.

59. $3\overrightarrow{AB}$

60. $\overrightarrow{AB} + \overrightarrow{CD}$

61. $|\overrightarrow{AB} + \overrightarrow{CD}|$

62. $|\overrightarrow{CD} - \overrightarrow{AB}|$

In Exercises 63 and 64, express vector $\overrightarrow{AB}$ as a linear combination of $\mathbf{i} = (1, 0)$ and $\mathbf{j} = (0, 1)$.

63. $A = (4, 0)$, $B = (2, 1)$

64. $A = (3, 1)$, $B = (5, 1)$

In Exercises 65 and 66, find the angle each vector makes with the positive x-axis and determine the angle between the vectors.

65. $\mathbf{u} = (4, 3)$, $\mathbf{v} = (2, 5)$

66. $\mathbf{u} = (-2, 4)$, $\mathbf{v} = (6, 4)$

67. Let $\mathbf{r} = (1, 2)$ and $\mathbf{s} = (2, -1)$. Show that *any* vector $\mathbf{v} = (x, y)$ can be expressed as a linear combination of $\mathbf{r}$ and $\mathbf{s}$.

68. A plane is flying on a bearing $10°$ east of south at 460 mph. A tail wind is blowing in the direction $20°$ west of south at 80 mph. Express the actual velocity of the plane as a vector. Also determine the plane's actual speed and direction.

69. A plane is flying on a bearing $10°$ east of south at 460 mph.

A 30-mph head wind is blowing in the direction $20°$ east of north. Express the actual velocity of the plane as a vector and determine the plane's actual speed and direction.

70. A sailboat under auxiliary power is proceeding on a bearing $25°$ north of west at 6.25 mph in still water. Then a tail wind blowing 15 mph in the direction $35°$ south of west alters the course of the sailboat. Express the actual velocity of the sailboat as a vector and determine the boat's actual speed and direction.

OMAR KHAYYAM

In the West, Omar Khayyam (ca. 1050–1122) is famous mostly as a romantic poet. In his native Iran, Eastern Europe, and Asia, however, he is regarded as a great mathematician and scientist. Although Khayyam did not come from a wealthy family, Islamic rulers of his time encouraged scholars, regardless of social background, to pursue studies of science and mathematics.

After Khayyam left his native city of Nishapur, he went to Samarkand, now in Uzbekistan, where he made astrological predictions for the Shah of Samarkand. He did not believe in astrology, but he pretended to do so because that gave him access to an excellent observatory for his work in astronomy.

Khayyam also did much original work in algebra and geometry. In algebra, he studied quadratic and third-degree equations and described general methods for solving them. Much of his work in geometry centered on Euclid's fifth postulate.

Parametric Equations and Polar Coordinates

9.1

Parametric Equations and Graphs

Objective

Students will be able to define parametric equations and graph curves parametrically.

Key Ideas

Curve
Parametric equation
Parameter
Parametrization of a curve
Endpoint of a curve
Curve of quickest descent

In the Exploration in Section 6.3, you were asked to set a graphing utility to "param mode" and to key in two functions; for example, $X_1(t) = \cos t$ and $Y_1(t) = \sin t$. You were using a method of describing a curve that is known as parametric graphing.

In those Explorations, once you pressed the **graph** key you watched the curve being generated over an interval of t-values. For each value of t, the coordinates of a point (x, y) were calculated. Because the position of the point (x, y) varies with t, the variable t is called the parameter. Both x and y are dependent variables and are functions of t.

Before we formally define parametric equations, experiment in the following Exploration.

475

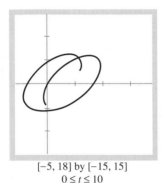

[−5, 18] by [−15, 15]
$0 \le t \le 10$

Figure 9.1 $X(T) = 5\cos T + \sin T + .5T$ and $Y(T) = 5\sin T + 2\cos T$.

🔍 EXPLORE WITH A GRAPHING UTILITY

Set a graphing calculator as follows:

- Rad mode,   Param mode
- Range: Tmin = 0, Tmax = 10, Tstep = 0.1
- Viewing rectangle: [−5, 18] by [−15, 15]
- Define: $X_1(T) = 5\cos T + \sin T + 0.5T$ and $Y_1(T) = 5\sin T + 2\cos T$

Describe what happens when you press the graph key (see Fig. 9.1).

1. How is the curve different if Tmax is changed to Tmax = 15? Or to Tmax = 25? Or to Tmax = 30?

2. As you move the trace cursor along the curve, how does the value of T relate to the values of X and Y on the grapher screen? How are these values related to the defining equations?

In this Exploration, you experienced a curve that was defined by parametric equations. You saw that as t varies, the corresponding point (x, y) also varies. You also saw that when Tmax is increased, a longer curve is drawn. You have experienced the concepts defined in Definition 9.1.

Definition 9.1   Parametric Equations

Let f and g be two functions defined on an interval I. The relation of all ordered pairs $\big(f(t), g(t)\big)$ for t in I is called a **curve** C. The equations

$$x = f(t) \qquad \text{and} \qquad y = g(t) \qquad \text{(for } t \text{ in } I)$$

are called **parametric equations** for C, and the variable t is the **parameter**.

If a curve C is defined by a pair of parametric equations, such as $x = f(t)$ and $y = g(t)$, with t in I, we say that the equations give a **parametrization of** C in terms of t.

Figure 9.2 shows several curves that can be defined by parametric equations.

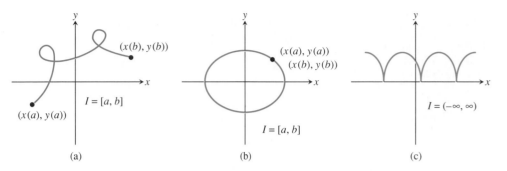

Figure 9.2 Several curves that can be defined by parametric equations.

The curves in Fig. 9.2(a, b) are defined for t in an interval $[a, b]$. The points $\big((x(a), y(a)\big)$ and $\big((x(b), y(b)\big)$ are called endpoints of the curve C. It is possible for a curve described parametrically to intersect itself, as shown in Fig. 9.2(a). This means that more than one value of t produces the same point in the coordinate plane.

EXAMPLE 1   Finding Endpoints of a Curve

Find the endpoints of the curve C defined by $x(t) = 1 - 2t$ and $y(t) = 2 - t$ for t in the interval $I = [0, 1]$. Then describe the nature of the curve as you move from one endpoint to the next.

Solution   To find the endpoints of the curve, evaluate the parametric equations at the endpoints of the interval I, that is, at $t = 0$ and $t = 1$: $x(0) = 1$, $y(0) = 2$, $x(1) = -1$, and $y(1) = 1$. Therefore, as t increases from 0 to 1, the curve goes from endpoint $(1, 2)$ to the endpoint $(-1, 1)$ (see Fig. 9.3).

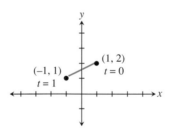

Figure 9.3   Curve C goes from $(1, 2)$ to $(-1, 1)$ as t goes from 0 to 1.

To describe the shape of the curve, consider that $x(t)$ and $y(t)$ are both linear functions. So x and y both decrease at a constant rate, suggesting that the curve is the line between the points $(1, 2)$ and $(-1, 1)$.   ≡

Eliminating the Parameter from a Parametric Representation

Sometimes a curve defined parametrically can also be described as the graph of a function. It is often possible in that event to describe the curve in terms of only the variables x and y. This process is called **eliminating the parameter**. Here is how to eliminate the parameter for the curve in Example 1.

$$x = 1 - 2t \quad \text{or} \quad t = 0.5(1 - x)$$

$$y = 2 - t \quad \text{or} \quad t = 2 - y$$

Notes on Examples

Graphing a linear system
of equations, such as
those in Example 1, is an
excellent way to begin the
approach to graphing curves
parametrically.

Because each equation has been solved for t, the two equations can be set
equal to each other.

$$2 - y = 0.5(1 - x)$$
$$4 - 2y = 1 - x$$
$$2y - x = 3$$

The fact that this final equation is an equation of a straight line provides further
evidence that the curve in Example 1 is part of a line. The extent of the line
segment is determined by the interval I from which the parameter is chosen.

**COMMON PRACTICE WITH
PARAMETRIC NOTATION**

In Example 1, curve C is
defined by $x(t) = 1 - 2t$
and $y(t) = 2 - t$. The
notation $x(t)$ and $y(t)$ is used
to emphasize the fact that
coordinates x and y are each
a function of t. For certain
situations, such as when
eliminating the parameter t, it
is convenient simply to write
$x = 1 - 2t$ and $y = 2 - t$.

E X A M P L E   2   Eliminating the Parameter

Eliminate the parameter from the curve C defined by $x(t) = t - 1$ and $y(t) = t^2$
for t in the interval $I = [0, 3]$. Describe this curve.

Solution   Solve the first parametric equation for t and substitute in the second
equation.

$$t = x + 1$$
$$y = t^2 = (x + 1)^2 = x^2 + 2x + 1 \quad \text{Square } t \text{ so that we can substitute for } t^2 = y.$$

Because $(x(0), y(0)) = (-1, 0)$ and $(x(3), y(3)) = (2, 9)$, we see that the
curve C is that part of the parabola $y = x^2 + 2x + 1$ that lies between $(-1, 0)$ and
$(2, 9)$ (see Fig. 9.4). ▪

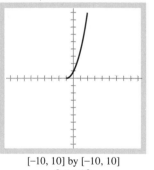

[−10, 10] by [−10, 10]
$0 \le t \le 3$

Figure 9.4   The curve C
is the part of the parabola
$y = x^2 + 2x + 1$ from $(-1, 0)$ to
$(2, 9)$.

Describing Any Function $y = f(x)$ with Parametric Equations

It is not always possible to eliminate the parameter t as was done in Example 2.
However, if you begin with a function $y = f(x)$, its graph can always be described
parametrically using the parametric equations

$$x = t$$
$$y = f(t).$$

E X A M P L E   3   Describing a Function as Parametric Equations

Describe the graph of $y = 3x^4 + 7x^3 - 8x^2 - 3$ with parametric equations.

Solution   The graph of this equation can be described by the parametric equations

$$x = t$$
$$y = 3t^4 + 7t^3 - 8t^2 - 3.$$

■

E X A M P L E   4   Finding Parametric Equations for a Circle

Find parametric equations for the circle with its center at the origin and a radius of 2.

Solution   It can be shown using trigonometry that if a point P on the unit circle determines an angle in standard position, then the coordinates of P are $(\cos\theta, \sin\theta)$ (see Fig. 9.5).

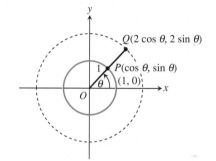

Figure 9.5   Coordinates of a point P on a unit circle and of a point Q on a circle with a radius of 2.

Parametric equations with parameter θ for a circle of radius 2 can be obtained by multiplying each coordinate function by 2 to obtain

$$x = 2\cos\theta$$
$$y = 2\sin\theta.$$

We obtain a complete circle if θ varies through any interval with a length of 2π, say $[0, 2\pi]$ (see Fig. 9.5).

■

Recall that for a relation (x, y), the set of all x-values is the domain of the relation and the set of all y-values is the range of the relation. In Example 4, the domain of the relation is $-2 \le x \le 2$ and the range of the relation is $-2 \le y \le 2$. Notice that the domain and range of a curve are not the same set as the domain for the parameter. In Example 4, the domain for the parameter is $[0, 2\pi]$.

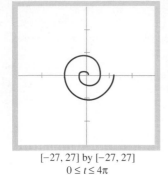

[−27, 27] by [−27, 27]
$0 \le t \le 4\pi$

Figure 9.6 $x = t \cos t$, $y = t \sin t$.

Notes on Exercises

Ex. 1–8 and 19–24 help students focus on the characteristics of the graphical representation. Be sure to emphasize domain and range and have students predict the shape of graphs before they use the graphing utility.

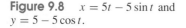

E X A M P L E   5   Graphing a Curve Defined Parametrically

Find the complete graph of the curve defined by $x = t \cos t$ and $y = t \sin t$ for $0 \le t \le 4\pi$.

Solution   The graph shown in Fig. 9.6 can be found with a graphing utility.   ≡

E X A M P L E   6   Graphing a Curve Defined Parametrically

Find the complete graph of the curve defined by $x(t) = 5 \sin t$ and $y(t) = 5t \cos t$ for $0 \le t \le 20.5$.

Solution   Notice that $-5 \le x \le 5$; therefore use the viewing rectangle $[-10, 10]$ by $[-100, 100]$.

A graph of this curve appears in Fig. 9.7.

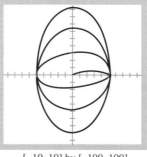

[−10, 10] by [−100, 100]
$0 \le t \le 20.5$

Figure 9.7 $x(t) = 5 \sin t$ and $y(t) = 5t \cos t$. (See Example 6.)   ≡

E X A M P L E   7   Graphing a Curve Defined Parametrically

Find a complete graph of the curve defined by the parametric equations $x = 5t - 5 \sin t$ and $y = 5 - 5 \cos t$.

Solution   Consider the range of values for x and y and note that y is a periodic function of t. Because $-5 \cos t$ varies from -5 to 5, observe that

$$0 \le y \le 10.$$

Likewise, $-5 \sin t$ varies from -5 to 5. However, since $5t$ is an increasing function, x is not a periodic function. So

$$-\infty \le x \le \infty \quad \text{as } t \text{ varies from } -\infty \text{ to } \infty.$$

A complete graph of this curve for t in $[-20, 20]$ is seen in Fig. 9.8.   ≡

[−100, 100] by [−5, 15]
$-20 \le t \le 20$

Figure 9.8 $x = 5t - 5 \sin t$ and $y = 5 - 5 \cos t$.

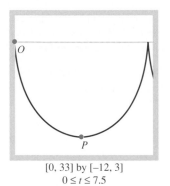

[0, 33] by [−12, 3]
$$0 \le t \le 7.5$$

Figure 9.9 $x(t) = 5t - 5\sin t$
and $y(t) = -5 + 5\cos t$.

The graph of the parametric equations $x = a(t - \sin t)$ and $y = a(1 - \cos t)$ for any real number a and for all values of t is called a **cycloid**. The graph in Fig. 9.8 is a cycloid.

Suppose one arc of a cycloid is reflected through the x-axis (see Fig. 9.9). Then, out of all possible curves connecting the origin to the point P (the first local minimum to the right of $x = 0$), a cycloid is the one that a ball will roll down, ignoring friction, in the *least* amount of time. Consequently, a physicist sometimes calls a cycloid a **curve of quickest descent**.

Assignment Guide

Day 1: Ex. 2, 5, 8, 10, 13, 15, 18–21, 23, 25–28
Day 2: Ex. 30, 31, 34, 35, 37, 40
Day 3: Ex. 41–48

Exercises for Section 9.1

In Exercises 1–8, find the endpoints of the curve C defined by the given pair of parametic equations and sketch a graph of the curve without using a graphing utility. Then support your work with a graphing utility.

1. $x(t) = 1 + t$ and $y(t) = t$ for t in $[0, 5]$

2. $x(t) = 1 + t$ and $y(t) = t$ for t in $[-3, 2]$

3. $x(t) = 2 - 3t$ and $y(t) = 5 + t$ for t in $[-1, 3]$

4. $x(t) = 2t - 3$ and $y(t) = 9 - 4t$ for t in $[3, 5]$

5. $x(t) = t$ and $y(t) = 2/t$ for t in $(0, 5]$

6. $x(t) = t + 2$ and $y(t) = 2/t$ for t in $[-3, 0) \cup (0, 3]$

7. $x(t) = 3\cos t$ and $y(t) = 3\sin t$ for t in $[0, 4\pi]$

8. $x(t) = 4\sin t$ and $y(t) = 4\cos t$ for t in $[-\pi/2, \pi/2]$

In Exercises 9–14, eliminate the parameter from the curve C defined by the given parametric equations and then describe the curve.

9. $x(t) = 2 - 3t$ and $y(t) = 5 + t$ for t in $(-\infty, \infty)$

10. $x(t) = 5 - 3t$ and $y(t) = 2 + t$ for t in $[-2, 3]$

11. $x(t) = 2t$ and $y(t) = t^2$ for t in $[-3, 1]$

12. $x(t) = 3t - 1$ and $y(t) = t^2 + 2$ for t in $[-1, 4]$

13. $x(t) = 3\cos t$ and $y(t) = 3\sin t$ for t in $[0, \pi]$ (*Hint:* Use a trig identity.)

14. $x(t) = 2\sin t$ and $y(t) = 2\cos t$ for t in $[0, 3\pi/2]$

In Exercises 15–18, use parametric equations to describe a curve that is the graph of the given functions.

15. $y = 3x^2 - 4x + 5$

16. $y = 9x^3 - 3x + 4$

17. $y = 7 + x^2 - 3x$

18. $y = \sin x + 4x^2$

19. Which of the curves in Exercises 1–8 are graphs of functions in x?

20. Which of the curves in Exercises 1–8 are graphs of one-to-one functions in x?

In Exercises 21–24, find the domain and range of each curve from the stated Exercise.

21. Exercise 2

22. Exercise 3

23. Exercise 5

24. Exercise 7

In Exercises 25–28, find parametric equations and sketch the graph of each circle with the given center and radius.

25. (0,0), 5

26. (10, 0), 4

27. (0, 10), 6

28. (a, b), r

In Exercises 29–36, draw a complete graph of the curve defined by the given pair of parametric equations.

29. $x(t) = 4t - 2$ and $y = 8t^2$ for t in $(-\infty, \infty)$

30. $x(t) = 4 \cos t$ and $y = 8 \sin t$ for t in $(-2\pi, 2\pi)$

31. $x(t) = 1 + 1/t$ and $y = t - 1/t$ for t in $(0, 20]$

32. $x(t) = 2t - 2 \sin t$ and $y(t) = 2 - 2 \cos t$ for t in $[0, 40]$

33. $x(t) = 4t - 4 \sin t$ and $y(t) = 4 - 4 \cos t$ for t in $[0, 40]$

34. $x(t) = 6 \cos t - 4 \cos \left(\frac{3}{2}t\right)$ and $y(t) = 6 \sin t - 4 \sin \left(\frac{3}{2}t\right)$ for t in $[0, 2\pi]$. This is an *epicycloid*, the path of a point on a circle rolling on another circle.

35. $x(t) = 5 \cos^3 t$ and $y(t) = 5 \sin^3 t$ for t in $[0, 2\pi]$. This is a *hypocycloid* of four cusps.

36. $x(t) = 4(\cos t + t \sin t)$ and $y(t) = 4(\sin t - t \cos t)$ for t in $[0, 50]$. This is the *involute* of a circle. (*Hint:* Use a large viewing rectangle.)

37. Find the domain and range of the curve in Exercise 35.

38. Find the domain and range of the curve in Exercise 36.

39. Let $f(x) = 3 - x^2$. Show that the curve defined parametrically by $x(t) = at + b$ and $y(t) = f(t)$ for t in the interval $I = [c, d]$ is a *function* $y = g(x)$. Find a rule for $g(x)$. What is the domain of g?

40. Find the maximum value (if any) of $y(t)$ for the curve defined parametrically by $x(t) = 2t - 1$ and $y(t) = 4 - t^2$.

41. Find a complete graph of $x(t) = a(t - \sin t)$ and $y(t) = a(1 - \cos t)$ for $a = 1, 2, 3, 4$, and 5. Use zoom-in and determine the period of the curve and the maximum y-coordinate with the least positive x-coordinate.

42. Based on the results of Exercise 41, make a conjecture

for the *period* and the maximum y-value with the least positive x-coordinate of the cycloid $x(t) = a(t - \sin t)$ and $y(t) = a(1 - \cos t)$ for an arbitrary value of a.

Exercises 43 to 48 refer to the following situation. Points $P_1(x_1, y_1)$ and $P_2(x_2, y_2)$ are given fixed points in a coordinate plane. Consider the parametric equations $x = tx_2 + (1 - t)x_1$ and $y = ty_2 + (1 - t)y_1$.

43. Show that when the parameter has value $t = 0$, the pair (x, y) is the point $P_1(x_1, y_1)$. Also show that when the parameter has value $t = 1$, the pair (x, y) is the point $P_2(x_2, y_2)$.

44. Show that when the parameter has value $t = 0.5$, the point (x, y) is the midpoint of the segment with endpoints (x_1, y_1) and (x_2, y_2).

45. Suppose $0 < t < 1$ and $P_t(x, y)$ is the point determined by the parametric equations $x = tx_2 + (1 - t)x_1$ and $y = ty_2 + (1 - t)y_1$. Show that $P_1 P_t / P_1 P_2 = t$. In other words show that P_t divides the segment with endpoints P_1 and P_2 into two segments whose lengths have the ratio $t/(1 - t)$.

46. Consider the points $P(-2, 1)$ and $Q(2, 4)$. Write the parametric equations that determine the points on the segment $\overline{PQ}$.

47. Consider the points $P(-3, -2)$ and $Q(3, 5)$. Find the point that is 0.25 of the distance from P and Q.

48. Consider the points $P(-4, 5)$ and $Q(2, 1)$. Find the point that is 0.42 of the distance from P and Q.

9.2 ───── Polar Coordinates and Graphs

Up to this point, we have used a rectangular coordinate system. However, some curves cannot be described easily in this type of system. In this section, we introduce a second coordinate system called the **polar coordinate system**. In the polar coordinate system, the first coordinate describes the distance from the origin and the second coordinate specifies an angle. Either degree measure or radian measure can be used to specify the second coordinate. Each number and angle pair determines a unique point, that is, no other point has these same polar coordinates.

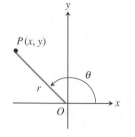

Figure 9.10 A point P with rectangular coordinates (x, y) and polar coordinates (r, θ).

Key Ideas

Polar coordinates
Conversion equations
Polar graph
Cardioid
Polar equations in parametric
    form

Teaching Note

The nonuniqueness of polar coordinates must be emphasized to students. Because of their extensive use of the Cartesian coordinate system, students have difficulty with this fact.

Let $P(x, y)$ be a point in the rectangular coordinate system (see Fig. 9.10). Point P lies on the terminal side of an angle θ in standard position and at a distance r from the origin O.

The pair of numbers (r, θ) are called **polar coordinates** of point P.

EXAMPLE 1   Plotting Points with Polar Coordinates

Graph the points whose polar coordinates are the following:

a) $P(2, \pi/3)$

b) $Q(1, 3\pi/4)$

c) $R(3, -45°)$

Solution   (See Fig. 9.11.)

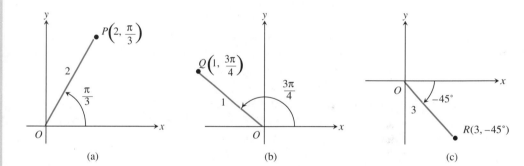

Figure 9.11   Three points and their polar coordinates.

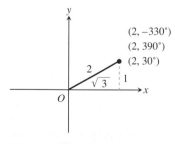

Figure 9.12   The point $P(\sqrt{3}, 1)$ with three different polar coordinates.

Example 1 illustrates that once a pair of polar coordinates is selected, a unique point has been determined. However, the converse is not true. The polar coordinates of a point $P(x, y)$ are *not unique*. For example, the polar coordinates $(2, 30°)$, $(2, 390°)$, and $(2, -330°)$ all name the same point with rectangular coordinates $P(\sqrt{3}, 1)$ (see Fig. 9.12).

Fig. 9.11(c) shows a point with a negative second coordinate. The first coordinate also can be negative, with the following interpretation. Suppose that $r > 0$. Figure 9.13 shows the relationship between the points whose polar coordinates are (r, θ) and $(-r, \theta)$. The distance of the point $P(-r, \theta)$ from the origin is r but in the direction opposite to the direction given by the terminal side of θ. So the polar coordinates $(-r, \theta)$ and $(r, \theta + \pi)$ name the same point.

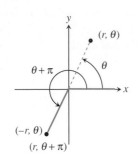

Figure 9.13 An interpretation for a negative value of r.

> **Theorem 9.1 Polar Coordinates of Points**
>
> Let (r, θ) be polar coordinates for a point P. Then, P also has the following for polar coordinates:
>
> $(r, \theta + 2n\pi)$ (where n is any integer)
>
> $(-r, \theta + \pi + 2n\pi) = (-r, \theta + (2n+1)\pi)$ (where n is any integer)

The origin is assumed to have polar coordinates $(0, \theta)$ where θ can be any angle.

E X A M P L E 2   Finding Polar Coordinates of Points

Find three polar coordinates, using radian measure, for points whose rectangular coordinates are the following:

a) $P(-\sqrt{3}, 1)$

b) $Q(-3, -3)$

Solution   (See Fig. 9.14.)

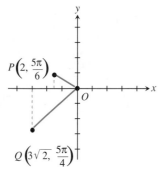

Figure 9.14 Reference triangles for the two points in Example 2.

a) The reference triangle for the point with rectangular coordinates $(-\sqrt{3}, 1)$ is a $30°$-$60°$ right triangle. Three of the many possible polar coordinates for point P are

$$\left(2, \frac{5\pi}{6}\right), \qquad \left(-2, \frac{-\pi}{6}\right), \qquad \text{and} \quad \left(-2, \frac{11\pi}{6}\right).$$

b) The reference triangle for the point with rectangular coordinates $(-3, -3)$ is a $45°$-$45°$ right triangle. Several polar coordinates for point Q are

$$\left(3\sqrt{2}, \frac{5\pi}{4}\right), \qquad \left(3\sqrt{2}, \frac{13\pi}{4}\right), \qquad \text{and} \quad \left(-3\sqrt{2}, \frac{\pi}{4}\right). \qquad ≡$$

Conversion Equations

It is convenient for both computational and theoretical reasons to be able to change the coordinates of a point from polar coordinates to rectangular coordinates and conversely. Figure 9.15 shows a point with rectangular coordinates $P(x, y)$ and polar coordinates $P(r, \theta)$ where r is positive. Using the reference triangle in this figure, we obtain the following relationships:

$$x = r \cos \theta, \qquad r^2 = x^2 + y^2,$$

$$y = r \sin \theta, \qquad \tan \theta = \frac{y}{x} \ (x \neq 0).$$

It can be shown that these equations remain true when r is negative.

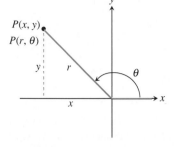

Figure 9.15 A point with polar coordinates (r, θ) and rectangular coordinates (x, y).

Notes on Examples

Theorem 9.2 serves the purpose of specifying how to convert between rectangular and polar coordinate systems in naming a given point.

> **Theorem 9.2 Polar ↔ Cartesian Conversion Equations**
>
> Let a point P have rectangular coordinates (x, y) and polar coordinates (r, θ). Then
>
> $$x = r \cos \theta, \qquad y = r \sin \theta, \qquad \textbf{(1)}$$
>
> $$r^2 = x^2 + y^2, \qquad \tan \theta = \frac{y}{x} \ (x \neq 0). \qquad \textbf{(2)}$$

Example 3 gives the polar coordinates of points and asks you to find the rectangular coordinates.

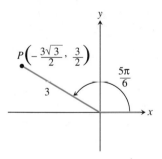

Figure 9.16 Converting $P(3, 5\pi/6)$ to rectangular coordinates.

E X A M P L E 3   Converting to Rectangular Coordinates

Find the rectangular coordinates of the points with the following polar coordinates:

a) $(3, 5\pi/6)$

b) $(2, -200°)$

Solution

a) Use Eq. (1) from Theorem 9.2 (see Fig. 9.16).

$$x = r \cos \theta \qquad\qquad y = r \sin \theta$$

$$= 3 \cos \frac{5\pi}{6} \qquad\qquad = 3 \sin \frac{5\pi}{6}$$

$$= -\frac{3\sqrt{3}}{2} \qquad\qquad = 3 \left(\frac{1}{2} \right) = \frac{3}{2}$$

The rectangular coordinates are $(-3\sqrt{3}/2, \frac{3}{2})$.

b) (See Fig. 9.17.)

$$x = r \cos \theta \qquad\qquad y = r \sin \theta$$
$$= 2 \cos(-200°) \qquad\qquad = 2 \sin(-200°)$$
$$= -1.88 \qquad\qquad = 0.68$$

The rectangular coordinates are $(-1.88, 0.68)$.

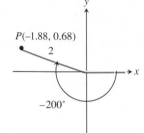

Figure 9.17   Converting
$P(2, -200°)$ to rectangular
coordinates.

In Example 4, the reverse situation occurs. The rectangular coordinates are
given and some polar coordinates are found. (Notice that we cannot say *the* coordinates because the polar coordinates are not unique.)

Suppose we are given a point $P(x, y)$ in rectangular coordinates and want
to find polar coordinates (r, θ) for P. For the first coordinate, we can use the
relation $r = \sqrt{x^2 + y^2}$ (Theorem 9.2). Finding the second coordinate θ is not
always so straightforward. Recall that the inverse tangent on a calculator always
returns values in radians between $-\pi/2$ and $\pi/2$. If P is in quadrants I or IV,
the coordinates $(r, \tan^{-1}(y/x))$ are polar coordinates for P. However, if P is
in quadrants II or III, the polar coordinates must be chosen more carefully, as
illustrated in Example 4.

E X A M P L E 4   Converting to Polar Coordinates

Find polar coordinates for the points with the following rectangular coordinates:

a) $(3, -5)$

b) $(-1, 1)$

c) $(-2, -3)$

Solution

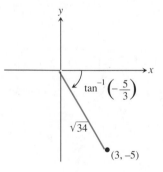

Figure 9.18   Converting
$P(3, -5)$ to polar coordinates.

a) Point $P(3, -5)$ is in quadrant IV (see Fig. 9.18).

$$r = \sqrt{3^2 + (-5)^2} = \sqrt{34}$$

$$\tan \theta = -\frac{5}{3} \quad \text{or} \quad \theta = \tan^{-1}\left(-\frac{5}{3}\right) = -1.03$$

Polar coordinates of point P are $(\sqrt{34}, -1.03)$ or $(5.83, -1.03)$.

b) Point $P(-1, 1)$ is in quadrant II (see Fig. 9.19).

$$r = \sqrt{(-1)^2 + 1^2} = \sqrt{2}$$

$$\tan \theta = -1$$

Because $\tan^{-1}(-1) = -\pi/4$ is in the fourth quadrant, use the angle $\pi + \tan^{-1}(-1) = 3\pi/4$ in the second quadrant whose tangent is also -1.

Polar coordinates of P are $(\sqrt{2}, \dfrac{3\pi}{4})$.

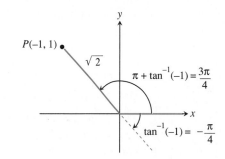

Figure 9.19 Converting $P(-1, 1)$ to polar coordinates.

c) The point $P(-2, -3)$ is in quadrant III (see Fig. 9.20).

$$r = \sqrt{(-2)^2 + (-3)^2} = \sqrt{13}$$

$$\tan\theta = \frac{-3}{-2}$$

Because $\tan^{-1}\left(\frac{3}{2}\right) = 0.98$ is in the first quadrant, use the angle $\pi + \tan^{-1}\left(\frac{3}{2}\right) = 4.12$ in the third quadrant, whose tangent is also $\frac{3}{2}$.

So polar coordinates of $(-2, -3)$ are $(\sqrt{13}, 4.12)$.   ▤

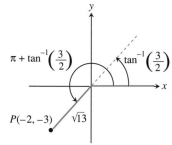

Figure 9.20 Converting $P(-2, -3)$ to polar coordinates.

The solution methods used in Example 4 can be generalized, as stated in Theorem 9.3.

Theorem 9.3   Polar Coordinates from Rectangular Coordinates

Let P have rectangular coordinates (x, y). Then the following are polar coordinates for P, where $r = \sqrt{x^2 + y^2}$:

a) $\left(r, \tan^{-1}\left(\dfrac{y}{x}\right)\right)$, if (x, y) is in the first or fourth quadrant

b) $\left(r, \pi + \tan^{-1}\left(\dfrac{y}{x}\right)\right)$ if (x, y) is in the second or third quadrant

Theorems 9.2 and 9.3 provide techniques for converting rectangular coordinates to polar coordinates and vice versa.

Polar Graphs

A function $r = f(\theta)$ where r and θ are polar coordinate variables can be graphed in the polar coordinate plane. The following definition applies to relations in addition to functions.

Definition 9.2   Graph of a Polar Equation

The **graph of a polar equation in the variables r and θ** is the set of all points (r, θ) in the polar coordinate plane, where (r, θ) is a solution to the equation.

The equations considered in the next example are the simplest kind of polar equations. They are constant functions analogous to the rectangular coordinate equations of the form $x = a$ and $y = b$.

E X A M P L E   5   Drawing Graphs of Simple Polar Equations

Draw a complete graph of each polar equation:

a) $r = 2$

b) $\theta = 2$

Solution

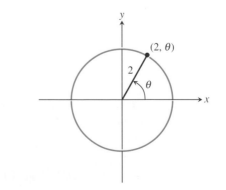

Figure 9.21   Polar coordinate graph of $r = 2$.

a) The graph of $r = 2$ consists of all points whose polar coordinates are $(2, \theta)$ where θ is any angle (see Fig. 9.21). The set of all such points is the circle with the center at the origin and a radius of 2.

b) The graph of $\theta = 2$ consists of all points with polar coordinates $(r, 2)$ where r is any real number (see Fig. 9.22). The set of all such points is a straight line through the origin. ▤

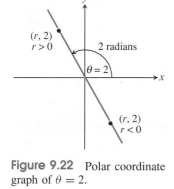

Figure 9.22 Polar coordinate graph of $\theta = 2$.

To draw the graph of a polar coordinate equation like $r = 3 + 3 \sin \theta$, consider all pairs (r, θ) as θ varies throughout an interval.

EXAMPLE 6   Graphing a Cardioid

Sketch a complete graph of the polar equation $r = 3 + 3 \sin \theta$ for $0 \leq \theta \leq 2\pi$.

Solution   If $0 \leq \theta \leq \pi/2$, then $0 \leq \sin \theta \leq 1$. Multiply all sides of this second inequality by 3 and add 3 to all sides to get

$$3 \leq 3 + 3 \sin \theta \leq 6.$$

Therefore, as θ increases from 0 to $\pi/2$, the variable r increases from 3 to 6. In like manner, you can determine the behavior of r for $\pi/2 \leq \theta \leq 2\pi$. The results are summarized here.

θ	$\sin \theta$	$r = 3 + 3 \sin \theta$
$0 \leq \theta \leq \dfrac{\pi}{2}$	Increases from 0 to 1	Increases from 3 to 6
$\dfrac{\pi}{2} \leq \theta \leq \pi$	Decreases from 1 to 0	Decreases from 6 to 3
$\pi \leq \theta \leq \dfrac{3\pi}{2}$	Decreases from 0 to -1	Decreases from 3 to 0
$\dfrac{3\pi}{2} \leq \theta \leq 2\pi$	Increases from -1 to 0	Increases from 0 to 3

Notice that the graph in Fig. 9.23 is traced in a counterclockwise fashion through the four quadrants beginning at $(3, 0)$, moving to $(6, \pi/2)$, then to $(3, \pi)$, followed by $\left(0, \frac{3\pi}{2}\right)$, and ending with $(3, 2\pi)$. ▤

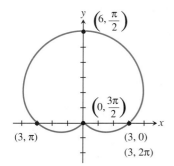

Figure 9.23 $y = 3 + 3 \sin \theta$.

The graph in Fig. 9.23 is called a **cardioid** because of its heartlike shape. All the graphs of polar equations of the form $r = a \pm a \sin \theta$ or $r = a \pm a \cos \theta$ (where a is any nonzero constant real number) are cardioids.

Because the function $r = 3 + 3\sin\theta$ in Example 6 is a periodic function with a period of 2π, a complete graph was found restricting the domain to the interval $[0, 2\pi]$. If θ varies from 2π to 4π, the curve is completely retraced. In fact, the graph in Fig. 9.23 is also the graph of $r = 3 + 3\sin\theta$ for θ in $(-\infty, \infty)$. These facts can be supported using a parametric grapher, as illustrated in Example 7, later in this section.

Converting from Polar to Rectangular Form

The polar equation $r = 3 + 3\sin\theta$ in Example 6 can be converted to a rectangular equation in x and y by using the conversion equations from Theorem 9.2. For example,

$$r = 3 + 3\sin\theta$$

$$r = 3 + 3\frac{y}{r}$$

$$r^2 = 3r + 3y$$

$$x^2 + y^2 = 3\sqrt{x^2 + y^2} + 3y.$$

Notice that this equation in x and y is a relation that does not define a function $y = f(x)$. This confirms the observation that the vertical line test, a test that applies only to the rectangular coordinate system, is not satisfied for the graph in Fig. 9.23. This also demonstrates a reason for using the polar coordinate system. Curves that cannot be described very easily in rectangular form often have simple polar equations. The cardioid is one such curve.

Graphing Polar Equations

Most modern graphing utilities can graph polar coordinate equations using the parametric graphing option illustrated in the next example.

Theorem 9.4 Polar Equations in Parametric Form

The graph of the polar equation $r = f(\theta)$ is the curve defined by the parametric equations

$$x(t) = f(t)\cos t$$

$$y(t) = f(t)\sin t.$$

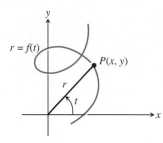

Figure 9.24 Consider a point $P(x, y)$ on a curve defined by a polar equation.

Proof Let (x, y) be the rectangular coordinates of a point P on the polar graph $r = f(t)$ (see Fig. 9.24). By definition of the trig functions, we have

$$\sin t = \frac{y}{r}, \qquad \cos t = \frac{x}{r}.$$

Thus $x = r \cos t = f(t) \cos t$, and $y = r \sin t = f(t) \sin t$. Notice that we have replaced θ with t in $r = f(\theta)$. ∎

E X A M P L E 7 Finding the Parametric Form

Determine parametric equations for the curve $r = 7 \sin 2\theta$. Find a complete graph of the curve.

Solution By Theorem 9.4, the parametric equations are

$$x(t) = 7 \sin 2t \cos t$$

$$y(t) = 7 \sin 2t \sin t.$$

A complete graph for $0 \le t \le 2\pi$ is shown in Fig. 9.25. ∎

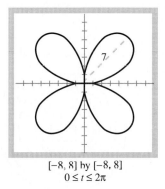

$[-8, 8]$ by $[-8, 8]$
$0 \le t \le 2\pi$

Figure 9.25 The four-leafed rose $r = 7 \sin 2\theta$.

Notice that the length of each loop or petal of the graph in Fig. 9.25 is 7 units—a result of the fact that the maximum value of $r = 7 \sin 2\theta$ is 7. In general, the graph of equations in the form $r = a \sin 2\theta$ (where a is a constant real number) is a **four-petaled rose** whose petals have a length of $|a|$ units.

The graph in the next example is called a **spiral of Archimedes.**

E X A M P L E 8 Graphing the Spiral of Archimedes

Find the graph of $r = \theta$ as θ varies over the following intervals. Complete a separate graph for each interval.

a) $[0, 2\pi]$, $[0, 4\pi]$, $[0, 8\pi]$

b) $[-4\pi, 0]$, $[-8\pi, 0]$, $[-8\pi, 8\pi]$

Solution To find this graph use the parametric equations $x = t \cos t$ and $y = t \sin t$.

a) Notice that the three curves in Fig. 9.26 are all counterclockwise spirals.

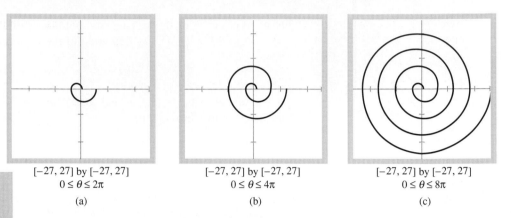

[−27, 27] by [−27, 27]
$0 \le \theta \le 2\pi$

(a)

[−27, 27] by [−27, 27]
$0 \le \theta \le 4\pi$

(b)

[−27, 27] by [−27, 27]
$0 \le \theta \le 8\pi$

(c)

Notes on Exercises

In Ex. 47–50, students are being asked to find the distance from the pole to the furthest point of a petal, not arc length.

In Ex. 58–61, caution students that the t-range settings must be appropriate in order for them to see a complete graph.

Figure 9.26 $r = \theta$ (spiral of Archimedes) for $\theta \ge 0$ over three intervals.

b) The two curves in Fig. 9.27(a,b) are clockwise spirals from the pole. Comparison of Figs. 9.26 and Fig. 9.27(a,b) shows that the spirals are counterclockwise as θ varies from zero to positive values and clockwise as θ varies from zero to negative values. The interval in Fig. 9.27(c) includes both positive and negative values, so both clockwise and counterclockwise spirals are present.

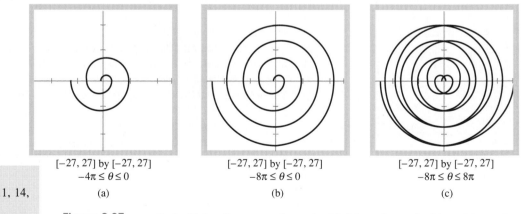

[−27, 27] by [−27, 27]
$-4\pi \le \theta \le 0$

(a)

[−27, 27] by [−27, 27]
$-8\pi \le \theta \le 0$

(b)

[−27, 27] by [−27, 27]
$-8\pi \le \theta \le 8\pi$

(c)

Assignment Guide

Day 1: Ex. 1–5, 7, 10, 11, 14, 17–19, 22, 25, 28

Day 2: Ex. 29, 32, 34, 35, 37, 40, 42, 44, 45, 48, 49

Day 3: Ex. 51–54, 56–58, 61

Figure 9.27 $r = \theta$: (a, b) $\theta \le 0$ over two intervals; (c) θ in an interval with positive and negative values.

Exercises for Section 9.2

In Exercises 1–4, compute the polar coordinates (r, θ) for $0 \le \theta < 360°$ for each point with the given rectangular coordinates.

1. $(1, 1)$

2. $(-10, 0)$

3. $(-3, 4)$

4. $(-2, 5)$

In Exercises 5–10, compute the rectangular coordinates for each point with the given polar coordinates.

5. $(0, \pi)$

6. $(2, 30°)$

7. $(-3, 135°)$

8. $(-2, -120°)$

9. $\left(3, \dfrac{3\pi}{4}\right)$

10. $\left(2, \dfrac{\pi}{12}\right)$

In Exercises 11–18, find three polar coordinate pairs for the point with given rectangular coordinates. Include at least one pair with a negative first coordinate.

11. $(-2, 2)$

12. $(2\sqrt{3}, -2)$

13. $(-\sqrt{3}, 1)$

14. $(-4, -4)$

15. $(2, -3)$

16. $(1, 4)$

17. $(-3, -1)$

18. $(-2, 3)$

In Exercises 19–28, sketch a complete graph of each equation without using a graphing utility. If possible, support your answer with a polar graphing utility.

19. $r = 3$

20. $\theta = \dfrac{\pi}{4}$

21. $\theta = -\dfrac{\pi}{2}$

22. $r = -1$

23. $r = 3\cos\theta$

24. $r = 2\sin\theta$

25. $r = 2 + 2\cos\theta$

26. $r = 4 + 4\sin\theta$

27. $r = -1 + 2\cos\theta$

28. $r = \sin 2\theta$

In Exercises 29–32, write the polar equation in parametric form for θ in the given interval. Graph *both* the parametric and polar forms with an appropriate graphing utility.

29. $r = 3\theta$, $[0, 10\pi]$

30. $r = 5\sin\theta$, $[-\pi, \pi]$

31. $r = 5\sin 2\theta$, $[0, \pi]$

32. $r = 5\sin 3\theta$, $[0, 2\pi]$

In Exercises 33–36, use a graphing utility to find the polar graph for θ in each of the given intervals.

33. $r = 3\theta$
   a) $0 \le \theta \le \dfrac{\pi}{2}$
   b) $0 \le \theta \le \pi$
   c) $0 \le \theta \le 2\pi$
   d) $-\pi \le \theta \le \pi$
   e) $0 \le \theta \le 4\pi$

34. $r = 5\sin\theta$
   a) $0 \le \theta \le \dfrac{\pi}{2}$
   b) $0 \le \theta \le \pi$
   c) $0 \le \theta \le 2\pi$
   d) $-\pi \le \theta \le \pi$
   e) $0 \le \theta \le 4\pi$

35. $r = 5\cos 3\theta$
   a) $0 \le \theta \le \dfrac{\pi}{2}$
   b) $0 \le \theta \le \pi$
   c) $0 \le \theta \le 2\pi$
   d) $-\pi \le \theta \le \pi$
   e) $0 \le \theta \le 4\pi$

36. $r = 5\sin 2\theta$
   a) $0 \le \theta \le \dfrac{\pi}{2}$
   b) $0 \le \theta \le \pi$
   c) $0 \le \theta \le 2\pi$
   d) $-\pi \le \theta \le \pi$
   e) $0 \le \theta \le 4\pi$

In Exercises 37–42, use a polar graphing utility to find a complete graph of the polar equation. Specify an interval $a \le \theta \le b$ of smallest length that gives a complete graph.

37. $r = 2\theta$

38. $r = 0.275\theta$

39. $r = 2 - 3\cos\theta$

40. $r = 2\sin 3\theta$

41. $r = 3\sin^2\theta$

42. $r = 2 - 2\sin\theta$

43. Use a polar graphing utility to find a complete graph of $r = 5\sin n\theta$ for $n = 1, 2, 3, 4, 5$, and 6. Explain how the number of petals is related to n.

44. Use a polar graphing utility to find a complete graph of $r = 5\cos n\theta$ for $n = 1, 2, 3, 4, 5$, and 6. Explain how the number of petals is related to n.

For Exercises 45 and 46, use a polar graphing utility to find a complete graph of the given equation for $a = 1, 2, 3, 4, 5$, and 10. What is the effect of the parameter a on the graph?

45. $r = a\sin 2\theta$

46. $r = a\cos 3\theta$

In Exercises 47–50, determine the length of *one* petal.

47. $r = 5\sin 3\theta$

48. $r = 5\sin 2\theta$

49. $r = 8\sin 5\theta$

50. $r = 8\sin 4\theta$

For Exercises 51–54, draw the graph of $r = 5\sin 2\theta$ and $r = 5\sin 3\theta$ in the standard viewing rectangle.

51. Predict what the graph of $r = 5\sin 2.5\theta$ looks like. Then draw a complete graph of $r = 5\sin 2.5\theta$ with a polar graphing utility.

52. Give an algebraic argument for the appearance of the graph in Exercise 51.

53. Predict the graph of $r = 5\sin 3.5\theta$. Confirm your graph using a polar graphing utility.

54. Predict the graph of

$$r = a\sin\left(\dfrac{m}{n}\theta\right),$$

where m and n are *relatively prime* positive integers, that is, m and n have no common factors.

55. Find polar coordinates of the vertices of a square with a side length of a whose center is at the origin. Assume the square is positioned with two sides parallel to the x-axis.

56. Find polar coordinates of the vertices of a regular pentagon if the center is at the origin, one vertex is on the positive x-axis, and the distance from the center to a vertex is a.

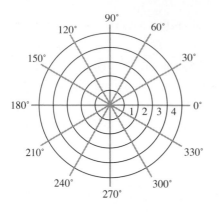

For Exercise 57.

57. Before computers could draw polar graphs, such graphs often were sketched using polar graph paper like that shown in the figure. Use polar graph paper (you can make it yourself using a ruler, protractor, and compass) to sketch a complete graph of $r = 3 \sin 2\theta$ for $0 \leq \theta \leq 360°$. Plot points for $\theta = 0 + k \cdot 30°$, where $k = 0, 1, 2, \ldots, 12$.

In Exercises 58–61, use the parametric equation method of Example 7 to find the points of intersection of these graphs.

58. $\begin{cases} r = 5 \sin \theta \\ r = 5 \cos \theta \end{cases}$

59. $\begin{cases} r = 5 \sin \theta \\ r = 5 \sin 3\theta \end{cases}$

60. $\begin{cases} r = \theta/\pi \\ r = 3 \sin \theta \end{cases}$

61. $\begin{cases} r = 2 \cos \theta - 1 \\ r = 3 \sin 2\theta \end{cases}$

62. Show that if P_1 and P_2 have polar coordinates (r_1, θ_1) and (r_2, θ_2) respectively, then the distance between P_1 and P_2 is $d(P_1, P_2) = \sqrt{r_1^2 + r_2^2 - 2r_1r_2 \cos(\theta_2 - \theta_1)}$.

9.3 — Motion Problems and Parametric Equations

Objective
Students will be able to solve application problems using parametric equations.

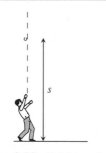

Figure 9.28 Height s of a ball thrown straight up in the air.

Key Ideas
Motion along a line
Motion in the plane

Motion Along a Line

In Section 2.5, we studied the projectile-motion problem situation in which a ball was thrown straight up in the air. The ball was imagined to be moving up and down along a vertical line (see Fig. 9.28).

In this section, we consider other kinds of linear and nonlinear motion problem situations and see how parametric equations can be used to represent them. Having briefly reviewed vertical motion, we now consider the case in which an object is moving along a horizontal line such as the x-axis.

Motion-on-a-line Problem Situation

The position (x-coordinate) of a particle moving on the x-axis is given by

$$x = s(t) = 2t^3 - 13t^2 + 22t - 5,$$

where t is time in seconds. Describe the motion of the particle. When does it speed up? When does it slow down? When does it change direction?

See how many of these questions you can answer while completing the following Exploration.

Frank says, "Linear and nonlinear motion problem situations can be beautifully represented using the parametric graphing utility. The foreshadowing of the calculus using the methods of this section is tremendously important for students."

🔍 EXPLORE WITH A GRAPHING UTILITY

Set a graphing calculator as follows:

- Rad mode,   Param mode
- Range: $T\text{min} = 0$, $T\text{max} = 5$, $T\text{step} = 0.05$
- Viewing rectangle: $[-10, 30]$ by $[-2, 6]$
- Define: $X_1(T) = 2T^3 - 13T^2 + 22T - 5$ and $Y_1(T) = 2$

1. Press the GRAPH key. After the curve has been drawn, press the TRACE key. Use the left arrow key to move the trace cursor until the parameter has the value $T = 0$. You are now ready to do a simulation.
2. Hold down the right arrow key and watch the cursor move. Write a description of this motion.
3. For what values of t does the particle change directions? What is the x position where it changes direction?
4. For an alternate visualization, let $X_2(T) = 2T^3 - 13T^2 + 22T - 5$ and $Y_2(T) = T$. Set the calculator to Simul mode. Now press GRAPH. (Ensure that all four functions $X_1(T)$, $Y_1(T)$, $X_2(T)$, and $Y_2(T)$ are selected.)

This Exploration gave you experience with motion along a line. Next, we use parametric equations to study motion in the plane.

Motion in the Plane

Recall the projectile-motion problem situation explored in Section 2.5. When an object is thrown straight up (that is, 90° from the horizontal) from a point s_0 feet above level ground with an initial velocity of v_0, then

$$s = -16t^2 + v_0t + s_0, \tag{1}$$

where s is the distance of the object above ground level t seconds after it is thrown.

Using parametric equations, this problem situation can be extended to the case where the ball is thrown at an angle other than 90° from the horizontal.

Suppose a ball is thrown with an initial velocity of 50 ft/sec at an angle of 60° with the x-axis and assume that gravity is the only force acting on the object.

Figure 9.29 Vector representing the velocity of a thrown ball.

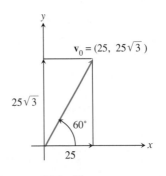

Figure 9.30 Vector components for the initial velocity of a ball thrown at an angle.

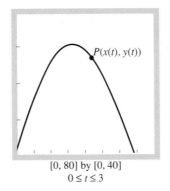

[0, 80] by [0, 40]
$0 \le t \le 3$

Figure 9.31 Complete graph of the problem situation in Example 1.

Velocity is a vector quantity, so this throw can be represented by a vector as shown in Fig. 9.29.

If v_0 is the initial velocity vector, the horizontal component is $50 \cos 60°$ and the vertical component is $50 \sin 60°$. Thus $v_0 = (25, 25\sqrt{3})$ (see Fig. 9.30). The horizontal direction of 25 ft/sec means that after t seconds, the ball has traveled $x = 25t$ ft horizontally.

The vertical component of $25\sqrt{3}$ ft/sec represents the initial velocity in a 90° direction from the horizontal—precisely the projectile motion described generally by Eq. (1). So the y-coordinate after t seconds is $y = -16t^2 + 25\sqrt{3}t$. Therefore the position $P(x(t), y(t))$ of the ball t seconds after it is thrown is described by the parametric equations

$$x = 25t \tag{2a}$$

$$y = -16t^2 + 25\sqrt{3}t. \tag{2b}$$

Notice that in using Eq. (1) to arrive at Eq. (2b), we let $s_0 = 0$ even though the ball leaves the thrower's hand a few feet above ground level. In the next example, we will continue to make this assumption; doing this will not alter the results significantly but will simplify the calculations. In later examples, we will consider problem situations in which $s_0 \ne 0$.

E X A M P L E   1   APPLICATION: Examining the Projectile Motion of a Baseball

A baseball is thrown with an initial velocity of 50 ft/sec at an angle of 60° with the positive x-axis. Assume that gravity is the only force acting on the baseball.

a) When will the ball hit the ground?

b) How far does the ball travel in the horizontal direction?

c) What is the maximum height attained by the ball?

Solution   The curve of the thrown ball is described by the parametric equations $x = 25t$ and $y = -16t^2 + 25\sqrt{3}t$. Find the complete graph shown in Fig. 9.31 and use TRACE to estimate that the ball hits the ground about 2.7 sec after it is thrown and it lands about 67.5 ft away. Its maximum height is about 29.3 ft.

a) The ball will hit the ground when $y(t) = 0$.

$$y(t) = -16t^2 + 25\sqrt{3}t = 0$$

$$t(25\sqrt{3} - 16t) = 0$$

Now, $y(t) = 0$ when $t = 0$ or $t = 25\sqrt{3}/16$. Thus the object will hit the ground when $t = 25\sqrt{3}/16 = 2.7063294$, or approximately 2.71 sec after launch.

b) The horizontal distance when it hits the ground is $x(t)$ when $t = 2.7063294$. That is,

$$x(2.7063294) = 25(2.7063294) = 67.658235.$$

The ball will travel 67.66 ft.

c) The path the ball travels is a parabola that opens down (see Fig. 9.31). Because its zeros occur when $t = 0$ or $t = 25\sqrt{3}/16$, the ball will reach its maximum height when t is halfway between, at $t = 25\sqrt{3}/32 = 1.3531647$. The maximum height will be

$$y\left(\frac{25\sqrt{3}}{32}\right) = 29.30 \text{ ft.}$$

In Example 1, the path of a thrown ball was described using parametric equations. With only a slight change in point of view, we can describe this path with vector notation.

The position of the ball at any time t of the object in vector form is

$$\mathbf{p} = (25t, 25\sqrt{3}t - 16t^2).$$

We can rewrite $\mathbf{p}$ as follows:

$$\mathbf{p} = (25t, 25\sqrt{3}t - 16t^2)$$
$$= (25t, 25\sqrt{3}t) + (0, -16t^2)$$
$$= t(25, 25\sqrt{3}) + (0, -16t^2)$$

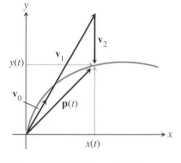

Figure 9.32 The path of an object is $\mathbf{p}(t) = \mathbf{v}_1 + \mathbf{v}_2$ where $\mathbf{v}_1 = t\mathbf{v}_0$.

So $\mathbf{p} = \mathbf{v}_1 + \mathbf{v}_2$ where $\mathbf{v}_1 = t\mathbf{v}_0 = t(25, 25\sqrt{3})$ and $\mathbf{v}_2 = (0, -16t^2)$. See Fig. 9.32. The position of $\mathbf{p}$ at time t thus depends on the two vectors $\mathbf{v}_0$ and $\mathbf{v}_2$, where $\mathbf{v}_2$ depends only on gravity.

Using Vector and Parametric Equations to Solve Motion Problems

The next four examples illustrate application problems that can be solved using vector and parametric equations. Without the use of powerful graphing tools, we could not easily solve these problems.

E X A M P L E 2   APPLICATION: Hitting a Baseball into the Wind

A baseball is hit when the ball is 3 ft above the ground and leaves the bat with initial velocity of 150 ft/sec and at an angle of elevation of 20°. A 6-mph wind is blowing in the horizontal direction against the batter. A 20-ft-high fence is 400 ft from home plate. Will the hit go over the fence and be a home run?

Notes on Examples

Each of the problem situations in the examples of this section is rich in terms of the mathematics to be learned. Exploring problem situations takes time, so it is suggested that some of the examples be examined in depth and that varying amounts of time be devoted to exploration. Stress the importance of using multiple representations in the examples by showing students mathematical modeling techniques. The use of algebra as a mathematical model can be taught effectively in this section.

Solution   Find a vector equation that describes the position of the ball at any time t. The time t that the ball is 400 ft from home plate can be used to determine the height of the ball.

Set up a coordinate system with the origin at home plate. Assume that the batter hits the ball in the direction of the positive x-axis. Fig. 9.33 diagrams this problem situation.

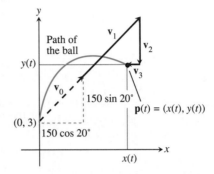

Figure 9.33   Vector diagram for Example 2. $\mathbf{p}(t)$ is the position of the ball at time t.

Initial Velocity v_0 $\mathbf{v_0} = (150\cos 20°, 150\sin 20°)$. Ignoring gravity and the wind, we see that the effect on the ball due to the hit at any time t is given by $\mathbf{v_1}$, where

$$\mathbf{v_1} = (150t\cos 20°, 150t\sin 20°).$$

Gravity Vector v_2   We have seen when studying the projectile-motion problem situation that the vertical position at time t is given by $y(t) = -16t^2 + 150t\sin 20° + 3$. In other words,

$$\mathbf{v_2} = (0, 3 - 16t^2).$$

Notice that this vector takes into account that the ball was hit 3 ft off the ground.

Wind Vector v_3   Because 60 mph is the same as 88 ft/sec, 6 mph is the same as 8.8 ft/sec. The wind vector is thus

$$\mathbf{v_3} = (-8.8t, 0).$$

The first coordinate is negative because the direction of the wind is opposite that of the hit ball, which is along the positive x-axis.

So the ball's path is described by $\mathbf{p}(t)$, where

$$\mathbf{p}(t) = \mathbf{v_1} + \mathbf{v_2} + \mathbf{v_3}$$

$$= (150t\cos 20°, 150t\sin 20°) + (0, 3 - 16t^2) + (-8.8t, 0)$$

$$= (150t\cos 20° - 8.8t, 150t\sin 20° + 3 - 16t^2).$$

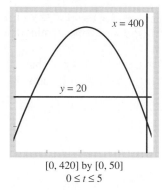

[0, 420] by [0, 50]
$0 \le t \le 5$

Figure 9.34 The curve described by the system of equations $x(t) = 150t \cos 20° - 8.8t$ and $y(t) = 150t \sin 20° + 3 - 16t^2$.

It now follows that

$$x(t) = 150t \cos 20° - 8.8t \qquad \text{and} \qquad y(t) = 150t \sin 20° + 3 - 16t^2 .$$

Figure 9.34 shows a graph of this pair of parametric equations. Clearly, when $x = 400$, $y < 20$ and the ball will *not* clear the fence. Therefore the hit will not be a home run. ≡

EXAMPLE 3   APPLICATION: Hitting a Baseball into the Wind

Suppose the ball of Example 2 leaves the bat at an angle of elevation of 30° but all other conditions of the example remain the same. Will this hit be a home run?

Solution   As only the angle value changes from the previous example, the parametric equations for the ball's path in this case are

$$x(t) = 150t \cos 30° - 8.8t$$

$$y(t) = 150t \sin 30° + 3 - 16t^2 .$$

Fig. 9.35 shows the ball's path. It is evident that at a distance of 400 ft, the ball is high enough to clear a 20-ft fence. Therefore the hit will be a home run. ≡

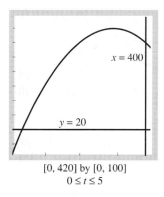

[0, 420] by [0, 100]
$0 \le t \le 5$

Figure 9.35 The ball clears the fence ($y = 20$) at 400 ft.

Now we use the information gained in previous examples to help solve the following problem posed by Neal Koblitz in the March 1988 issue of *The American Mathematical Monthly*.

EXAMPLE 4   APPLICATION: Throwing a Ball at a Ferris Wheel

Eric is standing on the ground at point D, a distance of 75 ft from the bottom of a ferris wheel that has a 20-ft radius (see Fig. 9.36). His arm is at the same height as the bottom of the ferris wheel. Janice is on the ferris wheel, which makes one revolution counterclockwise every 12 sec. At the instant she is at point A, Eric throws a ball to her at 60 ft/sec at an angle of 60° above the horizontal. Assume that $g = 32$ ft/sec² (force due to gravity) and neglect air resistance. Will the ball reach Janice?

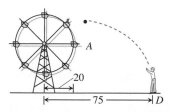

Figure 9.36 A person standing on the ground throws a ball to a person on a rotating ferris wheel.

Solution   To solve this problem, find parametric equations that describe the path of both the ferris wheel and the thrown ball. By graphing these two paths simultaneously, you will simulate the situation.

Place a coordinate system on the problem situation diagram so that the center of the ferris wheel is at point $(0, 20)$ and the ball is thrown from point $(75, 0)$.

Path of Ferris Wheel (Path A) The parametric equations for the position of a point P on a circle with center $C(0, 20)$ and radius 20 are

$$x_A = 20 \cos \theta$$

$$y_A = 20 + 20 \sin \theta,$$

where θ is the angle that radius CP makes with the positive x-axis (see Fig. 9.37).

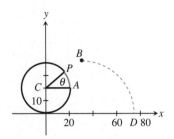

Figure 9.37 Diagram for Example 4.

The ferris wheel makes one revolution every 12 sec. Therefore θ as a function of t in seconds is

$$\theta(t) = \frac{2\pi}{12} t = \frac{\pi}{6} t.$$

Therefore the parametric equations that describe the position P of the friend on the ferris wheel are

$$x_A(t) = 20 \cos \frac{\pi}{6} t$$

$$y_A(t) = 20 + 20 \sin \frac{\pi}{6} t.$$

Path of Thrown Ball (Path B) The ball is thrown towards the friend at a $60°$ angle with the ground, which makes an angle of $120°$ with respect to the positive x-axis. Therefore the parametric equations that describe the path of the thrown ball are

$$x_B(t) = 75 + (60 \cos 120°)t = 75 - 30t$$

$$y_B(t) = -16t^2 + (60 \sin 120°)t = -16t^2 + 30\sqrt{3}\, t.$$

Graphing the two curves simultaneously shows that the ball comes close to Janice but does not reach her. Using the **TRACE** key, we learn that the minimum distance between Janice and the ball occurs between 2.1 and 2.3 sec. ▤

In Example 5, we use a graphing utility to investigate more closely the minimum distance between the ball and the person on the ferris wheel.

Notes on Exercises

Ex. 1–12 provide a revisitation of the basic ideas in Section 9.1 of this chapter.

In Ex. 7–40, there are eight different problem-situation sections. You might want to give each of these problem situations to small groups of students to solve and present in class. Some of these problems will be quite challenging.

Assignment Guide

Day 1: Ex. 1, 2, 4, 5, 7–14
Day 2: Ex. 15–20, 23–30
Day 3: Ex. 31–36, 42, 43

E X A M P L E 5   APPLICATION: Calculating Minimum Distance

Find the minimum distance between the ball and the friend on the ferris wheel and the time at which the minimum occurs.

Solution   Earlier we saw that $x_A(t) = 20\cos(\pi t/6)$ and $y_A(t) = 20+20\sin(\pi t/6)$ are parametric equations for the path of the friend on the ferris wheel and that $x_B(t) = 75 - 30t$ and $y_B(t) = 30\sqrt{3}\,t - 16t^2$ are parametric equations for the path of the ball.

The distance between the friend and the ball is given by

$$D(t) = \sqrt{[x_A(t) - x_B(t)]^2 + [y_A(t) - y_B(t)]^2}$$

$$= \sqrt{[20\cos\left(\frac{\pi}{6}t\right) - (75 - 30t)]^2 + [20 + 20\sin\left(\frac{\pi}{6}t\right) - (30\sqrt{3}t - 16t^2)]^2}.$$

Fig. 9.38 shows the graph of $D(t)$ for t over the interval $[0, 4]$. Note that the ball hits the ground in a little over 3 sec.

Using zoom-in, we determine that the minimum distance is 1.58 ft with an error of at most 0.01 and that this minimum occurs when $t = 2.19$ sec.   ▤

[0, 4] by [0, 20]

Figure 9.38   The distance between the friend and the ball is given by $D(t)$.

Exercises for Section 9.3

In Exercises 1–4, find a complete graph of the curve defined by the parametric equations.

1. $x(t) = 2t - 6\sin t$ and $y(t) = 2 - 6\cos t$

2. $x(t) = \cos^2 t$ and $y(t) = (1 - \cos t)\sin t$

3. $x(t) = 2 - 5\cos t$ and $y(t) = -2 + 4\cos t$

4. $x(t) = \dfrac{4}{1 + \sin t}\cos t$ and $y(t) = \dfrac{4}{1 + \sin t}\sin t$

5. An arrow is shot with an initial velocity of 205 ft/sec and at an angle of elevation of 48°. Find when and where the arrow will strike the ground.

6. With what initial velocity must a ball be thrown from the ground at an angle of 35° from the horizontal in order to travel a horizontal distance of 255 ft?

Exercises 7–12 refer to the following **problem situation** : A dart is thrown upward with an initial velocity of 58 ft/sec at an angle of elevation of 41°. Consider the position of the dart at any time t, where $t = 0$ when the dart is thrown. Neglect air resistance.

7. Find parametric equations that model the problem situation.

8. Draw a complete graph of the model.

9. What portion of the graph represents the problem situation?

10. When will the dart hit the ground?

11. Find the maximum height of the dart. At what time will this occur?

12. How far does the dart travel in the horizontal direction? Neglect air resistance.

Exercises 13 and 14 refer to the following **problem situation** : A golfer hits a ball with an initial velocity of 133 ft/sec and at an angle of 36° from the horizontal.

13. Find when and where the ball will hit the ground.

14. Will the ball in Exercise 13 clear a fence 9 ft high that is at a distance of 275 ft from the golfer?

Exercises 15–20 refer to the following **problem situation** : Chris and Linda are standing 78 ft apart. Simultaneously, each throws a softball toward the other. Linda throws her ball with an initial velocity of 45 ft/sec with an angle of inclination of 44°. Chris throws her ball with an initial velocity of 41 ft/sec with an angle of inclination of 39°.

15. Find two sets of parametric equations that represent a model of the problem situation.

16. Find complete graphs of both sets of parametric equations in the same viewing rectangle.

17. What values of t make sense in this problem situation?

18. Find the maximum height of each ball. How far does each travel in the horizontal direction and when does it hit the ground? Whose ball hits first?

19. By choosing t_{max} carefully (guess and check), estimate how close to each other the two balls get and the time when they are closest (minimum distance). Be careful to use square windows.

20. Use the distance formula and zoom-in to find the minimum distance (and the time at which it occurs) with an error of at most 0.01.

Exercises 21 and 22 refer to the following **problem situation**: A river boat's paddle wheel has a diameter of 26 ft and at full speed, makes one revolution clockwise in 2 sec.

For Exercises 21 and 22.

21. Write parametric equations describing the position of a point A on the paddle. Assume that at $t = 0$, A is at the very top of the wheel.

22. How far will point A, which is fixed on the wheel, move in 1 min?

Exercises 23–26 refer to the ferris-wheel-and-ball **problem situation** described in Example 4.

23. Solve the problem if the radius of the ferris wheel is 26 ft and the ball is thrown with an initial velocity 76 ft/sec at an angle of 52° with the horizontal and from a distance of 62 ft from the bottom of the ferris wheel. Use parametric simulation and estimate your answer. (Assume the ferris wheel revolves at a constant rate.)

24. Use the distance formula and zoom in to find the minimum distance between Janice and the ball.

25. Using the parametric simulation, vary Eric's position from 62 ft to see how close he can get the ball to Janice on the ferris wheel. Use a guess and check method. (Keep all other values the same.)

26. Using the parametric simulation, vary the angle of elevation from 52° to see how close you can get Janice and the ball. Use a guess and check method. (Keep all other values the same.)

Exercises 27–30 refer to the following **problem situation**: An NFL punter at the 15-yd line kicks a football downfield with an initial velocity of 85 ft/sec at an angle of elevation of 56°.

27. Find a complete graph of the problem situation.

28. How far downfield will the football first hit the ground?

29. Determine the ball's maximum height above the field.

30. What is the "hang time" (the total time the football is in the air)?

Exercises 31–36 refer to the following **problem situation**: A major league baseball player hits a ball with an initial velocity of 103 ft/sec in the direction of a 10-ft fence that is 300 ft from home plate. For each exercise, draw a complete graph of the problem situation and determine whether the hit is a home run (clears the fence). In Exercises 31–33, assume that gravity is the only force affecting the ball's path. In all exercises in this group, disregard air resistance.

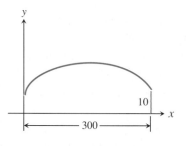

For Exercises 31 to 36.

31. The hit is at an angle of elevation of 35°.

32. The hit is at an angle of elevation of 43°.

33. The hit is at an angle of elevation of 49°.

34. The hit is at an angle of elevation of 41° and the wind is blowing at 22 ft/sec in the same direction as the horizontal path of the ball.

35. The hit is at an angle of elevation of 41° and the wind is blowing at 22 ft/sec in the direction opposite to the horizontal path of the ball.

36. The hit is at an angle of elevation of 41° and the wind is blowing at 22 ft/sec in the direction opposite to the horizontal path of the ball and at an angle of depression of 12°.

Exercises 37–40 refer to the following **problem situation** regarding projectile motion with air resistance: For many non-spinning projectiles, the main effect on the path of the projectile, other than gravity, is a slowing-down influence due to air resistance. This effect is called *drag force* and it acts in a direction opposite to the projectile's velocity. For projectiles moving through the air at low speeds, the drag force can be assumed to be linear and directly proportional to the speed. Using a linear drag model, it can be shown that the projectile's position at time t is given by

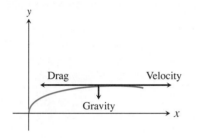

For Exercises 37 to 40.

$$x(t) = \frac{v_0}{k} \cos \alpha (1 - e^{-kt})$$

$$y(t) = \frac{v_0}{k} \sin \alpha (1 - e^{-kt}) + \frac{32}{k^2}(1 - kt - e^{-kt}),$$

where k is a constant representing the air density, v_0 is the initial velocity, and α is the initial angle of elevation. Suppose that $v_0 = 95$ ft/sec and $\alpha = 42°$.

37. Graph in the same viewing rectangle the motion of the projectile assuming no air resistance and the motion of the projectile assuming air resistance with $k = 0.3$.

38. Determine with and without drag force the maximum height of the projectile and the time this occurs.

39. Determine with and without drag force the range (horizontal distance) the projectile travels and the time of impact.

40. Solve all parts of this problem for $k = 0.01$.

41. Show that the path of a fixed point P on a circle with radius a and center C that rolls on a straight line is the *cycloid* described by

$$x(\theta) = a(\theta - \sin \theta)$$

$$y(\theta) = a(1 - \cos \theta),$$

where θ represents the angle the radius CP has turned through. Assume the starting position is at the origin and that θ is a positive angle even though the rotation is clockwise.

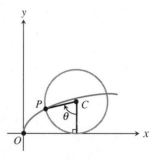

For Exercise 41.

42. Design an experiment in physics to show the following remarkable property of cycloids. If Q is the lowest point of an arch of an inverted cycloid, then the time it takes for a frictionless bead to slide down the curve to point Q is independent of the starting point. In other words, if A and B are dropped at the same time they will land at Q at the same time.

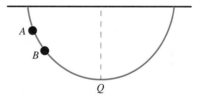

For Exercise 42.

43. **Writing to Learn** Write a paragraph that explains why Figs. 9.36 and 9.37 are actually distorted. How would you change the viewing rectangle to see the shape of the actual path?

44. **Writing to Learn** Explain how these parametric equations are related to Example 2.

$$x_1(t) = 400$$

$$y_1(t) = Y_{max}(T/T_{max})$$

$$x_2(t) = X_{max}(T/T_{max})$$

$$y_2(t) = 20$$

9.4 _____

Conic Sections

Imagine two lines intersecting in three-dimensional space. When one line is fixed and the other rotates around the first, the resulting set of points is called a **cone**. The cone has two parts, called **nappes**, each of which looks like an ice cream cone. If the two lines maintain a constant angle between them during the rotation, then the result is a **double-napped right circular cone**.

A **conic section** (or simply a **conic**) is any curve obtained by intersecting a double-napped right circular cone with a plane. The three basic conic sections are the **parabola**, the **ellipse**, and the **hyperbola** (see Fig. 9.39). A circle is a special case of an ellipse. When a plane slices through only one nappe of the cone, either a parabola or an ellipse is formed, depending on the angle of the plane (see Fig. 9.39a,b). If the plane slices through both nappes of the cone, the result is a hyperbola (see Fig. 9.39c).

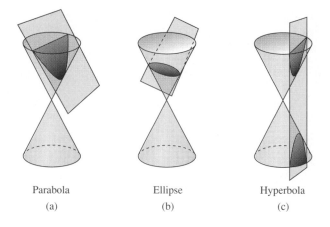

Parabola   Ellipse   Hyperbola
(a)   (b)   (c)

Figure 9.39 How the three basic conic sections are derived from the intersection of a right circular cone and a plane.

You were introduced informally to two of these conic sections—the parabola and the ellipse—in earlier chapters. In the material that follows, we shall analyze all three conic sections from an algebraic point of view. We will give a formal definition and then develop standard forms for equations of each.

Parabolas

In Section 1.5, you learned that the graph of any quadratic function $y = ax^2 + bx + c$ can be obtained from the graph of $y = x^2$ by using vertical stretching or shrinking, reflection through the x-axis, vertical shifting, and horizontal shifting. The name *parabola* was used to refer to graphs of these functions. Here, finally, is a definition of a parabola.

Teaching Note

It is assumed that students have substantial knowledge about parabolas and circles from prior courses and from the material in previous sections of this course. The focus of this section is the use of geometric approaches to conics, using the distance formula as a means to this end.

Definition 9.3 Parabola

Let L be any line and $F(x_0, y_0)$ any point not on L. The set of all points $P(x, y)$ in the plane that are equidistant from L and F is called a **parabola** (see Fig. 9.40). The point F is the **focus of the parabola**, and the line L is the **directrix of the parabola**. The line perpendicular to L through the focus is the **line of symmetry of the parabola**, and the point common to the parabola and its line of symmetry is the **vertex of the parabola.**

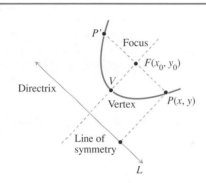

Figure 9.40 Terms used in defining a parabola.

Example 1 shows how you can find the equation of a parabola if the focus is on one axis and the directrix is parallel to the other axis.

E X A M P L E   1   Finding the Equation of a Parabola

Determine an equation for the parabola with focus $(0, a)$ and directrix $y = -a$ where $a > 0$ (see Fig. 9.41).

Solution   In Fig. 9.41, the distance from $P(x, y)$ to the line $y = -a$ is $|y + a|$, and the distance from $P(x, y)$ to $(0, a)$ is $\sqrt{(x - 0)^2 + (y - a)^2}$. (Why?) Apply the definition of a parabola to obtain

$$\sqrt{x^2 + (y - a)^2} = |y + a|. \tag{1}$$

Next, square each side of Eq. (1) and simplify.

$$\left(\sqrt{x^2 + (y - a)^2}\right)^2 = |y + a|^2$$

$$x^2 + (y - a)^2 = (y + a)^2$$

$$x^2 + y^2 - 2ay + a^2 = y^2 + 2ay + a^2$$

$$x^2 = 4ay \tag{2}$$

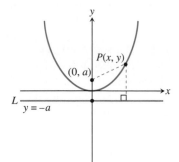

Figure 9.41 $x^2 = 4ay$, where $a > 0$. The distance from P to the focus $(0, a)$ is equal to the distance from P to L.

Thus every point on the parabola is a solution to Eq. (2). Because the steps used to derive Eq. (2) from Eq. (1) are reversible, every solution to Eq. (2) corresponds to a point on the parabola. We can rewrite Eq. (2) in the form

$$y = \frac{1}{4a}x^2$$

to see that the graph can be obtained by starting with $y = x^2$ and applying a vertical stretch or shrink by the factor $1/4a$. The coordinates of the vertex are $(0, 0)$. Notice that a is half the distance from the focus to the directrix.   ≡

Notice that the parabola developed in Example 1, whose equation is $x^2 = 4ay$, where $a > 0$, opens upward (see Fig. 9.41). If $a < 0$, the parabola $x^2 = 4ay$ opens downward (see Fig. 9.42a).

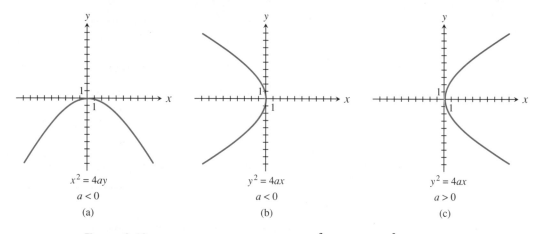

$$x^2 = 4ay$$
$$a < 0$$
(a)

$$y^2 = 4ax$$
$$a < 0$$
(b)

$$y^2 = 4ax$$
$$a > 0$$
(c)

Figure 9.42   Parabolas whose equations are $x^2 = 4ay$ and $y^2 = 4ax$. The fourth case is shown in Fig. 9.41.

By repeating the algebra carried out in Example 1, we can establish that the parabola opens right or left in the case of $y^2 = 4ax$ (see Fig. 9.42b,c).

Consider a parabola with vertex (h, k) and focus $F(h, k+a)$ where $a > 0$ (see Fig. 9.43). Then the equation of the directrix would be $y = k - a$. The equation of the parabola can be obtained from $x^2 = 4ay$ by a horizontal and vertical shift, as the following development confirms.

$$\sqrt{(x - h)^2 + [y - (k + a)]^2} = |y - (k - a)|$$
$$(x - h)^2 + [(y - k) - a]^2 = [(y - k) + a]^2$$
$$(x - h)^2 + (y - k)^2 - 2a(y - k) + a^2 = (y - k)^2 + 2a(y - k) + a^2$$
$$(x - h)^2 = 4a(y - k)$$

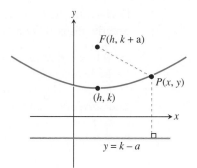

Figure 9.43 Vertex (h, k) and focus $F(h, k + a)$.

Theorem 9.5 Equations of Parabolas in Standard Form

The graph of each of the following equations is a parabola with vertex (h, k):

1. $(x - h)^2 = 4a(y - k)$: The focus is $(h, k + a)$, the directrix is $y = k - a$, and the line of symmetry is $x = h$.
2. $(y - k)^2 = 4a(x - h)$: The focus is $(h + a, k)$, the directrix is $x = h - a$, and the line of symmetry is $y = k$.

These equations are called **standard forms for the parabola**.

Once an equation for a parabola is in standard form, Theorem 9.5 can be used to find the important characteristics of the parabola.

E X A M P L E 2 *Using the Standard Form of a Parabola*

Find the vertex, line of symmetry, focus, and directrix of the parabola $(x + 1)^2 = -12(y - 3)$. Also find the x- and y-intercepts and sketch a complete graph.

Solution Because this equation is in standard form, use Theorem 9.5 to draw the following conclusions:

Vertex An equation in standard form has vertex (h, k), so the vertex of this parabola is the point $(-1, 3)$.

Line of Symmetry The second degree variable is x, therefore the parabola opens upward or downward and the line of symmetry is the line $x = -1$.

Focus   Because $4a = -12$, $a = -3$. Therefore the parabola opens downward and the focus is the point $(-1, 3 - 3) = (-1, 0)$.

Directrix   Because $a = -3$, the directrix is 3 units above the vertex and therefore is the line $y = 6$.

To find the x- and y-intercepts, substitute $(a, 0)$ and $(0, b)$ into the equation and solve for a and b.

$$(a + 1)^2 = -12(0 - 3)$$

$$(a + 1)^2 = 36$$

$$a + 1 = \pm 6$$

Therefore the points $(-7, 0)$ and $(5, 0)$ are x-intercepts. Similarly,

$$(0 + 1)^2 = -12(b - 3)$$

$$b - 3 = -\frac{1}{12}$$

$$b = -\frac{1}{12} + 3 = \frac{35}{12}.$$

The point $(0, 35/12)$ is the y-intercept.

With this information, sketch the graph shown in Fig. 9.44. Support your answer with a grapher.   ▤

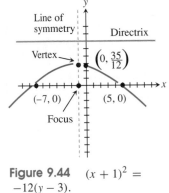

Figure 9.44 $(x + 1)^2 = -12(y - 3)$.

E X A M P L E   3   Finding the Standard Form of a Parabola

Find the standard form, vertex, focus, directrix, and line of symmetry of the parabola $y^2 - 6x + 2y + 13 = 0$.

Solution   Because this equation is quadratic in the variable y, complete the square on terms involving y.

$$y^2 - 6x + 2y + 13 = 0$$

$$y^2 + 2y = 6x - 13$$

$$y^2 + 2y + 1 = 6x - 13 + 1$$

$$(y + 1)^2 = 6x - 12$$

$$(y + 1)^2 = 6(x - 2)$$

This parabola opens to the right and has a vertex $(2, -1)$, and a line of symmetry $y = -1$. Because $4a = 6$, $a = 1.5$. The focus is $(3.5, -1)$, and the directrix is $x = 0.5$.   ▤

E X A M P L E 4   Finding a Complete Graph of a Parabola

Find a complete graph of the parabola $y^2 - 6x + 2y + 13 = 0$ with a graphing utility.

Solution   To sketch this parabola, it is helpful first to change the equation to standard form as illustrated in Example 3. However, to find a complete graph on a graphing utility it is necessary to solve this original equation for y. Rewrite the original equation in the form $y^2 + 2y + (13 - 6x) = 0$ and use the quadratic formula to solve for y.

$$y^2 + 2y + (13 - 6x) = 0$$

$$y = \frac{-2 \pm \sqrt{4 - 4(13 - 6x)}}{2}$$

$$= -1 \pm \sqrt{1 - (13 - 6x)} \quad \text{(Why?)}$$

$$= -1 \pm \sqrt{6x - 12}$$

To get a complete graph of the parabola (see Fig. 9.45), we must graph both $y = -1 + \sqrt{6x - 12}$ and $y = -1 - \sqrt{6x - 12}$.   ≡

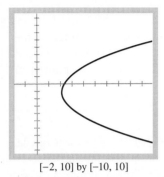

[−2, 10] by [−10, 10]

Figure 9.45 $y^2 - 6x + 2y + 13 = 0$.

In the real world, the reflective properties of objects with parabolic surfaces enable the parabola to be used in many applications. For example, if a source of light is placed at the focus of a parabolic surface, light will be reflected off the surface in lines parallel to its line of symmetry. This principle is used in the design of car headlights (see Fig. 9.46). Conversely, sound waves traveling *into* the opening of a parabola parallel to its line of symmetry will be reflected through the focus of the parabola. Parabolic microphones use this principle.

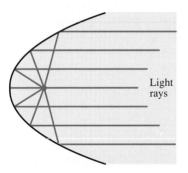

Light rays

Figure 9.46   Light rays from a source at the focus are reflected off a parabolic mirror in parallel directions.

E X A M P L E 5   APPLICATION: Placing the Sound Receiver in a Parabolic Microphone

A parabolic microphone is formed by revolving the portion of the parabola $15y = x^2$ between $x = -9$ and $x = 9$ about its line of symmetry. Where should the sound receiver be placed?

Solution   The receiver should be placed at the focus of the parabola $15y = x^2$. Because $4a = 15$, $a = 3.75$. The focus is thus $(0, 3.75)$. The receiver should be placed 3.75 units from the vertex along its line of symmetry.   ≡

Ellipses

When a plane slices through just one nappe of a cone to form a closed curve, the resulting shape is called an ellipse. An algebraic description of an ellipse also can be given in a manner somewhat the same as for a parabola. One difference, however, is that a parabola has one focal point whereas the ellipse has two.

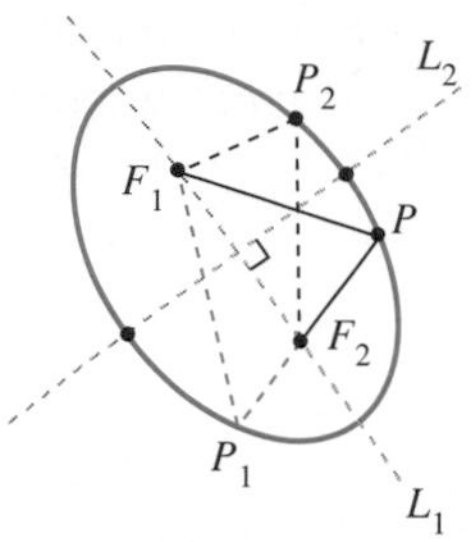

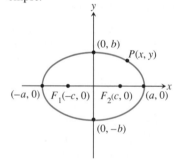

Figure 9.47 $|F_1P| + |PF_2|$ is a constant for all points P on the ellipse.

Definition 9.4   Ellipse

Let F_1 and F_2 be any two points in the coordinate plane. An **ellipse** is the set of all points P in the plane such that the sum of the distances PF_1 and PF_2 is a constant (see Fig. 9.47). F_1 and F_2 are the **foci (focal points) of the ellipse.**

Notice that an ellipse can be thought of as a locus of points, just as a parabola.

L_1, the line through the focal points, and L_2, the perpendicular bisector of line segment F_1F_2, are both lines of symmetry for any ellipse determined by points F_1 and F_2. For now, we will consider only ellipses whose lines of symmetry are parallel to the coordinate axes; later in Section 9.4, we will examine ellipses not parallel to the coordinate axes.

E X A M P L E   6   Finding a Standard Equation of an Ellipse

Let $F_1 = (-c, 0)$ and $F_2 = (c, 0)$ be the two foci of the ellipse consisting of the points P where $d(P, F_1) + d(P, F_2) = 2a$ such that $a > c > 0$. Determine an equation for the ellipse and the coordinates of the x-intercepts, the y-intercepts, and the center of the ellipse (see Fig. 9.48).

Figure 9.48 An ellipse with focal points F_1 and F_2 on the x-axis.

Solution   Use the distance formula to obtain an equation for the ellipse:

$$d(P, F_1) + d(P, F_2) = 2a$$

$$\sqrt{(x+c)^2 + y^2} + \sqrt{(x-c)^2 + y^2} = 2a$$

Next, obtain an alternate form of this equation that does not involve radicals.

$$\sqrt{(x+c)^2 + y^2} + \sqrt{(x-c)^2 + y^2} = 2a$$

$$\sqrt{(x+c)^2 + y^2} = 2a - \sqrt{(x-c)^2 + y^2}$$

$$(x+c)^2 + y^2 = 4a^2 - 4a\sqrt{(x-c)^2 + y^2} + (x-c)^2 + y^2$$

$$x^2 + 2cx + c^2 + y^2 = 4a^2 - 4a\sqrt{(x-c)^2 + y^2} + x^2 - 2cx + c^2 + y^2$$

$$2cx = 4a^2 - 4a\sqrt{(x-c)^2 + y^2} - 2cx$$

$$4a\sqrt{(x-c)^2 + y^2} = 4a^2 - 4cx$$

$$a\sqrt{(x-c)^2 + y^2} = a^2 - cx \quad \text{Square each side and collect terms}$$

$$(a^2 - c^2)x^2 + a^2 y^2 = a^2(a^2 - c^2) \tag{3}$$

Notes on Examples

Help students to view ellipses as generalized forms of circles. Point out that there are only two cases for which an ellipse is oriented parallel to the coordinate axes, whereas there are four cases for the parabola.

Divide each side of Eq. (3) by $a^2(a^2 - c^2)$ to obtain the following:

$$\frac{x^2}{a^2} + \frac{y^2}{a^2 - c^2} = 1. \tag{4}$$

Because $a > c$, you can define a positive number b such that $b^2 = a^2 - c^2$ and rewrite Eq. (4) in the form

$$\frac{x^2}{a^2} + \frac{y^2}{b^2} = 1. \tag{5}$$

Thus every point on the ellipse is a solution to Eq. (5). It also can be shown that every solution of Eq. (5) corresponds to a point on the ellipse. This form is called the standard form for the equation of an ellipse. The x-intercepts are $\pm a$, and the y-intercepts are $\pm b$. (Why?) The center of the ellipse is $(0, 0)$. ≡

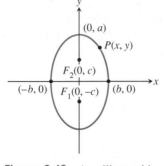

The **major axis** of the ellipse in Example 6 is the line segment from $(-a, 0)$ to $(a, 0)$, and the **minor axis** is the line segment from $(0, -b)$ to $(0, b)$. Note that regardless of orientation, the foci always lie on the major axis.

Now consider an ellipse with foci $F_1(0, -c)$ and $F_2(0, c)$ on the y-axis (see Fig. 9.49). It can be shown that the standard form for the equation of this ellipse, which satisfies $d(P, F_1) + d(P, F_2) = 2a$ such that $a > c > 0$, is

$$\frac{x^2}{b^2} + \frac{y^2}{a^2} = 1, \tag{6}$$

where $b^2 = a^2 - c^2$. This equation is the same as Eq. (5) except that the number under x^2 is smaller than the number under y^2.

Equations (5) and (6) are the standard forms for an ellipse centered at the origin with axes parallel to the coordinate axes. If the center of an ellipse is the point (h, k), the standard forms of its equations are as stated in this theorem.

Figure 9.49 An ellipse with foci F_1 and F_2 on the y-axis.

Theorem 9.6 Equations of Ellipses in Standard Form

Let $a > b$ and $c = \sqrt{a^2 - b^2}$. The graph of each of the following equations is an ellipse with center (h, k) and lines of symmetry $x = h$ and $y = k$ (see Fig. 9.50).

1. $\dfrac{(x - h)^2}{a^2} + \dfrac{(y - k)^2}{b^2} = 1$ with foci $(h \pm c, k)$.

2. $\dfrac{(x - h)^2}{b^2} + \dfrac{(y - k)^2}{a^2} = 1$ with foci $(h, k \pm c)$.

These equations are the **standard forms for an ellipse.**

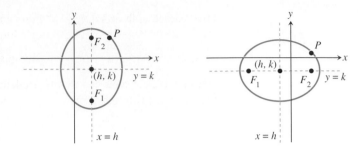

Figure 9.50 Ellipses with center (h, k) and axes parallel to the coordinate axes.

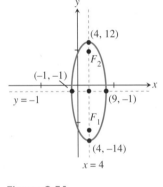

Figure 9.51
$$\frac{(x-4)^2}{25} + \frac{(y+1)^2}{169} = 1.$$

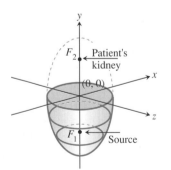

Figure 9.52 A lithotripter is an ellipsoidal device emitting UHF shock waves at one focus to break up kidney stones at the other focus.

E X A M P L E 7   Using the Standard Form of an Ellipse

Find the center and the foci, lines of symmetry, and endpoints of the major and minor axes and sketch a complete graph of

$$\frac{(x-4)^2}{25} + \frac{(y+1)^2}{169} = 1.$$

Solution   Because the equation is in standard form, use Theorem 9.6 to conclude that the lines of symmetry are $x = 4$ and $y = -1$. Therefore the center of the ellipse is $(4, -1)$.

Now $a = 13$ and $b = 5$, so $c = 12$. The foci are $(4, -13)$ and $(4, 11)$. The endpoints of the major axis are $(4, -14)$ and $(4, 12)$, and the endpoints of the minor axis are $(-1, -1)$ and $(9, -1)$. Figure 9.51 shows a complete graph of the ellipse. ≡

Ellipses, like parabolas, have reflective properties. Light or sound emitted at one focus of an ellipse is reflected by the ellipse through the other focus. One example of a real-world application is a lithotripter, a special device that produces ultra-high-frequency (UHF) shock waves moving through water to break up kidney stones. This device is in the shape of an **ellipsoid**. An ellipsoid is a three-dimensional solid that is formed by rotating an ellipse about its major axis. The foci of an ellipsoid are the same as for the ellipse used to generate it.

To use a lithotripter, doctors make careful measurements of the patient's kidney stones. The lithotripter is positioned so that the source producing the shock waves is at one focus of the ellipsoid and the kidney stones are at the other focus. The shock waves reflect off the inner surface of the ellipsoid and pass through the second focus to break up the kidney stones (see Fig. 9.52).

E X A M P L E 8   APPLICATION: Using a Lithotripter

Consider the following problem situation. Assume that the center of an ellipse rotated to form the ellipsoid of the lithotripter is $(0, 0)$. The two ends of the major axis are $(-6, 0)$ and $(6, 0)$, and one end of the minor axis is $(0, -2.5)$. Determine the coordinates of the lithotripter's foci.

Solution From the given information, note that $a = 6$ and $b = 2.5$. Thus the equation for the ellipse that is rotated is $\dfrac{x^2}{6^2} + \dfrac{y^2}{2.5^2} = 1$. Now $c^2 = a^2 - b^2$, so $c^2 = 29.75$ and $c = 5.45$. Therefore the foci are $(-5.45, 0)$ and $(5.45, 0)$. The shock source (an underwater spark discharge) is placed at the negative focus and the patient's kidney at the positive focus. ▤

Hyperbolas

When a plane intersects two nappes of a cone, the resulting shape is a hyperbola.

Recall that an ellipse is defined algebraically in terms of the *sum* of two distances being a constant. A hyperbola, on the other hand, is defined in terms of the *difference* of two distances being a constant.

Definition 9.5   Hyperbola

Let F_1 and F_2 be any two points in the coordinate plane. A **hyperbola** is the set of all points P in the plane such that the absolute value of the difference between the distances PF_1 and PF_2 is a constant. F_1 and F_2 are the **foci (focal points)**.

A hyperbola can be described as the locus of points P such that $|PF_1 - PF_2|$ is a constant.

L_1, the line through the focal points, and line L_2, the perpendicular bisector of line segment F_1F_2, are both lines of symmetry for any hyperbola determined by points F_1 and F_2.

The line through the focal points, L_1, is called the **principal axis**, line L_2 is the **transverse axis**, and the intersection of these two axes is the **center of the hyperbola**.

The calculation completed for an ellipse in Example 6 can be repeated with minor alterations for the hyperbola. In Fig. 9.53, the focal points are $F_1(-c, 0)$ and $F_2(c, 0)$. The hyperbola consists of all points P where $|d(P, F_1) - d(P, F_2)| = 2a$ such that $0 < a < c$.

Completing this calculation as we did in Example 6 leads to the following equation for a hyperbola:

$$\frac{x^2}{a^2} - \frac{y^2}{b^2} = 1, \tag{7}$$

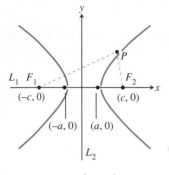

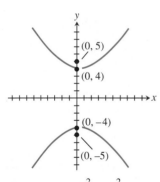

Figure 9.53 $\dfrac{x^2}{a^2} - \dfrac{y^2}{b^2} = 1.$

where $b^2 = c^2 - a^2$. The x-intercepts $(-a, 0)$ and $(a, 0)$ are called the **vertices of the hyperbola**.

Notice from Fig. 9.53 that Eq. (7) describes a hyperbola whose principal axis is the x-axis. The next example considers the equation of a hyperbola whose principal axis is the y-axis.

E X A M P L E 9　Finding an Equation of a Hyperbola

Find the equation of the hyperbola with focal points $(0, 5)$ and $(0, -5)$ and vertices $(0, 4)$ and $(0, -4)$ (see Fig. 9.54).

Solution　We see that $a = 4$, $c = 5$, and therefore $b = \sqrt{5^2 - 4^2} = 3$. Because the principal axis is the y-axis instead of the x-axis, the variables x and y must be reversed. Therefore the equation for this hyperbola is

$$\frac{y^2}{16} - \frac{x^2}{9} = 1. \qquad \blacksquare$$

Asymptotes of a Hyperbola

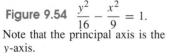

Figure 9.54 $\dfrac{y^2}{16} - \dfrac{x^2}{9} = 1.$
Note that the principal axis is the y-axis.

Equation (7) can be rewritten as follows:

$$\frac{x^2}{a^2} - \frac{y^2}{b^2} = 1$$

$$\frac{x^2}{a^2} - 1 = \frac{y^2}{b^2}$$

$$y^2 = \frac{b^2 x^2}{a^2} - b^2$$

$$y^2 = \frac{b^2 x^2}{a^2}\left(1 - \frac{a^2}{x^2}\right). \qquad (8)$$

From Eq. (8), it is apparent that as the absolute value of x gets very large, a^2/x^2 approaches zero. Therefore

$$y^2 \qquad \text{is very close to} \qquad \frac{b^2 x^2}{a^2}.$$

Hence the two lines $y = bx/a$ and $y = -bx/a$ are the **asymptotes of the hyperbola**.

Stan says, "The asymptotic behavior of hyperbolas is an extension of the ideas about asymptotes that students know from previous experiences with graphing. Be sure to allow time for students to explore hyperbolas in detail."

E X A M P L E 10　Finding the Graph of a Hyperbola

Find the asymptotes and sketch a complete graph of the hyperbola $\dfrac{x^2}{16} - \dfrac{y^2}{9} = 1$.

Solution   The given equation is in the form of Eq. (7). Its center is $(0, 0)$, its principal axis is the x-axis, and $a = 4$ and $b = 3$. Therefore the asymptotes are

$$y = \frac{3}{4}x \quad \text{and} \quad y = -\frac{3}{4}x.$$

The foci are $(-5, 0)$ and $(5, 0)$ and the vertices are $(-4, 0)$ and $(4, 0)$. A complete graph is shown in Fig. 9.55.   ▤

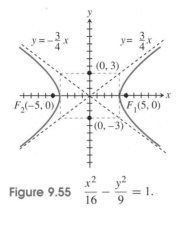

Figure 9.55 $\dfrac{x^2}{16} - \dfrac{y^2}{9} = 1.$

A horizontal and vertical shift moves a hyperbola centered at $(0, 0)$ to one centered at (h, k). The equations of the hyperbolas with axes parallel to the coordinate axes and centered at (h, k) are given in the following theorem.

Theorem 9.7   Equations of Hyperbolas in Standard Form

Let $c = \sqrt{a^2 + b^2}$. The graph of each of the following equations is a hyperbola with center (h, k) and lines of symmetry $x = h$ and $y = k$:

1. $\dfrac{(x - h)^2}{a^2} - \dfrac{(y - k)^2}{b^2} = 1$ with foci $(h \pm c, k)$ and vertices $(h \pm a, k)$; the asymptotes have a slope of $\pm b/a$.

2. $\dfrac{(y - k)^2}{a^2} - \dfrac{(x - h)^2}{b^2} = 1$ with foci $(h, k \pm c)$ and vertices $(h, k \pm a)$; the asymptotes have a slope of $\pm a/b$.

These equations are **standard forms for the hyperbola**.

E X A M P L E   11   *Changing to Standard Form*

Write $x^2 - 4y^2 + 2x - 24y = 39$ in standard form. Determine the center, foci, vertices, lines of symmetry, and asymptotes. Sketch a complete graph.

Solution   First, group the terms involving x and the terms involving y and then complete the square.

$$x^2 - 4y^2 + 2x - 24y = 39$$
$$(x^2 + 2x) - 4(y^2 + 6y) = 39$$
$$(x^2 + 2x + 1) - 4(y^2 + 6y + 9) = 39 + 1 - 4(9)$$

$$(x + 1)^2 - 4(y + 3)^2 = 4$$

$$\frac{(x + 1)^2}{4} - \frac{(y + 3)^2}{1} = 1$$

Use this standard form and refer to Theorem 9.7 to complete the example. The graph of this equation is a hyperbola whose center is $(-1, -3)$ and with lines of symmetry $x = -1$ and $y = -3$. Because $a = 2$ and $b = 1$, $c^2 = 5$. The vertices are $(-3, -3)$ and $(1, -3)$, and the foci are $(-1 - \sqrt{5}, -3) = (-3.24, -3)$ and $(-1 + \sqrt{5}, -3) = (1.24, -3)$. The asymptotes are $y + 3 = \pm\frac{1}{2}(x + 1)$ (see Fig. 9.56).

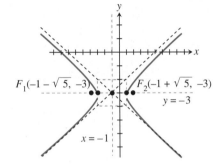

Figure 9.56   The asymptotes of $x^2 - 4y^2 + 2x - 24y = 39$ are $y + 3 = \pm\frac{1}{2}(x + 1)$.

E X A M P L E   12   Using a Graphing Utility

Use the quadratic formula and a graphing utility to find a complete graph of $x^2 - 4y^2 + 2x - 24y = 39$.

Solution   Solve this equation for y using the quadratic formula.

$$x^2 - 4y^2 + 2x - 24y = 39$$

$$-4y^2 - 24y + (x^2 + 2x - 39) = 0$$

$$4y^2 + 24y - (x^2 + 2x - 39) = 0$$

$$y = \frac{-24 \pm \sqrt{576 + 16(x^2 + 2x - 39)}}{8}$$

The graph of

$$y = \frac{-24 + \sqrt{576 + 16(x^2 + 2x - 39)}}{8}$$

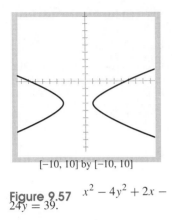

Figure 9.57 $x^2 - 4y^2 + 2x - 24y = 39$.

[−10, 10] by [−10, 10]

produces the portion of the hyperbola in Fig. 9.56 that is above the line $y = -3$, and the graph of

$$y = \frac{-24 - \sqrt{576 + 16(x^2 + 2x - 39)}}{8}$$

produces the portion below the line $y = -3$. A complete graph is shown in Fig. 9.57.

Reflective Properties of Hyperbolas

Hyperbolas have a reflective property that is important in the construction of lenses for cameras, glasses, and telescopes. A simple reflecting lens is created by coating one branch of a hyperbola with a substance that causes it to reflect light. A light ray aimed toward the focus F_1 behind that surface is reflected by the surface to the second focus F_2 (see Fig. 9.58). A telescope is constructed by using both parabolic and hyperbolic lenses (see Fig. 9.59). The main lens is parabolic with focus F_1 and vertex F_2, and the secondary lens is hyperbolic with foci F_1 and F_2. Thus one focus of the hyperbola is also the focus of the parabola. The eye is positioned at the point F_2.

This application will be referred to in Exercises 73 to 78.

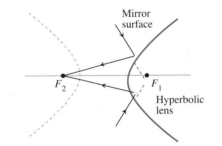

Figure 9.58 How a hyperbolic mirror works.

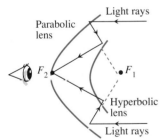

Figure 9.59 The reflective properties of the parabola and hyperbola combine in a telescope's lens system.

Graphing Equations in Quadratic Form

Next, we use a function graphing utility to determine graphs of equations of the form

$$Ax^2 + Bxy + Cy^2 + Dx + Ey + F = 0, \tag{9}$$

where A, B, and C are not all zero. This is a general second-degree equation. It is also called a **quadratic form in variables x and y**. A complete graph of Eq. (9) either is one of the three ordinary conic sections (parabola, ellipse, or hyperbola) or is a **degenerate conic**. Degenerate forms, created when a plane passes through the vertex of a double-napped cone, include a pair of intersecting lines, one line, or a single point. Other degenerate forms include a pair of parallel lines and no graph.

If $C \neq 0$, we can rewrite Eq. (9) as a quadratic in y^2 and using the quadratic formula results in the following two functions:

$$f(x) = \frac{-(Bx + E) + \sqrt{(Bx + E)^2 - 4C(Ax^2 + Dx + F)}}{2C} \qquad \textbf{(10a)}$$

$$g(x) = \frac{-(Bx + E) - \sqrt{(Bx + E)^2 - 4C(Ax^2 + Dx + F)}}{2C} \qquad \textbf{(10b)}$$

By graphing these two functions, we obtain a complete graph of the conic of Eq. (9).

E X A M P L E   13   Finding a Complete Graph

Use the quadratic formula and a function graphing utility to obtain a complete graph of the ellipse $5x^2 + 3y^2 - 20x + 6y = -8$.

Solution   First, rewrite the equation.

$$5x^2 + 3y^2 - 20x + 6y = -8$$

$$3y^2 + 6y + (5x^2 - 20x + 8) = 0$$

Next, use the quadratic formula to solve for y.

$$y = \frac{-6 \pm \sqrt{36 - 12(5x^2 - 20x + 8)}}{6}$$

It is *not* necessary to simplify this equation before using a graphing utility. The graph of

$$y = \frac{-6 + \sqrt{36 - 12(5x^2 - 20x + 8)}}{6}$$

produces the upper half of the complete graph of the ellipse in Fig. 9.60, and the graph of

$$y = \frac{-6 - \sqrt{36 - 12(5x^2 - 20x + 8)}}{6}$$

produces the lower half.   ≡

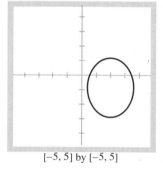

[−5, 5] by [−5, 5]

Figure 9.60 $5x^2 + 3y^2 - 20x + 6y = -8$.

Notes on Examples

Examples 14–16 concern conics given by the general equation on page 517, with the *rotational* term Bxy not necessarily zero. The intent here is for students to consider conics that have an axis of symmetry not parallel to a coordinate axis. The examples serve to integrate all of the ideas presented in this section.

E X A M P L E   14   Graphing a Conic in Quadratic Form

Use a graphing utility to determine a complete graph of the equation $x^2 + 4xy + 4y^2 - 30x - 90y + 450 = 0$.

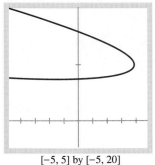

[−5, 5] by [−5, 20]

Figure 9.61 $x^2 + 4xy + 4y^2 - 30x - 90y + 450 = 0$.

Solution Notice that in this equation, $B = 4$. The functions of Eq. (10a,b) become

$$y = \frac{-(4x - 90) \pm \sqrt{(4x - 90)^2 - 16(x^2 - 30x + 450)}}{8}.$$

To make it easier to enter this expression into a graphing utility, simplify it to obtain the functions:

$$f(x) = \frac{45 - 2x + \sqrt{225 - 60x}}{4} \quad \text{and} \quad g(x) = \frac{45 - 2x - \sqrt{225 - 60x}}{4}.$$

Figure 9.61 shows a complete graph of each of these functions in the same viewing rectangle. The result is a parabola. ≡

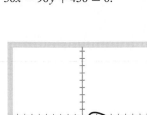

[−10, 10] by [−10, 10]

Figure 9.62 $3x^2 + 4xy + 3y^2 - 12x + 2y + 7 = 0$.

E X A M P L E 15 Graphing a Conic in Quadratic Form

Find a complete graph of $3x^2 + 4xy + 3y^2 - 12x + 2y + 7 = 0$.

Solution Equations (10a,b) become

$$f(x) = \frac{-(4x + 2) + \sqrt{(4x + 2)^2 - 12(3x^2 - 12x + 7)}}{6}$$

$$g(x) = \frac{-(4x + 2) - \sqrt{(4x + 2)^2 - 12(3x^2 - 12x + 7)}}{6}.$$

Figure 9.62 shows a complete graph of these functions. The result is an ellipse. ≡

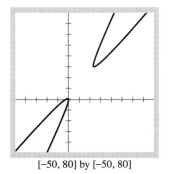

[−50, 80] by [−50, 80]

Figure 9.63
$4x^2 - 6xy + 2y^2 - 3x + 10y - 6 = 0$.

E X A M P L E 16 Graphing a Conic in Quadratic Form

Find a complete graph of $4x^2 - 6xy + 2y^2 - 3x + 10y - 6 = 0$.

Solution Equation (10a,b) becomes

$$f(x) = \frac{(6x - 10) + \sqrt{(-6x + 10)^2 - 8(4x^2 - 3x - 6)}}{4}$$

$$g(x) = \frac{(6x - 10) - \sqrt{(-6x + 10)^2 - 8(4x^2 - 3x - 6)}}{4}.$$

Figure 9.63 shows a complete graph of these functions. The result is a hyperbola. ≡

Teaching Note

Theorem 9.8 shows how the value of $B^2 - 4AC$ can be used to determine whether a parabola, ellipse, or hyperbola is obtained for a conic in general form.

Assignment Guide

Day 1: Ex. 3, 6, 8, 9, 12, 15, 16, 19, 21, 24, 25, 28, 29, 32, 33, 36, 38
Day 2: Ex. 39, 42, 43, 45, 48, 50, 51, 54, 55, 58, 61, 63, 66
Day 3: Ex. 67, 70, 72, 73–75, 79, 82, 85, 88
Day 4: Ex. 89, 91, 94, 97, 99, 100–102

Examples 14 through 16 have shown that a graph of a quadratic form can be any one of the conics. Theorem 9.8 formalizes this observation.

Theorem 9.8   Graphs of a Quadratic Form

The graph of the equation $Ax^2 + Bxy + Cy^2 + Dx + Ey + F = 0$ is (possibly degenerate)

1. a parabola if $B^2 - 4AC = 0$,
2. an ellipse if $B^2 - 4AC < 0$, or
3. a hyperbola if $B^2 - 4AC > 0$.

Exercises for Section 9.4

In Equations 1–8, find the vertex, line of symmetry, focus, and directrix of each parabola.

1. $24y = x^2$

2. $8x = y^2$

3. $12(y + 1) = (x - 3)^2$

4. $6(y - 3) = (x + 1)^2$

5. $(2 - y) = 16(x - 3)^2$

6. $22(x - 3) = (y + 5)^2$

7. $(x + 4)^2 = -6(y - 1)$

8. $(y - 7)^2 = 12x$

In Equations 9–12, list a sequence of transformations that when applied to the graph of $y^2 = x$ or $x^2 = y$ will result in the given parabola.

9. $12(x + 3) = (y + 4)^2$

10. $y = 18(x - 3)^2 + 7$

11. $(y - 2) = 6(x + 4)^2$

12. $x - 3 = 5(y + 6)^2 + 1$

In Exercises 13–15, find an equation, the vertex, and the line of symmetry of the parabola, determined by the given focus and directrix. Also sketch a complete graph.

13. $(2, -1)$, $y = 3$

14. $(2, -1)$, $y = -3$

15. $(2, -1)$, $x = 3$

In Exercises 16–20, find an equation, the focus, and the line of symmetry of the parabola, determined by the given vertex and directrix. Also sketch a complete graph.

16. $(2, -1)$, $x = -3$

17. $(3, 2)$, $x = 5$

18. $(3, 2)$, $x = -5$

19. $(3, 2)$, $y = 5$

20. $(3, 2)$, $y = -5$

In Exercises 21 and 22, find the standard form, vertex, focus, directrix, and line of symmetry of each parabola and sketch a complete graph without using a graphing utility. Then support your sketch with a graphing utility.

21. $x^2 + 2x - y + 3 = 0$

22. $3x^2 - 6x - 6y + 10 = 0$

23. Consider the graph of $y = 4x^2$. Explain how a horizontal shrink by a factor of $\frac{1}{2}$ applied to the graph of $y = x^2$ produces the same graph as applying a vertical stretch by a factor of 4 to $y = x^2$. [*Hint:* What happens to the point $(1, 1)$ on $y = x^2$?]

24. Consider the graph of $x = 4y^2$. Explain how a vertical shrink by a factor of $\frac{1}{2}$ applied to $x = y^2$ produces the same graph as applying a horizontal stretch by a factor of 4 to the graph of $x = y^2$.

In Exercises 25–28, for each ellipse give the endpoints of the major and minor axes, the coordinates of the center and foci, and the lines of symmetry. Sketch a complete graph without using a graphing utility.

25. $\dfrac{(x - 1)^2}{2} + \dfrac{(y + 3)^2}{4} = 1$

26. $(x + 3)^2 + 4(y - 1)^2 = 16$

27. $\dfrac{(x+2)^2}{5} + 2(y-1)^2 = 1$

28. $\dfrac{(x-4)^2}{16} + 16(y+4)^2 = 8$

In Exercises 29–32, draw a complete graph using only a graphing utility.

29. $2x^2 - 4x + y^2 - 6 = 0$

30. $3x^2 - 6x + 2y^2 + 8y + 5 = 0$

31. $2x^2 - y^2 + 4x + 6 = 0$

32. $3y^2 - 5x^2 + 2x - 6y - 9 = 0$

In Exercises 33–38, write the equation of the conic in standard form and give the endpoints of the major and minor axes, the coordinates of the center and foci, the lines of symmetry, and the asymptotes (if any). Draw a complete graph and support your sketch using a graphing utility.

33. $9x^2 + 4y^2 - 18x + 8y - 23 = 0$

34. $3x^2 + 5y^2 - 12x + 30y + 42 = 0$

35. $9x^2 - 4y^2 - 36x + 8y - 4 = 0$

36. $y^2 - 4y - 8x + 20 = 0$

37. $25y^2 - 9x^2 - 50y - 54x - 281 = 0$

38. $9x^2 + 16y^2 + 54x - 32y - 47 = 0$

In Exercises 39 and 40, determine the endpoints of the major and minor axes and the standard form of the equation of the ellipse. Also sketch a complete graph. Do not use a graphing utility.

39. Foci $(3, 0)$ and $(-3, 0)$; minor axis = 14

40. Foci $(0, 4)$ and $(0, -4)$; minor axis = 9

In Exercises 41–44, determine the endpoints of the major and minor axes and the standard form of the equation of the ellipse. Also sketch a complete graph. Do not use a graphing utility.

41. The sum of the distances from the foci is 9; the foci are $(-4, 0)$ and $(4, 0)$.

42. The sum of the distances from the foci is 11; the foci are $(-0.5, 0)$ and $(0.5, 0)$.

43. The foci are $(1, 3)$ and $(1, 9)$; the major axis has a length of 12.

44. The center is $(1, -4)$, the foci are $(1, -2)$ and $(1, -6)$; the minor axis has a length of 10.

In Exercises 45–48, determine the standard form of the equation and sketch a complete graph of the hyperbola. Do not use a graphing utility.

45. The difference between the distances from any point on the hyperbola to the foci is 4; the foci are $(-3, 0)$ and $(3, 0)$.

46. The difference between the distances from any point on the hyperbola to the foci is 8; the foci are $(0, -8)$ and $(0, 8)$.

47. The difference between the distances from any point on the hyperbola to the foci is 3; the foci are $(-3, 2)$ and $(3, 2)$.

48. The difference between the distances from any point on the hyperbola to the foci is 6; the foci are $(1, -2)$ and $(9, -2)$.

In Exercises 49–52, determine the coordinates of the center, the foci, and the vertices of each hyperbola, find the lines of symmetry and the asymptotes, and sketch a complete graph. Do not use a graphing utility.

49. $\dfrac{x^2}{4} - \dfrac{(y-3)^2}{5} = 1$

50. $\dfrac{(y-3)^2}{9} - \dfrac{(x+2)^2}{4} = 1$

51. $4(y-1)^2 - 9(x-3)^2 = 36$

52. $4(x-2)^2 - 9(y+4)^2 = 1$

In Exercises 53 and 54, find the points of intersection (if any) between the line and the parabola.

53. $y = 2x^2 - 6x + 7$, $2x + 3y - 6 = 0$

54. $x = 3y^2 - 2y + 6$, $x - 4y - 10 = 0$

In Exercises 55–62, determine whether the graph of each of the equations is a parabola, ellipse, or hyperbola. Find the vertices, lines of symmetry, and foci of each.

55. $y^2 - 2y + 4x - 12 = 0$

56. $4y^2 - 9x^2 - 18x - 8y - 41 = 0$

57. $4x^2 + y^2 - 32x + 16y + 124 = 0$

58. $9y^2 - 9x - 6y - 5 = 0$

59. $16x^2 - y^2 - 32x - 6y - 57 = 0$

60. $2x^2 + 3y^2 + 12x - 24y + 60 = 0$

61. $2x^2 - 6x + 5y - 13 = 0$

62. $9x^2 + 4y^2 - 18x + 16y - 11 = 0$

In Exercises 63 to 66, the graph of each system of parametric equations is a conic. Substitute specific values for a, b, h, and k and find a complete graph of the conic.

63. $\begin{cases} x = t + h \\ y = \dfrac{1}{4a}t^2 + k \end{cases}$ **64.** $\begin{cases} x = a\cos t + h \\ y = b\sin t + k \end{cases}$

65. $\begin{cases} x = \dfrac{a}{\cos t} + h \\ y = b\tan t + k \end{cases}$ **66.** $\begin{cases} x = \dfrac{1}{4a}t^2 + h \\ y = t + k \end{cases}$

67. A parabolic microphone is formed by revolving the portion of the parabola $10y = x^2$ between the lines $x = -7$ and $x = 7$ about its line of symmetry. Where should the sound receiver be placed for best reception?

68. A parabolic microphone is formed by revolving the portion of the parabola $18y = x^2$ between $x = -12$ and $x = 12$ about its line of symmetry. Where should the sound receiver be placed for best reception?

69. A parabolic headlight is formed by revolving the portion of the parabola $y^2 = 12x$ between the lines $y = -4$ and $y = 4$ about its line of symmetry. Where should the headlight bulb be placed for maximum illumination?

70. A parabolic headlight is formed by revolving the portion of the parabola $y^2 = 15x$ between the lines $y = -3$ and $y = 3$ about its line of symmetry. Where should the headlight bulb be placed for maximum illumination?

71. The form for a lithotripter derives from rotating the portion of an ellipse below its minor axis about its major axis. The *major diameter* (length of the major axis) is 26 in. and the maximum depth from the major axis is 10 in. Where should the shock-wave source and the patient be placed for maximum effect? Give the appropriate measurements.

72. There are elliptical pool tables that have been constructed with a single pocket at one of the foci. Suppose such a table has a major diameter of 6 ft and a minor diameter of 4 ft.

  a) Explain how a "pool shark" who knows conic geometry is at a great advantage over a "mark" who knows no conic geometry.

  b) How should the ball be hit so it bounces off the cushion directly into the pocket? Give specific measurements.

Exercises 73–75 refer to the parabolic-hyperbolic-lens **problem situation** represented here. Figure 9.59 in the text explains the arrangement of lenses.

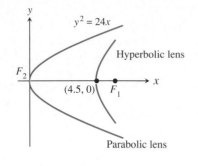

For Exercises 73 to 75.

73. What are the coordinates of the foci of the hyperbola?

74. Determine the standard form of the equation of the generating hyperbola.

75. Find complete graphs of both conics in the same viewing rectangle. Explain how this lens arrangement works.

Exercises 76–78 refer to the parabolic-hyperbolic-lens **problem situation** shown here. (See also Fig. 9.59.)

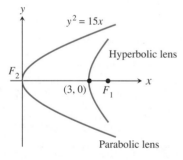

For Exercises 76 to 78.

76. What are the coordinates of the foci of the hyperbola?

77. Determine the standard form of the equation of the generating hyperbola.

78. Find complete graphs of both conics in the same viewing rectangle. Explain how this lens arrangement works.

In Exercises 79–88, find a complete graph and identify the conic section.

79. $2x^2 - xy + 3y^2 - 3x + 4y - 6 = 0$

80. $-x^2 + 3xy + 4y^2 - 5x - 10y - 20 = 0$

81. $2x^2 - 4xy + 8y^2 - 10x + 4y - 13 = 0$

82. $2x^2 - 4xy + 2y^2 - 5x + 6y - 15 = 0$

83. $10x^2 - 8xy + 6y^2 + x - 5y + 20 = 0$

84. $10x^2 - 8xy + 6y^2 + x - 5y - 30 = 0$

85. $3x^2 - 2xy - 5x + 6y - 10 = 0$

86. $5xy - 6y^2 + 10x - 17y + 20 = 0$

87. $-3x^2 + 7xy - 2y^2 - x + 20y - 15 = 0$

88. $-3x^2 + 7xy - 2y^2 - 2x + 3y - 10 = 0$

89. Determine an equation for the ellipse with foci $(-1, 1)$ and $(1, 2)$ such that 3 is the sum of the distances of its points from the foci.

90. Determine an equation for the hyperbola with foci $(-1, 1)$ and $(1, 2)$ such that 1 is the absolute value of the difference between the distances between its points from the foci.

In Exercises 91–97, graph each degenerate conic.

91. $x^2 = -1$

92. $x^2 = 0$

93. $x^2 = 1$

94. $x^2 - xy = 0$

95. $x^2 + y^2 = 0$

96. $x^2 - y^2 = 0$

97. $x^2 + y^2 + 1 = 0$

Exercises 98 and 99 are locus problems.

98. Find the locus of the point $P(x, y)$ if the sum of the squares of its distances from the two points $(-3, 2)$ and $(4, 1)$ is always 75.

99. Find the locus of the point $P(x, y)$ that is equidistant from the points $(-1, 1)$ and the origin.

A set of lines with a common property is called a **family of curves**. Exercises 100 and 101 ask you to identify a particular line in a family of lines.

100. A line that intersects a parabola in exactly one point is called a **tangent line of the parabola**. Which one line in the family of lines $y = 2x + k$, where k is a real number, is a tangent line to the parabola $y = x^2$.

101. Find two lines in the family $y = kx - 1$ that are tangent lines to the parabola $y = x^2$.

102. A line is a **normal line** to a conic at a point $P(x_0, y_0)$ if it passes through point P and is perpendicular to the tangent line at P. Show that the line $x + 4y - 18 = 0$ is a normal line to $y = x^2$ at the point $P(2, 4)$.

9.5 _____ Polar Equations of Conics

In Section 9.4, we defined a parabola as the set of all points P such that the perpendicular distance from P to a fixed line (directrix) equals the distance from P to a fixed point (focus) not on the directrix. Ellipses and hyperbolas also can be described in terms of a focus and a directrix when used with the concept of *eccentricity*—a concept also used in the polar form of the conics.

Eccentricity

Definition 9.6   Eccentricity

The **eccentricity** e of an ellipse or a hyperbola is defined by $e = c/a$, where $2c$ is the distance between foci F_1 and F_2 and $2a$ is the length of the major axis of an ellipse or the principal axis of a hyperbola. The eccentricity of a parabola is defined to be 1; that of a circle is 0.

Teaching Note

This is a difficult section for most students. Do not try to go through the examples too rapidly or you will lose many students. Connecting the concepts of conic sections and polar coordinates will require some time for students to understand.

Notes on Examples

Work through all of the examples in this section with students, allowing time for exploration. Consider having some students present a few of the examples in class.

Recall that for an ellipse, $c = \sqrt{a^2 - b^2}$, so $0 < c < a$ and $e = \dfrac{c}{a} < 1$. For a hyperbola, $c = \sqrt{a^2 + b^2}$, so $c > a > 0$ and $e = c/a > 1$.

Once the values for two of the constants $a, b,$ and c of a conic are known, the eccentricity of that conic can be found.

E X A M P L E   1   Finding the Eccentricity

Find the eccentricity of each conic.

a) $\dfrac{x^2}{16} + \dfrac{y^2}{9} = 1$

b) $\dfrac{y^2}{16} - \dfrac{x^2}{9} = 1$

c) $4x^2 + 9y^2 - 24x + 36y + 36 = 0$

Solution

a) In this case, $a > b$, so $a = 4$ and $b = 3$. Therefore, because $c = \sqrt{a^2 - b^2} = \sqrt{7}$, $e = c/a = \sqrt{7}/4$.

b) Here $a = 4$, $b = 3$, and $c = \sqrt{a^2 + b^2} = 5$; therefore $e = c/a = \frac{5}{4}$.

c) First complete the square for both variables x and y to find the standard form.

$$4x^2 + 9y^2 - 24x + 36y + 36 = 0$$

$$4x^2 - 24x + 9y^2 + 36y = -36$$

$$4(x^2 - 6x + 9) + 9(y^2 + 4y + 4) = 36$$

$$4(x - 3)^2 + 9(y + 2)^2 = 36$$

$$\frac{(x - 3)^2}{9} + \frac{(y + 2)^2}{4} = 1$$

From the standard form, it is evident that $a = 3, b = 2,$ and $c = \sqrt{a^2 - b^2} = \sqrt{5}$. Thus $e = c/a = \sqrt{5}/3$.   ▤

It turns out that if P is a point on either an ellipse or a hyperbola, the ratio $d(P, F)/d(P, Q)$ is equal to the eccentricity, where F is the closest focal point and $d(P, Q)$ is the perpendicular distance to the directrix. This fact is the subject of the next two theorems.

These theorems assume that the major axis (for an ellipse) or principal axis (for a hyperbola) is the x-axis. There are corresponding statements for cases where the major axis or the principal axis lies on the y-axis.

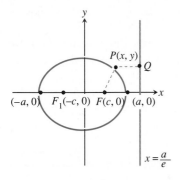

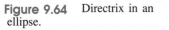

Figure 9.64   Directrix in an ellipse.

Theorem 9.9   An Ellipse and Its Eccentricity

Suppose $F(c, 0)$ (where $c > 0$) is a focus and $e = c/a$ is the eccentricity of the ellipse

$$\frac{x^2}{a^2} + \frac{y^2}{b^2} = 1.$$

Let ℓ be the directrix defined as $x = a/e$ (see Fig. 9.64). If P is any point of the ellipse and Q is a point on ℓ such that PQ is perpendicular to ℓ, then

$$\frac{d(P, F)}{d(P, Q)} = e.$$

Proof   Compute $d(P, F)$ where point $P(x, y)$ is on the ellipse.

$$d(P, F) = \sqrt{(x - c)^2 + y^2}$$

$$= \sqrt{(x - c)^2 + b^2 - \frac{b^2 x^2}{a^2}} \qquad \text{\small Solve the given equation for } y^2 \text{\small and substitute into the radical.}$$

$$= \sqrt{x^2 - 2cx + c^2 + b^2 - \frac{b^2 x^2}{a^2}}$$

$$= \sqrt{x^2 - 2cx + a^2 - b^2 + b^2 - \frac{b^2 x^2}{a^2}}$$

$$= \sqrt{x^2 \left(1 - \frac{b^2}{a^2}\right) - 2cx + a^2}$$

$$= \sqrt{x^2 \left(\frac{a^2 - b^2}{a^2}\right) - 2cx + a^2}$$

$$= \sqrt{\frac{c^2}{a^2} x^2 - 2cx + a^2}$$

$$= \sqrt{\frac{c^2}{a^2} \left(x^2 - \frac{2a^2}{c} x + \frac{a^4}{c^2}\right)}$$

$$= \sqrt{\frac{c^2}{a^2} \left(x - \frac{a^2}{c}\right)^2}$$

$$= \sqrt{\frac{c^2}{a^2} \left(x - \frac{a}{e}\right)^2}$$

$$= \frac{c}{a} \left|x - \frac{a}{e}\right|$$

$$= e\left|x - \frac{a}{e}\right|$$

$$= e \cdot d(P, Q) \qquad\qquad \blacksquare$$

Theorem 9.9 means that the vertical line $x = a/e$ is the directrix and the point $(c, 0)$ is the focus for an ellipse. In a similar way, it can be shown that $x = -a/e$ can be taken as a directrix with respect to the focus $(-c, 0)$.

Theorem 9.10 establishes that a hyperbola can also be defined in terms of a focus and a directrix. Notice that for a hyperbola, the directrix lies between the two foci.

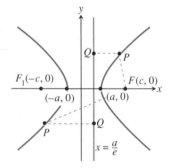

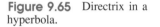

Figure 9.65   Directrix in a hyperbola.

Theorem 9.10   A Hyperbola and Its Eccentricity

Suppose $F(c, 0)$ (where $c > 0$) is a focus and $e = c/a$ is the eccentricity of the hyperbola

$$\frac{x^2}{a^2} - \frac{y^2}{b^2} = 1.$$

Let ℓ be the directrix defined as $x = a/e$ (see Fig. 9.65). If P is any point of the hyperbola and Q is a point on ℓ such that PQ is perpendicular to ℓ, then

$$\frac{d(P, F)}{d(P, Q)} = e.$$

The conclusions of Theorems 9.9 and 9.10 are true also in the event that the centers of the conics are at point (h, k) instead of the origin $(0, 0)$.

We now can give the following alternative definition of a conic.

Definition 9.7   Conics in Terms of Focus and Directrix

Let F be a fixed point and ℓ be a line *not* containing F. A **conic** is the locus of all points P with $d(P, F)/d(P, Q) = e$, where e is a positive constant called the eccentricity and Q is the point on ℓ such that PQ is perpendicular to ℓ. F is a **focus** and ℓ is the **directrix of the conic**.

The conic in Definition 9.7 is an ellipse if $e < 1$, a parabola if $e = 1$, and a hyperbola if $e > 1$.

Polar Equations of Conics

In Section 9.4, the standard forms in a rectangular coordinate system were developed for equations of conic sections. In the case of the ellipse and the hyperbola, these standard forms are described in terms of the center of the conic. The simplest situation is when the center is the origin.

This section develops equations for the conics in a polar coordinate system. For a polar coordinate system, it is most convenient to develop the equations assuming that a focus rather than the center of the conic is at the origin.

Consider a conic whose focus is the origin $(0, 0)$ and whose directrix is the line $x = -h$ (where $h > 0$). There are two cases to consider: whether a point P on the conic is on the same or opposite side of the directrix as the focus. Figure 9.66 shows the case where P is on the same side.

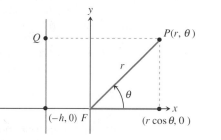

Figure 9.66 If $P(r, \theta)$ is a point on the conic and P and the focus F are on the same side of the directrix, then $d(P, F) = e \cdot d(P, Q)$. (*Note:* Conic not shown.)

Using the polar coordinates to describe these distances, we see that

$$d(P, F) = e \cdot d(P, Q)$$

$$r = e(h + r \cos \theta)$$

$$r - er \cos \theta = eh$$

$$r(1 - e \cos \theta) = eh$$

$$r = \frac{eh}{1 - e \cos \theta}.$$

Any point on the conic on the same side of the directrix as the focus has polar coordinates that satisfy Eq. (1).

On the other hand, suppose P is on the opposite side of the directrix from the focus (see Fig. 9.67). If (r, θ) is a polar coordinate of point P and $r < 0$, then $-r = d(P, F)$ and $d(P, Q) = -h - r \cos \theta$.

$$-r = e(-h - r \cos \theta)$$

$$r = eh + er \cos \theta$$

$$r = \frac{eh}{1 - e \cos \theta} \qquad \textbf{(1)}$$

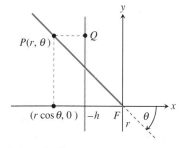

Figure 9.67 If P is a point on the conic and P and F are on opposite sides of the directrix, then $d(P, F) = e \cdot d(P, Q)$.

So if P is a point on a conic, then regardless of where the focus, directrix, and P are in relation to each other, the polar coordinates of P will satisfy an equation in the form of Eq. (1). It is important to observe that our development of Eq. (1) did not make any assertions about the eccentricity. If $0 < e < 1$, Eq. (1) is an ellipse, and if $1 < e$, Eq. (1) is a hyperbola.

A graphing utility can be used to find complete graphs of conics in polar form.

E X A M P L E 2   Finding a Complete Graph

Find complete graphs of

$$r = \frac{2e}{1 - e \cos \theta}$$

for e equal to 0.25, 0.5, 1.5, and 4.

Solution For each of these equations, the directrix is $x = -2$ and a focus is at the origin. See Fig. 9.68 for complete graphs. ≡

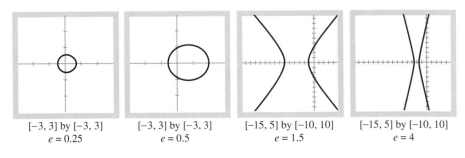

| $[-3, 3]$ by $[-3, 3]$ | $[-3, 3]$ by $[-3, 3]$ | $[-15, 5]$ by $[-10, 10]$ | $[-15, 5]$ by $[-10, 10]$ |
| $e = 0.25$ | $e = 0.5$ | $e = 1.5$ | $e = 4$ |

Figure 9.68 $r = 2e/(1 - e \cos \theta)$ for four values of e.

From Example 2, we see that when e is close to 0, the ellipse is close to being a circle and that as $e > 1$ gets larger, the hyperbola becomes more and more like a pair of parallel lines.

Notice that Eq. (1) applies when the directrix is on the negative side of the y-axis. Using a development that parallels the one above, it can be shown that when the directrix is on the positive side of the y-axis, the polar equation is

$$r = \frac{eh}{1 + e \cos \theta}. \tag{2}$$

If $\cos \theta$ is replaced in Eqs. (1) and (2) by $\sin \theta$, a conic is again obtained.

Theorem 9.11 summarizes the equations for the polar coordinates of conics.

Theorem 9.11   Polar Equations of Conics

A nondegenerate conic with eccentricity e has one of the following polar equations:

1. $r = \dfrac{eh}{1 + e \cos \theta}$; directrix $x = +h$

2. $r = \dfrac{eh}{1 - e \cos \theta}$; directrix $x = -h$

3. $r = \dfrac{eh}{1 + e \sin \theta}$; directrix $y = +h$

4. $r = \dfrac{eh}{1 - e \sin \theta}$; directrix $y = -h$

Moreover, the graph of any such equation is a conic. The conic is an ellipse if, and only if, $0 < e < 1$, a parabola if, and only if, $e = 1$; and a hyperbola if, and only if, $1 < e$.

The next three examples explore conics in polar form.

E X A M P L E   3   Examining a Conic in Polar Form

Find a complete graph of

$$r = \frac{6}{4 - 3 \cos \theta}.$$

Specify a directrix and a range for θ that produces a complete graph. Find the standard form for the equation of the conic.

Solution   Divide the numerator and denominator of this equation by 4 to put it into the form given in Theorem 9.11.

$$r = \frac{6}{4 - 3 \cos \theta}$$

$$r = \frac{1.5}{1 - 0.75 \cos \theta}$$

Then $eh = 1.5$ and $e = 0.75$, so $h = 2$. The conic is an ellipse whose directrix is $x = -2$. The range $0 \le \theta \le 2\pi$ produces a complete graph (see Fig. 9.69).

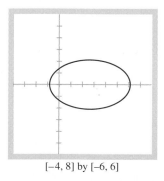

[−4, 8] by [−6, 6]

Figure 9.69 $r = \dfrac{6}{4 - 3 \cos \theta}$ and $0 \le \theta \le 2\pi$.

When $\theta = 0$, $r = 6$, and when $\theta = \pi$, $r = 0.86$. The major axis vertices are thus $(-0.86, 0)$ and $(6, 0)$, and the center of the ellipse is $((-0.86 + 6)/2, 0)$, or $(2.57, 0)$.

Because c is the distance from the center to a focus and one focus is $(0, 0)$, $c = 2.57$. The length of the major axis is $2a = 6 - (-0.86) = 6.86$, so $a = 3.43$.

$$b^2 = a^2 - c^2$$

$$b = \sqrt{(3.43)^2 - (2.57)^2} = 2.27$$

Therefore the standard form for the equation of this conic is

$$\frac{(x - 2.57)^2}{(3.43)^2} + \frac{y^2}{(2.27)^2} = 1.$$

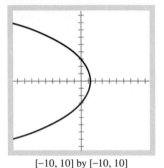

[−10, 10] by [−10, 10]

Figure 9.70 $r = \dfrac{3}{1 + \cos\theta}$ and $0 \le \theta \le 2\pi$.

EXAMPLE 4   Examining a Conic in Polar Form

Find a complete graph of

$$r = \frac{3}{1 + \cos\theta}.$$

Specify a directrix and a range for θ that produces a complete graph.

Solution   Notice from the form of the equation that $e = 1$ and $h = 3$. This conic is a parabola with directrix $x = 3$. A complete graph is obtained by letting $0 \le \theta \le 2\pi$ (see Fig. 9.70).

EXAMPLE 5   Examining a Conic in Polar Form

Find a complete graph of

$$r = \frac{4}{3 - 4\sin\theta}.$$

Specify a directrix and a range for θ that produces a complete graph.

Solution   Dividing numerator and denominator by 3 yields

$$r = \frac{\frac{4}{3}}{1 - \frac{4}{3}\sin\theta}.$$

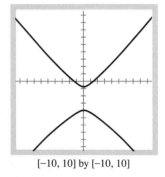

[−10, 10] by [−10, 10]

Figure 9.71 $r = \dfrac{4}{3 - 4\sin\theta}$ and $0 \le \theta \le 2\pi$.

Thus $e = \frac{4}{3}$, so this conic is a hyperbola. Also, because $h = (4/3)\left(\dfrac{1}{e}\right) = 1$, $y = -1$ is a directrix. A complete graph is obtained by allowing θ to vary from 0 to 2π (see Fig. 9.71).

Exercises for Section 9.5

In Exercises 1–6, determine the eccentricity of the conic and sketch a complete graph. Support your sketch using a graphing utility.

1. $4x^2 + y^2 = 4$

2. $4x^2 + 25y^2 = 100$

3. $9x^2 - 16y^2 = 144$

4. $4x^2 - y^2 = 4$

5. $2x^2 - 3y^2 - 4x + 6y - 10 = 0$

6. $-6x^2 + 8y^2 - 6x + 9y + 10 = 0$

In Exercises 7 and 8, determine the center and equations of the asymptotes.

7. The conic of Exercise 5   **8.** The conic of Exercise 6

In Exercises 9 and 10, estimate the eccentricity of the ellipse. (*Hint:* First estimate a and b. Then compute c and e.) Sketch a complete graph.

9. $2x^2 - 3xy + 6y^2 - 8x + 10y - 5 = 0$

10. $10x^2 + 52xy + 300y^2 - 100x + 250y - 500 = 0$

For Exercises 11–14, find complete graphs of the given conic for $e = 0.8$, 1, and 1.2 in the same viewing rectangle.

11. $r = \dfrac{2e}{1 - e \cos \theta}$

12. $r = \dfrac{2e}{1 + e \cos \theta}$

13. $r = \dfrac{2e}{1 - e \sin \theta}$

14. $r = \dfrac{2e}{1 + e \sin \theta}$

In Exercises 15–20, identify and draw a complete graph of the conic in polar form. Specify a directrix and a range for θ that produces a complete graph.

15. $r = \dfrac{5}{1 - \cos \theta}$

16. $r = \dfrac{4}{2 - 4 \sin \theta}$

17. $r = \dfrac{3}{4 - 2 \cos \theta}$

18. $r = \dfrac{4}{3 + 4 \sin \theta}$

19. $r = \dfrac{-8}{6 + 3 \cos \theta}$

20. $r = \dfrac{-4}{4 - 2 \sin \theta}$

In Exercises 21 and 22, draw a complete graph, identify the conic, and determine the standard form of the conic.

21. $r = \dfrac{16}{12 + 9 \cos \theta}$

22. $r = \dfrac{-4}{2 + 6 \sin \theta}$

23. Consider an ellipse in which the major axis has a length $2a = 8$ and the distance of a focus from the center is

$c = 2$. Compute the eccentricity and find graphs of r_1 and r_2 in the same viewing rectangle. What conclusions can you draw?

$$r_1 = \frac{a - ce}{1 + e \cos \theta} \qquad r_2 = \frac{ce - a}{1 - e \cos \theta}$$

24. Consider an ellipse in which the major axis has a length $2a = 12$ and the distance of a focus from the center is $c = 3.5$. Compute the eccentricity and find graphs of r_1 and r_2 in the same viewing rectangle. What conclusions can you draw?

$$r_1 = \frac{a - ce}{1 + e \cos \theta} \qquad r_2 = \frac{ce - a}{1 - e \cos \theta}$$

25. Consider the graph

$$r = \frac{eh}{-1 - e \cos \theta}$$

for $e = 7$ and $h = 2$. What type of conic is the graph? Why?

26. Show that if an ellipse has major axis length of $2a$ and minor axis length of $2b$, then a polar equation of the ellipse is

$$r = \frac{b^2}{a + \sqrt{a^2 - b^2} \, \cos \theta},$$

where a focus is at the origin and the directrix is $x = h$ ($h > 0$).

In Exercises 27–30, compute the eccentricity and sketch the polar graphs of each ellipse. Assume that $a = 4$ and $b = \sqrt{12}$.

27. $r_1 = \dfrac{b^2}{a + \sqrt{a^2 - b^2} \, \cos \theta}$

28. $r_2 = \dfrac{-b^2}{a + \sqrt{a^2 - b^2} \, \cos \theta}$

29. $r_3 = \dfrac{-b^2}{a - \sqrt{a^2 - b^2} \, \cos \theta}$

30. $r_4 = \dfrac{b^2}{a - \sqrt{a^2 - b^2} \, \cos \theta}$

31. Find the graphs of the conics from Exercises 27–30 in the same viewing rectangle. What conclusions can you draw?

Chapter 9 Review

KEY TERMS (The number following each key term indicates the page of its introduction.)

asymptotes of a hyperbola, 514
cardioid, 489
center of a hyperbola, 513
cone, 504
conic, 504
conic section, 504
curve of quickest descent, 481
cycloid, 481
degenerate conic, 517
directrix of a conic, 526
directrix of a parabola, 505
double-napped right
  circular cone, 504

eccentricity e, 523
eliminating the parameter, 477
ellipse, 504
endpoints, 477
focus of a conic, 526
focus of a parabola, 505
foci (focal points) of
  a hyperbola, 513
foci (focal points) of
  an ellipse, 510
four-petaled rose, 491
graph of a polar equation
  in variables r and θ, 488

hyperbola, 504
line of symmetry of
  a parabola, 505
major axis, 511
minor axis, 511
parabola, 504
parametric equations, 476
parametrization of C, 476
polar coordinates, 482
polar coordinate system, 482
principal axis, 513
quadratic form in variables
  x and y, 517

spiral of Archimedes, 491
standard form for
  an ellipse, 511
standard form for
  a hyperbola, 515
standard form for
  a parabola, 507
transverse axis, 513
vertex of a parabola, 505
vertices of a hyperbola, 514

REVIEW EXERCISES

In Exercises 1–3, describe each parametrically defined curve. Sketch the graph without using a graphing utility. Support your sketch using a parametric graphing utility.

1. $x(t) = 2 + 3t$, $y(t) = t$, t in $[0, 5]$

2. $x(t) = 2t$, $y(t) = 3/t$, t in $[1, 4]$

3. $x(t) = 4 \cos t$, $y(t) = 4 \sin t$, t in $[-2\pi, 2\pi]$

4. Which of the curves in Exercises 1–3 are functions?

5. What are the domains and ranges of the curves in Exercises 1–3?

In Exercises 6–10, find a complete graph.

6. $x(t) = 3t + 2$, $y = 6t^2$, t in $(-\infty, \infty)$

7. $x(t) = 2 - 1/t$ and $y = t + 1/t$, t in $(0, 20]$

8. $x(t) = \sin^2 t$, $y(t) = (1 - \sin t \cos t)$, t in $[0, 2\pi]$

9. $x(t) = t + \sin t$, $y(t) = 2 + \cos t$, t in $[-10, 10]$

10. $x(t) = 12 \cos t - 8 \cos \frac{3}{2}t$, $y(t) = 12 \sin t - 8 \sin \frac{3}{2}t$, t in $[0, 2\pi]$. (This is an *epicycloid*, the path of a point on a circle rolling on another circle.)

11. Find parametric equations and draw the graph of the circle centered at $(5, 0)$ with a radius of 4.

In Exercises 12–17, sketch a complete graph of each equation without using a graphing utility. Then support your answer with a graphing utility.

12. $r = 5$

13. $r = 2 \sin \theta$

14. $r = 3 + 3 \cos \theta$

15. $\theta = 78°$

16. $r = 1 - \sin \theta$

17. $r = \cos 3\theta$

In Exercises 18 and 19, use a graphing utility to draw the polar graph for θ over the given intervals.

18. $r = 4\theta$; $[0, \pi/2], [0, \pi], [0, 2\pi], [-\pi, \pi], [0, 4\pi]$

19. $r = 5 \sin 3\theta$; $[0, \pi/2], [0, \pi], [0, 2\pi], [-\pi, \pi], [0, 4\pi]$

In Exercises 20 and 21, use a graphing utility to draw a complete graph of each polar equation. Specify an interval $a \le \theta \le b$ of smallest length that gives a complete graph.

20. $r = 2 \cos 4\theta$

21. $r = 3\theta$

22. Determine the length of one rose "petal" for $r = 6 \sin 2\theta$.

In Exercises 23 and 24, use the parametric equation method to find the points of intersection of the graph of each system.

23. $\begin{cases} r = 3 \cos \theta \\ r = 3 \sin \theta \end{cases}$

24. $\begin{cases} r = 2 + \sin \theta \\ r = 1 - \cos \theta \end{cases}$

In Exercises 25 and 26, determine an equation for the parabola from the given focus and directrix. Give the coordinates of the vertex and the line of symmetry and draw a complete graph.

25. $(-2, 1)$, $y = -3$

26. $(-2, 1)$, $x = -3$

27. Determine an equation, the line of symmetry, and the focus of the parabola whose vertex is $(1, 4)$ and directrix is $x = 7$. Draw a complete graph.

28. Determine an equation, the line of symmetry, and the directrix of the parabola whose focus is $(0, 2)$ and vertex is $(0, -4)$. Draw a complete graph.

In Exercises 29 and 30, find the standard form, vertex, focus, directrix, and line of symmetry of each parabola. Find a complete graph without using a graphing utility and then support your graph with a graphing utility.

29. $x^2 + 4x - y + 6 = 0$

30. $6y^2 - 12y - 12x + 20 = 0$

In Exercises 31 and 32, give the endpoints of the major and minor axes, the coordinates of the center and foci, and the lines of symmetry of each ellipse. Draw a complete graph of the ellipse without using a graphing utility.

31. $\dfrac{(x + 3)^2}{4} + \dfrac{(y - 1)^2}{6} = 1$

32. $(x - 1)^2 + 8(y + 2)^2 = 16$

In Exercises 33–35, determine the endpoints of the major and minor axes and the standard form of the equation of each ellipse. Draw a complete graph without using a graphing utility.

33. The sum of the distances from the foci is 16; the foci are $(-5, 0)$ and $(5, 0)$.

34. The foci are $(2, 0)$ and $(-2, 0)$; the minor axis has a length of 10.

35. The foci are $(2, 3)$ and $(2, 9)$; the major axis has a length of 12.

In Exercises 36 and 37, write the equation of the ellipse in standard form. Give the endpoints of the major and minor axes, the coordinates of the center and foci, and the lines of symmetry. Draw a complete graph and support your answer using a graphing utility.

36. $4x^2 + 9y^2 + 8x - 36y + 4 = 0$

37. $4x^2 + 5y^2 - 24x + 10y + 21 = 0$

In Exercises 38 and 39, determine the coordinates of the center, foci, vertices, lines of symmetry, and asymptotes of each hyperbola. Draw a complete graph without using a graphing utility.

38. $\dfrac{x^2}{9} - \dfrac{(y + 2)^2}{6} = 1$

39. $3(y - 3)^2 - 5(x + 1)^2 = 15$

In Exercises 40–43, identify the conic and write the equation in standard form. Give the foci, vertices, center, axes of sym-

metry, and asymptotes (if any). Draw a complete graph and then support your answer using a graphing utility.

40. $25x^2 - 9y^2 - 50x - 36y - 236 = 0$

41. $y^2 - x + 6y + 9 = 0$

42. $5x^2 + 4y^2 + 30x - 8y + 29 = 0$

43. $4y^2 - 9x^2 + 24y + 35 = 0$

44. The parabola $y^2 - 4y + 2 = x$ does not define y as a function of x. How can it be graphed using only a function graphing utility?

45. Explain how a conic of the form $Ax^2 + Bxy + Cy^2 + Dx + Ey + F = 0$ can be graphed using a *function* graphing utility, not a special conic graphing utility.

In Exercises 46–49, find a complete graph and identify the conic.

46. $4x^2 + 3xy - y^2 - 10x - 5y - 20 = 0$

47. $2x^2 - 4xy + 2y^2 + 6x - 5y - 15 = 0$

48. $6x^2 - 8xy + 10y^2 - 5x + y - 30 = 0$

49. $-2x^2 + 7xy - 3y^2 + 3x - 2y - 10 = 0$

50. Estimate the coordinates of the center and the equations of the asymptotes of the hyperbola $2x^2 - 3xy + y^2 + 2x - 3y - 2 = 0$.

51. Estimate the coordinates of the center, the endpoints of the major and minor axes, and lines of symmetry of the ellipse $3x^2 - xy + 2y^2 + 6x + 2y - 10 = 0$.

52. Determine a pair of equations that give an end behavior model of the hyperbola $4x^2 + 8xy - 3y^2 + 5x - 10y - 30 = 0$. To support your answer, find graphs of the two equations and of the hyperbola in the same large viewing rectangle.

In Exercises 53 and 54, list a sequence of transformations that when applied to the graph of $y = x^2$, will result in the graph of the given parabola.

53. $6(y - 1) = (x - 3)^2$ **54.** $x^2 - 2x - 6 = 2y$

55. Explain what transformations applied first to a circle in standard form will result in the ellipse $9x^2 + 16y^2 = 144$.

56. What transformations applied first to a conic in standard form will result in the conic $4x^2 + 9y^2 + 16x - 18y - 11 = 0$?

In Exercises 57 and 58, determine the standard form of the equation and draw a complete graph of the hyperbola without using a graphing utility.

57. The absolute value of the difference of the distances from any point on the hyperbola to the foci is 2; the foci are $(-2, 0)$ and $(2, 0)$.

58. The absolute value of the difference of the distances from any point on the hyperbola to the foci is 10; the foci are $(0, -2)$ and $(0, 10)$.

59. Find the points of intersection of the line and the parabola (if any)

$$y = x^2 - 4x + 5 \qquad 2x + y - 7 = 0.$$

60. A parabola of the form $y = ax^2 + b$ passes through the points $(2, 4)$ and $(5, 8)$. Determine an equation of the parabola.

61. An ellipse has a major axis of length 10 along the x-axis and a minor axis of length 4 along the y-axis. Determine the foci and standard form of the equation and draw a complete graph without using a graphing utility.

62. A golfer hits a ball with an initial velocity of 125 ft/sec and at an angle of $33°$ from the horizontal. Find when and where the ball will hit the ground.

63. A batter hits a ball with an initial velocity of 100 ft/sec and at an angle of $50°$ from the horizontal. Find the maximum height attained and the distance traveled by the ball.

Exercises 64–68 refer to the following **problem situation**: A dart is thrown upward with an initial velocity of 60 ft/sec at an angle of elevation $35°$. Consider the position of the dart at any time t ($t = 0$ when the dart is thrown). Neglect air resistance.

64. Find parametric equations that model the problem situation.

65. Draw a complete graph of the model.

66. What portion of the graph represents the problem situation?

67. When will the dart hit the ground?

68. Find the maximum height of the dart. At what time will it reach maximum height?

Exercises 69–72 refer to the following **problem situation**: An NFL punter at the 20-yd line kicks a football downfield with an initial velocity of 80 ft/sec at an angle of elevation of $55°$.

69. Draw a complete graph of the problem situation.

70. How far downfield will the football first hit the ground?

71. Determine the ball's maximum height above the field.

72. What is the "hang time" (the total time the football is in the air)?

SONYA KOVALEVSKY

Although Sonya Kovalevsky (1850–1891) was the daughter of a Russian nobleman, her great achievements in science and mathematics were earned through hard work, determination, and sacrifice. As a nineteenth-century woman, she could not travel freely, could not attend public lectures, and had difficulty finding a job. She taught herself trigonometry and calculus as a young girl. At age 18 she left Russia to study in Germany, where she was tutored by several renowned mathematicians. She received her doctorate in 1874.

Later in her career, Kovalevsky worked with the famous mathematician Karl Weierstrass and received a life professorship at the University of Stockholm in Sweden. She is considered one of the greatest mathematical talents of the nineteenth century for her remarkable contributions in the areas of partial differential equations, Abelian integrals, and infinite series in mathematics, as well as in astronomy.

10

Matrices and Systems of Equations and Inequalities

10.1 _____ Solving Systems of Equations Algebraically

Sometimes a problem situation includes a set of conditions that must be satisfied simultaneously. Each condition results in an equation, and so the algebraic representation of the problem situation includes a set of equations, called a **system of equations.** For example, suppose we want to find the length L and width W of a rectangle with a perimeter of 100 units and an area of 300 square units (see Fig. 10.1). Then the algebraic representation of this problem situation is the following system of equations:

$$2L + 2W = 100$$

$$LW = 300$$

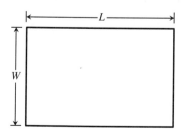

Figure 10.1 A rectangle with width W and length L.

A solution to the problem requires that we find a solution to this system of equations, a technique discussed in Section 3.7.

As another example, recall that in Example 6 of Section 2.2 the Quick Manufacturing Company wanted to find their break-even point. A comparison of revenue and cost led to the system of equations

$$y = 1.5x + 200{,}000$$

$$y = 4x.$$

(1)

535

A solution of the Quick Manufacturing Company system of equations is the number pair (a, b) that satisfies both equations of the system. For instance, Example 6 of Section 2.2 found that a solution to system (1) is the number pair (80,000, 320,000). We verify that claim here.

$$320{,}000 = 1.5(80{,}000) + 200{,}000$$

$$320{,}000 = 4(80{,}000)$$

Up to this point, we have used graphical methods to solve a system of equations; that is, we found a point (a, b), usually through zoom-in, at which the graphs of the equations intersected. The coordinates of that point represented a solution to the system.

In this section, we introduce two new algebraic methods, the **substitution method** and the **elimination method**.

Substitution Method

Consider the system

$$2y + 2x = 100$$

$$xy = 300.$$

To solve this system graphically, first solve each equation for y.

$$y = 50 - x$$

$$y = \frac{300}{x}$$

Then find complete graphs of both equations in the same viewing rectangle (see Fig. 10.2). Use zoom-in to find the points of intersection.

We can solve this system algebraically using a somewhat similar technique. First, solve one equation for y and then substitute that expression for y into the other equation. The resulting equation in the single variable x can be solved. To illustrate this method, Example 1 returns to the rectangle problem that opened this section.

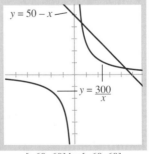

$y = 50 - x$

$y = \frac{300}{x}$

$[-60, 60]$ by $[-60, 60]$

Figure 10.2 Graphs of $y = 50 - x$ and $y = 300/x$.

E X A M P L E   1   Solving a System by Substitution

Solve the system

$$2y + 2x = 100$$

$$xy = 300.$$

Solution Solve the first equation for y to get $y = 50 - x$. Then substitute this value of y into the second equation.

$$xy = 300$$

$$x(50 - x) = 300$$

$$x^2 - 50x + 300 = 0$$

Next, use the quadratic formula to solve for x.

$$x = \frac{50 \pm \sqrt{(-50)^2 - 4(300)}}{2}$$

$$= \frac{50 \pm 10\sqrt{13}}{2}$$

$$= 25 \pm 5\sqrt{13}$$

Substituting these values into the equation $y = 50 - x$ results in the following:

$$y = 50 - (25 + 5\sqrt{13}) = 25 - 5\sqrt{13}$$

$$y = 50 - (25 - 5\sqrt{13}) = 25 + 5\sqrt{13}.$$

Thus the two solutions are

$$(25 + 5\sqrt{13}, 25 - 5\sqrt{13}) \qquad \text{and} \qquad (25 - 5\sqrt{13}, 25 + 5\sqrt{13}). \qquad \blacksquare$$

Frank says, "The graph of an equation in two variables is the set of solutions to the equation represented by points in the plane. Every solution to a system of equations in two variables is an *ordered pair*. Geometrically, an ordered pair is a point in the plane. A simultaneous solution to a system of equations is the point of intersection of the graph of each equation."

E X A M P L E 2   Solving the Rectangle Problem

Find the exact dimensions of a rectangle with a perimeter of 100 units and an area of 300 square units.

Solution The system of equations in Example 1 is the algebraic representation for this problem, and that system has the two solutions:

$$(25 + 5\sqrt{13}, 25 - 5\sqrt{13}) \qquad \text{and} \qquad (25 - 5\sqrt{13}, 25 + 5\sqrt{13}).$$

The first solution represents a rectangle with a length and width of

$$\ell = 25 + 5\sqrt{13} \qquad \text{and} \qquad w = 25 - 5\sqrt{13},$$

and the second solution represents a rectangle with a length and width of

$$\ell = 25 - 5\sqrt{13} \qquad \text{and} \qquad w = 25 + 5\sqrt{13}.$$

But these two rectangles are congruent (see Fig. 10.3); therefore there is a unique (one and only one) solution.

The dimensions of the rectangle are $25 - 5\sqrt{13}$ and $25 + 5\sqrt{13}$. In decimal form the dimensions are 6.97 and 43.03 with an error of at most 0.01. $\blacksquare$

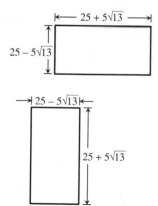

Figure 10.3 Rectangles corresponding to the two solutions of the system in Examples 1 and 2.

Elimination Method

A system of two linear equations with two unknowns may also be solved by the elimination method. In this method, you rewrite the pair of equations so that the coefficients of one of the variables, say x, have the same absolute value but are opposite in sign. Then add the two equations together. In the resulting equation, the coefficient of x becomes 0 and the variable x has been eliminated. You can then solve the resulting equation of one variable.

E X A M P L E   3   Solving a System by Elimination

Find the simultaneous solutions to

$$2x + 3y = 5$$

$$-3x + 5y = 21.$$

Solution   The least common multiple of the coefficients of x (2 and 3) is 6. So multiply the first equation by 3 and the second equation by 2.

$$6x + 9y = 15 \quad \text{Multiply the first equation by 3.}$$

$$-6x + 10y = 42 \quad \text{Multiply the second equation by 2.}$$

Then add the two equations together to obtain

$$6x + 9y + (-6x) + 10y = 15 + 42$$

$$19y = 57$$

$$y = 3.$$

Substituting $y = 3$ into either of the original equations gives $x = -2$.

Therefore the unique solution to this system is $x = -2, y = 3$, or more simply, $(-2, 3)$. ◼

Example 3 can also be solved graphically. First, write each equation in function form.

$$y = -\frac{2}{3}x + \frac{5}{3}$$

$$y = \frac{3}{5}x + \frac{21}{5}$$

Then find complete graphs of each and use zoom-in to find the point of intersection as shown in Fig. 10.4.

Notice a major difference between the system in Example 1 and that in Example 3. In Example 3, both equations in the system are linear, while in Example 1, the second equation $xy = 300$ is not linear. The elimination method used in

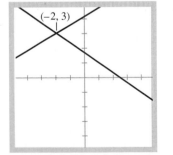

[-5, 5] by [-5, 5]

Figure 10.4 $y = -\frac{2}{3}x + \frac{5}{3}$ and $y = \frac{3}{5}x + \frac{21}{5}$.

Bert says, "Although the focus in this section is upon the algebraic methods of solving systems of linear equations, encourage students to use the graphing utility to verify their solutions graphically."

Example 3 can be used for any linear system of two equations in two variables. However, it may not work for nonlinear systems.

Types of Solutions to Systems of Equations

What types of solutions exist for systems of two linear equations with two variables? There are three possibilities. First, the graphs of the two equations may intersect in exactly one point, as in Example 3. Second, the two graphs may be parallel lines, in which case the system has no solutions. Finally, the two graphs may be the same line, in which case there are infinitely many solutions. Examples 4 and 5 illustrate these two possibilities.

E X A M P L E   4   Solving a System of Two Linear Equations

Find the simultaneous solutions to the system.

$$2x - 3y = 5$$
$$-6x + 9y = 10$$

Solution   Multiply each side of the first equation by 3 so that in both equations x has the same coefficient and opposite sign.

$$6x - 9y = 15$$
$$-6x + 9y = 10$$

Then add the two equations together and check the results.

$$6x - 6x - 9y + 9y = 25$$
$$0 = 25$$

The false statement $0 = 25$ tells us that the system of equations has no solutions; when we eliminate one variable, we also eliminate the other.   ≡

Figure 10.5 confirms that the graphs of the two equations in the system of Example 4 are a pair of parallel lines. Because these lines do not cross, the system has no solutions.

Still another way to see that the lines are parallel is to observe that the slope of each line is $\frac{2}{3}$, but they have different y-intercepts.

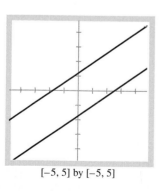

[−5, 5] by [−5, 5]

Figure 10.5 $2x - 3y = 5$ (lower) and $-6x + 9y = 10$ (upper).

E X A M P L E   5   Solving a System of Two Linear Equations

Find the simultaneous solutions to the system.

$$x - 3y = -2$$
$$-2x + 6y = 4$$

Solution   Multiply the first equation by 2 to obtain the following system.

$$2x - 6y = -4$$

$$-2x + 6y = 4$$

Adding the two equations in the system yields the equation

$$0 = 0.$$

This means that the graphs of the two equations are the same line and that each point on that line is a solution to each equation.   ▤

Extending the Elimination Method

The elimination method of solving a system of linear equations can be extended to systems of three linear equations in three variables, or even to larger systems.

The elimination method of solving a system of linear equations generalizes and can be applied to systems of m equations each with n variables, where n is not necessarily equal to m. Let $x_1, x_2, \ldots, x_n$ be variables, b be a real number, and $a_1, a_2, \ldots, a_n$ be real numbers not all of which are zero. Any equation of the form

$$a_1x_1 + a_2x_2 + \cdots + a_nx_n = b$$

is called a **linear equation in the variables $x_1, x_2, \ldots, x_n$.**

To solve a system of three or more linear equations, choose one variable to be eliminated. Then combine one equation with each of the remaining equations to eliminate the chosen variable from the system. Repeat this process with each of the remaining variables. Example 6 illustrates the method for three equations with three variables.

Notes on Examples

Example 6 illustrates the elimination method for solving three linear equations. After discussing this example, have students solve a system of equations using all three methods: substitution, elimination, and graphical. Then have them discuss strategies they can use to choose a method of solution. Exercise 13 should provide for a good discussion in this regard.

E X A M P L E   6   *Solving a System of Three Linear Equations*

Find the simultaneous solutions to the following system.

$$2x + 3y - z = -1 \tag{2a}$$

$$3x - 2y + 2z = 15 \tag{2b}$$

$$x + 4y - 3z = -8 \tag{2c}$$

Solution   Eliminate z by pairing the first equation with the other two.
Multiply Eq. (2a) by 2 and pair it with Eq. (2b) to obtain the system

$$4x + 6y - 2z = -2 \tag{3a}$$

$$3x - 2y + 2z = 15. \tag{3b}$$

Multiply Eq. (2a) by -3 and rewrite Eq. (2c) to obtain the system

$$-6x - 9y + 3z = 3 \qquad \text{(4a)}$$

$$x + 4y - 3z = -8. \qquad \text{(4b)}$$

Add Eqs. (3a) and (3b) to obtain $7x + 4y = 13$, and add Eqs. (4a) and (4b) to obtain $-5x - 5y = -5$ or $x + y = 1$.

So now consider the reduced system.

$$7x + 4y = 13 \qquad \text{(5a)}$$

$$x + \quad y = 1 \qquad \text{(5b)}$$

Multiply Eq. (5b) by -4 and add the resulting equation to Eq. (5a) to obtain

$$7x - 4x + 4y - 4y = 9 \qquad \text{(6a)}$$

$$3x = 9 \qquad \text{(6b)}$$

$$x = 3. \qquad \text{(6c)}$$

Substitute $x = 3$ into either Eq. (5a) or (5b) to obtain $y = -2$ and substitute $x = 3$ and $y = -2$ into Eq. (2a), (2b), or (2c) to obtain $z = 1$.

The solution of the original system of three equations with three variables is

$$x = 3, \qquad y = -2, \qquad z = 1. \qquad \blacksquare$$

It turns out that systems of three linear equations in three variables also have either one solution, no solutions, or infinitely many solutions. However, the geometry in this case is more complicated than for systems of two linear equations in two variables. In Section 10.2, you will see that the graph of a linear equation in three variables is a plane in three-dimensional space. By considering the possible ways that three planes can intersect, it becomes evident why only the three cases mentioned above occur for solutions of systems of three linear equations in three variables.

Notes on Exercises

Ex. 17 is similar to Example 2 and should be discussed in class since there is only one practical solution to the problem.
Ex. 22 is an interesting problem since the solutions to the system involving x and y are given, and students are required to create a system involving a and b.

Assignment Guide

Day 1: Ex. 2, 3, 6, 7, 10, 11, 15–17
Day 2: Ex. 20, 21, 23, 25–27, 29

Exercises for Section 10.1

In Exercises 1–6, use the substitution method to find the simultaneous solutions to each system of equations.

1. $\begin{cases} x = 3 \\ x - y = 20 \end{cases}$

2. $\begin{cases} 2x - 3y = -23 \\ x + y = 0 \end{cases}$

3. $\begin{cases} 3x + y = 20 \\ x - 2y = 10 \end{cases}$

4. $\begin{cases} y = x^2 \\ y - 9 = 0 \end{cases}$

5. $\begin{cases} x - 3y = 6 \\ 6y - 2x = 4 \end{cases}$

6. $\begin{cases} x = y + 3 \\ x - y^2 = 3y \end{cases}$

In Exercises 7–12, use the elimination method to solve each system of equations.

7. $\begin{cases} x - y = 10 \\ x + y = 6 \end{cases}$

8. $\begin{cases} 2x + y = 10 \\ x - 2y = -5 \end{cases}$

9. $\begin{cases} 3x - 2y = 8 \\ 5x + 4y = 28 \end{cases}$ **10.** $\begin{cases} 2x - 4y = 8 \\ -x + 2y = -4 \end{cases}$

11. $\begin{cases} 2x - y = 3 \\ -4x + 2y = 5 \end{cases}$ **12.** $\begin{cases} x^2 - 2y = -6 \\ x^2 + y = 4 \end{cases}$

13. Consider the system

$$2x + 2y = 200$$

$$xy = 500.$$

a) Which method, substitution or elimination, can be used to solve this system algebraically?

b) Find complete graphs of both equations in the same viewing rectangle.

c) How many solutions are there to the system?

d) Solve the system.

In Exercises 14 and 15, find the simultaneous solutions to each system.

14. $\begin{cases} x - y + z = 6 \\ 2x - z = 6 \\ 2x + 2y - z = 12 \end{cases}$ **15.** $\begin{cases} 2x + y - 2z = 1 \\ 6x + 2y - 4z = 3 \\ 4x - y + 3z = 5 \end{cases}$

16. A candy recipe calls for 2.25 times as much brown sugar as white sugar. If 23 oz of sugar are required, how much of each type is needed?

17. Find the dimensions of a rectangular cornfield with a perimeter of 220 yd and an area of 3000 sq yd.

18. Hank can row a boat 1 mi upstream (against the current) in 24 min. He can row the same distance downstream in 13 min. If both the rowing speed and current speed are constant, find Hank's rowing speed and the speed of the current.

19. Four hundred fifty-two tickets were sold for a high school basketball game. There were two ticket prices: student at $0.75 and nonstudent at $2.00. How many tickets of each type were sold if the total proceeds from the sale of the tickets were $429?

20. An airplane flying with the wind from Los Angeles to New York takes 3.75 hr. Flying against the wind, the plane takes 4.4 hr on the return trip. If the air distance between Los Angeles and New York is 2500 mi and the plane speed and wind speed are both constant, find the plane speed and the wind speed.

21. At a local convenience store, the total cost of one small, one medium, and one large soft drink is $2.34. The cost of a large soft drink is 1.75 times the cost of a small soft drink. The difference in cost between a medium soft drink and a small soft drink is $0.20. Determine the cost of each size of soft drink.

22. Determine a and b so that the graph of $y = ax + b$ contains the two points $(-1, 4)$ and $(2, 6)$.

23. Jessica invests $38,000, part at 7.5% simple interest and the remainder at 6% simple interest. If her annual interest income is $2600, how much does she have invested at each rate?

24. A 5-lb nut mixture is worth $2.80 per pound. The mixture contains peanuts worth $1.70 per pound and cashews worth $4.55 per pound. How many pounds of each type of nut are in the mixture?

25. Writing to Learn Write several paragraphs explaining whether you prefer finding a graphical solution or an algebraic solution to the system

$$y = 0.2x + 2$$

$$y = 0.18x + 1.9.$$

Give reasons for your choice.

Exercises 26 and 27 refer to the following economic **problem situation**: Economists have determined that supply curves are usually increasing (as the price increases, sellers increase production) and the demand curves are decreasing (as the price increases, consumers buy less). Suppose a certain single-commodity market situation is modeled by the following system:

Supply: $P = 17 + 0.1x$

Demand: $P = 25 - 0.03x$

26. Graph both the supply and the demand equations on the same coordinate system in the *first* quadrant where p represents the price (vertical axis) and x represents the number of units of the commodity produced.

27. Determine the equilibrium price, that is, the price where the supply is equal to the demand.

28. Repeat Exercises 26 and 27 for a single-commodity market situation modeled by the following system:

Supply: $P = 5 + 0.125x^2$

Demand: $P = 20 - 0.2x$

29. Use zoom-in on a graphing utility to support the solution to Example 1 graphically.

10.2 _____ Matrices and Systems of Equations

Objective

Students will be able to use matrix methods to solve and interpret systems of linear equations.

Key Ideas

Matrix
Elements of a matrix
Square matrix
Matrix of the system
Matrix model
Elementary row operations
Row echelon form
Inconsistent system
Matrix row operations
Matrix of the coefficients

In Section 10.1, we solved systems of equations algebraically using the elimination or substitution method. In this section, we streamline the elimination method of solving systems of linear equations by using matrices.

EXAMPLE 1   Solving a System of Equations

Solve the following system of equations:

$$x - 2y + z = -1 \tag{1a}$$

$$2x + 3y - 2z = -3 \tag{1b}$$

$$x + 3y - 2z = -2 \tag{1c}$$

Solution   Multiply Eq. (1a) by -2 and add the result to Eq. (1b) to obtain $7y - 4z = -1$. Replace Eq. (1b) by the new equation $7y - 4z = -1$ to obtain a new system equivalent to system (1).

$$x - 2y + z = -1 \tag{2a}$$

$$7y - 4z = -1 \tag{2b}$$

$$x + 3y - 2z = -2 \tag{2c}$$

Now repeat this process. Multiply Eq. (2a) by -1 and add the result to Eq. (2c) to obtain $5y - 3z = -1$.

$$x - 2y + z = -1 \tag{3a}$$

$$7y - 4z = -1 \tag{3b}$$

$$5y - 3z = -1 \tag{3c}$$

We have now eliminated the x-variable from two of the equations.

Next, multiply Eq. (3b) by -5 and Eq. (3c) by 7 and add the two new equations to obtain $-z = -2$. We obtain the following equivalent system of equations:

$$x - 2y + z = -1 \tag{4a}$$

$$7y - 4z = -1 \tag{4b}$$

$$-z = -2 \tag{4c}$$

Finally, multiply Eq. (4b) by $\frac{1}{7}$ and Eq. (4c) by -1:

$$x - 2y + z = -1 \tag{5a}$$

$$y - \tfrac{4}{7}z = -\tfrac{1}{7} \tag{5b}$$

$$z = 2 \tag{5c}$$

Substituting Eq. (5c) into Eq. (5b) gives $y = 1$. Then, substituting the values of y and z into Eq. (5a) yields $x = -1$. Thus the solution to system (1) is the ordered triple $(-1, 1, 2)$.

Using Matrices to Solve Systems of Equations

Example 1 provided a transition from the elimination method of Section 10.1 to a method that uses matrices. In solving Example 1, we added one equation to another to eliminate a variable and made the resulting equation a part of a new system of equations, equivalent to the preceding system. Notice that all the action took place with the coefficients. The primary role of the variables was to keep track of which columns the coefficients were written in.

This same solution is repeated below by arranging the coefficients into a rectangular array, called a **matrix**. A solution to the system in Example 1 begins with the following matrix:

$$\begin{pmatrix} 1 & -2 & 1 & -1 \\ 2 & 3 & -2 & -3 \\ 1 & 3 & -2 & -2 \end{pmatrix}$$

The first step in Example 1 was to multiply Eq. (1a) by -2 and add the result to Eq. (1b). When the system of equations is expressed as a matrix, that first step becomes "multiply row 1 by -2, add the result to row 2, and write the result in row 2." This step is symbolized as follows:

$-2r_1 + r_2.$ Multiply row 1 by -2 and add the result to row 2.

Here then is a solution to Example 1 using matrix notation. Notice that the goal of this solution process is to obtain zeros in the bottom left part of the matrix with the first nonzero entry of each row a 1.

$$\begin{pmatrix} 1 & -2 & 1 & -1 \\ 2 & 3 & -2 & -3 \\ 1 & 3 & -2 & -2 \end{pmatrix} \xrightarrow{-2r_1+r_2} \begin{pmatrix} 1 & -2 & 1 & -1 \\ 0 & 7 & -4 & -1 \\ 1 & 3 & -2 & -2 \end{pmatrix}$$

$$\xrightarrow{-r_1+r_3} \begin{pmatrix} 1 & -2 & 1 & -1 \\ 0 & 7 & -4 & -1 \\ 0 & 5 & -3 & -1 \end{pmatrix} \xrightarrow{-5r_2+7r_3} \begin{pmatrix} 1 & -2 & 1 & -1 \\ 0 & 7 & -4 & -1 \\ 0 & 0 & -1 & -2 \end{pmatrix}$$

$$\xrightarrow{\frac{1}{7}r_2, \ -r_3} \begin{pmatrix} 1 & -2 & 1 & -1 \\ 0 & 1 & -\frac{4}{7} & -\frac{1}{7} \\ 0 & 0 & 1 & 2 \end{pmatrix}$$

The expressions $-r_1 + r_3$, $-5r_2 + 7r_3$, and $\frac{1}{7}r_2$, $-r_3$ explain how the third, fourth, and fifth arrays were obtained.

Notice that the last matrix represents the following system of equations, the final system (5) from the Example 1 solution:

$$x - 2y + z = -1$$

$$y - \frac{4}{7}z = -\frac{1}{7}$$

$$z = 2.$$

As you might expect, modern technology can easily handle such matrix calculations.

We will use this matrix method in Example 2.

E X A M P L E   2   Using Matrices to Solve a System

Solve the system

$$x \qquad + z = 4$$

$$2x + 2y + 4z = 10$$

$$x + 6y + 8z = 4.$$

Solution   Perform operations on the rows of the matrix until 0s appear in the bottom left corner and the first nonzero entry of each row is 1. The leading 1s should move to the right as you look down the matrix.

$$\begin{pmatrix} 1 & 0 & 1 & 4 \\ 2 & 2 & 4 & 10 \\ 1 & 6 & 8 & 4 \end{pmatrix} \xrightarrow{-2r_1+r_2} \begin{pmatrix} 1 & 0 & 1 & 4 \\ 0 & 2 & 2 & 2 \\ 1 & 6 & 8 & 4 \end{pmatrix}$$

$$\xrightarrow{-r_1+r_3} \begin{pmatrix} 1 & 0 & 1 & 4 \\ 0 & 2 & 2 & 2 \\ 0 & 6 & 7 & 0 \end{pmatrix} \xrightarrow{-3r_2+r_3} \begin{pmatrix} 1 & 0 & 1 & 4 \\ 0 & 2 & 2 & 2 \\ 0 & 0 & 1 & -6 \end{pmatrix}$$

$$\xrightarrow{\frac{1}{2}r_2} \begin{pmatrix} 1 & 0 & 1 & 4 \\ 0 & 1 & 1 & 1 \\ 0 & 0 & 1 & -6 \end{pmatrix}$$

Note from the final matrix that $z = -6$. Substitute this value into the equation $y + z = 1$ to see that $y = 7$, and substitute these values into $x + z = 4$ to see that the solution for the system is

$$x = 10, \qquad y = 7, \qquad z = -6. \qquad \blacksquare$$

System of m Equations in n Variables

We now formalize some of the terminology and concepts used in the discussion for Examples 1 and 2.

Definition 10.1 $m \times n$ matrix

Let m and n be positive integers. An $m \times n$ **matrix** A is a rectangular array of numbers with m rows and n columns.

$$A = \begin{pmatrix} a_{11} & a_{12} & a_{13} & \cdots & a_{1n} \\ a_{21} & a_{22} & a_{23} & \cdots & a_{2n} \\ \vdots & \vdots & \vdots & & \vdots \\ a_{m1} & a_{m2} & a_{m3} & \cdots & a_{mn} \end{pmatrix}$$

The numbers a_{ij} are called the **elements of the matrix**. The subscript i indicates that element a_{ij} is in the ith row, and the subscript j indicates that a_{ij} is in the jth column. If $m = n$, the matrix is called a **square matrix**.

In general, capital letters denote the matrix as a whole and lowercase letters denote individual matrix elements.

It is common to refer to the unique matrix associated with a system of linear equations as the **matrix of the system**. For example,

$$A = \begin{pmatrix} 1 & 0 & 1 & 4 \\ 2 & 2 & 4 & 10 \\ 1 & 6 & 8 & 4 \end{pmatrix}$$

is the matrix of the system of equations in Example 2. The size of matrix A is determined by the fact that there were three equations with three variables. A is a 3×4 matrix. Notice that there is one more column than the number of variables.

Next, consider the general situation with this system of m linear equations in the n variables $x_1, x_2, \ldots, x_n$.

$$\begin{array}{ccccccccc} a_{11}x_1 & + & a_{12}x_2 & + & \cdots & + & a_{1n}x_n & = & b_1 \\ a_{21}x_1 & + & a_{22}x_2 & + & \cdots & + & a_{2n}x_n & = & b_2 \\ \vdots & & \vdots & & & & \vdots & & \vdots \\ a_{m1}x_1 & + & a_{m2}x_2 & + & \cdots & + & a_{mn}x_n & = & b_m \end{array} \qquad (6)$$

The $m \times (n + 1)$ matrix

$$\begin{pmatrix} a_{11} & a_{12} & \cdots & a_{1n} & b_1 \\ a_{21} & a_{22} & \cdots & a_{2n} & b_2 \\ \vdots & \vdots & & \vdots & \vdots \\ a_{m1} & a_{m2} & \cdots & a_{mn} & b_m \end{pmatrix} \qquad (7)$$

is a **matrix model** of the system, whereas the unique matrix determined by the particular system is the matrix of the system.

Matrix Row Operations

The matrix method used to solve Example 2 involved arithmetic operations on the rows of a matrix. This method can be extended to any system of m linear equations in n variables. The row operations are summarized in Theorem 10.1.

Teaching Note

Many graphing calculators and other scientific calculators have the capability of performing elementary row operations. Encourage students to use their calculators to perform matrix operations.

Theorem 10.1   Matrix Row Operations

When any one of the following **elementary row operations** is applied to a matrix, the system of equations of the resulting matrix is equivalent to the original system of equations. (Shorthand notation is given for each row operation.)

a) Interchanging any two rows: $r_{i,j}$.
b) Multiplying all elements of a row by the nonzero number k: kr_i.
c) Multiplying k times the elements in one row and adding the result to the corresponding elements in another row: $kr_i + r_j$.

Row Echelon Form

In Example 2, notice that after the elementary row operations were applied, the final matrix had 0's in the bottom left corner of the matrix. These 0's resulted in the last equation being $z = -6$ and made the calculations relatively simple when values were substituted back to find the values for x and y. The next definition describes the final form that should result when using the row operations.

Definition 10.2   Row Echelon Form

A matrix is said to be in **row echelon form** if the following conditions are satisfied:

a) The first nonzero entry in each row is a 1.
b) The index j of the column in which the first nonzero entry of a row occurs is less than the column index of the first nonzero entry of the next row.
c) Any rows consisting entirely of 0's occur at the bottom of the matrix.

E X A M P L E   3   Solving a System of Equations—Echelon Form

Solve the system

$$x - 2y + z \qquad = 7$$
$$2x + 3y \qquad - w = 0$$

$$y + 2z - 3w = 2$$
$$-x - y + 3z - w = 7. \tag{8}$$

Solution   Begin with the matrix of the system and use row operations to bring it to row echelon form.

$$
\begin{pmatrix}
1 & -2 & 1 & 0 & 7 \\
2 & 3 & 0 & -1 & 0 \\
0 & 1 & 2 & -3 & 2 \\
-1 & -1 & 3 & -1 & 7
\end{pmatrix}
\xrightarrow{-2r_1+r_2}
\begin{pmatrix}
1 & -2 & 1 & 0 & 7 \\
0 & 7 & -2 & -1 & -14 \\
0 & 1 & 2 & -3 & 2 \\
-1 & -1 & 3 & -1 & 7
\end{pmatrix}
$$

$$
\xrightarrow{r_1+r_4}
\begin{pmatrix}
1 & -2 & 1 & 0 & 7 \\
0 & 7 & -2 & -1 & -14 \\
0 & 1 & 2 & -3 & 2 \\
0 & -3 & 4 & -1 & 14
\end{pmatrix}
\xrightarrow{r_{2,3}}
\begin{pmatrix}
1 & -2 & 1 & 0 & 7 \\
0 & 1 & 2 & -3 & 2 \\
0 & 7 & -2 & -1 & -14 \\
0 & -3 & 4 & -1 & 14
\end{pmatrix}
$$

$$
\xrightarrow{-7r_2+r_3}
\begin{pmatrix}
1 & -2 & 1 & 0 & 7 \\
0 & 1 & 2 & -3 & 2 \\
0 & 0 & -16 & 20 & -28 \\
0 & -3 & 4 & -1 & 14
\end{pmatrix}
\xrightarrow{3r_2+r_4}
\begin{pmatrix}
1 & -2 & 1 & 0 & 7 \\
0 & 1 & 2 & -3 & 2 \\
0 & 0 & -16 & 20 & -28 \\
0 & 0 & 10 & -10 & 20
\end{pmatrix}
$$

$$
\xrightarrow{\frac{1}{10}r_4}
\begin{pmatrix}
1 & -2 & 1 & 0 & 7 \\
0 & 1 & 2 & -3 & 2 \\
0 & 0 & -16 & 20 & -28 \\
0 & 0 & 1 & -1 & 2
\end{pmatrix}
\xrightarrow{r_{3,4}}
\begin{pmatrix}
1 & -2 & 1 & 0 & 7 \\
0 & 1 & 2 & -3 & 2 \\
0 & 0 & 1 & -1 & 2 \\
0 & 0 & -16 & 20 & -28
\end{pmatrix}
$$

$$
\xrightarrow{16r_3+r_4}
\begin{pmatrix}
1 & -2 & 1 & 0 & 7 \\
0 & 1 & 2 & -3 & 2 \\
0 & 0 & 1 & -1 & 2 \\
0 & 0 & 0 & 4 & 4
\end{pmatrix}
\xrightarrow{\frac{1}{4}r_4}
\begin{pmatrix}
1 & -2 & 1 & 0 & 7 \\
0 & 1 & 2 & -3 & 2 \\
0 & 0 & 1 & -1 & 2 \\
0 & 0 & 0 & 1 & 1
\end{pmatrix}
$$

The last matrix corresponds to the following system of equations:

$$x - 2y + z = 7 \tag{9a}$$
$$y + 2z - 3w = 2 \tag{9b}$$
$$z - w = 2 \tag{9c}$$
$$w = 1 \tag{9d}$$

System (9) is equivalent to system (8).

Substituting Eq. (9d) into Eq. (9c) gives $z = 3$. Continuing substitution yields $y = -1$ and $x = 2$. You can check that the ordered 4-tuple $(2, -1, 3, 1)$ is a solution to each of the four original equations.   ≡

With practice, you will be able to perform two or more row operations on a matrix at the same time. We will often do this to save space.

Interpreting the Solution of a System of Linear Equations Geometrically

It can be shown that a complete graph of one linear equation with three variables, such as $x + y + 2z = -2$, is a plane in three-dimensional space. For a system of two equations and three variables, like the one considered in the next example, a solution exists if the two planes intersect in a line. In that case, the system has an infinite number of solutions (see Fig. 10.6).

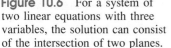

Figure 10.6 For a system of two linear equations with three variables, the solution can consist of the intersection of two planes.

E X A M P L E   4   Solving a System with an Infinite Number of Solutions

Solve the system

$$x + y + \ z = 3$$
$$2x + y + 4z = 8. \qquad\qquad \textbf{(10)}$$

Solution   First, reduce the matrix of the system to row echelon form.

$$\begin{pmatrix} 1 & 1 & 1 & 3 \\ 2 & 1 & 4 & 8 \end{pmatrix} \xrightarrow{-2r_1 + r_2} \begin{pmatrix} 1 & 1 & 1 & 3 \\ 0 & -1 & 2 & 2 \end{pmatrix} \xrightarrow{-r_2} \begin{pmatrix} 1 & 1 & 1 & 3 \\ 0 & 1 & -2 & -2 \end{pmatrix}$$

Next, simplify this system even more by completing the row operation $-r_2 + r_1$. The resulting matrix

$$\begin{pmatrix} 1 & 0 & 3 & 5 \\ 0 & 1 & -2 & -2 \end{pmatrix}$$

corresponds to the system

$$x + 3z = 5$$
$$y - 2z = -2.$$

Consequently, $x = -3z + 5$ and $y = 2z - 2$. The system consists of all ordered triples $(-3z + 5,\ 2z - 2,\ z)$, where z can be any real number. Vector addition as developed in Section 8.5 can be generalized to the three-dimensional case and ordered triples. Treat this triple as a vector and use vector addition to write this triple as

$$(-3z + 5, 2z - 2, z) = (5, -2, 0) + z(-3, 2, 1).$$

The set of solutions consists of points on the line L where L is obtained by translating the line of scalar multiples of the vector $(-3, 2, 1)$ by the vector $(5, -2, 0)$ (see Fig. 10.7). ▬

VECTOR EQUATIONS OF LINES

Notice that the equation $\mathbf{v}(z) = (5, -2, 0) + z(-3, 2, 1)$, where z is a real number, is a vector equation for the line that passes through the point $(5, -2, 0)$ and is parallel to the vector $(-3, 2, 1)$.

In general the equation

$$\mathbf{v}(t) = \mathbf{u} + t\mathbf{w}$$

is an equation of the line that passes through point u and is parallel to the vector $\mathbf{w}$.

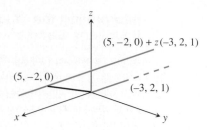

Figure 10.7 The line through the origin determined by $(-3, 2, 1)$ is translated by $(5, -2, 0)$.

A second possibility for a system of two equations with three variables is that the two planes representing the equations are parallel. In that case, the system has no solutions since the two planes have no points in common.

A system with no solutions is called an **inconsistent system** (see Fig. 10.8).

Example 5 illustrates a system of three equations with three variables in which there are no solutions.

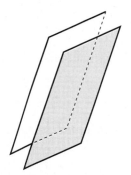

Figure 10.8 If a system of two equations and three variables is inconsistent, the planes representing the equations will be parallel.

E X A M P L E   5   Solving a System with No Solutions

Solve the system of equations

$$x + y - 2z = -2$$
$$2x - 3y + z = 1$$
$$2x + y - 3z = -2.$$

Solution   Use elementary row operations to reduce the matrix of the system to echelon form.

$$\begin{pmatrix} 1 & 1 & -2 & -2 \\ 2 & -3 & 1 & 1 \\ 2 & 1 & -3 & -2 \end{pmatrix} \xrightarrow[-2r_1+r_3]{-2r_1+r_2} \begin{pmatrix} 1 & 1 & -2 & -2 \\ 0 & -5 & 5 & 5 \\ 0 & -1 & 1 & 2 \end{pmatrix}$$

$$\xrightarrow{r_{2,3}} \begin{pmatrix} 1 & 1 & -2 & -2 \\ 0 & -1 & 1 & 2 \\ 0 & -5 & 5 & 5 \end{pmatrix} \xrightarrow{-5r_2+r_3} \begin{pmatrix} 1 & 1 & -2 & -2 \\ 0 & -1 & 1 & 2 \\ 0 & 0 & 0 & -5 \end{pmatrix}$$

$$\xrightarrow[-0.2r_3]{-r_2} \begin{pmatrix} 1 & 1 & -2 & -2 \\ 0 & 1 & -1 & -2 \\ 0 & 0 & 0 & 1 \end{pmatrix}$$

The third row of the last matrix corresponds to the false equation $0 = 1$. This means the system has no solutions.

Reduced Row Echelon Form

In Examples 3 and 5, the matrix of the system was reduced to row echelon form. The matrix of the system was simplified even further in Example 4 to what is called reduced row echelon form.

A matrix is in **reduced row echelon form** if it is in row echelon form and all other entries are 0 in each column that contains a leading 1. Which one of these matrices is in reduced row echelon form?

$$\begin{pmatrix} 1 & 2 & 0 & -6 \\ 0 & 1 & 0 & 1 \\ 0 & 0 & 1 & -2 \end{pmatrix} \quad \begin{pmatrix} 0 & 1 & 0 & 2 \\ 0 & 0 & 1 & -1 \\ 0 & 0 & 0 & 1 \end{pmatrix} \quad \begin{pmatrix} 1 & 0 & 0 & -2 \\ 0 & 1 & 0 & 7 \\ 0 & 0 & 1 & 4 \end{pmatrix}$$

Once the matrix of a system of equations is in reduced row echelon form, the solutions to the system are even more immediate. This method is illustrated in Example 6.

EXAMPLE 6   Solving a System-reduced Row Echelon Form

VECTOR EQUATIONS OF PLANES

Notice that the equation $\mathbf{v}(s, t) = (1, -1, 0) + s(-1, 2, 1) + t(-2, 1, 0)$, where s and t are real numbers, is a vector equation for the plane that passes through the point $(1, -1, 0)$ and is parallel to the plane determined by the vectors $(-1, 2, 1)$ and $(-2, 1, 0)$.

In general the equation

$$\mathbf{R}(s, t) = \mathbf{u} + s\mathbf{v} + t\mathbf{w}$$

is an equation for the plane that passes through point u and is parallel to the plane determined by the vectors $\mathbf{v}$ and $\mathbf{w}$.

Solve the system

$$x + 2y - 3z \quad\quad = -1$$
$$2x + 3y - 4z + w = -1$$
$$3x + 5y - 7z + w = -2.$$

Solution   First, bring the matrix of the system to reduced row echelon form:

$$\begin{pmatrix} 1 & 2 & -3 & 0 & -1 \\ 2 & 3 & -4 & 1 & -1 \\ 3 & 5 & -7 & 1 & -2 \end{pmatrix} \xrightarrow[-3r_1+r_3]{-2r_1+r_2} \begin{pmatrix} 1 & 2 & -3 & 0 & -1 \\ 0 & -1 & 2 & 1 & 1 \\ 0 & -1 & 2 & 1 & 1 \end{pmatrix}$$

$$\xrightarrow[-r_2+r_3]{2r_2+r_1} \begin{pmatrix} 1 & 0 & 1 & 2 & 1 \\ 0 & -1 & 2 & 1 & 1 \\ 0 & 0 & 0 & 0 & 0 \end{pmatrix} \xrightarrow{-r_2} \begin{pmatrix} 1 & 0 & 1 & 2 & 1 \\ 0 & 1 & -2 & -1 & -1 \\ 0 & 0 & 0 & 0 & 0 \end{pmatrix}$$

The following system of equations corresponds to the last matrix above:

$$x + \quad\quad z + 2w = \quad 1$$
$$y - 2z - \quad w = -1.$$

This is equivalent to

$$x = -z - 2w + 1$$
$$y = 2z + w - 1.$$

Thus the solutions to the system consist of all ordered 4-tuples of the form

$$(-z - 2w + 1,\ 2z + w - 1,\ z,\ w) = (1, -1, 0, 0) + z(-1, 2, 1, 0) + w(-2, 1, 0, 1),$$

where z and w can be any real numbers.   ≡

Because the two variables z and w can be chosen independently, these 4-tuples are called a **two-parameter family of solutions**.

Notes on Exercises

Ex. 31–36 use the concepts established in Example 7. Students should be able to translate problem situations into algebraic representations before applying matrix methods.

E X A M P L E   7   APPLICATION: Solving a Mixture Problem

A chemist wants to prepare a 60-l mixture that is 40% acid using three concentrations of acid. The first concentration is 15% acid, the second is 35% acid, and the third is 55% acid. Because of the amounts of acid solution on hand, the chemist wants to use twice as much of the 35% solution as the 55% solution. How much of each solution should be used?

Solution   Choose variables to represent the amounts of each solution to make a 60-l, 40% acid solution.

$$x : \text{number of liters of 15\% solution}$$

$$y : \text{number of liters of 35\% solution}$$

$$z : \text{number of liters of 55\% solution}$$

Then each of these equations must be satisfied.

$$x + \quad y + \quad z = 60$$

$$0.15x + 0.35y + 0.55z = 24 \qquad \text{40\% of 60 is 24.}$$

$$y - \quad 2z = \quad 0$$

Reduce the matrix of this system to row echelon form.

$$\begin{pmatrix} 1 & 1 & 1 & 60 \\ 0.15 & 0.35 & 0.55 & 24 \\ 0 & 1 & -2 & 0 \end{pmatrix} \xrightarrow{-0.15r_1 + r_2} \begin{pmatrix} 1 & 1 & 1 & 60 \\ 0 & 0.2 & 0.4 & 15 \\ 0 & 1 & -2 & 0 \end{pmatrix}$$

$$\xrightarrow{r_{2,3}} \begin{pmatrix} 1 & 1 & 1 & 60 \\ 0 & 1 & -2 & 0 \\ 0 & 0.2 & 0.4 & 15 \end{pmatrix} \xrightarrow{-0.2r_2 + r_3} \begin{pmatrix} 1 & 1 & 1 & 60 \\ 0 & 1 & -2 & 0 \\ 0 & 0 & 0.8 & 15 \end{pmatrix}$$

$$\xrightarrow{\frac{1}{0.8}r_3} \begin{pmatrix} 1 & 1 & 1 & 60 \\ 0 & 1 & -2 & 0 \\ 0 & 0 & 1 & 18.75 \end{pmatrix}$$

The last matrix is equivalent to the following system of equations:

$$x + \ y + \ z = 60$$

$$y - 2z = \ 0$$

$$z = 18.75.$$

Assignment Guide

Day 1: Ex. 8–14, 16, 17, 20, 23

Day 2: Ex. 25, 28, 29, 32–35, 37

Solving this system gives $z = 18.75$, $y = 37.50$, and $x = 3.75$. Consequently, 3.75 l of 15% acid, 37.50 l of 35%, and 18.75 l of 55% acid are needed to make a 60-l, 40% acid solution. ∎

Exercises for Section 10.2

In Exercises 1–7, consider the matrix

$$A = \begin{pmatrix} -1 & 2 & 3 & -1 \\ 4 & 0 & 1 & 4 \\ 2 & 6 & -2 & 3 \end{pmatrix}.$$

1. What is the size of the matrix? How many rows are there? How many columns are there?

2. What is the value of each of these entries in the matrix? $a_{21}, a_{33}, a_{23}, a_{14}, a_{24}$

3. Find the matrix that results from applying the elementary row operation $r_1 + r_2$ to A.

4. Find the matrix that results from applying the elementary row operation $2r_1 + r_3$ to A.

5. Find the matrix that results from applying the elementary row operation $4r_1 + r_2$ to A.

6. Find the row echelon form of A.

7. Find the reduced row echelon form of A.

In Exercises 8–14, consider the matrix

$$A = \begin{pmatrix} 3 & 0 & -9 & 2 & 1 \\ 1 & 1 & -1 & 0 & 3 \\ -6 & 0 & 18 & -4 & -2 \end{pmatrix}.$$

8. What is the size of the matrix? How many rows are there? How many columns are there?

9. What is the value of each of these entries in the matrix? $a_{13}, a_{32}, a_{23}, a_{25}, a_{34}$

10. Find the matrix that results from applying the elementary row operation $\frac{1}{3}r_1$ to A.

11. Find the matrix that results from applying the elementary row operation $2r_1 + r_3$ to A.

12. Find the matrix that results from applying the elementary row operation $-\frac{1}{3}r_1 + r_2$ to A.

13. Find the row echelon form of A.

14. Find the reduced row echelon form of A.

15. Determine the reduced row echelon form of the matrix of system (1) of Example 1.

16. Determine the reduced row echelon form of the matrix of system (10) of Example 3.

In Exercises 17–30, write a matrix model for each system of equations. Solve the system by reducing the matrix of the system to row echelon form or reduced row echelon form.

17. $\begin{cases} x - 3y = 6 \\ 2x + y = 19 \end{cases}$

18. $\begin{cases} 1.3x - 2y = 3.3 \\ 5x + 1.6y = 3.4 \end{cases}$

19. $\begin{cases} x - y + z = 6 \\ x + y + 2z = -2 \end{cases}$

20. $\begin{cases} x + y = 3 \\ x - y = 5 \\ 2x + y = -1 \end{cases}$

21. $\begin{cases} x - y + z = 0 \\ 2x - 3z = -1 \\ -x - y + 2z = -1 \end{cases}$

22. $\begin{cases} 2x - y = 0 \\ x + 3y - z = -3 \\ 3y + z = 8 \end{cases}$

23. $\begin{cases} x + y - 2z = 2 \\ 3x - y + z = 4 \\ -2x - 2y + 4z = 6 \end{cases}$

24. $\begin{cases} 2x - y = 10 \\ x - z = -1 \\ y + z = -9 \end{cases}$

25. $\begin{cases} x + y - 2z = 2 \\ 3x - y + z = 1 \\ -2x - 2y + 4z = -4 \end{cases}$

26. $\begin{cases} 1.25x \quad\quad + z = 2 \\ \quad\quad y - 5.5z = -2.75 \\ 3x - 1.5y \quad\quad = -6 \end{cases}$

27. $\begin{cases} x + y - z \quad\quad = 4 \\ \quad\quad y \quad + w = 4 \\ x - y \quad\quad = 1 \\ x \quad\quad + z + w = 4 \end{cases}$

28. $\begin{cases} \frac{1}{2}x - y + z - w = 1 \\ -x + y + z + 2w = -3 \\ x \quad\quad - z \quad\quad = 2 \\ \quad y \quad + w = 0 \end{cases}$

29. $\begin{cases} 2x + y + z + 2w = -3.5 \\ x + y + z + w = -1.5 \end{cases}$

30. $\begin{cases} 2x + y \quad\quad + 4w = 6 \\ x + y + z + w = 5 \end{cases}$

31. At a zoo in Pittsburgh, Pennsylvania, children ride a train for 25 cents, but adults must pay $1.00. On a given day, 1088 passengers paid a total of $545 for the rides. How many passengers were children? Adults?

32. One silver alloy is 42% silver, another is 30% silver. How many grams of each are required to produce 50 g of a new alloy that is 34% silver?

33. Jimenez inherits $20,000. He is advised to divide the money into three amounts and make three different investments. The first earns interest at 6% APR, the second

8% APR, and the third 10% APR. How much is invested at 6%, 8%, and 10% if the amount of the first investment is twice that of the second investment and the total annual interest received in the first year is $1640.00?

34. A scientist observes that data derived from an experiment seem to be parabolic when plotted on ordinary graph paper. Three of the observed data points are (1, 59), (5, 75), and (10, 50). Determine an algebraic representation of the parabola that is a model for this problem situation.

35. A scientist observes that data derived from an experiment seem to be parabolic when plotted on ordinary graph paper. Three of the observed data points are (1, 10), (2, 8), and (3, 4). Determine an algebraic representation of the parabola that is a model for this problem situation.

36. Joe has three employees of different abilities. They are assigned to work together on a task. Susan can do the task by herself in 6 hr, Jim and Bill working together can do the task in 1.2 hr, and Susan and Bill working together can do the task in 1.5 hr. How long will it take Susan, Bill, and Jim working together to complete the task?

37. Writing to Learn A certain investor has $50,000 to place and plans to place some into a long-term investment that earns 9% interest, some in a short-term investment that earns 7%, and some in a money market account that earns 5%. The investor expects her interest income to be $3750 for the first year. Write several paragraphs explaining how this problem is related to finding a solution to a system of linear equations.

10.3 _____

Solving a System of Equations with Matrix Multiplication

In Section 10.2, you learned a method for solving a system of linear equations that used row operations on the matrix of the system of equations. Reducing the matrix of a system to reduced row echelon form is a general method that applies to systems of any size.

A system of linear equations is a **square system** if there are the same number of equations as there are variables in the system. For example, the following system is a square system because there are three equations and three variables x, y, and z:

$$\begin{aligned} x - 2y + z &= -1 \\ 2x + 3y - 2z &= -3 \\ x + 3y - 2z &= -2 \end{aligned} \tag{1}$$

The method of solving a square system of linear equations discussed in this section is made much easier by modern technology because of increased ease in completing matrix operations.

Matrix Multiplication

We begin this section by defining matrix multiplication for 2×2 matrices, that is, matrices with two rows and two columns. Consider the two matrices A and B defined as follows:

$$A = \begin{pmatrix} a_{11} & a_{12} \\ a_{21} & a_{22} \end{pmatrix} \qquad B = \begin{pmatrix} b_{11} & b_{12} \\ b_{21} & b_{22} \end{pmatrix}.$$

The product of these two matrices, denoted AB, is defined as follows:

$$\begin{aligned} AB &= \begin{pmatrix} a_{11} & a_{12} \\ a_{21} & a_{22} \end{pmatrix} \cdot \begin{pmatrix} b_{11} & b_{12} \\ b_{21} & b_{22} \end{pmatrix} \\ &= \begin{pmatrix} a_{11}b_{11} + a_{12}b_{21} & a_{11}b_{12} + a_{12}b_{22} \\ a_{21}b_{11} + a_{22}b_{21} & a_{21}b_{12} + a_{22}b_{22} \end{pmatrix} \end{aligned} \tag{2}$$

EXAMPLE 1   Finding a Product of Two Matrices

Find the product $A \cdot B$ where

$$A = \begin{pmatrix} 2 & -1 \\ 1 & -3 \end{pmatrix} \qquad \text{and} \qquad B = \begin{pmatrix} 1 & -1 \\ -2 & 0 \end{pmatrix}.$$

Solution

$$\begin{pmatrix} 2 & -1 \\ 1 & -3 \end{pmatrix} \cdot \begin{pmatrix} 1 & -1 \\ -2 & 0 \end{pmatrix} = \begin{pmatrix} 2 \cdot 1 + (-1)(-2) & 2(-1) + (-1)0 \\ 1 \cdot 1 + (-3)(-2) & 1(-1) + (-3)0 \end{pmatrix}$$

$$= \begin{pmatrix} 4 & -2 \\ 7 & -1 \end{pmatrix} \qquad \blacksquare$$

It is important to observe that matrix multiplication is not commutative. Example 1 found the product AB. Verify for yourself that $BA \neq AB$.

In general, any two $n \times n$ matrices can be multiplied. Sometimes it is possible to multiply pairs of matrices that are not square. The product AB can be defined if the number of *columns* of A equals the number of *rows* of B.

Definition 10.3   Matrix Multiplication

Let A be an $m \times r$ matrix and B be an $r \times n$ matrix. Then the **matrix product**, AB, is the $m \times n$ matrix C whose entries are found as follows. To find the entry c_{ij} in row i and column j, pair row i of A with column j of B and add products of corresponding entries.

$$c_{ij} = (a_{i1} \quad a_{i2} \quad \cdots \quad a_{ir}) \cdot \begin{pmatrix} b_{1j} \\ b_{2j} \\ \vdots \\ b_{rj} \end{pmatrix} = a_{i1}b_{1j} + a_{i2}b_{2j} + \cdots + a_{ir}b_{rj}$$

Notice that the definition in Eq. (2) for the product of two 2×2 matrices agrees with the more general product in Definition 10.3.

Notes on Exercises

Ex. 1–6 can be used as a follow-up to Example 2.

E X A M P L E   2   Finding a Product of Two Matrices

Find the products AB and BA where

$$A = \begin{pmatrix} 2 & 1 & -3 \\ 0 & 1 & 2 \end{pmatrix} \quad \text{and} \quad B = \begin{pmatrix} 1 & -4 \\ 0 & 2 \\ 1 & 0 \end{pmatrix}.$$

Solution

$$AB = \begin{pmatrix} 2 & 1 & -3 \\ 0 & 1 & 2 \end{pmatrix} \cdot \begin{pmatrix} 1 & -4 \\ 0 & 2 \\ 1 & 0 \end{pmatrix} = \begin{pmatrix} -1 & -6 \\ 2 & 2 \end{pmatrix}$$

and

$$BA = \begin{pmatrix} 1 & -4 \\ 0 & 2 \\ 1 & 0 \end{pmatrix} \cdot \begin{pmatrix} 2 & 1 & -3 \\ 0 & 1 & 2 \end{pmatrix} = \begin{pmatrix} 2 & -3 & -11 \\ 0 & 2 & 4 \\ 2 & 1 & -3 \end{pmatrix}.$$

Two matrices are equal if each pair of corresponding elements are equal. The next example uses equality of matrices to show that a system of linear equations can be expressed as a product of two matrices.

Notes on Exercises

Ex. 7–10 can be used after discussing Example 3.

E X A M P L E   3   Solving a System of Equations as a Matrix Equation

Express the system of linear equations

$$2x + 3y = 7$$
$$-4x - y = 3 \qquad (3)$$

as a product of matrices.

Solution   Let A be the matrix $A = \begin{pmatrix} 2 & 3 \\ -4 & -1 \end{pmatrix}$. Then

$$A \begin{pmatrix} x \\ y \end{pmatrix} = \begin{pmatrix} 2 & 3 \\ -4 & -1 \end{pmatrix} \cdot \begin{pmatrix} x \\ y \end{pmatrix} = \begin{pmatrix} 2x + 3y \\ -4x - y \end{pmatrix}.$$

Therefore the matrix equation

$$\begin{pmatrix} 2 & 3 \\ -4 & -1 \end{pmatrix} \cdot \begin{pmatrix} x \\ y \end{pmatrix} = \begin{pmatrix} 7 \\ 3 \end{pmatrix}$$

is equivalent to system (3).   ■

In a similar manner, any square system of linear equations can be expressed as a **matrix equation**. For example, system (1) is equivalent to the following matrix equation:

$$\begin{pmatrix} 1 & -2 & 1 \\ 2 & 3 & -2 \\ 1 & 3 & -2 \end{pmatrix} \cdot \begin{pmatrix} x \\ y \\ z \end{pmatrix} = \begin{pmatrix} -1 \\ -3 \\ -2 \end{pmatrix}.$$

Identity Matrices and Inverse Matrices

The **identity matrix** I has the property that $AI = IA = A$ for all square matrices of a certain size. It can be shown that each of the following are identity matrices. The subscript identifies the size of the matrix.

$$I_2 = \begin{pmatrix} 1 & 0 \\ 0 & 1 \end{pmatrix}, \qquad I_3 = \begin{pmatrix} 1 & 0 & 0 \\ 0 & 1 & 0 \\ 0 & 0 & 1 \end{pmatrix}, \qquad I_4 = \begin{pmatrix} 1 & 0 & 0 & 0 \\ 0 & 1 & 0 & 0 \\ 0 & 0 & 1 & 0 \\ 0 & 0 & 0 & 1 \end{pmatrix}$$

It is reasonable to call these matrices identity matrices. They play the same role for matrix multiplication that the number 1 plays for multiplication of real numbers. When you multiply an $n \times n$ matrix A by the identity matrix I_n, the product is the original matrix A.

The real numbers $\frac{1}{3}$ and 3 are called multiplicative inverses because $\frac{1}{3} \cdot 3 = 3 \cdot \frac{1}{3} = 1$. Inverse matrices are defined in a similar manner.

Definition 10.4   Inverse Matrices

Two $n \times n$ matrices A and B are **inverse matrices** if

$$AB = BA = I_n.$$

We say that A is the inverse of B, denoted $A = B^{-1}$, and that B is the inverse of A, denoted $B = A^{-1}$.

Notes on Exercises

Ex. 11–14 provide additional practice after presenting Examples 4 and 5.

E X A M P L E   4   Verifying Inverse Matrices

Show that

$$A = \begin{pmatrix} 1 & 2 \\ 1 & 3 \end{pmatrix} \quad \text{and} \quad B = \begin{pmatrix} 3 & -2 \\ -1 & 1 \end{pmatrix}$$

are inverse matrices.

Solution

$$AB = \begin{pmatrix} 1 & 2 \\ 1 & 3 \end{pmatrix} \cdot \begin{pmatrix} 3 & -2 \\ -1 & 1 \end{pmatrix} = \begin{pmatrix} 1 & 0 \\ 0 & 1 \end{pmatrix}$$

and

$$BA = \begin{pmatrix} 3 & -2 \\ -1 & 1 \end{pmatrix} \cdot \begin{pmatrix} 1 & 2 \\ 1 & 3 \end{pmatrix} = \begin{pmatrix} 1 & 0 \\ 0 & 1 \end{pmatrix}.$$

Not all matrices have inverses. A matrix that has an inverse is called an **invertible matrix**. This fact raises an important question: When does a square matrix have an inverse, and how can you find the inverse of a matrix if it exists? Theorem 10.2 gives an answer for 2×2 matrices.

Theorem 10.2   Inverses of 2 × 2 Matrices

Consider the 2×2 matrix $A = \begin{pmatrix} a & b \\ c & d \end{pmatrix}$. If $ad - bc \neq 0$, then the inverse of matrix A exists and is given by

$$A^{-1} = \frac{1}{ad - bc} \begin{pmatrix} d & -b \\ -c & a \end{pmatrix}.$$

The expression $ad - bc$ is called the **determinant** of the 2×2 matrix $\begin{pmatrix} a & b \\ c & d \end{pmatrix}$ and is denoted by either

$$\det A \qquad \text{or} \qquad \begin{vmatrix} a & b \\ c & d \end{vmatrix}.$$

Theorem 10.2 tells us that a 2×2 matrix A has an inverse if $\det A \neq 0$. Although in this text we restrict our definition of the determinant to 2×2 matrices, it is possible also to calculate the determinant of larger square matrices. The situation described for 2×2 matrices generalizes to $n \times n$ matrices as described in this theorem.

Theorem 10.3   Inverse of an $n \times n$ Matrix

An $n \times n$ matrix has an inverse matrix if, and only if, $\det A \neq 0$.

We suggest using a calculator or a computer to find the determinant of a larger matrix.

E X A M P L E   5   Inverse Matrices

Determine whether each matrix is invertible. If so, find its inverse matrix.

a) $A = \begin{pmatrix} 3 & 1 \\ 4 & 2 \end{pmatrix}$

b) $B = \begin{pmatrix} -4 & 6 \\ -2 & 3 \end{pmatrix}$

Solution

a) Because $\det A = ad - bc = 3 \cdot 2 - 1 \cdot 4 = 2 \neq 0$, conclude that matrix A has an inverse and use Theorem 10.2 to find it.

$$A^{-1} = \frac{1}{2} \begin{pmatrix} 2 & -1 \\ -4 & 3 \end{pmatrix} = \begin{pmatrix} 1 & -\frac{1}{2} \\ -2 & \frac{3}{2} \end{pmatrix}$$

b) Because $\det B = ad - bc = -12 - (-12) = 0$, B does not have an inverse. ▧

Solving Systems of Linear Equations

Consider the system of equations

$$\begin{aligned} x + 2y &= 7 \\ x + 3y &= 3. \end{aligned} \tag{4}$$

GRAPHER SKILLS UPDATE

Calculators that have a matrix capability usually have a function DET that will calculate the determinant of a matrix. We suggest that you learn how to calculate the determinant of matrices with your calculator.

GRAPHER SKILLS UPDATE

Calculators that have a matrix capability will calculate the inverse of a matrix. If a matrix does not have an inverse, an error message should occur. Experiment with your calculator. Will it calculate inverse matrices?

The matrix

$$A = \begin{pmatrix} 1 & 2 \\ 1 & 3 \end{pmatrix}$$

is called the **matrix of coefficients** for system (4). We have seen in Example 3 that this system can be written as the matrix equation

$$A \begin{pmatrix} x \\ y \end{pmatrix} = \begin{pmatrix} 7 \\ 3 \end{pmatrix}. \qquad (5)$$

Example 4 shows that

$$A^{-1} = \begin{pmatrix} 3 & -2 \\ -1 & 1 \end{pmatrix}.$$

Multiplying both sides of equation (5) above by A^{-1} results in the following:

$$A^{-1}A \begin{pmatrix} x \\ y \end{pmatrix} = A^{-1} \begin{pmatrix} 7 \\ 3 \end{pmatrix}$$

$$\begin{pmatrix} x \\ y \end{pmatrix} = \begin{pmatrix} 3 & -2 \\ -1 & 1 \end{pmatrix} \cdot \begin{pmatrix} 7 \\ 3 \end{pmatrix}$$

$$= \begin{pmatrix} 15 \\ -4 \end{pmatrix}.$$

Therefore the solution to this system is

$$x = 15 \qquad \text{and} \qquad y = -4.$$

Once the inverse of matrix A is found (which can be done with a calculator or computer), the solution to the system is found by matrix multiplication. The following theorem is stated for n equations with n variables.

Common Errors

Some calculators use the x^{-1} as a reciprocal key in some applications and as an inverse key in others. Be sure students understand that this key may be used to denote the inverse of a matrix in entering the equation $X = A^{-1}B$.

Theorem 10.4   Solving a System of Equations

Suppose a system of n equations and n variables is written as the matrix equation $AX = B$, where A is the matrix of coefficients of the system and X and B are $n \times 1$ matrices of variables and constants, respectively.

If $\det A \neq 0$, then the system

$$AX = B$$

has the unique solution

$$X = A^{-1}B.$$

EXAMPLE 6   Solving a System of Equations

Use matrix multiplication to solve the system

$$3x - 3y + 6z = 20$$
$$x - 3y + 10z = 40$$
$$-x + 3y - 5z = 30.$$

Solution   Let

$$A = \begin{pmatrix} 3 & -3 & 6 \\ 1 & -3 & 10 \\ -1 & 3 & -5 \end{pmatrix}$$

be the matrix of coefficients of the system of equations. Then

$$A \cdot \begin{pmatrix} x \\ y \\ z \end{pmatrix} = \begin{pmatrix} 20 \\ 40 \\ 30 \end{pmatrix} \quad \text{and} \quad \begin{pmatrix} x \\ y \\ z \end{pmatrix} = A^{-1} \begin{pmatrix} 20 \\ 40 \\ 30 \end{pmatrix}.$$

The unique solution to this system is $x = 18$, $y = 39.33$, and $z = 14$.   ▤

In the following exercises, if the systems are larger than two equations and two variables, use a calculator or computer to find A^{-1} and to solve the system of equations.

Notes on Examples

Calculators with a matrix capability can calculate determinants and the inverse of a matrix. If a matrix does not have an inverse, an error message should occur. Experiment with your calculator to see if it has these capabilities.

Notes on Exercises

Ex. 27–38 provide an opportunity for students to look at some advanced topics with matrices. Encourage students to find the eigenvalues of a matrix in Ex. 29.

Assignment Guide

Day 1: Ex. 1, 4, 6, 7, 10, 12, 13, 15, 17, 20
Day 2: Ex. 21, 24, 25, 27–38

Exercises for Section 10.3

In Exercises 1–6, find the products $A \cdot B$ and $B \cdot A$ (if defined) for the given matrices.

1. $A = \begin{pmatrix} 2 & 3 \\ -1 & 5 \end{pmatrix}$; $B = \begin{pmatrix} 1 & -3 \\ -2 & -4 \end{pmatrix}$

2. $A = \begin{pmatrix} 1 & -4 \\ 2 & 6 \end{pmatrix}$; $B = \begin{pmatrix} 5 & 1 \\ -2 & -3 \end{pmatrix}$

3. $A = \begin{pmatrix} 2 & 0 & 1 \\ 1 & 4 & -3 \end{pmatrix}$; $B = \begin{pmatrix} 1 & 2 \\ -3 & 1 \\ 0 & -2 \end{pmatrix}$

4. $A = \begin{pmatrix} -1 & 4 \\ 0 & 6 \end{pmatrix}$; $B = \begin{pmatrix} 3 & -1 & 5 \\ 0 & -2 & 4 \end{pmatrix}$

5. $A = \begin{pmatrix} -1 & 0 & 2 \\ 4 & 1 & -1 \\ 2 & 0 & 1 \end{pmatrix}$; $B = \begin{pmatrix} 2 & 1 & 0 \\ -1 & 0 & 2 \\ 4 & -3 & -1 \end{pmatrix}$

6. $A = \begin{pmatrix} 1 & 0 & -2 & 3 \\ 2 & 1 & 4 & -1 \end{pmatrix}$; $B = \begin{pmatrix} 5 & -1 \\ 0 & 2 \\ -1 & 3 \\ 4 & 2 \end{pmatrix}$

In Exercises 7–10, write each system of equations as a matrix equation.

7. $\begin{cases} 2x + 5y = -3 \\ x - 2y = 1 \end{cases}$

8. $\begin{cases} x - 2y = 1 \\ 2x - 5y = 3 \end{cases}$

9. $\begin{cases} 5x - 7y + z = 2 \\ 2x - 3y - z = 3 \\ x + y + z = -3 \end{cases}$

10. $\begin{cases} 2x + 3y - z = 2 \\ 2x - 3y + 2z = -1 \\ -x - y + 3z = -4 \end{cases}$

In Exercises 11–14, write each matrix equation as a system of linear equations.

11. $\begin{pmatrix} 3 & -1 \\ 2 & 4 \end{pmatrix} \begin{pmatrix} x \\ y \end{pmatrix} = \begin{pmatrix} -1 \\ 3 \end{pmatrix}$

12. $\begin{pmatrix} 2 & 4 \\ -1 & -2 \end{pmatrix} \begin{pmatrix} x \\ y \end{pmatrix} = \begin{pmatrix} 5 \\ -2 \end{pmatrix}$

13. $\begin{pmatrix} 1 & 0 & -3 \\ 2 & -1 & 3 \\ -2 & 3 & -4 \end{pmatrix} \begin{pmatrix} x \\ y \\ z \end{pmatrix} = \begin{pmatrix} 3 \\ -1 \\ 2 \end{pmatrix}$

14. $\begin{pmatrix} 1 & -1 & 0 \\ 2 & 1 & -3 \\ -1 & 1 & 2 \end{pmatrix} \begin{pmatrix} x \\ y \\ z \end{pmatrix} = \begin{pmatrix} 3 \\ -1 \\ 4 \end{pmatrix}$

In Exercises 15–20, determine whether each matrix has an inverse and if so, find it. When the matrix is larger than 2×2, use a calculator or computer to answer the question.

15. $\begin{pmatrix} 2 & -3 \\ 4 & -1 \end{pmatrix}$

16. $\begin{pmatrix} 1 & -3 \\ 2 & 4 \end{pmatrix}$

17. $\begin{pmatrix} 6 & 3 \\ 4 & 2 \end{pmatrix}$

18. $\begin{pmatrix} -1 & -2 \\ -4 & -8 \end{pmatrix}$

19. $\begin{pmatrix} 2 & 0 & 1 \\ 4 & 1 & 2 \\ 2 & 0 & 4 \end{pmatrix}$

20. $\begin{pmatrix} 3 & -1 & 0 \\ 0 & 0 & 2 \\ -1 & 2 & 0 \end{pmatrix}$

In Exercises 21–26, solve each system of equations using matrix multiplication.

21. $\begin{cases} 2x - 3y = 7 \\ 4x + y = 2 \end{cases}$

22. $\begin{cases} x + 2y = 5 \\ -x + 3y = 6 \end{cases}$

23. $\begin{cases} 3x - 2y = 6 \\ x + y = 2 \end{cases}$

24. $\begin{cases} 7x - 5y = 12 \\ 2x + 3y = 4 \end{cases}$

25. $\begin{cases} x + 2y + z = -1 \\ x - 3y + 2z = 1 \\ 2x - 3y + z = 5 \end{cases}$

26. $\begin{cases} 2x + y + z - w = 1 \\ 2x - y + z + w = -2 \\ -x + y - z + w = -3 \\ x - 2y + z - w = 1 \end{cases}$

Exercises 27–31 refer to the following **problem situation**: For

the 2×2 matrix

$$A = \begin{pmatrix} a & b \\ c & d \end{pmatrix},$$

the polynomial

$$C(x) = \begin{vmatrix} a - x & b \\ c & d - x \end{vmatrix} = (a - x)(d - x) - bc$$

is called the *characteristic polynomial* of matrix A. Let

$$A = \begin{pmatrix} 3 & 2 \\ 1 & 5 \end{pmatrix} \quad \text{and} \quad B = \begin{pmatrix} 2 & -1 \\ -5 & 2 \end{pmatrix}.$$

27. Find the characteristic polynomials for matrices A and B.

28. For each characteristic polynomial from Exercise 27 find a complete graph.

29. The roots of the characteristic polynomial of a matrix A are called the *eigenvalues* of the matrix. Find the eigenvalues of the matrices A and B.

30. Compare det A and det B with the y-intercept of the characteristic polynomial of each. What is your conjecture?

31. Add the numbers on the main diagonal of A and B ($a_{11} + a_{22}$ and $b_{11} + b_{22}$). Compare this sum with the eigenvalues of each matrix. What is your conjecture?

Exercises 32–37 refer to this **problem situation**: Each of the following sets of points represents data from some problem situation. Determine a polynomial $f(x) = a_n x^n + a_{n-1} x^{n-1} + \ldots + a_1 x + a_0$ that contains the given points.

a) Write a system of linear equations that determine the coefficients of the polynomial.

b) Write a matrix equation $AX = B$ that is equivalent to the system in (a).

c) Determine the coefficients of the desired polynomial (by matrix methods $X = A^{-1}B$).

d) Graph the data points.

e) Support the solution in part (c) by overlaying the graph of the polynomial with the graph of the data points in (d).

32. (2, 3), (5, 8), (7, 2)

33. (2, 8), (6, 3), (9, 4)

34. (2, 3), (5, 8), (7, 2), (9, 4)

35. (2, 8), (4, 5), (6, 3), (9, 4)

36. (−2, −4), (1, 2), (3, 6), (4, −2), (7, 8)

37. (−1, 8), (1, 2), (4, −6), (7, 5), (8, 2)

38. Generalize The characteristic polynomial of an $n \times n$ matrix A is $C(x) = \det(A - xI_n)$. For the matrices in Exercises 19 and 20, graph $C(x)$ to find the eigenvalues.

10.4 _____

Matrices, Transformations, and Conic Sections

In this section, we will study the rotation transformation. We begin with an Exploration.

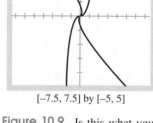

[−7.5, 7.5] by [−5, 5]

Figure 10.9 Is this what your grapher looks like?

⌕ EXPLORE WITH A GRAPHING UTILITY

Set a graphing calculator as follows:

- Deg mode, Param mode
- Range: $T\text{min} = -3, T\text{max} = 3, T\text{step} = 0.1$
- Viewing Rectangle: $[-7.5, 7.5]$ by $[-5, 5]$
- Define: $X_{1T}(T) = T, Y_{1T}(T) = T^3$ and
  $$X_{2T}(T) = X_{1T}\cos\theta - Y_{1T}\sin\theta, \quad Y_{2T}(T) = X_{1T}\sin\theta + Y_{1T}\cos\theta$$
- Enter (on the home screen) $45 \to \theta$

Describe what you see when you press the graph key. How does the second curve drawn compare to the first graph drawn? (See Fig. 10.9.)

1. How does the graph change when you enter $30 \to \theta$? Or $-30 \to \theta$?

2. What happens to the graph when you enter $120 \to \theta$?

Complete a Generalization How do the second set of parametric equations transform the curve generated by the first pair of equations?

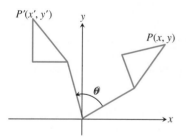

Figure 10.10 A rotation of angle θ.

This Exploration provides visual experience with rotation transformations. Figure 10.10 shows a triangle with vertex $P(x, y)$. When this triangle is rotated counterclockwise through angle θ, its image falls upon the triangle with vertex $P'(x', y')$.

The Rotation Equations

A **rotation about the origin through angle θ** is a transformation that turns the entire coordinate plane with the origin being the center of the turn. If θ is positive, the turn is counterclockwise, and if θ is negative, the turn is clockwise.

A rotation transformation can be defined in terms of the coordinates of the points in the coordinate plane. For example, in Fig. 10.11 point $P'(x', y')$ is the rotation image of point $P(x, y)$ about the origin through an angle θ. Applying

definition 6.3, we conclude that

$$x = r \cos \alpha \tag{1}$$

$$y = r \sin \alpha$$

$$x' = r \cos(\alpha + \theta) \tag{2}$$

$$y' = r \sin(\alpha + \theta)$$

Using the sum identities of Theorems 7.6 and 7.7, the second pair of equations can be rewritten as follows:

$$x' = r \cos(\alpha + \theta) = r(\cos \alpha \cos \theta - \sin \alpha \sin \theta) \tag{3}$$

$$y' = r \sin(\alpha + \theta) = r(\sin \alpha \cos \theta + \cos \alpha \sin \theta)$$

Substituting Eqs. (1) into Eqs. (3), we obtain

$$x' = x \cos \theta - y \sin \theta$$

$$y' = x \sin \theta + y \cos \theta.$$

We have verified Theorem 10.5.

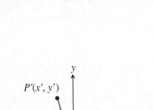

Figure 10.11 Point P and its rotation image P'.

Stan says, "The beauty in the mathematics of rotations of conics has been hidden in the prior emphasis placed upon the algebraic skills needed to produce a graph. With technology, students can see not only the beauty in the graphs, but also the ease with which transformations may be explored."

Theorem 10.5   Rotation through Angle θ

The rotation through angle θ maps each point $P(x, y)$ in the rectangular coordinate plane to the point $P'(x', y')$, where

$$x' = x \cos \theta - y \sin \theta$$

$$y' = x \sin \theta + y \cos \theta.$$

This next example shows how the experience of the exploration relates to Theorem 10.5.

E X A M P L E   1   Finding a Rotation

Suppose that curve C is the locus of points determined by the parametric equations

$$x = t \qquad -3 \le t \le 3$$

$$y = t^3.$$

Find the parametric equations of the curve C', which is a 30° counterclockwise rotation.

Solution The angle of rotation is $\theta = 30°$. Because $\cos 30° = \dfrac{\sqrt{3}}{2}$ and $\sin 30° = \dfrac{1}{2}$, the transformation equations used to find curve C' are

$$x' = \frac{\sqrt{3}}{2}x - \frac{1}{2}y$$

$$y' = \frac{1}{2}x + \frac{\sqrt{3}}{2}y.$$

By substituting the parametric equations $x = t$, $y = t^3$ into these equations, we obtain the parametric equations for curve C'. They are

$$x' = \frac{\sqrt{3}}{2}t - \frac{1}{2}t^3$$

$$y' = \frac{1}{2}t + \frac{\sqrt{3}}{2}t^3.$$

Figure 10.12 shows the curve C and its rotation image. Compare this figure to Fig. 10.9. ▤

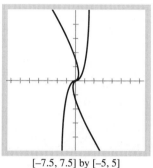

[–7.5, 7.5] by [–5, 5]

Figure 10.12 Compare this figure with the exploration.

Matrix Multiplication and Rotations

A rotation about the origin can be described by a matrix equation. The equations of Theorem 10.5,

$$x' = x \cos \theta - y \sin \theta$$

$$y' = x \sin \theta + y \cos \theta,$$

are equivalent to the following matrix product:

$$\begin{pmatrix} x' \\ y' \end{pmatrix} = \begin{pmatrix} \cos \theta & -\sin \theta \\ \sin \theta & \cos \theta \end{pmatrix} \cdot \begin{pmatrix} x \\ y \end{pmatrix}$$

Consequently, the rotation image of any curve can be expressed as the product of matrices, as illustrated in Example 2.

Notes on Exercises

Ex. 20–21 practice the skills in Example 2.

E X A M P L E 2 Using Matrices to Rotate

Suppose the curve C is the locus of points determined by the parametric equations

$$x = t^2 \qquad -3 \le t \le 3$$

$$y = t^3.$$

Write a matrix equation that can be used to find the 35° clockwise rotation of curve C. Use matrix multiplication to find parametric equations of curve C', a 35°

clockwise rotation of this curve. Find this curve's complete graph and its rotation image.

Solution   A 35° clockwise rotation is represented using the angle $\theta = -35°$. Therefore, by Theorem 10.5, the matrix to use is

$$\begin{pmatrix} \cos(-35°) & -\sin(-35°) \\ \sin(-35°) & \cos(-35°) \end{pmatrix}.$$

By using matrix multiplication, we find x' and y' in terms of x and y.

$$\begin{pmatrix} x' \\ y' \end{pmatrix} = \begin{pmatrix} \cos(-35°) & -\sin(-35°) \\ \sin(-35°) & \cos(-35°) \end{pmatrix} \cdot \begin{pmatrix} x \\ y \end{pmatrix}$$

$$\begin{pmatrix} x' \\ y' \end{pmatrix} = \begin{pmatrix} \cos(-35°)x - \sin(-35°)y \\ \sin(-35°)x + \cos(-35°)y \end{pmatrix}$$

Then substituting $x = t^2$ and $y = t^3$, we see that the parametric equations for C' are

$$x' = \cos(-35°)t^2 - \sin(-35°)t^3 \tag{4}$$

$$y' = \sin(-35°)t^2 + \cos(-35°)t^3$$

(see Fig.10.13).   ≡

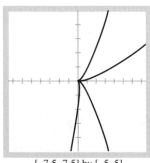

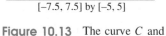
[−7.5, 7.5] by [−5, 5]

Figure 10.13   The curve C and its 35° clockwise rotation.

Rotations of Conic Sections

The techniques of the last subsection can be applied to any curve that is defined with parametric equations. We begin this section by verifying parametric equations that describe an ellipse.

GRAPHER REMINDER

Rather than use equations (4) in your grapher, you can use the structure of the earlier Exploration.

$X_{1T}(T) = T^2$

$Y_{1T}(T) = T^3$

$X_{2T}(T) = X_{1T} \cos\theta - Y_{1T} \sin\theta$

$Y_{2T}(T) = X_{1T} \sin\theta + Y_{1T} \cos\theta$

Then store $-35°$ as θ.

E X A M P L E   3   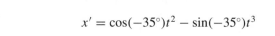 Parametric Equations for an Ellipse

Show that the points on the curve C determined by equations

$$x = a\cos t \qquad 0 \le t \le 2\pi$$

$$y = b\sin t$$

are on the ellipse

$$\frac{x^2}{a^2} + \frac{y^2}{b^2} = 1. \tag{5}$$

Common Errors

Students should again be cautioned to pay attention to the MODE in which their calculators are set. Many of the problems in this section involve arguments expressed in degrees, but some involve arguments expressed in radians.

Solution   We shall show that the points $x = a \cos t$ and $y = b \sin t$ that determine curve C satisfy equation (5).

$$\frac{x^2}{a^2} + \frac{y^2}{b^2} = 1$$

$$\frac{(a \cos t)^2}{a^2} + \frac{(b \sin t)^2}{b^2} = 1$$

$$\frac{a^2 \cos^2 t}{a^2} + \frac{b^2 \sin^2 t}{b^2} = 1$$

$$\cos^2 t + \sin^2 t = 1 \qquad \text{This equation is a basic trigonometric identity.}$$

Notice that the curve determined by the parametric equations

$$x = a \cos t + h \tag{6}$$

$$y = b \sin t + k$$

can be obtained from curve C by a horizontal shift of h units followed by a vertical shift of k units. In other words, the parametric equations (6) are the equations for the general ellipse with center (h, k).

E X A M P L E   4   A Rotation of an Ellipse

Find the $\dfrac{3\pi}{4}$ radian counterclockwise rotation image about the origin of the ellipse E defined by the parametric equations

$$x = 3.5 \cos t + 3$$

$$y = 2 \sin t - 2.$$

Solution   Describe the rotation as a product of matrices.

$$\begin{pmatrix} x' \\ y' \end{pmatrix} = \begin{pmatrix} \cos \dfrac{3\pi}{4} & -\sin \dfrac{3\pi}{4} \\ \sin \dfrac{3\pi}{4} & \cos \dfrac{3\pi}{4} \end{pmatrix} \cdot \begin{pmatrix} x \\ y \end{pmatrix}$$

The parametric equations for the rotation image of E are

$$x' = \cos \frac{3\pi}{4} x - \sin \frac{3\pi}{4} y$$

$$y' = \sin \frac{3\pi}{4} x + \cos \frac{3\pi}{4} y.$$

Figure 10.14 shows a complete graph of both E and its rotation image.

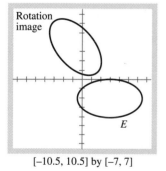

[–10.5, 10.5] by [–7, 7]

Figure 10.14   Ellipse E and its rotation image.

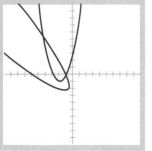

[−10, 10] by [−10, 10]

Figure 10.15 The parabola $x = t - 2$ and $y = t^2 - 1$ and its rotation image.

E X A M P L E 5   Using Parametric Equations for a Conic

Eliminate the parameter t to determine the conic that is defined by the parametric equations

$$x = t - 2$$
$$y = t^2 - 1.$$

Solution   Solve for t to obtain $t = x + 2$. Next, substitute t into the second equation to obtain

$$y = (x + 2)^2 - 1.$$

We see that this curve can be obtained from the graph of $y = x^2$ by a horizontal shift left 2 units followed by a vertical shift down 1 unit. It is a parabola. ▤

In this next example, we find the $45°$ counterclockwise rotation image of this parabola.

E X A M P L E 6   Rotating a Parabola

Find a complete graph of both the parabola P defined by

$$x = t - 2$$
$$y = t^2 - 1$$

and its image under a $45°$ counterclockwise rotation.

Solution   The parametric equations for the rotation image of P are

$$x' = \cos\frac{\pi}{4}x - \sin\frac{\pi}{4}y$$
$$y' = \sin\frac{3\pi}{4}x + \cos\frac{\pi}{4}y.$$

A complete graph is shown in Fig. 10.15. ▤

This same method can be used to find the rotation image of a hyperbola provided we can find parametric equations that describe the hyperbola. Exercises 28 through 31 relate to this task.

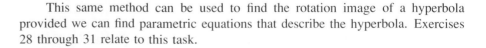

Exercises for Section 10.4

For Exercises 1–7, suppose C is the curve determined by the parametric equations $x = t^3$ and $y = t$. Use the equations found in Theorem 10.5 to write the parametric equations for a rotation about the origin with the specified angle of rotation.

1. A rotation about the origin whose angle is $30°$.

2. A rotation about the origin whose angle is $78°$.

3. A rotation about the origin whose angle is $120°$.

4. A rotation about the origin whose angle is $\pi/3$.

5. A rotation about the origin whose angle is $\pi/6$.

6. A rotation about the origin whose angle is $\pi/8$.

7. A rotation about the origin whose angle is $5\pi/12$.

For Exercises 8–17, write a matrix that you could use to describe the rotation for the given value of θ, the angle of rotation.

8. $\theta = 45°$ 9. $\theta = 30°$
10. $\theta = 120°$ 11. $\theta = -60°$
12. $\theta = -45°$ 13. $\theta = 21°$
14. $\theta = \pi/2$ 15. $\theta = 3\pi/4$
16. $\theta = 5\pi/6$ 17. $\theta = 2\pi/3$

18. Show that the points on the curve C determined by the parametric equations

$$x = 3t$$
$$y = t^2$$

are on a parabola. Specify the parabola.

19. Show that the points on the curve C determined by the parametric equations

$$x = t + 2$$
$$y = t^2 - 3$$

are on a parabola. Specify the parabola.

For Exercises 20 and 21 suppose that C' represents the rotation image for the rotation about the origin that is specified for the given curve. Write a matrix equation that describes curve C'.

20. Let C be the curve defined by $x = t^2$ and $y = (t - 3)/2$ and let the rotation be determined by angle $\theta = 45°$.

21. Let C be the curve defined by $x = t^2$ and $y = (t - 3)/2$ and let the rotation be determined by angle $\theta = 60°$.

For Exercises 22–25, find a complete graph of the rotation image for a rotation with angle θ of the curve C determined by the given parametric equations.

22. $\theta = \pi/3$ and C is determined by

$$x = 3\cos t$$
$$y = 2\sin t.$$

23. $\theta = 5\pi/6$ and C is determined by

$$x = 5\cos t$$
$$y = 3\sin t.$$

24. $\theta = 3\pi/8$ and C is determined by

$$x = \frac{t}{5}$$
$$y = t^2 - 3.$$

25. $\theta = 5\pi/6$ and C is determined by

$$x = 2t - 4$$
$$y = t^2 + 1.$$

26. Let f be the function defined by $f(x) = \dfrac{e^x + e^{-x}}{2}$. Find a complete graph of f and convince yourself that f is the same function that your calculator describes as $y = \cosh x$.

27. Let g be the function defined by $g(x) = \dfrac{e^x - e^{-x}}{2}$. Find a complete graph of g and convince yourself that g is the same function that your calculator describes as $y = \sinh x$.

28. Use the fact that $\dfrac{e^x + e^{-x}}{2} = \cosh x$ and $\dfrac{e^x - e^{-x}}{2} = \sinh x$ to show that the points on the curve determined by the parametric equations

$$x = \cosh t$$
$$y = \sinh t$$

all fall on the hyperbola $x^2 - y^2 = 1$.

29. Show that the points on the curve determined by the parametric equations

$$x = -\cosh t$$
$$y = \sinh t$$

all fall on the hyperbola $x^2 - y^2 = 1$.

30. Find a complete graph of the curve C determined by the parametric equations

$$x = \cosh t \quad -3 \le t \le 3$$

$$y = \sinh t$$

and the parametric equations

$$x = -\cosh t \quad -3 \le t \le 3$$

$$y = \sinh t.$$

31. If C is the curve determined by the parametric equations

$$x = \cosh t \quad -3 \le t \le 3$$

$$y = \sinh t,$$

find a complete graph of both C and its rotation image C' for a rotation about the origin with an angle of $\pi/3$ radians.

10.5 ———— Systems of Equations and Inequalities with Nonlinear Relations

Objective

Students will be able to use graphical methods to solve systems of equations and inequalities that are nonlinear.

Key Ideas

General second-degree equation

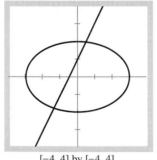

[−4, 4] by [−4, 4]

Figure 10.16 $4x^2 + 9y^2 = 36$ and $2x - y = -1$.

In Sections 10.1 through 10.3, the equations in the systems of equations considered were primarily linear. In this section, we consider systems of equations like

$$4x^2 + 9y^2 = 36$$

$$2x - y = 0,$$

whose first equation defines a relation that is not a function. A graphical method can be used to solve this type of system once you learn how to find the graph of the relation.

In Section 2.6, Example 7, the relation $4x^2 + 9y^2 = 36$ was graphed by finding the two functions

$$y = \left(\frac{1}{3}\sqrt{36 - 4x^2}\right) \quad \text{and} \quad y = -\left(\frac{1}{3}\sqrt{36 - 4x^2}\right)$$

in the same viewing rectangle. Example 1 uses this method.

E X A M P L E   1   Solving a Nonlinear System of Equations

Solve the system

$$4x^2 + 9y^2 = 36 \tag{1a}$$

$$2x - y = -1. \tag{1b}$$

Solution   Begin by graphing $y = \frac{1}{3}\sqrt{36 - 4x^2}$, $y = -\frac{1}{3}\sqrt{36 - 4x^2}$, and $y = 2x + 1$ in the viewing rectangle $[-4, 4]$ by $[-4, 4]$ (see Fig. 10.16).

Use zoom-in to find that the two solutions are $(0.487, 1.973)$ and $(-1.387, -1.773)$ with an error of at most 0.01. ▤

Finding the Graph of a Second-degree Equation

Equation (1a) in Example 1 is a special case of the **general second-degree equation** in two variables,

$$Ax^2 + Bxy + Cy^2 + Dx + Ey + F = 0,$$

where A, B, C, D, E, and F are real number constants. You learned in Theorem 9.8 that the graph of an equation of this form is called a conic section. To find the graph of this general second-degree equation, for $C \neq 0$, we can rewrite the equation as a quadratic in y^2 and use the quadratic formula as follows:

$$Cy^2 + (Bx + E)y + (Ax^2 + Dx + F) = 0$$

$$y = \frac{-(Bx + E) \pm \sqrt{(Bx + E)^2 - 4C(Ax^2 + Dx + F)}}{2C} \tag{2}$$

A complete graph of equation (2) can be obtained by finding the graphs of

$$f(x) = \frac{-(Bx + E) + \sqrt{(Bx + E)^2 - 4C(Ax^2 + Dx + F)}}{2C}$$

and

$$g(x) = \frac{-(Bx + E) - \sqrt{(Bx + E)^2 - 4C(Ax^2 + Dx + F)}}{2C}$$

in an appropriate viewing rectangle.

We can use these equations to complete a solution to Example 2.

E X A M P L E   2   Solving a Nonlinear System of Equations

Find the simultaneous solutions to the system

$$x^2 - 2xy + \quad y^2 - \quad 8x - \quad 8y + \quad 48 = 0 \tag{3a}$$

$$5x^2 + \quad xy + 6y^2 - 79x - 73y + 196 = 0. \tag{3b}$$

Solution   We begin by rewriting each equation in system (3) using the quadratic formula

$$y = \frac{2x + 8 \pm \sqrt{(-2x - 8)^2 - 4(x^2 - 8x + 48)}}{2}$$

$$y = \frac{73 - x \pm \sqrt{(-73 + x)^2 - 24(5x^2 - 79x + 196)}}{12}.$$

Notice that four functions are represented here. Two of them form the parabola, the other two form the ellipse (see Fig. 10.17). Use zoom-in to find that the solutions are (3.26, 11.74) and (15.05, 4.60).   ≡

Teaching Note

Students may need some review of algebra in preparation for finding graphs of second-degree equations in two variables. Demonstrate how to rewrite an equation as a quadratic in y using the quadratic formula. This will help students to review the algebraic skills needed to enter a general second-degree equation into a graphing calculator.

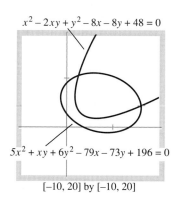

$x^2 - 2xy + y^2 - 8x - 8y + 48 = 0$

$5x^2 + xy + 6y^2 - 79x - 73y + 196 = 0$

[−10, 20] by [−10, 20]

Figure 10.17   System (3). The two equations yield four *functions* but when graphed, only two distinct curves.

Application: Locating a Position

Three observers are strategically located with synchronized watches. Person A is located relative to a large coordinate system at the origin $(0, 0)$; person B is at point $(0, 4000)$, and person C is at point $(7000, 0)$. Observer B hears the sound of a gun 2 sec before observer A, and observer C hears the sound 4 sec before observer A.

We want to find the point $P(x, y)$, where the gun is located (see Fig. 10.18).

Using the information given in the problem situation and the fact that sound travels at the rate of 1100 ft/sec, we draw the following conclusions. Why?

$$d(P, A) - d(P, B) = 2200 \quad \text{and} \quad d(P, A) - d(P, C) = 4400 \quad \textbf{(4)}$$

Using the distance formula, we obtain

$$d(P, A) - d(P, B) = 2200$$

$$\sqrt{(x - 0)^2 + (y - 0)^2} - \sqrt{(x - 0)^2 + (y - 4000)^2} = 2200. \quad \textbf{(5)}$$

By completing the rather long and tedious algebraic simplification, we conclude that Eq. (5) simplifies to

$$\frac{(y - 2000)^2}{(1100)^2} - \frac{x^2}{2,790,000} = 1.$$

In a similar fashion, the equation $d(P, A) - d(P, C) = 4400$ simplifies to

$$\frac{(x - 3500)^2}{(2200)^2} - \frac{y^2}{7,410,000} = 1.$$

The gun is located at a point that lies on both of these two hyperbolas, so we need to find the simultaneous solutions to this pair of equations.

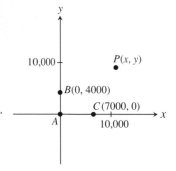

Figure 10.18 Observers are located at points A, B, and C. They want to find the location P of a gun.

E X A M P L E   3   Solving a Nonlinear System of Equations

Find the simultaneous solutions to the hyperbolas

$$\frac{(x - 3500)^2}{(2200)^2} - \frac{y^2}{7,410,000} = 1 \quad \textbf{(6a)}$$

$$\frac{(y - 2000)^2}{(1100)^2} - \frac{x^2}{2,790,000} = 1. \quad \textbf{(6b)}$$

Solution To find the graph of each of these equations, solve each one for y. Equation (6a) becomes

$$y = \pm \sqrt{7,410,000 \left[\frac{(x - 3500)^2}{(2200)^2} - 1 \right]},$$

and Eq. (6b) becomes

$$y = 2000 \pm 1100\sqrt{\frac{x^2}{2,790,000} + 1}.$$

From the graphs of these four functions (see Fig. 10.19), it is apparent there are four points of intersection and thus four simultaneous solutions. Use zoom-in to determine that the four solutions are the points (162.63, 3105.20), (1241.99, 629.24), (6414.37, −2365.08), and (11,714.31, 9792.52). ≡

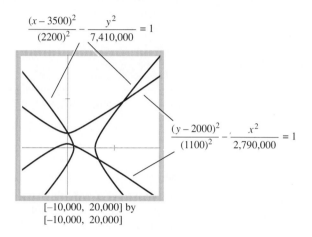

$$\frac{(x - 3500)^2}{(2200)^2} - \frac{y^2}{7,410,000} = 1$$

$$\frac{(y - 2000)^2}{(1100)^2} - \frac{x^2}{2,790,000} = 1$$

[−10,000, 20,000] by
[−10,000, 20,000]

Figure 10.19   System of equations (6). Note the four curves with four points of intersection.

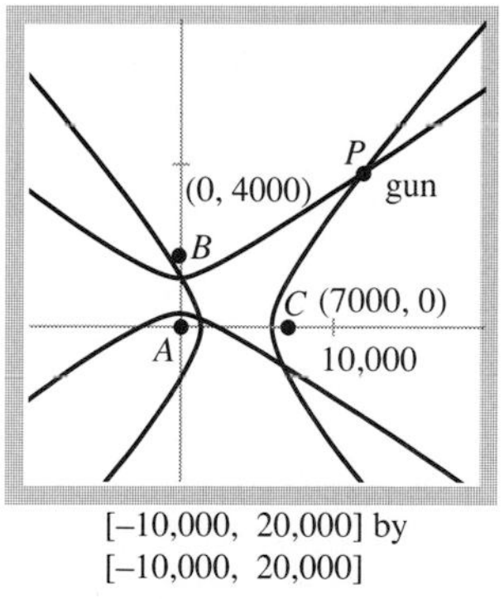

Figure 10.20   The gun is located at one of the four points of intersection of system (6).

[−10,000, 20,000] by
[−10,000, 20,000]

E X A M P L E   4   APPLICATION: Locating a Point

Determine the position of the gun in the problem that led to the system of equations solved in Example 3.

Solution   The calculations in Example 3 resulted in four possible locations for the gun.

Figure 10.20 shows the original locations A, B, and C of the observers (from Fig. 10.18) and the curves of system (6) and their four points of intersection (from Fig. 10.19). The point P must be one of these four points.

Because $d(P, A) - d(P, C) = 4400 > 0$ (see Eq. 4), P must be farther from A than C, which indicates that the gun location must be either (6414.37, −2365.08) or (11,714.31, 9792.52). Further, because $d(P, A) - d(P, B) = 2200 > 0$ (see Eq. 4), the point P also must be farther from A than B. This condition narrows the choice to the point (11,714.31, 9792.52).

The gun is located at point (11,714.31, 9792.52). ≡

This same technique can be used in navigation to determine location by using radio waves instead of sound waves.

Solving Systems of Inequalities

The technique for solving a system of inequalities is the same as that for solving equalities. First, sketch the graph of the solution set for each individual inequality. Then overlay these solution sets and identify the points that belong to each solution set of the system. This process is a graphical means of finding the intersection of the individual solution sets. This intersection is the solution to the system of inequalities.

E X A M P L E   5   Solving a System of Inequalities

Solve the system

$$y > x^2$$

$$2x + 3y < 4. \tag{7}$$

Solution   The graph of the equation $y = x^2$ is the boundary of the region that is the solution to the inequality $y > x^2$. Because $(0, 2)$ is a solution to the inequality, the solution must consist of all the points *inside* the parabola; the points on the boundary are not included. (Why?) (See Fig. 10.21a).

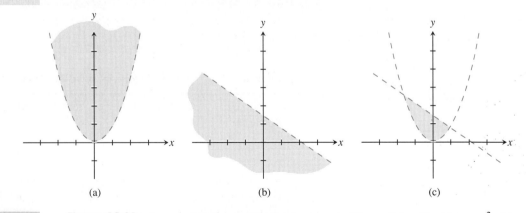

(a)   (b)   (c)

Figure 10.21   Steps in solving graphically the system of inequalities (7): (a) $y > x^2$; (b) $2x + 3y < 4$; (c) graph of the system. Dashed lines indicate that the points on the line and parabola are not part of the solution.

The solution to the inequality $2x + 3y < 4$ is the region below the line $2x + 3y = 4$. The points on this line are not included in the solution (see Fig. 10.21b).

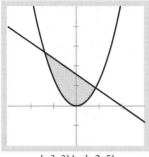

[−3, 3] by [−2, 5]

Figure 10.22
SHADE $(x^2, (1/3)(4 − 2x))$ is the solution to system (7).

Notes on Exercises

Ex. 21–23, 25, and 26 are problem situations that use concepts similar to those in Example 7.

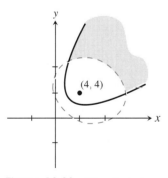

Figure 10.23   The shaded region is the solution to the system of inequalities (8).

Assignment Guide

Day 1: Ex. 1, 4, 7, 10, 11, 14, 15, 18, 19
Day 2: Ex. 21–26, 28

Therefore the solution is the region below the line $2x + 3y < 4$ and above the parabola $y = x^2$ (Fig. 10.21c). Figure 10.22 shows the result of finding the solution directly on a graphing utility with a shade function.   ≡

E X A M P L E   6   Solving a System of Inequalities

Solve the system

$$x^2 − 2xy + y^2 − 8x − 8y + 48 \le 0 \qquad \textbf{(8a)}$$

$$5x^2 + xy + 6y^2 − 79x − 73y + 196 > 0. \qquad \textbf{(8b)}$$

Solution   Complete graphs of $x^2 − 2xy + y^2 − 8x − 8y + 48 = 0$ and $5x^2 + xy + 6y^2 − 79x − 73y + 196 = 0$ were found in Example 2 (see Fig. 10.17). We complete the solution to system (8) by checking to see where the test point $(4, 4)$ falls.

Check $(4, 4)$ in the first inequality:

$$4^2 − 2(4)(4) + 4^2 − 8(4) − 8(4) + 48 = −16$$

$(4, 4)$ satisfies the first inequality, so all other points inside the parabola are also solutions.

Check $(4, 4)$ in the second inequality:

$$5(4^2) + (4)(4) + 6(4^2) − 79(4) − 73(4) + 196 = −220$$

$(4, 4)$ does not satisfy the second inequality, so points outside the ellipse are solutions.

Because the point $(4, 4)$ does not satisfy inequality (8b), the inside of the ellipse is not shaded. Therefore the solution is the region on or inside the parabola and outside the ellipse (see Fig. 10.23).   ≡

The final example of this section involves a system of inequalities as an algebraic representation of a problem situation.

E X A M P L E   7   APPLICATION: Determining Paint Mixtures

White paint at $1.00/cm^3$ is mixed with red paint that costs $0.20/cm^3$ in a container with a capacity of 1050 cm^3. If x is the number of cubic centimeters of red paint, and y is the number of cubic centimeters of white paint, color mixologists have determined that the product of x and y must be greater than 100,000 to produce eye-pleasing colors. What mixtures are possible that produce eye-pleasing colors and have a total cost less than or equal to $500?

a) Determine an algebraic representation of the problem situation.
b) Draw a graph of the algebraic representation and shade the portion of the graph that represents a solution to the problem situation.

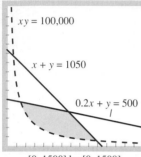
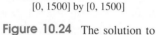

$[0, 1500]$ by $[0, 1500]$

Figure 10.24 The solution to the system of inequalities (9).

Solution

a) Choose variables as follows:

x : the number of cm^3 of red paint

y : the number of cm^3 of white paint

$x + y \leq 1050$	The capacity of the container is 1050 cm^3.	**(9a)**
$0.2x + y \leq 500$	The total cost must be no more than \$500.	**(9b)**
$xy > 100,000$	This condition ensures eye-pleasing colors.	**(9c)**

It is also true that $x \geq 0$ and $y \geq 0$. (Why?)

b) A complete graph of this system appears in Fig. 10.24. Notice that the shaded region includes the points on and below the lines $x + y = 1050$ and $0.2x + y = 500$ and above but not on the curve $xy = 100,000$. ▤

Exercises for Section 10.5

In Exercises 1–10, find the simultaneous solutions to each system.

1. $\begin{cases} 4x^2 + 9y^2 = 36 \\ x + 2y = 2 \end{cases}$

2. $\begin{cases} 4x^2 + 9y^2 = 36 \\ x - 2y = 2 \end{cases}$

3. $\begin{cases} 9x^2 - 4y^2 = 36 \\ x + 2y = 4 \end{cases}$

4. $\begin{cases} 9x^2 - 4y^2 = 36 \\ x - 2y = 4 \end{cases}$

5. $\begin{cases} 9x^2 - 4y^2 = 36 \\ 4x^2 + 9y^2 = 36 \end{cases}$

6. $\begin{cases} 9x^2 - 4y^2 = 36 \\ 9x^2 + 4y^2 = 36 \end{cases}$

7. $\begin{cases} x^2 + 4y^2 = 4 \\ y = 2x^2 - 3 \end{cases}$

8. $\begin{cases} x^2 - 4y^2 = 4 \\ y = 2x^2 - 3 \end{cases}$

9. $\begin{cases} x^2 - 4y^2 = 4 \\ x = 2y^2 - 3 \end{cases}$

10. $\begin{cases} x^2 + 4y^2 = 4 \\ x = 2y^2 - 3 \end{cases}$

In Exercises 11–14, draw a complete graph of each equation and find the simultaneous solutions to each system.

11. $\begin{cases} x^2 + xy + y^2 + x + y - 6 = 0 \\ 2x^2 - 3xy + 2y^2 + x + y - 8 = 0 \end{cases}$

12. $\begin{cases} 5x^2 - 40xy + 20y^2 - 17x + 25y + 50 = 0 \\ xy - 3x = 0 \end{cases}$

13. $\begin{cases} 2x^2 - 3xy + 2y^2 + x + y - 8 = 0 \\ 2x^2 - 8xy + 3y^2 + x + y - 10 = 0 \end{cases}$

14. $\begin{cases} 3x^2 - 5xy + 6y^2 - 7x + 5y - 9 = 0 \\ x^2 + y^2 - 2x - 6 = 0 \end{cases}$

In Exercises 15–20, draw a graph representing the solution to each system of inequalities.

15. $\begin{cases} y \geq x^2 \\ x^2 + y^2 \leq 4 \end{cases}$

16. $\begin{cases} y \leq x^2 + 4 \\ x^2 + y^2 \geq 4 \end{cases}$

17. $\begin{cases} x^2 - y^2 < 4 \\ x^2 + y^2 < 4 \end{cases}$

18. $\begin{cases} x^2 + 4y^2 - 2x + 4y - 6 > 0 \\ y < \frac{1}{2}x + 1 \end{cases}$

19. $\begin{cases} x^2 + 2xy - 5y^2 + 2x + 4y - 10 > 0 \\ 4x^2 + xy + 4y^2 - 5x + 5y - 15 < 0 \end{cases}$

20. $\begin{cases} x^2 + 2xy + 5y^2 + 2x + 4y - 10 > 0 \\ 5y^2 + 2x + 3y - 6 < 0 \end{cases}$

Exercises 21–23 refer to the following **problem situation.** Consider an acid solution of 84 oz that is 58% pure acid. How many ounces of pure acid (x) must be added to obtain a solution that is 70% to 80% acid? Let y be the concentration of acid in the final solution.

21. Find an algebraic representation of this problem situation consisting of a system of two inequalities and one equation whose simultaneous solution can be used to solve the problem situation.

22. Find a complete graph of the system of equations in Exercise 21.

23. Find the values of x and y that make sense in this problem situation and solve the system.

Exercise 24 refers to the following **problem situation.** Suppose three observers A, B, and C are listening for illegal dynamite explosions. They are positioned so the angle of the line between A and B and the line between B and C is $90°$. Further suppose that A and B are 6000 ft apart and that B and C are 2000 ft apart. A hears the explosion 4.06 sec before B, and C hears the explosion 0.63 sec before B.

24. The system of equations

$$\frac{(x - 3000)^2}{2233^2} - \frac{y^2}{4,013,711} = 1$$

$$\frac{(y - 1000)^2}{346.5^2} - \frac{x^2}{879,937.75} = 1$$

is an algebraic representation of this problem situation. Draw a complete graph of this situation. Find the location of the dynamite assuming it is located in the first quadrant.

Exercises 25 and 26 refer to the following **problem situation.** Hush Dog shoes are made from two types of leather, A and B. Type A costs \$0.25/sq in. and type B costs \$0.65/sq in. The Hush Dog Company must buy each day at least 3000 sq in. of A and 8600 sq in. of B, (or the supplier will not do business with the company). However, the total square inches of material provided daily by the supplier must be fewer than 30,000 sq in. Experience has shown the company's president that total leather costs must be less than \$10,000 daily or no one will buy Hush Dogs (because they would need to be priced too high).

25. Determine a system of inequalities whose solution models all possible combinations of the amounts of leather A and leather B that could be used in this problem situation.

26. Find the graph of the region that represents all possible combinations of the amounts of leather A and leather B that could be used in this problem situation.

In Exercises 27 and 28, use the shade feature of your grapher to solve the following systems:

27. $\begin{cases} y > x^3 - 4x \\ y < 6 - x^2 \end{cases}$

28. $\begin{cases} y > x + 2 \\ y < \sqrt{x + 3} \end{cases}$

Chapter 10 Review

KEY TERMS   (The number following each key term indicates the page of its introduction.)

REVIEW EXERCISES

In Exercises 1 and 2, use the substitution method to find the simultaneous solutions to each system of equations. Support your answer with a graphing utility.

1. $\begin{cases} 2x - y = 4 \\ y = 6 \end{cases}$

2. $\begin{cases} x + 5y = 16 \\ 2x + y = 5 \end{cases}$

In Exercises 3 and 4, use the elimination method to find the simultaneous solutions to each system of equations. Support your answer with a graphing utility.

3. $\begin{cases} 6x - 3y = 9 \\ -4x + 2y = -6 \end{cases}$

4. $\begin{cases} x^2 + 3y = 7 \\ x^2 - y = -1 \end{cases}$

5. Find the simultaneous solutions to the system

$$x + 2y + z = 1$$

$$2x - y + z = 2$$

$$x - 3y - z = 6.$$

In Exercises 6–11, write a matrix for each system of equations. Solve the system by reducing the matrix of the system to row echelon form or reduced row echelon form.

6. $\begin{cases} x + y = 3 \\ x - y = 1 \\ 2x + y = 1 \end{cases}$

7. $\begin{cases} x - y + z = 6 \\ x + y + 2z = -4 \end{cases}$

8. $\begin{cases} x - y + z = 0 \\ 2x - 3z = 4 \\ -x - y + 2z = -4 \end{cases}$

9. $\begin{cases} 4x - 2y = 0 \\ x + 3y - z = -3 \\ 6y + 2z = 16 \end{cases}$

10. $\begin{cases} x + y - 2z = -4 \\ 3x - y + z = -7 \\ -2x - 2y + 4z = 8 \end{cases}$

11. $\begin{cases} 2x - 4y = 10 \\ x - z = -1 \\ 2y - z = -5 \end{cases}$

For Exercises 12–16, use the following matrices and find the matrix products indicated.

$$A = \begin{pmatrix} 3 & -2 & 1 \\ 0 & -1 & 2 \\ 1 & -3 & 1 \end{pmatrix} \qquad B = \begin{pmatrix} -1 & 2 \\ 4 & 3 \end{pmatrix}$$

$$C = \begin{pmatrix} 1 & -3 & 2 \\ 0 & -2 & 4 \end{pmatrix} \qquad D = \begin{pmatrix} 1 & 0 \\ -3 & 4 \\ 1 & 1 \end{pmatrix}$$

12. $A \cdot D$

13. $C \cdot A$

14. $D \cdot B$

15. $B \cdot C$

16. $(D \cdot C) \cdot A$

In Equations 17–22, solve each system using matrix multiplication.

17. $\begin{cases} 2x - 3y = 2 \\ 4x + y = 5 \end{cases}$

18. $\begin{cases} 5x - 17y = 42 \\ 9x + 6y = 19 \end{cases}$

19. $\begin{cases} 0.32x + 1.8y = 3.2 \\ 2.4x - 0.08y = 5.2 \end{cases}$

20. $\begin{cases} 3x - 4y + 5z = 7 \\ 2x + y - z = 2 \\ -x + 2y + 2z = 5 \end{cases}$

21. $\begin{cases} 0.02x - 5.1y + 2z = 6 \\ x + 2.3y + 0.7z = -1 \\ x - 0.5y + 1.1z = 2 \end{cases}$

22. $\begin{cases} 2x - 3y + z - 1.5w = 3 \\ x + 2y - 3z + w = -2 \\ -x - 1.3y + 2.5z - 0.7w = 4 \\ 3x + 7y - 4.2z + 5w = 17 \end{cases}$

In Exercises 23–27, solve the given system of equations.

23. $\begin{cases} y = 115x - 3x^3 \\ y = 50\cos x \end{cases}$

24. $\begin{cases} 9x^2 + 25y^2 = 225 \\ x + 2y = 2 \end{cases}$

25. $\begin{cases} 9x^2 - 25y^2 = 225 \\ x + 2y = 4 \end{cases}$

26. $\begin{cases} 9x^2 - 25y^2 = 225 \\ 9x^2 + 25y^2 = 225 \end{cases}$

27. $\begin{cases} x^2 + 9y^2 = 9 \\ y = 2x^2 + 3 \end{cases}$

In Exercises 28 and 29, determine the number of simultaneous solutions to each system.

28. $\begin{cases} 2x^2 - 3xy + 2y^2 + x + y - 8 = 0 \\ 3x^2 - 8xy + 2y^2 + x + y - 10 = 0 \end{cases}$

29. $\begin{cases} 3x^2 - 5xy + 6y^2 - 7x + 5y - 9 = 0 \\ x^2 + y^2 - 2y - 6 = 0 \end{cases}$

30. Find the simultaneous solutions to the system in Exercise 28.

31. Find the simultaneous solutions to the system in Exercise 29.

In Exercises 32 and 33, draw a graph representing the solution to the system of inequalities.

32. $\begin{cases} x^2 + y^2 < 9 \\ y > x^2 \end{cases}$

33. $\begin{cases} x^2 - y^2 > 9 \\ x^2 + y^2 < 9 \end{cases}$

SRINIVASA RAMANUJAN

Born in southern India, Srinivasa Ramanujan (1887–1919) spent much of his life in poverty and obscurity. He never graduated from college. He worked at several menial jobs for low wages, and he spent most of his energies recording his mathematical discoveries derived by intuition in notebooks.

At the urging of friends, he eventually began to exchange letters with G. H. Hardy of Cambridge University. This was the beginning of a five-year collaboration that produced more than 30 papers on topics such as pi, infinite series, prime and composite numbers, integers as the sum of squares, function theory, and combinatorics. More than 70 years after his death, Ramanujan's work is still an important source of new mathematical ideas. In the 1970s it was used by computer programmers to calculate millions of digits in the decimal expansion of pi. His notebooks are just now being published, much to the delight of modern-day mathematicians.

11

Sequences, Series, and Three-Dimensional Geometry

11.1

Objective

Students will be able to define a sequence, either by formula or recursively, and identify the first several terms and the nth term of the sequence. They will be able to identify and solve problems about arithmetic and geometric sequences.

Key Ideas

Sequence
Recursively defined sequences
Arithmetic sequences
Geometric sequences

Sequences

The word *sequence* as used in ordinary English means "one thing after another." A set of events occurs in sequence if first one occurs, then the next, then the next, and so on. A **sequence of numbers** is a set of numbers with a specific order. For example,

$$1, \frac{1}{2}, \frac{1}{3}, \frac{1}{4}, \ldots, b_n, \ldots$$

is an example of a sequence in mathematics. For a sequence to be uniquely determined, there must be some rule that determines the nth term. The above sequence is uniquely defined by the rule

$$b_n = \frac{1}{n},$$

where it is understood that n has values $1, 2, 3, \ldots$

Bert says, "Students often think of a sequence as a set of numbers listed in some order. Recognizing that a sequence is a *function* whose domain is the set of positive integers is the secret to understanding sequences and their graphs."

A second sequence,

$$2, 5, 8, 11, \ldots, a_n, \ldots$$

is defined by the rule $a_n = 3n - 1$.

Any function $f(x)$ whose domain includes all positive integers defines the sequence

$$f(1), f(2), f(3), \ldots, f(n), \ldots .$$

In fact, the function concept is key to understanding the concept of sequence.

Definition 11.1   Sequence

An **infinite sequence** is a function whose domain is the set of all positive integers $\{1, 2, 3, \ldots, n, \ldots\}$.

A **finite sequence** is a function whose domain is some initial subset of positive integers $\{1, 2, 3, \ldots, n\}$ for some fixed positive integer n.

A sequence is typically expressed in list form $a_1, a_2, a_3, \ldots, a_n, \ldots$, and the term a_n is called the nth term of the sequence. We often use the notation $\{a_n\}$ to denote a sequence whose nth term is a_n.

E X A M P L E   1   Finding Terms in Sequences

List the first three terms and the 15th term of the following sequences:

a) $a_n = (n^2 - 1)/n$

b) $b_n = (-1)^n 2^n$

Solution

a)

$$a_1 = \frac{1^2 - 1}{1} = 0, \quad a_2 = \frac{2^2 - 1}{2} = \frac{3}{2}, \quad a_3 = \frac{3^2 - 1}{3} = \frac{8}{3},$$

and the 15th term is

$$a_{15} = \frac{15^2 - 1}{15} = \frac{224}{15}.$$

Therefore the sequence is

$$0, \frac{3}{2}, \frac{8}{3}, \ldots, \frac{224}{15}, \ldots, \frac{n^2 - 1}{n}, \ldots .$$

b) $b_1 = (-1) \cdot 2 = -2, \quad b_2 = (-1)^2 \cdot 2^2 = 4, \quad (-1)^3 \cdot 2^3 = -8,$ and the 15th term is $a_{15} = (-1)^{15} \cdot 2^{15} = -32{,}768.$ Therefore the sequence is

$$-2, 4, -8, \ldots, -32{,}768, \ldots, (-1)^n \cdot 2^n, \ldots .$$ ≡

Because a sequence is a function, the graph of the sequence is precisely its graph as a function. For example, to find the graph of the sequence $a_n = (n+1)/n$, simply plot the following set of points on the rectangular coordinate plane:

$$\{(1, 2), (2, 3/2), (3, 4/3), (4, 5/4), (5, 6/5), \ldots\}$$

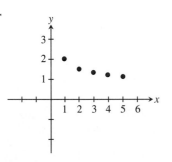

Figure 11.1 First five terms of $a_n = (n + 1)/n$.

(see Fig. 11.1).

A second approach to finding the graph of a sequence involves finding a complete graph $y = f(x)$ where f is the function rule that defines the sequence. After finding the complete graph, place dots on the graph for the points that represent the graph of the sequence. This latter approach will be used in Example 2.

EXAMPLE 2   Graphing a Sequence

Find the graph of the sequence $a_n = (n^2 - 1)/n$.

Solution   First, find the complete graph of $f(x) = (x^2 - 1)/x$. Then, place dots on the graph for points

$$(1, f(1)), (2, f(2)), (3, f(3)), (4, f(4)), (5, f(5)), \ldots .$$

This graph is shown in Fig. 11.2.   ≡

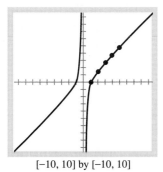

[−10, 10] by [−10, 10]

Figure 11.2 Graph of the sequence $a_n = (n^2 - 1)/n$.

The sequences in Example 1 were defined by stating an explicit formula that defined the nth term. A second method for specifying a sequence is to use a **recursive formula**. In a recursive formula, the first few terms are given and the nth term a_n is defined in terms of previous terms of the sequence. Here is an example of a recursive formula:

$$a_n = 3 + a_{n-1}, \quad a_1 = 2.$$

The first few terms of this sequence are

$$a_1 = 2, \quad a_2 = 3 + a_1 = 3 + 2 = 5, \quad a_3 = 3 + a_2 = 3 + 5 = 8, \ldots .$$

Stan says, "Find a complete graph of $a_n = \frac{(n^2-1)}{n}$ in a viewing window whose x-screen coordinates are integers. Notice how easily TRACE can be used to generate the sequence pairs (1, $f(1)$), (2, $f(2)$), . . . found in Example 2."

EXAMPLE 3   Using a Recursive Formula

Determine the first four terms and the eighth term for the sequence $\{a_n\}$ defined by $a_n = 3 + a_{n-1}$ for $n \geq 2$, where $a_1 = 4$.

Solution Use $a_n = 3 + a_{n-1}$ for $n = 2$ to find a_2, that is, $a_2 = 3 + a_1 = 3 + 4 = 7$. Also, $a_3 = 3 + a_2 = 3 + 7 = 10$, and $a_4 = 3 + a_3 = 3 + 10 = 13$.

When you know a term of a recursively defined sequence, you can find the next term, or you can work backwards to find an earlier term in the sequence.

$$a_8 = 3 + a_7$$
$$= 3 + (3 + a_6)$$
$$= 6 + (3 + a_5)$$
$$= 9 + (3 + a_4)$$
$$= 12 + 13 = 25$$

Notes on Examples

Examples 3 and 4 should be studied together. Example 3 uses a recursive definition to find the first several terms of a sequence. In Example 4, the same recursive rule is used to form a pattern and pattern recognition is used to describe the nth term in terms of variable n.

E X A M P L E   4   Finding the nth Term of a Recursive Formula

Discover a pattern and state a formula with variable n for the nth term of the sequence $\{a_n\}$ defined by $a_n = 3 + a_{n-1}$ for $n \geq 2$, where $a_1 = 4$.

Solution Compute terms of the sequence until a pattern is evident and you can conjecture a formula for the nth term.

$$a_2 = 3 + a_1$$
$$a_3 = 3 + a_2 = 2(3) + a_1$$
$$a_4 = 3 + a_3 = 3(3) + a_1$$
$$a_5 = 3 + a_4 = 4(3) + a_1$$
$$\vdots$$

The pattern seems to be clear. Conjecture that

$$a_n = 3(n-1) + a_1 = 3(n-1) + 4$$

and write $a_n = 3n + 1$.

In Section 11.5, we will introduce another method for verifying the nth term of a sequence, called mathematical induction. In the meantime use the "discover a pattern" method.

Example 4 illustrates that it is sometimes possible to express a_n as a formula with variable n. However, it is often difficult to determine such a formula. The next example introduces the **Fibonacci sequence**, a sequence that is easy to define recursively but difficult to express as a function of n.

Exercises 44 through 46 refer to a real-world occurrence of this sequence.

EXAMPLE 5 Finding Terms in the Fibonacci Sequence

Determine the first six terms of the sequence $\{a_n\}$ defined by $a_1 = a_2 = 1$ and $a_n = a_{n-1} + a_{n-2}$ for $n \geq 3$.

Solution Determine a_n for $n = 3, 4, 5$, and 6.

$$a_1 = 1$$
$$a_2 = 1$$
$$a_3 = a_2 + a_1 = 2$$
$$a_4 = a_3 + a_2 = 3$$
$$a_5 = a_4 + a_3 = 5$$
$$a_6 = a_5 + a_4 = 8$$

≡

Arithmetic Progression

Certain groups of sequences are identified as being of the same type because they all have an identical pattern. Arithmetic sequences are one such group.

Definition 11.2 Arithmetic Sequences

A sequence $\{a_n\}$ is called an **arithmetic progression** or an **arithmetic sequence** if there is a real number d such that

$$a_n = a_{n-1} + d$$

for every positive integer n. The number d is called the **common difference** of the arithmetic sequence.

Suppose that $\{a_n\}$ is an arithmetic sequence with a common difference d. The sequence satisfies the pattern

$$a_2 = a_1 + d$$
$$a_3 = a_2 + d = (a_1 + d) + d = a_1 + 2d$$
$$a_4 = a_3 + d = (a_1 + 2d) + d = a_1 + 3d$$
$$\vdots$$
$$a_n = a_{n-1} + d = a_1 + (n-2)d + d = a_1 + (n-1)d.$$

This pattern suggests the following theorem.

Theorem 11.1 Arithmetic Sequences

If $\{a_n\}$ is an arithmetic sequence with a common difference d, then

$$a_n = a_1 + (n-1)d$$

for every positive integer n.

Notice that the a_n can be described as a function of n if the constant difference d and the first term a_1 are known. Examples 6 and 7 illustrate that in fact if any two terms of an arithmetic sequence are known, a_1 and d can be found.

E X A M P L E   6   Finding an Arithmetic Sequence

The first two terms of an arithmetic progression are -8 and -2. Find the tenth term and a formula for the nth term.

Solution $a_1 = -8$, $a_2 = -2$, and $a_2 = a_1 + d$. Therefore

$$a_2 = a_1 + d$$

$$-2 = -8 + d \quad \text{Substitute values for } a_1 \text{ and } a_2.$$

$$d = 6.$$

So, $a_{10} = a_1 + 9d = -8 + 9(6) = 46$. The nth term is thus

$$a_n = a_1 + (n-1)d = -8 + 6(n-1) = 6n - 14. \qquad \equiv$$

E X A M P L E   7   Finding an Arithmetic Sequence

The third and eighth terms of an arithmetic progression are 13 and 3, respectively. Determine the first term and a formula for the nth term.

Solution   Use $a_n = a_1 + (n-1)d$ to establish the following pair of equations:

$$a_3 = a_1 + 2d = 13$$

$$a_8 = a_1 + 7d = 3$$

Subtracting the first equation from the second yields $5d = -10$, so $d = -2$ and $a_1 = 17$. Therefore

$$a_n = a_1 + (n-1)d = 17 - 2(n-1) = 19 - 2n. \qquad \equiv$$

Geometric Progression

In an arithmetic sequence, any term is obtained by adding a constant to the preceding term. The sequences known as geometric sequences satisfy the property that any given term is obtained from the preceding term by multiplying by a constant.

Definition 11.3 Geometric Sequences

A sequence $\{a_n\}$ is called a **geometric progression** or **geometric sequence** if there is a nonzero real number r such that

$$a_n = r \cdot a_{n-1}.$$

For every positive integer n, the number r is called the **common ratio** of the geometric sequence.

Suppose that $\{a_n\}$ is a geometric sequence with a common ratio r. Then the sequence satisfies the following pattern:

$$a_2 = a_1 r$$

$$a_3 = a_2 r = (a_1 r) \cdot r = a_1 r^2$$

$$a_4 = a_3 r = (a_1 r^2) \cdot r = a_1 r^3$$

$$\vdots$$

$$a_n = a_{n-1} r = (a_1 r^{n-2}) \cdot r = a_1 r^{n-1}.$$

This pattern suggests the following theorem.

Theorem 11.2 Geometric Sequences

If $\{a_n\}$ is a geometric sequence with a common ratio r, then

$$a_n = a_1 r^{n-1}$$

for every positive integer n.

Notice that if a_k and a_{k+1} are successive terms of a geometric sequence, the ratio a_{k+1}/a_k simplifies as follows:

$$\frac{a_{k+1}}{a_k} = \frac{a_1 r^k}{a_1 r^{k-1}}$$

$$= r.$$

This demonstrates why r is called the common ratio.

Teaching Note

Presented with a sequence of numbers such as 2, 6, 18, 54, . . . , students often have a difficult time deciding whether it is arithmetic, geometric, or neither. Teach them to find the difference d and the ratio r of the first two terms; for example,

$d = 6 - 2 = 4$ and $r = \frac{6}{2} = 3$.

Students should then ask themselves:
(1) Is the third term found by adding d to the second term? No. The sequence is not arithmetic.
(2) Is the third term found by multiplying the second term by r? Yes. The sequence is geometric.

E X A M P L E   8   Finding a Geometric Sequence

The second and third terms of a geometric sequence are -6 and 12, respectively. Determine the first term and a formula for the nth term.

Solution   Because r is the ratio of successive terms,

$$r = \frac{a_3}{a_2} = \frac{12}{-6} = -2.$$

Further, because $a_n = a_1 r^{n-1}$ where $r = -2$, it follows that $a_2 = (-2)a_1 = -6$. Therefore $a_1 = 3$ and

$$a_n = 3 \cdot (-2)^{n-1}.$$ ≡

Example 9 revisits compound interest to illustrate that geometric sequences find application in the business world.

Notes on Exercises

Example 5 introduced the Fibonacci sequence. Ex. 44–47 revisit this sequence from a different point of view. Ex. 51–53 are related to the concept *a sequence $\{a_n\}$ converges to a real number K.* Notice how this concept is related to a horizontal asymptote of a function.

Assignment Guide

Day 1: Ex. 1–20
Day 2: Ex. 21–35 odd, 37–41, 44–56

E X A M P L E   9   APPLICATION: Computing Compound Interest

Two hundred dollars are deposited in an account that pays 9% interest annually, compounded monthly. Let P_n be the amount in the account at the end of the nth month. Show that $P_1, P_2, P_3, \ldots$ is a geometric sequence. Find the common ratio and the value of P_1.

Solution   To find the monthly interest rate, divide the annual rate by 12: $0.09/12 = 0.0075$. Using the formula for compound interest studied in Section 5.2, note that

$$P_n = 200(1 + 0.0075)^n.$$

Therefore

$$\frac{P_n}{P_{n-1}} = \frac{200(1.0075)^n}{200(1.0075)^{n-1}} = 1.0075.$$

The ratio of successive terms is 1.0075, so this sequence is by definition a geometric sequence with a common ratio $r = 1.0075$. Therefore

$$P_n = P_1 \cdot r^{n-1} = 200(1.0075)^n = [200(1.0075)](1.0075)^{n-1}.$$

Consequently, the first term $P_1 = 200(1.0075)$.   ≡

Exercises for Section 11.1

In Exercises 1–8, determine the first four terms and the tenth term of the sequence.

1. $a_n = (-1)^{n+1}$

2. $a_n = 2n - 1$

3. $a_n = \dfrac{n+1}{n}$

4. $a_n = (-1)^n 2n$

5. $a_n = \dfrac{n^3 - 1}{n+1}$

6. $a_n = \dfrac{4}{n+2}$

7. $a_n = (-1)^n \left(1 + \dfrac{1}{n}\right)$

8. $a_n = \dfrac{1}{8} \cdot 3^n$

In Exercises 9–12, determine the first four terms and the eighth term of the sequence and a rule for the nth term; that is, determine a_n as an explicit function of n.

9. $a_1 = 3; a_n = a_{n-1} + 1, n = 2, 3, \ldots$

10. $a_1 = -2; a_n = a_{n-1} + 2, n = 2, 3, \ldots$

11. $a_1 = 2; a_n = 2a_{n-1}, n = 2, 3, \ldots$

12. $a_1 = \dfrac{3}{2}; a_n = \dfrac{1}{2}a_{n-1}, n = 2, 3, \ldots$

In Exercises 13–16, discover a pattern and find a formula with variable n for the nth term of each sequence $\{a_n\}$.

13. $a_1 = 3; a_n = a_{n-1} + 2$ for $n \geq 2$

14. $a_1 = -5; a_n = a_{n-1} + 7$ for $n \geq 2$

15. $a_1 = 2; a_n = 3a_{n-1}$ for $n \geq 2$

16. $a_1 = 3; a_n = (-2)a_{n-1}$ for $n \geq 2$

In Exercises 17–20, determine the common difference, the tenth term, and a formula for the nth term of each arithmetic sequence.

17. $6, 10, 14, 18, \ldots$

18. $-4, 1, 6, 11, \ldots$

19. $-5, -2, 1, 4, \ldots$

20. $-7, 4, 15, 26, \ldots$

In Exercises 21–24, determine the common ratio, the eighth term, and a formula for the nth term of each geometric sequence.

21. $2, 6, 18, 54, \ldots$

22. $1, -2, 4, -8, 16 \ldots$

23. $3, 6, 12, 24, \ldots$

24. $-2, 2, -2, 2, \ldots$

25. The fourth and seventh terms of an arithmetic progression are -8 and 4, respectively. Determine the first term and a formula for the nth term.

26. The fifth and ninth terms of an arithmetic progression are -5 and -17, respectively. Determine the first term and a formula for the nth term.

Which sequences in Exercises 27–32 could be arithmetic or geometric? For such sequences, state the appropriate common difference or ratio and determine a formula for the nth term.

27. $5, 10, 20, 40, \ldots$

28. $-0.25, 1, -4, 16, -64, \ldots$

29. $-16, -9, -2, 5, \ldots$

30. $10.1, 10.201, 10.30301, 10.4060401, \ldots$

31. $-2, 1, -\dfrac{1}{2}, \dfrac{1}{4}, \ldots$

32. $1, 5, 7, 11, 17, \ldots$

In Exercises 33–36, sketch a graph of each sequence.

33. $a_n = 2 - \dfrac{1}{n}$

34. $a_n = \sqrt{n} - 3$

35. $a_n = n^2 - 5$

36. $a_n = 3 + 2n$

37. Roberta had \$1250 in a savings account 3 yr ago. What will the value of her account be 2 yr from now, assuming no deposits or withdrawals are made and the account earns 6.5% interest compounded annually?

38. Ellen has \$12,876 in a savings account today. She made no deposits or withdrawals during the past 6 yr. What was the value of her account 6 yr ago? (Assume the account earned interest at 5.75% interest compounded monthly.)

Exercises 39–41 refer to the following **problem situation:** The half-life of a certain unstable radioactive substance is 1 wk. Suppose 1000 g of the substance exist today. Let n represent the number of weeks the substance exists.

39. Determine an infinite geometric sequence that models the amount of the substance at week n, where $n = 1, 2, 3, \ldots$. List the first 10 terms of the sequence. What is the common ratio?

40. When will there be only 0.05 g of the substance remaining?

41. Will the substance ever be reduced to nothing?

Exercises 42 and 43 refer to the following **problem situation:** The height of a certain fast-growing plant in a rain forest increases at the rate of 2.5% per month. Assume the plant is 15 in. in height today. Let n represent the number of months the plant grows and assume the plant dies in 10 mo.

42. Determine a finite geometric sequence that is a model for the height of the plant after n months. Write out all the terms of the sequence. What is the common ratio?

43. How long would the plant need to live in order to double in height?

Exercises 44–47 refer to the following rabbit population **problem situation**: Assume rabbits become fertile 1 mo after birth and each male-female pair of fertile rabbits produces one new male-female pair of rabbits each month. Further assume the rabbit colony begins with one newborn male-female pair of rabbits and no rabbits die for 12 mo. Let a_n represent the number of *pairs* of rabbits in the colony after $n - 1$ months.

44. Explain why $a_1 = 1$, $a_2 = 1$, and $a_3 = 2$.

45. Determine a_4, a_5, a_6, a_7, and a_8.

46. Explain why the sequence in Exercise 45 is a model for the size of the rabbit colony.

47. Compute the first seven terms of

$$a_n = \frac{1}{\sqrt{5}}\left(\frac{1 + \sqrt{5}}{2}\right)^n - \frac{1}{\sqrt{5}}\left(\frac{1 - \sqrt{5}}{2}\right)^n.$$

Do you recognize this sequence?

48. Investigate the value of $a_n = 10n\sin(\pi/n)$ for $n = 10$, 100, and 1,000. Compute the circumference of a circle with a radius of 5. What conclusions can you draw?

49. Sketch a graph of a_n for $1 \leq n \leq 100$ where a_n is defined

as in Exercise 48. How does your graph relate to Exercise 48?

50. Consider the sequence $a_1 = 1$ and $a_n = na_{n-1}$. Compute the first six terms of the sequence and determine a rule for a_n as an explicit function of n.

A sequence $\{a_n\}$ *converges to a real number* K, denoted by $a_n \to K$ as $n \to \infty$, if the line $y = K$ is a horizontal asymptote of the graph of a_n. For each sequence in Exercises 51–54, find a complete graph of a_n and a line $y = K$ in the same viewing rectangle to show that the sequence converges to a number K. Identify K.

51. $a_n = \dfrac{2n}{n + 1}$

52. $a_n = \left(1 + \dfrac{0.05}{n}\right)^n$

53. $a_n = 3 + \dfrac{(-1)^n}{n}$

54. $a_n = n\sin\dfrac{\pi}{2n}$

55. Let $f(x) = (\frac{1}{2})^x$ and define a sequence $a_n = f(a_{n-1})$, where $a_1 = f(1)$. Compute the first eight terms of this sequence. To what number does this sequence appear to converge?

56. Find complete graphs of $y = x$ and $f(x) = \left(\frac{1}{2}\right)^x$ in the same viewing rectangle. Zoom in to find a solution to the equation $f(x) = x$. How does this solution compare with the sequence in Exercise 55?

11.2   Finite and Infinite Series

Finite Series

In Section 11.1, we used the notation $\{a_n\}$ to denote a sequence. For example, the terms of the sequence $\{a_n\}$ where $a_n = (2n - 1)^2$ are

$$1^2, 3^2, 5^2, 7^2, 9^2, \ldots, (2n - 1)^2, \ldots.$$

Consider the following sums:

$$S_1 = 1^2$$

$$S_2 = 1^2 + 3^2$$

$$S_3 = 1^2 + 3^2 + 5^2$$

$$S_4 = 1^2 + 3^2 + 5^2 + 7^2$$

$$\vdots$$

$$S_n = 1^2 + 3^2 + \cdots + (2n - 1)^2.$$

Each sum S_n is called a **finite series** since there are a finite number of terms in each sum.

To illustrate the concept of *series* in general, consider the general sequence $\{a_n\}$. Then the **sequence of partial sums** of $\{a_n\}$ is the sequence $\{S_n\}$ shown here:

$$S_1 = a_1$$

$$S_2 = a_1 + a_2$$

$$S_3 = a_1 + a_2 + a_3$$

$$\vdots$$

$$S_n = a_1 + a_2 + \cdots + a_n$$

Again, each term S_k is an example of a finite series. The number S_n is called the **nth partial sum** of the sequence $\{a_n\}$.

The Greek capital letter Σ (sigma) is used as a symbol for sum.

Sigma Notation

The sum of n terms $a_1, a_2, \ldots, a_n$ is written as

$$S_n = \sum_{k=1}^{n} a_k = a_1 + a_2 + \cdots + a_n.$$

Read $\displaystyle\sum_{k=1}^{n} a_k$ as "the sum from $k = 1$ to n of a_k", where a_k is called the **kth term of the sum** and k is called the **index of summation.**

The sum $\displaystyle\sum_{k=1}^{n}$ is referred to as **sigma notation**, and the sum

$$a_1 + a_2 + \ldots + a_n$$

is referred to as **expanded notation**.

Example 1 illustrates how to convert from sigma notation to expanded form.

E X A M P L E   1   *Expanding a Sum*

Expand the sum $\displaystyle\sum_{k=1}^{n} (2k + 3)^3$.

Solution

$$\sum_{k=1}^{n}(2k+3)^3 = 5^3 + 7^3 + 9^3 + \cdots + (2n+3)^3$$

≡

Sigma notation is often a notational convenience. When the form of expressions involving sigma notation needs to be changed, properties about the summation notation given by Theorem 11.3, stated here without proof, are useful.

Theorem 11.3   Summation Properties

Let $\{a_n\}$ and $\{b_n\}$ be two sequences and n be a positive integer. Then

1. $\displaystyle\sum_{k=1}^{n}(a_k + b_k) = \sum_{k=1}^{n}a_k + \sum_{k=1}^{n}b_k,$

2. $\displaystyle\sum_{k=1}^{n}(a_k - b_k) = \sum_{k=1}^{n}a_k - \sum_{k=1}^{n}b_k,$ and

3. $\displaystyle\sum_{k=1}^{n}ca_k = c\sum_{k=1}^{n}a_k$ for every real number c.

E X A M P L E   2   Using Summation Properties

Use Theorem 11.3 to combine the following expression into a single sum:

$$\sum_{k=1}^{n}(5k^2 - 4) + \sum_{k=1}^{n}(7 - 2k + k^2)$$

Solution

$$\sum_{k=1}^{n}(5k^2 - 4) + \sum_{k=1}^{n}(7 - 2k + k^2) = \sum_{k=1}^{n}[(5k^2 - 4) + (7 - 2k + k^2)]$$

$$= \sum_{k=1}^{n}(6k^2 - 2k + 3)$$

≡

Summation Formulas

The next theorem gives several summation formulas. These formulas can be proven using mathematical induction, a method that will be discussed in Section 11.5. Using these formulas, a sum can be replaced with an expression in the variable n.

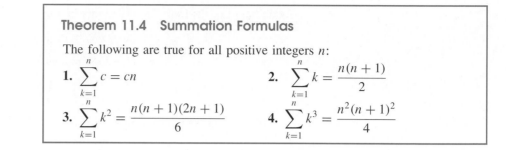

> **Theorem 11.4   Summation Formulas**
>
> The following are true for all positive integers n:
>
> **1.** $\displaystyle\sum_{k=1}^{n} c = cn$ **2.** $\displaystyle\sum_{k=1}^{n} k = \frac{n(n+1)}{2}$
>
> **3.** $\displaystyle\sum_{k=1}^{n} k^2 = \frac{n(n+1)(2n+1)}{6}$ **4.** $\displaystyle\sum_{k=1}^{n} k^3 = \frac{n^2(n+1)^2}{4}$

E X A M P L E   3   Simplifying Sums

Use Theorems 11.3 and 11.4 to simplify $\displaystyle\sum_{k=1}^{n}(6k^2 - 2k + 3)$.

Solution

$$\sum_{k=1}^{n}(6k^2 - 2k + 3) = \sum_{k=1}^{n} 6k^2 - \sum_{k=1}^{n} 2k + \sum_{k=1}^{n} 3$$

$$= 6\sum_{k=1}^{n} k^2 - 2\sum_{k=1}^{n} k + 3n$$

$$= 6\frac{n(n+1)(2n+1)}{6} - 2\frac{n(n+1)}{2} + 3n$$

$$= n(2n^2 + 3n + 1) - (n^2 + n) + 3n$$

$$= 2n^3 + 2n^2 + 3n$$

≡

Finite Geometric Series

A **finite geometric series** is a finite series determined by a geometric sequence $a_1, a_1 r, a_1 r^2, \ldots, a_1 r^{n-1}, \ldots$. That is, this sum is a finite geometric series.

$$S_n = \sum_{k=1}^{n} a_1 r^{k-1} = a_1 + a_1 r + \cdots + a_1 r^{n-1}$$

To develop a summation formula for a geometric series, start with expressions for S_n and rS_n.

$$S_n = \sum_{k=1}^{n} a_1 r^{k-1} = a_1 + a_1 r + \cdots + a_1 r^{n-1} \tag{1}$$

$$rS_n = \sum_{k=1}^{n} a_1 r^{k} = a_1 r + a_1 r^2 + \cdots + a_1 r^{n} \tag{2}$$

Subtracting Eq. (2) from Eq. (1) results in

$$S_n - rS_n = a_1 - a_1r^n$$

$$(1 - r)S_n = a_1(1 - r^n)$$

$$S_n = a_1 \frac{1 - r^n}{1 - r}.$$

We have proved the following theorem.

Teaching Note

Theorem 11.5 is the basis for the entire section. Make sure students have a firm understanding of this theorem. Develop several examples of your own like Example 4.

Theorem 11.5   Finite Geometric Series

If S_n is equal to a finite geometric series

$$S_n = \sum_{k=1}^{n} a_1 r^{k-1} = a_1 + a_1r + a_1r^2 + \cdots + a_1r^{n-1},$$

then

$$S_n = a_1 \frac{1 - r^n}{1 - r}.$$

E X A M P L E   4   Finding a Sum of a Finite Geometric Series

Determine the fourth partial sum of the geometric sequence $2, \frac{2}{3}, \frac{2}{9}, \ldots$.

Solution   To verify that these three terms are the beginning of a geometric sequence, check that the ratio of each pair of successive terms is the same.

$$r = \frac{\frac{2}{3}}{2} = \frac{\frac{2}{9}}{\frac{2}{3}} = \frac{1}{3}$$

Using Theorem 11.5 with $r = \frac{1}{3}$, $a_1 = 2$, and $n = 4$, we obtain

$$S_n = a_1 \frac{1 - r^n}{1 - r}$$

$$S_4 = 2 \frac{1 - (\frac{1}{3})^4}{1 - (\frac{1}{3})}$$

$$= \frac{2(1 - 1/81)}{\frac{2}{3}}$$

$$= \frac{240}{81}$$

$$= 2.9630.$$

Infinite Series

So far in this section, we have studied finite series, that is, the sums of the form

$$S_n = \sum_{k=1}^{n} a_k = a_1 + a_2 + \cdots + a_n.$$

If n is large, it may take a tremendous amount of computation time, but it is possible to actually add all the terms together.

We now introduce the concept of infinite series. An **infinite series** is denoted by the expression

$$\sum_{k=1}^{\infty} a_k = a_1 + a_2 + \cdots + a_n + \cdots. \tag{3}$$

Because an infinite number of terms cannot actually be added together, the sum of an infinite series cannot be found through direct addition.

To describe what the infinite sum of Eq. (3) means, consider the following sequence of partial sums:

$$S_1 = a_1$$

$$S_2 = a_1 + a_2$$

$$\vdots$$

$$S_n = a_1 + a_2 + \cdots + a_n$$

If there is a number S for which the line $y = S$ is a horizontal asymptote for the sequence $\{S_n\}$, then we say that "the sequence $\{S_n\}$ converges to S as n approaches infinity" and we call S the sum of the series. The notation for this is

$$S_n \to S \quad \text{as} \quad n \to \infty.$$

Definition 11.4   Sum of an Infinite Series

The **sum of an infinite series** $\displaystyle\sum_{k=1}^{\infty} a_k$, if it exists, is the number S where

$$\sum_{k=1}^{n} a_k \to S \quad \text{as} \quad n \to \infty.$$

The series is said to **converge** to S.

An infinite series of the form

$$\sum_{k=1}^{\infty} a_1 r^{k-1} = a_1 + a_1 r + a_1 r^2 + \cdots + a_1 r^{n-1} + \cdots$$

is called an **infinite geometric series**. Consider the following example.

Teaching Note

Students often fail to grasp the fact that an infinite series is not really a *sum*. Rather it is an infinite sequence, the sequence of partial sums. Highlight Examples 5 and 6. Notice that in Example 5 three specific terms—S_4, S_5, and S_6—of the sequence of partial sums are found. Then, in Example 6, a grapher is used to support the claim that this sequence converges to 3.

E X A M P L E   5   Finding Sums of an Infinite Geometric Series

Find S_4, S_5, and S_6 for the infinite geometric series $\sum_{k=1}^{\infty} 2 \left(\frac{1}{3}\right)^{k-1}$. Observe that $\{S_n\}$ appears to converge to 3.

Solution   Example 4 showed that $S_4 = 2.9630$. The other two sums are

$$S_5 = 2 \frac{1 - (\frac{1}{3})^5}{1 - (\frac{1}{3})}$$

$$= \frac{2(1 - \frac{1}{243})}{\frac{2}{3}}$$

$$= 2.9877,$$

$$S_6 = 2 \frac{1 - (\frac{1}{3})^6}{1 - (\frac{1}{3})}$$

$$= \frac{2(1 - \frac{1}{729})}{\frac{2}{3}}$$

$$= 2.9959.$$

The sequence of partial sums $\{S_n\} = \{2, 2.6667, 2.8889, 2.9630, 2.9877, 2.9959, \ldots\}$ appears to converge to 3.

Example 5 was based on direct computation. Now we develop a method for interpreting infinite geometric series graphically.

Theorem 11.5 states that the finite geometric series

$$\sum_{k=1}^{n} a_1 r^{k-1} \quad \text{is equal to} \quad a_1 \frac{1 - r^n}{1 - r}.$$

To obtain a visualization for the sum of the corresponding infinite series, consider the function

$$f(x) = a_1 \frac{1 - r^x}{1 - r}.$$

Then for each positive integer n, $f(n) = S_n$ is the nth partial sum of the infinite geometric series with the common ratio r. Consequently, the graph of $\{S_n\}$ consists of some points on the graph of f, and we can investigate the convergence of

$$\sum_{k=1}^{\infty} a_n r^{k-1}$$

by considering the graph of $f(x)$.

Example 6 provides a graphical method for finding the sum of the infinite series found in Example 5.

EXAMPLE 6   Visualizing an Infinite Geometric Series

Use a graphical method to find the sum of the series

$$\sum_{k=1}^{\infty} 2\left(\frac{1}{3}\right)^{k-1} = 2 + \frac{2}{3} + \frac{2}{3^2} + \cdots .$$

Solution   In this geometric series, $a_1 = 2$ and $r = \frac{1}{3}$. So consider the graph of

$$f(x) = 2 \cdot \frac{1 - r^x}{1 - r} = 2 \cdot \frac{1 - (\frac{1}{3})^x}{\frac{2}{3}} = 3 - 3\left(\frac{1}{3}\right)^x.$$

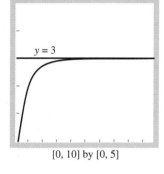

$y = 3$

[0, 10] by [0, 5]

Figure 11.3 $y = 3$ and $f(x) = 3 - 3(\frac{1}{3})^x.$

Figure 11.3 shows the graph of both $f(x)$ and $y = 3$. The line $y = 3$ is a horizontal asymptote for the graph of f. Because the graph of S_n consists of points on the graph of f, it follows that $S_n \to 3$ as $n \to \infty$; therefore the sum of this series is 3.   ≡

EXAMPLE 7   Visualizing an Infinite Geometric Series

Use a graphical method to find the sum of the series

$$\sum_{k=1}^{\infty} 4 \cdot 3^{k-1} = 4 + 4 \cdot 3 + 4 \cdot 3^2 + \cdots + 4 \cdot 3^{k-1} + \cdots .$$

Solution   In this geometric series, $a_1 = 4$ and $r = 3$. Consider the graph of

$$f(x) = 4 \cdot \frac{1 - 3^x}{1 - 3} = -2 + 2 \cdot 3^x.$$

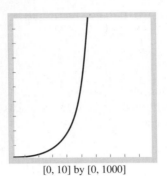

[0, 10] by [0, 1000]

Figure 11.4 $f(x) = -2 + 2 \cdot 3^x$.

Figure 11.4 shows a complete graph of f. As no horizontal asymptote exists, the sum of this infinite series does not exist.   ≡

When the sum of an infinite series does not exist, as illustrated in Example 7, the series is said to **diverge**.

Sum of an Infinite Geometric Series

We can see from Examples 6 and 7 that some infinite geometric series converge and some diverge. That raises the question, when does a geometric series converge and when does it diverge? The following Exploration addresses that question.

Frank says, "The grapher gives students a way to visualize the convergence of an infinite geometric series. This visual understanding was not available to us before the arrival of technology. Make sure students experience this Explore."

☌ EXPLORE WITH A GRAPHING UTILITY

Find a complete graph of the function

$$f(x) = \frac{1 - r^x}{1 - r}$$

for each of the following values of r:

a) $r = 0.8$

b) $r = 0.9$

c) $r = 1.1$

d) $r = 1.2$

For which of these values of r does the complete graph seem to have a horizontal asymptote?

Generalize   For what values of r does this geometric series converge?

$$\sum_{k=1}^{\infty} a_1 r^{k-1} = a_1 + a_1 r + a_1 r^2 + \cdots + a_1 r^{k-1} + \cdots$$

It can be shown that a geometric series $\displaystyle\sum_{k=1}^{\infty} a_1 r^{k-1}$ converges if $|r| < 1$. It can also be shown that if $|r| < 1$, the sum is $a_1 / (1 - r)$.

Theorem 11.6   Convergence of an Infinite Geometric Series

If $|r| < 1$, then the infinite geometric series

$$\sum_{k=1}^{\infty} a_1 r^{k-1} = a_1 + a_1 r + a_1 r^2 + \cdots + a_1 r^{k-1} + \cdots$$

has sum $a_1/(1 - r)$.

E X A M P L E   8   Finding the Sum of an Infinite Geometric Series

Find the sum of the infinite series

$$\sum_{k=1}^{\infty} 5\left(\frac{1}{2}\right)^{k-1} = 5 + \frac{5}{2} + \frac{5}{2^2} + \cdots + \frac{5}{2^{k-1}} + \cdots .$$

Solution   This series is a geometric series with $a_1 = 5$ and $r = 1/2$. Conclude from Theorem 11.6 that this series converges and has a sum of $\dfrac{5}{1 - (1/2)} = 10$.

≡

Annuities

Recall that an ordinary annuity is a sequence of equal periodic payments. Section 5.3 gave the following formula for the future value of an annuity.

Future Value of an Ordinary Annuity

The **future value S of an annuity** consisting of n equal payments of R dollars, each with interest rate i per period (payment interval), is given by

$$S = R\frac{(1 + i)^n - 1}{i}.$$

We shall show that this formula is the sum of a finite geometric series. Consider an annuity in which R dollars are deposited into an account at the end of each of n payment intervals. Suppose the account pays an interest rate i per period. The R dollars deposited at the end of the first payment interval earns interest for $n - 1$ intervals, so the value of the deposit at the end of the nth period is $R(1 + i)^{n-1}$ (see Fig. 11.5a).

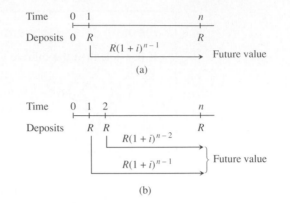

Figure 11.5 R is the amount deposited at the end of each payment interval, n is the number of payment intervals, and i is the interest per interval: (a) How much the first payment earns by the end of the nth interval; (b) How much the first and second payments earn by the end of the nth interval.

Teaching Note

The development of the future value of an annuity formula should be done carefully. Make sure students see the link between Fig. 11.5 and Eq. (4). Students often miss the fact that Eq. (4) is a finite geometric sum written in the reverse of the usual order.

Notes on Examples

Students are usually shocked by the result of Example 9. Consider expanding this example. Ask students to guess what the result would be using 600 months (50 years) instead of 240 months. Then verify that the result is $1,065,245. Do your students think it is within their reach to retire as a millionaire?

Notes on Exercises

Ex. 1–12 focus on sigma notation and related summation formulas. The remaining exercises focus primarily on some aspect of geometic series.

Assignment Guide

Day 1: Ex. 1–12, 13–35 odd
Day 2: Ex. 14–36 even, 39–47 odd, 48–53

Similarly, the value of the R dollars deposited at the end of the second payment interval will have a value of $R(1+i)^{n-2}$ at the end of the nth payment interval (see Fig. 11.5b).

Consequently, the value of the n deposits at the end of the nth period is

$$S_n = R(1+i)^{n-1} + R(1+i)^{n-2} + \cdots + R(1+i) + R. \tag{4}$$

Equation (4) is a finite geometric series where $a_1 = R$ and $r = (1+i)$. Theorem 11.5 states that this finite sum S_n is

$$S_n = a_1 \frac{1-r^n}{1-r} = R \frac{1-(1+i)^n}{1-(1+i)} = R \frac{(1+i)^n - 1}{i}.$$

E X A M P L E   9   APPLICATION: Finding the Future Value of an Annuity

Bob deposits $75 at the end of each month into an IRA (individual retirement account) that pays interest at 9.5% APR compounded monthly. Find the value of Bob's account at the end of 20 yr.

Solution   This is an example of an annuity with a monthly interest rate $i = 0.095/12$. Find the future value of this account when $n = 240$, $i = 0.095/12$ and $R = 75$.

$$S = R \frac{(1+i)^n - 1}{i}$$

$$= 75 \frac{(1 + 0.095/12)^{240} - 1}{0.095/12}$$

$$= 53,394.27 \qquad \blacksquare$$

Exercises for Section 11.2

In Exercises 1–6, expand each sum.

1. $\displaystyle\sum_{k=1}^{n}(2k)^2$ **2.** $\displaystyle\sum_{k=1}^{n}2k^2$ **3.** $\displaystyle\sum_{k=1}^{n}(3k-1)$

4. $\displaystyle\sum_{k=1}^{n}(k^2-1)$ **5.** $\displaystyle\sum_{k=1}^{n}3\left(\frac{1}{2}\right)^k$ **6.** $\displaystyle\sum_{k=1}^{n}\left(\frac{1}{4}\right)3^k$

In Exercises 7–12, use Theorems 11.3 and 11.4 to simplify each expression to obtain a formula with variable n.

7. $\displaystyle\sum_{k=1}^{n}(3k+2)$ **8.** $\displaystyle\sum_{k=1}^{n}(5-2k)$

9. $\displaystyle\sum_{k=1}^{n}(k^2-3k+4)$ **10.** $\displaystyle\sum_{k=1}^{n}(2k^2+5k-2)$

11. $\displaystyle\sum_{k=1}^{n}(k^3-1)$ **12.** $\displaystyle\sum_{k=1}^{n}(k^3+4k-5)$

In Exercises 13–20, write the nth partial sum S_n for each of these geometric series.

13. $a_1 = 3$ and $r = 2$ **14.** $a_1 = 4$ and $r = -2$

15. $a_1 = -2$ and $r = \frac{1}{2}$ **16.** $a_1 = 5$ and $r = \frac{1}{3}$

17. $a_1 = 10$ and $r = -\frac{1}{10}$ **18.** $a_1 = 8$ and $r = 0.4$

19. $a_1 = -2$ and $r = -0.5$ **20.** $a_1 = -7$ and $r = \frac{3}{2}$

In Exercises 21–24, use sigma notation to write each sum. Assume that the suggested patterns continue for infinite sums.

21. $2+5+8+11+\cdots+29$

22. $-1+2+7+14+23+\cdots+62$

23. $\dfrac{1}{4}+\dfrac{1}{16}+\dfrac{1}{64}+\cdots$

24. $-2+2-2+2-2+\cdots$

In Exercises 25–28, use sigma notation to write the nth partial sum of the sequence. Assume that the patterns continue as suggested.

25. $-8, -6, -4, \ldots$ **26.** $2, 5, 10, 17, 26, \ldots$

27. $3, -6, 9, -12, \ldots$ **28.** $\dfrac{5}{2}, \dfrac{5}{3}, \dfrac{5}{4}, \ldots$

In Exercises 29–32, assume the suggested pattern continues and determine a formula for the nth term of the sequence of partial sums. Which sequences could be arithmetic? Which geometric?

29. $1, 1+\dfrac{1}{2}, 1+\dfrac{1}{2}+\dfrac{1}{4}, 1+\dfrac{1}{2}+\dfrac{1}{4}+\dfrac{1}{8}, \ldots$

30. $1, 1+2, 1+2+3, 1+2+3+4, \ldots$

31. $-2, -2+1, -2+1-\dfrac{1}{2}, -1+1-\dfrac{1}{2}+\dfrac{1}{4}, \ldots$

32. $2, 2+4, 2+4+6, 2+4+6+8, \ldots$

For Exercises 33–36, find the partial sums $S_1, S_2, S_3, S_4, S_5,$ and S_6 for each of these infinite series. Decide whether each series converges.

33. $\dfrac{1}{8}+\dfrac{1}{16}+\dfrac{1}{32}+\dfrac{1}{64}+\cdots$

34. $-\dfrac{1}{4}-\dfrac{1}{8}-\dfrac{1}{16}-\dfrac{1}{32}-\cdots$

35. $\dfrac{3}{10}+\dfrac{3}{100}+\dfrac{3}{1000}+\cdots$

36. $\dfrac{1}{2}+\dfrac{1}{8}+\dfrac{1}{32}+\dfrac{1}{128}+\cdots$

37. Does a horizontal asymptote exist for the function

$$f(x) = 2\frac{1-1.05^x}{1-1.05}?$$

Explain how the answer relates to the infinite series

$$2 + 2.1 + 2.205 + 2.31525 + \cdots.$$

In Exercises 38–41, use a graphical method to find the sum of each series.

38. $1+\dfrac{1}{3}+\dfrac{1}{9}+\dfrac{1}{27}+\cdots$ **39.** $2+\dfrac{2}{5}+\dfrac{2}{25}+\dfrac{2}{125}+\cdots$

40. $3+2+\dfrac{4}{3}+\dfrac{8}{9}+\cdots$

41. $2+1.9+1.805+1.71475+\cdots$

In Exercises 42–47, determine whether the infinite geometric series converges. If so, find its sum.

42. $\displaystyle\sum_{k=1}^{\infty}2\left(\frac{1}{9}\right)^{k-1}$ **43.** $\displaystyle\sum_{k=1}^{\infty}3\left(-\frac{1}{4}\right)^{k-1}$

44. $\displaystyle\sum_{k=1}^{\infty}\frac{1}{16}(2^k)$ **45.** $\displaystyle\sum_{k=1}^{\infty}(0.1)^k$

46. $\displaystyle\sum_{k=1}^{\infty}\frac{4^{k-1}}{3^k}$ **47.** $\displaystyle\sum_{k=1}^{\infty}\frac{3^{k-1}}{4^k}$

Exercises 48–50 refer to this **problem situation**: A $150 monthly annuity has an APR of 7% compounded monthly.

Payments are made each month for 15 yr. All annuity payments are assumed to be made at the end of the payment period.

48. Find the value at the end of 15 yr (the future value) of the following:

a) The first month's payment.

b) The second month's payment.

c) The third month's payment.

49. Represent S_n, the future value of the annuity at the end of the nth month, using sigma notation.

50. Determine the future value of the annuity in 15 yr.

Exercises 51–53 refer to this **problem situation**: A $250

monthly annuity has an APR of 9% compounded monthly. Payments are made each month for 20 yr. All annuity payments are assumed to be made at the end of the payment period.

51. Find the value at the end of 20 yr (the future value) of the following:

a) The first month's payment.

b) The second month's payment.

c) The third month's payment.

52. Represent S_n, the future value of the annuity at the end of the nth month, using sigma notation.

53. Determine the future value of the annuity in 20 yr.

11.3   Binomial Theorem

Objective

Students will be able to apply the binomial theorem to expand a binomial expression of the form $(a + b)^n$. Students also will be able to find the coefficient of any given term of a binomial expansion and relate this coefficient to Pascal's triangle.

Key Ideas

Binomial
Pascal's triangle
Binomial coefficient n
    choose r
Binomial theorem

Section 11.2 introduced the concept of *finite series*. A finite sum also occurs when an expression like $a + b$, called a **binomial** since it has two terms, is raised to a power. Using multiplication, each of these formulas can be verified.

$$(a + b)^2 = a^2 + 2ab + b^2$$
$$(a + b)^3 = a^3 + 3a^2b + 3ab^2 + b^3$$
$$(a + b)^4 = a^4 + 4a^3b + 6a^2b^2 + 4ab^3 + b^4$$
$$(a + b)^5 = a^5 + 5a^4b + 10a^3b^2 + 10a^2b^3 + 5ab^4 + b^5$$

Observing the patterns that occur in this set of equations will make it easier to learn to find $(a + b)^n$.

1. Notice that in each expansion there are $n + 1$ terms.

2. There is symmetry in each expansion, with respect to both the coefficients and the exponents of each term.

3. The sum of the exponents in each term is n. From left to right in each expansion, the degree of a decreases by 1 as the degree of b increases by 1.

4. The first and last terms are a^n and b^n, respectively.

5. The second and second-to-last terms are $na^{n-1}b$ and nab^{n-1}, respectively.

Notice that the pattern of the expansion is as follows:

$$(a + b)^n = a^n + \underline{}\ a^{n-1}b + \underline{}\ a^{n-2}b^2 + \ldots + \underline{}\ ab^{n-1} + b^n.$$

We need only learn what numerical coefficient is needed for each blank. The pattern of numbers known as Pascal's triangle can be used to find these coefficients.

Pascal's Triangle

Consider the following array of numbers known as **Pascal's triangle**. What numbers should replace the dots in the next row of the array?

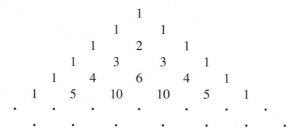

There are several important observations about the pattern in this array.

1. Each row begins and ends with a 1.

2. Each entry of a row is the sum of the two numbers directly above it in the preceding row.

3. The numbers of the nth row of this array are precisely the coefficients of the terms in the binomial expansion of $(a + b)^n$. Consider the 1 at the top of the triangle to be row zero.

Observe that row 6 of the array is

$$1 \quad 6 \quad 15 \quad 20 \quad 15 \quad 6 \quad 1.$$

E X A M P L E 1   Finding a Binomial Expansion

Expand $(a + b)^6$.

Solution   The sixth row of Pascal's triangle contains the coefficients of the terms in the expression.

$$(a + b)^6 = a^6 + \underline{\ ?\ }\ a^5b + \underline{\ ?\ }\ a^4b^2 + \underline{\ ?\ }\ a^3b^3 + \underline{\ ?\ }\ a^2b^4 + \underline{\ ?\ }\ ab^5 + b^6$$

Therefore the correct expansion is

$$(a + b)^6 = a^6 + 6a^5b + 15a^4b^2 + 20a^3b^3 + 15a^2b^4 + 6ab^5 + b^6.$$ ≡

Common Errors

Students often make errors in sign when they apply the binomial theorem to an expression of the form $(a - b)^n$. Notice that this expression must be interpreted as $(a + (-b))^n$. Example 2 illustrates a correct method.

E X A M P L E   2   Finding a Binomial Expansion

Expand $(2x - 3y^2)^4$.

Solution   The binomial $2x - 3y^2$ can be rewritten as $[2x + (-3y^2)]$. Let $a = 2x$ and $b = -3y^2$ and substitute into the expression $(a + b)^4$.

$$(a + b)^4 = a^4 + 4a^3b + 6a^2b^2 + 4ab^3 + b^4$$

$$(2x - 3y^2)^4 = (2x)^4 + 4(2x)^3(-3y^2) + 6(2x)^2(-3y^2)^2 + 4(2x)(-3y^2)^3 + (-3y^2)^4$$

$$= 16x^4 + 4(-24)x^3y^2 + 6(36)x^2(y^2)^2 + 4(-54)x(y^2)^3 + 81(y^2)^4$$

$$= 16x^4 - 96x^3y^2 + 216x^2y^4 - 216xy^6 + 81y^8 \qquad \blacksquare$$

Binomial Coefficients $\binom{n}{r}$

Teaching Note

Point out to students that the binomial coefficient n choose r is denoted on most graphers as $_nC_r$.

While Pascal's triangle is handy to use if n is fairly small, it is not adequate for large values of n or in a situation when a formula is needed (for example, when writing a computer program) for the coefficient of the rth term in $(a + b)^n$.

Before introducing such a formula, we must introduce the concept of *factorial*. For any positive integer n, we define n **factorial**, denoted $n!$, by

$$n! = 1 \cdot 2 \cdot \; \cdots \; \cdot n.$$

Notice that $n!$ is the product of all the positive integers starting with 1 and ending with n. For example,

$$3! = 1 \cdot 2 \cdot 3 = 6 \qquad \text{and} \qquad 5! = 1 \cdot 2 \cdot 3 \cdot 4 \cdot 5 = 120.$$

We also define $0! = 1$.

Definition 11.5   Binomial Coefficient $\binom{n}{r}$

If n and r are two nonnegative integers, the number called the **binomial coefficent n choose r**, denoted $\binom{n}{r}$, is defined by

$$\binom{n}{r} = \frac{n!}{r!(n-r)!}.$$

Example 3 illustrates how to compute binomial coefficients.

EXAMPLE 3   Finding Binomial Coefficients

Find each of these constants:

   a) $\binom{6}{4}$

   b) $\binom{9}{5}$

Solution

a) $\binom{6}{4} = \dfrac{6!}{4!(6-4)!} = \dfrac{6 \cdot 5}{2 \cdot 1} = 15$

b) $\binom{9}{5} = \dfrac{9!}{5!(9-5)!} = \dfrac{9 \cdot 8 \cdot 7 \cdot 6}{4 \cdot 3 \cdot 2 \cdot 1} = 126$

   Notice that $\binom{n}{r}$ is the rth entry in the nth row of Pascal's triangle, where the first entry is considered the zeroth entry. Evaluate each of these binomial coefficients and convince yourself that this array is really Pascal's triangle.

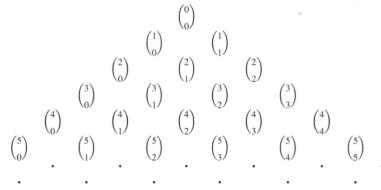

We have seen earlier in this section that Pascal's triangle can be used to find the coefficients in a binomial expansion. This relationship is formalized in the **binomial theorem**.

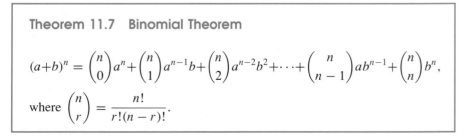

Theorem 11.7   Binomial Theorem

$$(a+b)^n = \binom{n}{0}a^n + \binom{n}{1}a^{n-1}b + \binom{n}{2}a^{n-2}b^2 + \cdots + \binom{n}{n-1}ab^{n-1} + \binom{n}{n}b^n,$$

where $\binom{n}{r} = \dfrac{n!}{r!(n-r)!}$.

Ex. 21–26 encourage students to use a grapher to develop a visual understanding of binomial expressions. Notice that Ex. 35 is the Pascal triangle relation. It says that each term of the triangle is the sum of the two terms immediately above it.

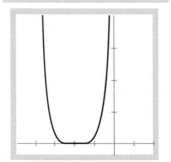

[–5, 2] by [–5, 40]

Figure 11.6 $f(x) = (x+2)^6$ overlaid with $y = x^6 + 12x^5 + 60x^4 + 160x^3 + 240x^2 + 192x + 64$.

Assignment Guide

Day 1: Ex. 1–31 odd, 33–35

EXAMPLE 4   Using the Binomial Theorem

Find the twelfth term of $(2x + 3y)^{15}$, where the leading term is considered the zeroth term.

Solution   From the binomial theorem, note that the twelfth term of $(2x + 3y)^{15}$ is

$$\binom{15}{12}(2x)^{15-12}(3y)^{12} = \left(\frac{15!}{12!3!}\right)(2x)^3(3y)^{12}$$

$$= \frac{15 \cdot 14 \cdot 13}{3 \cdot 2 \cdot 1} \cdot 8x^3 \cdot 3^{12}y^{12}$$

$$= 5 \cdot 7 \cdot 13 \cdot 8x^3(531{,}441) \cdot y^{12}$$

$$= 1{,}934{,}445{,}240\, x^3 y^{12}.$$ ▄

EXAMPLE 5   Using the Binomial Theorem

Use the binomial theorem to represent $f(x) = (x + 2)^6$ in the usual polynomial form.

Solution   (See Fig. 11.6.)

$$f(x) = (x + 2)^6$$

$$= x^6 + \binom{6}{1}x^5(2) + \binom{6}{2}x^4(2^2) + \binom{6}{3}x^3(2^3) + \binom{6}{4}x^2(2^4) + \binom{6}{5}x(2^5) + 2^6$$

$$= x^6 + 6 \cdot 2x^5 + 15 \cdot 4x^4 + 20 \cdot 8x^3 + 15 \cdot 16x^2 + 6 \cdot 32x + 64$$

$$= x^6 + 12x^5 + 60x^4 + 160x^3 + 240x^2 + 192x + 64$$ ▄

Exercises for Section 11.3

In Exercises 1 and 2, find the indicated row of Pascal's triangle.

1. The eighth row

2. The tenth row

In Exercises 3–10, expand each expression. Use Pascal's triangle to find the coefficients of each term.

3. $(a + b)^5$ **4.** $(a - b)^6$ **5.** $(x + 2y)^4$

6. $(3x - y)^7$ **7.** $(x + y)^6$ **8.** $(a - 2b^2)^4$

9. $(x - y)^6$ **10.** $(1 + 0.08)^5$

In Exercises 11–16, find each binomial coefficient.

11. $\binom{7}{3}$ **12.** $\binom{9}{4}$

13. $\binom{8}{4}$ **14.** $\binom{12}{3}$

15. $\binom{15}{11}$ **16.** $\binom{13}{7}$

In Exercises 17–20, find the coefficient of the given term in the binomial expansion.

17. x^5y^3 term, $(x + y)^8$ **18.** x^3y^5 term, $(x + y)^8$

19. x^4 term, $(x + 4)^6$ **20.** x^6 term, $(x - 3)^9$

In Exercises 21–26, use the binomial theorem to find a polynomial expansion for each function. Support your work with a graphing utility as for Figure 11.6.

21. $f(x) = (x - 2)^5$

22. $g(x) = (x + 3)^6$

23. $h(x) = (2x - 1)^7$

24. $f(x) = (3x + 4)^5$

25. $g(x) = (x + a)^4$

26. $h(x) = (3x - 4)^8$

In Exercises 27–32, use the binomial theorem to find an expansion for each expression.

27. $(2x + y)^5$

28. $(\sqrt{x} + 3)^4$

29. $(\sqrt{x} - \sqrt{y})^6$

30. $(2y - 3x)^{12}$

31. $(x^{-2} + 3)^5$

32. $(a - b)^{15}$

33. Show that $\binom{n}{r} = \binom{n}{n-r}$.

34. Show that $\binom{8}{3} + \binom{8}{4} = \binom{9}{4}$.

35. Show that, in general, $\binom{k}{r-1} + \binom{k}{r} = \binom{k+1}{r}$.

11.4 _____ Polynomial Approximations to Functions

Objective

Students will be able to demonstrate graphically that functions such as $\sin x$, $\cos x$, $\sqrt{1 + x}$ can be approximated on a given interval by a polynomial expression.

Key Ideas

Polynomial approximations
Power series
Power series representation of $\sin x$
Power series representation of $\frac{1}{(1-x)}$
Generalized binomial expansion
Power series representation of $\sqrt{1 + x}$

Polynomial functions are among the simplest functions discussed so far in this text. We are interested in exploring the extent to which a function can be approximated by a simpler function.

Consider the function $g(x) = (1 + x)^{10}$. Using the binomial theorem, this function can be expanded to the form

$$g(x) = 1 + \binom{10}{1}x + \binom{10}{2}x^2 + \binom{10}{3}x^3 + \cdots + x^{10}.$$

Counting the 1 as the first term, the first six terms of this expansion form the fifth-degree polynomial

$$f(x) = 1 + 10x + 45x^2 + 120x^3 + 210x^4 + 252x^5.$$

For what values of x does $f(x)$ approximate

$$g(x) = (1 + x)^{10}?$$

We begin by considering these functions for the specific value $x = 0.06$.

E X A M P L E   1   *Using a Binomial Expansion as Approximation*

Compare the value of the sum of the first six terms of the binomial expansion of $(1 + 0.06)^{10}$ with the value of $(1.06)^{10}$.

Solution   Use a calculator to find that $(1.06)^{10} = 1.790847697$, which is accurate to 10 digits. To complete the comparison, use the binomial theorem to expand $(1 + 0.06)^{10}$.

$$(1 + 0.06)^{10} = 1 + \binom{10}{1}(0.06) + \binom{10}{2}(0.06)^2 + \binom{10}{3}(0.06)^3$$

$$+ \binom{10}{4}(0.06)^4 + \binom{10}{5}(0.06)^5 + \cdots + (0.06)^{10}$$

We are interested in how the value of the sum of the first 6 terms compares to the exact value of the sum of all 11 terms. So we use a calculator to find that

$$1 + \binom{10}{1}(0.06) + \binom{10}{2}(0.06)^2 + \binom{10}{3}(0.06)^3$$

$$+ \binom{10}{4}(0.06)^4 + \binom{10}{5}(0.06)^5$$

$$= 1 + 10(0.06) + 45(0.06)^2 + 120(0.06)^3 + 210(0.06)^4 + 252(0.06)^5$$

$$= 1.790837555.$$

The two values 1.790847697 and 1.790837555 agree through the fourth decimal place. ▤

Notes on Examples

Suppose that $g(x) = (1 + x)^{10}$ and $f(x) = 1 + 10x + 45x^2 + 120x^3 + 210x^4 + 252x^5$. Example 1 shows numerically that $g(x)$ can be approximated by $f(x)$, and Example 2 shows graphically that $g(x)$ can be approximated by $f(x)$.

From this example, it is clear that the function $f(x)$ approximates $g(x)$ through the fourth decimal place, when $x = 0.06$. We ask, for what other values of x is this approximation close?

E X A M P L E   2   Finding an Interval of Values

Determine graphically an interval of values for x for which

$$f(x) = 1 + 10x + 45x^2 + 120x^3 + 210x^4 + 252x^5$$

is a good approximation to $g(x) = (1 + x)^{10}$.

Teaching Note

Figure 11.7 is important because it demonstrates that while a polynomial may approximate a function on one interval, it may not on another.

Solution   Figure 11.7(a) shows a graph of both f and g. It appears that f approximates g when $-0.2 \le x \le 0.2$.

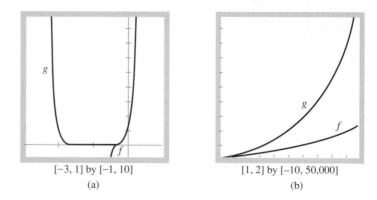

[−3, 1] by [−1, 10]	[1, 2] by [−10, 50,000]
(a)	(b)

Figure 11.7   Two views of $f(x) = 1 + 10x + 45x^2 + 120x^3 + 210x^4 + 252x^5$ and $g(x) = (1 + x)^{10}$: (a) f is a good approximation to g over $[-0.2, 0.2]$; (b) f is not a good approximation over $[1, 2]$.

We say that f is a good approximation to g over the interval $[-0.2, 0.2]$. ▣

Example 1 does not claim that the interval $[-0.2, 0.2]$ is the largest interval or the only interval on which the approximation is good. It also is of interest to observe in Fig. 11.7(b) that the approximation is not good on the interval $[1, 2]$. So an approximation may be good on some intervals and not on others.

Many functions can be approximated by polynomial functions. The intervals on which the approximation is good vary from one function to the next. The material that follows uses the two functions, $f(x) = \sin x$ and $g(x) = 1/(1 - x)$, to illustrate that polynomial functions can be used as approximations to these functions on certain intervals.

Power Series Representation of sin *x*

This section investigates when the partial sum of a power series is a good approximation to the function $f(x) = \sin x$. We begin with a definition.

Definition 11.6   Power Series

A **power series** in x is any series of the form

$$\sum_{k=0}^{\infty} a_k x^k = a_0 + a_1 x + a_2 x^2 + \cdots + a_n x^n + \cdots .$$

Notice that the partial sums of a power series are polynomials. Therefore a function that can be represented as a power series can be approximated by a polynomial function.

$$f_0(x) = a_0$$

$$f_1(x) = a_0 + a_1 x$$

$$f_2(x) = a_0 + a_1 x + a_2 x^2$$

$$\vdots$$

$$f_n(x) = a_0 + a_1 x + a_2 x^2 + \cdots + a_n x^n$$

Many of the functions studied earlier in this text can be represented by a power series. To do a complete study of all power series representations goes beyond the scope of this text. Here, we simply illustrate power series representations for a few specific functions. For example, the function $f(x) = \sin x$ can be represented as a power series, which means it can be approximated by polynomials.

Power Series Representation of sin x

It can be shown that for all real numbers x,

$$\sin x = x - \frac{x^3}{3!} + \frac{x^5}{5!} - \frac{x^7}{7!} + \cdots = \sum_{k=0}^{\infty} (-1)^k \frac{x^{2k+1}}{(2k+1)!}.$$

This power series representation of $f(x) = \sin x$ is valid for all real numbers. This means that for any real number a, there is a positive integer n such that the nth partial sum of the power series is a good approximation to $\sin a$. Said another way, the power series converges to $\sin a$ for all values of a.

But the $(n + 1)$th partial sum of this series is the degree-$(2n + 1)$ polynomial

$$f_n(x) = \sum_{k=0}^{n} (-1)^k \frac{x^{2k+1}}{(2k+1)!} = x - \frac{x^3}{3!} + \frac{x^5}{5!} - \frac{x^7}{7!} + \cdots + (-1)^n \frac{x^{2n+1}}{(2n+1)!}.$$

Therefore $f(x) = \sin x$ can be approximated by polynomial functions.

Every power series converges for some interval of values of x. This interval can be a single point. We call this interval the **interval of convergence**.

Example 3 provides visual evidence that as n increases, the interval of convergence gets larger as well.

E X A M P L E 3   Finding Intervals of Convergence

Determine graphically intervals on which $f(x) = \sin x$ is approximated by

$$f_n(x) = \sum_{k=0}^{n} (-1)^k \frac{x^{2k+1}}{(2k+1)!}$$

for $n = 1, 2,$ and 3.

Solution   We begin by finding a graph of $f(x) = \sin x$ in the same viewing rectangle with the graphs of

$$f_1(x) = x - \frac{1}{3!}x^3, \qquad f_2(x) = x - \frac{1}{3!}x^3 + \frac{1}{5!}x^5,$$

$$f_3(x) = x - \frac{1}{3!}x^3 + \frac{1}{5!}x^5 - \frac{1}{7!}x^7.$$

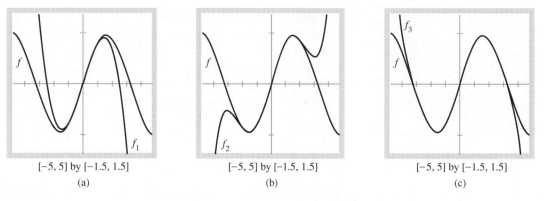

$$[-5, 5] \text{ by } [-1.5, 1.5]$$

(a)

$$[-5, 5] \text{ by } [-1.5, 1.5]$$

(b)

$$[-5, 5] \text{ by } [-1.5, 1.5]$$

(c)

Figure 11.8 $f(x) = \sin x$ plotted with three partial sums: (a) $f_1(x) = x - \frac{1}{6}x^3$; (b) $f_2(x) = x - \frac{1}{6}x^3 + \frac{1}{120}x^5$; and (c) $f_3(x) = x - \frac{1}{6}x^3 + \frac{1}{120}x^5 - \frac{1}{5040}x^7$.

Frank says, "The content of this section demonstrates the power of visualization as well as any topic in the entire text. You can *see* that many nonpolynomial functions can be approximated by polynomials over certain intervals."

a) Figure 11.8(a) shows that $f_1(x)$ gives a good approximation to $f(x) = \sin x$ over the interval $-1 \le x \le 1$.

b) Figure 11.8(b) shows that $f_2(x)$ gives a good approximation to $f(x) = \sin x$ over the interval $-2 \le x \le 2$.

c) Figure 11.8(c) shows that $f_3(x)$ gives a good approximation to $f(x) = \sin x$ over the interval $-3 \le x \le 3$. ▤

It appears in Example 3 that as n increases, the interval of values of x for which $f_n(x)$ is a good approximation to $f(x) = \sin x$ becomes larger. It can be shown that this is true for the power series representation for $f(x) = \sin x$. However, it is not true of power series representations of all functions.

Power Series Representation of $g(x) = 1/(1 - x)$

Theorem 11.6 in Section 11.2 tells us that the infinite geometric series with the common ratio x,

$$\sum_{k=0}^{\infty} x^k,$$

converges to $1/(1 - x)$ for all x in the interval $-1 < x < 1$. This gives a power series representation for $g(x) = 1/(1 - x)$.

Power Series Representation for $g(x) = 1/(1 - x)$

If x is a real number in the interval $(-1, 1)$, then

$$\frac{1}{1 - x} = 1 + x + x^2 + x^3 + \cdots = \sum_{k=0}^{\infty} x^k.$$

The partial-sum polynomials

$$g_n(x) = 1 + x + x^2 + \cdots + x^n$$

are approximations to $g(x) = 1/(1-x)$. Although in this example x must satisfy the condition $-1 < x < 1$, it is true here, as it was in Example 3, that the interval of values for which the approximation is good becomes larger as n increases. However, Example 4 will show that for a particular value of n, say $n = 5$, the error in the approximation is not the same for all values of x.

E X A M P L E 4   Finding the Error of Approximation

Determine an interval of values for which $g_5(x) = 1 + x + x^2 + x^3 + x^4 + x^5$ is a good approximation to $g(x) = 1/(1-x)$. Then estimate the error in approximating $g(0.2)$ by $g_5(0.2)$ and $g(0.3)$ by $g_5(0.3)$.

Solution Figure 11.9 shows a graph of both $g(x) = 1/(1-x)$ and $g_5(x)$. This figure shows that $g_5(x)$ gives a good approximation to $g(x)$ over the interval $-0.7 \leq x \leq 0.7$.

To find an estimate on the error of approximation to $g(0.2)$ and $g(0.3)$, graph both g and g_5 in the viewing rectangles $[0.1999, 0.2001]$ by $[1.249, 1.251]$ and $[0.299, 0.301]$ by $[1.42, 1.43]$.

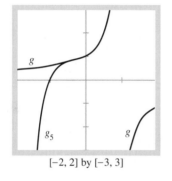

[−2, 2] by [−3, 3]

Figure 11.9 $g(x) = 1/(1-x)$ and $g_5(x)$.

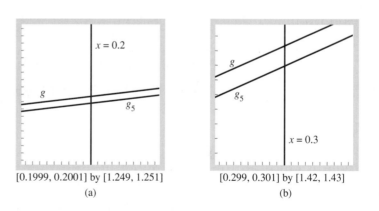

[0.1999, 0.2001] by [1.249, 1.251]
(a)

[0.299, 0.301] by [1.42, 1.43]
(b)

Figure 11.10 $g(x) = 1/(1-x)$ compared with a partial-sum approximation for two values of x: (a) $g(0.2)$ and $g_5(0.2)$; (b) $g(0.3)$ and $g_5(0.3)$.

Figure 11.10(a) shows that $g_5(0.2)$ underestimates $g(0.2)$ by about 0.0001. Similarly, $g_5(0.3)$ underestimates $g(0.3)$ by about 0.001. So we see that the approximation of g by g_5 is less accurate at the positive end of the interval of convergence. We would find the same pattern of decreasing accuracy if we tested smaller and smaller values of x at the negative end of this interval. ≡

Generalized Binomial Expansion

Using Theorem 11.7 of Section 11.3, we see that

$$f(x) = (1+x)^n = 1 + \binom{n}{1}x + \binom{n}{2}x^2 + \cdots + x^n. \tag{1}$$

In the remainder of this section, we will examine Eq. (1) for the case where $n = \frac{1}{2}$. This generalization results in a power series representation of an approximation for the function $f(x) = \sqrt{1+x}$.

Until now, binomial coefficients $\binom{n}{r}$ have been defined only for nonnegative integers. To state this power series representation requires that we first find an appropriate way to define $\binom{n}{r}$ in the case where n is a fraction. We use Definition 11.5 and divide by $(n-r)!$ to obtain a new representation for $\binom{n}{r}$.

$$\binom{n}{r} = \frac{n!}{r!(n-r)!} = \frac{n(n-1)\cdots(n-r+1)}{r!} \tag{2}$$

Using the right side of Eq. (2), where $n = \frac{1}{2}$, we obtain

$$\binom{0.5}{r} = \frac{0.5(0.5-1)\cdots(0.5-r+1)}{r!}.$$

For example,

$$\binom{0.5}{2} = \frac{0.5(0.5-1)}{2!} = -\frac{1}{8} \quad \text{and} \quad \binom{0.5}{3} = \frac{0.5(0.5-1)(0.5-2)}{3!} = \frac{1}{16}.$$

We can now state a power series representation for the function $f(x) = \sqrt{1+x}$.

Power Series Representation for $f(x) = \sqrt{1+x}$

It can be shown that for all real numbers x in the interval $[-1, 1]$,

$$\sqrt{1+x} = 1 + \binom{0.5}{1}x + \binom{0.5}{2}x^2 + \binom{0.5}{3}x^3 + \cdots = \sum_{k=0}^{\infty} \binom{0.5}{k}x^k. \tag{3}$$

Equation (3) is sometimes called a **generalized binomial expansion.**

As Examples 3 and 4 have illustrated, the polynomials that are the partial sums of an infinite series approximate the function described by the power series. Example 4 compared the error in the approximation for two values in the interval of convergence.

In Example 5, we compare the error in the approximation between the partial-sum polynomials $f_3(x)$ and $f_4(x)$ for the function $f(x) = \sqrt{1 + x}$.

E X A M P L E   5   Comparing Errors of Approximation

Let $f(x) = \sqrt{1 + x}$. Estimate graphically the error in approximating $\sqrt{1.5} = f(0.5)$ by $f_3(0.5)$ and by $f_4(0.5)$, where f_3 and f_4 are the following partial-sum approximations to the power series representation of $f(x)$.

$$f_3(x) = 1 + \frac{1}{2}x - \frac{1}{8}x^2 + \frac{1}{16}x^3$$

$$f_4(x) = 1 + \frac{1}{2}x - \frac{1}{8}x^2 + \frac{1}{16}x^3 - \frac{5}{128}x^4$$

Solution   Find the graphs of $f(x)$ and $f_3(x)$ in the viewing rectangle [0.499, 0.501] by [1.22, 1.23] and the graphs of $f(x)$ and $f_4(x)$ in the viewing rectangle [0.4999, 0.5001] by [1.224, 1.225].

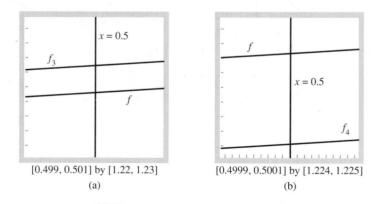

[0.499, 0.501] by [1.22, 1.23]
(a)

[0.4999, 0.5001] by [1.224, 1.225]
(b)

Figure 11.11 $f(x) = \sqrt{1 + x}$ for $x = 0.5$ compared with two of its partial-sum approximations: (a) $f(0.5)$ and $f_3(0.5)$; (b) $f(0.5)$ and $f_4(0.5)$.

Figure 11.11(a) shows that $f_3(0.5)$ is an overestimate of $f(0.5)$ by about 0.002. Similarly, Fig. 11.11(b) shows that $f_4(0.5)$ is an underestimate of $f(0.5)$ by about 0.0006.

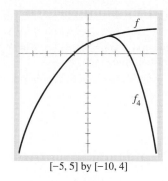

[−5, 5] by [−10, 4]

Figure 11.12 $f_4(x)$ and $f(x) = \sqrt{1+x}$. The approximation is good for x in $[-1, 1]$. ☰

Example 5 found the error in the approximation to $f(x) = \sqrt{1+x}$ by $f_4(x)$, where

$$f_4(x) = 1 + \frac{1}{2}x - \frac{1}{8}x^2 + \frac{1}{16}x^3 - \frac{5}{128}x^4$$

for $x = 0.5$. Figure 11.12 shows the graph of $f_4(x)$ overlaid onto the graph of f. From this global view, it is clear that the approximation is good for x in the interval $[-1, 1]$.

This section ends with a summary of some of the conclusions that come from the examples in this section.

Summary

Suppose that $f(x)$ has a power series representation

$$\sum_{k=0}^{\infty} a_k x^k$$

over some interval (a, b).

1. As n increases, the interval of values for which the partial-sum polynomial $f_n(x)$ is a good approximation to f becomes larger.
2. The approximation to $f(x)$ by $f_n(x)$ may not be as good near the ends of the interval of convergence as toward the midpoint of the interval.
3. For a particular value $x = c$, the approximation $f_n(c)$ to $f(c)$ gets better as n increases.

Notes on Exercises

Notice that the correct conjecture for Ex. 8 is the number $e = 2.71828\ldots$. Ex. 9–36 provide power series representations for many of the nonpolynomial functions studied previously in this course. A few of them are $f(x) = \frac{1}{(1+x)}$, $f(x) = e^x$, $f(x) = \cos x$, and $f(x) = \tan^{-1} x$. The fact that polynomials can be used to provide good approximations to many other types of functions is one of the reasons that polynomial functions are important.

Assignment Guide

Day 1: Ex. 1–12
Day 2: Ex. 13–23, 31

Exercises for Section 11.4

Exercises 1–6 refer to the function $f(x) = (1 + x)^6$ whose partial-sum approximations are the following:

$$f_1(x) = 1 + 6x$$

$$f_2(x) = 1 + 6x + 15x^2$$

$$f_3(x) = 1 + 6x + 15x^2 + 20x^3$$

$$f_4(x) = 1 + 6x + 15x^2 + 20x^3 + 15x^4$$

$$f_5(x) = 1 + 6x + 15x^2 + 20x^3 + 15x^4 + 6x^5$$

$$f_6(x) = 1 + 6x + 15x^2 + 20x^3 + 15x^4 + 6x^5 + x^6$$

1. Verify that $f(x) = (1 + x)^6 = f_6(x)$.

2. Find complete graphs of $f_1, f_2, \ldots, f_6$ in the same viewing rectangle.

3. Use a graphing method to determine the error in using f_4 to approximate the value of $f(x) = (1 + x)^6$ when $x = 1$.

4. Compute $|f(1) - f_4(1)|$ and compare the result with the error estimate determined in Exercise 3.

5. Is $f_4(1)$ an underestimate or overestimate of $f(1)$? Why?

6. Repeat Exercises 3 to 5 using f_5 to approximate $f(x) = (1 + x)^6$ when $x = 1$.

For Exercises 7 and 8 consider the infinite series

$$1 + \frac{1}{1!} + \frac{1}{2!} + \frac{1}{3!} + \cdots .$$

Assume the pattern continues as suggested.

7. Compute the first 10 partial sums of the series.

8. It can be shown the series converges to a number ℓ. Make a conjecture about the value of ℓ.

Exercises 9–12 refer to the following functions:

$$f_1(x) = 1 - x$$

$$f_2(x) = 1 - x + x^2$$

$$f_3(x) = 1 - x + x^2 - x^3$$

$$\vdots$$

$$f_n(x) = 1 - x + x^2 - x^3 + \cdots + (-1)^n x^n$$

9. Graph $f_1, f_2, f_3, f_4, f_5, f_6, f_7$, and f_8 together in the viewing rectangle $[-2, 2]$ by $[-5, 5]$.

10. Find a complete graph of $f(x) = 1/(x + 1)$. Compare this graph with those in Exercise 9 to conclude that the power series $1 - x + x^2 - x^3 + \cdots + (-1)^n x^n + \cdots$ represents $f(x) = 1/(x + 1)$ for some values of x.

11. Based on Exercises 9 and 10, make a conjecture about the interval of convergence of the power series expansion

$$\frac{1}{x + 1} = 1 - x + x^2 - x^3 + \cdots + (-1)^n x^n + \cdots .$$

12. Relate the power series expansion of $1/(x + 1)$ to a geometric series.

Exercises 13–18 refer to the following functions:

$$f_1(x) = 1 + x$$

$$f_2(x) = 1 + x + \frac{x^2}{2!}$$

$$f_3(x) = 1 + x + \frac{x^2}{2!} + \frac{x^3}{3!}$$

$$\vdots$$

$$f_n(x) = 1 + x + \frac{x^2}{2!} + \frac{x^3}{3!} + \cdots + \frac{x^n}{n!}$$

13. Find complete graphs of the six polynomial functions $f_1, f_2, \ldots, f_6$ in the viewing rectangle $[-2, 2]$ by $[-3, 8]$.

14. Find a complete graph of $f(x) = e^x$ and compare it with the graphs found in Exercise 13. Find complete graphs of $f(x) = e^x$ and f_6 in the same viewing rectangle.

15. Is $f_6(-8)$ a good estimate for e^{-8}? Why?

16. Is $f_6(3)$ a good estimate for e^3? Why?

17. On what interval is $f_6(x)$ a good approximation to e^x?

18. Use a graphing method to determine the error in using $f_3(3)$ to approximate e^3. Repeat using $f_4(3)$, $f_5(3)$, and $f_6(3)$. Are they underestimates or overestimates?

Exercises 19–23 refer to the following functions:

$$f_1(x) = 1 - \frac{x^2}{2!}$$

$$f_2(x) = 1 - \frac{x^2}{2!} + \frac{x^4}{4!}$$

$$f_3(x) = 1 - \frac{x^2}{2!} + \frac{x^4}{4!} - \frac{x^6}{6!}$$

$$\vdots$$

$$f_n(x) = 1 - \frac{x^2}{2!} + \frac{x^4}{4!} - \frac{x^6}{6!} + \cdots + \frac{(-1)^n x^{2n}}{(2n)!}$$

19. Find complete graphs of the six polynomial functions f_1, $f_2, \ldots, f_6$ in the viewing rectangle $[-6, 6]$ by $[-3, 3]$.

20. Find a complete graph of $f(x) = \cos x$ and compare it with the graphs found in Exercise 19. Find complete graphs of $f(x) = \cos x$ and f_6 in the same viewing rectangle.

21. Is $f_6(-8)$ a good estimate for $\cos(-8)$? Why?

22. Is $f_6(3)$ a good estimate for $\cos 3$? Why?

23. Explain why $f_{10}(3)$ is a very good estimate for $\cos 3$. What is the error in the estimate? Is $f_{10}(3)$ an underestimate or overestimate?

Exercises 24–28 refer to the family of functions $f_n(x)$ defined as follows:

$$f_n(x) = 1 + \frac{1}{3}x + \frac{\frac{1}{3}\left(-\frac{2}{3}\right)}{2!}x^2 + \frac{\frac{1}{3}\left(-\frac{2}{3}\right)\left(-\frac{5}{3}\right)}{3!}x^3$$

$$+ \cdots + \frac{\frac{1}{3}\left(-\frac{2}{3}\right)\left(-\frac{5}{3}\right)\cdots\left(\frac{1}{3}-n+1\right)}{n!}x^n$$

24. Determine f_1, f_2, f_3.

25. Find complete graphs of the polynomial functions f_1, f_2, f_3 in the same viewing rectangle and compare them with the graph of $f(x) = \sqrt[3]{1+x}$.

26. On what interval do you think $f_n(x)$, for large values of n, closely approximates $f(x) = \sqrt[3]{1+x}$?

27. Use a graphing method to estimate the error in using $f_3(-0.25)$ as an estimate for $\sqrt[3]{0.75}$.

28. Use a graphing method to estimate the error in using $f_6(-0.25)$ as an estimate for $\sqrt[3]{0.75}$.

Exercises 29–31 give power series representations for several other functions. Use a graphing argument to complete each.

29. Give an argument that

$$\tan^{-1}(x) = x - \frac{x^3}{3} + \frac{x^5}{5} - \frac{x^7}{7} + \cdots + \frac{(-1)^n x^{2n+1}}{2n+1} + \cdots$$

for $-1 < x < 1$ by graphing $f(x) = \tan^{-1}(x)$ and the polynomial function consisting of the first eight terms of the power series $x - x^3/3 + x^5/5 - \cdots$ in the same viewing rectangle.

30. Give an argument that

$$\ln(1+x) = x - \frac{x^2}{2} + \frac{x^3}{3} + \cdots + (-1)^{n+1}\left(\frac{x^n}{n}\right) + \cdots$$

for $0 < x < 1$ by graphing $g(x) = \ln(1+x)$ and the polynomial function consisting of the first eight terms of the power series $x - x^2/2 + x^3/3 - x^4/4 + \cdots$ in the same viewing rectangle.

31. Give an argument that

$$\frac{e^x - e^{-x}}{2} = x + \frac{x^3}{3!} + \frac{x^5}{5!} + \cdots + \frac{x^{2n-1}}{(2n-1)!} + \cdots$$

for $-2 < x < 2$ by graphing $g(x) = (e^x - e^{-x})/2$ and the polynomial function equal to the first eight terms of the power series

$$x + \frac{x^3}{3!} + \frac{x^5}{5!} + \cdots + \frac{x^{2n-1}}{(2n-1)!} + \cdots$$

in the same viewing rectangle.

Exercises 32 and 33 refer to the *hyperbolic cosine* (cosh) function, which is defined as

$$\cosh x = \frac{e^x + e^{-x}}{2}.$$

32. Use the fact that $e^x = 1 + x + x^2/2! + x^3/3! + \cdots$ for all real number values of x (see Exercises 13–18) to show that

$$\cosh x = 1 + \frac{x^2}{2!} + \frac{x^4}{4!} + \frac{x^6}{6!} + \cdots$$

for all real x. Assume that an infinite series can be added term by term.

33. Graph $g(x) = \cosh x$ and the polynomial function f consisting of the first four terms (up to the 6th power) of the power series $1 + x^2/2! + x^4/4! + \cdots$ in the same viewing rectangle. What conclusions can you draw? For what values of x is f a good approximation of $g(x) = \cosh x$?

Exercises 34–36 refer to the power series representation of $\sin x$,

$$\sin x = x - \frac{x^3}{3!} + \frac{x^5}{5!} - \frac{x^7}{7!} + \cdots .$$

34. How many terms of this power series must be used to obtain a good polynomial approximation to $f(x) = \sin x$ over $[-\pi/2, \pi/2]$? Use a graphing argument.

35. How many terms of this power series must be used to obtain a good polynomial approximation to $f(x) = \sin x$ over $[-\pi, \pi]$? Use a graphing argument.

36. How many terms of this power series must be used to obtain a good polynomial approximation to $f(x) = \sin x$ over $[-2\pi, 2\pi]$? Use a graphing argument.

It has been established in the previous exercises that

$$\sin x = x - \frac{x^3}{3!} + \frac{x^5}{5!} - \cdots$$

$$\cos x = 1 - \frac{x^2}{2!} + \frac{x^4}{4!} - \cdots$$

$$e^x = 1 + x + \frac{x^2}{2!} + \frac{x^3}{3!} + \cdots .$$

Use these power series to complete Exercises 37 and 38.

37. Show that $\cos x + i \sin x = e^{ix}$ (*Euler's formula*) by expanding e^u where $u = ix$. Assume infinite series can be added term by term.

38. The five most important constants in algebra are $0, 1, e, \pi,$ and i. Use the results from Exercise 37 to link these five constants by showing that

$$e^{\pi i} + 1 = 0.$$

11.5   Mathematical Induction

This section introduces a method of proof called the **principle of mathematical induction**. Let us consider the general sequence of partial sums.

$$S_1 = 1^2$$

$$S_2 = 1^2 + 2^2$$

$$S_3 = 1^2 + 2^2 + 3^2$$

$$S_4 = 1^2 + 2^2 + 3^2 + 4^2$$

$$\vdots$$

$$S_n = 1^2 + 2^2 + \cdots + n^2$$

In Theorem 11.4, the following equality was stated without proof:

$$S_n = 1^2 + 2^2 + \cdots + n^2 = \frac{n(n+1)(2n+1)}{6}.$$

For what values of n is this equation true? The implication is that the equation is true for all positive integers n. How can this be proven?

We begin with an Exploration.

Discuss in a Small Group

Let S be a set of numbers that have the following properties:

1. 1 is an element of S.
2. If k is in S, then $k + 1$ is also in S.

It is agreed that each element of S is obtained by applying these two properties.

What numbers are in the set S? Debate this question with members of your small group. Can the question be answered without additional information? Does your group agree?

Follow this reasoning: Because 1 is in S (by property 1), let $k = 1$ and conclude from property 2 that $1 + 1 = 2$ is also in S. Knowing that 2 is in S, apply property 2 again to conclude that $2 + 1 = 3$ is in S. Continuing this reasoning leads to the conclusion that all the positive integers are in S.

It is common for groups of students to conclude in this Exploration that S consists of the set of all integers $\{\ldots -3, -2, -1, 0, 1, 2, 3, \ldots\}$. This conclusion results from the following *false* reasoning:

> 1 is in S, so if $k + 1 = 1$ is in S, then $k = 0$ must also be in S.

Continuing this reasoning, you could conclude that -1 is in S, and so forth. What is false about this reasoning? It assumes that because you accept the statement

$$\text{if } k \text{ is in } S, \text{ then } k + 1 \text{ is also in } S, \qquad (1)$$

you can also accept the statement

$$\text{if } k + 1 \text{ is in } S, \text{ then } k \text{ is also in } S. \qquad (2)$$

Statement (2) is the converse of the first. It is not correct to accept the converse of a statement.

One way to remember that statement 1 and its converse are not both true is to visualize these falling dominoes.

Figure 11.13   Imagine a row of dominoes "sitting" on the real number line. Dominoes that have fallen or will eventually fall represent integers in the set S as a result of properties 1 and 2.

When the first domino, P_1, falls it does so in only one direction. And then all the dominoes in that direction fall. The others remain standing.

This discussion leads us to the following axiom.

Axiom of Induction

Let S be a set of the positive integers with the following two properties:

1. S contains the integer 1.
2. S contains the integer $k + 1$ whenever S contains the integer k.

Then S is the entire set of positive integers.

EXAMPLE 1 Using the Axiom of Induction

Consider the sequence $\{a_n\}$ defined recursively by $a_1 = 4$ and $a_n = 3 + a_{n-1}$. Prove that for all positive integers n, $a_n = 3n + 1$.

Solution Let S be the set of all positive integers for which the statement $a_n = 3n + 1$ is true. The goal is to show that this set S satisfies the two properties of the axiom of induction.

1. Demonstrate that 1 is in S: $a_1 = 4$ (given) and 4 is the value for $3n + 1$ when $n = 1$. Therefore the formula $a_n = 3n + 1$ is satisfied for $n = 1$. So 1 is in S.

2. Demonstrate that if k is in S, then $k + 1$ is in S: If k is in S, then $a_k = 3k + 1$.

$$a_{k+1} = 3 + a_k \qquad \text{This is true by the recursive definition of } a_n = 3 + a_{n-1}.$$

$$= 3 + (3k + 1) \qquad \text{Replace } a_k \text{ with its equivalent value } 3k + 1.$$

$$= (3k + 3) + 1$$

$$= 3(k + 1) + 1$$

Therefore $a_n = 3n + 1$ is true for $n = k + 1$. This shows that $k + 1$ is in S.

Both properties of the axiom of induction are satisfied, therefore you can conclude that S is the set of all positive integers. Further, because S is defined to be the set of all positive integers for which $a_n = 3n + 1$ is true, you can conclude that $a_n = 3n + 1$ is true for all positive integers n. ≡

Example 1 illustrates how to use the principle stated in the next theorem.

Theorem 11.8 Principle of Mathematical Induction

Let P_n be a statement that is defined for each positive integer n. All the statements P_n are true, provided the following two conditions are satisfied:

1. P_1 is true.
2. P_{k+1} is true whenever P_k is true.

Using the Principle of Mathematical Induction

Examples 2 through 4 show how to use Theorem 11.8 to prove that a statement P_n is true for all positive integers n.

E X A M P L E   2   *Using Mathematical Induction*

Use mathematical induction to prove that

$$1^2 + 3^2 + 5^2 + \cdots + (2n - 1)^2 = \frac{n(2n - 1)(2n + 1)}{3}$$

is true for every positive integer n.

Solution   Let P_n be the statement

$$1^2 + 3^2 + 5^2 + \cdots + (2n - 1)^2 = \frac{n(2n - 1)(2n + 1)}{3}. \tag{3}$$

For $n = 1$, the left-hand side of Eq. (3) is $1^2 = 1$ and the right-hand side is $1(2 - 1)(2 + 1)/3 = 1$. Thus P_1 is true.

Next, show that if P_k is true, then P_{k+1} is true. If P_k is true, then

$$1^2 + 3^2 + 5^2 + \cdots + (2k - 1)^2 = \frac{k(2k - 1)(2k + 1)}{3}$$

is true. To show that P_{k+1} is true means showing that the following is also:

$$1^2 + 3^2 + 5^2 + \cdots + (2(k + 1) - 1)^2 = \frac{(k + 1)(2(k + 1) - 1)(2(k + 1) + 1)}{3}$$

$$1^2 + 3^2 + 5^2 + \cdots + (2k + 1)^2 = \frac{(k + 1)(2k + 1)(2k + 3)}{3}. \tag{4}$$

The second-to-last term on the left-hand side of Eq. (4) is $(2k - 1)^2$. (Why?) Use this fact to rewrite the left-hand side of Eq. (4).

$$1^2 + 3^2 + 5^2 + \cdots + (2k + 1)^2 = 1^2 + 3^2 + 5^2 + \cdots + (2k - 1)^2 + (2k + 1)^2$$

$$= \frac{k(2k - 1)(2k + 1)}{3} + (2k + 1)^2$$

$$= (2k + 1)\left[\frac{k(2k - 1)}{3} + 2k + 1\right]$$

$$= (2k + 1)\left(\frac{2k^2 - k + 6k + 3}{3}\right)$$

$$= \frac{(2k + 1)(2k^2 + 5k + 3)}{3}$$

$$= \frac{(2k + 1)(k + 1)(2k + 3)}{3}$$

$$= \frac{(k + 1)(2k + 1)(2k + 3)}{3}$$

Thus Eq. (4) is true, so P_{k+1} must be true. By Theorem 11.8,

$$1^2 + 3^2 + 5^2 + \cdots + (2n - 1)^2 = \frac{n(2n - 1)(2n + 1)}{3}$$

for all positive integers n.   ▤

Example 3 proves one of the summation formulas stated without proof in Theorem 11.4.

E X A M P L E 3   Using Mathematical Induction

Use mathematical induction to prove that

$$1^3 + 2^3 + \cdots + n^3 = \frac{n^2(n + 1)^2}{4}. \tag{5}$$

Solution   Let P_n be the statement of Eq. (5). For $n = 1$, the left-hand side of Eq. (5) is $1^3 = 1$ and the right-hand side is $1^2(1 + 1)^2/4 = 1$. Thus P_1 is true.

Next, show that if P_k is true, then P_{k+1} is true. That is, show that the following is true:

$$1^3 + 2^3 + \cdots + (k + 1)^3 = \frac{(k + 1)^2(k + 1 + 1)^2}{4}$$

$$= \frac{(k + 1)^2(k + 2)^2}{4} \tag{6}$$

The second-to-last term on the left-hand side of Eq. (6) is k^3. Therefore

$$1^3 + 2^3 + \cdots + k^3 + (k+1)^3 = \frac{k^2(k+1)^2}{4} + (k+1)^3$$

$$= (k+1)^2 \left[\frac{k^2}{4} + (k+1) \right]$$

$$= (k+1)^2 \left[\frac{k^2 + 4k + 4}{4} \right]$$

$$= \frac{(k+1)^2(k+2)^2}{4}.$$

Thus

$$1^3 + 2^3 + \cdots + (k+1)^3 = \frac{(k+1)^2(k+2)^2}{4},$$

so P_{k+1} is true. By Theorem 11.8,

$$1^3 + 2^3 + \cdots + n^3 = \frac{n^2(n+1)^2}{4}$$

is true for all positive integers n.   ≡

Suppose A dollars are invested in an account that compounds interest. If i is the interest rate paid per compounding period, then, as shown in Section 5.2, the amount in the account after n periods is $A(1+i)^n$. For example, if \$200 is invested in an account that pays 9% annual interest compounded monthly, then $i = 0.09/12$ and the amount in the account after n months is

$$A(1+i)^n = 200 \left(1 + \frac{0.09}{12} \right)^n.$$

EXAMPLE 4   APPLICATION: Computing Compound Interest

Suppose A dollars are invested in a compound-interest-bearing account that pays interest rate i per compounding period. Use mathematical induction to prove that the amount in the account at the end of the nth period is $A(1+i)^n$.

Solution   Let P_n be the statement that the amount in the account at the end of the nth period is $A(1+i)^n$. First, show that P_1 is true. At the end of the first period, the amount in the account is the initial investment of A dollars plus the interest Ai earned during the first period. Thus

$$A + Ai = A(1+i).$$

Next, show that if P_k is true, then P_{k+1} is true. Assuming that P_k is true means that the amount in the account at the end of the kth period is $A(1+i)^k$. You must show that the amount in the account at the end of the $(k+1)$st period is $A(1+i)^{k+1}$. The interest earned during the $(k+1)$st period on this investment

Teaching Note

The fundamental counting principle refers to breaking a procedure into n successive ordered stages. Encourage students to identify what the stages are and to mentally go through the procedure. For example, in Example 3 the procedure involves choosing two letters and three digits. Draw a strip of 5 squares to begin the procedure. Then have students choose a letter and ask in how many ways this can be done (26). Next, they should choose a different letter. In how many ways can this be done? (25) Continue in this way.

Fundamental Counting Principle

Suppose that a certain procedure P can be broken into n successive ordered stages, $S_1, S_2, \ldots, S_n$, and suppose that

$$S_1 \text{ can occur in } r_1 \text{ ways,}$$
$$S_2 \text{ can occur in } r_2 \text{ ways,}$$
$$\vdots$$
$$S_n \text{ can occur in } r_n \text{ ways.}$$

Then the number of ways procedure P can occur is

$$r_1 \cdot r_2 \cdots r_n.$$

This fundamental counting principle is sometimes called the **multiplication principle.**

Using the Fundamental Counting Principle

When applying the fundamental counting principle, it is important to identify correctly the aspects of the particular counting problem under consideration that model the stages $S_1, S_2, \ldots, S_n$. Once these stages are correctly identified, the counting process proceeds according to the fundamental counting principle.

E X A M P L E   3   Using the Counting Principle

Suppose the license plates in a certain state begin with two letters followed by three digits. How many different license plates are possible? Assume that none of the characters (letters or numbers) repeats within each license plate.

Solution   Let P be the procedure of writing down a license plate number, S_1 be the stage of selecting the first letter, and S_2 the stage of selecting the second letter, and so forth.

Teaching Note

Encourage students to ask the following questions prior to actually doing any counting.
1. What is the process that is being completed? Does order matter?
2. What is the first stage? How many ways can it be completed?
3. What is the second stage? How many ways can it be completed? And so on.

Notice that S_2 can be completed in 25 ways since a letter cannot be repeated. The total number of license plates meeting the conditions of this problem is

$$26 \cdot 25 \cdot 10 \cdot 9 \cdot 8 = 468{,}000.$$

E X A M P L E   4   Using the Counting Principle

There are eight sprinters in a 100-m final dash. How many different outcomes are possible to this race?

Solution   This problem is equivalent to the following one.

In how many different ways can eight different names be written down in order? This second problem can be thought of as the procedure of actually writing down the eight names of the eight contestants: S_1 is the act of writing the first name, S_2 is the act of writing the second name, and so on.

Stages	S_1	S_2	S_3	S_4	S_5	S_6	S_7	S_8
Number of ways to complete each stage	8	7	6	5	4	3	2	1

Using the fundamental counting principle, conclude that the total number of outcomes to the race is

$$8 \cdot 7 \cdot 6 \cdot 5 \cdot 4 \cdot 3 \cdot 2 \cdot 1 = 8! = 40,320.$$

TRY THIS

Most graphing calculators have the capability of calculating factorials. Refer to your manual and use your calculator to confirm the results of Example 4.

Permutations

Examples 3 and 4 used the multiplication principle to find the number of different arrangements for letters, numbers, and names. We were in fact studying what is called a permutation.

Definition 11.7   Permutation

A **permutation** of n elements is an ordering of the n elements in a row.

Example 4 showed that the number of different permutations of eight objects is $8! = 40,320$. In general, the number of different permutations of n elements is $n!$.

Theorem 11.9   Number of Permutations of n Elements

There are $n!$ different permutations of n elements.

Some situations dealing with permutations of elements contain only a subset of the elements arranged in any one order. One such situation is considered in Example 5.

E X A M P L E 5   Finding the Number of Permutations

Suppose the 15 members of a club are to select three officers—president, vice president, and secretary-treasurer. In how many different ways can these offices be filled?

Solution   This problem is equivalent to asking how many ways three names, selected from the 15 persons, can be written in a specific order.

Using the multiplication principle, we obtain the answer

$$15 \cdot 14 \cdot 13 = 2730.$$ ≡

Example 5 can be generalized to the situation of ordering r names selected from n names where $r \leq n$. We call this generalization an r-permutation.

Definition 11.8 r-Permutation

An **r-permutation** of a set of n elements is an ordered selection of r elements taken from the set of n elements. The number of r-permutations of a set of n elements is denoted $_nP_r$.

This next theorem states how $_nP_r$ can be calculated using the concept of *factorial*.

Theorem 11.10   A Formula for $_nP_r$

If n and r are integers and $0 \leq r \leq n$, then

$$_nP_r = \frac{n!}{(n-r)!}.$$

Proof   Apply the fundamental counting principle to conclude that

$$_nP_r = n \cdot (n-1) \cdots (n-r+1)$$

$$= n \cdot (n-1) \cdots (n-r+1) \frac{(n-r) \cdots 2 \cdot 1}{(n-r) \cdots 2 \cdot 1} \quad \text{Multiply the numerator and denominator by the same factors.}$$

$$= \frac{n!}{(n-r)!}.$$ ≡

E X A M P L E   6   Counting Permutations

How many different five-letter code words are there if no letter is repeated?

TRY THIS

Most graphing calculators have the capability of calculating $_nP_r$. Refer to your manual and use your calculator to confirm the results of Example 6.

Solution This problem asks us to count 5-permutations where $n = 26$.

$$_{26}P_5 = \frac{26!}{(26-5)!}$$

$$= \frac{26!}{21!}$$

$$= \frac{26(25)(24)(23)(22) \cdot 21!}{21!}$$

$$= 26 \cdot 25 \cdot 24 \cdot 23 \cdot 22$$

$$= 7,893,600$$

Teaching Note

The solution to Example 6 does some paper and pencil simplifications of the factorial expression 26!/21!. Of course, this paper and pencil work is not required if the $_nP_r$ function found on a grapher is used. However, the numbers in a counting problem may exceed the capacity of the grapher. In this case, it is important that students have the algebraic skills to solve the problem.

Combinations

The last counting situation we will discuss is the problem of finding the number of subsets of r elements that can be selected from n elements. This is called counting the **combinations of n things taken r at a time.**

E X A M P L E 7   Counting Subsets

How many committees of three can be selected from four people?

Solution To simplify the discussion of this problem, call the four people A, B, C, and D.

Example 1 gave a complete listing of all possible orderings of three of these four people. That list of 24 orderings is repeated here.

ABC	ABD	ACB	ACD	ADB	ADC
BAC	BAD	BCA	BCD	BDA	BDC
CAB	CAD	CBA	CBD	CDA	CDB
DAB	DAC	DBA	DBC	DCA	DCB

However, the committee consisting of persons A, B, and C, denoted $\{A, B, C\}$, is the same committee regardless of the order in which the three members are listed. For example, this committee can be listed in the following six ways:

ABC	ACB	BAC	BCA	CAB	CBA

Common Errors

Students find counting problems challenging. They often want to apply a formula blindly rather than to go through a thinking process.

The total number of different committees is thus equal to the number of different listings, $_4P_3$, divided by the 3! different orderings for each committee. Therefore the number of different committees is

$$_4P_3/3! = \frac{4!}{(4-3)!3!}$$

$$= \frac{4!}{1! \cdot 3!}.$$

Notice that when a counting problem is a combination problem, the order in which the elements are listed is not important.

The solution to Example 7 can be generalized as follows to find the number of committees of r persons selected from a larger group of n persons.

1. List all possible r-permutations. There are $n!/(n-r)!$ of them.

2. Each committee has been listed in all $r!$ of its different orderings.

3. The number of different committees is thus

$$\frac{n!}{r!(n-r)!}.$$

Notice that the number of combinations of n elements taken r at a time returns to the topic of binomial coefficients: The numbers $n!/[r!(n-r)!]$ are precisely the binomial coefficients studied in Section 11.3.

Theorem 11.11   Number of Combinations of n Elements Taken r at a Time

If n and r are nonnegative integers where $r \leq n$, then the number of combinations of n elements taken r at a time, denoted ${}_nC_r$, is

$$_nC_r = \frac{n!}{r!(n-r)!}.$$

Examples 8 and 9 show how Theorem 11.11 can be applied.

E X A M P L E   8   Finding the Number of Starting Teams

A certain basketball team has 12 players. How many different starting fives are possible?

Solution   This problem asks how many different five-person subsets can be selected from a set of 12 persons. That is, we must find $_nC_r$ when $n = 12$ and $r = 5$.

$$_{12}C_5 = \frac{12!}{5!(12-5)!}$$
$$= \frac{12 \cdot 11 \cdot 10 \cdot 9 \cdot 8}{5 \cdot 4 \cdot 3 \cdot 2 \cdot 1}$$
$$= 11 \cdot 9 \cdot 8 = 792$$

There are 792 different starting lineups.

TRY THIS

Most graphing calculators have the capability of calculating $_nC_r$. Refer to your manual and use your calculator to confirm the results of Examples 8 and 9.

The number of different subsets that can be chosen is often much greater than expected.

EXAMPLE 9   Counting Different Card Hands

In the game of bridge, each player is dealt a 13-card hand selected from a deck of 52 cards. How many different bridge hands are there?

Solution   The order in which cards are held in a hand is not important. That is, you are interested in counting different subsets, not different orderings, so this is a combination problem, not a permutation problem. We must find $_nC_r$ when $n = 52$ and $r = 13$.

$$_{52}C_{13} = \frac{52!}{13!(52 - 13)!}$$

$$= 635{,}013{,}559{,}600$$

The number of different bridge hands is 635,013,559,600. Suppose a bridge hand can be dealt in 1 sec. How many years would it take to deal all possible hands?

Notes on Exercises

Ex. 24 is important because of its relationship to probability problems that will be studied in Chapter 12.

Assignment Guide

Day 1: Ex. 1–12
Day 2: Ex. 13–28

Exercises for Section 11.6

For Exercises 1–6, either list all possibilities or draw a tree diagram to count all the possibilities.

1. How many different three-letter code words are there, with no repeated letters, using the letters A, T, and X?

2. How many different three-letter code words are there, with no repeated letters, selecting from the letters R, S, T, and V?

3. There are three roads from town A to town B and four roads from town B to town C. How many different routes are there from A to C?

4. Using the information in Exercise 3, how many different routes are there from A to C and back to A?

5. There are four candidates for homecoming queen and three candidates for king. How many king-and-queen pairs are possible?

6. When ordering an airline ticket you can request first class, business class, or coach. You can also choose a window, aisle, or middle seat. How many different ways can you order a ticket?

For Exercises 7–12, use the fundamental counting principle.

7. How many seven-digit telephone numbers are there? (A number may not begin with a 0 or a 1. Why?)

8. A social security number has the format XXX-XX-XXXX. How many different social security numbers are there?

9. How many different license plates are there that begin with two digits, followed by two letters, and then three digits if no letter or digit repeats?

10. How many different license plates are there that consist of five symbols, either digit or letter? (Repetitions are allowed.)

11. A die is one of a pair of dice. Suppose that one red die and one green die are rolled. How many different outcomes are there?

12. How many different sequences of heads and tails are there if a coin is tossed 10 times?

In Exercises 13–20, calculate each value.

13. $_8P_5$ **14.** $_{12}P_7$ **15.** $_{18}P_6$ **16.** $_{11}P_7$

17. $_{14}C_5$ **18.** $_8C_6$ **19.** $_{24}C_{15}$ **20.** $_{18}C_8$

21. A three-person committee is to be elected from an organization's membership of 25 persons. How many different committees can be elected?

22. How many different six-card hands can be dealt from a deck of 20 cards?

23. Jon has money to buy only three distinct titles from among the 48 discs available. How many different purchases can he make if he purchases no repeated titles?

24. A coin is tossed 20 times and the heads and tails sequences are recorded. From among all possible sequences of heads and tails, how many have exactly seven heads?

25. How many different 13-card hands are there that include

the ace and king of spades? (See Example 9.)

26. An employer interviews eight people for three openings. How many different groups of three can be employed?

27. Verify algebraically that $_nC_r + _nC_{r+1} = _{n+1}C_{r+1}$, where n and r are positive integers and $r + 1 \leq n$.

28. Writing to Learn   Suppose a chain letter is sent to five people the first week of the year. Each of these five people sends a copy of the letter to five more people during the second week of the year, and this pattern is continued each week of the year. Explain how you know with certainty that you will receive a second copy of this letter later in the year. Why is it illegal to participate in "pyramid schemes," like this chain letter, which involve money?

11.7 _____ Three Dimensional Geometry

Objective

Students will be able to locate points in three dimensional geometry by rectangular, cylindrical, and spherical coordinates. They will also be able to associate a surface in 3-space with an equation of up to three variables.

Key Ideas

Cartesian coordinates
Distance formula
Midpoint formula
Cylindrical coordinates
Spherical coordinates
Quadric surface
Ellipsoid
Trace of a surface
Length or norm of a vector
Parametric equation for a line
Cartesian form of equation for a line
Equation of a plane in 3-space

Cartesian Coordinates

In Chapter 1, we introduced the rectangular coordinate system for the two-dimensional plane. It consists of two number lines that intersect at the zero on each line. In much the same way, a **Cartesian coordinate system** can be introduced in three-dimensional space by placing three mutually perpendicular number lines (coordinate lines) in 3-space so that they intersect at the zero on each line (see Fig.11.15). The point O of intersection of the three lines is called the **origin**. The three lines are called **coordinate axes**.

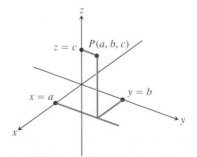

Figure 11.15   For point $P(a, b, c)$, we call $a, b,$ and c the **x-coordinate**, **y-coordinate**, and **z-coordinate**, respectively.

The Cartesian coordinate system is called **right-handed** because a counterclockwise rotation of the positive x-axis in the xy-plane will cause a right-handed screw to advance along the positive z-axis. (A right-handed screw tightens under a clockwise turn and loosens under a counterclockwise turn.)

The three coordinate planes divide 3-space into eight *octants*. The first octant is above the first quadrant of the xy-plane, the second octant is above the second quadrant, etc; the fifth octant is below the first quadrant of the xy-plane, the sixth is below the second quadrant etc., in order.

Each ordered triple (x, y, z) of real numbers is associated with a unique point P of 3-space in such a way that the plane through $P(a, b, c)$ parallel to the yz-plane intersects the x-axis in the point $x = a$. A similar property holds for the second and third coordinates b and c. The points $(x, y, 0)$, $(x, 0, z)$, and $(0, y, z)$ are sometimes called the **projection of the point (x, y, z) onto the xy-plane, xz-plane, and the yz-plane**, respectively.

The set of triples $P(x, y, z)$ that satisfy the equation $x = 3$ consists of all those points on the plane through $(3, 0, 0)$ that is parallel to the yz-plane (see Fig. 11.16).

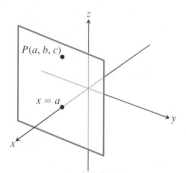

Figure 11.16 The graph of the equation $x = a$.

EXAMPLE 1   Planes Parallel to the Coordinate Planes

Find equations of the three planes that pass through the point $P(2, -5, 3)$ and are parallel to the three coordinate planes.

Solution

$x = 2$	The set of points satisfying this equation are the triples $(2, y, z)$–the plane parallel to the yz-plane.
$y = -5$	The set of points satisfying this equation are the triples $(x, -5, z)$–the plane parallel to the xz-plane.
$z = 3$	The set of points satisfying this equation are the triples $(x, y, 3)$–the plane parallel to the xy-plane.

The distance formula for 3-space can be derived from the distance formula for the plane. The derivation will be completed in Exercise 46.

The Distance Formula

The distance $d(P, Q)$, between points $P(x_1, y_1, z_1)$ and $Q(x_2, y_2, z_2)$ in 3-space is

$$d(P, Q) = \sqrt{(x_2 - x_1)^2 + (y_2 - y_1)^2 + (z_2 - z_1)^2}.$$

In a similar fashion, the formula for finding the midpoint between two points in the plane can be generalized to the following formula for 3-space.

The Midpoint Formula

The coordinates of the midpoint M of the segment $\overline{PQ}$ for points $P(x_1, y_1, z_1)$ and $Q(x_2, y_2, z_2)$ is the point with coordinates

$$\left(\frac{x_1 + x_2}{2}, \frac{y_1 + y_2}{2}, \frac{z_1 + z_2}{2} \right).$$

E X A M P L E   2   Finding the Distance between Points

Find the distance between the points $P(-2, 3, -1)$ and $Q(4, -1, 5)$ and the coordinates of the midpoint of the line segment PQ.

Solution   The distance formula gives

$$d(P, Q) = \sqrt{(4 + 2)^2 + (-1 - 3)^2 + (5 + 1)^2}$$

$$= \sqrt{88}$$

$$= 9.38.$$

The coordinates of the midpoint of the line segment $\overline{PQ}$ are

$$\left(\frac{-2 + 4}{2}, \frac{3 - 1}{2}, \frac{-1 + 5}{2} \right) = (1, 1, 2).$$

≣

Cylindrical Coordinates and Spherical Coordinates

There are two other coordinate systems frequently used to locate points in 3-space: cylindrical coordinates and spherical coordinates. Cylindrical coordinates are often used when a three-dimensional graph has a line of symmetry; spherical coordinates are used when the graph has a point of symmetry.

The **cylindrical coordinates** of a point P in 3-space is the triple (r, θ, z) where r and θ are essentially the polar coordinates of the projection of P to the xy-plane and z is the usual third coordinate of the Cartesian coordinate system.

The (x, y, z) coordinates and (r, θ, z) coordinates are related in the following way (see Fig. 11.17a):

$$x = r \cos \theta \qquad r^2 = x^2 + y^2$$

$$y = r \sin \theta \qquad \tan \theta = \frac{y}{x}.$$

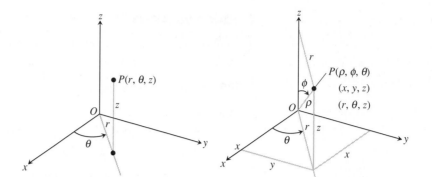

Figure 11.17 (a) The cylindrical coordinates of P; (b) the spherical coordinates of P.

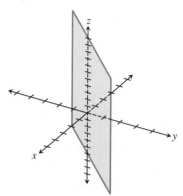

Figure 11.18 The plane $y = x$ and $\theta = 45°$.

The **spherical coordinates** of a point P in 3-space is the triple (ρ, ϕ, θ), where ρ is the distance from P to the origin and ϕ and θ are angles as shown in Fig. 11.17(b).

E X A M P L E   3   Drawing Graphs of Equations

Sketch a complete graph of each equation:

a) $y = x$

b) $\theta = 45°$

c) $r = 2$

d) $\rho = 2$

Figure 11.19 The cylinder $r = 2$ or $x^2 + y^2 = 4$.

Solution

a) The graph of $y = x$ in the xy-plane is the 45° line L passing through the origin and in the first and third quadrants. So the graph of $y = x$ in 3-space is the plane passing through L that is perpendicular to the xy-plane (see Fig. 11.18).

b) Think of θ as the second coordinate of the cylindrical coordinate system. The graph of $\theta = 45°$ in the xy-plane is the same 45° line referred to in part (a). So the graph of $\theta = 45°$ in 3-space is the same as the plane shown in Fig. 11.18.

c) In the xy-plane, the graph of $r = 2$ is a circle with radius 2 centered at the origin. In each plane $z = c$ parallel to the xy-plane, the graph of $r = 2$ is a circle with radius 2 centered at $(0, 0, c)$. Thus the graph of $r = 2$ is a right circular cylinder with the z-axis as the axis of symmetry (see Fig. 11.19).

d) The graph of $\rho = 2$ consists of all points in 3-space at a distance 2 from the origin. This is a sphere of radius 2 with its center at the origin (see Fig. 11.20).

Quadric Surfaces

The equation $r = 2$ of Example 3(c) also can be written in the form $x^2 + y^2 = 4$. (Why?) Similarly, the equation $\rho = 2$ of Example 3(d) can be written in the form $x^2 + y^2 + z^2 = 4$. (Why?) (See Fig. 11.19 and 11.20.) These equations are quadratic in the variables involved. Their graphs are called quadric surfaces.

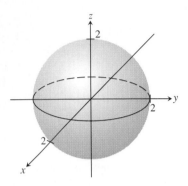

Figure 11.20 The sphere $\rho = 2$ or $x^2 + y^2 + z^2 = 4$.

Definition 11.9   Quadric Surfaces

A surface that is a graph of an equation quadratic in the variables x, y, and z is called a **quadric surface**.

In the next examples, you should notice the similarities between graphs of quadratic equations in x and y and graphs of quadric surfaces in x, y, and z.

The graph of $\dfrac{x^2}{a^2} + \dfrac{y^2}{b^2} = 1$ is called an **elliptic cylinder**. Notice that if $a = b$ in $\dfrac{x^2}{a^2} + \dfrac{y^2}{b^2} = 1$, then the surface is a right circular cylinder similar to that in Fig. 11.19. This is the subject of Example 4.

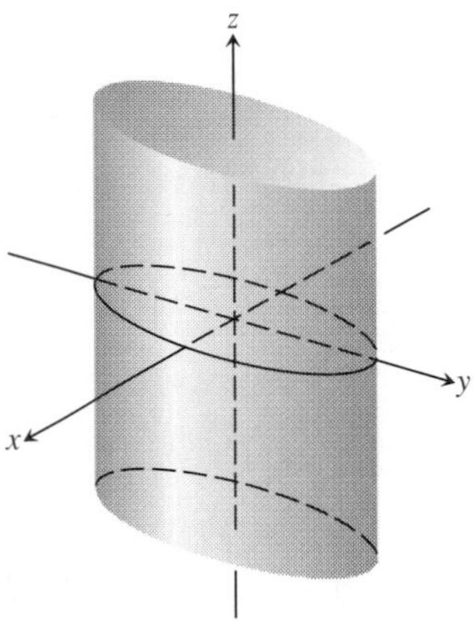

EXAMPLE 4   Graphing an Elliptic Cylinder

Sketch a complete graph of $\dfrac{x^2}{4} + \dfrac{y^2}{9} = 1$.

Solution   In the xy-plane, the graph of $\dfrac{x^2}{4} + \dfrac{y^2}{9} = 1$ is an ellipse with its center at the origin, major axis of length 6, and minor axis of length 4. The graph in any plane $z = c$ is an ellipse with a center at $(0, 0, c)$, major axis of length 6, and minor axis of length 4. The endpoints of the major axis are $(0, \pm 3, c)$ and the endpoints of the minor axis are $(\pm 2, 0, c)$. The complete graph is the cylinder shown in Fig. 11.21. ▰

The graph of $\dfrac{x^2}{a^2} + \dfrac{y^2}{b^2} + \dfrac{z^2}{c^2} = 1$ is called an **ellipsoid**. This is the subject of Example 5.

EXAMPLE 5   Graphing an Ellipsoid

Confirm that a complete graph of $\dfrac{x^2}{a^2} + \dfrac{y^2}{b^2} + \dfrac{z^2}{c^2} = 1$ is the quadric surface shown in Fig. 11.22.

Figure 11.21   An elliptic cylinder.

Common Errors

Students often think that an equation like the one in Example 4 has a graph in the xy-plane since no z-variable occurs in the equation. In fact, the graph is in 3-space and the z-coordinate of points on the graph has no restrictions.

Solution Notice that the x-intercepts are $(\pm a, 0, 0)$, the y-intercepts are $(0, \pm b, 0)$, and the z-intercepts are $(0, 0, \pm c)$. We can assume that $a > 0, b > 0$, and $c > 0$. If $|x| > a$, then $\dfrac{x^2}{a^2} > 1$ and there are no solutions to the equation

$$\frac{x^2}{a^2} + \frac{y^2}{b^2} + \frac{z^2}{c^2} = 1. \tag{1}$$

(Why?) Similarly, if $|y| > b$ or $|z| > c$, there are no solutions. So the graph of this quadric surface is contained in the box determined by the planes $x = \pm a$, $y = \pm b$, and $z = \pm c$.

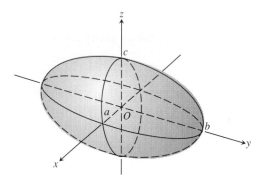

Figure 11.22 An ellipsoid.

If we set $z = 0$ in equation (1), the equation reduces to

$$\frac{x^2}{a^2} + \frac{y^2}{b^2} = 1,$$

whose graph is a cross section ellipse in the xy-plane. In a similar fashion, setting $y = 0$ or $x = 0$ we can confirm that the cross sections in the xz and yz-planes are each an ellipse. These observations support the claim that the surface in Fig. 11.22 is the graph of Eq. (1).

In Example 5, we found it useful to consider the intersection of the surface and the coordinate planes. We were using the concept of **trace** that we now define.

Teaching Note

Point out to students that the concept of *trace* in Definition 11.10 is not the same concept as TRACE on a grapher.

Definition 11.10 Trace of a Surface in a Plane

The **trace of a surface in a plane** is the intersection of the surface with the plane; that is, it is the set of points common to both the surface and the plane.

It is often helpful to analyze the trace of a surface in a plane.

E X A M P L E   6   Analyzing a Trace

Confirm that a complete graph of $\dfrac{z}{c} = \dfrac{y^2}{b^2} - \dfrac{x^2}{a^2}$ is the quadric surface shown in Fig. 11.23.

Solution   The trace of

$$\frac{z}{c} = \frac{y^2}{b^2} - \frac{x^2}{a^2} \tag{2}$$

in the xz-plane can be found by letting $y = 0$ in Eq (2). The resulting equation $\dfrac{z}{c} = -\dfrac{x^2}{a^2}$ or $z = -\left(\dfrac{c}{a^2}\right)x^2$ is a parabola in the xz-plane. Similarly, the trace in the yz-plane is the parabola $z = \left(\dfrac{c}{b^2}\right)y^2$. Notice that if $c > 0$, the first parabola opens down and the second opens up. These observations are consistent with the surface shown in Fig. 11.23.

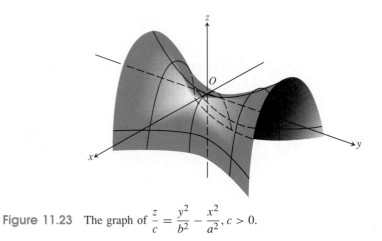

Figure 11.23   The graph of $\dfrac{z}{c} = \dfrac{y^2}{b^2} - \dfrac{x^2}{a^2}, c > 0.$

The trace of the surface in the xy-plane ($z = 0$) is the pair of intersecting lines $y = \pm\dfrac{b}{a}x$. On the other hand, the trace in the plane $z = d$ is the hyperbola

$$\frac{d}{c} = \frac{y^2}{b^2} - \frac{x^2}{a^2}, \quad \text{or} \quad 1 = \frac{y^2}{\dfrac{db^2}{c}} - \frac{x^2}{\dfrac{da^2}{c}}.$$

The asymptotes are the lines $y = \pm\dfrac{b}{a}x$ in the plane $z = d$ parallel to the xy-plane. (Why?) If $\dfrac{d}{c} > 0$, the vertices are

$$\left(0, \pm\sqrt{\frac{db^2}{c}}, d\right),$$

and if $\dfrac{d}{c} < 0$, then the vertices are

$$\left(\pm\sqrt{\frac{-da^2}{c}}, 0, d\right). \qquad\qquad \blacksquare$$

The graph in Fig. 11.23 appears to resemble a *saddle* near the origin. Consequently, it is often called a saddle surface. In general, the graph of

$$\frac{z}{c} = \frac{y^2}{b^2} - \frac{x^2}{a^2}$$

is called an **hyperbolic paraboloid**.

By using an analysis similar to the one of Example 6, we can confirm that the graph of $\dfrac{z}{c} = \dfrac{x^2}{a^2} + \dfrac{y^2}{b^2}$ is the **elliptic paraboloid** shown in Fig. 11.24.

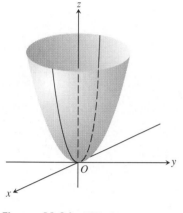

Figure 11.24 Elliptic paraboloid $\dfrac{z}{c} = \dfrac{x^2}{a^2} + \dfrac{y^2}{b^2}$.

Vectors in 3-Space

The study in Chapter 8 of vectors in two dimensions can be easily extended to vectors in 3-space. Two vectors are said to be **equal** if either vector can be obtained from the other by a shift parallel to the x-axis, the y-axis, the z-axis, or some combination of these transformations. Every vector in 3-space is equal to a unique vector of the form $\mathbf{v} = (x, y, z)$ with initial point the origin $(0, 0, 0)$ and terminal point the point (x, y, z). We call this representation the **standard form** for the vector. Similar to Chapter 8, we use 3-tuples to represent points and vectors, with the context clarifying a particular use of the given representation.

Addition, subtraction, and scalar multiplication of vectors in 3-space can be accomplished by performing the operation on each component.

$$(x_1, y_1, z_1) \pm (x_2, y_2, z_2) = (x_1 \pm x_2, y_1 \pm y_2, z_1 \pm z_2)$$

$$c(x, y, z) = (cx, cy, cz)$$

The **length**, or **norm**, of the vector $\mathbf{v} = (x, y, z)$ is defined by

$$|\mathbf{v}| = \sqrt{x^2 + y^2 + z^2}.$$

Two vectors in 3-space are **parallel** if, and only if, in standard form one vector is a scalar multiple of the other.

EXAMPLE 7 Finding Vectors

VECTOR CROSS PRODUCT

In mechanics another operation applied to vectors, **the cross product**, is used to compute moments of force. There are also applications of cross product in electricity and magnetism. See Ex. 47 to 51.

Find the standard form and length of the vector with initial point $A = (-1, 2, 3)$ and terminal point $B = (2, -3, 4)$.

Solution   The procedure used in Chapter 8 also works in 3-space. Let $\mathbf{v}$ be the standard form for the vector $\overrightarrow{AB}$, and $\overrightarrow{OA}$ and $\overrightarrow{OB}$ be the vectors with initial point O and terminal point A and B, respectively. Then

$$\mathbf{v} = \overrightarrow{OB} - \overrightarrow{OA}$$

$$= (2, -3, 4) - (-1, 2, 3)$$

$$= (3, -5, 1).$$ ≡

Equations of a Line in 3-Space

Suppose that a vector $\mathbf{v} = (A, B, C)$ and a point $P(x_0, y_0, z_0)$ are given and that ℓ is the line through P parallel to $\mathbf{v}$ (see Fig. 11.25).

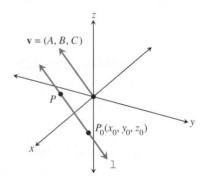

Figure 11.25   Line ℓ is parallel to $\mathbf{v}$.

If $P(x, y, z)$ is any point on ℓ, then $\overrightarrow{OP} - \overrightarrow{OP_0}$ is a vector in standard form that is parallel to $\mathbf{v}$. Consequently, it is a scalar multiple of $\mathbf{v}$. Therefore,

$$\overrightarrow{OP} - \overrightarrow{OP_0} = t\mathbf{v},$$

$$(x - x_0, y - y_0, z - z_0) = t(A, B, C).$$

So if we know a point on the line and a vector $\mathbf{v}$ that determines the direction of the line, we can write an equation that determines the line. This development leads to this definition.

Definition 11.11   Parametric Equations for a Line

Let $P = (x_0, y_0, z_0)$ be a point in 3-space and $\mathbf{v} = (A, B, C)$ be any vector. A **parametric form** for the equation of the line through P and parallel to $\mathbf{v}$ is given by

$$x = tA + x_0, \quad y = tB + y_0, \quad \text{and} \quad z = tC + z_0,$$

where the parameter t can be any real number.

If we take each parametric equation in Definition 11.11 and solve for t, we obtain

$$t = \frac{x - x_0}{A}, \qquad t = \frac{y - y_0}{B}, \quad \text{and} \quad t = \frac{z - z_0}{C}.$$

This leads to the next definition.

Definition 11.12   Cartesian Form for a Line

Let $P = (x_0, y_0, z_0)$ be a point in 3-space and $\mathbf{v} = (A, B, C)$ be any vector. A **Cartesian form** for the equation of the line through P and parallel to $\mathbf{v}$ is given by

$$\frac{x - x_0}{A} = \frac{y - y_0}{B} = \frac{z - z_0}{C}.$$

E X A M P L E   8   Finding Equations for a Line

Let ℓ be the line through the point $P(4, 3, -1)$ and parallel to the vector $\mathbf{v} = (-2, 2, 7)$. Find the following:

a) Parametric equations for line ℓ

b) A Cartesian equation form for line ℓ

Solution

a) From Definition 11.11, we see that

$$x = 4 - 2t, \quad y = 3 + 2t, \quad \text{and} \quad z = -1 + 7t.$$

b) Solving each equation in (a) for t, we obtain

$$\frac{x-4}{-2} = t, \quad \frac{y-3}{2} = t, \quad \text{and} \quad \frac{z+1}{7} = t.$$

Therefore

$$\frac{x-4}{-2} = \frac{y-3}{2} = \frac{z+1}{7}.$$

≡

E X A M P L E   9   Finding Equations for a Line

Find Cartesian and parametric forms for an equation of the line determined by the points $A = (3, 0, -2)$ and $B = (-1, 2, -5)$.

Solution   The line is parallel to the vector $\mathbf{v} = \overrightarrow{OB} - \overrightarrow{OA} = (-4, 2, -3)$. We can use the point A and $\mathbf{v}$ to obtain a Cartesian form.

$$\frac{x-3}{-4} = \frac{y}{2} = \frac{z+2}{-3}$$

The parametric form for the equation of this line is

$$x = 3 - 4t, \qquad y = 2t, \quad \text{and} \quad z = -2 - 3t.$$

≡

Notice that the solution to Example 9 is not unique. There are other forms that are also correct. For example, multiplying the equation by -1 we obtain

$$\frac{x-3}{4} = \frac{y}{-2} = \frac{z+2}{3}.$$

If as the fixed point, B is used instead of A, then we obtain

$$\frac{x+1}{-4} = \frac{y-2}{2} = \frac{z+5}{-3}.$$

Equations of a Plane in 3-Space

It is known that the graph of any equation of the form $ax + by + cz = d$, where not all of $a, b,$ and c are zero, is a plane. Moreover, every plane in 3-space is the graph of such an equation.

E X A M P L E   10   Drawing a Graph of a Plane

Draw a graph of the planes $x + y + z = 3$ and $2x + y + 4z = 8$.

Solution   To find the x-intercept for the plane $x + y + z = 3$, let $y = 0$ and $z = 0$, then solve for x. We conclude that the x-intercept is 3; similarly, the y-

and z-intercepts also are 3. Figure 11.26 shows the portion of the plane in the first octant.

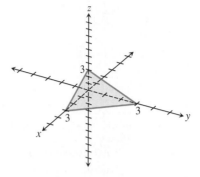

Figure 11.26   Plane $x + y + z = 3$.

The plane $2x + y + 4z = 8$ has an x-intercept of 4, a y-intercept of 8, and a z-intercept of 2. Figure 11.27 shows the portion of the plane in the first octant.

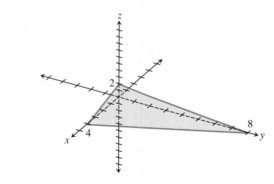

Figure 11.27   The plane $2x + y + 4z = 8$.

Notes on Exercises

For Ex. 23–26 it is assumed that a rectangular, cylindrical, and spherical coordinate system are superimposed upon one another. Notice that the *cross product* referred to in Ex. 47–51 is a concept that does not exist in the xy-plane.

Assignment Guide

Day 1: Ex. 1–25 odd
Day 2: Ex. 28 33, 35 39 odd, 43, 45

Exercises for Section 11.7

For Exercises 1–4 plot the given point in 3-space.

1. $(3, 4, 2)$ **2.** $(2, -3, 6)$
3. $(1, -2, -4)$ **4.** $(-2, 3, -5)$

For Exercises 5–8 name the octant of the given point.

5. $(-3, 5, 2)$ **6.** $(1, -4, -2)$
7. $(-2, -1, -5)$ **8.** $(3, -1, -3)$

9. Compute the distance between the points $(-1, 2, 5)$ and $(3, -4, 6)$.

10. Compute the distance between the points $(2, -1, -8)$ and $(6, -3, 4)$.

11. Determine the midpoint of the line segment between the points $(-1, 2, 5)$ and $(3, -4, 6)$.

12. Determine the midpoint of the line segment between the points $(2, -1, -8)$ and $(6, -3, 4)$.

13. Let $\mathbf{v} = (2, 4, 3)$ and $P = (4, 6, 1)$. Find both a parametric form and a Cartesian form for the line ℓ containing P parallel to $\mathbf{v}$.

14. Let $\mathbf{v} = (1, 5, 3)$ and $P = (2, 3, 6)$. Find both a parametric form and a Cartesian form for the line ℓ containing P parallel to $\mathbf{v}$.

For Exercises 15–18, compute the cylindrical coordinates of the points with the given Cartesian coordinates.

15. $(4, 0, 0)$ 16. $(0, 0, 5)$

17. $(4, 4, 3)$ 18. $(2, 2, -3)$

For Exercises 19–22, compute the spherical coordinates of the points with the given Cartesian coordinates.

19. $(1, 0, 0)$ 20. $(0, 0, -1)$

21. $(1, 1, \sqrt{2})$ 22. $(1, -1, -\sqrt{2})$

For Exercises 23–26, describe a complete graph in 3-space for each of the following equations.

23. $y = 2x$ 24. $\theta = 30°$

25. $r = 3$ 26. $\rho = 4$

27. Given that (ρ, ϕ, θ) are the spherical coordinates of a point with Cartesian coordinates (x, y, z), show that $\rho^2 = x^2 + y^2 + z^2$.

For Exercises 28–33, describe and sketch the quadric surface.

28. $4x^2 + 9y^2 = 36$ 29. $x^2 + 9z^2 = 9$

30. $16x^2 + y^2 = 16z$ 31. $x^2 + y^2 + z^2 = 64$

32. $9x^2 + 4y^2 = 36z$ 33. $144x^2 + 64y^2 + 36z^2 = 576$

For Exercises 34–39, sketch a graph of the plane.

34. $x + y + 3z = 9$ 35. $x + y - 2z = 8$

36. $x + z = 3$ 37. $2y + z = 6$

38. $x - 3y = 6$ 39. $x = 3$

40. Let $A = (x_1, y_1, z_1)$ and $B = (x_2, y_2, z_2)$ be two distinct points in 3-space. Show that the equation of the line through AB in *Cartesian* equation form is

$$\frac{x - x_1}{x_2 - x_1} = \frac{y - y_1}{y_2 - y_1} = \frac{z - z_1}{z_2 - z_1}.$$

41. Let $A = (x_1, y_1, z_1)$ and $B = (x_2, y_2, z_2)$ be two distinct points in 3-space. Write the equation of the line through AB in parametric form.

For Exercises 42–45, use the result of Exercises 40 and 41 to find two forms of the equation of the given line. Let $A = (-1, 2, 4)$, $B = (0, 6, -3)$, and $C = (2, -4, 1)$.

42. Write an equation for the line through points A and B in
   a) Cartesian form.
   b) parametric form.

43. Write an equation for the line through points A and C in
   a) Cartesian form.
   b) parametric form.

44. Write an equation for the line through points B and C in
   a) Cartesian form.
   b) parametric form.

45. Write an equation for the line through points M and A where M is the midpoint of line segment BC in
   a) Cartesian form.
   b) parametric form.

46. Use the distance formula for the plane to verify the distance formula $d(P, Q) = \sqrt{(x_2 - x_1)^2 + (y_2 - y_1)^2 + (z_2 - z_1)^2}$ for 3-space.

Exercises 47 to 51 refer to the **cross product** of vectors defined here. Given vectors $\mathbf{u} = (a_1, a_2, a_3)$ and $\mathbf{v} = (b_1, b_2, b_3)$, the cross product $\mathbf{u} \times \mathbf{v}$ is defined by:

$\mathbf{u} \times \mathbf{v} = (a_2 b_3 - a_3 b_2)(1, 0, 0) + (a_3 b_1 - a_1 b_3)(0, 1, 0)$

$$+ (a_1 b_2 - a_2 b_1)(0, 0, 1)$$

47. Find the cross product $\mathbf{u} \times \mathbf{v}$ where $\mathbf{u} = (1, -2, 3)$ and $\mathbf{v} = (-2, 1, -1)$.

48. Find the cross product $\mathbf{u} \times \mathbf{v}$ where $\mathbf{u} = (4, -1, 2)$ and $\mathbf{v} = (1, -3, 2)$.

49. Let $\mathbf{u} = (1, -2, 3)$ and $\mathbf{v} = (-2, 1, -1)$. Find the dot products $\mathbf{u} \times \mathbf{v} \cdot \mathbf{u}$ and $\mathbf{u} \times \mathbf{v} \cdot \mathbf{v}$. (See Exercises 57 and 58 of Section 8.5.)

50. Let $\mathbf{u} = (4, -1, 2)$ and $\mathbf{v} = (1, -3, 2)$. Find the dot products $\mathbf{u} \times \mathbf{v} \cdot \mathbf{u}$ and $\mathbf{u} \times \mathbf{v} \cdot \mathbf{v}$.

51. Based upon the experience of Exercises 49 and 50 conjecture a generalization about the perpendicularity of $\mathbf{u} \times \mathbf{v}$ compared to $\mathbf{u}$ and $\mathbf{v}$.

11.8 —————

Graphs of Functions of Two Variables

A function consists of a domain and a rule that assigns each element of the domain with a unique element of the range. If the domain D of a function f is a subset of the xy-plane and the range R is a subset of the real numbers, then f is called a function of two variables. Figure 11.28 shows a diagram of the domain and range.

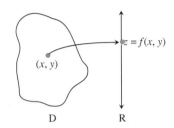

Figure 11.28 D represents the domain and R represents the range of f.

If (x, y) is an ordered pair in D, then the real number $z = f(x, y)$ is the corresponding range element of f. Here, x and y are called independent variables and z the dependent variable.

Definition 11.13 Function of Two Variables

A **function of two variables** defined on a domain D in the xy-plane is a rule that assigns each point (x, y) in D to a unique real number $f(x, y)$.

E X A M P L E 1 Finding the Domain and Range of a Function

Determine the domain of the function $f(x, y) = \dfrac{x + y}{\sqrt{4 - x^2 - y^2}}$.

Solution Because an expression under a square root radical must be non-negative and further, because a denominator cannot be zero, we conclude that we must have $4 - x^2 - y^2 > 0$ in order for $f(x, y)$ to be defined.

$$4 - x^2 - y^2 > 0$$
$$\iff \quad x^2 + y^2 < 4$$

Therefore the domain of f is the interior of the circle with center $(0, 0)$ and radius 2.
≡

The understood domain of a function of two variables is the largest subset of the xy-plane for which the function rule is defined.

E X A M P L E   2   Finding the Domain and Range of a Function

Determine the domain and the range of the function $f(x, y) = x^2 + y^2$.

Solution   Because $x^2 + y^2$ is defined for all pairs of real numbers (x, y), the domain of f is the entire xy-plane. For each pair (x, y) of real numbers, $x^2 + y^2 \geq 0$. This means the range of f is $[0, \infty)$.   ≡

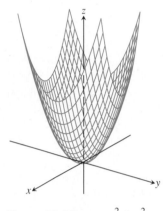

Figure 11.29 $z = x^2 + y^2$.

We use the expression R^2 to denote the xy-plane. Thus we can say that the domain of the function $f(x, y) = x^2 + y^2$ of Example 2 is R^2. Actually, any plane in three-dimensional space is a copy of the xy-plane and is often referred to as R^2. The symbol R^3 is often used to denote three-dimensional space.

Recall that in Section 11.7, you learned that the complete graph of the equation $\dfrac{z}{c} = \dfrac{x^2}{a^2} + \dfrac{y^2}{b^2}$ is an elliptic paraboloid. By setting $a = b = c = 1$, this equation becomes $z = x^2 + y^2$, the function of Example 2. Figure 11.29 shows a complete graph of f.

Definition 11.14   Graph of a Function of Two Variables

Let $z = f(x, y)$ be a function of two variables with domain D. The **graph of the function f** is the set of all points (x, y, z) in 3-space, where (x, y) is in D and $z = f(x, y)$. A graph of such a function is called a **surface**.

Three-dimensional Graphing Utilities

A graphing utility that produces graphs of a function $y = f(x)$ of one variable shows only a rectangular portion of the plane known as a viewing rectangle. Similarly, a 3-D graphing utility shows a rectangular parallelepiped portion of 3-space called a **viewing box**. If the viewing box consists of all triples (x, y, z) satisfying

$$A \leq x \leq B, \quad C \leq y \leq D, \quad E \leq z \leq F,$$

we describe this region as the **Viewing Box [A,B] by [C,D] by [E,F]**. The computer graphs drawn in this section will specify the viewing box.

x and y Sections or Cuts

We shall use the graph of $z = x^2 + y^2$ in Fig. 11.29 to illustrate principles that can be used to find the complete graph.

The plane $x = 2$ intersects the surface $z = x^2 + y^2$ to form a curve. By letting $x = 2$, we see that this curve is $z = 4 + y^2$, which is a parabola. Recall that we also call this curve the trace of $z = x^2 + y^2$ in the plane $x = 2$.

In a similar manner, we can find the traces in planes $x = 0$, $x = 4$, and $x = 6$ to be the parabolas $z = y^2$, $z = 16 + y^2$, and $z = 36 + y^2$. Figure 11.30 shows these four traces of $z = x^2 + y^2$.

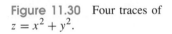

Figure 11.30 Four traces of $z = x^2 + y^2$.

EXAMPLE 3 Finding Traces

Describe and sketch the traces of $z = x^2 + y^2$ in the following planes:

a) $y = 0$

b) $y = 2$

c) $y = 4$

d) $y = 6$

Solution We find the trace $z = x^2 + y^2$ in each plane.

a) The trace in plane $y = 0$ is the parabola $z = x^2 + 0^2$.

b) The trace in plane $y = 2$ is the parabola $z = x^2 + 2^2$.

c) The trace in plane $y = 4$ is the parabola $z = x^2 + 4^2$.

d) The trace in plane $y = 6$ is the parabola $z = x^2 + 6^2$.

Figure 11.31 shows a sketch of these traces. ≣

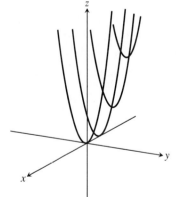

Figure 11.31 Four traces of $z = x^2 + y^2$.

A trace in the plane $x = a$ is often called an **x-section** or an **x-cut**. Similarly, a trace in the plane $y = b$ is often called a **y-section** or a **y-cut**. Observe that Fig. 11.29 shows both the x-cuts and the y-cuts—a method for showing a sketch of the surface $z = x^2 + y^2$.

Maximum and Minimum Values

The notions about local extrema and absolute extrema of a function of one variable extend naturally to functions of two variables. We begin with two definitions.

Definition 11.15   Maximum Values

The function $z = f(x, y)$ has a **local maximum value,** or **local maximum,** at (x_0, y_0) if there is a rectangular region $a \leq x \leq b, c \leq y \leq d$ in the xy-plane containing (x_0, y_0) such that $f(x, y) \leq f(x_0, y_0)$ for all (x, y) in the rectangular region that are in the domain of f. The local maximum value is $z_0 = f(x_0, y_0)$. If $f(x, y) \leq f(x_0, y_0)$ for all (x, y) in the domain of f, then f has $z_0 = f(x_0, y_0)$ as an **absolute maximum value.**

Definition 11.16   Minimum Values

The function $z = f(x, y)$ has a **local minimum value,** or **local minimum,** at (x_0, y_0) if there is a rectangular region $a \leq x \leq b, c \leq y \leq d$ in the xy-plane containing (x_0, y_0) such that $f(x, y) \geq f(x_0, y_0)$ for all (x, y) in the rectangular region that are in the domain of f. The local minimum value is $z_0 = f(x_0, y_0)$. If $f(x, y) \geq f(x_0, y_0)$ for all (x, y) in the domain of f, then f has $z_0 = f(x_0, y_0)$ as an **absolute minimum value.**

Notes on Examples

In Example 4, students do not need calculus or a grapher to find the maximum value of the function. Help students to understand that the maximum value occurs when $x = 0$ and $y = 0$. Point out that for any other values either z is not defined or $64 - x^2 - y^2$ is less than 64.

E X A M P L E   4   *Analyzing a Function of Two Variables*

Determine the domain, the range, and any local extrema of $z = \sqrt{64 - x^2 - y^2}$ and draw its complete graph.

Solution   Recall that a complete graph of $x^2 + y^2 + z^2 = 64$ is a sphere centered at the origin with radius 8. Suppose we solve this equation for z.

$$x^2 + y^2 + z^2 = 64$$
$$z^2 = 64 - x^2 - y^2$$
$$z = \sqrt{64 - x^2 - y^2}$$

We expect the graph of z to be the top hemisphere.

a) Consider the x-section $x = a$.

This section is the graph of the equation $z = \sqrt{64 - a^2 - y^2}$, the upper half circle centered at $(a, 0, 0)$ with radius $\sqrt{64 - a^2}$.

b) Consider the y-section $y = b$.

This section is the graph of the equation $z = \sqrt{64 - x^2 - b^2}$, the upper half circle centered at $(0, b, 0)$ with radius $\sqrt{64 - b^2}$.

A graph of this surface is shown in Fig. 11.32.

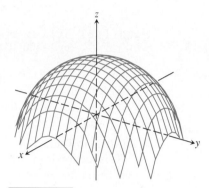

Figure 11.32 $z = \sqrt{64 - x^2 - y^2}$ in the viewing box $[-20, 20]$ by $[-20, 20]$ by $[-20, 20]$.

Domain of z:

The domain is the disk $x^2 + y^2 \leq 64$ that consists of all the interior points and the boundary of the circle with center at $(0, 0)$ and radius 8.

Local and Absolute Extrema:

The value of z is 8 at $(0, 0)$ and 8 is both a local maximum and an absolute maximum of z. Similarly, 0 is both a local minimum and an absolute minimum of z. The value 0 occurs at any point on the circle $x^2 + y^2 = 64$ in the xy-plane. The range of z is $[0, 8]$. ≣

Stan says, "Determining local extrema, as discussed in Example 5, would be beyond the level of this course without the use of technology."

E X A M P L E 5   *Analyzing a Function of Two Variables*

Determine the domain, the range, and any local extrema of $z = y^2 - x^2$ and draw its complete graph.

Solution

Domain and Range of z:

The domain of the function $z = y^2 - x^2$ is R^2, and the range is $(-\infty, \infty)$. (Why?)

To understand the graph of this function consider these sections or cuts.

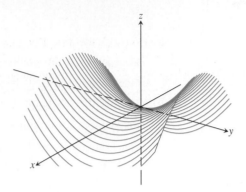

Teaching Note

Figures 11.33, 11.34, and 11.35 are very instructive in showing how computer images of surfaces are constructed. These figures are also valuable in using visual methods to determine important characteristics of functions of two variables.

Figure 11.33 The x-sections of $z = y^2 - x^2$ in the viewing box $[-10, 10]$ by $[-10, 10]$ by $[-100, 100]$.

1. x-sections $x = a$

These sections are the graphs of the equations $z = y^2 - a^2$, parabolas that open upward with vertices $(a, 0, -a^2)$ (see Fig. 11.33).

2. Section in the xz-plane

This section is the parabola $z = -x^2$. Notice that the vertices of the x-sections $x = a$ lie on this parabola $z = -x^2$.

3. y-sections $y = b$

These sections are the graphs of the equations $z = b^2 - x^2$, which are parabolas with vertices $(0, b, b^2)$ that open downward (see Fig. 11.34).

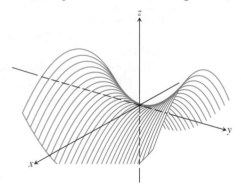

Figure 11.34 The y-sections of $z = b^2 - x^2$ in the viewing box $[-10, 10]$ by $[-10, 10]$ by $[-100, 100]$.

4. Section in the yz-plane

This section is the parabola $z = y^2$. Notice that the vertices of the y-sections $y = b$ lie on this parabola $z = y^2$.

Local and Absolute Extrema

The parabola of the xz-plane suggests there is a local maximum at $(0, 0)$. However, the yz-plane suggests there is a local minimum at $(0, 0)$. In fact, there are neither. This function has no local extrema or absolute extrema. We say that the function has a "saddle point" at the origin (see Fig. 11.35).

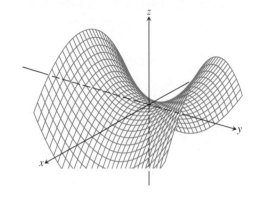

Figure 11.35 The graph of $z = y^2 - x^2$ in the viewing box $[-10, 10]$ by $[-10, 10]$ by $[-10, 10]$.

Using a 3-D grapher, you can experiment with other functions of two variables. For example, consider this fairly complicated function

$$f(x, y) = \frac{30}{(x^2 + 1)[(y + 3)^2 + 4]} - \frac{20}{(x^2 + 1)[(y - 4)^2 + 4]}.$$

The denominators of the fractions in the rule for the function f are different from zero for any element in R^2. Thus the domain of f is R^2. The graph of this function can be seen in Fig. 11.36

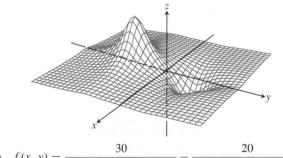

Figure 11.36 $f(x, y) = \dfrac{30}{(x^2 + 1)[(y + 3)^2 + 4]} - \dfrac{20}{(x^2 + 1)[(y - 4)^2 + 4]}$ in the viewing box $[-5, 5]$ by $[-10, 10]$ by $[-10, 10]$.

By zooming in, you can determine that this graph is not complete. This is because f has a local maximum value of about 0.0012 when x is 0 and y is 52.3 in addition to the local high point above the xy-plane and the local low point below the xy-plane that you see in Fig. 11.36.

Notice how the values of f get closer to zero as the distance of the domain elements (x, y) from $(0, 0)$ increases. This means an end behavior model of f is the xy-plane.

Application

We complete this section with an application.

Suppose we want to construct a box with a top that has capacity 300 cubic units. For economic reasons, we would like to use the least amount of material to construct the box. This means we want to find the minimum possible value of the surface area of the box.

Let the dimensions of the base be x by y and the height be h (see Fig. 11.37).

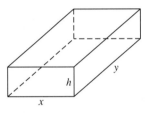

Figure 11.37   A box with dimensions x, y, and h.

The surface area S of the box is $2xy + 2xh + 2yh$. (Why?) We have $xyh = 300$ because the box is to have volume 300. If we solve $xyh = 300$ for h and substitute into the expression for the surface area of the box, we obtain

$$S = 2xy + 2xh + 2yh$$

$$= 2xy + 2x\left(\frac{300}{xy}\right) + 2y\left(\frac{300}{xy}\right)$$

$$= 2xy + \frac{600}{y} + \frac{600}{x}.$$

Thus the surface area S of such a box is a function of two variables x and y, the dimensions of the base of the box.

E X A M P L E   6   Finding Minimum Surface Area

Determine the dimensions of a box with top that has capacity 300 cubic units and has minimum possible surface area.

Solution   Let S be the surface area of the box. By the discussion preceding the example, we know that

$$S = 2xy + \frac{600}{y} + \frac{600}{x},$$

where x and y are the dimensions of the base of the box. Both x and y must be positive. We need to determine the minimum value of S in the first octant.

A grapher image of this surface is shown in Fig. 11.38. We can see from this graph that there is a local minimum value between $z = 265$ and $z = 270$. By zooming in, we estimate that this local low point occurs about $(6.7, 6.7, 269)$

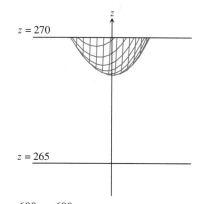

Figure 11.38 $S = 2xy + \dfrac{600}{y} + \dfrac{600}{x}$ in the viewing box $[5, 9]$ by $[5, 9]$ by $[265, 270]$.

This means that the dimensions of a box with a top and volume 300 cubic units that has minimum surface area is a *cube* of side length about 6.7 units. (Why?) The corresponding value of the surface area is about 269 square units.   ▬

It is very tedious to use a surface graphing utility and zoom-in to obtain accurate solutions of extrema problems. We will be content to estimate the values. Your estimates for the solution to Example 6 may differ from ours by a sizable margin.

Notes on Exercises

Ex. 1–10 provide practice in methods that can be used in later exercises to help determine extreme values and saddle points.
Graphs of Ex. 19–28 are difficult to imagine. A 3-*D* graphing program is highly recommended.
Ex. 49 and 50 are provided for those students who have access to the necessary graphing technology.

Assignment Guide

Day 1: Ex. 1–12
Day 2: Ex. 17, 21, 33–40

Exercises for Section 11.8

For Exercises 1–12, let $f(x, y) = 9 - x^2 - y^2$ and $g(x, y) = x^2 + y^2 - 9$.

1. Determine the domain of f and of g.

2. Determine the domain of $z = \sqrt{f(x, y)}$.

3. Determine the domain of $z = \sqrt{g(x, y)}$.

4. Determine the domain of $z = \dfrac{1}{f(x, y)}$.

5. Draw the trace of f in the planes $z = 0$, $z = 3$, and $z = 9$.

6. Draw the trace of g in the planes $z = 0$, $z = 3$, and $z = 9$.

7. Draw the trace of f in the planes $x = 0$, $x = 1$, $x = 2$, and $x = 3$.

8. Draw the trace of g in the planes $x = 0, x = 1, x = 2$, and $x = 3$.

9. Draw the trace of f in the planes $y = 0, y = 1, y = 2$, and $y = 3$.

10. Draw the trace of g in the planes $y = 0, y = 1, y = 2$, and $y = 3$.

11. Draw a complete graph of f. Identify all extrema.

12. Draw a complete graph of g. Identify all extrema.

For Exercises 13–16, $f(x, y) = 9 - (x - 2)^2 - (y - 3)^2$ and $g(x, y) = \sqrt{9 - (x - 2)^2 - (y - 3)^2}$.

13. Determine the domain of f.

14. Determine the domain of g. (*Hint:* When will $(x - 2)^2 + (y - 3)^2 \le 9$?)

15. Draw a complete graph of f and identify all extrema.

16. Draw a complete graph of g and identify all extrema.

For Exercises 17–28, determine the domain and draw a complete graph of the function. Estimate the coordinates (a, b, c) of any extrema. Identify any *saddle* points.

17. $z = 25 - x^2 - y^2$

18. $z = \sqrt{25 - x^2 - y^2}$

19. $z = \sqrt{25 - x^2 + 4x - y^2 + 9y + 13}$

20. $z = 5\sin(\frac{1}{2}x)$

21. $z = \dfrac{1}{\sqrt{9 - x^2 - y^2}}$

22. $z = \dfrac{1}{9 - x^2 - y^2}$

23. $z = y^2 - 8y - x^2 + 16$

24. $z = \ln(\cos x) - \ln(\cos y)$

25. $z = 24y^2 + 1.6y^3 - y^4 - 32x^2$

26. $z = 20xe^{-x^2 - y^2}$

27. $z = 0.7y^2 + 0.05y^3 - 0.03y^4 - x^2 + 5$

28. $z = 1 + \dfrac{1}{x} + \dfrac{1}{y}$

For Exercises 29–32, consider the function $z = 4xy + \dfrac{300}{x} + \dfrac{100}{y}$.

29. Let $x = 2, 4, 6$, and 8. For each x, compute z for $y = 1, 3, 7$, and 10.

30. What happens to the values of z as $x \to \infty$ and $y \to \infty$?

31. Graph z in the region $0 \le x \le 10, 0 \le y \le 10$, and $100 \le z \le 200$. Use $(0, 0, 150)$ as your aiming point.

32. Estimate the local minimum value and where it occurs.

Exercises 33–36 refer to this **problem situation:** The front and back of a box are constructed from material that costs \$1/sq in. The material for the top and bottom of the box costs \$2/sq in. and the material for the two sides costs \$3/sq in. Let the box have length x, width y, and height h.

33. Determine an algebraic representation in terms of x and y that gives the total cost C of the materials used in construction of a box with volume 50 cu in.

34. What values of x and y make sense in this problem situation?

35. Draw a complete graph of $z = C(x, y)$ in the first octant.

36. Estimate the minimum cost of a box with volume 50 cu in. and the dimensions of such a box of minimum cost.

Exercises 37–40 refer to this **problem situation**: A box with a top is constructed having a volume of 500 cu in., length x, width y, and height h.

37. Determine an algebraic representation of the total surface area S of the box in terms of x and y.

38. What values of x and y make sense in this problem situation?

39. Draw a complete graph of $z = S(x, y)$ in the first octant.

40. What are the dimensions of the box with minimum surface area? What is the minimum surface area?

Exercises 41–44 refer to this **problem situation:** A box with *no* top is constructed having a volume of 500 cu in., length x, width y, and height h.

41. Determine an algebraic representation of the total surface area S of the box in terms of x and y.

42. What values of x and y make sense in this problem situation?

43. Draw a complete graph of $z = S(x, y)$ in the first octant.

44. What are the dimensions of the box with minimum surface area? What is the minimum surface area?

For Exercises 45–48, determine an end behavior model for the function $z = f(x, y)$. Use a graphical argument.

45. $f(x, y) = xe^{-|x+y|}$

46. $f(x, y) = \dfrac{1}{2 + x^2 + y^2}$

47. $f(x, y) = 2x \sin \dfrac{1}{x}$

48. $f(x, y) = 3 + \dfrac{2}{x} + \dfrac{4}{y}$

Any function $y = f(x)$ can be used to define a surface in 3-space that provides an important useful *geometric representation of the complex zeros* of the function. In Exercises 49 and 50, consider the following definition and theorem, both stated without proof.

 Definition: Let f be any function of one complex variable. The **modulus surface** of f is the surface defined by $z = g(x, y) = |f(u)|$, where u is the complex number $x + iy$. The function g is called the **modulus function** of f.

 Theorem: Let $u = x + yi$ and let f be a function with domain and range that are subsets of the complex numbers. The complex number $a + bi$ is a zero of f if, and only if, the modulus surface $z = g(x, y) = |f(u)|$ touches the complex plane at $a + bi$. That is, $a + bi$ is a zero of f if, and only if, $(a, b, 0)$ is a point on the graph of $z = g(x, y) = |f(u)|$.

 Recall that $|a + bi| = \sqrt{a^2 + b^2}$.

49. a) Let $f(t) = t^2 + 1$ and $u = x + yi$. Show that the algebraic representation for the modulus surface of $y = f(t)$ is

$$z = g(x, y) = |f(u)| = |f(x + yi)|$$

$$= \sqrt{(x^2 - y^2 + 1)^2 + 4x^2 y^2}.$$

 b) Use a 3-*D* grapher and draw a complete graph of the modulus surface in part (a). (See the following figure.)

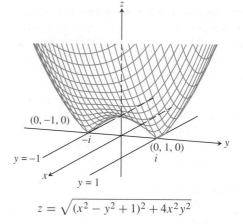

$$z = \sqrt{(x^2 - y^2 + 1)^2 + 4x^2 y^2}$$

 c) Explain how the surface in this figure relates to the complex zeros of $f(t) = t^2 + 1$.

50. a) Let $f(t) = t^3 - t^2 + t - 1$. Determine an algebraic representation of the modulus surface of f. (*Hint:* Use the binomial theorem to expand $(x + yi)^3$.)

 b) Use a 3-*D* grapher and draw a complete graph of the modulus surface in part (a). (See the figure below.)

 c) Use the rational zeros theorem and completely factor $f(t) = t^3 - t^2 + t - 1$ into 3 linear factors.

 d) Explain what this figure has to do with part (c).

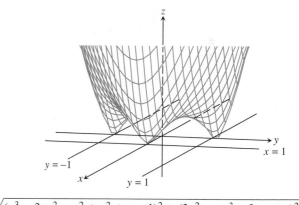

$$z = \sqrt{(x^3 - 3xy^2 - x^2 + y^2 + x - 1)^2 + (3x^2 y - y^3 - 2xy + y)^2}$$

Chapter 11 Review

KEY TERMS (The number following each key term indicates the page of its introduction.)

absolute maximum value, 648
absolute minimum value, 648
arithmetic progression, 585
arithmetic sequence, 585
axiom of induction, 620
binomial, 602
binomial coefficient
  n choose r, 604

REVIEW EXERCISES

In Exercises 1–8, determine the first four terms and the tenth term of the sequence. Draw a graph of the sequence.

1. $a_n = (-1)^{n+1}(n-1)$ **2.** $a_n = 2n^2 - 1$

3. $a_n = 2^n - 1$ **4.** $a_n = \cos \pi n$

5. $a_n = a_{n-1} + 3, a_1 = 2$ **6.** $a_n = 5 - a_{n-1}, a_1 = 7$

7. $a_n = 3 \cdot a_{n-1}, a_1 = 5$ **8.** $a_n = (-2) \cdot a_{n-1}, a_1 = 1$

9. The fourth and seventh terms of an arithmetic progression are -8 and 16, respectively. Determine the first term and a formula for the nth term.

10. The fifth and ninth terms of an arithmetic progression are -5 and 13, respectively. Determine the first term and a formula for the nth term.

In Exercises 11–14, expand each summation for $n = 8$.

11. $\sum_{k=1}^{n} (3k + 1)$ **12.** $\sum_{k=1}^{n} 3k^2$

13. $\sum_{k=1}^{n} (2^k - 1)$ **14.** $\sum_{k=1}^{n} (k^2 - 2k + 1)$

In Exercises 15 and 16, use summation notation to write the sum. Assume the patterns continue as suggested.

15. $-2 - 5 - 8 - 11 - \cdots - 29$

16. $4 + 16 + 64 + 256 + \cdots$

In Exercises 17 and 18, use summation notation to write the nth partial sum of the sequence. Assume the patterns continue as suggested.

17. $8, 6, 4, \ldots$ **18.** $-3, 6, -9, 12, \ldots$

In Exercises 19–22, use summation formulas to evaluate each expression.

19. $\sum_{k=1}^{25} (k^2 - 3k + 4)$ **20.** $\sum_{k=1}^{100} (k^3 - 2k)$

21. $\sum_{k=1}^{175} (3k^2 - 5k + 1)$ **22.** $\sum_{k=1}^{75} (k^3 - k^2 + 2)$

In Exercises 23–26, determine whether each infinite geometric series converges. If so, find its sum.

23. $\sum_{k=1}^{\infty} 3(0.5)^k$ **24.** $\sum_{k=1}^{\infty} (1.2)^k$

25. $\displaystyle\sum_{k=1}^{\infty} 2(0.01)^k$

26. $\displaystyle\sum_{k=1}^{\infty} \left(\frac{1}{1.05}\right)^k$

In Exercises 27–32, expand each expression.

27. $(2x + y)^5$

28. $(4a - 3b)^7$

29. $(3x^2 + y^3)^5$

30. $\left(1 + \frac{1}{x}\right)^6$

31. $(2a^3 - b^2)^9$

32. $(x^{-2} + y^{-1})^4$

Exercises 33–38, refer to the following sequence of polynomial functions:

$$f_1(x) = 1 - x$$

$$f_2(x) = 1 - x + \frac{x^2}{2!}$$

$$f_3(x) = 1 - x + \frac{x^2}{2!} - \frac{x^3}{3!}$$

$$\vdots$$

$$f_n(x) = 1 - x + \frac{x^2}{2!} - \frac{x^3}{3!} + \cdots + (-1)^n \frac{x^n}{n!}$$

33. Find complete graphs of the six polynomial functions f_1, f_2, $\ldots$, f_6 in the same viewing rectangle.

34. Find a complete graph of $f(x) = e^{-x}$ and compare it with the graphs in Exercise 33.

35. Is $f_6(-2)$ a good estimate for e^2? Why?

36. Is $f_6(4)$ a good estimate for e^{-4}? Why?

37. On what interval is $f_6(x)$ a good approximation for e^{-x}?

38. Use a graphing method to determine the error in using $f_3(-2)$ to approximate e^2. Repeat using $f_4(-2)$, $f_5(-2)$, and $f_6(-2)$. Are they underestimates or overestimates?

In Exercises 39–43, use mathematical induction to prove that each of the following is true for all positive integer values of n.

39. $1 + 3 + 6 + \cdots + \frac{n(n+1)}{2} = \frac{n(n+1)(n+2)}{6}$

40. $1 \cdot 2 + 2 \cdot 3 + 3 \cdot 4 + \cdots + n(n+1) = \frac{n(n+1)(n+2)}{3}$

41. $\displaystyle\sum_{i=1}^{n} \left(\frac{1}{2}\right)^i = 1 - \frac{1}{2^n}$

42. $2^{n-1} \le n!$

43. $n^3 + 2n$ is divisible by 3.

44. Find $_{12}P_7$.

45. Find $_{15}P_8$.

46. Find $_{18}C_{12}$.

47. Find $_{35}C_{28}$.

48. How many license plates are there that begin with AT followed by one letter and four digits?

49. How many five-character code words are there if the first character is always a letter and the other characters are letters and/or digits?

50. A travel agent is trying to schedule a client's trip from city A to city B. There are three direct flights, three flights from A to a connecting city C, and four flights from this connecting city C to city B. How many trips are possible?

51. How many license plates are there that begin with two letters followed by four digits or that begin with three digits followed by three letters? Assume no letters or digits are repeated.

52. A club has 45 members and the membership committee has three members. How many different membership committees are possible?

53. How many bridge hands are there that include the ace, king, and queen of spades?

54. How many bridge hands include all four aces and exactly one king?

55. Suppose a coin is tossed five times. How many different outcomes include at least two heads?

56. A certain small business has 35 employees—21 women and 14 men. How many different employee representative committees are there if the committee must consist of two women and two men?

57. Show algebraically that

$$_nP_k \times _{n-k}P_j = _nP_{k+j}.$$

For Exercises 58 and 59, sketch a graph of the plane.

58. $2x + y + 3z = 12$

59. $x + z = 5$

60. Compute the distance between the points $(1, 2, -5)$ and $(-3, 4, 6)$.

61. Determine the midpoint of the line segment between the points $(-1, 2, 5)$ and $(5, -2, 7)$.

62. Write an equation for the line through points $(2, 1, 5)$ and $(6, 4, 8)$ in

a) Cartesian form.

b) parametric form.

63. Write an equation for the line through points $(3, -4, 6)$ and $(2, 8, -2)$ in

a) Cartesian form.

b) parametric form.

For Exercises 64–67, describe and sketch the given quadric surface.

64. $9x^2 + y = 9$

65. $4x^2 + 9z^2 = 36$

66. $16x^2 + 64y^2 + z^2 = 64$

67. $x^2 + y^2 + z^2 = 81$

For Exercises 68–73, let $f(x, y) = 16 - x^2 - y^2$ and $g(x, y) = x^2 + y^2 - 16$.

68. Determine the domain of f and of g.

69. Determine the domain of $z = \sqrt{f(x, y)}$.

70. Determine the domain of $z = \sqrt{g(x, y)}$.

71. Determine the domain of $z = \dfrac{1}{f(x, y)}$.

72. Draw the trace of f in the planes $x = 0, x = 1, x = 2, x = 3$, and $x = 4$.

73. Draw the trace of g in the planes $y = 0, y = 1, y = 2, y = 3$, and $y = 4$.

For Exercises 74–77, determine the domain and draw a complete graph of the function. Estimate the coordinates (a, b, c) of any extrema. Identify any *saddle* points.

74. $z = 9 - x^2 - y^2$

75. $z = \sqrt{9 - x^2 - y^2}$

76. $z = \sqrt{5 - x^2 + y^2 + 2x - 6y}$

77. $z = 0.25(x - 3)y$

SEKI KOWA

Seki Kowa (1642–1708) was born into a samurai warrior family in Tokyo, Japan, but he was adopted by the family of an accountant. He showed his great mathematical aptitude as a child prodigy. In 1674, Seki Kowa published solutions to 15 problems thought to be unsolvable up to that time. This gave him a reputation throughout Japan as a brilliant mathematician. He was also an inspirational teacher.

Among his contributions were an improved method of solving higher-degree equations, the use of determinants in solving simultaneous equations, and a seventeenth-century form of calculus native to Japan, known as *yenri*. It is difficult to know the entire extent of his work, because the samurai code demanded great modesty. Seki Kowa is credited with awakening the scientific spirit in Japan that continues to thrive today.

CHAPTER 7

12

Probability and Statistics

12.1

Probability

In today's complex society, many situations have uncertain outcomes; consequently, we are concerned about the probability that these outcomes will occur.

What are the chances it will rain today? What are the chances I will get the job I just applied for? A manufacturer wonders, what is the probability that a randomly selected item coming off the production line will be defective? An airline's booking agent asks, what is the probability that a passenger with a reservation will be a no-show? The scientists at the Centers for Disease Control ask, what is the probability that someone exposed to a certain virus will contract the disease?

Each of these situations, and many more, can be thought of as an **experiment**. The various possible results of an experiment are called **outcomes** of the experiment, and the set of all possible outcomes is called the **sample space** of the experiment.

It is important to be able to identify the sample space of a probability experiment and to determine the number of elements in the sample space.

E X A M P L E 1   Finding the Sample Space

An experiment consists of tossing a red die and a green die. List all the elements in sample space S.

Figure 12.1 Consider the experiment of rolling dice.

Solution List all the pairs in sample space S. The first number of the pair represents the outcome of the red die and the second number the outcome of the green die.

(1, 1)	(1, 2)	(1, 3)	(1, 4)	(1, 5)	(1, 6)
(2, 1)	(2, 2)	(2, 3)	(2, 4)	(2, 5)	(2, 6)
(3, 1)	(3, 2)	(3, 3)	(3, 4)	(3, 5)	(3, 6)
(4, 1)	(4, 2)	(4, 3)	(4, 4)	(4, 5)	(4, 6)
(5, 1)	(5, 2)	(5, 3)	(5, 4)	(5, 5)	(5, 6)
(6, 1)	(6, 2)	(6, 3)	(6, 4)	(6, 5)	(6, 6)

There are 36 outcomes in the sample space of this experiment. ≡

A subset of a sample space is called an **event**. For example, in the experiment of Example 1, we can identify event D, "rolling doubles," or event N, "sum of 9," or event F, "sum less than 4." These events include the following outcomes:

D	(1, 1)	(2, 2)	(3, 3)	(4, 4)	(5, 5)	(6, 6)
N	(3, 6)	(5, 4)	(4, 5)	(6, 3)		
F	(1, 1)	(1, 2)	(2, 1)			

Bert says, "Probability theory got its start as an analysis of games of chance. It has grown to include many industrial and scientific applications."

E X A M P L E 2   Finding the Sample Space

An experiment consists of tossing a coin three times in succession and recording the outcomes of heads (H) and tails (T). List all the outcomes of sample space S and all the outcomes in event E, "at least two heads."

Solution The sample space is the set

$$S = \{HHH, HHT, HTH, THH, HTT, THT, TTH, TTT\},$$

and event E is the set

$$E = \{HHH, HHT, HTH, THH\}.$$ ≡

Figure 12.2 Consider the experiment of tossing a coin.

In the experiments considered so far, we assumed that no particular outcome is any more likely than any other. We therefore say that the outcomes in the sample space are **equally likely**. This condition is described by saying the dice are "fair," or the coin is "fair."

When any probability experiment with equally likely outcomes is conducted, any given event has a certain likelihood of occurring, called the event's theoretical probability. The **theoretical probability** for an experiment of equally likely outcomes is the ratio of the number of elements in the event divided by the number of outcomes in the sample space.

We use the notation $n(S)$ to represent the number of outcomes in sample space S and $n(E)$ to represent the number of outcomes in event E.

> **Definition 12.1   Probability of an Event**
>
> In an experiment with a finite number of equally likely outcomes, the **probability** (or theoretical probability) of an event E, denoted $P(E)$, is
>
> $$P(E) = \frac{n(E)}{n(S)}.$$

Notice that because $n(E)$ and $n(S)$ are positive integers where $n(E) \leq n(S)$, $0 \leq P(E) \leq 1$ for any event E. An event A is impossible if and only if $P(A) = 0$, and an event B is certain if and only if $P(B) = 1$.

Teaching Note

When students are faced with a new probability experiment, ask them to list a few elements of the sample space. It will help them to understand the problem. Only after students fully understand the problem should they begin calculating the probability of an event.

Finding the Probability of an Event

Finding the probability of an event requires calculating the quotient $n(E)/n(S)$ introduced in Definition 12.1. Consequently, in any probability experiment it is important to identify the number of elements in the sample space and the number of elements in the event. Each of these tasks is a counting problem.

E X A M P L E   3   Finding the Probability of an Event

Find the probability of event E, "rolling a sum of 5," in the experiment of Example 1.

Solution   From Example 1, $n(S) = 36$. Because E consists of the set of outcomes $\{(1, 4), (2, 3), (3, 2), (4, 1)\}$, $n(E) = 4$. Therefore

$$P(E) = \frac{n(E)}{n(S)} - \frac{4}{36} - \frac{1}{9}.$$

To find the theoretical probability, it is often necessary to use the counting methods discussed in Section 11.6. The fundamental counting principal is used in Example 4 to find both $n(S)$ and $n(E)$.

E X A M P L E   4   Finding the Probability of an Event

What is the probability that a five-digit telephone number chosen at random has no repeated digits?

Solution   Let E be the event "no repeated digits" and S represent the entire sample space. Use the fundamental counting principle to calculate

$$n(E) = 8 \cdot 9 \cdot 8 \cdot 7 \cdot 6 = 24{,}192.$$

(Recall that a telephone number cannot begin with a 0 or a 1.) Also,

$$n(S) = 8 \cdot 10 \cdot 10 \cdot 10 \cdot 10 = 80,000.$$

Therefore

$$P(E) = \frac{24,192}{80,000} = 0.3024.$$

≡

Two events A and B from the same sample space are called **mutually exclusive events** if they have no outcomes in common. In the experiment of rolling a set of dice, discussed in Example 1, the events "rolling a sum of 5" and "rolling doubles" are mutually exclusive. By comparison, the events "rolling a sum of 5" and "red die is a 1" are not mutually exclusive events since the outcome $(1, 4)$ occurs in both events.

If two events are mutually exclusive, then the following probability formula applies.

Probability of Mutually Exclusive Events

If A and B are mutually exclusive events for the same experiment, then

$$P(A \text{ or } B) = P(A) + P(B). \tag{1}$$

Stan says, "I find that when students are faced with a new probability experiment they tend to begin manipulating probability formulas before they fully understand the experiment."

When analyzing a probability experiment, it is important to be able to recognize that two events are mutually exclusive. By developing a thorough understanding of the experiment, you will learn to do this.

E X A M P L E 5   Probability of Mutually Exclusive Events

Suppose one card is drawn from a deck of 20 cards consisting of one ace, king, queen, jack, and 10 in each of the four suits of spades, hearts, diamonds, and clubs. What is the probability of drawing a 10 or a face card (king, queen, or jack)?

Solution   Let A be the event "draw a 10" and B be the event "draw a face card." Because a 10 is not a face card, events A and B are mutually exclusive. Then $P(A) = 4/20 = 1/5$ and $P(B) = 12/20 = 3/5$, and

$$P(A \text{ or } B) = P(A) + A(B) = \frac{1}{5} + \frac{3}{5} = \frac{4}{5}.$$

≡

The **complement of an event** A, denoted A^c, is the set of all outcomes in the sample space that are not in event A. Notice that events A and A^c taken together consist of the entire sample space. Therefore $P(A \text{ or } A^c) = 1$.

Events A and A^c are mutually exclusive by definition. Therefore

$$P(A \text{ or } A^c) = P(A) + P(A^c)$$

$$1 = P(A) + P(A^c)$$

$$P(A^c) = 1 - P(A).$$

Sometimes it is easier to find the probability of the complement of an event than it is to find the probability of the event itself. Example 6 is an example of such a situation.

E X A M P L E   6   Using the Complement of an Event

A 13-card hand is dealt from a standard 52-card deck. Find the probability that the hand has at least one heart.

Solution   Let H be the event "at least one heart." However, it is easier to determine the probability of H^c, that is, the event "no hearts." Because the number of hands with no hearts is $_{39}C_{13}$,

$$P(H^c) = \frac{_{39}C_{13}}{_{52}C_{13}}$$

$$= \frac{39!39!}{26!52!}$$

$$= 0.01279.$$

Therefore

$$P(H) = 1 - P(H^c) = 1 - 0.01279 = 0.98721.$$

Independent Events

Two events are **independent events** if the occurrence of one has no effect on the occurrence of the other. For example, when tossing a fair coin, the event "head on the second toss" is independent of the event "tail on the first toss." The coin is not influenced by what happens on previous tosses.

When two events are known to be independent, we can use the following formula to find the probability of both occurring.

Probability of Independent Events

If A and B are independent events, then the probability that both will occur is

$$P(A \text{ and } B) = P(A) \cdot P(B). \tag{2}$$

E X A M P L E   7   Finding the Probability of Independent Events

Find the probability of the event "HTH" when a coin is tossed three times.

Solution   Suppose E is the event "HTH". Let H_1 be the event "head on the first toss," T_2 the event "tail on the second toss," and H_3 the event "head on the third toss." In other words, the event E occurs if and only if all three events $H_1, T_2,$ and H_3 occur. These three events are independent. Therefore

$$P(E) = P(H_1 \text{ and } T_2 \text{ and } H_3)$$
$$= P(H_1) \cdot P(T_2) \cdot P(H_3)$$
$$= \frac{1}{2} \cdot \frac{1}{2} \cdot \frac{1}{2}$$
$$= \frac{1}{8}.$$

≡

Experimental Probability

Throughout this section, we have been calculating *theoretical* probability. For example, suppose a fair coin is tossed. The theoretical probability of obtaining a head on a single toss of a fair coin is $1/2$.

On the other hand, suppose this fair coin is physically tossed 100 times and 48 heads occur. Then the **experimental probability** of heads is 0.48.

E X A M P L E   8   Finding Experimental Probability

A fair coin is tossed 55 times and heads occur 26 times. What is the experimental probability of the event "tossing a tail" in this experiment?

Solution

Number of heads: 26

Number of tails: 29

Therefore the experimental probability of tossing a tail is $\dfrac{29}{55} = 0.527$.

≡

Notes on Exercises

The focus of Ex. 32 is that
the sum of probabilities of
mutually exclusive events of a
probability experiment cannot
exceed 1. You might want to
ask students to describe the
probability experiment that
is being referred to in this
exercise.

Assignment Guide

Day 1: Ex. 1–20
Day 2: Ex. 21–33

(*Note:* In the next section, you will learn how to find the theoretical probability for the event in Example 8.)

When a physical experiment is conducted, the experimental probability should be consistent with, though not identical to, the theoretical probability. As the number of repetitions of this experiment is increased, the experimental probability and the theoretical probability approach one another in the following sense. Suppose a fair coin is tossed n times and H_n is the number of times a head occurs. Because the mathematical probability of obtaining a head on a given toss is $1/2$, we should expect that $H_n/n \to 1/2$ as $n \to \infty$. In other words, the observed probability of a given event should be a good approximation of the mathematical probability if a large number of repetitions of the experiment are performed.

Exercises for Section 12.1

In Exercises 1–8, list the elements of the sample space for each experiment.

1. A single die is rolled.

2. A single fair coin is tossed.

3. A penny and dime are tossed simultaneously.

4. A penny, nickel, and dime are tossed simultaneously.

5. A fair coin is tossed and a single die is rolled.

6. A game spinner, numbered 1 through 8, is spun.

7. Ten balls numbered 1 through 10 are in an urn. One ball is selected.

8. Five balls numbered 1 through 5 are in an urn. Two balls are selected; the second ball is drawn before the first has been returned.

Exercises 9–20 refer to the following **problem situation**: A red die and a green die are rolled. List the outcomes in each of the following events:

9. The sum is 9.   10. The sum is even.

11. The number on one die is one more than that on the other die.

12. Both dice are even.

13. The sum is less than 10.

14. Both dice are odd.

15. Find the probability of the event in Exercise 9.

16. Find the probability of the event in Exercise 10.

17. Find the probability of the event in Exercise 11.

18. Find the probability of the event in Exercise 12.

19. Find the probability of the event in Exercise 13.

20. Find the probability of the event in Exercise 14.

In Exercises 21–24, state the complement of each event for the experiment of rolling a red die and a green die.

21. Rolling a sum greater than 1

22. Rolling an even sum

23. Rolling doubles

24. Rolling a sum less than 11

25. A coin is tossed 110 times and heads occurred exactly 53 times. What is the experimental probability of this event?

26. A red die and a green die are rolled 65 times. On eight rolls, the sum of the dice was "less than five." What is the experimental probability of this event?

Exercises 27–31 refer to the following **problem situation**: A deck of 20 cards consists of one ace, king, queen, jack, and 10 in each of the four suits of spades, hearts, diamonds, and clubs. An experiment consists of dealing a hand of five cards.

27. Find the probability that the hand consists of the five spades.

28. Find the probability that all five cards are of the same suit.

29. Find the probability that the hand includes four aces.

30. Find the probability that the hand contains only aces and face cards (king, queen, or jack).

31. Find the probability that the hand contains the king of clubs.

32. Explain why the following statement cannot be true: The probabilities that a computer salesperson will sell zero, one, two, or three computers in any one day are 0.12, 0.45, 0.38, and 0.15, respectively.

33. Writing to Learn   During July in a certain city, the probability of at least 1 hr of sunshine a day is 0.78, the probability of at least 30 min of rain a day is 0.44, and the probability it will be cloudy all day is 0.22. Write a paragraph explaining whether this statement could be true.

12.2

Objective

Students will be able to find the probability of any experiment that consists of repeated independent trials in which there are only two outcomes.

Key Ideas

Independent repeated trials
Binomial probability
    theorem
Monte Carlo methods

Teaching Note

Cooperative learning groups can be used to complete experimental probability activities. For example, have each group toss a coin 100 times and record the number of *heads* and *tails*. Then have the groups calculate their experimental probability of obtaining heads. Compare the results of each group. How do the results compare with using a table of random digits, or the random number generator of a grapher to simulate the experiment?

The Binomial Theorem and Probability

Example 7 in Section 12.1 refers to the experiment of tossing a coin three times. Because each toss of the coin is an independent event, we can think of tossing a coin three times as three repetitions of tossing a coin. This is an example of **independent repeated trials**.

Independent Repeated Trials, Each with Exactly Two Possible Outcomes

Many examples of independent repeated trials exist among the many applications of probability theory. Following are a few examples:

a) A coin is tossed 100 times.

b) A quality control inspector checks for defects among randomly selected products.

c) Drivers are selected at random to check if they are wearing seat belts.

d) The IRS inspects a tax return checking for errors.

In each of these examples, there are only two possible outcomes for each trial. A coin toss results in either heads or tails; an inspection results in a defect or no defect; a driver is either wearing or not wearing a seat belt; an IRS inspector finds either an error or no error.

Whenever you have independent repeated trials in which each trial has only two possible outcomes, the methods given in the next several examples will apply.

E X A M P L E   1   Finding the Probability of a Defect

Suppose there is a random check for defects every hour as items come off an assembly line. If the probability of no defect on each trial is 0.98, what is the probability of no defect on the second, but not the first or third of three successive checks?

Solution   Let D be the event "defect" and N be the event "no defect." We are to find $P(NDN)$ on three successive trials. Because N and D are complementary events, we have

$$P(N) = 0.98 \quad \text{and} \quad P(D) = 1 - P(N) = 0.02.$$

The trials are independent events, therefore

$$P(NDN) = P(N) \cdot P(D) \cdot P(N)$$
$$= (0.98)(0.02)(0.98)$$
$$= 0.019.$$

The method of Example 1 can be used in any situation that consists of a sequence of independent events. In the following examples, coin-tossing experiments will be considered. Coin-tossing experiments can be used to model many different probability applications.

E X A M P L E   2   Finding the Probability in Coin Tossing

Find the probability of the event "HHTTH" in the experiment of tossing a coin five times.

Solution   Because each coin toss is an independent event, the letters H and T refer to the events of tossing a head or a tail, respectively. So, by Eq. (2),

$$P(HHTTH) = P(H) \cdot P(H) \cdot P(T) \cdot P(T) \cdot P(H)$$
$$= \left(\frac{1}{2}\right)^5 = \frac{1}{32} = 0.03125.$$

In Example 2, the specific sequence of heads and tails—$HHTTH$—was given. Determining the probability of tossing exactly three heads in five tosses is more complex. Example 3 relates to this situation.

E X A M P L E   3   Finding the Probability in Coin Tossing

Find the probability of tossing exactly one head in three tosses of a fair coin.

Solution   The event "exactly one head" can occur in three ways: HTT, THT, or TTH. These are mutually exclusive events, each consisting of three repetitions of independent events; therefore

$$P(HTT \text{ or } THT \text{ or } TTH) = P(HTT) + P(THT) + P(TTH)$$
$$= \left(\frac{1}{2}\right)^3 + \left(\frac{1}{2}\right)^3 + \left(\frac{1}{2}\right)^3$$
$$= \frac{3}{8}.$$

Using the Binomial Theorem

The method used in Example 3 required that we list all the ways exactly one head could occur in three tosses. However, creating such a listing is tedious and unreasonable when the number of repeated tosses is large, so we next consider another way to analyze this situation.

Suppose a coin is tossed four times. Apply the Binomial theorem developed in Section 11.3 to the expression $(H + T)^4$ to obtain:

$$(H + T)^4 = \binom{4}{0} H^4 + \binom{4}{1} H^3 T + \binom{4}{2} H^2 T^2 + \binom{4}{3} HT^3 + \binom{4}{4} T^4 \quad (1)$$

The coefficient $\binom{4}{2}$ of $H^2 T^2$ represents the number of ways to toss two heads and two tails in four tosses. Likewise the coefficient $\binom{4}{3}$ of HT^3 represents the number of ways to toss one head and three tails in four tosses. These observations can be used to find probabilities.

E X A M P L E   4   Finding the Probability in Coin Tossing

Find the probability of the event E "tossing exactly two heads in four tosses of a fair coin."

Solution   The event $HHTT$ is one of the outcomes in the event E and

$$P(HHTT) = \left(\frac{1}{2}\right)^4 = \frac{1}{16}.$$

From the term $\binom{4}{2} H^2 T^2$ in the binomial expansion of $(H + T)^4$, we see that there are $\binom{4}{2} = {}_4C_2$ ways for two heads to occur in four tosses. Therefore

$$P(E) = {}_4C_2 \cdot \frac{1}{16}$$

$$= 6 \cdot \frac{1}{16} = 0.375.$$

∎

We can generalize Example 4 as follows: Each coefficient $\binom{4}{i}$ gives the number of ways i tails can occur when a coin is tossed four times; that is, $\binom{4}{i} = {}_4C_i$.

If we let $H = T = 1/2$, the probability of a head or a tail on one toss of a single coin, then we can rewrite the right-hand side of Eq. (1) as follows:

$$1 = \left(\frac{1}{2} + \frac{1}{2}\right)^4 = \frac{{}_4C_0}{16} + \frac{{}_4C_1}{16} + \frac{{}_4C_2}{16} + \frac{{}_4C_3}{16} + \frac{{}_4C_4}{16}$$

$$= \frac{1}{16} + \frac{4}{16} + \frac{6}{16} + \frac{4}{16} + \frac{1}{16}$$

Each term ${}_4C_i/16$ represents the probability of the event consisting of exactly i tails occurring on a toss of four coins. This generalizes to tosses of any number of fair coins, as illustrated in the next example. We note that a toss of n fair coins and n tosses of one fair coin are equivalent regarding events about the number of heads or tails that occur.

EXAMPLE 5   Finding the Probability of Independent Events

A fair coin is tossed seven times. Find the probability of obtaining exactly three tails on those seven tosses.

Solution   According to the discussion preceding this example, the desired probability is given by substituting $H = T = 1/2$ in ${}_7C_3H^4T^3$. Thus the probability of exactly three tails occurring on seven tosses of a fair coin is

$$_7C_3\left(\frac{1}{2}\right)^4\left(\frac{1}{2}\right)^3 = \frac{7!}{4!3!}\frac{1}{2^7} = \frac{35}{128}.$$

Frank says, "The solution to Example 5 reflects a paper and pencil solution. I'd encourage keying the expression $_7C_3(.5)\text{^}4(.5)\text{^}3$ directly into the grapher."

The analysis in Example 5 applies to any experiment that meets the following two conditions: (a) There are two outcomes to the experiment, one called H and the other T, each with probability of $1/2$ of occurring; and (b) Repetitions of the experiment are independent events. For example, a coin can be tossed n times with outcomes of heads (H) and tails (T). Or n light bulbs can be checked as they come off the assembly line with outcomes of defective (H) or nondefective (T). A basketball player can shoot n foul shots with outcomes of made (H) or missed (T). In each case, if $P(H) = k$ such that $0 \le k \le 1$, then $P(T) = 1 - k$.

Theorem 12.1   Binomial Probability Theorem

Suppose a certain experiment has only the two outcomes, H and T, and suppose repetitions of the experiment are independent events. If $P(H) = k$ and E is the event "T occurs exactly r times in n repetitions," then

$$P(E) = {}_nC_r P(H)^{n-r}P(T)^r$$

$$= \frac{n!}{r!(n-r)!}k^{n-r}(1-k)^r.$$

In Example 6, the Binomial Probability theorem will be applied to a typical problem that occurs in a manufacturing situation.

E X A M P L E 6   APPLICATION: Detecting Defective Chips

Suppose it is known that one out of 1000 of a certain brand of computer chip is defective. Five computer chips are selected at random. What is the probability that a lot of five will contain one defective chip?

Solution   Let H represent the outcome that a selected computer chip is good and T that it is defective. Then $P(H) = 999/1000 = 0.999$ and $P(T) = 1/1000 = 0.001$. Event E with exactly one defective computer chip can be represented by $\binom{5}{1} H^4 T$. The probability of this event is

$$P(E) = {}_5C_1 P(H)^4 P(T) = 5(0.999)^4(0.001)$$
$$= 0.00498003.$$

Thus the probability that one of the five selected computer chips is defective is about 0.00498. ≡

Notes on Examples

Notice that Example 8 on page 671 does not employ any technique not already discussed in other exercises. Its main purpose is to demonstrate that as the number of trials is increased, the experimental probability and theoretical probability become closer to each other.

GRAPHER UPDATE

Some graphers include a random number generator. Such a grapher can be used instead of a table of random digits to simulate a probability experiment.

Monte Carlo Methods

Simulation techniques are often applied in problem-solving endeavors. When the process being simulated involves an element of chance, it is referred to as a **Monte Carlo method.** Using these simulations often saves the cost of building and operating an expensive piece of equipment.

It often is expedient when conducting a Monte Carlo simulation to use tables of random digits in place of rolling dice or tossing coins to generate the outcomes of a simulated probability experiment. Figure 12.3 shows a table of 250 random digits.

We can use this table of random digits to simulate tossing a coin by letting 0, 2, 4, 6, 8 represent heads and 1, 3, 5, 7, 9 represent tails.

E X A M P L E 7   Using Random Digit Tables

Use the first 100 digits of the table in Fig. 12.3 to simulate tossing a fair coin 100 times. Find the experimental probability of heads for this simulation.

Solution   The one hundredth digit in Fig. 12.3 is the last digit of the fifth row of digits. Let 0, 2, 4, 6, 8 represent heads and 1, 3, 5, 7, 9 represent tails. By counting these digits, we see the following:

Heads: 57

Tails: 43

$$P(H) = \frac{57}{100} = 0.57$$ ≡

17057	03620	70700	82713
01340	76684	68426	07161
92416	41655	86773	50613
25431	17424	40065	27244
32215	33067	90841	58844
22076	24858	58376	07007
17341	14221	44651	43803
54540	35358	35461	13050
59383	47111	48921	80653
36377	43636	52091	99960
52045	67095	06496	94386
61976	52958	92485	12751
03034	61589		

Figure 12.3   A table of 250 random digits.

EXAMPLE 8   Using Random Digit Tables

Use the table in Fig. 12.3 to simulate tossing a fair coin 250 times. Find the experimental probability of heads for this simulation.

Solution   Let $0, 2, 4, 6, 8$ represent heads and $1, 3, 5, 7, 9$ represent tails. By counting these digits, we see the following:

Heads: 129

Tails: 121

$$P(H) = \frac{129}{250} = 0.516$$

◼

Comparing Examples 7 and 8, we can observe that as the number of trials increases, the experimental probability and the theoretical probability more nearly agree.

Notes on Exercises

Ex. 19–21 assume that certain subsets of a table of random digits can be used to obtain a table of random digits. Clearly, not any subset chosen will be a set of random digits. For example, the subset of all 1's would fail.

Assignment Guide

Day 1: Ex. 1–16
Day 2: Ex. 17–30

Exercises for Section 12.2

In Exercises 1–4, assume a fair coin is being tossed.

1. Find the probability of HTT occurring in three tosses of the coin.

2. Find the probability of $HHTH$ occurring in four tosses of the coin.

3. Find the probability of at least one head occurring in four tosses of the coin.

4. Find the probability of eight heads occurring in eight tosses of the coin.

In Exercises 5–8, assume N is the event "no defect" and D is the event "defect" in an inspection experiment. Assume $P(N) = 0.99$.

5. What is the probability of the event DND in three trials?

6. What is the probability of the event $NDNN$ in four trials?

7. What is the probability of the event NNN in three trials?

8. What is the probability of the event $DDNN$ in four trials?

In Exercises 9–16, assume a fair coin is being tossed five times. Use the binomial expansion of $(H + T)^5$ to answer the following questions:

9. In how many ways can exactly one head occur?

10. In how many ways can exactly three heads occur?

11. In how many ways can either one or two heads occur?

12. In how many ways can either four or five heads occur?

13. Find the probability of exactly two heads occurring.

14. Find the probability of exactly three heads occurring.

15. Find the probability of exactly four heads occurring.

16. Find the probability of either two or three heads occurring.

In Exercises 17–22, use the random digits in Fig. 12.3 to complete the indicated Monte Carlo simulations.

17. Use the last 100 digits of the random digit table as a simulation of tossing a fair coin 100 times.

18. Use the digits of the random digit table as a simulation of rolling a die 25 times. Let 1, 2, 3, 4, 5, and 6 represent rolling a 1 through 6. Ignore the digits 0, 7, 8, and 9.

19. Use the first 10 digits in the random digit table that are a 1 to 6 to simulate rolling a red die 10 times and use the last 10 digits starting from the end that are a 1 to 6 to simulate rolling a green die. Use these data as a simulation of rolling a set of dice 10 times. List the 10 outcomes from this simulated experiment.

20. Use the digits in the first column of the random digit table to simulate tossing one coin 13 times and the second column of digits to simulate tossing a second coin 13 times.

Find the experimental probability of tossing an HH in this experiment.

21. Pairing the first and last columns and the second and next to last columns of random digits, simulate tossing a pair of coins 25 times. Find the experimental probability of tossing an HT on these 25 tosses.

22. **Writing to Learn** Write a paragraph that explains how to use the table of random digits to simulate rolling a set of dice 20 times.

In Exercises 23–27, refer to the experiment of tossing a fair coin 10 times and find the following probabilities:

23. "Heads on the second toss"

24. "Heads on the first and last toss"

25. "Heads on all 10 tosses"

26. "Exactly two heads"

27. "At least one head"

28. A factory's management knows that an item coming off an assembly line has a probability of 0.015 of being defective. If three items are selected at random during the course of a work day, find the probability that no item is defective.

29. A car agency has 25 cars available for rental—12 compact and 13 intermediate sized. If two cars are selected at random, what is the probability they are both compact?

30. In a game of *Yahtzee*, five dice are tossed simultaneously. Find the probability of rolling five of a kind on a single roll.

12.3 _____ Stem-and-leaf Tables, Histograms, and Line Graphs

Objective

Students will be able to use a set of data to construct an appropriate stem-and-leaf table. Using the stem-and-leaf table, they will be able to construct a frequency table and a histogram. They will also be able to calculate and interpret the mean, median, and mode.

TABLE 12.1 Salaries for ABC Company for the year 1992 (in Thousands of Dollars)

28.3, 29.7, 28.2, 31.7, 29.8, 31.6, 21.7, 22.3, 23.4, 25.1, 24.5, 26.8, 27.8, 22.9, 23.7, 28.2, 25.6, 29.1, 28.5, 25.7, 29.5, 29.6, 29.7, 28.3, 21.4, 27.4, 23.4, 30.6, 23.8, 31.5, 27.9, 28.6, 25.6, 28.8, 31.7, 22.8, 24.8, 26.7, 30.8, 28.7, 31.6, 21.8

In this section, we continue showing that numbers are used to report quantitative data. Sometimes, whole numbers are sufficient to describe the data; other times, rational or real numbers are required. The set of numbers used in reporting the data depends on the nature of the data.

Data are easier to understand and interpret when they are translated into a visual representation. Whether a bar graph or a pie chart is chosen depends on the nature of the data: A bar graph always reports the frequency of occurrence of a given data value; a pie chart reports fractions of the whole for some quantity. So both the set of numbers used and the type of visual representation depends on the nature of the data being studied. In this section, you will learn about stem-and-leaf tables, histograms, and line graphs.

Stem-and-leaf Tables

In a **stem-and-leaf table**, sometimes called a stem-and-leaf **plot**, the data are arranged into a specific table format. In this format, a reader can make observations about the data more easily than when viewing the listing of raw data.

The raw data shown in Table 12.1 are salaries for the year 1992 for employees of the ABC Company. To complete a stem-and-leaf plot of these data, the first two digits of each data item, reading from left to right, are considered a **stem** and the last digit is a **leaf**. Write the stems in order down the first column, and for each data item, considered one by one, write the leaf digit in the appropriate stem row.

TABLE 12.2   Life Expectancy Data

	Male	Female
Austria	72.1	78.6
Belgium	76.0	76.8
Cyprus	73.8	77.8
Czechoslovakia	67.3	74.7
Denmark	71.8	77.6
Finland	70.5	78.2
France	72.0	80.3
Germany	71.8	78.4
Greece	72.2	76.4
Hungary	66.2	74.0
Ireland	70.1	75.3
Italy	72.0	78.4
Netherlands	72.9	79.1
Norway	72.8	79.2
Poland	67.1	75.7
Portugal	68.4	75.7
Spain	72.5	78.1
Sweden	74.2	79.6
Switzerland	73.9	80.2
United Kingdom	71.9	77.3
Yugoslavia	68.4	75.0

Key Ideas

Stem-and-leaf table
Frequency table
Histogram
Mean
Median
Mode
Line graphs

Teaching Note

Notice that each leaf is a single digit. A stem is usually more than one digit and represents a certain interval.

Salaries for the ABC Company for 1992

Stem	Leaf
21	7 4 8
22	3 9 8
23	4 7 4 8
24	5 8
25	1 6 7 6
26	8 7
27	8 4 9
28	3 2 2 5 3 6 8 7
29	7 8 1 5 6 7
30	6 8
31	7 6 5 7 6

By a quick glance at the stem-and-leaf table, we can determine the following, none of which is immediately obvious from the raw data:

- There are three salaries between $21,000 and $22,000.
- There are four salaries between $23,000 and $24,000.
- There are more salaries between $28,000 and $29,000 than between any other interval.

E X A M P L E   1   Completing a Stem-and-leaf Table

Using data shown in Table 12.2, construct a stem-and-leaf table for the life expectancy of males.

Solution   Because each data item is a three-digit number, each stem consists of two digits and each leaf a single digit.

List the stems in order from 66 to 76.

Stem	Leaf
66	2
67	3 1
68	4 4
69	
70	5 1
71	8 8 9
72	1 0 2 0 9 8 5
73	8 9
74	2
75	
76	0

There are no leaves for stem 69 and 75 because there are no data items with those stems.

One strength of stem-and-leaf plots is that each data item in the set of raw data appears in the stem-and-leaf table. The data are not condensed, rounded, or otherwise changed. They simply have been rearranged and presented in a different format.

Frequency Tables

Each stem of a stem-and-leaf table represents a certain range of data values. Useful information about the data can be obtained by counting the number of leaves for each stem. This number, called the **frequency** of the leaf, is often recorded in a **frequency table**. Table 12.3 shows a frequency table for the ABC Company salary stem-and-leaf plot given in Table 12.1.

SALARIES FOR THE ABC COMPANY FOR 1992

Stem	Leaf
21	7 4 8
22	3 9 8
23	4 7 4 8
24	5 8
25	1 6 7 6
26	8 7
27	8 4 9
28	3 2 2 5 3 6 8 7
29	7 8 1 5 6 7
30	6 8
31	7 6 5 7 6

TABLE 12.3   Frequency Table of Salary Data

Interval	Frequency
21,000 to 21,999	3
22,000 to 22,999	3
23,000 to 23,999	4
24,000 to 24,999	2
25,000 to 25,999	4
26,000 to 26,999	2
27,000 to 27,999	3
28,000 to 28,999	8
29,000 to 29,999	6
30,000 to 30,999	2
31,000 to 31,999	5

Completing a Histogram

A **histogram** is a bar graph that visually represents the information in a frequency table. Figure 12.4 shows a histogram of the information in Table 12.3 where each bar corresponds to a range in the table and the height of each bar represents the frequency of a data item in the corresponding range. (*Note:* In a histogram, the bars usually touch each other because there are no gaps between the ranges.)

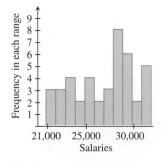

Figure 12.4   Histogram of salaries for ABC Company.

E X A M P L E   2   Completing a Histogram

Draw a histogram for the male life expectancy data given in Table 12.2.

Solution   Begin by completing a frequency table for the data in the stem-and-leaf plot in Example 1.

Interval	Frequency
66.0 to 66.9	1
67.0 to 67.9	2
68.0 to 68.9	2
69.0 to 69.9	0
70.0 to 70.9	2
71.0 to 71.9	3
72.0 to 72.9	7
73.0 to 73.9	2
74.0 to 74.9	1
75.0 to 75.9	0
76.0 to 76.9	1

Following is a completed histogram.

We will end this section by completing a line graph of the salary data for the ABC Company given in Table 12.1. In so doing, we will illustrate differences between data that can be represented with a histogram and those that can be represented by a line graph. However, we first must digress to discuss measures of central tendency.

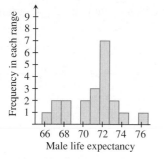

Figure 12.5   Histogram of male life expectancy.

Measures of Central Tendency

The salary data of the ABC Company (see Table 12.1) reflect the salaries for 1 yr. But suppose salary trends are studied over several years. A department manager would like a single number that reflects the "typical" or "average" salary in any particular year. This single number would be the **central tendency** of the complete set of salary data. The **mean** is one measure of central tendency for a set of data.

The concept of mean is defined by the following formula.

Definition 12.2   Mean

The **mean**, denoted by $\overline{x}$, of a set of data is the sum of all data numbers divided by the total number of items. That is,

$$\overline{x} = \frac{x_1 + x_2 + \cdots + x_n}{n}.$$

The mean is commonly called "the average value" of a data set.

NOTE ABOUT MATHEMATICAL
SYMBOLS
A letter used to represent a number is called a **variable**. In the definition of mean, $\overline{x}$ is a variable, as are the letters $x_1, x_2, \ldots, x_n$. Subscripts are used to indicate that x_1 represents the first data item, x_2 the second data item, etc. Variables are inherent in solving problems algebraically.

EXAMPLE 3   Finding the Mean

Find the mean of the data set $\{23, 48, 39, 42, 28, 33\}$ using the formula in Definition 12.2. Support your calculation using the statistical package of your calculator.

Solution

$$\overline{x} = \frac{x_1 + x_2 + \cdots + x_n}{n}$$

$$= \frac{23 + 48 + 39 + 42 + 28 + 33}{6}$$

$$= 35.5$$

Figure 12.6 shows the first three data items as they are keyed into a grapher DATA menu. The grapher supports that the mean is 35.5.

```
DATA
x1 = 23
y1 = 1
x2 = 48
y2 = 1
x3 = 39
y3 = 1
```

Figure 12.6 The first three of six data items keyed into a DATA menu on the grapher we use.

Two additional frequently used measures of central tendency are the **median** and the **mode**.

Definition 12.3   Median and Mode

The **median** of a set of data listed in order from the smallest to largest is

a) the middle number if there is an odd number of data items, or
b) the mean of the two middle numbers if there is an even number of data items.

The **mode** of a data set is the number that occurs most frequently in the set.

EXAMPLE 4   Finding the Median and the Mode

Find the median and the mode of the data set $\{21, 18, 28, 21, 27, 24, 21, 26, 21, 28, 27, 29\}$.

Solution   Begin by reordering the numbers from smallest to largest.

$$\{18, 21, 21, 21, 21, 24, 26, 27, 27, 28, 28, 29\}$$

Median: $\dfrac{24 + 26}{2} = 25$ The two middle numbers are 24 and 26.

Mode: 21   21 occurs on the list four times.

Depending on the nature of the data, the three measures of central tendency—mean, median, and mode—might all be equal or no two might be equal. Generally, there is no correlation among these three numbers.

Finding the Mean from a Frequency Table

When there are repeated values in a data set, it can be helpful to complete a frequency table first and then calculate the mean using the following formula, which is equivalent to that in Definition 12.2.

$$
\begin{array}{ll}
\text{Items of data:} & x_1 \quad x_2 \quad \cdots \quad x_n \\
\text{Frequency:} & f_1 \quad f_2 \quad \cdots \quad f_n
\end{array}
$$

Calculating the Mean from a Frequency Table

If data are arranged in a frequency table like the one above, the **mean** $\overline{x}$ can be found using the following equation:

$$
\overline{x} = \frac{x_1 f_1 + x_2 f_2 + \cdots + x_n f_n}{f_1 + f_2 + \cdots + f_n} = \frac{\displaystyle\sum_{i=1}^{n} x_i f_i}{\displaystyle\sum_{i=1}^{n} f_i}
$$

E X A M P L E   5   Finding the Mean

Find the mean of the following data:

$$
\begin{array}{ll}
\text{Data:} & 5 \quad 6 \quad 9 \quad 11 \\
\text{Frequency:} & 2 \quad 3 \quad 5 \quad 2
\end{array}
$$

Solution

$$
\begin{aligned}
\overline{x} &= \frac{(5)(2) + (6)(3) + (9)(5) + (11)(2)}{2 + 3 + 5 + 2} \\
&= \frac{95}{12} = 7.91\overline{6}
\end{aligned}
$$

Line Graphs

The histogram in Fig. 12.4 shows the salaries of the ABC Company in 1992. The mean of these salaries is 27.1 thousand dollars. Mean salaries for other years are

TABLE 12.4 Average Salary for ABC Company for the Past 6 yr

Year	Salary (in thousands)
1987	21.4
1988	22.7
1989	24.6
1990	25.8
1991	26.3
1992	27.1

Notes on Examples

The scale on the horizontal axis on the line graph in Example 6 represents time. When time is one of the varying factors, a line graph is usually the proper way to represent the data.

TABLE 12.5 Life Expectancy in the United States

Year	Male	Female
1920	53.6	54.6
1930	58.1	61.6
1940	60.8	65.2
1950	65.6	71.1
1960	66.6	73.1
1970	67.1	74.7
1980	70.0	77.5
1990	71.8	78.5

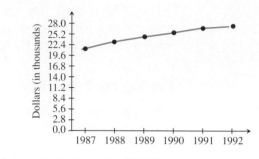

Figure 12.7 Line graph of salaries for ABC Company.

shown in Table 12.4. The line graph in Fig. 12.7 provides a visual understanding of how the data vary over time. Notice that the horizontal scale is *time* in years and the vertical scale is *dollars*.

To complete this line graph, first draw a pair of perpendicular lines, then draw the time scale on the horizontal axis and the dollar scale on the vertical axis. The vertical scale must begin at zero. For each year, plot a point on the graph the appropriate distance above the horizontal scale as indicated by the data. These data points are joined by a line to obtain the line graph.

Compare the difference between a histogram and a line graph. In a histogram, the data groupings, as indicated in the stem-and-leaf table, determine the width of the bars. Often this scale is horizontal. In a line graph, the vertical scale relates to the data values and the horizontal scale shows the time.

E X A M P L E 6 Completing a Line Graph

Using the data given in Table 12.5, complete a line graph showing the life expectancy trend for males in the United States from 1920 to 1990.

Solution (See Fig. 12.8.) Draw and label the perpendicular axes and plot the data points.

Figure 12.8 Male life expectancy in the United States.

Exercises for Section 12.3

In Exercises 1–3, refer to Table 12.1 and the stem-and-leaf table constructed for that data. In each case, determine if it is easier to answer the question by referring to Table 12.1 or to the stem-and-leaf table.

1. What is the lowest salary for the ABC Company?

2. How many pairs of employees of the ABC Company have identical salaries?

3. What is the highest salary for the ABC Company?

4. Complete a stem-and-leaf table for the female life expectancy data in Table 12.2.

5. Which country listed in Table 12.2 has the greatest life expectancy for females? Is it easier to answer this question by looking at Table 12.2 or at the stem-and-leaf table completed in Exercise 4?

In Exercises 6–11, refer to the following data. The salaries of the workers in one department of Smith Brothers Company (given in thousands of dollars) are as follows:

33.5, 35.3, 33.8, 29.3, 36.7, 32.8, 31.7, 36.3, 33.5, 28.2,

34.8, 33.5, 35.3, 29.7, 38.5, 32.7, 34.8, 34.2, 31.6, 35.4

6. Complete a stem-and-leaf table for this data set.

7. After completing Exercise 6, create a frequency table for the data.

8. After completing Exercises 6 and 7, complete, by hand, a histogram for the data.

9. What viewing window would you choose to generate this histogram on a grapher?

10. Find the histogram on a grapher for these data.

11. Why does a line graph not work well for this data?

In Exercises 12–17, refer to the following data. The average wind speeds for 1 yr at 44 climatic data centers around the United States were as follows:

9.0, 6.9, 9.1, 9.2, 10.2, 12.5, 12.0, 11.2, 12.9, 10.3, 10.6,

10.9, 8.7, 10.3, 11.0, 7.7, 11.4, 7.9, 9.6, 8.0, 10.7, 9.3,

7.9, 6.2, 8.3, 8.9, 9.3, 11.6, 10.6, 9.0, 8.2, 9.4, 10.6,

9.5, 6.3, 9.1, 7.9, 9.7, 8.8, 6.9, 8.7, 9.0, 8.9, 9.3

12. Use the stems 6, 7, 8, 9, 10, 11, and 12 to complete a stem-and-leaf table for these data.

13. After completing Exercise 12, create a frequency table for the data.

14. After completing Exercises 12 and 13, build, by hand, a histogram for the data.

15. What viewing window would you choose to generate this histogram on a grapher?

16. Find the histogram on a grapher for these data.

17. Why does a line graph not work well for this data?

In Exercises 18–29, find the mean and mode of the specified data set. Unless told otherwise, use a grapher.

18. Calculate by hand the mean of the data set {12, 23, 15, 48, 36}.

19. Calculate by hand the mean of the data set {4, 8, 11, 6, 21}.

20. Find the mean of the data set {32.4, 48.1, 85.3, 67.2, 72.4, 55.3}.

21. Find the mean of the data set in the stem-and-leaf table in Example 1.

22. Find the mean salary for the salaries given in Table 12.1.

23. Find the mean of the data in the following frequency table:

Data values:	15	12	8	21
Frequency:	3	2	5	4

24. Find the mean of the data in the following frequency table:

Data values:	33	29.3	41	28.9
Frequency:	2	4	8	5

25. Find the median of the data set in Exercise 18.

26. Find the median of the data set in Exercise 20.

27. Find the median of the data set represented by the stem-and-leaf table in Example 1.

28. Find the mode of the data set in Exercise 23.

29. Find the mode of the data set in Exercise 24.

In Exercises 30–32, complete line graphs of the given data.

30. Use the data in Table 12.5 to complete a line graph showing the life expectancy trends for females in the United States.

31. A certain company has had profits (in millions) as indicated in the following table:

Year	1986	1987	1988	1989	1990	1991
Profit	12.3	14.5	18.2	16.4	19.2	21.3

Complete a line graph that shows this company's profit trend.

32. The total cost (tuition, room, and board) of attending a certain private university is given in the following table:

Year	1986	1987	1988	1989	1990	1991
Costs	8400	8900	9525	9900	10,400	11,020

Complete a line graph that shows the cost trends of attending this university.

33. Writing to Learn Write a paragraph that explains how the frequency table in Table 12.3 differs from the frequency table in Example 5. Why do the data in one table lend themselves to a histogram while the others do not?

34. List a set of data for which the

$$mode < median < mean.$$

35. List a set of data for which the

$$median < mean < mode.$$

36. List a set of data for which the

$$mean < mode < median.$$

12.4

Box-and-whisker Plots

In Section 12.3, stem-and-leaf tables, histograms, and line graphs were introduced as methods for providing visual interpretation of sets of data. In this section, we introduce another method called a **box-and-whisker plot**.

The stem-and-leaf table that follows reflects the actual scores on a math test in a Calculus II course.

Scores on a Math Test

Stem	Leaf
5	1 1 2 8
6	1 2 3 6
7	2 4 4 9
8	0 3 4 8
9	4 8
10	0 5
11	0

From this stem-and-leaf table, we see that there were 21 students in the class and the median score was $M = 74$.

Completing a Box-and-whisker Plot

To complete a box-and-whisker plot, we first need to find the upper and lower quartile. The **upper quartile**, Q_U, is the median of those scores greater than the median 74, and the **lower quartile**, Q_L, is the median of those scores less than the median 74.

For these test scores, $Q_L = 61.5$ and $Q_U = 91$. Figure 12.9 shows these values plotted above a number line.

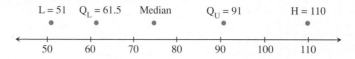

Figure 12.9 The first step in completing a box-and-whisker plot.

GRAPHER UPDATE

Some graphers are capable of producing box-and-whisker plots. Consult your grapher's owner's manual to determine the capability of your grapher.

To complete a box-and-whisker plot, draw a box extending from the lower quartile to the upper quartile, with a vertical line through the median. Then draw the whiskers from the box to L and H. (See Fig. 12.10.)

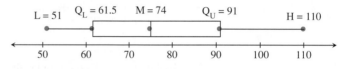

Figure 12.10 A completed box-and-whisker plot.

The difference between the highest and lowest values, $H - L$, is called the **range** of the data. Notice that about half of the data falls in the range of the box. The data greater than Q_U or less than Q_L are **outliers** and the values represented by the box constitute the **interquartile range.**

EXAMPLE 1 Completing a Box-and-whisker Plot

MALE LIFE EXPECTANCY DATA

Stem	*Leaf*
66	2
67	3 1
68	4 4
69	
70	5 1
71	8 8 9
72	1 0 2 0 9 8 5
73	8 9
74	2
75	
76	0

Complete a box-and-whisker plot for the male life expectancy data shown in the stem-and-leaf table in Example 1, Section 12.3, and repeated here.

Solution Using the data from the stem-and-leaf plot, we can conclude that $L = 66.2$, $H = 76.0$, $M = 72.0$, $Q_L = 69.25$, and $Q_U = 72.85$. Figure 12.11 shows a completed plot.

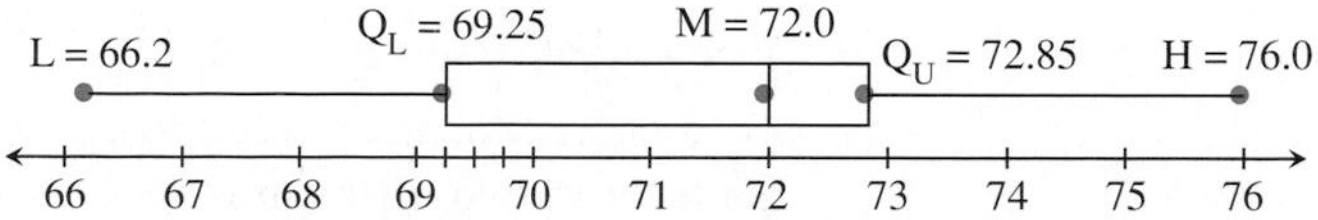

Figure 12.11 A box-and-whisker plot for the male life expectancy data.

Comparing Box-and-whisker Plots

When the box-and-whisker plots of two related data sets are compared, information is immediately visible. For example, in Fig. 12.12 the life expectancy data are compared for the 21 countries listed in Table 12.2.

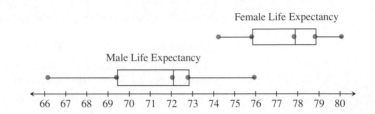

Figure 12.12 A comparison between the box-and-whisker plots for male and female life expectancy data listed in Table 12.2.

The following comparisons can be made through a visual inspection of these plots:

For the list of countries in Table 12.2,

- the male life expectancy range is nearly 10 yr, whereas the female's is about 6 yr;
- the male life expectancy interquartile range spans nearly 4 yr, while that for females spans about 3 yr;
- there is no overlap of the interquartile boxes for males and for females;
- the male life expectancy outliers have a greater range than that for females; and
- the male highest life expectancy is about 2 yr greater than the female lowest life expectancy.

None of these observations are obvious by a visual inspection of the raw data in Table 12.2 in Section 12.3.

E X A M P L E 2   Comparing Box-and-whisker Plots

Students in two different classes are given a nationally standardized test. The top 20 percentile scores in each class are recorded on the following stem-and-leaf table. Complete box-and-whisker plots for each class.

Leaf of Class 1	Stem	Leaf of Class 2
7 8	9	4 3 1
4 6 7 8 9 9	8	9 9 8 7 7 6 4 3
3 4 7 8	7	9 8 7 4
2 3 4 5 6 9	6	8 5 4
7 1	5	9 5

Solution

For Class 1: $L = 51$, $H = 98$, $M = 75.5$, $Q_L = 64.5$, and $Q_U = 87.5$

For Class 2: $L = 55$, $H = 94$, $M = 83.5$, $Q_L = 71$, and $Q_U = 88.5$

Figure 12.13 shows the completed plots.

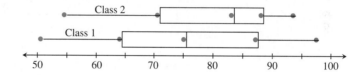

Figure 12.13   Box-and-whisker plots of test scores.

It often is easy to make comparisons between groups by inspecting their box-and-whisker plots. Example 3 asks for comparisons between Class 1 and Class 2 using the plots in Fig. 12.13.

E X A M P L E   3   Comparing Box-and-whisker Plots

Examine Fig. 12.13 to answer the following questions:

1. Which class has the student with the highest percentile score?
2. Which class has the student with the lowest percentile score?
3. Which class has the most students above the 70th percentile?
4. Which class would you say has the strongest students?
5. Which class has the greatest diversity in scores?

Solution

1. Class 1
2. Class 1
3. Class 2. The top three quarters, or 15 students, are above the 70th percentile.

4. Class 2. We can see that the top half of the students are above the 83rd percentile and the top three quarters are above the 70th percentile. For Class 1, both groups have lower percentiles.

5. Class 1

≡

Teaching Note

Make sure students understand the difference between the concepts of *range* and *spread* for a set of data. Figure 12.14 can be used to discuss the difference between these concepts. In this figure the range is equal but the spread is not. The box-and-whisker plots permit visual comparisons of spread. The standard deviation is a numerical measure of spread.

Measuring the Spread of Data

Figure 12.14 shows two data sets with equal ranges and medians. It is visually clear that the data in Data Set 2 are more dispersed; therefore we can say that Data Set 2 has a **greater spread** than does Data Set 1. Correspondingly, Data Set 1 has a **smaller spread** than does Data Set 2.

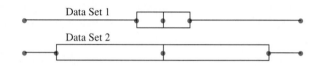

Data Set 1

Data Set 2

Figure 12.14 Two data sets with equal ranges and medians.

A **statistic** is a number that describes some characteristic of a data set, for example, the mean and median. The goal of this section is to develop a statistic that measures the spread of a data set. We need a way to measure the extent to which Data Set 2 in Fig. 12.15 has a greater spread than does Data Set 1.

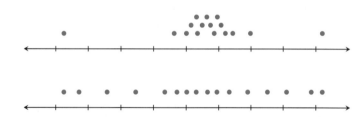

Figure 12.15 Two data sets with equal ranges but different spreads.

Two statistics that measure this spread are the **variance** and the **standard deviation**. To find the variance, calculate the distance that each data point is from the mean. For example, in Fig. 12.16 the difference $\bar{x} - x_7$ is shown. This difference is calculated for each data point.

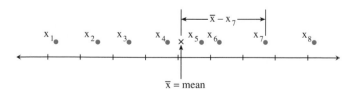

Figure 12.16 The difference between the mean and a data point.

Notice that the difference $\bar{x} - x_7$ is a negative number. In fact, for the eight numbers $x_i - \bar{x}$, $1 \le i \le 8$, four are positive and four are negative. Their sum therefore would be nearly zero, which would indicate, falsely, that there is not much spread. To prevent this false conclusion, we use the *square* $(x_i - \bar{x})^2$ of each difference. Dividing this sum of squares by n results in the measure called the **variance**. The **standard deviation** of a data set is the square root of the variance.

Definition 12.4   Variance and Standard Deviation

The **variance**, denoted σ^2, and the **standard deviation**, denoted σ, of a data set $\{x_1, x_2, \cdots, x_n\}$ are defined as follows:

$$\text{Variance} = \sigma^2 = \frac{(x_1 - \bar{x})^2 + (x_2 - \bar{x})^2 + \cdots + (x_n - \bar{x})^2}{n}$$

$$\text{Standard deviation} = \sigma = \sqrt{\text{variance}} = \sqrt{\frac{\displaystyle\sum_{i=1}^{n}(x_i - \bar{x})^2}{n}}$$

Calculating the variance and standard deviation by hand is tedious. It is natural to use a calculator or a grapher for this calculation, which we do in the next example.

E X A M P L E   4   Calculating the Variance and Standard Deviation

Use a grapher to find the mean and standard deviation of the data set $\{12, 17, 18, 24, 18, 25, 19, 23\}$.

Solution   Use the STAT menu on a grapher to enter the data set and find the 1-Variable statistics. On the grapher we use, the output display is as shown in Fig. 12.17.

```
1-Var
x̄ = 19.5
Σx = 156
Σx² = 3172
Sₓ = 4.3094589037
σₓ = 4.031128874
n = 8
```

Figure 12.17   Grapher output.

Assignment Guide

Day 1: Ex. 1–17
Day 2: Ex. 18–33

Line 1, $\bar{x} = 19.5$, and line 5, $\sigma = 4.031128874$, are the values to be found. The mean of the data set is $\bar{x} = 19.5$ and the standard deviation is $\sigma_x = 4.031128874$.

≡

Exercises for Section 12.4

In Exercises 1–6, refer to the ABC Company salaries listed in Section 12.3. The stem-and-leaf table for these data is repeated here.

Stem	Leaf
21	7 4 8
22	3 9 8
23	4 7 4 8
24	5 8
25	1 6 7 6
26	8 7
27	8 4 9
28	3 2 2 5 6 8 7 3
29	7 9 1 5 6 7
30	6 8
31	7 6 5 7 6

1. Find the high, low, median, and upper and lower quartiles—H, L, M, Q_L, and Q_U—for these data.

2. Complete a box-and-whisker plot for these data.

3. What is the range of these data?

4. What is the interquartile range of these data?

5. Is there a greater spread among the upper outliers or the lower outliers?

6. Is there a greater difference between the top salary and median salary, or between the median salary and lowest salary?

7. Do the outliers account for more than half, less than half, or about half of the range?

In Exercises 8–11, refer to the Smith Brothers salary data, repeated below, first introduced in Exercises 6–11 in Section 12.3.

33.5, 35.3, 33.8, 29.3, 36.7, 32.8, 31.7, 36.3, 33.5, 28.2,

34.8, 33.5, 35.3, 29.7, 38.5, 32.7, 34.8, 34.2, 31.6, 35.4

8. Find the high, low, median, and upper and lower quartiles—H, L, M, Q_L, and Q_U—for these data.

9. Complete a box-and-whisker plot for these data.

10. What is the range of these data?

11. Do the outliers account for more than half or less than half of the range?

In Exercises 12–17, refer to the average wind speeds at 44 climatic data centers around the United States that were first introduced in Exercises 12–17 in Section 12.3.

9.0, 6.9, 9.1, 9.2, 10.2, 12.5, 12.0, 11.2, 12.9, 10.3, 10.6,

10.9, 8.7, 10.3, 11.0, 7.7, 11.4, 7.9, 9.6, 8.0, 10.7, 9.3,

7.9, 6.2, 8.3, 8.9, 9.3, 11.6, 10.6, 9.0, 8.2, 9.4, 10.6,

9.5, 6.3, 9.1, 7.9, 9.7, 8.8, 6.9, 8.7, 9.0, 8.9, 9.3

12. Find the high, low, median, and upper and lower quartiles—H, L, M, Q_L, and Q_U—for these data.

13. Complete a box-and-whisker plot for these data.

14. What is the range of these data?

15. Do the outliers account for more than half or less than half of the range?

16. Some wind turbine generators, to be efficient generators of power, require average wind speeds of at least 10.5 mph. Approximately what fraction of the climatic centers are suited for these wind turbine generators?

17. If technology improves the efficiency of the wind turbines so that they are efficient in winds that average at least 7.5 mph, approximately what fraction of the climatic centers are suited for these improved wind turbine generators?

In Exercises 18–22, refer to the data in the following stem-and-leaf table. The top 20 students in two additional classes took the nationally standardized test referred to in Example 2.

Leaf of Class 3	Stem	Leaf of Class 4
	90	3
1 2 4 6	80	4 7
1 4 5	70	1 8
1 4 5 6 6	60	2 3 5 7
3 4 8 9	50	1 4 6 7 8
3 4 8 9	40	2 3 3 5 7 8

18. Find the high, low, median, and upper and lower quartiles—H, L, M, Q_L, and Q_U—for both data sets.

19. Complete box-and-whisker plots for both of these classes.

20. Which class has the greater range?

21. Which class has the larger interquartile range?

22. Writing to Learn Write a paragraph explaining which of these two classes you think is the stronger.

In Exercises 23–25, refer to the following box-and-whisker plots:

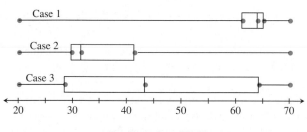

For Exercises 23–25.

23. Which plot is most likely to describe the salaries for the 100 employees of a manufacturing plant? Defend your answer.

24. Which plot is most likely to describe the salaries for the attendees of a 25-yr college class reunion? Defend your answer.

25. Which plot is most likely to describe the salaries for all employees of an urban school system? Defend your answer.

Exercises 26–28 refer to the following box-and-whisker plots:

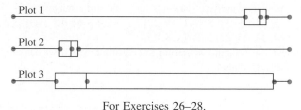

For Exercises 26–28.

26. Write a data set whose box-and-whisker plot closely resembles Plot 1.

27. Write a data set whose box-and-whisker plot closely resembles Plot 2.

28. Write a data set whose box-and-whisker plot closely resembles Plot 3.

29. Box-and-whisker plots also can be shown vertically. If you were the coach of a softball team, would you prefer your team's batting averages to look like Plot A, B, or C?

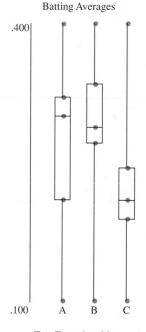

For Exercise 29.

30. Find the standard deviation of the data set $\{23, 45, 29, 34, 39, 41, 19, 22\}$.

31. Find the standard deviation of the data set $\{28, 84, 67, 71, 92, 37, 45, 32, 74, 96\}$.

32. The following two data sets have the same range:
{23, 45, 29, 34, 39, 41, 19, 22}
{19, 23, 25, 24, 29, 27, 28, 45}
Use the standard deviation to determine which set has the greatest spread.

33. The following two data sets have the same range:
{61, 72, 28, 59, 63, 96, 68, 57, 62, 71}
{28, 84, 67, 71, 92, 37, 45, 32, 74, 96}
Use the standard deviation to determine which set has the greatest spread.

12.5 _____ Scatter Plots and Least-squares Lines

Research workers often investigate whether there is a relationship between two given quantities; that is, as one quantity changes in value, does the other one change in some predictable way?

We have seen in Sections 12.3 and 12.4 that when graphs are used, relationships become evident visually. In this section, a graph known as a **scatter plot** is used to determine whether there is a relationship between two quantities.

Scatter Plots

A group of children had their heights measured and then recorded with their age in months, as follows:

Age (in months)	6	9	11	15	25	33	27	46	52	48
Height (in inches)	23	26	17	23	33	36	32	44	43	39

To determine whether there is a relationship between these two quantities, we plot the data on an age/height graph. The horizontal axis represents age; the vertical axis represents height. Figure 12.18 shows the ordered pairs (age, height) plotted on a rectangular coordinate system.

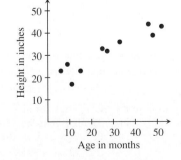

Figure 12.18 A scatter plot of the age/height data.

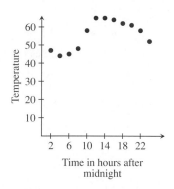

Figure 12.19 Scatter plot of time/temperature relationship at one weather station.

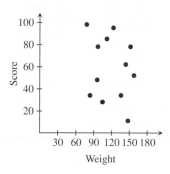

Figure 12.20 Scatter plot of weight/percentile score.

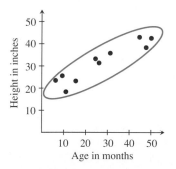

Figure 12.21 Scatter plot showing a linear correlation.

The graph shown in Fig. 12.18 is called a **scatter plot**. It is evident these data form a trend: As the age of a child increases, his or her height also increases.

E X A M P L E 1   Completing a Scatter Plot

The following data compare the number of hours after midnight and the air temperature at one weather station. Complete a scatter plot of this data.

Hours after midnight	2	4	6	8	10	12	14	16	18	20	22	24
Air temperature	47	44	45	48	58	65	65	64	62	61	58	52

Solution   The horizontal axis represents time, and the vertical axis represents temperature in degrees F. A completed scatter plot appears in Fig. 12.19.   ≡

We see in Fig. 12.19 that the time/air temperature correlation appears to be approximately sinusoidal, whereas the age/height correlation shown in Fig. 12.18 appears closer to a linear correlation. Obviously, then, there are different types of correlations or trends. Example 2 illustrates that sometimes there is very little correlation between two quantities.

E X A M P L E 2   Completing a Scatter Plot

The following data compare the weight for 12 randomly chosen students and their percentile score on a nationally standardized test. Complete a scatter plot of this data.

Weight	78	84	96	97	105	112	123	136	144	148	152	158
Percentile score	98	34	48	78	28	85	95	34	62	11	78	52

Solution   (See Fig. 12.20.)   ≡

Finding Linear Correlation in Scatter Plots

To determine whether there is a linear correlation between the quantities graphed in a scatter plot, draw an oval around the data as shown in Fig. 12.21. Draw the oval as narrowly as possible; the narrower the oval, the stronger the linear correlation. Because the oval in Fig. 12.21 has a positive slope, we say that the data have a **positive linear correlation.** When the oval has a negative slope, we say that the data have a **negative linear correlation.**

Fig. 12.22 shows several scatter plots and identifies the types of linear correlation that can occur.

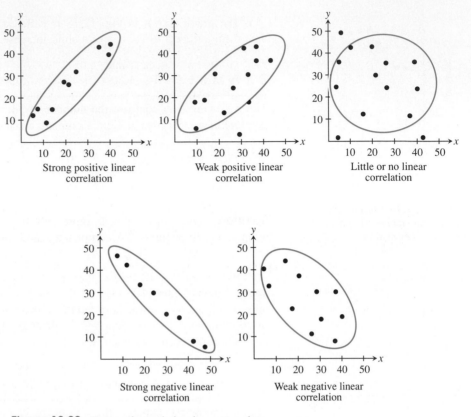

Figure 12.22 Types of correlation in scatter plots.

Example 3 asks that a data set be examined to determine the type of linear correlation, if any, that exists.

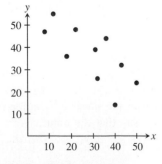

Figure 12.23 Scatter plot of data in Example 3.

E X A M P L E   3   Determining the Type of Linear Correlation

Determine by visual inspection the type of linear correlation, if any, that exists between the x and y quantities in the following data:

x	8	12	18	22	31	32	36	40	43	50
y	47	55	36	48	39	26	44	14	32	24

Solution   The scatter plot, graphed in Fig. 12.23, shows there is a weak negative linear correlation. ≡

The Line of Best Fit

When a scatter plot has a strong linear correlation, either positive or negative, we naturally want to seek a line that models the correlation. If the data points are the result of collecting data in an experiment, it is likely that experimental error explains the discrepancy between the obtained scatter plot and an actual line.

Consequently, we are interested in finding the straight line that is the best choice to represent this relationship, that is, the **line of best fit**. Figure 12.24 shows a scatter plot with a strong negative correlation and a line that has been drawn through the data points that is a candidate for the line of best fit.

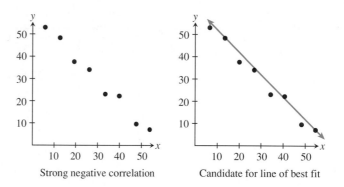

Strong negative correlation   Candidate for line of best fit

Figure 12.24 A scatter plot and a line through the data points.

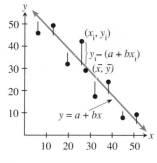

Figure 12.25 A graph of the line $y = a + bx$.

For any given set of data $\{(x_1, y_1), \cdots, (x_i, y_i), \cdots, (x_n, y_n)\}$, there is a unique line that is the line of best fit. How is it found?

Finding the Line of Best Fit

Suppose a scatter plot of a data set $\{(x_1, y_1), \cdots, (x_i, y_i), \cdots, (x_n, y_n)\}$ is drawn. We can use a theory called **linear regression** to find the straight line that is the best fit for the given data. The first requirement of the line of best fit is that it must pass through the point $(\overline{x}, \overline{y})$ whose coordinates are the means of the x- and y-coordinates, respectively, of the data points. Figure 12.25 shows a data set and a line through this point.

Many lines pass through the point $(\overline{x}, \overline{y})$. The line of best fit is the one line that minimizes the extent to which the data points deviate from this line.

If $y = a + bx$ is the equation of the line of best fit, then the difference

$$y_i - (a + bx_i),$$

sometimes called the **residual**, represents the extent to which the one data point (x_i, y_i) deviates from the line. To ensure the quantity is positive, use the square

$[y_i - (a + bx_i)]^2$ and consider the sum

$$\sum_{i=1}^{n} [y_i - (a + bx_i)]^2. \tag{1}$$

The line of best fit is the line whose slope makes the sum (1) as small as possible. Consequently, it is also called the **least-squares line**. In other words, the least-squares line is the line $y = a + bx$, where a and b are the values that make the sum (1) as small as possible.

It can be shown that the line of best fit, that is, the least-squares line, is the line described as follows.

Teaching Note

The solution to Example 4 shows how to find the line of best fit by paper and pencil methods. It is important for students to see how tedious the method is even for three data points. Ordinarily, we would expect students to use a grapher to find the line of best fit as illustrated in Example 5.

Least-squares Line

The **least-squares line** or the **line of best fit** for a set of n data pairs $\{(x_1, y_1), \cdots, (x_i, y_i), \cdots, (x_n, y_n)\}$ is the line described as follows:

1. The line passes through the point $(\overline{x}, \overline{y})$.
2. The slope b of the line is

$$b = \frac{\overline{xy} - \overline{x} \cdot \overline{y}}{\sigma_x^2} \qquad \text{or} \qquad b = \frac{(\sum x_i y_i)/n - \overline{x} \cdot \overline{y}}{\sigma_x^2},$$

where $\overline{xy}$ is the mean of the data set $\{x_1 y_1, x_2 y_2, \cdots, x_n y_n\}$ and σ_x^2 is the variance of the data set $\{x_1, x_2, \cdots, x_n\}$.

Common Errors

The paper and pencil method of Example 4 requires using the point-slope form of the line. However, the grapher output in Fig. 12.28 provides values a and b that are the constants in the slope-intercept form of the line. Students are sometimes confused by this difference.

In Example 4, the equation for the line of best fit is found for a simple set of data points.

EXAMPLE 4   Finding the Line of Best Fit

Find an equation of the line of best fit for the data points $\{(1, 2), (3, 4), (5, 9)\}$ and support the result with a grapher.

Solution   Begin by finding $\overline{x}$ and $\overline{y}$ for the data sets $x = \{1, 3, 5\}$ and $y = \{2, 4, 9\}$.

$$\overline{x} = \frac{1 + 3 + 5}{3} = 3 \qquad \overline{y} = \frac{2 + 4 + 9}{3} = 5$$

The line of best fit passes through the point $(3, 5)$. To find the slope b of this line, we first calculate $\overline{xy}$.

$$\overline{xy} = \frac{1 \cdot 2 + 3 \cdot 4 + 5 \cdot 9}{3} = \frac{59}{3}$$

REMINDER

When calculating the slope of the line of best fit, find the value of σ_x^2 on a statistics menu rather than use a rounded value.

Using a calculator, we find that $\sigma_x^2 = 2\frac{2}{3}$. Therefore the slope of the line of best fit is

$$b = \frac{\overline{xy} - \overline{x} \cdot \overline{y}}{\sigma_x^2} = \frac{59/3 - 3 \cdot 5}{2\frac{2}{3}} = 1.75.$$

Using the point-slope form, an equation of the line of best fit is

$$y - 5 = 1.75(x - 3).$$

See Fig. 12.26. ▤

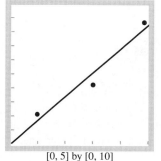

[0, 5] by [0, 10]

Figure 12.26 Equation of line of best fit: $y - 5 = 1.75(x - 3)$.

The calculations for the line of best fit are tedious. It is natural to use a grapher for these calculations. On the grapher we use, the menus in Fig. 12.27 are submenus of the VARS key and are helpful for these statistical calculations. Read your owner's manual to learn how statistical functions work on your grapher.

XY
1: n
2: $\overline{x}$
3: Sx
4: σx
5: $\overline{y}$
6: Sy
7: σy

Σ
1: Σ x
2: Σ x²
3: Σ y
4: Σ y²
5: Σ xy

LR
1: a
2: b
3: r
4: ReqEQ

Figure 12.27 Three grapher statistical menus that are helpful. Learn what each entry means.

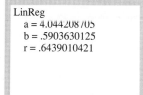

LinReg
a = 4.044208705
b = .5903630125
r = .6439010421

Figure 12.28 The number r is called the **correlation coefficient**. It is a measure of "the goodness of fit," that is, the closer r is to 1 or to -1, the better the fit.

In Example 4, it was most natural to use the point-slope form of the equation of the line of best fit because of the formal description of the least-squares line. On the other hand, information like that shown in Fig. 12.28 gives values for the slope and the y-intercept of the line $y = a + bx$. The value r is called the **correlation coefficient** and is a measure of the "goodness of fit," that is, the closer the r is to 1 or -1, the better the fit.

E X A M P L E   5   Finding the Line of Best Fit

For the following data set, use a grapher to
a) find an equation for the line of best fit for the data points, and
b) plot both the points and the line of best fit for the data points.

$\{(5, 4), (6, 8), (12, 15), (2, 5), (3, 6), (1.8, 6), (4.2, 6.8), (2.3, 7.9), (9, 5.2), (6.8, 7.3)\}$

Solution Use a grapher as follows:

1. Enter all the data points.
2. Find the linear regression menus as shown in Fig. 12.28. The equation $y = a + bx$ is the line of best fit when a and b have the values given in Fig. 12.28.

The equation $y = 4.044208705 + .5903630125x$ is an equation of the line of best fit. The grapher output is shown in Fig. 12.29.

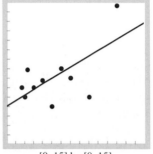

[0, 15] by [0, 15]

Figure 12.29 A scatter plot and line of best fit for the data in Example 5.

Logarithmic, Exponential, and Power Curves of Best Fit

If the linear correlation for the data is reasonably strong, as shown in Fig. 12.22, then the curve that best models the data is probably a straight line. However, suppose the scatter plot of the data points look like one of the plots in Fig. 12.30. In these cases, a nonlinear curve would be a better fit than would a linear one.

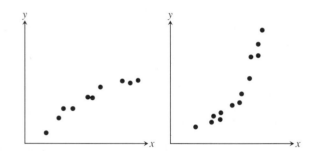

Figure 12.30 Two scatter plots whose curves of best fit are nonlinear.

Just as the linear regression facility on a grapher can be used to find the line of best fit (see Fig. 12.28), the grapher we use also allows us to find the logarithmic, exponential, or power function that models the data. We call this function the **curve of best fit**.

Model	Formula	Restrictions
Logarithmic	$y = a + b \ln x$	All $x > 0$
Exponential	$y = a \cdot b^x$	All $y > 0$
Power	$y = ax^b$	All $x > 0$ and $y > 0$

E X A M P L E 6 Finding the Curve of Best Fit

For the data set $\{(1, 0.5), (3, 2), (5, 6), (6, 12), (8, 15)\}$, use a grapher to find the equation of the curve of best fit for the following models:

a) Linear

b) Logarithmic

c) Exponential

d) Power

Solution Use a grapher to find the four regression models shown in Fig. 12.31.

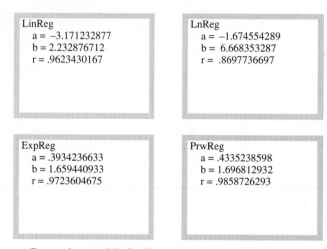

```
LinReg
  a = −3.171232877
  b = 2.232876712
  r = .9623430167
```

```
LnReg
  a = −1.674554289
  b = 6.668353287
  r = .8697736697
```

```
ExpReg
  a = .3934236633
  b = 1.659440933
  r = .9723604675
```

```
PrwReg
  a = .4335238598
  b = 1.696812932
  r = .9858726293
```

Figure 12.31 Regression models for linear, logarithmic, exponential, and power curves.

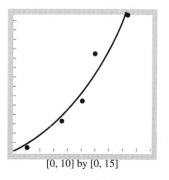

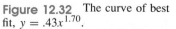
[0, 10] by [0, 15]

Figure 12.32 The curve of best fit, $y = .43x^{1.70}$.

Using the a and b values rounded to hundredths from the grapher output for each model, we find that the following functions are the curves of best fit for each of the four models:

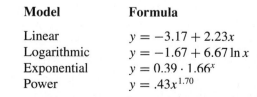

Model	Formula
Linear	$y = -3.17 + 2.23x$
Logarithmic	$y = -1.67 + 6.67 \ln x$
Exponential	$y = 0.39 \cdot 1.66^x$
Power	$y = .43x^{1.70}$

Of the four models shown, the correlation coefficient for the power model, $r = 0.9858726293$, is the closest to 1. We conclude that the power model curve, shown in Fig. 12.32, is the curve of best fit.

Relationship between Nonlinear and Linear Regression Models

Data pairs (x, y) that can be modeled as logarithmic, exponential, and power curves can be transformed into data pairs that satisfy a linear model. The relationships between the nonlinear data and corresponding linear data are as follows:

$$\text{logarithmic pair } (x, y) \longleftrightarrow \text{linear pair } (\ln x, y)$$

$$\text{exponential pair } (x, y) \longleftrightarrow \text{linear pair } (x, \ln y)$$

$$\text{power pair } (x, y) \longleftrightarrow \text{linear pair } (\ln x, \ln y)$$

For example, suppose a data set is modeled by the exponential curve $y = a \cdot b^x$. Observe the following result of taking the logarithm of both sides of this equation:

$$\ln y = \ln(a \cdot b^x)$$

$$\ln y = \ln a + \ln(b^x)$$

$$\ln y = \ln a + x \ln b$$

So the data pairs $(x, \ln y)$ satisfy the linear model $y = (\ln b)x + \ln a$.

Notes on Exercises

Ex. 9–14 do not require mathematical skills. They encourage students to develop judgments concerning when quantities may be correlated and when they may not.

Assignment Guide

Day 1: Ex. 1–14
Day 2: Ex. 15–22, 26–35

E X A M P L E   7   Converting to Linear Data

Transform the data of Example 6 to obtain a data set whose curve of best fit is a linear model.

Solution   Example 6 shows that the power model curve is the curve of best fit for the data set $\{(1, 0.5), (3, 2), (5, 6), (6, 12), (8, 15)\}$. Correspondingly, the curve of best fit for the data set

$$\{(\ln 1, \ln 0.5), (\ln 3, \ln 2), (\ln 5, \ln 6), (\ln 6, \ln 12), (\ln 8, \ln 15)\}$$

is a linear model.   ☰

Exercises for Section 12.5

In Exercises 1 and 2, complete a scatter plot of the given data.

1. A group of male children was weighed. Their individual ages and weights are recorded in the following data table. Complete a scatter plot of these data.

Age (in months)	18	20	24	26	27	29	34	39	42	48
Weight (in pounds)	23	25	24	32	33	29	35	39	44	42

2. A group of female children was weighed. Their individual ages and weights are recorded in the following data table. Complete a scatter plot of these data.

Age (in months)	19	21	24	27	29	31	34	38	43	47
Weight (in pounds)	22	23	25	28	31	28	32	34	39	41

Think Visually   In Exercises 3–5, state for the given scatter plot whether the correlation is positive or negative.

3.

4.

5.

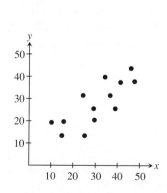

In Exercises 6–8, for the given scatter plot, identify the correlation as strong positive, weak positive, strong negative, weak negative, or little or no correlation.

6.

7.

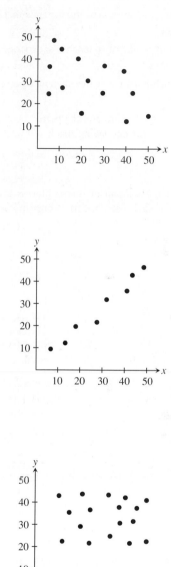

8.

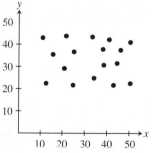

Writing to Learn   In Exercises 9–14, write a few sentences that explain why you think the correlation for each pair of variables will be strong, weak, or nonexistent.

9. The weight of a person and that person's running speed

10. A person's running speed and the distance that person can broad jump

11. The number of years of a teacher's experience and the teacher's salary

12. A student's hand span and the student's percentile score on the SAT exam

13. The length of a pendulum and the period of the pendulum (that is, the time for one swing, back and forth)

14. The force with which a ball is thrown and the speed of the ball

In Exercises 15 and 16, a pair of scatter plots is given. Determine which plot of each pair has the strongest correlation.

15.

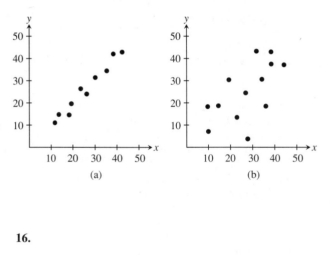

(a)   (b)

16.

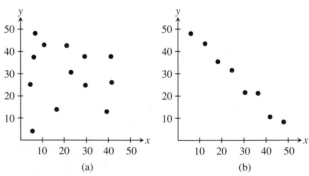

(a)   (b)

In Exercises 17–19, refer to the data set $\{(1, 3), (2, 4), (5, 9)\}$. Complete these exercises without using the STAT menu of a grapher.

17. Find the means $\bar{x}$ and $\bar{y}$ for the given data set.

18. Find the slope of the line of best fit for the given data set.

19. Find the equation for the line of best fit for the given data set.

In Exercises 20–22, refer to the data set $\{(2, 8), (3, 6), (5, 9), (6, 8), (8, 11), (10, 15), (12, 14), (15, 14)\}$ and use a grapher and its linear regression capabilities to do the following:

20. Find a scatter plot of these data in the viewing rectangle $[0, 15]$ by $[0, 15]$.

21. Find an equation of the line of best fit for these data.

22. Graph in the same viewing rectangle both the line of best fit found in Exercise 21 and the scatter plot.

In Exercises 23–25, refer to the data set $\{(1, 18), (3, 12), (5, 13), (6, 10), (8, 8), (10, 3), (12, 1), (15, 4)\}$ and use a grapher and its linear regression capabilities to do the following:

23. Find a scatter plot of these data in the viewing rectangle $[0, 15]$ by $[0, 20]$.

24. Find an equation of the line of best fit for these data.

25. Graph in the same viewing rectangle both the line of best fit found in Exercise 24 and the scatter plot.

In Exercises 26–35, refer to the data set $\{(.5, 1.3), (1.2, 3.5), (3.4, 5.2), (3.8, 4.1), (4.2, 6.5), (5.5, 7.3), (7.2, 9.1), (8.2, 6.9), (9.1, 8.7), (9.8, 9.3)\}$ and use a grapher and its regression capabilities to do the following:

26. Find a scatter plot of these data in the viewing rectangle $[0, 10]$ by $[0, 10]$.

27. Find the linear equation model (the line of best fit) for these data.

28. Find in the same viewing rectangle the graph of both this line of best fit and the scatter plot.

29. Find the logarithmic equation model (the logarithmic curve of best fit) for these data.

30. Find in the same viewing rectangle the graph of the logarithmic model and the scatter plot.

31. Find the exponential equation model (the exponential curve of best fit) for these data.

32. Find in the same viewing rectangle the graph of the exponential model and the scatter plot.

33. Find the power equation model for these data.

34. Find in the same viewing rectangle the graph of the power equation model and the scatter plot.

35. Determine whether the curve of best fit for these data is linear, logarithmic, exponential, or power.

In Exercises 36–38, refer to the data set {(1, 2.9), (2, 7.6), (3, 18.1), (4, 47.2)}.

36. Are these data best modeled by a linear, logarithmic, exponential, or power curve of best fit?

37. Graph in the same viewing window a scatter plot of these data and the curve of best fit.

38. Transform the given data to a related data set whose curve of best fit is a straight line.

Chapter 12 Review

KEY TERMS   (The number following each key term indicates the page of its introduction.)

Binomial Probability
  theorem, 669
box-and-whisker plot, 680
complement of an event, 663
correlation coefficient, 693
curve of best fit, 694
equally likely events, 660
event, 660
experiment, 659
experimental probability, 664
exponential model, 694

frequency table, 674
histogram, 674
independent events, 663
independent repeated
  trials, 666
interquartile range, 681
least-squares line, 692
line of best fit, 691
line graph, 678
linear correlation,
  positive and negative, 689

linear regression, 691
logarithmic model, 694
mean, 675
median, 676
mode, 676
Monte Carlo method, 670
mutually exclusive events, 662
outcomes, 659
outliers, 681
power model, 694
probability of an event, 661

quartile,
  upper and lower, 680
range, 681
residual, 691
sample space, 659
scatter plot, 688
standard deviation, 684
statistic, 684
stem-and-leaf table, 672
theoretical probability, 660
variance, 685

REVIEW EXERCISES

In Exercises 1–4, list the elements of the sample space for the probability experiment that is described.

1. A game spinner numbered 1 through 6 is spun.

2. A red die and a green die are rolled.

3. A two-digit code is selected from the digits {1, 3, 6}, where no digits are to be repeated.

4. A product is inspected as it comes off the production line and is classified as either defective or nondefective.

In Exercises 5–9, refer to the probability experiment of tossing a penny, nickel, and dime.

5. List all possible outcomes of this experiment.

6. List all outcomes in the event "two heads or two tails."

7. List all outcomes in the complement of the event in Exercise 6.

8. Find the probability of tossing three heads.

9. Find the probability of tossing at least one head.

10. If a fair coin is tossed six times, find the probability of the event "HHTHTT."

11. If a fair coin is tossed four times, find the probability of obtaining one head and three tails.

12. If a fair coin is tossed five times, find the probability of obtaining two heads and three tails.

13. In a random check of one item on an assembly line, the probability of finding a defective item is 0.003. Find the probability of a nondefective item occurring 10 times in a row.

14. A probability experiment has only two possible outcomes—success (S) and failure (F)—and repetitions of the experiment are independent events. If P(S) = 0.5, find the probability of obtaining three successes and one failure in four repetitions of the experiment.

15. In the experiment in Exercise 14, explain why the probability of one success and three failures is equal to the probability of three successes and one failure.

In Exercises 16–19, refer to the following experiment. A probability experiment has only two possible outcomes—success (S) and failure (F)—and repetitions of the experiment are independent events. The probability of success is 0.4.

16. Find the probability of SF on two repetitions of the experiment.

17. Find the probability of SFS on three repetitions of the experiment.

18. Find the probability of at least one success on two repetitions of the experiment.

19. Explain why the probability of one success and three failures is not equal to the probability of three successes and one failure.

20. Explain how the random number table on page 670 can be used to simulate tossing a dime and a nickel simultaneously.

21. Use the random number table on page 670 to simulate tossing a pair of coins 50 times. Use this simulation to find the experimental probability of HH when two coins are tossed simultaneously.

22. Use the random number generator on a grapher to simulate tossing a coin 50 times. Find the experimental probability of tossing a head. (Use 50 repeated trials.)

23. Use the random number generator on a grapher to find the experimental probability of rolling a six with a die. (Use 50 repeated trials.)

In Exercises 24–27, find the theoretical probabilities of the following events. Assume all the coins are fair and have two distinct sides.

24. Find the probability of obtaining exactly two heads in a toss of four coins.

25. Find the probability of obtaining exactly three heads in a toss of five coins.

26. Find the probability of obtaining exactly four tails in a toss of nine coins.

27. Find the probability of obtaining exactly two heads in a toss of nine coins.

28. Suppose a fair coin with two distinct sides is tossed 50 times and each time it shows a head. What is the probability that the 51st toss will result in a head? How likely is obtaining no tails (all heads) on a toss of 50 coins? Be specific.

29. Suppose the probability of producing a defective bat is 0.02. Four bats are selected at random. What is the probability that the lot of four bats contains the following:
a) No defective bats?
b) One defective bat?

30. Suppose the probability of producing a defective light bulb is 0.0004. Ten light bulbs are selected at random. What is the probability that the lot of 10 contains the following:
a) No defective light bulbs?
b) Two defective light bulbs?

In Exercises 31–35, refer to the data set {12, 16, 25, 28, 15, 13, 15, 12, 15, 19, 17, 18, 21, 14, 13, 15} and find the following:

31. The mean of these data

32. The mode of these data

33. The median of these data

34. The variance of these data

35. The standard deviation of these data

In Exercises 36–40, refer to the data in the following frequency table:

Data:	7	12	15	18
Frequency:	4	2	3	3

Find the following:

36. The mean of these data

37. The median of these data

38. The mode of these data

39. The variance for these data

40. The standard deviation for these data

In Exercises 41–44, use the following salary data (given in thousands) for the employees of a certain company:

$$\{28.7, 29.2, 31.5, 36.7, 30.5, 31.7, 29.4,$$

$$31.2, 29.8, 32.1, 31.7, 29.9, 31.2, 30.7,$$

$$32.6, 33.5, 34.8, 33.9, 32.4, 28.9, 30.4, 31.8,$$

$$30.7, 30.8, 32.9\}$$

41. Complete a stem-and-leaf table for these data.

42. Complete a histogram for these data.

43. Complete a box-and-whisker plot for these data.

44. Suppose each person receives a 6% salary increase. Complete a stem-and-leaf table for these new data.

45. A weather station daily records the temperature every 15 min. If you are to graph the data for a 24-hr period, would you use a histogram, a line graph, or box-and-whisker graph? Explain your choice.

46. Complete a stem-and-leaf table for the salary data in Table 12.1.

47. Writing to Learn   Write a paragraph comparing the performance of two classes on a test. Which class seems to have the strongest second quartile? The strongest third quartile? Which class would you choose to send to an academic competition?

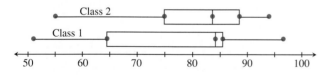

In Exercises 48–54, refer to the following data pertaining to a group of children:

Height (in inches)	24	23	26	27	24	31	29	33	34
Weight (in pounds)	35	38	37	48	58	65	69	64	62
Age (in months)	15	18	20	23	27	32	36	42	48

48. Complete a (height, weight) scatter plot.

49. Complete a (age, height) scatter plot.

50. Complete a (weight, age) scatter plot.

51. Find the least-squares line for the scatter plot in Exercise 48.

52. Find the least-squares line for the scatter plot in Exercise 49.

53. Find the least-squares line for the scatter plot in Exercise 50.

54. Is the curve of best fit for the data in Exercise 50 a linear, logarithmic, exponential, or power curve?

In Exercises 55–58, refer to the data set $\{(1, 1.3), (2, 4), (3, 8.2), (4, 14.3)\}$.

55. Is the curve of best fit for these data a logarithmic, exponential, or power curve?

56. Transform the data to a data set whose curve of best fit is a least-squares line.

57. Suppose $y = 2.3(4.1)^x$ is the curve of best fit for a data set (x, y). Find the equation for the line of best fit for the data $(x, \log y)$.

58. Suppose $y = 1.4x^3$ is the curve of best fit for a data set (x, y). Find the equation for the line of best fit for the data $(\log x, \log y)$.

DAVID BLACKWELL

David Blackwell (b. 1919) is a noted professor of statistics at the University of California, Berkeley. He graduated from high school in Centralia, Illinois, at age 16 and then entered the University of Illinois. Within six years he had not only received his baccalaureate degree, but his Ph.D. in mathematics as well, and had become the first African-American to receive a fellowship to the Institute for Advanced Study at Princeton. Professor Blackwell has made contributions to Bayesian statistics, game theory, set theory, information theory, probability, and dynamic programming.

Review of Basic Concepts

A.1   Graphing $y = x^n$ and Properties of Exponents

Complete an experiment: Graph both of the following equations, one after another, on the same coordinate system:

$$y = (x + 3)(x - 2) \quad \text{and} \quad y = x^2 + x - 6.$$

Notice that the grapher seems to be drawing only one graph. As it draws the second graph, it retraces the graph of the first equation. This exercise demonstrates visually that the two equations may be equivalent.

Complete the following Exploration and formulate generalizations about equations that may be equivalent.

🔎 EXPLORE WITH A GRAPHING UTILITY

On the same coordinate system and viewing rectangle, graph each of the following pairs of equations:

- $y = x(x^2)$ and $y = x^3$
- $y = (x^3)(x^2)$ and $y = x^5$
- $y = x^5 \div x^3$ and $y = x^2$

Study the patterns and generalize. Write some equations of your own that you think might be equivalent and check your guess with a graphing utility.

Questions

1. Which pairs of equations appear to be equivalent?
2. What generalizations can you make regarding exponents?

Exploring with the graphing utility can show visually some relationships about exponents in equations, for example, as in $(x^2)(x^3) = x^5$. On the other hand, it is important to realize that the properties of exponents discussed in this section are not true because of evidence produced by the grapher; they are true for logical reasons, as discussed in this section. The grapher is not automatically intelligent; it is programmed by engineers to reflect mathematical properties. Furthermore, although two graphs may appear to be the same, they are not necessarily the same.

Positive Integer Exponents

An exponent is a notational convenience—a type of mathematical shorthand. It is conventional to agree that when n is a positive integer, a^n means that n factors of a have been multiplied together.

Definition A.1 Exponential Notation

Let n be a positive integer. Then we read a^n as **a to the nth power** and

$$a^n = \underbrace{a \cdot a \cdot \cdots \cdot a}_{n \text{ factors}},$$

where n is called the **exponent** and a is called the **base**.

We use this definition to find an equivalent form for an expression like $(a^3)(a^5)$.

$$(a^3)(a^5) = \underbrace{a \cdot a \cdot a}_{3 \text{ factors}} \cdot \underbrace{a \cdot a \cdot a \cdot a \cdot a}_{5 \text{ factors}} = \underbrace{a \cdot a \cdot a \cdot a \cdot a \cdot a \cdot a \cdot a}_{8 \text{ factors}} = a^8$$

Verify this relationship for positive-integer exponents other than 3 and 5 and discover the following generalization:

$$(a^n)(a^m) = a^{n+m}.$$

Notice that for the following equation to be true, a^0 must be interpreted as 1:

$$a^{n+0} = a^n a^0.$$

Likewise, each of the following properties of exponents can be shown to be true.

Properties of Exponents

Let a, b, x, and y be real numbers and n and m be positive integers. Then the following properties are true:

Properties	**Example**
1. $a^m a^n = a^{m+n}$	$5^3 5^6 = 5^{3+6} = 5^9$
2. $\dfrac{a^n}{a^m} = a^{n-m}$	$\dfrac{x^9}{x^4} = x^{9-4} = x^5$
3. $a^0 = 1$	$8^0 = 1$
4. $(ab)^m = a^m b^m$	$(2y)^5 = 2^5 y^5 = 32y^5$
5. $(a^m)^n = a^{mn}$	$(u^2)^3 = u^{2\cdot3} = u^6$
6. $\left(\dfrac{a}{b}\right)^m = \dfrac{a^m}{b^m}$	$\left(\dfrac{x}{y}\right)^7 = \dfrac{x^7}{y^7}$

E X A M P L E 1   Simplifying Expressions with Positive Exponents

Simplify the following expressions:

a) $(2ab^3)(5a^2 b^5)$

b) $\left(\dfrac{x}{2}\right)^2 (2y)^3$

c) $\dfrac{(u^2v)^3}{v^2}$

d) $\left(\dfrac{x^3}{y^2}\right)^4$

Solution

a) $(2ab^3)(5a^2b^5) = 10\,(a \cdot a^2)(b^3 \cdot b^5) = 10a^3b^8$

b) $\left(\dfrac{x}{2}\right)^2(2y)^3 = \dfrac{x^2}{2^2}(2^3y^3) = 2x^2y^3$

c) $\dfrac{(u^2v)^3}{v^2} = \dfrac{(u^2)^3v^3}{v^2} = \dfrac{u^6v^3}{v^2} = u^6v$

d) $\left(\dfrac{x^3}{y^2}\right)^4 = \dfrac{(x^3)^4}{(y^2)^4} = \dfrac{x^{12}}{y^8}$ ∎

Negative-integer Exponents

If we agree that $a^{-n} = 1/a^n$ when n is a positive integer, then the six properties of exponents remain true when the exponent is a negative integer. The next example illustrates this fact.

E X A M P L E   2   Simplifying Expressions with Negative Exponents

Simplify the following expressions:

a) $\dfrac{x^2}{x^{-3}}$

b) $\dfrac{u^2v^{-2}}{u^{-1}v^3}$

c) $[(4)^{-2}]^3$

d) $\left(\dfrac{x^2}{2}\right)^{-3}$

Solution

a) $\dfrac{x^2}{x^{-3}} = x^2 \cdot \dfrac{1}{x^{-3}} = x^2 \cdot x^3 = x^{2+3} = x^5$

b) $\dfrac{u^2v^{-2}}{u^{-1}v^3} = \dfrac{u^2u^1}{v^2v^3} = \dfrac{u^{2+1}}{v^{2+3}} = \dfrac{u^3}{v^5}$

c) $[(4)^{-2}]^3 = 4^{(-2)(3)} = 4^{-6} = \dfrac{1}{4^6} = \dfrac{1}{4096}$

d) $\left(\dfrac{x^2}{2}\right)^{-3} = \dfrac{(x^2)^{-3}}{2^{-3}} = \dfrac{x^{-6}}{2^{-3}} = \dfrac{2^3}{x^6} = \dfrac{8}{x^6}$

≡

Rational-number Exponents

What does the symbol $a^{1/2}$ mean? This notation has been defined so that the properties of exponents remain true. For example, according to property 5,

$$(a^{1/2})^2 = a^{(1/2)(2)} = a^1 = a.$$

So $a^{1/2}$ is a solution of the equation $x^2 = a$, which means $a^{1/2} = \sqrt{a}$ or $a^{1/2} = -\sqrt{a}$. By convention, $a^{1/2} = \sqrt{a}$, the **principal square root**.

In general, if a is a real number and n is a positive integer, then

$$(a^{1/n})^n = a^{(1/n)n} = a^1 = a.$$

So $a^{1/n}$, called the **principal nth root** of a, is a solution to the equation $x^n = a$. In particular, $a^{1/2}$ is called the principal square root of a (as just shown), and $a^{1/3}$ is known as the **principal cube root** of a. Similarly, $a^{1/4}$, $a^{1/5}$, ... are the principal fourth and fifth roots of a, and so on.

🔍 EXPLORE WITH A GRAPHING UTILITY

Using the viewing rectangle $[-100, 100]$ by $[-10, 10]$, graph the following pairs of equations on the same coordinate system:

- $y = \sqrt{x}$ and $y = x^{1/2}$
- $y = \sqrt[3]{x}$ and $y = x^{1/3}$

Questions

1. Do the equations in each pair appear to be equivalent? Does your experiment support the meaning we have given to fractional exponents?

2. Why did the graph of the first pair fall entirely to the right of the y-axis, whereas the graph of the second pair appeared on both sides of the y-axis?

3. What generalization can you make for $x^{1/n}$ that depends on whether n is odd or even?

Recall that the square root of a negative number, for example $\sqrt{-3}$, does not represent a real number since any number times itself must be positive. In other words, in the expressions $\sqrt{x}$ and $x^{1/2}$, x cannot be a negative number. The same is true for $x^{1/n}$ when n is any even integer. This observation gives a partial answer to Question 2 of the Exploration.

In contrast, $\sqrt[3]{-8}$ is defined. Why? Because $\sqrt[3]{-8} = -2$, since $(-2)^3 = -8$. So when n is odd and x is negative, $x^{1/n}$ is defined and is negative. This finding is summarized in the following definition.

Definition A.2 $a^{1/n}$ nth Root of a Number

Let n be a positive integer and a a real number. Then

$$a^{1/n} = \begin{cases} \text{the } n\text{th root of } a \text{ if } n \text{ is odd;} \\ \text{the nonnegative } n\text{th root of } a \text{ if } n \text{ is even and } a \geq 0. \end{cases}$$

We next need to decide what symbols like $x^{2/3}$ mean. It can be shown that whenever m is an integer and n is a positive integer, then

$$\left(a^{1/n}\right)^m = \left(a^m\right)^{1/n},$$

where a is a nonzero real number for which the indicated powers exist. It makes sense to replace each expression in this equation with the single symbol $a^{m/n}$.

GRAPHER SKILLS UPDATE

When you want to graph $y = x^{m/n}$, use one of the two forms from Definition A.3: $y = \left(x^{1/n}\right)^m$ or $y = \left(x^m\right)^{1/n}$. For example, if you key $y = x^{2/3}$, into a grapher you might not get the correct graph. The graphing utility does not store the definition of $x^{2/3}$; you have to tell it the definition.

Definition A.3 $a^{m/n}$ Rational Number Exponents

For all integers m and all positive integers n, m/n is in lowest terms, and for all nonzero real numbers a for which all the indicated powers exist, then

$$a^{m/n} = \left(a^{1/n}\right)^m = \left(a^m\right)^{1/n}.$$

Our development of the concept *exponent* began with positive-integer exponents. We then extended that meaning to include negative-integer exponents, exponents of the form $1/n$, and finally exponents of the form m/n. This discussion leads to the following theorem.

Theorem A.1 Properties of Rational Exponents

Exponent properties 1 through 6 listed earlier in this appendix are true when the exponents represent rational numbers.

We use this theorem to complete the following example.

E X A M P L E 3 Simplifying Expressions with Fractional Exponents

Simplify the following expressions:

COMMON ERROR

Students often make the mistake of assuming that $(\sqrt{x^2}) = x$. In fact, the correct equation is $(\sqrt{x^2}) = |x|$. Notice that $\sqrt{(-3)^2} \neq -3$.

a) $8^{2/3}$

b) $x^{1/4}x^{1/3}$

c) $(x^3y^{2/3})^{1/2}$

d) $\left(\dfrac{x^2}{y^3}\right)^{-1/2}$

Solution

a) $8^{2/3} = (8^{1/3})^2 = 2^2 = 4$

b) $x^{1/4}x^{1/3} = x^{(1/3+1/4)} = x^{7/12}$

c) $(x^3y^{2/3})^{1/2} = (x^3)^{1/2}(y^{2/3})^{1/2} = x^{3/2}y^{1/3}$

d) $\left(\dfrac{x^2}{y^3}\right)^{-1/2} = \left(\dfrac{y^3}{x^2}\right)^{1/2} = \dfrac{(y^3)^{1/2}}{(x^2)^{1/2}} = \dfrac{y^{3/2}}{|x|}.$

■

Exercises for Section A.1

In Exercises 1–14, evaluate the numerical expressions.

1. $(2^3) \cdot (3^2)$

2. 3^{-1}

3. $3^4 \cdot 3^{-6}$

4. $\left(\dfrac{4}{9}\right)^{1/2}$

5. $7^{-2} \cdot 7^5$

6. $\dfrac{4 \cdot 5^{-4}}{5^{-2} \cdot 4^{-2}}$

7. $9^{1/2}$

8. $16^{1/4}$

9. $8^{2/3}$

10. $16^{3/4}$

11. $16^{-1/2}$

12. $8^{-1/3}$

13. $27^{2/3}$

14. $32^{-3/5}$

In Exercises 15 to 18, use a calculator to evaluate the expressions.

15. $(3.2)^{1/3}$

16. $(0.015)^{3.125}$

17. $(1.25)^{-2/3}$

18. $(3.14)^{3.1}$

In Exercises 19 to 21, use a calculator.

19. Which number is larger, $2^{2/3}$ or $3^{3/4}$?

20. Which number is smaller, $4^{-2/3}$ or $3^{-3/4}$?

21. Order these numbers from smallest to largest: $3^{3/4}$, $4^{5/8}$, $12^{3/10}$.

22. As you completed Exercises 9–14, which form did you find easier to use: $y = (x^m)^{1/n}$ or $y = (x^{1/n})^m$?

23. Graph these equations on a grapher: $y = (x^2)^{1/3}$ and $y = (x^{1/3})^2$. Do the results show that these equations are equivalent?

24. Graph these equations on a grapher: $y = (x^5)^{1/3}$ and $y = (x^{1/3})^5$. Do the results show that these equations are equivalent?

25. Graph these equations on a grapher: $y = (x^3)^{1/4}$ and $y = (x^{1/4})^3$. Do the results show that these equations are equivalent?

26. Graph these equations on a grapher: $y = x^{2/3}$ and $y = (x^{1/3})^2$. Are the graphs identical? Which allows negative values of x?

27. Experiment with a grapher and find a generalization. Under what conditions does the graph of $y = (x^{1/n})^m$, where m and n are integers, lie entirely in the first quadrant? (*Note:* Not all graphers produce identical results.)

In Exercises 28–43, simplify the expressions.

28. $\dfrac{x^4 \cdot y^3}{x^2 \cdot y^5}$

29. $\dfrac{(u \cdot v^2)^3}{v^2 \cdot u^3}$

30. $\dfrac{(3x^2)^2 \cdot y^4}{3y^2}$

31. $\left(\dfrac{4}{x^2}\right)^2$

32. $\dfrac{(2x^2y)^{-1}}{xy^2}$

33. $(3x^2y^3)^{-2}$

34. $\dfrac{x^{-3}y^3}{x^{-5}y^2}$

35. $\left(\dfrac{2}{xy}\right)^{-3}$

36. $\left(\dfrac{1}{x} + \dfrac{1}{y}\right)(x + y)^{-1}$

37. $\dfrac{(x + y)^{-1}}{(x - y)^{-1}}$

38. $\left(x^2y^4\right)^{1/2}$

39. $\left(x^{2/3}\right)^{1/2}$

40. $\left(\dfrac{x^{1/2}}{y^{2/3}}\right)^6$

41. $\left(x^2y^3\right)^{3/4}$

42. $\dfrac{\left(p^2q^4\right)^{1/2}}{\left(27q^3p^6\right)^{1/3}}$

43. $\dfrac{\left(x^{-3}y^2\right)^{-4}}{\left(y^6x^{-4}\right)^{-2}}$

Exercises 44 and 45 refer to the following **problem situation**: The amount A after t years in a savings account earning an annual interest rate r compounded n times per year is

$$A = P\left(1 + \frac{r}{n}\right)^{nt},$$

where P is the **principal** (the original amount saved).

44. If $5500 is deposited in a savings account that earns 8% per year compounded quarterly, how much will be in the account after 5 yr.?

45. Which is the better deal, a savings account that pays 8.25% compounded annually or one that pays 8% compounded monthly?

46. Suppose you place a single penny on a checkerboard's corner square, two pennies on the next square, four pennies on the next square, and so on. How many pennies will there be on the 64th square?

47. **Estimate Mentally**   In Exercise 46, approximately how high will the stack of pennies on the last square of the checkerboard be?
   a) The height of a four-story building
   b) The height of a skyscraper
   c) The height of a space needle reaching to the moon
   d) A height greater than all of the above

48. Use the definition of exponential notation to show that $a^7/a^2 = a^{7-2}$.

49. Use the definition of exponential notation to show that $(a^3)^4 = a^{3 \cdot 4}$.

A.2 _____ Algebraic Expressions

In algebra, we often use letters to represent numbers. Sometimes we use letters to represent fixed numbers (called **constants**) like $\pi \approx 3.14$ or $e \approx 2.718$. But more often, we let letters stand for a collection of numbers, as in $y = .25x + 15$, in which case we call the letters **variables**. An **algebraic expression** is a collection of variables and constants; these are combined by using addition, subtraction, multiplication, division, and radicals or fractional exponents. Following are some examples:

$$3x^2 + 4x - 5, \qquad \frac{(x - 3)^{1/2}}{x^2}, \qquad \sqrt{x^2 + y^2}, \qquad \text{and} \qquad 3x^2y + 4x + 7y.$$

A **term** in an algebraic expression is a number or the product of a number and one or more variables. For example, $3x^2 + 4x - 5$ has three terms: $3x^2$, $4x$, and 5. The real numbers in each term are called the **coefficients** of the term. The coefficient of the second term of $3x^2 + 4x - 5$ is 4.

To **evaluate** an algebraic expression means to substitute a specific number value for each variable and then calculate the result.

E X A M P L E   1   Evaluating an Expression

The volume and surface area of a box with length l, height h, and width w are given by the expressions

$$\text{volume} = lwh \qquad \text{and} \qquad \text{surface area} = 2lw + 2lh + 2wh.$$

Find the volume and surface area of a box with $l = 18$ cm, $w = 9$ cm, and $h = 5$ cm.

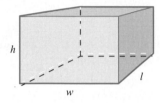
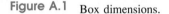

Figure A.1   Box dimensions.

Solution   (See Fig. A.1.)

$$\text{Volume} = lwh$$

$$= 18 \cdot 9 \cdot 5 \qquad \text{Substitute } 18, 9, \text{ and } 5 \text{ for } l, w, \text{ and } h, \text{ respectively.}$$

$$= 810$$

$$\text{Surface area} = 2lw + 2lh + 2wh$$

$$= 2 \cdot 18 \cdot 9 + 2 \cdot 18 \cdot 5 + 2 \cdot 9 \cdot 5 \qquad \begin{array}{l}\text{Substitute } 18, 9, \text{ and } 5 \text{ for} \\ l, w, \text{ and } h, \text{ respectively.}\end{array}$$

$$= 594$$

The volume is 810 cm^3 and the surface area is 594 cm^2.   ≡

Two expressions are **equivalent** when they yield the same numerical value for all values of the variables.

A **polynomial** is an algebraic expression defined as the sum of a finite number of terms with only nonnegative integer exponents permitted. Some examples are

$$2x^3 + 2x - 3, \qquad y^4 + 3y^2 + y + 5, \qquad \text{and} \qquad 3x^2y + xy^3 - 14.$$

The first is a polynomial in x, the second is a polynomial in y, and the third is a polynomial in x and y. A polynomial containing exactly three terms is called a **trinomial**, and one with exactly two terms is called a **binomial**.

Following is a formal definition.

Definition A.4   Polynomial

A **polynomial in** x is an expression of the form

$$a_n x^n + a_{n-1} x^{n-1} + \cdots + a_1 x^1 + a_0,$$

where a_n, a_{n-1}, ..., and a_0 are real numbers and n is an integer such that $n \geq 0$. n is the **degree** of the polynomial, and a_n (read "a sub n") is called the **leading coefficient**.

Operations with Polynomials

Polynomials are added and subtracted by using commutative, associative, and distributive properties and by combining like terms. Study this example.

E X A M P L E   2   Adding and Subtracting Polynomials

Perform the indicated operations and simplify.

a) $(2x^3 - 3x^2 + 4x - 1) + (x^3 + 2x^2 - 5x + 3)$

b) $(4x^2 + 3x - 4) - (2x^3 + x^2 - x + 2)$

Solution

a) $(2x^3 - 3x^2 + 4x - 1) + (x^3 + 2x^2 - 5x + 3)$

$\qquad = (2x^3 + x^3) + (-3x^2 + 2x^2) + (4x - 5x) + (-1 + 3)$ Group like terms.

$\qquad = 3x^3 + (-1)x^2 + (-1)x + 2$ Combine like terms.

$\qquad = 3x^3 - x^2 - x + 2$

b) $(4x^2 + 3x - 4) - (2x^3 + x^2 - x + 2)$

$\qquad = -2x^3 + (4x^2 - x^2) + (3x - (-1x)) + (-4 - 2)$ Group like terms.

$\qquad = -2x^3 + 3x^2 + (3 - (-1))x + (-6)$ Combine like terms.

$\qquad = -2x^3 + 3x^2 + 4x - 6$ ≡

To find the **product** of two polynomials, use the distributive properties and the properties of exponents. For example, the product of the two binomials $x^2 + 3$ and $x - 7$ can be completed as follows:

$$(x^2 + 3)(x - 7) = x^2(x - 7) + 3(x - 7) \qquad \text{Use the distributive property.}$$

product of first terms   product of outer terms   product of inner terms   product of last terms

$$= x^2 \cdot x - 7x^2 + 3x - 21 \qquad \text{Use the distributive property again.}$$

$$= x^3 - 7x^2 + 3x - 21.$$

Notice that after using the distributive property the second time, there are four terms. The first term is the product of the first terms of the two binomials; the second is the product of the outer terms of the two binomials; and the third and fourth terms are the products of the inner and last terms of the two binomials, respectively. This procedure is often called the FOIL method of multiplying two binomials, where FOIL stands for *F*irst, *O*uter, *I*nner, and *L*ast.

E X A M P L E 3 Multiplying Two Binomials with the FOIL Method

Find the product of $x^7 + 3x^2$ and $x^3 - 5x$.

Solution

$$(x^7 + 3x^2)(x^3 - 5x) = x^7 \cdot x^3 + x^7 \cdot (-5x) + 3x^2 \cdot x^3 + 3x^2 \cdot (-5x)$$

$$= x^{10} - 5x^8 + 3x^5 - 15x^3 \qquad \equiv$$

Sometimes the product of the outer and inner terms have the same degree. In that case, these two terms are combined into a single term called the **middle term**. With practice, these two terms can be combined mentally to find the middle term.

E X A M P L E 4 Multiplying Binomials Yielding One Middle Term

Find the product of $x + 4$ and $x - 5$.

Solution

$$(x + 4)(x - 5) = x^2 + ? - 20 \qquad \text{Think: } -5x + 4x = -x \text{ is the middle term.}$$

$$= x^2 - x - 20 \qquad \equiv$$

The FOIL method also can be used to multiply nonpolynomial expressions that each have two terms.

E X A M P L E 5 Multiplying Expressions with the FOIL Method

Find the product of $1/x^2 + 3$ and $x + 6/x^3$ and express the answer using positive exponents.

Solution

$$\left(\frac{1}{x^2} + 3\right)\left(x + \frac{6}{x^3}\right) = (x^{-2} + 3)(x + 6x^{-3})$$

$$= x^{-2}x + x^{-2} \cdot 6x^{-3} + 3x + 3 \cdot 6x^{-3}$$

$$= x^{-1} + 6x^{-5} + 3x + 18x^{-3}$$

$$= \frac{1}{x} + \frac{6}{x^5} + 3x + \frac{18}{x^3}$$

■

$y = (x^2 - 3)(x + 4)$

$y = x^3 + 4x^2 - 12$

[−5, 5] by [−15, 15]

Figure A.2 Graphs of $y = x^3 + 4x^2 - 12$ and $y = (x^2 - 3)(x + 4)$.

A grapher can be used to support that the algebraic manipulations are correct. Attempts to graph the equations $y = (x^2 + 3)(x - 7)$ and $y = x^3 - 7x^2 + 3x - 21$ in the same viewing rectangle will produce identical graphs. Note that this method is *not a proof;* it is merely suggestive. What happens when $y = x$ and $y = x + 0.01$ are graphed in the standard viewing rectangle?

A grapher also can be used to determine that two expressions have been multiplied incorrectly. For example, to show that the statement

$$(x^2 - 3)(x + 4) = x^3 + 4x^2 - 12$$

is false, graph $y = (x^2 - 3)(x + 4)$ and $y = x^3 + 4x^2 - 12$ and observe that these graphs are not identical (see Fig. A.2).

🔍 **EXPLORE WITH A GRAPHING UTILITY**

Use a grapher to find which of the following statements is false:

1. $(x + 2)^2 = x^2 + 4$
2. $(x - 3)^2 = x^2 - 6x - 9$
3. $(x - 1)(x + 4) = x^2 + 3x - 4$
4. $x^4 - 1 = (x - 1)^4$

For each false statement write several other false statements that have the same type of error. Support your work with a grapher.

In Examples 3 and 4, two binomial expressions were multiplied together to obtain an expression. How can this process be reversed? That is, given a polynomial, how can we find two factors whose product is that given polynomial? The process of writing one polynomial as a product of two polynomials is called **factoring**.

Notice that $x^2 + x - 6 = (x + 3)(x - 2)$. We say that $x^2 + x - 6$ is a **factored polynomial** and that $x + 3$ and $x - 2$ are **factors** of $x^2 + x - 6$. It is easy to observe that $y = (x + 3)(x - 2)$ has a value of 0 when $x = -3$ or $x = 2$. It is not as obvious when this same polynomial is written in the form $y = x^2 + x - 6$. Important properties of a polynomial are thus sometimes more observable in factored form.

Common Factors

Often it is not obvious how to factor a polynomial. However, there are patterns of coefficients in certain polynomials that make factors more recognizable. One pattern involves sharing a common factor. In an expression of the form $ab + ac$, the terms ab and ac have a common factor of a. So

$$ab + ac = a(b + c).$$

E X A M P L E   6   Removing Common Factors

Factor the following:

a) $3x^4 + 6x^2 - 24x$
b) $4x^2 y^3 + 12x^4 y^2$

Solution

a) $3x^4 + 6x^2 - 24x = 3x(x^3 + 2x - 8)$
b) $4x^2 y^3 + 12x^4 y^2 = 4x^2 y^2(y + 3x^2)$ ≡

Difference of Squares

A second pattern called the **difference of two squares** has a factorization that is easily recognizable. For example,

$$u^2 - v^2 = (u - v)(u + v).$$

The correctness of this pattern can be verified by multiplying the factors together. We agree not to factor an expression that requires the introduction of irrational numbers. For example, even though $x^2 - 2 = (x - \sqrt{2})(x + \sqrt{2})$, we will not factor $x^2 - 2$.

Sometimes after an expression has been factored, one or both factors can be factored again. Once each new factor has been factored as often as possible, the expression has been **factored completely**.

E X A M P L E   7   Factoring the Difference of Squares

Factor the following as completely as possible without introducing radicals:

a) $25x^2 - 36$

b) $4y^4 - 64$

Solution

a) $25x^2 - 36 = (5x - 6)(5x + 6)$ Think: $25x^2 = (5x)^2$

b) $4y^4 - 64 = 4(y^2 - 4)(y^2 + 4)$ Remove common factors.

$$= 4(y - 2)(y + 2)(y^2 + 4)$$

≡

Factoring Trinomials into Binomial Factors

REMINDER

Use your grapher to support algebraic work.

1. Graph the problem and solution in the same viewing rectangle.

2. If they match, your algebraic manipulation is probably correct.

3. If they do not match, you have likely made an algebraic error.

The final method of factoring requires reversing the thought process used when multiplying by the FOIL method. First, recall the FOIL method used to multiply $(x + 2)(x - 3)$.

$(x + 2)(x - 3) = x^2 +$? $+ (-6)$ Think: Outer is $-3x$, inner is $2x$ for a sum of $-x$, and last is -6.

$$= x^2 - x - 6$$

To factor $x^2 - x - 6$, reverse the thought process.

$x^2 - x - 6 = (x + ?)(x + ?)$ Think: Find two numbers with the sum of -1 and the product of -6.

$= (x - 3)(x + 2)$ Is it -2 and 3, or 2 and -3? It is -3 and 2.

E X A M P L E   8   Factoring Trinomials

Factor the following trinomials:

a) $x^2 + 3x - 10$

b) $x^2 + 11xy + 28y^2$

c) $6x^2 - x - 12$

Solution

a) $x^2 + 3x - 10 = (x + ?)(x + ?)$ Think: factors of -10 whose sum is 3

$$= (x + 5)(x - 2)$$

b) $x^2 + 11xy + 28y^2 = (x + ?y)(x + ?y)$ Think: factors of 28
whose sum is 11

$$= (x + 4y)(x + 7y)$$

c) $6x^2 - x - 12 = (3x + ?)(2x + ?)$ Think: factors of -12
that lead to a middle term of $-x$

$$= (2x - 3)(3x + 4)$$

≡

Exercises for Section A.2

1. Write three different expressions using all of the following expressions in each: x^2, $\sqrt{x - 1}$, and $9x$.

2. Is $\sqrt{2x^3 + 4x^2 - 5x + 7}$ a polynomial? Why?

3. Write an expression that is not a polynomial and state why it fails to be a polynomial.

4. An object in free fall (ignoring air friction) falls approximately D feet in t seconds where $D = 16t^2$. How far does a skydiver fall during the first 5 sec of free fall?

5. Jon found that his cost C for a telephone call was \$0.85/min, so the cost for t minutes can be found using the relationship $C = 0.85t$. What is the cost of a 45-min call?

6. The volume V of a cylinder can be found using $V = \pi r^2 h$, where r is the radius of the base and h is the height. Find the volume of a soup can that is 12 cm tall with a radius of 4 cm.

For Exercise 6.

7. What is the degree and leading coefficient of the polynomial $3x^4 - 5x + 3$?

8. What is the degree and leading coefficient of the polynomial $3x^2 + 5x^3 - 2x + 1$?

In Exercises 9–18, perform the indicated operations and simplify.

9. $(3x^2 + 4) + (5x - 2)$

10. $(x^2 - 3x + 7) + (3x^2 + 5x - 3)$

11. $(-3x^2 - 5) - (x^2 + 7x + 12)$

12. $(4x^3 - x^2 + 3x) - (x^3 + 12x - 3)$

13. $(2x + x^2 - 3) + (5x^2 + 3x + 4)$

14. $3x - [7 - (3x + 5)]$

15. $(12x^2 - 5x + 2) - (-x^2 - 2x + 3)$

16. $3x + (x^2 + 4x - 3) - (x^2 + 7)$

17. $(7 - 3x + x^2) - (4x + 2x^2 - 8)$

18. $(4x - 7 + x^2) - (-3x^2 + 4 - 7x)$

In Exercises 19–36, find the given product (check with a grapher).

19. $(x + 3)(x + 5)$ 20. $(x - 2)(x + 5)$

21. $(x + 7)(x - 2)$ 22. $(2x - 1)(x + 3)$

23. $(x - 3)(x - 1)$ 24. $(x - 9)(x - 3)$

25. $(2x + 3)(4x + 1)$ 26. $(3x - 5)(x + 2)$

27. $(5 - x)(2 + x)$ 28. $2x(x + 3)$

29. $(x^? + 1)(x^? - 3)$ 30. $(x + 4)(x^2 + 2)$

31. $(x^2 + 1)(x^3 + 4)$ 32. $x^5(x^2 - 3)$

33. $(x^{-2} + 1)(x^{-1} + 3)$ 34. $(x^{-2} + 4)(x^3 - 3)$

35. $(x^2 + 2x + 3)(x + 4)$

36. $(x^2 + x - 3)(x^2 + 3x + 1)$

For Exercises 37–43, use a grapher to determine which equations are false.

37. $(x + 4)^2 = x^2 + 16$

38. $(x - 1)^3 = x^3 - 1$

39. $(x - 3)(x - 1) = x^2 - 3x + 3$

40. $(x^2 + 2)(x^2 - 2) = x^4 - 4^2$

41. $(x^2 + 3)(x - 1) = (x^2 + 2x - 3)$

42. $x^3 - 1 = (x - 1)(x^2 - 1)$

43. $x^3 - 1 = (x - 1)(x^2 + 1)$

Use your grapher for Exercises 44–46, then confirm the results by multiplying or expanding.

44. Graph the (a) sum, (b) difference, and (c) product of $x^2 - 3$ and $x + 2$.

45. Which is the correct factorization of $12x^2 + x - 6$?
   a) $(3x + 1)(x - 6)$ **b)** $(3x - 2)(4x + 3)$

46. Which is the correct factorization of $x^3 + 2x^2 + 4x + 8$?
   a) $(x^2 + 4)(x + 2)$ **b)** $(x^2 + 2)(x + 4)$

Factor Exercises 47–71 completely (support with a grapher when possible).

47. $4x - 64$ **48.** $2x^3 - 4x^2$

49. $x^3 - 5x^2 + 7x$ **50.** $y^4 - 3y^3 + y$

51. $(x - 3)(x + 2)^2 + (x - 3)^2(x + 2)$

52. $x^2 - 16$

53. $y^2 - 4x^2$

54. $x^2y^2 - 25$

55. $x^4 - 81$

56. $(x + 9)^2 - 16$

57. $(x - 1)^4 - 9$

58. $x^2 - 6x + 9$

59. $x^2 + 6x + 9$

60. $x^2 - 3x - 4$

61. $x^2 + x - 6$

62. $x^2 + 3x - 4$

63. $x^2 + 8x + 12$

64. $x^2 - 8x + 12$

65. $x^2 + 5xy + 6y^2$

66. $2x^2 - x - 6$

67. $8x^2 - 8x - 6$

68. $x^2 + 3xy - 10y^2$

69. $4x^2 - 2x - 12$

70. $x^4 - x^2 - 12$

71. $x^4 - 14x^2 - 32$

72. Writing to Learn   Write a few paragraphs that explain your answer to the following question. In the Exploration in Section A.2, you used a grapher to decide that two expressions were not equal. When you want to conclude that two expressions are equal, which is more convincing, using an algebraic argument or a grapher?

A.3 _____ Fractional Expressions

A **fractional expression** is a quotient of two algebraic expressions. If the expression is written in the fraction form a/b, then a is the **numerator** and b is the **denominator**. Following are several examples:

$$\frac{\sqrt{x^2 + 1}}{x + 2}, \qquad \frac{4x^3 - 3x^2 + 5}{x^2 + 1}, \qquad \text{and} \qquad \frac{5}{\sqrt{x^3 - 3x}}. \tag{1}$$

If both numerator and denominator are polynomials, then the fractional expression is called a **rational expression**. The middle expression in line (1) above is a rational expression. The first one, however, fails to be rational because the numerator has a radical, while the third one is not rational because of the radical in the denominator. Following are several other rational expressions:

$$\frac{x^4 + 3x^3 - x + 5}{x^2 - 7}, \qquad \frac{x + 3}{x^2 - 4x + 3}, \tag{2}$$

$$\frac{x^{17} - 5x^4 + 1}{x^2 - 2x + 1}, \qquad \text{and} \qquad \frac{x^4 - 3x^2 + 2}{x^7 - 5}.$$

Recall that a fraction is undefined if the denominator equals zero; therefore you cannot substitute a value of x into the expression that makes the denominator have the value of zero. For example, in the first expression in line (1), -2 cannot replace x, while in the first expression in line (2), $\sqrt{7}$ cannot replace x.

Simplifying Rational Expressions

🔍 EXPLORE WITH A GRAPHING UTILITY

Use a grapher to find which of the two expressions might be equivalent to the given expression.

1. Is $\dfrac{2x+8}{4}$ equivalent to $\dfrac{x+4}{2}$ or $\dfrac{x+8}{2}$?

2. Is $\dfrac{x^2+x}{x}$ equivalent to $x+1$ or x^2+1?

3. Is $\dfrac{x^2+x-2}{x-1}$ equivalent to $x+2$ or $-(x^2-2)$?

Conjecture a Generalization   For each expression write an equivalent expression and support your solution with a graphing utility:

$$\frac{x^2+x}{x^2}, \qquad \frac{4x^2+3x}{x}, \qquad \text{and} \qquad \frac{x^2-1}{x+1}.$$

What algebraic property can be used to confirm the answers in this Exploration? Recall that numerical fractions can be simplified by dividing out common factors and **reducing to lowest terms**, as follows:

$$\frac{ac}{bc} = \frac{a}{b} \qquad (b, c \neq 0).$$

This same property can be used when a, b, and c represent algebraic expressions. We illustrate this property in the next two examples.

EXAMPLE 1   Reducing to Lowest Terms

Reduce to lowest terms the rational expression

$$\frac{3x^3 - 9x^2 + 27x}{6x}.$$

Solution

$$\frac{3x^3 - 9x^2 + 27x}{6x} = \frac{3x(x^2 - 3x + 9)}{2 \cdot 3x}$$

$$= \frac{x^2 - 3x + 9}{2}$$

$$= \frac{1}{2}x^2 - \frac{3}{2}x + \frac{9}{2} \qquad \equiv$$

AGREEMENT

When we write

$$\frac{x^2 + x}{x} = x + 1,$$

it is understood that $x \neq 0$.

Often, before an algebraic expression can be simplified, it must be factored completely, as the next example illustrates.

E X A M P L E 2 Simplifying Rational Expressions

Reduce to lowest terms the rational expression

$$\frac{(x^2 - 9)(x^2 + 2x + 1)}{(x + 3)(x + 1)^2}.$$

Solution

$$\frac{(x^2 - 9)(x^2 + 2x + 1)}{(x + 3)(x + 1)^2} = \frac{(x - 3)(x + 3)(x + 1)^2}{(x + 3)(x + 1)^2} \qquad \text{Factor the numerator.}$$

$$= x - 3, \ (x \neq -3), \ (x \neq -1) \qquad \text{Cancel common factors.} \ \blacksquare$$

You can use a grapher to support the algebraic conclusion of the above example. Graph both

$$y = \frac{(x^2 - 9)(x^2 + 2x + 1)}{(x + 3)(x + 1)^2} \qquad \text{and} \qquad y = x - 3$$

in $[-5, 5]$ by $[-5, 5]$. What do you observe?

Multiplying and Dividing Rational Expressions

Algebraic rational expressions are multiplied and divided in the same way as are numerical rational expressions. For example, if $a, b, c,$ and d represent polynomial expressions, the **product of a/b and c/d** is given by

$$\frac{a}{b} \cdot \frac{c}{d} = \frac{a \cdot c}{b \cdot d} = \frac{ac}{bd}.$$

REMINDER

Even though the graphs of

$$y = \frac{(x^2 - 9)(x^2 + 2x + 1)}{(x + 3)(x + 1)^2}$$

and $y = x - 3$ appear to be identical on graphing utilities, it is important to observe that they are not identical. If $x = -3$ or -1, the denominator of the rational expression is zero and hence undefined. For all other values of x, the rational expression is equal to $y = x - 3$.

To explain the quotient of rational expressions, recall that a quotient of whole numbers can be thought of as a **missing factor**. For example, $8 \div 2 = x$, where x is the missing factor in $8 = 2 \cdot x$. In the same way,

$$\frac{a}{b} \div \frac{c}{d} = x, \qquad \text{where} \qquad \frac{a}{b} = \frac{c}{d} \cdot x.$$

Solving for x yields the **quotient of a/b and c/d**:

$$\frac{a}{b} \div \frac{c}{d} = \frac{a}{b} \cdot \frac{d}{c}.$$

This definition for division is sometimes described by the phrase "invert and multiply."

EXAMPLE 3   Dividing Rational Expressions

Perform the indicated division and simplify.

$$\frac{x^2 - y^2}{2y} \div \frac{x + y}{4}$$

Solution

$$\frac{x^2 - y^2}{2y} \div \frac{x + y}{4} = \frac{(x - y)(x + y)}{2y} \cdot \frac{4}{x + y}$$ Factor expressions and invert and multiply.

$$= \frac{4(x - y)(x + y)}{2y(x + y)}$$ Complete the multiplication and simplify.

$$= \frac{2(x - y)}{y}$$ ≡

Notice that in Examples 1 through 3, the rational expressions were simplified by factoring the numerator and/or denominator and dividing out common factors. If multiplication is changed to addition or subtraction, there are no common factors to divide out.

Warning   In general,

$$\frac{a + c}{b + c} \neq \frac{a}{b}.$$

🔍 **EXPLORE WITH A GRAPHING UTILITY**

Use a grapher to find which of the two expressions might be equivalent to the given expression.

1. Is $\dfrac{x}{2} + \dfrac{1}{4}$ equivalent to $\dfrac{2x + 1}{4}$ or $\dfrac{x + 2}{4}$?

2. Is $\dfrac{1}{x} + \dfrac{1}{4}$ equivalent to $\dfrac{x + 4}{4x}$ or $\dfrac{1}{4 + x}$?

3. Is $\dfrac{2}{x} + \dfrac{4}{3x}$ equivalent to $\dfrac{10}{3x}$ or $\dfrac{6}{4x}$?

Conjecture a Generalization   For each expression, write an equivalent expression and check your solution with a graphing utility:

$$\frac{x}{3} + \frac{1}{5}, \qquad \frac{1}{x} + \frac{2}{7}, \qquad \text{and} \qquad \frac{1}{4x} + \frac{1}{6x}.$$

Adding and Subtracting Rational Expressions

To add or subtract two rational expressions, apply the same definition used for numerical fractions; that is, if the denominators are identical, simply add or subtract the numerators. If a, b, and c represent expressions, then

$$\frac{a}{c} + \frac{b}{c} = \frac{a+b}{c} \qquad \text{and} \qquad \frac{a}{c} - \frac{b}{c} = \frac{a-b}{c}. \qquad \textbf{(3)}$$

E X A M P L E   4   Adding Expressions with Like Denominators

Perform the addition

$$\frac{(x^2 - 6)}{x - 2} + \frac{x}{x - 2}$$

and simplify. Support using a grapher.

Solution

$$\frac{(x^2 - 6)}{x - 2} + \frac{x}{x - 2} = \frac{x^2 + x - 6}{x - 2} \qquad \text{Add the numerators.}$$

$$= \frac{(x - 2)(x + 3)}{x - 2} \qquad \text{Factor and cancel the common factor.}$$

$$= x + 3, \qquad (x \neq 2)$$

The check using a grapher is shown in Fig. A.3.   ▤

REMINDER

The two graphs in Fig. A.3 appear identical on graphing utilities. However, it is important to observe that one expression is undefined for $x = 2$, so the graphs are not really identical.

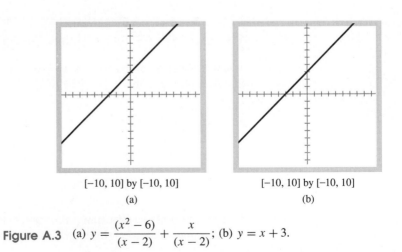

[−10, 10] by [−10, 10]   [−10, 10] by [−10, 10]
(a)   (b)

Figure A.3   (a) $y = \dfrac{(x^2 - 6)}{(x - 2)} + \dfrac{x}{(x - 2)}$; (b) $y = x + 3$.

If the denominators are not identical, there are two strategies you could follow: Either multiply the numerator and denominator of each term by the appropriate expression to obtain the **lowest common denominator (LCD)** or use the following definition:

$$\frac{a}{b} + \frac{c}{d} = \frac{ad + bc}{bd} \qquad \text{and} \qquad \frac{a}{b} - \frac{c}{d} = \frac{ad - bc}{bd}. \tag{4}$$

Recall that the LCD of several fractions consists of the product of all prime factors in the denominators, with each factor given the highest power of its occurrence in any denominator.

E X A M P L E   5   Finding the LCD and Subtracting

Perform the indicated subtraction and simplify.

$$\frac{5}{2x} - \frac{2x - 3}{3x} = \frac{3 \cdot 5}{3 \cdot 2x} - \frac{2(2x - 3)}{2 \cdot 3x}$$

Multiply numerator and denominator by the appropriate factor to obtain the common denominator $6x$.

$$= \frac{15 - 2(2x - 3)}{6x}$$

$$= \frac{15 - 4x + 6}{6x}$$

Subtract numerators, use the distributive property, and combine terms.

$$= \frac{21 - 4x}{6x}$$

E X A M P L E   6   Adding and Simplifying Fractional Expressions

Perform the indicated addition and simplify.

$$\frac{3}{x - 3} + \frac{4}{x + 2}$$

Solution

$$\frac{3}{x - 3} + \frac{4}{x + 2} = \frac{3(x + 2) + 4(x - 3)}{(x - 3)(x + 2)}$$

Use the definition in Eq. (4).

$$= \frac{3x + 6 + 4x - 12}{(x - 3)(x + 2)}$$

Use the distributive property.

$$= \frac{7x - 6}{(x - 3)(x + 2)}$$

Combine terms.

The results of Examples 5 and 6 can be supported using a grapher.

Negative exponents have the same meaning with algebraic expressions as they do with numerical expressions. Study the following example.

EXAMPLE 7   Using Negative Exponents

Perform the indicated addition and simplify.

$$3 + x^{-1} + x$$

Solution

$$3 + x^{-1} + x = 3 + \frac{1}{x} + x \qquad \text{Note that } 3 + x^{-1} \neq (3 + x)^{-1}$$
$$\text{and } x^{-1} + x \neq x^0.$$

$$= \frac{3x + 1 + x^2}{x}$$

$$= \frac{x^2 + 3x + 1}{x}$$

∎

Exercises for Section A.3

In Exercises 1–6, find the missing numerator or denominator so that the two rational expressions are equal.

1. $\dfrac{2}{3x} = \dfrac{?}{6x^2}$

2. $\dfrac{4y}{3x} = \dfrac{12xy}{?}$

3. $\dfrac{9x^2}{27xy} = \dfrac{?}{3y}$

4. $\dfrac{14x^3y^2}{2x^2y} = \dfrac{7xy}{?}$

5. $\dfrac{x^2 - 3x}{x^3} = \dfrac{?}{x^2}$

6. $\dfrac{(x-1)(x+3)}{x^2 + 5x + 6} = \dfrac{x-1}{?}$

In Exercises 7–16, reduce each to lowest terms. Support your work with a graphing utility.

7. $\dfrac{4x^2}{12x}$

8. $\dfrac{15xy^2}{3xy}$

9. $\dfrac{(x-1)(3x+2)}{(x-1)^2}$

10. $\dfrac{(x-1)(x+3)}{x^2 - 2x + 1}$

11. $\dfrac{(x+2)(x-3)}{x^2 + x - 2}$

12. $\dfrac{x^2 + 3x - 4}{(x-1)(x-2)}$

13. $\dfrac{x^3 - x^2}{x - 1}$

14. $\dfrac{9 - x^2}{x - 3}$

15. $\dfrac{x^2 - y^2}{3x + 3y}$

16. $\dfrac{x^2 - x - 1}{x^2 + 3x + 2}$

In Exercises 17–20, use a grapher to identify the correct equivalency and then confirm the results algebraically.

17. Is $\dfrac{x^2 - 1}{x - 1}$ equivalent to

a) x, or

b) $x + 1$?

18. Is $\dfrac{3x^3 + 12x}{9x}$ equivalent to

a) $\dfrac{x^2 + 4}{3}$, or

b) $x + 4$?

19. Is $\dfrac{x^3 - 2x^2 + 2x - 1}{x - 1}$ equivalent to

a) $x^2 + 2$, or

b) $x^2 - x + 1$?

20. Is $\dfrac{x^2 - 3}{x^2 + 2\sqrt{3}x + 3}$ equivalent to

a) $\dfrac{x - \sqrt{3}}{x + \sqrt{3}}$, or

b) $\dfrac{-3}{2\sqrt{3} + 3}$?

In Exercises 21–34, perform the indicated operations and simplify. Support your work with a grapher.

21. $\dfrac{3}{x - 1} \cdot \dfrac{x^2 - 1}{9}$

22. $\dfrac{x + 3}{7} \cdot \dfrac{14}{2x + 6}$

23. $\dfrac{x + 3}{x - 1} \cdot \dfrac{1 - x}{x^2 - 9}$

24. $\dfrac{1}{2x} \div \dfrac{1}{4}$

25. $\dfrac{18x^2 - 3x}{3xy} \cdot \dfrac{12y^2}{6x - 1}$

26. $\dfrac{4x}{y} \div \dfrac{8y}{x}$

27. $\dfrac{x^2 - 3x}{14y} \div \dfrac{2xy}{3y^2}$

28. $\dfrac{7(x - y)}{4} \div \dfrac{14(x - y)}{3}$

29. $\dfrac{2x^2 y/(x-3)^2}{8xy/(x-3)}$

30. $\dfrac{(x+2)^2(x-1)/(x+4)^2}{(x-1)^3/(x+2)^2(x+4)}$

31. $\dfrac{x^2+3x+2}{x^2-1} \cdot \dfrac{x^2+2x+1}{x^2+x-2}$

32. $\dfrac{x^2-y^2}{2xy} \cdot \dfrac{4x^2 y}{y^2-x^2}$

33. $\dfrac{(3x-2)(x+5)}{2x-1} \div \dfrac{(x+5)(2x-1)}{(3x-2)^2}$

34. $\dfrac{2x^2+3x-3}{x+4} \div \dfrac{x^2+x-2}{x^2+5x+4}$

In Exercises 35–38, use a grapher to identify the correct equivalency and then confirm the results algebraically:

35. Is $\dfrac{x}{7} + \dfrac{1}{x}$ equivalent to

a) $\dfrac{x+1}{7+x}$, or **b)** $\dfrac{x^2+7}{7x}$?

36. Is $\dfrac{1}{3x} + \dfrac{1}{4x}$ equivalent to

a) $\dfrac{1}{12x^2}$, or **b)** $\dfrac{7}{12x}$?

37. Is $\dfrac{x}{3} + \dfrac{x}{7}$ equivalent to

a) $\dfrac{2x}{10}$, or **b)** $\dfrac{10x}{21}$?

38. Is $x^{-1} + 4$ equivalent to

a) $\dfrac{4x+1}{x}$, or **b)** $(x+4)^{-1}$?

In Exercises 39–44, perform the indicated operations and simplify.

39. $\dfrac{1}{x} + \dfrac{2}{5}$

40. $\dfrac{1}{2x} - \dfrac{1}{y}$

41. $\dfrac{4}{x-1} + \dfrac{3}{x-2}$

42. $\dfrac{7}{x+3} + \dfrac{1}{x-2}$

43. $\dfrac{5x}{x+5} - \dfrac{3}{x+2}$

44. $\dfrac{3x}{x-1} + \dfrac{x}{x+3}$

In Exercises 45–48, perform the indicated operations and simplify. Support your work with a grapher if possible.

45. $\dfrac{4}{(x-1)(x+2)} - \dfrac{7}{(x-1)(x+4)}$

46. $\dfrac{2}{(x-7)(x+3)} + \dfrac{5x}{(x-2)(x+3)}$

47. $\dfrac{5x}{x^2+x-2} + \dfrac{2x}{x^2+3x+2}$

48. $\dfrac{x-3}{x^2-2x-8} + \dfrac{x+2}{x^2-x-6}$

In Exercises 49–54, perform the indicated operations and simplify. When exponents are needed, write answers with positive-integer exponents.

49. $\dfrac{2^{-1}+4^{-1}}{8^{-1}}$

50. $\dfrac{3^{-1}+6^{-1}}{5^{-1}}$

51. $x^{-1}+y^{-1}$

52. $\left(x^{-1}+y^{-1}\right)^{-1}$

53. $\dfrac{1}{(x+1)^{-1}} + \dfrac{1}{(x-1)^{-1}}$

54. $\dfrac{x^{-1}}{x+y} + \dfrac{y^{-1}}{x+y}$

55. Use the problem-solving strategies of "Use reasoning" and "Consider a special case" to show that the expressions $(x^2-1)/(x+1)$ and $x-1$ are equal for all values of x except one.

Glossary

A

ABS   Built-in absolute value function on computers and graphing calculators.

absolute extremum   Either an absolute maximum or an absolute minimum.

absolute maximum   A value, $f(c)$, of a function such that $f(c) \geq f(x)$ for all x in the domain.

absolute minimum   A value, $f(b)$, of a function such that $f(b) \leq f(x)$ for all x in the domain.

absolute value   Denoted by $|c|$, represents the distance of c from zero on the real-number line.

acute angle   An angle whose measure is between $0°$ and $90°$.

acute triangle   A triangle in which all angles measure less than $90°$.

algebraic expression   A collection of variables and constants that are combined by using addition, subtraction, multiplication, division, and radicals or fractional exponents.

algebraic methods   Methods used to solve an equation algebraically, such as addition, subtraction, multiplication, division, factoring, taking roots, . . .

algebraic representation   See mathematical model.

algorithm   A systematic procedure which, if followed, accomplishes a particular task.

altitude   The line segment through one vertex of a triangle perpendicular to the opposite side.

analytic trigonometry   Aspects of trigonometry which are important in calculus.

angle   Union of two rays with a common endpoint called the vertex of the angle. The beginning ray, called the initial side of the angle, is rotated about its endpoint. The final position is called the terminal side of the angle.

angular speed   The angle swept out in one second by a line segment from the center of a wheel to a point on the wheel's circumference.

area   Number of square units in a geometric figure.

arithmetic operations The four basic operations for numbers: addition, subtraction, multiplication, and division.

arithmetic progression or arithmetic sequence A sequence $\{a_n\}$ of the form $a_n = a_{n-1} + d$ for every positive integer n. The number d is called the common difference of the arithmetic sequence.

associative property of addition $a + (b + c) = (a + b) + c$

associative property of multiplication $a(bc) = (ab)c$

asymptote A straight line continually approached by a given curve but never met.

B

base See exponential function and logarithmic function.

bearing Measure of the clockwise angle that the line of travel makes with due north.

binary operations Operations where two terms are combined to form another term. Two examples of binary operations are addition and multiplication.

binomial A polynomial that contains two terms.

box-and-whisker plot A visual interpretation of sets of data.

C

Cartesian coordinate system An ordered set of numbers in a plane that defines the position of a point.

center A point at equal distance from the sides.

central angle An angle whose vertex is the center of a circle.

circle A set of points equally distant from a fixed point called the center.

circular functions Another name for trigonometric functions.

closure property of addition $a + b$ is real.

closure property of multiplication ab is real.

coefficients The real numbers in each term are called the coefficients of the term.

combination Any in a set of possible arrangements of a certain number or all elements of a set, in which order is not important.

common difference See arithmetic progression.

common logarithmic function The logarithmic function with base 10.

common ratio See infinite geometric series.

commutative property of addition $a + b = b + a$

commutative property of multiplication $ab = ba$

complement of an event The set of outcomes in the sample space not in the event.

complementary angles Two angles of positive measure whose sum is $90°$.

complete graph A graph of an equation that suggests all points of the graph and all the important features of the graph.

complex conjugates The complex numbers $a + bi$ and $a - bi$ are complex conjugates.

complex numbers For real numbers a and b, the expression $a + bi$ is called a complex number. The real number a is called the real part and the real number b is called the imaginary part.

complex plane When the coordinate plane is viewed as a representation of the complex numbers, it is called the complex plane or Gaussian plane. The x-axis of the complex plane is called the real axis and the y-axis is called the imaginary axis.

conditional equation An equation in x that is true for only certain values of x.

cone If two lines intersect, then the set of points obtained when one line is rotated about the other fixed line is called a cone.

congruent angles Angles that have the same measure.

conic or conic section Any curve obtained by intersecting a double-napped right circular cone with a plane.

constant A letter that stands for a specific number.

continuous function A function that is continuous on its entire domain.

continuous on an interval J A function f is continuous on the interval J if for all a and b in J, it is possible to trace the graph of the function between a and b without lifting the pencil from the paper.

convergence   When the sum of an infinite series exists.

correlation coefficient   The measure of the strength of the linear relationship between two variables.

coterminal angles   Two angles are coterminal if they have initial side and the same terminal side.

cubic function   A degree three polynomial.

curve   Set of points whose coordinates are determined by an equation.

cut   The curve formed when a plane intersects a surface.

D

decreasing function   A function that decreases over its entire domain.

decreasing on an interval J A function whose values $f(x)$ decrease as x increases in J.

degenerate conic   Conics created when a plane passes through the vertex of a double-napped cone.

degree   Unit of measurement for angles or arcs, equal to 1/360 of the circumference of a circle.

degree mode   A mode found on a calculator to find a trigonometric function in terms of degrees. (Also see radian mode.)

degree of a polynomial   The largest exponent on the variable in any of the terms of the polynomial.

denominator   The expression below the line in a fraction.

determinant   A number that is associated with a square matrix.

discontinuous on an interval J A function that is not continuous on an interval. See continuous on an interval J.

discontinuous at a point $x = a$ A function that is not defined at $x = a$, or has a break at $x = a$, or has a hole in it at $x = a$.

discriminant   For the function $f(x) = ax^2 + bx + c$, the discriminant is $b^2 - 4ac$.

distributive property $a(b + c) = ab + ac$

diverge   When a series does not have a sum.

domain (of a function)   The input values for a function.

dot product   The number found when the corresponding components of two vectors are multiplied and then summed.

double-napped right circular cone   A cone in which the two lines maintain a constant angle between them during the rotation.

E

elementary row operations   One of the following three row operations: multiply a row by a nonzero constant; interchange two rows; and replace a row by the sum of it and a multiple of another row.

elements of a matrix   The numbers in the matrix.

ellipse   The set of all points in the plane such that the sum of the distances from a pair of fixed points, called foci, is a constant.

end behavior   What happens to the graph of a function on the extreme right and left ends of the x-axis.

equal vectors   Two vectors are equal if and only if they have the same magnitude and direction.

equally likely events   When each outcome of an experiment has the same chance of occurring.

equivalent equations   Equations that have the same set of solutions.

equivalent expression   Expressions that are interchangeable.

equivalent statements   Statements that are interchangeable.

error   The difference between an estimated solution and the exact solution.

event   A subset of a sample space of an experiment.

experiment   An activity with observable results.

experimental probability   The probability associated with an actual experiment.

exponential function   A function of the form $f(x) = a^x$, where a is a positive real number other than 1.

extraneous solution   Any solution to the resulting equation that is not a solution of the original equation.

F

factor   Any of the numbers or algebraic expressions which, when multiplied together, form a product.

factorial   For any positive integer n, n factorial, denoted by $n!$, is given by the formula $n! = 1 \cdot 2 \cdot \ldots \cdot n$.

finite sequence   A function whose domain is some initial subset of positive integers $\{1, 2, 3, \ldots, n\}$ for some fixed integer n.

finite series   Sum of a finite number of terms.

foci, focus   One or two points used in determining an ellipse, parabola, or hyperbola.

FOIL method   First-Outer-Inner-Last method used in multiplying two binomials.

fractional expression   Quotient of two algebraic expressions.

frequency   Ratio of the number of times an event occurs to the total number of possible occurrences.

function   A relation that associates each value in a set called the domain with exactly one value.

G

gaps in screen coordinates   See x-gap and y-gap.

Gaussian plane   See complex plane.

grapher or graphing utility   Graphing calculator or a desktop computer with graphing software.

graphical method   A method to find the solution of a function using a graphing utility.

greater than   Order relation $a > b$ if $a - b$ is positive.

greater than or equal to   Order relation $a \geq b$ if $a - b$ is nonnegative.

greatest integer function   The greatest integer function of x is the greatest integer less than or equal to x.

H

half-life   The half-life of a radioactive substance is the amount of time it takes for half of the substance to decay.

histogram   A bar graph that visually represents the information in a frequency table.

horizontal asymptote   An asymptote that is a horizontal line.

horizontal line   A line whose slope is zero.

horizontal line test   A graphical test to determine if a function is one-to-one.

horizontal shift   A transformation of a function obtained by adding a constant to each input of the domain.

Horner's algorithm   A procedure to divide a polynomial by the linear factor, $x - a$.

hyperbola   Set of points in a plane the difference of whose distances from two fixed points, the foci, is a constant.

hypotenuse   Side opposite the right angle in a right triangle.

I

identity   An equation that is always true.

identity matrix   A square matrix with 1 down the main diagonal and 0 elsewhere.

identity property of addition $a + 0 = 0 + a = a$

identity property of multiplication $a1 = 1a = a$

image of x under f   Output value from the function f for input value x.

imaginary axis   See complex plane.

imaginary part of a complex number   See complex number.

inconsistent system   A system of equations with no solution.

increasing function   A function that increases over its entire domain.

increasing on an interval J   A function whose values $f(x)$ increase as x increases in J.

independent events   Two events are independent events if the occurrence of one has no effect on the occurrence of the other.

index   See radical.

index of summation   See sigma notation.

infinite geometric series   A series of the form $\sum_{k=1}^{\infty} ar^{k-1}$ is called an infinite geometric series, and r is called the common ratio.

infinite sequence   A function whose domain is the set of all positive integers $\{1, 2, 3, \ldots, n, \ldots\}$.

initial side of an angle   See angle.

integers   The numbers $\ldots, -3, -2, -1, 0, 1, 2, 3, \ldots$.

intercept(s) of a graph   Points where a line crosses the x and y axes.

intercepted arc   The arc formed when a central angle intersects the circle.

intermediate value property   If a function is continuous on $[a, b]$, then f assumes every value between $f(a)$ and $f(b)$.

interquartile range   The values between the lower quartile and upper quartile.

interval   Continuous subset of the real number line.

interval notation   Notation used to describe intervals.

interval of convergence   The set of all values for which a series converges.

Inverse function   An inverse relation of a function that is also a function.

inverse of a matrix   The inverse of matrix A, if it exists, is a matrix B, such that $AB = BA = I$.

inverse property of addition $a + (-a) = (-a) + a = 0$.

inverse property of multiplication $a \left(\dfrac{1}{a} \right) = \left(\dfrac{1}{a} \right) a = 1, a \neq 0$.

inverse relation (of R)   Denoted as R^{-1}, consists of all ordered pairs (b, a) for which (a, b) belongs to R.

irrational numbers   Real numbers that are not rational.

irrational zeros   Zeros or roots of a function that are irrational numbers.

L

leading coefficient   The coefficient of the leading term.

leading term   The term having the highest degree in a polynomial.

leaf   See stem-and-leaf table.

less than   The order relation: $a < b$ if $b - a$ is positive.

less than or equal to   The order relation: $a \leq b$ if $b - a$ is nonnegative.

line of symmetry   A line over which one half of the graph is a mirror image of the other half.

line of travel   Direction an object travels.

linear correlation   A procedure to determine if there is a linear relationship between two variables.

linear equation   Any equation in which the exponents on the variables are 1, and no term contains a product or quotient of two variables.

linear function   A function that can be written in the form $f(x) = mx + b$, where m and b are real numbers.

linear inequality (in two variables x and y)   An inequality that can be written in one of the following forms: $y < mx + b$, $y \leq mx + b$, $y > mx + b$, or $y \geq mx + b$.

linear regression   Procedure to find the straight line that is the best fit for the data.

linear speed   Distance an object travels per unit of time.

local extrema   A local maximum or a local minimum.

local maximum   A value $f(a)$ is a local maximum of f if there is an open interval (c, d) containing a such that $f(x) \leq f(a)$ for all values, x in (c, d).

local minimum   A value $f(b)$ is a local minimum of f if there is an open interval (c, d) containing b such that $f(x) \geq f(b)$ for all values, x in (c, d).

logarithmic function The inverse of the exponential function $y = a^x$ is called the logarithmic function with base a and is denoted by $y = \log_a x$.

lower bound (for real zeros) A number d is lower bound for the set of real zeros of f if $f(x) \neq 0$ whenever $x < d$.

lower quartile The median of the scores less than the median.

M

magnitude of a vector Length of the vector.

mathematical model An equation or set of equations that describes a problem.

matrix A rectangular array of numbers.

mean (of a set of data) The sum of all data numbers divided by the total number of items.

median (on a data set) The median of a set of data list in order from smallest to largest is (a) the middle number if there is an odd number of data items, or (b) the mean of the two middle numbers if there is an even number of data items.

median (on a triangle) The line segment through one vertex of a triangle to the midpoint of the opposite side.

minute 1/60 of a degree.

mode (of a data set) The number that occurs most frequently in the set.

modulus The absolute value of the complex number $z = a + bi$ is given by $\sqrt{a^2 + b^2}$. Also the length of the segment from the origin to z in the complex plane.

mutually exclusive events Events that have nothing in common.

N

$n!$ See factorial.

natural exponential function The function $f(x) = e^x$.

natural logarithmic function The logarithmic function with base e.

norm Length of a vector.

numerator The expression above the line in a fraction.

O

oblique triangle A triangle that is not a right triangle.

obtuse angle An angle whose measure is between $90°$ and $180°$.

obtuse triangle A triangle in which one angle is greater than $90°$.

octants Any of the eight parts into which a space is divided into by the perpendicular coordinate axes.

ordinary annuity A sequence of equal regular periodic payments to be made in the future.

origin The point where the number lines cross in the Cartesian coordinate system.

outcomes The various possible results of an experiment.

outliers Scores outside the interquartile range.

P

parabola Set of points in a plane that are equidistant from a fixed point, focus, and a fixed line, directrix.

parallel lines Two lines that are either both vertical or have equal slopes.

parallelogram A quadrilateral whose opposite sides are parallel.

parameter The variable used as input in each parametric equation of a relation.

parametric equations A method for defining a relation that employs a function for each x and y.

partial fractions The process of splitting a fraction into a sum of fractions.

periodic function A function f is said to be a periodic function if there is a positive real number h such that $f(x + h) = f(x)$ for every value x in the domain of f.

permutation Any in a set of possible arrangements of a certain number or all elements of a set, in which order is important.

perpendicular lines   Two lines that are at right angles to each other.

piecewise defined function   A function whose domain is divided into several parts and a different function rule is applied to each part.

point-slope form (of a line)   An equation of the line through the point (x_1, y_1) with slope m, $y - y_1 = m(x - x_1)$.

polar coordinate system   A coordinate system whose ordered pair is based on the distance from the origin and the angle measured from the positive x-axis.

polynomial function   A function that can be written in the form $f(x) = a_n x^n + a_{n-1} x^{n-1} + \cdots + a_1 x + a_0$, where n is a nonnegative integer and the coefficients $a_0, a_1, \cdots, a_n$ are real numbers.

power function   A function of the form $f(x) = ax^m$, where a is a positive constant and x is a positive independent variable.

power series   A power series in x is any series of the form $\sum_{k=0}^{\infty} a_k x^k = a_0 + a_1 x + a_2 x^2 + \cdots + a_n x^n + \cdots$.

probability   The relative frequency with which an event can be expected to occur.

Pythagorean theorem   In a right triangle, the square of the hypotenuse is equal to the sum of the squares of the other two sides.

Q

quadrant   Any of the four parts into which a plane is divided by the perpendicular coordinate axes.

quadrantal angle   An angle in standard position whose terminal side coincides with a coordinate axis.

quadratic equations (in x)   An equation that can be written in the form $ax^2 + bx + c = 0$.

quadratic formula   The formula $x = \dfrac{-b \pm \sqrt{b^2 - 4ac}}{2a}$ used to solve $ax^2 + bx + c = 0$.

quadratic function   A function that can be written in the form $f(x) = ax^2 + bx + c$, where a, b, and c are real numbers, and $a \neq 0$.

quadric surface   A surface that is a graph of an equation quadratic in the variables x, y, and z.

R

radian   One radian is the measure of a central angle whose intercepted arc has a length equal to the circle's radius.

radian mode   A mode found on a calculator to find a trigonometric function in terms of radians.

radical   Any expression of the form $\sqrt[n]{a}$ where n is an integer greater than one. The integer n is called the index of the radical and a is called the radicand.

radicand   See radical.

range (of a function)   Set of all images of elements in the domain.

range screen   A special function of a graphing calculator that allows the user to set the viewing box.

rational function   Functions of the form $\dfrac{P(x)}{Q(x)}$, where $P(x)$ and $Q(x)$ are polynomials, $Q(x)$ is not the zero polynomial.

rational numbers   Numbers that are of the form a/b, where a and b are integers, and b is not zero.

rational zeros   Zeros or roots of a function that are rational numbers.

real axis   See complex plane.

real-number properties of equality   The properties reflexive, symmetric, and transitive.

real numbers   The numbers that are associated with points on a number line.

real part of a complex number   See complex numbers.

reciprocal function   The function $f(x) = 1/x$.

rectangular coordinate system   See Cartesian coordinate system.

recursive formula   A formula in which the first few terms are given and the nth term a_n is defined in terms of previous terms of the sequence.

reduced row echelon form   A standard form to express a matrix that all matrices can be placed into.

reference angle See reference triangle.

reference triangle For an angle θ in standard position, the reference triangle is the triangle formed by the terminal side of angle θ, the x-axis, and a perpendicular dropped from a point on the terminal side to the x-axis. The angle at the origin is the reference angle.

reflection A transformation of a function obtained by multiplying the function by -1.

reflectional symmetry One half of the graph is a mirror image of the other half.

reflexive property of equality $a = a$

relation Any set of ordered pairs in the coordinate plane.

relatively prime Two expressions are relatively prime if they have no common factors.

Richter scale Logarithmic scale used in measuring the intensity of an earthquake.

right angle A $90°$ angle.

right triangle A triangle with a $90°$ angle.

root (of a function) A value of the domain that makes the function zero.

rotation equations Equations to transform every point in the rectangular coordinate plane to a new plane by a rotation through the origin.

row operations See elementary row operations.

S

sample space Set of all possible outcomes of an experiment.

scalar A real number.

scalar product See dot product.

scatter plot A plot of all the ordered pairs of two variable data on a coordinate axis.

screen coordinates The coordinates of a point in a viewing rectangle of a grapher.

second 1/60 of a minute, 1/360 of a degree.

sequence A set of numbers with a specific order.

series A finite or infinite sum of terms.

sigma notation The sum $\sum\limits_{k=1}^{n}$ is called sigma notation and k is called the index of summation.

similar triangles Triangles whose corresponding angles are congruent.

slope $\dfrac{\text{Ratio change in } y}{\text{change in } x}$.

slope-intercept form (of a line) An equation of the line with slope m and y intercept b; $y = mx + b$.

SQR or SQRT Built-in square root function on computers and graphing calculators.

square matrix A matrix whose number of rows equals the number of columns.

square system A system of equations where the number of variables equals the number of equations.

standard deviation A measure of how a data set is spread.

standard position An angle is called an angle in standard position when it is positioned on a rectangular coordinate system with its vertex at the origin and its initial side on the positive x-axis.

statistic A number that describes some characteristic of a data set.

stem See stem-and-leaf table.

stem-and-leaf table An arrangement of a numerical data set into a specific table format. Each piece of data is split into a stem (the higher order part of the data) and a leaf (the lower order part of the data).

step function A function whose complete graph looks like a series of steps.

supplementary angles Two angles of positive measure whose sum is $180°$.

surface The graph of a function of two variables.

symmetric property of equality If $a = b$, then $b = a$.

symmetry The exact correspondence of one half of a graph with the other half of the graph.

synthetic division See Horner's algorithm.

system of equations Set of equations.

T

term   Parts of an expression separated by plus or minus signs.

terminal side of an angle   See angle.

theoretical probability   Probability that is based on reasonable probability assignments.

trace of a surface in a plane   The intersection of the surface with the plane.

transitive property   If $a = b$ and $b = c$, then $a = c$.

tree diagram   One way to organize a procedure in which every possibility is diagramed at each stage in the experiment.

triangle   A polygon with three sides and three angles.

trigonometric function   Any of the functions sine, cosine, tangent, secant, cotangent, or cosecant that deal with the relations between sides and angles of triangles. Also related to the ordered pairs of points on the unit circle.

trigonometric identity   An equation that includes trigonometric functions and is true for all permissible values of the variable.

U

unit circle   A circle with radius 1.

unit vectors   Vector of length one.

upper bound (for real zeros)   A number c is upper bound for the set of real zeros of f if $f(x) \neq 0$ whenever $x > c$.

upper quartile   The median of the scores greater than the median.

V

variable   A letter that stands for any number.

variance   A measure of how a data set is spread.

vector   A directed line segment.

vertex of a parabola   The lowest or highest point of the parabola.

vertex of an angle   See angle.

vertical asymptote   An asymptote that is a vertical line.

vertical line   A line with NO slope.

vertical line test (for a function)   If every vertical line intersects the graph of a relation in at most one point, then the relation is a function.

vertical shift (slide)   A transformation of a function obtained by adding a constant.

vertical shrink or stretch   A transformation obtained by multiplying the function by a positive constant.

viewing box   A rectangular parallel-piped portion of 3-space shown by a 3-D graphing utility.

viewing rectangle   The portion that is displayed at any one time on a graphing calculator or computer.

X

x-axis   Usually the horizontal coordinate line with positive direction to the right.

x-component of a vector   See vector.

x-coordinate   The directed distance from the y-axis to the point P in a plane.

x-cut   Trace in the plane $x = a$.

x-gap   The width between x-coordinates of points in a viewing rectangle.

x-intercept   A point that lies on both the graph and the x-axis.

X-max   The value of the right side of the viewing rectangle.

X-min   The value of the left side of the viewing rectangle.

X-scl   The scale of the x-coordinate's tick mark on a graphing calculator or computer.

x-section   See x-cut.

xy-plane   Plane found by letting $z = 0$.

xz-plane   Plane found by letting $y = 0$.

Y

y-axis   Usually the vertical coordinate line with positive direction up.

y-component of a vector   See vector.

y-**coordinate**   The directed distance from the x-axis to the point P in a plane.

y-**cut**   Trace in the plane $y = a$.

y-**gap**   The width between y-coordinates of points in a viewing rectangle.

y-**intercept**   A point that lies on both the graph and the y-axis.

Y-**max**   The value of the top of the viewing rectangle.

Y-**min**   The value of the bottom of the viewing rectangle.

Y-**scl**   The scale of the y-coordinate's tick mark on a graphing calculator or computer.

y-**section**   See y-cut.

yz-**plane**   Plane found by letting $x = 0$.

Z

z-**axis**   In a three dimensional coordinate system, usually the vertical coordinate line with positive direction up.

zero or root   A value of the domain that makes the function zero.

zoom-in   A procedure of a graphing utility used to find solutions to a high degree of accuracy.

zoom-out   A procedure of a graphing utility used to find the end behavior of a function.

Teacher's Answer Section

CHAPTER 1

SECTION 1.1

1. Associative property of addition

2. Distributive property

3. Commutative property of addition

4. Associative property of multiplication

5.

6.

7.

8.

9.

10.

11. 8

12. 5

13. $5 - \sqrt{3}$

14. $\sqrt{5} - 2$

15. 12

16. 7

17. 4

18. 8

19. $|x - 4|$ or $|4 - x|$

20. $|x + 3|$ or $|-3 - x|$

21. **(a)** QI **(b)** y-axis
(c) x-axis **(d)** QIII
(e) QII **(f)** QIII

[-8,8] by [-8,8]

22. $X\text{scl} = Y\text{scl} = 5$

23. $X\text{scl} = Y\text{scl} = 1$

24. $X\text{scl} = Y\text{scl} = 10$

25. $X\text{scl} = 5; Y\text{scl} = 1$

26. 5

27. 3.16

28. 9

29. 6.71

30. $d[(1, 3), (4, 7)] = d[(4, 7), (8, 4)]$ so the three points are the vertices of an isosceles triangle.

31. $(2, 6)$

32. $\left(\dfrac{1}{2}, -2\right)$

33. $\left(\dfrac{9}{2}, \dfrac{\sqrt{2} + 2}{2}\right)$

34. $\left(\dfrac{a + 3}{2}, \dfrac{b - 6}{2}\right)$

35. lengths of the sides of $\triangle ABC$: $d(A, B) = \sqrt{(-1-3)^2 + (2-0)^2} = \sqrt{20} = 2\sqrt{5}$, $d(B, C) = \sqrt{(5-(-1))^2 + (4-2)^2} = \sqrt{40} = 2\sqrt{10}$, $d(A, C) = \sqrt{(5-3)^2 + (4-0)^2} = \sqrt{20} = 2\sqrt{5}$; thus two sides have the same length. $\triangle ABC$ is isosceles but not equilateral.

36. The midpoint of the hypotenuse AB is $M\left(\dfrac{0+5}{2}, \dfrac{7+0}{2}\right) = \left(\dfrac{5}{2}, \dfrac{7}{2}\right)$, $d(A, M) = \sqrt{\left(\dfrac{5}{2}\right)^2 + \left(\dfrac{7}{2}\right)^2} = \sqrt{\dfrac{74}{4}} = \dfrac{1}{2}\sqrt{74}$, $d(B, M) = \sqrt{\left(\dfrac{5}{2}\right)^2 + \left(\dfrac{7}{2}\right)^2} = \sqrt{\dfrac{74}{4}} = \dfrac{1}{2}\sqrt{74}$, $d(C, M) = \sqrt{\left(\dfrac{5}{2}\right)^2 + \left(\dfrac{7}{2}\right)^2} = \sqrt{\dfrac{74}{4}} = \dfrac{1}{2}\sqrt{74}$; therefore the midpoint $M\left(\dfrac{5}{2}, \dfrac{7}{2}\right)$ is equidistant from all three vertices.

37. Consider the right triangle with vertices $A = (0, a)$, $B = (b, 0)$, and $C = (0, 0)$. The midpoint of the hypotenuse is $D = \left(\dfrac{b}{2}, \dfrac{a}{2}\right)$. Compute the distance from D to each vertex: $d(A, D) = d(B, D) = d(C, D) = \dfrac{1}{2}\sqrt{a^2 + b^2}$

38. $\left(\dfrac{7}{3}, 4\right), \left(\dfrac{17}{3}, 5\right)$

39. $\left(4, \dfrac{9}{2}\right), \left(\dfrac{3}{2}, \dfrac{15}{4}\right), \left(\dfrac{13}{2}, \dfrac{21}{4}\right)$

40. $\sqrt{(a-2)^2 + (b-3)^2}$

41. $\sqrt{16 + y^2}$

42. $\sqrt{16 + y^2}$

43. $\sqrt{x^2 + 2x + 10}$

44. $x = 7$ or $x = -5$

45. $x = 8$ or $x = -4$

46. $x = 75$, $y = 43$. Answers may vary depending on grapher model.

47. Answers depend on the grapher model

48. Answers depend on the grapher model

49. If $a \geq 0$, $|a| = a \geq 0$ because a is a positive number. If $a < 0$, $|a| = -a \geq 0$ because $-a$ is a positive number. Thus $|a| \geq 0$.

50. If $a = 0$ or $b = 0$, $|a| \cdot |b| = 0 = |a \cdot b|$. If $a > 0$ and $b > 0$, then $a \cdot b > 0$. Thus $|a| \cdot |b| = a \cdot b = |a \cdot b|$. If $a < 0$ and $b < 0$, then $a \cdot b > 0$. Thus $|a| \cdot |b| = (-a) \cdot (-b) = a \cdot b = |a \cdot b|$. If $a < 0$ and $b > 0$, then $a \cdot b < 0$. Thus $|a| \cdot |b| = (-a) \cdot b = -(a \cdot b) = |a \cdot b|$. Similarly, for $a > 0$ and $b < 0$, $|a| \cdot |b| = a \cdot (-b) = -(a \cdot b) = |a \cdot b|$.

51. Responses should mention that math coordinates are points with no area, but screen coordinates have length and width. There are infinitely many math coordinates in a viewing rectangle but only a finite number of pixels to represent screen coordinates.

SECTION 1.2

1.

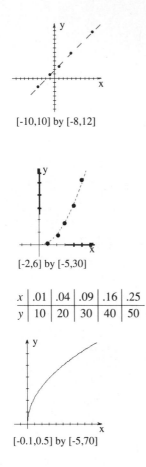

[-10,10] by [-8,12]

2.

[-2,6] by [-5,30]

3.

x	.01	.04	.09	.16	.25
y	10	20	30	40	50

[-0.1,0.5] by [-5,70]

4. The ordered pairs given satisfy $y = x + 2$.

5. The ordered pairs given satisfy $y = x^2$.

6. (b) $y = x^2 + x - 2$

7. (a) $-\$10.75$ (b) $\$8.00$
(c) $\$26.75$ (d) $\$70.50$

8. $y = 0.25x - 35.75$

9.

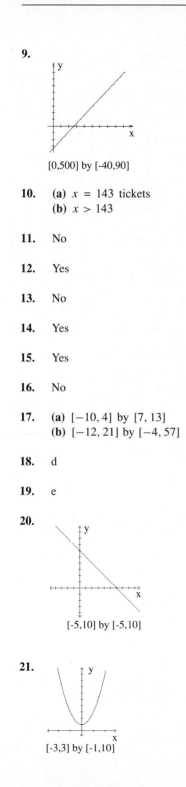

[0,500] by [-40,90]

10. **(a)** $x = 143$ tickets
(b) $x > 143$

11. No

12. Yes

13. No

14. Yes

15. Yes

16. No

17. **(a)** $[-10, 4]$ by $[7, 13]$
(b) $[-12, 21]$ by $[-4, 57]$

18. d

19. e

20.

[-5,10] by [-5,10]

21.

[-3,3] by [-1,10]

22.

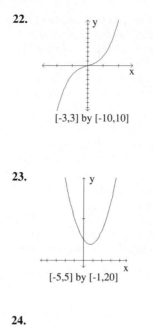

[-3,3] by [-10,10]

23.

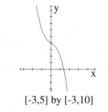

[-5,5] by [-1,20]

24.

[-3,5] by [-3,10]

25.

[-40,40] by [-10,100]

26.

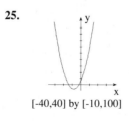

[-2,2] by [-200,200]

27.

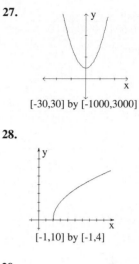

[-30,30] by [-1000,3000]

28.

[-1,10] by [-1,4]

29.

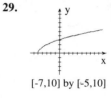

[-7,10] by [-5,10]

30.

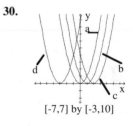

[-7,7] by [-3,10]

31.

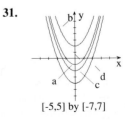

[-5,5] by [-7,7]

32.

x	-3	-2	-1	0	1
y	1	1.41	1.73	2	2.24

x	2	3	4	5	6
y	2.45	2.65	2.83	3	3.16

33.

x	0.25	0.5	1	1.5
y	−0.60	−0.30	0	0.18

x	2	3	4	5
y	0.30	0.48	0.60	0.70

34.

x	0	0.52	1.05	1.57	2.09	2.62
y	0	0.50	0.87	1.00	0.87	0.50

x	3.14	3.67	4.19	4.71
y	0.00	−0.50	−0.87	−1.00

35.

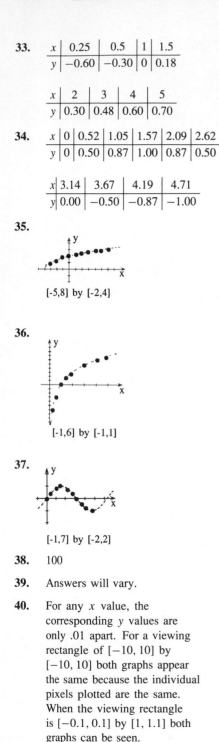

[-5,8] by [-2,4]

36.

[-1,6] by [-1,1]

37.

[-1,7] by [-2,2]

38. 100

39. Answers will vary.

40. For any x value, the corresponding y values are only .01 apart. For a viewing rectangle of $[-10, 10]$ by $[-10, 10]$ both graphs appear the same because the individual pixels plotted are the same. When the viewing rectangle is $[-0.1, 0.1]$ by $[1, 1.1]$ both graphs can be seen.

41. The line $y = 0.5x$ appears "jagged" because a grapher graphs functions by plotting and connecting a finite number of consecutive points. Each curve is drawn by a series of line segments, which gives any curve a "jagged" appearance.

42. Certain viewing rectangles do not show a significant portion of the graph or do not properly indicate the behavior of the graph.

SECTION 1.3

1. $f(0) = -1$, $f(1) = 0$

2. $f(3) = 8$, $f(-5) = 24$

3. $g(0) = 1$, $g(1) = \dfrac{1}{2}$

4. $g(3) = \dfrac{1}{4}$, $g(-5) = -\dfrac{1}{4}$

5. $\dfrac{t}{2 + t}$

6. $x^2 + 4x + 3$

7. $f(0) = 2$; $f(-1) = -3$; $f(4) = 2$

8. $-\dfrac{1}{2}, 4\dfrac{1}{2}$

9. $-\dfrac{1}{4}, 4\dfrac{1}{4}$

10. $-2, 6$

11. **(a)** It is a function. **(b)** It is not a function.

12. **(a)** It is a function. **(b)** It is not a function.

13. **(a)** It is not a function. **(b)** It is not a function.

14. Domain $= [0, \infty)$; range $= [0, \infty)$

15. Domain $= (-\infty, 3) \cup (3, \infty)$; range $= (-\infty, 0) \cup (0, \infty)$

16. Domain $= (-\infty, \infty)$; range $= [-3, \infty)$

17. Domain $= (-\infty, \infty)$; range $= [0, \infty)$

18. Domain $= (-\infty, -4) \cup (-4, \infty)$; range $= (-\infty, 1)$

19. Domain $= (-\infty, 8]$; range $= [0, \infty)$

20. Domain $= (-\infty, \infty)$; range $= [-0.5, 0.1]$

21. Domain $= (-1, \infty)$; range $= (0, \infty)$

22. Domain $= [-\sqrt{8}, 0] \cup [\sqrt{8}, \infty)$; range $= [0, \infty)$

23. Domain $= (-\infty, \infty)$; range $= [-16.25, \infty)$

24. (d) or (e)

25. (d)

26. (c)

27.

[-5,10] by [-3,3]

28.

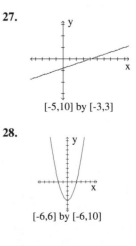

[-6,6] by [-6,10]

29.

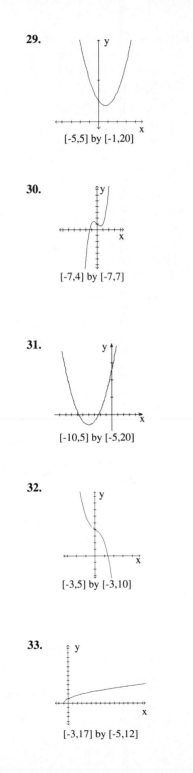

[-5,5] by [-1,20]

30.

[-7,4] by [-7,7]

31.

[-10,5] by [-5,20]

32.

[-3,5] by [-3,10]

33.

[-3,17] by [-5,12]

34.

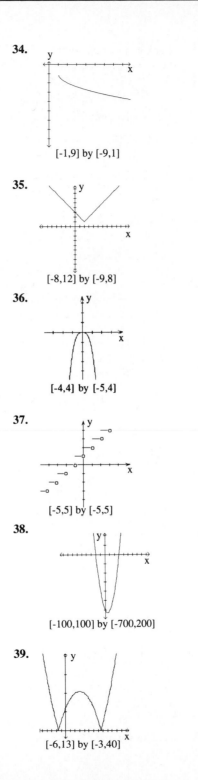

[-1,9] by [-9,1]

35.

[-8,12] by [-9,8]

36.

[-4,4] by [-5,4]

37.

[-5,5] by [-5,5]

38.

[-100,100] by [-700,200]

39.

[-6,13] by [-3,40]

40.

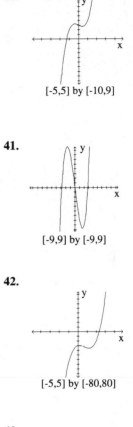

[-5,5] by [-10,9]

41.

[-9,9] by [-9,9]

42.

[-5,5] by [-80,80]

43.

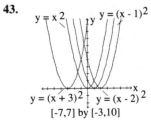

$y = x^2$ $y = (x - 1)^2$

$y = (x + 3)^2$ $y = (x - 2)^2$

[-7,7] by [-3,10]

44.

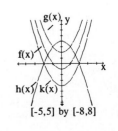

$g(x)$

$f(x)$

$h(x)$ $k(x)$

[-5,5] by [-8,8]

45.

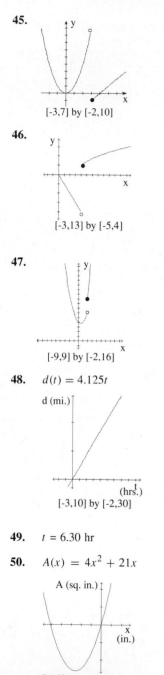

[-3,7] by [-2,10]

46.

y

x

[-3,13] by [-5,4]

47.

y

x

[-9,9] by [-2,16]

48. $d(t) = 4.125t$

d (mi.)

(hrs.)
t

[-3,10] by [-2,30]

49. $t = 6.30$ hr

50. $A(x) = 4x^2 + 21x$

A (sq. in.)

x
(in.)

[-6,3] by [-30,30]

51. Only positive real numbers

52. $x = 0.82$ in.

53. $A(x) = \left(\dfrac{x}{4}\right)^2 + \left(\dfrac{80-x}{4}\right)^2$;
values between 0 and 40 ft since x is smaller perimeter

54. $x = 11.72$ ft

A (sq. ft.)

x
(ft.)

[-50,150] by [-20,500]

55. $y = C(t) = 0.48 + 0.28(\text{INT}(t))$
if t is positive and not an integer.
$y = C(t) = 0.48 + 0.28 \times (\text{INT}(t-1)) = 0.48 + 0.28(t-1)$
if t is a positive integer. Range:
$\{0.48, 0.76, 1.04, \ldots\}$

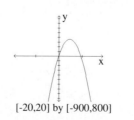

C

t

[0,5] by [0,3]

56. All values of t greater than 0

57. $5 < x \le 6$ min

58. $s = f(t) = -16t^2 + 150t + 10$

y

x

[-20,20] by [-900,800]

59. Graph in QI

60. $t = 2.73$ sec and $t = 6.65$ sec

SECTION 1.4

1. $y = 3x$ rises, $y = \dfrac{1}{3}x$ rises,
$y = -3x$ falls

2. $y = -3x$ falls, $y = 2x$ rises,
$y = -\dfrac{1}{4}x$ falls

3. $y = -2x - 1$ falls, $y = -4x + 3$
falls, $y = 2x + 5$ rises

4. $y = 3 - 2x$ falls, $y = -2x + 3$
falls, $y = 5 + 3x$ rises

5. $-\dfrac{8}{3}$

6. $-\dfrac{13}{4}$

7. $\dfrac{4}{5}$

8. -3

9. $m = 2;\ b = -4$

10. $m = -3;\ b = 1$

11. $m = -\dfrac{2}{3};\ b = \dfrac{2}{3}$

12. $m = \dfrac{1}{3};\ b = -\dfrac{2}{3}$

13. $m = -\dfrac{1}{2};\ b = \dfrac{3}{2}$

14. $m = -\dfrac{1}{3};\ b = -\dfrac{8}{3}$

15.

y

P(1,2)

x

[-4,4] by [-4,5]

16.

y

x

P(2,-1)

[-5,5] by [-5,5]

17.

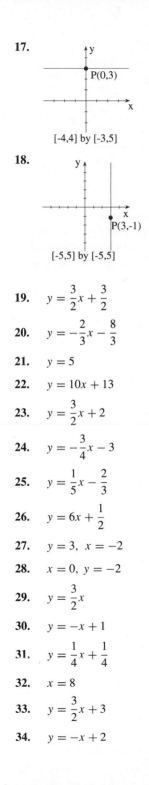

P(0,3)

[-4,4] by [-3,5]

18.

P(3,-1)

[-5,5] by [-5,5]

19. $y = \dfrac{3}{2}x + \dfrac{3}{2}$

20. $y = -\dfrac{2}{3}x - \dfrac{8}{3}$

21. $y = 5$

22. $y = 10x + 13$

23. $y = \dfrac{3}{2}x + 2$

24. $y = -\dfrac{3}{4}x - 3$

25. $y = \dfrac{1}{5}x - \dfrac{2}{3}$

26. $y = 6x + \dfrac{1}{2}$

27. $y = 3, \; x = -2$

28. $x = 0, \; y = -2$

29. $y = \dfrac{3}{2}x$

30. $y = -x + 1$

31. $y = \dfrac{1}{4}x + \dfrac{1}{4}$

32. $x = 8$

33. $y = \dfrac{3}{2}x + 3$

34. $y = -x + 2$

35. $y = 6.5x + 300$

36. $y = \dfrac{85}{3}x + \dfrac{175}{3}$

37. Possible solution:
$X\min = -4.8, Y\min = -3.2; X\max = 4.7, Y\max = 3.1; X\text{scl} = 1, Y\text{scl} = 1$

38. $y = 3x + 1$ appears steeper; actually, $y = 5x - 1$ has the greater slope

(a)

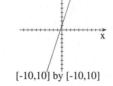

[-10,10] by [-10,10]

(b)

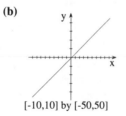

[-10,10] by [-50,50]

39. (a) appears steeper; however, the slopes are the same

(a)

[-10,10] by [-10,10]

(b)

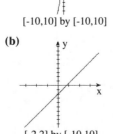

[-2,2] by [-10,10]

40. (b) appears steeper; however, the slopes are the same

(a)

[-10,10] by [-10,10]

(b)

[-25,25] by [-10,10]

41. Use $[-2, 2]$ by $[-10, 10]$ to get an identical graph

42. $y = \dfrac{3}{5}x + \dfrac{17}{5}$

43. $y = x - 2$

44. $y = -\dfrac{1}{2}x + 1$

45. $y = \dfrac{1}{4}x + \dfrac{9}{2}$

46. $x - -2$

47. $y = 2$

48. 4.5 yr

49. Show two sides of $\triangle ABC$ are perpendicular; $m_{AB} = \dfrac{-2 - 2}{-6 - (-1)} = \dfrac{4}{5}$; $m_{BC} = \dfrac{-12 - (-2)}{2 - (-6)} = -\dfrac{5}{4}$; thus $m_{AB} = \dfrac{-1}{m_{BC}}$ and $\triangle ABC$ is a right triangle

50. a represents the x-intercept and b represents the y-intercept

51. $m_{AB} = \dfrac{y_2 - y_1}{x_2 - x_1} =$
$$\dfrac{-1(-y_2 + y_1)}{-1(-x_2 + x_1)} = \dfrac{y_1 - y_2}{x_1 - x_2} = m_{BA}$$

52. $y \le x$

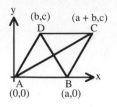

[0,600] by [0,600]

53. Not collinear

54. $D_1(5, 4),\ D_2(-3, 0),\ D_3(9, 10)$

55. The sides do not form a rectangle.

56. $M_{BC} = (2, 3);\ m_{AM} = -3$
and $m_{BC} = \dfrac{1}{3}$. Thus the lines
are perpendicular because
$$m_{AM} = \dfrac{-1}{m_{BC}}.$$

57. For A and M_{BC}: $y = \dfrac{6}{7}x$; for B
and M_{AC}: $y = -12x + 30$; for
C and M_{AB}: $y = -\dfrac{3}{4}x + \dfrac{15}{4}$

58. $A = \left(\dfrac{b}{2}, \dfrac{c}{2}\right)$,
$B = \left(\dfrac{b+a}{2}, \dfrac{c}{2}\right)$. Because
$m_{AB} = 0$, the line AB is
$y = \dfrac{c}{2}$, which is parallel to the
x-axis or the third side of the
triangle.

59. The four midpoints
are: $M_1 = \left(\dfrac{a}{2}, 0\right)$,
$M_2 = \left(\dfrac{b+a}{2}, \dfrac{c}{2}\right)$,
$M_3 = \left(\dfrac{b+d}{2}, \dfrac{e+c}{2}\right)$,

and $M_4 = \left(\dfrac{d}{2}, \dfrac{e}{2}\right)$.
The slopes of the lines
connecting the midpoints
are: $m_{1,2} = \dfrac{c}{b} = m_{3,4}$ and
$m_{2,3} = \dfrac{e}{d - a} = m_{4,1}$. Show by
calculating slopes that opposite
sides of $M_1 M_2 M_3 M_4$ are
parallel. Thus $M_1 M_2 M_3 M_4$ is a
parallelogram.

60. Let the vertices of the
quadrilateral be $(0, 0)$, $(0, b)$,
(c_1, c_2), and (d_1, d_2). Then
the midpoints of each are
$\left(0, \dfrac{b}{2}\right)$; $\left(\dfrac{c_1}{2}, \dfrac{b + c_2}{2}\right)$;
$\left(\dfrac{c_1 + d_1}{2}, \dfrac{c_2 + d_2}{2}\right)$; and
$\left(\dfrac{d_1}{2}, \dfrac{d_2}{2}\right)$. The midpoint
of the segment joining the
first and third points is
$\left(\dfrac{c_1 + d_1}{4}, \dfrac{b + c_2 + d_2}{4}\right)$. The
midpoint of the segment joining
the second and fourth points is
$\left(\dfrac{c_1 + d_1}{4}, \dfrac{b + c_2 + d_2}{4}\right)$. The
line segments share a midpoint,
thus they bisect each other.

61. Label the vertices of the
rectangle as: $A = (0, 0)$,
$B = (0, b)$, $C = (c, b)$, and
$D = (c, 0)$. Then $d(A, C) =$
$\sqrt{c^2 + b^2} = d(B, D)$.

62. Let the vertices of the rectangle
be $(0, 0)$, $(b, 0)$, (b, d), and
$(0, d)$. Then the slope from
$(0, 0)$ to (b, d) is $m_1 = \dfrac{d}{b}$ and
the slope from $(b, 0)$ to $(0, d)$ is
$m_2 = -\dfrac{d}{b}$; $m_1 = -\dfrac{1}{m_2}$ only if
$d = b$, then the rectangle is a
square.

63. Let $ABCD$ be a rhombus with
all sides of length a, as shown:

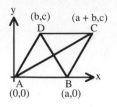

$d(A, B) = d(A, D)$ so
$a^2 - b^2 = c^2$. $ABCD$ is
a parallelogram, so the
coordinates of C are $(a + b, c)$.
The slopes of the diagonals AC
and BD are $m_1 = \dfrac{c}{a + b}$ and
$m_2 = \dfrac{-c}{a - b}$. Then $m_1 \cdot m_2 =$
$\dfrac{-c^2}{a^2 - b^2} = \dfrac{-c^2}{c^2} = -1$.
Therefore the diagonals are
perpendicular.

64. $y = \dfrac{3}{5}x + \dfrac{9}{5}$

65. $y = -\dfrac{1}{11}x + \dfrac{35}{11}$

66. $y = \dfrac{1}{7}x + \dfrac{19}{7}$

67. ℓ_1: $y = \dfrac{c - a}{b}(a) = \dfrac{a(c - a)}{b}$,
ℓ_2: $x = a$ is clearly
satisfied by $\left(a, \dfrac{a(c - a)}{b}\right)$,
ℓ_3: $y = -\dfrac{a}{b}(a) + \dfrac{ac}{b} =$
$\dfrac{-a^2 + ac}{b} = \dfrac{a(c - a)}{b}$

68. Solve each equation for y
to get: $y = -\dfrac{a}{b}x + \dfrac{c}{b}$ and
$y = -\dfrac{a}{b}x + \dfrac{d}{b}$. If $b \ne 0$, the
two lines have the same slope
and hence are parallel. If $b = 0$,
the lines are both vertical, and
therefore parallel.

69. $V = 3187.5t + 42,000$ where
$t = 0$ corresponds to time of
purchase

70. 15.06 yr

71. 15.06 yr after purchase

72. $I = 0.03A + 900$

73.
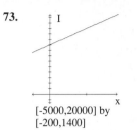
[-5000,20000] by [-200,1400]

74. $4000 at 8%

75. Year 2: 4%; Year 3: 3.85%;
Year 4: 3.70%; Year 5: 3.57%

76.

[-5,5] by [-7,5]

77.
[-4,5] by [-6,4]

78.
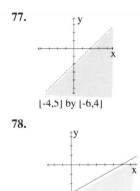
[-3,8] by [-4,4]

79.
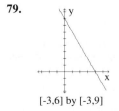
[-3,6] by [-3,9]

80.
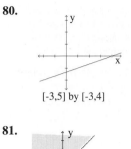
[-3,5] by [-3,4]

81.
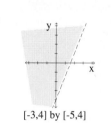
[-4,4] by -4,4]

82.
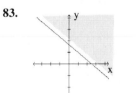
[-3,4] by [-5,4]

83.

[-4,4] by [-3,4]

84. (a) $y = \dfrac{B}{A}x + \left(y_0 - \dfrac{B}{A}x_0 \right)$

(b) $Q(x_1, y_1) =$
$\left(\dfrac{B^2 x_0 - ABy_0 - AC}{A^2 + B^2}, \right.$
$\left. \dfrac{A^2 y_0 - ABx_0 - BC}{A^2 + B^2} \right)$

(c) $d = \sqrt{\dfrac{(Ax_0 + By_0 + C)^2}{A^2 + B^2}}$

(d) From part (c),
$d = \dfrac{|Ax_0 + By_0 + C|}{\sqrt{A^2 + B^2}}$

85. From Exercise 84,
$d = \dfrac{|Ax_0 + By_0 + C|}{\sqrt{A^2 + B^2}}$. From A:

$d = 5.37$. From B: $d = 5.27$.
From C: $d = 6.52$.

86. Vertical lines cannot be graphed;
$x = a$ cannot be written in the
form $y = mx + D$

87. $2.50x + 3.75y > 1500$

88. Answers will vary.

89. Answers will vary.

90. The solution for $0 < 3x + 5$ is
an interval on the number line,
$\left(-\dfrac{5}{3}, \infty \right)$. The solution for
$y < 3x + 5$ is a set of ordered
pairs (x, y) below the line
$y = 3x + 5$.

91. Answers will vary.

SECTION 1.5

1.
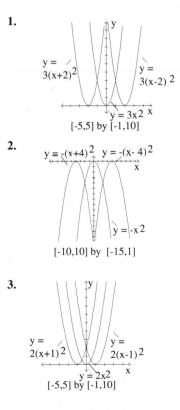
[-5,5] by [-1,10]

2.
[-10,10] by [-15,1]

3.
[-5,5] by [-1,10]

4.

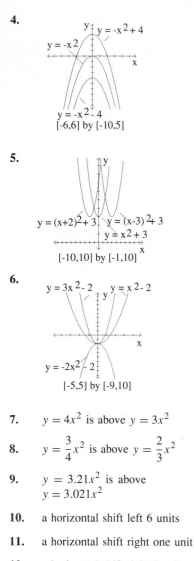

$y = -x^2 + 4$
$y = -x^2$
$y = -x^2 - 4$
[-6,6] by [-10,5]

5.

$y = (x+2)^2 + 3$ $y = (x-3)^2 + 3$
$y = x^2 + 3$
[-10,10] by [-1,10]

6.

$y = 3x^2 - 2$ $y = x^2 - 2$
$y = -2x^2 - 2$
[-5,5] by [-9,10]

7. $y = 4x^2$ is above $y = 3x^2$

8. $y = \dfrac{3}{4}x^2$ is above $y = \dfrac{2}{3}x^2$

9. $y = 3.21x^2$ is above $y = 3.021x^2$

10. a horizontal shift left 6 units

11. a horizontal shift right one unit

12. a horizontal shift right 3 units

13. a horizontal shift right 3 units

14. a vertical stretch by a factor of 4

15. a vertical stretch by a factor of 3, followed by a reflection through the x-axis

16. a horizontal shift right 5 units

17. a horizontal shift left 1 unit

18. a vertical stretch by a factor of 2, followed by a vertical shift down 3 units

19. a vertical stretch by a factor of 3, followed by a reflection through the x-axis then a vertical shift up 2 units

20. Vertex: $(1, 5)$; line of symmetry: $x = 1$

21. Vertex: $(-2, -3)$; line of symmetry: $x = -2$

22. Vertex: $(3, -7)$; line of symmetry: $x = 3$

23. Vertex: $(\sqrt{3}, 4)$; line of symmetry: $x = \sqrt{3}$

24. Vertex: $(5, \sqrt{2})$; line of symmetry: $x = 5$

25. Vertex: $(-4, 3)$; line of symmetry: $x = -4$

26.

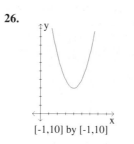

[-1,10] by [-1,10]

27.

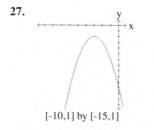

[-10,1] by [-15,1]

28.

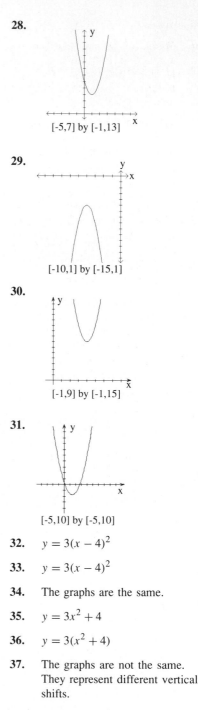

[-5,7] by [-1,13]

29.

[-10,1] by [-15,1]

30.

[-1,9] by [-1,15]

31.

[-5,10] by [-5,10]

32. $y = 3(x - 4)^2$

33. $y = 3(x - 4)^2$

34. The graphs are the same.

35. $y = 3x^2 + 4$

36. $y = 3(x^2 + 4)$

37. The graphs are not the same. They represent different vertical shifts.

38. $y = 3(x + 2)^2 - 4$

39. $y = -2(x-4)^2 + 3$

40. Vertex: $(2, 2)$; line of symmetry: $x = 2$; transformations: a horizontal shift right 2 units followed by a vertical shift up 2 units

41. Vertex: $(3, 3)$; line of symmetry: $x = 3$; transformations: a horizontal shift right 3 units, followed by a vertical shift up 3 units

42. Vertex: $(2, 12)$: line of symmetry: $x = 2$; transformations: a horizontal shift right 2 units followed by a vertical stretch by a factor of 2, then a vertical shift up 12 units

43. Vertex: $(-8, 74)$; line of symmetry: $x = -8$; transformations: a horizontal shift left 8 units followed by a reflection across the x-axis, followed by a vertical shift up 74 units

44. 2 real zeros

45. 1 real zero

46. no real zeros

47. 2 real zeros

48. 2 real zeros

49. no real zeros

50. $(3, 0)$

51. $\left(\dfrac{1}{2}, 0\right)$

52. $f(x) = x^2 - (a+b)x + ab$ to fit into the form $f(x) = Ax^2 + Bx + C$, $A = 1$, $B = -(a+b)$, $C = ab$; line of symmetry is $x = -\dfrac{B}{2A} = -\dfrac{-(a+b)}{2 \cdot 1}$ so $x = \dfrac{a+b}{2}$

53. $\left(\dfrac{a+b}{2}, \dfrac{2ab - a^2 - b^2}{4}\right)$

54. $(-3, -10)$, $(-1, 2)$, $(3, 2)$

55. $f(x-2) = |x-2|$; the graph is obtained by a horizontal shift right 2 units

56. $3f(x) = 3|x|$: the graph can be obtained by a vertical stretch by a factor of 3

57. $2f(x+3) - 1 = 2|x+3| - 1$; the graph is obtained by a horizontal shift left 3 units, then a vertical stretch by a factor of 2, followed by a vertical shift down 1 unit

58. 12 ft by 15 ft

59. $x = 3.5$ ft

60. 25 ft by 25 ft

61. 9 ft and 11 ft

62. 2.50 in.

63. 31,250 sq ft

64. Answers will vary.

65. $y = x^2$ in the standard VR appears steeper, but $y = 3x^2$ is really steeper when both graphs are sketched in the same VR.

SECTION 1.6

1. $(f+g)(x) = x^2 + 2x - 1$, domain $= (-\infty, \infty)$; $(f-g)(x) = -x^2 + 2x - 1$, domain $= (-\infty, \infty)$; $(fg)(x) = 2x^3 - x^2$, domain $= (-\infty, \infty)$; $\left(\dfrac{f}{g}\right)(x) = \dfrac{2x-1}{x^2}$, domain $= (-\infty, 0) \cup (0, \infty)$

2. $(f+g)(x) = x^2 - 3x + 4$, domain $= (-\infty, \infty)$;

$(f-g)(x) = x^2 - x - 2$, domain $= (-\infty, \infty)$; $(fg)(x) = -x^3 + 5x^2 - 7x + 3$, domain $= (-\infty, \infty)$; $\left(\dfrac{f}{g}\right)(x) = \dfrac{(x-1)^2}{3-x}$, domain $= (-\infty, 3) \cup (3, \infty)$

3. $(f+g)(x) = x^2 + 2x$, domain $= (-\infty, \infty)$; $(f-g)(x) = x^2 - 2x$, domain $= (-\infty, \infty)$; $(fg)(x) = 2x^3$, domain $= (-\infty, \infty)$; $\left(\dfrac{f}{g}\right)(x) = \dfrac{x}{2}$, domain $= (-\infty, 0) \cup (0, \infty)$

4. $(f+g)(x) = \sqrt{x} + x - 2$, domain $= [0, \infty)$; $(f-g)(x) = \sqrt{x} - x + 2$, domain $= [0, \infty)$; $(fg)(x) = x^{3/2} - 2x^{1/2}$, domain $= [0, \infty)$; $\left(\dfrac{f}{g}\right)(x) = \dfrac{\sqrt{x}}{x-2}$, domain $= [0, 2) \cup (2, \infty)$

5. $(f+g)(x) = \dfrac{5x+8}{3}$, domain $= (-\infty, \infty)$; $(f-g)(x) = \dfrac{x+10}{3}$, domain $= (-\infty, \infty)$; $(fg)(x) = \dfrac{2x^2 + 5x - 3}{3}$, domain $= (-\infty, \infty)$; $\left(\dfrac{f}{g}\right)(x) = \dfrac{3x+9}{2x-1}$, domain $= \left(-\infty, \dfrac{1}{2}\right) \cup \left(\dfrac{1}{2}, \infty\right)$

6. Domain $= (-\infty, 5) \cup (5, \infty)$; range $= \{-1, 1\}$

[-5,10] by [-5,10]

7. Domain = $(-\infty, -4) \cup (-4, \infty)$; range = $\{-1, 1\}$

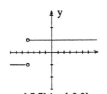

[-7,7] by [-3,3]

8. $(f \circ g)(3) = 5$, $(g \circ f)(-2) = -6$

9. $(f \circ g)(3) = 8$, $(g \circ f)(-2) = 3$

10. $(f \circ g)(3) = 2$, $(g \circ f)(-2) = \sqrt{3}$

11. $(f \circ g)(3) = 9$, $(g \circ f)(-2) = 66$

12. $(f \circ g)(x) = 3x - 1$, $(g \circ f)(x) = 3x + 1$; domain of $f \circ g = (-\infty, \infty)$, range of $f \circ g = (-\infty, \infty)$; domain of $g \circ f = (-\infty, \infty)$, range of $g \circ f = (-\infty, \infty)$

13. $(f \circ g)(x) = \left(\dfrac{1}{x-1}\right)^2 - 1$, $(g \circ f)(x) = \dfrac{1}{x^2 - 2}$; domain of $f \circ g = (-\infty, 1) \cup (1, \infty)$, range of $f \circ g = (-1, \infty)$; domain of $g \circ f = (-\infty, -\sqrt{2}) \cup (-\sqrt{2}, \sqrt{2}) \cup (\sqrt{2}, \infty)$, range of $g \circ f = \left(-\infty, -\dfrac{1}{2}\right) \cup (0, \infty)$

14. $(f \circ g)(x) = x - 2$, $(g \circ f)(x) = x - 1$; domain of $f \circ g = (-\infty, \infty)$, range of $f \circ g = (-\infty, \infty)$; domain of

$g \circ f = (-\infty, \infty)$, range of $g \circ f = (-\infty, \infty)$

15. $(f \circ g)(x) = x - 1$, $(g \circ f)(x) = \sqrt{x^2 - 1}$; domain of $f \circ g = [-1, \infty)$, range of $f \circ g = [-2, \infty)$; domain of $g \circ f = (-\infty, -1] \cup [1, \infty)$, range of $g \circ f = [0, \infty)$

16. $(f \circ g)(x) = \dfrac{1}{x(x+2)}$, $(g \circ f)(x) = \left(\dfrac{x}{x-1}\right)^2$; domain of $f \circ g = (-\infty, -2) \cup (-2, 0) \cup (0, \infty)$, range of $f \circ g = (-\infty, -1] \cup (0, \infty)$; domain of $g \circ f = (-\infty, 1) \cup (1, \infty)$, range of $g \circ f = [0, \infty)$

17. $(f \circ g)(x) = x - 1$, $(g \circ f)(x) = \sqrt{x^2 - 1}$; domain of $f \circ g = [-2, \infty)$, range of $f \circ g = [-3, \infty)$; domain of $g \circ f = (-\infty, -1] \cup [1, \infty)$, range of $g \circ f = [0, \infty)$

18. $(f \circ g)(x) = x$; $(g \circ f)(x) = x$; domain of f is $(-\infty, \infty)$, range of f is $(-\infty, \infty)$; domain of g is $(-\infty, \infty)$, range of g is $(-\infty, \infty)$; domain of $f \circ g$ is $(-\infty, \infty)$, range of $f \circ g$ is $(-\infty, \infty)$; domain of $g \circ f$ is $(-\infty, \infty)$, range of $g \circ f$ is $(-\infty, \infty)$

19. Solve $A(t) \geq 120$ or $12 + 7t + t^2 \geq 120$. Determine where $t^2 + 7t - 108 = 0$: $t = \dfrac{-7 \pm \sqrt{481}}{2}$. From Example 10, we know the solution is the points to the right of the positive zero, so $t \geq \dfrac{-7 + \sqrt{481}}{2}$

$g \circ f = (-\infty, \infty)$, range of $g \circ f = (-\infty, \infty)$

15. $(f \circ g)(x) = x - 1$, $(g \circ f)(x) = \sqrt{x^2 - 1}$; domain of $f \circ g = [-1, \infty)$, range of $f \circ g = [-2, \infty)$; domain of $g \circ f = (-\infty, -1] \cup [1, \infty)$, range of $g \circ f = [0, \infty)$

20. symmetry about $y = x$

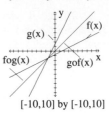

[-10,10] by [-10,10]

21. symmetry about $y = x$

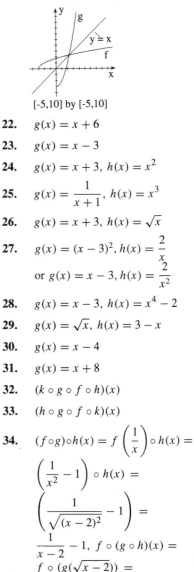

[-5,10] by [-5,10]

22. $g(x) = x + 6$

23. $g(x) = x - 3$

24. $g(x) = x + 3$, $h(x) = x^2$

25. $g(x) = \dfrac{1}{x+1}$, $h(x) = x^3$

26. $g(x) = x + 3$, $h(x) = \sqrt{x}$

27. $g(x) = (x-3)^2$, $h(x) = \dfrac{2}{x}$ or $g(x) = x - 3$, $h(x) = \dfrac{2}{x^2}$

28. $g(x) = x - 3$, $h(x) = x^4 - 2$

29. $g(x) = \sqrt{x}$, $h(x) = 3 - x$

30. $g(x) = x - 4$

31. $g(x) = x + 8$

32. $(k \circ g \circ f \circ h)(x)$

33. $(h \circ g \circ f \circ k)(x)$

34. $(f \circ g) \circ h(x) = f\left(\dfrac{1}{x}\right) \circ h(x) = \left(\dfrac{1}{x^2} - 1\right) \circ h(x) = \left(\dfrac{1}{\sqrt{(x-2)^2}} - 1\right) = \dfrac{1}{x-2} - 1$, $f \circ (g \circ h)(x) = f \circ (g(\sqrt{x-2})) =$

$$f\left(\frac{1}{\sqrt{x-2}}\right) =$$

$$\left(\frac{1}{\sqrt{x-2}}\right)^2 - 1 = \frac{1}{x-2} - 1$$

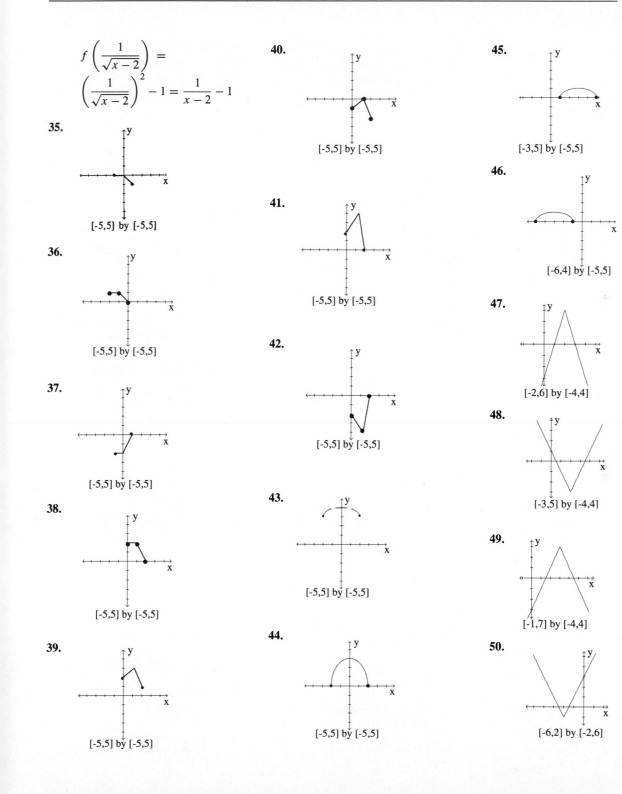

35.

[-5,5] by [-5,5]

36.

[-5,5] by [-5,5]

37.

[-5,5] by [-5,5]

38.

[-5,5] by [-5,5]

39.

[-5,5] by [-5,5]

40.

[-5,5] by [-5,5]

41.

[-5,5] by [-5,5]

42.

[-5,5] by [-5,5]

43.

[-5,5] by [-5,5]

44.

[-5,5] by [-5,5]

45.

[-3,5] by [-5,5]

46.

[-6,4] by [-5,5]

47.

[-2,6] by [-4,4]

48.

[-3,5] by [-4,4]

49.

[-1,7] by [-4,4]

50.

[-6,2] by [-2,6]

51.

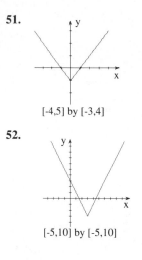

[-4,5] by [-3,4]

52.

[-5,10] by [-5,10]

53. A horizontal shift right 3 units

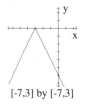

[-2,7] by [-2,6]

54. A horizontal shift left 3 units, followed by a reflection through the x-axis, then a vertical stretch by a factor of 2

[-7,3] by [-7,3]

55. A horizontal shift right 4 units, followed by a vertical stretch by a factor of 3, then a vertical shift up 2 units

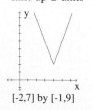

[-2,7] by [-1,9]

56. Horizontal shift right 2 units, followed by a reflection through the x-axis, then a vertical shift up 3 units

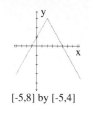

[-5,8] by [-5,4]

57. $(3, 8)$

58. $(3, 16)$

59. $(3, 6)$

60. $(3, -14)$

61. $b = 3$

62. $b = 5$

63. $(4, 5)$, $(2, 3)$

64. $(2, 7)$, $(0, 5)$

65. $(4, -15)$, $(2, -9)$

66. $(4, 13)$, $(2, 9)$

67. $t = 6.68$ sec

68. $D = \sqrt{144 + \left(\dfrac{48}{7}t\right)^2}$; when $t = 14.5$ sec, $D = 100$ ft

69. $t \geq 3.63$ sec

70. $y = 4t$; when $t = 5$ sec, $y = 20$ ft

71. $V = 36\pi t^3$; when $t = 3$, $V = 3053.63$ cu in.

72. $A = 5.29\pi t^2$; when $t = 6$, $A = 598.28$ sq ft

73. $y = 4\pi(1.6 - 0.0027t)^2$, $t = 9.88$ min

74. $t \geq 1.62$ sec

75. $C(x) = $
$\begin{cases} 0.72|1 - [\![1 - x]\!]|, 0 < x < 5 \\ 0.63|1 - [\![4 - x]\!]| + 2.34, 5 \leq x < 15 \\ 0.51|1 - [\![14 - x]\!]| + 8.88, 15 \leq x \end{cases}$

76. Domain $= (0, \infty)$; range $= \{0.72, 1.44, \ldots, 3.60\} \cup \{4.23, 4.86, \ldots, 9.90\} \cup \{9.90 + 0.51k | k$ is a positive integer$\}$

77.

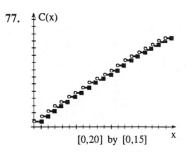

[0,20] by [0,15]

Chapter 1 Review

1. Commutative property of addition; distributive property

2. Associative property of addition; commutative property of addition

3.

4.

5.

6.

7.

8.

9. 2.5

10. $\dfrac{1}{8}$

11. 9

12. $\dfrac{1}{4}$

13. 3

14. 4

15. 8

16. 8192

17. (a) $\sqrt{7} - 2.6$ (b) $\sqrt{7} - 2.6$
(c) $\pi - 3$ (d) $x - 5$

18. $(3, 4)$

19. $(-2, -1)$

20. (a) 6 (b) 7 (c) $\pi - 1$

21. $\sqrt{13}$

22. $\sqrt{146}$

23. $\sqrt{5}$

24. $\sqrt{13}$

25. $\sqrt{13 - 6y + y^2}$

26. $\sqrt{x^2 + 16}$

27. (a) $d(AB) = \sqrt{74}$,
$d(BC) = \sqrt{74}$, $d(AC) = 2\sqrt{2}$;
therefore $\triangle ABC$ is isosceles
(b) $m(AC) = 1$, m(midpoint
segment) $= -1$; slopes are
negative reciprocals, therefore
perpendicular

28. $y = 3x - 10$, $y = 0$,
$y = -\dfrac{3}{4}x + \dfrac{5}{2}$; three lines all
contain the point $\left(\dfrac{10}{3}, 0\right)$, so
they intersect in one point

29. Look at the square with vertices
$(0, 0)$, $(0, a)$, $(a, 0)$, and (a, a).
The slope of the segment
between $(0, 0)$ and (a, a) is
1. The slope of the segment
between $(0, a)$ and $(a, 0)$ is -1.
Since their slopes are negative
reciprocals of each other, the
diagonals are perpendicular.

30. $y = 7x - 17$

31. $y = -\dfrac{1}{7}x + \dfrac{6}{7}$

32. $y = -\dfrac{1}{7}x + \dfrac{6}{7}$

33. (a) $y = x$

34. (b) $y = 5x$

35. $X\text{scl} = 1 = Y\text{scl}$

[-13,7] by [-10,10]

36. $X\text{scl} = 1 = Y\text{scl}$

[-10,10] by [-10,10]

37. $X\text{scl} = 0.1$; $Y\text{scl} = 50$

[-1,1] by [-200,200]

38. $X\text{scl} = 5$; $Y\text{scl} = 100$

[-20,20] by [-500,1500]

39. d

40. d

41. $[-10, 4]$ by $[-9, 21]$

42. $[-12, 21]$ by $[-12, 160]$

43.

[-4,4] by [-7,9]

44.

[-10,10] by [-10,10]

45.

[-6,6] by [-70,70]

46.

[-25,10] by [-10,10]

47.

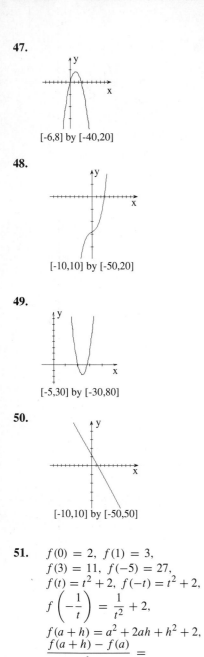

[-6,8] by [-40,20]

48.

[-10,10] by [-50,20]

49.

[-5,30] by [-30,80]

50.

[-10,10] by [-50,50]

51. $f(0) = 2,\ f(1) = 3,$
$f(3) = 11,\ f(-5) = 27,$
$f(t) = t^2 + 2,\ f(-t) = t^2 + 2,$
$f\left(-\dfrac{1}{t}\right) = \dfrac{1}{t^2} + 2,$
$f(a+h) = a^2 + 2ah + h^2 + 2,$
$\dfrac{f(a+h) - f(a)}{h} =$
$\dfrac{2ah + h^2}{h} = 2a + h$

52. $g(2)$ undefined,
$g(0) = \dfrac{1}{2},\ g(-2) =$

$\dfrac{1}{4},\ g(a) = \dfrac{1}{2-a},\ g\left(\dfrac{1}{a}\right) =$
$\dfrac{a}{2a-1},\ g(a+h) = \dfrac{1}{2-a-h}$

53.   **(a)** $f(0) = 0,\ f(-1) = -2,$
$f(2) = 4,$ **(b)** $f(x) = 0,$
$x = 0$ **(c)** $f(x) = 2,$
$x = 0.3$ or 2.70

54.

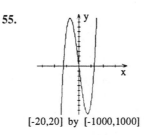

[-10,10] by [-50,20]

55.

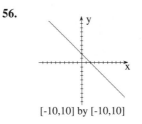

[-20,20] by [-1000,1000]

56.

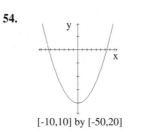

[-10,10] by [-10,10]

57.

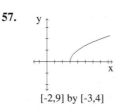

[-2,9] by [-3,4]

58.

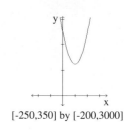

[-250,350] by [-200,3000]

59.

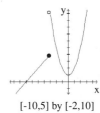

[-5,6] by [-6,2]

60.   Domain $= [3, \infty)$, range $=$
$[0, \infty)$

61.   Domain $= (-\infty, \infty)$, range $=$
$[1250, \infty)$

62.   Domain $= [-3, \infty)$, range $=$
$(-\infty, -2]$

63.   Domain $= (-\infty, \infty)$, range $=$
$(-\infty, \infty)$

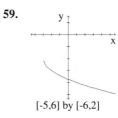

[-10,5] by [-2,10]

64.   Domain $= (-\infty, \infty)$, range $=$
$(-\infty, \infty)$

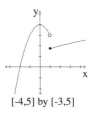

[-4,5] by [-3,5]

65. Domain = $(-\infty, \infty)$, range = $(-2, \infty)$

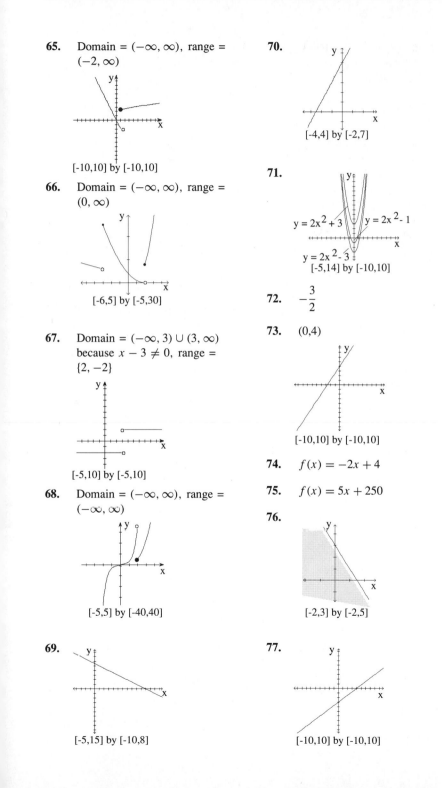

[-10,10] by [-10,10]

66. Domain = $(-\infty, \infty)$, range = $(0, \infty)$

[-6,5] by [-5,30]

67. Domain = $(-\infty, 3) \cup (3, \infty)$ because $x - 3 \neq 0$, range = $\{2, -2\}$

[-5,10] by [-5,10]

68. Domain = $(-\infty, \infty)$, range = $(-\infty, \infty)$

[-5,5] by [-40,40]

69.

[-5,15] by [-10,8]

70.

[-4,4] by [-2,7]

71.

$y = 2x^2 + 3$ $y = 2x^2 - 1$

$y = 2x^2 - 3$

[-5,14] by [-10,10]

72. $-\dfrac{3}{2}$

73. $(0,4)$

[-10,10] by [-10,10]

74. $f(x) = -2x + 4$

75. $f(x) = 5x + 250$

76.

[-2,3] by [-2,5]

77.

[-10,10] by [-10,10]

78.

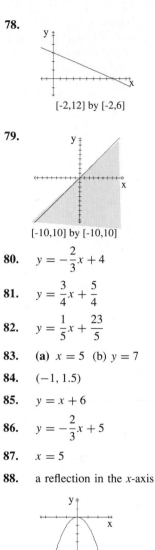

[-2,12] by [-2,6]

79.

[-10,10] by [-10,10]

80. $y = -\dfrac{2}{3}x + 4$

81. $y = \dfrac{3}{4}x + \dfrac{5}{4}$

82. $y = \dfrac{1}{5}x + \dfrac{23}{5}$

83. **(a)** $x = 5$ **(b)** $y = 7$

84. $(-1, 1.5)$

85. $y = x + 6$

86. $y = -\dfrac{2}{3}x + 5$

87. $x = 5$

88. a reflection in the x-axis

[-4,4] by [-7,2]

89. a horizontal shift right 3 units

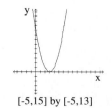

[-5,15] by [-5,13]

90. a reflection in the *x*-axis, followed by vertical stretch by a factor of 2, horizontal shift left 3 units, then a vertical shift up 4 units

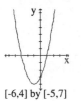

[-7,2] by [-4,5]

91. a horizontal shift left 1 unit then vertical shift down 4 units

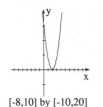

[-6,4] by [-5,7]

92. a horizontal shift right 2 units, followed by a vertical stretch by a factor of 4

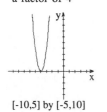

[-8,10] by [-10,20]

93. a horizontal shift left 4 units, followed by a vertical stretch by a factor of 4

[-10,5] by [-5,10]

94. a horizontal shift left 3 units, followed by a vertical stretch by

a factor of 2 then a reflection in the *x*-axis

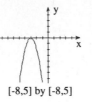

[-8,5] by [-8,5]

95. a horizontal shift right 3 units, followed by a vertical stretch by a factor of 2

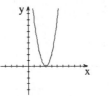

[-5,10] by [-5,10]

96. a horizontal shift right 3 units, followed by a vertical stretch by a factor of 2

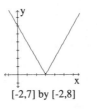

[-2,7] by [-2,8]

97. a horizontal shift left 4 units, then a reflection through the *x*-axis followed by a vertical shift down 3 units

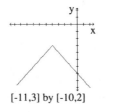

[-11,3] by [-10,2]

98. $b = 3$

99. (a) 2 (b) 0 (c) −6 (d) 2

100. (a) $y = 2x^2 + 1$

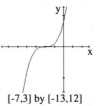

[-3,3] by [-2,7]

(b) $y = 2(x^2 + 1)$ (c) no, the line is shifted vertically 2 units instead of 1

101. Vertical stretch by a factor of 2, then horizontal shift right 3 units, then vertical shift down 14 units

102. a reflection in the *x*-axis, then horizontal shift 3 units left, then vertical shift up 23 units

103. a horizontal shift left 2 units

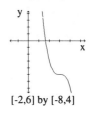

[-7,3] by [-13,12]

104. Reflection in *x*-axis; vertical stretch by a factor of 2; horizontal shift right 3 units; vertical shift down 5 units

[-2,6] by [-8,4]

105. (a) $f(3) > -3$ (b) $g(4) > 1$

106. (a) $f(-2) > g(-2)$
(b) $f(2) < g(2)$

107. $x = -1$ is a solution to (c)

108. a horizontal shift left 3 units, followed by a vertical shift up 2

units

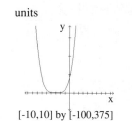

[-10,10] by [-100,375]

109. $(f + g)(x) = \dfrac{x^3 + 3x^2 - 3}{x - 1}$,
domain $= (-\infty, 1) \cup (1, \infty)$;
$(f - g)(x) = \dfrac{-x^3 - 3x^2 + 5}{x - 1}$,
domain $= (-\infty, 1) \cup (1, \infty)$;
$(fg)(x) = \dfrac{(x + 2)^2}{x - 1}$, domain $=$
$(-\infty, 1) \cup (1, \infty)$; $\left(\dfrac{f}{g}\right)(x) =$
$\dfrac{1}{(x - 1)(x + 2)^2}$, domain $=$
$(-\infty, -2) \cup (-2, 1) \cup (1, \infty)$

110. $(f \circ g)(-3) = -28$, $(g \circ f)(2) = 13$

111. $(f \circ g)(-3) = 8$, $(g \circ f)(2)$ is undefined

112. $(f \circ g)(x) = \dfrac{x^2 + 6x + 17}{4}$, $(g \circ f)(x) = \dfrac{x^2 + 5}{2}$

113. Domain of $f = (-\infty, \infty)$,
range of $f = (-\infty, \infty)$;
domain of $g = (-\infty, \infty)$, range
of $g = [-2, \infty)$; domain of
$f \circ g = (-\infty, \infty)$, range of
$f \circ g = [-1.75, \infty)$; domain
of $g \circ f = (-\infty, \infty)$, range of
$g \circ f = [-2, \infty)$

114. (a) $g(x) = x - 5$
(b) $g(x) = x + 4$

115. (a) $g(x) = x + 5$
(b) $g(x) = x - 6$

116. $y = (k \circ g \circ f \circ h)(x)$

117. $g(x) = x + 5$, $h(x) = x^2$

118. $g(x) = (x - 2)^2, h(x) = \dfrac{3}{x}$ or
$g(x) = (x - 2), h(x) = \dfrac{3}{x^2}$

119.

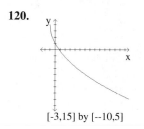

[-10,10] by [-10,10]

120.

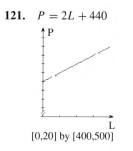

[-3,15] by [--10,5]

121. $P = 2L + 440$

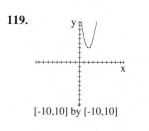

[0,20] by [400,500]

122. $A = 125w$

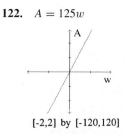

[-2,2] by [-120,120]

123. $d = 4.25t$ $t = 6.12$ hr

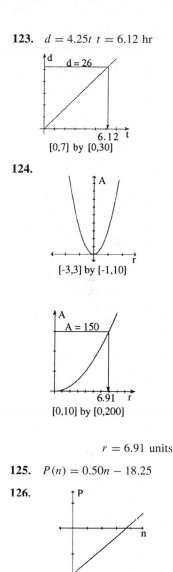

[0,7] by [0,30]

124.

[-3,3] by [-1,10]

[0,10] by [0,200]

$r = 6.91$ units

125. $P(n) = 0.50n - 18.25$

126.

[-10,50] by [-20,15]

The problem situation is
represented by those points
on the graph corresponding
to nonnegative integer values
for n.

127. For the algebraic representation,
$D = R = (-\infty, \infty)$.
For the problem situation,

$D = \{0, 1, 2, 3, \ldots\}$, $R =$ $\{-18.25, -17.75, -17.25, -16.75,$ $\ldots, 0, 0.25, \ldots\}$

128. 37 tickets sold yields a profit $(P > 0)$.

129. $C(x) = 85x + 75,000$

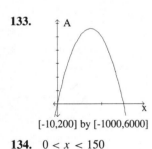

[0,1000] by [0,160000]

130. The problem situation is represented by the portion of the graph in QI that corresponds to integer values of x.

131. 800 bikes

132. Let $x =$ length and $w =$ width. If $x + w = 150$, then $w = 150 - x$. Thus $A = \ell \cdot w = x(150 - x)$.

133.

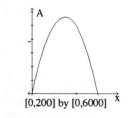

[-10,200] by [-1000,6000]

134. $0 < x < 150$

[0,200] by [0,6000]

135. $P(x) = 0.15x - 83,000$

136. 553,334 jars

137. 620,000 jars

138. $d = 48t$

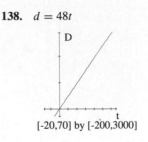

[-20,70] by [-200,3000]

139. The portion of the graph in QI represents the problem situation (i.e., $t \geq 0$).

140. $t = 25$ hr

141. $0.40x + 0.15y > 1200$

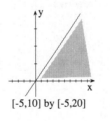

[-500,4000] by [-2000, 9000]

142. $.75x + 1.10y \leq .999(x + y)$ where $x \geq 0, y \geq 0$

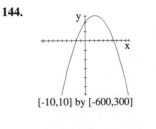

[-5,10] by [-5,20]

143. $s(t) = -16t^2 + 70t + 200$

144.

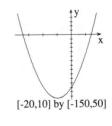

[-10,10] by [-600,300]

145. first quadrant

146. 0.39 sec and 3.98 sec

147. 0.39 sec $< t <$ 3.98 sec

CHAPTER 2

SECTION 2.1

1. 3, -2

2. $\pm\sqrt{14}$

3. 2, 1

4. There are no real solutions.

5. 0, ± 1

6. 8, -4

7. 2 solutions

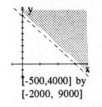

[-4,6] by [-14,5]

8. 2 solutions

[-20,10] by [-150,50]

9. 2 solutions

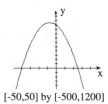

[-50,50] by [-500,1200]

10. 2 solutions

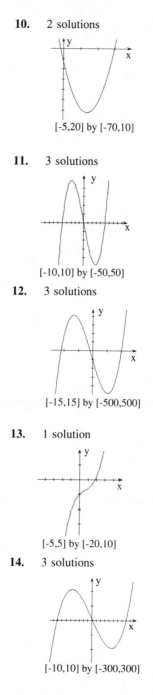

[-5,20] by [-70,10]

11. 3 solutions

[-10,10] by [-50,50]

12. 3 solutions

[-15,15] by [-500,500]

13. 1 solution

[-5,5] by [-20,10]

14. 3 solutions

[-10,10] by [-300,300]

15. The large range required for a complete graph of

$y = \frac{1}{8}x^4 - 5x^2 + 2$ makes it difficult to determine the zeros between -1 and 1.

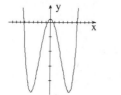

[-10,10] by [-50,10]

16. Graph $y = \frac{1}{2}x^3 - 7x^2 + 3$. The large range required for a complete graph of $y = \frac{1}{2}x^3 - 7x^2 + 3$ makes it difficult to determine the zeros between -1 and 1.

[-5,15] by [-250,50]

17. 0.01

18. 0.1

19. 0.5

20. 0.001

21. 0.01

22. 0.001

23. 0.326

24. 0.05

25. Answers will vary. One sequence is [1, 2] by [−1, 1]; [1.5, 1.6] by [−0.1, 0.1]; [1.57, 1.58] by [−0.01, 0.01]; [1.574, 1.575] by [−0.001, 0.001]

26. Answers will vary. One sequence is: [−2, −1] by

[−1, 1], [−1.9, −1.8] by [−0.1, 0.1], [−1.90, −1.89] by [−0.01, 0.01], [−1.894, −1.893] by [−0.001, 0.001]

27. between 1.52 and 1.53

28. between 10.88 and 10.89

29. between 10.19 and 10.20

30. between 0.31 and 0.32

31. 4.39, 0.61

32. 0.43

33. 5.00, 15.00, 40.00

34. −0.44, 4.94

35. −1.5, 4.5

36. 3

37. $\sqrt[3]{10}$ Approximations may vary.

38. $\sqrt[3]{-4}$ Approximations may vary.

39. 3.39, 0.70

40. −2.00, 2.68, 9.32

41. 1.86, 5.00, 8.14 radians

42. There is no solution.

43. 2858 cones

44. Let x = length. Then $P = 2x + 2w = 320$ implies $2w = 320 - 2x$ or $w = 160 - x$. Both length and width must be positive. Therefore $160 - x > 0$ and $0 < x < 160$ in. for the problem situation.

45. $A = x(160 - x)$

46.

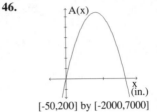

[-50,200] by [-2000,7000]

47.   If $x = -40$, $A(x) = y = -40(160 - (-40)) = -8000$. Thus $(-40, -8000)$ is a point on the graph, but neither length nor width can be negative. The coordinates have no meaning in the problem situation.

48.   67.88

49.   0.003 cm

50. $A(x) = (11 - 2x)(8.5 - 2x)$

51.   1.28 in.

52.   \$12.24, 552 units

53.   \$78.40, 255 units

54.   Answers will vary.

55.   **(a)**

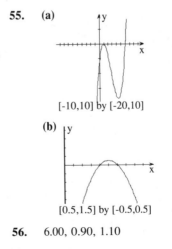

[-10,10] by [-20,10]

     (b)

[0.5,1.5] by [-0.5,0.5]

56.   6.00, 0.90, 1.10

57. $t = 7.47$ sec

SECTION 2.2

1.

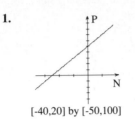

[-40,20] by [-50,100]

2.

[-3,3] by [-10,20]

3.

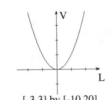

[-25,25] by [-2000,4000]

4.

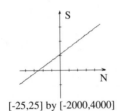

[-3,3] by [-100,1000]

5. $A = 50L$

6. $A = 200W$

7.

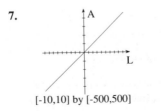

[-10,10] by [-500,500]

8.

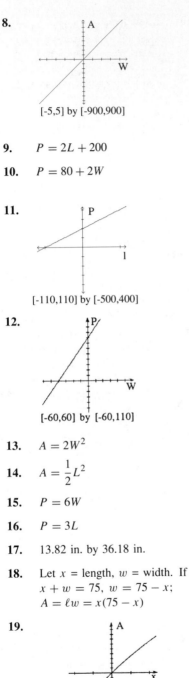

[-5,5] by [-900,900]

9. $P = 2L + 200$

10. $P = 80 + 2W$

11.

[-110,110] by [-500,400]

12.

[-60,60] by [-60,110]

13. $A = 2W^2$

14. $A = \dfrac{1}{2}L^2$

15. $P = 6W$

16. $P = 3L$

17.   13.82 in. by 36.18 in.

18.   Let x = length, w = width. If $x + w = 75$, $w = 75 - x$; $A = \ell w = x(75 - x)$

19.

x = -10
y = -850

[-12,12] by [-1000,1000]

20. It is a complete graph of the algebraic representation, not the problem situation, since it includes domain values that don't apply to the problem.

21. $A = x(180 - x)$

22. $0 < x < 180$

23.

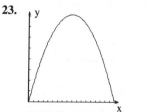

[0,180] by [0,8100]

24. The graph of the algebraic representation will include all real values for x: the graph of the problem situation will include only positive values for x.

25. $625 = LW$

26. $L = \dfrac{625}{W}$

[-1000,1000] by [-10,10]

27. The portion in QI represents the problem situation since neither width nor area can be negative.

28. $y = \dfrac{6000}{x}$

29.

[-20,100] by [-200,1000]

30. 133.33 ft

31. $I = 0.07x + 0.085(12{,}000 - x)$

32.

[-10000,10000] by [-1000,2000]

33. $0 < x < 12{,}000$

34. $8000 at 7% and $4000 at 8.5%

35. $C = 30x + 100{,}000$

36.

[-1000,1000] by [-200000,150000]

37. 8000 pairs of shoes

38. $R = 50x$

39. 5000 pairs of shoes

40. Answers will vary.

41. Answers will vary.

42. $11,400 is invested at 6.75% and $8600 is invested at 8.6%

43. $8000 is invested at 7%, $4000 is invested at 8.5%

SECTION 2.3

1. conditional equation

2. identity

3. identity

4. conditional equation

5. conditional equation

6. identity

7. 2 is a solution

8. 3 is not a solution

9. **(a)** -3 is a solution **(b)** -2 is not a solution **(c)** -1 is a solution

10. **(a)** 1 is not a solution **(b)** -3 is a solution **(c)** 2 is a solution

11. 1

12. -8

13. $-\dfrac{46}{9}$

14. $-\dfrac{4}{5}$

15. $\dfrac{8}{19}$

16. 29

17. 3 or -5

18. there is no solution

19. $\dfrac{7}{4}$ or $\dfrac{5}{4}$

20. $-\dfrac{1}{5}$ or $\dfrac{7}{5}$

21. -4 or $\dfrac{2}{3}$

22. -2 or 1

23. 3 or 2

24. 5 or −4

25. 3 or 1

26. $\frac{1}{2}$ or −3

27. $\frac{1}{2}$ or $\frac{3}{2}$

28. 3 or 5

29. −5 or 1

30. $x = \dfrac{-7 \pm \sqrt{57}}{2}$

31. $x = -8,\ 2$

32. $x = 2 \pm \sqrt{11}$

33. $x = \dfrac{5 \pm \sqrt{37}}{2}$

34. $\dfrac{-1 \pm \sqrt{5}}{2}$

35. $2 \pm \sqrt{2}$

36. $-4 \pm 3\sqrt{2}$

37. 1 or $\dfrac{1}{2}$

38. $1 \pm 2\sqrt{2}$

39. 4 or −1

40. $-4 \pm \sqrt{21}$

41. $\dfrac{\sqrt{3} \pm \sqrt{23}}{2}$

42. 8 or −3

43. 0

44. ±2

45. ±1

46. 2 real solutions

47. no real solutions

48. 2 real solutions

49. 2 real solutions

50. 37°C

51. $\dfrac{5}{9}(F - 32) = C$

52. $2(20) + 2l = 360$

53. 3200 sq in.

54. $\ell = 120$ in. and $w = 60$ in.

55. $y = \dfrac{180}{x}$

56. 7.5 cm

57. 15 cm

58. 15ℓ of 20% acid and 10ℓ of 35% acid should be used

59. 0.75ℓ of distilled water should be added

60. $\dfrac{S - P}{Pr} = n$

61. 25 yr

62. $\dfrac{2A - hb_2}{h} = b_1$

63. 11.98 units

64. $15

65. 440 units

66. $149, 42.43 or 43 units

67. 4 or 17 listings yield $300 profit.

68. Maximum profit is $520 when $x = 10$.

69. As the number increases to 10, profits increase. Beyond 10 listings, profits decrease. More agents might be needed to handle the additional listings.

SECTION 2.4

1. Any x in $(-\infty, 2)$ is a solution.

2. Any x in $(-\infty, -2)$ is a solution.

3. Any x in $[2, 8]$ is a solution.

4. Any x in $\left(\dfrac{9}{4}, \dfrac{11}{2}\right]$ is a solution.

5. $-3 \le x < 5$

6. $-3 < x < 7$

7. $x < 4$

8. $x \ge 3$

9.

10.

11.

12.

13. $(-1, 1)$

14. $(0, 3]$

15. $(-1, \infty)$

16. $[-1, 0] \cup (1, \infty)$

17. $(-\infty, 0] \cup (2.5, \infty)$

18.

19.

20.

21.

22. $|x| < 2$

23. $|x - 1| \ge 1$

24. $-2 > x$ or $(-\infty, -2)$

25. $x \le \dfrac{34}{7}$ or $\left(-\infty, \dfrac{34}{7}\right]$

26. $(-1, \infty)$

27. $\left(-\infty, \dfrac{33}{13}\right)$

28. $\left(-\infty, -\dfrac{11}{7}\right)$

29. $[-4, 3)$

30. $\dfrac{1}{3} < x < 3$ or $\left(\dfrac{1}{3}, 3\right)$

31. $\dfrac{17}{2} \geq x \geq -\dfrac{1}{2}$ or $\left[-\dfrac{1}{2}, \dfrac{17}{2}\right]$

32. $\left(2, \dfrac{18}{5}\right)$

33. $[-3, 2)$

34. $(-1, 0)$

35. $[7, \infty)$

36. $(2, 5], \ 2 < x \leq 5$

37. $(3, 7), \ 3 < x < 7$

38. $|x - 3| < 2$

39. $|x| < 4$

40.
-2 0 2

41.
-3 0 3

42.
-1 1 3

43.
-3 2 7

44. $(1, 5)$

45. $[-8, 2]$

46. $\left(\dfrac{20}{3}, \infty\right)$

47. $\left(-\infty, -\dfrac{4}{3}\right) \cup \left(\dfrac{4}{3}, \infty\right)$

48. $(0, 2) \cup (2, \infty)$

49. $(-\infty, -2) \cup (-2, 3)$

50. $\left(-\infty, -\dfrac{1}{3}\right) \cup \left(\dfrac{1}{3}, \infty\right)$

51. $\left(-\infty, \dfrac{3}{2}\right)$

52. $x < 5$

[-3,10] by [-5,8]

53. $x > 3.5$

[-2,7] by [-2,8]

54. $\left(1, \dfrac{18}{5}\right]$

55. $(2, \infty)$

56. $\left(-13, \dfrac{9}{2}\right)$

57. $\left(-\dfrac{33}{13}, \infty\right)$

58. $(-\infty, 1)$

59. $\left(-\dfrac{14}{3}, \infty\right)$

60. $7.50(x + 1) + 5.75 \leq 45$

61. $x \leq 4.2\overline{3}$

62. 4 friends

63. $r \geq 52.5$ mph

64. less than $4\dfrac{1}{4}$ hr

65. $0 < w < 34$ in.

66. 100,001 candy bars

67. $10 \leq P \leq 20$

68. more than $100,000

69. $a > b$ if $b - a$ is a negative number

70. $\dfrac{9}{5}C + 32 \geq 212$

71. $C \geq 100$

72. $0.42x - 20,000$

73. $0.42x - 20,000 > 0$

74. produce at least 47,620 lb of candy

75. $(4, \infty)$

76. $(-\infty, -1.50)$

77. $(-2.50, \infty)$

78. **(a)** $(1.9, 2.1)$ **(b)** $(0.7, 1.3)$ on the y-axis

79.
1.9 2.1

80.
1.3
1.0
0.7

81. $0 < d < 0.00\overline{3}$

82. Part 1: If $a < b$, then $a + c < b + c$.

- $a < b$ implies that $b - a > 0$
  Definition of less than

- $b - a + (c - c) > 0$
  Additive property of zero

- $b + c - a - c > 0$
  Commutative property of addition

- $(b + c) - (a + c) > 0$
  Distributive property and

Associative property of addition

- $a + c < b + c$
  Definition of less than

  Part 2: If $a < b$, and $c > 0$, then $ac < bc$.

- $a < b$ implies that $b - a > 0$
  Definition of less than
  If $c > 0$, $(b - a)c > 0$
  The product of 2 positive numbers is a positive number

- $bc - ac > 0$
  Distributive property

- $ac < bc$
  Definition of less than

  Part 4: If $a < b$, and $b < c$, then $a < c$. $a < b$ implies that $b - a > 0$
  Definition of less than
  If $b < c$, $c - b > 0$
  Definition of less than

- $b - a + c - b > 0$
  The sum of 2 positive numbers is a positive number

- $b - b + c - a > 0$
  Commutative property of addition

- $c - a > 0$
  Additive property of zero

- $a < c$
  Definition of less than

83. For Method 1, graph
$y = [3(x - 1) + 2] - [5x + 6] = -2x - 7$ and find the values of x for which the graph lies below the x-axis and the x-coordinates of the x-intercept. For Method 2, graph $y = 3(x - 1) + 2$ and $y = 5x + 6$ in the same viewing rectangle. Find the values of x for which the graph of $y = 3(x - 1) + 2$ is above or intersects the graph of $y = 5x + 6$.

SECTION 2.5

1. 3 sec

2. $-16t^2 + 80t = 0$

3. (a) $-16t^2 + 80t > 10$
(b) $-16t^2 + 80t \geq 10$

4. $s = -16t^2 + 48t$

5. $s = -16t^2 + 32t + 120$

6. $(-\infty, 2)$, $(-2, 3)$, $(3, \infty)$

7. $(-\infty, -6)$, $(-6, -5)$, $(-5, \infty)$

8. $(-\infty, -3)$, $(-3, -1)$, $(-1, \infty)$

9. $(-\infty, -5)$, $(-5, -3)$, $(-3, 2)$, $(2, \infty)$

10. $(-\infty, -2)$, $(-2, 3)$, $(3, 4)$, $(4, \infty)$

11. $(-\infty, 2)$, $(2, 4)$, $(4, \infty)$

12. $(-\infty, -2)$, $(-2, 3)$, $(3, 5)$, $(5, \infty)$

13. $(-\infty, -2)$, $(-2, -1)$, $(-1, 1)$, $(1, 2)$, $(2, \infty)$

14. $(-\infty, -3)$, $(-3, 2)$, $(2, \infty)$

15. $(-\infty, -1 - \sqrt{2})$, $(-1 - \sqrt{2}, -1 + \sqrt{2})$, $(-1 + \sqrt{2}, 3)$, $(3, \infty)$

16. $y = x^2 - 3x + 5$

17. $y = 7x^3 - 2x^2 - 5x + 3$

18. $y = x^2 - 3x - 18$

19. $y = x^5 - 6x^2 - x - 3$

20. $(-2, 1)$

21. $[0, 3]$

22. $(-\infty, 2] \cup [3, \infty)$

23. $(-3, 7)$

24. $(-\infty, -3] \cup \left[\dfrac{1}{2}, \infty\right)$

25. $(0, 1)$

26. $(-\infty, -3] \cup (1, \infty)$

27. $[-2, 0] \cup [4, \infty)$

28. $(-\infty, -1] \cup [0.67, 3]$

29. $(-\infty, -1.98) \cup (3.94, 7.04)$

30. $[2.55, \infty)$

31. $[0, 0.41) \cup (2.73, 6.69) \cup (9.01, 10]$ using radians

32. $(2, \infty)$

33. $(-\infty, -4) \cup (-1, 2)$

34. $(-1, 1)$

35. $(-3, 1)$

36. $(-\infty, 0.70) \cup (4.30, \infty)$

37. $[0.28, 2.39]$

38. $[2.15, \infty)$

39. $(-\infty, -3.66)$

40. $(-\infty, -5) \cup (2, \infty)$

41. $(-\infty, 2)$

42. $(3, \infty)$

43. $(-1, 1] \cup (3, \infty)$

44. $S(t) = -16t^2 + 100t + 200 = 0$

45. The complete graph of the algebraic representation is shown. The graph of the problem situation is that portion of $y = -16t^2 + 100t + 200$ in QI.

[-10,10] by [-100,400]

46. 7.84 sec

47. $t = 7.84$ sec

48. For 44, $-16t^2 + 275t + 155 = 0$.
For 45, graph
$y = -16t^2 + 275t + 155$. The
complete graph of the algebraic
representation is shown. The
graph of the problem situation
is that portion in QI.

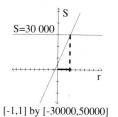

[-5,25] by [-200,1500]

For 46, 17.73 sec in the air. For
47, $t = 17.73$ sec

49. $A(x) = 4x^2 + 100x$

50. $200 < 4x^2 + 100x < 360$

51. $1.86 < x < 3.19$ ft

52. $30,000 \le 5000(1 + 18r)$

53. $[0.28, \infty)$

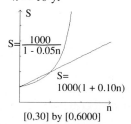

[-1,1] by [-30000,50000]

54. $0 < x < 11$ in.

55. $n = 10$ yr

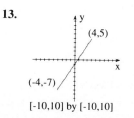

[0,30] by [0,6000]

56. after ten yr

57. $|x + 5| \le |x| + |5|$ for all values
of x

58. true for all values of x

59. true for all x

60. $|a + b| \le |a| + |b|$ for all values
of a and b

SECTION 2.6

1. $(3, 8)$ is in the relation; $(4, 2)$ is
not in the relation; $2(3) - 8 = -2\checkmark$
$2(4) - 2 = 6 \ne -2$

2. $(2, 5)$ is in the relation; $(5, 2)$ is
not in the relation; $2(2)^2 - 5 = 3\checkmark$
$2(5)^2 - 2 = 48 \ne 3$

3. $(-3, 6)$ is in the relation;
$(-2, 4)$ is not in the relation;
$-3 + 6^2 = 33\checkmark$
$-2 + 4^2 = 14 \ne 33$

4. $(1, 2)$ is in the relation; $(3, 2)$ is
not in the relation; $5(1)^3 - 2(2) = 1\checkmark$
$5(3)^3 - 2(2) = 131 \ne 1$

5. $(3, 5)$ is not in the relation

6. $(7, 2)$ is in the relation

7. $(1, 2)$ is not in the relation

8. $(3, 1)$ is not in the relation

9. $(2, 3)$ is not in the relation

10. $(3, 4)$ is in the relation

11. $(2, \sqrt{2})$ is in the relation

12. $(\sqrt{3}, 2)$ is not in the relation

13.

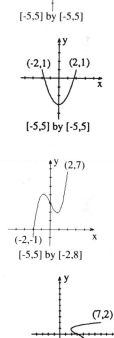

[-10,10] by [-10,10]

14.

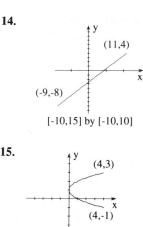

[-10,15] by [-10,10]

15.

[-5,5] by [-5,5]

16.

[-5,5] by [-5,5]

17.

[-5,5] by [-2,8]

18.

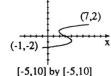

[-5,10] by [-5,10]

19. $(2, 0)$ and $(-2, 0)$ are
x-intercepts; $(0, 1)$ and $(0, -1)$
are y-intercepts

20. $(-3, 0)$ and $(3, 0)$ are
x-intercepts; $(0, 2)$ and $(0, -2)$
are y-intercepts

21. $(1, 0)$ and $(-1, 0)$ are x-intercepts; there are no y-intercepts

22. x-intercept is $(4, 0)$; there is no y-intercept

23. $(\sqrt{5}, 0)$ and $(-\sqrt{5}, 0)$ are x-intercepts; $(0, \sqrt{3})$ and $(0, -\sqrt{3})$ are y-intercepts

24. x-intercept is $(-5, 0)$; there is no y-intercept

25. Center $(3, -4)$, radius 4

26. Center $(1, -3)$, radius 7

27. Center $(2, -3)$, radius $\sqrt{2}$

28. Center $(7, 4)$, radius $\sqrt{19}$

29. $(x - 1)^2 + (y - 2)^2 = 25$

30. $(x + 3)^2 + (y - 2)^2 = 1$

31. $(x + 1)^2 + (y + 4)^2 = 9$

32. $(x - 5)^2 + (y + 3)^2 = 64$

33. $(h, k) = (3, 1)$ and $r = 6$

34. $(h, k) = (-4, 2)$ and $r = 11$

35. **(a)** $(2, -3)$ **(b)** $(-2, 3)$
(c) $(-2, -3)$

36. **(a)** (a, b) **(b)** $(-a, -b)$
(c) $(-a, b)$

37. **(a)** symmetric with respect to y-axis   **(b)** symmetric with respect to origin

38. **(a)** symmetric with respect to origin
(b) symmetric with respect to x-axis

39.

[-3,3] by [-3,3]

40.

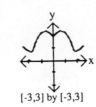

[-3,3] by [-3,3]

41.

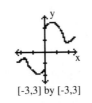

[-3,3] by [-3,3]

42.

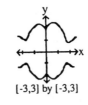

[-3,3] by [-3,3]

43. symmetric about the y-axis

44. symmetric about the x-axis

45. symmetric about the origin

46. symmetric about the y-axis

47. symmetric with respect to the y-axis, x-axis, and origin

48. no symmetry

49. symmetric about the y-axis

50. no symmetry

51. symmetric with respect to the origin

52. y-axis symmetry

53. The equation of the reflectional line of symmetry is $x = 2$.

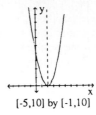

[-5,10] by [-1,10]

54. symmetric about the point $(0, 2)$

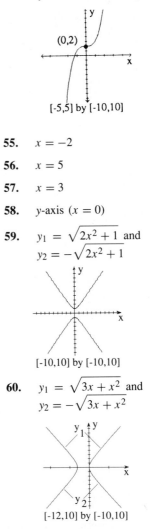

(0,2)

[-5,5] by [-10,10]

55. $x = -2$

56. $x = 5$

57. $x = 3$

58. y-axis $(x = 0)$

59. $y_1 = \sqrt{2x^2 + 1}$ and $y_2 = -\sqrt{2x^2 + 1}$

[-10,10] by [-10,10]

60. $y_1 = \sqrt{3x + x^2}$ and $y_2 = -\sqrt{3x + x^2}$

y_1
y_2

[-12,10] by [-10,10]

61. $y_1 = \sqrt{x^4 + 3x^2}$ and
$y_2 = -\sqrt{x^4 + 3x^2}$

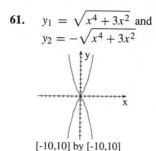

[-10,10] by [-10,10]

SECTION 2.7

1. Inverse: $x = t^2 - 2$ and
$y = 2t - 3$

2. Inverse: $x = 2t$ and $y = t^3 - 2$

3. Inverse: $x = t$ and $y = 3^t$

4. Inverse: $x = t^3$ and $y = t^2$

5. Inverse: $x = t^2$ and $y = 2t$

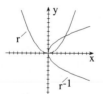

[-10,10] by [-10,10]

6. Inverse: $x = 2t - 3$
and $y = t^2 - 3t + 2$

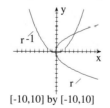

[-10,10] by [-10,10]

7. Inverse: $x = t^3 + 3$
and $y = t^2 - 1$

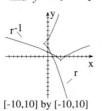

[-10,10] by [-10,10]

8. Inverse: $x = 2t$ and $y = t - t^3$

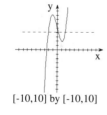

[-10,10] by [-10,10]

9. Suppose (a, b) is in the relation $y = x^3$. Then $b = a^3$. This equation shows that (b, a) satisfies $x = y^3$.

10. Suppose (a, b) is in the relation $y = x^2 + 1$. Then $b = a^2 + 1$. Thus (b, a) satisfies the equation $x = y^2 + 1$.

11. Suppose (a, b) is in the relation $y = x^2 - 4$. Then $b = a^2 - 4$. This equation shows that (b, a) satisfies the equation $x = y^2 - 4$.

12. Suppose (a, b) is in the relation $y = x^3 + x^2 - 6x$. Then $b = a^3 + a^2 - 6a$. Thus (b, a) satisfies the equation $x = y^3 + y^2 - 6y$.

13. $f(g(x)) = 3\left[\dfrac{1}{3}(x + 2)\right] - 2 - x + 2 - 2 = x, \ g(f(x)) = \dfrac{1}{3}(3x - 2 + 2) = \dfrac{1}{3}(3x) = x$

14. $f(g(x)) = \dfrac{(4x - 3) + 3}{4} = \dfrac{4x}{4} = x, \ g(f(x)) = 4\left(\dfrac{x + 3}{4}\right) - 3 = x + 3 - 3 = x$

15. $f(g(x)) = \left[(x - 1)^{1/3}\right]^3 + 1 = x - 1 + 1 = x, \ g[f(x)] = (x^3 + 1 - 1)^{1/3} = (x^3)^{1/3} = x$

16. $f(g(x)) = \dfrac{1}{1/x} = x$, for $x \neq 0$,
$g(f(x)) = \dfrac{1}{1/x} = x$, for $x \neq 0$

17. not one-to-one

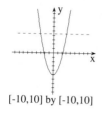

[-10,10] by [-10,10]

18. not one-to-one

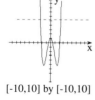

[-10,10] by [-10,10]

19. not one-to-one

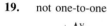

[-10,10] by [-10,10]

20. one-to-one

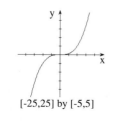

[-25,25] by [-5,5]

21. one-to-one

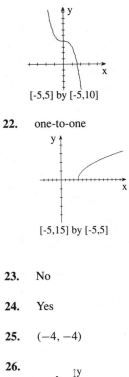

[-5,5] by [-5,10]

22. one-to-one

[-5,15] by [-5,5]

23. No

24. Yes

25. $(-4, -4)$

26.

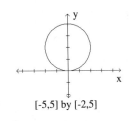

[-3,4] by [-4,4]

27.

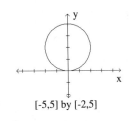

[-5,5] by [-2,5]

28. $f^{-1}(x) = -\sqrt{x}$

29. $f^{-1}(x) = x^2 + 2$, $x \geq 0$; for f, domain = $[2, \infty)$ and range = $[0, \infty)$; for f^{-1}, domain = $[0, \infty)$ and range = $[2, \infty)$

30. The graph shows f is one-to-one; $f^{-1}(x) = \dfrac{x+6}{3} = \dfrac{1}{3}x + 2$

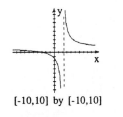

[-10,10] by [-10,10]

31. The graph shows $f(x) = 2x + 5$ is one-to-one; $f^{-1}(x) = \dfrac{1}{2}x - \dfrac{5}{2}$

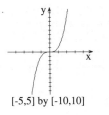

[-10,10] by [-10,10]

32. The graph shows f is one-to-one; $f^{-1}(x) = \sqrt[3]{x}$

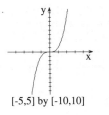

[-5,5] by [-10,10]

33. The graph shows $f(x) = \dfrac{x+3}{x-2}$ is one-to-one; $f^{-1}(x) = \dfrac{2x+3}{x-1}$

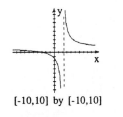

[-10,10] by [-10,10]

34. The graph shows f is one-to-one; $f^{-1}(x) = \dfrac{-x-3}{x-2}$

35. The graph shows $f(x) = \sqrt{x+2}$ is one-to-one; $f^{-1}(x) = x^2 - 2$ for $x \geq 0$

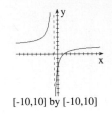

[-6,10] by [-8,8]

36.

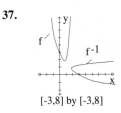

[-10,10] by [-10,10]

37.

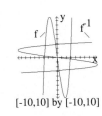

[-3,8] by [-3,8]

38.

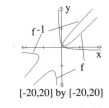

[-20,20] by [-20,20]

39.

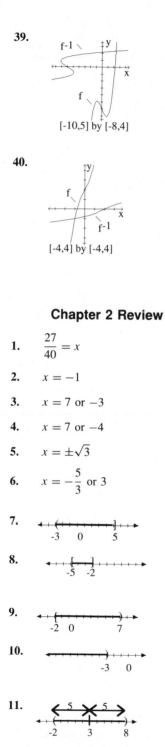

[-10,5] by [-8,4]

40.

[-4,4] by [-4,4]

Chapter 2 Review

1. $\dfrac{27}{40} = x$

2. $x = -1$

3. $x = 7$ or -3

4. $x = 7$ or -4

5. $x = \pm\sqrt{3}$

6. $x = -\dfrac{5}{3}$ or 3

7.

-3   0   5

8.

-5 -2

9.

-2 0   7

10.

-3   0

11.

-2   3   8

12.

-3   4   11

13.

1   1.5   2

14.

$\frac{-5}{3}$ $\frac{-2}{3}$ $\frac{1}{3}$

15.

-6   -2 0

16.

-1 $\frac{3}{2}$ 4

17.

-2/5 0   4/5

18.

-0.46 $\frac{\sqrt{3}}{2}$ 2.19

19. $\pm\sqrt{6}$

20. $x = 4$ or -2

21. $\dfrac{3 \pm \sqrt{65}}{4}$

22. $\dfrac{-3 \pm \sqrt{41}}{2}$

23. $-0.44,\ 1.69$

24. $-1.93,\ 2.59$

25. 2.11

26. $-1.75,\ 7.42$

27. No real solutions because there is no x-intercept.

28. -1.95

29. $-1.00,\ 0.50$

30. 1.84

31. 2.23

32. $0.06,\ -5.60,\ 5.54$

33. ± 2.28

34. $0,\ 3.53$

35. 0.85

36. $-16.50,\ -0.49,\ 1.99$

37. $\left(-\infty, -\dfrac{9}{2}\right]$

38. $(17, \infty)$

39. $\left[1, \dfrac{23}{4}\right)$

40. $\left[-\infty, -\dfrac{5}{3}\right] \cup [3, \infty)$

41. No real solutions

42. $[-2, -1) \cup (-1, \infty)$

43. $(-\infty, -1] \cup [4, \infty)$

44. $(-3, 2)$

45. $(-\infty, -7] \cup [-3, \infty)$

46. $(-\infty, -5) \cup (8, \infty)$

47. $(-\infty, 2] \cup [3, \infty)$

48. No real solutions

49. $(-2, 0) \cup \left(\dfrac{3}{2}, \infty\right)$

50. $(-\infty, -5) \cup (-2, \infty)$

51. $(-\infty, -2) \cup \left(\dfrac{5}{2}, \infty\right)$

52. $(-\infty, 2]$

53. $(-5, 2)$

54. $\left(1, \dfrac{5}{3}\right]$

55. $(-\infty, 0.23] \cup [1.43, \infty)$

56. $(-\infty, -1) \cup (0.70, 4.30)$

57. $[-0.91, \infty)$

58. $(-0.30, \infty)$

59. $[0, 1]$ by $[-1, 1]$; $[0.3, 0.4]$ by $[-0.1, 0.1]$; $[0.33, 0.34]$ by $[-0.01, 0.01]$; $[0.333, 0.334]$ by $[-0.001, 0.001]$

60. 215 craft kits

61. $y = \dfrac{10500}{25x}$; 35 cm

62. 210 adult tickets and 115 children's tickets

63. 6 pennies, 5 nickels, 12 dimes

64. Sandy's swimming speed is 0.08 miles per minute or 4.83 miles per hour, and the speed of the current is 0.03 miles per minute or 1.83 miles per hour

65. $0 < $ width $ < 72.5$ in.

66. 8

67. $P = 2x + 2w = 530$ yields $w = 265 - x$. Then $A = \ell \cdot w = x(265 - x)$.

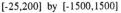

[-100,400] by [-15000,20000]

68. When the length of the rectangle is 100 yards, the area of the rectangle is 16,500 square yards. $100(265 - 100) = 100(165) = 16500$

69. Thus $(-100, -36,500)$ is a point on the graph, but neither length nor width can be negative. The coordinates have no meaning in the problem situation. $-100[265 - (-100)]$ $= -100(365) = -36500$

70. $A(x) = 4x^2 - 72x + 320$

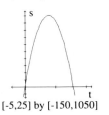

[-10,20] by [-20,400]

71. $0 < x < 8$ in.

72. Width is 1.03 in.

73. $\dfrac{x+1}{2}(8.50) \le 36$

74. Chris can invite 7 friends.

75. $P = 15.25x - [(x - 100)^2 + 400]$

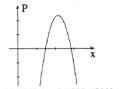

[-25,200] by [-1500,1500]

76. The problem situation is the portion of the graph corresponding to positive integer values of x.

77. 74 books will avoid a loss, but not more than 142.

78. $P = 5.15x - (1.26x + 34,000)$

79. $5.15x - (1.26x + 34,000) \ge 42,000,\ x \ge 19,537.28$

80. at least 19,538 lb of candy

81. $A = 4x^2 + 710x$

82. $2900\ \text{ft}^2 < A < 3900\ \text{ft}^2$; thus $2900\ \text{ft}^2 < 4x^2 + 710x < 3900\ \text{ft}^2$

83. $(3.99, 5.33)$

84. $s(t) = -16t^2 + 250t + 30$

85. Problem situation is represented by the portion of the graph in QI.

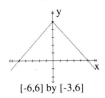

[-5,25] by [-150,1050]

86. 2.47 sec and 13.15 sec after shot

87. $x < 3$

88. symmetric about the y-axis

89. symmetric about the x-axis

90. no symmetry

91. symmetric about the origin

92. symmetric about the y-axis, the x-axis, and the origin

93. symmetric about the origin

94. line of symmetry is $x = -5$

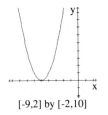

[-9,2] by [-2,10]

95. Domain $= (-\infty, \infty)$, range $= [0, \infty)$; line of symmetry: $x = 3$; y is a function of x

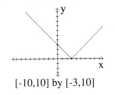

[-10,10] by [-3,10]

96. Domain $= (-\infty, \infty)$, range $= (-\infty, 5]$; line of symmetry: $x = 0$; f is a function of x

[-6,6] by [-3,6]

97. Domain $= [0, \infty)$, range $= (-\infty, \infty)$; line of symmetry:

$y = 4$; y is not a function of x

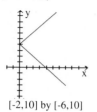

[-2,10] by [-6,10]

98. Domain = $(-\infty, \infty)$, range = $[0, \infty)$; line of symmetry: $x = 0$; f is a function of x

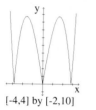

[-4,4] by [-2,10]

99. Domain = $(0, \infty)$, range = $(-\infty, 0) \cup (0, \infty)$; line of symmetry: $y = 0$; y is not a function of x

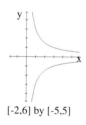

[-2,6] by [-5,5]

100. Domain = $(-\infty, 2) \cup (2, \infty)$ because $x - 2 \neq 0$, range = $(0, \infty)$; line of symmetry: $x = 2$; f is a function of x

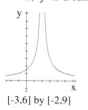

[-3,6] by [-2,9]

101. (a)

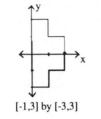

[-1,3] by [-3,3]

(b)

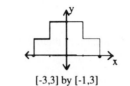

[-3,3] by [-1,3]

(c)

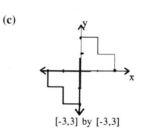

[-3,3] by [-3,3]

102. $f^{-1}(x) = -\dfrac{1}{3}x + \dfrac{5}{3}$ is a function

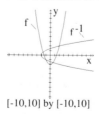

[-2,6] by [-4,6]

103. $y^2 = x + 2$; the inverse relation is not a function of x

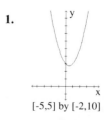

[-10,10] by [-10,10]

104. f is not one-to-one; $x = (y + 2)^2$: the inverse relation is not a function of x

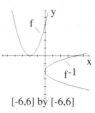

[-6,6] by [-6,6]

105. f is one-to-one; $y = \dfrac{x^2}{4} + 4$, $x \geq 0$; the inverse relation is a function of x.

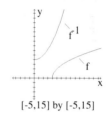

[-5,15] by [-5,15]

106. Domain of f is $(-\infty, \infty)$, range of the inverse is $(-\infty, \infty)$; range of f is $[0, \infty)$, domain of the inverse is $[0, \infty)$

107. (a) symmetric about the origin
(b) symmetric about the y-axis

CHAPTER 3

SECTION 3.1

1.

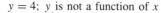

[-5,5] by [-2,10]

2.

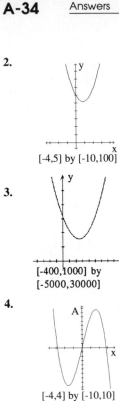

[-4,5] by [-10,100]

3.

[-400,1000] by
[-5000,30000]

4.

[-4,4] by [-10,10]

5.

[-5,30] by [-5000,5000]

6.

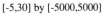

[-2,4] by [-10,5]

7.

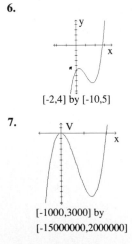

[-1000,3000] by
[-15000000,2000000]

8.

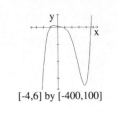

[-4,6] by [-400,100]

9.   local minimum of 1 at $x = 2$

10.   local minimum of $\dfrac{831}{28}$ when

$x = \dfrac{3}{14}$

11.   local minimum of 0 when $x = 2$
or $x = 5$ and a local maximum
of 2.25 when $x = 3.5$

12.   local maximum of 3 when
$x = -\dfrac{7}{4}$

13. $\pm\sqrt{10}$ (zeros), local maximum
of 10 when $x = 0$

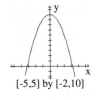

[-5,5] by [-2,10]

14.   no x-intercepts, local minimum
of 5.5 when $x = 1.5$

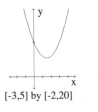

[-3,5] by [-2,20]

15.   Zeros: $x = 0$, $x = 10$; local
maximum of 0 when $x = 0$;
local minimum of -148.15

when $x = 6.67$

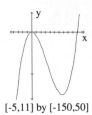

[-5,11] by [-150,50]

16.   no local extrema; zero:
$x = 1.30$

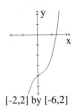

[-2,2] by [-6,2]

17.   no x-intercepts; local maximum
of 12.59 at $x = -0.41$; local
minima of 11.91 when
$x = -1.16$ and 3.51 when
$x = 1.57$

[-3,3] by [-3,20]

18.   Zero: $x = 2.10$; local maximum
of 4.10 when $x = 1.19$;
local minima of 2.93 when
$x = 0.14$ and 0 when $x = 2.10$.

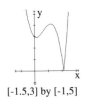

[-1.5,3] by [-1,5]

19.   Zero: $x = 3.81$; local maximum
of -29.85 when $x = 0.33$; local

minimum of -30 when $x = 1$

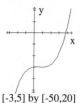

[-3,5] by [-50,20]

20. Zeros at $x = 0$, $x = 17$, $x = 26.5$; local maximum of 5465.81 when $x = 6.75$; local minimum of -1985.81 when $x = 22.25$

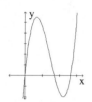

[-10,32] by [-2000,6000]

21. Zero: $x = -0.93$; local maximum of 8.08 when $x = -0.44$; local minima of 0 when $x = -0.93$ and 2.59 when $x = 0.38$

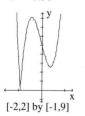

[-2,2] by [-1,9]

22. Zeros: $x = 0$, $x = 4$, $x = 11$; local maximum at 145.74 when $x = 1.79$; local minimum of -385.74 when $x = 8.21$

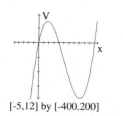

[-5,12] by [-400,200]

23. Zero: $x = 2.36$; local maximum of 12.10 when $x = 0.79$; local minimum of 11.90 when $x = 0.21$

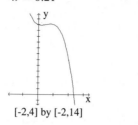

[-2,4] by [-2,14]

24. Zero: $x = 1.53$; local maximum of -5.201 when $x = 0.639$; local minimum of -5.202 when $x = 0.699$

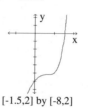

[-1.5,2] by [-8,2]

25. Increasing on $(1.5, \infty)$; decreasing on $(-\infty, 1.5)$

26. Increasing on $(-\infty, -1.79)$, $(1.12, \infty)$; decreasing on $(-1.79, 1.12)$

27. Increasing on $(-\infty, 0.67)$, $(2, \infty)$; decreasing on $(0.67, 2)$

28. Increasing on $(-\infty, -4.10)$, $(-0.57, \infty)$; decreasing on $(-4.10, -0.57)$

29. Increasing on $(1.38, \infty)$; decreasing on $(-\infty, 1.38)$

30. Increasing on $(-\infty, -1.33)$, $(0.47, 1.61)$; decreasing on $(-1.33, 0.47)$, $(1.61, \infty)$

31. Increasing on $(-\infty, 0.33)$, $(1, \infty)$; decreasing on $(0.33, 1)$

32. Increasing on $(-\infty, -0.07)$, $(0, \infty)$; decreasing on $(-0.07, 0)$

33. reflection through the x-axis, followed by a vertical stretch by a factor of 2, a horizontal shift left 3 units, then a vertical shift up 1 unit

34. decreases on $(-\infty, \infty)$

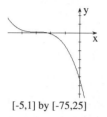

[-5,1] by [-75,25]

35. increasing on $(-\infty, \infty)$ when $a > 0$, decreasing on $(-\infty, \infty)$ when $a < 0$

36. $y = x^3 - 4x^2 - 4x + 16$ has both increasing and decreasing intervals, but a transformed image of $y = x^3$ is either always increasing or always decreasing

37. Dimensions: 262.5 ft by 525 ft; maximum area: $137{,}812.5$ ft^2

38. $s(t) = -16t^2 + 40t + 300$

39. maximum height of 325 ft at $t = 1.25$ sec

40. $t = 5.76$ sec

41. The maximum area is 171.13 in.^{2} when $x = 9.25$ in.

42. **(a)** $450 - 15x$, x an integer
(b) $1900 + 20x$

43. $R(x) = (1900 + 20x)(450 - 15x)$, domain $= 0 \le x < 30$, x an integer

44. \$450

45. $A(x) = \left(\dfrac{x}{4}\right)^2 + \left(\dfrac{300 - x}{4}\right)^2$

46.

[-200,400] by
[-200,10000]

47. first quadrant

48. each wire is 150 in.

49. Total area maximized by making one square use as much of the 300 in. as possible. Max. area approaches 5625 in^2.

50. $V(x) = x^2(100 - x) + \frac{1}{2}x(30 - x)(100 - x)$

51.

[-40,120] by
[-2500,120000]

52. $0 < x < 30$

53. maximum volume approaches 63,000 ft^3, base approaches 30 ft by 70 ft, top vanishingly small

54. $R(x) = xp = x(2 + 0.002x - 0.0001x^2)$

55. $0 \le x \le 151$

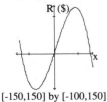

[-150,150] by [-100,150]

56. maximum daily revenue is $123.35, 89 glasses

57. Answers will vary.

SECTION 3.2

For Ex. 1–8, intervals may vary. Examples are given.

1. continuous

2. discontinuous; continuous on $[-5, -4]$ and $[0, \infty)$; discontinuous on $[-5, 5]$ and $[-4, -2]$

3. continuous

4. discontinuous; continuous on $(1, 2)$ and $(-2, 1)$; discontinuous on $(0, 2)$ and $(-5, 4)$

5. discontinuous; continuous on $(-3, 0)$ and $(2, \infty)$; discontinuous on $(0, 4)$ and $(-5, 5)$

6. discontinuous; continuous on $(-2, 2)$ and $(7, 10)$; discontinuous on $(-5, -3)$ and $(3, 5)$

7. discontinuous; continuous on $(-2, 0)$ and $(5, \infty)$; discontinuous on $(1, 3)$ and $(3, 5)$

8. continuous

9. discontinuous at every integer

10. no discontinuities

11. no discontinuities

12. $x = 0$

13. $x = 3$

14. $x = -1$

15. $c = \frac{17}{3}$

16. $c = \frac{7}{4}$

17. $c = 4.57$

18. $c = \frac{L - b}{a}$

19. $f(x) \to -\infty$

20. $f(x) \to \infty$

21. $f(x) \to 0$

22. $f(x) \to 0$

23. $(\swarrow, \nearrow)$

24. $(\nwarrow, \nearrow)$

25. $(\nwarrow, \searrow)$

26. $(\swarrow, \nearrow)$

27. $(\swarrow, \nearrow)$

28. $(\nwarrow, \searrow)$

29. EB model: $g(x) = 3x^4$

30. EB model: $g(x) = 0.05x^7$

31. EB model: $g(x) = 4x^3$

32. EB model: $g(x) = 2x^5$

33. $[-30, 30]$ by $[-5000, 5000]$ (Answers will vary.)

34. $[-30, 30]$ by $[-10,000, 10,000]$ (Answers will vary.)

35. $0 \le x < 1019.62$

36. increases on the interval $(0, 676.52)$

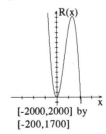

[-2000,2000] by
[-200,1700]

37. maximum revenue is $1615.81 when 676.52 lbs are sold

38. maximum volume is 663.85 cu in.; side $x = 3.28$ in.

39. largest possible area is 0.87 ft^2 when $x = 1.73$ ft

40.

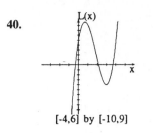

[-4,6] by [-10,9]

41. 5

42. Answers will vary.

43. Answers will vary. Example: velocity and time

SECTION 3.3

1. $q(x) = x - 1$, $r(x) = 2$

2. $q(x) = x^2 - x + 1$, $r(x) = -2$

3. $q(x) = 2x^2 - 5x + \dfrac{7}{2}$, $r(x) = -\dfrac{9}{2}$

4. $q(x) = x^2 - 4x + 12$, $r(x) = -32x + 18$

5. 3

6. -4

7. 5

8. 23

9. is a factor

10. is a factor

11. is a factor

12. is not a factor

13. $x = 3$, $x = 2$

14. $x = \dfrac{2}{3}$, $x = -2$

15. $x = 0$, $x = -3$, $x = 3$

16. $x = 1$

17. $x = \pm\sqrt{\dfrac{5}{2}}$

18. $x = 1$, $x = \pm\sqrt{2}$

19. $x = -3$, $x = \dfrac{17}{5}$

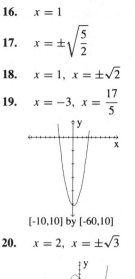

[-10,10] by [-60,10]

20. $x = 2$, $x = \pm\sqrt{3}$

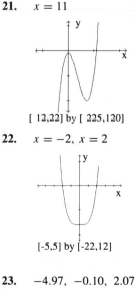

[-5,5] by [-10,10]

21. $x = 11$

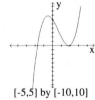

[12,22] by [225,120]

22. $x = -2$, $x = 2$

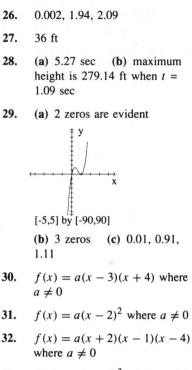

[-5,5] by [-22,12]

23. -4.97, -0.10, 2.07

24. 0.15

25. -0.002

26. 0.002, 1.94, 2.09

27. 36 ft

28. **(a)** 5.27 sec **(b)** maximum height is 279.14 ft when $t = 1.09$ sec

29. **(a)** 2 zeros are evident

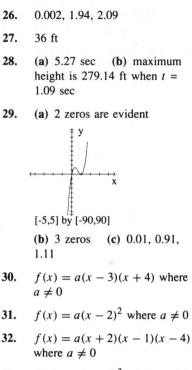

[-5,5] by [-90,90]

(b) 3 zeros **(c)** 0.01, 0.91, 1.11

30. $f(x) = a(x - 3)(x + 4)$ where $a \neq 0$

31. $f(x) = a(x - 2)^2$ where $a \neq 0$

32. $f(x) = a(x + 2)(x - 1)(x - 4)$ where $a \neq 0$

33. $f(x) = a(x + 1)^3$ where $a \neq 0$ or $f(x) = k(x+1)(ax^2+bx+c)$ where $k \neq 0$ and $b^2 < 4ac$

34. production level is 36,270 units; equilibrium price is \$53.70

35. production level is 28,127; equilibrium price is \$61.00

36. $k = -4$

37. $(x + 1)$, $(x - 3)$

38. **(a)** $r(x) = -2$
(b) $r(x) = -16$

39. **(a)** Area $= \dfrac{1}{2}xh = 50$; solve for $h = \dfrac{100}{x}$; from Pythagorean theorem, $x^2 + h^2 = (x + 2)^2$. Substituting for h, we get:
$$x^2 + \left(\dfrac{100}{x}\right)^2 = (x + 2)^2;$$
$$x^4 + 100^2 = x^2(x + 2)^2; \text{ thus}$$

$10,000 - 4x^3 - 4x^2 = 0$
(b) 13.25 in.

40. $0 = x^4 - 16x^3 + 500x^2 - 8000x + 32,000$

SECTION 3.4

1. $\dfrac{p}{q}$: $\pm 1, \pm 2, \pm 4$

2. $\dfrac{p}{q}$: $\pm 1, \pm \dfrac{1}{2}$

3. $\dfrac{p}{q}$: $\pm 1, \pm \dfrac{1}{2}, \pm 2, \pm 5, \pm \dfrac{5}{2}, \pm 10$

4. $\dfrac{p}{q}$: $\pm 1, \pm \dfrac{1}{2}, \pm \dfrac{1}{4}, \pm 2, \pm 3, \pm \dfrac{3}{2},$ $\pm \dfrac{3}{4}, \pm 6$

5. $f(1) = 1^3 + 2 \cdot 1^2 + 1 - 1 = 3 \neq 0;$ $f(-1) = (-1)^3 + 2(-1)^2 + (-1) - 1 = -1 \neq 0$

6. $f(1) = 2; \; f(-1) = 2;$ $f(2) = 8; \; f(-2) = -4$

7. $f(1) = 2; \; f\left(\dfrac{1}{2}\right) = 1;$ $f(-1) = 4; \; f\left(-\dfrac{1}{2}\right) = 1.25$

8. $f(1) = 3; \; f\left(\dfrac{1}{3}\right) = \dfrac{5}{9};$ $f(-1) = 1; \; f\left(-\dfrac{1}{3}\right) = \dfrac{5}{3}$

9.
$$\begin{array}{r|rrrr} 2 & 1 & 1 & -10 & 8 \\ & & 2 & 6 & -8 \\ \hline & 1 & 3 & -4 & 0 \end{array}$$
$(x - 2)(x + 4)(x - 1)$

10.
$$\begin{array}{r|rrrrr} -3 & 1 & 3 & -4 & -11 & 3 \\ & & -3 & 0 & 12 & -3 \\ \hline & 1 & 0 & -4 & 1 & 0 \end{array}$$
$(x + 3)(x^3 - 4x + 1)$

11.
$$\begin{array}{r|rrr} -\dfrac{1}{2} & 2 & 1 & -6 & -3 \\ & & -1 & 0 & 3 \\ \hline & 2 & 0 & -6 & 0 \end{array}$$
$(2x + 1)(x - \sqrt{3})(x + \sqrt{3})$

12.
$$\begin{array}{r|rrrr} \dfrac{4}{3} & 3 & -7 & 10 & -8 \\ & & 4 & -4 & 8 \\ \hline & 3 & -3 & 6 & 0 \end{array}$$
$(3x - 4)(x^2 - x + 2)$

13. $q(x) = x^2 - 3; \; r(x) = 7$

14. $q(x) = -2x^2 - 3x - 3;$ $r(x) = -7$

15. $q(x) = x^3 - 3x^2 + 4x - 5;$ $r(x) = 7$

16. $q(x) = 2x^3 - 6x^2 + 15x - 44;$ $r(x) = 137$

17. no real zeros

18. $0, \; -2, \; 2$

19. 0

20. 0

21.

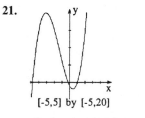
[-5,5] by [-5,20]

Rational zero: $x = 1$; irrational zeros: $x = \dfrac{-5 \pm \sqrt{21}}{2}$

22. Rational zero: $x = 1$; irrational zeros: $x = \dfrac{-3 \pm \sqrt{3}}{6}$

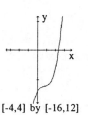
[-1.5,2] by [-3,1.5]

23. Rational zero: $x = \dfrac{7}{3}$

[-4,4] by [-16,12]

24. Rational zero: $x = \dfrac{3}{2}$; irrational zeros: $x = \dfrac{-1 \pm \sqrt{13}}{2}$

[-4,4] by [-3,20]

25. Rational zeros: $x = 4, \; -\dfrac{1}{2}$; no irrational zeros

26. Rational zero: $x = \dfrac{2}{3}$; irrational zero: $x = -0.68$

27. Rational zeros: $x = -3, \; 3, \; \dfrac{1}{2}$; no irrational zeros

28. Rational zero: $x = \dfrac{1}{3}$; no irrational zeros

29. Rational zeros: $x = \pm 2$; no irrational zeros

30. no real zeros

31. 4 is an upper bound; -2 is a lower bound; answers may vary

32. 2 is an upper bound; -1 is a lower bound; answers may vary

33. 2 is an upper bound; -3 is a lower bound; answers may vary

34. 3 is an upper bound; -4 is a lower bound

35. irrational zeros:
$x = -1.11, \; 0.86$

36. 7

37. There are one positive and two or no negative real zeros possible. A complete graph of $y = f(x)$ shows one positive real zero.

38. There are one positive and two or no negative real zeros possible. A complete graph of $y = f(x)$ shows one positive real zero.

39. There are three or one positive and no negative real zeros possible. A complete graph of $y = f(x)$ shows one positive real zero.

40. There are no positive and three or one negative real zeros possible. A complete graph of $y = f(x)$ shows one negative real zero.

41. domain = $(-\infty, \infty)$;
$0 \le d \le 172$ for the problem situation

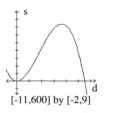
[-11,600] by [-2,9]

42. $d = 95.78$ ft

43. 3.35 ft when $x = 172$ ft

44.

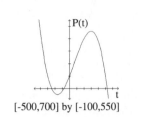
[-500,700] by [-100,550]

45. 460 turkeys after 300 days

46. 523.22 days

47. Answers will vary.

48. vertical stretch by a factor of 2, then vertical shift up 1 unit

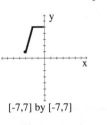
[-7,7] by [-7,7]

49. reflection in the x-axis

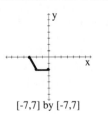
[-7,7] by [-7,7]

50. reflection in the x-axis, then a vertical shrink by a factor of $\dfrac{1}{2}$

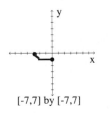
[-7,7] by [-7,7]

51. reflection in the y-axis, a vertical stretch by a factor of 2, then a vertical shift down 1 unit

[-7,7] by [-7,7]

52. reflection in the x-axis, a vertical stretch by a factor of 2, then a horizontal shift right 2 units

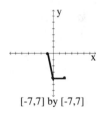
[-7,7] by [-7,7]

53. reflection in the x-axis, a horizontal shift left 1 unit, then a vertical shift up 2 units

[-7,7] by [-7,7]

54. reflection in the y-axis

[-7,7] by [-7,7]

55. reflection in the *x*-axis

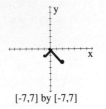

[-7,7] by [-7,7]

56. reflection in the *x*-axis, then a horizontal shift right 3 units

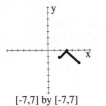

[-7,7] by [-7,7]

57. reflection in the *x*-axis, a reflection in the *y*-axis, then a horizontal shift right 3 units

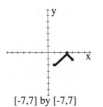

[-7,7] by [-7,7]

58. reflection in the *x*-axis, a vertical stretch by a factor of 3, then a horizontal shift left 1 unit, and a vertical shift up 2 units

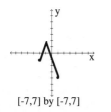

[-7,7] by [-7,7]

59. reflection in the *y*-axis, a vertical stretch by a factor of 2, then a horizontal shift right 1 unit, and

a vertical shift down 1 unit

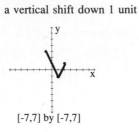

[-7,7] by [-7,7]

60. The only candidates for rational roots are ± 1, but, $f(1) = -1 \neq 0$ and $f(-1) = -3 \neq 0$

61.

[0.65,0.66] by [-0.005,0.01]

62. $x = 0.77$ m

63. $x = 0.57$ m

64. Answers may vary. $g(x) = 3f(x)$, so both functions have the same zeros. A polynomial with rational coefficients can be converted to a polynomial with integer coefficients by multiplying the polynomial by the least common denominator. The roots of this new polynomial will be the same as the original polynomial.

SECTION 3.5

1. $8 - 3i$

2. $8 - 7i$

3. $7 + 4i$

4. $5 + 5i$

5. $1 + i$

6. $20 + 3i$

7. $9 + 8i$

8. $-2 - 2i$

9. $2 + 3i$

10. $6i$

11. $-3 - 4i$

12. $-1 + \sqrt{2}i$

13. $\dfrac{2}{5} - \dfrac{1}{5}i$

14. $-\dfrac{1}{5} + \dfrac{2}{5}i$

15. $\dfrac{3}{5} + \dfrac{4}{5}i$

16. $\dfrac{1}{3} - \dfrac{2}{3}i$

17. $\dfrac{1}{2} - \dfrac{7}{2}i$

18. $\dfrac{26}{29} + \dfrac{7}{29}i$

19. two nonreal zeros

20. three real zeros

21. two nonreal zeros and one real zero

22. two nonreal zeros and two real zeros

23. two nonreal zeros and two real zeros

24. four nonreal zeros and one real zero

25. 1 is an integer zero; $-\dfrac{1}{2} + \dfrac{\sqrt{19}}{2}i$ and $-\dfrac{1}{2} - \dfrac{\sqrt{19}}{2}i$ are nonreal complex zeros.

26. 3 is an integer zero; $\dfrac{7}{2} + \dfrac{\sqrt{43}}{2}i$ and $\dfrac{7}{2} - \dfrac{\sqrt{43}}{2}i$ are nonreal complex zeros

27. 1 and -1 are integer zeros; $-\dfrac{1}{2} + \dfrac{\sqrt{23}}{2}i$ and $-\dfrac{1}{2} - \dfrac{\sqrt{23}}{2}i$ are nonreal complex zeros

28. -2 is an integer zero; $\dfrac{1}{3}$ is a noninteger rational zero; $-\dfrac{1}{2} + \dfrac{\sqrt{3}}{2}i$ and $-\dfrac{1}{2} - \dfrac{\sqrt{3}}{2}i$ are nonreal complex zeros

29. $f(x) = (3x - 1)(x - (1 + i))(x - (1 - i))$; real zero: $x = \dfrac{1}{3}$; nonreal complex zeros: $1 + i,\ 1 - i$

30. $f(x) = (x^2 - 2)(x - (3 - 2i))(x - (3 + 2i))$; real zeros: $x = \pm\sqrt{2}$; nonreal complex zeros: $x = 3 - 2i,\ x = 3 + 2i$

31. $(x - 1)(x + i)(x - i)$

32. $(x + 1)(x - 1)(x - \sqrt{5})(x + \sqrt{5})$

33. $f(x) = (x - 1)\left(x - \left(-\dfrac{1}{4} + \dfrac{\sqrt{31}i}{4}\right)\right)\left(x - \left(-\dfrac{1}{4} - \dfrac{\sqrt{31}i}{4}\right)\right)$

34. $(x + 3)(x + 3)(x + \sqrt{2})(x - \sqrt{2})$

35. $f(x) = a(x^2 - 4x + 13),\ a \neq 0$

36. $f(x) = a(x^3 - x^2 + x - 1),\ a \neq 0$

37. $f(x) = a(x^3 - 5x^2 + 8x - 6),\ a \neq 0$

38. $f(x) = a(x^4 + 2x^3 - x^2 - 2x + 10),\ a \neq 0$

39. $f(x) = k \cdot (x + 2)(x - (a + bi))(x - (a - bi))$ for $k,\ a,\ b \neq 0$

40. No

41. $f(x) = a(x^4 + 4x^3 - x^2 - 6x + 18),\ a \neq 0$

42. $f(x) = \dfrac{1}{18}(x^4 + 4x^3 - x^2 - 6x + 18)$

43. $h = 3.78$ ft

44. $h = 6.51$ ft

45. No solution because $h = -5.26 < 0$.

46. $x = -1, 1$

47. $x = 2,\ -1 + \sqrt{3}i,\ -1 - \sqrt{3}i$

48. $z + \bar{z} = (a - bi) + (a + bi) = 2a$, which is a real number

49. $z \cdot \bar{z} = (a - bi)(a + bi) = a^2 - b^2 \cdot i^2 = a^2 + b^2$

50. $f(x) = x^3 - ix^2 + 2ix + 2$, $f(i) = i^3 - i^3 + 2i^2 + 2 = -2 + 2 = 0$

SECTION 3.6

1. $(-\sqrt{2}, 0) \cup (0, \sqrt{2})$

2. $(-\sqrt{3}, 0) \cup (\sqrt{3}, \infty)$

3. $(-1, 0) \cup (0, 1)$

4. $[-\sqrt{5}, \sqrt{5}]$

5. $(-\infty, 0) \cup (0, 1)$

6. $(-1, 0) \cup (1, \infty)$

7. $(-1, 0) \cup (1, \infty)$

8. $\left(-\infty, -\dfrac{1}{\sqrt{2}}\right) \cup \left(0, \dfrac{1}{\sqrt{2}}\right)$

9. $(-2, \infty)$

10. $(-\infty, -2.89] \cup [0.13, 2.76]$

11. $(-1, \infty)$

12. $[1.75, \infty)$

13. $(-\infty, 2)$

14. $(-\infty, \infty)$

15. $(1, \infty)$

16. $(-\infty, -3] \cup [0, 1]$

17. $(-1, 0) \cup (1, \infty)$

18. $(-1, 0) \cup (2, \infty)$

19. $[-1, 1] \cup [4, \infty)$

20. $x = 0,\ \pm\sqrt{8}$

21.
$[-5,10]$ by $[-20,20]$

22.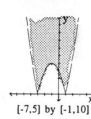
$[-7,5]$ by $[-1,10]$

23. $(0, 0.55] \cup [6.79, 7.50)$, x in in. (If $x = 0$ or $x = 7.50$, there is no box.)

24. $[0.93, 3.64]$, x in cm

25. $0 < x < 21.50$

26. $A(x) = x(335 - 2x)$

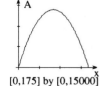
$335 - 2x$

27. $0 < x < 167.50$

$[0,175]$ by $[0,15000]$

28. $(0, 50] \cup [117.5, 167.50)$, x in ft

29. From 30 to 586 customers

30. 201 or 429 customers

31. **(a)**

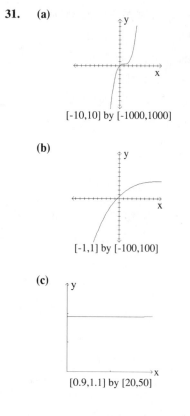

[-10,10] by [-1000,1000]

(b)

[-1,1] by [-100,100]

(c)

[0.9,1.1] by [20,50]

32. [0.99, 1.03] by [38.9997, 39.0001]

33. No local maximum or minimum; real zero: $x = 0.72$; decreasing on $(-\infty, \infty)$

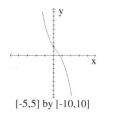

[-5,5] by [-10,10]

34. Local maximum of 0.03 when $x = 0.60$; local minimum of 0 when $x = 1$; real zeros: $x = 0,\ 1$; increasing on $(-\infty, 0.60),\ (1, \infty)$; decreasing

on $(0.60, 1)$

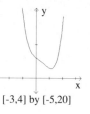

[-1,2] by [-0.2,0.2]

35. Local minimum of 2.84 when $x = 1.15$; no real zeros; decreasing on $(-\infty, 1.15)$; increasing on $(1.15, \infty)$

[-3,4] by [-5,20]

36. No local maximum or minimum; real zero: $x = 0.77$; increasing on $(-\infty, \infty)$

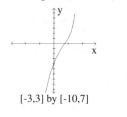

[-3,3] by [-10,7]

SECTION 3.7

1. E_1 and E_2 are parallel lines. There is no solution to the system E_1 and E_2.

2. E_1 and E_3 yield the same line. Any point on the line $y = \frac{1}{2}x - 2$ is a solution to the system E_1 and E_3.

3. $(6, 2)$

4. $(9, -5)$

5. $(10, 10)$

6. $(0, 0),\ (-\sqrt{2}, -\sqrt{2}),\ (\sqrt{2}, \sqrt{2})$

7. maximum error in y of at most 0.01; solution: $(3.1427, 15.713)$; VR: $[3.1425, 3.1429]$ by $[15.71, 15.72]$

8. $(-2.8907, -2.356),$ $(2.4907, -0.204)$

9. $(-2.236, 1.00),\ (2.236, 1.00)$

10. $(1.3146, 2.272)$

11. $(-0.103, -9.668),$ $(-1.5386, -0.650),$ $(3.142, 0.318)$

12. $(0.6159, 9.621),$ $(4.2261, -7.860),$ $(-3.842, -4.761)$

13. $(-7.64261, -63.695)$ $(-0.5106, 8.718),$ $(1.153, 10.977)$

14. $(1.062, 0.873),\ (-1.728,$ $-0.988)$

15. $(1.173, 1.612)$

16. $(0, 0),\ (1.317, 0.968),$ $(-1.317, -0.968)$

17. Solutions for $x < 5$ are: $(0.538, 1.859),$ $(2.130, 0.469),\ (3.876, 0.258)$

18. $H(x) = \frac{1}{3}x - \frac{11}{3}$

19. $T(x) = \frac{11}{3}x^2 - \frac{32}{3}x + 5$

20. \$5494.51 at 5%, \$1098.90 at 6%, \$6593.41 at 10%

21. 5 nickels, 2 dimes, 10 quarters

22. 79

23. Complete graphs of E_1 and E_2 show that there are 2 simultaneous solutions to the system

24. $(-1.646, 3.646)$,
$(3.646, -1.646)$

25. $(1 + \sqrt{7}, 1 - \sqrt{7})$,
$(1 - \sqrt{7}, 1 + \sqrt{7})$

26. Complete graphs of E_1 and E_2 show that there are 4 simultaneous solutions to the system.

27. $(-4.306, 2.541)$,
$(4.306, 2.541)$,
$(-3.5297, -3.541)$,
$(3.5297, -3.541)$

28. $\left(-4.306, \dfrac{-1 + \sqrt{37}}{2}\right)$,
$\left(4.306, \dfrac{-1 + \sqrt{37}}{2}\right)$,
$\left(-3.530, \dfrac{-1 - \sqrt{37}}{2}\right)$,
$\left(3.530, \dfrac{-1 - \sqrt{37}}{2}\right)$

29. $V(x) = x(30 - 2x)(40 - 2x)$

30. $y = 1200$; $y = x(30 - 2x)(40 - 2x) = 4x^3 - 140x^2 + 1200x$; there are 2 solutions for the problem situation

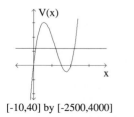

[-10,40] by [-2500,4000]

31. $x = 1.1490$ with an error of at most 0.0001; $x = 11.8890$ with an error of at most 0.0001

32.

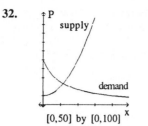

[0,50] by [0,100]

33. $23.25

34. **(a)**

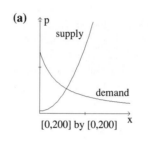

[0,200] by [0,200]

(b) $53.64

35.

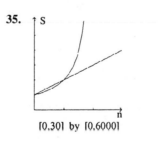

[0,30] by [0,6000]

36. in 10 years

37. As n approaches 20, the denominator in the simple discount formula approaches 0 and the future value increases without bound. The simple discount model is used only for short periods because of the asymptotic behavior for large n. **(a)** $20,000 **(b)** $40,000 **(c)** $200,000 **(d)** $2,000,000 **(e)** undefined

Chapter 3 Review

1.

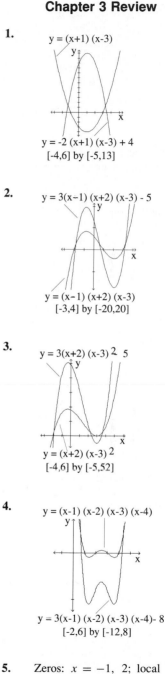

$y = (x+1)(x-3)$

$y = -2(x+1)(x-3) + 4$
[-4,6] by [-5,13]

2. $y = 3(x-1)(x+2)(x-3) - 5$

$y = (x-1)(x+2)(x-3)$
[-3,4] by [-20,20]

3. $y = 3(x+2)(x-3)^2 - 5$

$y = (x+2)(x-3)^2$
[-4,6] by [-5,52]

4. $y = (x-1)(x-2)(x-3)(x-4)$

$y = 3(x-1)(x-2)(x-3)(x-4) - 8$
[-2,6] by [-12,8]

5. Zeros: $x = -1$, 2; local maximum of 0 when $x = -1$; local minimum of -4 when

$x = 1$

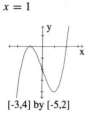

[-3,4] by [-5,2]

6.   Zero: $x = -0.10$; local maximum of 5 when $x = 1$; local minimum of 1 when $x = 3$

[-2,5] by [-2,6]

7.   No real zeros; local minimum of 2 when $x = 1$

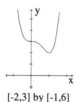

[-2,3] by [-1,6]

8.   Zeros: $x = -20.91$, ± 0.19, 12.91; local maximum of 10 when $x = 0$; local minimum of $-37{,}115$ when $x = -15$ and -9467 when $x = 9$

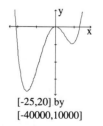

[-25,20] by [-40000,10000]

9.   Real zeros: $x = -3.60$, -0.88, 1.63; local maximum of 74.12

when $x = -2.85$ and -2.52 when $x = 0.37$; local minimum of -3.43 when $x = -0.32$ and -4.33 when $x = 1.19$

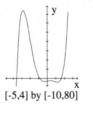

[-5,4] by [-10,80]

10.   Real zeros: $x = -3$, 1.10, 1.12; local maximum of $10{,}285.28$ when $x = -1.63$; local minimum of -0.41 when $x = 1.11$

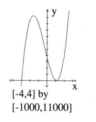

[-4,4] by [-1000,11000]

11.   Real zeros: $x = -1.26$, 5; local maximum of -9.49 when $x = 0.39$; local minimum of -10.47 when $x = -0.35$ and -68.45 when $x = 3.71$

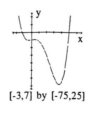

[-3,7] by [-75,25]

12.   Real zero: $x = -1.39$; local maximum of 779.86 when $x = -0.84$ and 451.31 when $x = 1.25$; local minimum of 408.42 when $x = 0.40$ and

451.17 when $x = 1.35$

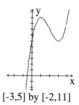

[-3,3] by [-150,850]

13. $6 < c < 10$ (answers for c may vary)

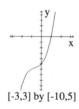

[-3,5] by [-2,11]

14.   No values for c work.

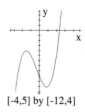

[-3,3] by [-10,5]

15. $-11.06 < c < -4.12$ (answers for c may vary)

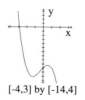

[-4,5] by [-12,4]

16. $-12.11 < c < -9.37$ (answers for c may vary)

[-4,3] by [-14,4]

17. $2 < c < 2.35$ (answers for c may vary)

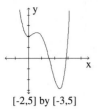

[-2,5] by [-3,5]

18. $-5 < c < -4$ (answers for c may vary)

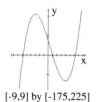

[-3,3] by [-5,3]

19. Real zeros: $x = -5.81$, 1.04, 6.27; local maximum of 211.00 when $x = -3.00$; local minimum of -132.00 when $x = 4.00$

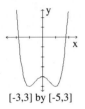

[-9,9] by [-175,225]

20. Real zeros: $x = 1$, 2.62; local maximum of 13 when $x = 0$; local minimum of 8 when $x = -1$ and -19 when $x = 2$

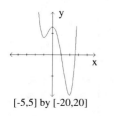

[-5,5] by [-20,20]

21. Real zero: $x = -3.63$; local maximum of 80.12 when $x = -2.85$ and 3.48 when $x = 0.37$; local minimum of 2.57 when $x = -0.32$ and 1.67 when $x = 1.19$

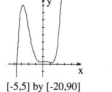

[-5,5] by [-20,90]

22. Domain $= (-\infty, \infty)$; range $= [-1, \infty)$; no points of discontinuity

23. Domain $= (-\infty, \infty)$; range $= (-\infty, \infty)$; no points of discontinuity

24. Domain $= (-\infty, -5) \cup (-5, \infty)$; range $= (-\infty, 0) \cup (0, \infty)$; discontinuous at $x = -5$

25. Domain $= (-\infty, \infty)$; range $=$ the integers; discontinuous when x is an integer

26. Domain $= (-\infty, -0.5) \cup (-0.5, \infty)$; range $= (-\infty, \infty)$; discontinuous at $x = -0.5$

27. Domain $= (-\infty, \infty)$; range $= (-1, \infty)$; discontinuous at $x = -2, 0$

28.

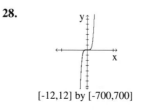

[-12,12] by [-700,700]

29. $f(x) \to -\infty$

30. $f(x) \to \infty$

31. $f(x) \to 2$

32. $f(x) \to 2$

33. $q(x) = 3x - 8$; $r(x) = 23$

34. $q(x) = 2x^2 - \frac{5}{3}x + \frac{37}{9}$; $r(x) = -\frac{65}{9}$

35. $r(x) = 18$

36. 27

37. $x - 3$ is a factor

38. $(-2, -4)$

39. $x = 0, \pm\sqrt{8}$

40. $x = \pm 2, \pm 3$

41. $x = 2$

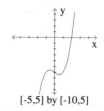

[-5,5] by [-10,5]

42. Real zero: $x = 2$; nonreal complex zeros: $x = -1 \pm \sqrt{2}i$

43. $x = 5$

44. **(a)** two zeros **(b)** three real zeros **(c)** $x = -1, -0.025, 0.05$

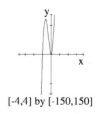

[-4,4] by [-150,150]

45. no real zeros

46. **(a)** $f(x)$ increases on $(-\infty, \infty)$ **(b)** no local extrema **(c)** $f(x)$ has two

local extrema

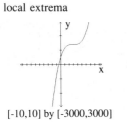

[-10,10] by [-3000,3000]

47. **(a)** For $m = -3$: 3 real zeros; 2 local extrema; increasing on $(-\infty, -0.5)$, $(0.5, \infty)$; decreasing on $(-0.5, 0.5)$

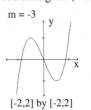

m = -3

[-2,2] by [-2,2]

(b) For $m = -2$: 3 real zeros; 2 local extrema; increasing on $(-\infty, -0.41)$, $(0.41, \infty)$; decreasing on $(-0.41, 0.41)$

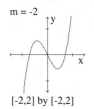

m = -2

[-2,2] by [-2,2]

(c) For $m = 0$: 1 real zero; no local extrema; increasing on $(-\infty, \infty)$

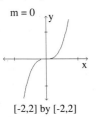

m = 0

[-2,2] by [-2,2]

(d) For $m = 2$: 1 real zero; no

local extrema; increasing on $(-\infty, \infty)$

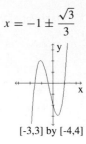

m = 2

[-2,2] by [-2,2]

(e) For $m = 4$: 1 real zero; no local extrema; increasing on $(-\infty, \infty)$

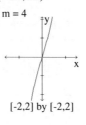

m = 4

[-2,2] by [-2,2]

48. **(a)** $a(x - 2)(x + 3)(x - c)$, $a \neq 0$ **(b)** $a(x - 2)^2(x + 3)$ or $a(x - 2)(x + 3)^2$, $a \neq 0$ **(c)** $a(x - 2)(x^2 - 4x + 5)$, $a \neq 0$

49. $k = -2$

50. Integer zero: $x = -3$; irrational zeros: $x = \pm\sqrt{5}$

51. Integer zeros: $x = \pm 3$

52. $\dfrac{p}{q}$: ± 1, ± 2, ± 5, ± 10; real zero: $x = 5$

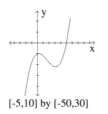

[-5,10] by [-50,30]

53. $\dfrac{p}{q}$: $\pm\dfrac{1}{3}$, $\pm\dfrac{2}{3}$, ± 1, ± 2; rational zero: $x = 1$; irrational zeros:

$x = -1 \pm \dfrac{\sqrt{3}}{3}$

[-3,3] by [-4,4]

54. $q(x) = 2x^2 - 9x + 41$; $r(x) = -200$

55. Rational zero: $x = -2$; irrational zeros: $x = \dfrac{-5 \pm \sqrt{41}}{2}$

[-7,3] by [-12,20]

56. Rational zeros: $x = \pm 1$; irrational zeros: $x = \dfrac{-3 \pm \sqrt{17}}{2}$

57. 1 is an upper bound; -1 is a lower bound

58. 2 is an upper bound; -2 is a lower bound

59. $g(-3) = (-3)^6 - 729 = 0$

60. $3 - 7i$

61. $2 + 11i$

62. $\dfrac{19}{41} - \dfrac{7}{41}i$

63. $1 + 0i$

64. one real zero and two nonreal complex zeros

65. one real zero and two nonreal complex zeros

66. integer zeros: $x = 4$, -1; nonreal complex zeros: $x = \pm 2i$

67. $a(x^2 - 2x + 5)$, $a \neq 0$

68. $a(x^3 - 6x^2 + 37x - 58)$, $a \neq 0$

69. $a(x^4 - 10x^3 + 38x^2 - 64x + 40)$, $a \neq 0$

70. $f(x) = \dfrac{1}{40}(x^4 - 10x^3 + 38x^2 - 64x + 40)$

71. Domain $= (-\infty, \infty)$, $0 \le t \le 24$ for the problem situation

72.

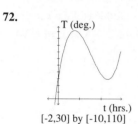

[-2,30] by [-10,110]

73. $t = 0.76$, 18.37 hr (at 6:46 a.m. and 12:22 a.m.)

74. Maximum temperature of $101.79°$ occurs at 1:38 p.m.

75. Minimum temperature of $29°$ occurs at 5:01 a.m.

76.

[-2,2] by [-2,2]

77.

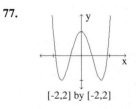

[-2,2] by [-2,2]

78.

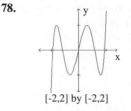

[-2,2] by [-2,2]

79. Exercise 76: local maximum of 1 when $x = -0.5$; local minimum of -1 when $x = 0.5$. Exercise 77: local maximum of 1 when $x = 0$; local minimum of -1 when $x = \pm 0.71$. Exercise 78: local maximum of 1 when $x = -0.81$ and $x = 0.31$; local minimum of -1 when $x = -0.31$ and $x = 0.81$

80. You can generate Chebyshev polynomials by the following recursive algorithm:
$$C_{n+1}(x) = 2xC_n(x) - C_{n-1}(x)$$

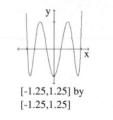

[-1.25,1.25] by [-1.25,1.25]

81. **(a)**

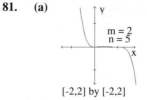

m = 2
n = 5

[-2,2] by [-2,2]

(b)

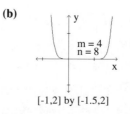

m = 4
n = 8

[-1,2] by [-1.5,2]

(c)

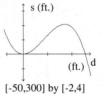

n = 10
m = 3

[-1.5,2] by [-2,2]

82. Domain $= (-\infty, \infty)$, $0 \le x \le 255$ for the problem situation

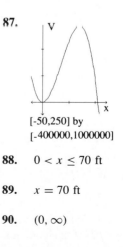

[-50,300] by [-2,4]

83. $d = 82.61$ ft or $d = 233.40$ ft

84. Maximum deflection is 2.09 ft, $d = 170$ ft

85. The beam is not of uniform density.

86. $V = \pi x^2(140 - 2x) + \dfrac{4}{3}\pi x^3$

87.

[-50,250] by [-400000,1000000]

88. $0 < x \le 70$ ft

89. $x = 70$ ft

90. $(0, \infty)$

91.

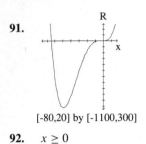

[-80,20] by [-1100,300]

92. $x \geq 0$

93.   no maximum revenue; the function is always increasing

94.

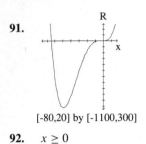

[-200,300] by [-100,300]

95.   182 pheasants in 111.65 days

96.   210.30 days

97.   Answers may vary. One scenario could be that the population increases until there are too many pheasants for the food supply. They then begin to die as the food disappears.

98. $(1.214, 0.142)$ or $\left(\dfrac{17}{14}, \dfrac{1}{7}\right)$

99. $(3.142, -1.546)$

100. $(2.2128, 1.3844)$

101.   There is no solution.

102. $(-0.67, 3.67), (2, 9)$

103. $(-0.2944, 5.632),$
$(6.7944, 14.493)$

104. $(-2.3273, 12.363),$
$(16.3273, 105.637)$

105. $(-3.2623, 5.358)$

106. $(-2.6711, -1.135),$
$(2.9364, -2.622)$

107. $(-6.3970, 49.676)$
$(0.4018, 46.017),$
$(5.9730, 47.613)$

CHAPTER 4

SECTION 4.1

1. $(-\infty, -1) \cup (-1, 2) \cup (2, \infty)$

2. $(-\infty, -1) \cup (-1, 1) \cup (1, \infty)$

3. $(-\infty, -3) \cup (-3, 1) \cup (1, \infty)$

4. $(-\infty, -3) \cup (-3, -1) \cup (-1, \infty)$

5. $(-\infty, -1) \cup (-1, 1) \cup (1, \infty)$

6. $(-\infty, -1) \cup (-1, 1.5) \cup (1.5, \infty)$

7. $(-\infty, 2 - \sqrt{5}) \cup (2 - \sqrt{5}, 2 + \sqrt{5}) \cup (2 + \sqrt{5}, \infty)$

8. $(-\infty, \infty)$

9. $(-\infty, -1) \cup (-1, 0) \cup (0, 1) \cup (1, \infty)$

10. $(-\infty, 0) \cup (0, 1) \cup (1, \infty)$

11. $y = 0$

12. $y = 3$

13. $y = -4$

14. $y = 2$

15. $y = -12$

16. $y = 17$

17.   HA at $y = 0$; VA at $x = -1$

18.   HA at $y = 0$; VA at $x = 2$

19.   HA at $y = 4$; VA at $x = -1$

20.   HA at $y - 2$; VA at $x = -3$

21.   HA at $y = 0$; VA at $x = 4$

22.   HA at $y = -1$; VA at $x = -1$

23.   horizontal shift right 3 units to $y = \dfrac{1}{x}$; HA at $y = 0$; VA at $x = 3$

[-4,8] by [-6,6]

24.   horizontal shift left 2 units to $y = \dfrac{1}{x}$, followed by a vertical stretch of 2; HA at $y = 0$; VA at $x = -2$

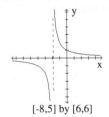

[-8,5] by [6,6]

25.   vertical stretch by a factor of 2 to $y = \dfrac{1}{x}$, then a reflection in the x-axis, followed by a horizontal shift left 5 units; HA at $y = 0$; VA at $x = -5$

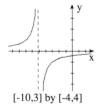

[-10,3] by [-4,4]

26.   horizontal shift left 3 units to $y = \dfrac{1}{x}$; HA at $y = 0$; VA at $x = -3$

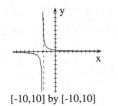

[-10,10] by [-10,10]

27. horizontal shift left 1 unit to $y = \dfrac{1}{x}$, followed by a vertical shift down 3 units; HA at $y = -3$; VA at $x = -1$

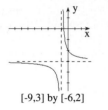

[-9,3] by [-6,2]

28. horizontal shift left 3 units to $y = \dfrac{1}{x}$, followed by a vertical shift down 2 units; HA at $y = -2$; VA at $x = -3$

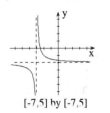

[-7,5] by [-7,5]

29. vertical stretch by a factor of 5 to $y = \dfrac{1}{x}$, followed by a reflection in the x-axis, then a horizontal shift right 1 unit; HA at $y = 0$; VA at $x = 1$

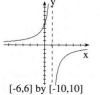

[-6,6] by [-10,10]

30. horizontal shift left 1 unit to $y = \dfrac{1}{x}$, followed by a reflection across the x-axis, then a vertical shift down 1 unit; HA at $y = -1$; VA at $x = -1$

[-5,5] by [-5,5]

31. HA at $y = 3$; VA at $x = -2$

32. HA at $y = 4$; VA at $x = 2$

33. HA at $y = 2$; VA at $x = 3$

34. HA at $y = \dfrac{1}{2}$; VA at $x = -2.5$

35. HA at $y = 1$; VA at $x = -4$

36. HA at $y = 2$; VA at $x = -2$

37. $f(x) \to 0$

38. $f(x) \to \infty$

39. $g(x) \to 4$

40. $g(x) \to -\infty$

41. Domain $= (-\infty, 1) \cup (1, \infty)$; range $= \left\{ \dfrac{1}{2} \right\}$; HA at $y = \dfrac{1}{2}$; no VA. There is a removable discontinuity at $x = 1$

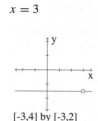

[-3,3] by [-3,3]

42. Domain $= (-\infty, 3) \cup (3, \infty)$; range $= \left\{ -\dfrac{3}{2} \right\}$; HA at $y = -\dfrac{3}{2}$; no VA. There is a removable discontinuity at

$x = 3$

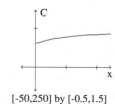

[-3,4] by [-3,2]

43. vertical stretch by a factor of 2, followed by a shift right 1 unit, then a reflection across the x-axis and a vertical shift up 3 units

44. horizontal shift right h units, then a vertical stretch by a factor of r units, followed by a vertical shift up k units

45. vertical stretch by a factor of $\dfrac{bc - ad}{c^2}$, then a horizontal shift of $-\dfrac{d}{c}$ units, followed by a vertical shift up $\dfrac{a}{c}$ units

46. $L = \dfrac{300}{x}$

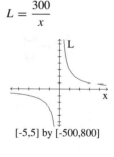

[-5,5] by [-500,800]

.15 units

47. $x \geq 80$ oz

48. **(a)**

[-50,250] by [-0.5,1.5]

(b) $x \geq 169.12$ oz

49. 60 oz of 60% solution, 40 oz of 10% solution

50. Domain = $(-\infty, h) \cup (h, \infty)$; range = $(-\infty, k) \cup (k, \infty)$

51. Domain = $\left(-\infty, -\dfrac{d}{c}\right) \cup$ $\left(-\dfrac{d}{c}, \infty\right)$; range = $\left(-\infty, \dfrac{a}{c}\right) \cup \left(\dfrac{a}{c}, \infty\right)$ provided by $c \neq 0$

SECTION 4.2

1. $(-\infty, -1) \cup (-1, 1) \cup (1, \infty)$

2. $(-\infty, 2 - \sqrt{5}) \cup (2 - \sqrt{5}, 2 + \sqrt{5}) \cup (2 + \sqrt{5}, \infty)$

3. $(-\infty, 0) \cup (0, \infty)$

4. $(-\infty, 0) \cup (0, \infty)$

5. $(-\infty, -1) \cup (-1, 1) \cup (1, \infty)$

6. $(-\infty, \infty)$

7. $(-\infty, -1) \cup \left(-1, \dfrac{3}{2}\right) \cup \left(\dfrac{3}{2}, \infty\right)$

8. $(-\infty, \infty)$

9. $g(x) = \dfrac{1}{x}$

10. $g(x) = \dfrac{2}{x}$

11. $g(x) = 3$

12. $g(x) = 4$

13. $g(x) = x$

14. $g(x) = x$

15. VA at $x = 3$; HA at $y = 0$; EB model is $g(x) = \dfrac{2}{x}$; EB

asymptote is $y = 0$

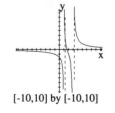

[-10,10] by [-10,10]

16. VA at $x = 3$, $x = -1$; HA at $y = 0$; EB model is $f(x) = \dfrac{1}{x}$; EB asymptote is $y = 0$

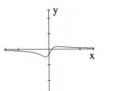

[-10, 10] by [-10,8]

17. VA does not exist; HA at $y = 0$; EB model is $g(x) = \dfrac{1}{x}$; EB asymptote is $y = 0$

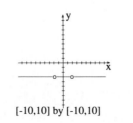

[-15,15] by [-3,3]

18. No vertical asymptotes; HA at $y = -3$; EB model is $y = -3$; EB asymptote is $y = -3$

[-10,10] by [-10,10]

19.

x	-1	-10	-100	-1000
$f(x)$	.50	-0.11	-0.01	-0.001

x	$-10,000$
$f(x)$	-0.0001

20.

x	3	2.90	2.99
$f(x)$	und	-3.88	-39.88

x	2.999	2.9999
$f(x)$	-399.88	-3999.88

21. $f(x) \to 0$

22. $f(x) \to -\infty$

23. Domain = $(-\infty, 0) \cup (0, \infty)$; range = $[-0.25, \infty)$; HA at $y = 0$; VA at $x = 0$; EB model is $g(x) = \dfrac{1}{x}$

[-10,10] by [-3,6]

24. Domain = $(-\infty, 1) \cup (1, 3) \cup (3, \infty)$; range = $(-\infty, \infty)$; HA at $y = 0$; VA at $x = 1$, $x = 3$; EB model is $g(x) = \dfrac{3}{x}$

[-10,10] by [-10,10]

25. Domain = $(-\infty, -1) \cup (-1, 1) \cup (1, \infty)$; range = $\left(-\infty, -\dfrac{1}{2}\right) \cup \left(-\dfrac{1}{2}, 0\right) \cup (0, \infty)$; removable discontinuity at $x = -1$; HA at $y = 0$; VA at

$x = 1$; EB model is $f(x) = \dfrac{1}{x}$

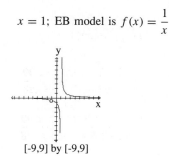

[-9,9] by [-9,9]

26. Domain $= (-\infty, 3) \cup (3, \infty)$; range $= (-\infty, \infty)$; HA at $y = 0$; VA at $x = 3$; EB model is $g(x) = \dfrac{2}{x}$

[-10,10] by [-10,8]

27. Domain $= (-\infty, -3.30) \cup (-3.30, 0.30) \cup (0.03, \infty)$; range $= (-\infty, 0.12] \cup [0.65, \infty)$; HA at $y = 0$; VA at $x = -3.30$, 0.30; EB model is $g(x) = \dfrac{1}{x}$

[-10,10] by [-5,5]

28. Domain $= (-\infty, -2) \cup (-2, 3) \cup (3, \infty)$; range $= (-\infty, \infty)$; HA at $y = 0$; VA at $x = -2$, $x = 3$; EB model is $f(x) = \dfrac{2}{x}$

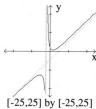

[-10,10] by [-10,8]

29.

x	3	2.1	2.01	2.001
$f(x)$	21	38.82	308.08	3008.01
$g(x)$	3	30	300	3000

x	2.0001
$f(x)$	30,008.00
$g(x)$	30,000

30.

x	1	1.9	1.99
$f(x)$	-1	-22.78	-292.08
$g(x)$	-3	-30	-300

x	1.999	1.9999
$f(x)$	-2992.01	$-29,992$
$g(x)$	-3000	$-30,000$

31.

x	1	10	100
$f(x)$	-1	200.38	20,000.03
$g(x)$	-3	0.38	0.03

x	1000	10,000
$f(x)$	2,000,000.00	200,000,000
$g(x)$	0.003	0.0003

32.

x	-1	-10	-100
$f(x)$	1	199.75	19,999.97
$g(x)$	-1	-0.25	-0.03

x	-1000
$f(x)$	1,999,999.997
$g(x)$	-0.003

x	$-10,000$
$f(x)$	199,999,999.997
$g(x)$	-0.0003

33. $f(x) \to \infty$

34. $f(x) \to -\infty$

35. $f(x) \to \infty$

36. $f(x) \to \infty$

37. EB asymptote is $y = x - 4$; VA at $x = -2$

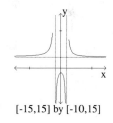

[-25,25] by [-25,25]

38. EB asymptote is $y = 3$; VA at $x = -2$, $x = 2$

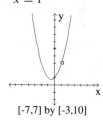

[-15,15] by [-10,15]

39. EB asymptote is $y = x^2 + x + 1$; removable discontinuity at $x = 1$

[-7,7] by [-3,10]

40. EB asymptote is $y = x$; no VA exists

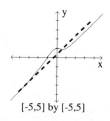

[-5,5] by [-5,5]

41. EB asymptote is $y = x - 6$; VA at $x = -3$

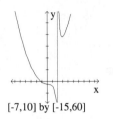

[-30,20] by [-30,20]

42. EB asymptote is $y = 2x^2 + 2x + 3$; VA at $x = 2$

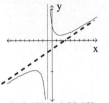

[-7,10] by [-15,60]

43. EB asymptote is $y = x^3 + 2x^2 - 3x + 1$; removable discontinuity at $x = \dfrac{5}{2}$

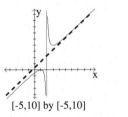

[-5,5] by [-10,35]

44. EB asymptote is $y = 2x^4 + 2x^3 - x^2 - x + 1$; VA at $x = 1$

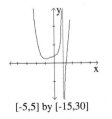

[-5,5] by [-15,30]

45. Domain = $(-\infty, -1) \cup (-1, \infty)$; range = $(-\infty, -9.66] \cup$

$[1.66, \infty)$; VA at $x = -1$; EB model is $g(x) = \dfrac{x^2}{x} = x$; EB asymptote is $y = x - 3$

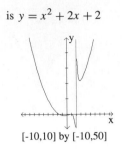

[-10,10] by [-20,10]

46. Domain = $(-\infty, 2) \cup (2, \infty)$; range = $(-\infty, 0] \cup [4, \infty)$; VA at $x = 2$; EB model is $f(x) = \dfrac{x^2}{x} = x$; EB asymptote is $y = x$

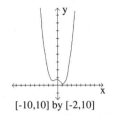

[-5,10] by [-5,10]

47. Domain = $(-\infty, -\sqrt[3]{3}) \cup (-\sqrt[3]{3}, \infty)$; range = $(-\infty, 2) \cup (2, \infty)$; VA at $x = -\sqrt[3]{3}$; HA at $y = 2$; EB model is $g(x) = \dfrac{2x^3}{x^3} = 2$; EB asymptote is $y = 2$

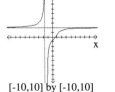

[-10,10] by [-10,10]

48. Domain = $(-\infty, 2) \cup (2, \infty)$; range = $(-\infty, -\infty)$; VA at $x = 2$, EB model is $f(x) = \dfrac{x^3}{x} = x^2$; EB asymptote

is $y = x^2 + 2x + 2$

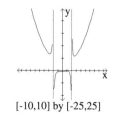

[-10,10] by [-10,50]

49. Domain = $(-\infty, \infty)$; range = $[0.18, \infty)$; no HA; no VA; EB model is $g(x) = \dfrac{x^4}{x^2} = x^2$; EB asymptote is $y = x^2 - 6$

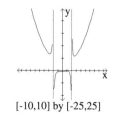

[-10,10] by [-2,10]

50. Domain = $(-\infty, -2) \cup (-2, 2) \cup (2, \infty)$; range = $(-\infty, -0.33] \cup [11.78, \infty)$; VA at $x = -2$, $x = 2$; EB model is $f(x) = \dfrac{x^4}{x^2} = x^2$; EB asymptote is $y = x^2 + 2$

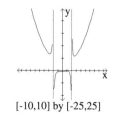

[-10,10] by [-25,25]

51. Increasing on $(-\infty, -1.00)$, $(1.00, \infty)$; decreasing on $(-1.00, 0)$, $(0, 1.00)$

52. Increasing on $(-\infty, -0.62)$, $(0.69, \infty)$; decreasing on $(-0.62, 0.69)$

53. Increasing on $(-2.27, -1.67)$, $(-1.67, -0.84)$, $(0.61, \infty)$; decreasing on $(-\infty, -2.27)$, $(-0.84, 0.61)$

54. Increasing on $(-\infty, 2)$, $(2, 2.94)$; decreasing on $(2.94, \infty)$

55. Local minimum of 0.92 when $x = -0.81$; no local maximum; increasing on $(-0.18, 0.5)$, $(0.5, \infty)$; decreasing on $(-\infty, -0.18)$; zero at $x = 1.75$; EB model is $g(x) = \dfrac{x^3}{2x} = \dfrac{1}{2}x^2$

56. Local minimum of -2.81 when $x = -2.62$; local maximum of -14.30 when $x = -5.40$; increasing on $(-\infty, -5.40)$, $(-2.62, 3)$, $(3, \infty)$; decreasing on $(-5.40, -4)$, $(-4, -2.62)$; zeros at $x = -3.46$, $x = -0.50$, $x = 3.46$; EB model is $g(x) = \dfrac{2x^3}{x^2} = 2x$

57. Local minimum of 0.10 when $x = -0.69$ and $x = 0$; local maximum of -5.28 when $x = -3.42$ and 0.10 when $x = -0.40$; increasing on $(-\infty, -3.42)$, $(-0.69, -0.40)$, $(0, \infty)$; decreasing on $(-3.42, -2.42)$, $(-2.42, -0.69)$, $(-0.40, 0)$; no zeros; EB model is $g(x) = \dfrac{2x^4}{x^3} = 2x$

58. Local minimum of 1.89 when $x = 0.79$; no local maximum; increasing on $(0.79, \infty)$; decreasing on $(-\infty, 0)$, $(0, 0.79)$; zero at $x = -1$; EB model is $g(x) = \dfrac{x^3}{x} = x^2$

59. Local minimum of 3.32 when $x = -1.50$ and 180.28 when $x = 5.83$; local maximum of -0.92 when $x = 0.75$; increasing on $(-1.50, -1)$, $(-1, 0.75)$, $(5.83, \infty)$; decreasing on $(-\infty, -1.50)$, $(0.75, 4)$, $(4, 5.83)$; no zeros; EB model is $g(x) = \dfrac{2x^4}{x^2} = 2x^2$

60. **(a)** No VA apparent in standard viewing rectangle.

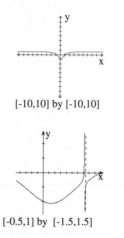

[-10,10] by [-10,10]

[-0.5,1] by [-1.5,1.5]

(b) the VA is visible in the new viewing rectangle.

61. EB asymptote is $y = 3x^3 + 7x^2 + 3x - 30$; EB model is $g(x) = 3x^3$

[-20,20] by [-5000,5000]

62. $P(x) = 2x + 2\left(\dfrac{182}{x}\right) = 2x + \dfrac{364}{x}$; 13.49 units by 13.49 units

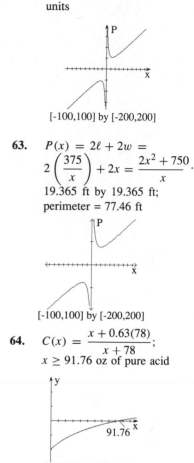

[-100,100] by [-200,200]

63. $P(x) = 2\ell + 2w = 2\left(\dfrac{375}{x}\right) + 2x = \dfrac{2x^2 + 750}{x}$. 19.365 ft by 19.365 ft; perimeter $= 77.46$ ft

[-100,100] by [-200,200]

64. $C(x) = \dfrac{x + 0.63(78)}{x + 78}$; $x \geq 91.76$ oz of pure acid

[0,120] by [-0.3,0.3]

65. 102.16 oz of 100% barium

SECTION 4.3

1. $-\dfrac{7}{2}$

2. $x = 3 \pm \sqrt{2}$

3. $x = \dfrac{11 \pm \sqrt{73}}{8}$

4. $x = \dfrac{13 \pm \sqrt{105}}{16}$

5. $x = 1$

6. $x = \dfrac{-3 \pm \sqrt{31}}{2}$

7. $x = 0, -1, 1$

8. $x = \pm\dfrac{1}{\sqrt{2}} = \pm 0.71$

9. $x = \dfrac{1 \pm \sqrt{13}}{6}$

10. $x = 5$

11. $\left(-\infty, \dfrac{11}{4}\right) \cup (3, +\infty)$

12. $(-\infty, -6] \cup (-5, \infty)$

13. $\left(-\infty, \dfrac{7}{2}\right) \cup \left(\dfrac{38}{9}, \infty\right)$

14. $(-4, 1)$

15. $\left(-\dfrac{13}{2}, -4\right)$

16. $(-\infty, -2) \cup (1, 2)$

17. $(-\infty, -1] \cup [1, \infty)$

18. $[-0.56, 2) \cup (2, 3.56]$

19. $(-\infty, -2) \cup (-1, 5)$

20. $x = -1.94, \; x = 1.94$

21. $x = -2.04, \; x = 1.44$

22. $(-1.71, 2.22)$

23. $[-1.66, -1.00] \cup (3, \infty)$

24. Zero at $x = -1.89$; y-intercept at $y = f(0) = \dfrac{3}{4}$; no VA; EB asymptote is $g(x) = x + 1$

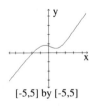

[-5,5] by [-5,5]

25. Zeros at $x = \pm 1$; y-intercept at $y = f(0) = -1$; VA at

$x = -1.62, \; x = 0.62$; EB asymptote is $g(x) = x^2 - x$

[-4,5] by [-10,20]

26. Zeros at $x = -0.93$, $x = 1.18$; y-intercept at $y = f(0) = -1$; VA at $x = 4$, $x = 1$; EB asymptote is $g(x) = x^2 + 3x + 14$

[-4,4] by [-20,20]

[-6,10] by [-150,200]

27. Zero at $x = 1.46$; y-intercept at $y = f(0) = 5$; VA at $x = -0.30$ and $x = 3.30$; EB asymptote is $y = x^3 + 3x^2 + 9x + 30$; more than one view required for a complete graph:

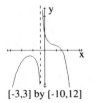

[-3,3] by [-10,12]

[-6,7] by [-220,450]

28. $T(x) = \dfrac{17}{x} + \dfrac{53}{x + 43}$

29. HA at $y = 0$; VA at $x = 0$, $x = -43$; zero at $x = -10.44$

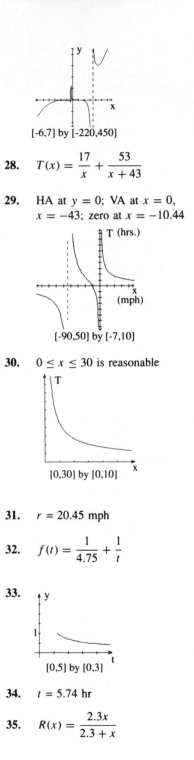

[-90,50] by [-7,10]

30. $0 \le x \le 30$ is reasonable

[0,30] by [0,10]

31. $r = 20.45$ mph

32. $f(t) = \dfrac{1}{4.75} + \dfrac{1}{t}$

33.

[0,5] by [0,3]

34. $t = 5.74$ hr

35. $R(x) = \dfrac{2.3x}{2.3 + x}$

36.

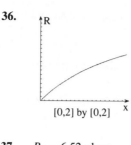

[0,2] by [0,2]

37. $R_2 = 6.52$ ohms

38. $S(r) = 2\pi r^2 + \dfrac{710}{r}$

39. $r > 0$

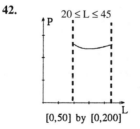

[0,10] by [0,800]

40. $r = 3.84$ cm and $h = 7.67$ cm;
$S = 277.55$ cm^2

41. $P(L) = \dfrac{1800}{L} + 2L$

42.

[0,50] by [0,200]

43. The minimum perimeter is
120 ft when $L = 30.00$ ft

44. $A = 1, B = -1$

45. $A = \dfrac{3}{5}, B = -\dfrac{3}{5}$

46. $A = \dfrac{1}{2}, B = \dfrac{1}{2}$

47. $A = \dfrac{1}{4}, B = \dfrac{7}{4}$

SECTION 4.4

1. $(1, \infty)$

2. $(1, \infty)$

3. $(0, 16)$

4. $(4096, \infty)$

5. horizontal shift left 3 units to
the graph of $y = \sqrt{x}$, then
apply a vertical stretch of 2 and
reflect through the x-axis to get
$y = -2\sqrt{x + 3}$

6. horizontal shift left 2 units to
the graph of $y = x^{1/2}$, followed
by a vertical shift up 4 units to
get $y = 4 + (x + 2)^{1/2}$

7. horizontal shift right 5 units to
the graph of $y = \sqrt{x}$, then a
vertical shift down 3 units to get
$y = -3 + \sqrt{x - 5}$

8. horizontal shift right 3 units to
the graph of $y = x^{1/3}$, followed
by a vertical stretch by a factor
of 2, then a reflection across the
x-axis, and a vertical shift up 3
units to get $y = 3 - 2(x - 3)^{1/3}$

9. horizontal shift right $\dfrac{5}{3}$ units
to the graph of $y = \sqrt[3]{x}$, then
apply a vertical stretch by a
factor of 2 and a horizontal
stretch by a factor of 3 to get
$y = 2\sqrt[3]{3x - 5} = 2\sqrt[3]{3(x - \dfrac{5}{3})}$

10. horizontal shift right $\dfrac{2}{5}$ units
to the graph of $y = x^{1/3}$,
followed by a vertical stretch
by a factor of $3\sqrt[3]{5}$ and a
reflection across the x-axis, then
a vertical shift down 2 units to
get $y = -2 + 3(2 - 5x)^{1/3} =$
$-2 - 3\sqrt[3]{5}\left(x - \dfrac{2}{5}\right)^{1/3}$

11.

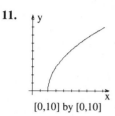

[0,10] by [0,10]

12.

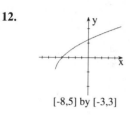

[-8,5] by [-3,3]

13.

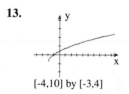

[-4,10] by [-3,4]

14.

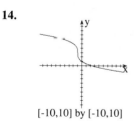

[-10,10] by [-10,10]

15.

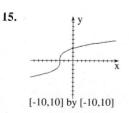

[-10,10] by [-10,10]

16.

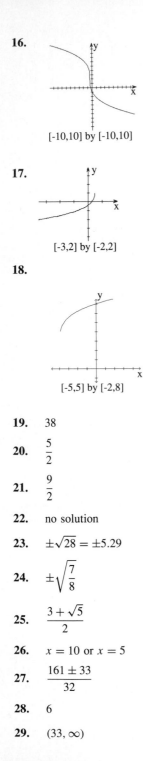

[-10,10] by [-10,10]

17.

[-3,2] by [-2,2]

18.

[-5,5] by [-2,8]

19. 38

20. $\dfrac{5}{2}$

21. $\dfrac{9}{2}$

22. no solution

23. $\pm\sqrt{28} = \pm 5.29$

24. $\pm\sqrt{\dfrac{7}{8}}$

25. $\dfrac{3+\sqrt{5}}{2}$

26. $x = 10$ or $x = 5$

27. $\dfrac{161 \pm 33}{32}$

28. 6

29. $(33, \infty)$

30. $\left(-\infty, -\dfrac{9}{8}\right)$

31. $x = \sqrt[3]{23} = 2.84$

32. $x = 0.27$

33. $(-1.30, 1.30)$

34. $[1.33, 15.57)$

35. The boundary is not in the solution

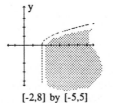

[-2,8] by [-5,5]

36. The boundary is not in the solution

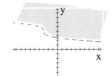

[-10,10] by [-8,10]

37. Domain = $(-\infty, 0) \cup (0, \infty)$;
range = $y = \pm 1$

[-3,3] by [-3,3]

38. The number of real solutions of $\sqrt[3]{x-1} = 4 - x^2$ will be the number of times the graphs of f and g intersect. A quick sketch will reveal two solutions.

39. EB model is $\sqrt{x^2}$ or $|x|$

40. EB model is $\sqrt{a}\sqrt{x^2}$ or $\sqrt{a}|x|$

41. $x = 4 - 2\sqrt{3} = 0.54$ is a solution

42. 21.21 ft above the ground

43. minimum distance is 21 units when $x = 14$

44. $x = 14.98$ mi, $T(x) = 1.73$ hr

45. $T(x) = \dfrac{\sqrt{x^2+400}}{30} + \dfrac{60-x}{50}$;
$T(0) = 1.87$ hr; this is the time elapsed if Penny goes to B by boat and drives 60 miles.
$T(60) = 2.11$ hr; this is the time elapsed if she goes by boat directly to the city at C

46. $0 \le x \le 60$

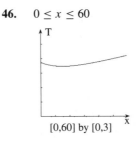

[0,60] by [0,3]

47. 14.94 mi from B; $t = 1.73$ hr

48. $s = \pi r\sqrt{r^2+441}$

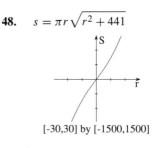

[-30,30] by [-1500,1500]

49. $r = 2.34$ ft

50. $S(r) = \pi r \sqrt{r^2 + \left(\dfrac{1140}{\pi r^2}\right)^2}$

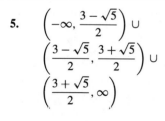

[-10,10] by [-2000,2000]

51. radius 6.35 ft, height 9.00 ft

Chapter 4 Review

1. $(-\infty, -2) \cup (-2, 1) \cup (1, \infty)$

2. $(-\infty, -1) \cup (-1, \infty)$

3. $(-\infty, -1) \cup (-1, 3) \cup (3, \infty)$

4. $(-\infty, -3) \cup (-3, 2) \cup (2, \infty)$

5. $\left(-\infty, \dfrac{3 - \sqrt{5}}{2}\right) \cup$ $\left(\dfrac{3 - \sqrt{5}}{2}, \dfrac{3 + \sqrt{5}}{2}\right) \cup$ $\left(\dfrac{3 + \sqrt{5}}{2}, \infty\right)$

6. $(-\infty, \infty)$

7. HA at $y = 0$; VA at $x = -5$

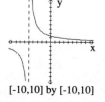

[-10,10] by [-10,10]

8. HA at $y = -2$; no VA; removable discontinuities at

$x = \pm\sqrt{3}$

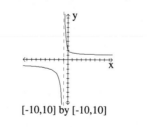

[-10,10] by [-10,10]

9. HA at $y = 5$; VA at $x = 3$

[-15,15] by [-10,20]

10. HA at $y = -1$ (for $x < 0$) and $y = 1$ (for $x \geq 0$); VA at $x = -1$ (when $x + 1 = 0$)

[-10,10] by [-10,10]

11. Domain $= (-\infty, 3) \cup (3, \infty)$; range $= (-\infty, 5) \cup (5, \infty)$

12. $y = \begin{cases} \dfrac{2 - x}{x + 1} & \text{for } x < 0, \ x \neq -1 \\[2mm] \dfrac{2 + x}{x + 1} & \text{for } x \geq 0 \end{cases}$

13. $f(x) \to 1$

14. $f(x) \to -\infty$

15. Apply a horizontal shift right of 2 to the graph of $y = \dfrac{1}{x}$. followed by a vertical stretch of

5, to get $f(x) = \dfrac{5}{x - 2}$; HA at $y = 0$; VA at $x = 2$

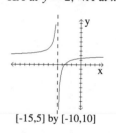

[-10,10] by [-10,10]

16. horizontal shift 5 left to the graph of $y = \dfrac{1}{x}$, then a vertical stretch by a factor of 3, then a reflection through the x-axis, followed by a vertical shift up 2 units to get $f(x) = 2 - \dfrac{3}{x + 5}$; HA at $y = 2$; VA at $x = -5$

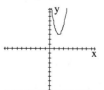

[-15,5] by [-10,10]

17. horizontal shift 2 units right to the graph of $y = x^2$, followed by a vertical stretch by a factor of 2, then a vertical shift up 3 units to get $g(x) = 2(x - 2)^2 + 3$; there are no asymptotes

[-10,10] by [-10,10]

18. horizontal shift left 1 unit to the graph of $y = \sqrt{x}$, followed by a vertical stretch by a factor of 3 and a reflection across the x-axis, then a vertical shift up 4 unts to get $h(x) = 4 - 3\sqrt{x + 1}$;

there are no asymptotes

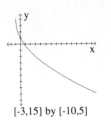

[-3,15] by [-10,5]

19. Domain = $(-\infty, -4) \cup (-4, 5) \cup (5, \infty)$; range = $(-\infty, 0] \cup (3, \infty)$ (from the graph); HA at $y = 3$; VA at $x = 5, -4$; zero at $x = 0$

[-10,12] by [-10,12]

20. Domain = $(-\infty, \infty)$; range = $[-1, 1)$ from the graph); HA at $y = 1$; no VA; zeros at $x = \pm 2$

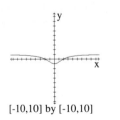

[-10,10] by [-10,10]

21. Domain = $(-\infty, -\sqrt[3]{2}) \cup (-\sqrt[3]{2}, \infty)$; range = $(-\infty, \infty)$; HA at $y = 2$; VA at $x = -\sqrt[3]{2}$; zeros at $x = -2.59, -0.16, 1.24$

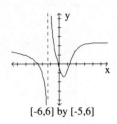

[-6,6] by [-5,6]

22. Increasing on $(-\infty, 0), (1.26, \infty)$, decreasing on $(0, 1.26)$

23. Increasing on $(-\infty, 0), (0, \infty)$

24. Local minimum of 10.95 at $x = 1.83$; local maximum of -10.95 at $x = -1.83$

[-30,30] by [-60,60]

25. Local minimum of 3.47 at $x = -0.89$; no local maxima

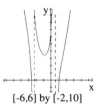

[-6,6] by [-2,10]

26. EB model is $g(x) = \dfrac{x^3}{x} = x$; no HA; VA at $x = \pm 3$; zero at $x = -1$; local minimum of 7.85 at $x = 5.23$; local maximum of -7.74 at $x = -5.16$; increasing on $(-\infty, -5.16), (5.23, \infty)$; decreasing on $(-5.16, -3)$,

$(-3, 3), (3, 5.23)$

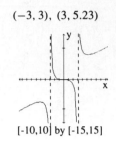

[-10,10] by [-15,15]

27. EB model is $g(x) = \dfrac{x^2}{x} = x$; no HA; VA at $x = -2$; no zeros; local minimum of 2 at $x = 3$; local maximum of -18 at $x = -7$; increasing on $(-\infty, -7), (3, \infty)$; decreasing on $(-7, -2), (-2, 3)$

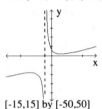

[-15,15] by [-50,50]

28. EB model is $g(x) = x^2$; EB asymptote is $y = x^2 - 7x + 7$

29. EB model is $g(x) = x^2$; EB asymptote is $y = x^2 + 2x + 4$

30. EB model is $g(x) = \dfrac{x^2}{x^3} = \dfrac{1}{x}$; EB asymptote is $y = 0$

31. EB model is $g(x) = \dfrac{x^4}{x} = x^3$; EB asymptote is $y = x^3 - 3x^2 - 5x + 7$

32.

x	-0.9	-0.99
$g(x)$	50	500
$f(x)$	50.81	500.9801

x	-0.999	-0.9999
$g(x)$	5000	50,000
$f(x)$	5000.998001	50,000.9998

33. $f(x) \to \infty$

34. $f(x) \to 0$

35. $f(x) \to -\infty$

36. 10

37. $\dfrac{6 \pm \sqrt{6}}{2}$

38. 0

39. No real solutions

40. $(-5, 2)$

41. $(-4, -2] \cup [2, \infty)$

42. $\left(\dfrac{8}{3}, 3\right)$

43. $(-\infty, 2) \cup \left(\dfrac{13}{4}, \infty\right)$

44. $-\dfrac{3}{2}, 2$

45. No real solutions

46. $(-0.43, 0.70)$

47. $(-1.91, 2.39)$

48. Zeros at $x = -0.93$, $x = 1.18$; y-intercept at $\dfrac{2}{7}$; VA at $x = 7$, $x = -2$; EB asymptote is $y = x^2 + 3x + 32$

[-20,30] by [-100,500]

[-4,6] by [-4,1]

49. 2

50. $\dfrac{-3 \pm \sqrt{37}}{2}$

51. 3

52. $-1 \pm \sqrt{129}$

53. Observe that the graph of y_1 is below the graph of y_2 for $x < 5$, so the inequality is not satisfied; $y_1 > y_2$ on $(5.35, \infty)$

54. horizontal shift right 3 units to the graph of $y = \sqrt[3]{x}$ to get $y = \sqrt[3]{x - 3}$

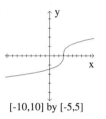

[-10,10] by [-5,5]

55. Reflect $y = \sqrt{x}$ through the y-axis, then apply a vertical stretch by a factor of 2, a horizontal shift right 5 units, and a vertical shift down 3 units to get $y = -3 + 2\sqrt{5 - x}$

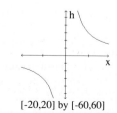

[-10,10] by [-8,10]

56. (a) $h = \dfrac{300}{x}$

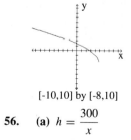

[-20,20] by [-60,60]

57. (a) $P(x) = 2x + \dfrac{1000}{x}$

(b) vertical asymptote at $x = 0$; EB asymptote is $y = 2x$

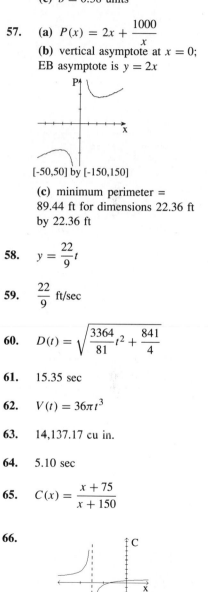

[-50,50] by [-150,150]

(c) minimum perimeter = 89.44 ft for dimensions 22.36 ft by 22.36 ft

58. $y = \dfrac{22}{9}t$

59. $\dfrac{22}{9}$ ft/sec

60. $D(t) = \sqrt{\dfrac{3364}{81}t^2 + \dfrac{841}{4}}$

61. 15.35 sec

62. $V(t) = 36\pi t^3$

63. 14,137.17 cu in.

64. 5.10 sec

65. $C(x) = \dfrac{x + 75}{x + 150}$

66.

[-300,100] by [-10,10]

(b) Domain = $(-\infty, 0) \cup (0, \infty)$; range = $(-\infty, 0) \cup (0, \infty)$; $x > 0$

(c) $b = 0.38$ units

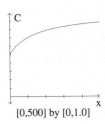

[0,500] by [0,1.0]

67. $x \geq 190.91$ oz

68. 27.13 oz of 72% solution;
57.87 oz of 25% solution

69. $T(x) = \dfrac{\sqrt{x^2 + 15^2}}{25} + \dfrac{55 - x}{40}$;
$T(0) = 1.975$ hr to go to
B by boat then drive to C;
$T(55) = 2.28$ hr to go to C by
boat only

70.

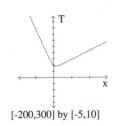

[-200,300] by [-5,10]

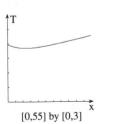

[0,55] by [0,3]

71. Eric should go to a point 12.01
mi from B

72. 1.84 hr is the least time

73. $T(x) = \dfrac{11}{x} + \dfrac{45}{x + 41}$

74. $0 < x \leq 30$

75. $x = 34.40$ mph

CHAPTER 5

SECTION 5.1

1.

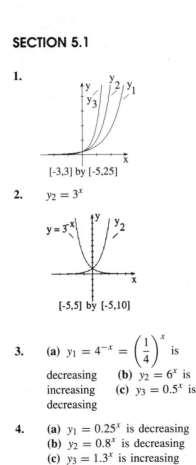

[-3,3] by [-5,25]

2. $y_2 = 3^x$

[-5,5] by [-5,10]

3. **(a)** $y_1 = 4^{-x} = \left(\dfrac{1}{4}\right)^x$ is
decreasing **(b)** $y_2 = 6^x$ is
increasing **(c)** $y_3 = 0.5^x$ is
decreasing

4. **(a)** $y_1 = 0.25^x$ is decreasing
(b) $y_2 = 0.8^x$ is decreasing
(c) $y_3 = 1.3^x$ is increasing

5. vertical shift down 4 units

6. vertical stretch by a factor of 3

7. horizontal shift right 3 units

8. reflection about the x-axis
followed by a reflection about
the y-axis

9. horizontal shift left 1 unit, then
a vertical shift up 7 units

10. a reflection about the x-axis,
followed by a vertical stretch by
a factor of 2, a horizontal shift
left 2 units, then a vertical shift
down 1 unit

11. $x < 0$

12. $x < 0$

13. $x > 0$

14. $x > 0$

15. **(a)** and **(c)**

16. **(b)** and **(c)**

17. **(b)** and **(c)**

18. **(a)** and **(b)**

19.

[-10,10] by [-1,10]

20.

[-10,10] by [-10,3]

21.

[-10,10] by [1,10]

22.

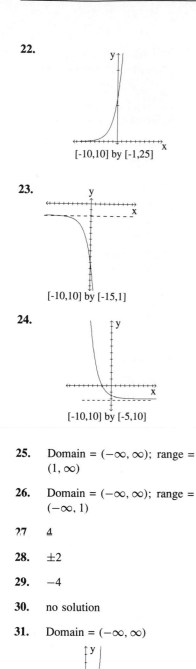

[-10,10] by [-1,25]

23.

[-10,10] by [-15,1]

24.

[-10,10] by [-5,10]

25. Domain = $(-\infty, \infty)$; range = $(1, \infty)$

26. Domain = $(-\infty, \infty)$; range = $(-\infty, 1)$

27. 4

28. ± 2

29. -4

30. no solution

31. Domain = $(-\infty, \infty)$

[-5,5] by [-1,10]

32. Domain = $(-\infty, \infty)$

[-5,5] by [-1,10]

33. Domain = $(-\infty, \infty)$

[-10,5] by [-2,10]

34. Domain = $(-\infty, \infty)$

[-5,10] by [-10,3]

35. Increasing on $(-\infty, \infty)$; no local extrema; range = $(0, \infty)$

36. Increasing on $(0, \infty)$; decreasing on $(-\infty, 0)$; local minimum of 0.5 when $x = 0$; range = $[0.5, \infty)$

37. Decreasing on $(-\infty, -0.91)$; increasing on $(-0.91, \infty)$; local minimum of -0.33 when $x = -0.91$; range = $[-0.33, \infty)$

38. Increasing on $(-\infty, 1)$; decreasing on $(1, \infty)$; local maximum of 0.37 when $x = 1$; range = $(-\infty, 0.37]$

39. Increasing on $(-0.85, 0.85)$; decreasing on $(-\infty, -0.85)$,

$(0.85, \infty)$; local minimum of -0.52 when $x = -0.85$; local maximum of 0.52 when $x = 0.85$; range = $[-0.52, 0.52]$

40. Decreasing on $(-\infty, \infty)$; no local extrema; range = $(-\infty, \infty)$

41.

[-10,10] by [-10,10]

42.

[-10,10] by [-10,10]

43. $(-0.77, 0.59)$, $(2.00, 4.00)$, $(4.00, 16.00)$

44. $(2.48, 15.25)$, $(3, 27)$

45. HA: $y = e^2 \approx 7.39$

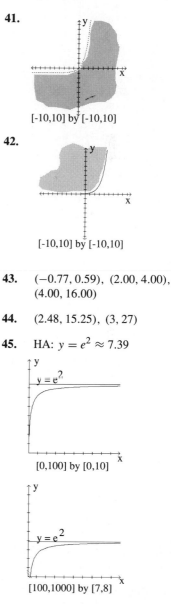

$y = e^2$

[0,100] by [0,10]

$y = e^2$

[100,1000] by [7,8]

46. HA: $y = e^3 \approx 20.09$

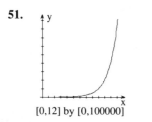

[0,100] by [0,25]

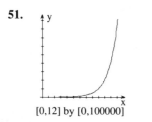

[0,1000] by [19,21]

47. (a) 5000 bacteria (b) 10,000 bacteria (c) $2500(2)^{t/3}$ bacteria

48.

[0,25] by [0,600000]

49. 15.97 hrs

50. After 1 year, 81,920 rabbits; after 5 years, 2.31×10^{19} rabbits

51.

[0,12] by [0,100000]

52. 8.97 mo

53. Answers will vary, but should mention food and space limits.

54. $P(t) = 475,000(1 + 0.0375)^t$; $t = 20.22$ yr

55. 1915, 15,689 people; 1940, 39,381 people

56. $P(t) = 123,000(1 - 0.02375)^t$; $t = 37.45$ yr

57. $P(t) = 6.58(0.5)^{t/14}$; after 38.05 days

58. $P(t) = 3.5\left(\dfrac{1}{2}\right)^{t/65}$, after 117.48 days

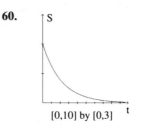

[-20,200] by [0,5]

59. $P(1.5) = 0.5Sg$;
$P(3) = 0.25Sg$;
$P(t) = S(2^{-1})^{t/1.5}g$

60.

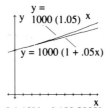

[0,10] by [0,3]

61. $2^{40}g$

SECTION 5.2

1. $200 interest earned; $1450

2. $250 remains

3. $641.13

4. 2.86 yr

5. $981.79

6. $2391.08

7. $33,492.13

8. $7429.74

9. $36,374.39

10. $11,149.60

11. $10,382.24

12. 11.1% compounded monthly is slightly better

13. 8% compounded annually is slightly better

14. Use y-max = 2000

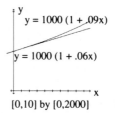

[0,10] by [0,2000]

15. Use y-max = 2000 (Answers will vary.)

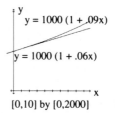

[-1,10] by [-100,2000]

16. For $y = 1000(1 + 0.06x)$, APR is 6%; for $y = 1000(1 + 0.09x)$, APR is 9%

17. Simple interest:
$y = 1000(1 + 0.05x)$;
compound interest:
$y = 1000(1.05)^x$; APR = 5%

18. Use y-max = 2500

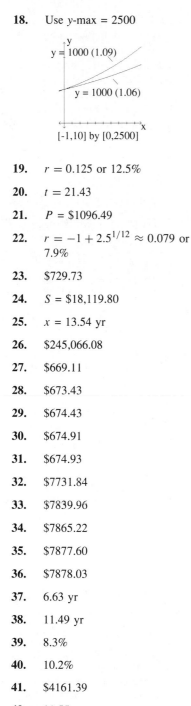

$[-1,10]$ by $[0,2500]$

19. $r = 0.125$ or 12.5%

20. $t = 21.43$

21. $P = \$1096.49$

22. $r = -1 + 2.5^{1/12} \approx 0.079$ or 7.9%

23. $729.73

24. $S = \$18,119.80$

25. $x = 13.54$ yr

26. $245,066.08

27. $669.11

28. $673.43

29. $674.43

30. $674.91

31. $674.93

32. $7731.84

33. $7839.96

34. $7865.22

35. $7877.60

36. $7878.03

37. 6.63 yr

38. 11.49 yr

39. 8.3%

40. 10.2%

41. $4161.39

42. 11.55 yr

43. $r = 4.3\%$; a \$1500 investment earns interest compounded continuously; determine the annual interest rate if the value of the investment is \$2300 after ten years

44. 12.14 yr

45. 17.62 yr

46. $0.74

47. $0.46

48. $0.25

49. $94,260.33

50. For simple interest, only the principal earns interest; for compound interest, both principal and previously paid interest earn interest; for simple interest, $S = P(1 + 10r)$ after ten years; for compound interest,
$$S = P\left(1 + \frac{0.08}{4}\right)^{4(10)};$$
12.08%

SECTION 5.3

1. The 6% investment yields the greatest return.

2. The 8.25% investment yields the greatest return.

3. The 7.20% investment is the better investment.

4. The 8.5% investment yields the greatest return.

5. $36,013.70

6. $13,937.28

7. $749.35

8. 8%, 12 payments per year

9. 9%, 4 payments per year

10. For $i = 0.05 = \dfrac{r}{k}$, any of the following are correct: one payment per year at 5% annual rate ($r = 0.05, k = 1$); two payments per year at 10% APR ($r = 0.10, k = 2$); four payments per year at 20% APR ($r = 0.20, k = 4$); etc.

11. For $i = 0.01 = \dfrac{r}{k}$, any of the following are correct: one payment per year at 1% APR ($r = 0.01, k = 1$); two payments per year at 2% APR ($r = 0.02, k = 2$); four payments per year at 4% APR ($r = 0.04, k = 4$); 12 payments per year at 12% APR ($r = 0.12, k = 12$); etc.

12.

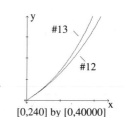

$[0,240]$ by $[0,40000]$

13. Refer to the graph in Exercise 12.

14. The values of y in Exercise 12 correspond to the future value of an annuity after x payments of \$100 per month at 6% APR. The values of y in Exercise 13 correspond to the future value of an annuity after x payments of \$100 per month at 8% APR.

15. $R = \$230.43$

16. $R = \$151.62$

17. $R = \$884.61$

18. $R = \$1032.14$

19. $R = \$1028.61$

20. 14 yr and 4 mo

21. 22 yr and 2 mo

22. 8%, 12 payments per year

23. 11%, 4 payments per year

24. Answers will vary; APR of 1% with 1 payment per year; APR of 12% with 12 payments per year; APR of 4% with 4 payments per year, etc.

25. Answers will vary; APR of 9% with 12 payments per year, since $\dfrac{0.09}{12} = 0.0075$

26.

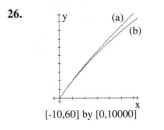

[-10,60] by [0,10000]

27. The values of y in (a) and (b) represent the present value of the loan for x monthly payments of $200 at 8% APR and 12% APR, respectively.

28. (a) $R = \dfrac{860}{1 - (1.01)^{-x}}$

(b)

[0,360] by [0,6000]

(c) The values of R represent monthly payments for an $86,000 loan at 12% APR to be paid in x installments.

29. $t = 69.66 \approx 70$ means a $10,000 loan at 12% APR with

monthly payments of $200 will be paid off in 70 months.

30. (Let $i = x$) $x = 0.0125$; 1.25% is the monthly interest rate required to pay off a $10,000 car loan in 5 years with monthly payments of $238

31. $r = 0.3571$ or 35.71%

32.

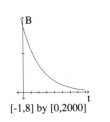

[-1,8] by [0,2000]

33. $0 \leq t \leq 6$

34. $2600.51

35.

B st. line method
const.pct. method

[0,6] by [0,20000]

36. $B(t) =$
$$\begin{cases} \dfrac{3800t}{3} + 22{,}600, & 0 \leq t \leq 3 \\ -5425(t - 7) + 4700, & 3 < t \leq 7 \end{cases}$$

37.

B

[0,10] by [0,30000]

38. $12,837.50

39. Since $r < 0$, let $r = -a$, where $a > 0$. Then

$S = C(1 - (-a)^n) = C(1 + a)^n$, where $a > 0$. Replace a by r for $r > 0$. Since $r > 0$, $1 + r > 1$ and $(1 + r)^n > 1^n$, since $n > 0$. Because C is multiplied by a number greater than 1 to get S, then $S > C$ and the asset has appreciated over time.

40.

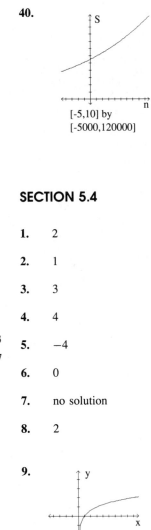

[-5,10] by [-5000,120000]

SECTION 5.4

1. 2

2. 1

3. 3

4. 4

5. −4

6. 0

7. no solution

8. 2

9.

[-5,10] by [-5,5]

10.

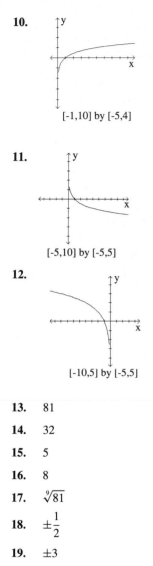

[-1,10] by [-5,4]

11.

[-5,10] by [-5,5]

12.

[-10,5] by [-5,5]

13. 81

14. 32

15. 5

16. 8

17. $\sqrt[9]{81}$

18. $\pm\dfrac{1}{2}$

19. ± 3

20. 0 or 2

21. $\log_7 y = x$

22. $4 = \log_3 xy$

23. $\log_2(x + y) = 8$

24. $\log_{(1+r)} P = n$

25. $3^5 = x$

26. $2^y = x$

27. $3^{-2} = \dfrac{x}{y}$

28. $P = (1 + r)^n$

29. $3\log_2 x + 2\log_2 y$ for $x > 0$, $y > 0$

30. $\ln x + 3\ln y$ for $x > 0$, $y > 0$

31. $2\log_a x - 3\log_a y$ for $x > 0$, $y > 0$

32. $3 + 4\log_{10} x$ for $x > 0$

33. $\log 5000 + \log x + 360\log(1+r)$ for $r > -1$, $x > 0$

34. $\dfrac{3}{5}(\log 6 + \log z)$ for $z > 0$

35.

y=1

[-5,10] by [-1,4]

36.

[-2,6] by [-5,3]

37.

[-2,20] by [-0.5,2]

38.

[-2,10] by [-10,10]

39. As $x \to \infty$, $f(x) \to 1$; as $x \to -\infty$, $f(x) \to 1$

40. As $x \to \infty$, $f(x) \to 0$; as $x \to -\infty$, $f(x) \to 0$

41. As $x \to \infty$, $f(x) \to 1$

42. As $x \to \infty$, $f(x) \to \infty$

43. 20.22 yr

44. 37.45 yr

45. 15,689 people in 1915; 39,381 people in 1940

46. $t > 38.05$ days

47. $t > 117.48$ days

48. $y = 2.75x^5$

49. Power rule for the first set of data is $y = 8.10x^{-0.1145}$; for the second set, the data, when plotted, are not linear.

50. 11.91

51. 5.45

52. 15.52

53. The equations have the form $S = P(1 + r)^t$ where S is the total value of the account, P the principal, r the annual interest rate. This is the formula for an investment with interest compounded annually for t yr.

54. $\dfrac{\ln S - \ln P}{\ln(1 + r)}$

55. 9.93 yr

56. 15.74 yr

57.

$\ln x$	18.3438	17.3990	18.0217
$\ln P$	5.8999	4.4773	5.4161

$\ln x$	18.7692	19.9955	20.6023
$\ln P$	6.5323	8.3733	9.2827

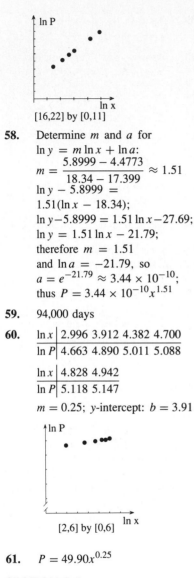

[16,22] by [0,11]

58. Determine m and a for
$\ln y = m \ln x + \ln a$:
$m = \dfrac{5.8999 - 4.4773}{18.34 - 17.399} \approx 1.51$
$\ln y - 5.8999 =$
$1.51(\ln x - 18.34)$;
$\ln y - 5.8999 = 1.51 \ln x - 27.69$;
$\ln y = 1.51 \ln x - 21.79$;
therefore $m = 1.51$
and $\ln a = -21.79$, so
$a = e^{-21.79} \approx 3.44 \times 10^{-10}$;
thus $P = 3.44 \times 10^{-10} x^{1.51}$

59. 94,000 days

60.

$\ln x$	2.996	3.912	4.382	4.700
$\ln P$	4.663	4.890	5.011	5.088

$\ln x$	4.828	4.942
$\ln P$	5.118	5.147

$m = 0.25$; y-intercept: $b = 3.91$

[2,6] by [0,6]

61. $P = 49.90 x^{0.25}$

SECTION 5.5

1. Domain $= (0, \infty)$; range $= (-\infty, \infty)$

[-2,10] by [-3,3]

2. Domain $= (0, \infty)$; range $= (-\infty, \infty)$

[-2,10] by [-2,2]

3. Domain $= (0, \infty)$; range $= (-\infty, \infty)$

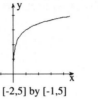

[-2,5] by [-1,5]

4. Domain $= (0, \infty)$; range $= (-\infty, \infty)$

[-1,20] by [-15,2]

5. Domain $= (0, \infty)$; range $= (-\infty, \infty)$

[-1,5] by [-1,9]

6. Domain $= (0, \infty)$; range $= (-\infty, \infty)$

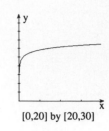

[0,20] by [20,30]

7.

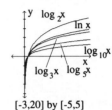

[-3,20] by [-5,5]

8. Horizontal shift right 2 units; VA at $x = 2$

9. Horizontal shift left 5 units, vertical shift down 1 unit; VA at $x = -5$

10. Reflection in the y-axis, horizontal shift right 5 units; VA at $x = 5$

11. Reflection in the y-axis, horizontal shift right 2 units; VA at $x = 2$

12. Horizontal shift right $\dfrac{4}{3}$ units, vertical shift up $\log_2 3 (\approx 1.58)$ units; VA at $x = \dfrac{4}{3}$

13. Horizontal shift left $\dfrac{3}{2}$ units, vertical shift up $\ln 2 + 2$ units; VA at $x = -\dfrac{3}{2}$

14. Reflection in the x-axis, horizontal shift left 3 units, vertical shift down 2 units; VA at $x = -3$

15. Reflection in the y-axis, horizontal shift right 4 units, reflection in the x-axis, vertical shift down 2 units; VA at $x = 4$

16. Domain = $(-3, \infty)$; range = $(-\infty, \infty)$

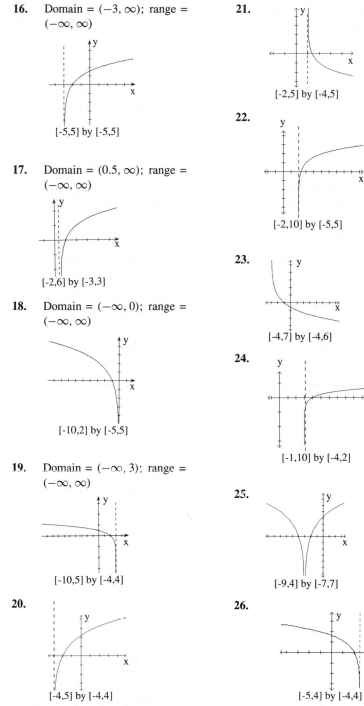

[-5,5] by [-5,5]

17. Domain = $(0.5, \infty)$; range = $(-\infty, \infty)$

[-2,6] by [-3,3]

18. Domain = $(-\infty, 0)$; range = $(-\infty, \infty)$

[-10,2] by [-5,5]

19. Domain = $(-\infty, 3)$; range = $(-\infty, \infty)$

[-10,5] by [-4,4]

20.

[-4,5] by [-4,4]

21.

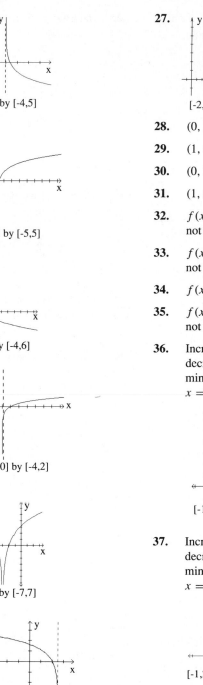

[-2,5] by [-4,5]

22.

[-2,10] by [-5,5]

23.

[-4,7] by [-4,6]

24.

[-1,10] by [-4,2]

25.

[-9,4] by [-7,7]

26.

[-5,4] by [-4,4]

27.

[-2,15] by [-2,7]

28. $(0, 1)$

29. $(1, \infty)$

30. $(0, 1)$

31. $(1, \infty)$

32. $f(x) \to \infty$ as $x \to \infty$; $f(x)$ is not defined for $x \in (-\infty, 0]$

33. $f(x) \to \infty$ as $x \to \infty$; $f(x)$ is not defined in $(-\infty, 0]$

34. $f(x) \to \infty$ as $|x| \to \infty$

35. $f(x) \to 0$ as $x \to \infty$; $f(x)$ is not defined in $(-\infty, 0]$

36. Increasing on $(0.37, \infty)$; decreasing on $(0, 0.37)$; local minimum of -0.37 when $x = 0.37$

[-1,6] by [-2,7]

37. Increasing on $(0.61, \infty)$; decreasing on $(0, 0.61)$; local minimum of -0.18 when $x = 0.61$

[-1,2] by [-2,7]

38. Increasing on $(-0.61, 0)$, $(0.61, \infty)$; decreasing on

$(-\infty, -0.61)$, $(0, 0.61)$; local minimum of -0.18 when $x = \pm 0.61$

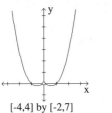

[-4,4] by [-2,7]

39. Increasing on $(0, 2.72)$; decreasing on $(2.72, \infty)$; local maximum of 0.37 when $x = 2.72$

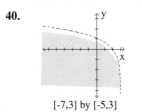

[-2,6] by [-5,2]

40.

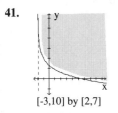

[-7,3] by [-5,3]

41.

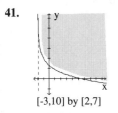

[-3,10] by [2,7]

42. $(4.93, 1.07)$

43. $(-2.02, 2.09)$ and $(2.23, 2.98)$

44. The graphs of $y_1 = \dfrac{1}{2}(e^x - e^{-x})$ and $y_2 = \ln(x + \sqrt{x^2 + 1})$ are reflections of each other about

the line $y = x$

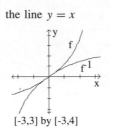

[-3,3] by [-3,4]

45. $f(x) > g(x)$ in $(-\infty, 2.48) \cup (3, \infty)$; $f(x) = g(x)$ when $x = 2.48$ or $x = 3$; $f(x) < g(x)$ in $(2.48, 3)$

46. $f(x) > g(x)$ in $(-\infty, 2.72) \cup (2.72, \infty)$; $f(x) = g(x)$ when $x = 2.72$

47. $f(x) > g(x)$ in $(0, 10)$; $f(x) = g(x)$ when $x = 10$; $f(x) < g(x)$ in $(10, \infty)$

48. $f(x) > g(x)$ in $[1, \infty)$

49. $f(x) > g(x)$ in $(1, \infty)$

50. $f(x) > g(x)$ in $(6.41, 93.35)$; $f(x) = g(x)$ when $x = 6.41$ or $x = 93.35$; $f(x) < g(x)$ in $(0, 6.41) \cup (93.35, \infty)$

51. Graph $y = \dfrac{\ln 4}{\ln x}$ or $y = \dfrac{\log 4}{\log x}$

52. 6.35

53. $6 = \log\left(\dfrac{a_1}{T}\right) + B = \log a_1 - \log T + B$; $4 = \log\left(\dfrac{a_2}{T}\right) + B = \log a_2 - \log T + B$; subtracting, we get $2 = \log a_1 - \log a_2 = \log\left(\dfrac{a_1}{a_2}\right)$. Then $10^2 = \dfrac{a_1}{a_2}$ and $100 a_2 = a_1$. So the first is 100 times more intense than the second.

54. $a > 0$

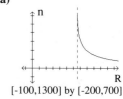

[-100,500] by [-2,7]

55. $I = 12 \cdot 10^{-0.00235x}$

56.

[-15,60] by [-15,25]

57. 10.20 lumens

58. 5.06 lumens

59.

[-1,10] by [-1,6]

60. $a = 3.47$ with $\log a = 0.54 = y_0$ and $b = 3.02$ with $\log b = 0.48 = m$

61. Given $f(x) = ab^x$, then $\log(f(x)) = \log(ab^x) = \log a + x \log b = (\log b)x + \log a$; thus $m = \log b$ and $y_0 = \log a$

62. $f(x) = (3.47)(3.02)^x$

63. (a)

[-100,1300] by [-200,700]

(b) $708.14   **(c)** 231 payments (19 yr, 3 mo)

64. (a)

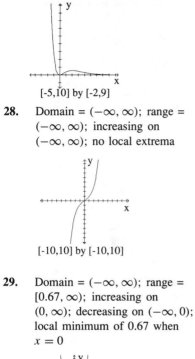

[-100,1400] by [-100,800]

(b) $923.17   **(c)** 252 payments (21 yr)

SECTION 5.6

1. 10,000

2. $\dfrac{1}{e}$

3. 5.25

4. $\dfrac{1}{2}$

5. 97

6. 1001

7. 16

8. 531,434

9. $x = 3$

10. 2.52

11. $x = \dfrac{e^3}{2}$

12. $x = 80$

13. $x = 1$

14. $\dfrac{1}{3}$

15. $D_1 = (0, \infty)$; $R_1 = (-\infty, \infty)$; $D_2 = (-\infty, 0) \cup (0, \infty)$; $R_2 = (-\infty, \infty)$

16. $D_1 = (0, \infty)$; $R_1 = (-\infty, \infty)$; $D_2 = (-\infty, -1) \cup (0, \infty)$; $R_2 = (-\infty, \infty)$

17. $D_1 = (0, \infty)$; $R_1 = (-\infty, \infty)$; $D_2 = (0, \infty)$; $R_2 = (-\infty, \infty)$

18. $D_1 = (0, \infty)$; $R_1 = (-\infty, 0)$; $D_2 = (-\infty, -1) \cup (0, \infty)$; $R_2 = (-\infty, 0) \cup (0, \infty)$

19. $\dfrac{\ln 4.1}{\ln 1.06}$

20. $\dfrac{\log 1.6}{\log 0.98}$

21. $\dfrac{\log 18.4}{\log 1.09}$

22. $\dfrac{\log 3.2}{\log 1.12}$

23. $\dfrac{\log(6 + \sqrt{37})}{\log 2}$

24. $\dfrac{\log(3 \pm 2\sqrt{2})}{\log 2}$

25. **(a)** ≈ 24.22 **(b)** ≈ -23.26
     (c) ≈ 33.79 **(d)** ≈ 10.26
     (e) ≈ 3.59 **(f)** $x \approx \pm 2.54$

26. Domain = $(-\infty, \infty)$; range = $[-0.37, \infty)$; increasing on $(1, \infty)$; decreasing on $(-\infty, -1)$; local minimum of -0.37 when $x = -1.00$

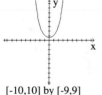

[-5,5] by [-5,5]

27. Domain = $(-\infty, \infty)$; range = $[0, \infty)$; increasing on $(0, 2.00)$; decreasing on $(-\infty, 0)$, $(2.00, \infty)$; local minimum of 0 when $x = 0$; local maximum of

0.54 when $x = 2.00$

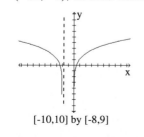

[-5,10] by [-2,9]

28. Domain = $(-\infty, \infty)$; range = $(-\infty, \infty)$; increasing on $(-\infty, \infty)$; no local extrema

[-10,10] by [-10,10]

29. Domain = $(-\infty, \infty)$; range = $[0.67, \infty)$; increasing on $(0, \infty)$; decreasing on $(-\infty, 0)$; local minimum of 0.67 when $x = 0$

[-10,10] by [-9,9]

30. Domain = $(-\infty, -2) \cup (0, \infty)$; range = $(-\infty, \infty)$; increasing on $(0, \infty)$; decreasing on $(-\infty, -2)$; no local extrema

[-10,10] by [-8,9]

31. Domain = $(-\infty, -2) \cup (0, \infty)$; range = $(-\infty, 0) \cup (0, \infty)$;

increasing on $(-\infty, -2)$,
$(0, \infty)$; no local extrema

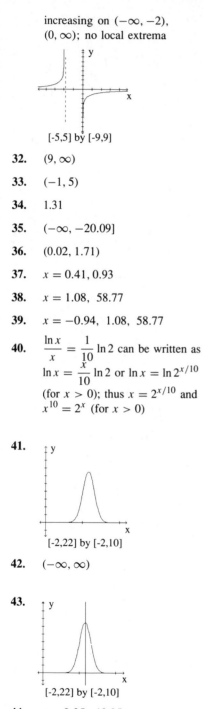

[-5,5] by [-9,9]

32. $(9, \infty)$

33. $(-1, 5)$

34. 1.31

35. $(-\infty, -20.09]$

36. $(0.02, 1.71)$

37. $x = 0.41, 0.93$

38. $x = 1.08, \ 58.77$

39. $x = -0.94, \ 1.08, \ 58.77$

40. $\dfrac{\ln x}{x} = \dfrac{1}{10} \ln 2$ can be written as
$\ln x = \dfrac{x}{10} \ln 2$ or $\ln x = \ln 2^{x/10}$
(for $x > 0$); thus $x = 2^{x/10}$ and
$x^{10} = 2^x$ (for $x > 0$)

41.

[-2,22] by [-2,10]

42. $(-\infty, \infty)$

43.

[-2,22] by [-2,10]

44. $x = 9.25, \ 13.95$

45. 9.69×10^{-4} lb/in^2

46. $(-\infty, 0) \cup (0, \infty)$

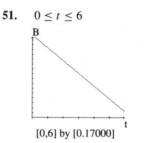

[-20,20] by [-10,30]

47. Values of r greater than zero
make sense.

48. Relative maximum of 2.38
when $r = 1.73$

49. 9.89% APR

50. 10.6% APR

51. $0 \le t \le 6$

[0,6] by [0,17000]

52. $6295.11

53. We are given $f_n(x) =$
$\dfrac{x^n}{n!} + \dfrac{x^{n-1}}{(n-1)!} + \cdots + x + 1$;
then $f_{n-1}(x) =$
$\dfrac{x^{n-1}}{(n-1)!} + \dfrac{x^{n-2}}{(n-2)!} + \cdots + x + 1$
and $f_n(x) = \dfrac{x^n}{n!} +$
$\left[\dfrac{x^{n-1}}{(n-1)!} + \cdots + x + 1 \right] =$
$\dfrac{x^n}{n!} + f_{n-1}(x)$

54.

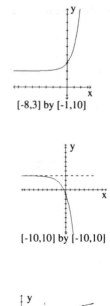

[-5,5] by [-25,25]

55. $(-1.32, 1.17)$

56. $(-2.13, 1.94)$

Chapter 5 Review

1.

[-8,3] by [-1,10]

2.

[-10,10] by [-10,10]

3.

[-2,10] by [4,5]

4.

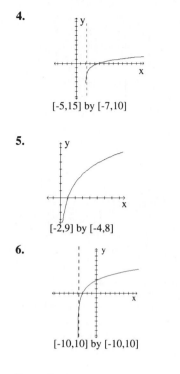

[-5,15] by [-7,10]

5.

[-2,9] by [-4,8]

6.

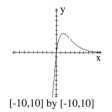

[-10,10] by [-10,10]

7.   5

8.   Domain = $(-\infty, \infty)$; range =
$(-\infty, 4.25]$; relative maximum
of 4.25 when $x = 1.44$

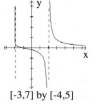

[-10,10] by [-10,10]

9.   Domain = $(-2, 2) \cup (2, \infty)$;
range = $(-\infty, \infty)$; no local
extrema

[-3,7] by [-4,5]

10.   Domain = $(-\infty, 5) \cup (5, \infty)$;
range = $(-\infty, \infty)$; no local
extrema

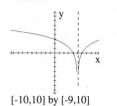

[-10,10] by [-9,10]

11.   Domain = $(-\infty, 0) \cup (3, \infty)$;
range = $(-\infty, 0) \cup (0, \infty)$; no
local extrema

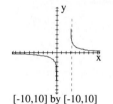

[-10,10] by [-10,10]

12.   Domain = $(-\infty, \infty)$; range =
$\left[\dfrac{2}{3}, \infty\right)$; local minimum of $\dfrac{2}{3}$
when $x = 0$

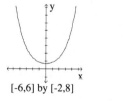

[-6,6] by [-2,8]

13.   Decreasing on $(-\infty, \infty)$; no
local extrema

14.   Increasing on $(-\infty, 1.44)$;
decreasing on $(1.44, \infty)$;
local maximum of 4.25 when
$x = 1.44$

15.   3

16.   No solution because 3 to any
power is positive.

17.   3

18.   64

19. $12 \log(1 + r) - \log r, \ r > 0$

20. $\dfrac{\log z}{\log(23 + x)}$

21.   as $x \to \infty, \ f(x) \to \infty$

22.   as $x \to 2^{+}, \ g(x) \to 0$

23.   as $x \to 1^{-}, \ f(x) \to -\infty$

24. $(0, 1)$

25.   as $x \to \infty, \ f(x) \to \infty; \ f(x)$
is not defined for $x \le 0$

26.   Increasing on $(0.72, \infty)$;
decreasing on $(0, 0.72)$; local
minimum of -0.12 when
$x = 0.72$

[-3,15] by [-3,15]

27. $\sqrt[5]{9}$ or $3^{2/5}$

28. $x = 5$

29. $D_1 = (0, \infty); \ R_1 = (-\infty, 0);$
$D_2 = (-\infty, -2) \cup (0, \infty);$
$R_2 = (-\infty, 0) \cup (0, \infty)$

30. $\dfrac{\log 0.90}{\log 1.5}$

31. ≈ -0.26

32. $x > 3.52$ or $(3.52, \infty)$

33. $x = 2$

34. $(0.72, \infty)$

35.   1.52

36. $x = 0.12, \ 1.13$

37. $(-\infty, 0.69]$

38. $(0.88, 1.63)$

39. $x = -0.77, 2, 4$; $2^x > x^2$ on $(-0.77, 2) \cup (4, \infty)$; $x^2 > 2^x$ on $(-\infty, -0.77) \cup (2, 4)$

40. $x = 2.48, 3$; $3^x > x^3$ on $(-\infty, 2.48) \cup (3, \infty)$; $x^3 > 3^x$ on $(2.48, 3)$

41. $701.28

42. $707.39

43. $708.81

44. $709.51

45. $705.30

46. $709.53

47. 6 yr

48. 11.18 yr

49.

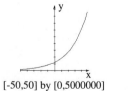

[0,10] by [0,1100]

50. $P(t) = 625{,}000(1 + 0.0405)^t$

51. Negative and positive values of t near 0 (recent past and near future) make sense given the current rate of growth.

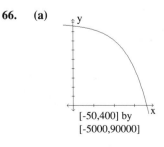

[-50,50] by [0,5000000]

52. 11.84 yr

53. The difference between the two functions is less than 0.01 in the interval $(0, 0.14)$

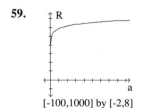

[-0.1,0.9] by [-0.2,1]

54. Let $S = 2P$. Then $2P = P(1 + r)^n$ where P is the principal and n is the number of years. Solve for n: $2 = (1+r)^n$; $\ln 2 = \ln(1 + r)^n = n \ln(1 + r)$; $n = \dfrac{\ln 2}{\ln(1 + r)}$; if $0.72 \approx \ln 2$, then $n \approx \dfrac{0.72}{\ln(1 + r)}$. From Ex. 53, $r \approx \ln(1 + r)$ when $0 < r < 0.14$, so if $r < 14\%$, r is a good approximation for $\ln(1 + r)$. Thus $n \approx \dfrac{0.72}{r}$ or $n \approx \dfrac{72}{100r}$.

55. $A(t) = 4.62(0.5)^{t/21}$

56.

[-5,70] by [-2,6]

57. $t > 46.37$ days

58. 6.29

59.

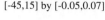

[-100,1000] by [-2,8]

60.

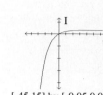

[-45,15] by [-0.05,0.07]

61.

[-5,22] by [-2,5]

62. $1729.20

63. $i = 0.014$ or 1.4%

64. This is the present value formula with $(360 - n)$ payments left to be made (30 yr with 12 payments per year).

65. $\log f(x) = \log a + \log b^x = \log a + x \log b$, or $\log f(x) = (\log b)x + \log a$; then $\log f(x_1) = (\log b)x_1 + \log a$; $\log f(x_2) = (\log b)x_2 + \log a$; so $\dfrac{\log f(x_2) - \log f(x_1)}{x_2 - x_1} = \dfrac{[(\log b)x_2 + \log a] - [(\log b)x_1 + \log a]}{x_2 - x_1} = \dfrac{(\log b)x_2 - (\log b)x_1}{x_2 - x_1} = \dfrac{(\log b)(x_2 - x_1)}{x_2 - x_1} = \log b$

66. (a)

[-50,400] by [-5000,90000]

(b) The y value is the outstanding balance after n of the 360 payments have been made.

67. **(a)** \$702.06   **(b)** 27.63 yr

CHAPTER 6

SECTION 6.1

1. QIII

2. QIV

3. QII

4. QIII

5. QI

6. QIII

7. 408°

8. −250°

9. 345°

10. 470°

11. $\alpha_1 = 415°$; $\alpha_2 = 775°$; $\beta_1 = -305°$; $\beta_2 = -665°$

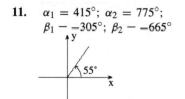

12. $\alpha_1 = 338°$; $\alpha_2 = 698°$; $\beta_1 = -382°$; $\beta_2 = -742°$

13. $\alpha_1 = 50°$; $\alpha_2 = 770°$; $\beta_1 = -310°$; $\beta_2 = -670°$

14. $\alpha_1 = 210°$; $\alpha_2 = 570°$; $\beta_1 = -510°$; $\beta_2 = -870°$

15. $\alpha_1 = \dfrac{9\pi}{4}$; $\alpha_2 = \dfrac{17\pi}{4}$; $\beta_1 = -\dfrac{7\pi}{4}$; $\beta_2 = -\dfrac{15\pi}{4}$

16. $\alpha_1 = \dfrac{7\pi}{2}$; $\alpha_2 = \dfrac{11\pi}{2}$; $\beta_1 = -\dfrac{\pi}{2}$; $\beta_2 = -\dfrac{5\pi}{2}$

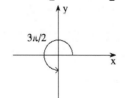

17. $\alpha_1 = \dfrac{13\pi}{6}$; $\alpha_2 = \dfrac{25\pi}{6}$; $\beta_1 = -\dfrac{11\pi}{6}$; $\beta_2 = -\dfrac{23\pi}{6}$

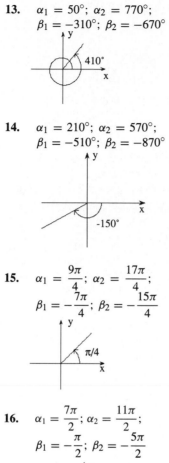

18. $\alpha_1 = \dfrac{17\pi}{6}$; $\alpha_2 = \dfrac{29\pi}{6}$; $\beta_1 = -\dfrac{7\pi}{6}$; $\beta_2 = -\dfrac{19\pi}{6}$

19. $\alpha_1 = \dfrac{5\pi}{3}$; $\alpha_2 = \dfrac{11\pi}{3}$; $\beta_1 = -\dfrac{7\pi}{3}$; $\beta_2 = -\dfrac{13\pi}{3}$

20. $\alpha_1 = \dfrac{\pi}{4}$; $\alpha_2 = \dfrac{9\pi}{4}$; $\beta_1 = -\dfrac{15\pi}{4}$; $\beta_2 = -\dfrac{23\pi}{4}$

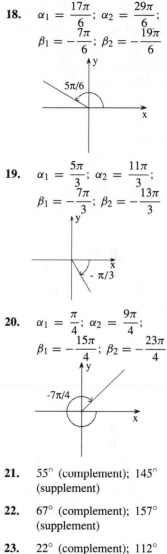

21. 55° (complement); 145° (supplement)

22. 67° (complement); 157° (supplement)

23. 22° (complement); 112° (supplement)

24. 78° (complement); 168° (supplement)

25. $\dfrac{\pi}{6}$ (complement); $\dfrac{2\pi}{3}$ (supplement)

26. $\dfrac{5\pi}{12}$ (complement); $\dfrac{11\pi}{12}$ (supplement)

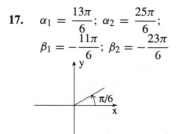

27. $\dfrac{3\pi}{26}$ (complement); $\dfrac{8\pi}{13}$ (supplement)

28. $\dfrac{\pi}{14}$ (complement); $\dfrac{4\pi}{7}$ (supplement)

29. $\theta = 180°$ or $\theta = \pi$

30. $\theta = 90°$ or $\theta = \dfrac{\pi}{2}$

31. $\theta = 45°$ or $\dfrac{\pi}{4}$

32. $\theta = 135°$ or $\dfrac{3\pi}{4}$

33. $\theta = 315°$ or $\dfrac{7\pi}{4}$

34. $\theta = 0°$ or $360°$; in radian measure, $\theta = 0$ or 2π

35. $180°$ or π

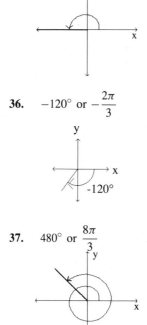

36. $-120°$ or $-\dfrac{2\pi}{3}$

37. $480°$ or $\dfrac{8\pi}{3}$

38. $-600°$ or $-\dfrac{10\pi}{3}$

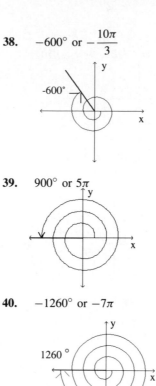

39. $900°$ or 5π

40. $-1260°$ or -7π

41. $\dfrac{\pi}{4}$

42. $\dfrac{\pi}{3}$

43. $\dfrac{3\pi}{4}$

44. $\dfrac{7\pi}{6}$

45. $\dfrac{2\pi}{3}$

46. $\dfrac{11\pi}{6}$

47. 0.40

48. 2.06

49. 1.26

50. 4.35

51. $75°$

52. $257.83°$

53. $214.86°$

54. $739.12°$

55. $\dfrac{\pi}{3}$

56. 7.98

57. 15.58

58. 25.82

59. 5.76 in.

60. 0.293 ft

61. 6.125 cm

62. 1.34 in.

63.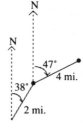

64. **(a)** $\theta = 45°$ **(b)** $\theta = 22.5°$
(c) $\theta = 67.5°$

65. Course correction: $(d) = 10°$ or $10°$ south of current course

66. ≈ 295.75 statute mi

67.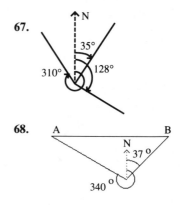

68.

69. (a)

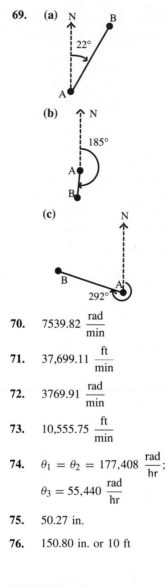

(b)

(c)

70. $7539.82 \dfrac{\text{rad}}{\text{min}}$

71. $37{,}699.11 \dfrac{\text{ft}}{\text{min}}$

72. $3769.91 \dfrac{\text{rad}}{\text{min}}$

73. $10{,}555.75 \dfrac{\text{ft}}{\text{min}}$

74. $\theta_1 = \theta_2 = 177{,}408 \dfrac{\text{rad}}{\text{hr}}$;

$\theta_3 = 55{,}440 \dfrac{\text{rad}}{\text{hr}}$

75. 50.27 in.

76. 150.80 in. or 10 ft

SECTION 6.2

1. $(0, 1)$

2. $\left(\dfrac{\sqrt{2}}{2}, \dfrac{\sqrt{2}}{2}\right)$

3. $\left(\dfrac{\sqrt{2}}{2}, -\dfrac{\sqrt{2}}{2}\right)$

4. $\left(-\dfrac{\sqrt{3}}{2}, \dfrac{1}{2}\right)$

5. $(0, 1)$

6. $\left(\dfrac{1}{2}, -\dfrac{\sqrt{3}}{2}\right)$

7. $\left(-\dfrac{\sqrt{2}}{2}, -\dfrac{\sqrt{2}}{2}\right)$

8. $(-1, 0)$

9. $\left(\dfrac{\sqrt{3}}{2}, -\dfrac{1}{2}\right)$

10. $\left(-\dfrac{\sqrt{2}}{2}, \dfrac{\sqrt{2}}{2}\right)$

11. $\left(-\dfrac{1}{2}, \dfrac{\sqrt{3}}{2}\right)$

12. $\left(-\dfrac{\sqrt{3}}{2}, -\dfrac{1}{2}\right)$

13. $\sin \dfrac{\pi}{6} = \dfrac{1}{2}$; $\cos \dfrac{\pi}{6} = \dfrac{\sqrt{3}}{2}$

14. $\sin \dfrac{\pi}{4} = \dfrac{\sqrt{2}}{2}$; $\cos \dfrac{\pi}{4} = \dfrac{\sqrt{2}}{2}$

15. $\sin \dfrac{\pi}{3} = \dfrac{\sqrt{3}}{2}$; $\cos \dfrac{\pi}{3} = \dfrac{1}{2}$

16. $\sin \dfrac{2\pi}{3} = \dfrac{\sqrt{3}}{2}$; $\cos \dfrac{2\pi}{3} = -\dfrac{1}{2}$

17. $\sin \dfrac{3\pi}{4} = \dfrac{\sqrt{2}}{2}$; $\cos \dfrac{3\pi}{4} = -\dfrac{\sqrt{2}}{2}$

18. $\sin \dfrac{5\pi}{6} = \dfrac{1}{2}$; $\cos \dfrac{5\pi}{6} = -\dfrac{\sqrt{3}}{2}$

19. $\sin \left(-\dfrac{2\pi}{3}\right) = -\dfrac{\sqrt{3}}{2}$;

$\cos \left(-\dfrac{2\pi}{3}\right) = -\dfrac{1}{2}$

20. $\sin \left(-\dfrac{\pi}{6}\right) = -\dfrac{1}{2}$;

$\cos \left(-\dfrac{\pi}{6}\right) = \dfrac{\sqrt{3}}{2}$

21. $\tan \dfrac{\pi}{6} = \dfrac{1}{\sqrt{3}}$; $\cot \dfrac{\pi}{6} = \sqrt{3}$;

$\sec \dfrac{\pi}{6} = \dfrac{2}{\sqrt{3}}$; $\csc \dfrac{\pi}{6} = 2$

22. $\tan \dfrac{\pi}{4} = 1$; $\cot \dfrac{\pi}{4} = 1$;

$\sec \dfrac{\pi}{4} = \dfrac{2}{\sqrt{2}}$ $\csc \dfrac{\pi}{4} = \dfrac{2}{\sqrt{2}}$

23. $\tan \dfrac{\pi}{3} = \sqrt{3}$; $\cot \dfrac{\pi}{3} = \dfrac{1}{\sqrt{3}}$;

$\sec \dfrac{\pi}{3} = 2$; $\csc \dfrac{\pi}{3} = \dfrac{2}{\sqrt{3}}$

24. $\tan \dfrac{2\pi}{3} = -\sqrt{3}$;

$\cot \dfrac{2\pi}{3} = -\dfrac{1}{\sqrt{3}}$; $\sec \dfrac{2\pi}{3} = -2$;

$\csc \dfrac{2\pi}{3} = \dfrac{2}{\sqrt{3}}$

25. $\tan \dfrac{3\pi}{4} = -1$; $\cot \dfrac{3\pi}{4} = -1$;

$\sec \dfrac{3\pi}{4} = -\dfrac{2}{\sqrt{2}}$; $\csc \dfrac{3\pi}{4} = \dfrac{2}{\sqrt{2}}$

26. $\tan \dfrac{5\pi}{6} = -\dfrac{1}{\sqrt{3}}$;

$\cot \dfrac{5\pi}{6} = -\sqrt{3}$;

$\sec \dfrac{5\pi}{6} = -\dfrac{2}{\sqrt{3}}$; $\csc \dfrac{5\pi}{6} = 2$

27. $\tan \left(-\dfrac{2\pi}{3}\right) = \sqrt{3}$;

$\cot \left(-\dfrac{2\pi}{3}\right) = \dfrac{1}{\sqrt{3}}$;

$\sec \left(-\dfrac{2\pi}{3}\right) = -2$;

$\csc \left(-\dfrac{2\pi}{3}\right) = -\dfrac{2}{\sqrt{3}}$

28. $\tan \left(-\dfrac{\pi}{6}\right) = -\dfrac{1}{\sqrt{3}}$;

$\cot \left(-\dfrac{\pi}{6}\right) = -\sqrt{3}$;

$\sec \left(-\dfrac{\pi}{6}\right) = \dfrac{2}{\sqrt{3}}$;

$\csc \left(-\dfrac{\pi}{6}\right) = -2$

29. $\sin 21° = 0.3583679495$;
$\cos 21° = 0.9335804265$;
$\tan 21° = 0.383864035$

30. $\sin 49° = 0.75$; $\cos 49° = 0.66$;
$\tan 49° = 1.15$

31. $\sin 1.23 = 0.94$;
$\cos 1.23 = 0.33$; $\tan 1.23 = 2.82$

32. $\sin 0.78 = 0.70$;
$\cos 0.78 = 0.71$; $\tan 0.78 = 0.99$

33. $\sin 82° = 0.99$; $\cos 82° = 0.14$;
$\tan 82° = 7.12$

34. $\sin 19° = 0.33$; $\cos 19° = 0.95$;
$\tan 19° = 0.34$

35. $\sin 0.27 = 0.27$;
$\cos 0.27 = 0.96$; $\tan 0.27 = 0.28$

36. $\sin 0.95 = 0.81$;
$\cos 0.95 = 0.58$; $\tan 0.95 = 1.40$

37. $\cot 38° = 1.28$; $\sec 38° = 1.27$;
$\csc 38° = 1.62$

38. $\cot 72° = 0.32$; $\sec 72° = 3.24$;
$\csc 72° = 1.05$

39. $\cot 0.83 = 0.91$;
$\sec 0.83 = 1.48$; $\csc 0.83 = 1.36$

40. $\cot 0.12 = 8.29$;
$\sec 0.12 = 1.01$; $\csc 0.12 = 8.35$

41. $\cot 46° = 0.97$; $\sec 46° = 1.44$;
$\csc 46° = 1.39$

42. $\cot 62° = 0.53$; $\sec 62° = 2.13$;
$\csc 62° = 1.13$

43. $\cot 1.35 = 0.22$;
$\sec 1.35 = 4.57$; $\csc 1.35 = 1.02$

44. $\cot 1.03 = 0.60$;
$\sec 1.03 = 1.94$; $\csc 1.03 = 1.17$

45. $\sin\left(\dfrac{13\pi}{4}\right) = -\dfrac{\sqrt{2}}{2}$;
$\cos\left(\dfrac{13\pi}{4}\right) = -\dfrac{\sqrt{2}}{2}$

46. $\sin\left(\dfrac{13\pi}{6}\right) = \dfrac{1}{2}$;
$\cos\left(\dfrac{13\pi}{6}\right) = \dfrac{\sqrt{3}}{2}$

47. $\sin\left(\dfrac{8\pi}{3}\right) = \dfrac{\sqrt{3}}{2}$;
$\cos\left(\dfrac{8\pi}{3}\right) = -\dfrac{1}{2}$

48. $\sin\left(\dfrac{23\pi}{6}\right) = -\dfrac{1}{2}$;
$\cos\left(\dfrac{23\pi}{6}\right) = \dfrac{\sqrt{3}}{2}$

49. $\sin^2\theta + \cos^2\theta = 1 \Rightarrow$
$\dfrac{\sin^2\theta}{\cos^2\theta} + \dfrac{\cos^2\theta}{\cos^2\theta} = \dfrac{1}{\cos^2\theta} \Rightarrow$
$\left(\dfrac{\sin\theta}{\cos\theta}\right)^2 + \left(\dfrac{\cos\theta}{\cos\theta}\right)^2 =$
$\left(\dfrac{1}{\cos\theta}\right)^2$. Thus
$\tan^2\theta + 1 = \sec^2\theta$.

50. $\sin^2\theta + \cos^2\theta = 1 \Rightarrow$
$\dfrac{\sin^2\theta}{\sin^2\theta} + \dfrac{\cos^2\theta}{\sin^2\theta} = \dfrac{1}{\sin^2\theta} \Rightarrow$
$\left(\dfrac{\sin\theta}{\sin\theta}\right)^2 + \left(\dfrac{\cos\theta}{\sin\theta}\right)^2 =$
$\left(\dfrac{1}{\sin\theta}\right)^2$. Thus
$1 + \cot^2\theta = \csc^2\theta$.

$\sin x \quad \cos x \quad \tan x$

51. $+ \quad + \quad +$
52. $+ \quad - \quad -$
53. $- \quad - \quad +$
54. $- \quad + \quad -$

55. $\sin\dfrac{7\pi}{3} = \dfrac{\sqrt{3}}{2}$; $\cos\dfrac{7\pi}{3} = \dfrac{1}{2}$;
$\tan\dfrac{7\pi}{3} = \sqrt{3}$

56. $\sin\dfrac{31\pi}{6} = -\dfrac{1}{2}$;
$\cos\dfrac{31\pi}{6} = -\dfrac{\sqrt{3}}{2}$;
$\tan\dfrac{31\pi}{6} = \dfrac{1}{\sqrt{3}}$

57. $\sin\dfrac{5\pi}{12} = 0.97$; $\cos\dfrac{5\pi}{12} = 0.26$;
$\tan\dfrac{5\pi}{12} = 3.73$

58. $\sin\dfrac{8\pi}{12} = \dfrac{\sqrt{3}}{2}$; $\cos\dfrac{8\pi}{12} = -\dfrac{1}{2}$;
$\tan\dfrac{8\pi}{12} = -\sqrt{3}$

59. $\sin\dfrac{11\pi}{24} = 0.99$;
$\cos\dfrac{11\pi}{24} = 0.13$;
$\tan\dfrac{11\pi}{24} = 7.60$

60. $\sin\dfrac{7\pi}{36} = 0.57$; $\cos\dfrac{7\pi}{36} = 0.82$;
$\tan\dfrac{7\pi}{36} = 0.70$

61. $\cos^2 t - \sin^2 t =$
$\cos^2 t - (1 - \cos^2 t) = 2\cos^2 t - 1$

62. $\sec^2 t - \tan^2 t =$
$(1 + \tan^2 t) - \tan^2 t = 1$

63. $1 + \dfrac{\sin^2 t}{\cos^2 t} = 1 + \left(\dfrac{\sin t}{\cos t}\right)^2 =$
$1 + \tan^2 t = \sec^2 t$

SECTION 6.3

1. amplitude $= |4| = 4$; maximum
value: $y = 4$

$[-5,10]$ by $[-5,5]$

2. amplitude $= |1| = 1$; maximum
value: $y = 1$

$[-1,6]$ by $[-2,2]$

3. amplitude $= |15| = 15$;
maximum value: $y = 15$

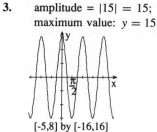

[-5,8] by [-16,16]

$$\frac{2\pi}{|b|} = \frac{2\pi}{|3|} = \frac{2\pi}{3}$$

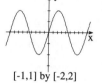

[-4,7] by [-1.5,2]

8. Domain $= (-\infty, \infty)$; range $=$ $[-1, 1]$; period: $\dfrac{2\pi}{|7|} = \dfrac{2\pi}{7}$

[-1,1] by [-2,2]

9. Domain $= (-\infty, \infty)$; range $=$ $[-4, 4]$; period: $\dfrac{2\pi}{5}$

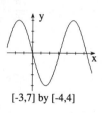

[-4,7] by [-5,5]

4. amplitude $= |-3| = 3$:
maximum value: $y = 3$

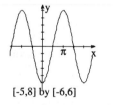

[-3,7] by [-4,4]

5. amplitude $= |-5| = 5$;
maximum value: $y = 5$

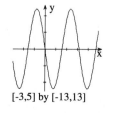

[-5,8] by [-6,6]

10. Domain $= (-\infty, \infty)$; range $=$ $[-2, 2]$; period: $\dfrac{2\pi}{9}$

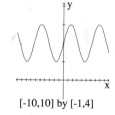

[-1,2] by [-5,5]

6. amplitude $= |-12| = 12$;
maximum value: $y = 12$

[-3,5] by [-13,13]

11. Domain $= (-\infty, \infty)$; range $=$ $[-3, 3]$; period: $\dfrac{2\pi}{2} = \pi$

[-4,7] by [-4,4]

7. Domain $= (-\infty, \infty)$;
range $= [-1, 1]$; period:

12. Domain $= (-\infty, \infty)$; range $=$ $[-6, 6]$; period: $\dfrac{2\pi}{9}$

[-0.5,1] by [-7,7]

13. Domain $= (-\infty, \infty)$; range $=$ $[-1, 1]$; period: $\dfrac{2\pi}{3}$

[-5,5] by [-2,2]

14. Domain $= (-\infty, \infty)$; range $=$ $[1, 3]$; period: $\dfrac{2\pi}{1} = 2\pi$

[-10,10] by [-1,4]

15. Domain $= (-\infty, \infty)$; range $=$ $[1, 3]$; period: $\dfrac{2\pi}{1} = 2\pi$

[-10,10] by [-4,4]

16. Domain $= (-\infty, \infty)$; range $=$ $[-2, 0]$; period: $\dfrac{2\pi}{1} = 2\pi$

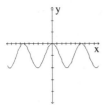

[-10,10] by [-6,3]

17. Domain $= (-\infty, \infty)$; range $=$ $[-4, 2]$; period: $\dfrac{2\pi}{1} = 2\pi$

[-4,9] by [-5,4]

18. Domain $= (-\infty, \infty)$; range $=$ $[0, 4]$; period: $\dfrac{2\pi}{1} = 2\pi$

[-6,6] by [-1,5]

19. Domain $= (-\infty, \infty)$; range $=$ $[-4, 2]$; period: $\dfrac{2\pi}{1} = 2\pi$

[-10,10] by [-6,6]

20. Domain $= (-\infty, \infty)$; range $=$ $[0, 4]$; period: $\dfrac{2\pi}{1} = 2\pi$

[-10,10] by [-1,5]

21. $[0, \pi]$ by $[-2, 2]$

22. $[0, 4\pi]$ by $[-2, 2]$

23. $[0, 4\pi]$ by $[-3, 3]$

24. $[0, 6\pi]$ by $[-2, 2]$

25. $[0, 10\pi]$ by $[-4, 4]$

26. $[0, 8\pi]$ by $[-4, 4]$

27. 0.52, 2.62

28. 3.79, 5.64

29. 0.93, 2.21

30. 0.30, 2.84

31. 0.93, 5.36

32. 1.98, 4.30

33. 0.60, 2.54

34. 1.45

35. 0.79, 3.93

36. 1.05, 5.24

37. 1.70, 3.14, 4.60

38. 2.36, 5.49

39. $0 \le x < 0.64$, $2.50 < x < 2\pi$

40. $0 \le x < 4.98$

41. $0 \le x < 2.28$, $4.00 < x < 2\pi$

42. $0.05 < x < 3.09$

43. $\tan(-x) = \dfrac{\sin(-x)}{\cos(-x)} = \dfrac{-\sin x}{\cos x} = -\tan x$

44. A vertical stretch of factor 2 yields $y = 2\sin x$, then a

vertical shift up 3 units produces $y = 3 + 2\sin x$. However, beginning with a vertical shift up 3 units yields $y = 3 + \sin x$, then a vertical stretch produces $y = 2(3 + \sin x) = 6 + 2\sin x$.

45. $y = 2\sin x + 3$, $y = 2(\sin x + 3)$

$y = 2(\sin x + 3)$

$y = 2\sin x + 3$
[-7,7] by [-2,9]

46. After the 1st, 3rd, and 5th rings, there is a gap of $\dfrac{2\pi}{3} = 2.09$ sec. After the 2nd, 4th, and 6th rings, there is a gap of $\dfrac{4\pi}{3} = 4.19$ sec.

47. 0.80 min

48. Answers may vary depending on student's interpretation. One possible explanation: If the distance varies as a function of time, the graph of $y = a\sin(bt)$ would suggest that distance becomes negative as time increases. It may be possible that distance is negative in a "Black Hole," but in our world, it is always positive.

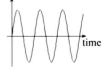

distance / time

SECTION 6.4

1. The graphs appear the same because each shows a complete graph of one period of the tangent function.

(a)

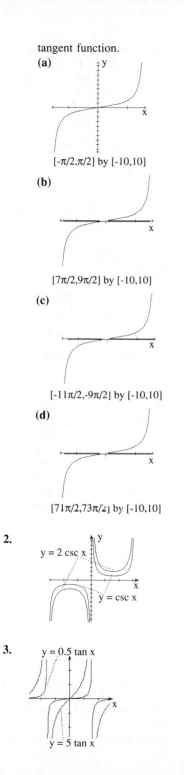

[-π/2,π/2] by [-10,10]

(b)

[7π/2,9π/2] by [-10,10]

(c)

[-11π/2,-9π/2] by [-10,10]

(d)

[71π/2,73π/2] by [-10,10]

2.

$y = 2 \csc x$

$y = \csc x$

3.

$y = 0.5 \tan x$

$y = 5 \tan x$

4. Four possibilities are: $[0, \pi]$ by $[-10, 10]$; $[3\pi, 4\pi]$ by $[-10, 10]$; $[-3\pi, -2\pi]$ by $[-10, 10]$; $[-\pi, 0]$ by $[-10, 10]$

5. Four possibilities are:
$\left[-\dfrac{\pi}{2}, \dfrac{3\pi}{2} \right]$ by $[-10, 10]$;

$\left[\dfrac{7\pi}{2}, \dfrac{11\pi}{2} \right]$ by $[-10, 10]$;

$\left[\dfrac{51\pi}{2}, \dfrac{55\pi}{2} \right]$ by $[-10, 10]$;

$\left[-\dfrac{9\pi}{2}, -\dfrac{5\pi}{2} \right]$ by $[-10, 10]$

6. **(a)** $y = \tan x$ and $y = \sec x$

7. period: $\dfrac{\pi}{|2|} = \dfrac{\pi}{2}$

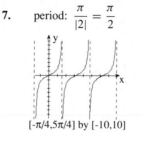

[-π/4,5π/4] by [-10,10]

8. period: $\dfrac{\pi}{|3|} = \dfrac{\pi}{3}$

[0,π] by [-10,10]

9. period: $\dfrac{2\pi}{|2|} = \pi$

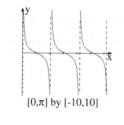

[-π/4,11π/4] by [-10,10]

10. period: $\dfrac{2\pi}{|1/2|} = 4\pi$

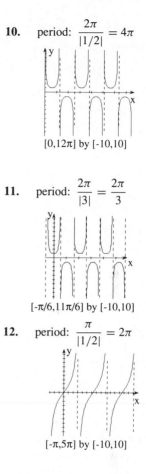

[0,12π] by [-10,10]

11. period: $\dfrac{2\pi}{|3|} = \dfrac{2\pi}{3}$

[-π/6,11π/6] by [-10,10]

12. period: $\dfrac{\pi}{|1/2|} = 2\pi$

[-π,5π] by [-10,10]

13. Domain = all reals except $x = \dfrac{\pi}{2} + k\pi$, k any integer; range = $(-\infty, \infty)$; period: $\dfrac{\pi}{1} = \pi$; VA at $x = \dfrac{\pi}{2} + k\pi$, k any integer

[-5,5] by [-6,6]

14. Domain = all reals except $x = \dfrac{\pi}{2} + k\pi$, k any integer; range = $(-\infty, \infty)$; period:

$\dfrac{\pi}{1} = \pi$; VA at $x = \dfrac{\pi}{2} + k\pi$, k any integer

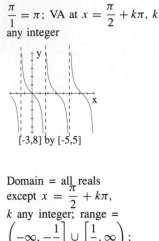

[-3,8] by [-5,5]

integer

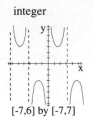

[-7,6] by [-7,7]

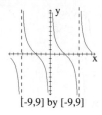

[-9,9] by [-9,9]

15. Domain = all reals except $x = \dfrac{\pi}{2} + k\pi$, k any integer; range = $\left(-\infty, -\dfrac{1}{2}\right] \cup \left[\dfrac{1}{2}, \infty\right)$; period: $\dfrac{2\pi}{1} = 2\pi$; VA at $x = \dfrac{\pi}{2} + k\pi$, k any integer

[-5,8] by [-4,4]

16. Domain = all reals except $x = \dfrac{\pi}{2} + k\pi$, k any integer; range = $(-\infty, -1] \cup [1, \infty)$; period: $\dfrac{2\pi}{1} = 2\pi$; VA at $x = \dfrac{\pi}{2} + k\pi$, k any integer

[-4,7] by [-5,5]

17. Domain = all reals except $x = k\pi$, k any integer; range = $(-\infty, -3] \cup [3, \infty)$; period: $\dfrac{\pi}{1} = \pi$; VA at $x = k\pi$, k any

18. Domain = all reals except $x = \dfrac{\pi}{2} + k\pi$, k any integer; range = $(-\infty, \infty)$; period: $\dfrac{\pi}{1} = \pi$; VA at $x = \dfrac{\pi}{2} + k\pi$, k any integer

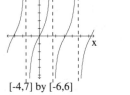

[-4,7] by [-6,6]

19. Domain = all reals except $x = \pi + 2k\pi$, k any integer; range = $(-\infty, \infty)$; period: $\dfrac{\pi}{|1/2|} = 2\pi$; VA at $x = \pi + 2k\pi$, k any integer

[-9,9] by [-9,9]

20. Domain = all reals except $x = 2k\pi$, k any integer; range = $(-\infty, \infty)$; period: $\dfrac{\pi}{|1/2|} = 2\pi$; VA at $x = 2k\pi$, k any integer

21. Domain = all reals except $x = k\pi$, k any integer; range = $(-\infty, -2] \cup [2, \infty)$; period: $\dfrac{2\pi}{1} = 2\pi$; VA at $x = k\pi$, k any integer

[-5,7] by [-6,5]

22. Domain = all reals except $x = \pi + 2k\pi$, k any integer; range = $(-\infty, -2] \cup [2, \infty)$; period: $\dfrac{2\pi}{|1/2|} = 4\pi$; VA at $x = \pi + 2k\pi$, k any integer

[-10,10] by [-10,10]

23. Domain = all reals except $x = \dfrac{\pi}{6} + \dfrac{k\pi}{3}$, k any integer; range = $(-\infty, \infty)$; period: $\dfrac{\pi}{|3|} = \dfrac{\pi}{3}$; VA at $x = \dfrac{\pi}{6} + \dfrac{k\pi}{3}$, k any integer

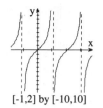

[-1,2] by [-10,10]

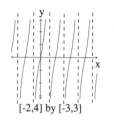

[-2,4] by [-3,3]

34. 0.79

35. ±0.90

36. 0.50

37. $\left(0.21, \dfrac{\pi}{2}\right)$

38. $\left(-\dfrac{\pi}{2}, -0.79\right) \cup (0, 0.78)$

39. $\left[0.67, \dfrac{\pi}{2}\right)$

40. $(0, 0.78)$

41. 0.85, 2.29

24. Domain = all reals except $x = \pi + 2k\pi$, k any integer; range = $(-\infty, -1] \cup [1, \infty)$; period: $\dfrac{2\pi}{|-1/2|} = 4\pi$; VA at $x = \pi + 2k\pi$, k any integer

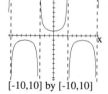

[-10,10] by [-10,10]

27. Domain = all reals except $x = \dfrac{\pi}{4} + k \cdot \dfrac{\pi}{2}$, k any integer; range = $(-\infty, -3] \cup [3, \infty)$; period: $\dfrac{2\pi}{|2|} = \pi$; VA at $x = \dfrac{\pi}{4} + k \cdot \dfrac{\pi}{2}$, k any integer

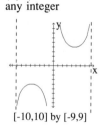

[-4,5] by [-6,6]

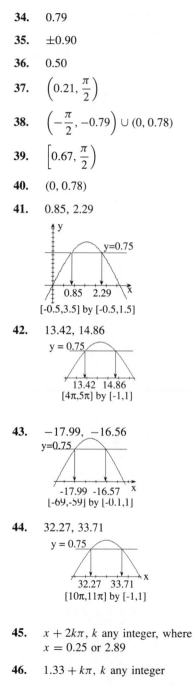

y=0.75

0.85 2.29

[-0.5,3.5] by [-0.5,1.5]

42. 13.42, 14.86

y = 0.75

13.42 14.86

[4π,5π] by [-1,1]

43. −17.99, −16.56

y=0.75

-17.99 -16.57

[-69,-59] by [-0.1,1]

44. 32.27, 33.71

y = 0.75

32.27 33.71

[10π,11π] by [-1,1]

25. Domain = all reals except $x = 1 + 2k$, k any integer; range = $(-\infty, \infty)$; period: $\dfrac{\pi}{|\pi/2|} = 2$; VA at $x = 1 + 2k$, k any integer

[-4,4] by [-4,4]

28. Domain = all reals except $x = 3k\pi$, k any integer; range = $(-\infty, -4] \cup [4, \infty)$; period: $\dfrac{2\pi}{|1/3|} = 6\pi$; VA at $x = 3k\pi$, k any integer

[-10,10] by [-9,9]

29. 1.27, 4.41, −1.87, −5.01

30. −0.18, −3.32, 2.96, 6.10

31. ±0.59, ±5.69

32. −0.33, 5.95, 3.47, −2.81

33. 0.67

26. Domain = all reals except $x = \dfrac{1}{2} + k$, k any integer; range = $(-\infty, \infty)$; period: $\dfrac{\pi}{|\pi|} = 1$; VA at $x = \dfrac{1}{2} + k$, k any integer

45. $x + 2k\pi$, k any integer, where $x = 0.25$ or 2.89

46. $1.33 + k\pi$, k any integer

47. $x + 2k\pi$, k any integer, where $x = 1.14$ or 5.15

48. $x + 2k\pi$, k any integer, where $x = 1.23$ or 5.05

49. $x + 2k\pi$, k any integer, where $-3.29 < x < 0.15$

50. $x + k\pi$, k any integer, where $1.11 < x < \dfrac{\pi}{2}$

51. $x + \dfrac{2k\pi}{3}$, k any integer, where $x = 0.19$ or 0.85

52. $x + k\pi$, k any integer, where $x = 0.28$ or 2.86

53. $x + 4k\pi$, k any integer, where $0.51 \le x \le 5.78$ or $2\pi < x < 4\pi$

54. $x + 4k\pi$, k any integer, where $0.88 < x < 5.40$

SECTION 6.5

1. $\dfrac{8}{\sqrt{113}}$

2. $\dfrac{7}{\sqrt{113}}$

3. $\dfrac{8}{7}$

4. $\dfrac{7}{8}$

5. $\dfrac{\sqrt{113}}{7}$

6. $\dfrac{\sqrt{113}}{8}$

7. $\sin B = \dfrac{7}{\sqrt{113}}$; $\cos B = \dfrac{8}{\sqrt{113}}$; $\tan B = \dfrac{7}{8}$; $\cot B = \dfrac{8}{7}$; $\sec B = \dfrac{\sqrt{113}}{8}$; $\csc B = \dfrac{\sqrt{113}}{7}$

8. $\dfrac{11}{15}$

9. $\dfrac{2\sqrt{26}}{15}$

10. $\dfrac{11}{2\sqrt{26}}$

11. $\dfrac{2\sqrt{26}}{11}$

12. $\dfrac{15}{2\sqrt{26}}$

13. $\dfrac{15}{11}$

14. $\sin B = \dfrac{2\sqrt{26}}{15}$; $\cos B = \dfrac{11}{15}$; $\tan B = \dfrac{2\sqrt{26}}{11}$; $\cot B = \dfrac{11}{2\sqrt{26}}$; $\sec B = \dfrac{15}{11}$; $\csc B = \dfrac{15}{2\sqrt{26}}$

15. $\sin 30° = \dfrac{1}{2}$; $\cos 30° = \dfrac{\sqrt{3}}{2}$; $\tan 30° = \dfrac{1}{\sqrt{3}}$; $\cot 30° = \dfrac{\sqrt{3}}{1} = \sqrt{3}$; $\sec 30° = \dfrac{2}{\sqrt{3}}$; $\csc 30° = \dfrac{2}{1} = 2$

16. $\sin 45° = \dfrac{1}{\sqrt{2}} = \dfrac{\sqrt{2}}{2}$; $\cos 45° = \dfrac{1}{\sqrt{2}} = \dfrac{\sqrt{2}}{2}$; $\tan 45° = 1$; $\cot 45° = 1$; $\sec 45° = \dfrac{\sqrt{2}}{1} = \sqrt{2}$; $\csc 45° = \dfrac{\sqrt{2}}{1} = \sqrt{2}$

17. $\sin \dfrac{\pi}{3} = \dfrac{\sqrt{3}}{2}$; $\cos \dfrac{\pi}{3} = \dfrac{1}{2}$; $\tan \dfrac{\pi}{3} = \dfrac{\sqrt{3}}{1}$; $\cot \dfrac{\pi}{3} = \dfrac{1}{\sqrt{3}}$; $\sec \dfrac{\pi}{3} = \dfrac{2}{1} = 2$; $\csc \dfrac{\pi}{3} = \dfrac{2}{\sqrt{3}}$

18. $\theta = \dfrac{\pi}{6} = 30°$; see Exercise 15

19. $\theta = 60° = \dfrac{\pi}{3}$; see Exercise 17

20. $\theta = \dfrac{\pi}{4} = 45°$; see Exercise 16

21. $\cos \theta = \dfrac{4}{5}$; $\tan \theta = \dfrac{3}{4}$; $\cot \theta = \dfrac{4}{3}$; $\sec \theta = \dfrac{5}{4}$; $\csc \theta = \dfrac{5}{3}$

22. $\sin \theta = \dfrac{1}{\sqrt{10}}$; $\cos \theta = \dfrac{3}{\sqrt{10}}$; $\cot \theta = \dfrac{3}{1} = 3$; $\sec \theta = \dfrac{\sqrt{10}}{3}$; $\csc \theta = \dfrac{\sqrt{10}}{1} = \sqrt{10}$

23. $\sin \theta = \dfrac{1}{2}$; $\tan \theta = \dfrac{1}{\sqrt{3}}$; $\cot \theta = \dfrac{\sqrt{3}}{1} = \sqrt{3}$; $\sec \theta = \dfrac{2}{\sqrt{3}}$; $\csc \theta = \dfrac{2}{1} = 2$

24. $\cos \theta = \dfrac{5}{13}$; $\tan \theta = \dfrac{12}{5}$; $\cot \theta = \dfrac{5}{12}$; $\sec \theta = \dfrac{13}{5}$; $\csc \theta = \dfrac{13}{12}$

25. 0.52 or 30°

26. 0.97 or 55.54°

27. 0.64 or 36.87°

28. 1.45 or 82.82°

29. 0.79 or 45°

30. 1.25 or 71.57°

31. 0.40 or 22.93°

32. 1.23 or 70.35°

33. 0.52 or 30°

34. 1.31 or 74.78°

35. 1.03 or 59.04°

36. 0.49 or 28.07°

37. 0.73 or 41.81°

38. 1.11 or 63.43°

39. $\angle B = 70°$; $c = 35.96$; $b = 33.79$

40. $c = 5$; $\angle B = 53.13°$; $\angle A = 36.87°$

41. $\angle B = 49°$; $b = 7.55$; $a = 6.56$

42. $\angle B = 35°$; $c = 27.16$; $a = 22.25$

43. $a = 4.90$; $\angle A = 44.42°$; $\angle B = 45.58°$

44. $b = 46.56$; $\angle B = 66.54°$; $\angle A = 23.46°$

45. $c = 9.46$; $\angle A = 12.20°$; $\angle B = 77.8°$

46. $\angle A = 31°$; $c = 9.71$; $b = 8.32$

47. $\angle A = 77.45°$; $b = 2.80$; $a = 12.58$

48. $\angle B = 79.8°$; $a = 2.57$; $b = 14.27$

49. $\sin \theta = 0$; $\cos \theta = -1$; $\tan \theta = 0$; $\csc \theta$ is undefined; $\sec \theta = -1$; $\cot \theta$ is undefined; $\theta = 180°$, $\theta' = 0°$

50. $\sin \theta = 1$; $\cos \theta = 0$; $\tan \theta$ is undefined; $\csc \theta = 1$; $\sec \theta$ is undefined; $\cot \theta = 0$; $\theta = 90° = \theta'$

51. $\sin \theta = \dfrac{3}{5}$; $\cos \theta = \dfrac{4}{5}$; $\tan \theta = \dfrac{3}{4}$; $\csc \theta = \dfrac{5}{3}$; $\sec \theta = \dfrac{5}{4}$; $\cot \theta = \dfrac{4}{3}$; $\theta = \theta' = 36.87°$

52. $\sin \theta = \dfrac{2}{\sqrt{5}}$; $\cos \theta = \dfrac{-1}{\sqrt{5}}$; $\tan \theta = \dfrac{2}{-1} = -2$; $\csc \theta = \dfrac{\sqrt{5}}{2}$; $\sec \theta = \dfrac{\sqrt{5}}{-1} = -\sqrt{5}$;

$\cot \theta = \dfrac{-1}{2} = -\dfrac{1}{2}$; $\theta' = 63.43°$; $\theta = 116.57°$

53. $\sin \theta = -\dfrac{3}{\sqrt{13}}$; $\cos \theta = -\dfrac{2}{\sqrt{13}}$; $\tan \theta = \dfrac{3}{2}$; $\csc \theta = -\dfrac{\sqrt{13}}{3}$; $\sec \theta = -\dfrac{\sqrt{13}}{2}$; $\cot \theta = \dfrac{2}{3}$; $\theta' = 56.31°$; $\theta = 236.31°$

54. $\sin \theta = -\dfrac{2}{\sqrt{29}}$; $\cos \theta = \dfrac{5}{\sqrt{29}}$; $\tan \theta = \dfrac{-2}{5} = -\dfrac{2}{5}$; $\csc \theta = -\dfrac{\sqrt{29}}{2}$; $\sec \theta = \dfrac{\sqrt{29}}{5}$; $\cot \theta = -\dfrac{5}{2}$; $\theta' = 21.80°$; $\theta = 338.20°$

55. $\sin \theta = -\dfrac{1}{\sqrt{2}}$; $\cos \theta = \dfrac{1}{\sqrt{2}}$; $\tan \theta = -1$; $\csc \theta = -\sqrt{2}$; $\sec \theta = \sqrt{2}$; $\cot \theta = -1$; $\theta' = 45°$; $\theta = 315°$

56. $\sin \theta = -\dfrac{1}{\sqrt{65}}$; $\cos \theta = -\dfrac{8}{\sqrt{65}}$; $\tan \theta = \dfrac{1}{8}$; $\csc \theta = -\sqrt{65}$; $\sec \theta = -\dfrac{\sqrt{65}}{8}$; $\cot \theta = 8$; $\theta' = 7.13°$; $\theta = 187.13°$

57. $\sin \theta = \sin \theta' = 0.41$; $\cos \theta = -\cos \theta' = -0.91$; $\tan \theta = -\tan \theta' = -0.45$; $\csc \theta = \csc \theta' = 2.46$; $\sec \theta = -\sec \theta' = -1.09$; $\cot \theta = -\cot \theta' = -2.25$; $\theta = 156°$; $\theta' = 24°$

[-10,10] by [-10,10]

58. $\sin \theta = \sin \theta' = 0.82$; $\cos \theta = \cos \theta' = 0.57$; $\tan \theta = \tan \theta' = 1.43$; $\csc \theta = \csc \theta' = 1.22$; $\sec \theta = \sec \theta' = 1.74$; $\cot \theta = \cot \theta' = 0.70$; $\theta = -305°$; $\theta' = 55°$

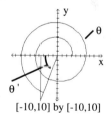

[-5,5] by [-5,5]

59. $\sin \theta = -\sin \theta' = -0.96$; $\cos \theta = -\cos \theta' = -0.28$; $\tan \theta = \tan \theta' = 3.49$; $\csc \theta = -\csc \theta' = -1.04$; $\sec \theta = -\sec \theta' = -3.63$; $\cot \theta = \cot \theta' = 0.29$; $\theta = 614°$; $\theta' = 74°$

[-10,10] by [-10,10]

60. $\sin \theta = -\sin \theta' = -0.54$; $\cos \theta = -\cos \theta' = -0.84$; $\tan \theta = \tan \theta' = 0.65$; $\csc \theta = -\csc \theta' = -1.84$; $\sec \theta = -\sec \theta' = -1.19$; $\cot \theta = \cot \theta' = 1.54$; $\theta = 213°$; $\theta' = 33°$

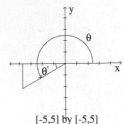

[-5,5] by [-5,5]

61. $\tan \dfrac{\pi}{2}$ and $\sec \dfrac{\pi}{2}$ are undefined

62. $\cot \pi$ and $\csc \pi$ are undefined

63. $\tan \dfrac{3\pi}{2}$ and $\sec \dfrac{3\pi}{2}$ are undefined

64. $\cot 5\pi$ and $\csc 5\pi$ are undefined

65. $\tan \dfrac{5\pi}{2}$ and $\sec \dfrac{5\pi}{2}$ are undefined

66. $\tan \dfrac{7\pi}{2}$ and $\sec \dfrac{7\pi}{2}$ are undefined

SECTION 6.6

1. $\sin \theta = \dfrac{-1}{\sqrt{2}}$; $\csc \theta = -\sqrt{2}$;
$\cos \theta = \dfrac{-1}{\sqrt{2}}$; $\sec \theta = -\sqrt{2}$;
$\tan \theta = 1$; $\cot \theta = 1$

2. $\sin \theta = \dfrac{2}{\sqrt{5}}$; $\csc \theta = \dfrac{\sqrt{5}}{2}$;
$\cos \theta = \dfrac{1}{\sqrt{5}}$; $\sec \theta = \sqrt{5}$;
$\tan \theta = 2$; $\cot \theta = \dfrac{1}{2}$

3. $\sin \theta = \dfrac{1}{\sqrt{2}}$; $\csc \theta = \sqrt{2}$;
$\cos \theta = \dfrac{1}{\sqrt{2}}$; $\sec \theta = \sqrt{2}$;
$\tan \theta = 1$; $\cot \theta = 1$

4. $\sin \theta = \dfrac{-2}{\sqrt{5}}$; $\csc \theta = \dfrac{\sqrt{5}}{-2}$;
$\cos \theta = \dfrac{-1}{\sqrt{5}}$; $\sec \theta = -\sqrt{5}$;
$\tan \theta = 2$; $\cot \theta = \dfrac{1}{2}$

5. $\sin \theta = \dfrac{3}{5}$; $\csc \theta = \dfrac{5}{3}$;
$\cos \theta = \dfrac{4}{5}$; $\sec \theta = \dfrac{5}{4}$;
$\tan \theta = \dfrac{3}{4}$; $\cot \theta = \dfrac{4}{3}$

6. $\sin \theta = \dfrac{1}{\sqrt{2}}$; $\csc \theta = \sqrt{2}$;
$\cos \theta = \dfrac{-1}{\sqrt{2}}$; $\sec \theta = -\sqrt{2}$;
$\tan \theta = -1$; $\cot \theta = -1$

7. $\sin \theta = \dfrac{-2}{\sqrt{13}}$; $\csc \theta = \dfrac{\sqrt{13}}{-2}$;
$\cos \theta = \dfrac{-3}{\sqrt{13}}$; $\sec \theta = \dfrac{\sqrt{13}}{-3}$;
$\tan \theta = \dfrac{2}{3}$; $\cot \theta = \dfrac{3}{2}$

8. $\sin \theta = \dfrac{-3}{\sqrt{34}}$; $\csc \theta = \dfrac{\sqrt{34}}{-3}$;
$\cos \theta = \dfrac{5}{\sqrt{34}}$; $\sec \theta = \dfrac{\sqrt{34}}{5}$;
$\tan \theta = -\dfrac{3}{5}$; $\cot \theta = -\dfrac{5}{3}$

9. $63.43°$

10. $33.69°$ in QI

11. The acute angle is $71.57°$

12. $30.96°$ in QI

13. QIII

14. QIV

15. QIII

16. QII

17. 100 ft

18. 219.80 ft

19. 680.55 ft

20. 1908.11 ft

21. 106.12 ft

22. $\alpha = 80.41°$ (angle between guy wire and horizontal); $\beta = 9.59°$ (angle between guy wire and antenna); $h = 29.58$ m

23. 101.57 ft

24. 30.09 ft high, 34.08 ft long

25. 4.28 mi

26. 3290.53 ft high

27. $12.80°$

28. 52.35 ft

29. 219.22 ft

30. $d(A, C) \approx 84.85$ mi; bearing from A to $C = 140°$

31. $d(A, C) \approx 178.89$ mi; bearing from A to $C = 128.43°$

32. $19.03°$

33. 3136.38 ft

34. 637.85 ft long

35. 3.67 mi

36. 9.78 units high

37. 8271.02 ft away

38. 2931.09 ft from shore

39. 6.16 mph

40. To show Definitions 6.2 and 6.6 are consistent, let t be the length on the unit circle subtended by angle θ and let $r = 1$. To show Definitions 6.5 and 6.6 are consistent, observe that, for an acute angle θ in standard position, a point $P(x, y)$ on the terminal side of θ determines an acute right triangle whose adjacent side, opposite side, and hypotenuse have lengths x, y, and r, respectively. The consistency of Definitions 6.2 and 6.5 follows from their mutual consistency with Definition 6.6.

41. The discrepancy in the hundredths place of the two solutions for h in Example 9 occurs because, in the second calculation, the measure of angle α was rounded from $82.33774434°$ to $82.34°$ BEFORE the value of h was computed. If the full calculator approximation for α is finished before rounding by calculating

$$h = 10\tan\left(\cos^{-1}\left(\frac{10}{75}\right)\right),$$

then $h = 74.33'$, the same value for h obtained in Example 9 when using the Pythagorean theorem.

Chapter 6 Review

1. QII; $\dfrac{3\pi}{4} = 2.36$

2. QIV; $-\dfrac{\pi}{4}$

3. on the positive y-axis; $450°$

4. QII; $135°$

5. QI; $\dfrac{13\pi}{30} = 1.36$

6. QII; $\dfrac{28\pi}{45} = 1.96$

7. QI; $15°$

8. QII; $126°$

9. $270°$; $\dfrac{3\pi}{2}$

10. $900°$; 5π

11. 1.11 or $63.43°$

12. $135°$ or 2.36 or $\dfrac{3\pi}{4}$

13. $120°$ or 2.09

14. $225°$ or 3.93 or $\dfrac{5\pi}{4}$

15. $296.57°$ or 5.18

16. $30°$ or $\dfrac{\pi}{6}$

17. $\dfrac{1}{2}$

18. 0

19. -1

20. $\dfrac{1}{2}$

21. $\dfrac{2}{\sqrt{3}}$

22. $-\sqrt{2}$

23. $\dfrac{2}{\sqrt{3}}$

24. 2

25. $-\dfrac{1}{2}$

26. $\dfrac{2}{\sqrt{3}}$

27. -1

28. 1

29. $\sin\dfrac{\pi}{9} = 0.34$

30. $\cos\dfrac{\pi}{5} = 0.81$

31. $\tan\dfrac{\pi}{6} = 0.58$

32. $\sin\dfrac{\pi}{8} = 0.38$

33. $\cot\dfrac{\pi}{5} = 1.38$

34. $\sec\dfrac{\pi}{12} = 1.04$

35. $\sin 121° = 0.86$

36. $\cos 72° = 0.31$

37. $\tan 83° = 8.14$

38. $\sin 19° = 0.33$

39. amplitude $= 1$; period $= \dfrac{2\pi}{3}$

40. amplitude $= 5$; period $= \pi$

41. amplitude is undefined for the tangent; period $= \dfrac{\pi}{2}$

42. amplitude $= 6$; period $= \dfrac{\pi}{2}$

43. amplitude is undefined for the secant; period $= \dfrac{2\pi}{3}$

44. amplitude $= 2$; period $= \dfrac{2}{3}$

45. $\sin 0° = 0$; $\cos 0° = 1$; $\tan 0° = 0$; $\csc 0°$ is undefined; $\sec 0° = 1$; $\cot 0°$ is undefined

46. $\sin\left(-\dfrac{\pi}{6}\right) = -\dfrac{1}{2}$;
$\cos\left(-\dfrac{\pi}{6}\right) = \dfrac{\sqrt{3}}{2}$;
$\tan\left(-\dfrac{\pi}{6}\right) = -\dfrac{1}{\sqrt{3}}$;
$\cot\left(-\dfrac{\pi}{6}\right) = -\sqrt{3}$;
$\sec\left(-\dfrac{\pi}{6}\right) = \dfrac{2}{\sqrt{3}}$;
$\csc\left(-\dfrac{\pi}{6}\right) = -2$;

47. $\sin\left(\dfrac{3\pi}{4}\right) = \dfrac{1}{\sqrt{2}}$;
$\cos\left(\dfrac{3\pi}{4}\right) = -\dfrac{1}{\sqrt{2}}$;
$\tan\left(\dfrac{3\pi}{4}\right) = -1$;

$\cot\left(\dfrac{3\pi}{4}\right) = -1;$

$\sec\left(\dfrac{3\pi}{4}\right) = -\sqrt{2};$

$\csc\left(\dfrac{3\pi}{4}\right) = \sqrt{2}$

48. $\sin 60° = \dfrac{\sqrt{3}}{2};\ \cos 60° = \dfrac{1}{2};$

$\tan 60° = \sqrt{3};\ \cot 60° = \dfrac{1}{\sqrt{3}};$

$\sec 60° = 2;\ \csc 60° = \dfrac{2}{\sqrt{3}};$

49. $\sin(-135°) = -\dfrac{1}{\sqrt{2}};$

$\cos(-135°) = -\dfrac{1}{\sqrt{2}};$

$\tan(-135°) = 1;$

$\cot(-135°) = 1;$

$\sec(-135°) = -\sqrt{2};$

$\csc(-135°) = -\sqrt{2};$

50. $\sin 300° = -\dfrac{\sqrt{3}}{2};$

$\cos 300° = \dfrac{1}{2};\ \tan 300° = -\sqrt{3};$

$\cot 300° = -\dfrac{1}{\sqrt{3}};\ \sec 300° = 2;$

$\csc 300° = -\dfrac{2}{\sqrt{3}};$

51.

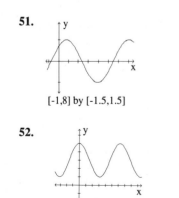

[-1,8] by [-1.5,1.5]

52.

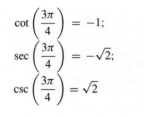

[-4,10] by [-2,7]

53.

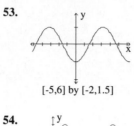

[-5,6] by [-2,1.5]

54.

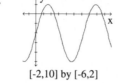

[-2,10] by [-6,2]

55.

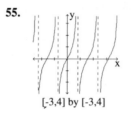

[-3,4] by [-3,4]

56.

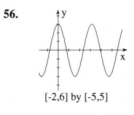

[-2,6] by [-5,5]

57.

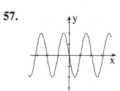

[-4,4] by [-4,4]

58.

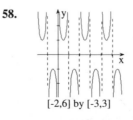

[-2,6] by [-3,3]

59. $\sin\angle A = \dfrac{12}{13};\ \csc\angle A = \dfrac{13}{12};$

$\cos\angle A = \dfrac{5}{13};\ \sec\angle A = \dfrac{13}{5};$

$\tan\angle A = \dfrac{12}{5};\ \cot\angle A = \dfrac{5}{12}$

60. $\sin\theta = \dfrac{2\sqrt{6}}{7};\ \csc\theta = \dfrac{7}{2\sqrt{6}};$

$\cos\theta = \dfrac{5}{7};\ \sec\theta = \dfrac{7}{5};$

$\tan\theta = \dfrac{2\sqrt{6}}{5};\ \cot\theta = \dfrac{5}{2\sqrt{6}}$

61. $\sin\theta = \dfrac{5}{13};\ \csc\theta = \dfrac{13}{5};$

$\cos\theta = \dfrac{12}{13};\ \sec\theta = \dfrac{13}{12};$

$\tan\theta = \dfrac{5}{12};\ \cot\theta = \dfrac{12}{5}$

62. $64.62°$

63. $71.57°$

64. 0.93

65. 0.22

66. $\angle B = 55°;\ a = 8.60;\ b = 12.29$

67. $a = 6;\ \angle A = 36.87°;$
$\angle B = 53.13°$

68. $\angle A = 42°;\ b = 7.77;\ c = 10.46$

69. $\angle B = 62°;\ a = 3.76;\ b = 7.06$

70. $a = 4.90;\ \angle A = 44.42°;$
$\angle B = 45.58°$

71. $c = 7.72;\ \angle A = 18.90°;$
$\angle B = 71.10°$

72. QIII

73. QII

74. QIV

75. QII

76. $\cot\theta = \dfrac{x}{y},\ \cos\theta = \dfrac{x}{1},$ and
$\sin\theta = \dfrac{y}{1}$ by Definition 7.3.

Thus $\dfrac{\cos\theta}{\sin\theta} = \dfrac{x/1}{y/1} = \dfrac{x}{1}\cdot\dfrac{1}{y} = \dfrac{x}{y} = \cot\theta$

77. 0.71 in QI; -0.71 in QII

78. 0.70 in QI; -0.70 in QIV

79. $\sin\theta = \dfrac{2}{\sqrt{5}}$; $\cos\theta = -\dfrac{1}{\sqrt{5}}$;

$\tan\theta = -2$; $\csc\theta = \dfrac{\sqrt{5}}{2}$;

$\sec\theta = -\sqrt{5}$; $\cot\theta = -\dfrac{1}{2}$

80. $\sin\theta = \dfrac{7}{\sqrt{193}}$; $\cos\theta = \dfrac{12}{\sqrt{193}}$;

$\tan\theta = \dfrac{7}{12}$; $\csc\theta = \dfrac{\sqrt{193}}{7}$;

$\sec\theta = \dfrac{\sqrt{193}}{12}$; $\cot\theta = \dfrac{12}{7}$

81. $\sin\theta = -\dfrac{3}{\sqrt{34}}$; $\cos\theta = -\dfrac{5}{\sqrt{34}}$;

$\tan\theta = \dfrac{3}{5}$; $\csc\theta = -\dfrac{\sqrt{34}}{3}$;

$\sec\theta = -\dfrac{\sqrt{34}}{5}$; $\cot\theta = \dfrac{5}{3}$

82. $\sin\theta = \dfrac{9}{\sqrt{97}}$; $\cos\theta = \dfrac{4}{\sqrt{97}}$;

$\tan\theta = \dfrac{9}{4}$; $\csc\theta = \dfrac{\sqrt{97}}{9}$;

$\sec\theta = \dfrac{\sqrt{97}}{4}$; $\cot\theta = \dfrac{4}{9}$

83. $f(x) = \dfrac{1}{x}\sin\dfrac{1}{x}$ diverges as $|x|\to\infty$

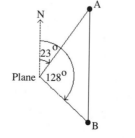

[-0.5,0.5] by [-100,100]

84. $|y|\to\infty$ as $|x|\to 0$

[-0.5,0.5] by [-10,10]

85. $50°$ or 0.87

86. $28.62°$ or 0.50

87. $60°$ or 1.05 or $\dfrac{\pi}{3}$

88. $45°$ or 0.79 or $\dfrac{\pi}{4}$

89. 0.25

90. 1.37

91. undefined

92. -1.43

93. $(-3.99, -1.36)\cup(0.55, \infty)$

94. 470.46 m high

95. 23.78 ft tall

96. 295.06 ft

97. Point P is 1.62 mi from Q

98.

99. 4669.58 ft from shore

CHAPTER 7

SECTION 7.1

1. transformations to $y = \sin x$: a vertical stretch by a factor of 4, followed by a reflection in the x-axis, a horizontal shrink by a factor of $\dfrac{1}{2}$, a vertical shift up 3 units, then a horizontal shift right $\dfrac{\pi}{2}$ units to obtain $y = 3 - 4\sin(2x - \pi)$

2. transformations to $y = \sin x$: a vertical stretch by a factor of 3, followed by a horizontal stretch by a factor of 2, a vertical shift up 2 units, then a horizontal shift left π units to obtain $y = 2 + 3\sin\left(\dfrac{x}{2} + \dfrac{\pi}{2}\right)$

3. transformations to $y = \tan x$: a reflection in the x-axis, followed by a vertical shift up 2 units, and a horizontal shift right π units to obtain $y = 2 - \tan(x - \pi)$

4. transformations to $y = \tan x$: a vertical stretch by a factor of 2, followed by a horizontal shrink by a factor of $\dfrac{1}{2}$, a vertical shift up 1 unit, then a horizontal shift right $\dfrac{\pi}{2}$ units to obtain $y = 1 + 2\tan(2x - \pi)$

5.

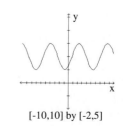

[-10,10] by [-2,5]

6.

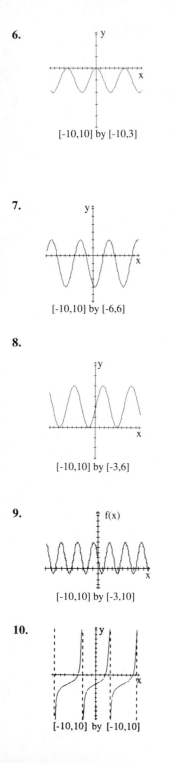

[-10,10] by [-10,3]

7.

[-10,10] by [-6,6]

8.

[-10,10] by [-3,6]

9.

f(x)

[-10,10] by [-3,10]

10.

[-10,10] by [-10,10]

11.

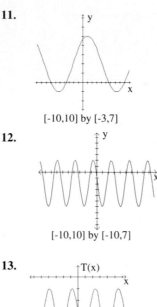

[-10,10] by [-3,7]

12.

[-10,10] by [-10,7]

13.

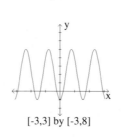

T(x)

[-6,6] by [-6,1]

14.

[-3,3] by [-3,8]

15. Domain = all reals except $x = \dfrac{\pi}{2} + k\pi$, k any intger; range = $(-\infty, \infty)$; period: $\dfrac{x}{|1|} = \pi$; VA: $x = \dfrac{\pi}{2} + k\pi$, k any integer

16. Domain = $(-\infty, \infty)$; range: [1, 3] (from a vertical shift up 2 units); period: $\dfrac{2\pi}{1} = 2\pi$

17. Domain = $(-\infty, \infty)$; range = $[-2, 0]$ (from a vertical shift down 1 unit); period: $\dfrac{2\pi}{1} = 2\pi$

18. Domain = $(-\infty, \infty)$; range: [0, 4] (from a vertical stretch of 2 and a vertical shift up 2 units); period: $\dfrac{2\pi}{1} = 2\pi$

19. Period: π; phase shift: $\dfrac{\pi}{2}$

20. Period: $\dfrac{2\pi}{5}$; phase shift: $\dfrac{\pi}{10}$

21. Period: π; phase shift: $\dfrac{\pi}{4}$

22. Period: $\dfrac{\pi}{3}$; phase shift: $\dfrac{1}{2}$

23. Period: π; phase shift: 0.25

24. Period: 4π; phase shift: 6

25. $y = 3\sin(2x - \pi)$

[-6,6] by [-4,5]

26. $y = \dfrac{1}{3}\sin\left(\dfrac{\pi x}{2} - \pi\right)$

[-10,10] by [-1,1]

27. $y = 2\sin\left(\dfrac{x}{2} + \dfrac{\pi}{8}\right)$

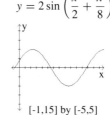

[-1,15] by [-5,5]

28. $y = 5\sin(2\pi x + 2\pi)$

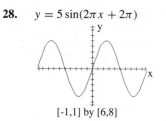

[-1,1] by [6,8]

2π; phase shift: $-\dfrac{\pi}{2}$

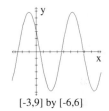

[-6,6] by [-5,5]

43. $y \approx 3.61\sin(x + 0.59)$

[-10,10] by [-8,8]

29.   Domain $= (-\infty, \infty)$; range $= [-3, 3]$; period: π; phase shift: 0

[-2π,2π] by [-4,4]

33.   Domain $= (-\infty, \infty)$; range $= [-1, 5]$; period: $\dfrac{\pi}{2}$; phase shift: $\dfrac{\pi}{4}$

[-π,π] by [-2,6]

44. $y \approx -5.83\sin(x - 0.54)$

[-3,9] by [-6,6]

30.   Domain $= (-\infty, \infty)$; range $= [-1, 3]$; period: 2π; phase shift: π

[-6,6] by [-2,4]

34.   Domain $= (-\infty, \infty)$; range $= [-5, -1]$; period: 6π; phase shift: $\dfrac{\pi}{2}$

[-12,12] by [-8,3]

45. $y \approx 1.41\sin(x + 0.79)$

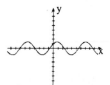

[-10,10] by [-8,8]

31.   Domain – all reals except $x = \pi k$, k any integer; range $= (-\infty, \infty)$; period: π; phase shift: π

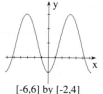

[-2π,2π] by [-4,4]

35. $x = \dfrac{x}{18} + \dfrac{2\pi}{3}k, \; \dfrac{5\pi}{18} + \dfrac{2\pi}{3}k$ for k any integer

36.   no solution

37. $x = \pm 2.66 + 4k\pi$, k any integer

38. $(-3.98, -3.75) \cup (-1.39, 0) \cup (1.39, 3.75) \cup (3.98, \infty)$

39.   0, 3.14

40.   0, 1.15, 1.99, 3.14, 4.29

41.   (0.69, 3.83)

42.   (1.22, 1.92) ∪ (4.36, 5]

46. $y \approx 5\sin(2x - 0.64)$

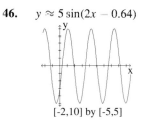

[-2,10] by [-5,5]

47.   Domain $= (-\infty, \infty)$; range $= [-2, 1.13]$; period: 2π; absolute maximum of 1.13 when $x = 0.25$ or 2.89; absolute minimum of -2 when $x = 4.71$; local minimum of 0

32.   Domain $=$ all reals except $x = \pi k$, k any integer; range $= (-\infty, -2] \cup [2, \infty)$; period:

when $x = 1.57$

[-10,10] by [-4,5]

48. Domain $= (-\infty, \infty)$; range $= [-5, 3.17]$; period: 2π; absolute maximum of 3.17 when $x = 0.17$ or 2.97; absolute minimum of -5 when $x = 4.71$; local minimum of -1 when $x = 1.57$

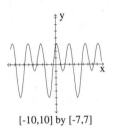

[-10,10] by [-7,7]

49. Domain $= (-\infty, \infty)$; range $= [-1.88, 1.88]$; period: 2π; absolute maximum of 1.88 when $x = 0.47$; local maxima of 0.15 when $x = 2.56$ and 1.06 when $x = 4.82$; absolute minimum of -1.88 when $x = 3.61$; local minima of -1.06 when $x = 1.68$ and -0.15 when $x = 5.70$

[-10,10] by [-5,5]

50. Domain $= (-\infty, \infty)$; range $= [-3.73, 3.73]$; period: 2π; absolute maximum of 3.73 when $x = 3.87$; local maximum of 2.31 when $x = 0.85$; absolute

minimum of -3.73 when $x = 5.55$; local minimum of -2.31 when $x = 2.29$

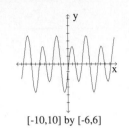

[-10,10] by [-6,6]

51. Domain $= (-\infty, \infty)$; range $= [-5.39, 5.39]$; period: 2π; absolute maximum of 5.39 when $x = 0.38$; absolute minimum of -5.39 when $x = 3.52$

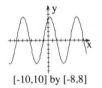

[-10,10] by [-8,8]

52. Domain $= (-\infty, \infty)$; range $= [-1.91, 1.91]$; period: 12π; absolute maximum of 1.91 when $x = 3.62$; absolute minimum of 0 when $x = 0$

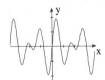

[-12π,12π] by [-3,3]

53. $P = \dfrac{2\pi}{3}$; $f = \dfrac{3}{2\pi}$

54. $P = 4\pi$; $f = \dfrac{1}{4\pi}$

55. $P = 4$; $f = \dfrac{1}{4}$

56. $P = 1$; $f = 1$

57. 217.22 in. long

58. 164.47 in. long

59. $\theta = \tan^{-1}\left(\dfrac{200t}{2055}\right) = 0.97$ or $55.59°$

60.

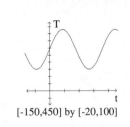

[-150,450] by [-20,100]

61. Hottest: 86° on June 30 ($t = 90$); coldest: 30° on December 27 ($t = 270$)

62. $A = -15$; a cycle is 4 seconds long; $B = \dfrac{\pi}{2}$; $y = -15\cos\left(\dfrac{\pi t}{2}\right)$

63. $t > 0$ for the problem situation

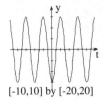

[-10,10] by [-20,20]

64. $t = 1.22 + 4k$, $2.78 + 4k$, $0.78 + 4k$, $3.22 + 4k$ for k any positive integer

65. $y = -28\cos 6\pi t$, $t = 0.093 + \dfrac{k}{3}$, $0.24 + \dfrac{k}{3}$, $0.074 + \dfrac{k}{3}$, $0.26 + \dfrac{k}{3}$ for k any positive integer

66. The maximum volume is 19,347.18 cu. units when $\theta = 0.62$. Then $r = 20\cos(0.62) = 16.28$ units and $h = 40\sin(0.62) = 23.24$

units

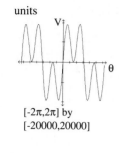

[-2π,2π] by
[-20000,20000]

67.

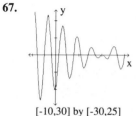

[-10,30] by [-30,25]

68. The portion of the graph in
Exercise 67 for $x < 20$ could be
a model for the motion of an
object on a spring. Friction will
cause the object to oscillate with
smaller and smaller amplitudes
until it stops.

69. A linear combination of sine
and cosine functions is a
sinusoid when both have the
same period.

SECTION 7.2

1. Domain $= \left[-\dfrac{1}{3}, \dfrac{1}{3}\right]$; range
$= \left[-\dfrac{\pi}{2}, \dfrac{\pi}{2}\right]$

[-1,1] by [-2,2]

2. Domain $= [-1, 1]$; range
$= \left[-\dfrac{\pi}{2}, \dfrac{\pi}{2}\right]$

[-3,3] by [-3,3]

3. Domain $= [-2, 0]$; range
$= \left[-\dfrac{\pi}{2}, \dfrac{\pi}{2}\right]$

[-3,1] by [-2,2]

4. Domain $= [-3, 3]$; range
$= [-\pi, \pi]$

[-4,4] by [-4,4]

5. Domain $= [1.5, 2.5]$; range
$= [0, 3\pi]$

[-1,4] by [-1,10]

6. Domain $= (-\infty, \infty)$; range
$= \left(\dfrac{\pi}{2}, \dfrac{3\pi}{2}\right)$

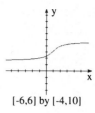

[-6,6] by [-4,10]

7. transformations to $y = \arcsin x$:
a reflection in the x-axis
followed by a vertical shift up 1
unit to obtain $y = 1 - \arcsin x$;
domain $= [-1, 1]$; range
$= [-0.57, 2.57]$

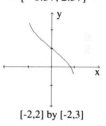

[-2,2] by [-2,3]

8. transformations to $y = \cos^{-1} x$:
a vertical shift up 3 units, then a
horizontal shift right 2 units to
obtain $y = 3 + \cos^{-1}(x - 2)$;
domain $= [1, 3]$; range
$= [3, 6.14]$

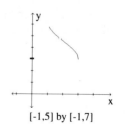

[-1,5] by [-1,7]

9. transformations to $y = \tan^{-1} x$:
a reflection in the x-axis, a
vertical shrink by a factor
of 0.25, then a horizontal
shift right π units to obtain
$y = -0.25 \tan^{-1}(x - \pi)$;
domain $= (-\infty, \infty)$; range

$= \left(-\dfrac{\pi}{8}, \dfrac{\pi}{8}\right);$

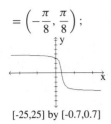

[-25,25] by [-0.7,0.7]

10. transformations to $y = \arccos x$: a vertical stretch by a factor of 3, a horizontal stretch by a factor of 2, then a horizontal shift right 2π units to obtain $g(x) = 3\arccos\left(\dfrac{x}{2} - \pi\right);$ domain $= [4.28, 8.28];$ range $= [0, 9.42]$

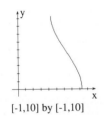

[-1,10] by [-1,10]

11. $y = \sin(\sin^{-1} x) = x$ for $-1 \le x \le 1;$ domain $= [-1, 1];$ range $= [-1, 1]$

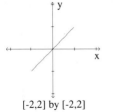

[-2,2] by [-2,2]

12. $-1 \le \sin x \le 1$ for all real x and $\sin^{-1}(t)$ is defined for t in $[-1, 1].$ Thus $y = \sin^{-1}(\sin x)$ is defined for all $x;$ domain $= (-\infty, \infty);$

range $= \left[-\dfrac{\pi}{2}, \dfrac{\pi}{2}\right]$

[-10,10] by [-3,3]

13. $21.22°$

14. undefined

15. $7.13°$

16. $-70.35°$

17. 0.48

18. 2.59

19. 1.17

20. -1.53

21. $\dfrac{\pi}{2}$

22. $\dfrac{\pi}{3}$

23. $\dfrac{\pi}{4}$

24. $\dfrac{5\pi}{6}$

25. $-\dfrac{\pi}{3}$

26. $-\dfrac{\pi}{2}$

27. 0.36

28. 0.82

29. 0.42

30. 0.99

31. undefined

32. 1.2

33. undefined

34. 1.14

35. 0.74

36. 3

37. $\dfrac{\sqrt{3}}{2}$

38. $\dfrac{\sqrt{2}}{2}$

39. $\dfrac{1}{2}$

40. $\dfrac{1}{2}$

41. 0.8

42. $\dfrac{2}{\sqrt{5}}$

43. 1

44. infinitely many solutions of the form $x = \pm 1 + 2\pi k$ where k is any integer

45. 0.48

46. -1.56

47. $\dfrac{x}{\sqrt{x^2 + 1}}$

48. $\dfrac{1}{\sqrt{x^2 + 1}}$

49. $\dfrac{x}{\sqrt{1 - x^2}}$

50. $\dfrac{x}{\sqrt{1 - x^2}}$

51. $x = 1.16 + k\pi$ for k any integer

52. $x = -0.85 + 2k\pi,$ or $x = 3.99 + 2k\pi$ for k any integer

53. $x = \pm 1.23 + 2k\pi$ for k any integer

54. $x = -0.20 + k\pi$ for k any integer

55. $[-\pi, \pi]$

56. $[-0.5, 0) \cup (0, 0.5]$

57. Let $|x| \leq 1$ so that $\sin^{-1}(-x)$
and $\sin^{-1}(x)$ are defined.
Let $y = \sin^{-1}(-x)$. Then
$-x = \sin y$ or $x = -\sin y$.
Now $-\sin y = \sin(-y)$
for y in $\left[-\dfrac{\pi}{2}, \dfrac{\pi}{2}\right]$. Thus
$x = -\sin(y) = \sin(-y)$
and $-y = \sin^{-1} x$. Therefore
$y = -\sin^{-1}(x)$ and
$\sin^{-1}(-x) = -\sin^{-1}(x)$ for
$|x| \leq 1$.

58. Let $\theta = \sin^{-1} x$. Then $\sin \theta = x$
and $\tan \theta = \dfrac{x}{\sqrt{1 - x^2}}$. Thus
$\theta = \tan^{-1}\left(\dfrac{x}{\sqrt{1 - x^2}}\right)$ and
$\sin^{-1} x = \tan^{-1}\left(\dfrac{x}{\sqrt{1 - x^2}}\right)$.

59. Let $u = \arccos x$, $v = \arcsin x$.
Then $\cos u = \sin v = x$.
Since $\cos u = \sin(90° - u)$,
$\sin(90° - u) = \sin v$. Therefore
$90° - u = v$ and $u + v = 90°$.
Thus $\arccos x + \arcsin x = 90°$.

60. Let $\theta = \sin^{-1} x$. Then
$\sin \theta = x$ and $\cos \theta = \sqrt{1 - x^2}$
(see Example 4). Thus
$\cos(\sin^{-1} x) = \sqrt{1 - x^2}$

61. $\theta = \tan^{-1}\left(\dfrac{50}{L}\right)$. The portion
of the graph for $L > 0$
represents the problem situation.

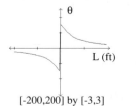

[-200,200] by [-3,3]

62. $\theta = \tan^{-1}\left(\dfrac{x}{3}\right)$. The portion of
the graph for $x > 0$ represents
the problem situation.

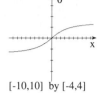

[-10,10] by [-4,4]

63. Let h be the altitude from
C to D, the midpoint of the base.

Then $\sin\left(\dfrac{\theta}{2}\right) = \dfrac{c/2}{a} = \dfrac{c}{2a}$

and $\cos\left(\dfrac{\theta}{2}\right) = \dfrac{h}{a}$.

Then $c = 2a \sin\left(\dfrac{\theta}{2}\right)$

and $h = a \cos\left(\dfrac{\theta}{2}\right)$.

The area is $A = \dfrac{1}{2}ch =$

$\dfrac{1}{2}\left(2a \sin\left(\dfrac{\theta}{2}\right)\right)\left(a \cos\left(\dfrac{\theta}{2}\right)\right) =$

$a^2 \sqrt{\dfrac{1 - \cos \theta}{2}} \sqrt{\dfrac{1 + \cos \theta}{2}} =$

$\dfrac{1}{2}a^2 \sin \theta$.

64. $-1, 1$

65. The equation $\sin y = 0.5$ has
an infinite number of solutions
because any angle y whose
sine value is 0.5 satisfies
the equation. The value of
$\sin^{-1}(0.5)$ is $\dfrac{\pi}{6}$ which is the
principal value of inverse sine
function.

SECTION 7.3

1. $t = 4.71 + 2k\pi$ for k any integer

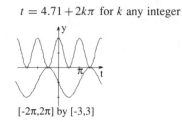

[-2π,2π] by [-3,3]

2. $t = 4.71 + 2k\pi$ for k any intger

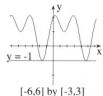

[-6,6] by [-3,3]

3. $x = 1.01 + 2k\pi$, $2.13 + 2k\pi$ for
k any integer

[-6,6] by [-4,4]

4. $x = 0.50 + 2k\pi$, $1.25 + 2k\pi$,
$1.57 + 2k\pi$, $1.89 + 2k\pi$,
$2.65 + 2k\pi$, and $4.71 + 2k\pi$ for
k any integer

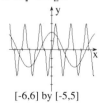

[-6,6] by [-5,5]

5. $x = 0.78 + 2k\pi$, $2.37 + 2k\pi$ for
k any integer

6. $x = \pm 0.45 + 2k\pi$ for k any
integer

7. no solution

8. $x = 1.22 + \pi k$ for k any integer

9. $\dfrac{\pi}{4}, \dfrac{5\pi}{4}$

10. $\dfrac{\pi}{6}, \dfrac{5\pi}{6}, \dfrac{3\pi}{2}$

11. $\dfrac{\pi}{6}, \dfrac{5\pi}{6}$

12. $\dfrac{\pi}{18}, \dfrac{5\pi}{18}, \dfrac{13\pi}{18}, \dfrac{17\pi}{18}, \dfrac{25\pi}{18}, \dfrac{29\pi}{18}$

13. 0.62, 2.53, 3.76, 5.67

14. 0, 1.05, 3.14, 5.24

15. $0, \pi$

16. $0, \dfrac{3\pi}{4}, \pi, \dfrac{7\pi}{4}$

17. $\dfrac{\pi}{2}, \dfrac{3\pi}{2}$

18. $\dfrac{\pi}{2}, \dfrac{3\pi}{2}$

19. $\theta = 0 + k\pi = k\pi$ for k any integer

20. $t = \pm\dfrac{\pi}{6} + k\pi$ for k any integer

21. $t = 0.52 + 2k\pi,\ 2.62 + 2k\pi$ for k any integer

22. $x = 0 + k\pi = k\pi$ for k any integer

23. $x = 1.57 + 2k\pi$ for k any integer

24. $t = -1.05 + 2k\pi,\ 0 + k\pi,\ 1.05 + 2k\pi$ for k any integer

25. $x = \dfrac{\pi}{6} + 2k\pi,\ \dfrac{5\pi}{6} + 2k\pi$ for k any integer

26. $x = \pm\dfrac{\pi}{3} + 2k\pi,\ \pi + 2k\pi$ for k any integer

27. $x = \pm 1.05 + 2k\pi,\ 3.14 + 2\pi k$

28. no solution

29. $\left[0, \dfrac{\pi}{6}\right) \cup \left(\dfrac{5\pi}{6}, 2\pi\right]$

30. $\left[0, \dfrac{\pi}{6}\right) \cup \left(\dfrac{5\pi}{6}, \dfrac{7\pi}{6}\right) \cup \left(\dfrac{11\pi}{6}, 2\pi\right]$

31. $(1.96 + k\pi, 3.53 + k\pi)$ for k any integer

32. $\left(k\pi, \dfrac{\pi}{2} + k\pi\right)$ for k any integer

33. $0, \pm 2.28$

34. 0

35. 0, 1.28

36. $(2.36 + 2k\pi, 5.50 + 2k\pi)$ for k any integer

37. $(1.57 + 2k\pi, 2.84 + 2k\pi) \cup (4.71 + 2k\pi, 6.59 + 2k\pi)$ for k any integer

38. $[-9.32, -6.44] \cup [-2.77, -1.11] \cup [1.11, 2.77] \cup [6.44, 9.32]$

39. $1.18 + 2k\pi,\ 2.36 + 2k\pi,\ 2.75 + 2k\pi,\ 4.32 + 2k\pi,\ 5.50 + 2k\pi,\ 5.89 + 2k\pi$ for k any integer

40. $0.17 + 2k\pi,\ 1.57 + k\pi,\ 2.97 + 2k\pi$ for k any integer

41. Local maximum of 5 when $x = 0$; local minimum of -1 when $x = 2\pi$

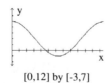

[0,12] by [-3,7]

42. Local maximum of 3.95 when $x = 2.29$; local minimum of

-3.95 when $x = -2.29$

[-π,π] by [-5,4]

43. Local maximum of 1 when $x = 0.64$ or 2.64; local minimum of -3 when $x = 1.64$

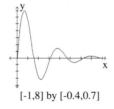

[0,3] by [-4,5]

44. Local maximum of 0.70 when $x = 0.66$, 0.14 when $x = 3.80$, and 0.03 when $x = 6.95$; local minimum of -0.32 when $x = 2.23$ and -0.07 when $x = 5.38$

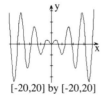

[-1,8] by [-0.4,0.7]

45. does not repeat in cycles

[-20,20] by [-20,20]

46. does not repeat in cycles, note that function is always positive as $x \to +\infty$ but always negative as $x \to -\infty$

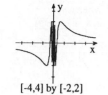

[-4,4] by [-2,2]

47. $0 \le \theta \le \dfrac{\pi}{2}$ represents the problem situation

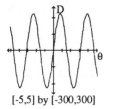

[-0.5,2] by [-500,8500]

48. 0.15 or 8.8°; 1.42 or 81.2°

49. 7812.5 ft

50. $0 \le \theta \le \dfrac{\pi}{2}$ represents the problem situation because the angle of elevation is acute or right

[-5,5] by [-300,300]

51. 36° or 54°

52. 45°

53. The graphs of
$f(x) = x \sin\left(\dfrac{1}{x}\right)$ below
show that the end behavior of
$f(x) = x \sin\left(\dfrac{1}{x}\right)$ is $g(x) = 1$.

As $|x| \to \infty$, $f(x) \to 1$

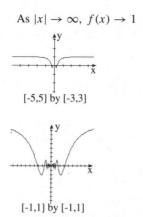

[-5,5] by [-3,3]

[-1,1] by [-1,1]

54. The graph of $f(x) = x^{\sin x}$ shows that this function does not have an end behavior. The graph consists of increasingly larger oscillations as $x \to \infty$

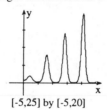

[-5,25] by [-5,20]

55. One method for solving $f(x) \le g(x)$ graphically is to graph $y = f(x)$ and $y = g(x)$ and find x values for which the graph of $y = f(x)$ is below the graph of $y = g(x)$ or intersects it. Another method for solving $f(x) \le g(x)$ graphically is to transform the equation to $f(x) - g(x) \le 0$. Then graph $y = f(x) - g(x)$ and find the x-values for which the function is at or below the x-axis.

56. The period of $\sin x$ is 2π. The right endpoint in $[0, 2\pi)$ is excluded because $x = 2\pi$ is the first x-value in the next period of the sine function.

SECTION 7.4

1. $\sin\theta$

2. 1

3. 1

4. $\cos\theta$

5. $\tan^2 x$

6. $\sin\theta$

7. 2

8. 1

9. $\tan x$

10. $\csc x$

11. $\dfrac{1}{\cos x \sin x}$

12. $2\sin\theta$

13. $\dfrac{1 + 2\sin t \cos t}{\sin^3 t \cos t}$

14. $\cos\theta - \sin\theta$

15. 1

16. $\cos x \sin x$

17. The graphs appear to be identical

18. not an identity

19. The graphs appear to be identical

20. The graphs appear to be identical

21. $1 + \cot^2\theta = 1 + \dfrac{\cos^2\theta}{\sin^2\theta} = \dfrac{\sin^2\theta + \cos^2\theta}{\sin^2\theta} = \dfrac{1}{\sin^2\theta} = \csc^2\theta$

22. $1 - 2\sin^2 x = 1 - 2(1 - \cos^2 x) = 1 - 2 + 2\cos^2 x = 2\cos^2 x - 1$

23. $2\cos^2\theta + \sin^2\theta = \cos^2\theta + (\cos^2\theta + \sin^2\theta) = \cos^2\theta + 1$

24. $\sin\theta + \cos\theta\cot\theta = \sin\theta +$
$\cos\theta \cdot \dfrac{\cos\theta}{\sin\theta} = \sin\theta + \dfrac{\cos^2\theta}{\sin\theta} =$
$\dfrac{\sin^2\theta + \cos^2\theta}{\sin\theta} = \dfrac{1}{\sin\theta} = \csc\theta$

25. $\dfrac{\sin x}{\tan x} = \dfrac{\sin x}{\sin x/\cos x} = \cos x$

26. $\sec^2\theta(1 - \sin^2\theta) =$
$\sec^2\theta \cdot \cos^2\theta = 1$

27. $(\cos t - \sin t)^2 + (\cos t + \sin t)^2 =$
$\cos^2 t - 2\cos t\sin t + \sin^2 t +$
$\cos^2 t + 2\cos t\sin t + \sin^2 t =$
$2\cos^2 t + 2\sin^2 t =$
$2(\cos^2 t + \sin^2 t) = 2$

28. $\sin^2\alpha - \cos^2\alpha = (1 - \cos^2\alpha) -$
$\cos^2\alpha = 1 - 2\cos^2\alpha$

29. $\dfrac{1 + \tan^2 x}{\sin^2 x + \cos^2 x} = \dfrac{\sec^2 x}{1} =$
$\sec^2 x$

30. $\dfrac{1}{\tan\beta} + \tan\beta = \dfrac{1 + \tan^2\beta}{\tan\beta} =$
$\dfrac{\sec^2\beta}{\tan\beta} = \dfrac{1/\cos^2\beta}{\sin\beta/\cos\beta} =$
$\dfrac{1}{\cos\beta\sin\beta} = \sec\beta\csc\beta$

31. $\dfrac{1 - \cos\theta}{\sin\theta} =$
$\dfrac{(1 + \cos\theta)}{(1 + \cos\theta)} \cdot \dfrac{(1 - \cos\theta)}{\sin\theta} =$
$\dfrac{1 - \cos^2\theta}{(1 + \cos\theta)\sin\theta} =$
$\dfrac{\sin^2\theta}{(1 + \cos\theta)\sin\theta} = \dfrac{\sin\theta}{1 + \cos\theta}$

32. $\dfrac{\tan x}{\sec x - 1} = \dfrac{\tan x}{\sec x - 1} \cdot$
$\dfrac{\sec x + 1}{\sec x + 1} = \dfrac{\tan x(\sec x + 1)}{\sec^2 x - 1} =$
$\dfrac{\tan x(\sec x + 1)}{\tan^2 x} = \dfrac{\sec x + 1}{\tan x}$

33. $\dfrac{\sin t - \cos t}{\sin t + \cos t} =$
$\dfrac{(\sin t - \cos t)}{(\sin t + \cos t)} \cdot \dfrac{(\sin t + \cos t)}{(\sin t + \cos t)} =$

34. $\dfrac{\sin^2 t - \cos^2 t}{\sin^2 t + 2\sin t\cos t + \cos^2 t} =$
$\dfrac{\sin^2 t - (1 - \sin^2 t)}{1 + 2\sin t\cos t} =$
$\dfrac{2\sin^2 t - 1}{1 + 2\sin t\cos t}$

34. $\dfrac{\sec x + 1}{\sec x - 1} = \dfrac{(1/\cos x) + 1}{(1/\cos x) - 1} =$
$\dfrac{(1 + \cos x)/\cos x}{(1 - \cos x)/\cos x} = \dfrac{1 + \cos x}{1 - \cos x}$

35. $(x\sin\alpha + y\cos\alpha)^2 +$
$(x\cos\alpha - y\sin\alpha)^2 =$
$x^2\sin^2\alpha + 2xy\sin\alpha\cos\alpha +$
$y^2\cos^2\alpha + x^2\cos^2\alpha -$
$2xy\cos\alpha\sin\alpha + y^2\sin^2\alpha =$
$x^2\sin^2\alpha + x^2\cos^2\alpha + y^2\sin^2\alpha +$
$y^2\cos^2\alpha = x^2(\sin^2\alpha + \cos^2\alpha) +$
$y^2(\sin^2\alpha + \cos^2\alpha) = x^2 + y^2$

36. $\dfrac{\sin t}{1 - \cos t} + \dfrac{1 + \cos t}{\sin t} =$
$\dfrac{\sin^2 t + 1 - \cos^2 t}{(1 - \cos t)(\sin t)} =$
$\dfrac{(1 - \cos^2 t) + 1 - \cos^2 t}{(1 - \cos t)\sin t} =$
$\dfrac{2 - 2\cos^2 t}{(1 - \cos t)\sin t} =$
$\dfrac{2(1 - \cos t)(1 + \cos t)}{(1 - \cos t)\sin t} =$
$\dfrac{2(1 + \cos t)}{\sin t}$

37. $\dfrac{\sin\theta}{1 + \cos\theta} + \dfrac{1 + \cos\theta}{\sin\theta} =$
$\dfrac{\sin^2\theta + 1 + 2\cos\theta + \cos^2\theta}{(1 + \cos\theta)\sin\theta} =$
$\dfrac{2 + 2\cos\theta}{(1 + \cos\theta)\sin\theta} =$
$\dfrac{2(1 + \cos\theta)}{(1 + \cos\theta)\sin\theta} = \dfrac{2}{\sin\theta} =$
$2\csc\theta$

38. $\dfrac{\tan A + \tan B}{1 - \tan A\tan B} =$
$\dfrac{(\sin A/\cos A) + (\sin B/\cos B)}{1 - (\sin A\sin B)/(\cos A\cos B)} =$

$\dfrac{\frac{(\sin A\cos B + \cos A\sin B)}{(\cos A\cos B)}}{\frac{(\cos A\cos B - \sin A\sin B)}{(\cos A\cos B)}} =$
$\dfrac{\sin A\cos B + \cos A\sin B}{\cos A\cos B - \sin A\sin B}$

39. $\sin^2 x = 1 - \cos^2 x$ implies
$\sin x = \pm\sqrt{1 - \cos^2 x}$,
depending on the sign of $\sin x$.
For example, $\sin\dfrac{3\pi}{2} = -1$
and $\cos\dfrac{3\pi}{2} = 0$. Then
$\sin^2\left(\dfrac{3\pi}{2}\right) = (-1)^2 =$
$1 - 0^2 = 1 - \cos^2\left(\dfrac{3\pi}{2}\right)$; but,
$\sin\left(\dfrac{3\pi}{2}\right) = -1 \neq \sqrt{1 - 0^2} =$
1. So $\sin x = \sqrt{1 - \cos^2 x}$ is
not an identity

40. The graphs of $y = \sin x$ and
$y = x$ appear to be identical in
the viewing rectangle given
below. This is not a proof that
$\sin x = x$, however. Further
investigation shows the two
functions differ at other points.
For example, $\sin\pi \neq \pi$.

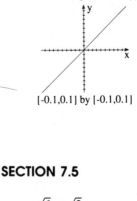

[−0.1, 0.1] by [−0.1, 0.1]

SECTION 7.5

1. $\dfrac{\sqrt{6} - \sqrt{2}}{4}$

2. $\dfrac{3 - \sqrt{3}}{3 + \sqrt{3}}$

3. $\dfrac{\sqrt{6}+\sqrt{2}}{4}$

4. $\dfrac{3+\sqrt{3}}{3-\sqrt{3}}$

5. $\dfrac{\sqrt{2}+\sqrt{6}}{4}$

6. $\dfrac{\sqrt{6}+\sqrt{2}}{4}$

7. $\dfrac{3+\sqrt{3}}{3-\sqrt{3}}$

8. $\dfrac{3-\sqrt{3}}{3+\sqrt{3}}$

9. $\cos x$

10. $\tan x$

11. $-\sin x$

12. $\cot x$

13. $\sin\left(\theta+\dfrac{\pi}{2}\right)=$
$\sin\theta\cos\dfrac{\pi}{2}+\cos\theta\sin\dfrac{\pi}{2}=$
$\sin\theta\cdot0+\cos\theta\cdot1=\cos\theta$

14. $\cos\left(\theta-\dfrac{\pi}{4}\right)=$
$\cos\theta\cos\dfrac{\pi}{4}+\sin\theta\sin\dfrac{\pi}{4}=$
$\cos\theta\cdot\dfrac{\sqrt{2}}{2}+\sin\theta\cdot\dfrac{\sqrt{2}}{2}=$
$\dfrac{\sqrt{2}}{2}(\cos\theta+\sin\theta)$

15. $\tan\left(\theta+\dfrac{\pi}{4}\right)=$
$\dfrac{\tan\theta+\tan(\pi/4)}{1-\tan\theta\tan(\pi/4)}=$
$\dfrac{\tan\theta+1}{1-\tan\theta\cdot1}=\dfrac{1+\tan\theta}{1-\tan\theta}$

16. $\cos\left(\theta+\dfrac{\pi}{2}\right)=$
$\cos\theta\cos\dfrac{\pi}{2}-\sin\theta\sin\dfrac{\pi}{2}=$
$\cos\theta\cdot0-\sin\theta\cdot1=-\sin\theta$

17. $2\sin\theta\cos\theta+\cos\theta$

18. $2\sin\theta\cos\theta+\cos^2\theta-\sin^2\theta$
(Answers may vary.)

19. $\cos^3\theta-3\sin^2\theta\cos\theta+$
$2\sin\theta\cos\theta$ (Answers may vary.)

20. $3\sin\theta\cos^2\theta+\cos^2\theta-\sin^3\theta-$
$\sin^2\theta$ (Answers may vary.)

21. $(4\sin\theta+1)(\cos\theta)(\cos^2\theta-$
$\sin^2\theta)-2\sin^2\theta\cos\theta$

22. $\dfrac{\sin\theta(3\cos^2\theta-\sin^2\theta)}{\cos\theta(\cos^2\theta-\sin^2\theta)}$

23. Amplitude: 2.24; period: $\dfrac{2\pi}{3}$;
phase shift: 0.15 units to the left

[-π,π] by [-3,3]

24. Amplitude: 3.6; period: π;
phase shift: 1.08 units to the left

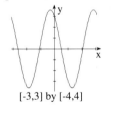

[-3,3] by [-4,4]

25. Amplitude: 4.64; period: 2π;
phase shift: 1.37 units to the left

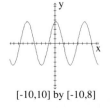

[-10,10] by [-10,8]

26. Amplitude: 3.79; period: π;
phase shift: 1.32 units to the right

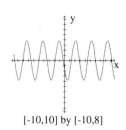

[-10,10] by [-10,8]

27. Domain = all reals except
$x=\pm\sqrt{\dfrac{\pi}{2}+k\pi}$ for k any
nonnegative integer; range
$=(-\infty,-1]\cup[1,\infty)$

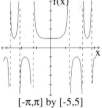

[-π,π] by [-5,5]

28. Domain $=(-\infty,\infty)$; range
$=(-\infty,\infty)$

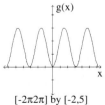

[-20,20] by [-100,100]

29. Domain $=(-\infty,\infty)$; range
$=[0,3]$

g(x)

[-2π,2π] by [-2,5]

30. Domain = all reals except $x = 0$ and $x = \frac{\pi}{2} + k\pi$ for k any integer; range $= (-\infty, \infty)$

[-5,5] by [-5,4]

31. (a) $-\dfrac{\sqrt{5}}{3}$ (b) $\dfrac{3}{2}$

32. (a) $-4\sqrt{5}$ (b) $\sqrt{\dfrac{3-\sqrt{5}}{6}} \approx 0.36$

33. -2

34. $-\dfrac{\sqrt{3}}{2}$

35. $-\sqrt{3}$

36. $-\dfrac{1}{2}$

37. $\dfrac{\pi}{6}, \dfrac{5\pi}{6}$

38. $\dfrac{\pi}{2}, \dfrac{3\pi}{2}$

39. $\dfrac{\pi}{3}, \dfrac{2\pi}{3}$

40. $\dfrac{\pi}{4}, \dfrac{5\pi}{4}$

41. $\dfrac{3\pi}{2}, \dfrac{\pi}{2}$

42. $\dfrac{\pi}{6}, \dfrac{5\pi}{6}, \dfrac{3\pi}{2}$

43. $0, \pi$

44. $0.94, 2.20, 3.46, 4.71, 5.97$

45. $\sin\left(\dfrac{\pi}{2} - \alpha\right) =$
$\sin\dfrac{\pi}{2}\cos\alpha - \cos\dfrac{\pi}{2}\sin\alpha =$
$1 \cdot \cos\alpha - 0 \cdot \sin\alpha = \cos\alpha$

46. $\sin(\alpha - \beta) = \sin(\alpha + (-\beta)) =$
$\sin\alpha\cos(-\beta) + \cos\alpha\sin(-\beta) =$
$\sin\alpha\cos\beta - \cos\alpha\sin\beta$ (Recall that $\cos(-\beta) = \cos\beta$ and $\sin(-\beta) = -\sin\beta$.)

47. $\tan(\alpha + \beta) = \dfrac{\sin(\alpha + \beta)}{\cos(\alpha + \beta)} =$
$\dfrac{\sin\alpha\cos\beta + \cos\alpha\sin\beta}{\cos\alpha\cos\beta - \sin\alpha\sin\beta} =$
$\dfrac{\frac{(\sin\alpha\cos\beta)}{(\cos\alpha\cos\beta)} + \frac{(\cos\alpha\sin\beta)}{(\cos\alpha\cos\beta)}}{\frac{(\cos\alpha\cos\beta)}{(\cos\alpha\cos\beta)} - \frac{(\sin\alpha\sin\beta)}{(\cos\alpha\cos\beta)}} =$
$\dfrac{\tan\alpha + \tan\beta}{1 - \tan\alpha\tan\beta}$

48. $\tan(\alpha - \beta) = \tan(\alpha + (-\beta)) =$
$\dfrac{\tan\alpha + \tan(-\beta)}{1 - \tan\alpha\tan(-\beta)} =$
$\dfrac{\tan\alpha - \tan\beta}{1 + \tan\alpha\tan\beta}$

49. $\cos 2\theta = \cos(\theta + \theta) =$
$\cos\theta\cos\theta - \sin\theta\sin\theta =$
$\cos^2\theta - \sin^2\theta$

50. (a) $\cos 2\theta = \cos^2\theta - \sin^2\theta =$
$(1 - \sin^2\theta) - \sin^2\theta = 1 - 2\sin^2\theta$
(b) $\cos 2\theta = \cos^2\theta - \sin^2\theta =$
$\cos^2\theta - (1 - \cos^2\theta) =$
$2\cos^2\theta - 1$

51. $\tan 2\theta = \tan(\theta + \theta) =$
$\dfrac{\tan\theta + \tan\theta}{1 - \tan\theta\tan\theta} = \dfrac{2\tan\theta}{1 - \tan^2\theta}$

52. $\cos 2\alpha = 1 - 2\sin^2\alpha$
or $\sin^2\alpha = \dfrac{1 - \cos 2\alpha}{2}$.
Let $\theta = 2\alpha$, so that
$\alpha = \dfrac{\theta}{2}$. Substitution yields
$\sin^2\left(\dfrac{\theta}{2}\right) = \dfrac{1 - \cos\theta}{2}$ or
$\sin\left(\dfrac{\theta}{2}\right) = \pm\sqrt{\dfrac{1 - \cos\theta}{2}}$

53. $\tan\left(\dfrac{\theta}{2}\right) = \dfrac{\sin(\theta/2)}{\cos(\theta/2)} =$
$\dfrac{\pm\sqrt{(1 - \cos\theta)/2}}{\pm\sqrt{(1 + \cos\theta)/2}} =$

$\pm\dfrac{\sqrt{1 - \cos\theta}}{\sqrt{1 + \cos\theta}} \cdot \dfrac{\sqrt{1 + \cos\theta}}{\sqrt{1 + \cos\theta}} =$
$\pm\dfrac{\sqrt{1 - \cos^2\theta}}{1 + \cos\theta} = \dfrac{\sin\theta}{1 + \cos\theta}$

54. $2\sin x\cos x = \sin 2x$ which is sinusoidal. $1 - 2\sin^2 x = \cos 2x = \sin\left(2x + \dfrac{\pi}{2}\right)$, which also has the form of a sinusoidal function.

55. The graph of $y = \cos(x + 4)$ can be obtained from the graph of $y = \cos x$ by a horizontal shift left 4 units. The graph of $y = \cos x + 4$ can be obtained from the graph of $y = \cos x$ by a vertical shift up 4 units. The two transformations have different effects on the graph of $y = \cos x$. Thus $\cos(x + 4) \neq \cos x + 4$.

56. If $\cos\left(\dfrac{\pi}{2} - \theta\right) = \sin\theta$, then $\sin\left(\dfrac{\pi}{2} - \alpha\right) = \sin\theta = \cos\left(\dfrac{\pi}{2} - \theta\right) = \cos\alpha$

57. $\sin(\alpha - \beta) =$
$\cos\left(\dfrac{\pi}{2} - (\alpha - \beta)\right) =$
$\cos\left(\left(\dfrac{\pi}{2} - \alpha\right) + \beta\right) =$
$\cos\left(\dfrac{\pi}{2} - \alpha\right)\cos\beta -$
$\sin\left(\dfrac{\pi}{2} - \alpha\right)\sin\beta =$
$\sin\alpha\cos\beta - \cos\alpha\sin\beta$ (from Exercise 56)

SECTION 7.6

1. $0.30, 2.84$

2. $0.72, 5.56$

3. 0.98

4. no solution

5. $\dfrac{\pi}{6}, \dfrac{5\pi}{6}$

6. $\dfrac{\pi}{8}$

7. $\dfrac{\pi}{6}, \dfrac{5\pi}{6}, 1.88, 4.41$

8. $0, \pi, 3.34, 6.08$

9. $2.50, 3.79, \dfrac{\pi}{3}, \dfrac{5\pi}{3}$

10. $1.47, 4.81$

11. $0, \dfrac{\pi}{4}, \pi, \dfrac{5\pi}{4}$

12. $0.98, 4.12, \dfrac{\pi}{3}, \dfrac{7\pi}{6}, \dfrac{5\pi}{3}, \dfrac{11\pi}{6}$

13. $0, \dfrac{\pi}{2}, \dfrac{3\pi}{2}$

14. $0, \dfrac{\pi}{3}, \pi, \dfrac{5\pi}{3}$

15. $\dfrac{\pi}{3}, \pi, \dfrac{5\pi}{3}$

16. $\dfrac{\pi}{3}, \dfrac{2\pi}{3}, \dfrac{4\pi}{3}, \dfrac{5\pi}{3}$

17. 0

18. $\dfrac{\pi}{6}, \dfrac{\pi}{2}, \dfrac{5\pi}{6}, \dfrac{3\pi}{2}$

19. $\dfrac{2\pi}{3}, \dfrac{4\pi}{3}$

20. $\dfrac{\pi}{2}, \dfrac{7\pi}{6}, \dfrac{11\pi}{6}$

21. $\dfrac{\pi}{2}, \dfrac{7\pi}{6}, \dfrac{11\pi}{6}$

22. $0.31, 2.83, 4.08, 5.34$

23. $\dfrac{\pi}{3}, \pi, \dfrac{5\pi}{3}$

24. $A = \dfrac{1}{2}ch = \dfrac{1}{2}(x_1 + x_2)h$
$= \dfrac{1}{2}x_1 h + \dfrac{1}{2}x_2 h$

25. $x_1 = b\sin\theta_1, \; x_2 = a\sin\theta_2,$
$h = b\cos\theta_1,$ and $h = a\cos\theta_2.$
From Exercise 24,

$A = \dfrac{1}{2}x_1 h + \dfrac{1}{2}x_2 h =$
$\dfrac{1}{2}(b\sin\theta_1)(a\cos\theta_2) +$
$\dfrac{1}{2}(a\sin\theta_2)(b\cos\theta_1) =$
$\dfrac{1}{2}ab\sin\theta_1\cos\theta_2 +$
$\dfrac{1}{2}ab\cos\theta_1\sin\theta_2$

26. $A = \dfrac{1}{2}ab\sin\theta_1\cos\theta_2 +$
$\dfrac{1}{2}ab\cos\theta_1\sin\theta_2 =$
$\dfrac{1}{2}ab(\sin\theta_1\cos\theta_2 +$
$\cos\theta_1\sin\theta_2) =$
$\dfrac{1}{2}ab\sin(\theta_1 + \theta_2) = \dfrac{1}{2}ab\sin C$

27. $\dfrac{\sin\alpha}{\sin\beta} = 2.42;$ thus
$\sin\beta = \dfrac{\sin\alpha}{2.42} = 0.413\sin\alpha$ and
$\beta = \sin^{-1}(0.413\sin\alpha);$ domain
$= \left[0, \dfrac{\pi}{2}\right]$

$[0,\pi/2]$ by $[-1,1]$

28. $\beta = 0.43 = 24.4°$

29. $\alpha = 1.11 = 63.9°$

30. $\beta = 0.296 = 16.98°$

31. If $\tan x = \sec x$, then
$\dfrac{\sin x}{\cos x} = \dfrac{1}{\cos x}$ and $\sin x = 1$.
Thus $x = \dfrac{\pi}{2} + 2k\pi$ for k any
integer; but $\sec x$ is undefined
for these values of x, so there
is no solution. The graphical
solution is misleading. The
graphs of $y = \tan x$ and
$y = \sec x$ appear to intersect
near $x = \dfrac{\pi}{2} + 2k\pi$. Similarly,

the graph of $y = \tan x - \sec x$
appears to intersect the x-axis at
these points.

Chapter 7 Review

1. Amplitude: 2; period: $\dfrac{2\pi}{3}$;
phase shift: 0; domain
$= (-\infty, \infty)$; range $= [-2, 2]$

2. Amplitude: 3; period:
$\dfrac{2\pi}{4} = \dfrac{\pi}{2}$; phase shift: 0;
domain $= (-\infty, \infty)$; range
$= [-3, 3]$

3. Amplitude: 1.5; period:
$\dfrac{2\pi}{2} = \pi$; phase shift:
$\dfrac{\pi/4}{2} = \dfrac{\pi}{8}$; domain
$= (-\infty, \infty)$; range
$= [-1.5, 1.5]$

4. Amplitude: 2; period: $\dfrac{2\pi}{3}$;
phase shift: $-\dfrac{1}{3}$; domain
$= (-\infty, \infty)$; range $= [-2, 2]$

5. Amplitude: 4; period: $\dfrac{2\pi}{2} = \pi$;
phase shift: $\dfrac{1}{2}$; domain
$= (-\infty, \infty)$; range $= [-4, 4]$

6. Amplitude: 2; period: $\dfrac{2\pi}{3}$;
phase shift: $\dfrac{\pi/3}{3} = \dfrac{\pi}{9}$; domain
$= (-\infty, \infty)$; range $= [-2, 2]$

7. A vertical stretch by a factor
of 3, followed by a horizontal
shrink by a factor of $\dfrac{1}{2}$, a
horizontal shift right $\dfrac{1}{2}$ unit,
then a vertical shift up 7 units
to obtain $y = 3\sin(2x - 1) + 7$.

8. A reflection in the x-axis,
followed by a vertical stretch by

a factor of 2, a horizontal shrink by a factor of $\frac{1}{2}$, a horizontal shift right 0.25 units, then a vertical shift down 3 units to obtain $y = -2\sin(2x - 0.5) - 3$.

9. A vertical stretch by a factor of 5, followed by a horizontal shrink by a factor of $\frac{1}{3}$, a horizontal shift left 0.5 units, then a vertical shift up 2.3 units to obtain $y = 5\sin(3x + 1.5) + 2.3$

10. A reflection in the x-axis, followed by a vertical stretch by a factor of 3, a horizontal shrink by a factor of $\frac{1}{\pi}$, a horizontal shift right $\frac{2}{\pi}$ units, then a vertical shift up 4 units to obtain $y = -3\sin(\pi x - 2) + 4$

11. A vertical stretch by a factor of 2, followed by a horizontal shrink by a factor of $\frac{1}{2\pi}$, a horizontal shift left $\frac{3}{2\pi}$ units, then a vertical shift down 2 units to obtain $y = 2\sin(2\pi x + 3) - 2$

12. A reflection in the x-axis, followed by a horizontal shift right $\frac{\pi}{2}$ units, then a vertical shift down 1 unit to obtain $y = -\sin\left(x - \frac{\pi}{2}\right) - 1$

13.

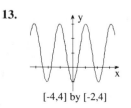

[-4,4] by [-2,4]

14.

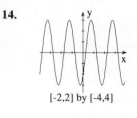

[-2,2] by [-4,4]

15.

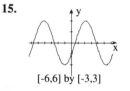

[-6,6] by [-3,3]

16.

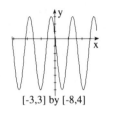

[-3,3] by [-8,4]

17. $y \approx 4.47\sin(x - 1.11)$

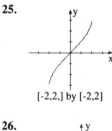

[-10,10] by [-5,5]

18. $y \approx 3.61\sin(2x + 2.16)$

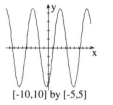

[-3,3] by [-4,4]

19. $y \approx 7.53\sin(2x - 1.45)$

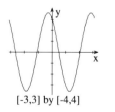

[-6,6,] by [-10,10]

20. $y \approx 3.83\sin(3x + 0.11)$

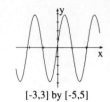

[-3,3] by [-5,5]

21. $\approx 3.61\sin(2x + 0.59)$

22. $\approx 5\sin(3x + 0.93)$

23. $\approx 5.83\sin(2x - 0.54)$

24. $\approx 8.06\sin(3x + 1.05)$

25.

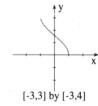

[-2,2,] by [-2,2]

26.

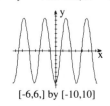

[-3,3] by [-3,4]

27. Apply a horizontal shrink by a factor of $\frac{1}{2}$ to the graph of $y = \sin^{-1} x$ to obtain $y = \sin^{-1} 2x$

[-2,2,] by [-2,2]

28.

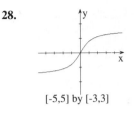

[-5,5] by [-3,3]

29. Apply a horizontal shrink by a factor of $\frac{1}{2}$ to the graph of $y = \tan^{-1} x$ to obtain $y = \tan^{-1} 2x$

[-10,10] by [-4,4]

30. apply a horizontal shrink by a factor of $\frac{1}{2}$ to the graph of $y = \sin^{-1} x$, followed by a horizontal shift right $\frac{1}{2}$ units to obtain $y = \sin^{-1}(2x - 1)$

[-3,3] by [-3,3]

31. Apply a horizontal shrink by a factor of $\frac{1}{3}$ to the graph of $y = \sin^{-1} x$, followed by a horizontal shift right $\frac{1}{3}$ units and a vertical shift up 2 units to obtain $y = \sin^{-1}(3x - 1) + 2$

32. Apply a horizontal shrink by a factor of $\frac{1}{2}$ to the graph of $y = \cos^{-1} x$, followed by a horizontal shift left $\frac{1}{2}$ units and a vertical shift down 3 units to obtain $y = \cos^{-1}(2x + 1) - 3$

[-2,2] by [-4,2]

33. $\dfrac{\pi}{12} + k\pi, \ \dfrac{5\pi}{12} + k\pi$

34. $\pm\dfrac{\pi}{6} + 2k\pi$

35. $-\dfrac{\pi}{4} + k\pi$

36. $0.78 + 2k\pi, \ 2.37 + 2k\pi$

37. $\pm 0.72 + k\pi$

38. $0.59 + k\pi$

39. $0.30 + 2k\pi, \ 2.84 + 2k\pi$

40. $\pm 0.45 + \dfrac{2k\pi}{3}$

41. 0.65

42. 1.56

43. Domain $= (-\infty, \infty)$; range $= (-\infty, \infty)$

[-20,30] by [-60,60]

44. Domain $= (-\infty, \infty)$; range $= [-5.75, 7]$

[-10,10] by [-6,9]

45. Domain $=$ all reals except $x = \dfrac{k\pi}{2}$ for k any integer; range $= (-\infty, \infty)$

[-10,10] by [-10,10]

46. Domain $= (-\infty, \infty)$; range $= (-\infty, \infty)$

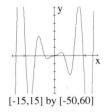

[-15,15] by [-50,60]

47. Domain $= (-\infty, \infty)$; range $= [0, 5]$

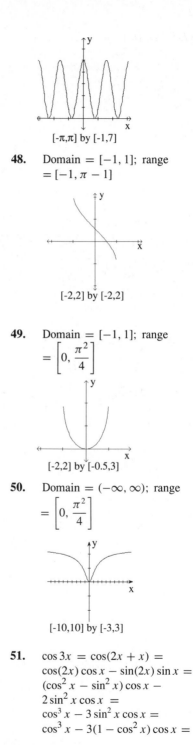

[-π,π] by [-1,7]

48. Domain $= [-1, 1]$; range $= [-1, \pi - 1]$

[-2,2] by [-2,2]

49. Domain $= [-1, 1]$; range $= \left[0, \dfrac{\pi^2}{4}\right]$

[-2,2] by [-0.5,3]

50. Domain $= (-\infty, \infty)$; range $= \left[0, \dfrac{\pi^2}{4}\right]$

[-10,10] by [-3,3]

51. $\cos 3x = \cos(2x + x) =$
$\cos(2x)\cos x - \sin(2x)\sin x =$
$(\cos^2 x - \sin^2 x)\cos x -$
$2\sin^2 x \cos x =$
$\cos^3 x - 3\sin^2 x \cos x =$
$\cos^3 x - 3(1 - \cos^2 x)\cos x =$
$\cos^3 x - 3\cos x + 3\cos^3 x =$
$4\cos^3 x - 3\cos x$

52. $\sin^2 x - \sin^2 2x = (1 - \cos^2 x) -$
$(1 - \cos^2 2x) = \cos^2 2x - \cos^2 x$

53. $\tan^2 x - \sin^2 x = \dfrac{\sin^2 x}{\cos^2 x} -$
$\dfrac{\sin^2 x}{1} = \dfrac{\sin^2 x - \sin^2 x \cos^2 x}{\cos^2 x} =$
$\dfrac{\sin^2 x(1 - \cos^2 x)}{\cos^2 x}$
$\dfrac{\sin^2 x \cdot \sin^2 x}{\cos^2 x} = \sin^2 x \tan^2 x$

54. $2\sin\theta \cos^3\theta + 2\sin^3\theta \cos\theta =$
$2\sin\theta \cos\theta(\cos^2\theta + \sin^2\theta) =$
$2\sin\theta \cos\theta(1) = \sin 2\theta$

55. $\csc x - \cos x \cot x =$
$\dfrac{1}{\sin x} - \dfrac{\cos x}{1} \cdot \dfrac{\cos x}{\sin x} =$
$\dfrac{1 - \cos^2 x}{\sin x} = \dfrac{\sin^2 x}{\sin x} = \sin x$

56. $\dfrac{\tan\theta + \sin\theta}{2\tan\theta} =$
$\dfrac{(\sin\theta/\cos\theta) + (\sin\theta/1)}{(2\sin\theta)/(\cos\theta)} =$
$\dfrac{(\sin\theta + \sin\theta\cos\theta)/\cos\theta}{(2\sin\theta)/(\cos\theta)} =$
$\dfrac{\sin\theta(1 + \cos\theta)}{2\sin\theta} = \dfrac{1}{2}(1 +$
$\cos\theta) = \left[\pm\sqrt{\dfrac{1}{2}(1 + \cos\theta)}\right]^2 =$
$\cos^2\left(\dfrac{\theta}{2}\right)$

57. $\dfrac{1 + \tan\theta}{1 - \tan\theta} + \dfrac{1 + \cot\theta}{1 - \cot\theta} =$
$\dfrac{1 + (\sin\theta/\cos\theta)}{1 - (\sin\theta/\cos\theta)} +$
$\dfrac{1 + (\cos\theta/\sin\theta)}{1 - (\cos\theta/\sin\theta)} =$
$\dfrac{(\cos\theta + \sin\theta)/\cos\theta}{(\cos\theta - \sin\theta)/\cos\theta} +$
$\dfrac{(\sin\theta + \cos\theta)/\sin\theta}{(\sin\theta - \cos\theta)/\sin\theta} =$
$\dfrac{\cos\theta + \sin\theta}{\cos\theta - \sin\theta} +$

$\dfrac{\sin\theta + \cos\theta}{-1(\cos\theta - \sin\theta)} =$
$\dfrac{\cos\theta + \sin\theta}{\cos\theta - \sin\theta} - \dfrac{\cos\theta + \sin\theta}{\cos\theta - \sin\theta} =$
0

58. $\sin 3\theta = \sin(2\theta + \theta) =$
$\sin 2\theta \cos\theta + \cos 2\theta \sin\theta =$
$2\sin\theta \cos\theta \cos\theta +$
$(\cos^2\theta - \sin^2\theta)\sin\theta =$
$3\cos^2\theta \sin\theta - \sin^3\theta$

59. not an identity; at $\dfrac{\pi}{2}$, left side is undefined but right side $= 0$

60. identical

61. $(2\sin x \cos x)(\cos x - \sin x) +$
$(\cos^2 x - \sin^2 x)(\sin x + \cos x)$

62. $\cos^3 x - 3\sin^2 x \cos x +$
$2\sin x \cos x$

63. $\cos^4 x - 2\cos^2 x \sin^2 x +$
$\sin^4 x - 2\sin x \cos x$

64. $3\sin x \cos^2 x - \sin^3 x -$
$6\sin x \cos x$

65. $\sqrt{1 - x^2}$

66. $\dfrac{x}{\sqrt{x^2 + 1}}$

67. $\sqrt{1 - x^2}$

68. $2x\sqrt{1 - x^2}$

69. $1.12 + 2k\pi,\ 5.16 + 2k\pi$

70. $3.79 + 2k\pi,\ 6.42 + 2k\pi$

71. ± 1.15

72. $-0.23,\ 1.85,\ 3.59$

73. $\dfrac{\pi}{12},\ \dfrac{5\pi}{12},\ \dfrac{13\pi}{12},\ \dfrac{17\pi}{12}$

74. $\dfrac{\pi}{3},\ \dfrac{5\pi}{3}$

75. $0,\ \pi,\ \dfrac{\pi}{4},\ \dfrac{3\pi}{4},\ \dfrac{5\pi}{4},\ \dfrac{7\pi}{4}$

76. $\dfrac{3\pi}{2}$

77. $0,\ \dfrac{2\pi}{3},\ \dfrac{4\pi}{3}$

78. no solution

79. $\left[0, \dfrac{\pi}{6}\right) \cup \left(\dfrac{5\pi}{6}, \dfrac{7\pi}{6}\right) \cup$
$\left(\dfrac{11\pi}{6}, 2\pi\right)$

80. $\left(\dfrac{\pi}{2}, \dfrac{3\pi}{2}\right)$

81. $\left(\dfrac{\pi}{3}, \dfrac{5\pi}{3}\right)$

82. $\left(-\dfrac{\pi}{2}, 0\right)$

83. As $|x| \to \infty$, $y \to 0$
Note: function oscillates wildly
near $x = 0$

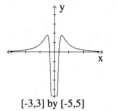

[-3,3] by [-5,5]

84. As $|x| \to \infty$, the graph
consists of increasingly larger
oscillations.

[-40,40] by [-50000,50000]

85.

[-10,10] by [-5,13]

86. $0 < \theta < \pi$ or $0° < \theta < 180°$

87. $\theta = 0.955 = 54.72°$

88. 7.72 in.^{2}

CHAPTER 8

SECTION 8.1

1. Answers will vary. Let $a = 11.2$
cm; $b = \dfrac{11.2 \sin 75°}{\sin 32°} = 20.42$
cm and $c = 20.21$ cm. Let
$a = 10$ in.; $b = 18.23$ in., and
$c = 18.05$ in.

2. Answers will vary. Let $a = 50$
m; $c = 17.04$ m and $b = 58.08$
m; let $a = 16.30$ ft; $c = 5.55$ ft
and $b = 18.93$ ft

3. $\gamma = 110°$, $a = 12.86$, $c = 18.79$

4. No triangle is possible

5. $\alpha = 90°$, $\gamma = 60°$, $c = 10.39$

6. $\gamma = 68°$, $b = 4.61$, $c = 4.84$

7. $\gamma = 68°$, $a = 3.88$, $c = 6.61$

8. $\alpha = 41.62°$, $\gamma = 53.38°$,
$c = 4.83$

9. Two triangles possible:
$\beta = 73.25°$, $\gamma = 56.75°$,
$c = 4.37$; $\beta = 106.75°$,
$\gamma = 23.25°$, $c = 2.06$

10. $\alpha = 29.51°$, $\gamma = 112.49°$,
$c = 30.01$

11. Two triangles possible:
$\alpha = 55.17°$, $\gamma = 86.83°$,
$c = 19.46$; $\alpha = 124.83°$,
$\gamma = 17.17°$, $c = 5.75$

12. $\alpha = 49.51°$, $\gamma = 14.49°$,
$c = 3.62$

13. no triangle exists

14. no triangle exists

15. $\alpha = 22.06°$, $\gamma = 5.94°$,
$c = 2.20$

16. $\alpha = 61°$, $b = 3.39$, $a = 10.77$

17. **(a)** There are two values
for α when $6.69 < b < 10$.
(b) There is one value for
α when $b = 6.69$ or $b \geq 10$.
(c) α has no value when
$0 < b < 6.69$.

18. **(a)** 54.60 ft   **(b)** 51.93 ft
apart

19. 19.70 miles from B, 15.05 mi
from A, 11.86 mi from the road

20. 24.93 ft high

21. 0.72 mi high

22. 1.255 ft long

23. $c = 3.85$, $\beta = 128.89°$,
$\alpha = 29.11°$

SECTION 8.2

1. no triangle exists

2. no triangle exists

3. $\alpha = 24.56°$, $\gamma = 56.23°$,
$\beta = 99.22°$

4. $\gamma = 142°$, $a = 8.73$, $b = 7.12$

5. $a = 9.83$, $\beta = 89.32°$,
$\gamma = 35.68°$

6. $b = 59.01$, $\alpha = 20.31°$,
$\gamma = 34.69°$

7. no triangle exists

8. Two triangles possible:
$\beta = 45.54°$, $\gamma = 98.46°$,
$c = 23.56$; $\beta = 134.46°$,
$\gamma = 9.54°$, $c = 3.95$

9. no triangle exists

10. $\alpha = 44.42°$, $\beta = 78.46°$,
$\gamma = 57.12°$

11. (a) There are two values for α when $6.78 < b < 8$. (b) $b = 6.78$ or $b \geq 8$ results in a unique triangle (c) α has no value when $0 < b < 6.78$

12. (a) $6.36 < b < 12$ (b) $b = 6.36$ or $b > 12$ (c) $0 < b < 6.36$

13. 17.55 sq units

14. 115.95 sq units

15. 110.35 sq units

16. 12.11 sq units

17. 5.56 sq units

18. 298.66 sq units

19. 841.22 ft apart

20. 66.78 ft

21. 92.82°

22. 42.18°

23. 42.50 ft

24. 1.45

25. 41.56 ft

26. 108.91 ft high

27. distance from $A = 893.64$ ft., $h = 1123.56$ ft

28. 5.80 ft

29. 61.73 ft high

30. Let O (located at the origin) be the intersection of lines ℓ_1 and ℓ_2 and P and Q be the intersections of the vertical line $x = 1$ with ℓ_1 and ℓ_2, respectively. The coordinates of P and Q are $(1, m_1)$ and $(1, m_2)$, respectively. Let α be the angle of $\triangle OPQ$ between OP and OQ. Either α or $180° - \alpha$ is the angle between ℓ_1 and ℓ_2. $d(P, Q) = |m_1 - m_2|$;

$d(O, P) = \sqrt{1 + m_1^2}$;

$d(O, Q) = \sqrt{1 + m_2^2}$; by the law of cosines: $d^2(P, Q) = d^2(O, P) + d^2(O, Q) - 2d(O, P)d(O, Q)\cos\alpha$;

$|m_1 - m_2|^2 = 1 + m_1^2 + 1 + m_2^2 - 2\sqrt{\left(1 + m_1^2\right)\left(1 + m_2^2\right)}\cos\alpha$;

$\cos\alpha = \dfrac{1 + m_1 m_2}{\sqrt{\left(1 + m_1^2\right)\left(1 + m_2^2\right)}}$

yields $\alpha =$

$\cos^{-1}\left[\dfrac{1 + m_1 m_2}{\sqrt{\left(1 + m_1^2\right)\left(1 + m_2^2\right)}}\right].$

If $\cos\alpha > 0$, α is acute. If $\cos\alpha < 0$, then $180° - \alpha$ is acute.

31. $\dfrac{A_{\text{sector}}}{A_{\text{circle}}} = \dfrac{\theta}{2\pi}$ yields $A_{\text{sector}} = A_{\text{circle}} \cdot \dfrac{\theta}{2\pi} = \pi r^2 \cdot \dfrac{\theta}{2\pi} = \dfrac{r^2\theta}{2}$

32. $0.0625 P^2$

33. From Exercise 32, $A = P^2\left[\dfrac{\theta}{2(2 + \theta)^2}\right]$ is the area of the sector; for P constant, A is maximum when $f(x) = \dfrac{x}{2(2 + x)^2}$ is maximum.

SECTION 8.3

1. $2 + 7i$

2. $6 - 9i$

3. $14 - 8i$

4. $-3 - 4i$

5. i

6. $\dfrac{-7}{41} - \dfrac{22i}{41}$

7. $a = b = 2$, $|z_1| = 2\sqrt{2}$

8. $a = 3$, $b = -4$, $|z_2| = 5$

9. $2\sqrt{2}\left(\cos\dfrac{\pi}{4} + i\sin\dfrac{\pi}{4}\right)$

10. $5(\cos 5.36 + i\sin 5.36)$

11.

$(3,0)$ x

[-5,5] by [-5,5]

12.

$(0,3)$

[-5,5] by [-5,5]

13.

$(4,-4)$

[-5,5] by [-5,5]

14.

$(-4,3)$

[-5,5] by [-5,5]

15.

$(\sqrt{3},1)$

[-5,5] by [-5,5]

16.

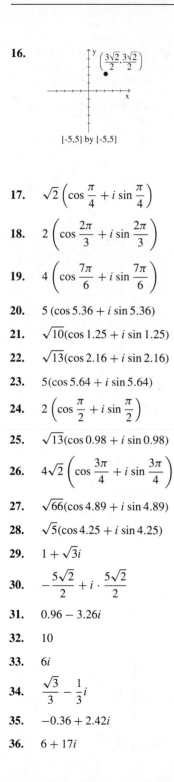

$\left(\frac{3\sqrt{2}}{2}, \frac{3\sqrt{2}}{2}\right)$

[-5,5] by [-5,5]

17. $\sqrt{2}\left(\cos\frac{\pi}{4} + i\sin\frac{\pi}{4}\right)$

18. $2\left(\cos\frac{2\pi}{3} + i\sin\frac{2\pi}{3}\right)$

19. $4\left(\cos\frac{7\pi}{6} + i\sin\frac{7\pi}{6}\right)$

20. $5(\cos 5.36 + i\sin 5.36)$

21. $\sqrt{10}(\cos 1.25 + i\sin 1.25)$

22. $\sqrt{13}(\cos 2.16 + i\sin 2.16)$

23. $5(\cos 5.64 + i\sin 5.64)$

24. $2\left(\cos\frac{\pi}{2} + i\sin\frac{\pi}{2}\right)$

25. $\sqrt{13}(\cos 0.98 + i\sin 0.98)$

26. $4\sqrt{2}\left(\cos\frac{3\pi}{4} + i\sin\frac{3\pi}{4}\right)$

27. $\sqrt{66}(\cos 4.89 + i\sin 4.89)$

28. $\sqrt{5}(\cos 4.25 + i\sin 4.25)$

29. $1 + \sqrt{3}i$

30. $-\frac{5\sqrt{2}}{2} + i \cdot \frac{5\sqrt{2}}{2}$

31. $0.96 - 3.26i$

32. 10

33. $6i$

34. $\frac{\sqrt{3}}{3} - \frac{1}{3}i$

35. $-0.36 + 2.42i$

36. $6 + 17i$

37. $z_1 z_2 = 0.26 + 0.97i,$
$z_1/z_2 = 0.97 + 0.26i$

38. $z_1 z_2 = 0.97 + 0.26i,$
$z_1/z_2 = -0.26 - 0.97i$

39. $z_1 z_2 = -2.20 + 8.20i;$
$z_1/z_2 = -2.05 + 0.55i$

40. $z_1 z_2 = 9.93 + 1.20i;$
$z_1/z_2 = 0.98 + 2.30i$

41. $1 + 7i$

42. $8 - 20i$

43. $11 + 16i$

44. $\sqrt{26}$

45. $5\sqrt{13}$

46. $-\frac{14}{29} + \frac{23}{29}i$

47. $5 + 7i$

y

(5,7)

(1,3)

(4,4)

x

[-1,6] by [-1,8]

48. $-7 - i$

y

(4,2)

(-3,1)

x

(-7,-1)

[-7,5] by [-2,4]

SECTION 8.4

1. $5(\cos 5.36 + i\sin 5.36)$

2. $2\sqrt{2}\left(\cos\frac{3\pi}{4} + i\sin\frac{3\pi}{4}\right)$

3. $\sqrt{34}(\cos 0.54 + i\sin 0.54)$

4. $\sqrt{89}(\cos 5.72 + i\sin 5.72)$

5. $2\left(\cos\frac{3\pi}{2} + i\sin\frac{3\pi}{2}\right)$

6. $10(\cos\pi + i\sin\pi)$

7. $1^3\left[\cos\left(3 \cdot \frac{\pi}{4}\right) + i\sin\left(3 \cdot \frac{\pi}{4}\right)\right] =$
$-\frac{\sqrt{2}}{2} + \frac{\sqrt{2}}{2}i$

8. $1^4\left[\cos\left(4 \cdot \frac{\pi}{3}\right) + i\sin\left(4 \cdot \frac{\pi}{3}\right)\right] =$
$-\frac{1}{2} - \frac{\sqrt{3}}{2}i$

9. $2[\cos(6\pi) + i\sin(6\pi)] = 2$

10. $3\left(\cos\frac{15\pi}{2} + i\sin\frac{15\pi}{2}\right) =$
$-3i$

11. $\sqrt{2}^5\left(\cos\frac{5\pi}{4} + i\sin\frac{5\pi}{4}\right) =$
$-4 - 4i$

12. $5^{20}[\cos(20(0.927)) +$
$i\sin(20(0.927))] =$
$5^{20}(0.95 - 0.30i)$

13. $2^3\left[\cos\left(3 \cdot \frac{5\pi}{3}\right) +$
$i\sin\left(3 \cdot \frac{5\pi}{3}\right)\right] = -8$

14. $-\left[1^4\left(\cos\left(4 \cdot \frac{\pi}{4}\right) +$
$i\sin\left(4 \cdot \frac{\pi}{4}\right)\right)\right] = 1$

15. $1^3\left[\cos\left(3 \cdot \frac{\pi}{3}\right) +$
$i\sin\left(3 \cdot \frac{\pi}{3}\right)\right] = -1$

16. $5^5 \cdot [\cos(5 \cdot 5.36) +$
$i\sin(5 \cdot 5.36)] = -237 + 3116i$

17. **(a)** Let $z = 2 + i$. Then
$r = \sqrt{2^2 + 1^2} = \sqrt{5}$ and
$\theta = \sin^{-1}\left(\frac{1}{\sqrt{5}}\right) = 0.4636;$
$(2 + i)^2 = [\sqrt{5}(\cos 0.4636 +$
$i\sin 0.4636)]^2 =$
$(\sqrt{5})^2(\cos 0.9273 +$

$i \sin 0.9273) = 5(0.6 + 0.8i) =$
$3 + 4i$ **(b)** $(2+i)(2+i) = 4 +$
$2i + 2i + i^2 = 4 + 4i - 1 = 3 + 4i$

18.

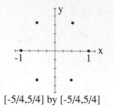

[-2,2] by [-2,2]

$1 + 0i$, $-0.5 + 0.87i$, and
$-0.5 - 0.87i$

19.

[-2,2] by [-2,2]

$1 + 0i$, $0.31 + 0.95i$,
$-0.81 + 0.59i$, $-0.81 - 0.59i$,
and $0.31 - 0.95i$

20. $z^2 = \cos\dfrac{\pi}{2} + i \sin\dfrac{\pi}{2}$,

$z^4 = \cos\pi + i \sin\pi$,

$z^5 = \cos\dfrac{5\pi}{4} + i \sin\dfrac{5\pi}{4}$,

$z^7 = \cos\dfrac{7\pi}{4} + i \sin\dfrac{7\pi}{4}$

21. $\left(\dfrac{1}{\sqrt{2}} + \dfrac{1}{\sqrt{2}}i\right)^8 =$

$\left[1\left(\cos\dfrac{\pi}{4} + i \sin\dfrac{\pi}{4}\right)\right]^8 =$

$1^8 \cdot \left(\cos\dfrac{8\pi}{4} + i \sin\dfrac{8\pi}{4}\right) =$

$1(1 + 0i) = 1$;

$\left(-\dfrac{1}{\sqrt{2}} + \dfrac{1}{\sqrt{2}}i\right)^8 =$

$\left[1\left(\cos\dfrac{3\pi}{4} + i \sin\dfrac{3\pi}{4}\right)\right]^8 =$

$1^8 \cdot \left(\cos\dfrac{24\pi}{4} + i \sin\dfrac{24\pi}{4}\right) =$

$1(1 + 0i) = 1$;

$\left(-\dfrac{1}{\sqrt{2}} - \dfrac{1}{\sqrt{2}}i\right)^8 =$

$\left[1\left(\cos\dfrac{5\pi}{4} + i \sin\dfrac{5\pi}{4}\right)\right]^8 =$

$1^8 \cdot \left(\cos\dfrac{40\pi}{4} + i \sin\dfrac{40\pi}{4}\right) =$

$1(1 + 0i) = 1$;

$\left(\dfrac{1}{\sqrt{2}} - \dfrac{1}{\sqrt{2}}i\right)^8 =$

$\left[1\left(\cos\dfrac{7\pi}{4} + i \sin\dfrac{7\pi}{4}\right)\right]^8 =$

$1^8 \cdot \left(\cos\dfrac{56\pi}{4} + i \sin\dfrac{56\pi}{4}\right) =$

$1(1 + 0i) = 1$

22. $z_0 = 1$, $z_1 = \dfrac{1}{2} + \dfrac{\sqrt{3}}{2}i$,

$z_2 = -\dfrac{1}{2} + \dfrac{\sqrt{3}}{2}i$, $z_3 = -1$,

$z_4 = -\dfrac{1}{2} - \dfrac{\sqrt{3}}{2}i$, $z_5 = \dfrac{1}{2} - \dfrac{\sqrt{3}}{2}i$

23. $z_0 = 1$, $z_1 = \dfrac{\sqrt{2}}{2} + \dfrac{\sqrt{2}}{2}i$,

$z_2 = i$, $z_3 = -\dfrac{\sqrt{2}}{2} + \dfrac{\sqrt{2}}{2}i$,

$z_4 = -1$, $z_5 = -\dfrac{\sqrt{2}}{2} - \dfrac{\sqrt{2}}{2}i$,

$z_6 = -i$, $z_7 = \dfrac{\sqrt{2}}{2} - \dfrac{\sqrt{2}}{2}i$

24. $z_0 = 1$, $z_1 = -\dfrac{1}{2} + \dfrac{\sqrt{3}}{2}i$,

$z_2 = -\dfrac{1}{2} - \dfrac{\sqrt{3}}{2}i$

25. $z_0 = 1$, $z_1 = i$ $z_2 = -1$,
$z_3 = -i$

[-5,5] by [-5,5]

26. Use the sixth roots of unity
from Exercise 22.

[-5/4,5/4] by [-5/4,5/4]

27. $z_0 = \sqrt[8]{8}\left(\cos\dfrac{\pi}{16} + i \sin\dfrac{\pi}{16}\right)$,

$z_1 = \sqrt[8]{8}\left(\cos\dfrac{9\pi}{16} + i \sin\dfrac{9\pi}{16}\right)$,

$z_2 = \sqrt[8]{8}\left(\cos\dfrac{17\pi}{16} + i \sin\dfrac{17\pi}{16}\right)$,

$z_3 = \sqrt[8]{8}\left(\cos\dfrac{25\pi}{16} + i \sin\dfrac{25\pi}{16}\right)$

[-1.5,1.5] by [-1.5,1.5]

28. $z_0 = \sqrt[8]{8}\left(\cos\dfrac{3\pi}{16} + i \sin\dfrac{3\pi}{16}\right)$,

$z_1 = \sqrt[8]{8}\left(\cos\dfrac{11\pi}{16} + i \sin\dfrac{11\pi}{16}\right)$,

$z_2 = \sqrt[8]{8}\left(\cos\dfrac{19\pi}{16} + i \sin\dfrac{19\pi}{16}\right)$,

$z_3 = \sqrt[8]{8}\left(\cos\dfrac{27\pi}{16} + i \sin\dfrac{27\pi}{16}\right)$

[-5/4,5/4] by [-5/4,5/4]

29. $z_0 = \sqrt[12]{8}\left(\cos\dfrac{\pi}{8} + i \sin\dfrac{\pi}{8}\right)$,

$z_1 = \sqrt[12]{8}\left(\cos\dfrac{11\pi}{24} + i\sin\dfrac{11\pi}{24}\right),$

$z_2 = \sqrt[12]{8}\left(\cos\dfrac{19\pi}{24} + i\sin\dfrac{19\pi}{24}\right),$

$z_3 = \sqrt[12]{8}\left(\cos\dfrac{9\pi}{8} + i\sin\dfrac{9\pi}{8}\right),$

$z_4 = \sqrt[12]{8}\left(\cos\dfrac{35\pi}{24} + i\sin\dfrac{35\pi}{24}\right),$

$z_5 = \sqrt[12]{8}\left(\cos\dfrac{43\pi}{24} + i\sin\dfrac{43\pi}{24}\right)$

[-5/4,5/4] by [-5/4,5/4]

30. $z_0 = 2(\cos 0 + i\sin 0) = 2,$

$z_1 = 2\left(\cos\dfrac{2\pi}{5} + i\sin\dfrac{2\pi}{5}\right),$

$z_2 = 2\left(\cos\dfrac{4\pi}{5} + i\sin\dfrac{4\pi}{5}\right),$

$z_3 = 2\left(\cos\dfrac{6\pi}{5} + i\sin\dfrac{6\pi}{5}\right),$

$z_4 = 2\left(\cos\dfrac{8\pi}{5} + i\sin\dfrac{8\pi}{5}\right)$

[-3,3] by [-3,3]

31. $z_0 = 1 + \sqrt{3}i, z_1 = -2,$
$z_2 = 1 - \sqrt{3}i, z = -8$

$1 + \sqrt{3}\,i$

$1 - \sqrt{3}\,i$

[-3,3] by [-3,4]

32. $z_1 = 2 - 2i, z_2 = -2 + 2i,$
$z_3 = 2 + 2i, z = -64$

[-3,3] by [-3,3]

33. $z_0 = \sqrt[8]{50}\left(\cos\dfrac{7\pi}{16} + i\sin\dfrac{7\pi}{16}\right),$

$z_1 = \sqrt[8]{50}\left(\cos\dfrac{15\pi}{16} + i\sin\dfrac{15\pi}{16}\right),$

$z_2 = \sqrt[8]{50}\left(\cos\dfrac{23\pi}{16} + i\sin\dfrac{23\pi}{16}\right),$

$z_3 = \sqrt[8]{50}\left(\cos\dfrac{31\pi}{16} + i\sin\dfrac{31\pi}{16}\right)$

[-3,3] by [-3,4]

34. The roots are 60° apart;
$z_0 = \sqrt[6]{2}(\cos 15° + i\sin 15°);$
the remaining angles are $\dfrac{5\pi}{12},$
$\dfrac{3\pi}{4}, \dfrac{13\pi}{12}, \dfrac{17\pi}{12}, \dfrac{7\pi}{4}$

[-5/4,5/4] by [-5/4,5/4]

35. $(-1 + i)^{12} =$
$\left[\sqrt{2}\left(\cos\dfrac{3\pi}{4} + i\sin\dfrac{3\pi}{4}\right)\right]^{12} =$
$(\sqrt{2})^{12}\left(\cos\left(\dfrac{12\cdot 3\pi}{4}\right) + \right.$

$\left. i\sin\left(\dfrac{12\cdot 3\pi}{4}\right)\right) =$
$64(-1 + 0i) = -64$

36. $z_0 = \sqrt{2}\left(\cos\dfrac{\pi}{12} + i\sin\dfrac{\pi}{12}\right);$
the remaining angles are: $\dfrac{\pi}{4},$
$\dfrac{5\pi}{12}, \dfrac{7\pi}{12}, \dfrac{3\pi}{4}, \dfrac{11\pi}{12}, \dfrac{13\pi}{12}, \dfrac{5\pi}{4},$
$\dfrac{17\pi}{12}, \dfrac{19\pi}{12}, \dfrac{7\pi}{4},$ and $\dfrac{23\pi}{12}$

[-3,3] by [-3,3]

SECTION 8.5

1. $3\sqrt{2}, 45°$

2. $\sqrt{26}, 191.31°$

3. $\sqrt{10}, -18.43°$ or $341.57°$

4. $\sqrt{5}, 63.43°$

5. $(1, 7)$

6. $(-3, -1)$

7. $(-3, 8)$

8. $(6, 12)$

9. $(4, -9)$

10. $(-10, -10)$

11. $5\sqrt{2}$

12. $\sqrt{10}$

13. **(a)** $(4, 9)$

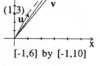

[-1,6] by [-1,10]

(b) $(-2, -3)$

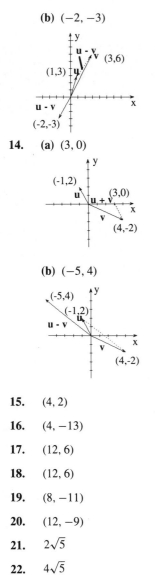

14. **(a)** $(3, 0)$

(b) $(-5, 4)$

15. $(4, 2)$

16. $(4, -13)$

17. $(12, 6)$

18. $(12, 6)$

19. $(8, -11)$

20. $(12, -9)$

21. $2\sqrt{5}$

22. $4\sqrt{5}$

23. 0

24. $\sqrt{185}$

25.

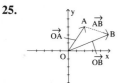

26.

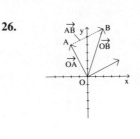

27. $-3\mathbf{i} + 6\mathbf{j}$

28. $3\mathbf{i} - \mathbf{j}$

29. $-\mathbf{i} + 8\mathbf{j}$

30. $-\mathbf{i} - 6\mathbf{j}$

31. $\alpha = 51.13°$, $\beta = 0°$, angle between $= 51.13°$

32. $\alpha = 116.57°$, $\beta = 33.69°$, angle between $= 82.88°$

33. $\alpha = 116.57°$, $\beta = 306.87°$, angle between $= 169.70°$

34. $\alpha = 303.69°$, $\beta = 239.04°$, angle between $= 64.65°$

35. $\dfrac{19}{5}\mathbf{r} + \dfrac{3}{5}\mathbf{s}$; geometrically, this means $\mathbf{v}$ is the sum of scalar multiples of vectors $\mathbf{r}$ and $\mathbf{s}$

36. $\dfrac{x + 2y}{5}\mathbf{r} + \dfrac{2x - y}{5}\mathbf{s}$

37. $(x, y) = t(a, b) + u(c, d) = (ta + uc, tb + ud)$; $t = \dfrac{dx - cy}{ad - bc}$

$u = \dfrac{ay - bx}{ad - bc}$ for $ad - bc \neq 0$

38. $(-223.99, 480.34)$

39. $(79.88, -453.01)$

40. $\mathbf{r} = (-136.87, 336.04)$; $|\mathbf{r}| = \sqrt{(-136.87)^2 + (336.04)^2} = 362.84$; $\theta = 180° + \tan^{-1}\left(\dfrac{336.04}{-136.87}\right) = 112.16°$ (in QII); the plane actually flies $22.16°$ west of north at 362.84 mph

41. $\mathbf{r} = (52.52, -528.19)$; the plane actually flies $5.68°$ east of south at 530.79 mph

42. $\mathbf{r} = (-17.95, -5.96)$; the boat actually sails $18.37°$ south of west at 18.92 mph

43. The pilot should fly on a bearing of $2.72°$ north of west at a speed of 433.49 mph

44. The pilot should fly on a bearing of $1.05°$ east of north at a speed of 758.91 mph

45. Answers will vary with the choice of graphing utility.

46. $\mathbf{v}_1 = (12.31, 33.83)$, $\mathbf{v}_{2.5} = (30.78, 84.57)$, $\mathbf{v}_4 = (49.25, 135.32)$

47. $\mathbf{v}_t = (36t\cos 70°, 36t\sin 70°)$

48. In the DEGREE mode, graph $X(t) = 36t\cos 70°$ and $Y(t) = 36t\sin 70°$ in the VR: $[0, 300]$ by $[0, 200]$; let $0 \leq t \leq 5$ with Tstep $= 0.25$

49. The description of the problem situation is not a realistic model because it neglects gravity and air resistance. In this model, the ball will never return to the ground. Its vertical distance $b = 36t\sin 70°$ increases by $36\sin 70°$ ft every second.

50. $\mathbf{v}_t = (0, 220 - 16t^2)$, $\mathbf{v}_2 = (0, 156)$, $\mathbf{v}_3 = (0, 76)$, $\mathbf{v}_4 = (0, -36)$

51. 3.71 sec

52. $(36t\cos 70°, 36t\sin 70° - 16t^2)$

53. 2.11 sec

54. maximum height is 17.88 ft when $t = 1.06$ sec; horizontal distance when the ball stops ($t = 2.11$ sec) is 26.03 ft

55. $\cos\beta = \dfrac{|\mathbf{u}|^2 + |\mathbf{v}|^2 - d^2}{2|\mathbf{u}||\mathbf{v}|}$
from the law of cosines; thus
$\beta = \cos^{-1}\left(\dfrac{|\mathbf{u}|^2 + |\mathbf{v}|^2 - d^2}{2|\mathbf{u}||\mathbf{v}|}\right)$

56. For each pair $\mathbf{u}$ and $\mathbf{v}$, apply the law of cosines:
(a) $(2+3)^2 + (3-2)^2 =$
$(\sqrt{13})^2 + (\sqrt{13})^2 -$
$2\sqrt{13}\sqrt{13}\cos(\beta-\alpha);$
$0 = -26\cos(\beta-\alpha)$ Thus
$\beta - \alpha = \cos^{-1}(0) = 90°$ and
the vectors are perpendicular
(b) $(1+1)^2 + (-1+1)^2 =$
$(\sqrt{2})^2 + (\sqrt{2})^2 -$
$2\sqrt{2}\sqrt{2}\cos(\beta-\alpha);$
$0 = -4\cos(\beta-\alpha)$ Thus
$\beta - \alpha = \cos^{-1}(0) = 90°$ and
the vectors are perpendicular.
(c) $(5+3)^2 + (-3+5)^2 =$
$(\sqrt{34})^2 + (\sqrt{34})^2 -$
$2\sqrt{34}\sqrt{34}\cos(\beta-\alpha);$
$0 = -68\cos(\beta-\alpha);$ Thus
$\beta - \alpha = \cos^{-1}(0) = 90°$ and
the vectors are perpendicular.

57. The dot product is zero for each vector pair. Generalization: If two vectors are perpendicular, then their dot product is zero.

58. Answers will vary. Three possible choices are $(1, 4)$, $(-2, -8)$, and $\left(\dfrac{1}{2}, 2\right)$

Chapter 8 Review

1. $\gamma = 68°$, $b = 3.88$, $c = 6.61$

2. $\alpha = 35.97°$, $\gamma = 34.03°$, $c = 4.76$

3. $\beta = 113.50°$, $c = 18.16$, $b = 27.55$

4. no triangle exists

5. $\gamma = 72°$, $a = 2.94$, $b = 5.05$

6. $\beta = 102°$, $a = 19.45$, $c = 48.90$

7. $\alpha = 44.42°$, $\beta = 78.46°$, $\gamma = 57.12°$

8. $\beta = 41.62°$, $\gamma = 53.38°$, $c = 4.83$

9. **(a)** $5.63 < b < 12$
(b) $b = 5.63$ or $b \geq 12$
(c) $0 < b < 5.63$

10. 7.48 sq units

11. 22.98 sq units

12. **(a)** 102.54 ft **(b)** 96.35 ft

13. 0.61 mi

14. 849.77 ft

15. 1.25

16. $2+4i$

17. $-2 - 5i$

18. $5 - i$

19. $10 - 10i$

20. $-3 + i$

21. $-\dfrac{17}{58} - \dfrac{i}{58}$

22.

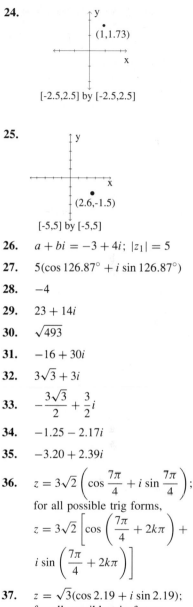

(-5. 2)

[-6,4] by [-4,5]

23.

(4,-3)

[-5,5] by [-5,5]

24.

(1,1.73)

[-2.5,2.5] by [-2.5,2.5]

25.

(2.6,-1.5)

[-5,5] by [-5,5]

26. $a + bi = -3 + 4i$; $|z_1| = 5$

27. $5(\cos 126.87° + i \sin 126.87°)$

28. -4

29. $23 + 14i$

30. $\sqrt{493}$

31. $-16 + 30i$

32. $3\sqrt{3} + 3i$

33. $-\dfrac{3\sqrt{3}}{2} + \dfrac{3}{2}i$

34. $-1.25 - 2.17i$

35. $-3.20 + 2.39i$

36. $z = 3\sqrt{2}\left(\cos\dfrac{7\pi}{4} + i\sin\dfrac{7\pi}{4}\right)$;
for all possible trig forms,
$z = 3\sqrt{2}\left[\cos\left(\dfrac{7\pi}{4} + 2k\pi\right) + i\sin\left(\dfrac{7\pi}{4} + 2k\pi\right)\right]$

37. $z = \sqrt{3}(\cos 2.19 + i\sin 2.19)$;
for all possible trig forms,
$z = \sqrt{3}[\cos(2.19 + 2k\pi) + i\sin(2.19 + 2k\pi)]$ where k is any integer

38. $z = \sqrt{34}(\cos 5.25 + i\sin 5.25)$;
for all possible trig forms,
$z = \sqrt{34}[\cos(5.25 + 2k\pi) +$

$i \sin(5.25 + 2k\pi)]$ where k is any integer

39. $z = 2\sqrt{2}\left(\cos\dfrac{5\pi}{4} + i\sin\dfrac{5\pi}{4}\right);$

for all possible trig forms,

$z = 2\sqrt{2}\left[\cos\left(\dfrac{5\pi}{4} + 2k\pi\right) + i\sin\left(\dfrac{5\pi}{4} + 2k\pi\right)\right]$ where k is any integer

40. $12i$

41. $5 + 5\sqrt{3}i$

42. $\dfrac{3\sqrt{3}}{8} - \dfrac{3}{8}i$

43. $\dfrac{5\sqrt{2}}{4} + \dfrac{5\sqrt{2}}{4}i$

44. $\dfrac{3\left(\frac{\sqrt{3}}{2} + \frac{1}{2}i\right)}{4\left(\frac{1}{2} + \frac{\sqrt{3}}{2}i\right)} = \dfrac{3\sqrt{3}}{8} - \dfrac{3}{8}i$

45. $\dfrac{5(0.259 + 0.966i)}{2\left(\frac{\sqrt{3}}{2} + \frac{1}{2}i\right)} =$

$1.77 - 1.77i \approx \dfrac{5\sqrt{2}}{4} + \dfrac{5\sqrt{2}}{4}i$

46. $-171.83 - 171.83i$

47. $-128 + 221.70i$

48. -125

49. $83{,}190.41 + 83{,}190.41i$

50. $z_0 = \sqrt[8]{18}\left(\cos\dfrac{\pi}{16} + i\sin\dfrac{\pi}{16}\right),$

$z_1 = \sqrt[8]{18}\left(\cos\dfrac{9\pi}{16} + i\sin\dfrac{9\pi}{16}\right),$

$z_2 = \sqrt[8]{18}\left(\cos\dfrac{17\pi}{16} + i\sin\dfrac{17\pi}{16}\right),$

$z_3 = \sqrt[8]{18}\left(\cos\dfrac{25\pi}{16} + i\sin\dfrac{25\pi}{16}\right)$

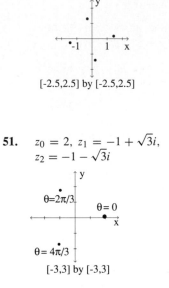

[-2.5,2.5] by [-2.5,2.5]

51. $z_0 = 2,\ z_1 = -1 + \sqrt{3}i,$
$z_2 = -1 - \sqrt{3}i$

$\theta = 2\pi/3$ $\theta = 0$
$\theta = 4\pi/3$

[-3,3] by [-3,3]

52. $z_2 = 2\sqrt{2}(\cos 165° + i\sin 165°),$
$z_3 = 2\sqrt{2}(\cos 285° + i\sin 285°),$
$z = -16 + 16i$

53. $z_0 = 0.30 + 1.51i,$
$z_1 = -1.51 + 0.30i,$
$z_2 = -0.30 - 1.51i,$
$z_3 = 1.51 - 0.30i$

[-2,2] by [-2,2]

54. $(-2, -3)$

55. $(1, 7)$

56. $\sqrt{37}$

57. $\sqrt{10}$

58. (a) $\mathbf{u} + \mathbf{v} = (1, 7)$

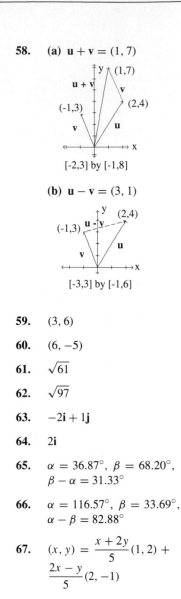

[-2,3] by [-1,8]

(b) $\mathbf{u} - \mathbf{v} = (3, 1)$

[-3,3] by [-1,6]

59. $(3, 6)$

60. $(6, -5)$

61. $\sqrt{61}$

62. $\sqrt{97}$

63. $-2\mathbf{i} + 1\mathbf{j}$

64. $2\mathbf{i}$

65. $\alpha = 36.87°,\ \beta = 68.20°,$
$\beta - \alpha = 31.33°$

66. $\alpha = 116.57°,\ \beta = 33.69°,$
$\alpha - \beta = 82.88°$

67. $(x, y) = \dfrac{x + 2y}{5}(1, 2) + \dfrac{2x - y}{5}(2, -1)$

68. $\mathbf{r} = (52.52, -528.19);$ speed $=$ 530.79 mph; direction: 5.68° east of south

69. $\mathbf{r} = (90.14, -424.82);$ the plane actually flies 11.98° east of south at 434.28 mph

70. See Exercise 42, Section 8.5 for the solution. Velocity:
$\mathbf{r} = (-17.95, -5.96);$ speed $=$

18.92 mph; direction: 18.37°
south of west

CHAPTER 9

SECTION 9.1

1. (1, 0) to (6, 5)

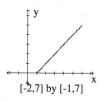

[-2,7] by [-1,7]

2. (−2, −3) to (3, 2)

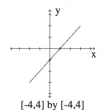

[-4,4] by [-4,4]

3. (5, 4) to (−7, 8)

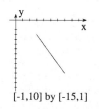

[-8,6] by [-2,10]

4. (3, −3) to (7, −11)

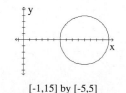

[-1,10] by [-15,1]

5. C is part of a hyperbola in QI with a VA at $x = 0$

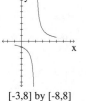

[-2,7] by [-1,10]

6. C is part of a hyperbola with a VA at $x = 2$

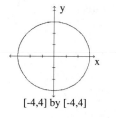

[-3,8] by [-8,8]

7. C begins and ends at (3, 0); circle

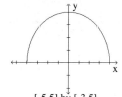

[-4,4] by [-4,4]

8. (−4, 0) to (4, 0)

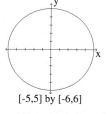

[-5,5] by [-3,5]

9. line $y = \dfrac{17}{3} - \dfrac{x}{3}$

10. line segment (11, 0) to (−4, 5) on $y = -\dfrac{1}{3}x + \dfrac{11}{3}$

11. part of a parabola $y = \dfrac{1}{4}x^2$ from (−6, 9) to (2, 1) with vertex (0, 0)

12. part of a parabola $y = \dfrac{1}{9}(x+1)^2 + 2$ from (−4, 3) to (11, 18) with vertex (−1, 2)

13. semicircle $y = \sqrt{9 - x^2}$ in QI and QII with radius 3 and center (0, 0)

14. three-quarters of a circle $x^2 + y^2 = 4$ in QI, QIV, and QIII with radius 2 and center (0, 0)

15. $x(t) = t,\ y(t) = 3t^2 - 4t + 5$

16. $x(t) = t,\ y(t) = 9t^3 - 3t + 4$

17. $x(t) = t,\ y(t) = 7 + t^2 - 3t$

18. $x(t) = t,\ y(t) = \sin t + 4t^2$

19. 1, 2, 3, 4, 5, 6, and 8

20. 1, 2, 3, 4, 5, and 6

21. Domain = [−2, 3]; range = [−3, 2]

22. Domain = [−7, 5]; range = [4, 8]

23. Domain = (0, 5]; range $= \left[\dfrac{2}{5}, \infty\right)$

24. Domain = [−3, 3]; range = [−3, 3]

25. $x(t) = 5\cos t,\ y(t) = 5\sin t,\ t$ in $[0, 2\pi]$

[-5,5] by [-6,6]

26. $x(t) = 10 + 4\cos t,$ $y(t) = 4\sin t,\ t$ in $[0, 2\pi]$

[-1,15] by [-5,5]

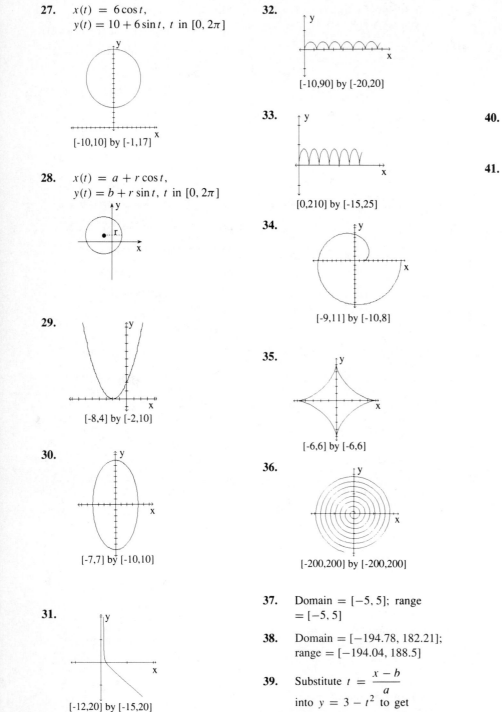

27. $x(t) = 6\cos t,$
$y(t) = 10 + 6\sin t,$ t in $[0, 2\pi]$

[-10,10] by [-1,17]

28. $x(t) = a + r\cos t,$
$y(t) = b + r\sin t,$ t in $[0, 2\pi]$

29.

[-8,4] by [-2,10]

30.

[-7,7] by [-10,10]

31.

[-12,20] by [-15,20]

32.

[-10,90] by [-20,20]

33.

[0,210] by [-15,25]

34.

[-9,11] by [-10,8]

35.

[-6,6] by [-6,6]

36.

[-200,200] by [-200,200]

37. Domain $= [-5, 5]$; range $= [-5, 5]$

38. Domain $= [-194.78, 182.21]$; range $= [-194.04, 188.5]$

39. Substitute $t = \dfrac{x - b}{a}$ into $y = 3 - t^2$ to get

$y = 3 - \left(\dfrac{x - b}{a}\right)^2 =$
$-\dfrac{1}{a^2}(x - b)^2 + 3$; domain
$= [ac + b, ad + b]$ for t in $[c, d]$

40. $y = 4$

41. **(a)** $a = 1$; period: 2π;
maximum: $y = 2$

$a = 1$

[-5,12] by [-3,4]

(b) $a = 2$; period: 4π;
maximum: $y = 4$

$a = 2$

[-5,20] by [-3,7]

(c) $a = 3$; period: 6π;
maximum: $y = 6$

$a = 3$

[-5,30] by [-10,15]

(d) $a = 4$; period: 8π;
maximum: $y = 8$

$a = 4$

[-5,32] by [-10,15]

(e) $a = 5$; period: 10π;
maximum: $y = 10$

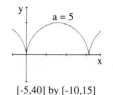

[-5,40] by [-10,15]

42. Conjecture: A cycloid
$x(t) = a(t - \sin t)$ and
$y(t) = a(1 - \cos t)$ has a period
of $2a\pi$ and a maximum value
of $y = 2a$ for an arbitrary value
of a.

43. When $t = 0$,
$x = 0 \cdot x_2 + (1 - 0)x_1 = x_1$ and
$y = 0 \cdot y_2 + (1-0)y_1 = y_1$. When
$t = 1$, $x = 1 \cdot x_2 + (1-1)x_1 = x_2$
and $y = 1 \cdot y_2 + (1 - 1)y_1 = y_2$

44. When $t = 0.5$,
$x = 0.5x_2 + 0.5x_1$ and
$y = 0.5y_2 + 0.5y_1$.
$M = \left(\dfrac{x_1 + x_2}{2}, \dfrac{y_1 + y_2}{2} \right) =$
$\left(\dfrac{1}{2}x_1 + \dfrac{1}{2}x_2, \dfrac{1}{2}y_1 + \dfrac{1}{2}y_2 \right) =$
(x, y).

45. $x_t - x_1 = t(x_2 - x_1)$;
$y_t - y_1 = t(y_2 - y_1)$; $P_1 P_t =$
$\sqrt{(x_t - x_1)^2 + (y_t - y_1)^2} =$
$t\sqrt{(x_2 - x_1)^2 + (y_2 - y_1)^2}$.
Then $\dfrac{P_1 P_t}{P_1 P_2} =$
$\dfrac{t\sqrt{(x_2 - x_1)^2 + (y_2 - y_1)^2}}{\sqrt{(x_2 - x_1)^2 + (y_2 - y_1)^2}} = t$

46. $x = 4t - 2$; $y = 3t + 1$

47. $P(-1.5, -0.25)$

48. $P(-1.48, 3.32)$

SECTION 9.2

1. $(\sqrt{2}, 45°)$

2. $(10, 180°)$

3. $(5, 126.87°)$

4. $(\sqrt{29}, 111.80°)$

5. $(0, 0)$

6. $(\sqrt{3}, 1)$

7. $\left(\dfrac{3\sqrt{2}}{2}, \dfrac{-3\sqrt{2}}{2} \right)$

8. $(1, \sqrt{3})$

9. $\left(\dfrac{-3\sqrt{2}}{2}, \dfrac{3\sqrt{2}}{2} \right)$

10. $(1.93, 0.52)$
Answers will vary for Exercises
11–18.

11. $(2\sqrt{2}, 135°)$, $\left(2\sqrt{2}, \dfrac{11\pi}{4} \right)$,
$(-2\sqrt{2}, -45°)$

12. $(4, -30°)$, $(4, 330°)$, $(-4, 150°)$

13. $(2, 150°)$, $(2, 510°)$, $(-2, -30°)$

14. $(5.66, 225°)$, $(5.66, -135°)$,
$(-5.66, 45°)$

15. $(\sqrt{13}, -56.31°)$,
$(\sqrt{13}, 303.69°)$
$(-\sqrt{13}, 123.69°)$

16. $(\sqrt{17}, 75.96°)$, $(\sqrt{17}, 435.96°)$,
$(-\sqrt{17}, 255.96°)$

17. $(\sqrt{10}, 198.43°)$,
$(\sqrt{10}, -161.57°)$,
$(-\sqrt{10}, 18.43°)$

18. $(3.61, 123.69°)$,
$(3.61, 483.69°)$,
$(-3.61, 303.69°)$

19.

[-5,5] by [-3,5]

20.

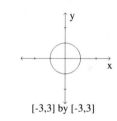

[-10,10] by [-10,10]

21. The graph of $\theta = -\dfrac{\pi}{2}$ is the
y-axis.

22.

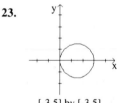

[-3,3] by [-3,3]

23.

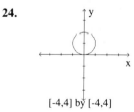

[-3,5] by [-3,5]

24.

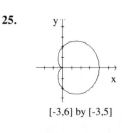

[-4,4] by [-4,4]

25.

[-3,6] by [-3,5]

26.

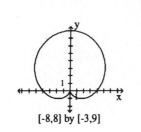

[-8,8] by [-3,9]

27.

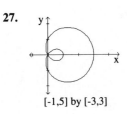

[-1,5] by [-3,3]

28.

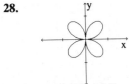

[-2,2] by [-2,2]

29. $x(t) = 3t \cos t,\ y(t) = 3t \sin t,\ t$ in $[0, 10\pi]$

[-150,150] by [-150,150]

30. $x(t) = 5 \sin t \cos t,$
$y(t) = 5 \sin t \sin t,\ t$ in $[-\pi, \pi]$

[-4,4] by [-1,7]

31. $x(t) = 5 \sin 2t \cos t,$
$y(t) = 5 \sin 2t \sin t,\ t$ in $[0, \pi]$

[-5,5] by [-5,5]

32. $x(t) = 5 \sin 3t \cos t,$
$y(t) = 5 \sin 3t \sin t,\ t$ in $[0, 2\pi]$

[-6,6] by [-6,6]

33. **(a)**

[-3,4] by [-1,5]

(b)

[-11,5] by [-3,8]

(c)

[-20,20] by [-20,20]

(d)

[-20,20] by [-20,20]

(e)

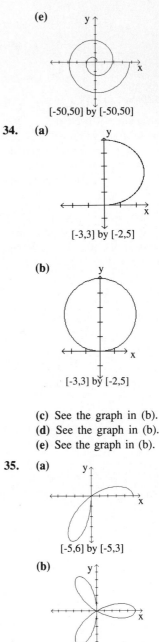

[-50,50] by [-50,50]

34. **(a)**

[-3,3] by [-2,5]

(b)

[-3,3] by [-2,5]

(c) See the graph in (b).
(d) See the graph in (b).
(e) See the graph in (b).

35. **(a)**

[-5,6] by [-5,3]

(b)

[-6,6] by [-6,6]

(c) See the graph in (b).
(d) See the graph in (b).
(e) See the graph in (b).

36. **(a)**

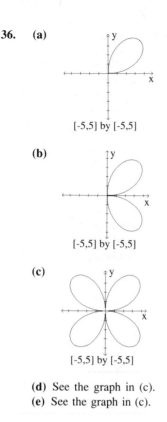

[-5,5] by [-5,5]

(b)

[-5,5] by [-5,5]

(c)

[-5,5] by [-5,5]

(d) See the graph in (c).
(e) See the graph in (c).

37. No interval gives a "complete" graph.

[-30,30] by [-30,30]

38. No interval gives a "complete" graph.

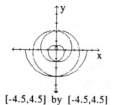

[-4.5,4.5] by [-4.5,4.5]

39. $0 \le \theta \le 2\pi$

[-6,2] by [-5,4]

40. $0 \le \theta \le \pi$

[-3,3] by [-3,3]

41. $0 \le \theta \le 2\pi$

[-3,3] by [-4,4]

42. $0 \le \theta \le 2\pi$

[-5,5] by [-5,5]

43. $2n$ petals for even n and n petals for odd n

44. $2n$ petals for even n and n petals for odd n

45. All of the graphs are 4-petaled roses; the length of the petals is determined by a

46. All of the graphs are 3-petaled roses; the length of the petals is determined by a

47. 5 units

48. 5 units

49. 8 units

50. 8 units

51.

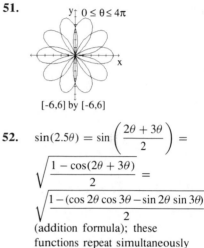

$0 \le \theta \le 4\pi$

[-6,6] by [-6,6]

52. $\sin(2.5\theta) = \sin\left(\dfrac{2\theta + 3\theta}{2}\right) =$

$\sqrt{\dfrac{1 - \cos(2\theta + 3\theta)}{2}} =$

$\sqrt{\dfrac{1 - (\cos 2\theta \cos 3\theta - \sin 2\theta \sin 3\theta)}{2}}$

(addition formula); these functions repeat simultaneously after a period of 4π

53.

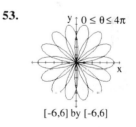

$0 \le \theta \le 4\pi$

[-6,6] by [-6,6]

54. $r = a \sin\left(\dfrac{m}{n}\theta\right)$ has m petals and a period of $n\pi$; if either m or n is even, the graph has $2m$ petals and a period of $2n\pi$

55. $\left(\dfrac{a\sqrt{2}}{2}, 45°\right)$, $\left(\dfrac{a\sqrt{2}}{2}, 135°\right)$,

$\left(\dfrac{a\sqrt{2}}{2}, 225°\right)$, and

$\left(\dfrac{a\sqrt{2}}{2}, 315°\right)$

56. $(a, 0)$, $(a, 72°)$, $(a, 144°)$, $(a, 216°)$, and $(a, 288°)$

57.

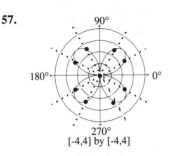

[-4,4] by [-4,4]

58. $(0, 0)$, $(2.5, 2.5)$

59. $(0, 0)$, $(2.5, 2.5)$, $(-2.5, 2.5)$

60. $(0, 0)$, $(1.48, 1.72)$, $(0.86, 0.27)$, $(1.24, 2.35)$, $(-1.48, 1.72)$ $(-1.24, 2.35)$, $(-0.86, 0.27)$

61. $(0, 0)$, $(0.95, 0.17)$, $(0.95, -0.17)$, $(2.27, 1.43)$, $(2.27, -1.43)$, $(0.43, 1.49)$, $(0.43, -1.49)$, $(-0.096, 0.71)$, $(-0.096, -0.71)$

62. $d(P_1, P_2) =$
$\sqrt{(x_2 - x_1)^2 + (y_2 - y_1)^2} =$
$[x_2^2 - 2x_1x_2 + x_1^2 +$
$y_2^2 - 2y_1y_2 + y_1^2]^{1/2} =$
$[(r_2^2 \cos^2 \theta_2 - 2r_1 \cos \theta_1 r_2 \cos \theta_2 +$
$r_1^2 \cos^2 \theta_1) + (r_2^2 \sin^2 \theta_2 -$
$2r_1 \sin \theta_1 r_2 \sin \theta_2 +$
$r_1^2 \sin^2 \theta_1)]^{1/2} =$
$[r_2^2 + r_1^2 - 2r_1r_2(\cos \theta_1 \cos \theta_2 +$
$\sin \theta_1 \sin \theta_2)]^{1/2} =$
$\sqrt{r_1^2 + r_2^2 - 2r_1r_2 \cos(\theta_2 - \theta_1)}$

SECTION 9.3

1.

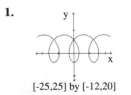

[-25,25] by [-12,20]

2.

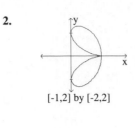

[-1,2] by [-2,2]

3.

[-5,9] by [-6,4]

4.

[-300,300] by [-3000,100]

5. The arrow hits in 9.52 sec at a horizontal distance of 1306.08 ft.

6. 93.19 ft/sec

7. $x(t) = 58t \cos 41°$, $y(t) = -16t^2 + 58t \sin 41°$

8.

[-100,200] by [-80,50]

9. QI

10. 2.38 sec

11. 22.62 ft at 1.19 sec

12. 104.10 ft

13. 525.73 ft in 4.89 sec

14. clears the fence

15. Place Chris at the origin; then Linda is at $(78, 0)$. Chris: $x_1(t) = 41t \cos 39°$; $y_1(t) = -16t^2 + 41t \sin 39°$. Linda: $x_2(t) = 78 - 45t \cos 44°$; $y_2(t) = -16t^2 + 45t \sin 44°$

16.

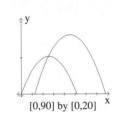

[0,90] by [0,20]

17. $0 \le t \le 1.95$ sec

18. Chris: The ball reaches a maximum height of 10.4 ft; it travels a horizontal distance of 51.38 ft in 1.61 sec. Linda: The ball reaches a maximum height of 15.27 ft; it travels a horizontal distance of 63.24 ft in 1.95 sec. Chris's ball hits first.

19. Answers will vary.

20. 6.60 ft apart at 1.2 sec

21. $x(t) = 13 \cos \left(\dfrac{\pi}{2} - \pi t \right)$ and $y(t) = 13 \sin \left(\dfrac{\pi}{2} - \pi t \right) + 13$

22. 2450.44 ft

23. Ferris wheel: $x_A(t) = 26 \cos \dfrac{\pi}{6} t$; $y_A(t) = 26 + 26 \sin \dfrac{\pi}{6} t$; Ball: $x_B(t) = 62 + (76 \cos 128°)t = 62 - 46.79t$; $y_B(t) = -16t^2 + (76 \sin 128°)t = -16t^2 + 59.89t$ (Answers will vary.)

24. The ball comes within 1.36 ft of Janice.

25. Answers will vary. It appears that a distance of 59 ft brings the ball closest to Janice.

26. It appears that an angle of about 50° brings the ball closest to Janice.

27.

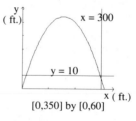

[-50,320] by [-40,120]

28. The ball lands about 70 yd downfield from the punter, on the opponents' 15-yard line.

29. The maximum height is 77.59 ft.

30. "hang time" is 4.4 sec

31. The hit is not a home run.

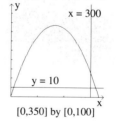

[0,350] by [0,60]

32. The hit is a home run.

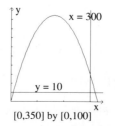

[0,350] by [0,100]

33. The hit is a home run.

[0,350] by [0,100]

34. The hit is a home run.

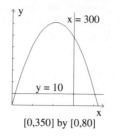

[0,350] by [0,80]

35. The hit is not a home run.

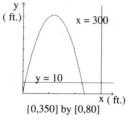

{0,350] by [0,80]

36. The hit is not a home run.

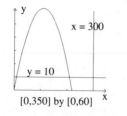

[0,350] by [0,60]

37.

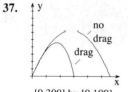

[0,300] by [0,100]

38. With drag: 45.7 ft at 1.6 sec; without drag: 63.1 ft at 2 sec

39. With drag: 150.47 ft at 3.4 sec; without drag: 280.49 ft at 4 sec

40. (a) With drag: $x_1(t) = \frac{95}{0.01}(\cos 42°)(1 - e^{-0.01t})$;

$y_1(t) = \frac{95}{0.01}(\sin 42°)(1 -$

$e^{-0.01t}) + \frac{32}{0.01^2}(1 -$

$0.01t - e^{-0.01t})$ Without drag: $x_2(t) = 95t \cos 42°$; $y_2(t) = -16t^2 + 95t \sin 42°$

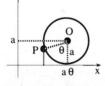

[0,300] by [0,100]

(b) With drag: maximum height of 62.3 ft at 2 sec; horizontal distance of 273 ft at 3.9 sec. Without drag: maximum height of 63.1 ft at 2 sec; horizontal distance of 280 ft at 4 sec.

41. coordinates of point P are $(a\theta - a \sin \theta, a - a \cos \theta)$; the parametric equations are $x(\theta) = a(\theta - \sin \theta)$ and $y(\theta) = a(1 - \cos \theta)$

42. Answers will vary.

43. Answers will vary.

44. $x_1(t) = 400$ represents the horizontal distance of the fence from home plate; $y_2(t) = 20$ represents the height of the fence; $\frac{T}{T_{max}}$ is the portion of the total time available that has elapsed at time T. Thus $y_1(t)$ represents the corresponding proportion of the maximum vertical distance and $x_2(t)$ represents the corresponding proportion of the maximum horizontal distance for the ball's flight.

SECTION 9.4

1. vertex: $(0, 0)$; focus: $(0, 6)$; directrix: $y = -6$; line of sym.: $x = 0$

2. vertex: $(0, 0)$; focus: $(2, 0)$; directrix: $x = -2$; line of sym.: $y = 0$

3. vertex: $(3, -1)$; focus: $(3, 2)$; directrix: $y = -4$; line of sym.: $x = 3$

4. vertex: $(-1, 3)$; focus: $(-1, 4.5)$; directrix: $y = 1.5$; line of sym.: $x = -1$

5. vertex: $(3, 2)$; focus: $\left(3, 2 - \dfrac{1}{64}\right) = \left(3, \dfrac{127}{64}\right)$; directrix: $y = 2 + \dfrac{1}{64} = \dfrac{129}{64}$; line of sym.: $x = 3$

6. vertex: $(3, -5)$; focus: $(8.5, -5)$; directrix: $x = -2.5$; line of sym.: $y = -5$

7. vertex: $(-4, 1)$; focus: $(-4, -0.5)$; directrix: $y = 2.5$; line of sym.: $x = -4$

8. vertex: $(0, 7)$; focus: $(3, 7)$; directrix: $x = -3$; line of sym.: $y = 7$

9. transformations to $y^2 = x$: a horizontal shrink by a factor of $\dfrac{1}{12}$, followed by a horizontal shift left 3, then a vertical shift down 4

10. transformations to $x^2 = y$: a vertical stretch by a factor of 18, followed by a horizontal shift right 3, then a vertical shift up 7

11. transformations to $x^2 = y$: a vertical stretch by a factor of 6, followed by a horizontal shift left 4, then a vertical shift up 2

12. transformations to $y^2 = x$: a horizontal stretch by a factor of 5, followed by a horizontal shift right 4, then a vertical shift down 6

13. Equation: $(x - 2)^2 = -8(y - 1)$; vertex: $(2, 1)$; line of sym.: $x = 2$

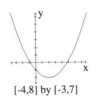

[-7,10] by [-7,2]

14. Equation: $(x - 2)^2 = 4(y + 2)$; vertex: $(2, -2)$; line of sym.: $x = 2$

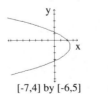

[-4,8] by [-3,7]

15. Equation: $(y + 1)^2 = -2(x - 2.5)$; vertex: $(2.5, -1)$; line of sym.: $y = -1$

[-7,4] by [-6,5]

16. Equation: $(y + 1)^2 = 20(x - 2)$; focus: $(7, -1)$; line of sym.:

$y = -1$

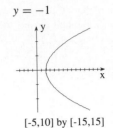

[-5,10] by [-15,15]

17. Equation: $(y - 2)^2 = -8(x - 3)$; focus: $(1, 2)$; line of sym.: $y = 2$

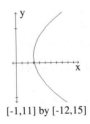

[-8,6] by [- 6,12]

18. Equation: $(y - 2)^2 = 32(x - 3)$; focus: $(11, 2)$; line of sym.: $y = 2$

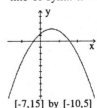

[-1,11] by [-12,15]

19. Equation: $(x - 3)^2 = -12(y - 2)$; focus: $(3, -1)$; line of sym.: $x = 3$

[-7,15] by [-10,5]

20. Equation: $(x - 3)^2 = 28(y - 2)$; focus: $(3, 9)$; line of sym.:

$x = 3$

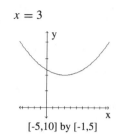

[-5,10] by [-1,5]

21. Std. form: $(x + 1)^2 =$ $4\left(\dfrac{1}{4}\right)(y - 2)$; vertex: $(-1, 2)$;

focus: $\left(-1, \dfrac{9}{4}\right)$; directrix:

$y = \dfrac{7}{4}$; line of sym.: $x = -1$

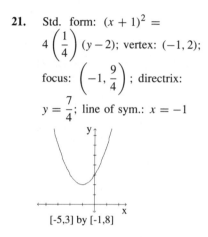

[-5,3] by [-1,8]

22. Std. form: $(x - 1)^2 =$ $2\left(y - \dfrac{7}{6}\right)$; vertex: $\left(1, \dfrac{7}{6}\right)$;

focus: $\left(1, \dfrac{5}{3}\right)$; directrix:

$y = \dfrac{2}{3}$; line of sym.: $x = 1$

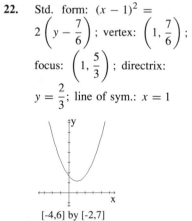

[-4,6] by [-2,7]

23. Apply a vertical stretch by a factor of 4 to $y = x^2$ to get $y = 4x^2$; apply a horizontal shrink by a factor of $\dfrac{1}{2}$ to $y = x^2$ to get $y = (2x)^2$. But

$y = (2x)^2 = 4x^2$, so they are equivalent.

24. Apply a vertical shrink by a factor of $\dfrac{1}{2}$ to $x = y^2$ to get $x = (2y)^2$; apply a horizontal stretch by a factor of 4 to $x = y^2$ to get $x = 4y^2$. But $x = 4y^2 = (2x)^2$, so they are equivalent.

25. Major axis: $(1, -1)$, $(1, -5)$; minor axis: $(1 - \sqrt{2}, -3)$, $(1 + \sqrt{2}, -3)$; center: $(1, -3)$; foci: $(1, \sqrt{2} - 3)$, $(1, -\sqrt{2} - 3)$; lines of sym.: $x = 1$, $y = -3$

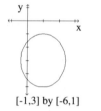

[-1,3] by [-6,1]

26. Major axis: $(-7, 1)$, $(1, 1)$; minor axis: $(-3, -1)$, $(-3, 3)$; center: $(-3, 1)$; foci: $(-3 - 2\sqrt{3}, 1)$, $(-3 + 2\sqrt{3}, 1)$; lines of sym.: $x = -3$, $y = 1$

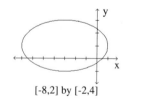

[-8,2] by [-2,4]

27. Major axis: $(-2 - \sqrt{5}, 1)$, $(-2 + \sqrt{5}, 1)$; minor axis: $\left(-2, 1 + \dfrac{1}{\sqrt{2}}\right)$, $\left(-2, 1 - \dfrac{1}{\sqrt{2}}\right)$; center: $(-2, 1)$; foci: $\left(-2 - \dfrac{3\sqrt{2}}{2}, 1\right)$,

$\left(-2 + \dfrac{3\sqrt{2}}{2}, 1\right)$; lines of sym.: $x = -2$, $y = 1$

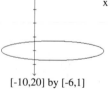

[-5,1] by [-1,3]

28. Major axis: $(4 - 8\sqrt{2}, -4)$, $(4 + 8\sqrt{2}, -4)$; minor axis: $\left(4, -4 - \dfrac{\sqrt{2}}{2}\right)$,

$\left(4, -4 + \dfrac{\sqrt{2}}{2}\right)$; center: $(4, -4)$; foci:

$\left(4 - \dfrac{\sqrt{510}}{2}, -4\right)$,

$\left(4 + \dfrac{\sqrt{510}}{2}, -4\right)$; lines of sym.: $x = 4$, $y = -4$

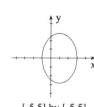

[-10,20] by [-6,1]

29.

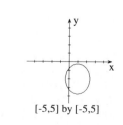

[-5,5] by [-5,5]

30.

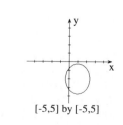

[-5,5] by [-5,5]

31.

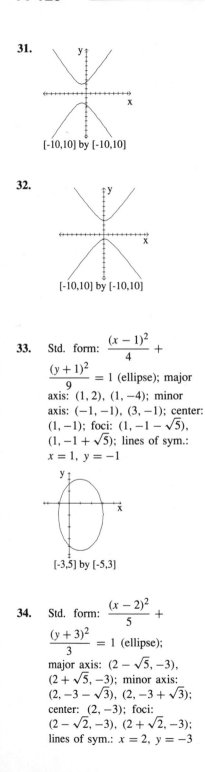

[-10,10] by [-10,10]

[-2,5] by [-6,1]

asymptotes: $y - 1 = \pm\dfrac{3}{5}(x + 3)$

[-10,4] by [-5,7]

32.

[-10,10] by [-10,10]

33. Std. form: $\dfrac{(x-1)^2}{4} + \dfrac{(y+1)^2}{9} = 1$ (ellipse); major axis: $(1, 2)$, $(1, -4)$; minor axis: $(-1, -1)$, $(3, -1)$; center: $(1, -1)$; foci: $(1, -1 - \sqrt{5})$, $(1, -1 + \sqrt{5})$; lines of sym.: $x = 1$, $y = -1$

[-3,5] by [-5,3]

34. Std. form: $\dfrac{(x-2)^2}{5} + \dfrac{(y+3)^2}{3} = 1$ (ellipse); major axis: $(2 - \sqrt{5}, -3)$, $(2 + \sqrt{5}, -3)$; minor axis: $(2, -3 - \sqrt{3})$, $(2, -3 + \sqrt{3})$; center: $(2, -3)$; foci: $(2 - \sqrt{2}, -3)$, $(2 + \sqrt{2}, -3)$; lines of sym.: $x = 2$, $y = -3$

35. Std. form: $\dfrac{(x-2)^2}{4} - \dfrac{(y-1)^2}{9} = 1$ (hyperbola); center: $(2, 1)$; foci: $(2 - \sqrt{13}, 1)$, $(2 + \sqrt{13}, 1)$; vertices: $(4, 1)$, $(0, 1)$; lines of sym.: $x = 2$, $y = 1$; asymptotes: $y - 1 = \pm\dfrac{3}{2}(x - 2)$

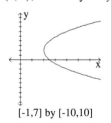

[-8,10] by [-8,9]

36. Std. form: $(y - 2)^2 = 8(x - 2)$ (parabola); vertex: $(2, 2)$; focus: $(4, 2)$; line of sym.: $y = 2$

[-1,7] by [-10,10]

37. Std. form: $\dfrac{(y-1)^2}{9} - \dfrac{(x+3)^2}{25} = 1$ (hyperbola); center: $(-3, 1)$; foci: $(-3, 1 - \sqrt{34})$, $(-3, 1 + \sqrt{34})$; vertices: $(-3, 4)$, $(-3, -2)$; lines of sym.: $x = -3$, $y = 1$;

38. Std. form: $\dfrac{(x+3)^2}{16} + \dfrac{(y-1)^2}{9} = 1$ (ellipse); center: $(-3, 1)$; foci: $(-3 - \sqrt{7}, 1)$, $(-3 + \sqrt{7}, 1)$; major axis: $(-7, 1)$, $(1, 1)$; minor axis: $(-3, -2)$, $(-3, 4)$; lines of sym.: $x = -3$, $y = 1$

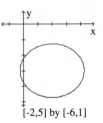

[-9,3] by [-3,5]

39. Major axis: $(-\sqrt{58}, 0)$, $(\sqrt{58}, 0)$; minor axis: $(0, -7)$, $(0, 7)$; std. form: $\dfrac{x^2}{58} + \dfrac{y^2}{49} = 1$

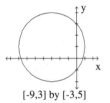

[-9,9] by [-9,9]

40. Major axis: $\left(0, -\dfrac{\sqrt{145}}{2}\right)$, $\left(0, \dfrac{\sqrt{145}}{2}\right)$; minor axis: $\left(-\dfrac{9}{2}, 0\right)$, $\left(\dfrac{9}{2}, 0\right)$; std. form: $\dfrac{x^2}{81/4} + \dfrac{y^2}{145/4} = 1$

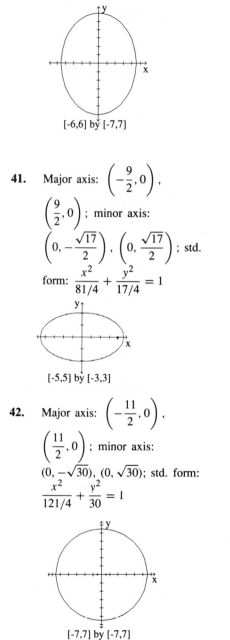

[-6,6] by [-7,7]

41. Major axis: $\left(-\dfrac{9}{2}, 0\right)$, $\left(\dfrac{9}{2}, 0\right)$; minor axis: $\left(0, -\dfrac{\sqrt{17}}{2}\right)$, $\left(0, \dfrac{\sqrt{17}}{2}\right)$; std. form: $\dfrac{x^2}{81/4} + \dfrac{y^2}{17/4} = 1$

[-5,5] by [-3,3]

42. Major axis: $\left(-\dfrac{11}{2}, 0\right)$, $\left(\dfrac{11}{2}, 0\right)$; minor axis: $(0, -\sqrt{30})$, $(0, \sqrt{30})$; std. form: $\dfrac{x^2}{121/4} + \dfrac{y^2}{30} = 1$

[-7,7] by [-7,7]

43. Major axis: $(1, 0)$, $(1, 12)$; minor axis: $(1 - 3\sqrt{3}, 6)$, $(1 + 3\sqrt{3}, 6)$; std. form:

$\dfrac{(x-1)^2}{27} + \dfrac{(y-6)^2}{36} = 1$

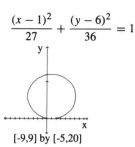

[-9,9] by [-5,20]

44. Major axis: $(1, -4 - \sqrt{29})$, $(1, -4 + \sqrt{29})$; minor axis: $(-4, -4)$, $(6, -4)$; std form: $\dfrac{(x-1)^2}{25} + \dfrac{(y+4)^2}{29} = 1$

[-5,7] by [-10,2]

45. Std. form: $\dfrac{x^2}{4} - \dfrac{y^2}{5} = 1$

[-9,9] by [-8,8]

46. Std. form: $\dfrac{y^2}{16} - \dfrac{x^2}{48} = 1$

[-7,8] by [-9,7]

47. Std. form: $\dfrac{x^2}{9/4} - \dfrac{(y-2)^2}{27/4} = 1$

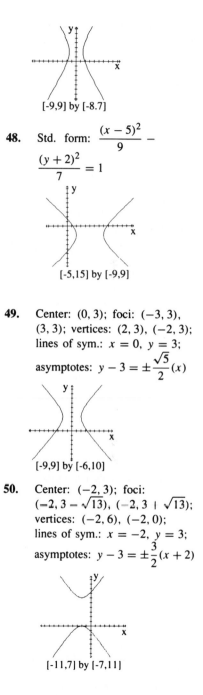

[-9,9] by [-8,7]

48. Std. form: $\dfrac{(x-5)^2}{9} - \dfrac{(y+2)^2}{7} = 1$

[-5,15] by [-9,9]

49. Center: $(0, 3)$; foci: $(-3, 3)$, $(3, 3)$; vertices: $(2, 3)$, $(-2, 3)$; lines of sym.: $x = 0$, $y = 3$; asymptotes: $y - 3 = \pm\dfrac{\sqrt{5}}{2}(x)$

[-9,9] by [-6,10]

50. Center: $(-2, 3)$; foci: $(-2, 3 - \sqrt{13})$, $(-2, 3 + \sqrt{13})$; vertices: $(-2, 6)$, $(-2, 0)$; lines of sym.: $x = -2$, $y = 3$; asymptotes: $y - 3 = \pm\dfrac{3}{2}(x + 2)$

[-11,7] by [-7,11]

51. Center: $(3, 1)$; foci: $(3, 1 - \sqrt{13})$, $(3, 1 + \sqrt{13})$;

vertices: $(3, 4)$, $(3, -2)$; lines of sym.: $x = 3$, $y = 1$; asymptotes: $y - 1 = \pm\dfrac{3}{2}(x - 3)$

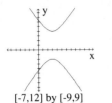

[-7,12] by [-9,9]

52. Center: $(2, -4)$; foci:
$$\left(2 - \frac{\sqrt{13}}{6}, -4\right),$$
$$\left(2 + \frac{\sqrt{13}}{6}, -4\right); \text{ vertices:}$$
$$\left(2\frac{1}{2}, -4\right), \left(1\frac{1}{2}, -4\right); \text{ lines}$$
of sym.: $x = 2$, $y = -4$;
asymptotes: $y + 4 = \pm\dfrac{2}{3}(x - 2)$

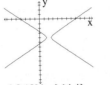

[-8,12] by [-14,4]

53. The graphs do not intersect.

54. $(7.89, -0.53)$, $(20.11, 2.53)$

55. parabola; vertex: $\left(\dfrac{13}{4}, 1\right)$;
line of sym.: $y = 1$; focus:
$\left(\dfrac{9}{4}, 1\right)$

56. hyperbola; center: $(-1, 1)$;
vertices: $(-1, 4)$, $(-1, -2)$;
lines of sym.: $x = -1$,
$y = 1$; foci: $(-1, 1 + \sqrt{13})$,
$(-1, 1 - \sqrt{13})$

57. ellipse; center: $(4, -8)$;
vertices: $(5, -8)$, $(3, -8)$,
$(4, -10)$, $(4, -6)$; lines of

sym.: $x = 4$, $y = -8$; foci:
$(4, -8 - \sqrt{3})$, $(4, -8 + \sqrt{3})$

58. parabola: vertex: $\left(-\dfrac{2}{3}, \dfrac{1}{3}\right)$;
line of sym.: $y = \dfrac{1}{3}$; focus:
$\left(-\dfrac{5}{12}, \dfrac{1}{3}\right)$

59. hyperbola; center: $(1, -3)$;
vertices: $(-1, -3)$, $(3, -3)$;
lines of sym.: $x = 1$,
$y = -3$; foci: $(1 - \sqrt{68}, -3)$,
$(1 + \sqrt{68}, -3)$

60. ellipse; center: $(-3, 4)$;
vertices: $(-3 - \sqrt{3}, 4)$,
$(-3 + \sqrt{3}, 4)$, $(-3, 4 - \sqrt{2})$,
$(-3, 4 + \sqrt{2})$; foci: $(-4, 4)$,
$(-2, 4)$; lines of sym.: $x = -3$,
$y = 4$

61. parabola; vertex: $\left(\dfrac{3}{2}, \dfrac{7}{2}\right)$;
line of sym.: $x = \dfrac{3}{2}$; focus:
$\left(\dfrac{3}{2}, \dfrac{23}{8}\right)$

62. ellipse: center: $(1, -2)$;
vertices: $(-1, -2)$, $(3, -2)$,
$(1, -5)$, $(1, 1)$; lines of
sym.: $x = 1$, $y = -2$; foci:
$(1, -2 - \sqrt{5})$, $(1, -2 + \sqrt{5})$

63. Let $h = 1$, $k = 1$, $a = 1$.

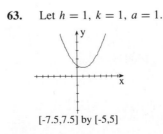

[-7.5,7.5] by [-5,5]

64. Let $a = 1$, $h = 0$, $b = 1$, $k = 0$.

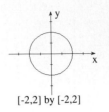

[-2,2] by [-2,2]

65. Let $a = 2$, $h = 1$, $b = 2$, $k = 1$.

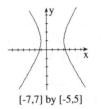

[-7,7] by [-5,5]

66. Let $a = 1$, $h = 0$, $k = 0$.

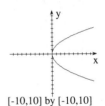

[-10,10] by [-10,10]

67. At the focus $\left(0, \dfrac{5}{2}\right)$, about 2.5
units from the vertex along the
line of symmetry

68. At the focus $\left(0, \dfrac{9}{2}\right)$, about 4.5
units from the vertex along the
line of symmetry

69. At the focus $(3, 0)$, about 3
units from the vertex along the
line of symmetry

70. At the focus $\left(\dfrac{15}{4}, 0\right)$, about
3.75 units from the vertex along
the line of symmetry

71. At the focus below the minor
axis, $\sqrt{69}$ in. below the minor
axis on the major axis

72. **(a)** The pool shark will always aim for the second focus, since a ball passing through this point will be reflected to the other focus (pocket). **(b)** If the pocket is at $(-\sqrt{5}, 0)$, the ball should be aimed at $(\sqrt{5}, 0)$

73. foci: $(0, 0)$ and $(6, 0)$

74. $\dfrac{(x-3)^2}{9/4} - \dfrac{y^2}{27/4} = 1$

75. Light is reflected from the parabolic surface toward the focus common to both conics. Then it is reflected from the hyperbolic surface toward the eyepiece at the second focus of the hyperbola.

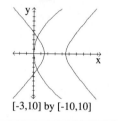

$[-3,10]$ by $[-10,10]$

76. foci: $(0, 0)$ and $\left(\dfrac{15}{4}, 0\right)$

77. $\dfrac{(x - 15/8)^2}{81/64} - \dfrac{y^2}{9/4} = 1$

78. See explanation for Exercise 75.

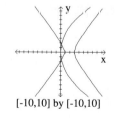

$[-10,10]$ by $[-10,10]$

79. Ellipse

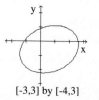

$[-3,3]$ by $[-4,3]$

80. Hyperbola

$[-10,10]$ by $[-10,10]$

81. Ellipse

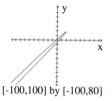

$[-8,12]$ by $[-10,7]$

82. Parabola

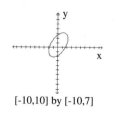

$[-100,100]$ by $[-100,80]$

83. Degenerate ellipse; no graph possible

84. Ellipse

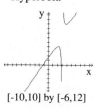

$[-10,10]$ by $[-10,7]$

85. Hyperbola

$[-10,10]$ by $[-6,12]$

86. Hyperbola

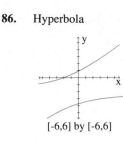

$[-6,6]$ by $[-6,6]$

87. Hyperbola

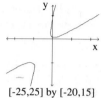

$[-25,25]$ by $[-20,15]$

88. Hyperbola

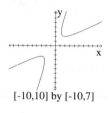

$[-10,10]$ by $[-10,7]$

89. $5x^2 - 4xy + 8y^2 + 6x - 24y + 9 = 0$

90. $3x^2 + 4xy - 6x - 1 = 0$

91. No graph is possible for $x^2 = -1$.

92. The line $x = 0$ is the graph of $x^2 = 0$.

93. The pair of lines $x = 1$ and $x = -1$ forms the graph of $x^2 = 1$.

94. The pair of lines $x = 0$ and $y = x$ forms the graph.

95. The point $(0, 0)$ is the graph of $x^2 + y^2 = 0$.

96. The pair of lines $y = x$ and $y = -x$ forms the graph.

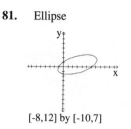

97. No graph is possible for $x^2 + y^2 = -1$.

98. a circle with center $\left(\frac{1}{2}, \frac{3}{2}\right)$ and radius 5

99. the line $y = x + 1$

100. $k = -1$; $y = 2x - 1$ is the tangent line

101. $k = \pm 2$; $y = 2x - 1$ and $y = -2x - 1$ are the tangent lines

102. Verify the line $x + 4y - 18 = 0$ passes through $P(2, 4)$: The slope of the tangent to $y = x^2$ at $P(2, 4)$ has slope $m_1 = 4$ and the slope of the normal line will be $m_2 = -\frac{1}{4}$. Since the slope of $x + 4y - 18 = 0$ is $m = -\frac{1}{4}$, it is the normal line.

SECTION 9.5

1. $e = \dfrac{\sqrt{3}}{2}$

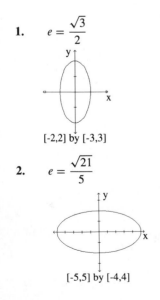

[-2,2] by [-3,3]

2. $e = \dfrac{\sqrt{21}}{5}$

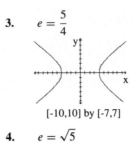

[-5,5] by [-4,4]

3. $e = \dfrac{5}{4}$

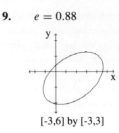

[-10,10] by [-7,7]

4. $e = \sqrt{5}$

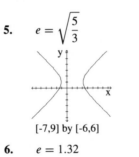

[-3,4] by [-4,4]

5. $e = \sqrt{\dfrac{5}{3}}$

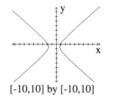

[-7,9] by [-6,6]

6. $e = 1.32$

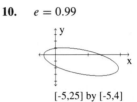

[-10,10] by [-10,10]

7. Center: $(1, 1)$ asymptotes:
$$y - 1 = \pm\sqrt{\frac{2}{3}}(x - 1)$$

8. Center: $\left(-\dfrac{1}{2}, -\dfrac{9}{16}\right)$
asymptotes:
$$y + \frac{9}{16} = \pm 0.87\left(x + \frac{1}{2}\right)$$

9. $e = 0.88$

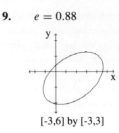

[-3,6] by [-3,3]

10. $e = 0.99$

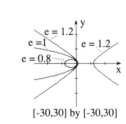

[-5,25] by [-5,4]

11.

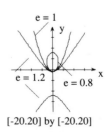

[-20,20] by [-20,20]

12.

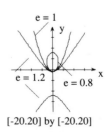

[-30,30] by [-30,30]

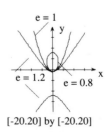

[-20,20] by [-20,20]

14.

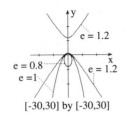

$e = 1.2$

$e = 0.8$ $e = 1.2$
$e = 1$

[-30,30] by [-30,30]

15. parabola; directrix: $x = -5$;
$0 \le \theta \le 2\pi$

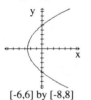

[-6,6] by [-8,8]

16. hyperbola; directrix: $y = -1$;
$0 \le \theta \le 2\pi$

[-6,6] by [-5,4]

17. ellipse; directrix: $x = -\dfrac{3}{2}$;
$0 \le \theta \le 2\pi$

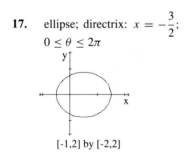

[-1,2] by [-2,2]

18. hyperbola; directrix: $y = 1$;
$0 \le \theta \le 2\pi$

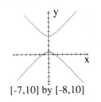

[-7,10] by [-8,10]

19. ellipse; directrix: $x = -\dfrac{8}{3}$;
$0 \le \theta \le 2\pi$

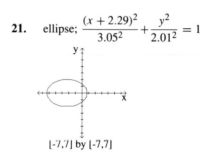

[-4,4] by [-4,4]

20. ellipse; directrix: $y = 2$;
$0 \le \theta \le 2\pi$

[-3,3] by [-3,3]

21. ellipse; $\dfrac{(x + 2.29)^2}{3.05^2} + \dfrac{y^2}{2.01^2} = 1$

[-7,7] by [-7,7]

22. hyperbola; $\dfrac{(y + 3/4)^2}{(1/4)^2} - \dfrac{x^2}{1/2} = 1$

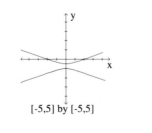

[-5,5] by [-5,5]

23. $e = \dfrac{1}{2}$; the graphs are the same

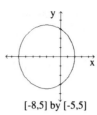

[-8,5] by [-5,5]

24. $e = 0.583$; the graphs are the
same

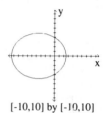

[-10,10] by [-10,10]

25. This is a hyperbola since $e > 1$.

[-10,5] by [-20,20]

26. $r = \dfrac{eh}{1 + e \cos\theta} = \dfrac{e[(a - ce)/e]}{1 + e \cos\theta} = \dfrac{a - ce}{1 + e \cos\theta}$;
substitute $e = \dfrac{c}{a}$ to get
$r = \dfrac{a - c(c/a)}{1 + (c/a) \cos\theta} = \dfrac{a^2 - c^2}{a + c \cos\theta}$; since $c^2 = a^2 - b^2$,
we get $c = \sqrt{a^2 - b^2}$,
$a^2 - c^2 = b^2$. Substituting
yields $r = \dfrac{b^2}{a + \sqrt{a^2 - b^2} \cos\theta}$

27. $e = \dfrac{2}{4} = \dfrac{1}{2}$; see Exercise 31 for
the graph

28. $e = \dfrac{1}{2}$; see Exercise 31 for the
graph

29. $e = \dfrac{1}{2}$; see Exercise 31 for the graph

30. $e = \dfrac{1}{2}$; see Exercise 31 for the graph

31. The graphs of the equations in Exercises 27 and 29 are the same, as are the graphs of the equations in Exercises 28 and 30.

27 & 29   28 & 30

[-10,10] by [-10,10]

Chapter 9 Review

1. C is the line segment from $(2, 0)$ to $(17, 5)$

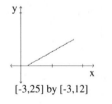

[-3,25] by [-3,12]

2. C is part of the hyperbola $y = \dfrac{6}{x}$ from $(2, 3)$ to $\left(8, \dfrac{3}{4}\right)$

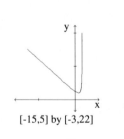

[-1,9] by [-2,4]

3. C is a circle centered at $(0, 0)$ with radius 4

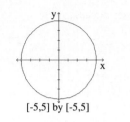

[-5,5] by [-5,5]

4. Exercises 1 and 2 are functions

5. For Exercise 1, domain $= [2, 17]$; range $= [0, 5]$; for Exercise 2, domain $= [2, 8)$; range $= \left[\dfrac{3}{4}, 3\right]$; for Exercise 3, domain $= [-4, 4]$; range $= [-4, 4]$

6.

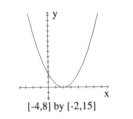

[-4,8] by [-2,15]

7.

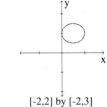

[-15,5] by [-3,22]

8.

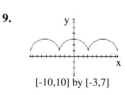

[-2,2] by [-2,3]

9.

[-10,10] by [-3,7]

10.

[-20,20] by [-20,15]

11. $x(t) = 5 + 4\cos t, \; y(t) = 4\sin t$

[-1,10] by [-4,5]

12.

[-6,6] by [-6,6]

13.

[-2,2] by [-0.5,3]

14.

[-2,7] by [-5,5]

15.

78°

[-5,5] by [-5,5]

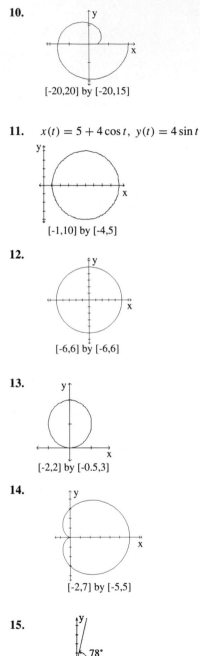

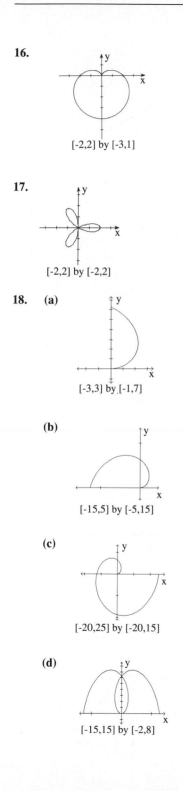

16.

[-2,2] by [-3,1]

17.

[-2,2] by [-2,2]

18. **(a)**

[-3,3] by [-1,7]

(b)

[-15,5] by [-5,15]

(c)

[-20,25] by [-20,15]

(d)

[-15,15] by [-2,8]

(e)

[-50,60] by [-50,40]

19. **(a)**

[0, π/2]

[-5,6] by [-6,4]

(b)

[0, π]

[-6,6] by [-6,4]

(c) See the graph in (b).
(d) See the graph in (b).
(e) See the graph in (b).

20. $0 \le \theta \le 2\pi$

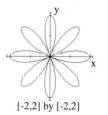

[-2,2] by [-2,2]

21. The graph does not repeat for
$-4\pi \le \theta \le 4\pi$; there is no
smallest length

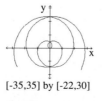

[-35,35] by [-22,30]

22. 6 units

23. $(0, 0)$, $(1.5, 1.5)$

24. $(-2, 0)$, $(0, -1)$

25. Eqn.: $(x + 2)^2 = 8(y + 1)$;
vertex: $(-2, -1)$; line of sym.:
$x = -2$

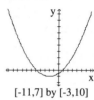

[-11,7] by [-3,10]

26. Eqn.: $(y - 1)^2 = 2\left(x + \dfrac{5}{2}\right)$;

vertex: $\left(-\dfrac{5}{2}, 1\right)$; line of

sym.: $y = 1$

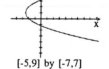

[-5,9] by [-7,7]

27. Eqn.: $(y - 4)^2 = -24(x - 1)$;
focus: $(-5, 4)$; line of sym.:
$y = 4$

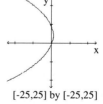

[-25,25] by [-25,25]

28. Eqn.: $x^2 = 24(y + 4)$; line
of sym.: $x = 0$; directrix:

$y = -10$

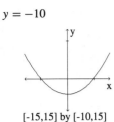

[-15,15] by [-10,15]

29. Std. form: $(x + 2)^2 = 4\left(\dfrac{1}{4}\right)(y - 2)$; vertex: $(-2, 2)$; focus: $(-2, 2.25)$; directrix: $y = 1.75$; line of sym.: $x = -2$

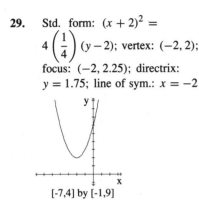

[-7,4] by [-1,9]

30. Std. form: $(y - 1)^2 = 4\left(\dfrac{1}{2}\right)\left(x - \dfrac{7}{6}\right)$; vertex: $\left(\dfrac{7}{6}, 1\right)$; focus: $\left(\dfrac{5}{3}, 1\right)$; directrix: $x = \dfrac{2}{3}$; line of sym.: $y = 1$

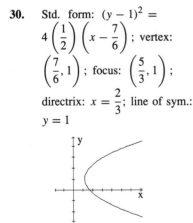

[-2,7] by [-3,5]

31. Major axis: $(-3, 1 - \sqrt{6})$, $(-3, 1 + \sqrt{6})$; minor axis: $(-5, 1)$, $(-1, 1)$; center: $(-3, 1)$; foci: $(-3, 1 - \sqrt{2})$, $(-3, 1 + \sqrt{2})$; lines of sym.:

$x = -3$, $y = 1$

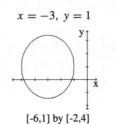

[-6,1] by [-2,4]

32. Major axis: $(-3, -2)$, $(5, -2)$; minor axis: $(1, -2 - \sqrt{2})$, $(1, -2 + \sqrt{2})$; center: $(1, -2)$; foci: $(1 - \sqrt{14}, -2)$, $(1 + \sqrt{14}, -2)$; lines of sym.: $x = 1$, $y = -2$

[-4,5] by [-4,1]

33. Major axis: $(-8, 0)$, $(8, 0)$; minor axis: $(0, -\sqrt{39})$, $(0, \sqrt{39})$; std. form: $\dfrac{x^2}{64} + \dfrac{y^2}{39} = 1$

[-9,9] by [-7,8]

34. Major axis: $(-\sqrt{29}, 0)$, $(\sqrt{29}, 0)$; minor axis: $(0, -5)$, $(0, 5)$; std form: $\dfrac{x^2}{29} + \dfrac{y^2}{25} = 1$

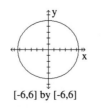

[-6,6] by [-6,6]

35. Major axis: $(2, 0)$, $(2, 12)$; minor axis: $(2 - \sqrt{27}, 6)$,

$(2 + \sqrt{27}, 6)$; std. form: $\dfrac{(x - 2)^2}{27} + \dfrac{(y - 6)^2}{36} = 1$

[-5,10] by [-2,14]

36. Std. form: $\dfrac{(x + 1)^2}{9} + \dfrac{(y - 2)^2}{4} = 1$; major axis: $(-4, 2)$, $(2, 2)$; minor axis: $(-1, 0)$, $(-1, 4)$; center: $(-1, 2)$; foci: $(-1 - \sqrt{5}, 2)$, $(-1 + \sqrt{5}, 2)$; lines of sym.: $x = -1$, $y = 2$

[-5,3] by [-1,5]

37. Std. form: $\dfrac{(x - 3)^2}{5} + \dfrac{(y + 1)^2}{4} = 1$; major axis: $(3 - \sqrt{5}, -1)$, $(3 + \sqrt{5}, -1)$; minor axis: $(3, -3)$, $(3, 1)$; center: $(3, -1)$; foci: $(2, -1)$, $(4, -1)$; lines of sym.: $x = 3$, $y = -1$

[-1,6] by [-3,2]

38. Center: $(0, -2)$; foci: $(-\sqrt{15}, -2)$, $(\sqrt{15}, -2)$; vertices: $(-3, -2)$, $(3, -2)$; lines of sym.: $x = 0$, $y = -2$;

asymptotes: $y + 2 = \pm\dfrac{\sqrt{6}}{3}(x)$

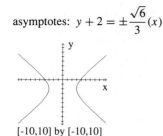

[-10,10] by [-10,10]

39. Center: $(-1, 3)$; vertices: $(-1, 3 - \sqrt{5})$, $(-1, 3 + \sqrt{5})$; foci: $(-1, 3 - \sqrt{8})$, $(-1, 3 + \sqrt{8})$; lines of sym.: $x = -1$, $y = 3$; asymptotes:

$$y - 3 = \pm\sqrt{\dfrac{5}{3}}(x + 1)$$

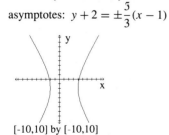

[-8,6] by [-5,10]

40. Std. form: $\dfrac{(x - 1)^2}{9} - \dfrac{(y + 2)^2}{25} = 1$; hyperbola; center: $(1, -2)$; foci: $(1 - \sqrt{34}, -2)$, $(1 + \sqrt{34}, -2)$; vertices: $(-2, -2)$, $(4, -2)$; lines of sym.: $x = 1$, $y = -2$; asymptotes: $y + 2 = \pm\dfrac{5}{3}(x - 1)$

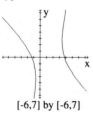

[-10,10] by [-10,10]

41. Std. form: $(y + 3)^2 = x$; parabola; focus: $\left(\dfrac{1}{4}, -3\right)$; vertex: $(0, -3)$; line of sym.:

$y = -3$

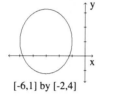

[-3,10] by [-8,3]

42. Std. form: $\dfrac{(x + 3)^2}{4} + \dfrac{(y - 1)^2}{5} = 1$; ellipse; center: $(-3, 1)$; foci: $(-3, 0)$, $(-3, 2)$; vertices: $(-3, 1 - \sqrt{5})$, $(-3, 1 + \sqrt{5})$, $(-5, 1)$, $(-1, 1)$; lines of sym.: $x = -3$, $y = 1$

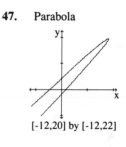

[-6,1] by [-2,4]

43. Std. form: $\dfrac{(y + 3)^2}{1/4} - \dfrac{x^2}{1/9} = 1$; hyperbola; center: $(0, -3)$; foci: $\left(0, -3 - \dfrac{\sqrt{13}}{6}\right)$, $\left(0, -3 + \dfrac{\sqrt{13}}{6}\right)$; vertices: $\left(0, -\dfrac{7}{2}\right)$, $\left(0, -\dfrac{5}{2}\right)$; axis of sym.: $x = 0$, $y = -3$; asymptotes: $y + 3 = \pm\dfrac{3}{2}x$

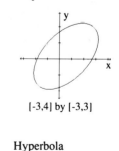

[-10,10] by [-10,10]

44. $y = 2 \pm \sqrt{x + 2}$; then graph $y_1 = 2 + \sqrt{x + 2}$ and

$y_2 = 2 - \sqrt{x + 2}$ in the same window

45. Solve for y: $y = \dfrac{-(Bx + E)\pm}{2C}$
$\dfrac{\sqrt{(Bx+E)^2 - 4C(Ax^2 + Dx + F)}}{2C}$
then graph both functions in the same viewing rectangle.

46. Hyperbola

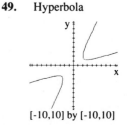

[-6,7] by [-6,7]

47. Parabola

[-12,20] by [-12,22]

48. Ellipse

[-3,4] by [-3,3]

49. Hyperbola

[-10,10] by [-10,10]

50. Center: $(-5.05, -6.11)$; asymptotes: (estimated) $y = 2x + 4$, $y = x - 1$

51. Center: $(-1.06, -0.86)$; major axis: $(0.03, 1.78)$, $(-2.12, -3.41)$; minor axis: $(-3.09, -0.02)$, $(0.79, -1.63)$; lines of sym.: $y = -0.41x - 1.32$, $y = 2.41x + 1.7$

52. $y = -0.44x - 1$, $y = 3.3x - 3$

53. transformations to $y = x^2$: a vertical shrink by a factor of $\frac{1}{6}$, followed by a horizontal shift right 3, then a vertical shift up 1 to get $y = \frac{1}{6}(x - 3)^2 + 1$

54. transformations to $y = x^2$: a vertical shrink by a factor of $\frac{1}{2}$, followed by a horizontal shift right 1, then a vertical shift down $\frac{7}{2}$ to get $y = \frac{1}{2}(x - 1)^2 - \frac{7}{2}$

55. a horizontal stretch by a factor of 4, then a vertical stretch by a factor of 3

56. Starting with $\frac{x^2}{9} + \frac{y^2}{4} = 1$, a horizontal shift left 2, then a vertical shift up 1

57. Std. form: $\frac{x^2}{1} - \frac{y^2}{3} = 1$

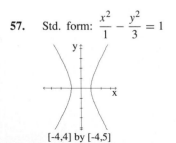

[-4,4] by [-4,5]

58. Std. form: $\frac{(y - 4)^2}{25} - \frac{x^2}{11} = 1$

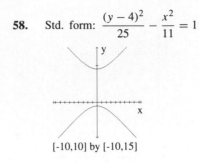

[-10,10] by [-10,15]

59. $(2.73, 1.54)$, $(-0.73, 8.46)$

60. $y = \frac{4}{21}x^2 + \frac{68}{21}$

61. Std. form: $\frac{x^2}{25} + \frac{y^2}{4} = 1$; foci: $(-\sqrt{21}, 0)$, $(\sqrt{21}, 0)$

[-6,6] by [-4,4]

62. The ball hits 446 ft downrange in 4.25 sec.

63. The maximum height is 91.69 ft at $t = 2.39$ sec. The ball goes 307.8 ft horizontally in 4.79 sec.

64. $x(t) = 60t \cos 35°$, $y(t) = -16t^2 + 60t \sin 35°$

65.

[-30,150] by [-50,35]

66. QI

67. 2.15 sec

68. Maximum height is 18.5 ft at $t = 1.08$ sec.

69.

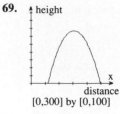

[0,300] by [0,100]

70. The ball first hits 188 ft downfield from the punter's 20-yard line, around the opponent's 17-yard line.

71. 67.1 ft

72. 4.1 sec

CHAPTER 10

SECTION 10.1

1. $(3, -17)$

2. $\left(-\frac{23}{5}, \frac{23}{5}\right)$

3. $\left(\frac{50}{7}, -\frac{10}{7}\right)$

4. $(-3, 9)$, $(3, 9)$

5. no solution

6. $(0, -3)$, $(4, 1)$

7. $(8, -2)$

8. $(3, 4)$

9. $(4, 2)$

10. $\left(x, \frac{1}{2}x - 2\right)$

11. no solution

12. $\left(-\sqrt{\frac{2}{3}}, \frac{10}{3}\right)$, $\left(\sqrt{\frac{2}{3}}, \frac{10}{3}\right)$

13. (a) Use substitution

(b)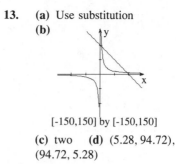

[-150,150] by [-150,150]

(c) two (d) (5.28, 94.72), (94.72, 5.28)

14. $(5, 3, 4)$

15. $\left(\dfrac{1}{2}, 6, 3\right)$

16. 7.08 oz of white sugar, 15.92 oz of brown sugar

17. 60 yd by 50 yd

18. Hank rows 3.56 mph and the current flows at a rate of 1.06 mph

19. 380 student tickets, 72 nonstudent tickets

20. The plane flies 617.42 mph against a 49.24 mph wind

21. The costs for small, medium, and large drinks are $0.57, $0.77, and $1.00, respectively.

22. $y = \dfrac{2}{3}x + \dfrac{14}{3}$

23. $21,333.33 at 7.5%, $16,666.67 at 6%

24. 3.07 lb of peanuts, 1.93 lb of cashews

25. Answers will vary.

26.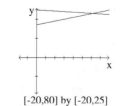

[-20,80] by [-20,25]

27. $23.15

28. $17.96

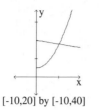

[-10,20] by [-10,40]

29. (43.03, 6.79) (6.97, 43.03)

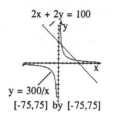

$2x + 2y = 100$

$y = 300/x$

[-75,75] by [-75,75]

SECTION 10.2

1. A is a 3×4 matrix with 3 rows and 4 columns

2. $a_{21} = 4$, $a_{33} = -2$, $a_{23} = 1$, $a_{14} = -1$, $a_{24} = 4$

3. $\begin{pmatrix} -1 & 2 & 3 & -1 \\ 3 & 2 & 4 & 3 \\ 2 & 6 & -2 & 3 \end{pmatrix}$

4. $\begin{pmatrix} -1 & 2 & 3 & -1 \\ 4 & 0 & 1 & 4 \\ 0 & 10 & 4 & 1 \end{pmatrix}$

5. $\begin{pmatrix} -1 & 2 & 3 & -1 \\ 0 & 8 & 13 & 0 \\ 2 & 6 & -2 & 3 \end{pmatrix}$

6. $\begin{pmatrix} 1 & -2 & -3 & 1 \\ 0 & 1 & 13/8 & 0 \\ 0 & 0 & 1 & -4/49 \end{pmatrix}$

7. $\begin{pmatrix} 1 & 0 & 0 & 50/49 \\ 0 & 1 & 0 & 13/98 \\ 0 & 0 & 1 & -4/49 \end{pmatrix}$

8. A is a 3×5 matrix with 3 rows and 5 columns

9. $a_{13} = -9$, $a_{32} = 0$, $a_{23} = -1$, $a_{25} = 3$, $a_{34} = -4$

10. $\begin{pmatrix} 1 & 0 & -3 & 2/3 & 1/3 \\ 1 & 1 & -1 & 0 & 3 \\ -6 & 0 & 18 & -4 & -2 \end{pmatrix}$

11. $\begin{pmatrix} 3 & 0 & -9 & 2 & 1 \\ 1 & 1 & -1 & 0 & 3 \\ 0 & 0 & 0 & 0 & 0 \end{pmatrix}$

12. $\begin{pmatrix} 3 & 0 & -9 & 2 & 1 \\ 0 & 1 & 2 & -2/3 & 8/3 \\ -6 & 0 & 18 & -4 & -2 \end{pmatrix}$

13. $\begin{pmatrix} 1 & 0 & -3 & 2/3 & 1/3 \\ 0 & 1 & 2 & -2/3 & 8/3 \\ 0 & 0 & 0 & 0 & 0 \end{pmatrix}$

14. The result in Exercise 13 is already in reduced row echelon form. It cannot be reduced further.

15. $\begin{pmatrix} 1 & 0 & 0 & -1 \\ 0 & 1 & 0 & 1 \\ 0 & 0 & 1 & 2 \end{pmatrix}$

16. $\begin{pmatrix} 1 & 0 & 0 & 0 & 2 \\ 0 & 1 & 0 & 0 & -1 \\ 0 & 0 & 1 & 0 & 3 \\ 0 & 0 & 0 & 1 & 1 \end{pmatrix}$

17. $\begin{pmatrix} 1 & -3 & 6 \\ 2 & 1 & 19 \end{pmatrix}$; $(9, 1)$

18. $\begin{pmatrix} 1.3 & -2 & 3.3 \\ 5 & 1.6 & 3.4 \end{pmatrix}$; $(1, -1)$

19. $\begin{pmatrix} 1 & -1 & 1 & 6 \\ 1 & 1 & 2 & -2 \end{pmatrix}$ $(x, y, z) = \left(-\dfrac{3}{2}z + 2, -\dfrac{1}{2}z - 4, z\right)$ for any real z

20. $\begin{pmatrix} 1 & 1 & 3 \\ 1 & -1 & 5 \\ 2 & 1 & -1 \end{pmatrix}$; no solution

21. $\begin{pmatrix} 1 & -1 & 1 & 0 \\ 2 & 0 & -3 & -1 \\ -1 & -1 & 2 & -1 \end{pmatrix}$;
$(1, 2, 1)$

22. $\begin{pmatrix} 2 & -1 & 0 & 0 \\ 1 & 3 & -1 & -3 \\ 0 & 3 & 1 & 8 \end{pmatrix}$;
$\left(\dfrac{5}{13}, \dfrac{10}{13}, \dfrac{74}{13} \right)$

23. $\begin{pmatrix} 1 & 1 & -2 & 2 \\ 3 & -1 & 1 & 4 \\ -2 & -2 & 4 & 6 \end{pmatrix}$; no
solution

24. $\begin{pmatrix} 2 & -1 & 0 & 10 \\ 1 & 0 & -1 & -1 \\ 0 & 1 & 1 & -9 \end{pmatrix}$;
$(0, -10, 1)$

25. $\begin{pmatrix} 1 & 1 & -2 & 2 \\ 3 & -1 & 1 & 1 \\ -2 & -2 & 4 & -4 \end{pmatrix}$;
$\left(\dfrac{1}{4}z + \dfrac{3}{4}, \dfrac{7}{4}z + \dfrac{5}{4}, z \right)$ for any
real z

26. $\begin{pmatrix} 1.25 & 0 & 1 & 2 \\ 0 & 1 & -5.5 & -2.75 \\ 3 & -1.5 & 0 & -6 \end{pmatrix}$;
$(0.4789, 4.9577, 1.4014)$

27. $\begin{pmatrix} 1 & 1 & -1 & 0 & 4 \\ 0 & 1 & 0 & 1 & 4 \\ 1 & -1 & 0 & 0 & 1 \\ 1 & 0 & 1 & 1 & 4 \end{pmatrix}$;
$(2, 1, -1, 3)$

28. $\begin{pmatrix} 1/2 & -1 & 1 & -1 & 1 \\ -1 & 1 & 1 & 2 & -3 \\ 1 & 0 & -1 & 0 & 2 \\ 0 & 1 & 0 & 1 & 0 \end{pmatrix}$;
$(2, 1, 0, -1)$

29. $\begin{pmatrix} 2 & 1 & 1 & 2 & -3.5 \\ 1 & 1 & 1 & 1 & -1.5 \end{pmatrix}$;
there are infinitely many

solutions; from row 2,
$y + z = 0.5$ so $y = -z + 0.5$;
from row 1, $x + w = -2$ so
$x = -w - 2$; solution:
$(-w - 2, -z + 0.5, w, z)$ for
any real w, z

30. $\begin{pmatrix} 2 & 1 & 0 & 4 & 6 \\ 1 & 1 & 1 & 1 & 5 \end{pmatrix}$; solution:
$(z - 3w + 1, 2w - 2z + 4, z, w)$
for any real z, w

31. 364 adults and 724 children

32. $16\dfrac{2}{3}$ g of 42% alloy and $33\dfrac{1}{3}$ g
of 30% alloy

33. $7200 at 6%, $3600 at 8%,
$9200 at 10%

34. $y = -x^2 + 10x + 50$

35. $y = -x^2 + x + 10$

36. It takes one hour if all 3 work
together.

37. Answers will vary.

SECTION 10.3

1. $A \cdot B = \begin{pmatrix} -4 & -18 \\ -11 & -17 \end{pmatrix}$;
$B \cdot A = \begin{pmatrix} 5 & -12 \\ 0 & -26 \end{pmatrix}$

2. $A \cdot B = \begin{pmatrix} 13 & 13 \\ -2 & -16 \end{pmatrix}$;
$B \cdot A = \begin{pmatrix} 7 & -14 \\ -8 & -10 \end{pmatrix}$

3. $A \cdot B = \begin{pmatrix} 2 & 2 \\ -11 & 12 \end{pmatrix}$;
$B \cdot A = \begin{pmatrix} 4 & 8 & -5 \\ -5 & 4 & -6 \\ -2 & -8 & 6 \end{pmatrix}$

4. $A \cdot B = \begin{pmatrix} -3 & -7 & 11 \\ 0 & -12 & 24 \end{pmatrix}$;
$B \cdot A$ is undefined

5. $A \cdot B = \begin{pmatrix} 6 & -7 & -2 \\ 3 & 7 & 3 \\ 8 & -1 & -1 \end{pmatrix}$;
$B \cdot A = \begin{pmatrix} 2 & 1 & 3 \\ 5 & 0 & 0 \\ -18 & -3 & 10 \end{pmatrix}$

6. $A \cdot B = \begin{pmatrix} 19 & -1 \\ 2 & 10 \end{pmatrix}$;
$B \cdot A = \begin{pmatrix} 3 & -1 & -14 & 16 \\ 4 & 2 & 8 & -2 \\ 5 & 3 & 14 & -6 \\ 8 & 2 & 0 & 10 \end{pmatrix}$

7. $\begin{pmatrix} 2 & 5 \\ 1 & -2 \end{pmatrix} \begin{pmatrix} x \\ y \end{pmatrix} = \begin{pmatrix} -3 \\ 1 \end{pmatrix}$

8. $\begin{pmatrix} 1 & -2 \\ 2 & -5 \end{pmatrix} \begin{pmatrix} x \\ y \end{pmatrix} = \begin{pmatrix} 1 \\ 3 \end{pmatrix}$

9. $\begin{pmatrix} 5 & -7 & 1 \\ 2 & -3 & -1 \\ 1 & 1 & 1 \end{pmatrix} \begin{pmatrix} x \\ y \\ z \end{pmatrix} =$
$\begin{pmatrix} 2 \\ 3 \\ -3 \end{pmatrix}$

10. $\begin{pmatrix} 2 & 3 & -1 \\ 2 & -3 & 2 \\ -1 & -1 & 3 \end{pmatrix} \begin{pmatrix} x \\ y \\ z \end{pmatrix} =$
$\begin{pmatrix} 2 \\ -1 \\ -4 \end{pmatrix}$

11. $3x - y = -1$
$2x + 4y = 3$

12. $2x + 4y = 5$
$-x - 2y = -2$

13. $x - 3z = 3$
$2x - y + 3z = -1$
$-2x + 3y - 4z = 2$

14. $x - y = 3$
$2x + y - 3z = -1$
$-x + y + 2z = 4$

15. $\begin{pmatrix} -1/10 & 3/10 \\ -2/5 & 1/5 \end{pmatrix}$

16. $\begin{pmatrix} 2/5 & 3/10 \\ -1/5 & 1/10 \end{pmatrix}$

17. A does not have an inverse

18. A does not have an inverse

19. $A^{-1} = \begin{pmatrix} 2/3 & 0 & -1/6 \\ -2 & 1 & 0 \\ -1/3 & 0 & 1/3 \end{pmatrix}$

20. $A^{-1} = \begin{pmatrix} 0.4 & 0 & 0.2 \\ 0.2 & 0 & 0.6 \\ 0 & 0.5 & 0 \end{pmatrix}$

21. $x = \dfrac{13}{14}, \; y = -\dfrac{12}{7}$

22. $x = \dfrac{3}{5}, \; y = \dfrac{11}{5}$

23. $x = 2, \; y = 0$

24. $x = \dfrac{56}{31}, \; y = \dfrac{4}{31}$

25. $x = 2.25, \; y = -0.75,$
$z = -1.75$

26. $x = -6, \; y = 2, \; z = 11.5,$
$w = 0.5$

27. For A, $C(x) = x^2 - 8x + 13$;
for B, $C(x) = x^2 - 4x - 1$

28. (a)

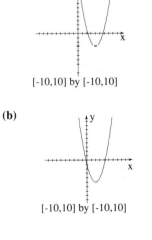

[-10,10] by [-10,10]

(b)

[-10,10] by [-10,10]

29. Eigenvalues for
$A: x = 2.27, \; 5.73$; eigenvalues
for $B: x = -0.24, \; 4.24$

30. The determinant of each matrix
is the same as the y- intercept of
its characteristic polynomial.

31. Conjecture: The sum of the
numbers on the main diagonal
of a 2×2 matrix is equal to the
sum of the eigenvalues of the
matrix.

32. (a) $\begin{aligned} 4a + 2b + c &= 3 \\ 25a + 5b + c &= 8 \\ 49a + 7b + c &= 2 \end{aligned}$

(b) $\begin{pmatrix} 4 & 2 & 1 \\ 25 & 5 & 1 \\ 49 & 7 & 1 \end{pmatrix} \begin{pmatrix} a \\ b \\ c \end{pmatrix} =$
$\begin{pmatrix} 3 \\ 8 \\ 2 \end{pmatrix}$ (c) $f(x) =$
$-0.93x^2 + 8.20x - 9.67$

(d) See the graph in (e).

(e)

[-5,10] by [-5,10]

33. (a) $\begin{aligned} 4a + 2b + c &= 8 \\ 36a + 6b + c &= 3 \\ 81a + 9b + c &= 4 \end{aligned}$

(b) $\begin{pmatrix} 4 & 2 & 1 \\ 36 & 6 & 1 \\ 81 & 9 & 1 \end{pmatrix} \begin{pmatrix} a \\ b \\ c \end{pmatrix} =$
$\begin{pmatrix} 8 \\ 3 \\ 4 \end{pmatrix}$ (c) $f(x) =$
$0.23x^2 - 3.06x + 13.21$

(d) See the graph in (e).

(e)

[-2,16] by [-2,16]

34. (a) $\begin{aligned} 8a + 4b + 2c + d &= 3 \\ 125a + 25b + 5c + d &= 8 \\ 343a + 49b + 7c + d &= 2 \\ 729a + 81b + 9c + d &= 4 \end{aligned}$

(b) $\begin{pmatrix} 8 & 4 & 2 & 1 \\ 125 & 25 & 5 & 1 \\ 343 & 49 & 7 & 1 \\ 729 & 81 & 9 & 1 \end{pmatrix}$

$\begin{pmatrix} a \\ b \\ c \\ d \end{pmatrix} = \begin{pmatrix} 3 \\ 8 \\ 2 \\ 4 \end{pmatrix}$ (c) $f(x) =$

$0.28x^3 - 4.8x^2 + 24.50x - 29$

(d) See the graph in (e).

(e)

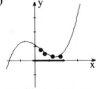

[-2,12] by [-2,12]

35. (a) $\begin{aligned} 8a + 4b + 2c + d &= 8 \\ 64a + 16b + 4c + d &= 5 \\ 216a + 36b + 6c + d &= 3 \\ 729a + 81b + 9c + d &= 4 \end{aligned}$

(b) $\begin{pmatrix} 8 & 4 & 2 & 1 \\ 64 & 16 & 4 & 1 \\ 216 & 36 & 6 & 1 \\ 729 & 81 & 9 & 1 \end{pmatrix}$

$\begin{pmatrix} a \\ b \\ c \\ d \end{pmatrix} = \begin{pmatrix} 8 \\ 5 \\ 3 \\ 4 \end{pmatrix}$ (c) $f(x) =$

$0.02x^3 - 0.12x^2 - 1.36x + 11.03$

(d) See the graph in (e).

(e)

[-10,20] by [-20,40]

36. **(a)**
$$16a - 8b + 4c - 2d + e = -4$$
$$a + b + c + d + e = 2$$
$$81a + 27b + 9c + 3d + e = 6$$
$$256a + 64b + 16c + 4d + e = -2$$
$$2401a + 343b + 49c + 7d + e = 8$$

(b)
$$\begin{pmatrix} 16 & -8 & 4 & -2 & 1 \\ 1 & 1 & 1 & 1 & 1 \\ 81 & 27 & 9 & 3 & 1 \\ 256 & 64 & 16 & 4 & 1 \\ 2401 & 343 & 49 & 7 & 1 \end{pmatrix}$$

$$\begin{pmatrix} a \\ b \\ c \\ d \\ e \end{pmatrix} = \begin{pmatrix} -4 \\ 2 \\ 6 \\ -2 \\ 8 \end{pmatrix}$$

(c) $f(x) = 0.18x^4 - 1.61x^3 + 1.64x^2 + 9.35x - 7.56$

(d) See the graph in (e).

(e)

[-5,10] by [-15,10]

37. **(a)**
$$a - b + c - d + e = 8$$
$$a + b + c + d + e = 2$$
$$256a + 64b + 16c + 4d + e = -6$$
$$2401a + 343b + 49c + 7d + e = 5$$
$$4096a + 512b + 64c + 8d + e = 2$$

(b)
$$\begin{pmatrix} 1 & -1 & 1 & -1 & 1 \\ 1 & 1 & 1 & 1 & 1 \\ 256 & 64 & 16 & 4 & 1 \\ 2401 & 343 & 49 & 7 & 1 \\ 4096 & 512 & 64 & 8 & 1 \end{pmatrix}$$

$$\begin{pmatrix} a \\ b \\ c \\ d \\ e \end{pmatrix} = \begin{pmatrix} 8 \\ 2 \\ -6 \\ 5 \\ 2 \end{pmatrix}$$

(c) $f(x) = -0.0569x^4 + 0.75x^3 - 1.9653x^2 - 3.75x + 7.0222$

(d) See the graph in (e).

(e)

[-10,20] by [-20,25]

38. **(a)** From Exercise 19:
$x = 1, \ 1.27, \ 4.73$
(b) From Exercise 20:
$x = 1.60, \ 3.29, \ -1.90$

SECTION 10.4

1. $x' = t^3 \cos 30° - t \sin 30° = \dfrac{\sqrt{3}}{2}t^3 - \dfrac{1}{2}t;$
$y' = t^3 \sin 30° + t \cos 30° = \dfrac{1}{2}t^3 + \dfrac{\sqrt{3}}{2}t$

2. $x' = t^3 \cos 78° - t \sin 78° = 0.21t^3 - 0.98t;$
$y' = t^3 \sin 78° + t \cos 78° = 0.98t^3 + 0.21t$

3. $x' = t^3 \cos 120° - t \sin 120° = -\dfrac{1}{2}t^3 - \dfrac{\sqrt{3}}{2}t;$
$y' = t^3 \sin 120° + t \cos 120° = \dfrac{\sqrt{3}}{2}t^3 - \dfrac{1}{2}t$

4. $x' = t^3 \cos \dfrac{\pi}{3} - t \sin \dfrac{\pi}{3} = \dfrac{1}{2}t^3 - \dfrac{\sqrt{3}}{2}t; \ y' = t^3 \sin \dfrac{\pi}{3} + t \cos \dfrac{\pi}{3} = \dfrac{\sqrt{3}}{2}t^3 + \dfrac{1}{2}t$

5. $x' = t^3 \cos \dfrac{\pi}{6} - t \sin \dfrac{\pi}{6} = \dfrac{\sqrt{3}}{2}t^3 - \dfrac{1}{2}t; \ y' = t^3 \sin \dfrac{\pi}{6} + t \cos \dfrac{\pi}{6} = \dfrac{1}{2}t^3 + \dfrac{\sqrt{3}}{2}t$

6. $x' = t^3 \cos \dfrac{\pi}{8} - t \sin \dfrac{\pi}{8} = 0.92t^3 - 0.38t;$

$y' = t^3 \sin \dfrac{\pi}{8} + t \cos \dfrac{\pi}{8} = 0.38t^3 + 0.92t$

7. $x' = t^3 \cos \dfrac{5\pi}{12} - t \sin \dfrac{5\pi}{12} = 0.26t^3 - 0.97t;$
$y' = t^3 \sin \dfrac{5\pi}{12} + t \cos \dfrac{5\pi}{12} = 0.97t^3 + 0.26t$

8. $\begin{pmatrix} \cos 45° & -\sin 45° \\ \sin 45° & \cos 45° \end{pmatrix}$ or $\begin{pmatrix} \dfrac{\sqrt{2}}{2} & -\dfrac{\sqrt{2}}{2} \\ \dfrac{\sqrt{2}}{2} & \dfrac{\sqrt{2}}{2} \end{pmatrix}$

9. $\begin{pmatrix} \cos 30° & -\sin 30° \\ \sin 30° & \cos 30° \end{pmatrix}$ or $\begin{pmatrix} \dfrac{\sqrt{3}}{2} & -\dfrac{1}{2} \\ \dfrac{1}{2} & \dfrac{\sqrt{3}}{2} \end{pmatrix}$

10. $\begin{pmatrix} \cos 120° & -\sin 120° \\ \sin 120° & \cos 120° \end{pmatrix}$ or $\begin{pmatrix} -\dfrac{1}{2} & -\dfrac{\sqrt{3}}{2} \\ \dfrac{\sqrt{3}}{2} & -\dfrac{1}{2} \end{pmatrix}$

11. $\begin{pmatrix} \cos(-60°) & -\sin(-60°) \\ \sin(-60°) & \cos(-60°) \end{pmatrix}$ or $\begin{pmatrix} \dfrac{1}{2} & \dfrac{\sqrt{3}}{2} \\ -\dfrac{\sqrt{3}}{2} & \dfrac{1}{2} \end{pmatrix}$

12. $\begin{pmatrix} \cos(-45°) & -\sin(-45°) \\ \sin(-45°) & \cos(-45°) \end{pmatrix}$ or $\begin{pmatrix} \dfrac{\sqrt{2}}{2} & \dfrac{\sqrt{2}}{2} \\ -\dfrac{\sqrt{2}}{2} & \dfrac{\sqrt{2}}{2} \end{pmatrix}$

13. $\begin{pmatrix} \cos 21° & -\sin 21° \\ \sin 21° & \cos 21° \end{pmatrix}$ or $\begin{pmatrix} 0.93 & -0.36 \\ 0.36 & 0.93 \end{pmatrix}$

14. $\begin{pmatrix} \cos \dfrac{\pi}{2} & -\sin \dfrac{\pi}{2} \\ \sin \dfrac{\pi}{2} & \cos \dfrac{\pi}{2} \end{pmatrix}$ or $\begin{pmatrix} 0 & -1 \\ 1 & 0 \end{pmatrix}$

19.

$4x^2 + xy + 4y^2 - 5x + 5v - 15 = 0$

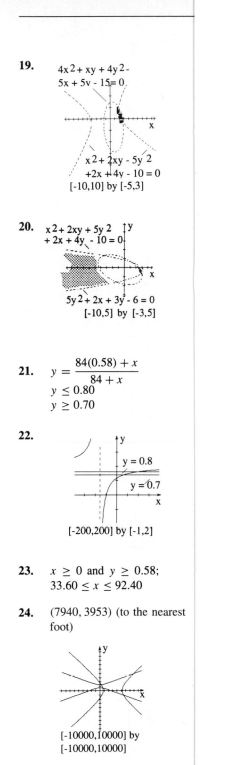

$x^2 + 2xy - 5y^2 + 2x + 4y - 10 = 0$

[-10,10] by [-5,3]

20.

$x^2 + 2xy + 5y^2 + 2x + 4y - 10 = 0$

$5y^2 + 2x + 3y - 6 = 0$

[-10,5] by [-3,5]

21. $y = \dfrac{84(0.58) + x}{84 + x}$

$y \le 0.80$

$y \ge 0.70$

22.

$y = 0.8$

$y = 0.7$

[-200,200] by [-1,2]

23. $x \ge 0$ and $y \ge 0.58$;

$33.60 \le x \le 92.40$

24. (7940, 3953) (to the nearest foot)

[-10000,10000] by [-10000,10000]

15. $\begin{pmatrix} \cos\dfrac{3\pi}{4} & -\sin\dfrac{3\pi}{4} \\ \sin\dfrac{3\pi}{4} & \cos\dfrac{3\pi}{4} \end{pmatrix}$ or $\begin{pmatrix} -\dfrac{\sqrt{2}}{2} & -\dfrac{\sqrt{2}}{2} \\ \dfrac{\sqrt{2}}{2} & -\dfrac{\sqrt{2}}{2} \end{pmatrix}$

16. $\begin{pmatrix} \cos\dfrac{5\pi}{6} & -\sin\dfrac{5\pi}{6} \\ \sin\dfrac{5\pi}{6} & \cos\dfrac{5\pi}{6} \end{pmatrix}$ or $\begin{pmatrix} -\dfrac{\sqrt{3}}{2} & -\dfrac{1}{2} \\ \dfrac{1}{2} & -\dfrac{\sqrt{3}}{2} \end{pmatrix}$

17. $\begin{pmatrix} \cos\dfrac{2\pi}{3} & -\sin\dfrac{2\pi}{3} \\ \sin\dfrac{2\pi}{3} & \cos\dfrac{2\pi}{3} \end{pmatrix}$ or $\begin{pmatrix} -\dfrac{1}{2} & -\dfrac{\sqrt{3}}{2} \\ \dfrac{\sqrt{3}}{2} & -\dfrac{1}{2} \end{pmatrix}$

18. Substitute $t = \dfrac{x}{3}$ to get $y = \dfrac{x^2}{9}$.

19. Substitute $t = x - 2$ to get $y = x^2 - 4x + 1$.

20. $\begin{pmatrix} x' \\ y' \end{pmatrix} =$

$\begin{pmatrix} \cos 45° & -\sin 45° \\ \sin 45° & \cos 45° \end{pmatrix}\begin{pmatrix} \dfrac{t^2}{2} - 3 \\ 2 \end{pmatrix} =$

$\begin{pmatrix} \dfrac{\sqrt{2}}{2} & -\dfrac{\sqrt{2}}{2} \\ \dfrac{\sqrt{2}}{2} & \dfrac{\sqrt{2}}{2} \end{pmatrix}\begin{pmatrix} \dfrac{t^2}{2} - 3 \end{pmatrix}$

21. $\begin{pmatrix} x' \\ y' \end{pmatrix} =$

$\begin{pmatrix} \cos 60° & -\sin 60° \\ \sin 60° & \cos 60° \end{pmatrix}\begin{pmatrix} \dfrac{t^2}{2} - 3 \\ 2 \end{pmatrix} =$

$\begin{pmatrix} \dfrac{1}{2} & -\dfrac{\sqrt{3}}{2} \\ \dfrac{\sqrt{3}}{2} & \dfrac{1}{2} \end{pmatrix}\begin{pmatrix} \dfrac{t^2}{2} - 3 \end{pmatrix}$

22. $x' = 3\cos t \cos\dfrac{\pi}{3} - 2\sin t \sin\dfrac{\pi}{3}$;

$y' = 3\cos t \sin\dfrac{\pi}{3} + 2\sin t \cos\dfrac{\pi}{3}$

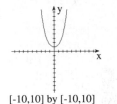

[-10,10] by [-10,10]

23. $x' = 5\cos t \cos\dfrac{5\pi}{6} - 3\sin t \sin\dfrac{5\pi}{6}$; $y' = 5\cos t \sin\dfrac{5\pi}{6} + 3\sin t \cos\dfrac{5\pi}{6}$

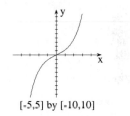

[-10,10] by [-10,10]

24. $x' = \dfrac{t}{5}\cos\dfrac{3\pi}{8} - (t^2 - 3)\sin\dfrac{3\pi}{8}$;

$y' = \dfrac{t}{5}\sin\dfrac{3\pi}{8} + (t^2 - 3)\cos\dfrac{3\pi}{8}$

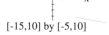

[-15,10] by [-5,10]

25. $x' = (2t - 4)\cos\dfrac{5\pi}{6} - (t^2 + 1)\sin\dfrac{5\pi}{6}$; $y' = (2t - 4)\sin\dfrac{5\pi}{6} + (t^2 + 1)\cos\dfrac{5\pi}{6}$

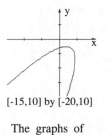

[-15,10] by [-20,10]

26. The graphs of

$y = f(x) = \dfrac{e^x + e^{-x}}{2}$ and

$y = \cosh x$ appear to coincide.

[-10,10] by [-10,10]

27. The graphs of

$y = g(x) = \dfrac{e^x - e^{-x}}{2}$ and

$y = \sinh x$ appear to coincide.

[-5,5] by [-10,10]

28. $x^2 - y^2 = \left(\dfrac{e^t + e^{-t}}{2}\right)^2 - \left(\dfrac{e^t - e^{-t}}{2}\right)^2 =$

$\dfrac{e^{2t} + 2e^t e^{-t} + e^{-2t}}{4} - \dfrac{e^{2t} - 2e^t e^{-t} + e^{-2t}}{4} = 1$

29. $x^2 - y^2 = (-\cosh t)^2 - (\sinh t)^2 = \cosh^2 t - \sinh^2 t = 1$ (from Exercise 28)

30.

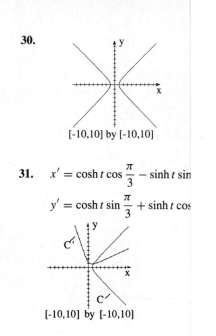

[-10,10] by [-10,10]

31. $x' = \cosh t \cos \dfrac{\pi}{3} - \sinh t \sin$

$y' = \cosh t \sin \dfrac{\pi}{3} + \sinh t \cos$

[-10,10] by [-10,10]

SECTION 10.5

1. $(-1.4796, 1.7398)$ and $(2.9196, -0.4598)$

2. $(2.9196, 0.4598)$ and $(-1.4796, -1.7398)$

3. $(-3.0981, 3.5490)$ and $(2.0981, 0.951)$

4. $(-3.0981, -3.5490)$ and $(2.0981, -0.95096)$

5. $(-2.1965, -1.362)$, $(-2.1965, 1.362)$, $(2.1965, -1.362)$, and $(2.1965, 1.362)$

6. $(-2, 0)$ and $(2, 0)$

7. $(-1.3657, 0.7305)$, $(1.3657, 0.7305)$, $(-1.0355, -0.8555)$, and $(1.0355, -0.8555)$

8. no real solution

9. $(4.3166, -1.913)$, $(4.3166, 1.913)$,

15. $B \cdot C = \begin{pmatrix} -1 & -1 & 6 \\ 4 & -18 & 20 \end{pmatrix}$

16. $(D \cdot C) \cdot A = \begin{pmatrix} 5 & -5 & -3 \\ 1 & -25 & 9 \\ 9 & -15 & -3 \end{pmatrix}$

17. $\begin{pmatrix} 17/14 \\ 1/7 \end{pmatrix}$

18. $\begin{pmatrix} 575/183 \\ -283/183 \end{pmatrix}$

19. $\begin{pmatrix} 2.2128 \\ 1.3844 \end{pmatrix}$

20. $\begin{pmatrix} 1.2041 \\ 1.3469 \\ 1.7551 \end{pmatrix}$

21. $\begin{pmatrix} 1.0479 \\ -1.0136 \\ 0.4049 \end{pmatrix}$

22. $\begin{pmatrix} 8.8983 \\ 17.1017 \\ 8.8983 \\ -18.4068 \end{pmatrix}$

23. $(-6.3970, 49.676)$, $(0.4018, 46.017)$, $(5.9730, 47.613)$

24. $(-2.8934, 2.447)$, $(4.5327, -1.266)$

25. $(-23.2622, 13.6311)$, $(5.0804, -0.5402)$

26. $(5, 0)$, $(-5, 0)$

27. no solution

28. There are 4 simultaneous solutions to the system E_1 and E_2

29. There are 2 simultaneous solutions to the system E_1 and E_2.

30. $(-2.6053, -0.3675)$, $(-0.2381, 1.6320)$, $(1.8436, 0.1517)$, $(0.1890, -2.0784)$

31. $(-1.2313, -1.342)$, $(2.4807, 1.920)$

32.

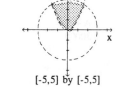

[-5,5] by [-5,5]

33. no solution

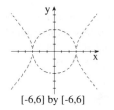

[-6,6] by [-6,6]

CHAPTER 11

SECTION 11.1

1. $a_1 = 1$, $a_2 = -1$, $a_3 = 1$, $a_4 = -1$, $a_{10} = -1$

2. $a_1 = 1$, $a_2 = 3$, $a_3 = 5$, $a_4 = 7$, $a_{10} = 19$

3. $a_1 = 2$, $a_2 = \dfrac{3}{2}$, $a_3 = \dfrac{4}{3}$, $a_4 = \dfrac{5}{4}$, $a_{10} = \dfrac{11}{10}$

4. $a_1 = -2$, $a_2 = 4$, $a_3 = -6$, $a_4 = 8$, $a_{10} = 20$

5. $a_1 = 0$, $a_2 = \dfrac{7}{3}$, $a_3 = \dfrac{13}{2}$, $a_4 = \dfrac{63}{5}$, $a_{10} = \dfrac{999}{11}$

6. $a_1 = \dfrac{4}{3}$, $a_2 = 1$, $a_3 = \dfrac{4}{5}$, $a_4 = \dfrac{2}{3}$, $a_{10} = \dfrac{1}{3}$

7. $a_1 = -2$, $a_2 = \dfrac{3}{2}$, $a_3 = -\dfrac{4}{3}$, $a_4 = \dfrac{5}{4}$, $a_{10} = \dfrac{11}{10}$

8. $a_1 = \dfrac{3}{8}$, $a_2 = \dfrac{9}{8}$, $a_3 = \dfrac{27}{8}$, $a_4 = \dfrac{81}{8}$, $a_{10} = \dfrac{59,049}{8}$

9. $a_1 = 3$, $a_2 = 4$, $a_3 = 5$, $a_4 = 6$, $a_8 = 10$; $a_n = 2 + n$

10. $a_1 = -2$, $a_2 = 0$, $a_3 = 2$, $a_4 = 4$, $a_8 = 12$; $a_n = -4 + 2n$

11. $a_1 = 2$, $a_2 = 4$, $a_3 = 8$, $a_4 = 16$, $a_8 = 256$; $a_n = 2^n$

12. $a_1 = \dfrac{3}{2}$, $a_2 = \dfrac{3}{4}$, $a_3 = \dfrac{3}{8}$, $a_4 = \dfrac{3}{16}$, $a_8 = \dfrac{3}{256}$; $a_n = \left(\dfrac{1}{2}\right)^n \cdot 3$

13. $a_n = 2n + 1$

14. $a_n = -12 + 7n$

15. $a_n = 2 \cdot 3^{n-1}$

16. $a_n = (-2)^{n-1} \cdot 3$

17. $d = 4$, $a_{10} = 42$; $a_n = 2 + 4n$

18. $d = 5$, $a_{10} = 41$; $a_n = -9 + 5n$

19. $d = 3$, $a_{10} = 22$; $a_n = -8 + 3n$

20. $d = 11$, $a_{10} = 92$; $a_n = -18 + 11n$

21. $r = 3$, $a_8 = 4374$; $a_n = 2 \cdot (3)^{n-1}$

22. $r = -2$, $a_8 = -128$; $a_n = 1 \cdot (-2)^{n-1}$

23. $r = 2$, $a_8 = 384$; $a_n = 3 \cdot (2)^{n-1}$

24. $r = -1$, $a_8 = 2$; $a_n = (-2)(-1)^{n-1}$

25. $a_1 = -20$; $a_n = -24 + 4n$

26. $a_1 = 7$; $a_n = 10 - 3n$

27. geometric, $r = 2$; $a_n = 5(2)^{n-1}$

28. geometric, $r = -4$;
$a_n = (-0.25)(-4)^{n-1}$

29. arithmetic, $d = 7$;
$a_n = -23 + 7n$

30. geometric, $r = 1.01$;
$a_n = 10.1(1.01)^{n-1}$

31. geometric, $r = -\dfrac{1}{2}$;
$a_n = -2\left(-\dfrac{1}{2}\right)^{n-1}$

32. neither arithmetic nor geometric

33.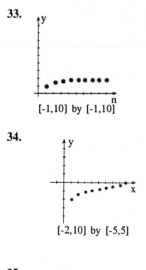
[-1,10] by [-1,10]

34.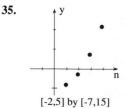
[-2,10] by [-5,5]

35.
[-2,5] by [-7,15]

36.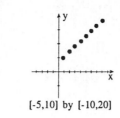
[-5,10] by [-10,20]

37. $1712.61

38. $9126.56

39. $a_1 = 500$, $a_2 = 250$, $a_3 = 125$,
$a_4 = 62.5$, $a_5 = 31.25$,
$a_6 = 15.625$, $a_7 = 7.8125$,
$a_8 = 3.90625$, $a_9 = 1.953125$,
$a_{10} = 0.9765625$; $r = \dfrac{1}{2}$,
$a_n = 1000\left(\dfrac{1}{2}\right)^n$

40. 14.29 weeks

41. No

42. $a_1 = 15.375$, $a_2 = 15.759$,
$a_3 = 16.153$, $a_4 = 16.557$,
$a_5 = 16.971$, $a_6 = 17.395$,
$a_7 = 17.830$, $a_8 = 18.276$,
$a_9 = 18.733$, $a_{10} = 19.201$;
after ten months, the plant dies;
$r = 1.025$

43. 28.07 mo

44. $a_1 = 1$ because at 0 mo, we
begin with one pair of rabbits;
$a_2 = 1$ because at 1 mo, the
original female becomes fertile
but has not produced offspring;
$a_3 = 2$ because at 2 mo, the
original female will have given
birth to a second pair of rabbits

45. $u_4 = 3$, $u_5 = 5$, $u_6 = 8$,
$a_7 = 13$, $a_8 = 21$

46. The sequence in Exercise 45
models the growth of the
colony; $P_n = P_{n-1} + F_{n-1}$,
$F_n = N_{n-1}$, and

$N_n = P_n$, where P_n is the
number of pairs producing
offspring, F_n is the number of
newly fertile females, and N_n is
the number of newborn pairs.
Then $a_n = P_n + F_n + N_n$

47. 1, 1, 2, 3, 5, 8, 13 ... is the
Fibonacci sequence

48. $a_{10} = 30.90$, $a_{100} = 31.41$,
$a_{1000} = 31.42$; circumference of
a circle = 31.42; as $n \to \infty$,
$a_n \to 10\pi$, which is the
circumference of a circle with
radius 5

49. if $x \to \infty$, a_n approaches 31.42
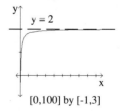
[0,100] by [0,40]

50. $a_1 = 1$, $a_2 = 2$, $a_3 = 6$,
$a_4 = 24$, $a_5 = 120$, $a_6 = 720$;
$a_n = n!$

51. $K = 2$

[0,100] by [-1,3]

52. $K = e^{0.05} = 1.05$

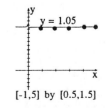
[-1,5] by [0.5,1.5]

53. $K = 3$

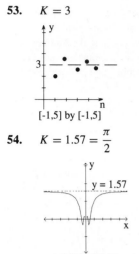

[-1,5] by [-1,5]

54. $K = 1.57 = \dfrac{\pi}{2}$

y = 1.57

[-5,5] by [-2,3]

55. $a_1 = 0.5$, $a_2 = 0.7071$,
$a_3 = 0.6125$, $a_4 = 0.6540$,
$a_5 = 0.6355$, $a_6 = 0.6437$,
$a_7 = 0.6401$, $a_8 = 0.6417$;
converge to $0.6411857445\ldots$

56. $x = 0.6411857445\ldots$, the
number to which the sequence
in Exercise 55 converges

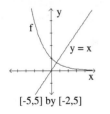

[-5,5] by [-2,5]

SECTION 11.2

1. $4 + 16 + 36 + 64 + \cdots + 4n^2$

2. $2 + 8 + 18 + 32 + \cdots + 2n^2$

3. $2 + 5 + 8 + 11 + \cdots + (3n - 1)$

4. $0 + 3 + 8 + 15 + \cdots + (n^2 - 1)$

5. $\dfrac{3}{2} + \dfrac{3}{4} + \dfrac{3}{8} + \dfrac{3}{16} + \cdots + \dfrac{3}{2^n}$

6. $\dfrac{3}{4} + \dfrac{9}{4} + \dfrac{27}{4} + \dfrac{81}{4} + \cdots + \dfrac{3^n}{4}$

7. $\dfrac{3n^2 + 7n}{2}$

8. $4n - n^2$

9. $\dfrac{n^3 - 3n^2 + 8n}{3}$

10. $\dfrac{4n^3 + 21n^2 + 5n}{6}$

11. $\dfrac{n^4 + 2n^3 + n^2 - 4n}{4}$

12. $\dfrac{n^4 + 2n^3 + 9n^2 - 12n}{4}$

13. $3(2^n - 1)$

14. $\dfrac{4}{3}(1 - (-2)^n)$

15. $(-4)\left(1 - \left(\dfrac{1}{2}\right)^n\right)$

16. $\left(\dfrac{15}{2}\right)\left(1 - \left(\dfrac{1}{3}\right)^n\right)$

17. $\left(\dfrac{100}{11}\right)\left(1 - \left(-\dfrac{1}{10}\right)^n\right)$

18. $\dfrac{40}{3}(1 - (0.4)^n)$

19. $-2\left(\dfrac{1 - (-0.5)^n}{1.5}\right)$

20. $14\left(1 - \left(\dfrac{3}{2}\right)^n\right)$

21. $\displaystyle\sum_{k=1}^{10}(3k - 1)$

22. $\displaystyle\sum_{k=1}^{8}(k^2 - 2)$

23. $\displaystyle\sum_{k=1}^{\infty}\left(\dfrac{1}{4}\right)^k$

24. $\displaystyle\sum_{k=1}^{\infty}2(-1)^k$

25. $\displaystyle\sum_{k=1}^{n}(2k - 10)$

26. $\displaystyle\sum_{k=1}^{n}(k^2 + 1)$

27. $\displaystyle\sum_{k=1}^{n}3k(-1)^{k+1}$

28. $\displaystyle\sum_{k=1}^{n}\dfrac{5}{k + 1}$

29. geometric, $S_n = 2\left(1 - \left(\dfrac{1}{2}\right)^n\right)$

30. arithmetic, $S_n = \dfrac{n(n + 1)}{2}$

31. geometric, $S_n = -\dfrac{4}{3}\left(1 - \left(-\dfrac{1}{2}\right)^n\right)$

32. arithmetic, $S_n = n(n + 1)$

33. $S_1 = \dfrac{1}{8}$, $S_2 = \dfrac{3}{16}$, $S_3 = \dfrac{7}{32}$,
$S_4 = \dfrac{15}{64}$, $S_5 = \dfrac{31}{128}$, $S_6 = \dfrac{63}{256}$;
the series converges to 0.25

34. $S_1 = -\dfrac{1}{4}$, $S_2 = -\dfrac{3}{8}$,
$S_3 = -\dfrac{7}{16}$, $S_4 = -\dfrac{15}{32}$,
$S_5 = -\dfrac{31}{64}$, $S_6 = -\dfrac{63}{128}$; the
series converges to -0.5

35. $S_1 = \dfrac{3}{10}$, $S_2 = \dfrac{33}{100}$,
$S_3 = \dfrac{333}{1000}$, $S_4 = \dfrac{3333}{10,000}$,
$S_5 = \dfrac{33,333}{100,000}$,
$S_6 = \dfrac{333,333}{1,000,000}$; the series
converges to $\dfrac{1}{3}$

36. $S_1 = \dfrac{1}{2}$, $S_2 = \dfrac{5}{8}$, $S_3 = \dfrac{21}{32}$,
$S_4 = \dfrac{85}{128}$, $S_5 = \dfrac{341}{512}$,
$S_6 = \dfrac{1365}{2048}$; the series converges
to $\dfrac{2}{3}$

37. The graph of
$f(x) = 2\left(\dfrac{1 - 1.05^x}{1 - 1.05}\right)$ shows
as $n \to \infty$, $S_n \to \infty$, so the
series does not converge. The
function, $f(x)$, has a horizontal
asymptote at $y = -40$.

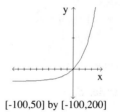

[-100,50] by [-100,200]

38. $S_n \to 1.5$ as $n \to \infty$

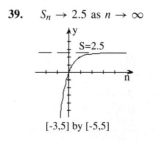

[-5,15] by [-10,5]

39. $S_n \to 2.5$ as $n \to \infty$

S=2.5

[-3,5] by [-5,5]

40. $S_n \to 9$ as $n \to \infty$

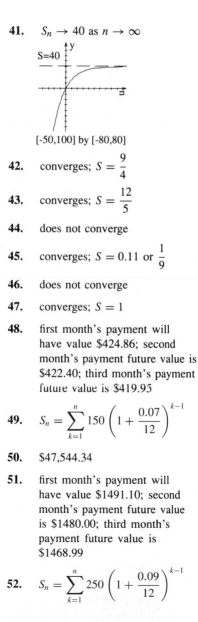

y = 9

[-5,10] by [-10,15]

41. $S_n \to 40$ as $n \to \infty$

S=40

[-50,100] by [-80,80]

42. converges; $S = \dfrac{9}{4}$

43. converges; $S = \dfrac{12}{5}$

44. does not converge

45. converges; $S = 0.11$ or $\dfrac{1}{9}$

46. does not converge

47. converges; $S = 1$

48. first month's payment will
have value \$424.86; second
month's payment future value is
\$422.40; third month's payment
future value is \$419.95

49. $S_n = \displaystyle\sum_{k=1}^{n} 150\left(1 + \dfrac{0.07}{12}\right)^{k-1}$

50. \$47,544.34

51. first month's payment will
have value \$1491.10; second
month's payment future value
is \$1480.00; third month's
payment future value is
\$1468.99

52. $S_n = \displaystyle\sum_{k=1}^{n} 250\left(1 + \dfrac{0.09}{12}\right)^{k-1}$

53. \$166,971.72

SECTION 11.3

1. 1, 8, 28, 56, 70, 56, 28, 8, 1

2. 1, 10, 45, 120, 210, 252, 210,
120, 45, 10, 1

3. $a^5 + 5a^4b + 10a^3b^2 + 10a^2b^3 + 5ab^4 + b^5$

4. $a^6 - 6a^5b + 15a^4b^2 - 20a^3b^3 + 15a^2b^4 - 6ab^5 + b^6$

5. $x^4 + 8x^3y + 24x^2y^2 + 32xy^3 + 16y^4$

6. $2187x^7 - 5103x^6y + 5103x^5y^2 - 2835x^4y^3 + 945x^3y^4 - 189x^2y^5 + 21xy^6 - y^7$

7. $x^6 + 6x^5y + 15x^4y^2 + 20x^3y^3 + 15x^2y^4 + 6xy^5 + y^6$

8. $a^4 - 8a^3b^2 + 24a^2b^4 - 32ab^6 + 16b^8$

9. $x^6 - 6x^5y + 15x^4y^2 - 20x^3y^3 + 15x^2y^4 - 6xy^5 + y^6$

10. $1 + 0.4 + 0.064 + 0.00512 + 0.0002048 + 0.0000032768 = 1.469328077$

11. 35

12. 126

13. 70

14. 220

15. 1365

16. 1716

17. 56

18. 56

19. 240

20. -2268

21. $x^5 - 10x^4 + 40x^3 - 80x^2 + 80x - 32$

22. $x^6 + 18x^5 + 135x^4 + 540x^3 + 1215x^2 + 1458x + 729$

23. $128x^7 - 448x^6 + 672x^5 - 560x^4 + 280x^3 - 84x^2 + 14x - 1$

24. $243x^5 + 1620x^4 + 4320x^3 + 5760x^2 + 3840x + 1024$

25. $x^4 + 4ax^3 + 6a^2x^2 + 4a^3x + a^4$

26. $6561x^8 - 69{,}984x^7 + 326{,}592x^6 - 870{,}912x^5 + 1{,}451{,}520x^4 - 1{,}548{,}288x^3 + 1{,}032{,}192x^2 - 393{,}216x + 65{,}536$

27. $32x^5 + 80x^4y + 80x^3y^2 + 40x^2y^3 + 10xy^4 + y^5$

28. $x^2 + 12x\sqrt{x} + 54x + 108\sqrt{x} + 81$

29. $x^3 - 6x^{5/2}y^{1/2} + 15x^2y - 20x^{3/2}y^{3/2} + 15xy^2 - 6x^{1/2}y^{5/2} + y^3$

30. $4096y^{12} - 73{,}728y^{11}x + 608{,}256y^{10}x^2 - 3{,}041{,}280y^9x^3 + 10{,}264{,}320y^8x^4 - 24{,}634{,}368y^7x^5 + 43{,}110{,}144y^6x^6 - 55{,}427{,}328y^5x^7 + 51{,}963{,}120y^4x^8 - 34{,}642{,}080y^3x^9 + 15{,}588{,}936y^2x^{10} - 4{,}251{,}528yx^{11} + 531{,}441x^{12}$

31. $x^{-10} + 15x^{-8} + 90x^{-6} + 270x^{-4} + 405x^{-2} + 243$

32. $a^{15} - 15a^{14}b + 105a^{13}b^2 - 455a^{12}b^3 + 1365a^{11}b^4 - 3003a^{10}b^5 + 5005a^9b^6 - 6435a^8b^7 + 6435a^7b^8 - 5005a^6b^9 + 3003a^5b^{10} - 1365a^4b^{11} + 455a^3b^{12} - 105a^2b^{13} + 15ab^{14} - b^{15}$

33. $\binom{n}{n-r} = \dfrac{n!}{(n-r)!(n-(n-r))!} = \dfrac{n!}{(n-r)!(n-n+r)!} =$

$\dfrac{n!}{(n-r)!r!} = \dfrac{n!}{r!(n-r)!} = \binom{n}{r}$

34. $\binom{8}{3} + \binom{8}{4} = \dfrac{8!}{3!5!} + \dfrac{8!}{4!4!} = \dfrac{4 \cdot 8! + 5 \cdot 8!}{5!4!} = \dfrac{9 \cdot 8!}{5!4!} = \dfrac{9!}{5!4!} = \binom{9}{4}$

35. $\binom{k}{r-1} + \binom{k}{r} = \dfrac{k!}{(r-1)!(k-(r-1))!} + \dfrac{k!}{r!(k-r)!} = \dfrac{r \cdot k! + (k-r+1)k!}{r!(k-r+1)!} = \dfrac{k!(r+k-r+1)}{r!(k-r+1)!} = \dfrac{k!(k+1)}{r!(k-r+1)} = \dfrac{(k+1)!}{r!(k+1-r)!} = \binom{k+1}{r}$

SECTION 11.4

1. From the binomial theorem, $(1+x)^6 = 1 + 6x + 15x^2 + 20x^3 + 15x^4 + 6x^5 + x^6 = f_6(x)$

2.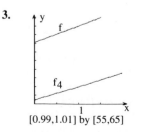
[-3,1] by [-5,5]

3.

[0.99,1.01] by [55,65]

error is approximately 7

4. 7; this agrees with the error estimate in Exercise 3

5. underestimate since $f_4(1) < f(1)$

6.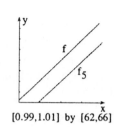
[0.99,1.01] by [62,66]

error is approximately 1; $|f(1) - f_5(1)| = 1$; $f_5(1) < f(1)$, so $f_5(1)$ is an underestimate

7. $S_1 = 1$, $S_2 = 2$, $S_3 = 2.5$, $S_4 = 2.67$, $S_5 = 2.7083$, $S_6 = 2.7167$, $S_7 = 2.7181$, $S_8 = 2.71825$, $S_9 = 2.718279$, $S_{10} = 2.718282$

8. $e = 2.718281828$, so the conjecture is $\ell = e$

9.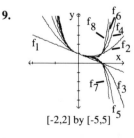
[-2,2] by [-5,5]

10. The graph of $f(x) = \dfrac{1}{x+1}$ appears to coincide with the graph of f_8 for some values of

x near $x = 0$

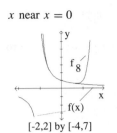

[-2,2] by [-4,7]

11. The interval of convergence is approximately $(-0.5, 0.5)$

12. $\dfrac{1}{x+1} = \displaystyle\sum_{k=1}^{\infty} (-x)^{k-1}$

13.

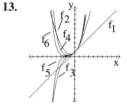

[-10,10] by [-10,10]

14. The graph of $f(x) = e^x$ appears to coincide with the graph of f_6 for some values of x near $x = 0$

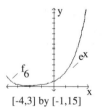

[-4,3] by [-1,15]

15. $|e^{-8} - f_6(-8)| = 201.36$; therefore $f_6(-8)$ is not a good estimate for e^8

16. $|e^3 - f_6(3)| = 0.67$; therefore $f_6(3)$ is a good estimate for e^3

17. $(-1.5, 1.5)$

18. **(a)** The error is about 7 units.

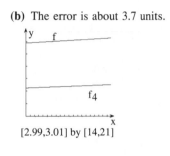

[2.99,3.01] by [11,22]

(b) The error is about 3.7 units.

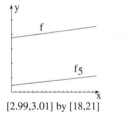

[2.99,3.01] by [14,21]

(c) The error is about 1.7 units.

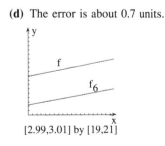

[2.99,3.01] by [18,21]

(d) The error is about 0.7 units.

[2.99,3.01] by [19,21]

All approximations above are underestimates for $y = e^3$.

19.

[-10,10] by [-10,10]

20. The graph of $f(x) = \cos x$ appears to coincide with the graph of f_6 values of x on $(-4.5, 4.5)$.

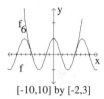

[-10,10] by [-2,3]

21. $|\cos(-8) - f_6(-8)| = 39.39$; therefore $f_6(-8)$ is not a good estimate of $\cos(-8)$

22. $|\cos(3) - f_6(3)| = 5.29 \times 10^{-5}$; therefore $f_6(3)$ is a good estimate of $\cos(3)$

23. $|f_{10}(3) - \cos(3)| < 2.75 \times 10^{-11}$ therefore $f_{10}(3)$ is a very good estimate for $\cos(3)$; $f_{10}(3)$ is an overestimate

24. $f_1(x) = 1 + \dfrac{1}{3}x$,

$f_2(x) = 1 + \dfrac{1}{3}x - \dfrac{1}{9}x^2$,

$f_3(x) = 1 + \dfrac{1}{3}x - \dfrac{1}{9}x^2 + \dfrac{5}{81}x^3$

25. The graphs of $f(x) = \sqrt[3]{1+x}$ and f_3 appear to coincide for values of x near $x = 0$.

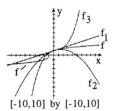

[-10,10] by [-10,10]

26. $(-0.7, 1)$

27. The error is approximately 0.00020 units.

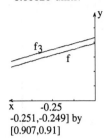

x   -0.25
-0.251,-0.249] by
[0.907,0.91]

28. The error is approximately 1×10^{-6} units.

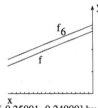

x
[-0.25001,-0.24999] by
[-0.90857,-0.90855]

29. appear to coincide on $(-1, 1)$

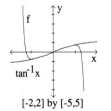

$\tan^{-1}x$

[-2,2] by [-5,5]

30. appear to coincide on $(0, 1)$

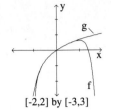

[-2,2] by [-3,3]

31. appear to coincide on $(-2, 2)$

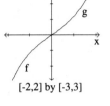

[-2,2] by [-3,3]

32. Given $e^x = 1 + x + \dfrac{x^2}{2!} + \dfrac{x^3}{3!} + \dfrac{x^4}{4!} + \dfrac{x^5}{5!} + \cdots$, substituting $-x$ for x yields: $e^{-x} = 1 - x + \dfrac{x^2}{2!} - \dfrac{x^3}{3!} + \dfrac{x^4}{4!} - \dfrac{x^5}{5!} + \cdots$; combine these two equations to get: $\dfrac{e^x + e^{-x}}{2} = 1 + \dfrac{x^2}{2!} + \dfrac{x^4}{4!} + \cdots = \cosh(x)$

33. appear to coincide on the interval $(-2.8, 2.8)$; f is a good approximation of $g(x) = \cosh x$ to within 0.01 on the interval

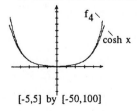

f_4

cosh x

[-5,5] by [-50,100]

34. Three terms of the power series approximate $f(x) = \sin x$ within 0.01 units on $\left[-\dfrac{\pi}{2}, \dfrac{\pi}{2}\right]$

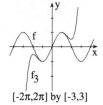

f

f_3

[-2π,2π] by [-3,3]

35. Five terms of the power series approximate $f(x) = \sin x$ within 0.01 units on $[-\pi, \pi]$.

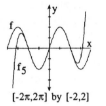

f

f_5

[-2π,2π] by [-2,2]

36. Nine terms of the power series are needed to approximate $f(x) = \sin x$ within 0.01 units on $[-2\pi, 2\pi]$.

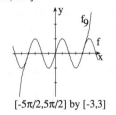

f_9

f

[-5π/2,5π/2] by [-3,3]

37. Let $u = ix$; $e^u = e^{ix} = 1 + (ix) + \dfrac{(ix)^2}{2!} + \dfrac{(ix)^3}{3!} + \cdots =$

$$\left(1 - \frac{x^2}{2!} + \frac{x^4}{4!} - \frac{x^6}{6!} + \cdots\right) +$$
$$i \cdot \left(x - \frac{x^3}{3!} + \frac{x^5}{5!} - \frac{x^7}{7!} + \cdots\right) =$$
$$\cos x + i \sin x$$

38. $e^{\pi i} + 1 = (\cos \pi + i \sin \pi) + 1 = (-1 + 0) + 1 = 0$

SECTION 11.5

1. $a_n = 3 + (n-1)5 = 5n - 2$.
Proof by induction. Let P_n be the statement $a_n = 5n - 2$. For $n = 1$, $a_1 = 3$ and $5(1) - 2 = 3$. Thus P_1 is true. Suppose that P_k is true; then $a_k = 5k - 2$. Now consider P_{k+1}: $a_{k+1} = a_k + 5 = (5k - 2) + 5 = 5(k + 1) - 2$. So P_{k+1} is true. Thus $a_n = 5n - 2$ for all positive integers n.

2. $a_n = 7 + (n-1)2 = 2n + 5$.
Proof by induction: Let P_n be the statement $a_n = 2n + 5$. For $n = 1$, $a_1 = 7$ and $2(1) + 5 = 7$. Thus P_1 is true. Suppose that P_k is true; then $a_k = 2k + 5$. Now consider P_{k+1}: $a_{k+1} = a_k + 2 = (2k + 5) + 2 = 2(k + 1) + 5$. So P_{k+1} is true. Thus $a_n = 2n + 5$ for all positive integers n.

3. $a_n = 2 \cdot 3^{n-1}$. Proof by induction: For $n = 1$, $a_1 = 2$ and $3^{(1-1)} \cdot 2 = 3^0 \cdot 2 = 2$. Thus P_1 is true. Suppose that P_k is true; then $a_k = 2 \cdot 3^{(k-1)}$. Now consider P_{k+1}: $a_{k+1} = 3 \cdot a_k = 3(2 \cdot 3^{k-1}) = 3^k \cdot 2 = 2 \cdot 3^k$. So P_{k+1} is true. Thus $a_n = 2 \cdot 3^{(n-1)}$ for all positive integers n.

4. $a_n = 3 \cdot 5^{n-1}$. Proof by induction: For $n = 1$, $a_1 = 3$ and $3 \cdot 5^{(1-1)} = 3$. Thus P_1 is true. Suppose that P_k is

true; then $a_k = 3 \cdot 5^{(k-1)}$. Now consider P_{k+1}: $a_{k+1} = 5 \cdot a_k = 5(3 \cdot 5^{k-1}) = 3 \cdot 5^k$. So P_{k+1} is true. Thus $a_n = 3 \cdot 5^{(n-1)}$ for all positive integers n.

5. $P(1)$: $1 = \dfrac{1(1+1)}{2}$; $P(k)$:
$1 + 2 + 3 + \cdots + k = \dfrac{k(k+1)}{2}$;
$P(k+1)$: $1 + 2 + 3 + \cdots + k + (k+1) = \dfrac{(k+1)(k+2)}{2}$

6. $P(1)$: $1^2 = \dfrac{1(1+1)(2\cdot1+1)}{6}$;
$P(k)$: $1^2 + 2^2 + 3^2 + \cdots + k^2 = \dfrac{k(k+1)(2k+1)}{6}$; $P(k+1)$:
$1^2 + 2^2 + 3^2 + \cdots + k^2 + (k+1)^2 = \dfrac{(k+1)(k+2)(2k+3)}{6}$

7. $P(1)$: $\dfrac{1}{1 \cdot 2} = \dfrac{1}{1+1}$;
$P(k)$: $\dfrac{1}{1 \cdot 2} + \dfrac{1}{2 \cdot 3} + \cdots + \dfrac{1}{k(k+1)} = \dfrac{k}{k+1}$; $P(k+1)$:
$\dfrac{1}{1 \cdot 2} + \dfrac{1}{2 \cdot 3} + \cdots + \dfrac{1}{k(k+1)} + \dfrac{1}{(k+1)(k+2)} = \dfrac{k+1}{k+2}$

8. $P(1)$: $\displaystyle\sum_{i=1}^{1} i^4 = \dfrac{1(2)(3)(5)}{30}$;
$P(k)$: $\displaystyle\sum_{i=1}^{k} i^4 = \dfrac{k(k+1)(2k+1)(3k^2+3k-1)}{30}$;
$P(k+1)$: $\displaystyle\sum_{i=1}^{k+1} i^4 = \dfrac{1}{30}\Big[(k+1)(k+2)(2k+3)\cdot (3(k+1)^2 + 3(k+1) - 1)\Big]$

9. $P(1)$: $(a + b)^1 = \displaystyle\sum_{i=0}^{1} a^i b^{1-i}$;
$P(k)$: $(a + b)^k = \displaystyle\sum_{i=0}^{k} a^i b^{k-i}$;
$P(k+1)$: $(a + b)^{k+1} = \displaystyle\sum_{i=0}^{k+1} a^i b^{k+1-i}$

10. $P(1)$: $2^1 < 3!$; $P(k)$:
$2^k < (k+2)!$; $P(k+1)$:
$2^{k+1} < (k+3)!$

11. $P(1)$: $\dbinom{1}{0} + \dbinom{1}{1} = 2$;
$P(k)$: $\dbinom{k}{0} + \dbinom{k}{1} + \dbinom{k}{2} + \cdots + \dbinom{k}{k} = 2^k$; $P(k+1)$:
$\dbinom{k+1}{0} + \dbinom{k+1}{1} + \dbinom{k+1}{2} + \cdots + \dbinom{k+1}{k+1} = 2^{k+1}$

12. Let P_n be the statement $a_n = a_1 + (n-1)d$. For $n = 1$, $a_1 = a_1 + (1-1)d = a_1$. Thus P_1 is true. Suppose P_k is true; then $a_k = a_1 + (k-1)d$. Consider P_{k+1}: $a_{k+1} = a_k + d = a_1 + (k-1)d + d = a_1 + k \cdot d$. Thus P_n is true for all positive integers n.

13. For $n = 1$, $1 = \dfrac{1(1+1)}{2} = \dfrac{2}{2} = 1$. Thus P_1 is true. Suppose P_k is true; then
$1 + 2 + 3 + \cdots + k = \dfrac{k(k+1)}{2}$.
Consider P_{k+1}:
$1 + 2 + 3 + \cdots + k + (k+1) = \dfrac{k(k+1)}{2} + (k+1) = \dfrac{k(k+1)}{2} + \dfrac{2(k+1)}{2}$. Thus
$1 + 2 + 3 + \cdots + k + k + 1 =$

$\dfrac{(k+2)(k+1)}{2}$. So P_n is true for all positive integers n.

14. For $n = 1$, $1^3 = 1^2\dfrac{(1+1)^2}{4}$

or $1 = \dfrac{4}{4}$. Thus P_1 is true.
Suppose P_k is true; then
$1^3 + 2^3 + 3^3 + \cdots + k^3 = \dfrac{k^2(k+1)^2}{4}$. Consider P_{k+1}:
$1^3 + 2^3 + 3^3 + \cdots + k^3 + (k+1)^3 = \dfrac{k^2(k+1)^2}{4} + (k+1)^3 = \dfrac{(k+1)^2(k^2+4k+4)}{4} = \dfrac{(k+1)^2(k+2)^2}{4}$. Thus P_n is true for all positive integers n.

15. For $n = 1$, $1 = 1(1)$. Thus P_1 is true. Suppose P_k is true; then $1 + 3 + 5 + \cdots + (2k-1) = k^2$. Consider P_{k+1}:
$1 + 3 + 5 + \cdots + (2k-1) + (2(k+1)-1) = k^2 + (2(k+1)-1) = k^2 + 2k + 1 = (k+1)^2$. Thus P_n is true for all positive integers n.

16. For $n = 1$, $1 = 1(1)$. Thus P_1 is true. Suppose P_k is true; then $1 + 5 + 9 + \cdots + (4k-3) = k(2k-1)$. Consider P_{k+1}:
$1 + 5 + 9 + \cdots + (4k-3) + (4(k+1)-3) = k(2k-1) + (4(k+1)-3) = 2k^2 + 3k + 1 = (2k+1)(k+1) = (2(k+1)-1)(k+1)$. Thus P_n is true for all positive integers n.

17. For $n = 1$, $2^{1-1} = 2^2 - 1$ or $1 = 1$. Thus P_1 is true. Suppose P_k is true; then $1 + 2 + 2^2 + 2^3 \cdots + 2^{k-1} = 2^k - 1$. Consider P_{k+1}:
$1 + 2 + 2^2 + 2^3 + \cdots + 2^{k-1} + 2^k = 2^k - 1 + 2^k = 2(2^k) - 1 = 2^{k+1} - 1$. Thus P_n is true for all positive integers n.

18. For $n = 1$, $\dfrac{1}{1 \cdot 2} = \dfrac{1}{1+1}$

or $\dfrac{1}{2} = \dfrac{1}{2}$. Thus P_1 is true.
Suppose P_k is true; then
$\dfrac{1}{1 \cdot 2} + \dfrac{1}{2 \cdot 3} + \cdots + \dfrac{1}{k(k+1)} = \dfrac{k}{k+1}$. Consider P_{k+1}:
$\dfrac{1}{1 \cdot 2} + \dfrac{1}{2 \cdot 3} + \cdots + \dfrac{1}{k(k+1)} + \dfrac{1}{(k+1)(k+2)} = \dfrac{k}{k+1} + \dfrac{1}{(k+1)(k+2)} = \dfrac{k(k+2)+1}{(k+1)(k+2)} = \dfrac{k^2 + 2k + 1}{(k+1)(k+2)} = \dfrac{(k+1)^2}{(k+1)(k+2)} = \dfrac{k+1}{k+2}$. Thus P_n is true for all positive integers n.

19. 125,250

20. 5,239,625

21. $\dfrac{n(n+1)}{2} - 6$

22. 8,122,500

23. 33,554,431

24. 14,400

SECTION 11.6

1. 6

2. 24

3. 12

4. 144

5. 12

6. 9

7. 8,000,000

8. 10^9

9. 19,656,000

10. 60,466,176

11. 36

12. 1024

13. 6720

14. 3,991,680

15. 13,366,080

16. 1,663,200

17. 2002

18. 28

19. 1,307,504

20. 43,758

21. 2300

22. 38,760

23. 17,296

24. 77,520

25. 37,353,738,800

26. 56

27. $_nC_r + {_nC_{r+1}} = \dfrac{n!}{r!(n-r)!} + \dfrac{n!}{(r+1)!(n-r-1)!} = \dfrac{(r+1)n! + (n-r)n!}{(r+1)!(n-r)!} = \dfrac{n!(r+1+n-r)}{(r+1)!(n-4)!} = \dfrac{n!(n+1)}{(r+1)!(n-r)!} = \dfrac{(n+1)!}{(r+1)!(n+1-r-1)!} = {_{n+1}C_{r+1}}$

28. Answers will vary. Using counting principles, we see that before the end of the year, the number of copies of the letter mailed out during one week would exceed the entire population. So, you would be guaranteed to receive a copy in that mailing.

SECTION 11.7

1.

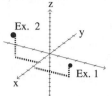

[-6,6] by [-6,6] by [-10,10]

2. See graph in Exercise 1.

3.

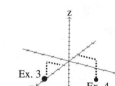

[-6,6] by [-6,6] by [-10,10]

4. See graph in Exercise 3.

5. Octant II

6. Octant VIII

7. Octant VII

8. Octant VIII

9. $\sqrt{53}$

10. $2\sqrt{41}$

11. $\left(1, -1, \dfrac{11}{2}\right)$

12. $(4, -2, -2)$

13. (a) $x = 2t + 4$,
$y = 4t + 6$, $z = 3t + 1$
(b) $\dfrac{x-4}{2} = \dfrac{y-6}{4} = \dfrac{z-1}{3}$

14. (a) $x = t + 2$,
$y = 5t + 3$, $z = 3t + 6$
(b) $\dfrac{x-2}{1} = \dfrac{y-3}{5} = \dfrac{z-6}{3}$

15. $(4, 0, 0)$

16. $(0, \text{any}, 5)$

17. $\left(4\sqrt{2}, \dfrac{\pi}{4}, 3\right)$

18. $\left(2\sqrt{2}, \dfrac{\pi}{4}, -3\right)$

19. $\left(1, \dfrac{\pi}{2}, 0\right)$

20. $(1, \pi, 0)$

21. $\left(2, \dfrac{\pi}{4}, \dfrac{\pi}{4}\right)$

22. $\left(2, \dfrac{3\pi}{4}, \dfrac{7\pi}{4}\right)$

23. The graph is the plane containing L and perpendicular to the xy-plane.

24. The graph is the plane containing L and perpendicular to the xy-plane where L is the $30°$ line in QI and QIII that passes through the origin.

25. The graph is the right circular cylinder with the z-axis as its central axis and radius 3.

26. The graph is a sphere centered at the origin with radius 4.

27. $\rho = d(P, O) =$
$\sqrt{(x-0)^2 + (y-0)^2 + (z-0)^2} =$
$\sqrt{x^2 + y^2 + z^2}$ so
$\rho^2 = x^2 + y^2 + z^2$

28. The graph is an elliptical cylinder with a major axis of 6 and a minor axis of 4. In the plane $z = c$, the trace is an ellipse with center $(0, 0, c)$ and endpoints $(0, \pm 2, c)$ and $(\pm 3, 0, c)$.

29. The graph is an elliptical cylinder with a major axis of 6 and a minor axis of 2. In the plane $y = c$, the trace is an ellipse with center $(0, c, 0)$ and endpoints $(\pm 3, c, 0)$ and $(0, c, \pm 1)$.

30. The graph is an elliptical paraboloid. The trace in the plane $y = 0$ is $z = x^2$. The trace in the plane $x = 0$ is $z = \dfrac{1}{16}y^2$.

31. The graph is a sphere centered at the origin with radius 8.

32. The graph is an elliptic paraboloid. The trace in the plane $x = 0$ is $z = \dfrac{1}{9}y^2$. The trace in the plane $y = 0$ is $z = \dfrac{1}{4}x^2$.

33. The graph is an ellipsoid with axes intercepts of $(\pm 2, 0, 0)$, $(0, \pm 3, 0)$, and $(0, 0, \pm 4)$.

34.

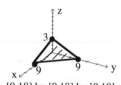

[0,18] by [0,18] by [0,10]

35.

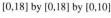

[0,12] by [0,12] by [0,12]

36.

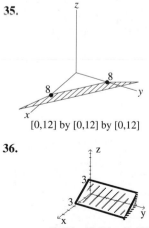

[0,6] by [0,6] by [0,10]

37.

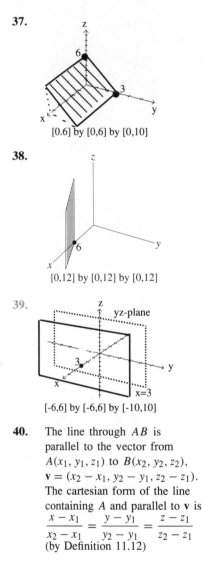

[0.6] by [0,6] by [0,10]

38.

[0,12] by [0,12] by [0,12]

39. yz-plane

x=3

[-6,6] by [-6,6] by [-10,10]

40. The line through AB is parallel to the vector from $A(x_1, y_1, z_1)$ to $B(x_2, y_2, z_2)$, $\mathbf{v} = (x_2 - x_1, y_2 - y_1, z_2 - z_1)$. The cartesian form of the line containing A and parallel to $\mathbf{v}$ is

$$\frac{x - x_1}{x_2 - x_1} = \frac{y - y_1}{y_2 - y_1} = \frac{z - z_1}{z_2 - z_1}$$

(by Definition 11.12)

41. $x = t(x_2 - x_1) + x_1$;
$y = t(y_2 - y_1) + y_1$;
$z = t(z_2 - z_1) + z_1$

42. **(a)** $\dfrac{x + 1}{1} = \dfrac{y - 2}{4} = \dfrac{z - 4}{-7}$
(b) $x = t - 1$; $y = 4t + 2$;
$z = -7t + 4$

43. **(a)** $\dfrac{x + 1}{3} = \dfrac{y - 2}{-6} = \dfrac{z - 4}{-3}$
(b) $x = 3t - 1$; $y = -6t + 2$;
$z = -3t + 4$

44. **(a)** $\dfrac{x}{2} = \dfrac{y - 6}{-10} = \dfrac{z + 3}{4}$
(b) $x = 2t$; $y = -10t + 6$;
$z = 4t - 3$

45. **(a)** $\dfrac{x + 1}{2} = \dfrac{y - 2}{-1} = \dfrac{z - 4}{-5}$
(b) $x = 2t - 1$; $y = -t + 2$;
$z = -5t + 4$

46. Let $P = (x_1, y_1, z_1)$ and $Q = (x_2, y_2, z_2)$

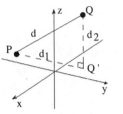

Let $Q' = (x_2, y_2, z_1)$. The distance between P and Q' in the plane $z = z_1$ is $d_1 = \sqrt{(x_2 - x_1)^2 + (y_2 - y_1)^2}$. The distance between Q and Q' in the plane $x = x_2$ is $d_2 = \sqrt{(y_2 - y_2)^2 + (z_2 - z_1)^2} = |z_2 - z_1|$. By the Pythagorean theorem, $d^2 = d_1^2 + d_2^2 = (x_2 - x_1)^2 + (y_2 - y_1)^2 + (z_2 - z_1)^2$. Therefore $d = \sqrt{(x_2 - x_1)^2 + (y_2 - y_1)^2 + (z_2 - z_1)^2}$.

47. $(-1, -5, -3)$

48. $(4, -6, -11)$

49. $\mathbf{u} \times \mathbf{v} \cdot \mathbf{u} = 0$; $\mathbf{u} \times \mathbf{v} \cdot \mathbf{v} = 0$

50. $\mathbf{u} \times \mathbf{v} \cdot \mathbf{u} = 0$; $\mathbf{u} \times \mathbf{v} \cdot \mathbf{v} = 0$

51. Generalization: The cross product $\mathbf{u} \times \mathbf{v}$ is perpendicular to both $\mathbf{u}$ and $\mathbf{v}$ in 3-space.

SECTION 11.8

1. Domain of f is all ordered pairs in the xy-plane. Domain of g is all ordered pairs in the xy-plane

2. Domain of $z = \sqrt{9 - x^2 - y^2}$ is the circle $x^2 + y^2 = 9$ and its interior in the xy-plane

3. Domain of $z = \sqrt{x^2 + y^2 - 9}$ is the circle $x^2 + y^2 = 9$ and its exterior in the xy-plane

4. Domain of $z = \dfrac{1}{9 - x^2 - y^2}$ is all ordered pairs in the xy-plane except those on the circle $x^2 + y^2 = 9$

5.

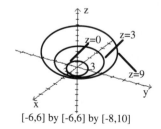

[-6,6] by [-6,6] by [-10,10]

6.

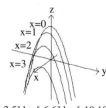

[-6,6] by [-6,6] by [-8,10]

7.

[-2,5] by [-6,6] by [-10,10]

8.

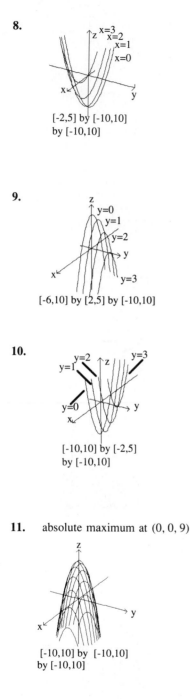

[-2,5] by [-10,10]
by [-10,10]

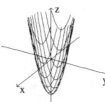

[-10,10] by [-10,10]
by [-10,10]

maximum at $(0, 0, 25)$

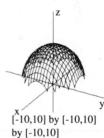

x [-10,10] by [-10,10]
by [-10,30]

9.

[-6,10] by [2,5] by [-10,10]

10.

[-10,10] by [-2,5]
by [-10,10]

11. absolute maximum at $(0, 0, 9)$

[-10,10] by [-10,10]
by [-10,10]

12. absolute minimum at $(0, 0, -9)$

13. Domain of f is all ordered pairs in the xy-plane

14. Domain of g is all points on the circle $(x - 2)^2 + (y - 3)^2 = 9$ and in its interior

15. absolute maximum at $(2, 3, 9)$

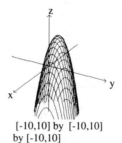

[-10,10] by [-10,10]
by [-10,10]

16. absolute maximum at $(2, 3, 3)$; absolute minimum value of 0 at all points on the circle $(x - 2)^2 + (y - 3)^2 = 9$, in the x-y plane

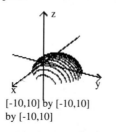

[-10,10] by [-10,10]
by [-10,10]

17. Domain of z is all pairs of real numbers (x, y); absolute

18. Domain of z is all points on the circle $x^2 + y^2 = 25$ and in its interior; absolute maximum at $(0, 0, 5)$; absolute minimum at $(x, y, 0)$ for all (x, y) on the circle $x^2 + y^2 = 25$

[-10,10] by [-10,10]
by [-10,10]

19. Domain of z is all points on the circle
$$(x - 2)^2 + \left(y - \frac{9}{2}\right)^2 = 62.25$$
and in its interior; absolute maximum at $\left(2, \frac{9}{2}, 7.9\right)$; absolute minimum at $(x, y, 0)$ for all (x, y) on the circle
$$(x - 2)^2 + \left(y - \frac{9}{2}\right)^2 = 62.25$$

[-5,10] by [-5,10] by [-5,10]

20. Domain is all pairs of real numbers (x, y); absolute maximum at $(x, y, 5)$ where $x = \pi + 4k\pi$; absolute minimum at $(x, y, -5)$ where $x = 3\pi + 4k\pi$, for any integer k.

[–6,15] by [–6,15] by [–10,20]

21. Domain is all points (x, y) in the interior of the circle $x^2 + y^2 = 9$; absolute minimum at $\left(0, 0, \dfrac{1}{3}\right)$

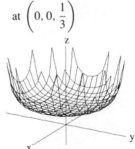

[-6,6] by [-6,6] by [-5,10]

22. Domain is all pairs of real numbers (x, y) except those on the circle $x^2 + y^2 = 9$; local minimum at $\left(0, 0, \dfrac{1}{9}\right)$

[-6,6] by [-6,6] by [-6,6]

23. Domain is all pairs of real numbers (x, y); saddle point at

$(0, 4, 0)$

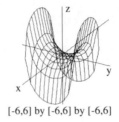

[-5,5] by [-5,10] by [-5,5]

24. Domain is all pairs of real numbers (x, y) where $\cos x > 0$ and $\cos y > 0$; saddle point at $(2\pi k, 2\pi n, 0)$ where k and n are integers

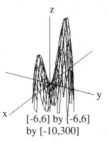

[-6,6] by [-6,6] by [-6,6]

25. Domain is all pairs of real numbers (x, y); absolute maximum at $(0, 4.1, 231.2)$; local maximum at $(0, -2.9, 92.1)$; saddle point at $(0, 0, 0)$

[-6,6] by [-6,6] by [-10,300]

26. Domain is all pairs of real numbers (x, y); absolute maximum at $(0.7, 0, 8.6)$; absolute minimum at

$(-0.7, 0, -8.6)$

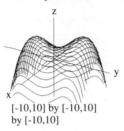

[-4,4] by [-4,4] by [-15,15]

27. Domain is all pairs of real numbers (x, y); absolute maximum at $(0, 4.1, 11.7)$; local maximum at $(0, -2.8, 7.5)$; saddle point at $(0, 0, 5)$

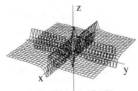

[-10,10] by [-10,10] by [-10,10]

28. Domain is all pairs of real numbers (x, y) except for $x = 0$ or $y = 0$; no local extrema or saddle points

[-10,10] by [-10,10] by [-10,10]

29.

	$x = 2$	$x = 4$	$x = 6$
$y = 1$	258	191	174
$y = 3$	207.3	156.3	155.3
$y = 7$	220.3	201.3	232.3
$y = 10$	240	245	300

	$x = 8$
$y = 1$	169.5
$y = 3$	166.8
$y = 7$	275.8
$y = 10$	367.5

30. As both $x \to \infty$ and $y \to \infty$, $z \to \infty$

31.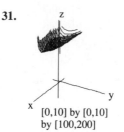
[0,10] by [0,10] by [100,200]

32. minimum at (6.1, 2.2, 148.0)

33. $C(x, y) = \dfrac{300}{x} + 4xy + \dfrac{100}{y}$

34. $0 < x < 50$ and $0 < y < 50$ are reasonable dimensions

35.
[0,12] by [0,12] by [0,385]

36. minimum cost of $148 when $x = 6.1$ in., $y = 2.0$ in., $h = 4.2$ in.

37. $S(x, y) = 2xy + \dfrac{1000}{x} + \dfrac{1000}{y}$

38. any $x > 0$ and $y > 0$ are *possible* dimensions mathematically

39.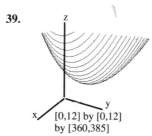
[0,12] by [0,12] by [360,385]

40. minimum surface area is 378 sq in. when the box has dimensions 7.9 in. by 7.9 in. by 7.9 in.

41. $S(x, y) = xy + \dfrac{1000}{y} + \dfrac{1000}{x}$

42. any $x > 0$ and $y > 0$ are *possible* dimensions mathematically

43.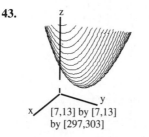
[7,13] by [7,13] by [297,303]

44. minimum surface area is 300 sq in. when the box has dimensions 10 in. by 10 in. by 5 in.

45. end behavior model is the plane $z = 0$

46. end behavior model is the plane $z = 0$

47. end behavior model is the plane $z = 2$

48. end behavior model is the plane $z = 3$

49. (a) $f(u) = (x + yi)^2 + 1 = (x^2 - y^2 + 1) + (2xy)i$; then $z = |f(u)| = \sqrt{(x^2 - y^2 + 1)^2 + (2xy)^2} = \sqrt{(x^2 - y^2 + 1)^2 + 4x^2y^2}$ (b) See the figure in the text for the graph. (c) Observe that $t^2 + 1 = (t + i)(t - i) = 0$ when $t = \pm i$. The graph in part (b) touches the complex plane at $(0, 1, 0)$ and $(0, -1, 0)$, corresponding to zeros at $\pm i$.

50. (a) $f(u) = (x + yi)^3 - (x + yi)^2 + (x + yi) - 1 = (x^3 - 3xy^2 - x^2 + y^2 + x - 1) + (3x^2y - y^3 - 2xy + y)i$.

Then $z = |f(u)| = \sqrt{\begin{array}{l}(x^3 - 3xy^2 - x^2 + y^2 + x - 1)^2 \\ + (3x^2y - y^3 - 2xy + y)^2\end{array}}$
(b) See the figure from part (d) in the text for the graph.
(c) $f(t) = (t - 1)(t + i)(t - i)$
(d) $f(t) = 0$ when $t = 1$, $t = i$, $t = -i$. The graph in part (d) shows the modulus surface touches the complex plane at $(1, 0, 0)$, $(0, 1, 0)$, and $(0, -1, 0)$, corresponding to the zeros of $f(t)$ at $t = 1$, i, and $-i$, respectively.

Chapter 11 Review

1. $a_1 = 0$, $a_2 = -1$, $a_3 = 2$, $a_4 = -3$, $a_{10} = -9$
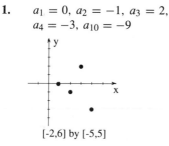
[-2,6] by [-5,5]

2. $a_1 = 1$, $a_2 = 7$, $a_3 = 17$, $a_4 = 31$, $a_{10} = 199$
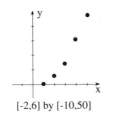
[-2,6] by [-10,50]

3. $a_1 = 1$, $a_2 = 3$, $a_3 = 7$, $a_4 = 15$, $a_{10} = 1023$
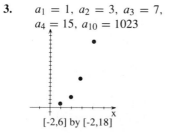
[-2,6] by [-2,18]

4. $a_1 = -1$, $a_2 = 1$, $a_3 = -1$, $a_4 = 1$, $a_{10} = 1$

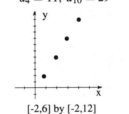

[-2,6] by [-3,3]

5. $a_1 = 2$, $a_2 = 5$, $a_3 = 8$, $a_4 = 11$, $a_{10} = 29$

[-2,6] by [-2,12]

6. $a_1 = 7$, $a_2 = -2$, $a_3 = 7$, $a_4 = -2$, $a_{10} = -2$

[-2,6] by [-3,8]

7. $a_1 = 5$, $a_2 = 15$, $a_3 = 45$, $a_4 = 135$, $a_{10} = 98,415$

[-2,6] by [-10,140]

8. $a_1 = 1$, $a_2 = -2$, $a_3 = 4$, $a_4 = -8$, $a_{10} = -512$

[-2,6] by [-10,10]

9. $a_1 = -32$; $a_n = 8n - 40$

10. $a_1 = -23$; $a_n = 4.5n - 27.5$

11. $4+7+10+13+16+19+22+25$

12. $3 + 12 + 27 + 48 + 75 + 108 + 147 + 192$

13. $1+3+7+15+31+63+127+255$

14. $0+1+4+9+16+25+36+49$

15. $\displaystyle\sum_{k=1}^{10}(1 - 3k)$

16. $\displaystyle\sum_{k=1}^{\infty}4^k$

17. $\displaystyle\sum_{k=1}^{n}(10 - 2k)$

18. $\displaystyle\sum_{k=1}^{n}(-1)^k 3 \cdot k$

19. 4650

20. 25,492,400

21. 5,328,575

22. 7,979,200

23. converges; $S = 3$

24. does not converge

25. converges; $S = \dfrac{2}{99}$

26. converges; $S = 20$

27. $32x^5 + 80x^4y + 80x^3y^2 + 40x^2y^3 + 10xy^4 + y^5$

28. $16,384a^7 - 86,016a^6b + 193,536a^5b^2 - 241,920a^4b^3 + 181,440a^3b^4 - 81,648a^2b^5 + 20,412ab^6 - 2187b^7$

29. $243x^{10} + 405x^8y^3 + 270x^6y^6 + 90x^4y^9 + 15x^2y^{12} + y^{15}$

30. $1+\dfrac{6}{x}+\dfrac{15}{x^2}+\dfrac{20}{x^3}+\dfrac{15}{x^4}+\dfrac{6}{x^5}+\dfrac{1}{x^6}$

31. $512a^{27} - 2304a^{24}b^2 + 4608a^{21}b^4 - 5376a^{18}b^6 + 4032a^{15}b^8 - 2016a^{12}b^{10} + 672a^9b^{12} - 144a^6b^{14} + 18a^3b^{16} - b^{18}$

32. $\dfrac{1}{x^8} + \dfrac{4}{x^6y} + \dfrac{6}{x^4y^2} + \dfrac{4}{x^2y^3} + \dfrac{1}{y^4}$

33.

[-10,10] by [-10,10]

34. The graphs of $f(x) = e^{-x}$ and $f_6(x)$ appear to agree on the interval $(-3, 3)$.

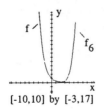

[-10,10] by [-3,17]

35. $|e^2 - f_6(-2)| = 0.0335$; therefore $f_6(-2)$ is a good estimate of e^2.

36. $|e^{-4} - f_6(4)| = 2.1372$; therefore $f_6(4)$ is not a good estimate of e^{-4}.

37. $(-1.5, 1.5)$

38. $f_3(x)$ approximates e^2 with an error of 1.06 at $x = -2$;

$f_4(x)$ approximates e^2 with an error of 0.39 at $x = -2$; $f_5(x)$ approximates e^2 with an error of 0.12 at $x = -2$; $f_6(x)$ approximates e^2 with an error of 0.03 at $x = -2$; all estimates are underestimates.

39. For $n = 1, 1 = \dfrac{1(2)(3)}{6}$

or $1 = \dfrac{6}{6}$, thus P_1 is true. Suppose P_k is true; then $1 + 3 + 6 + \cdots + \dfrac{k(k+1)}{2} = \dfrac{k(k+1)(k+2)}{6}$. Consider $P_{k+1}: 1 + 3 + 6 + \cdots + \dfrac{k(k+1)}{2} + \dfrac{(k+1)(k+2)}{2} = \dfrac{k(k+1)(k+2)}{6} + \dfrac{(k+1)(k+2)}{2} = \dfrac{k(k+1)(k+2) + 3(k+1)(k+2)}{6} = \dfrac{(k+1)(k+2)(k+3)}{6}$; so P_n is true for all positive integers n

40. For $n = 1, 1 \cdot 2 = \dfrac{1(2)(3)}{3}$

or $2 = \dfrac{6}{3}$, thus P_1 is true. Suppose P_k is true; then $1 \cdot 2 + 2 \cdot 3 + 3 \cdot 4 + \cdots + k(k+1) = \dfrac{k(k+1)(k+2)}{3}$. Consider $P_{k+1}: 1 \cdot 2 + 2 \cdot 3 + 3 \cdot 4 + \cdots + k(k+1) + (k+1)(k+2) = \dfrac{k(k+1)(k+2)}{3} + (k+1)(k+2) = \dfrac{k(k+1)(k+2) + 3(k+1)(k+2)}{3} = \dfrac{(k+1)(k+2)(k+3)}{3}$; so P_n is true for all positive integers n.

41. For $n = 1, \displaystyle\sum_{i=1}^{1}\left(\dfrac{1}{2}\right)^i = 1 - \dfrac{1}{2}$

or $\dfrac{1}{2} = 1 - \dfrac{1}{2}$, thus P_1 is

true. Suppose P_k is true; then $\displaystyle\sum_{i=1}^{k}\left(\dfrac{1}{2}\right)^i = 1 - \dfrac{1}{2^k}$. Consider

$P_{k+1}: \displaystyle\sum_{i=1}^{k+1}\left(\dfrac{1}{2}\right)^i = 1 - \dfrac{1}{2^k} + \left(\dfrac{1}{2}\right)^{k+1} = 1 - \dfrac{1}{2^k} + \dfrac{1}{2^{k+1}} = 1 - \dfrac{2}{2^{k+1}} + \dfrac{1}{2^{k+1}} = 1 - \dfrac{1}{2^{k+1}}$;

so P_n is true for all positive integers n.

42. For $n = 1, 2^{1-1} \le 1!$ or $2^0 \le 1$, thus P_1 is true. Suppose P_k is true; then $2^{k-1} \le k!$. Consider $P_{k+1}: 2^{(k+1)-1} \le 2 \cdot k! \le (k+1)k! \le (k+1)!$ So P_n is true for all positive integers n.

43. For $n = 1, 1^3 + 2(1) = 3$ which is divisible by 3, thus P_1 is true. Suppose P_k is true; that is, $k^3 + 2k$ is divisible by 3. Then $k^3 + 2k = 3t$ for some integer t. Consider $P_{k+1}: (k+1)^3 + 2(k+1) = k^3 + 3k^2 + 3k + 1 + 2k + 2 = (k^3 + 2k) + 3(k^2 + k + 1) = 3t + 3(k^2 + k + 1)$ (for some integer t) $= 3(t + k^2 + k + 1)$; the right hand side is divisible by 3, so P_{k+1} is true. Therefore P_n is true for all positive integers n.

44. 3,991,680
45. 259,459,200
46. 18,564
47. 6,724,520
48. 260,000
49. 43,670,016
50. 15
51. 14,508,000
52. 14,190
53. 8,217,822,536
54. 708,930,508
55. 26
56. 19,110

57. $_n P_k \cdot {_{n-k}}P_j = \dfrac{n!}{(n-k)!} \cdot \dfrac{(n-k)!}{(n-k-j)!} = \dfrac{n!}{(n-k-j)!} = \dfrac{n!}{(n-(k+j))!} = {_n}P_{k+j}$

58.

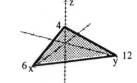

[-6,6] by [-12,12] by [-10,10]

59.

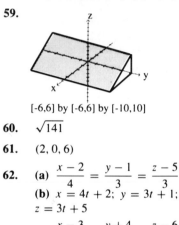

[-6,6] by [-6,6] by [-10,10]

60. $\sqrt{141}$

61. $(2, 0, 6)$

62. (a) $\dfrac{x-2}{4} = \dfrac{y-1}{3} = \dfrac{z-5}{3}$
(b) $x = 4t + 2; \ y = 3t + 1;$
$z = 3t + 5$

63. (a) $\dfrac{x-3}{-1} = \dfrac{y+4}{12} = \dfrac{z-6}{-8}$
(b) $x = -t + 3; \ y = 12t - 4;$
$z = -8t + 6$

64. $y - 9 - 9x^2$, a parabolic cylinder. The trace in plane $z = c$ is a parabola with vertex $(0, 9, c)$.

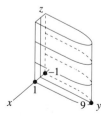

[–2,6] by [0,9] by [0,10]

65. $\dfrac{x^2}{9} + \dfrac{z^2}{4} = 1$, an elliptic cylinder parallel to the y-axis.

65. The trace in plane $y = c$ is an ellipse with major axis end points at $(\pm 3, c, 0)$ and minor axis end points at $(0, c, \pm 2)$.

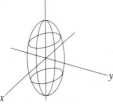

[−8,12] by [−8,12] by [−6,12]

66. $\dfrac{x^2}{4} + \dfrac{y^2}{1} + \dfrac{z^2}{64} = 1$, an ellipsoid centered at the origin and intercepting the axes at $(\pm 2, 0, 0)$, $(0, \pm 1, 0)$, and $(0, 0, \pm 8)$.

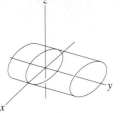

[−8,12] by [−8,12] by [−10,12]

67. sphere centered at the origin with radius 9

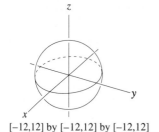

[−12,12] by [−12,12] by [−12,12]

68. Domain of f is all pairs of real numbers (x, y); domain of g is all pairs of real numbers (x, y).

69. Domain of $z = \sqrt{f(x, y)}$ is all points on the circle $x^2 + y^2 = 16$ and in its interior.

70. Domain of $z = \sqrt{g(x, y)}$ is all points on the circle $x^2 + y^2 = 16$ and in its exterior.

71. Domain of $z = \dfrac{1}{f(x, y)}$ is all pairs of real numbers (x, y) except those on the circle $x^2 + y^2 = 16$.

72.

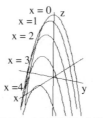

[-5,5] by [-5,5] by [-10,20]

73.

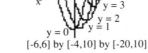

[-6,6] by [-4,10] by [-20,10]

74. Domain is all pairs of real numbers (x, y); absolute maximum at $(0, 0, 9)$

[-4,4] by [-4,4] by [-10,10]

75. Domain is all pairs (x, y) on the circle $x^2 + y^2 = 9$ and in its interior; absolute maximum at $(0, 0, 3)$; absolute minimum of 0 for all points on the circle

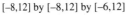

[-6,6] by [-6,6] by [-10,20]

76. Domain is all points (x, y) on the hyperbola $\dfrac{(y - 3)^2}{3} - \dfrac{(x - 1)^2}{3} = 1$ and in its exterior; absolute minimum of 0 at the hyperbola

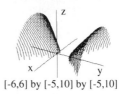

[-6,6] by [-5,10] by [-5,10]

77. Domain is all pairs of real numbers (x, y); saddle point at $(3, 0, 0)$

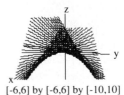

[-6,6] by [-6,6] by [-10,10]

CHAPTER 12

SECTION 12.1

1. $S = \{1, 2, 3, 4, 5, 6\}$

2. $S = \{H, T\}$

3. Order listed is penny, dime: $S = \{HH, HT, TH, TT\}$

4. Order listed is penny, nickel, dime: $S = \{HHH, HHT, HTH, THH, TTH, THT, HTT, TTT\}$

5. $S = \{H1, H2, H3, H4, H5, H6, T1, T2, T3, T4, T5, T6\}$

6. $S = \{1, 2, 3, 4, 5, 6, 7, 8\}$

7. $S = \{1, 2, 3, 4, 5, 6, 7, 8, 9, 10\}$

8. $S = \{(1, 2), (1, 3), (1, 4), (1, 5), (2, 1), (2, 3), (2, 4), (2, 5), (3, 1), (3, 2), (3, 4), (3, 5), (4, 1), (4, 2), (4, 3), (4, 5), (5, 1), (5, 2), (5, 3), (5, 4)\}$

In Exercises 9–20, order listed is red, green:

9. $E = \{(3, 6), (4, 5), (5, 4), (6, 3)\}$

10. $E = \{(1, 1), (1, 3), (1, 5), (2, 2), (2, 4), (2, 6), (3, 1), (3, 3), (3, 5), (4, 2), (4, 4), (4, 6), (5, 1), (5, 3), (5, 5), (6, 2), (6, 4), (6, 6)\}$

11. $E = \{(1, 2), (2, 1), (2, 3), (3, 2), (3, 4), (4, 3), (4, 5), (5, 4), (5, 6), (6, 5)\}$

12. $E = \{(2, 2), (2, 4), (2, 6), (4, 2), (4, 4), (4, 6), (6, 2), (6, 4), (6, 6)\}$

13. $E = \{(1, 1), (1, 2), (1, 3), (1, 4), (1, 5), (1, 6), (2, 1), (2, 2), (2, 3), (2, 4), (2, 5), (2, 6), (3, 1), (3, 2), (3, 3), (3, 4), (3, 5), (3, 6), (4, 1), (4, 2), (4, 3), (4, 4), (4, 5), (5, 1), (5, 2), (5, 3), (5, 4), (6, 1), (6, 2), (6.3)\}$

14. $E = \{(1, 1), (1, 3), (1, 5), (3, 1), (3, 3), (3, 5), (5, 1), (5, 3), (5, 5)\}$

15. $\dfrac{1}{9}$

16. $\dfrac{1}{2}$

17. $\dfrac{5}{18}$

18. $\dfrac{1}{4}$

19. $\dfrac{5}{6}$

20. $\dfrac{1}{4}$

21. Rolling a sum less than or equal to 1 (an impossible event)

22. Rolling an odd sum

23. Rolling different numbers on each die

24. Rolling a sum of 11 or 12

25. $\dfrac{53}{110}$

26. $\dfrac{8}{65}$

27. 0.0000645

28. 0.000258

29. 0.001032

30. 0.28173

31. 0.25

32. The sum of the probabilities is greater than one and the events are mutually exclusive.

33. Answers will vary. The probabilities of two mutually exclusive events, i.e., no sunshine (0.22) and at least one hour of sunshine (0.78) sum to 1.00 but do not account for all events. For example, there could be 30 minutes of sunshine. So the statement cannot be true.

SECTION 12.2

1. $\dfrac{1}{8}$

2. $\dfrac{1}{16}$

3. $\dfrac{15}{16}$

4. $\dfrac{1}{256}$

5. 0.000099

6. 0.009703

7. 0.9703

8. 0.000098

9. 5

10. 10

11. 15

12. 6

13. $\dfrac{5}{16}$

14. $\dfrac{5}{16}$

15. $\dfrac{5}{32}$

16. $\dfrac{5}{8}$

17. 45 heads (even digits) and 55 tails (odd digits); experimental probabilities are $P(H) = \dfrac{45}{100}$, $P(T) = \dfrac{55}{100}$.

18. 25 occurrences of 1, 2, 3, 4, 5, and 6 in first 45 digits; experimental probabilities: $P(1) = \dfrac{6}{25}$, $P(2) = \dfrac{4}{25}$, $P(3) = \dfrac{3}{25}$, $P(4) = \dfrac{4}{25}$, $P(5) = \dfrac{1}{25}$, $P(6) = \dfrac{7}{25}$.

19. Outcome order is (first, last) = (red, green). Outcomes: $(1, 5)$, $(5, 1)$, $(3, 6)$, $(6, 4)$, $(2, 3)$, $(2, 3)$, $(1, 1)$, $(3, 5)$, $(1, 2)$, $(3, 1)$

20. One occurrence of (even, even); $P(HH) = \dfrac{1}{13}$

21. Seven occurrences of (even, odd); $P(HT) = \dfrac{7}{25}$

22. Use the method from Exercise 18 with the first 20 digits from the table that are a 1 to 6 and the last 20 digits that are a 1 to 6. Ignore the digits 6, 7, 8, and 9.

23. $\dfrac{1}{2}$

24. $\dfrac{1}{4}$

25. $\dfrac{1}{1024} = 0.00098$

26. $\dfrac{45}{1024} = 0.0439$

27. $\dfrac{1023}{1024} = 0.99902$

28. 0.9557

29. 0.22

30. 0.000772

SECTION 12.3

1. 21.4 or \$21,400 (stem-and-leaf)

2. 7 pairs (stem-and-leaf)

3. 31.7 or \$31,700 (stem-and-leaf)

4.

Stem	Leaf
74	7 0
75	3 7 7 0
76	8 4
77	8 6 3
78	6 2 4 4 1
79	1 2 6
80	3 2

5. France (Table 12.2)

6.

Stem	Leaf
28	2
29	3 7
30	
31	7 6
32	8 7
33	5 8 5 5
34	8 8 2
35	3 3 4
36	7 3
37	
38	5

7.

Interval	Frequency
28,000 to 28,999	1
29,000 to 29,999	2
30,000 to 30,999	0
31,000 to 31,999	2
32,000 to 32,999	2
33,000 to 33,999	4
34,000 to 34,999	3
35,000 to 35,999	3
36,000 to 36,999	2
37,000 to 37,999	0
38,000 to 38,999	1

8.

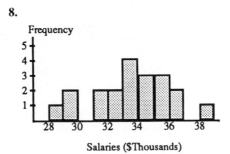

Salaries (\$Thousands)

9. Window: [27, 40] by [−1, 5]. Answers will vary. (Data are displayed in thousands of dollars.)

10. The histogram on your grapher should look like the one in Exercise 8.

11. A line graph displays pairs of data arrayed relative to two scales. Often it shows trends in data over time. These data vary in only one dimension (salary); there is no time factor to show changes over time.

12.

Stem	Leaf
6	9 2 3 9
7	7 9 9 9
8	7 0 3 9 2 8 7 9
9	0 1 2 6 3 3 0 4 5 1 7 0 3
10	2 3 6 9 3 7 6 6
11	2 0 4 6
12	5 0 9

13.

Interval	Frequency
6.0 to 6.9	4
7.0 to 7.9	4
8.0 to 8.9	8
9.0 to 9.9	13
10.0 to 10.9	8
11.0 to 11.9	4
12.0 to 12.9	3

14.

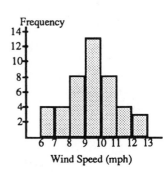

Wind Speed (mph)

15. Window: [5, 14] by [−1, 13]. (Answers will vary.)

16. The histogram on your grapher should look like the one in Exercise 14.

17. A line graph displays pairs of data arrayed relative to two scales. Often it shows trends in data over time. These data vary in only one dimension (wind speed); there is no time factor to show changes over time.

18. 26.8

19. 10

20. $60.11\overline{6}$

21. 71.33

22. $27.1\overline{3}$ or $27,133

23. 13.79

24. 34.51

25. 23

26. 61.25

27. 72.0

28. 8

29. 41

30.

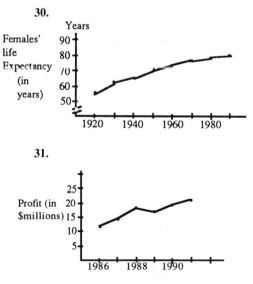

31.

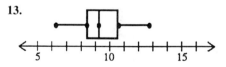

32.

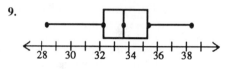

33. The frequency table in Table 12.3 displays intervals of data whereas the frequency table from Example 5 shows individual data points explicitly. Table 12.3 lends itself to a histogram because the data are grouped into nonoverlapping intervals. The data from Example 5 cannot be grouped easily into a series of intervals to generate a meaningful histogram.

34. Answers will vary. The data 1, 2, 2, 4, 5, 10 have mode = 2, median = 3, and mean = 4.

35. Answers will vary. The data 1, 2, 3, 7, 7 have median = 3, mean = 4, and mode = 7.

36. Answers will vary. The data 1, 13, 13, 14, 15, 16 have mean = 12, mode = 13, and median = 13.5.

SECTION 12.4

1. $H = 31.7$, $L = 21.4$, $M = 28.05$, $Q_L = 24.5$, $Q_U = 29.6$

2.

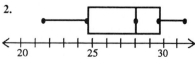

3. 10.3

4. 5.1

5. There is a greater spread among the lower outliers.

6. There is a greater difference between median salary and the lowest salary.

7. About half of the range

8. $H = 38.5$, $L = 28.2$, $M = 33.65$, $Q_L = 32.2$, $Q_U = 35.3$

9.

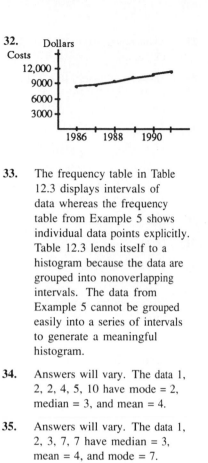

10. 10.3

11. More than half of the range

12. $H = 12.9$, $L = 6.2$, $M = 9.25$, $Q_L = 8.5$, $Q_U = 10.6$

13.

14. 6.7

15. More than half of the range

16. $\dfrac{12}{44}$ or about 27%

17. $\dfrac{40}{44}$ or about 91%

18. Section 3: $H = 86$, $L = 43$, $M = 64.5$, $Q_L = 53.5$, $Q_U = 74.5$; Section 4: $H = 93$, $L = 42$, $M = 57.5$, $Q_L = 47.5$, $Q_U = 69.0$

19.

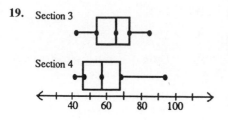

Section 3

Section 4

20. Section 4

21. Section 4

22. Section 3 is stronger; 70% of the students in Section 3 scored above 57.5 (the median for Section 4) compared to 50% of the students in Section 4.

23. Answers will vary. Case 2 could represent a plant where most of the workers earn between $30,000 and $40,000 with a few earning lower salaries and a few supervisors or managers with higher salaries.

24. Answers will vary. Case 1 would represent the salaries for the class members because most of the data are at the high end, above $60,000. College graduates who have been in the work force for 25 years would be more likely to have high salaries.

25. Answers will vary. One possibility is Case 3 because a school system would have large numbers of employees working at many levels (staff, teachers, administrators) and with a great deal of variation in seniority. Case 3 has the broadest range of salaries for the majority of the population.

26. Answers will vary: 10, 70, 76, 76, 77, 80

27. Answers will vary: 10, 13, 15, 15, 16, 60

28. Answers will vary: 10, 13, 14, 15, 16, 16, 17, 19, 25, 25, 25, 26

29. Answers will vary. A coach might choose B because $\frac{3}{4}$ of the players are batting above .250 and B has the highest upper quartile value.

30. 9.08

31. 23.99

32. The first set has the greater spread.

33. The second set has the greater spread.

SECTION 12.5

1.

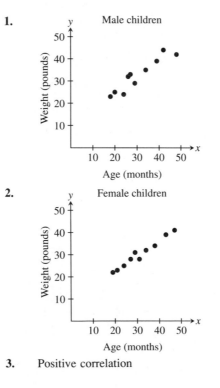

Male children

2.

Female children

3. Positive correlation

4. Negative correlation

5. Positive correlation

6. Weak negative correlation

7. Strong positive correlation

8. Little or no correlation

9. There is a weak correlation between weight and running speed. Although an overweight person probably runs much slower than a sprinter who is lean, the sprinter runs much faster than a child who weighs far less.

10. There is a strong correlation between running speed and distance in the broad jump because higher speeds result in greater forward velocity upon "takeoff" and carry the jumper farther.

11. There is a strong correlation because the pay scale is based on years of experience.

12. There is little or no correlation because physical attributes are not indicators of knowledge or test-taking ability.

13. There is a strong correlation because the length of a pendulum affects its period.

14. There is a strong correlation because the speed of the ball is directly related to the force with which it is thrown.

15. The plot in (a) has the stronger correlation.

16. The plot in (b) has the stronger correlation.

17. $\bar{x} = \frac{8}{3}, \bar{y} = \frac{16}{3}$

18. $b = \frac{20}{13}$

19. $y = \dfrac{16}{13} + \dfrac{20}{13}x$

20.

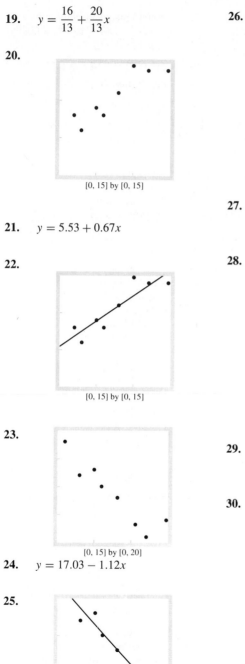

[0, 15] by [0, 15]

21. $y = 5.53 + 0.67x$

22.

[0, 15] by [0, 15]

23.

[0, 15] by [0, 20]

24. $y = 17.03 - 1.12x$

25.

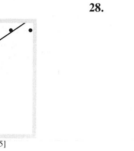

[0, 15] by [0, 15]

26.

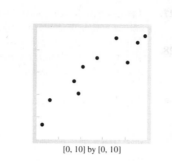

[0, 10] by [0, 10]

27. $y = 2.21 + 0.75x$

28.

[0, 10] by [0, 10]

29. $y = 2.67 + 2.56 \ln x$

30.

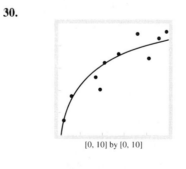

[0, 10] by [0, 10]

31. $y = (2.35)(1.17)^x$

32.

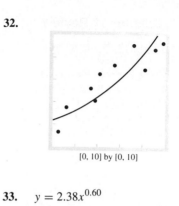

[0, 10] by [0, 10]

33. $y = 2.38x^{0.60}$

34.

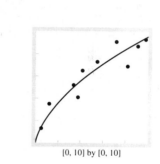

[0, 10] by [0, 10]

35. The power equation is the curve of best fit.

36. These data are best modeled by an exponential curve of best fit, $y = 1.16(2.52)^x$.

37.

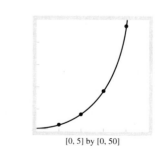

[0, 5] by [0, 50]

38. Transform the data from (x, y) to $(x, \ln y)$: {(1, 1.06), (2, 2.03), (3, 2.90), (4, 3.85)}

Chapter 12 Review

1. $S = \{1,\ 2,\ 3,\ 4,\ 5,\ 6\}$

2. Order listed is (red, green).
$S = \{(1,1),\ (1,2),\ (1,3),$
$(1,4),\ (1,5),\ (1,6),\ (2,1),$
$(2,2),\ (2,3),\ (2,4),\ (2,5),$
$(2,6),\ (3,1),\ (3,2),\ (3,3),$
$(3,4),\ (3,5),\ (3,6),\ (4,1),$
$(4,2),\ (4,3),\ (4,4),\ (4,5),$
$(4,6),\ (5,1),\ (5,2),\ (5,3),$
$(5,4),\ (5,5),\ (5,6),\ (6,1),$
$(6,2),\ (6,3),\ (6,4),\ (6,5),$
$(6,6)\}$

3. $S = \{13,\ 16,\ 31,\ 36,\ 61,\ 63\}$

4. $S = \{D, N\}$

5. Order listed is penny, nickel, dime: $S = \{HHH,\ HHT,$
$HTH,\ THH,\ HTT,\ THT,$
$TTH,\ TTT\}$

6. $E = \{HHT,\ HTH,\ THH,$
$TTH,\ THT,\ HTT\}$

7. $E = \{HHH,\ TTT\}$

8. $\dfrac{1}{8}$

9. $\dfrac{7}{8}$

10. $\dfrac{1}{64}$

11. $\dfrac{1}{4}$

12. $\dfrac{5}{16}$

13. 0.9704

14. $\dfrac{1}{4}$

15. The probability of success equals the probability of failure. Therefore the probability of one success and three failures is equal to the probability of three successes and one failure: $_4C_1 = {_4C_3}$ and $P(S) = P(F) = 0.5$.

16. 0.24

17. 0.096

18. 0.64

19. In this case the probability of success does not equal the probability of failure: $_4C_1 = {_4C_3}$ but $(0.4)^3(0.6) \neq (0.4)(0.6)^3$.

20. Let pairs of adjacent digits represent the outcome of flipping a dime and a nickel, where an even digit represents a head and an odd digit represents a tail.

21. For the first 50 pairs, $P(HH) = \dfrac{17}{50}$

22. Answers will vary, depending on the digits generated. In one simulation, $P(H) = \dfrac{23}{50} = 0.46$.

23. Answers will vary, depending on the digits generated. In one simulation, $P(6) = \dfrac{5}{50} = 0.10$.

24. 0.375

25. 0.3125

26. 0.2461

27. 0.0703

28. The probability that the 51st toss results in a head is $\dfrac{1}{2}$ because each toss is an independent event. The probability of no tails in 50 tosses is 8.88×10^{-16}, a highly unlikely event.

29. (a) 0.9224   (b) 0.0753

30. (a) 0.9960   (b) 7.177×10^{-6}

31. 16.75

32. 15

33. 15

34. 19.5625

35. 4.42295

36. $12.58\overline{3}$

37. 13.5

38. 7

39. 19.24

40. 4.39

41.
Stem	Leaf
28	7 9
29	2 4 8 9
30	5 7 4 7 8
31	5 7 2 7 2 8
32	1 6 4 9
33	5 9
34	8
35	
36	7

42.
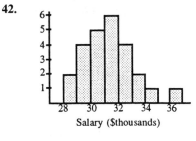
Salary ($thousands)

43.
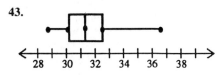

$H = 36.7$, $L = 28.7$,
$M = 31.2$, $Q_U = 32.5$,
$Q_L = 30.15$

44.

Stem	Leaf
30	4 6
31	0 2 6 7
32	3 5 2 5 6
33	4 6 1 6 1 7
34	0 6 3 9
35	5 9
36	9
37	
38	9

45. A line graph shows trends over time and would be appropriate to display changes in temperature over the course of the day.

46.

Stem	Leaf
21	7 4 8
22	3 9 8
23	4 7 4 8
24	5 8
25	1 6 7 6
26	8 7
27	8 4 9
28	3 2 2 5 3 6 8 7
29	7 8 1 5 6 7
30	6 8
31	7 6 5 7 6

47. Answers will vary. Class 2 has a greater second quartile range covering a broader range of scores. Class 1 has a compact second quartile, with a lower top score but a higher bottom score. Class 2 has a stronger third quartile because it is completely contained between 75 and 83, whereas the third quartile for Class 1 falls between 64 and

84. It is impossible to decide which class is stronger without the actual scores, but one might choose Class 2 because 75% of the students have scores above 75 while 75% of students in Class 1 scored above 64%.

48.

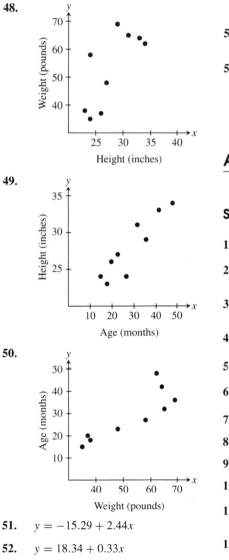

49.

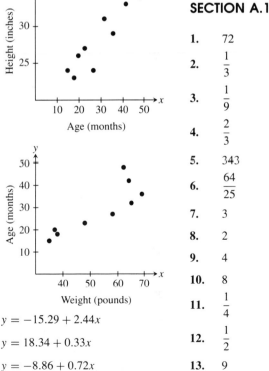

50.

51. $y = -15.29 + 2.44x$

52. $y = 18.34 + 0.33x$

53. $y = -8.86 + 0.72x$

54. The power curve $y = 0.143x^{1.33}$ is the best fit with $r = 0.911$.

55. The best fit is the power curve $y = 1.27x^{1.72}$ with $r = 0.9993$

56. Transform (x, y) to $(\ln x, \ln y)$: $\{(0, 0.26),$ $(0.69, 1.39), (1.10, 2.10),$ $(1.39, 2.66)\}$ The curve of best fit is $\ln y = 0.236 + 1.72 \ln x$.

57. $\ln y = (\ln(4.1))x + \ln(2.3)$ or $y' = 1.41x + 0.83$

58. $\ln y = \ln(1.4) + 3 \ln x$ or $y' = 0.34 + 3x'$

APPENDIX

SECTION A.1

1. 72

2. $\dfrac{1}{3}$

3. $\dfrac{1}{9}$

4. $\dfrac{2}{3}$

5. 343

6. $\dfrac{64}{25}$

7. 3

8. 2

9. 4

10. 8

11. $\dfrac{1}{4}$

12. $\dfrac{1}{2}$

13. 9

14. $\dfrac{1}{8}$

15. 1.47

16. 0.000002

17. 0.86

18. 34.71

19. $3^{3/4}$

20. $4^{-2/3}$

21. $12^{3/10}$, $3^{3/4}$, $4^{5/8}$

22. $y = (x^{1/n})^m$

23. Yes

[-8,8] by [-2, 8]

24. Yes

[-10,10] by [-10,10]

25. Yes

[-8,8] by [-2,8]

26. No, the graph of $y = (x^{1/3})^2$ allows negative values.

[-10,10] by [-10,10]

27. $(x^{1/n})^m$ lies entirely in the first quadrant if n is even and $n \neq 0$.

28. $\dfrac{x^2}{y^2}$

29. v^4

30. $3x^4 y^2$

31. $\dfrac{16}{x^4}$

32. $\dfrac{1}{2x^3 y^3}$

33. $\dfrac{1}{9x^4 y^6}$

34. $x^2 y$

35. $\dfrac{x^3 y^3}{8}$

36. $\dfrac{1}{xy}$

37. $\dfrac{x - y}{x + y}$

38. $y^2 |x|$

39. $|x|^{1/3}$

40. $\dfrac{x^3}{y^4}$

41. $|x|^{3/2} y^{9/4}$

42. $\dfrac{q}{3|p|}$

43. $x^4 y^4$

44. $8172.71

45. 8% compounded monthly is the better deal

46. 9.22×10^{18}

47. (d)

48. $\dfrac{a^7}{a^2} = \dfrac{a \cdot a \cdot a \cdot a \cdot a \cdot a \cdot a}{a \cdot a} = a \cdot a \cdot a \cdot a \cdot a = a^5 = a^{7-2}$

49. $(a^3)^4 = (a \cdot a \cdot a)(a \cdot a \cdot a)(a \cdot a \cdot a)(a \cdot a \cdot a) = a^{12} = a^{3 \cdot 4}$

SECTION A.2

1. Three possibilities are:
(a) $x^2 + 9x + \sqrt{x - 1}$
(b) $\dfrac{x^2(\sqrt{x - 1})}{9x}$
(c) $\dfrac{9x\sqrt{x - 1}}{x^2}$

2. No; the power of the leading term is not an integer.

3. Example: $\dfrac{1}{2x}$ is not a polynomial because the degree of x, -1, is not an integer greater than zero.

4. 400 ft

5. $38.25

6. 603.19 cm^3

7. Degree: 4; leading coefficient: 3

8. Degree: 3; leading coefficient: 5

9. $3x^2 + 5x + 2$

10. $4x^2 + 2x + 4$

11. $-4x^2 - 7x - 17$

12. $3x^3 - x^2 - 9x + 3$

13. $6x^2 + 5x + 1$

14. $6x - 2$

15. $13x^2 - 3x - 1$

16. $7x - 10$

17. $-x^2 - 7x + 15$

18. $4x^2 + 11x - 11$

19. $x^2 + 8x + 15$

20. $x^2 + 3x - 10$

21. $x^2 + 5x - 14$

22. $2x^2 + 5x - 3$

23. $x^2 - 4x + 3$

24. $x^2 - 12x + 27$

25. $8x^2 + 14x + 3$

26. $3x^2 + x - 10$

27. $-x^2 + 3x + 10$

28. $2x^2 + 6x$

29. $x^4 - 2x^2 - 3$

30. $x^3 + 4x^2 + 2x + 8$

31. $x^5 + x^3 + 4x^2 + 4$

32. $x^7 - 3x^5$

33. $\dfrac{1}{x^3} + \dfrac{3}{x^2} + \dfrac{1}{x} + 3$

34. $x - \dfrac{3}{x^2} + 4x^3 - 12$

35. $x^3 + 6x^2 + 11x + 12$

36. $x^4 + 4x^3 + x^2 - 8x - 3$

37. False

38. False

39. False

40. False

41. False

42. False

43. False

44. (a) $x^2 + x - 1$ (b) $x^2 - x - 5$
    (c) $x^3 + 2x^2 - 3x - 6$

[-10,10] by [-10,10]

45. b

46. a

47. $4(x - 16)$

48. $2x^2(x - 2)$

49. $x(x^2 - 5x + 7)$

50. $y(y^3 - 3y^2 + 1)$

51. $(x - 3)(x + 2)(2x - 1)$

52. $(x + 4)(x - 4)$

53. $(y + 2x)(y - 2x)$

54. $(xy + 5)(xy - 5)$

55. $(x^2 + 9)(x + 3)(x - 3)$

56. $(x + 13)(x + 5)$

57. $(x^2 - 2x + 4)(x^2 - 2x - 2)$

58. $(x - 3)^2$

59. $(x + 3)^2$

60. $(x - 4)(x + 1)$

61. $(x + 3)(x - 2)$

62. $(x + 4)(x - 1)$

63. $(x + 6)(x + 2)$

64. $(x - 6)(x - 2)$

65. $(x + 2y)(x + 3y)$

66. $(2x + 3)(x - 2)$

67. $2(2x + 1)(2x - 3)$

68. $(x + 5y)(x - 2y)$

69. $2(2x + 3)(x - 2)$

70. $(x + 2)(x - 2)(x^2 + 3)$

71. $(x + 4)(x - 4)(x^2 + 2)$

72. An algebraic argument is more convincing because, as we've seen, for some viewing rectangles different functions look the same.

SECTION A.3

1. $4x$

2. $9x^2$

3. x

4. 1

5. $x - 3$

6. $x + 2$

7. $\dfrac{x}{3}$

8. $5y$

9. $\dfrac{3x + 2}{x - 1}$

10. $\dfrac{x + 3}{x - 1}$

11. $\dfrac{x - 3}{x - 1}$

12. $\dfrac{x + 4}{x - 2}$

13. x^2

14. $-(3 + x)$

15. $\dfrac{x - y}{3}$

16. This is already in lowest terms.

17. (b)

18. (a)

19. (b)

20. (a)

21. $\dfrac{x + 1}{3}$

22. 1

23. $\dfrac{-1}{x - 3}$

24. $\dfrac{2}{x}$

25. $12y$

26. $\dfrac{x^2}{2y^2}$

27. $\dfrac{3(x - 3)}{28}$

28. $\dfrac{3}{8}$

29. $\dfrac{x}{4(x-3)}$

30. $\dfrac{(x+2)^4}{(x+4)(x-1)^2}$

31. $\dfrac{(x+1)^2}{(x-1)^2}$

32. $-2x$

33. $\dfrac{(3x-2)^3}{(2x-1)^2}$

34. $\dfrac{(2x^2+3x-3)(x+1)}{(x+2)(x-1)}$

35. (b)

36. (b)

37. (b)

38. (a)

39. $\dfrac{5+2x}{5x}$

40. $\dfrac{y-2x}{2xy}$

41. $\dfrac{7x-11}{(x-1)(x-2)}$

42. $\dfrac{8x-11}{(x+3)(x-2)}$

43. $\dfrac{5x^2+7x-15}{(x+5)(x+2)}$

44. $\dfrac{4x^2+8x}{(x-1)(x+3)}$

45. $\dfrac{-3x+2}{(x-1)(x+2)(x+4)}$

46. $\dfrac{5x^2-33x-4}{(x-7)(x+3)(x-2)}$

47. $\dfrac{7x^2+3x}{(x+2)(x-1)(x+1)}$

48. $\dfrac{2x^2-8x+1}{(x-4)(x+2)(x-3)}$

49. 6

50. $\dfrac{5}{2}$

51. $\dfrac{y+x}{xy}$

52. $\dfrac{xy}{y+x}$

53. $2x$

54. $\dfrac{1}{xy}$

55. $\dfrac{x^2-1}{x+1} = \dfrac{(x+1)(x-1)}{(x+1)} = x-1$ for all $x \neq -1$. If $x = -1$, $\dfrac{x^2-1}{x+1}$ is undefined because the denominator is 0.

Index